The Encyclopedia of Applied Animal Behaviour and Welfare

The Encyclopedia of Applied Animal Behaviour and Welfare

Editor-in-chief

Daniel S. Mills

University of Lincoln
Department of Biological Sciences
Riseholme, Lincoln, UK

Editors

Jeremy N. Marchant-Forde

USDA-ARS, LBRU
Purdue University
West Lafayette, Indiana, USA

Paul D. McGreevy

Faculty of Veterinary Science,
University of Sydney
NSW, Australia

David B. Morton

School of Biosciences
University of Birmingham
Edgbaston, Birmingham, UK

Christine J. Nicol

School of Clinical Veterinary Science
University of Bristol
Langford, UK

Clive J.C. Phillips

Centre for Animal Welfare and Ethics
The University of Queensland
Gatton, QLD, Australia

Peter Sandøe

Danish Centre for Bioethics and Risk Assessment
University of Copenhagen
Frederiksberg C, Denmark

Ronald R. Swaisgood

San Diego Zoo's Institute for Conservation Research
Escondido, California, USA

www.cabi.org

CABI is a trading name of CAB International

CABI Head Office
Nosworthy Way
Wallingford
Oxfordshire OX10 8DE
UK

Tel: +44 (0)1491 832111
Fax: +44 (0)1491 833508
E-mail: cabi@cabi.org
Website: www.cabi.org

CABI North American Office
875 Massachusetts Avenue
7th Floor
Cambridge, MA 02139
USA

Tel: +1 617 395 4056
Fax: +1 617 354 6875
E-mail: cabi-nao@cabi.org

A catalogue record for this book is available from the British Library, London, UK.

Library of Congress Cataloging-in-Publication Data

The encyclopedia of applied animal behaviour and welfare / editor-in-chief, Daniel S. Mills ; editors, Jeremy N. Marchant-Forde ... [et al.].

p. cm.

Includes bibliographical references and index.

ISBN 978-0-85199-724-7 (alk. paper)

1. Animal behavior--Encyclopedias. 2. Animal welfare--Encyclopedias. I. Mills, D. S. II. Marchant-Forde, Jeremy N. III. Title.

QL750.3.E526 2010
591.503--dc22

2009039900

ISBN-13: 978 0 85199 724 7

Commissioning editors: Rebecca Stubbs, Rachel Cutts, Claire Parfitt
Production editor: Tracy Head

Typeset by Columns Design Ltd, Reading, UK
Printed and bound in the UK by Cambridge University Press, Cambridge

Contributors

Alberts, Allison C., San Diego Zoo's Institute for Conservation Research, 15600 San Pasqual Valley Road, Escondido, CA 92027-7000, USA

Aldenhoven, Jaclyn, Department of Biology, University of Utah, 257 South 1400 East, Salt Lake City, UT 84112, USA

Anderson, Roland C., The Seattle Aquarium, 1483 Alaskan Way, Seattle, WA 98101, USA

Anthony, Raymond X., Department of Philosophy, University of Alaska Anchorage, 3211 Providence Drive, Anchorage, AK 99508, USA

Arney, David, Estonian University of Life Sciences Institute of Veterinary Medicine and Animal Sciences, Kreutzwaldi 46, 51006 Tartu, Estonia

Bailey, Natalie, Assistant Director, Bushmeat Crisis Task Force, 19K Hillside Road, Greenbelt, MD 20770, USA

Barnett, John†, formerly Animal Welfare Science Centre, Department of Primary Industries, 600 Sneydes Road, Werribee, Victoria 3030, Australia

Bashaw, Meredith, Department of Psychology, Franklin and Marshall College, PO Box 3003, Lancaster, PA 17604-3003, USA

Bercovitch, Fred B., San Diego Zoo's Institute for Conservation Research, 15600 San Pasqual Valley Road, Escondido, CA 92027-7000, USA

Berg, Elena C., Evolutionary Biology, Department Biologie II, University of Munich (LMU), Grosshaderner Str. 2, D-82152 Planegg-Martinsried, Germany

Blackwell, Emily, Animal Welfare and Behaviour Group, School of Clinical Veterinary Science, University of Bristol, Langford House, Langford BS40 5DU, UK

Bloomsmith, Mollie A., Director of Behavioral Management, Yerkes National Primate Center, 954 Gatewood Road, Atlanta, GA 30329, USA

Blumstein, Daniel T., University of California, Department of Ecology & Evolutionary Biology, 621 Charles E. Young Drive South, Los Angeles, CA 90095-1606, USA

Bonacic, Cristián, Fauna Australis Laboratory, School of Agriculture and Forestry Sciences, Pontificia Universidad Catolica de Chile, Casilla 306, Correo 22, Santiago, Chile

Boydston, Erin E., US Geological Survey, Western Ecological Research Center, 4727 Portola Parkway, Irvine, CA 92620, USA

Bradshaw, John W.S., Anthrozoology Institute, University of Bristol, Dolberry Building, Langford BS40 5DU, UK

Brain, Paul F., Department of Pure and Applied Ecology, School of the Environment and Society, Swansea University, Swansea SA2 8PP, UK

Broom, Donald M., Colleen Macleod Professor of Animal Welfare, Centre for Animal Welfare and Anthrozoology, Department of Veterinary Medicine, University of Cambridge, Madingley Road, Cambridge CB3 0ES, UK

Bshary, Redouan, Institute of Biology, University of Neuchâtel, Rue Emile-Argand 11, Case postale 158, CH-2009 Neuchâtel, Switzerland

Buchanan-Smith, Hannah M., Scottish Primate Research Group, Department of Psychology, University of Stirling, Stirling FK9 4LA, UK

Buckley, Ralf, International Centre for Ecotourism Research, Griffith University, Gold Coast, QLD 4222, Australia

Burghardt, Gordon M., Departments of Psychology and Ecology and Evolutionary Biology, 1404 Circle Drive, University of Tennessee, Knoxville, TN 37996-0900, USA

Burman, Oliver H.P., Animal Welfare and Behaviour Group, School of Clinical Veterinary Science, University of Bristol, Langford House, Langford BS40 5DU, UK

Burn, Charlotte C., Department of Veterinary Clinical Sciences, The Royal Veterinary College, Hawkshead Lane, Hatfield AL9 7TA, UK

Butterworth, Andrew, Animal Welfare and Behaviour Group, School of Clinical Veterinary Science, University of Bristol, Langford House, Langford BS40 5DU, UK

Caplen, Gina, Animal Welfare and Behaviour Group, School of Clinical Veterinary Science, University of Bristol, Langford House, Langford BS40 5DU, UK

Cheng, Heng-Wei, USDA-ARS, LBRU, Purdue University, Poultry Science Building, 125 S. Russell St, West Lafayette, IN 47907, USA

Clucas, Barbara, College of the Environment, University of Washington, Box 352100, Seattle, WA 98195, USA

Clutton-Brock, Juliet, South Barn, High Street, Fen Ditton, Cambridgeshire CB5 8ST, UK

Cooper, Jonathan, University of Lincoln, Department of Biological Sciences, Riseholme, Lincoln LN2 2LG, UK

Corridan, Claire L., University of Lincoln, Department of Biological Sciences, Riseholme, Lincoln LN2 2LG, UK

Cross, Nicki, Ministry of Agriculture and Forestry, Pastoral House, 25 The Terrace, Wellington, New Zealand

Curry, Mark R., University of Lincoln, Department of Biological Sciences, Riseholme, Lincoln LN2 2LG, UK

Custance, Deborah, Goldsmiths College, Department of Psychology, University of London, Lewisham Way, New Cross, London SE14 6NW, UK

Damm, Birgitte I., Department of Large Animal Sciences, Faculty of Life Sciences, University of Copenhagen, Denmark

Deeming, D. Charles, University of Lincoln, Department of Biological Sciences, Riseholme, Lincoln LN2 2LG, UK

De Keuster, Tiny, Oostveldkouter 222, 9920 Lovendegem, Belgium

delBarco-Trillo, Javier, Cornell University, Department of Psychology, Uris Hall, Ithaca, NY 14853, USA

† Deceased.

Dennis, Rachel, USDA-ARS, LBRU, Purdue University, Poultry Science Building, 125 S. Russell St, West Lafayette, IN 47907, USA

de Passillé, Anne Marie, Agriculture and Agri-Food Canada, Pacific Agri-Food Research Centre, Agassiz, BC VOM 1A0, Canada

de Vere, Natasha, National Botanic Garden of Wales, Llanarthne, Carmarthenshire SA32 8HG, UK

de Waal, Frans, Living Links, Yerkes Primate Center, Emory University, 954 N. Gatewood Road, Atlanta, GA 30322, USA

Diehl, Nancy K., Pennsylvania State Horse Racing Commission, Pennsylvania Department of Agriculture, Agriculture Building Room 304, 2301 North Cameron Street, Harrisburg, PA 17110, USA

Dielenberg, Robert, School of Biomedical Sciences (MSB 510), University of Newcastle, Newcastle, NSW 2308, Australia

Dixon, Greg J.P., Maes Glas Vet Clinic, Church Acre, Brackla, Bridgend CF31 2JT, UK

Dixon, Laura M., Poultry Behaviour and Welfare, Avian Science Research Centre, SAC, Auchincruive, Ayr KA6 5HW, UK

Dyson, Miranda, Department of Biological Sciences, The Open University, Walton Hall, Milton Keynes MK7 6AA, UK

Eady, Paul, University of Lincoln, Department of Biological Sciences, Riseholme, Lincoln LN2 2LG, UK

Edgar, Joanne L., Animal Welfare and Behaviour Group, School of Clinical Veterinary Science, University of Bristol, Langford House, Langford BS40 5DU, UK

Eicher, Susan, USDA-ARS, LBRU, Purdue University, Poultry Science Building, 125 S. Russell St, West Lafayette, IN 47907, USA

Elwood, Robert W., Queen's University Belfast, School of Biological Sciences, Medical Biology Centre, 97 Lisburn Road, Belfast BT9 7BL, UK

Eves, Heather E., College of Natural Resources – National Capital Region, Virginia Polytechnic and State University, Falls Church, VA 22043, USA

Fahey, Alan G., School of Agriculture, Food Science & Veterinary Medicine, Veterinary Science Centre, Belfield, Dublin 4, Ireland

Ferkin, Michael H., Department of Biology, The University of Memphis, Ellington Hall, Memphis, TN 38152, USA

Fernández-Juricic, Esteban, Lilly Hall G-420, 915 W. State Street, Department of Biological Sciences, Purdue University, West Lafayette, IN 47907, USA

Fisher, Heidi S., Harvard University, Museum of Comparative Zoology, 26 Oxford St, Cambridge, MA 02138, USA

Flower, Frances C., formerly Animal Welfare Program, University of British Columbia, 2357 Main Mall, Vancouver, BC V6T 1Z4, Canada

Forkman, Björn A., Department of Animal Science and Animal Health, University of Copenhagen, Grønnegårdsvej 8, 1870 Frederiksberg, Denmark

Fougere, Barbara, All Natural Vet Care, 292 Lyons Road, Russell Lea, NSW 2046, Australia

Freire, Rafael, School of Animal and Veterinary Sciences, Charles Sturt University, Wagga Wagga, NSW 2350, Australia

Garshelis, David L., Minnesota Department of Natural Resources, 1201 E. Hwy 2, Grand Rapids, MN 55744,USA

Gaughan, John, Animal Biometeorology, School of Animal Studies, The University of Queensland, Gatton, QLD 4343, Australia

Gilligan, Brenda, University of Lincoln, Department of Biological Sciences, Riseholme, Lincoln LN2 2LG, UK

Gomez, Juan Carlos, School of Psychology, University of St Andrews, St Andrews KY16 9JU, UK

Gonyou, Harold W., Prairie Swine Centre, Inc., PO Box 21057, 2105 8th St East, Saskatoon, Saskatchewan S7H 5N9, Canada

Grandin, Temple, Department of Animal Sciences, Colorado State University, Fort Collins, CO 80523, USA

Grether, Gregory F., University of California, Department of Ecology and Evolutionary Biology, 621 Charles E. Young Drive South, Los Angeles, CA 90095-1606, USA

Guo, Kun, University of Lincoln, Department of Psychology, Brayford Pool, Lincoln LN6 7TS, UK

Hall, Stephen J.G., University of Lincoln, Department of Biological Sciences, Riseholme, Lincoln LN2 2LG, UK

Hardiman, Jessica R., University of Lincoln, Department of Biological Sciences, Riseholme, Lincoln LN2 2LG, UK

Hare, Brian, Hominoid Psychology Research Group, Duke University, Durham, NC 27708, USA

Harris, Justin, School of Psychology, University of Sydney, NSW 2006, Australia

Harris, Moira, Animals Group, Harper Adams University College, Edgmond, Newport, Shropshire TF10 8NB, UK

Harris, Richard B., Department of Ecosystem and Conservation Sciences, University of Montana, Missoula, MT 59812, USA

Haslam, Sue M.†, formerly Animal Welfare and Behaviour Group, School of Clinical Veterinary Science, University of Bristol, Langford House, Langford BS40 5DU, UK

Hawkins, Penny, Research Animals Department, RSPCA, Wilberforce Way, Southwater, West Sussex RH13 9RS, UK

Held, Suzanne D.E., Animal Welfare and Behaviour Group, School of Clinical Veterinary Science, University of Bristol, Langford House, Langford BS40 5DU, UK

Hemsworth, Paul H., Animal Welfare Science Centre, The Melbourne School of Land and Environment, The University of Melbourne, Parkville, Victoria 3010, Australia

Higgs, Suzanne, School of Psychology, University of Birmingham, Edgbaston, Birmingham B15 2TT, UK

Horner, Victoria, Living Links, Yerkes National Primate Research Center, Emory University, 2409 Taylor Lane, Lawrenceville, GA 30043, USA

Hothersall, Becky, Animal Welfare and Behaviour Group, School of Clinical Veterinary Science, University of Bristol, Langford House, Langford BS40 5DU, UK

Houpt, Katherine Albro, Animal Behaviour Clinic, College of Veterinary Medicine, Cornell University, Ithaca, NY 14853-6401, USA

Inglis, Ian R., The Food and Environment Research Agency, Sand Hutton, York YO41 1LZ, UK

Insley, Stephen J., Department of Biology, University of Victoria, PO Box 3020 STN CSC, Victoria, BC V8W 3N5, Canada

Janczak, Andrew M., Norwegian School of Veterinary Science, Department of Production Animal Clinical Sciences, PO Box 8146 Dep./0033, Oslo, Norway

Jefferson, Laura, Department of Animal Science, Purdue University, Poultry Science Building, 125 S. Russell St, West Lafayette, IN 47907, USA

Jones, Bidda, RSPCA Australia, PO Box 265, Deakin West, ACT 2600, Australia

Kanaan, Vanessa, formerly Department of Animal Science, Purdue University, Poultry Science Building, 125 S. Russell St, West Lafayette, IN 47907, USA

Kerr, Annie, Animal Welfare and Behaviour Group, School of Clinical Veterinary Science, University of Bristol, Langford House, Langford BS40 5DU, UK

Kirkden, Richard D., Department of Veterinary Medicine, University of Cambridge, Madingley Road, Cambridge CB3 0ES, UK

Kitchener, Andrew C., Principal Curator of Mammals and Birds, Department of Natural Sciences, National Museums Scotland, Chambers Street, Edinburgh EH1 1JF, UK

Klimley, A. Peter, Department of Wildlife, Fish and Conservation Biology, Academic Surge, University of California Davis, CA 95616, USA

Knowles, Toby, Animal Welfare and Behaviour Group, School of Clinical Veterinary Science, University of Bristol, Langford House, Langford BS40 5DU, UK

Kornell, Nate, Department of Psychology, Williams College, 18 Hoxsey Street, Williamstown, MA 01267, USA

Kreger, Michael, Division of Bird Habitat Conservation, US Fish and Wildlife Service, 4401 North Fairfax Dr., MS-4075, Arlington, VA 22203, USA

Kuhar, Christopher W, Cleveland Metroparks Zoo, 3900 Wildlife Way, Cleveland, OH 44130, USA

Kyriazakis, Ilias, Veterinary Faculty, University of Thessaly, PO Box 199, 43100 Karditsa, Greece and School of Agriculture Food and Rural Development, Newcastle University, Newcastle-upon-Tyne NE1 7RU, UK

Lamb, David, School of Psychology, Animal Behaviour Unit, University of Southampton and Department of Bioethics, University of Birmingham, UK

Landsberg, Gary M., Doncaster Animal Clinic, Thornhill, Ontario, Canada

Law, Graham, Faculty of Biomedical and Life Sciences, Graham Kerr Building, University of Glasgow, Glasgow G12 8QQ, UK

Lay Jr, Donald C., USDA-ARS, LBRU, Purdue University, Poultry Science Building, 125 S. Russell St, West Lafayette, IN 47907, USA

Ligout, Séverine, Laboratoire d'Ecologie et Neuroéthologie Sensorielles, Faculté des Sciences et Techniques, Université Jean Monnet, 23 rue Paul Michelon, 42023 St Etienne, France

Lindburg, Donald G., San Diego Zoo's Institute for Conservation Research, 15600 San Pasqual Valley Road, Escondido, CA 92027-7000, USA

Littin, Kate, MAFBNZ Animal Welfare, Ministry of Agriculture and Forestry, PO Box 2526, Wellington 6140, New Zealand

Loomis, Caroline P., Marine Laboratory, Nicholas School of the Environment and Earth Sciences, Duke University, 135 Duke Marine Lab Rd, Beaufort, NC 28516, USA

Luddem, S. Tabitha, formerly Department of Animal Behaviour & Ecology, The School of Biology, Biology Building, The University of Nottingham, University Park, Nottingham NG7 2RD, UK

Majolo, Bonaventura, University of Lincoln, Department of Psychology, Brayford Pool, Lincoln LN6 7TS, UK

Malpass, Mark, School of the Environment and Natural Resources, Bangor University, Bangor LL57 2UW, UK

Marchant-Forde, Jeremy N., USDA-ARS, LBRU, Purdue University, Poultry Science Building, 125 S. Russell St., West Lafayette, IN 47907, USA

Marchant-Forde, Ruth, Purdue University, School of Veterinary Medicine, 625 Harrison Street, West Lafayette, IN 47907, USA

McGonigle, Brendan[†], formerly Laboratory for Cognitive Neuroscience, Level 8 Appleton Tower, University of Edinburgh, Edinburgh EH8 9JZ, UK

McGonigle, Margaret, Department of Psychology, University of Edinburgh, 7 George Square, Edinburgh EH8 9JZ, UK

McGreevy, Paul D., Faculty of Veterinary Science, Room 206, Gunn Building (B19), Regimental Crescent, University of Sydney, NSW 2006, Australia

McGregor, Peter K., Cornwall College Newquay, Newquay, Cornwall TR7 2LZ, UK

McLean, Andrew N., Australian Equine Behaviour Centre, 730 Clonbinane Road, Clonbinane, Victoria 3658, Australia

Melfi, Vicky, Whitley Wildlife Conservation Trust, Paignton Zoo Environmental Park, Totnes Road, Paignton, Devon TQ4 7EU, UK

Merrill, Ralph J.N., Waltham Centre for Pet Nutrition, Waltham-on-the-Wolds, Leicestershire LE14 4RT, UK

Mills, Daniel S., University of Lincoln, Department of Biological Sciences, Riseholme, Lincoln LN2 2LG, UK

Morgan, Bethan J., San Diego Zoo's Institute for Conservation Research, 15600 San Pasqual Valley Road, Escondido, CA 92027-7000, USA; Ebo Forest Research Project, BP 3055, Messa, Yaoundé, Cameroon; and Department of Psychology, University of Stirling, Stirling FK9 4LA, UK

Morgan, Kathleen N., Department of Psychology, Wheaton College, Norton, MA 02766, USA

Morris, Heather, formerly Biophysics Group, Howard Building, Silsoe Research Institute, Wrest Park, Silsoe MK45 4HS, UK

Morton, David B., University of Birmingham, School of Biosciences, Edgbaston, Birmingham B15 2TT, UK

Mullan, Siobhan, Animal Welfare and Behaviour Group, School of Clinical Veterinary Science, University of Bristol, Langford House, Langford BS40 5DU, UK

Nicholas, Frank W., Faculty of Veterinary Science, University of Sydney, NSW 2006, Australia

Nicol, Christine J., Animal Welfare and Behaviour Group, School of Clinical Veterinary Science, University of Bristol, Langford House, Langford BS40 5DU, UK

Nordby, J. Cully, UCLA, Institute of Environment and Department of Ecology and Evolutionary Biology, La Kretz Hall, Suite 300, Box 951496, Los Angeles, CA 90095-1496, USA

Nordstrom, Lisa A., San Diego Zoo's Institute for Conservation Research, 15600 San Pasqual Valley Road, Escondido, CA 92027-7000, USA

Olsson, I. Anna S., Laboratory Animal Science Group, IBMC – Instituto de Biologia Molecular e Celular, Universidade do Porto, Rua Campo Alegre 823, 4150-180 Porto, Portugal

Omed, Hussein, School of Environment and Natural Resources, Bangor University, Bangor LL57 2UW, UK

Ord, Terry J., School of Biological, Earth and Environmental Sciences, University of New South Wales, Sydney, NSW, Australia

Owen, Megan A., San Diego Zoo's Institute for Conservation Research, 15600 San Pasqual Valley Road, Escondido, CA 92027-7000, USA

Parker, Richard M.A., Animal Welfare and Behaviour Group, School of Clinical Veterinary Science, University of Bristol, Langford House, Langford BS40 5DU, UK

Paul, Elizabeth, Animal Welfare and Behaviour Group, School of Clinical Veterinary Science, University of Bristol, Langford House, Langford BS40 5DU, UK

Pellis, Sergio M., Canadian Centre for Behavioural Neuroscience, Department of Neuroscience, University of Lethbridge, Lethbridge, Alberta T1K 3M4, Canada

Pellis, Vivien C., Canadian Centre for Behavioural Neuroscience, Department of Neuroscience, University of Lethbridge, Lethbridge, Alberta T1K 3M4, Canada

Petherick, J. Carol, Queensland Primary Industries and Fisheries, Dept Employment, Economic Development and Innovation, Rockhampton, QLD 4072, Australia

Phillips, Clive J.C., Centre for Animal Welfare and Ethics, The University of Queensland, Gatton Campus, Gatton, QLD 4343, Australia

Pines, Mathew K., formerly Centre for Animal Welfare and Ethics, The University of Queensland, Gatton Campus, Gatton, QLD 4343, Australia

Plotnik, Joshua, Living Links, Yerkes Primate Center, Emory University, 954 N. Gatewood Road, Atlanta, GA 30322, USA

Plowman, Amy, Whitley Wildlife Conservation Trust, Paignton Zoo Environmental Park, Totnes Road, Paignton, Devon TQ4 7EU, UK

Poletto, Rosangela, Department of Animal Science, Purdue University, Poultry Science Building, 125 S. Russell St, West Lafayette, IN 47907, USA

Pullen, P. Kirsten, Whitley Wildlife Conservation Trust, Paignton Zoo Environmental Park, Totnes Road, Paignton, Devon TQ4 7EU, UK

Redgate, Sarah, University of Lincoln, Department of Biological Sciences, Riseholme, Lincoln LN2 2LG, UK

Righetti, Joanne, Delta Society Australia, 50 Carlton Crescent, Summer Hill, NSW 2130, Australia

Robinson, Paul A., Quercus Editorial Service, Bath, UK

Röcklinsberg, Helena, Teol dr., Centre for Theology and Religious Studies, Lund University, Allhelgona Kyrkogata 8, SE-223 62 Lund, Sweden

Rogers, Lesley J., Centre for Neuroscience and Animal Behaviour, School of Science and Technology, University of New England, Armidale, NSW 2351, Australia

Ruedisueli, Frank L., University of Lincoln, Department of Biological Sciences, Riseholme, Lincoln LN2 2LG, UK

Rushen, Jeffrey, Agriculture and Agri-Food Canada, Pacific Agri-Food Research Centre, Agassiz, BC VOM 1A0, Canada

Saltzman, Wendy, Department of Biology, University of California, Riverside, CA 92521, USA

Sandøe, Peter, Danish Centre for Bioethics and Risk Assessment, University of Copenhagen, Faculty of Life Sciences, Rolighedsvej 25, DK-1958, Frederiksberg C, Denmark

Schenck, Erin L., Department of Animal Science, Purdue University, Poultry Science Building, 125 S. Russell St, West Lafayette, IN 47907, USA

Schoener, Thomas W., Section of Evolution and Ecology, University of California, Davis, One Shields Avenue, Davis, CA 95616, USA

Schulte, Bruce A., Department of Biology, Western Kentucky University, 1906 College Heights Blvd #11080, Bowling Green, KY 42101-1080, USA

Seksel, Kersti, Sydney Animal Behaviour Service, 55 Ethel Street, Seaforth, NSW 2092, Australia

Sherwin, Chris, Animal Welfare and Behaviour Group, School of Clinical Veterinary Science, University of Bristol, Langford House, Langford BS40 5DU, UK

Shier, Debra M., San Diego Zoo's Institute for Conservation Research, 15600 San Pasqual Valley Road, Escondido, CA 92027-7000, USA

Slater, Peter J.B., School of Biology, University of St Andrews, Bute Building, St Andrews, Fife KY16 9TS, UK

Smith, Andrew, Department of Life Sciences, Anglia Ruskin University, East Road, Cambridge CB1 1PT, UK

Spinka, Marek, Ethology Group, Research Institute of Animal Production, 104 01 Prague – Uhříněves, Czech Republic

Stamps, Judy A., Section of Evolution and Ecology, University of California, Davis, One Shields Avenue, Davis, CA 95616, USA

Statham, Poppy, Animal Welfare and Behaviour Group, School of Clinical Veterinary Science, University of Bristol, Langford House, Langford BS40 5DU, UK

Stookey, Joseph M., Department of Large Animal Clinical Sciences, Western College of Veterinary Medicine, University of Saskatchewan, 52 Campus Dr., Saskatoon, SK S7N 5B4 Canada

Swaisgood, Ronald, San Diego Zoo's Institute for Conservation Research, 15600 San Pasqual Valley Road, Escondido, CA 92027-7000, USA

Tarou, Loraine R., Ohio Wesleyan University, Department of Psychology, Philips Hall 52, Delaware, OH 43015, USA

Taylor, Katy D. University of Lincoln, Department of Biological Sciences, Riseholme, Lincoln LN2 2LG, UK

Thompson, Katerina V., Behavior, Ecology, Evolution and Systematics Program, University of Maryland, College Park, Maryland, USA

Tinsley, Richard C., School of Biological Sciences, University of Bristol, Bristol BS8 1UG, UK

Tolkamp, Bert, Animal Health, SAC, Bush Estate, Midlothian EH26 0PH, UK

Toscano, Mike, Animal Welfare and Behaviour Group, School of Clinical Veterinary Science, University of Bristol, Langford House, Langford BS40 5DU, UK

Tucker, Cassandra B., University of California, Department of Animal Science, One Shields Avenue, Davis, CA 95616, USA

VanBuskirk, Richard W., Environmental Studies Program, Pacific University, 2043 College Way, Forest Grove, OR 97116, USA

van der Zee, Emile, University of Lincoln, Department of Psychology, Brayford Pool, Lincoln LN6 7TS, UK

van Dierendonck, Machteld C., Equus Research and Therapy, Tolnegenweg 39, 3776PT Stroe, Netherlands

Van Horn, Russell C., San Diego Zoo's Institute for Conservation Research, 15600 San Pasqual Valley Road, Escondido, CA 92027-7000, USA

Ventura, Raffaella, University of Abertay Dundee, School of Social and Health Sciences, Division of Psychology, 158 Marketgait, Dundee DD1 1NJ, UK

Warburton, Harriet J., National Centre for the Replacement, Refinement and Reduction of Animals in Research (NC3Rs), 20 Park Crescent, London W1B 1AL, UK

Warren-Smith, Amanda K., Millthorpe Equine Research Centre, 426 Forest Reefs Road, Millthorpe, NSW 2798, Australia

Warriss, Paul D., Animal Welfare and Behaviour Group, School of Clinical Veterinary Science, University of Bristol, Langford House, Langford BS40 5DU, UK

Watters, Jason V., Chicago Zoological Society – Brookfield Zoo, 3300 Golf Road, Brookfield, IL 60513, USA

Weary, Daniel M., Animal Welfare Program, Faculty of Land and Food Systems, The University of British Columbia, 2357 Main Mall, Vancouver V6T 1Z4, Canada

Weeks, Claire A., Animal Welfare and Behaviour Group, School of Clinical Veterinary Science, University of Bristol, Langford House, Langford BS40 5DU, UK

Wei, Cynthia A., School of Biological Sciences, University of Nebraska – Lincoln, Lincoln, NE 68588, USA

Whistance, Lindsay, Faculty of Agricultural Sciences, Department of Animal Health and Bioscience, University of Aarhus, Blichers Allé 20 – PO Box 50, DK-8830 Tjele, Denmark

Wildhaber, Mark L., US Geological Survey, Columbia Environmental Research Center, 4200 New Haven Road, Columbia, MO 65201, USA

Williams, Fiona, J., University of Lincoln, Department of Biological Sciences, Riseholme, Lincoln LN2 2LG, UK

Woods, Vanessa, Hominoid Psychology Research Group, Duke University, Durham, NC 27708, USA

Wright, Hannah Frances, University of Lincoln, Department of Biological Sciences, Riseholme, Lincoln LN2 2LG, UK

Würbel, Hanno, Justus-Liebig-Universität Gießen, Frankfurter Str. 104 – D-35392 Gießen, Germany

Zimmerman, Patrick H., Animal Welfare and Behaviour Group, School of Clinical Veterinary Science, University of Bristol, Langford House, Langford BS40 5DU, UK

Zulch, Helen, University of Lincoln, Department of Biological Sciences, Riseholme, Lincoln LN2 2LG, UK

Guide to Authors' Initials

AB	Butterworth, Andrew
ACA	Alberts, Allison C.
ACK	Kitchener, Andrew C.
AGF	Fahey, Alan G.
AK	Kerr, Annie
AM	McLean, Andrew N.
AMdP	de Passillé, Anne Marie
AMJ	Janczak, Andrew M.
AO	Olsson, Anna
AP	Plowman, Amy
APK	Klimley, A. Peter
AS	Smith, Andrew
AW-S	Warren-Smith, Amanda
BAF	Forkman, Björn A.
BAS	Schulte, Bruce A.
BC	Clucas, Barbara
BeH	Hothersall, Becky
BF	Fougere, Barbara
BG	Gilligan, Brenda
BH	Hare, Brian
BID	Damm, Birgitte I.
BJ	Jones, Bidda
BJM	Morgan, Bethan J.
BM	Majolo, Bonaventura
BMcG	McGonigle, Brendan
BT	Tolkamp, Bert
CAW	Wei, Cynthia A.
CB	Bonacic, Cristián
CBT	Tucker, Cassandra B.
CCB	Burn, Charlotte C.
CJCP	Phillips, Clive J.C.
CJN	Nicol, Christine J.
CLC	Corridan, Claire L.
CPL	Loomis, Caroline P.
CS	Sherwin, Chris
CW	Weeks, Claire A.
CWK	Kuhar, Christopher W.
DA	Arney, David
DBM	Morton, David B.
DC	Custance, Deborah
DCD	Deeming, D. Charles
DCL	Lay Jr, Donald C.
DGL	Lindburg, Donald G.
DL	Lamb, David
DLG	Garshelis, David L.
DMB	Broom, Donald M.
DMS	Shier, Debra M.
DSM	Mills, Daniel S.
DTB	Blumstein, Daniel T.
DW	Weary, Daniel M.
EB	Blackwell, Emily
ECB	Berg, Elena C.
EEB	Boydston, Erin E.
EF-J	Fernández-Juricic, Esteban
ELS	Schenck, Erin L.
EP	Paul, Elizabeth
EvdZ	van der Zee, Emile
FBB	Bercovitch, Fred B.
FdW	de Waal, Frans
FF	Flower, Frances C.
FJW	Williams, Fiona J.
FLR	Ruedisueli, Frank L.
FWN	Nicholas, Frank W.
GC	Caplen, Gina
GD	Dixon, Greg J.P.
GFG	Grether, Gregory F.
GL	Law, Graham
GMB	Burghardt, Gordon M.
GML	Landsberg, Gary M.
HEE	Eves, Heather E.
HFW	Wright, Hannah Frances
HJW	Warburton, Harriet J.
HM	Morris, Heather
HMB-S	Buchanan-Smith, Hannah M.
HO	Omed, Hussein
HR	Röcklinsberg, Helena
HSF	Fisher, Heidi S.
HW	Würbel, Hanno
H-WC	Cheng, Heng-Wei
HWG	Gonyou, Harold W.
HZ	Zulch, Helen
IK	Kyriazakis, Ilias
IRI	Inglis, Ian R.
JA	Aldenhoven, Jaclyn
JAS	Stamps, Judy A.
JB	Barnett, John
JC	Cooper, Jonathan
JC-B	Clutton-Brock, Juliet
JCN	Nordby, J. Cully
JCP	Petherick, J. Carol
JdB-T	delBarco-Trillo, Javier
JG	Gaughan, John
JH	Harris, Justin
JLE	Edgar, Joanne L.
JMS	Stookey, Joseph M.
JNM-F	Marchant-Forde, Jeremy N.
JP	Plotnik, Joshua
JR	Righetti, Joanne
JRH	Hardiman, Jessica R.
JRu	Rushen, Jeffrey

JVW	Watters, Jason V.
JWSB	Bradshaw, John W.S.
KAH	Houpt, Katherine Albro
KG	Guo, Kun
KL	Littin, Kate
KNM	Morgan, Kathleen N.
KP	Pullen, Kirsten
KS	Seksel, Kersti
KT	Taylor, Katy D.
KVT	Thompson, Katerina V.
LAN	Nordstrom, Lisa A.
LJ	Jefferson, Laura
LJR	Rogers, Lesley J.
LMD	Dixon, Laura M.
LRT	Tarou, Loraine R.
LW	Whistance, Lindsay
MAB	Bloomsmith, Mollie A.
MAO	Owen, Megan A.
MB	Bashaw, Meredith
MD	Dyson, Miranda
MH	Harris, Moira
MHF	Ferkin, Michael H.
MK	Kreger, Michael
MKP	Pines, Mathew K.
MLW	Wildhaber, Mark L.
MM	Malpass, Mark
MMcG	McGonigle, Margaret
MRC	Curry, Mark R.
MS	Spinka, Marek
MT	Toscano, Mike
MVD	van Dierendonck, Machteld C.
NB	Bailey, Natalie
NC	Cross, Nicki
NdV	de Vere, Natasha
NK	Kornell, Nate
NKD	Diehl, Nancy Kate
OHPB	Burman, Oliver H.P.
PAR	Robinson, Paul A.
PDM	McGreevy, Paul D.
PDW	Warriss, Paul D.
PE	Eady, Paul
PFB	Brain, Paul F.
PH	Hawkins, Penny
PHH	Hemsworth, Paul H.
PHZ	Zimmerman, Patrick H.
PJBS	Slater, Peter J.B.
PKM	McGregor, Peter K.
PoS	Statham, Poppy
PS	Sandøe, Peter
RA	Anthony, Raymond X.
RB	Bshary, Redouan
RBH	Harris, Richard B.
RCA	Anderson, Roland C.
RCB	Buckley, Ralf
RCT	Tinsley, Richard C.
RCVH	Van Horn, Russell C.
RD	Dennis, Rachel
RDK	Kirkden, Richard D.
RF	Freire, Rafael
RJNM	Merrill, Ralph J.N.
RMAP	Parker, Richard M.A.
RM-F	Marchant-Forde, Ruth
RoD	Dielenberg, Robert
RP	Poletto, Rosangela
RS	Swaisgood, Ronald
RV	Ventura, Raffaella
RWE	Elwood, Robert W.
RWVB	VanBuskirk, Richard W.
SDEH	Held, Suzanne
SE	Eicher, Susan
SH	Haslam, Sue M.
SJGH	Hall, Stephen J.G.
SJI	Insley, Stephen J.
SL	Ligout, Séverine
SM	Mullan, Siobhan
SMP	Pellis, Sergio M.
SR	Redgate, Sarah
SuH	Higgs, Suzanne
TDK	De Keuster, Tiny
TG	Grandin, Temple
TJO	Ord, Terry J.
TK	Knowles, Toby
TL	Luddem, S. Tabitha
TWS	Schoener, Thomas W.
VCP	Pellis, Vivien C.
VH	Horner, Victoria
VK	Kanaan, Vanessa
VM	Melfi, Vicky
VW	Woods, Vanessa
WS	Saltzman, Wendy

Preface

The importance of an appreciation of animal behaviour science and welfare is increasingly being recognized in a range of disciplines. Greater scientific attention to these subject areas over the last 50 years has increased both the quality of the research being undertaken and its application. While animal behaviour science has, for a long time, been seen as integral to the study of animal welfare, it is perhaps worth emphasizing that this text is not just for those with this interest. The principles of, and concepts explored within, animal behaviour are of relevance to those working in many other pure and applied fields such as the psychological sciences and conservation biology. Similarly, the importance of considering animal welfare by all those whose work might have an impact on other species is being more widely encouraged; it is no longer the domain of those concerned primarily with animals in captivity, such as in farms, laboratories, zoos or homes, but is also a concern for those who work to manage wild communities, be they feral, pest or under human protection for their conservation.

This concern is not simply a peripheral matter or distraction, but has a moral dimension too, and so it is important that we reflect and are guided by ethicists in our decisions when it comes to our interactions with other species. It is with this in mind that this text was developed. This publication has been an enormous undertaking, and would not have been possible without an excellent team of editors (Jeremy Marchant-Forde, Paul McGreevy, David Morton, Christine Nicol, Clive Phillips, Peter Sandøe and Ron Swaisgood) to whom I am particularly grateful, and the large number of academics who have so willingly given their time and shared their knowledge. I am struck by the generosity of so many in helping to ensure that this project became a reality and I believe this is largely because of a common passion for our disciplines and care for, and about, animals. I have learned much and it has been a genuine honour for me to work with so many distinguished scientists in compiling this text.

At this point I must acknowledge that any limitations to the text are probably mine (and at times difficult decisions have had to be made about the extent or format of some entries in the final run-through), and all credit must go to the many individuals who have contributed. Not only are the contributors outstanding scientists in their own right, but they have also worked hard to become clear and careful communicators of their subject. This is not always an easy task. The field is rapidly changing and, while new publications are constantly appearing, the objective of this text is to produce a reliable and thought-provoking point of reference for those working both directly and indirectly in the full spectrum of relevant fields. This includes higher-level students as well as established academics (especially those diversifying their interests as outlined above) and those working practically in the field. This text is for you.

Science is concerned with reducing uncertainty and, in the applied zoological sciences, we must often make inferences, whether concerning the function of a given behaviour or its psychological basis, because we can never know the mind of another. This has been a particular challenge for those working in the field of animal welfare for some time, but it is also faced by those working in a range of rapidly developing pure and applied fields such as comparative psychology and clinical animal behaviour. None the less, with careful study and experimentation, we have gone a long way towards helping to reduce this uncertainty, as explained in many of the entries in this text.

Opinions inevitably differ on many matters of judgement and, while one function of this text is to try to increase the consistency of use of terminology that can give rise to confusion (the first entry in this encyclopedia on 'abnormal behaviour' is a good example of this), we do not seek to preach a particular opinion. Thus, in addition to trying to clarify key concepts, issues for consideration are also highlighted in many of the entries. I am sure some may hope for more information or perhaps a greater emphasis on some aspects, and to this end we have tried to suggest further reading where appropriate. As already mentioned, this work is a collection derived from the efforts of many and, I very much hope, that you as the reader will share in this process. If, as you read it, you have comments then you can help us by contacting either me or the publishers so that we can consider these points in any future revision. Drawing boundaries has sometimes been a difficult but necessary editorial decision and additional contributors are always appreciated; I am especially grateful to those authors who have stepped in quite late to fill gaps that became apparent as we edited the entries and recognized additional cross-references we wanted to cover. Some I know personally, others I do not, but I will happily buy you a drink should we meet!

In order to try to limit the length of the text, there is substantial cross-referencing throughout, and I suspect that if you started with almost any of the substantive entries and then committed to read all the associated cross-references, you might well end up reading the whole book before you are finished! Another feature, aimed at trying to limit the length of the volume, is to consider some topics within related entries. For this reason, it is important that the reader uses the index when seeking information, rather than depending on being able to turn to the relevant letter in the main part of the text. The index, too, is cross-referenced.

This and many of the other production tasks have been undertaken by a dedicated team at CABI to whom I am also very grateful, especially Rachel Cutts, Quentin Scott, Tracy Head, Claire Parfitt and Rebecca Stubbs. I must thank my family, my wife Connie and my two young sons Stefan and Tomas. This project was conceived before Tomas was born, and I know they have themselves given up much to allow it to come to fruition. They have supported me throughout and put up with me locking myself away when I am sure they would have preferred we were together. I am forever grateful to you, and hope you appreciate that you too have played your part in producing something of a unique text, which is aimed at helping those who want to deepen their understanding of the science and philosophy of applied animal behaviour and welfare, for the betterment of others. One of the joys of educating others is that you never know what others may achieve with the knowledge you share, and so I would also like to thank all my tutors and colleagues who have helped me to become the person I am, and especially Sir Robert Hinde, whose guest lecture in Bristol on the interdependence of the behaviour sciences and subsequent correspondence over 20 years ago, convinced me of the importance of a text such as this.

In this vein I would like to conclude by thanking you, the readership. This text is dedicated to you, in the hope that you may be inspired and share our enthusiasm for science with others.

Daniel Mills
September 2009

Foreword

Every waking moment billions of photons stream into our eyes, our ear drums resonate with arrays of complex fluctuations in air pressure and data-streams from throughout our bodies pour through sensory nerves into our brains. Luckily for us, we (our conscious selves) do not have to deal with this ceaseless gushing of raw information as it sprays in as if from the flailing end of some severed broadband fire-hose. Within milliseconds, the tireless crews of our subconscious processing departments have it all neatly reviewed, sifted, assimilated, tidied up and summarized. And, aware of our limited capacities, they politely present the key points to us, colour coded for simplicity wherever possible, so that we immediately understand, for example, that '*there's a wasp on my apple*'. (For a fuller, proper, and very neat, account of conciousness, see within).

Progress in any subject is dependent on a similar process: information gathering followed by identification and elucidation of key patterns, higher-level conclusions and principles. Interest in animal behaviour – and all the other disciplines that inform assessments of animal welfare and practical measures to be taken to ensure or improve it – has increased dramatically in recent decades. This has resulted in a correspondingly huge proliferation of information, conclusions and opinions. Review, sifting, assimilation and summary are essential to facilitate understanding of developments across this large and varied field to make it widely accessible and to provide a basis for future progress.

It is clear from the pre-publication drafts that an excellent job has been done of distilling large amounts of complex information into easily assimilated packages and I look forward to reading the entries and essays that I have not yet seen. The contributions cover a wide field and are very informative, interesting and well and engagingly written. In producing this book, Daniel Mills, his editorial team and the contributors (numbering almost 200) have managed to overcome the daunting challenge posed by the scale and complexity of the task (there are some 1000 entries) and, in so doing, have provided a very valuable service for everyone interested in the field and for the future development of it.

James K. Kirkwood
UFAW
November 2009

A

Abandoned animal

An abandoned animal is any animal whose previous care by another has been relinquished. Animals abandoned by humans as opposed to their parents (**see: Mis-mothering**, for example) may be given over to the care of another, as occurs in the case of surrender to an **animal shelter** (**see also: Animal rescue**), or released into the wider general environment. In either case, the previous carer may be distressed by the decision to sever their relationship. Some relinquishers may consider the release (abandonment) of their animals into the general environment as a more humane option ('giving the animal a chance to live naturally'), even though legislation that prohibits this may exist, while others may consider this type of action irresponsible and seek another to take responsibility of the animal.

Studies throughout a range of countries suggest that behavioural problems are the most common reason cited for the surrender of animals to animal shelters in the industrialized world, even though it has been suggested that these may be under-reported by relinquishing owners in an attempt not to reduce the chances of the animal being rehomed. The most common reported reasons given for relinquishment for cats and dogs appear to be aggression, house-soiling, failure to integrate with other **pets**, destructiveness, hyperactivity and undesirable vocalization, most of which represent normal behaviours and reflect inadequate provision for the animal's **needs**.

Behavioural problems may also underlie other common reasons for surrender such as 'moving home', 'unanticipated demands' and 'changes in lifestyle'. In the case of horses, few data exist relating to the reasons for their surrender to shelters, but reasons tend to relate to owner circumstances (especially financial demands) and some sort of behavioural problem with the horse; a considerable number are released or abandoned on to open land, although requirements for mandatory registration may reduce this. Specific but widespread circumstances may result in bursts of abandonment for other reasons, for example, following natural disaster, such as the aftermath of Hurricane Katrina in 2005, which resulted in widespread animal abandonment as some rescue services would not take animals. Indeed there are some reports of people dying as they refused to evacuate without their pet. Other circumstances leading to an increase include: (i) media coverage over actual or potential zoonotic diseases such as rabies or avian influenza, resulting in relinquishment of dogs and parrots, respectively; and (ii) legislative changes in relation to the ownership of particular animals, e.g. the introduction of owner licensing fees or more stringent ownership requirements in relation to so-called 'dangerous' species or breeds. Abandonment is a serious cause for concern relating to the **welfare** of **companion animals**; for example, in the USA it is estimated that approximately one-quarter of dogs pass through a shelter at some time and that just under half of these are surrendered directly by their owner. The rest are obtained as strays, with around 15% reclaimed. It is estimated that about 60% of shelter dogs are euthanized in the USA (approximately 2 million per year). Long-term maintenance within a shelter is an obvious cause for concern but, even for those that are successfully rehomed, the experience of passing through a shelter is inevitably stressful and in some cases may result in long-term behavioural and emotional disturbance.

The abandonment of animals does not relate just to the relinquishment of care towards companion and domestic species, but may also relate to wild animals that come under the care of people to a greater or lesser degree and for variable periods of time. This includes the abandonment of animals within a wild **animal sanctuary** or managed reserve (perhaps as a result of local conflict or lack of funds), which may result in a resurgence of poaching, as has occurred in the case of mountain gorillas in Rwanda. The abandonment of rabbits and edible snails, neither of which is originally native to the UK, may be considered as the cause of their naturalization within this nation. They were originally brought to the UK by invaders (Normans and Romans, respectively), who used them as a convenient food source (**see also: Exotic species invasions**). Also pest species, which are caught in so-called 'humane traps' that do not kill the animal concerned, could be considered as coming under the temporary care of a human who becomes responsible (physically and possibly morally) for the consequences for this animal. The release of this animal 'into the wild' may then be considered an act of abandonment, especially since the domestic setting may be a more appropriate environment for this species than a field with predators, etc. (**see also: Pest control – ethics of; Pest species – welfare of**).

Many of the common medium- to large-sized companion animal species (i.e. cats, dogs, horses) that are abandoned can survive and may breed successfully if given the opportunity to live outside captivity, as long as they do not have traits dependent on human intervention (e.g. coats requiring extensive grooming, assistance with giving birth due to anatomical extremes, etc.), since they are tolerant of a wide range of environments. The abandonment of animals is therefore one reason why an animal may become **feral**. Smaller animals and ectothermic species may face problems, especially with **thermoregulation** at cooler times of year in regions with markedly different (wetter and/or cooler) weather patterns; and even abandoned indigenous species (e.g. mice in the UK) may fail to survive and be subject to a higher risk of predation, due to both physical (e.g. coat markings that inhibit camouflage) and behavioural (e.g. poor **predator avoidance** behaviour) characteristics.

Welfare problems also arise when animals are abandoned but confined, and so given less opportunity to adapt (**see also: Confinement**). Such confinement may be by means of some kind of tether or enclosure, such as a fence or the walls of a building, and typically will result in starvation or dehydration.

In many countries legislation in relation to abandoned animals exists. This legislation may relate directly to control over the act of abandonment or the control of animals who have escaped or been abandoned into the general environment (stray animals). The former emphasizes the responsibility of citizens to animals under their care, while the latter is often aimed at protecting citizens from the harm that may be caused by them (e.g. injury from a bite or road traffic accident). In some countries/regions abandoned animals are taken into care, under the responsibility of the police or a regional authority, for a short period before being euthanized if an owner cannot be found, but in others **euthanasia** may not be allowed and the animals must be cared for in the long term. This can be problematic because available space may become limited and the welfare of animals kept in such environments may in itself be seriously compromised. A key component to the enforcement of legislation relating to the abandonment of animals is the identification of the legal carer, and this too can be problematic without a requirement for legal identification of an animal and its ownership.

While abandonment is often considered a permanent or long-lasting act, the term 'abandoned animal' is also used occasionally in relation to a temporary absence of responsibility, e.g. the 'abandonment' of animals in cars during hot weather. (DSM)

Abnormal/abnormality

The concept of normality and abnormality is widely used in reference to animal behaviour and **welfare**; for example, in the **five freedoms** and the 'Codes of Recommendations for the Welfare of Livestock' (**see: Codes of practice**) it is stated that domestic animals should have the opportunity to perform most normal patterns of behaviour. However, it is less clear what this means and what is acceptable in practice.

While behaviours vary according to context, animals also have species-specific behaviour patterns that have an element of consistency, and abnormal behaviour may be considered that which deviates from this standard. This poses the problem of recognizing the standard being used, the behavioural measure (pattern of behaviours, individual form, intensity, functionality, etc.), its level of deviation, as well as its context must be defined. In many instances the term abnormal is simply used to describe behaviour that is rare (literally away from the statistical norm), but in these circumstances it must be recognized that there is not necessarily a link between the abnormality of a behaviour and animal suffering. In other instances the functionality of the behaviour may be used as the reference for normality, but in this case the term 'functional' is to be preferred to 'normal', as it is less likely to infer a link with an animal's welfare. For example, stereotypic behaviour (**see: Stereotypies**) might be considered abnormal because it is generally not seen in wild, free-roaming animals. However, if polar bears in captivity are considered as the reference point, then when using a statistical approach it might be argued that it is not abnormal because it occurs quite frequently in this context.

In psychology, the concept of 'social deviance' is sometimes applied to assess the abnormality of a behaviour. This involves assessing the impression of other people of the behaviour and the impact that behaviour might have on them. One major problem with this approach is that it depends on the subjective opinion of the person using the term. The philosopher Wittgenstein has argued that many concepts have a similar problem of definition. Things that belong within a given category have certain resemblances without necessarily sharing the same features, like members of a family. All of the underlying defining features that make up the concept of abnormality need to be identified, and then simple assumptions about the link between the concept and its implications questioned. Seligman *et al.* (2000) propose seven criteria that might be used when referring to something as abnormal. These may be adapted to the field of animal behaviour thus, but the different usages have different implications for assessing the welfare of the performer:

1. **Suffering.** Behaviours related to suffering may be divided into those that cause harm to the performer or another, those that are associated with an attempt to adapt or cope with a suboptimal environment and those associated with an inadequate or noxious state, e.g. signs of frustration.
2. **Maladaptiveness.** This refers to the failure of a behaviour to fulfil its goal at either an **appetitive** or **consummatory** level, e.g. a failure to achieve **homeostasis**. Maladaptiveness may also refer to suboptimal behaviour, e.g. the ingestion of a non-nutritive foodstuff.
3. Unconventionality. This refers to the statistical rareness of a behaviour in a given context (see definition above).
4. Unpredictability. It is often implied that behaviours that cannot be predicted are abnormal because there is no recognized control of the behaviour. However, spontaneous behaviour is a rare occurrence and more often is a reflection of the knowledge of the reporter of the behaviour. Once the occurrence of a behaviour can be explained, it may cease to appear abnormal. For example, an owner who cannot predict their dog's aggression may describe it as abnormal, although a clear stimulus may be discerned by a therapist.
5. Incomprehensibility. Similarly, if the nature of a behaviour cannot be understood it may be described as abnormal until it is explained.
6. Observer discomfort. If a behaviour causes concern to its observer for any reason, although the reason for this should be given and justified, then it may also be described as abnormal on this basis. For example, some people may consider homosexual behaviour offensive and so describe the behaviour as abnormal.
7. Violation of standards. Standards may be set which, if they are not met, may classify a behaviour as abnormal. For example, there may be an expectation for a given behaviour not to occur, such as aggression in horses towards people, so its occurrence is abnormal by definition, even if it can be understood.

Many of these criteria are subjective, but still they encapsulate the essence of how the term 'abnormality' may be used. Any behaviour described as abnormal may meet one or more of these criteria in a given context. It is important that any link to welfare is demonstrated logically and not inferred by virtue of the use of the term 'abnormality' of the behaviour alone. Just because a behaviour fulfils one of these criteria, it does not infer that it fulfils any other, i.e. abnormality is not necessarily a welfare problem. (DSM)

Reference

Seligman, M.E.P., Walker, E.F. and Rosenhan, D.L. (2000) *Abnormal Psychology*, 4th edn. W.W. Norton & Co., Inc., New York.

Acetylcholine

Acetylcholine (ACh) is a neurotransmitter that occurs widely in animals, being found throughout the somatic nervous system, but less ubiquitously within the **autonomic nervous system**. Acetylcholine is stored in vesicles within nerve endings and then released through exocytosis in response to an action potential. It is cleared from the synapse largely as a result of the enzyme acetylcholinestase. Smooth and skeletal muscle respond to ACh by contracting, while cardiac muscle will relax. Acetylcholine is released by the terminal nerves of the **parasympathetic nervous system**, while **epinephrine** is released by the terminal nerves of the **sympathetic nervous system**. It is therefore associated with derousal. Centrally, ACh release facilitates both learning and short-term memory at the cellular level, and it is involved in both top-down and bottom-up **attention** processes; it is also involved in the scheduling of REM **sleep**. Two general types of ACh receptor are recognized, muscarinic and nicotinic, on the basis of other chemicals that also stimulate these receptors (muscarine and nicotine). (DSM)

ACTH (adrenocorticotropic hormone)

ACTH is the abbreviation for adrenocorticotropic hormone. ACTH is a 39-amino-acid peptide hormone produced in the anterior **pituitary gland**. Production of this hormone is stimulated by corticotropin-releasing factor (CRF) which travels from the **hypothalamus** through the blood to bind to receptors on corticotropic cells of the anterior pituitary. Activation of these receptors then causes the production of **pro-opio-melanocortin**, referred to as POMC. Enzymatic cleavage of POMC in the anterior pituitary gland forms several hormones, including **beta-endorphin**, melanocyte-stimulating hormone and ACTH.

ACTH is released into the peripheral vasculature, which allows it to travel to the cortex of the adrenal gland to bind to receptors, causing the production of **glucocorticoids**. The predominant glucocorticoid in cattle, swine and sheep is cortisol, whereas in poultry and rodents the predominant glucocorticoid is corticosterone. The release of glucocorticoids causes the liver to produce glucose (gluconeogenesis).

ACTH production and excretion is controlled by the negative feedback effect of glucocorticoids on the anterior pituitary gland to directly decrease ACTH synthesis and excretion. Glucocorticoids also have a negative feedback effect on the hypothalamus to decrease CRF synthesis and excretion.

Because ACTH plays a dominant role in increasing energy availability in the form of glucose, it has been considered of major interest in terms of how animals cope with and respond to environmental challenges and **stress**. Stress research has long focused on the concept of the 'fight or flight' response which requires that the animal has rapid availability of energy (glucose) to mount a response to a threat. Therefore, researchers have used the measurement of ACTH and its associated activation path to measure a response to stress. One such test is the ACTH challenge test (ACTH stimulation test), which involves injecting ACTH to stimulate the adrenal gland to release glucocorticoid. Researchers can then measure the circulating concentrations. The reason that this test may show when an animal is stressed is because under chronic stress, the adrenal gland of animals enlarges due to hyperplasia and hypertrophy. This enlargement enables the animal to produce more glucocorticoids. When given ACTH the circulating concentrations of a chronically stressed animal should be greater than an animal that is not stressed. Thus, circulating glucocorticoids can potentially give us a quantitative measure of the level of stress an animal is experiencing, although there are limitations to its interpretation and usefulness. (DCL)

See also: Corticosteroids

Further reading

Charmandari, E., Tsigos, C. and Crousos, G. (2005) Endocrinology of the stress response. *Annual Review of Physiology* 67, 259–284.

Adaptability

Adaptability describes the extent to which the regulatory systems of an animal can facilitate physiological or behavioural responses that allow it to deal with changes in its environment. The adaptability of an individual depends not only on its health, age and status but also on the type and degree of that environmental change. (KT)

See also: Adaptive; Coping; General adaptation syndrome; Homeostasis

Adaptive

Behavioural and physiological traits that have been preserved by **selection**, because they offer a **fitness** advantage to individuals that carry the genes for their expression, may be said to be adaptive. This includes traits that increase the individual's ability to reproduce – for example, a successful mating strategy or attractiveness to the opposite sex. It also includes traits that increase the individual's ability to survive – for example, traits that enable an individual to find food or cope with changes to its environment.

Adaptive behaviours are not just those that have been selected for genetically during the evolutionary history of the species because they confer a fitness advantage to the individual in the environment in which the species evolved; behaviours may also be transmitted between generations as a result of cultural transmission (**see: Social learning**), for example, the deliberate action of moving into water or higher pastures by feral horses to escape airborne parasites. With time such traits may acquire a genetic basis.

The adaptive value is the increase in fitness that a physiological or behavioural trait confers on an individual, relative to the cost to the individual of having this trait. If a trait confers a large advantage on the individual in terms of survival or reproduction and at little cost in terms of energy or risk of reduced fitness, then that trait can be said to have a high adaptive value. Traits with high adaptive value are more likely to be preserved within the gene pool.

Adaptive intelligence is the ability to solve problems as a consequence of previous experience with similar problems. If a problem is encountered frequently enough the individual has more opportunity to learn the best solution and apply this to other, similar, situations. This ability to learn is an adaptive response because it helps maximize the animal's fitness. (KT)

See also: Coping; Evolution; Fitness; Maladaptive; Selection

Further reading

Keiper, R.R. and Berger, J. (1982) Refuge-seeking and pest avoidance by feral horses in desert and island environments. *Applied Animal Ethology* 9, 111–120.

Adoption

The young of many species depend upon the care provided by one or both parents to survive. In some species, individuals other than the genetic parents provide a portion of care, a phenomenon known as **alloparenting**. Adoption is an extreme form of alloparenting, in which an individual other than the genetic parent provides all (or nearly all) care for the young.

Adoption in nature is generally rare, although it has been reported anecdotally in hundreds of avian and mammalian species. The evolution of any behaviour depends upon how that behaviour ultimately affects an individual's ability to survive and reproduce (direct fitness) or have a positive impact on the survival and reproduction of its relatives (indirect fitness). In order for a behaviour to become widespread in a population, the fitness benefits of performing that behaviour must exceed its costs (**see also: Optimality**). Adoption involves many potential fitness costs to the adoptive parent, such as: (i) energy; (ii) increased exposure to predators; (iii) loss of the opportunity for producing its own young or decreased ability to care for existing descendent young; and (iv) the fitness benefits are usually minimal unless the adopted individual is a relative.

Usually, adoption appears to be the result of misdirected **parental behaviour**. While it might be theoretically possible for caregivers to discriminate fairly accurately between descendent and non-descendent young, any such discrimination mechanism would carry a risk that a parent would mistakenly deny care to its own offspring. In an evolutionary context, being responsive to offspring and exhibiting proper caregiving behaviours are so important that they often outweigh the cost of providing care to non-descendent young.

The value of a precise discrimination mechanism increases with the increasing likelihood that a caregiver will encounter non-descendent young. Thus colonial, synchronously breeding species with precocial (and therefore independently mobile) young tend to have more highly developed discrimination mechanisms than those that breed in isolation. In practice, rather than showing a precise ability to discriminate, most species use 'rules of thumb' in providing parental care. These rules of thumb are often based on spatial criteria ('provide care to any young in the nest') or familiarity ('provide care to the young that were encountered immediately after parturition').

One type of adoption, termed brood parasitism, has evolved many times in nature. In brood-parasitic species such as wood ducks, mothers provide care to any young that are in the nest and have limited ability to discriminate between individual young. This type of rearing system is vulnerable to exploitation by conspecific females that deposit eggs surreptitiously in the nest, leaving the nest owner to raise the hatchlings. Another form of adoption occurs in several colonial bird species (e.g. emperor penguins, kittiwakes, ring-billed gulls, Alpine swifts). In these species, the precocial young themselves move between nests, abandoning their natal nest to join one that is less infested with parasites or that contains younger, smaller nestlings that they can dominate.

In mammals, adoptions are very rare but it is possible to make some generalizations based upon anecdotes. Two factors seem to precipitate adoption events. First, orphaned or abandoned infants can be very persistent in soliciting care from other females that they encounter. Secondly, females that have given birth to stillborn young or whose young die shortly after birth will still experience the powerful surge of hormones that triggers the onset of parental care and facilitates parent–offspring **bonding**. This can cause them to behave maternally towards non-descendent young and, in some cases, leads to the adoption of an orphaned or abandoned infant. Most cases of adoption involve females that are already lactating, but the sensory stimulation associated with suckling can, in some species, induce lactation in a female that has not recently given birth. The highest likelihood of adoption is expected in species where females live in kin groups, because adoptive mothers would increase their inclusive fitness by rearing an orphaned relative.

While adoption is clearly not an artefact of **captivity**, it does appear to be more common in captive settings than in nature. This may simply be because reproduction and infant rearing are more closely observed in captivity. Additionally, specific aspects of the captive environment, such as enforced spatial proximity or enhanced reproductive synchrony, may facilitate adoption. Captive animals are generally more likely to show aberrant parental behaviours, which increases the chance that an infant will be abandoned or neglected by its genetic mother.

The existence of adoption in nature can have useful implications for animals in laboratories, **zoos** or other captive settings. The lack of a precise ability to discriminate between descendent and non-descendent young can enable the **fostering** of young between different females for reasons of experimental design or to equalize brood or litter sizes. Captive managers may also enhance the survival of offspring born to individuals with infanticidal tendencies, by removing them from their parents and fostering them to individuals with proven parenting skills. Additionally, it may be possible to increase reproductive rates in genetically valuable individuals by fostering their offspring to be raised by less genetically valuable individuals, thereby allowing the valuable individuals to re-breed sooner. It is important to recognize that among species (and among individuals within a given species) the degree of tolerance to non-descendent young can vary along a continuum from highly tolerant to infanticidal. For those that show tolerance of non-descendent young, adoption provides another option besides **hand-rearing** for orphaned or abandoned young. To capitalize on this opportunity, adopted young must be closely monitored to ensure that the adoptive mother is providing adequate care to both the adopted infant and any current dependent young. (KVT)

See also: Captive breeding programmes; Kinship; Recognition among animals; Sociobiology

Further reading

Reidman, M. (1982) The evolution of alloparental care and adoption in mammals and birds. *Quarterly Review of Biology* 57, 405–435.

Adrenal gland

The adrenal glands are a set of paired organs located beside the kidneys. They are comprised of an outer portion – the cortex – and an inner portion – the medulla. The adrenal cortex is comprised of three layers and produces **glucocorticoids** and mineralocorticoids. The adrenal medulla is a modified ganglion and, upon stimulation of the **sympathetic nervous system**, produces **epinephrine** (adrenaline) and **norepinephrine** (noradrenaline). (DCL)

See also: Corticosteroids; Stress; Stress hormones

Aetiology

Derived from the Greek word *aitios*, meaning cause, origin or reason, aetiology is the study of causation. As applied to animals, aetiology refers to the cause of a disease, disorder or behaviour. If the aetiology of a disease or disorder is unknown, the causal factor(s) have yet to be identified and may therefore be called idiopathic. (KT)

See also: Causal factor

Affiliation

The evaluation of affiliative behaviours has underpinned a large proportion of the research into the social complexity of many social species, particularly non-human primates. Affiliative behaviours are often termed 'positive interactions' and, within the primates, can include **grooming**, physical contact and infant care. Social grooming is one of the most widely used measures of affiliative interactions in the studies of social relationships.

However, other affiliative behaviours are exhibited by social species, e.g. the spotted hyena (*Crocuta crocuta*) demonstrates a number of behaviours that can be termed affiliative, including mother–infant attachment and social play. In addition, hyenas demonstrate ritualized greeting ceremonies as a means of advertising and reaffirming social relationships. As hyenas have a **fission–fusion society**, individual clan members may spend significantly long periods of time in isolation. Greeting ceremonies facilitate the constant transition from solitary to social status. Ritualized greeting behaviours usually occur at the communal den after a period of separation, although these behaviours have also been seen in other social circumstances. Animals of either sex will stand head to tail with legs lifted in reciprocal presentation of the genitalia and anal scent glands. Virilization of the female genitalia in the spotted hyena allows both sexes to engage in erect phallus display. In accordance with a strict social etiquette, subordinates will always lift their legs first.

Affiliative behaviours can apply to a variety of social models, including **reciprocity**, altruistic behaviours, reconciliation and consolation, and can be affected by kinship, dominance hierarchies, familiarity, etc.

Reciprocity

In many mammal species affiliative behaviour is directed towards particular classes of individuals. Although developmental research has shown that many preferences for partners in affiliative behaviour will emerge during infancy, these preferences may play an important role in determining group affiliation patterns at a later stage in life. Age, gender, kinship and social status can play important roles in the development of these group affiliations. Changes in female reproductive status have also been seen to override and alter some of the affiliative bonds in non-human primates.

In non-human primates kinship is recognized as a significant determinant of affiliative relationships, particularly in levels of grooming behaviour and proximity. In many female **philopatric** species female grooming is significantly more kin-oriented than male grooming. This kin bias for affiliative behaviours can assist in the establishment of **matrilines** (particularly in macaques, *Macaca* spp.). Matrilines can have a strong influence on group behaviour, including feeding patterns and group travel. Females with a strong matriline can rely on support from their kin in social contests and will often hold a dominant position within the group. Matrilines can also have a strong impact on the hierarchy positions of individual animals. Several species of Old World monkeys have dominance hierarchies in which individuals within a matriline occupy adjacent ranks, and their offspring often acquire a rank just below that of their mothers. This close mother–infant bond can then be maintained (and reaffirmed through affiliative behaviours) into later life stages.

Dominance hierarchies describe relationship patterns between animals that engage in repeated social encounters, particularly where status is not necessarily predictable from physical characteristics. Social rank and dominance hierarchies can be seen to interact strongly with kinship in the distribution of affiliative behaviour. High-ranking individuals generally groom at a far lower rate and receive grooming at an increased rate in comparison with lower-ranking individuals. Grooming may be most commonly seen between closely ranked and closely related females.

The distribution of grooming within a primate group can also be strongly influenced by the reproductive status of the females within that group. Seasonal breeders will become the focus of increased grooming behaviour from males (i.e. will receive higher levels of grooming) in the breeding season. Even in non-seasonal breeders the oestrous cycle and the formation of consortships can lead to higher levels of grooming behaviour. In this situation it has been suggested that increased grooming behaviour reduces the risk of **aggression**, allows the increased proximity necessary for mating to occur and can temporarily impact on social status within groups. This has been observed in hamadryas baboon harems, where females in oestrus give higher levels of grooming directed towards the alpha male while receiving higher levels from the male, and have been observed to 'leapfrog' positions within the female hierarchy.

Altruism

Social grooming can be seen to fit the definition of altruistic behaviour, as it appears to incur greater costs than benefits to the recipient. The costs involved in a grooming bout include expenditure of time and energy and the reduction of time available to engage in other necessary activities (i.e. feeding, vigilance behaviours). However, the benefits of grooming can be more difficult to define. In some cases it may appear that groomers may derive protection from the groomed animal, or it could be considered to be a reinforcement of dominance ranks between individuals. It seems likely that the benefit received by the groomer is social tolerance, so that a low-ranking animal grooming a higher-ranking animal is able to maintain proximity and thereby some level of protection or sharing of resources. However, there is a lack of easily interpretable data on a casual relationship between affiliative behaviours (such as grooming) and agonistic support, partly because these interactions often occur at a low rate and can be separated by long time intervals. Tolerance may not be the only benefit to a grooming exchange. Evidence of grooming exchange for sexual favours, food and access to infants has been identified.

Although grooming tends to be heavily biased towards kin selection, grooming between non-kin is often thought to be maintained through grooming reciprocity, where the reciprocation can be either in the same currency (reciprocity) or in a different currency (interchange). Early studies indicated that grooming between non-kin in stump-tailed macaques was usually reciprocated once grooming had been initiated, although conflict over which member of the dyad would initiate grooming was not unusual. However, most individuals did appear to attempt to maximize grooming received while performing as little grooming as possible. In addition, grooming levels between familiar and non-familiar animals can be seen to vary greatly. Interactions between familiar cage-mates demonstrated a high level of affiliative behaviours, whereas non-familiar individuals tended not to reciprocate. These interactions were categorized by a high level of grooming solicitations that were not followed by grooming bouts. This suggests that grooming between non-kin follows strategies of reciprocation such as 'tit for tat', in which the cooperative first move is followed by the individual either continuing or defecting, depending on their partner's behaviour.

In social systems conflicts of interest can arise between individuals within the group, most commonly competition for access to resources (i.e. food, mates). In these situations behavioural interactions may escalate into aggression which, in turn, may compromise the benefits of group living or future cooperation between the opponents. Behavioural mechanisms that may work to prevent aggressive escalation can be strongly selected for in social animals; however, there has been very little research focused on this particular area to date. In comparison there is a growing body of research on behavioural mechanisms following escalated aggressive incidences, particularly the use of affiliative behaviours for both 'reconciliation' and 'consolation'.

Reconciliation

Initial research into this phenomenon with captive chimpanzees indicated that opponents in an aggressive conflict spent more time in close proximity and exchanging friendly behaviour than before the aggressive incident. This has been termed 'reconciliation' and can be viewed as a mechanism to repair social bonds that have potentially been damaged by aggression. The vast majority of research into reconciliation has been undertaken with primates in captivity, where it has been argued that limited enclosure sizes can create an artificial social environment. This has led to the suggestion that reconciliatory behaviour may in fact be an artefact of the captive environment. However, research carried out in large enclosures where prolonged avoidance of individuals is possible and research with habituated wild primate groups both suggest that reconciliation does occur outside of the captive environment, although possibly at lower frequencies.

In species that tend to show high conciliatory tendencies, reconciliation can involve behaviour patterns that can be termed affiliative but are rarely seen in other contexts (explicit reconciliation). In other species regularly observed affiliative behaviours such as grooming or physical contact can be seen to have a reconciliatory function; for example, in a number of species (hyenas, *Crocuta* spp. and baboons, *Papio* spp.), ritualized greeting ceremonies (such as that described above for hyenas) are thought also to serve a reconciliatory function in conflict management. In this case the affiliative behaviour may provide a means of repairing social relationships and achieving a balance with aggressive behaviour.

The recognition of reconciliatory behavioural patterns within a range of species has assisted with the classification of 'dominance styles' within species; **'despotic' species** (rhesus macaques, *Macaca mulatta*) are relatively aggressive, have steep power gradients and reconcile fewer conflicts. In comparison, 'tolerant' species (Tonkean macaques, *Macaca tonkeana*) typically display low-intensity aggression and high levels of affiliation and reconciliation. Only the tolerant species have been seen to exhibit explicit gestures for reconciliation, although this has also been seen to vary between groups of conspecifics.

Consolation

'Consolation' has been defined as any increase in affiliative interaction between targets and third parties specific to post-conflict contexts, although it is perhaps best reserved for those interactions initiated by third parties. This raises a further question as to whether these interactions may then have been triggered by **'empathy'** on the part of the third party. The major function proposed for consolation focuses on the alleviation of stress in a post-aggressive context and, potentially, protection from further aggression.

Factors influencing development of affiliative behaviour

Many developmental psychologists believe that the parent–infant relationship has a unique role in the socialization of the infant; however, an alternative view is that social development is mainly affected by the experiences within a peer group. Therefore, the acquisition and development of affiliative behaviours can be paramount to an individual's development. Research has indicated that individuals that are deprived of social interaction with peer groups may exhibit reduced social competence in adult life.

Research conducted with non-human primates has demonstrated clear gender differences in affiliative behaviours (particularly grooming and play), emerging early on in life. Female macaques will be both giving and receiving higher levels of grooming interactions by 2 to 3 years of age. Female–female grooming increases steadily throughout the juvenile period to adulthood, with greatest emphasis being placed on the grooming of kin. Juvenile females will groom both older females and younger individuals until the onset of menstrual activity, when the grooming of males will begin to increase. Conversely, when macaque males are 3 to 4 years old, their grooming activity is significantly lower than that of the females and is not kin-biased. The majority of grooming interactions are directed towards the adult males; however, it is rarely reciprocated. Male–female grooming will increase with the onset of puberty and emigration from the natal group.

Gender differences are also prevalent in the amount of **play** and the types of play exhibited by young non-human primates. In most primate species after the first 6 months of life males will engage in higher levels of 'rough-and-tumble play' than most females. Juvenile males will also both initiate and accept invitations to play far more frequently than females. Juvenile and sub-adult males also demonstrate higher levels of social play and of a type that is more vigorous. Females will engage in higher levels of solitary play but, in addition, will also begin to demonstrate play parenting within the first years of life.

Affiliative play behaviour can also be influenced by kin effects. Infants with closely related and similarly ranked mothers will often engage in higher levels of play than those whose mothers are unrelated or have a large gap in rank.

Spotted hyenas have also been observed to engage in social interactions that can develop into play patterns from the early stages of life (as early as the second stage of life in captivity). By approximately 1 month of age cubs engage in more frequent bouts of rough-and-tumble play and can therefore be said to have a system of affiliation. This facilitates introduction to peers and clan members on transfer from the natal to the communal den. It has been seen that dominant hyena cubs will initiate play bouts more frequently that subordinates. During play bouts a dominant animal is more tolerant of contact (being knocked to the ground, gently mauled) from a subordinate.

In some primate species (i.e. gorillas, *Gorilla* spp., black howler monkeys, *Alouatta pigra*) strong affiliative relationships have been seen between infants and adult males. The infants will spend more time interacting with or being closer to the males than their own mothers and the males will hold, cuddle, nuzzle or groom the infants. These infants will often turn to the males in times of distress. The males are often very tolerant of infant play and will allow youngsters to climb and tumble on their backs. Baboon males have been seen to 'baby-sit' infants when the mother is absent, sometimes grooming or carrying the infant. Protection for the infant appears to be an important benefit for these affiliative interactions, with the adult males often acting as a buffer between the female–infant dyad and threats from conspecifics. Tolerance and food sharing between males and infants has also been seen in species exhibiting these strong affiliative bonds. In addition, for mountain gorillas it appears that male infants that establish a strong bond with the silverback (the mature male) and maintain it through adolescence are more likely to inherit leadership of the group, whereas those male infants that do not establish such a strong bond are more likely to emigrate from the group. These strong male–infant bonds are often accompanied by strong affiliative bonds between the male and the infant's mother, although they can also be influenced by male rank, dyadic preferences and infant needs.

Although most aspects of affiliative behaviours within non-human primates have been documented as intra-group social interactions, there are cases of inter-group affiliative interactions (play, grooming, touching). The primary proponents of these interactions are young males, although this is far from exclusive of the other age-classes. However, there is a lack of comparable data of the frequency of these interactions to be able to comment on this. (KP)

See also: Aggression; Altruism; Chimpanzee; Cooperation; Intraspecific behaviour; Kin selection; Social behaviour; Sociobiology

Further reading

Campbell, C.J., Fuentes, A., MacKinnon, K.C., Panger, M. and Bearder, S.K. (2007) *Primates in Perspective*. Oxford University Press, Oxford, UK.

Drea, C.M. and Frank, L.C. (2003) The social complexity of spotted hyenas. In: de Waal, F.B.M. and Tyack, P. (eds) *Animal Social Complexity*. Harvard University Press, Cambridge, Massachussetts, pp. 121–148.

Roney, J.R. and Maestripieri, D. (2003) Social development and affiliation. In: Maestripieri, D. (ed.) *Primate Psychology*. Harvard University Press, Cambridge, Massachussetts, pp. 171–204.

Ageing

An animal's behaviour may change when it becomes aged due to: (i) nerve cell loss, reduced connections between nerve cells and other cellular changes in the **brain**; (ii) deteriorating perceptual function, as in the case of impaired vision and hearing; and (iii) impaired physiological changes separate from those occurring in the nervous system (e.g. arthritis, cardiovascular dysfunction).

Behavioural testing of dogs has shown that ageing slows learning, particularly spatial memory, and alters activity cycles, but there is so much variation between species and even strains/breeds that it is difficult to ascertain what might be considered as old age. Individual differences are also marked. Diet and life history can affect ageing. Aged rats housed in an enriched environment, allowing social interaction with other members of their species, have more connections between the nerve cells in their cerebral cortex than do rats housed in an impoverished environment in isolation. To a large extent our knowledge of ageing in animals is based on those species kept as pets and on a limited number of laboratory species used in research on ageing (rats and some species of primates). Aged dogs, 9 years old or more, show cell death (apoptosis) and senile plaques in the brain, and a behavioural test is available for scoring dementia in dogs: the degree of dementia correlates with the extent of apoptosis but is unrelated to the number of senile plaques. Many species of birds are very long-lived compared with mammals of equivalent weight.

According to one theory of ageing, the oxidative stress theory, which states that free radicals cause the degenerative changes in the brain, the longevity of birds is unexpected since they have higher metabolic rates than mammals and hence should be at greater risk of degenerative effects. This has led to the suggestion that birds may have a special protective mechanism against ageing. Other theories of ageing include: (i) the neuroendocrine theory, with new evidence for a role of the cell growth and survival factor, IGF-1; (ii) the evolutionary theory of ageing, which considers the accumulation of late-acting DNA mutations; and (iii) the theory of telomere shortening, which concerns cell division. (LJR)

Further reading

Erwin, J.M. and Hof, P.R. (eds) (2002) Interdisciplinary Topics in Gerontology. Volume 31, *Aging in Nonhuman Primates*. Karger, Basel, Switzerland.

Holmes, D.J. and Austad, S.N. (1995) Birds as animal models for the comparative biology of aging: a prospectus. *Journal of Gerontology: Biological Sciences* 50A, B59–B66.

Milgram, N.W. (2003) Cognitive experience and its effect on age-dependent cognitive decline in beagle dogs. *Neurochemical Research* 28, 1677–1682.

Aggression

Aggression is derived from the Latin word '*aggredi*' meaning 'to attack'. S.A. Barnett described aggression as 'a word with many meanings, and a source of much confusion', largely because it is used in very different contexts by biologists, psychologists, psychiatrists, sociologists and the media. The term has had an unfortunate tendency to have different meanings, even within the course of a single scientific article or review. Aggression is generally applied to the threat and attack components of **agonistic behaviour** in animals. In addition, it is actually a somewhat 'loaded' epithet, often implying that there is something excessive, inappropriate or undesirable about the severity and/or nature of the animal's responses. The term is especially problematic when applied to **companion animals**, where a behaviour may be welcomed in one context and disapproved in another, e.g. a dog barking at a stranger but not at the owner's partner. The basic problem is that aggression (because of its almost ubiquitous occurrence in animals) clearly serves biological functions but is often uniformly regarded as reprehensible in society (this is called **'negative apperception'** by some social scientists). Such activities clearly cause, however, problems in captive and domestic animals.

To qualify for the label 'aggression', it has been suggested an activity should:

1. Have at least the potential to cause injury or damage to another animal (often a member of the same species), with the focus generally being on physical injury rather than psychological damage. Consequently, elements such as biting, pecking, kicking, pushing, stinging and goring are usually stressed in descriptions of such behaviour in non-human animals.
2. Be deliberate or intentional (rather than accidental). It is difficult to establish this with any certainty in most cases.
3. Be **aversive** to the victim of attack or threat (such that the animal tries to avoid it). This is to eliminate some sexual displays in certain species from inclusion.
4. Be accompanied by arousal (evidence of internal activation in the form of increased heart rate, altered tissue distribution of blood, release of 'stress' hormones or elevated rates of breathing) in the aggressor (this would exclude many forms of predatory behaviour from the definition, as they often lack this characteristic). Most studies attempting to assess arousal do not directly measure internal physiological changes but rely on 'body language' such as piloerection and positions of the ears in cats and dogs, as well as tooth exposure in primates.

None of these characteristics, on its own, is sufficient for firm characterization of a response as being 'aggression'. Conversely, there are many examples of activities with one of the above attributes that clearly do not constitute aggression in most schema.

The entire range of available animal senses may be employed in expressions of aggression. Visual examples include the wide use of characteristic postures, but also other bodily features, such as: (i) the red abdomen of the territorial male three-spined stickleback (such seasonal changes are not limited to this fish); (ii) the comb and wattles of the domestic chicken (their size and brightness reflect the animal's social status); (iii) the black patches on the chests of female golden hamsters (the larger, the more dominant the animal); and (iv) the antler displays of red deer (these annually produced structures increase in size and complexity with maturity).

Examples involving **olfaction** include the use of **urine** and faeces in territorial marking in dogs and cats (this can be regarded as a 'problem' behaviour), but also in many rodent and ungulate species (this is one reason why attention should be paid to the cleaning of cages and enclosures).

Examples involving audition (i.e. the production of particular sounds) include: (i) croaking by frogs and toads (pitch is often an indication of the animal's size); (ii) rattling by the rattlesnake; (iii) 'roaring' in rutting red deer stags; and (iv) a range of bird and primate calls. Certain species use sounds at frequencies below (infrasound) or above (ultrasound) human hearing. Other species may use other senses such as touch (including pressure waves in water by fish) and even electromagnetic communication (e.g. in some electric fish). Clearly, understanding aggressive displays implies also understanding the sensory systems employed by particular species. Most species use a range of sensory communication in their displays.

The above types of **communication** may enable the animal to succeed using threat alone, as overt fighting can be damaging (see later) and generally only occurs *in extremis* or if individuals are equally matched in competitive ability. Even if it does not result in injury, aggression is very energetically demanding. The constraints of captive housing may render threat a less effective means of avoiding conflict, as it is difficult for animals to remove themselves from the vicinity of the potential aggressor.

The term 'aggression' has been applied to a very heterogeneous set of phenomena, being used as an epithet for activities as diverse as 'spontaneous' fighting in wild and captive animals to subjects biting their handler or another species. This can be illustrated by the very wide ranges of rodent models of aggression (but other species, e.g. dogs and deer, also show similar diversity) repeatedly employed in the laboratory. They include:

1. Social conflict, where threat and attack (generally in encounters between members of the same species) are used to compete for a resource, the winning of which increases the victor's relative **fitness** (breeding potential). Such conflicts

often employ strategies ('**ritualization**') that minimize the potential for serious damage, e.g. by avoiding bites to vulnerable areas of the opponent's body (this is important as all animals are vulnerable to defeat at some stage in their lives). Social conflict is expressed differently in varied animal species (and one should not assume functions by extrapolation from other species). In some animals (e.g. red deer) the conflict is used by males to compete for mates (a harem). In a smaller number of species, females fight for males. In others, fighting and threat are used to delineate a defended area or territory (e.g. the European robin). In others (e.g. domestic chickens and many primate species) the behaviour determines an animal's status within a social hierarchy (e.g. a 'peck order'). Some individuals directly compete for food that is necessary to 'fuel' their reproductive efforts as well as behaviour in general.
2. Parental defence, where threat and attack between members of the same or two different species are used to protect either the attacker's young or nest sites from potentially destructive intruders. For example, lactating mammals will often attack intruders coming near their offspring.
3. Self-defence, where threat and attack between members of the same or two different species are used to protect the 'aggressor' from potential predators or attacking members of their own species. These behaviours are generally only seen *in extremis* if flight or escape is precluded (as is frequently the case when caging is used or if an animal is suddenly cornered). They involve no injury-limiting ritualization strategies and attack can be especially damaging.
4. **Infanticide** involves the killing of (generally) unrelated young. In males, this strategy seems to be a method of increasing the individual's reproductive fitness (as it facilitates the attacker's reproduction), whereas in females of some species it may be a device for reducing litter or clutch size to manageable proportions.
5. **Predation** may occur within a species (cannibalism) but often involves different species. It often is characterized by an efficient killing strategy (prey can be dangerous to the attacker) and is generally followed by feeding activity. In laboratory rodents, mouse-killing behaviour by rats and insect killing by mice and hamsters have been the most studied examples of such phenomena. Many workers regard predation as being very different from more accepted types of aggression (notably because of the general lack of emotional arousal), but killing (e.g. by domestic cats of garden birds) can be regarded as problematic 'aggression'.

Structure of aggression

Aggression seen in different situations clearly serves a variety of functions, particularly offence, defence and predation. Having said this, the idea that an individual's behaviour can be classified as 'purely' offence or defence is not supported by experience. An individual's motivation generally changes over the course of an encounter (e.g. from dominant to submissive in the case of a defeated animal). The diversity of 'aggressions' in humans is even greater (largely because of the inclusion of their verbal and written responses). Although many of the human examples are unlikely to have parallels in non-human animals, there is a tendency in the media to assume that aggression has similar meanings in the biological and social sciences. As noted earlier, this causes further confusion.

All forms of aggression are products of complex interplay between:

1. Biological factors such as genes, particular neural circuits, neurotransmitters and hormones, etc.
2. Situational determinants, namely the context in which the animal finds itself (e.g. encountering a competitor in its home cage).
3. Experiential influences acquired as a result of the animal's accumulated positive and negative experiences in previous conflict situations.

It is not easy to separate these elements. The diversity of the concept means, however, that there will be no single physiology of aggression, i.e. all forms of threat and attack cannot be linked, for example, to the Y chromosome or to **androgens** (the view that aggression is a 'male' attribute is mistaken). Hoping to find simple 'cures' by **castration** or the use of psychoactive drugs for problem behaviours is unduly optimistic. Situations even within the same species have to be considered on a case-by-case basis.

Factors affecting aggression

Genotype has a strong impact on social aggression (an expression of threat and attack that is important in animal **welfare** terms). Different lines or strains of animals (from mice to dogs) may show considerable variation in their aggressiveness (the ease with which they display threat and attack).

Individual housing (**see also: Isolation**) is generally accompanied by increased aggressiveness in the males of a variety of laboratory rodent species, including many mouse lines, rats, the guinea pig, the deer mouse (*Peromyscus* sp.) and the Mongolian gerbil. The majority of social organizations recorded within groups of male mice in small cages are despotic dominance relationships, where one animal is dominant to a group of equally ranked subordinates (this may actually suggest that male mice are not notably social animals). There are, however, as in the peck order of chickens, linear hierarchical organizations in some social species. Attempts to 'environmentally enrich' captive animals often involve providing opportunities for social interaction with conspecifics. Having said this, the aggression seen in some groups (especially if the species is not notably social, e.g. polar bears) may be a source of concern or even be linked to **stereotypies**. Social grouping is not uniformly beneficial and each situation has to be considered on a case-by-case basis. Even stable groups need consistent monitoring for conflict.

Breeding experience (with separation from other mature males) induces a propensity to attack in males of some species (e.g. mice). This may reflect the common use of territorial behaviour for breeding activity by males. There is substantial evidence that the presence of females increases aggressiveness in males and that pregnancy/lactation may be associated with reduced tolerance by females of strangers.

The practice of replacing the sawdust substrate regenerates fighting in males of established social groups of mice. This temporary increase may be a consequence of disrupting the patterns of odour marking used in this species. Obviously, cleaning of captive animals can have important effects on their displayed aggression. Some small primates may be prone to similar problems.

Familiarity with the test situation (e.g. a 'home cage') in many animals may increase the probability of seeing fighting. Furthermore, familiarity with particular partners tends to reduce aggression ('strangers' are most likely to be attacked). Repeated experience of defeat and success in fights generally, respectively, reduces and increases subsequent threat and attack.

Cage design (e.g. the siting of feeding and watering locations) can have profound effects on fighting behaviour in captive animals. Dominants may exclude subordinates from these essential resources and cause them serious difficulties.

Consequences of aggression

Obviously, overt aggression can cause serious tissue damage. As well as reducing or suppressing a defeated or subordinated individual's reproductive activity (there is evidence in groups of some species that most of the breeding is performed by dominant or territorial individuals), the 'stressful' effects of fighting (and especially defeat) have other deleterious actions. Fighting associated with 'crowding' increases the retention of introduced endoparasitic larvae in mice. Furthermore, the hormonal correlates associated with 'social stress' may slow the antibody response to challenge with standard antigens. Such changes may account for the often-recorded reductions in disease resistance in defeated or subordinated rodents. The 'stresses' generated by fighting may also have detrimental effects on the functioning of particular organ systems, e.g. these can produce chronic and irreversible hypertension accompanied later by arteriosclerosis and nephritis. It is interesting, in this respect, to note that **ACTH** damages the Bowman's capsules in the kidney. Blood urea levels (indicative of kidney damage) are elevated in 'free range' territory holders and subordinates compared with dominant and subordinate mice from stable groups.

Consequently, the generation of **social stress** by aggression is a complex phenomenon with wide-ranging repercussions. Although the traditional association between hormones and fighting has been emphasized, other aspects of physiology should also be related to much wider descriptions of behaviour. Although the stress generated by conflict can acutely cause considerable problems leading to injury (e.g. wounding, breaking of limbs, blinding, etc.), failure to gain weight at a normal rate and reduced fitness in laboratory rodents, there are husbandry steps that one can take to ameliorate the effects of social conflict. Similar steps can be taken for farm and **zoo** animals, but the precise programme would depend on the species, the strain, the age of the subjects and their prior history.

Hormonal variations (however produced) clearly modulate the behavioural characteristics of animals and change the individual's production and detection of social signals. The relation of these factors to aggression is clearly very complex, involving actions at different stages of the animal's life.

The experience gained with rats and mice might well help guide the necessary pioneering investigations on the effects of social stress in farm animals by workers on both chickens and pigs. Control of aggression is also something that is essential for persons involved in captive breeding programmes in zoos and elsewhere. We have to accept that these 'undesirable' behaviours are rooted in the animal's basic biology, having had survival value in the wild progenitors of captive animals. Understanding the origins of and controlling the factors involved in aggression are the best means of generating solutions to the deleterious consequences of this activity. (PFB)

See also: Castration; Dominance; Feedlot; Hierarchy

Aggressive behaviour

Behaviour may be referred to as being aggressive or a form of **aggression** if it is considered by an observer that it may cause harm to another. This has led to considerable confusion over the definition of aggression, with some authors classifying elements of behaviours such as **play** and **predation** alongside components of agonistic interaction and responses to **frustration** as different 'forms' of aggression (play aggression, predatory aggression, competitive aggression, etc.). However, play and predation relate to qualitatively different functional processes and types of **arousal**. This has led some to describe the hostile behaviour that occurs in association with arousal within an **aversive** situation (such as in response to frustration or perceived threat) as affective aggression, in order to distinguish it from the states related to play and predation. It can be argued that the use of the adjectival derivative is preferable in relation to play (i.e. 'aggressive play') and that the term aggression be deleted in relation to predation in order to avoid inadvertently implying that these behaviours are mechanistically related to each other or to affective aggression. (DSM)

Agonistic behaviour

Agonistic behaviour is classically defined as being any behaviour that is associated with conflict and competition – including attack, escape, threat, defence and **appeasement**. **Aggression** or **aggressive behaviour** is an integral part of agonistic behaviour, but the two terms should not be used synonymously. Most animals kept by man as companion, laboratory or farm animals are social animals – ordinarily existing in **social groups**. Agonistic behaviour can occur between members of different groups and between members of the same group.

From an evolutionary standpoint, group living confers a number of benefits (as well as costs). A group is likely to be more efficient at detecting, acquiring and defending food (or other resource). It can detect and defend against predators and, when confronted by a predator, the chance of an individual within a group being predated is much less than if the individual were confronted alone. A group can be better at surviving cold weather and group members may assist each other in the rearing of offspring. In terms of disadvantages, a group may be more conspicuous than an individual, thereby attracting a predator. Once food (or other resource) is found, there is competition for access. Contagious illness is spread more rapidly and effectively through groups.

Within free-living animals, the 'normal' size of the group will depend on the species and the distribution of resources. Groups are likely to be quite dynamic in nature, not necessarily through the active choice of animals leaving or joining a group – though this can occur – but because there will be mortality and birth. Even though a group may have a dynamic nature, there will be mechanisms contributing to maintaining group cohesion. Each animal within the group is usually striving to gain maximum advantage for itself and some will be more successful than others. Consequently, group members will compete against each other for resource access and, through social interactions, a dominance hierarchy will become apparent. Many of the social interactions involved in hierarchy formation are overtly aggressive, such as attacking and biting, but individuals may appease without being threatened and, once established, the stability of the hierarchy is evidenced from behaviour that is more subtle and less obviously aggressive, such as threats with corresponding withdrawal or retreat, or even by withdrawal or retreat alone.

The precise **ethogram** of agonistic behaviour will be species-specific and, for nearly all species, a full cataloguing and understanding of the behaviours carried out during an agonistic encounter are lacking. Considering the importance placed upon the aggressive components of agonistic interactions and their impact on an individual's welfare, this gap in knowledge is worthy of further study. The major difficulty for ethologists looking in from the outside is to record the behaviours in sufficient detail and also to correctly assign them. Some authors have attempted to describe specific postures, movements and visual signals, etc., but there is still the risk of misidentification or misinterpretation. Major challenges are the speed at which they may be carried out and the fact that observable behaviours may well be accompanied by both olfactory and auditory cues that may act to modify 'meaning'. As stated above, once dominance is established, the subtlety of behaviours needed to maintain social status also makes recording difficult.

Within applied **ethology**, relatively few studies have been carried out to investigate agonistic interactions in detail and to attempt to develop a definitive ethogram of agonistic behaviour. Often, the behavioural category 'agonistic behaviour' may be used in more generalized study of the behaviour of animals being subjected to experimental treatments. Also, most agonistic interactions are very brief, meaning that unless continual direct observation techniques are being used, true quantification of agonistic interaction incidence is unlikely. Scan sampling, in particular, is a poor method for recording the often short-duration behaviours that are performed during agonistic interactions.

A thorough example of an ethogram of agonistic behaviours is that of male horses, described by McDonnell and Haviland (1995). Following a review of the literature, comments from peers and detailed breakdown of 50 hours of video recordings, a final ethogram of 49 behavioural elements, including five distinct **vocalizations**, was described. Three complex behavioural sequences were also described. The individual elements have been assigned names (with a cross-reference to names used to describe the same behaviour in other literature), with a full definition or description of the behaviour given and accompanying comments providing extra information about the behaviour. (JNM-F)

See also: Castration; Dominance

Further reading and references

Fatjo, J., Feddersen-Petersen, D., Ruiz de la Torre, J.L., Amat, M., Mets, M., Braus, B. and Manteca, X. (2007) Ambivalent signals during agonistic interactions in a captive wolf pack. *Applied Animal Behaviour Science* 105, 274–283.

Jensen, P. (1980) An ethogram of social interaction patterns in group-housed dry sows. *Applied Animal Ethology* 6, 341–350.

McDonnell, S.M. and Haviland, J.C.S. (1995) Agonistic ethogram of the equid bachelor band. *Applied Animal Behaviour Science* 43, 147–188.

McGlone, J.J. (1986) Agonistic behaviour in food animals: review of research and techniques. *Journal of Animal Science* 62, 1130–1139.

Alarm reaction

The term 'alarm reaction' is used to describe two different behavioural responses. It can refer to the first part of the **general adaptation syndrome**, described by Selye as the general response to a stressor; alternatively, it may refer to a range of behaviours that alert others to perceived danger (warning behaviour), for example following detection of a predator.

During the alarm phase of the general adaption syndrome, activation of the **sympathetic nervous system** results in the production of **epinephrine** (adrenaline) and **norepinephrine** (noradrenaline) from the adrenal medulla, which helps to prepare the animal for a burst of activity. Physiological changes include an increase in **heart rate** and respiratory rate, an increase in blood sugar (related to a decrease in the production of and reduction in sensitivity to insulin), diversion of blood away from the gastrointestinal system and an increase in flow to the skeletal muscles, together with a reduction in anabolic hormone levels, including a fall in thyroid levels that suppresses the immune and reproductive systems. Many of these changes are used during the assessment of animal **welfare (see: Welfare measurement)**.

In response to signs of danger, animals may exhibit an alarm reaction that alerts others to the threat. Warning signals may be transmitted through a variety of sensory media and may involve overt and specific signals emitted before taking flight, such as alarm calls, signals associated with the act of taking flight (e.g. the white flashing on the underside of the rabbit tail) and signals that are apparent even after flight has taken place, such as chemical markers (**pheromones**), which may indicate previous danger in an area.

Alarm reactions that alert others may represent a form of **altruism** if they increase the risk to the signaller, but many alarm signals appear to minimize the risk to the individual. For example, alarm calls are often relatively low-frequency sounds with little change in pitch, which are more difficult for a predator to localize. These signals also travel further than high-pitched sounds and may serve to encourage others to emit similar calls or undertake behaviour that might reduce the individual risk further: for example, the simultaneous flight occurring in response to a call might confuse a predator, or individuals might be recruited to attack a predator, increasing the defensive potential of the group. Even if the warning behaviour does increase the risk to the individual it may still be beneficial from an inclusive **fitness** perpective, for example by encouraging related individuals such as offspring to take cover.

Some alarm calls may simply communicate to the predator that they have been detected and so deter **predation**, but others clearly have greater informational value for conspecifics, since in vervet monkeys for example, different types of call may be used in response to different threats. Flight responses serving as alarm signals include both the appearance of stimuli not normally present, such as the flashing on the white underside of the tail of the rabbit in flight, and the absence of normal stimuli, such as the elimination of intention signals before taking flight, which normally suppress the flight of others in flocks of birds. (DSM)

Further reading

Cheney, D.L. and Seafarth, R.M. (1982) How vervet monkeys perceive their grunts: field playback experiments. *Animal Behaviour* 30, 739–751.

Marler, P. (1955) Characteristics of some animal calls. *Nature* 176, 6–8.

Moberg, G.P. and Mench, J.A. (2000) *The Biology of Animal Stress. Basic Principles and Implications for Animal Welfare*. CAB International, Wallingford, UK.

Allodynia

Allodynia is an exaggerated pain response to normally innocuous stimuli and, although the mechanisms are unknown, allodynia probably arises in the structures of the limbic system of the **brain**, such as the amygdala and periaqueductal grey, which are associated with the processing of emotions. Allodynia may explain some apparently excessive aggressive responses triggered by touch. (DSM)

Allogrooming

Allogrooming (mammals) or allopreening (birds) refers to the care given to the pelage of one individual by another (in contrast to autogrooming, in which an individual grooms itself). Allogrooming serves a hygienic function, given that grooming tends to be directed at sites that are inaccessible during autogrooming and is often directed at offspring that are incapable of autogrooming. However, allogrooming is also important in maintaining social cohesion and **dominance hierarchies**. Among non-human primate societies, allogrooming functions in conflict reconciliation (**see also: appeasement**) and in the building and maintenance of social alliances. It has also been proposed that allogrooming enables individuals to assess their own positions within such societies. Within some primate and non-primate species allogrooming has been associated with **mate choice**, with males increasing their grooming of females around the time of oestrus. For example, in crab-eating macaques a female is more likely to engage in sexual activity with a male that has recently groomed her. (PE)

Alloparenting

Alloparenting describes the giving of parental care to non-descendent kin. This can occur interspecifically, as in the case of a cuckoo chick being fed by its host parents (interspecific brood parasitism) or, more commonly, intraspecifically when conspecifics give care to the offspring of others. In species that practice cooperative breeding, alloparenting is often directed at the young of close relatives, so that the alloparent enhances its inclusive **fitness** through the production of non-descendent kin. However, not all cooperative breeding systems are characterized by **kin selection** and, in these cases, the alloparent is thought to benefit directly through having increased chances of future reproduction via a host of mechanisms including territory inheritance, mate inheritance and coalition formation.

Alloparenting may also arise when parents are tricked into caring for non-kin. In some species of birds, females are known to lay eggs in the nests of conspecifics (intraspecific brood parasitism), while in others females are known to engage in extra-pair copulations that result in extra-pair young that are fed by the genetically unrelated father. (PE)

Allostasis

Allostasis refers to a model of physiological regulation that produces stability through change. It differs importantly from the concept of **homeostasis**, which is a system that focuses on the response to change aimed at returning the animal to a set point, by integrating predictive systems that bring about reactions in anticipation of change. For example, a rise in cortisol (**see: Corticosteroids**) from a homeostatic perspective is seen as a response to stress that helps the animal to cope, whereas from an allostatic perspective it is a reaction in anticipation of further stress that minimizes the impact of the anticipated change in circumstances. From an allostatic perspective it is not the level of cortisol that is so important as a measure of stress, but rather the functioning of the system. Thus, a failure of an appropriate rise may be just as important as an extreme response.

Within an allostatic context the systems are dynamic and may maintain a degree of baseline activity reflecting minimal phylogenetic levels of anticipated challenge; therefore, if an animal is kept in circumstances where that minimal level of challenge is not encountered, then the system may actually be harmful (e.g. in the absence of normal levels of immunological challenge, autoimmune problems may develop). Another feature of allostatic models is the potential importance of previous experience in shaping the response to given challenges. It has been suggested that allostatic models may be useful in evaluating **welfare**, since they explain why understimulation may be as significant a welfare problem as overstimulation. By contrast, a model of welfare based on homeostasis tends to predict that understimulation will maintain homeostasis and therefore minimize **stress**. (DSM)

Further reading

Korte, S.M., Olivier, B. and Koolhaas, J.M. (2007) A new welfare concept based on allostasis. *Physiology and Behaviour* 92, 422–428.

Altricial

Developmental maturity is recognized as varying between **reptiles**, birds and mammals, with neonatal animals differing in their ability to survive without parental supervision. Neonatal reptiles, and some birds (e.g. the Megapodidae and kiwis), are fully independent of parental care and exhibit 'super-precocial' maturity. **Precocial** offspring are fully mobile but dependent on their parents for food (as is the case in mammals), thermoregulation or protection from predators (as is often the case in birds). Semi-precocial neonates are active and well covered with down or fur, but require feeding by their parents because usually they are incapable of securing their food. Semi-altricial neonates may be well covered with down or fur but are inactive, blind and require brooding and feeding. At the opposite extreme of this spectrum of developmental maturity are altricial neonates of birds and mammals, which are naked, often blind and totally dependent on their parents for maintenance of bodily functions. These neonates are incapable of effective thermoregulation or movement and require brooding and feeding for an extended period of time after hatching or birth.

Altricial development is often associated with relatively short periods of incubation or gestation. Adoption of this strategy represents a shift from **parental investment** in an extended period of the reproductive cycle that is required to produce precocial offspring to a strategy where the adult is able to provision its free-living offspring more effectively. In birds, eggs producing altricial neonates have a lower energy content and a higher water content, and have a relatively shorter incubation period than those producing precocial offspring. The energy-poor eggs minimize female investment and the short incubation period minimizes predation risk during incubation. Most of the reproductive effort is shifted to the post-hatching period, when the altricial offspring have to be maintained within a nest until they are able to develop bodily functions akin to independent life. (DCD)

See also: Development of behaviour

Altruism

In its everyday sense, the word altruism refers to the conferring of some benefit on another person at some cost to oneself. It is not usually applied within a family, for it seems obvious that actions such as feeding or protecting one's children are beneficial to one's own interests. Biologists have put this idea more formally, following two very influential papers by William Hamilton, who put forward what has come to be known as Hamilton's rule (Hamilton, 1964). This states that, for altruism to pay, the following must hold true:

Benefits to recipient > $1/r$ Costs to donor

where r is the coefficient of relatedness between the two individuals involved, the probability that the two share a gene by common descent. For identical twins this is 1.0 and for siblings it is 0.5, as it is also for parents and offspring; for first cousins it is 0.125. Thus, in genetic terms, it pays to give aid to a full sib if the benefit to them is more than twice the cost to oneself. At an extreme, it may be worth saving a very distant relative from drowning if the risk to oneself is very slight.

Hamilton's rule helps to explain why relatives may help each other, but one could dispute whether this is altruism at all, for the basis of it is that the individual is furthering the interest of their own genes. Is that then all there is to it? While it is certainly true that many examples of apparent altruism are in groups of closely related animals, and may therefore be explained by kinship, there are reasons why animals might also behave altruistically towards non-relatives.

Robert Trivers (1971) put forward the theory of reciprocal altruism. In this he suggested that it might pay an individual to help another if there was a good chance the favour would be returned in due course. For such an arrangement to evolve it was necessary that animals should live in fairly stable groups so that the same individuals would meet repeatedly. It was also important that they would be able to recognize each other, both so that they could form reciprocal relationships and so that 'cheats' (individuals that accepted favours without returning them) could be ostracized. Given these conditions, Trivers argued that altruism might arise in groups of animals.

Reciprocation is obviously a rather important phenomenon among humans: if someone invites us out we tend to invite them in turn. But is there evidence for this among animals? There is, but so far it is quite limited; the theory is rather far ahead of the data. Animals often show cooperation so that two or more individuals derive immediate benefits from their relationship, but this does not involve altruism. For altruism one requires evidence that unrelated animals may help one another without any immediate gain.

The most oft-quoted case of reciprocal altruism is the food sharing of vampire bats (*Desmodus rotundus*). These animals fly out from their roosts at night in search of blood meals. Some fail to find food and might starve if this occurred repeatedly. On the other hand some return to the roost gorged, having found a good source of blood. Under these circumstances animals will share blood with one another even if unrelated and individuals that received blood on one occasion will tend to reciprocate on another. To date this remains the best example of reciprocal altruism among animals in the wild. Another that may be best viewed in this way, and certainly does not involve kinship as different species are concerned, is the relationship between cleaner fish and the larger fish that are their clients. Some of these larger fish are carnivores and could easily consume the cleaner; the cleaners could also go beyond their normal scavenging for parasites and take valuable scales or mucus from the host. But both of them forgo these short-term gains in favour of their long-term relationship.

Other cases may be more properly viewed as **mutualism**, both individuals gaining from cooperating without the need for reciprocation. Pairs of male yellow baboons (*Papio cynocephalus*) may cooperate, one to distract the dominant male in their group while the other mates with his female. This was originally described as reciprocal altruism, but the evidence for reciprocation is weak; instead, it seems most likely that the two combine because both gain a chance of mating and this would be absent were they not to cooperate.

Bird alarm calls have also been suggested as altruistic; an individual sees a predator calling and so warns the others in its group, which will reciprocate when they are the first to spot the hawk. Thus the costs are spread out and all gain in the long term. But it is hard to see how cheating could be avoided in this case, for it would involve the difficult task of identifying an individual that saw a hawk and failed to call. Many more plausible theories of alarm calling, including advantage to kin, have been put forward.

Most cases where animals behave altruistically towards one another may be explained because they are kin, and the altruist gains genetic benefits from the action. In other cases the altruism may be more apparent than real and both individuals may gain, either in the short term or, in the small number of cases that have been described, through later reciprocation. Genuine altruism, where an animal selflessly bears costs for the benefit of another with no possibility of payback, seems unlikely to occur and has not so far been described in animals. (PJBS)

Further reading

Hamilton, W.D. (1964) The genetical evolution of social behaviour II. *Journal of Theoretical Biology* 7, 17–52.

Slater, P.J.B. (1994) Kinship and altruism. In: Slater, P.J.B. and Halliday, T.R. (eds) *Behaviour and Evolution*. Cambridge University Press, Cambridge, UK, pp. 193–222.

Trivers, R.L. (1971) The evolution of reciprocal altruism. *Quarterly Review of Biology* 46, 35–57.

Wilkinson, G.A. (1984) Reciprocal food sharing in the vampire bat. *Nature* 308, 181–184.

Ambivalent behaviour

Ambivalent behaviour is a form of **conflict behaviour** in which the condition of contradictory motivational states results in behaviour that either oscillates between the motivational states or takes an intermediate form.

An animal is said to be in a motivational conflict when two motivational tendencies are at a similar and high level. For example, a conflict between hunger and fear can arise in the situation where an animal reaches a preferred feeding location and notices the threatening presence of a predator. When the animal is very hungry and does not want to flee, an approach-avoidance conflict occurs. This approach-avoidance conflict could be expressed by the animal rapidly and repeatedly switching between feeding and retreating. In the scenario that the animal is too frightened to feed but at the same time too hungry to avoid the predator the conflict remains unresolved, and this is likely to result in a state of **frustration** in the animal.

A three-way motivational conflict between sexual interest, defence and offence takes place in many species during courtship. An example of such a conflict is the courtship of the cockerel. 'Circling' behaviour is part of this courtship, and during this circling (signalling sexual interest) the wing facing the female is held low (signalling defensive fear) while the other wing is held high (signalling offence). (PHZ)

See also: Conflict behaviour; Displacement behaviour

Amphibian

Amphibians (Class Amphibia) are a major and diverse group of tetrapods that represent the living relatives of the first vertebrates to venture on to land around the Late Devonian (~360 million YBP). Amphibians were dominant during the Cambrian and Permian epochs (354–248 million YBP), until the evolution of reptiles. Modern amphibian forms arose in the Triassic and Jurassic and so have persisted for at least 144 million years.

Today, amphibians are divided into three main groups: caecilians (Order Gymnophiona), newts and salamanders (Order Caudata) and frogs and toads (Order Anura). There are around 5400 species, dominated by diverse species of frogs and toads (~4700 species); newts and salamanders are much less diverse (~470 species). There are fewer than 200 species of the eel-like caecilians.

Characterized by a moist, glandular skin, amphibians are found on all continents with the exception of the polar regions. No amphibians have adapted to life in seawater. At high latitudes cold temperatures drive amphibians to seek refuges on land or water to overwinter in a state of torpor. Some species are tolerant to freezing. Locomotion on land is through walking or hopping, but in water undulations of the body are used by newts and salamanders to swim. By contrast, anurans have shortened vertebrae and elongated legs that combine to produce a hopping action on land, and swimming is by simultaneous thrusts of the legs. Communication is via visual, aural and chemical means.

A key feature of amphibians lies in their pattern of reproduction. Despite living in terrestrial environments for much of their juvenile and adult life, most amphibians are tied to water for embryonic development. There is a clear distinction in reproductive strategy between caudates and anurans. Newts and salamanders often exhibit elaborate **courtship** and **display behaviour**, leading up to production of a spermatophore by the male that is then taken up by the female into her reproductive tract, which allows internal fertilization. Typically, females of this group then produce relatively few eggs (100–200 per season), laid individually and deliberately hidden. Egg laying often takes place over several weeks. By contrast, frogs and toads tend to return to water *en masse* and sperm is released into the water when the thousands of eggs are released. Egg laying by an individual occurs only once. The adults then leave the breeding pond to resume their terrestrial habits.

Embryonic development proceeds via aquatic larval stages and, again, newts and frogs differ. The larval newt is very similar to the adult in appearance for much of its time in the water. Frog tadpoles are quite different from the adult for much of the aquatic stage: for instance, they possess a tail. Development proceeds to a final metamorphosis, producing an air-breathing juvenile that will typically leave the breeding pond for a terrestrial habitat. Amphibians are carnivores in their adult forms although in many species the larval species are herbivorous or omnivorous.

In some salamanders the final stage of metamorphosis is never completed – often through a lack of iodine in the environment, which is essential for the formation of thyroxine, a key hormone controlling metamorphosis (**see: Thyroid hormone**). In these cases the juvenile continues to grow in the water and becomes sexually mature and active. Such neotenous development is most typically associated with the axolotl (*Ambystoma mexicanum*).

Amphibians are found in captivity as pets, for commercial production and as subjects for scientific research. The numbers of species kept under such circumstances are relatively few but pet frogs, toads and salamanders, particularly the axolotl, are popular and becoming increasingly so. Commercial exploitation of amphibians often lies with their production of meat, which in some parts of the world can be very significant. Research is directed at a few species, such as the common frog (*Rana temporaria*) or the clawed toad (***Xenopus** laevis*), which are kept in laboratory conditions. Amphibians of all types are the subject of research in their natural habitats all around the world.

One species, the cane toad (*Bufo marinus*), which was introduced to some regions (such as Australia) as a biological agricultural pest control measure, has been very successful biologically (though less so as a pest control measure) and has caused considerable ecological damage as it is highly toxic to native predators (**see: Exotic species invasion**). However, there occurred a worldwide decline in amphibian numbers and diversity during the 20th century, which appears to be accelerating in the 21st century. There are a variety of factors causing this decline, including pollution, exposure to excessive amounts of UV light, habitat loss, introduction of non-native predators and competitors, and commercial exploitation. One key problem is the spread of the parasitic chytrid fungus on a global scale that is threatening many species with extinction in the wild. These problems mean that amphibians are increasingly being brought into captivity as a conservation measure.

Welfare issues with amphibians lie primarily with the poor understanding of their specialized needs in captivity. Handling should be minimized to prevent damage to the skin, but also to prevent excessive rises in temperature in the smaller species. The correct environment and housing need to be provided to ensure that the appropriate temperature and humidity conditions are maintained. Appropriate nutrition and health care are also causes for concern with captive amphibians. (DCD)

Further reading

Phillips, B.L. and Shine, R. (2006) The morphology, and hence impact, of an invasive species (the cane toad, *Bufo marinus*): changes with time since colonisation. *Animal Conservation* 8, 407–413.

Stuart, S.N., Chanson, J.S., Cox, N.A., Young, B.E., Rodrigues, A.S.L., Fischman, D.L. and Waller, R.W. (2004) Status and trends of amphibian declines and extinctions worldwide. *Science* 306, 1783–1786.

Anaesthesia

A state of not being able to feel; often applied to a whole animal being 'anaesthetized' (general anaesthesia – GA), but can also apply to a part of the body (local anaesthesia – LA). GA can be induced by means of a liquid chemical substance (or mixture of chemicals) that is injected into an animal, or the chemical may be vaporized and inhaled (e.g. ether, isoflurane); occasionally it may be given orally. GAs work by depressing the **central nervous system**, but sometimes a mixture of chemicals is used to provide the three major aspects of GA, i.e. **restraint**, analgesia and muscle relaxation (so-called balanced anaesthesia). Balanced anaesthesia is normally safer than general anaesthesia, as none of the individual chemicals used approaches its toxic level.

Local anaesthetics work, on the other hand, by blocking nerve conduction to produce a numbness in a particular area of the body by either: (i) local infiltration affecting only the nerves in that area; (ii) blocking the nerve that supplies that region (nerve block) at a site away from the area of interest; or (iii) injection into the vertebral column affecting the spinal cord, from which the nerves that supply that region originate.

Anaesthesia can be induced in some species by immersion when chemicals are absorbed through the skin (e.g. amphibians) or via the gills (fish).

During anaesthesia the autonomic controls and reflexes – for example, to low oxygen and low body temperature – are reduced or removed and so it is important to monitor physiological states and maintain **homeostasis** artificially through monitoring of blood gases and body temperature and compensating when necessary.

The behavioural effects of anaesthesia on animals vary according to the species but, in general, as an animal becomes anaesthetized it loses control of its motor functions such as its righting reflex, and it will fall about uncontrollably. As it becomes deeper so it will lose various protective responses, e.g. the blink reflex to touching the cornea of the eye, the withdrawal reflex to a potentially painful pinch to the foot and, at this point, it is normally safe and humane to carry out surgical interventions. As an animal recovers from surgical anaesthesia so these reflexes are regained in the order in which they were lost. If, however, the animal goes deeper, the central depression may progress to one where it loses its respiratory drive, and the lack of oxygen to the heart and **brain** will lead to a loss of respiratory and cardiac functions and, ultimately, **death**.

The behavioural effects of anaesthesia are not just confined to the period of anaesthesia but also affect the animal's **circadian rhythm** for several days afterwards. Depending on the type and duration of anaesthesia, as well as post-anaesthetic care, the heart rate, temperature and mobility can all be affected for 4 days or more. Moreover, some anaesthetics render a second anaesthetic less predictable in terms of duration and depth, particularly those anaesthetics that must be metabolized for recovery, e.g. barbiturates. Those anaesthetics that rely on exhalation for recovery are better tolerated. (DBM)

See also: Analgesia

Analgesia

A state of not being able to experience pain. This is part of **anaesthesia** but it is a state that can also be specifically induced with chemicals, e.g. morphine, and with physical forces, e.g. rubbing or exposure to cold. Ideally, treatment of pain, e.g. after surgery, should be preventative (so-called *pre-emptive analgesia*) rather than after pain has been experienced by the patient, as the feeling of pain can result in a condition known as 'wind-up'. Wind-up sensitizes the spinal cord, resulting in a local spread in the area where pain can be felt, as well as an increased intensity of pain felt at the site (**hyperalgesia**); furthermore, previously non-painful stimuli become painful (**allodynia**).

Stress and heavy physical exertion (e.g. exercise) can induce analgesia by acting centrally to promote the release of natural opioids. Analgesics can work centrally (e.g. opioids like morphine) or peripherally at the site of pain by acting on the inflammatory response to injury (e.g. non-steroidal anti-inflammatory drugs (NSAIDs), affecting prostaglandin metabolism) or by blocking impulse transmission along afferent sensory nerves (local anaesthetics). Paracetamol (Tylenol®) is another centrally acting analgesic, but its mechanism of action is not clear. Centrally acting opiates bind to the mu (μ) agonist, which has the advantage that it lasts for several hours – even up to 8 h in some species (dog and cat).

The behavioural effects of an analgesic will depend on the drug being used. Some drugs stimulate the appetite whereas others may cause respiratory depression to the point of cyanosis (blue coloration of the mucous membranes), although this is far less profound in animals than in humans and so drugs such as the opiates are safer to use. (DBM)

Analogy

Two structures are said to be analogous if they have the same or similar function although their evolutionary pathway is different (this process is known as **convergent evolution**). A classic example of analogy is the wings of insects and birds: they evolved independently but have a similar function. (BM)

See also: Homology

Androgens

Androgens are a family of **steroid hormones** principally concerned with the development and maintenance of the male sex characteristics and gamete production. The term androgen is also used to encompass other molecules with androgen receptor-binding properties, including steroidal and non-steroidal synthetic compounds able to act as androgen receptor agonists. Androgen molecules have 19 carbon atoms (C-19) arranged in a basic four-carbon (cyclopentanoperhydrophenanthrene) ring structure common to a wide range of steroid hormones. A complex biosynthetic network utilizes cholesterol to first make pregnenolone, which acts as a general precursor for a number of steroid hormones including, among the androgens, relatively low-activity molecules such as dehydroepiandrosterone (DHEA) and androstenedione, which subsequently undergo further processing to make the more active **testosterone** and dihydrotestosterone (DHT).

Although studies on **castration** described from the early 1700s onwards had shown that a testicular factor was responsible for expression of the masculine appearance and behaviour, it was not until 1936 that testosterone was identified and characterized by Leopold Ruzicjka. The primary source of androgens in vertebrates is the testes; however, androgen production is maintained at very low levels in juvenile males and increases significantly only at the time of puberty, when testosterone production is initiated in the interstitial Leydig cells by the action of pituitary **gonadotropins**, particularly **luteinizing hormone** (LH). Post-pubertal testosterone production is regulated by a series of feedback mechanisms operating within the hypothalamic–pituitary–gonadal endocrine axis. Increased circulating androgen levels at puberty initiate spermatogenesis in the testes and are responsible for acquisition of, and thereafter maintenance of, the secondary male sex characteristics and other androgen-dependent structures such as the accessory glands.

Prenatally, androgens also play a crucial role in fetal development. The fetal testes are active in testosterone production from an early stage in development and it is testosterone that signals development of the male internal and external genitalia; in the absence of this androgenic support the default female morphology is expressed. Fetal androgens also play an important role in gender-specific differentiation of the brain, acting during fetal life for the promotion of development and maintenance of sex-specific behaviour. Sexually dimorphic behaviour patterns in postnatal development are also dependent to some degree on circulating levels of gonadal steroids.

Androgens are known to exert profound effects on the organization and function of the **central nervous system**, a range of androgen receptors has been identified in different brain regions and there is increasing evidence in support of the general hypothesis that testosterone levels influence aspects of **cognitive** function. There has been an extensive amount of work investigating the effects of circulating androgen levels on the promotion of **sexual behaviour**, but also considering other types of behaviour patterns and particularly in connection with **aggression**.

In addition to the testes, the cortex of the **adrenal gland** is a secondary source of low-affinity androgens, in both males and females. In the adult male these low-activity forms can be converted to testosterone and DHT in a range of peripheral tissues, but the amount of testosterone produced via this route is probably not physiologically significant, amounting to only around 5% of total circulating androgen levels. However, during fetal development in the prepubertal male and in the female, the adrenal gland may be an important source of androgen. Although androgens have been considered as the definitive male hormone they do have essential physiological roles, including maintenance of muscle and bone and behavioural effects in the female, where production is divided fairly evenly between the ovaries and the adrenal glands. In addition, androgens are vital within the female endocrine system as they provide an essential precursor that undergoes a process of aromatization to produce **oestrogens**, which may be considered as the female counterpart to the male androgens.

Testosterone has been shown to have potent anabolic effects on the musculoskeletal system, including an increase in lean body mass, a dose-related hypertrophy of muscle fibres and an increase in muscle strength; androgens can also inhibit protein catabolism and fat deposition. Synthetic androgen derivatives such as trenbolone, a steroid with a threefold higher affinity for the androgen receptor than testosterone, have been extensively used as growth promoters within the livestock production industry. The effects of supraphysiologic doses of exogenous androgens on muscle mass and strength have led to their continued use and abuse as performance enhancers.

(MRC)

Animal abuse

Animal abuse or **animal cruelty** involves causing significant harm to animals, when a person either intentionally and actively harms an animal or fails to act appropriately towards an animal for which he/she is responsible. In most legal texts, harm to animals is not considered cruelty or abuse if it takes place within accepted agricultural, management or research practice and if the harm is not greater than considered necessary to obtain the purpose. (AO)

Animal activist

An animal activist is a person who engages in different actions to bring about changes to the treatment and/or status of animals. This activism can include a wide variety of actions and positions, but people referred to (by themselves or others) as animal activists tend to hold more radical positions. (AO)

See also: Animal rights

Animal-assisted therapy

In recent years, increasing attention has been paid to the potential therapeutic benefits that interactions with animals can have in the lives of ill, disabled and disadvantaged individuals. The expression 'animal-assisted therapy' has been coined to refer to a variety of treatment programmes that employ animals, usually in association with a handler, in formalized therapeutic interventions with patients or clients. Related expressions include '**pet**-facilitated therapy' and 'animal-facilitated therapy'. Hospital visitation schemes are a common form of animal-assisted therapy programme. These involve a **companion animal** (usually a dog) and its handler visiting hospital patients, particularly children, elderly and psychiatric patients, on a regular basis.

Similar schemes are also employed to bring nursing home residents into contact with animals. The aim of such schemes is to provide interest, entertainment and distraction from pain, distress, anxiety or depression, and to offer patients opportunities for interaction and conversation (e.g. see Bernstein *et al.*, 2000; Marr *et al.*, 2000). Such programmes are thought to have therapeutic effects via a number of different routes. For example, the opportunity for tactile contact with another living being may confer feelings of relaxation and well-being to some patients; for others, animals may serve as a facilitator, enabling interaction with the handler and other patients. The unconditional affection that many animals offer also provides additional support to those for whom disability, deformity or infirmity limits successful interactions with other people. Such schemes are not a modern invention: as long ago as the 18th century, a York retreat in England included animals within its environs. Interest in such therapeutic interventions, however, has grown substantially in recent years, with several umbrella organizations supporting such work emerging in a number of countries across the world.

Animal-assisted therapies among prisoner populations take a wide variety of forms. Some are similar to hospital visitation programmes, with volunteer handlers and dogs coming into prisons to visit particular inmates within the system. Others are on-site programmes, in which prisoners care for and raise livestock, train dogs or horses, rehabilitate wildlife, socialize guide dogs, etc. At a less formal level, many prisons also now allow inmates to keep and care for companion animals individually, in the hope that such experience may help to instil compassion and empathy, as well as practical caregiving skills. Such programmes, however, can give rise to concerns about the welfare of the animals involved. Educational interventions, and supervision of training and caregiving activities, can help to ameliorate such risks.

Therapeutic horse-riding and -driving programmes aim to benefit disabled children and adults physically, emotionally and cognitively. Often, over a period of many years, riders and drivers are taught horsemanship skills, helped to develop balance and strength and may also experience boosts in self-confidence, social skills, the ability to communicate with others and follow directions given by the instructor. Organizations dedicated to equestrian-based therapy (hippotherapy) programmes include the Riding for the Disabled Associations of the UK, Ireland, New Zealand and Australia, the North American Riding for the Handicapped Association and the French Association Nationale Handi-Cheval.

Therapies involving wild animals such as dolphins have increased dramatically in popularity in recent years. However, such schemes are usually considerably more expensive to operate than volunteer-based visitation programmes, given the extensive overhead costs of accommodation, feeding, husbandry and training of the animals concerned. Waiting lists are common, and therapeutic contact is frequently limited to just one or two relatively brief sessions. Despite this, such programmes attract high levels of public support, with the potential benefits that interactions with dolphins may confer on children with autistic spectrum disorders generating considerable popular attention (Smith, 1988).

Despite the increasing popularity of animal-assisted therapies, there has been relatively little empirical research into the efficacy of such programmes. In many of the studies that have been conducted, problems have arisen with respect to small sample sizes, lack of adequate control groups and outcome assessments being made by observers who are not blind to the treatments assigned. In addition to such methodological problems, the nature of any potential benefits of animal-assisted therapy is likely to be multifaceted, often subtle, and frequently difficult to measure in an objective fashion. Effects are also likely to vary considerably from one user to another, depending on their previous experience with and attitudes towards the animal species concerned. Regardless of these problems, however, those providing animal-assisted therapy are clearly encouraged by the simple enjoyment that many people derive from the contact they experience with animals. In addition, the interest that such schemes generate in patients and staff alike, and the sometimes dramatic improvements that appear to be associated with the initiation of treatment in some individuals, encourage many to believe that animal-assisted therapy often provides a beneficial experience.

On the negative side, animal-assisted therapy programmes also present potential health risks to patients and clients, particularly in the form of infection and injury. The majority of programmes therefore employ strict guidelines regarding animal selection, cleanliness and safety precautions in order to minimize such hazards. Another potentially negative issue within animal-assisted therapy concerns the welfare of the therapeutic animals used. With domestic species such as dogs, selection procedures can help to alleviate this by excluding animals that are temperamentally unsuited to the task at hand (e.g. those that show anxiety in the presence of unfamiliar adults or children). With wild animals such as dolphins, additional and more rigorous safeguards, including appropriate amounts of 'time out' from therapeutic sessions, need to be built into programmes.

Animal-assisted therapies are distinct from the use of assistance or **service animals** (such as guide dogs for the blind and hearing dogs for the deaf), in that their primary purpose is therapeutic rather than practical. Nevertheless, there are many overlaps between the uses of therapeutic and assistance animals. For example, many prison-based animal therapy programmes involve the training of dogs to be used as assistance animals for the disabled. In these instances, the organizations are specifically structured so that the same dogs are acting as both therapeutic and assistance animals (for the trainers and end-users, respectively). At a more informal level, guide dog owners frequently report social and emotional benefits, as well as more specific practical benefits, arising from the relationships they form with their dogs. (EP)

See also: Assistance animal

References

Bernstein, P.L., Friedmann, E. and Malaspina, A. (2000) Animal-assisted therapy increases resident social interaction and initiation in long-term care facilities. *Anthrozoos* 13, 213–224.

Marr, C.A., French, L., Thompson, D., Drum, L., Greening, G., Mormon, J., Henderson, I. and Hughes, C.W. (2000) Animal-

assisted therapy in psychiatric rehabilitation. *Anthrozoos* 13, 43–47.
Smith, B.A. (1988) The autistic person experiences Atlantic bottlenose dolphins as therapy. *National Aquatics Journal* 4, 5–7.

Animal cruelty

Cruelty to, or abuse of, animals involves causing these animals significant harm. A person who acts cruelly to an animal can either intentionally and actively harm the animal or fail to act appropriately in relation to an animal for which he/she is responsible. This failure to act usually involves neglecting to provide the animal with wholesome food, water, appropriate shelter or necessary veterinary care.

Legal protection of animals against cruelty dates back to at least the 18th century BC, with some elements in the Codex Hammurabi of ancient Babylon and the Atharva Veda of Hinduism. The first Western law against cruelty to animals was contained in the Massachusetts Bay Colony Body of Liberties in 1641; however, it did not remain part of the laws of the colony or the later state of Massachusetts. The first enduring parliamentary animal protection law was established in the UK in 1822, when what was later to be termed The Martin's Act (after its promoter Richard Martin) made it a crime to wantonly and cruelly beat or ill-treat any 'horse, mare, gelding, mule, ass, ox, cow, heifer, steer, sheep or other cattle'. This law was followed by similar legislative texts in other northern European countries and North American states during the following decades, including the New York State anti-cruelty law (1828 for cattle, amended in 1866 to include all animals), the Saxonian criminal law (1838) and the French *Loi Grammont* (1850). Parallel to the establishment of legislation in the 19th century, and instigated by the same public debate, the various associations for the prevention of cruelty to animals were developed. The best known of these associations is probably the British **Royal Society for the Prevention of Cruelty to Animals**, established in 1824.

Under earlier legislation, mistreatment of animals was considered illegal only to the extent that it harmed the owner of the animals by harming his property. However, even though The Martin's Act extended the legal protection for animals, it covered only cruel acts towards an animal when committed by those persons other than the animal's legal owner. The protection was further limited by the fact that this Act, as well as most other legislation against animal cruelty in the 19th and early 20th centuries, only considered instances where animals were subject to deliberate and unjustified harm. What determined whether an act was cruel was not the harm as such, but the context in which it was inflicted and whether there was any objective purpose behind the action.

During the second half of the 20th century, animal protection legislation was developed not only to ban abnormal practice (wanton cruelty) but also to regulate normal practice, in particular in farm animal production, **transport** and **slaughter** and in animal research. In many countries, there is now specific legislation laying down the conditions for the housing, handling and husbandry of these animals; however, enforcement varies enormously between countries, and sometimes within countries. Such legislation aims to protect the **welfare** of the animals involved and may, for example, demand that **analgesia** is always used in painful procedures, that animals are guaranteed basic levels of comfort and freedom of movement and may even prohibit certain specific practices (such as the tethering or crating of pregnant sows, already prohibited in several European countries and regulated through a European Directive affecting all member states by 2013). Whether or not harm within regulated practices in animal production and research is ethically justifiable, however, remains a matter of debate. In many cases, animal protection and **animal rights** associations refer to harm done to animals also within such circumstances as cruelty or abuse.

In Europe, legal animal protection is increasingly regulated at the European Community (EC) level. The first transnational regulations were the Council of Europe Conventions covering international transport of animals (ETS 65 from 1968), farm animals (ETS 87 from 1976), slaughter (ETS 102 from 1979), laboratory animals (ETS 123 from 1986) and companion animals (ETS 125 from 1987). In most cases, a convention has later been followed by a legally binding European directive. From the EC, there has also been an effort to bring animal welfare into international trade agreements. The OIE (World Organisation for Animal Health) has identified animal **welfare** as a priority, with guiding principles laid out in the Terrestrial Animal Health Code 2004. Welfare standards have now been produced for animals during transport and slaughter.

Altogether, the process of expanding legislation reflects an increasing concern over the welfare of animals – but essentially a concern that does not question the use of animals as a principle. A more demanding view on legislation has been increasingly defended during the last two decades by some **animal activists**, including philosophers and lawyers, who claim that animals deserve respect as such and that they have a right to life, and that legislation should grant them such rights. One example is the Great Ape Project (http://www.greatapeproject.org), which aims 'to include the non-human great apes within the community of equals by granting them the basic moral and legal protection that only human beings currently enjoy', and consequently strongly restricts the situations when these animals could be harmed. (AO)

See also: Animal abuse

Further reading

Knierim, U. and Jackson, W.T. (1997) Legislation. In: Appleby, M.C. and Hughes, B.O. (eds) *Animal Welfare*. CAB International, Wallingford, UK.
Tannenbaum, J. (1995) Animals and the law: property, cruelty, rights – in the company of animals. *Social Research* (autumn). Available at: http://www.findarticles.com/p/articles/mi_m2267/is_n3_v62/ai_17909880 (accessed 25 September 2009).

Animal enclosure

Animals have been kept by man for a variety of reasons including religion, food, entertainment and as symbols of wealth and power. The earliest date of an animal housed in a **zoo**-type situation comes in the form of an illustration going as far back as 2495 BC.

In more modern times cages are normally associated with animals in zoos, research laboratories or battery farms. Cages exist in all shapes and sizes, from those designed to hold insects to those constructed to house large mammal species such as lion. When considering the type of environment in which an animal is going to be held, it is important to have some idea of its natural behaviour. Provision within the cage of an opportunity to explore and deal with challenges may be essential, if the creature is to be prevented from becoming bored and apathetic or frustrated (**see: Boredom; Frustration**). It has been shown that rats kept in an unstimulating environment, even when given complex motor tasks to perform, have smaller brains than those individuals living in enriched, complex cages. Cage **enrichment** should not be thought of as the exclusive preserve of complex mammal species. Research carried out on the behaviour of spiders has shown that they are also affected by the complexity of the environment in which they are caged.

It is generally perceived that the larger the cage the better. However, it was Heini Hediger who suggested that it was not the quantity of space that an animal was provided with that matters, but the quality. Hediger's suggestion was that a small but high-quality environment was better for the animal than merely providing it with space that either it may not feel secure in or was an inappropriate situation for its behavioural repertoire to be expressed. Hediger also discussed the human concept of what a caged animal was lacking in its life, when compared with the freedom of its wild counterpart.

An animal in the wild has a territory of a certain size, as this area provides it with the resources it requires. The wild animal does not wander without restrictions for the simple sake of expressing its freedom to travel. Primarily it travels to protect its territorial resources, namely food, water and a mate.

Cages have been constructed from a wide combination and variety of materials. Regardless of the materials used, an animal cage should be designed to fit the specific needs of the intended occupant. It should keep the inhabitant secure, have places for hiding and, if outdoors, should provide shelter from the elements; and, in the case of zoos, it should not be open to viewing by the public on all sides. An animal cage should be designed, where necessary, to afford protection to the occupant from any predators that may exist in the external area.

Depending on the species involved it may be essential to equip the cage with different gauges of perching, so that the animal, bird or insect can utilize the maximum height of the area involved. Using shelves, branches and ledges around a cage increases the usable surface area. Certain species like to rest above the ground, and safely secured, suitably sized supports should be provided for this purpose. Some species (e.g. marmosets) benefit from the perches in their cages being able to move, as the branches of a tree would in the wild, while others – specifically bird species – benefit from the perches being a specific diameter. The floor and walls of a cage, depending on the species involved and the purpose for which it is being kept, can be utilized to maximize the cage's potential. In the case of large zoo cages, the animal involved may benefit from some of the enclosure walls being clad in rough timber planking, which can be clawed or rubbed against. Wood in general is a preferable surface for animals to make comfortable contact with than the metal bars or brick walls that may make up the cage structure. However, for some species the bars of a cage can offer a worthwhile, useful surface for climbing on.

The **floor** of any cage is extremely important and should reflect the needs of the animal and the purpose for which it is being held. For instance, the laboratory animal may need to be kept in a sterile environment, whereas the zoo or farm animal may not be subjected to such strictures. Substrates can be used to cover the floor and make it more user-friendly for the creature involved. Substrates add a whole new dimension to a cage and have many different benefits. For some species soft substrates can be kinder to the feet and legs, allow for play behaviour and digging and provide a surface where food can be hidden to encourage foraging. Appropriate substrates have also been shown to be self-sterilizing, allowing helpful bacteria to grow to the detriment of others. Deep woodchip flooring has also been associated with a reduction in parasite levels in captive carnivores.

A number of countries have opted for a mandatory code of practice that dictates minimum cage sizes for captive animals in zoos, laboratories and farms. But many have an ad hoc system where some categories of animals are covered while others are not. Standardized legislation is, however, proposed for all laboratory and zoo animals within the European Community (EC) countries.

Laboratory rodent cages are predominantly manufactured from polycarbonate, formed into boxes and fitted with a stainless steel lid. For convenience these cages are usually slid into a mobile rack that holds a large number of cages. The cage base and lid are made from high-quality materials that can withstand the rigours of being autoclaved and sterilized. The lid may be fitted with one or two wire food hoppers and space to accommodate a water bottle. Many different styles of **laboratory animal** caging are available for housing. A small selection is highlighted here to give a general feel for how many animals are allowed to be housed in each style mentioned, in compliance with the Home Office Animals Scientific Procedures Act 1986, currently in place in the UK, by way of example of what are generally considered to be high standards for animals in this environment.

Three styles of standard mouse cages are commonly used in laboratories. One of the smallest, referred to as a short box by the care staff, can hold up to three adult mice. They measure approximately 33 × 15 × 13 cm (internal size 330 cm^2 × 12 cm). The second standard mouse box is generally referred to as a long box, measuring approximately 48 × 15 × 13 cm (internal size 490 cm^2 × 12 cm), and is deemed suitable for the accommodation of up to four adult mice. The third type of standard cage is considerably larger, measuring 45 × 28 × 13 cm (internal size 960 cm^2 × 12 cm). This type of cage can be used to house nine adult mice.

As one would expect for laboratory rats, the space requirements are greater (see Table A.1). One cage type in vogue comprises a base of moulded polypropylene with a high wire lid fitted on top to give the animals extra height. This cage measures 56 × 38 × 25 cm (internal size 1575 cm^2 × 25 cm),

Table A.1. Housing space allowances for laboratory mice, rats and rabbits in the UK.

Species	Typical weight (g)	Floor space (cm^2)		Height (cm)
		Single	Group	
Mouse	> 30	200	100	12 (min)
Rat	350–450		300	20
Rabbit	4000–6000		3300	45

and can be used to house five rats of the weight listed in the table.

Rabbits, in many establishments, are now held in floor pens rather than in individual cages. Home Office legislation in the UK also requires that animals are provided with appropriate cage **enrichment**, and this usually comprises bedding materials, cardboard fun tunnels and chewing blocks, depending on the species concerned.

Laboratory cats are usually kept in large rooms. Cats can be held in small social groups of up to 12. Various forms of caging are used but, regardless of the style, minimum floor space and shelving must be provided. Proposed UK legislation dictates that one cat in a room should have 1.5 m^2 of floor space, 0.5 m^2 of shelving and the cage height must not be less than 2.0 m. For each additional cat added to the room, 0.75m^2 of floor must be provided and 0.25 m^2 of shelving. For cats litter trays (300 × 400 mm) should be provided at a rate of one for every two cats. Toys and food puzzles should also be provided, and trays of grass for them to chew are recommended.

Additional legislation is in place to cover the accommodation in cages for breeding groups of laboratory animals, and it is essential to check with the appropriate authorities for up-to-date information.

Until the 1950s most chickens kept for egg laying were kept in a free-range situation during the day and locked up for security at night in a barn or courtyard. From the early 1960s the number of free-range chicken farms in many industrialized countries began to decline in favour of battery systems. Today, the minimum cage sizes used for battery hens are a point of contention in Europe and it is generally acknowledged that the welfare standards for chickens held in such conditions are poor (**see: Laying hen housing**). As the birds do not have much space to move around their bones become brittle and their muscles weak. Several countries have called for a ban on battery caging on the grounds that the birds held in such systems experience acute **welfare** problems. In the European Union (EU) battery cages allow for between 450 and 500 cm^2 of cage space per bird. It has been estimated that, on average, a resting chicken needs 600 cm^2 of space. At the moment 90% of the EU's egg-laying hens are housed in battery cages. However, the EU has agreed to ban battery-caging systems by 2012.

For zoo animals, countries such as Switzerland have in place regulations that cover minimum cage sizes for dozens of species. These regulations include an extensive list of the animals' special requirements, such as 'muddy pool – possibility for pigs to wallow and root' and 'screens – possibility to withdraw and hide'. For example, for a group of three gorillas the outdoor enclosure surface area should be 50 m^2 and the volume of the area should be 200 m^3. The indoor enclosure should have a surface of 50 m^2 and a volume of 200 m^3. For a pair of rhinos, the outside enclosure should have a surface area of 500 m^2 and the indoor enclosure should have a surface area of no less than 25 m^2 per animal. Regulations for the keeping of a number of marine mammals are supplied. As an interesting example, to keep two killer, white or pilot whales, the surface area of the pool should be 400 m^2 and have a minimum depth of 4.0 m. The pool volume should be no less than 1600 m^3. Special requirements for keeping whale species include an overall circulation rate (of the total water volume) through the filters in a maximum of 4 h, and the pool should contain salt water. It is also required that an adjoining pool and separating pool be provided, at least one offering the possibility of an independent water supply.

More recently, Petherick and Phillips (2009) have proposed a number of formulae based on allometric equations to calculate space allowances for animals in different contexts, below which there is evidence of adverse effects on welfare. These can be applied cross-species and this approach may be fruitful for deriving recommended minimal size requirements for a wide range of animal enclosures. (GL)

See also: Confinement; Enrichment; Space; Zoo

Reference and further reading

Broom, D.M. and Fraser, A.F. (2007) *Domestic Animal Behaviour and Welfare*, 4th edn. CAB International, Wallingford, UK.

Chamove, A.S., Anderson, J.R., Morgan-Jones, S.C. and Jones, S.P. (1982) Deep woodchip litter: hygiene, feeding and behavioural enhancement in eight primate species. *International Journal for the Study of Animal Problems* 3, 308–318.

Draft Appendix A of the European Convention for the Protection of Vertebrate Animals used for Experimental and other Scientific Purposes (ETS No. 123). Guidelines for Accommodation and Care of Animals (Article 5 of the Convention) to be Approved by the Multilateral Consultation Council of Europe. European Commission, Brussels.

Hediger, H. (1970) *Man and Animal in the Zoo, Zoo Biology*. Routledge & Kegan Paul Ltd, London.

Home Office Animals (Scientific Procedures) Act (1986) *Code of Practice for The Housing and Care of Animals used in Scientific Procedures*. Her Majesty's Stationery Office, London.

Petherick, J.C. and Phillips, C.J.C. (2009) Space allowances for confined livestock and their determination from allometric principles. *Applied Animal Behaviour Science* 117, 1–12.

Animal Ethics Committees

Animal Ethics Committees (AECs) are established to protect the **welfare** of animals used in research, teaching and training and, in some countries, they are required by law. AECs operate in a range of organizations, including universities and research institutes, research sponsors, scientific journals, scientific societies and animal trainers. AECs sometimes publish guidelines on how animals should or should not be treated, and what is best practice for the care of the animals and the quality of science, teaching or training. It is the Ethics Committee's duty to ensure that animal use complies with the spirit and intention of any law and guidelines. When seeking ethical approval, animal users must show that the guidelines have been, or will be, adhered to.

When assessing the ethical acceptability of animal use, most AECs adopt a **utilitarian** framework. The principle of this framework is that it is ethically justifiable for animals to be 'used', as long as there are benefits and any costs involved are proportionate to the expected gains. Benefits are considered to be improvements in the quality of life for humans, other animals and the environment; costs can include animal welfare concerns, reduced survivability, a reduced potential for breeding success or damage to the environment. AECs most frequently use a cost–benefit analysis, in which the benefits are balanced against the costs; however, the overall aim is to minimize the costs, not simply to outweigh the benefits. In science, great costs can be 'offset' in the cost–benefit analysis by achieving a high quality of research and/or answering very important questions, as summarized in Bateson's Ethical Decision Cube (Bateson, 1986). AECs consider that welfare should be assessed at the level of the individual animal, so the numbers of animals used should be minimized without compromising the power and validity of the research. The composition of AECs varies, but they usually include scientists (both expert and non-expert in the subject), veterinarians, statisticians, ethicists, other experts and lay persons (often a theologian). (CS)

See also: Ethics

Reference and further reading

Anon. (2006) Guidelines for the treatment of animals in behavioural research and teaching. *Animal Behaviour* 71, 245–253. Available at: http://asab.nottingham.ac.uk/downloads/guidelines2006.pdf (accessed 9 October 2009).

Bateson, P. (1986) When to experiment on animals. *New Scientist* 109, 30–32.

Animal experiments

Research that involves the use of living animals, usually vertebrates but also some invertebrates (e.g. in the UK this applies only to *Octopus vulgaris*), that is carried out with the intention of providing new information in the area of study, is often subject to some form of legal control, e.g. licensing of the person doing the work, as well as of the research programme and the premises.

The procedures used may be invasive, such as transplant surgery, or non-invasive, such as observing normal maternal behaviour. Even if a minor procedure is being used, such as taking a blood sample or plucking a feather, if the procedure has the potential to cause **pain**, **suffering**, **distress** or lasting harm, the work may have to be covered by the relevant licences. Some areas may involve wildlife research as opposed to animals in a laboratory and this can also be invasive, through the use of, e.g. location devices or those that monitor internal physiological performance (heart rate, blood pressure, etc.) or position (fish depth in the sea, the position of migrating birds by satellite tracking, etc.). The number of procedures carried out on an animal as part of a research protocol will vary according to the particular experiment.

Animals are often used as models in experiments, and the main aims of these 'animal experiments' are generally to:

1. Acquire new knowledge that may prove useful in the treatment or understanding of normal functioning or disease in humans and other animals.
2. Elucidate any toxicity that a chemical may cause, as well as to determine the target organ and the dosage level at which it does this (safety tests) according to how it is to be used (different hazards can arise depending on a chemical mode of action and method of administration, e.g. a chemical released into the environment in a spray may affect wildlife as well as human operators; a medical treatment for a person may have side effects and the margin of safety is that area between its therapeutic and its adverse effects).
3. Ensure the safety of medical and biological products such as vaccines and saline drips.
4. Make a diagnosis using animals in bioassays, e.g. for shellfish toxins, to protect the environment or to provide information on an endangered species.

The level of animal suffering during an experiment will be influenced by the species, housing and husbandry of the animals, as they are often kept in cages or pens for all of their lives, in addition to the effects of the scientific procedures. More often than not, any animal suffering is not caused deliberately but is an unwanted side effect (consider transplant research where the objective is to test a new immunosuppressant drug in a model of kidney transplantation – the surgical model will produce some pain, but the aim of the experiment is to develop better anti-rejection drugs). On some occasions, the research may be into a particular type of suffering, e.g. pain. In this sort of work the aim particularly has to be to cause only the minimum degree of pain to achieve the scientific objective, although this should be the aim in all areas of research.

In some countries the level of pain involved in an experiment is limited to a notional level (mild, moderate and substantial are permitted categories in the UK, and any level of suffering has to be justified) according to the perceived worth of the research project. Compare research into a life-threatening disease in humans like AIDS with wildlife research. In the former a higher level of pain may be permitted than in the latter.

Since 1990 there has been a dramatic increase in the creation and use of **transgenic** animals, notably mice, replacing the more traditional use of animals. Transgenic animals may suffer depending on the exact nature of the genetic modification, and it may range from none at all to substantial; in some cases there may even be a welfare benefit. In most developed countries animal research is carefully controlled through laws that aim: (i) to replace living animals altogether (e.g. cell cultures); (ii) to reduce the number of animals used to the minimum; and (iii) not to cause more suffering than is necessary to achieve the scientific objective (**see: Three Rs principle**). There are several ways in which legal controls are implemented, either through a national competent authority, a local ethics committee or a mixture of both. In the UK it is through a national licensing system under the Animals (Scientific Procedures) Act 1986 that is administered by the Home Office, coupled with a local process of ethical review; similar frameworks exist in many other countries. In the UK, there are three types of licence: (i) the research premises are approved as being suitable for the research and are inspected by the Home Office in order to be registered and granted a *Certificate of Designation* as either a Breeding, a Supplying or a Scientific Procedure establishment; (ii) the programme of research is approved through a *Project Licence*; and (iii) the person carrying out the procedures on the animals is granted a *Personal Licence* in order to gain competence through training and supervision. All three types of licence aim to produce good scientific data with the minimum of animal suffering.

The number of animals used in research varies depending on the country, but in the UK it was about 3.2 million in 2007, whereas it is estimated to be around 80 million in the USA (they do not routinely count all the animal species used, unlike in the EU). Of these, 83% involved mice, rats and other rodents, 10% were fish and 4 % birds. Four thousand primates were used; of this number in the UK, 21% were used for the research and development of human medicines, 5% for veterinary medicines, 31% for fundamental research and 13% for safety testing. About 36% of the total animals used were genetically modified (93% mice, 6% fish). (DBM)

See also: Laboratory animal

Reference

Home Office (UK) Homepage. Available at: http://www.homeoffice.gov.uk

Animal Health and Welfare Officer

This is a person who is identified in the **Animals (Scientific Procedures) Act 1986** as the Named Person in Day-to-Day Charge (unofficially renamed as the Named Animal Care and Welfare Officer (NACWO)). His or her role is to ensure the day-to-day care of all animals in the research establishment that is part of an institute's animal research programme. S/he has certain responsibilities, including drawing the attention of scientists licensed to carry out research to any experimental animal whose welfare gives rise to concern, and to deal with any other animal that is not the responsibility of a licensed person, e.g. a stock animal that may need care. Animals that give rise to concern include those that may have exceeded the permitted level of severity that has been designated in the project licence protocol. The NACWO may report directly to the Home Office Inspector, if necessary. (DBM)

Animal hoarding

Animal hoarding describes the collection of animals (of one or more species) by an individual to the extent that it exceeds their capacity to care for the animals to an acceptable level; this lack of care is not usually acknowledged or recognized by the individual concerned, leading to the suggestion that this condition may reflect a mental health disorder. Animals are usually kept in unsanitary conditions and are often undernourished, with a significant proportion requiring veterinary attention. The situation not only results in severely compromised welfare for the animals, but often a serious public health problem as well. There have been few scientific studies into animal hoarding, although it is recognized as occurring by relevant agencies (e.g. local government authorities and larger **animal rescue** organizations) in the industrialized world. Scientific literature suggests that hoarding of cats may be more common than of dogs, both of which are markedly more common than the hoarding of other species. (DSM)

Further reading

Patronek, G.J. (1999) Hoarding of animals: an under-recognized public health problem in a difficult to study population. *Public Health Reports* 114, 81–87.

The Hoarding of Animals Research Consortium. Available at: http://www.tufts.edu/vet/cfa/hoarding (accessed 25 September 2009).

Animal integrity

Rapid and radical modifications of animals brought about by conventional breeding and (especially) advanced biotechnologies sometimes raise public concerns. **Bioethicists** have coined the phrase 'animal integrity' to express one such concern. The basic idea is that we should not violate the integrity of animal species. 'Integrity' here means *wholeness* or *intactness*.

This idea seems to assume that animals have a relatively fixed nature and that human interference with this is morally (or perhaps aesthetically – it is sometimes hard to tell what is at stake) objectionable. There are several difficulties with this attitude. Animal species evolve over time, so it is not easy to identify the true nature of a species. Again, farm animals have been altered substantially by conventional breeding methods, largely without comment. Why should the fact that this process has recently accelerated cause concern? Some people will argue that, as long as the animals involved will not be caused to suffer poor levels of welfare, there is no problem about changing animals. For example, there is in principle no problem with breeding featherless chickens, as poultry breeders in Israel managed to do some time ago, as long as the chickens involved will be kept in a hot environment where they will not need their feathers to keep warm.

However, these people may also wish to stress the need for *precaution* in this area (**see: Precautionary principle**). In practice it may be difficult accurately to foresee effects on modified animals. Those who are essentially pro-modification may therefore interpret the idea of animal integrity as a precautionary attitude to rapid and radical animal modifications. Others may believe that, although alterations of animals through breeding are inevitable, there is a limit to how far we, as humans, may go in changing our fellow creatures. This view may be based on the ethical claim that we should respect nature in its own right (**see also: Environmental ethics**). An obvious issue here is whether *domesticated* animals can sensibly be considered part of nature worth protecting for its own sake. (PS)

Further reading

Sandøe, P. and Christiansen, S.B. (2008) *Ethics of Animal Use* (Chapter 9). Blackwell, Oxford, UK.

Vorstenbosch, J. (1993) The concept of integrity. Its significance for the ethical discussion on biotechnology and animals. *Livestock Production Science* 36, 109–112.

Animal rescue

Animal rescue refers to the removal of an animal from a place of potential danger, ideally to a place where its welfare is better, although rescued animals may be euthanized in some circumstances. Rescue may involve the removal of an animal from an acute crisis, e.g. a cow that slips over a cliff, or from a chronic situation, e.g. the rescue of a collection of animals from a keeper who is unable to provide for them adequately, or the acceptance of **abandoned animals**, including those surrendered by their owners. *In situ* rescue is also possible, for example by providing additional resources for animals within a given environment, such as the provision of hay to sheep stranded by flooding. Animals may be rescued by organizations or individuals.

Animal rescue organizations are not the same as animal welfare organizations, despite the popular claim by the former to be the latter and the well-meaning intention of those involved in animal rescue in general. For example a not uncommon priority for those operating animal rescue is a focus on saving life rather than preserving or ensuring a high **quality of life.** This can easily lead to animals being 'rescued' into situations where their welfare may be seriously compromised due to inadequate resources and a desire to save them from dying. The tendency of individuals or organizations to exceed their financial capacity is not uncommon, and larger rescue organizations may be called upon to take over other rescue organizations or populations as a consequence, potentially compromising their own strategic plans for animal welfare as a result. At the extremes, what may begin as rescuing a few individuals may develop into an obsessive situation or result in **animal hoarding.** (DSM)

Animal rights

Animal rights refers to the claim that animals have moral rights

This concept is primarily associated with the ethical theory first defended by **Tom Regan**, American philosopher and author of the book *The Case for Animal Rights* (1984), of which the main message is that sentient animals ('living subjects of a life', in the author's own definition) possess individual rights that ought to be respected.

In the language of ethics, a right confers protection to an individual, primarily by setting limits to how others can treat him or her. Consequently, if an individual has a right, s/he is justified in making corresponding claims on others. In that sense, rights have been described as a sort of trump, a claim that overrules any other argument made over the issue. Rights can be moral or legal, or a combination of both. A distinction is also sometimes made between *active rights* (the rights to do things) and *passive rights* (the rights to have things done for or to one), and between *negative rights* (the right *not* to have something done to one) and *positive rights* (the right to have something done for or to one).

In ethics, rights is a controversial topic: there is disagreement about whether they exist at all, how extensive they are and who can hold them. Who can be a rights-holder is an essential question when discussing animal rights. Some philosophers argue that animals cannot have rights, because rights are always coupled with duties and responsibility, so that only an individual that can be held responsible can have a right. This position is, however, not compatible with the view of our society, in which newborn babies and mentally disabled persons are also granted certain rights. Even though there is a coupling of rights and duties, the right and its corresponding duty are not necessarily held by the same individual, and the distinction between active and passive rights is useful in understanding this. If a right to a certain type of action is granted to an individual (an active right), it seems reasonable that a corresponding responsibility is demanded. But if the right is passive, a right to have something done for one, the duty or responsibility falls on those that are to do something for the rights-holder. If children have rights, their adult guardians have duties towards them. The issue of animal rights can be seen in the same way: if animals have rights, then humans have duties towards animals.

Regan builds his theory of animal rights on the work of German 18th-century philosopher **Immanuel Kant**. The essence of Kant's approach to rights is that humans have 'an intrinsic worth, i.e. dignity', and that therefore they ought to be treated 'always as an end and never as a means only'. That an individual is to be treated always as an end and never as a means can be understood in the way that it can never be justified to sacrifice an individual, even if this will lead to overall beneficial consequences. Tom Regan (1989) argues that the principle of dignity must be extended beyond the human community:

> But attempts to limit its scope to humans only can be shown to be rationally defective. Animals, it is true, lack many of the abilities humans possess. They can't read, do higher mathematics, build a bookcase or make *baba ghanoush*. Neither can many human beings, however, and yet we don't (and shouldn't) say that they (these humans) therefore have less inherent value, less of a right to be treated with respect, than do others. It is the *similarities* between those human beings who most clearly, most noncontroversially have such value (the people reading this, for example), not our differences that matter most. And the really crucial, basic similarity is simply this: we are each of us the experiencing subject of a life, a conscious creature having an individual welfare that has importance to us whatever our usefulness to others. We want and prefer things, believe and feel things, recall and expect things. And all these dimensions of our life, including our pleasure and pain, our enjoyment and suffering, our satisfaction and frustration, our continued existence or our untimely death – all make a difference to the quality of our life as lived, as experienced, by us as individuals. As the same is true of those animals that concern us (the ones that are eaten and trapped, for example), they too must be viewed as the experiencing subjects of a life, with inherent value of their own.

But animal rights are not necessarily defined as extensively as does Regan. A much more moderate position can still grant animals rights, albeit more limited, such as for example the right to be protected from extreme suffering or the right not to be killed for a trivial purpose. In legal terms, many countries' animal **welfare** legislation may be seen as a protection of such more moderate moral rights of animals.

Animal rights is also an umbrella term for the more radical section of the animal protection movement

The animal rights movement draws upon the writings of both **Peter Singer** and Tom Regan's philosophy, although only the latter is a rights theorist in the philosophical terminology. A distinction is often made between 'animal rights' organizations and 'animal welfare' (or in the Germanic languages 'animal protection') organizations. Animal welfare/protection advocates accept the human use of animals as long as animals are guaranteed acceptable welfare. On the contrary, the animal rights position that animals' individual rights should be respected is generally incompatible with most, if not all, human use of animals. Consequently, organizations with an animal welfare ideology are typically reformist, accepting improvements within current practices, whereas animal rights organizations are typically abolitionists, demanding the end of animal farming, animal-based research and most other ongoing practices where non-human animals are used in the interest of humans. A small fraction of the animal rights movement maintains that the animal cause justifies the use of illegal and/or violent actions, but both Singer and Regan have dissociated themselves from that position. (AO)

Further reading

Almond, B. (1991) Rights. In: Singer, P. (ed.) *A Companion to Ethics*. Blackwell Publishers, Oxford, UK.

Regan, T. (1984, rev. 2004) *The Case for Animal Rights*. University of California Press, Berkeley, California.

Regan, T. (1989) The case for animal rights. In: Regan, T. and Singer, P. (eds) *Animal Rights and Human Obligations*. Prentice Hall, Englewood Cliffs, New Jersey.

Animal sanctuary

An animal sanctuary is an enclosure or area which houses a group of orphaned or rescued animals; typically, the animals are kept either permanently or for an extended period of time prior to release back into their natural environment. By contrast, an animal shelter aims to provide for an animal in the short term with a view to rehoming it (**see: Animal shelter** for details relating to the rescue and rehoming of domestic animal species).

Animal rescue is not synonymous with animal welfare, but sanctuaries may provide part of the solution to future research on a range of captive species, given the rising costs of keeping animals in captivity and the concern over the use of laboratory animals. This may be particularly the case for work on captive great apes. The increasing ethical dilemma and expense of working with animals, especially apes, in biomedical facilities has prompted a need for alternative research sites. Working with animals in African sanctuaries benefits the animals, who have a better quality of life, and supports conservation efforts. It also benefits the researchers as the animals have a more natural upbringing and there are more test subjects, leading to higher-quality data. A similar argument could be applied to other exotic species living in sanctuaries, but this article focuses on the case relating to great apes, as a diverse range of benefits relating to welfare and conservation from both pure and applied research can be demonstrated very clearly.

The benefit of animal sanctuaries as a place of animal research

Historically, most of what we know about ape cognition has been discovered in biomedical laboratories from tests on **chimpanzees**. However, largely due to expense, most **great ape** laboratories outside the USA have been closed. Remaining US centres house only chimpanzees (no bonobos). The three chimpanzee subspecies have not been maintained as separate populations, and so subspecies comparisons are not possible. Laboratory populations consist almost entirely of adults, who are ageing rapidly. There is no younger generation, as the National Institute of Health (NIH) banned breeding chimpanzees for all NIH-funded projects. The majority of chimpanzees in laboratories were reared and live in simple concrete cages that were designed before any research on wild chimpanzees had been conducted. Therefore, current laboratory facilities do not provide the rich and complex environment in which chimpanzees normally develop and for which they have evolved to live.

At the same time, the remaining unprotected habitats in Africa are being destroyed and the apes living within that area are killed. The only surviving apes from such populations are infants that are taken from dead mothers to be sold as pets within Africa or on the international black market (including Europe and the USA). The lucrative trade in these infants fuels the demise of other wild ape populations, as some try to profit by selling more infants obtained from the wild.

Most infants die within days of capture, but those that survive face a difficult life. Apes are many times stronger than their human caregivers and can not be tamed – inevitably leading to psychological and physical abuse when owners can no longer manage them.

Governments in Africa have created laws banning the capture and sale of great apes. Sanctuaries were created to offer lifelong care to those individuals confiscated as a result of such laws. Without sanctuaries, laws protecting wild apes from capture could not be enforced and thus would do little to stop the trade of live infants. The alternative would require euthanizing confiscated apes. Today there are 17 sanctuaries in 12 countries, and together they provide lifelong care for over 1000 wild-born African apes and economic benefit to many Africans.

Ape sanctuaries offer an ideal opportunity for researchers to conduct innovative research on semi-free-ranging ape behaviour, intelligence, genetics, disease/health, morphology and physiology. Sanctuaries manage apes at a fraction of the cost of laboratories. This means that non-invasive work requires only small grants (invasive work may require licensing under current legislation and so is currently less feasible in many sanctuaries). It is easy to remodel workspace for short- or long-term projects, and the cost of living in Africa is affordable compared with other sites of research. All ape species (including the largest population of captive bonobos in the world, with over 50 individuals) are represented in sanctuaries. Subspecies are maintained as separate populations, so all three chimpanzee subspecies can be compared. While breeding is controlled, the majority of individuals in sanctuaries are infants and juveniles.

Day enclosures for large social groups are areas of tropical forest that can be as large as 40 ha (100 acres). This means that the apes are reared and live in environments similar to those for which apes evolved to thrive mentally and physically – an environment that allows most animals to quickly recover from their early trauma.

Why should sanctuaries cooperate with researchers?

Researchers can make various valuable contributions to sanctuaries. In addition to helping with funds and enrichment of sanctuary animals, researchers can work with sanctuaries towards a common goal, primarily raising awareness of the species and providing material for education programmes. Research fees become another source of income for sanctuaries, helping towards management costs or improvements for research. Researchers may also donate equipment like computers and veterinary supplies. Furthermore, researchers may have expertise in areas like disease screening that could benefit the sanctuary animals. Non-invasive research, which is painless and stress free, may potentially even provide enrichment for captive populations. Many forms of primate behavioural research involve presenting non-human apes with problem-solving tasks to study their intelligence. Apes enjoy problem solving and readily volunteer to participate in such games, since these typically involve obtaining food rewards. People respect and then protect what they understand. The results of studies on non-human apes add to our overall knowledge of great ape species and thus could lead to more respect and protection for great apes. In addition, non-invasive research in sanctuaries could set a new ethical standard for research with non-human apes, by offering researchers an alternative to working on **laboratory animals**.

For example, researchers at the Max Planck Institute for Evolutionary Anthropology and Harvard University are carrying out successful research programmes at three sanctuaries in Uganda, the People's Republic of Congo and the Democratic Republic of Congo.

The aim of this research is to identify precisely what it is about human psychology that can explain why, in many cases, humans behave so differently from other animals. Comparisons between the intelligence of bonobos, chimpanzees and human children are believed to be the only way to accomplish this goal. In trying to find out what makes humans different from other apes, we must also understand in what ways we are similar to our closest living relatives. Therefore, this type of research often identifies surprising similarities as often as it does differences between our species.

Sanctuaries recognize the importance of learning about their captive populations through non-harmful techniques that can enrich or improve the lives of their animals. Continued success will depend on whether future researchers approach sanctuaries respectfully, with a collaborative spirit and genuine interest in ape welfare and conservation. If they do, sanctuaries will become the world's finest resource for improving our understanding of the great apes, and potentially other species, in the decades to come – to the benefit of all. (VW, BH)

Further reading

Gagneux, P., Moore, J. and Varki, A. (2005) The ethics of research on great apes. *Nature* 437, 27–29.

Hominoid Psychology Research Group. Available at: http://www.eva.mpg.de/3chimps

ILAR, National Research Council (1997) *Chimpanzees in Research: Strategies for their Ethical Care, Management and Use.* National Academy Press, Washington, DC.

Melis, A.P., Hare, B. and Tomasello, M. (2006) Chimpanzees recruit the best collaborators. *Science* 313, 1297–1300.

Pan African Sanctuary Alliance. Available at: http://www.panafricanprimates.org

VandeBerg, J.L. and Zola, S.M. (2005) A unique biomedical resource at risk. *Nature* 437, 30–32.

Animal shelter

An animal shelter is typically a place of temporary refuge (**see: Animal sanctuary** by contrast) and is often associated with **animal rescue** and **abandoned animals**. The aim of a shelter is to provide for the animal until it can be placed in more appropriate long-term accommodation. In some countries and regions, local laws prevent the **euthanasia** of healthy animals and some shelter organizations also adopt a 'no-kill' policy; in this case these shelters are in effect animal sanctuaries. While the euthanasia of healthy animals is a cause for potential ethical concern (**see: Ethics**), the maintenance of animals in a restricted environment, e.g. dogs in kennels, is a cause of **welfare** concern. This has led to a growth in research, largely sponsored by animal welfare organizations, into methods of environmental **enrichment** for companion animals in **confinement**.

Most shelters seek to rehome animals and many invest in efforts to match an owner with a potential **pet**, although there has been little scientific validation of the efficacy of matching processes or in-shelter temperament testing. One recent report (Diesal *et al.*, 2008) suggests that the majority of unsuccessful canine rehomings are due to behavioural problems in the dog. The attendance at training classes and realistic expectation of the work involved in owning a dog also reduce the risk of return. Thus the importance of personality matching as opposed to more general owner education on the animal husbandry may be questioned. (DSM)

Reference and further reading

Diesal, G., Pfeiffer, D.U. and Brodbelt, D. (2008) Factors affecting the success of rehoming dogs in the UK during 2005. *Preventive Veterinary Medicine* 84, 228–241.

Mellen, J.D. and Shepherdson, D.J. (1997) Environmental enrichment for felids: an integrated approach. *International Zoo Yearbook* 35, 191–197.

Taylor, K.D. and Mills, D.S. (2006) The development and assessment of temperament tests for adult companion dogs. *Journal of Veterinary Behaviour: Clinical Applications and Research* 1, 94–108.

Taylor, K.D. and Mills, D.S. (2007) The effect of the kennel environment on canine welfare: a critical review of experimental studies. *Animal Welfare* 16, 435–447.

Animal trade

Animal products form an important part of global trade. For example, meat chickens (broilers) are the most numerous farmed terrestrial animal, with an estimated annual production of 40 billion birds – and a large proportion of this huge production is shipped around the world (in 1 million t; 1 t of poultry meat can be produced from approximately 1200 chickens):

- Exports: USA, 2.3; Brazil, 1.1; EU, 0.65; Thailand, 0.4.
- Imports: Russia, 1.3; Japan, 0.7; EU, 0.6; Saudi Arabia, 0.35; Mexico, 0.3; China, 0.3.

Global trade is influenced by national government decisions. For example, there may be legislation concerning hygiene standards or animal **welfare** standards, and individual governments may provide subsidies to producers. Produce sold on the international market must usually meet global standards – for example, to control livestock diseases or to prevent trade in endangered species.

The General Agreement on Tariffs and Trade (GATT) was drawn up by 23 countries in 1947. GATT was formed after World War II as a result of the 'Bretton-Woods Order', which also created the International Trade Organization, the International Monetary Fund and the World Bank. This Order was designed to provide an international forum that encouraged free trade between member states. The Marrakech Agreement established the World Trade Organization (WTO) on 1 January 1995. This Agreement restated the General Agreement on Tariffs and Trade (GATT) (1947 and 1994). GATT had 128 signatories by the end of 1994 and, in 2008, WTO had 153 member countries. The WTO provides a forum for negotiation of trade relations between its members, promotes free trade and adjudicates in disputes between its members. A number of WTO side-agreements have the potential to affect animal welfare and related marketing issues: The Agreement on Technical Barriers to Trade (1994), The Agreement on Agriculture (1994) and The Agreement on the Application of Sanitary and Phytosanitary Measures (1994).

In practical terms, the WTO facilitates trade agreements between countries (which usually apply to particular products), assists enforcement of bans on trade of certain wild animals and their products, assists in the application of disease control standards (for example foot-and-mouth disease) and mediates the application of trade concessions to countries in the developing world.

The interpretation and application of these articles depends upon comparison of the '*likeness*' of products; '*like*' products may not be discriminated against or protected. The issue of likeness is at the core of the potential for animal product marketing problems as the GATT dispute panels have determined that, when determining whether two products are '*like*' one another, they may only consider the '*end product*' and not '*the way in which it was produced*'. Some products – for instance, those organically produced – are demonstrably '*unlike*' conventionally farmed animal products, because they meet a distinct legal standard requirement (IFOAM, UK Compendium or local legal standards) and can be marketed as a distinct product type. To demonstrate that a product meets a certain separate standard, 'certification' is usually required to discriminate this type of 'unlike' product from a conventionally farmed product.

WTO agreements may hamper local efforts to introduce legislation to protect animals, as the inability to differentiate systems of production may create disincentives to producers to adopt measures designed to improve animal welfare. Under WTO rules a country is unable to ban imports of products on ethical grounds or to insist that laws designed to protect animals or the environment must apply to imported products as well as those produced domestically. WTO prevents restrictions on imports on the basis of production method, and so countries may be powerless to stop domestic markets being undermined by imports. Additionally, EU Directive 2000/13/EC prohibits the attribution to any foodstuff of the property of preventing, treating or curing a human disease, or reference to such properties; this can impinge on animal welfare-based marketing claims, as claims of human health benefits for products emanating from, for example, free range or organic production systems may not be made. The European Commission has recognized the potentially damaging effects of WTO agreements on global moves towards animal welfare improvement and has stated its intention to 'ensure that the trend toward liberalisation does not lead to competitive reduction in animal welfare standards'.

Many countries have developed **farm assurance schemes**, and these have the potential to improve local animal welfare by offering financial (and 'ethical') incentives for producers to rear animals to raised 'standards'. These standards are sometimes significantly above the lowest denominator set by the law, and may allow local differentiation of the animal product into '*welfare friendly*'. Internationally, it seems likely that farm assurance work may be undermined by the inability of systems of animal rearing offering improved animal welfare to visibly (through labelling or through trade restriction) differentiate their products in the global market.

Many animal welfare organizations and other stakeholders see the WTO agreements as having the potential to block progress on animal protection. These organizations propose that reform of the WTO rules could help to safeguard standards of animal welfare. Some of their proposed reforms include: (i) that animal welfare be considered as a legitimate 'non-trade concern' under WTO; (ii) a generalized alteration in the legal status of animals from trade products (property) to sentient beings; (iii) allowance of trade-related measures, such as import bans on genuine animal welfare grounds; and (iv) promotion of labelling schemes to distinguish between otherwise like products that are produced to different animal welfare standards (e.g. battery versus free range eggs). (AB)

Further reading

Garcés, L. (ed.) (2002) *The Detrimental Impacts of Industrial Animal Agriculture*. Compassion in World Farming Trust, Godalming, UK.

Stevenson, P. (2002) The World Trade Organization rules: a legal analysis of their adverse impact on animal welfare. *Northwestern School of Law, Lewis & Clark College Animal Law* 8, 107–141.

http://www.wto.org

http://ec.europa.eu/food/animal/welfare/international/index

Animals (Scientific Procedures) Act 1986

For many years, live animals were used in scientific experimentation in the UK with no legal safeguards for their **welfare**, until the Cruelty to Animals Act 1876. By this Act, there were to be no prosecutions for cruelty to animals used for experiments, but they were required to be anaesthetized, used only once and then killed. The flaws were that **anaesthesia** was potentially inadequate and need not be used if it 'frustrated the object of the experiment'; there could be more than one procedure within one experiment and animals need not be killed if, again, the object of the experiment would be defeated.

Over 100 years later the Animal (Scientific Procedures) Act 1986 came into being, and it is recognized as one of the most advanced pieces of legislation safeguarding the welfare of animals used in science, and a model for other countries, although opponents of animal experimentation claim it is a 'vivisectionist's charter'. It also controls the breeding and supply of animals to be used for research and the welfare of such animals.

Control is through the granting of licences for research projects, which must be meritorious (following a cost–benefit analysis (**see: Ethics**)); the environment in which the research is to be conducted must have adequate facilities to meet animal welfare needs; and individuals involved in the research must have suitable training and qualifications. Licences are granted based on criteria, i.e. there must be no alternative to the use of animals, suffering must be minimal, with anaesthetics and painkillers being used, and the animals must be used once and killed humanely. A major feature is the cost–benefit analysis, where the cost of the animal's suffering (mild, moderate or severe) is weighed against the benefit of the research, which must have a clear objective. Mostly small rodents are used, with larger animals and primates only being allowed in exceptional circumstances (**see: Three Rs principle**). Control and licensing is by the Home Office, through the Animals in Scientific Procedures Inspectorate and the Animal Procedures Committee. (BG)

Anthropocentrism

Anthropocentrism (from the Greek *anthropos*, human being) refers to the worldwide view and philosophical theories that consider human beings as the centre of moral considerations. The distinction is sometimes made between strong (*only* human beings have moral standing) and weak (human beings are central and other beings peripheral to moral considerations) anthropocentrism.

The traditional approaches to morality in Western society are essentially anthropocentric. This is usually seen as the outcome of the Judaeo–Christian view of human beings as having been created in the image of God and the only possessors of an immortal soul, in combination with the focus on rationality (originating in Aristotle's ethics and further emphasized during the Enlightenment).

The first seriously considered challenge to this strong anthropocentrism came with the English 18th-century writer **Jeremy Bentham**. Bentham argued that whether an individual should be considered a moral subject depended on its capacity for having subjective experiences or, in other words, to feel pleasure or pain. Such an approach to ethics is sometimes termed *sentiocentric*, and the capacity for subjective experiences ***sentience***. Several modern philosophers, nevertheless, disagree with the idea that sentience provides the complete answer to the question of who should be granted moral consideration. An alternative approach is suggested by Paul Taylor, who argues that all living beings – not only sentient animals but also plants and microorganisms – have an inherent value. This is because, Taylor argues, all these beings can be seen as having 'a good of their own' which they are – consciously or not – pursuing. Other philosophers go further and criticize the focus on individuals inherent in both the sentiocentric view and in Taylor's *biocentric* view. For many environmental ethicists there is a specific value also in species, ecosystems and even in the evolutionary process itself, as well as in non-living entities. Approaches which place the emphasis on other aspects than individual beings are usually referred to as *holistic* (in contrast to individualistic theories) or *ecocentric* (in contrast to sentiocentric and biocentric theories). (AO)

Further reading

Palmer, C. (2003) An overview of environmental ethics. In: Light, A. and Rolston III, H. (eds) *Environmental Ethics*. Blackwell Publishing, Oxford, UK.

Russow, L.M. (2002) Environment. In: Comstock, G.L. (ed.) *Life Science Ethics*. Iowa State University Press, Ames, Iowa.

Anthropomorphism

In the context of animal behaviour, 'anthropomorphism' is the assumption that animals share the same thoughts, feelings and motivations as humans. In everyday life, animal companions are (and probably always have been) very often treated with extreme anthropomorphism; that is, they are treated as almost human and sometimes even as substitute children. Frans de Waal calls this 'naive behaviourism', or taking an anthropocentric attitude to animals (Waal, 2001). Most ethologists today would probably claim that they avoid **anthropocentrism** and try to understand the animals they are studying and interpret their behaviour on the animals' terms. However, the attitude of these ethologists to the use of anthropomorphism in the interpretation of the way they study animals behave will depend on their cultural as well as their scientific training.

In parallel with the development of ethology as a science, anthropomorphism has undergone three alternating phases – acceptance, rebuttal and re-acceptance. In the 1930s, when **Konrad Lorenz** (1903–1989) began his studies of animal behaviour, which can be truly called the beginnings of **ethology**, he had no compunction in relating the ways his jackdaws behaved to the ways of people, and he used the most personal and anthropocentric language. Lorenz was followed by **Tinbergen** and Thorpe, who, among others, were more rigorous in their direct observations of animal behaviour and led the subject into a scientific discipline. Then, from around the middle of the 20th century, **behaviourism** began to make an impact when, as Daniel Dennett wrote in the *Times Literary Supplement* (16 May 1997, p. 5), 'Icy blasts of reinforcement extinguished all thought, all feeling, all mentality, and the world was a dead mechanism'. In this second phase of attitudes, anthropomorphism became a pejorative term applied to anyone who dared to believe that animals were capable of conscious thinking.

There were, however, opponents to anthropomorphism who were not as single-minded as the radical behaviourists who appeared to have inherited the beliefs of Descartes. Over 50 years, J.S. Kennedy never wavered from his published view that, 'If the study of animal behaviour is to mature as a science, the process of liberation from the delusions of anthropomorphism must go on'. Kennedy wrote about what he saw as the unwitting anthropomorphism of the **neobehaviourists**, who he defined as modern behaviourists who differed from their radical forbears in not discounting internal processes in the causation of behaviour and in not excluding some measure of cognitive activity in their study animals. He pointed to some examples of less subjective terms, such as 'scanning' rather than 'searching' and 'resisted mating' instead of 'rape', that can be usefully used by ethologists to avoid overt anthropomorphisms.

Anthropomorphism is not just an out-of-date concept – discussion about it is important because the language that is used in everyday life reflects the culture, religion and education of the society that uses it. Those who believe that anthropomorphic language is to be avoided (apart from such obvious examples as those quoted above) and that the behaviour of animals is on a totally separate level from that of human thoughts and consciousness may be considered inheritors of the Judaeo–Christian tradition, which has taught over the centuries that only humans have a soul. It is hard to escape from this cultural and religious edict, which led to the association of the supernatural attribute of a soul with consciousness, and therefore to a general denial of consciousness in animals.

Towards the end of the 20th century, with the great expansion of knowledge about the behaviour of humans and animals, as well as an increased understanding of evolutionary theory, has come a renewed acceptance of anthropomorphism. Ethologists have been joined by psychologists and philosophers in research into the minds of animals. This multidisciplinary field of research, known as **cognitive ethology**, has resulted in a multitude of investigations into **consciousness, cognition, self-awareness** and **intelligence**, as well as on whether animals feel pain, anger, fear, love and have a **theory of mind**.

With the expanding research into animal minds has come the general realization that anthropomorphism does not disrupt scientific observation but supports the continuity between humans and animals. A strong supporter of this view is Frans de Waal who, in writing about attitudes to anthropomorphism, cites the example of his chimpanzee, Georgia, who regularly 'plays a trick' on visitors by taking a drink of water and then spraying them with it from her cheek pouches (Waal, 2001). Waal then discusses whether believing that the chimpanzee is consciously 'playing a trick' is an anthropomorphism; he argues that the most parsimonious conclusion is to assume that the chimpanzee and the humans have a cognitive similarity based on their close evolution, and he proposes the term 'anthropodenial' for the rejection of shared characteristics between humans and animals.

It is from this *milieu* that the study of the interactions between humans and animals is developing, and the investigators in this field of research may be truly called the new anthropomorphists, to whom improvements in animal welfare are of prime importance. (JC-B)

Reference and further reading

Kennedy, J.S. (1992) *The New Anthropomorphism.* Cambridge University Press, Cambridge, UK.

Waal, F. de (2001) *The Ape and the Sushi Master; Cultural Reflections by a Primatologist.* Allen Lane, Penguin Press, London.

Anthrozoology

Anthrozoology is the study of animal–human interactions. The combination of *anthro-*, meaning of humans, and *zoology* first appeared in 1987 as the title of the journal *Anthrozoös,* and was later used for the name of the International Society for Anthrozoology (although *anthropozoology* would be more correct etymologically). Applied literally, anthrozoology would therefore encompass much of conservation science, animal **welfare** and applied animal ecology, but it is normally applied to studies of associations, and especially relationships, between individual humans and individual animals, rather than to interactions that take place at the population level. Anthrozoology encompasses, but is not restricted to, study of the **human–animal 'bond'**. It should logically also include unsuccessful and dysfunctional interactions (although it rarely does).

The current tradition of research into human–animal interactions is usually considered to have begun with a paper by the psychiatrist Boris Levinson in 1962, which described the effects of interaction with a dog on a young patient of his. Two decades later, by which time several research centres had appeared, particularly in the USA, Levinson addressed one of the first major conferences in this field, in Philadelphia, where he suggested four areas on which future research might focus: (i) the role of **companion animals** in human cultures from prehistory to the present, including archaeozoology, anthropology and sociology; (ii) the effects of associations with animal companions on the development of personality, attitudes and other traits in humans, encompassing psychology and psychiatry; (iii) the therapeutic effects of companion animals; and (iv) the ethology of human–animal interactions. A further two decades on, knowledge has increased in all of these areas, and there have also been attempts to forge links between them to develop underlying theories of the human–animal 'bond', which will be covered in the latter part of this entry.

The roles of animals in human cultures have often been studied from an **anthropocentric** viewpoint and so fall outside the scope of this volume. Archaeozoology has made a valuable contribution towards our understanding of the behaviour and appearance of modern companion animals, by enabling us to trace the early course of their domestication. However, for the domestic **dog** there are currently serious discrepancies between the timing of domestication as indicated by archaeology and that suggested by the molecular biology of modern dogs and wolves. Moreover, the main focus of archaeology and historical studies has been on the contributions of animals to changes in human society, rather than on human–animal relationships. For example, the use of **horses** is cited as a major factor in the domination of the advanced societies of South America by a comparatively small number of Spaniards. At the population level, the **domestication** of animals provided opportunities for infectious agents of animals to transfer to the human host, creating such modern **diseases** as smallpox, influenza, tuberculosis, measles and cholera. Most of these appear to have originated in agricultural animals, one exception being the possible origin of whooping cough (pertussitis) in the dog.

However, relationships can occasionally be deduced from archaeological finds: the puppy found in a Mesolithic grave of 12,000 YBP, buried beneath the hand of an elderly human, was presumably a **pet**, whereas the knife marks found on some **cat** bones from mediaeval sites in Britain indicate that these animals had been killed to make cat-skin coats.

The idea that pet keeping has a positive benefit on child development is not a new one, although the underlying concepts have changed. For example, many Victorians believed that companion animals in the household could be used to teach children an abhorrence of cruelty and the value of gratitude and fidelity, emotions that pets were presumed to display towards their owners, and these attitudes have persisted in many societies up to the present day. More recently the research emphasis has switched to the effects of pet keeping on the development of **empathy**. Pet keeping in childhood seems to be reliably associated with positive attitudes towards animals in young adulthood, and there appears to be a link between empathy towards animals and empathy towards humans.

Empathy is a complex construct that is generally considered to have both cognitive ('I can imagine what you are thinking') and affective ('I can feel what you are feeling') components, and it is not yet clear precisely which aspects are most affected by pet keeping. The few studies of pets used as an intervention in educational settings have generally reported an increase in empathy, but associations between pet ownership and empathy in children appear to be more complex: for example, one study found slightly higher empathy in dog owners, but lower empathy in cat owners, compared with non-owners. Since most family pets are chosen by parents, the latter association could be due to a personality characteristic of parents that drives both a preference for cats as pets and an inhibition of the development of empathy in children, rather than a negative effect association that a cat has on the development of empathy. Whatever the mechanisms involved, it seems likely that different species of companion animals may have different effects on emotional development, and that classifying them all as 'pets' may be an oversimplification. Moreover, research in this field has overwhelmingly focused on the development of desirable traits, and the possibility that not all pets may be beneficial for all children has barely been considered, even though links have been established between abuse of animals and violence towards people.

A variety of benefits to health have been proposed for pet ownership, and for emotional attachments to animals, and there has been much discussion about the role of pets in enhancing various aspects of human quality of life. In societies where they are well accepted, dogs act as catalysts for social interactions, especially between owners and people they have not met before, thereby expanding existing social networks. Similar effects have been observed for trained assistance dogs, such as those used as guide dogs and for people with other disabilities, and it has been suggested that these benefits may be as important as those arising from the work that the dogs are trained to do. Comparable effects due to other pet species are less well documented. Physical contact with a friendly pet has often been demonstrated to have a relaxing effect, for example by lowering blood pressure, although the undoubted stress occasionally generated by pets, for example when they misbehave, has received little research attention.

More contentious has been the idea that pet keeping may be generally beneficial to mental and physical health, and specifically protect against cardiac illnesses. Such beliefs have apparently become widespread among Western societies, but supporting scientific evidence has often been equivocal or ambiguous. Several studies in different countries have reported fewer visits to doctors among pet owners than comparable non-owners, but these findings are generally correlational rather than cause and effect. Moreover, not all have factored out the effects of additional exercise that dog ownership normally incurs. Overall, it appears likely that pet ownership is associated with slightly better health, but the mechanisms involved are still far from clear, and in considering the possible health benefits that pets might provide it is important to allow for the likelihood that unhealthy people may be unlikely to acquire pets and more likely to relinquish ownership.

Levinson was originally inspired to consider the benefits of pet keeping by the effects of the presence of his dog on the behaviour of otherwise uncommunicative children who came to him for psychotherapy, and he proposed that the dog had acted as a 'bridge' for the establishment of rapport between child and therapist. Since his pioneering work, such effects have been widely observed (**see: Animal-assisted therapy**). Interaction with dogs, in particular, has a strong motivational effect on many children with both physical and learning disabilities, perhaps because interaction with the dog is experienced as non-judgemental and unconditionally positive.

Ethology, the study of the behaviour of animals in their natural habitat, has made only a limited contribution to anthrozoology, largely because in its modern form it specifically addresses, or at least implies, the evolutionary origins and adaptive functions of behaviour. These are not easy to unravel for domesticated animals whose behaviour has been both consciously and unconsciously selected by man. However, ethological concepts such as **dominance**, **territoriality** and **imprinting** have been widely used to interpret the behaviour of pet animals. For example, the behavioural disorders of pet cats are often divided into those of ethological origin, i.e. normal behaviour that would have been adaptive to the cat's wild ancestors but is objectionable to owners, such as spray-urination, and those that are abnormal, such as **pica**.

The concept of **socialization**, largely derived from the ethological theories of imprinting and sensitive periods, is the cornerstone of our understanding of how young companion animals learn to react to humans and other species, and how behavioural disorders can result from deficiencies in this process. In addition, ethological methods, involving the detailed observation of animals in their normal surroundings, with minimal interventions, have been used to provide detailed descriptions of interactions between pets and their owners. For example, they have been used to differentiate between the competitive play that takes place between two dogs, and the socially interactive play between dogs and people.

Theories of the human–animal 'bond'

Anthrozoology has usually been studied by specialists from one of the constituent disciplines referred to above, but at its core it is multidisciplinary, and attempts have been made to explain the keeping of pet animals, especially dogs and cats, using various theoretical frameworks, in an attempt to arrive at an explanation of the human–animal 'bond'. Some have adopted a biological perspective, for example those based upon traditional comparative zoology and ethology.

In the terminology of comparative zoology the relationship between owner and pet could potentially be classified as **commensalism**, **mutualism** or **parasitism**. Some domesticated species may have initially had a commensal relationship with man, probably as scavengers, including the wolf, the progenitor of the dog, and possibly the pig. Domestic animals primarily kept for their utilitarian value, such as dairy cattle or guarding dogs, have a mutualistic relationship with man in which costs and benefits are deliberately balanced out by the human partner. Pet keeping is undeniably costly for owners: for example, in the USA more than US$20 billion is spent annually on pet food and pet health care. Since pets return no immediate economic benefit to their owners, some authors have classified them as **social parasites**, analogous to the cuckoo chick in the nest of its foster-parents.

Supporters of this idea have pointed out that the foster-parents presumably find it rewarding to feed the cuckoo chick, even though they inadvertently incur substantial biological cost in the process. In this conceptualization, therefore, the emotional rewards of pet keeping experienced by the owner are no defence against the charge of parasitism. However, pet keeping has two distinct differences from the cuckoo-type of social parasitism. First, the majority of pets are deliberately obtained by their host species, although the stray cat that 'adopts' an owner could be cast in a cuckoo-like role. Secondly, pet owners do terminate the relationship with their pets if the costs become too high, especially social costs, such as the dog that bites, the cat that defaecates in the house, or the pet-induced allergies that can interfere with family relationships. Moreover, owners are aware, unlike the cuckoo's foster parents, that they can make such decisions at any time. The possession of this consciousness by one of the partners introduces complexities that may be difficult to assimilate into the conventional classifications of comparative zoology.

The ethological concept of **social releasers** has been used to explain the attractiveness of those pets that possess features redolent of those of human infants. Thus domestic cats, and some breeds of dog, have large, round, flat-faced heads and large forward-facing eyes that are superficially baby-like, and could trigger nurturant behaviour in humans. That such features are indeed preferred by humans is evident from their progressive appearance in teddy bears, which originally had bear-like proportions, and Mickey Mouse (who originally looked like a mouse). In some cases, such as the selection for flat-faced (brachycephalic) dogs such as the pug, human preference for such features has resulted in obstructions to normal breathing that constitute a serious welfare problem. However, this theory cannot account for dogs with pointed noses, or for the general preference for furry pets over hairless ones, even though the skin of the latter is more similar to that of a human infant.

Nevertheless, concepts from **sociobiology** and **evolutionary psychology** have been useful in probing the nature of the bond between owner and pet. Historically, one of the defining characteristics of the brain of *Homo sapiens sapiens* appears to be its capacity to attribute thought processes to animals. Evolving perhaps as recently as 40,000 YBP, this ability is proposed to underlie the emergence in man of complex hunting strategies, which were highly adaptive (in the evolutionary sense), and enabled both the colonization of new habitats and the displacement of Neanderthal man. Pet keeping may be a side effect of this ability, since it enables humans to impart human characteristics to animals and thereby form personal relationships with them, rather than conceiving of them simply as commodities. In this context it is interesting that pet keeping is widespread in some modern-day hunter–gatherer societies, and may possibly provide a functional benefit by providing opportunities for children to learn about animal behaviour and ultimately become more successful hunters.

Certainly the human mind has a propensity to over-attribute human thought processes to animals, and to pets in particular. Thus most pet owners express the belief that their pets are sensitive to their feelings and understand much of what they say to them, despite the current lack of supporting evidence for this from scientific study. The pet–owner relationship appears to be based upon emotions attributed by the owner to the pet, which are more consistently rewarding because they are unconditional and non-judgemental than equivalent relationships with humans, which can be both conditional and judgemental.

From a psychological perspective, the pet–owner bond is often addressed in terms of **attachment** theory. Owners usually express strong attachments towards their pets, and feelings of grief when they die that are as acute, if not so long lasting, as those felt on the death of a close relative. From studies of the ways in which people interact with their pets, and the language that they use when talking to them, parallels have been drawn with parent–child interactions (the nurturant or caregiving part of the relationship), and the benefits of reassurance obtained from the pet may be a substitute for companionship with other adult humans. However, the charge that relationships with pets are no more than a poor substitute for deficient relationships with other people has not been borne out by research, except in the extreme cases of animal 'collectors'. (JWSB)

See also: Farmed animals

Further reading

Archer, J. (1997) Why do people love their pets? *Evolution and Human Behaviour* 18, 237–259.

Beck, A.M. and Katcher, A. (1996) *Between Pets and People: the Importance of Animal Companionship*. Purdue University Press, West Lafayette, Indiana.

Fawcett, N.R. and Gullone, E. (2001) Cute and cuddly and a whole lot more? A call for empirical investigation into the therapeutic benefits of human–animal interaction for children. *Behaviour Change* 18, 124–133.

Serpell, J.A. (1996) *In the Company of Animals*. Cambridge University Press, Cambridge, UK (Canto).

Serpell, J.A. (2003) Anthropomorphism and anthropomorphic selection – beyond the 'cute response'. *Society and Animals* 11, 83–100.

Antidiuretic hormone

Antidiuretic hormone (ADH), also termed arginine vasopressin (AVP), is a nonapeptide hormone synthesized by hypothalamic neurones and secreted from the posterior **pituitary gland**. Principal functions are concerned with water homeostasis, release being triggered by increased plasma osmolarity activating hypothalamic osmoreceptors; ADH acts to insert membrane water channels (Aquaporin-2) into the kidney collecting duct, promoting reabsorption of water and secretion of a smaller volume of concentrated urine. There is also neurological release of ADH into the **brain** and there is evidence to indicate ADH effects on cognitive functions such as **memory**; it has also been implicated in the expression of a range of **social behaviour** patterns.

ADH shares a largely common structure with **oxytocin**, another nonapeptide pituitary hormone, and there is some cross-over between the actions of the two. (MRC)

Anti-predator training (for reintroduction)

Anti-predator training is the teaching of captive-reared animals to avoid predators. It is becoming a common component of pre-release conditioning prior to the reintroduction of captive-reared animals to the wild. It may be surprising that animals require experience for the development of effective anti-predator skills, since anti-predator behaviour must be functional when a predator is first encountered. However, the ability to recognize and avoid predators is often dependent on experience and has been demonstrated across taxa. Anti-predator training typically entails the use of classical conditioning techniques to link an aversive stimulus – e.g. unpleasant stimulus such as water squirts, or environmental cues of predator presence such as conspecific alarm calls – with the presentation of a live or model predator to which the individual initially has a neutral response. This linkage quickly generates an acquired fear response to the predator or model, generally in one or two exposures. For example, in fish, if the presentation of a predatory fish (which had previously elicited no alarm) is paired with a chemical alarm substance released from the damaged skin of conspecifics, individuals acquire an alarm response.

Anti-predator training can also involve improving the efficiency of responses to predators as a result of experience. For example, young Siberian polecats (*Mustela eversmanni*) trained with a model badger presented concomitantly with an aversive stimulus (being shot with rubber bands) reduced their escape times after a single exposure. To be most effective, anti-predator training should mimic the critical features of ontogenetic processes in the wild: the stimuli used, the developmental timing of the training and the social and physical context in which it occurs must all be appropriate.

For example, training animals in a social context has the potential to amplify the effects of this intervention. Laboratory-reared rhesus monkeys became fearful and avoided snakes after observing conspecifics interacting with snakes. Moreover, when trained animals are reintroduced to the wild, they may function as demonstrators for predator-naive individuals, which may then acquire effective anti-predator behaviour. Thus, social facilitation of acquired anti-predator skills can increase the effects of the initial investment in pre-release training by extending the benefits of training, over the short term, to a much larger number of animals and, over the long term, through the generations. (DMS)

See also: Reintroduction

Further reading

Griffin, A.S. (2004) Social learning about predators: a review and prospectus. *Learning and Behaviour* 32, 131–140.

Griffin, A.S., Blumstein, D.T. and Evans, C.S. (2000) Training captive-bred or translocated animals to avoid predators. *Conservation Biology* 14, 1317–1326.

Anxiety

Anxiety can be defined as an emotional response to stimuli associated with potential danger. **Fear** and acute anxiety may be very similar, both in terms of the observed behavioural and physiological responses shown and the cognitive processes involved. Anxiety, like fear, protects animals from dangers, and different types of anxiety may be classified according to the types of eliciting stimuli. The definition of anxiety may vary according to whether the approach taken is ethological (Marks and Nesse, 1994) or clinical (Dodman and Shuster, 1998). However, most animal behaviour researchers agree that fear responses are shown primarily to well-defined threats, whereas anxiety is shown in healthy animals when the source of the threat is less clearly defined.

Differences between fear- and anxiety-eliciting stimuli are matched by differences in the responses shown by an animal. Whereas fear is normally expressed as avoidance and flight in freely moving animals, anxiety is associated with arousal, observation and cautious assessment of the threat. Stimulus-specific anxiety can be exemplified by the behaviour of a rat exposed to the smell of a cat in the familiar home environment, a stimulus associated with potential but not actual danger. The rat may initially avoid this stimulus but will soon respond with cautious, investigative, risk-assessment behaviours, which stand in sharp contrast to the flight responses observed when a rat is exposed to the actual presence of a cat. This example illustrates how acute anxious responses may be considered adaptive: they facilitate caution, information collection and planning of appropriate responses. An underlying or basal level of anxiety in the absence of obvious signs of threat is also adaptive in that it increases observation and scanning of the environment for potential danger. Such a basal level of anxiety may be especially important for ensuring the survival of both wild and extensively farmed animals, and may be exemplified by a grazing deer that periodically scans the surroundings for signs of danger.

Types of anxiety

Anxiety may be viewed and studied as either a trait (trait anxiety) or a state (state anxiety). Trait anxiety is essentially the same as fearfulness, and may be considered to be a component of **personality (see also: Temperament)**. State anxiety refers to the emotional response that is elicited in a given situation by a specific set of stimuli associated with potential threat. There appear to be several subtypes of state anxiety in many species, because different threats are associated with different responses. The behavioural responses and cognitive processes necessary for coping with social isolation, such as escape from an enclosure and reinstatement of social contact, may be very different from the responses necessary for coping with exposure to a novel object in the home environment.

Whereas social isolation may induce **separation anxiety** in dogs, and result in attempts at social reinstatement and a number of other behavioural signs of distress, sudden reactions to an approaching stranger or exposure to a conspicuous object may induce **neophobic** responses observed as initial startle, vocalization and avoidance, followed by cautious investigation of the person or object. These examples illustrate that, although reactions to different types of stimuli may all be influenced by fearfulness, these stimuli may elicit very different responses.

The view that different types of state anxiety are independent of one another is also indicated by studies showing no correlation between different responses to potentially dangerous stimuli. This is found for laboratory rodents such as mice, where avoidance of a brightly lit novel arena may not be correlated with avoidance of an elevated open corridor, even though both situations clearly elicit state anxiety (Janczak *et al*., 2000, 2003). The fact that an anxiolytic substance may attenuate one form of anxiety but have no effect on another supports the idea that there are different subtypes of anxiety, and suggests that these different subtypes may be mediated by different receptor systems. Different types of stimulus-specific anxiety are the products of evolutionary processes (Marks and Nesse, 1994). Anxiety subtypes reflect evolutionary adaptations to different environments and should therefore vary independently of one another. This has the consequence that the level of anxiety elicited by different stimuli and the correlations between different responses may be species- or even breed-specific (Janczak *et al.*, 2000, 2003). The independence of anxiety subtypes allows an animal to adjust its responses in relationship to specific characteristics of the situation and thereby maximize survival in the face of potential danger.

Despite the fact that there are different types of anxiety, they are all responses to stimuli associated with potential danger and may all be affected by personality characteristics such as fearfulness, to a certain degree. Furthermore, different types of anxiety are also similar in that they are usually associated with a number of symptoms of physiological **stress** that may be detrimental from an animal **welfare** perspective. Whereas moderate anxiety may be beneficial for welfare by protecting the animal from damage, extreme or chronic anxiety may reduce welfare and productivity in domesticated animals. Certain forms of anxiety may also be a significant problem for household and sports animals and call for a combination of behavioural modification therapy and pharmacological treatment (see Dodman and Shuster, 1998). The component of anxiety that is influenced by fearfulness can also be reduced through genetic selection and by exposing animals to a predictable but varied range of stimuli, especially during the juvenile period.

Measuring anxiety

Methods for measuring anxiety, and the validity of these measures, are described elsewhere (**see also: Fear**) and will be mentioned only briefly here. Most models used for measuring anxiety have involved the observation of unconditioned avoidance of potentially dangerous stimuli such as novel objects, novel corridors, novel social partners and novel, brightly lit arenas (see Andersen *et al.*, 2000). Many test methods also include the registration of ethological parameters such as vocalization or other species-specific anxiety-motivated responses. The **approach test**, which may involve different stimuli, may be either a test of anxiety if the object is merely a novel object or a test of fear if the stimulus is something the animal has had uncomfortable experience with in the past. The **open-field test** is a typical behavioural model that induces a state of anxiety by exposing animals to a novel arena. (AMJ)

References and further reading

Andersen, I.L., Færevik, G., Bøe, K.E., Janczak, A.M. and Bakken, M. (2000) Behavioural evaluation of methods for assessing fear responses in weaned pigs. *Applied Animal Behaviour Science* 69, 227–240.

Dodman, N.H. and Shuster, L. (1998) *Psychopharmacology of Animal Behaviour Disorders*. Blackwell Science, Oxford, UK.

Janczak, A.M., Braastad, B. and Bakken, M. (2000) Fear-related behaviour in two mouse strains differing in litter size. *Animal Welfare* 9, 25–38.

Janczak, A.M. Sandem, A.I. and Bakken, M. (2003) Behavioural differences in sub-adult female mice exposed to an elevated plus-maze: correlated effects of selection for high litter size. *Applied Animal Behaviour Science* 80, 347–354.

Marks, I.M. and Nesse, R.M. (1994) Fear and fitness: an evolutionary analysis of anxiety disorders. *Ethology and Sociology* 15, 247–261.

Apathy

Apathy can be used to describe abnormal, maladaptive, reduced activity, lack of interest or concern (i.e. indifference), and a lack of feeling and emotion (impassiveness). Although the term is often used in animal behaviour work to refer to a state of general behavioural **depression** (or suppression), the relationship between apathy and depression in humans appears to be disease-specific and generally uncorrelated. Apathy is considered to be a neuropsychiatric syndrome that is clinically distinct from depression. Human apathy is mediated by frontal lobe systems, highly prevalent in neurodegenerative diseases and significantly correlated with cognitive impairment (Levy *et al.*, 1998).

Apathetic behaviour (i.e. few, if any, signs of vigour or excitement) tends to occur when animals are kept in environments that are inappropriate, offer few challenges or in which the animal has little control. Sows confined for long periods within small pens/stalls demonstrate inactivity and reduced levels of responsiveness to stimuli (Broom and Johnson, 1993). Animals that are unable to avoid an unpleasant stimulus (e.g. aversive conditions) initially try to adapt via the psychological process of **habituation**. If unsuccessful they may enter an apathetic state, described by some as **learned helplessness** (i.e. they learn that they have no control over the stimulus) or learned hopelessness (i.e. they predict that they cannot eliminate the stimulus), whereby they remain distressed but no longer attempt to make appropriate responses to improve their situation via escape or avoidance. Since animals exhibiting such behaviour provide a cause for concern, it is important to distinguish apathy from resting contentment.

It is possible that the tendency for an animal to become apathetic may be related to its personality, temperament or coping style, with those individuals identified as passive, or those that internalize their response to **frustration**, more likely to exhibit this behaviour. However, to date there has been little animal **welfare** research specifically investigating apathetic behaviour beyond that associated with the learned helplessness paradigm in psychology. (DSM, GC)

References and further reading

Broom, D.M. and Johnson, K.G. (1993) *Stress and Animal Welfare*. Springer, New York, 211 pp.

Levy, M.L., Cummings, J.L., Fairbanks, L.A., Masterman, D., Miller, B.L., Craig, A.H., Paulsen, J.S. and Litvan, I. (1998) Apathy is not depression. *The Journal of Neuropsychiatry and Clinical Neurosciences* 10, 314–319.

Meehan, C.L. and Mench, J.A. (2007) The challenge of challenge: can problem solving opportunities enhance animal welfare? *Applied Animal Behaviour Science* 102, 246–261.

Wood-Gush, D.G.M. and Vestergaard, K. (1989) Exploratory behaviour and the welfare of intensively kept animals. *Journal of Agricultural and Environmental Ethics* 2, 161–169.

Appeasement

Appeasement refers to the gestures used to reduce social tension in order to avoid violence. Traditionally, the term has been used to refer to what are commonly referred to as signs of **submissiveness** issued by lower-ranking animals towards others (often described as more dominant animals as a result) to discourage attacks or aggression when there is competition for access to a resource. As such, appeasement may be seen as integral to the maintenance of social hierarchy. However, it is now recognized that some species appear to display a wider range of behaviours that tend to have more general derousing effects in a greater variety of contexts, and that these behaviours may not necessarily be issued by the typically subordinate individual. In dogs these behaviours have been popularly referred to as 'calming signals', and include gestures such as head turning, blinking, yawning and slow movements; these can be shown in a range of situations involving stress in a social context, even if there is no resource dispute, e.g when an owner appears angry with their dog for not coming when called. Some authors prefer the term 'appeasement' to submission or **submissiveness**, because it can refer not only to a wider range of behaviours, but also because the latter can seem to imply that the animal has awareness of or is motivated by its social standing, as opposed to avoiding conflict, which can be determined objectively without inference about the cognitive state or capacity of the individuals concerned. (LMD, DSM)

See also: Agonistic behaviour; Dominance; Hierarchy

Appetitive behaviour

Appetitive behaviour is the variable, flexible, searching phase of a behavioural sequence, indicative of the need or desire to achieve a certain goal. Appetitive behaviour is usually followed by **consummatory acts**, which are more invariant or fixed and usually lead to termination of the behavioural sequence. Appetitive behaviour may function not just to obtain a goal, but also to obtain information about the location or availability of a resource and its quality for future requirements. It therefore serves an important role in information gathering.

A typical example of appetitive behaviour is the searching for food as a part of **foraging behaviour**. The consummatory phase of foraging behaviour is the ultimate eating that ends the foraging sequence. In this context appetitive behaviour and consummatory behaviour come in a variety of forms, depending on the species. Examples of appetitive behaviour include: (i) the hunting of feline and canine species; (ii) the ground-pecking and -scratching of poultry; and (iii) the rooting and nosing of pigs. Examples of consummatory eating behaviour include pecking, biting, chewing or swallowing whole.

Foraging activities are influenced by both internal factors, such as nutritional requirements, and external factors, such as social factors, and the distribution and availability of food. In general, hungry animals spend less time on appetitive behaviour and are less selective in their choice of food. Some farm animals show **contra-freeloading** (i.e. they continue to put effort into performing foraging activities even though food is freely available), but less so when they have been deprived of food. Social factors that influence appetitive behaviour include, for instance, competition for food and **social facilitation**. Social facilitation is commonly described as the increase in frequency or intensity of particular responses already in the animal's repertoire, when shown in the presence of others engaged in the same behaviour. It has been found that a satiated domestic chicken, when seeing another chicken feed, will show increased consummatory eating behaviour but an even more marked increase in appetitive pecking at the food (Keeling and Hurnik, 1996).

In farm animals food is usually readily available, either continuously or in discrete meals. Delivering food in meals can affect both quantity and quality of appetitive behaviours, through anticipation and frustration. Animals, for example, that can anticipate the delivery of food through external cues but experience a delay between cues and the actual arrival of food, often develop stereotypic behaviours that originate from appetitive behaviours (**see: Stereotypies**) (Rushen *et al.*, 1993). Furthermore, by giving animals like sows food in meals the full performance of appetitive behaviour is often thwarted, which can result in the performance of stereotypic behaviour after feeding.

Other examples of appetitive behaviour include courtship as part of **sexual behaviour** and nest building as part of **reproductive behaviour**.

Intensive husbandry conditions can put constraints on the performance of appetitive behaviour by not providing the relevant stimuli or commodities. How this affects the animal and its welfare depends on the actual causation of behaviour. For instance, some appetitive behaviour sequences are clearly goal-directed. A goal is defined as the external situation that may elicit the behaviour and the achievement of which will bring the sequence to an end. So, providing an animal with the goal (e.g. food, a nest) makes it unnecessary for that animal to perform the accompanying appetitive behaviour to achieve that goal. This probably has no negative impact on the welfare of the animal. In contrast to goal-directed appetitive behaviour, the performance of some other appetitive behaviour is strongly internally motivated and probably rewarding in itself. Thwarting of such appetitive behaviour is more likely to have a negative impact on welfare because the animal still has the need to perform it, irrespective of the consequences.

One example of such a behavioural need is **dust-bathing** behaviour in laying hens. The function of dust-bathing behaviour is to remove feather lipids and maintain the plumage condition. In the initial, appetitive phase of dust bathing, a hen tosses the dusty substrate into her feathers with vertical movements of her wings. In the next, consummatory, phase she rolls on her side and rubs the dust thoroughly through her feathers. The dust-bathing sequence ends with the hen standing up and shaking her body vigorously. An essential part of the release mechanism for dust bathing is the build-up of internal motivation in interaction with a pronounced diurnal rhythm. In the absence of a dust-bathing substrate or even a plumage, a hen will still perform the appetitive phases of dust bathing (even if in a 'vacuum' form (**see: Vacuum behaviour**)). Thus, complete prevention of dust bathing is likely to impair a hen's welfare. Whether the performance of vacuum dust bathing is less satisfying than dust bathing with a full substrate is less clear. Long-term deprivation also changes the quality of the behaviour; the time spent on the appetitive tossing phase decreases, while the consummatory rubbing phase becomes more predominant.

An example of how goal-directedness determines the termination of a behaviour is seen in the nest-building behaviour of sows (e.g. Jensen, 1993). The appetitive phase consists of nosing, rooting and pawing potential nesting material and is mainly elicited by an internal factor, the level of prolactin. In the subsequent consummatory phase the sow gathers and arranges the material to construct the nest. Sows that do not have nesting material at their disposal do not proceed to the consummatory phase but persist in performing appetitive behaviours. Thus, although the onset of the appetitive phase of nesting behaviour is mainly caused by an internal factor, externally derived factors – e.g. the presence/absence of substrate – certainly play a role in proceeding to the consummatory phase and termination of the nest-building sequence.

Thwarting of appetitive behaviour, especially foraging behaviour, can lead to stereotypic behaviour in several farm animal species. These stereotypic behaviours vary in form and time. Mason and Mendl (1997) pointed out that species differences in natural foraging behaviour might underlie these differences. Pigs and chickens under semi-natural conditions spend most of their day searching for food. Pigs perform rooting and nosing of the ground for up to 50% of their day, whereas chickens in the wild spend almost 50% of their time ground-pecking and -scratching. Both species have a very broad diet, usually with a patchy distribution. In such patch-feeders, feeding enhances foraging; bouts of appetitive behaviour will naturally be alternated with bouts of consummatory acts. When food is clumped, there is a clear advantage in increasing the intensity of foraging when food is discovered. Most carnivores also spend a lot of time foraging, which usually involves a large amount of physical effort (e.g. hunting in hyenas, cheetahs). However, a foraging sequence usually terminates once they have caught the prey. This might explain why pigs and chickens perform most stereotypic behaviours after feeding while carnivores, such as mink, perform them before feeding. Thus, it is not surprising that hunting animals tend to pace, whereas other species show oral stereotypies.

(PHZ)

See also: Consummatory acts; Contra-freeloading; Emancipation; Frustration; Lorenz, Konrad; Motivation; Social facilitation; Stereotypies; Vacuum behaviour

References

Jensen, P. (1993) Nest building in domestic sows: the role of external stimuli. *Animal Behaviour* 45, 351–358.

Keeling, L.J. and Hurnik, J.F. (1996) Social facilitation acts more on the appetitive than the consummatory phase of feeding behaviour in domestic fowl. *Animal Behaviour* 52, 11–15.

Mason, G. and Mendl, M. (1997) Do the stereotypies of pigs, chickens and mink reflect adaptive species differences in the control of foraging? *Applied Animal Behaviour Science* 53, 45–58.

Rushen, J., Lawrence, A.B. and Terlouw, E.M.C. (1993) The motivational basis of stereotypies. In: Lawrence, A.B. and Rushen, J. (eds) *Stereotypic Animal Behaviour – Fundamentals and Applications to Welfare*. CAB International, Wallingford, pp. 41–65.

Applied ethology

Applied ethology is the study of the behaviour of animals in relation to their management or **welfare**. Behavioural studies may be conducted with any animal used in some context by humans, for example in laboratory research, in entertainment, as companions or for food. Applied ethology is the application of our knowledge of the behaviour of such animals from these studies, usually in order to improve their management. (KT)

See also: Ethology; ISAE

Approach test

The approach test involves the observation of an animal's approach and avoidance of a clearly defined object and is used to quantify the forces favouring approach relative to the forces favouring avoidance. Miller's model of approach–avoidance conflict provides a conceptual framework for interpreting behaviour in the approach test (**see also: Fear**). Although the approach test has often been focused on quantifying fear or **anxiety**, similar methods can also be used to measure the strength of a number of **emotions** or **motivations**, depending on the design of the test. Modifications of the approach test can also be used to study cognitive processes such as learning (**see also: Cognition; Conditioning**). The strength of the approach test is that it is flexible and can be designed to measure a large number of different psychological characteristics. It also allows the experimenter to exert some control over the effects of variables that are potentially confounding.

Forces favouring approach include hunger, thirst, curiosity, attraction to familiar stimuli in an unfamiliar environment or attraction to mildly unfamiliar stimuli in a familiar environment. Forces that favour avoidance include **fear**, **anxiety**, **pain**, **frustration** and **boredom**. The original work of Miller involved conditioned behaviour in rats. Rats were first trained to perform an **operant response** in order to receive a food reward and then punished with an electric shock while performing the operant response. He showed that approach and avoidance tendencies could be independently manipulated. Varying the duration of food deprivation affected the approach tendency, and varying the strength of electric shock altered the avoidance tendency. This illustrates that the different factors influencing approach and avoidance tendencies can be consciously incorporated into the design of an approach test. This would increase the likelihood that the test actually measures what the experimenter is interested in measuring.

Different forms of the voluntary approach test are commonly used to quantify fear and anxiety in animals. This test involves the placement of a test animal into a corridor and the quantification of the time spent close to or in contact with a human or inanimate object at one end. It has high internal validity for measuring fear of humans and fear of an object associated with pain. It is also internally valid for measuring the anxiety induced by exposure to a novel object or area. Avoidance of dangerous and potentially dangerous stimuli is one of the most fundamental aspects of fear and anxiety, respectively. This is mainly indicated by the fact that aversive experience with an object increases subsequent avoidance of the object in most species and that most species also initially avoid novelty. It has also been shown that forced exposure to novelty and objects associated with pain induces behavioural signs of risk assessment and acute fear. Forced exposure also results in physiological and neuroendocrine responses characteristic of fear and anxiety. These symptoms of fear and anxiety subside when an animal is no longer exposed to the aversive stimuli.

Because the strength of fear and anxiety increases with increasing proximity to an aversive object, it is possible to validate independent measures in relation to this gradient. Ethological, physiological and neuroendocrine indicators of increased fear should increase with increasing proximity to the object or person, and decrease with increasing distance.

Measurement of avoidance as an indicator of fear also has external validity in many species because it is often predictive of fear responses to similar stimuli outside of the test situation. This has been shown for pigs, where animals that avoid humans in standard tests are subject to signs of chronic stress when they are regularly exposed to humans (Hemsworth *et al.*, 1986; Janczak *et al.*, 2003).

Fear of humans or novelty-induced anxiety causes avoidance, whereas appetitive motivation and curiosity cause approach. The time that the test animal spends at different distances from the stimulus can be used to quantify the strength of fear or anxiety relative to the strength of emotions favouring approach. Alternative measures are the latency to contact the stimulus and the time spent in contact with it. A frequently cited but easily solved problem with these tests is the concern that animals may simply ignore the stimulus. Although this is clearly a question of calibration in designing the test, it does not necessarily cause interpretational problems; increasing the strength of the approach tendency can solve this problem. Food can be offered to food-deprived animals or contact with a conspecific can be offered to highly social species. If the rewarding stimulus (food or social partner) is placed at the same end of the corridor as the avoided stimulus, it should be possible to quantify the balance between approach and avoidance tendencies in animals that would otherwise ignore the feared human or object by itself.

Because of this possibility for standardizing effects of social-reinstatement tendencies in the approach test, a properly designed approach test is not subject to the same interpretational problems that are inherent in the **open-field test**. In the latter, it is difficult or impossible to standardize such confounding effects on behaviour. It is interesting to note that variance in approach and avoidance tendencies is mainly a product of the strongest and most variable motivations. Because fear and anxiety are powerful emotions, they initially inhibit curiosity-motivated, hunger-motivated and socially motivated responses. Variance in approach and avoidance therefore normally reflects fear and anxiety in animals that are exposed to humans or novel objects. However, if animals are tested in the absence of fear- or anxiety-inducing stimuli, the influence of these emotions can be negated and modifications of the approach test can be used to measure the relative attractiveness of different stimuli that elicit approach responses (**see also: Preference**). This modification of the approach test has been used for measuring effects on social reinstatement responses in a number of species (Marin *et al.*, 2001). In conclusion, different forms of the approach test, if properly designed and validated, should be very useful for measuring a number of emotions and motivations in a range of different species. (AMJ)

References

Hemsworth, P.H., Barnett, J.L. and Hansen, C. (1986) The influence of handling by humans on the behaviour, reproduction and corticosteroids of male and female pigs. *Applied Animal Behaviour Science* 15, 303–314.

Janczak, A.M., Pedersen, L.J., Rydhmer, L. and Bakken, M. (2003) Relation between early fear- and anxiety-related behaviour and maternal ability in sows. *Applied Animal Behaviour Science* 82, 121–135.

Marin, R.H., Freytes, P., Guzman, D. and Jones, R.B. (2001) Effects of an acute stressor on fear and on the social reinstatement responses of domestic chicks to cagemates and strangers. *Applied Animal Behaviour Science* 71, 57–66.

Approach–avoidance conflict

Approach–avoidance conflict refers to the inferred motivational state of an animal exhibiting behavioural inhibition when presented with either a new stimulus of potential significance or a situation of ambiguous valence (desirable and undesirable characteristics). It is used in many experimental models to assess '**anxiety**'. For example, the time spent in approach–avoidance conflict is often recorded in tests such as the **elevated plus-maze**, the **light–dark test** or **open-field tests,** which may be used to assess the anxiolytic (anti-anxiety) effects of chemotherapeutic agents. Ambiguous situations can be used to assess the potential aversiveness or attractiveness of a given stimulus by combining it with stimuli of opposite valence and varying intensity, in order to determine at what point approach avoidance is induced and/or overcome (**see also: Choice test**). In response to approach–avoidance conflict, animals may express **ambivalent behaviour, displacement behaviour** or signs of emotional **frustration** (such as **aggression**). (DSM)

See also: Conflict behaviour

Further reading

Elliot, A.J. (ed.) (2008) *Handbook of Approach and Avoidance Motivation*. Psychology Press, Taylor & Francis Inc., London.

Aquinas, Thomas

The Aristotelian (384–322 BC) view that human beings are unique in their possession of the faculty of reason and that there is a natural hierarchy of life forms is amalgamated with the teachings of the Catholic Church in the works of leading Church father St Thomas Aquinas (AD 1225/1227–1274). While non-human animals occupy a higher status than mere vegetative life forms, Aquinas believed that they are morally inferior to human beings because they are without reason and, hence, do not possess free will. Furthermore, he believed that human beings are 'intellectual creatures [who] hold the highest place, because they approach nearest to the divine likeness'. Aquinas' view of 'dominion over Nature' has been interpreted as giving license to human rule over and subsequent exploitation of animals 'either by killing or in any other way whatever'.

Upon closer analysis, it is found that Aquinas did not think that we could *treat* animals as we wanted. While 'the love of charity' does not extend to 'irrational creatures', Aquinas believed that there were proper ways of interacting with animals. In his view, the mistreatment of animals did not constitute a harm to the animals themselves. Instead, the caveat against abuse or neglect reflected his fear that mistreatment of animals would result in cruelty to other human beings.

In contrast to Aquinas, other Christian thinkers like St Francis of Assisi have extended direct concern to animals. (RA)

See also: Christianity

Further reading

Aquinas, T. (1918) *Summa Theologica*. Benziger Brothers, Chicago, Illinois, Part II, question 64: article 1, questions 25: article 3.

Aquinas, T. (1928) *Summa contra Gentiles*. Benziger Brothers, Chicago, Illinois, Third book, Part II.
Linzey, A. (1991) *Christianity and the Rights of Animals*. Cross Roads, New York.

Arousal

Arousal is a state of psycho-physiological activation that determines the responsiveness of an individual to environmental stimuli. From a behavioural point of view, arousal is characterized by increased alertness, sensory sensitivity and readiness to respond to stimulation through faster cognitive processing and motor response. Arousal results from the activation of several neurobiological systems (i.e. the reticular activating system in the brain stem, the **autonomic nervous system** and the endocrine system), involving the action of various neurotransmitters such as **acetylcholine**, **norepinephrine**, **dopamine** and **serotonin** in order to meet the energy requirements for the behavioural response and to maintain **homeostasis**. From a physiological point of view, this activation leads to an increased heart rate, blood pressure and the production of **corticosteroid**. Arousal is closely related to other concepts such as **stress**, **anxiety**, **attention** and **motivation**.

Part of the response to stress is, for instance, elicited by the emotional arousal that arises from environmental stimulation. Emotional responses arise when the individual evaluates a challenging event and then influence its adaptive response through various ensuing arousal levels. The study of emotion in animals has recently been approached, for instance, in sheep through an original framework derived from appraisal theories developed in human cognitive psychology. It hypothesizes that particular emotions stem from a particular combination of a limited number of criteria (i.e. suddenness, familiarity, predictability, pleasantness, conformity with expectation, controllability). From a welfare point of view, acting on such criteria could, for instance, allow reducing negative emotions in animals and thus lowering of emotional arousal and stress.

Arousal is a major aspect of many learning theories. The Yerkes–Dodson law, for instance, predicts that there is an optimal level of arousal for learning to occur, depending of the nature of the task to be learned. A mild stress would increase attention and thus facilitate learning, but by contrast too low or too intense stress/arousal would have detrimental effects. Müller and Pilzecker have also proposed a 'perseveration–consolidation' hypothesis, which has been supported by several human and animal experiments showing that emotional arousal enhances long-term memory of immediately preceding neutral stimuli, through the activation of the amygdala. (SL)

Further reading

McGaugh, J.L. (2006) Make mild moments memorable: add a little arousal. *Trends in Cognitive Sciences* 10, 345–347.
Veissier, I. and Boissy, A. (2007) Stress and welfare: two complementary concepts that are intrinsically related to the animal's point of view. *Physiology and Behaviour* 92, 429–433.

Arrhythmia

A cardiac arrhythmia may be defined as any deviation from the normal cardiac rhythm or heart beat. Arrhythmias may present themselves as an increase (tachycardia) or decrease (bradycardia) in rate or as a chaotic motion (fibrillation) in the cardiac muscle. There are multiple causes of arrhythmias, and prognosis ranges from innocuous to fatal depending on the type and intensity of arrhythmia present and whether or not treatment is sought. (RM-F)

Artificial insemination

Artificial insemination (AI) involves the collection of semen from a male animal and its introduction by artificial means into the reproductive tract of a female, with the intention of achieving conception and a subsequent pregnancy. Artificial insemination was the first of what may be termed assisted-breeding technologies to be developed, with anecdotal accounts referring to its use with native Arab horses dating back as far as the 14th century. The first documented experiments in mammals were performed in the 1780s on dogs by the Italian physiologist Spallanzani, who reported the live birth of three pups born to a bitch he had artificially inseminated with a syringe 62 days previously. However, AI remained something of a scientific curiosity throughout most of the 19th century and it was not until the later part of that century and early 20th century, as part of a new scientific approach to agriculture, that AI began to be established as a technique for livestock **breeding** and improvement.

Successful artificial insemination is dependent on three distinct stages – collection of viable semen from the male, generally followed by a degree of semen processing and frequently by a short- or long-term holding period, and finally introduction of semen to an appropriate site within the female reproductive tract at an appropriate time in the reproductive cycle.

Ejaculated semen can be collected from many domestic livestock breeds by use of an artificial vagina (cattle, horses, sheep) or by manipulation (pigs, dogs, poultry, fish species), but these methods are restricted to species that are both safe to handle and are to some degree trainable. Excessive use of an individual may raise **welfare** concerns, as musculoskeletal injuries may occur as a result of excessive mounting and dismounting. For recalcitrant domestic or wild species semen collection can in many instances be accomplished under sedation or full **anaesthesia** by **electroejaculation** using rectal probes. Electroejaculation may also be used for valuable males whose reproductive function has been compromised by injury or illness, and in such cases pharmaceutical techniques have also been used to stimulate semen release. However, the routine use of electroejaculation – particularly in the absence of sedation – does raise welfare issues. Surgical and post-mortem collection has also been used, by taking sperm directly from the vas deferens or epididymis.

Following collection, procedures most often involve a dilution stage where the ejaculate is extended with a diluent designed to both maintain short-term viability and dilute the semen to allow greater efficiency of usage. Medium- or long-term storage of the semen may be undertaken at this stage either by chilling to temperatures above 0°C or by cryopreservation in liquid nitrogen, but the success of chilling or freezing processes is highly species-dependent and in many cases subject to individual variation within species. Deposition into the female tract must be timed to coincide with ovulation, either during the course of a natural cycle or by giving exogenous hormones to stimulate timed ovulation. The site of semen deposition is usually into the uterus, by canulation of the cervix.

Many domestic species will allow insemination when in oestrus, but wild species will require sedation or full anaesthesia to ensure safety and compliance. In some species, e.g. sheep, the morphology of the cervix makes canulation difficult and insemination has more commonly been performed using a surgical technique involving laparoscopy, which can be performed under local anaesthetic; this inevitably increases the level of training and experience required in the inseminator, increases costs and raises welfare issues for the animals concerned. Where the numbers of sperm available for insemination are low or semen is generally of poor quality, deep intra-uterine techniques have been developed in which a catheter is guided using either an endoscope or ultrasound to place reduced numbers of sperm adjacent to the utero–tubal junction at the top of the uterine horn.

Artificial insemination has four major advantages over natural covering: (i) it maximizes the use of superior males; (ii) ejaculates can be extended and divided to obtain the highest number of insemination doses from each ejaculate; (iii) the procedure overcomes geographic limitations, allowing semen to be shipped to females; and (iv) chilled and frozen semen also overcomes temporal restrictions as semen can be collected and stored on a short- or long-term basis before use. The increased availability of the highest-quality males has driven the dramatic increases in production seen, for example, in the dairy industry since the introduction of AI. AI has also had a major role in the reduction of venereal disease, by restricting direct contact between male and female animals and by allowing semen to be held while animals complete quarantine periods. However, semen is itself a route of infections, and international AI programmes are potentially also at risk of spreading venereal diseases if males are not fully screened.

Artificial insemination plays an increasing role in conservation programmes for endangered species, making breeding plans involving small, geographically isolated and often valuable captive populations simpler and more successful. The establishment of frozen semen banks can extend the reproductive life of individual males and thereby effectively increase breeding population size while also acting as an insurance against catastrophic loss. Frozen semen banks also have a therapeutic role in overcoming infertility and are of increasing importance in maintaining ever-expanding numbers of genetically modified animals, particularly laboratory rodents. The use of AI also enables breeders to avoid behavioural incompatibilities and to achieve pregnancies in species where correct sexual and mating behaviour is either insufficiently known or difficult to recreate under captive conditions, thereby preventing possible injuries consequent on attempting to pair animals in an inappropriate way or at an inappropriate time. (MRC)

See also: Captivity

Assistance animal

An assistance animal is one that is used to facilitate an activity by a human who is otherwise impaired in this action. Many assistance animals are trained to complete tasks in relation to a human disability. Such animals are provided to increase the safety, mobility and/or independence of their human companions. Examples of tasks include guidance when walking (guide dogs for the blind), warning of imminent seizures (seizure-alert dogs), informing the human of important sounds (hearing dogs for the deaf) and assistance with other, more general tasks related to mobility (**service** or assistance animals, typically dogs or monkeys). There is growing interest in the use of **companion animals** in the management of mental health problems, including autism and learning difficulty, where they may act as a catalyst for improved communication. (KT)

See also: Animal-assisted therapy

Further reading

Lane, D.R., McNicholas, J. and Collis, G.M. (1998) Dogs for the disabled: benefits to recipients and the welfare of the dog. *Applied Animal Behaviour Science* 59, 49–60.

Siegel, M.E. and Koplin, H.M. (1984) *More than a Friend: Dogs with a Purpose*. Walker & Co., New York.

Attachment

The theory of social attachment comes from psycho-ethology and provides a framework for the discussion of affectionate relationships. It was developed in the 1950s by Bowlby and then later on by Ainsworth in human babies, and refers to the strong emotional bond that develops between infant and caregiver and that provides the infant with emotional security. The concept of attachment is used to describe the mechanism that leads to the establishment of the relationship, but also most frequently the relationship itself. The result of the attachment process in mammals is thought to be the equivalent of the **imprinting** described by **Lorenz** in birds. Imprinting indeed occurs during sensitive developmental periods (i.e. **critical period**), and specific behaviours are then restricted and directed towards particular individuals only.

However, while imprinting seems to be almost instantaneous, the attachment mechanism in mammals appears to be slower. An individual can develop several attachments during a lifetime: a primary attachment towards the mother (or the father) and secondary attachments such as attachment between youngsters or adults (e.g. sexual partner), which are weaker than the previous one. Several definitions of attachment have been suggested since 1958, but the overall specific characteristic of the attachment mechanism is that it results in the establishment of a relationship which is emotional, selective and reassuring. The attachment figure (e.g. the mother) has an effect on the emotional reactivity of the subject. Furthermore, even if several attachments can exist, each of them is supposed to be unique in its expression. Finally, the attachment figure has been shown to act as a base of security from which the subject could gain reassurance during challenging situations and in the presence of which s/he will show a greater propensity to explore his/her environment. If attachment is described as an enduring **bond**, it is not clearly defined whether it will last a lifetime after a long separation period.

At birth, social attachment is based on innate behaviours displayed by the young (e.g. tendency to follow, vocalizations), which will be directed towards the mother or any caregiver and then enable the development of the affectional bond. According to the ethological point of view, the behaviour of the neonate can be understood in terms of adaptive value as it ensures that the young will get food, warmth and protection from its mother. The process of attachment is defined as being based on internal **reinforcement**.

Ethologists thus emphasize the biological and genetic basis of development, even though they do not dismiss the implication of some learning. The ethologists' attachment theory has been subjected to several controversies. From a behaviourist perspective for instance, learning and feeding would have a central role in the attachment process: the newborn would learn to prefer their mother as they associate her with the feeding activity (**see: Behaviourism**). This hypothesis has been rejected following the work of Harlow and colleagues on monkeys, who showed that attachment behaviours such as contact or security seeking occurred even without the provision of food by the figure of attachment. According to the behaviourists, **operant** conditioning based on the reciprocal responsiveness between the mother and the young could explain the attachment process. This implies that the attachment behaviours of the young are reinforced by food and positive interactions, and in contrast that they would be decreased by negative interactions. According to the operant conditioning theory, a young individual that had been mistreated by its caregiver should not exhibit attachment behaviours as they are not reinforced. This hypothesis has also been ruled out by studies in animals showing that young animals continued to seek physical contact from their mother even though she might reject them. Discomfort or **distress** states such as hunger, thirst or **fear** have been suggested to facilitate the expression of attachment behaviours.

Recognition of the partner is an essential prerequisite for attachment to occur. In the context of maternal attachment, for instance, the young individual needs to learn the characteristics of the figure of attachment to be able to recognize it and subsequently express a preference. Recognition is conditioned by exposure to the characteristics to be learned (i.e. contact), learning and memorization. Complex interactions between different hormones and neurotransmitters that are influenced by genetic and environmental factors underpin attachment through the promotion of contact seeking and social behaviours and the facilitation of sensory learning, social memorization and the establishment of preference. **Oxytocin**, vasopressin, opiates and **norepinephrine** have been proved to play major roles in this process. These neurochemicals have also been shown to be implied in the expression of attachment, such as separation-induced **vocalizations**.

A behaviour can be considered as being indicative of attachment if it is expressed more often and more intensively towards a particular individual than towards any other. Several criteria should be met to conclude the existence of an attachment relationship. The figure of attachment should elicit the expression of a preference compared with any other familiar individual. It should also elicit contact seeking, separation distress, appeasement following reunion and should be used as a secure base.

A measure of attachment combining all those elements has been developed by Mary Ainsworth in human infants. Her Strange Situation Test is carried out in a novel environment and based on a succession of eight periods consisting of separations from and reunions with the caregiver, combined with exposure to a stranger. The use of the Strange Situation Test is not restricted to the field of human psychology. The nature of the bond that can exist between domestic animals and humans can indeed also be studied using adapted versions of this test. Recent results in dogs based on this paradigm give support to the hypothesis of the dog–human attachment. Attachment theory has thus been suggested for use as a framework with regards to canine **separation anxiety**.

Other studies have also tried to assess the possibility of an attachment between farm animals and humans using, for instance, artificially reared lambs as a model. Test situations have been developed to assess exploration of a new environment (secure base hypothesis) and feeding activity in the presence/absence of the caregiver, as well as preference for and reaction to separation from and reunion with the caregiver. Treatment procedures use various levels of food provision and contact with the caregiver, and test situations have been carried out under different levels of satiety to help differentiate between a conditioning versus attachment hypothesis. Even if no definitive conclusion can be drawn, results suggest that attachment could indeed occur between farm animals and humans. (SL)

Further reading

Nelson, E.E. and Panksepp, J. (1998) Brain substrates of mother–infant attachment: contributions of opioids, oxytocin, and norepinephrine. *Neuroscience and Biobehavioural Reviews* 22, 437–452.

Topal, J., Miklosi, A., Csanyi, V. and Doka, A. (1998) Attachment behaviour in dogs: a new application of Ainsworth's Strange Situation Test. *Journal of Comparative Psychology* 112, 219–229.

Attention

Despite William James' assertion in his *Principles of Psychology* in 1890 that 'every one knows what attention is', measuring or enumerating attention is complex because it encompasses subjective mental states of consciousness and awareness. James stated: 'Focalization, concentration of concsiousness are of its essence. It implies withdrawal from some things in order to deal effectively with others.' Definitions usually centre around the idea that computational constraints prevent the brain from processing all incoming sensory information simultaneously; the purpose of attention is to permit prioritization of important information by restricting what input is noticed, processed and stored in memory.

Attention is a necessary condition for learning to take place and so underpins all sorts of processes vital to animal survival. In models of classical **conditioning**, unconditioned stimuli that are unexpected or surprising attract more attention and so have greater associative strength; when unconditioned stimuli are paired, an attention-grabbing element may block or overshadow learning about the other stimulus. Novelty and fast changes are responded to so readily because they denote new information that may require further analysis. Ability to pay attention will decrease with task difficulty or strong distracters. **Habituation** occurs when a stimulus no longer attracts attention. Habituation will be slow and distraction more difficult from highly motivated activities or stimuli that relate to fulfilling a behavioural need – such as finding food or materials to make a nest – because these are allocated a high priority for attention. Barren environments provide little input on which to focus attention and may cause maladaptive or redirected behaviour in captive animals. Rats, mice and macaques reared in impoverished or socially isolated conditions have demonstrated over-excitement or repetitious, unvarying responses towards test stimuli as well as failure to habituate to novel or irrelevant input. Environmental **enrichments** for **zoo**, laboratory or **pet** animals will be most successful where these continue to attract attention over a long period, stimulating **foraging** and **exploratory behaviours**.

Animal behaviour often appears to represent adaptations to a limited attentional capacity. Animals attend to certain important features or dimensions of the environment while ignoring others (selective attention), discriminate more accurately when able to focus on one rather than two stimulus dimensions (divided attention) and find ways to reduce the demands of tasks requiring prolonged concentration (sustained attention). Other adaptations such as camouflage and sit-and-wait predation work precisely by avoiding drawing attention. Exploring attention in humans has relied heavily on verbal reporting of experience, making the design of analogous tests for animals challenging. Attributing an attentional cause to behaviour generally relies on carefully controlled laboratory testing, but this can often make sense of behaviours in a natural context.

Selective attention

This is a mechanism for coping with limitations in processing ability by focusing on the most salient features. Its study generally measures the ability to focus on a stimulus while avoiding other, distracting information. Discrimination tasks indicate that animals direct their attention to certain features to form a 'search image' of prey or food items. Where prey is cryptic, the task is harder and may require a very specific search image that restricts selection to one prey type. This could mean that other, palatable food is ignored; indeed, attentional models of capture efficiency appear to best explain real foraging patterns and priming effects. Priming effects describe the increase in detection speed or accuracy seen when a cue indicates the timing or position of a target's appearance. Focusing attention on an area of space or time reduces distractions, resulting in improved performance compared with un-cued or miscued trials. Exposing pigs to task cues or a novel environment between learning and completing a maze task adversely affected recall, suggesting that distractions that redirect attention during the retention of learning can interfere with **memory** storage or retrieval.

Divided attention

This examines the capacity for, and constraints on, performing more than one task simultaneously. In nature, it is rare that an animal will be able to attend exclusively to one stimulus. Animals must find or select food while monitoring the presence of predators, group-mates or rivals. Divided attention has been most extensively explored in animals using delayed matching to sample tasks where subjects – usually birds – are taught to recall which of two stimuli matches a stimulus shown briefly beforehand (limiting the time available to process all the information). In separate sets of 'element' trials, the subject is taught to match a single dimension such as colour or shape. In compound trials, the stimulus to be matched presents a 'compound' of both learned shape and colour. As before, the comparison stimuli in these trials consist of only one dimension, and only one stimulus provides a match. Performance is consistently poorer on compound trials, where attention must be shared between the two features in order later to select the one to be matched. Reduced performance on divided attention tasks is reliably associated with cognitive changes related to ageing.

Sustained attention

This describes the ability to concentrate on one thing for a prolonged period. In pet dogs, it has been suggested that a deficit in attention span appears to cause behaviour problems analogous to those seen in human children with attention deficit hyperactivity disorder (ADHD). In wild animals, sustained attention is most widely studied in vigilance behaviour. Like selective attention, vigilance enhances information processing by directing alertness towards certain stimuli. In sustained attention, the focal stimulus is a rare and unpredictable event such as the appearance of a predator. The **central nervous system** cannot focus indefinitely on any one task: a decrement in detection ability is seen over time and occurs faster with more difficult tasks, such as spotting well-camouflaged predators or prey or learning new tasks. Recovery of vigilance requires rest and one advantage of group living is to increase the proportion of time individuals can spend engaged in resting, foraging and other behaviours. Time spent vigilant is commonly inversely proportional to group size. Ducks intermittently 'peek' during rest to check for predators, and individuals' rates of scanning increase as group size falls.

Attention is vital to higher cognitive processes such as **social learning**, where the presence or activity of a **conspecific** increases the salience of one stimulus over another, leading to its exploration or manipulation. Hens' maternal displays actively attract chicks' attention to profitable food and away from harmful or unprofitable items. If a novel behaviour is to be transmitted, it must attract observers' attention. In order to imitate, an observer must focus on another individual's movements throughout a demonstration in order to replicate its actions. Indeed, experimental findings show that marmosets watch for longer when a conspecific is engaged in a food-finding task rather than in exploration.

Most mathematical modelling of attention is based on the assumption of resource constraint but, even without it, recent work suggests that limiting attention can be an optimal strategy. Attending preferentially to stimuli whose predictive relationships are unknown is statistically efficient regardless of how much else can be processed. Modelling also suggests that restrictions on the field of view that can be effectively scanned may represent an optimal trade-off between time spent detecting predators (and so not foraging) and the metabolic costs of increasing cognitive capacity.

The evolutionary and ecological history of a species helps to determine what is attended to, complicating attempts to carry out comparative experiments. A number of studies have indicated that scatter-hoarding birds depend more heavily on spatial cues such as the arrangement of landmarks than on the visual appearance of goals, whereas closely related species that do not retrieve caches of food divide their attention more equally between cues. Manipulation of cues shows that cachers are capable of navigating using visual cues alone, suggesting that preferential use of visual cues may be due to an attentional bias rather than differential ability to learn or remember different types of cue. Attempts to compare attention in animals against human capacities may give a distorted picture, due to the prevalent use of tasks relying on the visual stimuli important to humans rather than on other modalities. (BeH)

Reference and further reading

Bushnell, P.P. (1998) Behavioural approaches to the assessment of attention in animals. *Psychopharmacology* 138, 231–259.

Deutsch, J.A. and Deutsch, D. (1963) Attention: some theoretical considerations. *Psychological Review* 70(1), 80–90.

James, W. (1890) *The Principles of Psychology*. Henry Holt and Co., reprinted (1950) by Dover Publications, New York, pp. 303–304.

Shettleworth, S.J. (1998) *Cognition, Evolution and Behaviour*. Oxford University Press, New York, pp. 49–94.

Zentall, T.R. (2005) Selected and divided attention in animals. *Behavioural Processes* 69, 1–15.

Attitude

The term 'attitude' is commonly used to describe a person's subjective opinion; in particular, whether they like something or not, but its true meaning is more complex. Attitude is a product of cognitive (information processing; reasoning), affective (emotional) and behavioural components. These are closely linked, so changes in one component can influence both the others and the resulting attitude. This dynamic interaction between reasoning, emotion and behaviour results in a value judgement that, together with social factors, produces what the subject believes are appropriate beliefs and behaviours.

There is a broad spectrum of attitudes towards animals. One extreme is the belief that animals are morally equal to humans, have rights and should never be used or harmed in any way (**see: Animal rights**). The opposite extreme is that animals have no intrinsic worth and can be used for any purpose, regardless of any suffering that this may cause. The beliefs of most people lie somewhere between these two extremes, often varying with the context of human–animal interactions. People variously regard animals as food, entertainment, commodities, status symbols, 'pests', icons, experimental tools, companions or as interesting or aesthetically pleasing beings. Perceptions of animals' intelligence and interpretations of their behaviour – although these can often be anthropomorphic or incorrect – also play a major role in shaping attitudes. All these economic, emotional, religious, cultural and social definitions determine attitudes towards animals and how they should be treated; for example, society as a whole sanctions suffering in 'pest' animals that would not be tolerated for companion animals.

Regulations and codes of practice that control animal use and husbandry both reflect and are influenced by the prevailing attitudes of relevant stakeholder groups and society as a whole. Some sectors also aim to ensure appropriate attitudes by screening new staff, ensuring that workers receive **ethics** training or setting up ethics or animal care and use committees. The importance of cultivating a positive attitude towards animals is also reflected in the national curriculum in some countries, e.g. the UK, which requires that children are encouraged to empathize with animals. (PH)

See also: Animal activist; Christianity; Codes of Practice; Hinduism; Humane societies; Humanism; Islam; Jainism; Judaism; Legislation to protect animals; Morality; Society and animal welfare; Speciesism; Stockperson; Utilitarianism

Audience effect

The audience effect refers to the influence that the presence or identity of conspecifics can have on the frequency, intensity or type of behaviour expressed by an animal. The feeding behaviour of domestic chickens is one well-studied example. Hens attract the attention of chicks to food with a complex behavioural display consisting of staccato calls and pecking movements directed to food on the ground. This display is not fixed. The sight of young chicks inhibits maternal feeding and prolongs the display. In natural conditions, the display of junglefowl hens is intensified when the chicks move too far away or fail to respond. Under experimental conditions, hens vocalize for longer and give more food calls when the chicks are visible but physically separated from the hen than when the chicks are free to interact with her. Nicol and Pope (1996) trained mother hens to expect that food of a particular colour was distasteful. When these hens saw their chicks eating food of the same colour, although there was no significant change in their vocalizations, other aspects of the display including food pecking, food dropping and food scratching were increased as a response to the chicks making apparent errors. Audience effects among hens are not limited to mothers and chicks: male domestic chickens give food-associated calls at higher rates in the presence of a hen than when isolated (Marler *et al.*, 1986). This occurs regardless of whether the hen is his mate or an unknown female; however, if the audience is another male chicken, the food-calling behaviour is suppressed (Marler *et al.*, 1986). Audience effects have been described in a wide variety of species, ranging from vervet monkeys (Cheney and Seyfarth, 1985) to Siamese fighting fish (Doutrelant *et al.*, 2001). (CS)

References

Cheney, D.L. and Seyfarth, R.M. (1985) Vervet monkey alarm calls – manipulation through shared information. *Behaviour* 94, 150–166.

Doutrelant, C., McGregor, P.K. and Oliveira, R.F. (2001) The effect of an audience on intrasexual communication in male Siamese fighting fish, *Betta splendens*. *Behavioural Ecology* 12, 283–286.

Marler, P., Duffy, A. and Pickert, R. (1986) Vocal communication in the domestic chicken: II. Is a sender sensitive to the presence and nature of a receiver? *Animal Behaviour* 34, 194–198.

Nicol, C.J. and Pope, S.J. (1996) The maternal feeding display of domestic hens is sensitive to perceived chick error. *Animal Behaviour* 52, 767–774.

Autonomic nervous system

The autonomic nervous system (ANS) is a very complex and highly differentiated constituent of animal physiology that innervates three types of cells – smooth muscle, cardiac muscle and glandular cells – and acts as an intermediate pathway between the **central nervous system** (CNS) and the body's organs. It regulates and coordinates many fundamental visceral functions, including gastrointestinal motility, cardiovascular reactivity, respiration, body temperature and blood pressure. It comprises two anatomically separate systems, the **parasympathetic nervous system** (PNS) and the **sympathetic nervous system** (SNS). (RM-F)

Autoshaping

The term 'autoshaping' was first used by experimental psychologists investigating animal learning to refer to the discovery that animals often responded to a stimulus such as a button or a lever, even though the response had never been associated with any reward. Hence, the prefix 'auto' refers to the animal's apparent shaping of the response by itself. For example, if a light is briefly illuminated a few seconds before food presentation, pigeons begin to peck the key even though no explicit response–reward contingency has been established. Since the arrival of food has been preceded by a novel event in the environment, the pigeons respond to the novel event as if it were food.

Autoshaping can be used in preliminary training to generate a reliable response which the experimenter can subsequently pair with a reward. For example, chickens will readily peck at a key introduced into their environment in an autoshaping procedure. Once the chicken reliably pecks the key, the experimenter may control the equipment so that there is a food reward after each peck. The advantage of using an autoshaping procedure rather than pairing pecks at the key with a food reward from the outset is that the former procedure is often faster in training the animal than the latter. An autoshaping procedure can also be less tedious and has the added advantage that the animal has less interaction with the experimenter than in other forms of training. (RF)

See also: Reinforcement

Aversion and aversives

An aversion is a reaction of avoidance. In this entry, innate aversions and conditioned aversions will be discussed with particular attention to conditioned taste aversion and the use of aversives to control the behaviour of captive animals. There are aversions that appear to be innate (instinctive or unlearned). These are responses to anything that is an extremely strong stimulus, such as a very loud sound or a very bright light. Many animals have **neophobia** – fear of anything new – and the degree to which an object is judged as new can vary, depending on the animal's boldness or lack thereof. Thunder and lightning are stimuli that most animals, including humans, find aversive. The sudden appearance of lightning and the loudness and low pitch of thunder appear to be fear provoking. This, however, is not the only reason why a dog may be afraid of thunderstorms. In addition to the initial sight and sound of an electrical storm, other frightening things may happen and be learned. For example, a tree struck by lightning may fall on the house. Many animals have an aversion to anything approaching them from above. This varies from the common complaint that a dog growls when petted on the head to a fear of hot-air balloons. Many animals are more likely to show aversions to things that startle or chase them, so a paper blowing across its path may startle a horse, and a cart dragging behind it will produce panic in an untrained horse.

All animals are averse to **pain**, but the stimuli that are associated with pain must be learned. An animal has to touch a flame or an electric fence before it will be avoided. Because an animal must learn to avoid pain, the use of remotely activated electronic **collars** may teach the animal to avoid something other than the intended action or object. For example, if a dog is shocked when it attacks a duck on a pond, it may learn to avoid the pond rather than ducks. Pain is probably the only stimulus quality that is routinely aversive. The relationship between perceptual strength and aversion is complex.

Acute perception is not related to acute aversion. Thus, just because dogs have a very acute sense of smell, it does not mean that it is easy to repel them with an odour at a lower concentration than would be perceived by humans. Citrus odours, such as citronella, are the ones most likely to repel dogs and cats, and are used to punish barking in some bark-activated devices – the sudden spurt of compressed air from the collar may contribute to its aversiveness. Some animals seem to learn to avoid aversive things by observing others (**see: Social learning**). Anecdotally, once one cow has been shocked by an electric fence, all the other cattle will avoid it.

Taste repellents are often not very effective in the artificial control of behaviour because – although aversive or non-preferenced – the aversion is only relative. A hungry cat will eat foul-smelling or bitter food, e.g. food treated with cinnaminic acid. Cats and dogs given a two-choice preference test between two garbage bags of foods, one of which had been sprayed with cinnaminic acid, would avoid that bag; however, in a field test, garbage bags sprayed with the repellent were torn open because that odour predicted the chicken inside. Many people spray repellents on the surfaces of fences and wooden buildings to discourage horses from wood chewing or cribbing. These probably discourage licking the surface because taste receptors are on the tongue, but there are no taste or pain receptors on the surface of the teeth. The most effective aversive agents are those that produce pain. Carbon tetrachloride, for example, stimulates trigeminal nerve receptors and will be avoided.

The use of negative reinforcement as a training method is based on aversion. Horse training is based on negative reinforcement. The horse finds being kicked in the flank aversive and quickly learns to move faster to escape. The horse learns to avoid being kicked by moving at the first touch of heel to skin. Gradually, subtler cues can be used to predict the kick. Soon, the horse will walk forward in response to the slightest squeeze of the rider's legs.

Conditioned taste aversion

The aversions to painful, bitter, loud or frightening stimuli are generally innate, although learning can strengthen them. The phenomenon of conditioned taste aversion refers to a specific form of classical conditioning associated with the development of an aversion to a given taste or smell. The unconditioned stimulus is gastrointestinal malaise. When an animal becomes ill after eating a novel food, it will avoid that food. Unlike most types of learning in which the unconditioned stimulus must be paired closely in time with the conditioned stimulus, a taste aversion can be formed when the unconditioned stimulus (nausea, vomiting) occurs hours after the conditioned stimulus of taste. Novelty is important because the animal will have learned that a more familiar food is safe, i.e. not followed by malaise.

Conditioned taste aversions were first discovered when, subsequent to irradiation, rats refused to drink flavoured water that they had drunk while being irradiated. Since then, use has been made of this phenomenon to produce a conditioned taste aversion to lamb in coyotes. Coyotes were fed lamb impregnated with lithium chloride – an emetic; after several meals that culminated in vomiting, the coyotes would not kill or eat lambs and might even vomit at the sight of a lamb. Application of this form of conditioned taste aversion has been used to reduce **predation** on sheep without killing the coyotes or interfering with their motivation to hunt other prey such as rabbits. Lamb carcasses were injected with lithium chloride and left in the sheep's home range for the coyotes to eat. One reason this method has not been more successful is that lithium chloride has a salty taste and is used as a salt substitute for people on a low-sodium diet. The coyotes were, in effect, becoming averse to salty lamb. Another problem is that coyotes may have eaten enough lamb before lithium-baited carcasses became available that lamb would not be a novel food.

Taste aversion has also been used to prevent plant poisoning in sheep and horses. Locoweed poisoning is common in many parts of the American West and South Africa. Locoweed does not produce malaise, but does contain toxins that accumulate to produce central nervous damage. For example, horses can be trained to avoid locoweed by pairing it with lithium chloride via stomach tube. (KAH)

See also: Reinforcement

Further reading

Pfister, J.A., Stegelmeier, B.L., Cheney, C.D., Ralphs, M.H. and Gardner, D.R. (2002) Conditioning taste aversions to locoweed in horses. *Journal of Animal Science* 80, 79–83.

Aversion learning

The term 'aversion learning' was first used by experimental psychologists to refer to the discovery that animals quickly and accurately learn to avoid certain stimuli. Initial work showed that when rats were made ill by independent means following ingestion of a previously preferred food, they quickly learned to avoid that food (Garcia *et al.*, 1955). Aversion learning can occur after just one trial, even if there is a long delay between the food **stimulus** and becoming ill. In this respect, aversion learning is established faster and perhaps more reliably than other conditioned responses. Aversion learning is sometimes called food aversion learning, to distinguish it from other forms of aversion learning such as colour aversion in birds learning to avoid noxious and distinctly marked prey.

Escape and avoidance conditioning are two forms of aversion learning. Escape conditioning occurs when an animal learns the response necessary to escape an aversive stimulus. Initially this is learned by trial and error and, with time, the animal quickly displays the correct response when the stimulus is presented. Avoidance conditioning occurs when an animal associates a neutral stimulus with a forthcoming presentation of an aversive stimulus. Hence if an animal is exposed to a light coming on prior to an electric shock it will, with time, learn to respond to the light in order to avoid the shock.

There are two main theories as to how aversion learning occurs: (i) the one-stage theory – the animal learns, by trial and error, to show the required response; and (ii) the two-stage theory – the subject initially learns to associate the warning stimulus with the aversive stimulus so that the warning stimulus elicits **fear**. With time, the animal learns to escape from the warning stimulus and avoid the resulting fear.

One striking property of aversion learning is that animals learn to form some associations better than others. For example, rats are capable of learning to avoid a nausea-inducing food if it is associated with a particular taste, but not if it is associated with a sound. Conversely, they can learn to crouch to avoid a shock when it is associated with a sound, but not if it is associated with a particular taste. These constraints on learning appear to be related to the natural behaviour of the animal, since a rat may expect a novel food to cause nausea and, similarly, the sound of a predator may be likely to cause pain (which in this case was brought about by the electric shock). Hence, it appears that natural selection has honed the animal's readiness to form certain associations.

One application of aversion learning techniques may be to train animals to avoid certain foods. For example, Pfister (2000) trained cattle to avoid eating a target plant (green pine needles) by making them ill through a lithium chloride injection subsequent to ingestion. Once released, cattle that were made ill following ingestion initially avoided the target plant more than controls. However, 16 days after release, trained and control cattle were eating similar amounts of pine needles. He found that trained cattle did not avoid pine litter initially (i.e. pine needles on the ground), and this failure of cattle to generalize to other forms of pine needles may have led to them losing their aversion to pine needles. One possibility is that if an animal accidentally ate the target plant while eating pine litter, and did not become ill, then the aversion conditioning would be weakened till eventually it would disappear. None the less, refinement in the training procedure should be possible to eliminate the rapid loss in aversion, and may yet prove to be a valuable method in the management of consumption of poisonous plants by livestock.

Aversion learning has also attracted attention from animal welfare scientists investigating the issue of animal suffering. From our own experience it could be argued that one function of **suffering** is to learn to avoid or generally try to escape from certain situations. It could also be argued from our own experiences that the greater the suffering then the greater our attempts to avoid situations that induce such feelings. Therefore by measuring the extent to which animals avoid certain situations, it may be possible to rank how much suffering is caused by each. For example, Rushen (1996) studied learning to avoid certain handling procedures in sheep by training them to run along a race to reach an area where the procedure was carried out. He measured the latency to run along the race in seven trials and assumed that longer latencies would indicate greater aversiveness. Using such an approach, he was able to rank the relative aversiveness of various handling procedures.

In some respects, an aversion learning technique is more useful than traditional behavioural measures of **welfare**, because the animal's immediate behaviour may be influenced by a large array of factors, not just the imminent handling procedure, making results difficult to replicate and interpret. Additionally, changes in physiological **stress** would probably not provide a good comparison of various handling procedures, as the stress response is seldom accurately related to the intensity of the **stressor**. In contrast, aversion learning techniques are sensitive to small variations in the intensity of stimuli, such as electric shock, noise and vibration.

However, aversion learning techniques are often not reliable. First, the natural response of the animal may interfere with the performance of the measurable avoidance behaviour. For example, chickens cannot be trained to move away from a predator-type stimulus because the natural response in this species is to freeze. Secondly, learning and **memory** ability tend to differ considerably, not only between animals of related species but also between individuals of the same species. Hence, negative results may easily arise because of problems with the method rather than from a lack of aversiveness of the treatment. The use of aversion learning techniques therefore requires an understanding of the memory and learning ability as well as a detailed knowledge of the behaviour of the species. (RF)

See also: Reinforcement

References

Garcia, J., Kimmeldorf, D.J. and Koelling, R.A. (1955) Conditioned aversion to saccharine resulting from exposure to gamma radiation. *Science* 122, 157–158.

Pfister, J.A. (2000) Food aversion learning to eliminate cattle consumption of pine needles. *Journal of Range Management* 53, 655–659.

Rushen, J. (1996) Using aversion learning techniques to assess the mental state, suffering and welfare of farm animals. *Journal of Animal Science* 74, 1990–1995.

Avoidance behaviour

Avoidance behaviour can be described as a sequence of actions involving retreat from, or lack of approach to, a perceived threat **stimulus**. For the purpose of scientific study, avoidance behaviour can be sub-divided into two main categories: unconditioned (unlearned) and conditioned (learned).

Unconditioned avoidance behaviour

Unconditioned behaviour is assumed to be entirely instinctive. In the parlance of **ethology** – the study of animals' natural action patterns (**see: Tinbergen, Niko; Lorenz, Konrad; von Frisch, Karl**) – unconditioned behaviours are those that are 'released' by 'triggers' or 'cues'. From a developmental point of view, it is impossible to rule out some form of learning, since from inception the organism is constantly being stimulated by the environment and so may react on levels as fundamental as the genome. The main point is that fixed-action patterns emerge as a function of the organism's particular structure and, having done so, appear to remain stable and resistant to change. By contrast, learned behaviour is more malleable and can change in response to environmental pressures. So learned behaviour is adaptive, whereas unlearned behaviour can be described as pre-adaptive.

The cockroach's (e.g. *Periplaneta americana*) aversion to light is a common example of unconditioned avoidance behaviour and clearly serves as a survival strategy. In the dark, the cockroach is less vulnerable to predation. So strong is its drive to seek darkness that it will run towards a dark escape hole in the wall even if a bright light exists directly over the hole. This is a good example of the non-malleability of unconditioned avoidance behaviour. Predator odours serve as another example. Cats are well-known predators of rats and mice. If a laboratory rat (e.g. *Rattus norvegicus*) is placed for the first time in an apparatus that has a worn cat collar attached at one end and a hide box at the other, it will perform a characteristic posture known as 'head out behaviour' (Dielenberg and McGregor, 2001). Descriptively, this involves the rat poking its head out of the hide box entrance hole while vigilantly orienting itself towards the cat collar. It performs this behaviour despite having never experienced a cat before, either directly or indirectly through its littermates. It also shows more generalized avoidance behaviour in that if it is exposed repeatedly to the cat collar, it will decrease its vigilant behaviour, but will not increase its approach to the collar. Interestingly, it will perform occasional forays up to the collar to investigate it. This is known as 'risk assessment', and is an important component of avoidance behaviour in that it allows the animal to determine the degree of risk that a threat stimulus holds.

Conditioned avoidance behaviour

Conditioned avoidance behaviours involve learning something aversive about a stimulus that is either intrinsically threatening to the species or has acquired some threatening aspect due to an association with an intrinsically threatening object or process.

Pavlov's famous experiments with dogs demonstrate the process of conditioning (**see: Conditioning**). Conditioned stimuli rely on being contextually associated with unconditioned responses. Unconditioned responses, in turn, are strongly linked to survival.

To condition an animal to avoid a place or an object, the experimenter must associate a conditioned stimulus with something the animal intrinsically fears (as opposed to food, which the animal likes). A classic example is the shuttle box experiment. The rat is placed in the apparatus, a box with two compartments, then occasionally given a mild foot shock irrespective of which compartment it is in. Importantly, just before delivering the shock, a tone is sounded. The animal quickly learns that when it hears the tone, it needs to get out of the compartment it is in. Soon the animal can be made to shuttle back and forth at the sound of the tone alone. In this way, a conditioned avoidance can be established. The animal has been conditioned to avoid the compartment in which it hears a tone.

A more subtle example has been shown in the case of rhesus monkeys (*Macaca mulatta*). Mineka and Cook (1988) demonstrated that juvenile rhesus monkeys learned to be fearful of snakes by watching video recordings of adults responding fearfully towards snakes. From this experiment, some researchers concluded that rhesus monkeys are not born with an innate fear of snakes; rather, they are born with a preparedness to fear snakes, which can be awakened given the appropriate social and or environmental cues (NB: this is a contentious hypothesis). Certainly, it makes evolutionary sense, since failure to learn fear about a truly dangerous stimulus could mean the difference between life and death. Interestingly, when the same monkeys were shown videos of adults responding fearfully to flowers (the adults were conditioned to fear the flowers), the juvenile monkeys did not form a response as strong as was observed with snakes. The reason for the weaker conditioned avoidance response is perhaps best explained by the fact that flowers do not normally present a survival threat to the monkey.

Another important phenomenon found with conditioned avoidance is the process of '**extinction**'. Fear reactions to intrinsically non-threatening objects can fade over time. For example, this occurs when an intrinsically non-threatening object, which was once paired with **pain**, is repeatedly exposed to the animal without pain. Gradually, the animal will learn that the stimulus is non-noxious. This is especially the case when the animal has been deprived of food and the non-threatening stimulus is a palatable food. The same process does not occur with a true threat stimulus, and once again reinforces the notion that each species possesses a predisposed fear towards certain threat objects relevant to its survival. (RoD)

See also: Predator avoidance

References

Dielenberg, R.A. and McGregor, I.S. (2001) Defensive behaviour in rats toward predatory odors: a review. *Neuroscience and Biobehavioural Reviews* 25, 597–609.

Mineka, S. and Cook, M. (1988) Social learning and the acquisition of snake fear in monkeys. In: Zentall, T. and Galef, B.G. (eds) *Social Learning: Psychological and Biological Perspectives.* Erlbaum, Hillsdale, New York, pp. 51–73.

Awareness

The use of the term 'awareness' can roughly be categorized into three instances: (i) as a synonym of **consciousness**; (ii) as a subjective state that is in some way differentiated from consciousness (perhaps being more 'basic' or elemental, or one of a number of 'types' of consciousness); or (iii) defined in purely neurological, behavioural or cognitive terms that do not necessarily imply the presence of any subjective element.

Awareness as a synonym of consciousness

Awareness is commonly used as one of a series of rough synonyms (Block, 1995), together with consciousness, **sentience** and so on, which point towards the elusive phenomena of private, **subjective experience**: what it is like to be either ourselves or another animal (after Nagel, 1974). Since many believe consciousness cannot be defined in a way that is non-circular, it is not surprising that definitions of one commonly cite the other in this manner. To say that an animal has awareness in this sense is to talk of a state that is arguably outside the limits of scientific measurement or proof, but is central to considerations of animal **welfare** and legislation.

Awareness as a type of consciousness

Griffin (1976) notes that as a means of describing animals' mental experiences, some find awareness more acceptable (i.e. less controversial) than consciousness. But how might awareness differ from consciousness? Some use the term to refer to a particular 'type' of consciousness (see also the section below for other ways in which the term is used). Examples of this in the scientific and philosophical literature are manifold, and sometimes contradictory, demonstrating that such semantics are very much in the mind of the beholder.

Lambie and Marcel (2002), for example, distinguish two levels of consciousness: phenomenology (which is 'what it is like' to have particular experiences, such as the sensation of pain, seeing a certain shade of green and so on) and awareness (which is 'a kind of knowing' about that phenomenology). The difference between these two levels of consciousness can be illustrated by asking the following question: 'How does your left foot feel?' Before being asked that question, did the feeling in your left foot exist? Unless you were aware of your left foot beforehand (e.g. maybe you were aware your new shoes were starting to pinch your left foot, or perhaps someone has just trodden on your toe and you were aware of the pain), it is likely that information (i.e. how your left foot feels) only became accessible to verbal report once you had been asked about it.

However, the feeling may, in some sense, have existed before then, i.e. phenomenology might occur separately from awareness. They note that the philosopher Block (1995) divides these mental phenomena along very similar lines, also distinguishing two types of consciousness: phenomenal consciousness (or P-consciousness, for short) and access-consciousness (A-consciousness). Block states that P-consciousness cannot be defined non-circularly, other than saying it is 'what it is like' to be in a particular experiential state (e.g. having sensations, feelings, perceptions, thoughts, desires, emotions and so on), and distinguishes this from A-consciousness, which is marked by 'the availability for use in reasoning and rationally guiding speech and action'. In contrast to Lambie and Marcel, though, he feels more comfortable equating P-consciousness with being aware and A-consciousness with being conscious (i.e. tends to label these phenomena the other way round; **see also: Attention**).

Awareness without (necessarily any) consciousness

The term awareness can be used without necessarily implying the presence of any subjective experience. Chalmers (1996) provides a good working definition of awareness in this context: 'Awareness can be broadly analyzed as a state wherein we have access to some information, and can use that information in the control of behaviour. One can be aware of an object in the environment, of a state of one's body, or of one's mental state, among other things.'

A definition such as this, expressed in terms of information processing and behavioural control, can be used to infer the status of an animal's *awareness* by observing particular stimuli and behavioural output, while remaining agnostic (i.e. non-committal) with regard to the existence of any private, subjective experiences. For example, an observer might state that an animal has become aware of a nearby predator by noting an acute, obvious change in behaviour (e.g. fleeing in the opposite direction) that is informed by, and an adaptive reaction to, the presence of that predator, without necessarily inferring the presence of any accompanying subjective element.

Of course, the lack of any apparent behavioural change does not necessarily mean an animal is unaware of a particular stimulus: in the above example, the animal may have been aware of the predator's approach but chose to stay still, or was unable to engage in an active behavioural response due to injury or illness. In keeping with this type of definition, Jennings (1998) defines awareness as the physical (neurological) processes associated with consciousness in those who are capable of such subjective experience, but suggests an animal might be aware without being conscious, i.e. an animal might undergo physical, neurobiological changes, indicating a change in the informational state of the brain, while essentially being an automaton, incapable of any subjective experience.

Chalmers' view of awareness is perhaps one of the most carefully considered and complete. Following his working definition, quoted above, he proceeds to define awareness as the 'psychological correlate of consciousness': the cognitive processes associated with the phenomenal, subjective property of the mind (**see also: Cognition**; **Mentation**). Consciousness does not occur without awareness, but he does not regard them as synonymous: the former refers to how the mind feels and the latter to what it (causally and functionally) does.

Whether awareness can occur without consciousness, though (as Jennings' definition allows), is a little more complex: Chalmers notes that he can be in some sense aware of a given fact – for example, the US President's identity – but there is a distinction between this fact being in his memory, accessible on request but not presently retrieved, and it actually being directly accessed so that he experiences it. As a result of these nuances, Chalmers defines awareness as requiring *direct access* (i.e. he defines it as a more faithful psychological correlate of consciousness), and he allows for the possibility of awareness occurring in non-linguistic species by characterizing the contents of awareness as being available for use in the global control of behavioural processes (rather than stating that the contents of awareness are solely characterized by their availability for verbal report). Although Chalmers' treatment of the term will not satisfy everyone's understanding of awareness, it is none the less an interesting example of how defining awareness in a cognitive sense can allow it to be variously decoupled and linked to subjective experience.

The content of awareness

Almost regardless of which of the above definitions one has in mind when talking about awareness, what an animal is aware of will depend on the type of information that is accessible and available. This will differ between individuals and across species due to differences in cognitive capabilities and the limitations of sensory apparati (**see also: Perception**). For example, olfactory awareness may be greater in an animal whose olfactory sense is particularly rich and well developed. Such an individual may be aware of differences in odour that another might be unable to distinguish, or become aware of the presence of an odour when it is at a much lower concentration in the air. Similarly, rats' hearing typically extends into the ultrasound range, i.e. to frequencies higher than those an (unaided) human can detect. Therefore, a rat kept in a laboratory, for example, might be aware of sounds to which a human is oblivious, possibly to the detriment of its welfare or to the detriment of experiments in which the rat is a subject.

In addition, Kendrick (1997) notes that the '**motivational state**' of an animal might be an important determinant of what information reaches awareness, a conclusion he draws from experiments indicating that when a ewe is not sexually receptive, she shows very little neurophysiological change when presented with the image of a ram; however, when she is in oestrus, she exhibits a substantial neurophysiological response (**see also: Motivation**).

Finally, the cognitive capabilities of an individual or species will be an important determinant of what one is, or ever can be, aware of. Although some animals are capable of forming simple abstract concepts, only humans are likely to be aware of highly abstract ideas (e.g. globalization, quantum mechanics, etc.); similarly, there are likely to be differences between individuals and species in the extent to which an animal can be aware of the state of another's mind, or be aware of aspects of 'self' (**see also: Concept formation; Self-awareness and self-recognition; Self-consciousness; Theory of mind**).

(RMAP)

References

Block, N. (1995) On a confusion about a function of consciousness. *Behavioural and Brain Sciences* 18, 227–287.

Chalmers, D.J. (1996) *The Conscious Mind*. Oxford University Press, Oxford, UK.

Griffin, D.R. (1976) *The Question of Animal Awareness*. Rockefeller University Press, New York.

Jennings, R.C. (1998) A philosophical consideration of awareness. *Applied Animal Behaviour Science* 57, 201–211.

Kendrick, K.M. (1997) Animal awareness. In: Forbes, J.M., Lawrence, T.L.J., Rodway, R.G. and Varley, M.A. (eds) *Animal Choices*. Occasional Publication No. 20, British Society of Animal Science, Edinburgh, UK.

Lambie, J.A. and Marcel, A.J. (2002) Consciousness and the varieties of emotion experience: a theoretical framework. *Psychological Review* 109, 219–259.

Nagel, T. (1974) What is it like to be a bat? *Philosophical Review* 4, 435–450.

B

Backtest

According to Hessing *et al.* (1993), a piglet's **coping** style can be obtained during the suckling period by assessing its response during a behavioural test, known as the 'backtest'.

Coping style is a concept in which **stress** responses shown by individuals within the same species may be divided into distinct categories. It is based on behavioural, physiological and psychological responses that are consistent across time and across situations, even if their overall level changes. Two distinct coping styles, 'active' and 'passive', have been shown to exist in rodents. Active individuals are characterized by high levels of **aggression**, greater activity levels and other physiological and neuroendocrinological characteristics predominant of a fight/flight response. On the other hand, passive individuals display the opposite pattern by showing conservation/withdrawal responses. A strong individual differentiation that might reflect different coping styles is present in several species (see Koolhaas *et al.*, 1999 for a review), but it is controversial in pigs.

In order to investigate the existence of coping styles in pigs, Hessing *et al.* (1993) introduced the backtest. Since its creation, the backtest has been widely used and there are slight variations as to how it is conducted. In general, a piglet is removed from its home pen, then placed on a flat surface (i.e. table) where it is manually restrained in the supine position for 1 min. To restrain a piglet, the handler places the palm of the left hand firmly over the piglet's head, and keeps its front legs between his/her fingers. The handler's right hand is placed loosely over the piglet's hind legs. Piglets' behavioural responses during the backtest range from uninterrupted immobility to continuous struggling, and show a unimodal distribution. The number of escape attempts, usually defined as a bout of wriggling, is used to assign a 'backtest score' to each piglet. Piglets that perform a low number of escape attempts are considered 'low resistant' (LR), whereas those that perform a high number of escape attempts are considered 'high resistant' (HR). The criteria used in defining the backtest score have varied among different investigators since the original description of the backtest. Nevertheless, it has been suggested that the backtest score can be predictive of piglets' responses to other stressful situations, and can consequently be used as an indicator of coping styles in piglets.

Several factors contribute to the popularity of the backtest among some researchers. The backtest is a quick and inexpensive way potentially to assess piglets' coping styles. It allows us to better understand the genetic influences on piglets' coping style, as individuals are young, naive and unlikely to be affected by previous experiences. The main argument for using the backtest as a measurement of piglets' coping style is that several studies have demonstrated that backtest scores correlate with piglets' responses in social and non-social tests, as well as with physiological measurements indicative of distinctive coping styles. For instance, animals classified as HR in the backtest have been shown to have a stronger fight/flight response pattern by being more aggressive in social tests, exploring a novel environment less, having lower cortisol levels and higher mean **heart rate**, **blood pressure** and metabolism than LR piglets, as would be expected for an 'active' individual. LR and HR also differ in apomorphine susceptibility, immune reactivity, post-mortem heart muscle alteration and stomach wall damage.

Despite its potential benefits, the backtest has been widely criticized. One of the main criticisms relates to its methodology, especially with regard to categorization of individuals based on the backtest score. Most studies that report the existence of distinct coping styles in pigs use only the extremes of the population when classifying individuals as HR or LR. This exclusion of any intermediate group results in the erroneous analysis of a unimodal distribution as a bimodal distribution. In other cases, the entire population is arbitrarily divided into two groups. It is generally accepted that domesticated pigs demonstrate large individual variation, but attempts to categorize animals on distinct coping style, as done for rodents, have been less successful and show conflicting results.

Other problems have been reported with the utilization of the backtest. Although some researchers find a correlation between the backtest and other behavioural and physiological measures, these results cannot always be replicated. One possible explanation is the lack of intra-test consistency for the backtest. Also, it is not clear at this time what are the appropriate behavioural tests or physiological measurements necessary to find correlations in order to define an individual's coping style. The lack of theoretical background makes the backtest difficult to interpret and evaluate.

There are several possible explanations for the discrepancy in the results reported by studies in pigs and why the patterns in other species, such as rodents, may not apply to domesticated animals. **Domestication** may play a role in the distribution of individual characteristics. Genetic selection of pigs that are more adapted to farm conditions may indirectly result in the selection for one type of coping style and a consequent reduction in individual variation. Another explanation is that wild ancestors of the domestic pig, unlike rodents, did not undergo selective pressures to evolve two distinct coping styles. At this time, the data on the existence of distinct coping style in pigs and the validity of the backtest are conflicting and inconclusive. Methods for assessing coping style, such as the backtest, need to be critically evaluated before being used routinely. (VK)

See also: Pig

References and further reading

Hessing, M.J.C., Hagelsø, A.M., van Beek, J.A.M., Wiepkema, P.R., Schouten, W.G.P. and Krukow, R. (1993) Individual behavioural characteristics in pigs. *Applied Animal Behavioural Science* 37, 285–295.

Jensen, P., Rushen, J. and Forkman, B. (1995) Behavioural strategies or just individual variation in behaviour? A lack of evidence for active and passive piglets. *Applied Animal Behavioural Science* 43, 135–139.

Koolhaas, J.M., Korte, S.M., De Boer, S.F., Van Der Vegt, B.J., Van Reenen, C.G., Hopster, H., De Jong, I.C., Ruis, M.A.W. and Blokhuis, H.J. (1999) Coping styles in animals: current status in behaviour and stress-physiology. *Neuroscience and Biobehavioural Reviews* 23, 925–935.

Baiting

Baiting is when an animal(s) is set on another animal, usually a wild animal. In the past this has included bears, bulls and badgers, but baiting of all these species is now illegal. As the animal that is baited is often stronger than the animals used to bait, the baited animal is handicapped in some way through being tethered or chained (bears), through fracturing a limb or jaw (badgers) or through being confined in a space from which it cannot escape. Most commonly dogs are used to bait, the traditional breeds being terriers (badgers) and bulldogs (bulls) or other, larger, dogs. These traditional countryside 'sports' were banned in the UK (e.g. Protection of Badgers Act 1992) but are still not uncommon in other countries. (DBM)

Bar biting

Bar biting is regarded as a stereotypic behaviour (**see: Stereotypies**) in which the animal carries out repetitive mouthing and biting of the metal bars of its environmental enclosure. It is commonly seen in rodents, calves and pigs housed in barren, confined environments and with restricted access to food. However, it has also been implicated as having a function in buffering gastric acidity in sows. (JNM-F)

See also: Repetitive behaviour

Beak trimming

The beaks of laying hens are commonly trimmed in the first 10 days of life. Trimming involves the removal of up to one-third of the upper and lower beaks. This is done by various means, the commonest being a hot, cauterizing blade. Beak trimming reduces feather loss and **cannibalism**-associated injury or deaths. It is considered of particular benefit in non-cage-housed flocks, where outbreaks of cannibalism can spread rapidly. There are other economic advantages to the procedure: trimmed birds tend to waste less food and have increased food conversion. The relative benefits of beak trimming vary among genetic stocks of birds.

There is evidence that the procedure is likely to be acutely painful, as the beak has many nociceptors. A reduction in feed pecking after trimming can be eliminated by the application of topical analgesia to the beak stump at trimming. There is anatomical evidence of neuroma formation following beak trimming in older pullets. There is also electrophysiological evidence of abnormal neural discharges, implicated in chronic **pain** syndromes, recorded from stumps up to 83 days after trimming. There is no evidence of neuroma formation in **turkeys**. Neuromas are less likely if layer pullets are conservatively trimmed at hatch. Behavioural evidence consistent with pain, including decreased pecking rates, has been seen in **chickens** and turkeys for several weeks after trimming. Behavioural changes were less long-lasting in chicks that were trimmed at 1 day of age when compared with those trimmed at an older age. The dermis of the beak normally contains many sensitive mechanoreceptors capable of fine tactile discrimination. These are absent from the scar tissue left after trimming, and this may account for some of the long-lasting behavioural changes seen. The procedure is frequently regarded as a **mutilation** and some governments discourage its routine use as a prophylaxis against cannibalism. (GD)

See also: Chicken; Feather; Laying hen housing; Mutilation; Pecking behaviour; Vices

Further reading

Glatz, P.C. (2000) Beak trimming methods – review. *Asian–Australasian Journal of Animal Sciences* 13, 1619–1637.

Hughes, B. and Gentle, M. (1995) Beak trimming of poultry: its implications for welfare. *World's Poultry Science Journal* 51, 51–59.

Bear

The bear family, Ursidae, in the Order Carnivora, is comprised of only eight living species, which range from the Arctic to the tropics. All but the polar bear are primarily forest dwellers, but they also inhabit grasslands, arid shrublands, tundra and semi-desert. Bears once existed on all continents except Australia and Antarctica, but prehistoric extinctions of several species, range collapse from habitat loss and over-hunting have resulted in their extirpation from Africa, most of the Middle East and Western Europe, Central America and all but the northern Andes of South America.

Descended from carnivorous canid ancestors, modern bears are mainly omnivorous. The polar bear, the most recently evolved species, is the only obligate carnivore, subsisting almost entirely on seals. Conversely, giant pandas, strict herbivores, subsist almost entirely on bamboo. Sloth bears, of the Indian subcontinent, are ant and termite specialists. All other bear species are known to eat insects, as well as numerous kinds of fruits and other plant matter. Sizes of bears (ranging from 30 to over 600 kg), reproductive rates, life history patterns, behaviour and social organization correspond largely with food supplies that vary geographically, seasonally and from year to year.

Three species of bears (brown, American black and Asiatic black) that occupy northern temperate environments hibernate when food supplies disappear during winter; they must then rely solely on stored body fat. In some parts of their geographic ranges, where food is available year-round, they may remain active. Debate still exists as to whether bears truly hibernate, as their body temperature drops only slightly; however, their heart rate and respirations profoundly decline, and they can endure periods of up to 7 months without eating, drinking, urinating or defecating, a feat matched by only a few species of ground squirrels (considered true hibernators).

Northern hemisphere bears (except pandas) give birth while hibernating. Thus, hibernating mothers must not only survive without nourishment, but also must provide nourishment (milk) to newborn young for up to 3 months before emerging from their dens and eating again. In non-hibernating species, females are also restricted to a den site before and after birthing, and may not eat for weeks or months, suggesting that they too possess an adaptation akin to hibernation.

Average litter sizes range from one to three cubs, varying by species and area; some individuals produce litters of four to six (rarely). Average age of first birthing ranges from 3 to 10 years, again varying between species and populations; in areas with abundant food (natural or human-related), bears grow faster and mature earlier.

Once mature, bears generally give birth every 2 or 3 years, although the inter-birth period is extended by 1 year or more in some brown bear populations. Females become reproductively senescent in their mid- to late 20s, but may live into their early to mid-30s.

Cubs are typically cared for by their mother until she mates again. However, mothers nurse cubs for only about 9 months – up to the first winter in temperate populations (but may resume lactation after hibernation in some cases). First year cub survival averages 60–75%, but varies widely.

In the few instances where family break-up has been witnessed in the wild, females actively drove off their cubs after males began to show interest in mating. It is possible that males are also actively involved in family break-up in some circumstances.

Mating occurs several (up to 8) months before birthing, although the fertilized egg does not implant in the uterus until 2 months prior to parturition. The period of delayed implantation may enable bears in poor bodily condition to easily abort a pregnancy. However, observations of captive pregnant bears in poor condition indicate that they are more apt to give birth and then immediately consume their young. Only the tropical sun bear appears not to have a distinct mating and birthing season.

Males mate with several (often three or more) females during a single breeding season, and females mate with multiple males. All mature females breed, whereas male breeding success is related to dominance status. Males are often at least 50% larger than females. Dominance among males is related to their size and age, and is established through both subtle and **agonistic** interactions, sometimes resulting in facial wounds and broken teeth.

Males may sequester females for mating and mate repeatedly for 1 or more days, possibly to stimulate ovulation. Breeding with multiple males sometimes results in litters of mixed paternity, as evidenced by genetic analysis. However, in species where coat colour varies from black to various shades of brown, litters of mixed colours do not necessarily signify mixed paternity.

Aside from mating and cub rearing, bears are generally solitary, although not asocial. They may congregate peaceably at concentrated food sources, such as rich berry patches, fruiting trees, salmon streams, insect aggregations or human-related areas such as garbage or agricultural fields. They also regularly communicate with each other through scent and physical marking of their environment (e.g. biting and rubbing on trees), occasional encounters and subtle **vocalizations**. Home ranges between and within sexes normally overlap, although this varies with food abundance. Home range size, which varies by four orders of magnitude across species and up to two orders of magnitude within species, is also related largely to food availability. Bears may make extensive seasonal movements, laterally or altitudinally, to locate better sources of food. Four of the eight species span an altitudinal range of more than 3500 m, and some individuals range over more than 1000 m of elevation.

Bears have few natural predators. Adult males may kill cubs, and fights between adult bears (e.g. females protecting their cubs from males) may result in death. In Asia, tigers and leopards are potential predators, although bears are generally adept and aggressive enough to chase them away. Sloth bear mothers carry their cubs on their backs, possibly as a defence against tiger **predation**. Young bears of most species climb trees as a refuge from predators. Andean bears of South America and sun bears of South-east Asia appear to be the most arboreal, but this behaviour is related more to their foods than threats of predation.

Humans are the main source of bear mortality. Four species are legally hunted in parts of their range. Most legally hunted populations are presently increasing, because **hunting** is limited and well enforced. Illegal hunting and habitat loss pose much greater risks, particularly in parts of Asia where deforestation and poaching of bears for their gall bladders are causing populations to fragment and decline.

The bile (found in the gall bladder) of bears contains unique compounds that may be instrumental in the hibernation process. Long ago, people in Eastern Asia discovered medicinal properties of these compounds, making bear bile a valuable commodity – this is problematic in today's world where humans are more plentiful and bears more rare. To supply this demand, thousands of bears in China and Vietnam are raised on farms so their bile can be regularly drained with a catheter inserted into the gall bladder. International animal welfare groups are working to rectify this situation.

Conservationists focus on reducing habitat loss, poaching and the removal of wild young bears for **pets**. Only two bear species, American black bears and brown bears, are not considered globally threatened (but many small, isolated brown bear populations are). (DLG)

See also: Game animal; Giant panda

Further reading

Garshelis, D.L. (2004) Variation in ursid life histories: is there an outlier? In: Lindburg, D. and Baragona, K. (eds) *Giant Pandas. Biology and Conservation*. University of California Press, Berkeley, California, pp. 53–73.

Garshelis, D.L., Stirling, I. and Zhi, L. (2001) Bear family. In: MacDonald, D. (ed.) *The New Encyclopedia of Mammals*. Oxford University Press, Oxford, UK, pp. 70–85.

Servheen, C., Herrero, S. and Peyton, B. (compilers) (1999) *Bears: Status Survey and Conservation Action Plan*. IUCN, Gland, Switzerland.

Stirling, I. (ed.) (1993) *Bears. Majestic Creatures of the Wild*. Rodale Press, Emmaus, Pennsylvania.

Stirling, I. and Derocher, A.E. (1990) Factors affecting the evolution and behavioural ecology of the modern bears. *International Conference on Bear Research and Management* 8, 189–204.

Behavioural ecology

Behavioural ecology is the study of animal behaviour within its evolutionary and ecological context – the integrative study of behaviour across **Tinbergen's four questions**, which address the **ultimate** and **proximate** causation of animal behaviour. Behavioural ecology thus integrates the evolutionary focus of traditionally field-based **ethology** with the methodological rigour of laboratory-based **comparative psychology**.

As an evolutionary science, behavioural ecology addresses the consequences of behaviour for an individual's evolutionary **fitness**. That is, how does the expression of the behaviour alter the likelihood that the individual's genes are replicated in future generations, compared with the replication of genes of individuals exhibiting other behaviours? The relative replication of an individual's genes in future generations through the individual's direct descendants is that individual's direct fitness, which is determined by the process of direct selection. Because an individual's genes are also present via shared co-ancestry in collateral kin (e.g. siblings, cousins), an individual's genes are also replicated in future generations in the descendants of collateral kin (e.g. nieces and nephews), although at a lower per capita efficiency than in direct descendants. The relative number of additional copies of genes in collateral kin is an individual's indirect fitness, and the process that determines an individual's indirect fitness is called kin selection. The sum of an individual's direct and indirect fitness is its inclusive fitness, the parameter altered by natural selection.

One of the great puzzles in behavioural ecology has been the existence of behaviours that appear altruistic. **Altruism** is the general term for behaviours that an individual performs at some cost to itself, to the benefit of another. Because of the cost of altruism, these behaviours appear to counter predictions from natural selection. The recognition that indirect fitness mattered during natural selection was a key advance in understanding the evolution of behaviours that otherwise appeared altruistic. That is, natural selection could result in the spread of an apparently altruistic behaviour if the individual performing the behaviour sufficiently increased its own inclusive fitness by helping kin. Under current understanding of behavioural evolution, altruism among non-kin is likely to evolve only under specific conditions, such as through a reiterated process of reciprocal altruism.

Because the indirect fitness benefits of behaviour are dependent on the kinship between individuals, natural selection has shaped mechanisms that produce discrimination of kin from non-kin, and those mechanisms can have an impact on management of captive and domestic populations, as well as conservation efforts. For example, the cues used by prairie falcons (a relatively abundant North American raptor) to identify and feed their own nestlings are sufficiently vague that these falcons will rear young peregrine falcons (an endangered raptor) placed into their nest, in a process called cross-fostering (**see also: Adoption**). If natural selection had resulted in the use of very precise cues by prairie falcons, cross-fostering would not have been an option for those seeking to rebuild peregrine falcon populations.

The discrimination of kin from non-kin has ramifications beyond altruistic behaviour. Inbreeding can have severe fitness consequences for an individual, and natural selection has shaped mechanisms of mate choice to reduce the risk of inbreeding. Possible cues of kin discrimination during mate choice include olfactory or other phenotypic cues, as well as learned associative cues. For example, striped mice avoid inbreeding by familiarity through prior association. The likelihood of successfully breeding striped mice, and other species that use familiarity as a cue to avoid inbreeding, would be increased by housing them separately during development.

Sexual selection is another topic in behavioural ecology that has applied implications. Sexual selection is the special case of natural selection that increases the likelihood of behaviours that improve reproductive success, at the cost of survival. In general, female reproduction is limited by access to resources, and male reproduction is limited by access to females. Sexual selection thus imposes selective pressures on males to compete for access to females. In rhinoceroses, male competition for females is based on size and aggression, so multiple male rhinoceroses cannot be kept in the same captive enclosures even with sufficient food and water.

Whether or not multiple individuals can be housed together may be influenced by aspects of a species' behavioural ecology other than its mating system. For example, the social units of wild spotted hyenas sometimes include over 100 individuals. However, **agonistic** encounters (i.e. win–lose encounters) in spotted hyenas are often resolved when the loser of an encounter temporarily leaves the vicinity of the winner. In captivity, a huge amount of space would be required for an individual hyena to separate itself from others; in the absence of enough room for spatial conflict resolution, aggression levels among captive hyenas rise to dangerous levels, so in practice these otherwise social animals cannot be kept in social groups. In contrast, many social birds such as monk parrots fare better when housed in groups; understanding a species' social behaviour can improve captive management plans.

Another aspect of an animal's behavioural ecology with clear applied consequences is the organism's mating system, which influences the amount of genetic diversity present in a population. An estimation of the number of individuals living in an area does not necessarily reflect the genetic diversity among those individuals, which represents the capacity of the population to respond to current and future evolutionary pressures. If sexual reproduction is highly skewed, as in polygamous mating systems where one male sires most of the offspring, then the existing genetic diversity in the population will be lower than if reproductive success were more evenly distributed, as in a monogamous mating system. In addition, **inbreeding** can accumulate more rapidly in mating systems with high reproductive skew, and an accumulation of inbreeding can reduce survival and reproductive success.

Behavioural ecology can also shed light on the impact of species interactions in contexts that have not previously existed. As an example, although the foraging behaviour of predators and the anti-predator behaviour of **prey** co-evolve in an 'evolutionary arms race', human activities may alter predator abundance in ways that have cascading impacts on prey. Predators have typically evolved to hunt more than one species, and have evolved under conditions in which they rarely face a superabundance of food. Consequently, some well-fed predators continue to hunt in response to cues from prey. **Predation** rates by individual feral and commensal cats are not tightly linked to the amount of food they receive from humans; although the demographic consequences of predation on prey populations are not always clear, subsidized predation by domestic cats often appears to outstrip the growth rate of small bird and mammal populations. (RCVH)

See also: Sociobiology

Further reading

Krebs, J.R. and Davies, N.B. (1993) *Introduction to Behavioural Ecology*, 3rd edn. Blackwell Publishing, Oxford, UK.

Krebs, J.R. and Davies, N.B. (eds) (1997) *Behavioural Ecology: an Evolutionary Approach*, 4th edn. Blackwell Publishing, Oxford, UK.

Behavioural elasticity

This is a term taken from economics referring to the behavioural effect of altering an animal's income (e.g. energy budget) or the cost of performing the behaviour. High elasticity means that small increases in cost, or decreases in income, cause large decreases in performance of the behaviour. Less valued behaviour may show higher elasticity. Conversely, inelastic behaviour is observed when consumption is maintained as price increases (or income decreases) and is indicative of more valued behaviour. (HJW)

See also: Choice test; Decision making; Economics of behaviour; Motivation; Preference

Behavioural need

A need is a requirement for something that is essential or very important. At any moment an individual has a variety of needs, some of greater urgency than others. The principle of a behavioural need is that *performance* of that behaviour is essential for psychological well-being. Thus, if performance is prevented, the animal may experience frustration and ultimately suffering. The phrase was first coined in a paper presented by W.H. Thorpe to the Brambell Committee in 1965, where he suggested that animals possess behavioural needs apart from the more obvious physiological needs. The phrase became popular and is still a key concept in discussions of animal welfare.

Animals generally have a complex and varied behavioural repertoire and it is unlikely that, in any given captive or managed environment, they need to perform all the behaviours in their repertoire. For example, it may not be beneficial for animals to show anti-predator behaviour in the absence of a predator. Providing predatory cues to produce the behaviour could cause stress to the animal. It is therefore essential that we consider which behaviours are important for captive animals to perform.

It is important to distinguish between ultimate and proximate needs. Dawkins (1983) described ultimate needs as being those that are necessary for life, such as food and water. There is an obvious daily need for animals to feed and drink, as denial of these would result in reproductive failure and ultimately death. So all animals have ultimate needs that must be met, but they may also have proximate needs. Failure to perform a behaviour that meets a proximate need may have less catastrophic or immediate consequences, but there may still be welfare implications, particularly if proximate needs are not met over an extended period of time. For example, a captive tiger may be highly motivated to perform hunting behaviour even when food is provided. Failure to hunt will not result in **death**; however, it could negatively impact on that animal's **welfare** through increased stress levels.

In the wild, ultimate and proximate needs will generally occur together. In the example above, a tiger will experience a proximate need for food, resulting in a motivation to hunt which will meet the ultimate need of avoiding starvation. In captivity, however, ultimate and proximate needs may become separated, making it difficult to determine which behaviours are actually needs. It is simple to understand why animals' ultimate needs must be met, but there is no easy way to tell whether, for example, a tiger needs to hunt, a chicken needs to dust bathe or a pig needs to forage. Therefore the underlying **motivation**, and consequences of prevention, of each behaviour must be thoroughly investigated.

Why might it be important to consider behavioural needs?

The idea of some behaviours being important to animals is embedded in several national and international Animal Welfare Codes, which state that animals have a need to express most normal patterns of behaviour. Yet the commercial conditions in which we keep farm and **laboratory animals** often deprive them of resources that they could naturally access. For instance they may have reduced or no access to mates, a lack of space, an inability to create social distance, limited foraging possibilities or no control over their environment. This means that they are being prevented from performing a number of their normal behaviour patterns.

For some behaviours the issue is quite straightforward, in that the conditions in which the animals are being kept do not enable them to meet their behavioural needs. For example, in the period prior to farrowing, sows have been shown to be highly motivated to perform nest-building behaviours. When they are kept in **farrowing crates** without any bedding they are restricted in their movements and have no means of performing this behaviour. This constraint results in increased **hypothalamic–pituitary–adrenal axis** activation, which is indicative of higher **stress** levels, and sows may also suffer higher levels of stillbirth. Thus it seems clear that sows have a behavioural need to perform nest-building behaviour prior to farrowing.

For other behaviours the issues are more complicated, since the behaviours that the animal is being prevented from performing would normally be carried out in order to obtain resources that are being readily supplied for them. For example, in the wild many animals would spend a large proportion of time foraging, and for pigs this may be around 50% of their time budget. Even though they may not be able to forage in the same way in commercial or laboratory conditions, concentrated high-energy food is in ready supply. This gives two possibilities, one being that the animals do not 'need' to perform foraging behaviour, they simply need to satisfy their energy and nutrient requirements and, since sufficient food is provided to meet these requirements, the animal will have no remaining motivation to forage. Thus, restricting the opportunity to forage will not reduce welfare. However, it is also possible that animals have a behavioural 'need' to perform foraging behaviour, whether or not food is freely available. Foraging behaviour may, for example, be motivated at an appetitive stage. Thus, an animal may desire to search or explore possible feeding sites prior to feeding. In this case the actual performance of foraging behaviour is important to the animal, as well as the outcome of the behaviour, i.e. the detection or consumption of food. These same principles can be applied to many other behaviours that animals kept in commercial or laboratory conditions are not able to perform.

Evidence for behavioural needs

The idea that animals have behavioural needs is intuitively appealing; however, it is important to consider the evidence. Sherwin *et al.* (2004) examined burrowing behaviour in laboratory mice. A comparison was made between the time mice spent nesting or burrowing when provided with a compartment containing peat and the same measure when they were supplied with previously constructed burrows. The mice spent an equal amount of time constructing new burrows in the presence of previously constructed ones as when no burrow was present. This suggests that mice need to perform burrowing behaviour as opposed to simply requiring a ready-formed burrow. An argument used against such evidence for behavioural needs is that the resource that is provided may be deficient in some way, and this is why the animal continues to show the behaviour. However, Sherwin's study ensured that the burrows had been dug by the same individual in a previous stage of the experiment. It is therefore difficult to argue that the burrows provided were in any way unsuitable for the mice.

If an animal has a need to perform a behaviour, then keeping it in conditions where it is unable to do so may result in **frustration** and reduced welfare. It is therefore important to distinguish which behaviours are needs.

Motivational aspects of behavioural needs

Some suggestions have been made as to the characteristics of behaviours that are likely to be needs. First, it has been proposed that the behavioural needs will be most likely when a candidate behaviour is primarily motivated by internal causal factors, e.g. changing hormone levels. This is logical, since a behaviour that has external causal factors would be triggered only in certain environments, whereas an internally motivated behaviour may be triggered in any conditions. For example, if the primary external causal factor for rooting behaviour in pigs was the presence of a suitable rooting material, then the pigs would not be motivated to root if they were kept in fully slatted pens without any bedding. However, if rooting was internally motivated by rising hunger levels, then the pigs would be stimulated to perform rooting in a fully slatted pen just as much as in a natural environment.

Externally motivated behaviours can also be behavioural needs, when animals are kept in conditions where the external causal factors are present. Normally, such environments would permit expression of the appropriate behaviour, thus returning motivation to its baseline levels. However, behavioural needs may be unmet where the external causal factors are present but the means to carry out the behaviour are not – e.g. a tethered pig where substrate is available. It is also possible that some behaviours may be motivated by a combination of internal and external factors. For example, Jensen *et al.* (1993) suggested that the onset of nest building in free-ranging sows was internally motivated, but that the course of nest-building and maternal-oriented behaviours was more likely to be externally motivated. Causal factors for the internally motivated aspects were changes in hormones such as **prolactin**, whereas the course of nest building was more influenced by the level of protection offered by the nest site and by the materials available. Thus it is essential that the motivation and causal factors for each behaviour are thoroughly understood, but even then it may not be clear whether the behaviour is a need or not.

Another potential characteristic of a behavioural need is that the motivation to do it should build when performance is being denied. This relates to the classic model of **motivation** proposed by **Konrad Lorenz**, who suggested that motivation may be an accumulation of action-specific energy that is released when the behaviour occurs. Rising motivation during thwarted behavioural performance may be significant if this is accompanied by feelings of increasing frustration. Denial of behaviours motivated in this way would therefore result in a welfare problem that would increase the longer that an animal was kept in the deprived conditions. However, if this definition were to be used it would exclude some behavioural needs simply because motivation was already at a consistently high level and could not increase further.

Thus behavioural needs are behaviours that are important for the animal to perform even if its physiological needs have already been met, i.e. it is the performance of the behaviour that is important to the animal and not the outcomes of that behaviour.

Experimental techniques to examine behavioural need

Observation of natural behaviour

One possibility when trying to establish which behaviours are needs is to study animals in their natural environment and to record all the behaviours that they perform. This enables a complete behavioural repertoire to be established, which seems to be a logical starting point. However, a behavioural need is not proved by the performance of the behaviour in natural conditions, since it may be dependent on the presence of external motivating stimuli. Also, many of the behaviours that are essential for survival in the wild would not be needed in captivity. Thus, this technique enables identification of the behaviours an animal performs in the wild but gives no indication of the relative importance of different elements of the repertoire, especially under captive conditions.

Observation of behaviour when restricted

In order to rule out the issue of behaviours being stimulated by external stimuli that may not be present in captivity, the animal can be observed under specifically structured conditions. It has been suggested that behaviours that are performed even when relevant external stimuli are absent must have particularly strong internal causal factors and that this is evidence of behavioural need (Hughes, 1980). It has also been suggested that animals performing increased levels of exploration when deprived of a resource are demonstrating strong internal motivation to perform the behaviour involved. Nicol and Guilford (1991) compared the time spent exploring by hens deprived of feed, litter or neither. The deprived hens spent more time exploring in a tunnel that had no prior association with the resources, suggesting that they were internally motivated to seek them out.

This technique demonstrates that animals can have strong internal motivation for a given behaviour, even in the absence of external stimuli, making a convincing argument for it being a need in any environment. However the technique is limited to situations where behaviour-specific external stimuli can be systematically removed. It is also restricted to cases where it is still possible for the animal to carry out the behaviour, despite external stimuli being absent. As a result this technique has limited application.

Recording physiological consequences
It has also been suggested that recording the physiological consequences of not performing a behaviour would give an indication as to whether it is a need or not. Classic stress responses could indicate that frustration was resulting from not being able to perform selected behaviours, as could the occurrence of **abnormal** behaviours. However, this technique has the same limitations as that of observing behaviour in restricted conditions, i.e. that it is difficult to restrict only one behaviour and, if this cannot be done, then it is not possible to ascertain which restriction is causing the stress response. Therefore it is possible to reach only limited conclusions about many behaviours using this technique.

Performance despite physiological needs being met
Another possibility is to provide animals with their physiological needs and then see if they continue to perform the behaviour. As previously described, this has proved a successful approach with laboratory mice (Sherwin *et al.*, 2004). A similar experiment was carried out with hens by Hughes *et al.* (1989), which showed that nest-building behaviour was not inhibited by the presence of a preformed nest. This suggests that the animals have a need to perform the behaviour as opposed to just requiring the outcome of the behaviour. However, in some instances the opposite is true. Gerbils' motivation to dig was noticeably reduced by the presence of a burrow (Wiedenmayer, 1997). In this case, provision of the outcome of the behaviour was sufficient for the animals, suggesting that digging behaviour was *not* a need. Overall, this technique seems to work well, but it remains restricted to goal-directed behaviours for which we are able to provide the outcome. It would, for example, be difficult to apply this method to examine whether there is a behavioural need for social contact. Secondly, we have to consider that this technique inevitably presents animals with external causal factors. Thus it is not possible to determine whether the animals would still be motivated to perform the behaviours in the absence of these stimuli. Ultimately, this method cannot tell us whether animals are suffering when they are kept in conditions without those external stimuli.

Consumer demand and behavioural resilience
The final method is to make animals work for the opportunity to perform different behaviours. By varying the amount of work that is required in order to gain the opportunity to perform specific behaviours, the strength of the animals' need for the behaviour can be determined. Alternatively, by measuring behavioural resilience or the inelasticity of demand, an assessment of the need for that behaviour or even a hierarchy of behavioural needs can be constructed (Young, 1999; and **see: Economics of behaviour**).

Thus, this technique has utility when examining individual behavioural needs. We can determine the relative strengths of an animal's need for a variety of behaviours at a particular point in time. However, it should be borne in mind that there will be variation in motivation at different times of the day, stages of life and so on. For example, sows will generally work hardest for food, except in the hours before farrowing when they are motivated to nest build, and will work equally hard for nesting material. Training the animals can be time consuming and outcomes have to be interpreted carefully, particularly since results are a continuum and a cut-off point must be defined in order to identify needs. Finally, as with the previous technique, the animal inevitably experiences the external stimuli, meaning that the conclusions about behavioural needs are limited when external stimuli are absent.

Alternatives to the concept of behavioural need
From this it can be seen that whether animals have behavioural needs or not is difficult to assess and therefore mention should be made of a different approach to the issue, which was raised by Hughes and Duncan (1988). They suggested that the problem for intensively farmed animals is not that they are unable to perform certain behavioural needs, but that they are unable to fill certain behavioural voids. So, for example, intensively kept animals have no need to find food, water or shelter as these are provided, and therefore their time budget is compacted. What may be important to the animal is how to fill the extra time available with the limited behaviours that they are able to perform within their situation. (PoS)

See also: Boredom; Causal factor; Economics of behaviour; Motivation

References

Dawkins, M.S. (1983) Battery hens and consumer demand theory. *Animal Behaviour* 31, 1195–1205.

Hughes, B.O. (1980) The assessment of behavioural needs. In: Moss, R. (ed.) *The Laying Hen and its Environment*. Martinus Nijhoff, The Hague, The Netherlands, pp. 149–160.

Hughes, B.O. and Duncan, I.J.H. (1988) The notion of ethological 'need', models of motivation and animal welfare. *Animal Behaviour* 36, 1696–1707.

Hughes, B.O., Duncan, I.J.H. and Brown, M.F. (1989) The performance of nest building by domestic hens: is it more important than the construction of a nest? *Animal Behaviour* 37, 210–214.

Jensen, P., Vestergaard, K. and Algers, B. (1993) Nestbuilding in free-ranging domestic sows. *Applied Animal Behaviour Science* 38, 245–255.

Nicol, C.J. and Guilford, T. (1991) Exploratory activity as a measure of motivation in deprived hens. *Animal Behaviour* 41, 333–341.

Sherwin, C.M., Haug, E., Terkelsen, N. and Vadgama, M. (2004) Studies on the motivation for burrowing by laboratory mice. *Applied Animal Behaviour Science* 88, 343–358.

Wiedenmayer, C. (1997) Causation of the ontogenetic development of stereotypic digging in gerbils. *Animal Behaviour* 53, 461–470.

Young, R.J. (1999) The behavioural requirements of farm animals for psychological well-being and survival. In: Dolins, F.L. (ed.) *Attitudes to Animals*. Cambridge University Press, Cambridge, UK, pp. 77–100.

Behavioural plasticity

Behavioural plasticity refers to the extent to which an individual can vary its response to a given situation. Plasticity may be used to refer to the level of change in a behaviour over time as a result of external feedback, i.e. the degree to which it is affected by learning. For example, fixed action patterns (**FAP**) are traditionally considered relatively non-plastic (or rigid), although it is now recognized that they do vary and adapt in response to **feedback**. This use is closely related to the use of the term plasticity in neurobiology in relation to synapses: synaptic plasticity refers to the extent to which the connection between two cells can vary, and so is closely linked to the processes of learning and **memory**. Alternatively, the term behavioural plasticity may refer to the ability of an individual to vary its response according to circumstances. For example, fight and flight are both responses to a potential threat, but one may be more appropriate than the other in a given circumstance. Some individuals may preferentially express one strategy over the other regardless of circumstances, and so their response is considered less plastic. In this context plasticity is closely linked to the study of **individual differences**. (DSM)

See also: Behavioural elasticity

Further reading

Briffa, M., Rundle, S.D. and Fryer, A. (2008) Comparing the strength of behavioural plasticity and consistency across situations: animal personalities in the hermit crab *Pagurus bernhardus*. *Proceedings of The Royal Society B* 275, 1305–1311.

Will, B., Dalrymple-Alford, J., Wolff, M. and Cassel, J.-C. (2008) Reflections on the use of the concept of plasticity in neurobiology [Translation and adaptation by Bruno Will, John Dalrymple-Alford, Mathieu Wolff and Jean-Christophe Cassel from J. Paillard (1976) Journal of Psychology 1, 33–47]. *Behavioural Brain Research* 192, 7–11.

Behavioural syndrome

The term behavioural syndrome derives from the use of the word 'syndrome' by evolutionary ecologists to describe packages of traits that go together (e.g. life history or migratory syndromes). For example, migratory syndromes occur in species where migrating individuals develop different groups of traits (or values of group traits) than non-migrants. A behavioural syndrome occurs when a population is comprised of multiple behavioural types. Individuals express behavioural types when they act consistently within and across functional contexts such as foraging or mating. The behavioural type is analogous to an individual's **personality** or **coping** style, and describes the manner in which the individual is likely to behave. For example, more aggressive individuals may engage in more agonistic interactions than less aggressive individuals, regardless of the circumstances.

This behavioural syndrome is marked by a correlation of behaviours within a single context over time or across contexts. For example, aggression and activity may correlate within a single context, with more aggressive individuals being more active and less aggressive individuals moving less. Across contexts correlations can occur if, for example, aggression in one context correlates with aggression in another. A key element of the behavioural syndrome is that individuals may occasionally act in a suboptimal manner. This suboptimality results because limited behavioural plasticity causes a carry-over of an individual's behavioural propensity from one context to another. The result is a **fitness** trade-off where individuals expressing one behavioural type fare well in one circumstance but worse in another. If individuals were unlimited in terms of **behavioural plasticity**, they would change their behaviour for each circumstance; this trade-off would not be observed and behavioural types would be uncommon.

However, behavioural types are noted in many species and the possibility for individuals to act in a manner that is more adaptive in some situations than in others thus occurs. For example, predators may easily notice the active individual that, even in the presence of predators, cannot reduce its activity level entirely. Thus, in situations where predators are common, active individuals may suffer higher mortality than less active individuals that are able to reduce their movements sufficiently to avoid being noticed. However, in an alternative situation where there are few predators, the active individual may locate and take advantage of food at a higher rate than the less active individual. In a behavioural syndrome then, the relative fitnesses of individuals vary within and between contexts.

In the applied sense, it may be useful to encourage the expression of behavioural syndromes in managed populations, be they wild or captive. Different behavioural types interact with their environment in different ways. The result of this differential interaction with the environment is that different behavioural types fill different portions of a species' **ecological niche**. For some individuals subtle changes in environmental conditions can result in behavioural change, while for others extrinsic factors have little effect on behaviour. Alternative behavioural types may also have variant physiological responses to pathogens. For example, individuals that are more proactive or bold appear to be more susceptible to autoimmune disease than their more shy, reactive counterparts. These physiological differences occur in a number of measures and result in different behavioural types being more or less susceptible to various pathogens than others. Because behavioural types have different responses to environmental challenge, it is likely that populations that express behavioural syndromes are resilient to environmental challenge, as their members express varied responses to the challenge. Some of these responses may be effective in dealing with the challenge, while others are not. Conversely, populations with only one behavioural type may respond inappropriately and falter in the face of environmental challenge.

Programmes that raise animals for release to a protected wild environment may benefit from considering the behavioural types of animals destined for release. In the wild setting, animals must face challenges in a number of functional contexts. They must avoid predators and exploit food and occasionally food types with which they have no experience, and they must interact with familiar and unfamiliar **conspecifics**. Thus, in preparing animals for release, it may be most beneficial to consider their responses under an array of conditions. Those individuals that perform poorly at one task may excel at another, and these differences may be predictable based on the behavioural types of individuals. Because of these different responses of alternative behavioural types to environmental challenge, populations that express behavioural syndromes may be more stable than those composed of individuals that are proficient at single, assumedly important tasks.

By promoting the expression of behavioural syndromes in managed populations managers can ensure that populations of concern are capable of dealing with a wide array of challenges. When animals live in social groups, alternative behavioural types are likely to fill specific roles within the group. Some individuals may be more likely to find and exploit new foods, while others may be more adept at locating predators and providing warning to conspecifics. Some behavioural types are more aggressive than others, and may be more likely to fight and gain dominant positions in the social hierarchy. In captive environments, the welfare of animals living in social groups may be enhanced if there is a mix of behavioural types that promotes decreased agonistic interactions in the group. (JVW)

See also: Captivity; Conservation; Ecological niche; Individual fitness; Optimality

Further reading

Gosling, S.D. (2001) From mice to men: what can we learn about personality from animal research? *Psychological Bulletin* 127, 45–86.

Koolhaas, J.M., Korte, S.M., De Boer, S.F., Van Der Vegt, B.J., Van Reenen, C.G., Hopster, H., De Jong, I.C., Ruis, M.A.W. and Blokhuis, H.J. (1999) Coping styles in animals: current status in behaviour and stress-physiology. *Neuroscience and Biobehavioural Reviews* 23, 925–935.

Sih, A., Bell, A.M., Johnson, J.C. and Ziemba, R.E. (2004) Behavioural syndromes: an integrative overview. *Quarterly Review of Biology* 79, 341–377.

Watters, J.V. and Meehan, C. (2007) Different strokes: can managing behavioural types increase post-release success? *Applied Animal Behaviour Science* 102, 364–379.

Behaviourism

Behaviourism (behaviorism) is a school in experimental psychology that focuses on the behaviour of the animal while avoiding the making of any inferences about the underlying processes. This is often called a 'black box' approach. The scientist studies the input, the stimulus, and the output, the behaviour or the response, without trying to understand what happens inside the animal. Because what is studied is the stimulus and the response, this has also been called an S–R approach to behaviour. The existence of internal processes is not denied but, since it was thought that they are impossible to investigate directly, they were excluded from study. This position has later been modified and the stance of behaviourism is today less rigid. Behaviouristic research focuses almost exclusively on the mechanisms of associative learning, especially operant conditioning.

Behaviourism as a school of thought was started in the USA in the early 1900s by the psychologist John B. Watson. It was founded as a reaction against the introspective approach of psychology, which relied on self-reports of mental experiences. The most famous behaviourist, **B.F. Skinner**, constructed a device in which the animal could perform a behaviour, e.g. pressing a lever in response to a signal. The pressing of the lever resulted in an event that could either be pleasant, e.g. food, or unpleasant, e.g. an electrical shock. The signal and the resulting event were controlled by a computer and the behaviour of the animal was registered automatically. This set-up came to be called a Skinner box, and has become a symbol for the behaviouristic approach to studying learning and behaviour. (BAF)

See also: Cognition; Conditioning

Further reading

Kennedy, J.S. (1992) *The New Anthropomorphism*. Cambridge University Press, Cambridge, UK.

Watson, J.B. (1913) Psychology as the behaviourist views it. *Psychological Review* 20, 158–177.

Belly nosing

Belly nosing is the term given to the distinctive, rhythmic up-and-down snout movements performed by one piglet on the belly or flank of another piglet. Belly nosing and belly sucking are considered to be **abnormal** or **problem behaviours**, and are considered by some researchers to be **stereotypic** behaviours, although controversy about their **aetiology** remains. In piglets reared in semi-natural environments and subjected to the natural weaning process (gradual weaning between 7 and 17 weeks of age), belly nosing is rarely seen. However, it is relatively common on commercial farms, where piglets are artificially weaned (abrupt weaning and separation from the sow usually between 2 and 4 weeks of age). Belly nosing begins 3–5 days after weaning and rises to a peak incidence around 2 weeks later, followed by a gradual decline. It is also present in piglets artificially reared on milk replacer (i.e. non-weaned, but maternally separated piglets), can persist through the grow–finish phase and has been observed in both gilts and sows.

Belly nosing is characterized by high individual variation. Some piglets do not belly nose, some spend very small amounts of time belly nosing, and others spend a relatively large amount (as much as 8%) of their time belly nosing. As belly nosing is temporally transient post-weaning, it has been suggested as being redirected suckling behaviour (**see: Redirected behaviour**). A number of factors are associated with increased belly nosing incidence: age, genetics, suckling behaviour, growth and feed intake, feeding/drinking system and environmental enrichment.

There have been a number of studies that have shown that the age at which piglets are weaned has an impact on the amount of belly nosing observed during the post-weaning period. For example, two studies have shown that piglets weaned at 2 weeks of age show more belly nosing than piglets weaned at 4 weeks of age. Comparisons have also been made between piglets weaned at 4 and 6 weeks, piglets weaned at 2 and 3 weeks and piglets weaned at 1, 2 and 4 weeks. In all cases the younger the weaning age, the more belly nosing is observed.

There is some evidence that the genetic line of the piglet increases the likelihood of belly nosing being observed post-weaning. A couple of studies have shown that Yorkshire/Landrace-type piglets show more belly nosing than Duroc-type piglets. Also, in one of these studies, it was shown that the individual sire within the breed also had a significant effect on belly nosing incidence. Many authors have proposed that belly nosing is closely related to suckling motivation, and it would therefore perhaps be expected that those piglets that show high levels of suckling behaviour pre-weaning would show correspondingly high levels of belly nosing post-weaning. In fact the opposite trend has been observed, with piglets that show low levels of suckling behaviour when with the sow showing more belly nosing post-weaning.

There appears to be no direct link between weaning weight and belly nosing, at least at the individual pig level, although it has been reported that lighter weight groups of piglets show more belly nosing than heavier weight groups, with weaning age being constant. However, a number of studies have demonstrated that piglets performing relatively high levels of belly nosing post-weaning have slower growth rates than their pen-mates. When piglets are abruptly weaned, they are usually transitioning from a milk diet to solid feed, although some piglets may have had access to solid feed prior to weaning (creep feeding). This usually results in a marked 'growth-check' as piglets struggle to adapt to the dietary change, and those piglets that are slow to adapt to the diet presumably become hungry. A number of studies have noted that belly nosing is higher in piglets that have low feed intake, and there are significant negative correlations between time spent belly nosing and time spent at the feeder and time spent lying, and a positive correlation between time spent belly nosing and time spent active.

As to whether lack of feeding is driving belly nosing or vice versa, there is inconclusive evidence. Studies to date appear to suggest that when hungry, some pigs redirect foraging behaviour to pen structures, increasing rooting and nosing of the floor, whereas others appear to redirect to pen-mates and increase belly nosing. Belly nosing does appear to be related to the piglet's ability to perform natural sucking behaviours, as illustrated by studies on feeder and drinker design. First, where piglets are given access to nipples attached to a feed trough or access to an artificial udder that allows massaging and sucking, belly nosing is reduced. Also, the design of the drinker can influence time spent belly nosing, with water provided in drinking bowls apparently allowing piglets to satisfy the motivation to 'suck' better than water provided via nipple drinkers, thereby reducing belly nosing. Presentation of feed in a liquid form results in piglets spending less time belly nosing and more time lying resting compared with piglets fed a dry, solid feed – again, presumably by allowing piglets to suck during drinking.

The pig is naturally an inquisitive and exploratory animal, spending large amounts of time engaged in foraging-related behaviour. Within a commercial setting, many pigs are housed in relatively barren environments without any foraging substrates. It has long been known that pigs in barren environments will spend significantly more time engaged in pen-mate-directed behaviours than pigs in enriched environments. Providing piglets with straw and/or other enrichment objects can reduce belly nosing.

In summary, although there have been a number of recent advances made in understanding the aetiology of belly nosing in piglets, the exact causation and functional significance of belly nosing remain to be elucidated. (ELS)

Further reading

Widowski, T.M., Torrey, S., Bench, C.J. and Gonyou, H.W. (2008) Development of ingestive behaviour and the relationship to belly nosing in early-weaned pigs. *Applied Animal Behaviour Science* 110, 109–127.

Bentham, Jeremy

Jeremy Bentham (1748–1832) was an English philosopher regarded as the founder of a school of philosophy known as **utilitarianism**. He was the son and grandson of lawyers and it was intended that he should also be a practising lawyer. But Bentham was so disenchanted with the current state of English law that he turned to philosophy and the study of the theoretical foundations of a perfect system of law, and consequently wrote manuscripts on economics, politics, law and jurisprudence. In order to establish the foundations of a perfect legal system Bentham developed the principle of utility, known as the 'greatest happiness' principle, which, he held, rested on the undisputed natural fact that all living beings prefer pleasurable experiences over painful ones.

In the opening paragraph of the *Introduction to the Principles of Morals and Legislation*, Bentham states that: 'Nature has placed mankind under the governance of two sovereign masters, *pain* and *pleasure*'. This is not merely a factual statement; it informs us about what we ought to do. The principle of utility indicates that the right law will be that which provides the greatest amount of pleasure and, in its negative form, states that the right law will be that which minimizes misery and suffering. Pleasure and pain, being the ultimate measures of value, provide Bentham with guidelines for the well-being of all sentient creatures. Generally speaking it is the negative aspect of the principle that has provided legislators with guidelines: it is easier to establish consensus with regard to actions that alleviate suffering than those aimed at the promotion of pleasure.

Sweeping aside theoretical and theological foundations of morality in Chapter XVIII, Sec. 1., of the Introduction, Bentham posed the question of moral obligation towards all sentient creatures as follows: 'The question is not, Can they reason? Nor can they talk? But, *Can they suffer*?' [author's italics]. This doctrine has obvious implications for animal welfare, offering a moral dimension for scientific studies on the likelihood of pain and suffering among animals in our care.

Bentham's principle of utility underwent several modifications in the light of criticism. The doctrine of psychological hedonism which, according to Bentham, admitted no qualitative distinction between various forms of pleasure, was modified by his 19th-century successor John Stuart Mill, who introduced a distinction between 'higher' and 'lower' pleasures corresponding to mental and physical characteristics, respectively. Classical utilitarianism was further modified to avoid the accusation that its appeal to the consequences of an action supported the questionable doctrine that 'the end justifies the means'. Notwithstanding these modifications, Bentham's utilitarianism continues to provide a viable framework for proposed legislation regarding animal welfare. (DL)

See also: Ethics

Reference

Bentham, J. (1970) *Introduction to the Principles of Morals and Legislation* (J.H. Burns and H.L.A. Hart, eds). Athlone Press, London.

Benzodiazepines

Benzodiazepines are a group of drugs that tend to be used for their inhibitory effect on the **central nervous system**, operating through the stimulation of benzodiazepine receptors on the gamma amino butyric acid A receptor (**GABA**^{A} receptor). Benzodiazepine receptors are abundant in the central amygdala, along the ventral amygdalofugal pathway and within the anterior and medial **hypothalamus**, *substantia nigra* and peri-aqueductal grey (PAG), and so are closely associated with the **fear** response. Benzodiazepines also stimulate appetite, inhibit **memory** and can result in the disinhibition of **aggression**. They are dependence forming, and many animals treated with benzodiazepines in a clinical setting relapse upon the withdrawal of medication, and so their use without behaviour modification is not generally to be advised, unless being used concurrently to manage an acute crisis. Commonly used benzodiazepines include diazepam, alprazolam and lorazepam. (DSM)

Further reading

Crowell-Davis, S.L. and Murray, T. (2006) *Veterinary Psychopharmacology*. Blackwell Science, Oxford, UK.

Bernard, Claude

An eminent French physiologist (1813–1878), whose work on living non-anaesthetized animals (**vivisection**) helped elucidate several physiological functions in the body. He started work as an aspiring writer but eventually was persuaded to study medicine. He then went to work with Francois Magendie, carrying out research into digestion and the pancreas, and eventually succeeded Magendie to the Chair of Experimental Physiology, where he was a proponent of experimental medicine on animals, so-called vivisection. He formulated the hypothesis of the internal **homeostasis** of the body (*milieu interieur*) that could be completely understood only in the whole animal. He was the first scientist in France to be granted a public funeral. (DBM)

Beta-blocker

The term beta-blocker is used to refer to a class of drugs that antagonize (block) the beta-noradrenergic receptors (receptors of **norepinephrine**), which are found in both the **central nervous system** (these tracts run from the *locus coeruleus* to the cortex and from the lateral tegmental area into the thalamic and hypothalamic regions and spinal cord) and peripherally in the **sympathetic nervous system**. Peripheral sympathetic arousal prepares the body for action by diverting blood from the viscera to the muscles and increasing cardiac output; centrally, noradrenergic activity results in changes in attention and memory that support this preparation process.

Blocking these receptors reduces the capacity to mount this response and has potential cognitive effects either directly or indirectly as a consequence. Alternative names for these drugs include beta-adrenergic blocking agents, beta-adrenergic antagonists, beta-adrenoceptor antagonists and beta-antagonists. The most commonly used beta-blocker is propranolol. They are used in animals not only in the management of heart failure, but also sometimes as part of the pharmacological management of fears in **companion animals** and may also be used (often illegally) in anxious animals at very low doses to improve performance. Their use for the management of fears is targeted at reducing the physiological response to the problem stimulus. Beta-blockers lower blood pressure because they relax both smooth (via beta-2 receptors) and cardiac (via beta-1 receptors) muscle, and this is a potentially serious side effect of its use in animals when administered for the management of fears. (DSM)

Further reading

Crowell-Davis, S.L. and Murray, T. (2006) *Veterinary Psychopharmacology*. Blackwell Science, Oxford, UK.

Beta-endorphin

Beta-(or β-)endorphin is an endogenous opioid neurotransmitter 31 amino acids long, resulting from the processing of preprohormone **pro-opio-melanocortin (POMC)**. β-endorphin has the following amino acid structure: Tyr-Gly-Gly-Phe-Met-Thr-Ser-Glu-Lys-Ser-Gln-Thr-Pro-Leu-Val-Thr-Leu-Phe-Lys-Asn-Ala-Ile-Ile-Lys-Asn-Ala-Tyr-Lys-Lys-Gly-Glu.

The first five amino acids are common among several endogenous opioid peptides. This neurotransmitter works through both the **peripheral** and **central nervous systems** and is secreted from neurons of the **hypothalamus**, in the arcuate nucleus and the adenohypophysis of the **pituitary gland** following stimulation by the hypothalamic hormone **corticotropin-releasing hormone** (CRH). β-endorphin acts via the opioid receptors and was initially described for its pain-suppressant effects. It acts naturally to reduce sensations of **pain** in animals over prolonged periods. Many narcotic painkillers work through the β-endorphin and other opioid receptors. β-endorphin response has also been shown to be initiated by other stressful and noxious stimuli such as exercise and foot shock. β-endorphins work primarily through the opioid receptor subtypes mu (μ) and delta (δ). Mu receptors were initially defined in relation to their morphine affinity. Along with β-endorphins, μ receptors also possess a high affinity for the opioid peptides **Dynorphin A** and the **enkephalins**. Delta receptors possess a high affinity for enkephalins and a slightly less high affinity for β-endorphins. Currently, δ receptors are less well defined than μ receptors.

Beta-endorphins are regulated by several neural and endocrine modulators. Activation of the **hypothalamic–pituitary–adrenal (HPA) axis** via stimulation of either **serotonin** (5HT)-1A or 5HT-2 receptors increases peripheral β-endorphin levels. However, antagonism of these receptors has not been shown to have an effect on plasma β-endorphin. Neuropeptide Y (NPY), another neural modulator of feeding behaviour, increases β-endorphin release from the basal hypothalamus.

Beta-endorphins have been shown to have a role in regulating **sexual behaviour** and **maternal behaviour**. A reduction in previously elevated β-endorphin levels in the brain just prior to parturition is suggested to induce the onset of maternal behaviour. Studies using morphine to block the β-endorphin pathway have disrupted normal maternal behaviour and maternal aggression in lactating females, altered pup-cleaning behaviour and prevented the eating of the placenta in females post-parturition. β-endorphins have also been shown to facilitate the **lordosis** behaviour in sexually mature females. β-endorphins are also involved in male **sexual behaviour** – male rats administered opiate antagonist displayed significantly fewer mounting attempts and reduced latencies to ejaculation.

Opioids such as β-endorphin also have a role in feed intake and **feeding** behaviour. Reduction in β-endorphin levels causes a reduction in appetite or food-seeking behaviour. Studies have shown that mice deficient in β-endorphin have less motivation to work for food as a reward than wild-type mice. Conversely, fasting has been shown to increase hypothalamic and peripheral β-endorphin in animals, including pigs, but not pituitary concentrations. However, elevated pituitary concentrations of β-endorphin have been shown in genetically obese mice and rats. Lack of feed or energy from feed (such as from low-energy feedstuffs) is suggested as being the major cause for this increase. Studies of bulimia in human females have also shown a reduced β-endorphin reactivity in women with bulimia. Additionally, these studies have shown a negative correlation between β-endorphin levels and the severity of bulimia. β-endorphins may also play a role in the preference of more palatable foods. Mice fed diets consisting of equivalent energy, but superior and inferior palatability, have also shown reduced and increased levels, respectively, of gene expression of endogenous opioids in the hypothalamus. Similarly, increased β-endorphin binding was found in the hypothalamus of animals following a palatable feed reward stimulus. These data suggest a possible role for β-endorphin in the reward pathway.

Many researchers suggest that opioids such as β-endorphins may provide a superior measure of animal well-being than traditional **glucocorticoids** alone. Studies have shown that **transport** stress in swine has a strong affect on β-endorphin reactivity. Following a single transport of just 2 h, plasma β-endorphin levels were twice the pre-transport levels. Two transports were conducted when pigs were 18 and 20 kg, respectively. Although an increase in peripheral β-endorphin levels was detected following both transports, the second transport elicited a significantly smaller response than the first, suggesting an acclimation to this **stress**. Similar studies in sheep revealed no effect of transportation on β-endorphin levels, however. β-endorphins have also been established as indicators of social stress. Increased β-endorphin levels have also been correlated with low social rank in sows. Decreased μ opioid receptors in the brain have been shown in sows kept in isolation, while sows maintained at high density had increased density of μ receptors.

β-endorphins have also been shown to have a regulatory effect on **sleep**. A clear diurnal rhythm has been illustrated in humans, with β-endorphins at their peak levels between 04.00 and 10.00 hours, and at the lowest between 22.00 and 03.30 hours. Additionally, in cats, administration of excess β-endorphin has been shown to produce insomnia, inhibiting deep, slow-wave sleep and completely suppressing rapid eye movement (REM) sleep. Following treatment with excess β-endorphin, intermittent episodes of light, slow-wave sleep became the most commonly noted sleep stage. β-endorphin treatment, following treatment with a β-endorphin antagonist, reversed the insomnia effect of the opioid and returned the two stages of slow-wave sleep to control levels. However, REM-stage sleep was still suppressed even in the presence of an antagonist, suggesting a role for multiple opioid receptors in the sleep–wakefulness cycle.

β-endorphins in the periphery are synthesized in peripheral blood mononuclear cells. The concentration within the mononuclear cells is independent of the plasma opioid concentration. Stress is one of the most potent influences on the concentration of β-endorphins in mononuclear cells, increasing drastically with stressful stimuli. Increased peripheral blood mononuclear cell concentration of β-endorphins has been shown to inhibit normal immune function in both humans and animals. β-endorphins also play a role in allergic response. Increased β-endorphin-like immunoreactivity has been shown in bronchial lavage sampling following grass pollen provocation in allergic subjects. (RD)

Further reading

Charmandari, E., Tsigos, C. and Crousos, G. (2005) Endocrinology of the stress response. *Annual Review of Physiology* 67, 259–284.

Beta-phenylethylamine

Beta-phenylethylamine is a monoamine synthesized in the body from phenylalanine and has mood-elevating effects in humans. It is a neuromodulator of dopaminergic systems in the **brain** (**see: Dopamine**) and is normally rapidly broken down by enzymes of the class monoamine oxidase-B. Primary interest in beta-phenylethylamine in applied animal behaviour relates to the use of the drug selegiline (l-deprenyl), which belongs to a class known as the monoamine oxidase-B inhibitors. It is a human antidepressant, also used in the treatment of Parkinson's disease, but is licensed in many countries for use in the treatment of animals with a range of emotional behaviour problems and **cognitive dysfunction**. Some of the effects of this drug may relate to the effects of phenylethylamine in the body. (DSM)

Further reading

Crowell-Davis, S.L. and Murray, T. (2006) *Veterinary Psychopharmacology*. Blackwell Science, Oxford, UK.

Biodiversity

The word biodiversity (shortened from biological diversity) began to be used by environmentalists and conservationists in the 1980s to explain the variety of life on Earth. It considers the variety of life at different levels of biological organization including genes, species and ecosystems.

Biodiversity is most often understood in terms of the number of species within an area, region or the Earth as a whole. As of the first decade of the 21st century, 1.75 million species have been identified, and estimates for the total number in existence vary from 3 to 100 million. Biodiversity also includes genetic diversity within species: these are the small differences within an organism's DNA that makes it look and behave differently from all other members of its species. Genetic diversity is important, as it is the raw material for evolution by natural selection. Finally, biodiversity includes ecosystems: this is where communities of species, including humans, interact with each other and the environment around them. The habitats of the world, including forests, deserts, savannahs, freshwater and marine environments are all composed of many ecosystems.

The importance of biodiversity was one of the key subjects of the 1992 World Summit held in Rio de Janeiro, Brazil. World leaders discussed a strategy for sustainable development, and the outcome of this meeting was the Convention on Biological Diversity (CBD). One hundred and fifty government leaders signed the convention, with its goals of 'the conservation of biological diversity, the sustainable use of its components, and the fair and equitable sharing of the benefits from the use of genetic resources'. It emphasizes that 'biological diversity is about more than plants, animals and micro-organisms and their ecosystems – it is about people and our need for food, security, medicines, fresh air and water, shelter, and a clean and healthy environment in which to live'.

Unsustainable human consumption and poverty have caused huge losses in biodiversity; these include the disruption of ecosystem processes, habitat destruction, species extinction and the erosion of genetic diversity within species. The IUCN Red List of Threatened Species compiled by the Species Survival Commission of the World Conservation Union (IUCN) provides a measure of the conservation status of species, and this provides a direct measure of the threat that biodiversity is under. The 2006 Red List states that one-quarter of mammals, one-third of all amphibians and one-quarter of all coniferous trees are threatened with extinction.

Biodiversity is not distributed evenly throughout the world: in 1988 Norman Myers introduced the concept of biodiversity hotspots; these are areas with exceptionally high numbers of species found nowhere else (endemic) and very high levels of habitat loss. A number of conservation organizations, such as Conservation International, state that these areas should be conservation priorities. There are currently 34 biodiversity hotspots recognized, each holding at least 1500 endemic plant species and having lost 70% of their original habitat area. The hotspots include: Madagascar and the Indian Ocean islands; the coastal forests of Eastern Africa; the Caribbean Islands; and the Mediterranean Basin. (NdV)

Further reading

Gaston, K.J. and Spicer, J.I. (2004) *Biodiversity: an Introduction*, 2nd edn. Blackwell Science, Oxford, UK.

Myers, N., Mittermeier, R.A., Mittermeier, C.G., da Fonseca, G.A.B. and Kent, J. (2000) Biodiversity hotspots for conservation priorities. *Nature* 403, 853–858.

Secretariat of the Convention on Biological Diversity (2001) *The Handbook of the Convention on Biological Diversity*. Earthscan, London.

IUCN (2006) Red List of Threatened Species. Available at: http://www.iucnredlist.org

Bioethics

Bioethics is a branch of applied **ethics** dealing with ethical issues arising from human interference with biological processes. Bioethics was developed after World War II, and initially focused on the use of human subjects for experimentation, but the field has since broadened to cover issues raised by other applications of biology and various uses of biological resources, including animal use. Writings on bioethical issues are often informed by the author's commitment to a certain ethical perspective. Both religion and law help to shape the issues as well. (PS)

Biophilia

Biophilia refers to the love of life, nature or living systems. Coined by psychologist Erich Fromm (1964) to capture the essence of humanitarian **ethics**, biophilia connotes a caring predisposition towards life forms that is manifest in risk taking to save an animal's life, for example, or in the positive emotion shown by adult mammals to certain traits of their young.

Two decades later, **E.O. Wilson**, an early and prominent advocate of the biological basis of all social behaviour (**see: Sociobiology**), published a small volume (Wilson, 1984) in which he postulated an innate human predisposition to love and care for nature that cannot be accounted for by science, money, power or culture. Rather, he argued, it is an emotion that appeared early in the human experience of being a part of nature and dependent upon it for survival. Humans arose out of nature. They gave names to its parts and assigned to them totemic value and supernatural power. Humans and other life forms are kin. In Wilson's words, an appreciation of nature's beauty can be said to lie in the genes of the beholder.

Modern humans' quest for open space in which to live, and the value assigned to a place with high vantage points and broken topography (i.e. an aesthetic view), are possible expressions of an ideal towards which they unconsciously strive. Cultivating ornamental gardens and placing flowers within our homes might give expression to an ancient attraction to a simpler and more spacious world. Biophilia cannot be scientifically proved, but many scientists who have focused on life and lifelike processes do not doubt its existence.

In the trammelled world of modern humans, biophilia is too infrequently evident. Yet, some argue it is humankind's best hope for its own future well-being, perhaps even its survival. Biophilia offers the best hope of conserving natural forms and systems that, in Darwin's phrase, offer a view of grandeur. The potential to respond to its inner force is there, as a substrate on which mental development can lead to love for nature. The time has come 'to invent moral reasoning of a new and more powerful kind' that reconnects us with nature and the values that will ensure that, in its grandest forms, it will always be with us. (DGL)

See also: Anthropomorphism; Anthrozoology; Conservation; Ethics

References

Fromm, E. (1964) *The Heart of Man*. Harper & Row, New York.

Wilson, E.O. (1984) *Biophilia: the Human Bond with Other Species*. Harvard University Press, Cambridge, Massachusetts.

Biostimulation

Biostimulation in relation to animal behaviour typically describes the positive effect of a male presence on female reproductive function. Several female reproductive variables can be affected – e.g. reduced time to sexual maturity, reduced postpartum or seasonal anoestrus, ovulation induction and improved sperm transport. Effects are most commonly mediated by olfactory cues through the release of what have been termed priming **pheromones**; however, other visual, tactile or auditory cues may also be effective. Biostimulation has been utilized in livestock husbandry to improve reproductive efficiency for a range of species. (MRC)

See also: Sexual behaviour

Biotelemetry

Biotelemetry is the remote detection and measurement of an animal function, activity or condition. This includes a range of techniques of varying invasiveness, including video monitoring, radio tracking and the use of remote data sampling systems. These sampling systems either transmit or store (log) physiological or behavioural data and may be mounted externally, e.g. using a harness, or internally by surgical implantation. Biotelemetry is used to obtain data such as heart rate, electrocardiogram, body temperature and blood pressure from experimental animals, commonly for fundamental physiology studies or in the process of pharmaceutical research and development or safety testing.

Biotelemetry can benefit both animal welfare and science. It can minimize or eliminate the need for external instrumentation and restraint, thereby reducing **stress**, and produce more relevant, better-quality data; this can in turn reduce the number of animals required for statistical significance. Telemetered data can also be used to provide evidence of animal well-being and to refine humane end points. However, there are a number of harms associated with biotelemetry: (i) surgical implantation or attachment procedures; (ii) the physical impact of the device on the animal; (iii) distress caused by housing animals individually, if devices are used that transmit at the same wavelength; and (iv) in studies using wild-caught animals, additional physiological and psychological stress and, if animals are released, reduced potential to monitor well-being and intervene if necessary.

There are a number of ways in which the impact of telemetry on experimental animals can be reduced. Devices can be selected and located so as to minimize their physical impact; good surgical technique and perioperative care, including analgesia, can significantly reduce postoperative discomfort or pain. Social species used in telemetry studies can be group-housed, either by using data loggers, transmitters that can be switched on and off *in situ* and used one at a time, or the 'buddy' system, where implanted animals are housed with companions without working devices. It is important that all proposed projects involving biotelemetry: (i) are subject to a comprehensive ethical review that questions the necessity of the data; (ii) ensure that the least invasive method has been chosen to obtain data; and (iii) ensure that all techniques and husbandry have been fully refined. For wild animal use and field studies, researchers should be able to demonstrate that they have access to adequate resources and expertise to ensure that animals' lives and **well-being** will not be avoidably compromised. (PH)

Further reading

Hawkins, P., Morton, D.B., Bevan, R., Heath, K., Kirkwood, J., Pearce, P., Scott, L., Whelan, G. and Webb, A. (2004) Husbandry refinements for rats, mice, dogs and non-human primates used in telemetry procedures. *Laboratory Animals* 38, 1–10.

Morton, D.B., Hawkins, P., Bevan, R., Heath, K., Kirkwood, J., Pearce, P., Scott, L., Whelan, G. and Webb, A. (2003) Refinements in telemetry procedures. *Laboratory Animals* 37, 261–299.

Bite prevention programmes

Owning a **dog** has been proved to have multiple benefits for humans. Dogs can provide love, companionship and acceptance for all family members, particularly for children. Having a good relationship with a dog can help a child to overcome grief and sadness, and there is evidence that dogs play a positive role in child development. In addition, children who have pets engage more in social interactions and show more empathy towards peers (**see: Companion animal; Petting**). Nevertheless, owning pets may also present potential hazards through dog bites, trauma or zoonotic transmission of disease. It is estimated that each year in the USA 4.7 million people are bitten by dogs, with 800,000 bites requiring medical attention and 370,000 bites severe enough to be treated in emergency departments (EDs). Comparable ratios are found in Europe, with 250,000 people reported bitten annually in the UK and 100,000 in Belgium, the latter figure representing 1% of the population being bitten annually, and this figure has been rising annually in recent years. It is estimated that 50% of dog bites remain unreported, so the true scale of the problem remains difficult to quantify.

According to hospital-based surveys, it seems most dog bites happen in familiar surroundings, involving young children bitten by their own pet or a familiar dog, and 80–90% of the bites appear to be triggered by an interaction from the child towards the pet dog; these bites may result in signs of post-traumatic stress syndrome. However, in veterinary caseloads and otherwise-sourced bite report surveys, adults are reported to be the victim of dog bites twice as often as children. The fact that dog bites that happen in public places receive a lot of media attention may contribute to the perception of 'dangerous dogs' and dangerous breeds.

In non-industrialized countries, rabies causes 55,000 human deaths per year. Surveys indicate that bite accidents with domestic dogs are the source of infection for the vast majority (> 95%) of cases. In the Western world, dog bite fatalities are rare. In The Netherlands each year approximately one person dies following a dog bite, while 11 fatalities occur due to sport and 23 following household accidents. Unfortunately, the majority of victims of serious dog bites and fatalities are young children, triggering media emphasis on 'aggressive breeds', with the result that many governments have developed breed-specific legislation as a preventive measure. Recently, the advice that 'families with young children should stop keeping dog as pets' and 'a total ban on dog keeping' has been suggested in the medical literature.

In human preventive medicine it is recognized that only sustained interventions have sustained effects. Sustainability is enhanced by focusing upon underlying causes and contexts of behaviour and embedding or integrating prevention and promotion initiatives into ongoing systems, institutions and organizations, so that they become a natural part of people's everyday lives.

A dog's tendency to bite depends on multiple interacting factors related to heredity, early experience, later socialization, training and education, medical health, behavioural health and victim behaviour; these must be considered in the context of human social and cultural factors that have been shown to influence human–dog interactions. In the past, extensive information regarding interactions between dogs and children has been published in the veterinary and medical literature, and 'Prevent a Bite' programmes have been established in primary schools and healthcare centres in various European countries. Most 'Prevent a Bite' programmes, however, are aimed at the age group of 7 years and older and focus mainly on public safety rules, like how to behave when encountering an unfamiliar dog. Programmes intended to prevent bite accidents in the home are very rare in Europe, although a CD-based training programme, focused on teaching appropriate behaviour towards dogs in the home for younger children (e.g. Blue dog; see De Keuster *et al.*, 2005), has recently been developed.

For the majority of risk situations, the message to the child is to encourage 'no' interaction, or 'leave your dog alone'.

In the past, dog bite prevention has generally been approached by teaching children skills on how to recognize the dog's body language. According to research, however, young children score badly in discriminating canine body language. Young children look mainly at the face of the dog in order to make their decisions. In addition, young children do not necessarily understand the dog's intentions: they will often confuse a fearful with a friendly dog. Therefore, rather than focusing on 'recognizing dog body language', it is potentially more useful to focus on teaching parents and children 'how to recognize and to assess situations that appear to trigger dog bites in a household situation'. Although there are commonly held beliefs and rules on safe play and safe petting of dogs, no data-based evidence exists on how to safely pet or how to safely play with the family dog. Thus it can be concluded that, for this issue, there are no rules applicable to 'the all-round family dog'. An alternative strategy is to improve parent awareness and knowledge of dog behaviour. (TDK)

Reference and further reading

Besser, R. (2007) Dog attacks: is it time for doctors to bite back? *British Medical Journal* 334, 425.

Bond, L.A. and Hauf, A.M. (2004) Taking stock and putting stock in primary prevention: characteristics of effective programs. *The Journal of Primary Prevention* 24, 199–221.

De Keuster, T., Moons, C. and De Cock, I. (2005) Dog bite prevention – how a Blue dog can help. *European Journal of Companion Animal Practice* 15, 137–139.

Ozanne-Smith, J., Ashby, K. and Stathakis, V.Z. (2001) Dog bite and injury prevention – analysis, critical review and research agenda. *Injury Prevention* 7, 321–326.

Bloat

Bloat, due to trapped gas in the stomach, can be observed as an acute swelling between the last rib and the hip on the left-hand side of a ruminant animal (in which it is also referred to as ruminal tympany), but may also affect certain monogastric species, such as the dog, when fed diets that produce a lot of gas. The animal is restless, finds lying uncomfortable and may eventually die of heart failure or suffocation.

Pasture bloat (also known as frothy bloat or primary bloat) in cattle is caused by stable foam in the rumen, which traps the gas that is normally released by eructation. The condition is often due to the rapid digestion of legumes in particular, although young, leafy grass that has recently received nitrogen fertilizer can also cause bloat. Lucerne is the most likely of all legumes to cause bloat, with cows sometimes dying within a few hours of entering a field for grazing. Some legumes have developed a chemical, tannin, which reduces the speed of protein digestion and probably discourages animals from grazing it. Tannins are present in sufficient quantities in birdsfoot trefoil to prevent the production of a stable foam, and the content in white clover increases sufficiently at flowering to make it safe to graze. If a mixed grass and clover sward has enough clover to cause bloat (probably more than 50% of the herbage by mass), it should not be grazed for long periods, but should either be conserved if there is sufficient mass or rested for a few weeks until the clover inflorescences appear, after which it can be grazed or conserved.

Cattle are most likely to become bloated in the late evening after a day's grazing, and also after a wet period when they avidly graze to make up for lost time. Wet grass reduces saliva production, and the saliva contains a mucin that disperses foam in the rumen. Herbage that has been frozen is particularly likely to cause bloat, as the rupture of plant cell walls releases the solutes that contain much potassium. Potassium-rich feeds, such as molasses, are well known for causing bloat, whereas grasses rich in sodium appear to be less likely to cause it. The precise mechanism has not yet been determined but may relate to the stimulation of saliva production by sodium-rich feeds and the foam-dispersing properties of the salivary mucin.

Forage supplements will usually slow down the rate of digestion and reduce bloat but, if there is adequate herbage, grazing supplements may not be eaten by some animals in sufficient quantities, particularly if they are based on straw or other low-quality forages. Mineral oils also help to disperse the foam and can be added to a concentrate feed, sprayed on to the pasture or, in the case of dairy cattle, applied to their flanks during milking to be licked off as needed; linseed oil is often used. A proprietary product, poloxalene, also breaks up the foam and can be used as a drench for clinical cases or included in feed blocks as a preventive measure. Often simply walking an animal from the field to the farm to receive medication may alleviate the swelling. It is important to keep a bloated animal on its feet if possible, as death can follow soon after recumbency.

There is a genetic component in the susceptibility to bloat, as well as reported breed differences in susceptibility; for example, Jersey cows are particularly prone to the disorder. Some species appear to be able to learn to tolerate feeds that are likely to cause bloat; this may be by altering their behaviour to spread their meals out more evenly over the day. Lactating cows are particularly susceptible due to their high intakes.

Pasture bloat remains a serious problem for farmers in countries like New Zealand, where the animals rely on pasture with little or no fertilizer applied and a high legume content. In Europe, the greater emphasis on controlling nitrogen emissions is encouraging farmers to use high-clover swards, potentially leading to more serious problems with bloat.

Obstructive bloat (also known as secondary bloat or free-gas bloat) occurs when gas cannot escape from the stomach due to physical obstruction and may, in theory, occur in any mammal. Often it is due to a foreign body (e.g. large piece of food such as a root vegetable) or twisting of the stomach (gastric torsion or volvulus), in which case it often requires surgical correction. The risk of twisting is increased in animals where the suspension of the stomach appears to be unstable, e.g. in certain deep-chested breeds of dog and animals that have undergone a period of starvation and so have an empty stomach. (CJCP)

See also: Feedlot

Blocking

Blocking refers to a form of interference with classical conditioning (**see: Conditioning – types of**) that arises as a result of prior experience with a contingent predictor (**see: Contingency**) of an event. Once an animal learns that a given **stimulus** predicts a certain event, it becomes a conditioned stimulus for that event. If this stimulus is then reliably paired with another stimulus, which was not previously associated with the event, then the animal may fail to learn the new association, i.e. the second stimulus does not become a conditional stimulus and learning has been blocked by the presence of the first conditional stimulus. This may be because the new stimulus does not provide additional information that helps to predict the event beyond that which is provided by the first conditional stimulus, and so it is not salient. This phenomenon can give rise to problems when training, but this may be resolved by ensuring that the potentially blocked stimulus is presented in advance of the blocking stimulus (i.e. the second stimulus is presented in advance of the already conditioned stimulus), so that it has predictive value and second order conditioning can occur. The phenomenon may also be useful because a conditional stimulus with a strong positive emotional association (e.g. a friendly owner) can be used to help block the learning of problematic behaviours associated with novel stimuli (e.g. a traumatic first experience at the veterinary clinic). (DSM)

Blood pressure

Blood pressure (BP) is the pressure exerted by the blood on the walls of the blood vessels. Usually, when reported, it is the arterial pressure that is measured and reported as two numbers – the first being the systolic pressure (the pressure when the heart contracts and pushes blood out into the pulmonary artery and the aorta) and the second being the diastolic pressure (the pressure when the heart dilates and fills with blood). Pressure is conventionally reported in millimetres of mercury (mmHg), and an example within the normal human range would be 'one-twenty over eighty', or 120/80 mmHg – meaning a systolic pressure of 120 mmHg coupled with a diastolic pressure of 80 mmHg.

Blood pressure can be measured either invasively (i.e. by inserting a measuring device into an artery) or, more commonly, non-invasively, by attaching a cuff externally and increasing pressure to occlude blood flow, before gradual release of the pressure to determine when blood is again flowing through the vessel of interest. Exact measurement may then be made by either listening through a stethoscope or using an oscillometer. For measurements in humans, the sphygmomanometer is still the most-used method, and comprises an inflatable cuff that is placed around the upper arm attached to an aneroid or mercury manometer. The cuff is inflated until the brachial artery is occluded and then pressure is released until turbulent blood flow can be heard through the artery. This pressure reading corresponds to the systolic pressure. Further pressure is released until turbulent flow can no longer be heard, with this reading corresponding to the diastolic pressure. Invasive techniques involve indwelling catheters or implanted radiotelemetry devices. For animals, the cuff technique can still be used, either on the tail or limbs, but for ambulatory studies invasive methods are more useful.

Blood pressure (mean arterial pressure (MAP)) can also be defined in terms of cardiac output (CO) and peripheral (or systemic) vascular resistance (SVR): BP = CO × SVR. Cardiac output is itself a product of stroke volume (SV) and heart rate (HR), so BP = SV × HR × SVR. Systemic vascular resistance is related to the radius of the blood vessels and, indeed, resistance is inversely proportional to the fourth power of the radius (i.e. r^4). Thus, if the radius of the blood vessel is decreased by 50%, the pressure within it will increase 2^4 times = 16-fold. Not surprisingly, it is by altering the vessel radius (vasoconstriction and vasodilation) that the body controls blood pressure most readily, although it can also bring about changes by altering cardiac output. Normal blood pressure is maintained by a number of mechanisms, acting over both the short and long terms.

Short-term control of blood pressure is mediated by the nervous system, chemicals and hormones through alteration of systemic resistance within seconds or minutes. The **autonomic nervous system** is the most rapidly responding regulator of blood pressure, and receives continuous information from the baroreceptors (pressure-sensitive nerve endings) situated in the carotid sinus and the aortic arch. Ordinarily, with blood pressure at 'normal', the baroreceptors are firing at a steady rate, which is exerting a tonic inhibitory influence on sympathetic outflow from the vasomotor centre in the medulla. If blood pressure increases, the rate of firing increases, resulting in a decrease in sympathetic activity and an increase in parasympathetic activity, which together will act to decrease cardiovascular output by increasing heart rate and to decrease peripheral resistance by vasodilation. Changes in the opposite direction occur if blood pressure falls below 'normal'.

Chemoreceptors situated in the aortic arch and carotid body can respond to changes in blood PO_2, PCO_2 and pH. A decrease in carotid body blood flow results in cellular hypoxia, hypercapnia and decreased pH, leading to an increase in receptor firing. The threshold PO_2 for activation in humans is about 80 mmHg (normal arterial PO_2 is about 95 mmHg). Any elevation of PCO_2 above a normal value of 40 mmHg, or a decrease in pH below 7.4, causes receptor firing. If respiratory activity is not allowed to change during chemoreceptor stimulation, then chemoreceptor activation causes bradycardia and coronary vasodilation (both via vagal activation) and systemic vasoconstriction (via sympathetic activation). If respiratory activity increases, then sympathetic activity stimulates both the heart and vasculature to increase arterial pressure.

Hormonal mechanisms exist for both lowering and raising of blood pressure. They act in various ways, including vasoconstriction, vasodilation and alteration of blood volume. The principal hormones affecting blood pressure are:

1. **Epinephrine** (adrenaline) and **norepinephrine** (noradrenaline) are secreted from the adrenal medulla in response to **sympathetic nervous system** stimulation. They increase cardiac output, causing vasoconstriction, and act very rapidly.
2. Atrial natriuretic peptide (ANP) is produced, stored and released by cardiac myocytes in the atria of the heart. It is released in response to atrial stretching. ANP binds to a specific set of receptors – ANP receptors. Receptor–agonist binding causes a reduction in blood volume and therefore a reduction in cardiac output and systemic blood pressure.

3. **Antidiuretic hormone** (ADH or arginine vasopressin), secreted from the posterior **pituitary gland**, increases the resistance of the peripheral vessels and increases arterial blood pressure.
4. Endothelial factors have an important role in the regulation of smooth muscle function. The three most important endothelial-derived substances are: nitric oxide (NO), endothelin (ET-1) and prostacyclin (PGI_2). NO and PGI_2 act as vasodilators, whereas ET-1 serves as a vasoconstrictor.

Over the longer term, blood pressure control is a function of renal mechanisms that regulate blood volume. Directly, any increase in blood volume and blood pressure will result in an increase in the rate of fluid removal by the kidneys, leading to decreased blood volume and decreased blood pressure. If blood volume and blood pressure decrease, the kidney will conserve and return more fluid, thereby increasing blood volume and raising blood pressure. The other major renal mechanism is the renin–angiotensin–aldosterone system (RAAS). When blood pressure drops, the juxtaglomerular cells of the kidneys secrete renin, which cleaves a peptide bond in angiotensinogen to create angiotensin I. This is then converted to angiotensin II in the lungs by angiotensin-converting enzyme (ACE). Angiotensin II is a strong vasoconstrictor which also stimulates the adrenal cortex to release aldosterone, which itself works to reduce the excretion of sodium and reduce water loss, thereby increasing blood volume and raising blood pressure.

As a **welfare** indicator, blood pressure has not been used extensively due to methodological limitations. It has been used during studies of **restraint** and **slaughter**, but not in animals in free-ranging situations. With the development of telemetric implants with blood pressure capability, blood pressure is likely to attract more research attention in the future.

(RM-F)

Body condition score

Body condition score (BCS) is a subjective, non-invasive assessment tool used to describe the level of body fat and muscle in animals. Generally, low body condition scores reflect emaciation while high values are associated with obesity. Body condition scores are used to monitor and manage nutrient requirements, to estimate body fat prior to **slaughter** and as an indicator of the level of feeding.

Scoring systems are available for a range of animals including **cattle**, **sheep**, **horses** and **chickens**. BCS is based on visual and/or tactile assessment of the fat/musculature in single or multiple regions of the body. The body regions included in the score differ between species and scoring systems, but BCS is often based on the amount of fat/muscle covering bony protuberances, such as the ribs. Scoring systems characterize body condition into discrete categories. The categories can be as few as 3, but 5-, 6- and 9- point integer scales are common. Although body condition scoring systems differ in methodology, there is often good agreement between the attributed scores in different systems, for example in dairy cows (Roche *et al.*, 2004).

Regardless of the scoring system used, the repeatability and validity of BCS affects its usefulness as an assessment tool. BCSs can be assigned consistently across observers (e.g. Edmonson *et al.*, 1989; Teixeira *et al.*, 1989), although training and experience can have a major influence. Body condition scores have been compared to more quantitative measurements of condition, including ultrasound, live weight and percentage of dissected fat, empty body weight, muscle weight and dewlap thickness (e.g. Teixera *et al.*, 1989; Domecq *et al.*, 1995, Gregory *et al.*, 1998). In addition to the relationship between BCS and other measures of condition, body condition affects aspects of biological function, such as **reproduction** (e.g. number of live offspring, time to conception). (CBT)

See also: Malnutrition

References

Domecq, J.J., Skidmore, A.L., Lloyd, J.W. and Kaneene, J.B. (1995) Validation of body condition scores with ultrasound measurements of subcutaneous fat of dairy cows. *Journal of Dairy Science* 78, 2308–2313.

Edmonson, A.J., Lean, I.J., Weaver, L.D., Farver, T. and Webster, G. (1989) A body condition scoring chart for Holstein dairy cows. *Journal of Dairy Science* 72, 68–78.

Gregory, N.G., Robins, J.K., Thomas, D.G. and Purchas, R.W. (1998) Relationship between body condition score and body composition in dairy cows. *New Zealand Journal of Agricultural Research* 41, 527–532.

Roche, J.R., Dillon, P.G., Stockdale, C.R., Baumgard, L.H. and VanBaale, M.J. (2004) Relationships among international body condition scoring systems. *Journal of Dairy Science* 87, 3076–3079.

Teixeira, A., Delfa, R. and Colomer, R.F. (1989) Relationships between fat depots and body condition score or tail fatness in the Rasa Aragonesa breed. *Animal Production* 49, 275–280.

Body weight – metabolic

Metabolic body weight (also known as metabolic weight) considers the relationship between body weight and metabolic rate in homeothermic species, and is used in the calculation and prediction of energy and other requirements related to the rate of metabolism of an animal. There is not a linear relationship between metabolism and body weight, since smaller animals are relatively speaking much more metabolically active than larger ones. This is mainly due to the relationship between surface area and volume, since much energy is lost as heat through the surface area of an animal, and the ratio of surface area to body volume decreases with increased size. A plot of the log of body weight against the log of heat production for various homeothermic species produces a straight line, and so the metabolic body weight of an animal is its body weight to the power of 0.75 ($BW^{0.75}$). It is calculated that the metabolic level of adult homeothermic mammals is approximately 70 kcal/$kg^{0.75}$. (DSM)

Bonding

Bonding relates to preferential social relationships that can be established between individuals. Bonds can develop between various types of partners, such as mother and young (**see: Bonding – parent–offspring**), mates (i.e. pair bonding), brothers and sisters, other non-kin conspecifics and even individuals from different species (e.g. the human–animal bond). Several elements are necessary for a bond to be established. The individual should first be motivated to approach and interact with another individual whom they should then subsequently be able to recognize, and selective positive interactions should then be reinforced. **Oxytocin** and vasopressin have been demonstrated to play a major role in all these stages.

The capacity/tendency to form social bonds stems from interactions between genetic and developmental processes. It can vary across species and may lead to a range of social organizations from solitary animals to extended family groups. Bonding is also influenced by gender, physiological state, previous experience or emotional context – for example, stress has been shown to facilitate social bonding.

The existence of preferential social relationships can be assessed through the use of behavioural observations. A selective approach in partner **preference** tests (**choice test**) is a widely used criterion. The animal is given a choice between different kinds of social partners (e.g. two familiar individuals or a familiar individual versus a stranger), and the time it spends in proximity with each partner is recorded. The maintenance of physical proximity or contact and positive social interactions (e.g. **allogrooming**) are also considered as being indicative of social affiliative relationships. Other parameters, such as the reactivity to separation from the partner, as well as its appeasing effect in a challenging context, can also be used to assess bonding. Social bonds can benefit the animals in numerous ways. The mother–young bond or the pair bond are of evident advantage in terms of survival and reproduction, and social bonds can more generally reduce anxiety, help recovery following challenging events (i.e. social buffering) and provide support during social interaction. The disruption, as well as the absence, of social bonds can thus be of much welfare concern. (SL)

See also: Attachment; Social behaviour

Further reading

Lim, M.M. and Young, L.J. (2006) Neuropeptidergic regulation of affiliative behaviour and social bonding in animals. *Hormones and Behaviour* 50, 506–517.

Veissier, I., Boissy, A., Nowak, R., Orgeur, P. and Poindron, P. (1998) Ontogeny of social awareness in domestic herbivores. *Applied Animal Behaviour Science* 57, 233–245.

Bonding – parent–offspring

In most animals it is the female parent, rather than the male, that normally becomes most closely bonded to the offspring, by virtue of her greater investment in the nurturing and birth of the offspring. Bonding develops soon after birth and, given the opportunity, will persist as a matriarchal family group. The initial bonding has the following characteristics. It develops during the **sensitive phase**, which for a precocious animal such as *Bos taurus* occurs in the immediate postnatal period. It is permanent and irreversible, but its strength will depend on the extent of other imprinted bonds the dam has. A cow with twins therefore develops a weaker bond with each calf than a cow with a single calf, and this encourages the twins to form sibling bonds. Within a semi-wild herd of cows and calves, the bond between dam and daughter will persist as the dominant social force, and the males disperse to form bachelor groups. Equal preference is generally given by the dam to her offspring, regardless of their age and sex.

However, a number of factors may affect the strength of the imprinting process. Multiparous cows form stronger bonds than primiparous cows, through greater contact and contact-seeking behaviour with their calves, and they are more disturbed by separation. Primiparous cows show more abnormal maternal behaviour but spend longer suckling their calves than multiparous cows, but this may simply be due to slow milk excretion rather than to more investment of time to enhance the bond. The creation of a strong maternal–filial bond is a learned characteristic, although much of the stimulation by the calf is innate. Intensely selected dairy breeds, such as the Holstein–Friesian, show weak dam–filial bonding. They are more easily cross-fostered (**see: Fostering**), and removal of the calf for artificial rearing has little effect on its temperament, whereas in less developed cattle breeds such as the Salers, artificial rearing reduces the social adaptability of calves.

The bond is naturally created at a time when the neonate is drawn to the dam by suckling motivation, and the chances of other bonds being created during the period of primary **socialization** are small. The bond is preserved by **grooming**, both during and after suckling, but other forms of communication like **vocalization** must be important, since calves that are muzzled to prevent suckling and grooming still develop a bond with their dams. The bond, and in particular suckling, maintains the postpartum anoestrus in cows for about 8 weeks to prevent early rebreeding.

The benefits of a strong mother–filial bond are many, particularly in wild animals, where the dam needs to teach the offspring how to survive. For the neonate, filial **imprinting**, which underpins the bonding process, may provide the basis upon which the young learn the characteristics of their species. There are many sources of reward provided through the bond: for example, reward may come in the form of protection and nutrition in the form of milk. Early licking of the perineal region stimulates the functioning of the gastrointestinal tract and excretion. Later, the juveniles learn dietary and other habits from their dams and may acquire a stable position in the group structure dependent partly on their dam's position. (CJCP)

Bone strength

Bone strength is a term used to refer to the bone density and the architecture of a given bone. Bone strength can be affected by a multitude of factors, including nutrition, stage of production, age and mechanical stress. To understand how bone strength is controlled, it is important to understand the mechanisms of bone degradation and rebuilding.

There are three main types of cells that make up the bone. Osteoblasts are bone-formation cells and osteoclasts are bone-remodelling cells. Osteocytes are the functional cells of bone and are responsible for responding to mechanical stimuli by initiating either bone formation or remodelling by chemical messengers. Osteoclasts and osteoblasts act concurrently in continually remodelling bone to keep the architecture of the bone sufficient to support the mechanical stress to which it is subjected. Constant remodelling also serves to help maintain plasma calcium concentrations. Several hormones are responsible for the initiation and regulation of both osteoclasts and osteoblasts. The sex hormone **oestrogen** stimulates osteoprotegerin (OPG) in both males and females to inhibit osteoclast activity. Thyroid-stimulating hormone (TSH) also plays a role in the stimulation of both osteoclasts and osteoblasts. In order for growth to occur, osteoblastic activity must be greater than the osteoclastic activity. Both **growth hormone** (GH) and **insulin-like growth factor**-1 (IGF-1) control the activity of osteoblasts during growth phases. Bone remodelling occurs throughout an animal's lifetime but slows with age.

As previously mentioned, there are several factors that affect bone strength. These factors are common among species and are described below.

Nutrition

Calcium is a vital component of bone, especially bone mineral content and bone density. Ninety-nine per cent of the body's calcium is stored in the bone as hydroxyapatite. Plasma calcium concentrations are controlled by the peptide **parathyroid hormone** (PTH) from the parathyroid glands. PTH controls calcium from the intestines and kidney and is capable of releasing calcium from bone. Adequate calcium intake throughout life, particularly during adolescence, is vital in developing and maintaining bone density. Inadequate calcium concentrations during development inhibit the skeletal system from reaching maximal bone density.

Production

At different stages of production, demands on the body are different. In gestating and lactating mammals and egg-laying birds, the calcium requirement is high and bone calcium reserves may be used to meet the demands of the developing fetus and milk production and eggshell production, respectively. In chronic calcium deficiency, bone calcium will be used to replenish plasma calcium concentrations. In the event that calcium is not replenished to the bone, long-term effects on bone strength will result. The use of bone calcium reserves will decrease bone density, increasing the females' susceptibility to fracture and other bone injuries. In most mammalian species, there is a short period between lactation and gestation of the next parity. This may not allow the female to replace the bone reabsorbed during the previous gestation or lactation, causing increasing bone loss. For egg-laying hens, egg production is continuous throughout their productive life from 18 to 22 weeks of age, and up to 60–70 weeks of age and longer for moulted birds. Hens at the end of lay are particularly susceptible to broken bones during removal from the laying house and transportation to the slaughter plant.

In animals raised for meat production, such as slaughter pigs, turkeys and broilers, high growth rates may not allow the animal ever to reach maximal bone density. If bone density is not sufficient, the risk of bone problems increases, thus increasing the risk of lame animals.

Mechanical stress

One of the primary ways in which bone is deposited and reabsorbed throughout an animal's life is in response to mechanical stress on the bone. Wolf's Law is a principal law in bone physiology, stating that bone responds to the forces placed upon it. In conditions of high mechanical stress, osteoblast activity will increase to deposit more bone to compensate for the added stress. In sedentary animals where mechanical stress is lacking, such as sows in stalls and hens in cages, osteoclast activity will increase, depleting the bone density. Short, dynamic exercise with rest periods between exercise bouts has been shown to significantly increase bone density. Animals that are kept in confinement may not be able to exercise enough in order to maintain or increase bone density.

Age

Bone modelling and remodelling occurs at a higher rate throughout adolescence until maturity. At maturity, bone activity tends to plateau and slowly decrease with age. As mentioned previously, the amount of bone deposited during adolescence and maintained during adulthood primarily depends on nutrition and mechanical stress on the bone. The sex hormone **oestrogen** also plays an essential role in bone formation during adolescence and bone maintenance in adult animals. Oestrogen is responsible for the closure of growth plates in adolescent males and females. In adult animals, oestrogen deficiencies can also cause considerable loss of bone, resulting in osteoporosis. This loss of bone may render the animal more susceptible to becoming lame.

Genetics

Like most traits, bone density is a genetically influenced trait. Bone strength can differ between breeds of the same species and between individuals of the same genetic line. Selective breeding is a common practice in which animals are bred to enhance a certain trait, such as muscle weight or height. With selective breeding of such traits, bone strength can be improved or negatively affected. With traits that increase growth rate or overall size of an animal, there is a risk of decreasing bone strength, not only by genetic factors but by the fact that the bone may be deposited at a slower rate than is muscle. This again can lead to leg and skeletal problems and increase the risk of **lameness**. (ELS)

See also: Chicken; Slaughter; Transport

Boredom

Boredom is a negative subjective state, defined by Webster (2005) as 'a sense of unease associated with the absence of any specific motivational need'. Although originally a phenomenon described in humans, boredom is increasingly being considered when assessing **welfare** in captive animals. Boredom is envisaged as resulting from a chronic lack of opportunity for active interaction with the environment (Wemelsfelder, 2005), and it has been suggested that animals housed in barren forms of farm, laboratory and **zoo** housing are at most risk of experiencing this unpleasant state (**see also: Allostasis**).

The impact of housing animals in a barren environment can be examined objectively by comparing their physiology and behaviour with animals housed in more complex environments. Behavioural changes associated with barren environments include more time spent sitting, lying and sleeping and more fearful or aggressive responses to novel or unpredictable stimuli. Other problems are also well documented. These range from **stereotypies** to self-injurious behaviour, **aggression** and even **cannibalism**. Another way of examining the impact of a barren environment is to determine whether animals will work to access more stimulating environments.

The phenomenon of **contra-freeloading** shows that animals will sometimes choose to work for food even when presented with 'free' food of the same value. Similarly, animals will explore novel objects that are of no biological relevance to them, suggesting that they do so to seek stimulation.

However, although such studies are useful, we must remember that barren environments may exert their effects in many ways (e.g. by increasing **stress** levels), and the appearance of **problem behaviours** is not, in itself, good evidence that animals feel bored. There may also be cognitive constraints on the extent to which animals can choose to be in a variable environment. Although hens can assess the future consequences of their choices this is generally done over very short time periods, indicating that animals may not always make the best choices for their long-term welfare.

Human studies have shown that the propensity to experience boredom is influenced by both internal and external factors, and that this tendency changes at different stages of a lifetime. External factors are thought to influence boredom largely when a discrepancy arises between an animal's current environment and the one to which it is evolutionarily adapted. A simple model was proposed by Webster (2005) in his adaptation of one of McFarland's motivational models (**see: Motivation**). In this model (Fig. B.1), the distance from the centre relates to the level of motivation associated with a given state, and at any given time an animal will be at a given point in this multidimensional state. As one need is satisfied, so its state regresses towards the centre for that dimension and another will take priority. For example, a very hungry animal that is quite tired could be placed somewhere between H″ and R in Fig. B.1, and will feed rather than rest. As it feeds, so hunger falls along the axis through H to H′ while exhaustion remains around the limit of R, and it will then rest. 'Limbo', a state caused by the absence of normal environmental challenges, was envisaged as the motivational state most likely to be associated with a subjective feeling of boredom in captive animals. Webster (2005) used an extension of the surface including H′–R′, which extends to the other motivational states (see Fig. B.1), to demonstrate how a cow in a solitary stall may enter the state of limbo. The cow has everything she needs and will not suffer from exhaustion, hunger or thirst, nor will she be exposed to any of the challenges around the sides of McFarland's 'egg'.

This line of reasoning is brought into focus when we compare the lives of domestic animals with those of their wild ancestors. Junglefowl (*Gallus gallus*), the wild ancestors of domestic chickens (*Gallus gallus domesticus*), for example, spend a large amount of time each day locating and consuming food, checking for predators, building nests and competing for resources within a complex social hierarchy. These behaviours take place within a complex jungle habitat, with varying temperature and light levels. In contrast, most domestic chickens are kept in flocks of the same age and genetic background, in predator- and weather-proof buildings and are provided with *ad libitum* high-quality feed. The behavioural void, or limbo, left by housing chickens in this way is likely to lead to reduced welfare, but whether this ultimately results in a feeling of boredom would depend on whether chickens have sufficient neural complexity to experience complex secondary emotions.

Like all subjective experiences, boredom is impossible to measure directly. Findings from human studies have revealed that boredom is an 'extremely unpleasant and distressing experience'. Although many believe that most vertebrates are indeed likely to be capable of experiencing this kind of negative state, it is salutary to consider that we know little about the psychological constructs required for the sensation of boredom. Presumably, a complex secondary emotion such as boredom depends on a certain level of emotional and cognitive complexity and possibly some degree of self-reflective **awareness** (or **sentience**). Whether an animal is considered as sentient might play a large part in whether it is capable of experiencing boredom. Humans might become bored because they consider that the situation they are currently in differs from an imagined situation that they might like to be in. Therefore, the degree of sentience and the ability to reflect on current status versus the alternatives might be limiting factors in whether or not an animal is capable of experiencing bore-

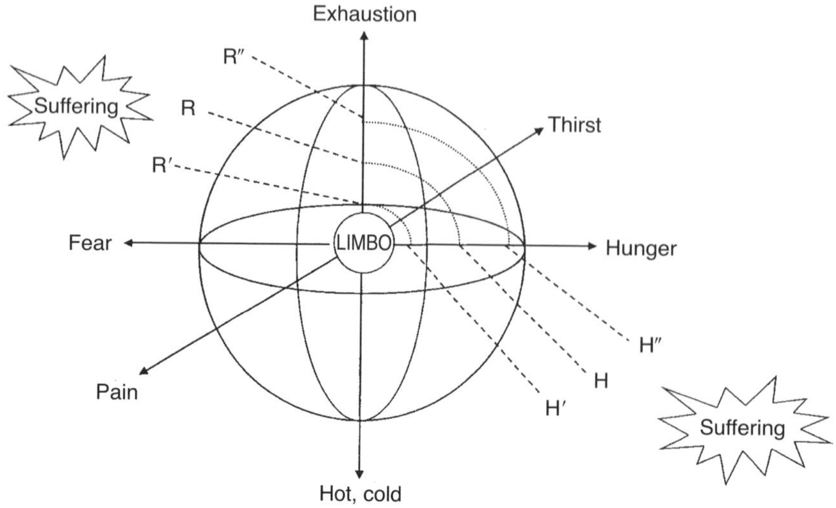

Fig. B.1. McFarland's 'egg', as adapted by Webster (2005).

dom, rather than a simpler, primary emotion, such as frustration.

Until the psychological constructs that support subjective states such as boredom are better defined, we may have to resort to drawing analogies between our experience as humans and that of animals. For example, we might get bored sitting in an empty room all day with nothing to do, and so might imagine that a chicken, with no species-specific stimulation, would also be bored. Some scientists believe that this method is **anthropomorphic** and inappropriate because of different cognitive abilities and requirements; others would claim that it at least gives the animal the 'benefit of the doubt'. A final approach is to consider whether boredom might have an adaptive function. Boredom could force animals to explore, learn and familiarize themselves with their environment and, through this, to adapt to challenges within it. Animals in different ecological niches, social animals or animals that have large home ranges could therefore experience different forms of boredom when confined in captivity. In addition, domesticated animals, or individuals housed in barren environments all their lives, might differ in their propensity to experience boredom.

In conclusion, the term boredom originated to describe our own human experience, but there has been little if any progress in identifying reliable physiological or behavioural correlates of these human feelings in non-human animals. Unlike feelings associated with emotions such as **frustration**, **fear** or **anxiety**, it is therefore difficult to identify which animals might share a feeling of boredom. It has been proposed that animals may feel bored when there is reduction in stimulation between what an animal expects from its environment (the environment it evolved in) and what we, as humans, provide for the animal (its captive environment). However, the use of the term animal boredom remains vague and testable predictions elusive. Further research and integration of objective measures could allow us to make inferences about whether an animal is bored, and might allow us to seek methods to alleviate that boredom and to improve welfare. (JLE)

See also: Exploratory behaviour; Stereotypies

References and further reading

Webster, J. (2005) *Animal Welfare: Limping Towards Eden*. UFAW Animal Welfare Series. Blackwell Publishing, Oxford, UK, pp. 74–75.

Wemelsfelder, F. (2005) Animal boredom: understanding the tedium of confined lives. In: McMillan, F. (ed.) *Mental Health and Well-Being in Animals*. Iowa State University Press, Ames, Iowa, pp. 79–93.

Wemelsfelder, F. and Birke, L.I.A. (1997) Environmental challenge. In: Appleby, M.C. and Hughes, B.O. (eds) *Animal Welfare*. CAB International, Wallingford, UK, pp. 35–47.

Brain

The brain is the organ that controls the **central nervous system**. It is closely connected to the sensory apparatus and is mainly responsible for behaviour. In vertebrates, the brain is differentiated into areas with specific functions (brain modularity). The neocortex is the part principally devoted to the higher cognitive processes in mammals. (BM)

Brambell Report

The Brambell Committee was a technical committee set up by the UK government to investigate the welfare of animals kept in intensive livestock production systems. They published their influential report in 1965 and this, along with the publication of **Ruth Harrison**'s *Animal Machines*, was thought to be one of the main factors responsible for the initiation of specific farm animal legislation in the UK. The report discussed many of the potential welfare concerns associated with modern farming, identified the most serious threats to farm animals and made recommendations as to which practices should or should not be allowed to continue.

(LMD)

Further reading

Brambell, F.W.R. (1965) *Report of the Technical Committee to Enquire into the Welfare of Animals Kept under Intensive Livestock Husbandry Systems*. HMSO, London, Command 2836.

Branding

Branding serves as a method for permanently identifying animals (usually livestock) in a way that is difficult to modify or falsify. Such a form of identification is important for proof of ownership and for trace-back, for example in the event of outbreaks of disease. For these reasons, in some countries branding is a legal requirement for certain classes of livestock, e.g. those over a specified liveweight being presented for sale.

Branding is normally conducted by one of two methods: heating (hot-iron branding) or cooling (freeze branding) of the skin. Brands are generally applied to the body of the animal (rump, ribs, shoulder), although with concerns for hide quality there has been a move to locate brands in positions that cause less damage to and/or loss of hides. Hot-iron brands are also used to mark the horns of animals such as stud sheep.

Hot-iron branding

Hot-iron branding is conducted by heating a branding iron, either by an internal electrical element or by placing it in a furnace or fire to a point where it glows a dull red (in excess of 500°C). The hot iron is then placed on the body of a suitably restrained animal for about 3–5 s. The visible brand that results is a hairless area of scar tissue. Performed correctly (i.e. the branding iron at the correct temperature, applied with the correct amount of force and for the correct length of time), hot-iron branding should burn the skin through to the dermis, producing a third-degree burn. Third-degree burns destroy the pain receptors in the burnt tissue, so that noxious sensory inputs are reduced. In humans, third-degree burns are reported to be much less painful than first- or second-degree burns that leave many pain receptors intact. However, at the periphery of third-degree burns will be first- and second-degree burns that probably cause the inflammatory responses and longer-term **pain**.

Freeze branding

For freeze branding the branding iron is chilled in a coolant, such as liquid nitrogen (–196°C) or dry ice in alcohol (–70°C) and it is then placed on a shaved area of the hide for 15–30 s until the skin is frozen. Another technique involves a stencil held over a shaved area of the hide and an aerosol spray of a chemical coolant (e.g. chlorodifluoromethane), which cools the skin to about –50°C. Freeze-brand marks take about 1 month to become visible; either the pigment-producing cells or the hair cells are destroyed and the hair grows white permanently or falls out, depending on whether the application time is short or long, respectively. If pigmentation changes are induced then, evidently, freeze branding is suitable only on dark-coloured animals. As a result of the increased restraint time and handling of animals to produce the brand and difficulties in obtaining and storing the coolant in remote areas, freeze branding is less popular in remote areas than hot-iron branding, particularly when large numbers of animals are being branded.

Research on branding in cattle has investigated changes in behaviour, **heart rate**, plasma cortisol, adrenaline and noradrenaline levels as indicators of pain and **stress**. Most studies have investigated only the short-term responses to branding and have compared hot-iron, freeze and sham (the branding iron at ambient temperature) branding. Based on the measures taken, both hot-iron and freeze branding procedures are stressful and painful, although the stress appears not to be severe and the pain relatively short-lived. Hot-iron branding is probably more painful than freeze branding, although some work indicates that the pain from freeze branding may last longer than that associated with hot-iron branding. Studies that have investigated the hide surface temperature in the vicinity of the brand demonstrated elevated temperatures for up to 7 days post-branding, but with higher temperatures at the freeze brand sites compared with the hot-iron brand sites in the first 2 days. One study found that the temperatures of sites for both methods peaked at 48 h post-branding and there was no difference between the two methods until 6 days post-branding, when the hot-iron brand sites were warmer than those of the freeze brand. Hot-iron brand sites were also warmer than control sites at 7 days post-branding but freeze brand sites were not.

These results have been interpreted as corresponding to the differences in damage caused by the methods of branding and the duration of the inflammatory response, with the **inflammation** giving an indirect 'measure' of pain and discomfort. Interestingly, another study found that there was no difference between hot-iron, freeze and sham branding in the sensitivity to touching the branding site 1 and 7 days post-branding.

The longer-term effects of hot-iron and freeze branding have also been investigated. Branding appeared not to be a sufficiently stressful procedure to induce illness or disease in the animals (as assessed by the level of antibiotic treatment of animals), and did not influence body weight gains over 10- and 28-day periods post-branding.

It follows that, as both hot-iron and freeze branding cause pain as a result of tissue damage, pain will be reduced by using small, single-piece brands compared with large and/or multiple-piece brands.

One interesting finding from some of these experiments relates to the **stress** induced by **handling** and restraining animals that are unaccustomed to it. In one study, plasma cortisol concentrations and heart rates of beef calves rose when handled and restrained, regardless of the branding method. In contrast, when dairy cows that were accustomed to being handled and restrained were branded, their behavioural and physiological responses were attenuated, although measurements still indicated that hot-iron branding was more painful and stressful than freeze branding, which in turn was more painful and stressful than sham branding.

The stencil and aerosol coolant method was investigated in a study with horses and produced only 'mild' reactions. However, the brands produced were of poor quality.

A third method of branding that has been used, although rarely, is the application of caustic chemicals. This method is banned in many countries on animal **welfare** grounds. However, it should be possible to use non-caustic chemicals to inhibit melanin synthesis in the pigment-producing cells. In one study a range of such chemicals was applied topically or by intradermal injection to **cattle**, but most were ineffective. Furthermore, there were problems with the chemicals causing necrosis of tissue. (JCP)

See also: Handling; Markers, animal

Breed

Variation between individuals of a species is a driving force of natural selection. Under natural situations factors, such as geographical isolation, can lead to natural selection producing populations of a species that exhibit distinctly different characteristics from other nearby populations of the same species. This variation is usually described in terms of **subspecies**.

Variation is also a characteristic of populations of domesticated animals, and artificial selection by humans has led to the development of distinct breeds. A breed is a group of animals that has been selected by man to possess a uniform appearance that is heritable and distinguishes it from other groups of animals within the same species. Small changes in the body plan, pelage or behaviour of an individual of a species can be considered desirable by humans. Such differences can have a functional aspect or can be simply aesthetic. Much variability in a population of animals is due to genetic mutations within individuals. Some kinds of genetic variation in a species are not discrete, and often over a wide geographic area different populations of a species interbreed to produce a **phenotypic cline**. Breeds, therefore, reflect human exploitation of the natural variability of domesticated species.

Breeds are typically associated with 'animal partners', e.g. **dogs**, **cattle**, **sheep** and **horses**, that have undergone complete domestication and changed in morphology and behaviour from their original progenitor species. In farm animals, breed development invariably enhances desirable production traits. In **companion animals**, e.g. **cats**, **rabbits** and **rodents**, breed development largely enhances aesthetic characteristics, producing breeds that exhibit particular colour combinations or pelage types. Humans have not exposed 'exploited captive' species, e.g. camels and elephants, to any significant selection pressure and so they tend not to exhibit breeds.

Numbers of breeds of cattle, equines, **goats**, **pigs** and sheep that are recognized to have arisen in Africa, Asia and Europe are significantly correlated with the size of the human population. Similar results were found for states within countries like Russia, India and China, but correlations were weaker for land areas. Therefore, conditions that favour human population growth also favour breed diversification. Geographical location is important in breed diversity; countries with the most breeds per million people are more remote. Even within the UK, Scotland and Wales have many more breeds of livestock per million head of the population than England. Breed diversification is, therefore, more likely in remote or difficult areas of the globe.

Development of breeds has a long history. For example, dogs that are recognizably hounds are depicted on Ancient Egyptian hieroglyphics. Although we have a similar hound breed (Pharaoh hound) in modern times, this does not suggest that there is a continuous bloodline for this breed from ancient times (indeed the breed is much more recent in its origins). Many other breeds that have been developed in recent times can be traced back to particular historical periods or people.

The Englishman Robert Bakewell (1725–1795) is credited with the development of modern breeding practices in agriculture; he bred shire horses, small white pigs, longhorn cattle and Leicester sheep. At a time when it was unknown, the young Bakewell travelled extensively about the country, learning agricultural techniques and studying the anatomy of animals. In 1760 he returned to work on his father's farm at Dishley, Leicestershire. Bakewell's innovation was to move away from the random breeding of stock to breeding 'in-and-in'. He separated male and female stock, only allowing mating between specific animals he selected. This process of 'inbreeding' allowed him to fix and exaggerate those traits he felt desirable.

Starting with old Leicestershire sheep, Bakewell bred the New Leicester, a big, fine-boned, barrel-bodied sheep with a good-quality fleece. The breed also had fatty forequarters, considered an important characteristic of sheep at that time. Bakewell also began hiring out his rams so that other farmers could improve their own stock. By the end of the 18th century the New Leicester had interbred with almost all other types of sheep, but it never replaced them. In 1783 Bakewell established the 'Dishley Society' to protect the purity of his stock. Matthew and George Culley were friends of Bakewell, and established a flock of New Leicesters in Northumberland, northern England, in 1767. Isolation from southern breeders meant that the Culleys interbred with local Teeswater sheep to produce the Border Leicester, which was recognized as a breed by the Highland and Agricultural Society in 1869. A breed society and flock book were established in 1898 in Edinburgh. Since 1850 the sale of crossbred sheep (Border Leicesters crossed with Cheviot or Blackface sheep) has been very important in British agriculture. The Border Leicester is still used to pass on large litter size and rapid growth in the progeny.

The process of breed differentiation has distinct steps. First, a type of animal is defined that is more useful and desirable than the ordinary type, but it is not differentiated in pedigree. This distinction may reflect recognition of a cline or the development of a **mutation**. Secondly, some of the best animals of that type are gathered into one or a few herds, which then cease to introduce much outside genetics. This population then undergoes **inbreeding** so that the animals become distinct from other animals in other captive populations. If this process is successful then the breed becomes more popular and more herds are established from the original inbred population. Marketing of the breed is crucial in becoming accepted and expanding its numbers. Once the breed becomes numerous, a herd book or studbook is established to record the numerous pedigrees. Finally, a breed society is formed to safeguard the breed and to advance the interest of the breeders.

There are around 5000 recognized breeds of domesticated herbivores around the world. For example, there are almost 1400 sheep breeds and almost 1300 cattle breeds, although there are only just over 100 breeds of ass. Not all breeds are thriving, because worldwide just over 700 breeds are considered extinct and a further 845 breeds are considered to be rare. Interest in old breeds of livestock remains high because of the historical value of these animals. However, breed differentiation provides genetic variability within the population of any species as a whole. Loss of any breed leads to a reduction in the genetic pool for the species. Maintenance of ancient breed types is critical to ensure that animals that are genetically relatively close to the wild progenitor are available for modern breeding programmes. (DCD)

See: Breeding; Evolution; Selection

Further reading

Hall, S.J.G. (2004) *Livestock Biodiversity. Animal Genetic Resources for the Farming of the Future*. Blackwell Publishing, Oxford, UK.

Breed societies

Breed societies are formed by animal stock-keepers for farmed, companion and sporting animals to promote the **breed** and to help ensure its survival and development. These societies in the UK and former colonies have an important influence on the development of a breed, as they set the 'breed standard' by which animals are judged. The breed standard in the past has sometimes been to the detriment of the individual animals concerned, as the standard tends to concentrate on the appearance of an animal rather than its function and fitness (in a Darwinian sense). (DBM)

Breeding

The term 'breeding' is associated with the propagation of a particular species. However, the term is extremely broad and may refer to any number of aspects of the process of sexual **reproduction**. For example, breeding may refer to the behaviours associated with courtship and the act of copulation or the physiology associated with reproduction. It may or may not also refer to the behaviours associated with the production of the offspring, including 'brooding', the old English term from which breeding is derived, and **gestation**. Alternatively, breeding may refer to a controlled process in which individuals are selectively paired to produce offspring with a particular characteristic. Here, breeding will be discussed within the context of breeding in the wild, **captive breeding** and special scenarios associated with captive breeding.

Breeding in the wild

From an evolutionary perspective, the goal of breeding individuals in the wild is to maximize their genetic contribution to future generations. In order to accomplish this goal each species has evolved a series of complex **signals** and behaviours to identify individuals that are ready to reproduce and to help individuals identify receptive mates. Because of the relatively low physiological costs of reproduction for males of most species, males tend to seek to reproduce as rapidly and as often as possible. Thus, they seek signals of readiness of a female to breed. These signals may include specific odours or behaviours that are influenced by hormone levels. These odours may be detected in the **urine** or faeces or from the anogenital regions of the female. The **flehmen** response seen in many mammalian species is thought to be a behaviour used to determine sexual readiness in females. Behavioural signals of readiness to breed include postures or particular combinations of behaviours that indicate that a female is prepared to mate, such as presenting for copulation. Generally, once males ascertain that a female may be physiologically ready to mate they begin courtship and attempts to copulate.

Alternatively, because of the high physiological costs associated with reproduction, females will often be selective in determining a male with which to mate. Given the costs and physiological constraints, females seek to determine whether they are getting the best male genes possible. As such, females tend to be selective and seek out certain characteristics in a mate. Females tend to seek out males that are physically robust, can provide adequate food, protection or living conditions, are socially dominant or whatever characteristics are most important, depending on the natural history of the species. Males may hold territories or physically clash with other males as a means of proving their worth; they may build nests or dens and they may engage in elaborate courtship displays to attract females. The idea is that by rating well on the required signal, a male can show his value relating to characteristics that are important to the species. Females use these signals, dependent on the criteria appropriate for their species, when selecting a mate.

Both the signals to the opposite sex and the actual act of copulation are important components of breeding. However, there are other components to breeding in some species. Egg-laying species may sit on the eggs to keep them warm or defend them from potential predators. For example, after copulation females of certain hornbill species nest in excavated cavities in trees. The female and her mate seal her inside the tree cavity with a combination of mud, food and faeces. A small hole is left open through which the male feeds her while she sits on the eggs. Once the eggs hatch the female breaks out and they reseal the tree cavity. This behaviour enables both the male and female to feed the chicks until the young are old enough to fledge. While this strategy helps to prevent egg and chick loss due to predators, it requires a great deal of investment from both parents. In this case, breeding involves much more than simple mate selection and copulation: it involves elaborate, coordinated behaviours of both the male and the female.

The principle of natural selection drives reproductive behaviours and cues. Individuals that are able to successfully produce more offspring spread a greater percentage of their genes into the population. Those behaviours and strategies with a genetic basis that are successful will become more prevalent in future generations because they comprise a greater percentage of the genes in the population. It is important to note that no single breeding strategy works for all species. The appropriate breeding signals and behaviours have developed over long periods of time and exist in response to the ecological pressures that face the individual species. As a result, each species, even closely related ones, may have developed different strategies and signals based on the techniques that have proved successful in the past.

Breeding in captivity

In contrast to breeding in the wild, where females and males make selections of appropriate mates based on their signals of worth, and those strategies, if effective, are passed on to future generations, breeding of animals in captivity often involves human intervention. Breeding of animals in a captive setting is often termed artificial selection, because it involves the human selection of animals to reproduce, not what would actually be successful according to the criteria of natural selection. Captive breedings are chosen to meet one of two criteria for artificial selection. The first selection type is based on the human desire to obtain some particular physical or behavioural characteristic in the offspring. For example, livestock have been bred to produce the largest individuals because this tends to make the animals more valuable for the amount of meat that they can provide. Animals such as dairy cows are often bred to maximize milk production, and fibre-producing animals such as alpaca are bred to produce certain colours or quality of fibre. Domestic animals may be bred for a certain behavioural characteristic such as chasing rodents, as in terriers, or retrieving hunted waterfowl, as in retrievers. Domestic animals are often bred for temperament, as the most easily handled animals are often bred to create offspring that are calm around people.

The goal of this type of artificial selection is to make a certain trait or characteristic more prevalent in the offspring. This may even involve a breeding process called line breeding, in which offspring are bred to their relatives to further increase the possibility of a desirable characteristic in future generations. This process of artificial selection greatly impacts the genetics of the species, as animals with less desirable traits (from the human perspective) are not provided with the opportunity to reproduce and, as a result, these traits become less common or may disappear from the captive population. Additionally, certain traits that occur at a variable level in the normal population become set in the captive population. That is, all individuals have that trait because there are no other genetic alternatives in the gene pool.

Another technique for breeding animals in captivity involves the selective breeding of individuals to maintain genetic diversity, but again without concern for which traits would be most adaptive in the wild condition. For example, in **zoos** and aquariums around the world, exotic animals are bred to maintain the largest variability of genes possible. By maintaining a greater diversity of genes, zoos and aquariums seek to avoid selectively breeding out characteristics that we currently feel are undesirable. The genetic population is nearer the naturally occurring genetic variability in the population than in populations that are bred for a certain characteristic. This effort is designed to prevent **inbreeding** and **genetic bottlenecks**, which may cause physical problems in later years. Additionally, it makes the genetic stock as similar as possible to the animals in the wild, given the chance that animals are able to be reintroduced or transferred to another region of the world.

Currently, there are zoo and aquarium associations in regions all over the world. Each organization attempts to manage the genetic diversity of individual species by regularly reviewing its population and conducting genetic, demographic and logistic analyses to produce a plan for individual breeding recommendations. The member zoos and aquariums of these organizations agree to breed or not breed their animals according to these recommendations in an effort to preserve long-term genetic diversity of the species and ensure their long-term success in captivity.

Breeding challenges in captivity

It is important to note the challenges associated with both captive breeding strategies. When animals are bred for a single characteristic there are often unwanted, unforeseen side effects. For example, extreme breeding for size in broiler chickens and other poultry may result in animals that cannot fly or may have difficulty supporting their weight. Additionally, purebred dogs may have physical ailments, such as bulldogs and pugs, which have been bred for shorter muzzles, and often have difficulty in breathing. Temperament characteristics may also be associated with physical characteristics. In attempts to breed a tamer fox, scientists found that the resulting offspring differed in coloration, tail placement and ear shape. In a sense, the new foxes did not look like foxes. Line breeding for a particular characteristic may result in a genetic composition that is quite different from the original species and may have many other characteristics.

Breeding for genetic diversity also has its challenges. When a single male is selected for a female that usually would select her own mate from many choices, it is always possible that the two selected individuals will not reproduce. Lack of reproduction in a particular pair causes many challenges when maintenance of genetic diversity is the goal. Zoos and aquariums often deal with small population sizes, and the amount of genetic diversity available is already limited. If particular individuals do not reproduce, they can have no genetic contribution to future generations and it makes it more difficult to maintain the genetic diversity goals into the future. Additionally, if some individuals of a species breed well while others do not, animal managers are potentially selecting for individuals that breed well within the captive environment, which may lead to the same problems associated with selectively breeding for a particular trait.

Finally, breeding for genetic diversity creates serious logistical challenges. As opposed to instances of line breeding in which it may be desirable for a single male to breed with dozens of females and their offspring because he has a desirable set of characteristics, breeding for genetic diversity calls for no particular individual to have an overabundance of offspring. This means that animals need to be moved from group to group or from institution to institution on a regular basis to produce the most genetically diverse pairings and offspring. This strategy requires that animals must be transported between institutions and they must be introduced to new groups or individuals. This is resource consuming and can be problematic when certain animals do not get along and will not reproduce.

Artificial breeding

One of the techniques that has been recently explored to help deal with some of the challenges associated with captive breeding is artificial breeding. Artificial breeding often involves the **artificial insemination** of females or embryo transfer. This process has been used in a number of zoo-housed animals, most notably elephants.

Artificial breeding is advantageous in a zoo setting because the challenges associated with frequently moving an animal are eliminated. This is particularly important when animals are susceptible to **transport**ation **stress** or when they are extremely difficult or dangerous to move. The ability to artificially inseminate elephants eliminates the need for long trips in a truck for the elephants, and allows institutions that do not have facilities for holding a bull elephant to be able to successfully breed their female elephants. Another potential advantage of artificial breeding is that sexual sorting of sperm can be carried out prior to the insemination technique. This can be helpful for species in which there is a large number of solitary males that are surplus to the reproductive population, such as gorillas or elephants. In these cases, male offspring are unlikely to be able to live in a social group after they reach reproductive maturity and, as a result, females are more desirable within the population for breeding purposes. While this option has been suggested for use within zoos and aquariums, it has yet to be perfected and tested on a large scale.

Artificial breeding within the captive context does have many advantages, but there are also some disadvantages. Chief among these is the fact that the technology associated with artificial insemination is quite expensive, and the process of collecting semen samples and the insemination process can be quite labour intensive. Many zoos and aquariums do not possess the funds to be able to invest in such a technology and, at the current time, it is cost prohibitive. However, as technology continues to move forward, it is not unlikely that we will begin to see more and more artificial breeding techniques being employed in the captive animal setting. (CWK)

See also: Captive breeding programmes; Farmed animals; Sexual behaviour

Further reading

Alcock, J. (1998) *Animal Behaviour: an Evolutionary Approach*, 6th edn. Sinauer Associates, Inc., Sunderland, Massachusetts.

Kleiman, D.G., Allen, M.E., Thompson, K.V. and Lumpkin, S. (eds) (1996) *Wild Mammals in Captivity: Principles and Techniques*. The University of Chicago Press, Chicago, Illinois.

Krebs, J.R. and Davies, N.B. (1997) *An Introduction to Behavioural Ecology*, 4th edn. Blackwell Scientific Publications, London.

Price, E.O. (2002) *Animal Domestication and Behavior*. CAB International, Wallingford, UK.

Buddhism

Buddhism has a strong position in Asia, but there is also a growing interest in Western cultures, partly thanks to its view of human relation to nature. Buddhism is perceived as building on a deeper understanding of human dependence on, and intimate relation to, animals than found in mainstream Judaeo–Christianity. Several central elements in Buddhism contribute to this. One fundament is the continuity of all life forms, expressed in the idea of reincarnation, i.e. every sentient being's rebirth in different life forms. In the popular Jataka stories (about Buddha's previous reincarnations), relations between animals and between animals and humans are important. An enlightened Buddhist will 'know' his or her previous lives, but also without this specific knowledge a Buddhist regards an animal as a possible future human, or even a former human. Tightly related to this is a second element, the idea of karma, i.e. that we will experience the effects of all our intentions in the present and the next life. A third element more explicitly influencing the human–animal relationship is the ideal of non-violence (ahimsa). This is expressed in the first of the Five Precepts 'Ye shall slay no living thing'. A fourth element is action guiding, the ideal of compassion, understood as the primary ethical value. On these grounds Buddha not only objected to animal sacrifice, but showed animals respect. These core elements are valuable incentives to treat animals well, or at least not intentionally to inflict harm.

According to the Lotus sutra, central in Mahayana Buddhism, animals also have Buddha nature, and can thus be fully enlightened. However, influential lines of Buddhism also include clear elements of hierarchy between humans and other animals. According to this only a human can attain enlightenment, and every human is more valuable than any other animal since their physical nature is a manifestation of moral status, and 'human nature is itself a product of moral evaluation' (Keown, 1995). In practice, neither the non-harming principle nor compassion has been strong enough to prevent captivity or instrumentalization of animals in Buddhist countries. Waldau (2002) concludes that both Buddhist ideals and practice to some extent are hierarchical, or speciesist. This tenet runs parallel to **Christianity**. (HR)

See also: Hinduism; Jainism

References and further reading

Chapple, C.K. (1993) *Nonviolence to Animals, Earth and Self in Asian Traditions*. State University of New York Press, Albany, New York.

Keown, D. (1995) *Buddhism and Bioethics*. Macmillan, London.

Smart, N. (1989) *The World's Religions*. Cambridge University Press, Cambridge, UK.

Waldau, P. (2002) *The Specter of Speciesism. Buddhist and Christian Views of Animals*. Oxford University Press, New York.

Buffalo

The African buffalo (*Syncerus caffer*) and the Asian buffalo (*Bubalus arnee*) are similar in appearance, both having typically angular horns in section compared with the circular or oval shape of the cattle horn. The former is a larger animal (examples exceed 1000 kg liveweight), with more massive horns that are joined at the base where they are heavily thickened. In addition, the African buffalo has not been successfully domesticated (though individual animals have been tamed). There are wild Asian buffaloes extant, but the Indian wild buffalo (*Bubalus arnee*) is the only one to have been domesticated, the domesticated species being named *Bubalus bubalis*. The largest of the domesticated buffalo weigh around 600 kg. The earliest evidence of **domestication** is derived from representations of tame buffalo in artefacts from around 2500 BC in Mesopotamia and the Indus valley. The wild Asian buffalo may have extended through Europe and Asia in the Pleistocene period, but the buffalo in Europe today was reintroduced as a domesticated species.

Their arrival as a domesticated animal has been attributed by various authors to Arab conquest, returning Crusaders or as gifts from Arab leaders to European kings. By whatever means, domesticated buffalo were to be found in Europe – in Italy and the Balkan states – by the end of the 13th century.

Buffalo are adapted to an environment that is warm and wet, though they are not particularly heat tolerant. The buffalo includes shade seeking and wallowing among behavioural strategies to combat **heat stress**. As they have few sweat glands, buffalo lose relatively little heat through sweating and, if they are denied access to shade or wallow resources, they can quickly become heat stressed in hot conditions. They will seek out water, but it is unclear whether allowing buffalo the opportunity to wallow, where it is often denied, is necessary for their optimal production in European conditions. If given the opportunity buffalo will spend long periods wallowing, but wallowing animals have been observed to graze for longer periods than those in the absence of wallows. The management of wallowing areas needs to be carefully regulated: if the water is not continually flowing, excreta and other waste will soon build up, compromising the welfare of the animals using it on health grounds. Whatever the breed, time of day or season, the body temperature of buffalo is always lower than that of other cattle, and this difference is greatest in the early morning or in the cool season, suggesting that their welfare may be more compromised than cattle in a cold environment.

The domesticated buffalo will breed with neither the African buffalo nor cattle to produce offspring (though copulation has been observed). They are broadly classified into river and swamp breeds, with river buffalo having higher milk yields. Between-breed hybridization has successfully bred productive animals.

The purposes of keeping buffalo worldwide include milk, meat, hide, work and entertainment/ritual. Milk yields average 2000 kg per lactation (commonly around 305 days). This low quantity compared with cattle is in part compensated for by the higher fat content (up to 10%), which, it is claimed, has a lower level of cholesterol. Individuals are productive for up to 20 lactations. Additionally, buffalo milk has less of an allergenic effect, allowing consumption by humans with dairy allergies. The milk is creamy in taste and has a pure white colour – the buffalo, unlike the cow, does not secrete beta-carotene in the milk. Irrespective of the diet, beta-carotene is converted more efficiently into vitamin A, which is colourless. Buffalo milk is used for the production of mozzarella-type cheese.

Meat is generally regarded as a by-product at the end of the dairy animal's life, and it can be of rather poor quality. However, if the buffalo is raised as a meat animal the product is similar in quality to beef. With a comparatively lower fat content than beef, and a higher protein content, there is a scope for development of buffalo meat production. They are reported to have good forage conversion rates and infrequent calving problems.

The hide is valued as it is not damaged by warble fly larvae, which cause widespread damage worldwide to the hides of cattle. It is used for the production of heavy leather articles such as shoes, belts and saddles. Their horns can be used to produce handicraft products on a cottage industry scale.

Both males and females are used for work – in the fields (ploughing and harrowing) and for threshing, milling, raising water from wells, transporting material by cart or by pack and for riding. Although less swift than cattle, they have greater muscular power and are particularly preferred for tasks such as ploughing heavy soils. The domesticated buffalo (*B. bubalis*) is a docile animal that is easily trained.

Buffalo have entertainment value and, in the UK, as an exotic experience: animal producers have added value by opening to the public and training animals to pull carts for rides, and so on. Buffalo racing is a sport in Indonesia and the Philippines, and entire males are kept as fighting animals – these are pushing matches, but may result in fatalities. The buffalo is also important in religious ceremonies and for ritual slaughter (making the blood of a buffalo flow is believed to increase rainfall and the rice harvest in parts of Indonesia, and decapitation with one blow of the khukri knife is believed to bring the promise of good harvest to some in Nepal).

Buffalo faeces are prized as a fertilizer in Europe and elsewhere, as a flooring medium and, when mixed with straw and chaff, a fuel. The last is particularly important in regions with scarce access to fuel. (DA)

See also: Religious slaughter

Bull-baiting

Baiting refers to the harassment of one animal by another and is a form of public entertainment that can be traced back thousands of years. In bull-baiting, the bull is typically tied with a 15-foot (5-m) rope around the base of its horns and is tethered to a stake in the middle of an arena. Trained dogs, generally bulldogs, surround the bull and are released one at a time to bait the bull. The dogs are taught to stay low and avoid the horns while trying to attack the bull, usually by seizing the bull by the nose and holding on in an attempt to immobilize it. This practice has long been criticized as unethical and was prohibited in the UK by an Act of Parliament in 1835. There are a number of animal **welfare** concerns associated with bull-baiting, including stress and injury of the bulls, **stress** and **injury** to the dogs and even injury to members of the audience and dog handlers. While bull-baiting is illegal in most developed counties, it continues as part of a cultural tradition in others and may still occur in areas where it is banned. (LMD)

See also: Sport – use of animals in

Buller–steer syndrome

Buller–steer syndrome is a behavioural problem characterized by repeated and persistent mounting of one steer (castrated bovines), termed the buller, usually by a group of steers, termed riders. The frequency of mounting can be very high, in some cases exceeding 60 mounts per hour, resulting in **stress** and exhaustion with associated minor or even major traumatic **injury** and, in extreme cases, **death** of the buller. The syndrome is most prevalent among feedlot steers, where the overall incidence has been reported to be between 2 and 4%, a level sufficient to raise significant economic and welfare concerns within the industry; it may also be occasionally observed with pasture-raised steers.

There has been some discussion as to whether this behaviour is initiated by the buller standing passively, or even in some way soliciting mounting by the riders, or whether it is the riders that target a particular steer for mounting. Removal of the buller animal has generally been seen to alleviate the problem and the attention of the riders is not automatically refocused on a new buller, which would seem to indicate that it is the buller that is the initiating factor. However, observational studies have shown that the majority of bullers, rather than standing voluntarily, do actively seek to avoid mounting by riders, suggesting that it may be the riders that are the initiators.

A number of variables have been investigated as possible causative factors for the syndrome, and there is evidence for both increased stocking density and the use of anabolic steroids such as growth promoters having an effect on incidence of the behaviour, while a range of other variables, including seasonal and environmental factors and entry weights, has been shown to play a much less clear-cut role. There is however, probably a relationship between the incidence of the syndrome and the establishment and maintenance of social hierarchies (**see: Hierarchy**), with an increased appearance of the behaviour when new hierarchies are being established or existing ones are disrupted by the movement of steers. (MRC)

See also: Feedlot; Sexual behaviour

Bullfighting

Bullfighting has its origins in the ancient Greek and Roman practice of bull sacrifice, but was introduced into Spain by the Moors, who established bullfights in the ancient Roman amphitheatres of Andalusia. The practice saw a resurgence in popularity in the 1970s and 1980s, despite mounting criticism from animal **welfare** activists. It is now legalized within the European Union by special exemption from **slaughter** regulations on the grounds of cultural heritage. It occurs mainly in the Iberian peninsula, south-west France and South America, with approximately 10,000 bulls killed annually. It has its greatest following in Spain, where there are just over 1000 bullrings hosting about 1500 fights per year, killing approximately 4000 bulls. There are also a number of practice fights with animals that are not thought suitable for a major event. A major industry has developed to support the sport, with approximately 200,000 people employed in rearing bulls, maintaining the bullrings, etc.

Over the years the practice has developed into a more elaborate spectacle, with glittering costumes and large prizes for successful matadors, and much celebration and fiesta. However, many young people have little interest in bullfighting and there is concern that the close-up pictures of the fight, and in particular the **death** of the bull, on television are bad for the moral upbringing of children and will desensitize them to violence. Children under 14 years of age are banned from attending fights in some regions of Spain, and other regions have banned bullfighting altogether.

The bullfight differs from ritual slaughter of animals, or even **cock-fighting** or fox **hunting**, because the bullfighters or matadors face considerable danger when fighting the animals. The contest between man and animal has been romanticized by leading authors, such as Ernest Hemingway, and artists, such as Pablo Picasso. The animals are bred to be brave and to respond rapidly and vehemently to the challenge by humans. Their courage is exemplified by a willingness to charge a man on horseback. The animals must be so reactive to provocation that they charge up to the point of death, and they must not cry out or retreat when wounded.

The breeding programme has led to the development of animals that are hypersensitive and nervous, quite different from cattle that are used for milk production, which have been bred for docility. The animals for fighting are selected at an early age by challenging them to assess their levels of aggression. They are kept on extensive pasturelands, so that they have little contact with humans, and their reaction to the matadors in the fight then appears all the more valiant if they have not been subjugated by humans before. Females are selected for breeding by testing their charge at horses when provoked with lances jabbed into their backs. For the fight the bulls must be lean, strong and not too heavy, or their charge is slow. To achieve this they mainly graze on natural pasture with little supplementation, to prevent them becoming obese. They are weaned at about 8 months and branded. Their slow growth makes them mature late and, typically, they fight at 4 years of age, by which time they weigh around 600 kg. They have little contact with humans and there are no practice fights before they enter the ring.

The bull must come to the fight well nourished, to give him enough energy to be able to charge with vigour to the last. He should be kept under as little **stress** as possible, otherwise his performance will suffer. A veterinarian is present at each fight and has to certify the animal's welfare before the fight.

There are regional variations to the bullfight, with a fight to the death being the norm in Spain and South America but not in France or Portugal, where the bull is professionally slaughtered immediately after the event. After the extreme stress suffered by bulls during the fight, the best course of action for the bull on welfare grounds is slaughter as soon as possible, regardless of whether it is in front of a crowd of spectators.

Each fight lasts for about 20 min, with four to six bulls per event. The matador normally fights two bulls at each event. He has assistance from two picadors – men carrying long pikes with 10 cm steel tips – and three bandilleros – men on horseback with decorated wooden sticks that end in a barbed harpoon. Initially, the picadors thrust the pikes two or three times into the bull's shoulder muscle as he charges, to weaken his tossing ability. Next, the bandilleros jab up to eight of their harpoons into the bull's back to further weaken it. A 'good' bull does not cry out when they are inserted and indeed shows no signs of pain throughout the event. Then the matador taunts the bull by tempting it to charge a large piece of cloth, coloured fuchsia and yellow, held outside his body. This stimulates the bull with its movement and takes its attention away from the matador. This taunting continues until the matador has complete control of the charging bull, which may take about 20 charges. During this period the bull may also charge the matador. Then, with his sword in his right hand, he attracts a final charge at a small bright red cape held directly in front of his body with his left hand. It seems likely that the bright red colour and the positioning of this cape are designed to extract the last vestiges of aggression in the bull. Red is commonly found to incite arousal in other species and this universal phenomenon is used to good effect by the bullfighters. As the bull charges, the sword is plunged between the bull's shoulder blades and, if correctly positioned, will cause instant collapse and death. If the bull fails to die, the matador will lance the neck with a short sword, causing immediate death. A bull is considered to be more courageous and honourable if it dies with its mouth closed. (CJCP)

See also: Sport – use of animals in

Bushmeat

In Africa, forest is often referred to as 'the bush'; wildlife and meat derived from forests is called 'bushmeat' (in French: *viande de brousse*). The term applies to all wildlife species – including those that are threatened or endangered – that are hunted for their meat. The rapid expansion of **hunting** for the commercial sale of wildlife meat has become the most immediate threat to the future of wildlife in the Congo Basin, and has already resulted in widespread local **extinctions** throughout West Africa (BCTF, 2000).

Wildlife has been hunted for food ever since humans first evolved, and is still viewed as a resource 'free' for the taking in many areas. Today in Africa, bushmeat continues to be an economically important food and trade item for thousands of poor rural and urban families. Bushmeat has also become a status symbol for urban elites trying to retain cultural links to 'the village', often commanding extremely high prices in city restaurants. The immediate loss of wildlife and the secondary loss of many plant species jeopardize the function and stability of natural habitats – including both forests and savannahs – threatening the long-term survival of ecosystems and the people dependent upon them.

The bushmeat crisis refers to the current unsustainable, illegal, commercial trade in wildlife in Central Africa and around the world. Problematic bushmeat hunting may take place with illegal methods (such as wire snares that capture and kill indiscriminately), may injure or kill protected species, may take place in protected areas or may remove too many individuals from a population (**see: Wildlife management**). Logging operations deep in the forest often transport hunters and traders to the forest and carry wildlife to urban markets. This commercial trade is markedly different from subsistence hunting. Subsistence bushmeat hunting may be sustainable where human population densities are very low and markets are inaccessible (Robinson and Bennett, 2000). However, growing human populations with few economic alternatives and little opposition from law enforcement officers will continue to demand bushmeat, both for its protein and for its cultural value.

The health of humans, wildlife and domestic animals are closely linked in the bushmeat trade, and disease transmission between species has been documented. Particularly distressing is the multiple transmission of immunodeficiency viruses between primates and humans: more than 26 species of primates carry a variety of simian immunodeficiency virus (SIV) (Wolfe *et al.*, 2000). Ebola, which results in death for approximately 80% of those who contract it, equally affects humans and great ape populations, and may be passed from one species to another in the preparation of bushmeat.

Though deforestation has an obvious impact on wildlife dependent on these habitats, hunting constitutes a comparable threat to the ecosystem itself, in that it often results in Empty Forest Syndrome: a forest filled with trees, but with few, if any, large animals (Redford, 1992). Such forests will, over the long term, suffer dramatic changes in structure and composition as the wildlife responsible for dispersing seeds is lost through overhunting.

The bushmeat crisis is a global problem, with effects on people, wildlife and landscapes around the world. In Central Africa, estimates suggest that 30 million people consume

2.5 million t of meat per year (Wilkie and Carpenter, 1999). In East and southern Africa bushmeat is a growing problem, and is frequently turned to during periods of economic hardship or when fast cash is needed. West Africa's wildlife has already been hunted to depletion: the main species available in today's West African bushmeat trade are rodents and small antelope.

The wildlife trade in other areas of the world is also of great concern. Numerous species of fish have been overharvested around the world to satisfy an unsustainable demand; overfishing off the coast of West Africa has been linked to increased demand for bushmeat in the region. In addition, nearly 60% of the world's population now lives in Asia; China's growing middle class drives an enormous demand for wildlife products used in medicine, food and cultural ornamentation. The Asian turtle trade is so great that it is measured in tonnes rather than in numbers of individuals.

From a biodiversity-conservation perspective, the commercial-scale bushmeat trade threatens the survival of many species, including **great apes**, **elephants**, antelopes and others, as well as the ecosystems in which they live. From a human development perspective, the commercial bushmeat trade threatens to remove a needed source of protein from rural, poor human communities that depend on bushmeat as their primary protein source. While momentum is building to legalize the commercial bushmeat trade in Central African countries, it is highly unlikely that a legal trade would be ecologically sustainable, nor would it alleviate poverty. A recent analysis of bushmeat data collected in Gabon concluded that implementation of a legal, sustainable bushmeat trade would be far too costly to regulate (not even a 25% tax on bushmeat sales would cover the costs for law enforcement and tax collection) (Wilkie *et al.*, 2006).

Using data from throughout Central Africa, Fa *et al.* (2003) concluded that maintaining current bushmeat offtake levels without protein alternatives would result in a reduction of 81% in bushmeat protein supply in less than 50 years. Four out of the five countries studied do not currently produce sufficient non-bushmeat protein to ensure that their citizens consume the minimum daily requirement of protein; if bushmeat off-take suddenly decreased to sustainable levels, only Gabon would be able to maintain its population's protein demands (Fa *et al.*, 2003). Clearly, development of alternative livelihood (protein as well as income) strategies is needed immediately to maintain both human and wildlife populations.

There is no panacea solution to the bushmeat crisis. Different countries, ecosystems, cultural groups, wildlife populations and economic systems require different approaches to assure the long-term security of biodiversity and the health of human populations. An example of an integrated approach to solving the bushmeat crisis is found in a collaboration between the government of the Republic of Congo, the *Congolaise Industrielle des Bois* logging company and the Wildlife Conservation Society in the Northern Republic of Congo. These stakeholders have teamed up to assure appropriate policy development, adequate law enforcement and capacity building, hunting regulations (including strict enforcement of zones, methods and species), providing alternative protein and income sources, awareness building and collaboration with the local community, and continued monitoring and evaluation (Bennett *et al.*, 2005).

Conservation professionals agreed in 1999 to create the Bushmeat Crisis Task Force (BCTF) to fulfil a mission of building a public, professional and government constituency aimed at identifying and supporting solutions that effectively respond to the bushmeat crisis in Africa and around the world. The BCTF goals include effective information management targeted at engaging key decision makers, facilitating education and training and raising public awareness. Resources for key decision makers, educators, journalists, scientists, communities, students and the public are available at http://www.bushmeat.org.

(NB, HEE)

References

Bennett, E.L., Ruggiero, R.G., Eves, H.E., Bailey, N.D. and Tobiason, A.W. (2005) Case study 8.2. The bushmeat crisis: approaches for conservation. In: Groom, M.J., Meffe, G.K. and Carroll, C.R. (eds) *Principles of Conservation Biology*, 3rd edn. Sinauer Associates, Sunderland, Massachusetts, pp. 280–283.

Bushmeat Crisis Task Force (BCTF) (2000) *Bushmeat: a Wildlife Crisis in West and Central Africa and around the World*. Bushmeat Crisis Task Force, Silver Springs, Maryland, 8 pp.

Fa, J.E., Currie, D. and Meeuwig, J. (2003) Bushmeat and food security in the Congo Basin: linkages between wildlife and people's future. *Environmental Conservation* 30, 71–78.

Redford, K. (1992) The empty forest. *Bioscience* 42, 412–422.

Robinson, J.G. and Bennett, E.L. (eds) (2000) *Hunting for Sustainability in Tropical Forests*. Columbia University Press, New York, 582 pp.

Wilkie, D.S. and Carpenter, J.F. (1999) Bushmeat hunting in the Congo Basin: an assessment of impacts and option for mitigation. *Biodiversity and Conservation* 8, 927–955.

Wilkie, D.S., Starkey, M., Bennett, E.L., Abernethy, K., Fotso, R., Maisels, F. and Elkan, P. (2006) Can taxation contribute to sustainable management of the bushmeat trade? Evidence from Gabon and Cameroon. *Journal of International Wildlife Law and Policy* 9(4), 335–349.

Wolfe, N.D., Eitel, M.N., Gockowski, J., Muchaal, P.K., Nolte, C., Prosser, A.T., Torimiro, J.N., Weise, S.F. and Burke, D.S. (2000) Deforestation, hunting and the ecology of microbial emergence. *Global Change and Human Health* 1, 10–25.

C

Caecotrophy (cecotrophy)

Caecotrophy is a particular form of **cophrophagia** (faeces eating), seen in certain species of hind gut-fermenting herbivores such as lagomorphs (including **rabbits**), whereby they eat faeces directly from the anus. Since these species digest fibre in an expansive caecal region, they have less opportunity to absorb some of the nutrients before their elimination. By re-ingesting these initial faecal pellets (caecotrophs), digestion is improved. In captive rabbits the use of surgical **collars** to prevent self-injury, e.g. following **castration**, can prevent this behaviour by inhibiting neck flexion and result in serious **malnutrition** if prolonged.

(DSM)

Camelids – South American

Basic biology

There are six types of Camelids: the domesticated; the one-humped dromedary of southern Asia and North Africa; the two-humped Bactrian camel, which is still found wild in the Mongolian steppes; the domesticated llama and alpaca; and the wild guanaco and vicuña found in South America (see Table C.1). The llama is among the world's oldest domestic animals. While the llama is used mostly as a beast of burden in the Andes of Peru, Bolivia, Chile and Argentina, carrying things for the native herdsmen, it also provides the herdsmen with meat, wool, hides for shelter, manure pellets for fuel and was used as a sacrificial offering to their gods. In North America, Europe and Australia llamas are used as **pet** animals, backpacking for trekking expeditions, tourism, cart pulling, public relations, showing, parades and fairs, guardians of other livestock (such as sheep), golf caddies, therapy and, less importantly, production herds (for their wool).

Llamas are the largest domesticated members of the South American camelids, four species that have been the product of a long evolutionary history, marked by an affinity towards arid and semi-arid environmental conditions and the development of a unique series of characteristics. A llama's lifespan is about 15–18 years; however, they can live longer with care. An adult llama weighs between 100 and 150 kg. Their stomach anatomy consists of three compartments; they possess high-altitude physiology adaptations that allow them to live up to 4600 m above sea level – such as elliptical blood cells and locomotory traits suitable for walking in arid environments (moveable toe pads). The split upper lip, the long, curved neck, lack of tensor skin between thigh and body are all, together with the fine fibre fleece, adapted for **thermoregulation** in extremes of cold and hot. Llamas are seasonal breeders and mate lying down on their chest – copulation last 10–20 min and induces ovulation. Gestation lasts 11 months and the mother gives birth to a single offspring in a standing position. Llamas neither lick the newborn nor eat the afterbirth; the newborn is active and follows its mother within 15–30 min of birth. The female comes into heat again 24 h after birth, but usually does not mate for 2 weeks following recovery of the uterus. The four South American camelids can interbreed, producing fertile offspring; they all have the same chromosome number ($n = 74$).

Evolution

The llama's ancestors originated in the Eocene, approximately 40–45 million YBP, in North America. *Hemiauchenia*, which originated in mid-Miocene, already presented many of the characteristics of the true camelids. This genus began spreading towards more southerly latitudes of North America. During the Pleistocene (3 million YBP), the formation of the Panama isthmus formed, allowing *Hemiauchenia* to reach and thrive in South America. *Hemiauchenia* evolved into *Paleolama* and *Lama* approximately 2 million YBP in the Andean Zone of South America. Both genera showed a reduction in the length of their extremities and dental changes, implying better adaptation to harsh grassland, accompanied by an increase in the level of browsing in the diet. In terms of comparison, the habitat of *Hemiauchenia* consisted of flat terrain with a **grazing** diet while the habitat of *Paleolama* was mountainous, favouring browsing and grazing behaviour. The genus *Vicugna* supposedly originated during the same time.

Table C.1. Camelids of South America.

Name	Order	Family	Genus	Species	Habitat
Llama	Artiodactyla	Camelidae	*Lama*	*Lama glama* Linnaeus 1758	Central Andes of Peru, W. Bolivia, N.E. Chile, N.W. Argentina
Guanaco	Artiodactyla	Camelidae	*Lama*	*Lama guanicoe* Muller 1776	Andean Foothills of Peru, Chile, Argentina, Patagonia
Alpaca	Artiodactyla	Camelidae	*Vicugna*	*Vicugna pacos* Linnaeus 1758	Central Andes Peru to W. Bolivia
Vicuña	Artiodactyla	Camelidae	*Vicugna*	*Vicugna vicugna* Molina 1758	High Andes of central Peru, W. Bolivia, N.E. Chile, N.W. Argentina

Towards the end of the Pleistocene, between 10,000 and 12,000 years ago both *Paleolama* and *Hemiauchenia* genera suffered extinction, as did the rest of the megafauna that had settled in South America. *Lama* and *Vicugna* have persisted until the present, being represented by two wild species, the guanaco and vicuña, whereas llamas and alpacas originated from the domestication of their wild counterparts, processes that began approximately 7000 years ago. Because the process of domestication is closely related to the cultural action of man and their livestock, the distribution and radiation of llamas and alpacas occurred in the Altiplano zones of the central Andes.

Domesticated camelids

The llama is the largest of the South American camelids, with a slender shape, and may be found in up to 50 different colours. The llama has elongated legs, neck and face, and may reach as high as 1.5–2.0 m from the ground to its head. Its long ears are erect and curve inward in a classic banana shape. There are two breeds of llama traditionally recognized – the woolier being called *Ch'aku* and that with less fibre on the neck and body known as *Q'ara*. Their fibre (technically it is 'fibre' and not 'wool') is less dense than in the alpaca, with an average diameter of 26 μm on the undercoat and 70 μm on the guard hair.

The alpaca presents two recognized breeds, the Huacaya and Suri. The alpaca is the primary South American camelid fibre producer of the Andean Highland.

The vicuña is the smallest of the Andean camelids and has the finest fibre coat, the overall colour of the soft woolly coat being ochre, light cinnamon or reddish brown, with the underparts, insides of the legs and underside of the head being a dirty white. Unique among living artiodactyls, the incisors of the vicuña grow throughout life. The guanaco has a similar silhouette to that of a llama, with a uniform coloration – a dark brown upper body, neck and limbs; whitish fibre on the underside of the neck and belly; and greyish to black face. The guanaco's fibre has a diameter of 18–24 μm.

Distribution and welfare aspects

Today there are approximately 21.5 million camelids in the world, with around 8 million in South America alone: llama, 3.7 million; alpaca, 3.3 million; guanaco, 875,000; vicuña, 250,000. Most alpacas and vicuñas are to be found in Perú, the majority of South American llamas are in Bolivia and nearly all the guanacos are in Argentina and Chile. Overall in South America the number of alpacas, vicuñas and guanacos is increasing due to the value of their fibre, whereas llama numbers are falling. In the USA and Canada there are an estimated 160,000–200,000 llamas and up to 60,000–90,000 alpacas.

People's attitudes towards llamas differ greatly between South America and North America, and this determines the kind of **welfare** problems that arise. Llamas comprise extensive herds grazing in open and poor grasslands with extreme temperature variation between day and night in the Andes. Poor, indigenous farmers had developed a traditional way of farming llamas even before the Inca empire and for more than 4000 years. Welfare problems such as freezing after shearing, predation, abandonment of crias (juveniles) and outbreaks of livestock diseases – such us enterotoxaemia, mange and rotavirus in crias – cause large mortalities and losses in Andean herds. Shearing also causes deep skin cuts and infections due to poor technique. In conclusion, **hunger** and freezing are the main animal welfare concerns in South American llamas. On the other hand, llama farming in North America faces problems of overfeeding, heat stress in humid and hot summers and behavioural problems in bottle-fed animals in close contact with humans. In addition, the isolation evidenced when llamas are kept as pets causes stress and abnormal behaviours.

Llamas and alpacas are becoming a new type of farm animal worldwide, and not simply the livestock of the traditional indigenous communities of the Andes. Herd practices and animal welfare recommendations should bear in mind their evolutionary adaptations to open and extreme environments moulded by cold weather and high altitude, and also the complex behavioural and physiological traits that make them a unique species. (CB)

Cannibalism

Cannibalism occurs when one individual consumes all or part of another individual of the same species. Cannibalism can be found in nature and may occur for a number of reasons. For example, sexual cannibalism occurs when, within a species, a member of one sex kills a member of the opposite sex (usually the female kills the male) before, during or after copulation; this occurs primarily in insects. Another well-known form of cannibalism is filial cannibalism, where the offspring of a species are killed and consumed by adults of the same species. A common example of this occurs after a new male kills and replaces the dominant male of a harem.

Cannibalism is also found in animal production systems. However, this type of cannibalism does not appear to have an evolutionary basis and is often described as **abnormal** or problem behaviour. It is thought that unnatural, intensive conditions lead to **stress** responses and **frustration**, and this may lead to the cannibalism. For example, chickens commonly **feather** peck, which is considered an abnormal or **stereotypic behaviour**, when kept in production systems. Feathers are removed during the process, which can lead to bleeding that stimulates increased pecking and cannibalism. Similarly, pigs kept in intensive production systems may **tail bite**, that is, they chew and consume tissue and blood from another pig's tail. Cannibalism is an ongoing problem in animal production and, after one incident occurs, others may follow, leading to an outbreak. There is no definitive solution to the problem, but the addition of **enrichments** to the environment, separation of injured animals and removal of dead animals can help control the problem.

Outbreaks of Bovine Spongiform Encephalopathy (BSE) or mad cow disease are believed to be linked to the feeding of infected cattle as meat- and bonemeal back to normally herbivorous cattle. This has lead to over 179,000 cattle becoming infected in the UK alone, with over 4.4 million cattle euthanized as a precaution. (LMD)

Further reading

Cloutier, S., Newberry, R.C. and Alldredge, J.R. (2002) Cannibalistic behaviour spread by social learning. *Animal Behaviour* 63, 1153–1162.

Elgar, M.A. and Crespi, B.J. (1992) *Cannibalism: Ecology and Evolution among Diverse Taxa*. Oxford University Press, Oxford, UK.

Captive bolt

A handheld gun containing a metal bolt that is used in the **slaughter** of animals. It renders animals unconscious immediately, through either penetrating the skull or by having a percussive effect that causes concussion and intracranial haemorrhage. The animal is subsequently killed by bleeding it out through severing the large blood vessels in the neck and upper part of the chest before it can regain consciousness.

(DBM)

Captive breeding programmes

Captive breeding refers to the managed propagation of wild animals in captivity, namely **zoos** and **conservation centres**. Systematic breeding programmes were established to create self-sustaining captive populations to prevent species extinctions in the wild. The long-term captive management of endangered species provides the time during which those factors leading to species extinction can be determined and hopefully ameliorated.

Many domesticated animals are selectively bred for desirable traits. For example, Robert Bakewell (1725–1795) bred the largest of his cattle together and, after a number of years, created a cow that had doubled in size (from 370 to 840 lb, or 815 to 1850 kg). Conversely, captive breeding programmes are set up to maximize genetic diversity and avoid **inbreeding**, thus avoiding selection for or against specific (desirable/undesirable) traits. This can be achieved through even production of offspring from all founders (wild-caught animals) and subsequent offspring, so that genetic integrity remains and is maintained within the captive population. The advantage associated with maintaining genetically diverse captive populations is that these will reduce the likelihood of animals becoming adapted to captivity and so they will represent their wild counterparts, thus improving the likelihood of surviving **reintroduction**, the long-term goal of much captive breeding. It is also reduces the likelihood of inbreeding.

Genetic selection occurs due to environmental pressures, both in the wild and in captivity, and results in animals that are able to survive and breed in their surroundings. It is difficult to avoid selection in animals in their temporary captive surroundings, as they may live in zoos for several generations. If wild animals do not breed in zoos their genetic contribution to the next generation will be lost, unless they have kin within the population. In these circumstances, against our best efforts, selection of animals that are able to breed in a captive environment occurs.

Another potential problem in captive breeding occurs when related animals are bred together – an activity employed to achieve 'pure blood' individuals, for example in pedigree dogs and cats. This process is called inbreeding and, although it results in animals that have some desirable qualities, as determined by the breeder, it is also associated with a range of problems (**see: Mutations**). Inbred animals have been found to be more susceptible to disease, and thus increased morbidity, and have physical abnormalities, reduced longevity and reproductive difficulties.

European bison were the first species to be systematically managed to prevent **extinction**. Heinz Heck managed the population and published his studbook in 1932. Studbooks are central to the management and propagation of many animals in captivity. Within the zoo community studbooks contain information about the population (living, historical and projected) and individuals (their age, location, genetic representation within the population, inbreeding status). Basic studbooks provide information on the number of births, deaths and transfers between zoos that occur yearly. Analysis of these data can provide important information about the demographic and genetic status of the population.

A key role for zoo associations worldwide is the coordinated captive breeding of animals. Captive breeding programmes

Table C.2. Summary of zoo captive breeding programmes for accredited zoos in Australasian, European and North American zoos (as of 16 November 2006).

Programme	AZA (American Association of Zoos and Aquaria)		ARAZPA (Australasian Regional Association of Zoo Parks and Aquaria)		EAZA (European Association of Zoos and Aquaria)	
Taxon advisory groups	46		16		40	
Captive breeding programmes	SSP: 111	PMP: 324	Conservation Plan: 33	PMP: 56	EEP: 165	ESB: 161
Collection of data: births/deaths/moves	✓	✓	✓	✓	✓	✓
Analysis of data	✓	✓	✓	✓	✓	✓
Production of studbook	✓	✓	✓	✓	✓	✓
Recommendation of moves/mates	✓	(✓)	✓	✓	✓	(✓)
Construction of long-term management plan	✓	(✓)	✓	✓	✓	(✓)
H & H guidelines	✓		✓	✓	✓	
Link with *in situ* conservation	(✓)		✓	(✓)	(✓)	

SSP, species survival programmes; PMP, population management plan; EEP, endangered species breeding programme; ESB, European studbook.

range from those requiring little input (low-level programmes) to those which are intensively managed (high-level programmes). The programmes implemented appear to differ between zoo associations; however, there are many similarities in the way they function (see Table C.2). Essentially, Taxon Advisory Groups (TAGs) coordinate breeding efforts for a particular group of animals; the number of species represented can vary from as few as two in an elephant TAG to the many species included in invertebrate and reptile TAGs. Discussions between TAG members prioritize which zoo-held species need to be managed and what level of management they will receive. A variety of factors will influence this decision, including: (i) *in situ* and *ex situ* population numbers; (ii) the role of the species in conservation, education, research or recreation; (iii) the level of expertise required to maintain the species; and (iv) how many people with appropriate expertise are available to maintain the animals, etc.

The simplest level of management programme involves keeping a tally of the number of animals in captivity each year. From this information the TAG can assess changes in the population over time and consider whether or not more intensive management is needed. High-level captive breeding programmes revolve around the information collected and analysed in the studbook, to assess the success and future plans for the programme. Recommendations are made to zoos holding the species to ensure that future breeding and moves between zoos results in a genetically diverse population with limited inbreeding. These programmes usually have long-term management plans and produce husbandry guidelines to facilitate these goals. Many of these high-level programmes are integrated with *in situ* conservation or reintroduction efforts.

Captive breeding programmes face many challenges. Frustratingly, animals recommended for breeding might not breed successfully yet offspring are produced from non-recommended pairings. This may be due to failed contraception, moves not occurring quickly enough, mismatched animals or simply bad luck. Analyses of genetic data may suggest unrealistic manipulations to maintain genetic diversity and limit inbreeding. For example, the number of individuals necessary to maintain 90% genetic diversity over 100 years may exceed that available within zoos. Logistical problems may in themselves make moves between zoos and breeding of 'ideal' animals difficult.

Many zoos worldwide, however, implement a variety of husbandry techniques, from single-sex housing and cessation of breeding and **artificial insemination** to environmental and social manipulations undertaken to achieve the recommendations made by studbook keepers. It is through these cooperative captive breeding programmes that many species have defeated the threat of extinction. Captive breeding successes include the Arabian oryx, golden lion tamarin, Mauritius kestrel and Round Island boa. (VBM)

See also: Breed; Breeding; Conservation; Endangered species

Further reading

Kleiman, D.G., Allen, M.E., Thompson, K.V. and Lumpkin, S. (eds) (1996) *Wild Mammals in Captivity: Principles and Techniques*. University of Chicago Press, Chicgo, Illinois.

Zoo Association websites, e.g. www.waza.org; www.eaza.net; www.aza.org

Captivity

Captive animal environments are diverse and include private homes, farms, ranches, **animal shelters**, laboratories, **zoos**, **circuses**, sanctuaries and others. While there is discussion on the **ethics** of keeping animals in captivity, a great deal has been gained and learned from captive animals. Some questions in which scientists are interested can best be answered by studying captive animals. For example, natural communication patterns among chimpanzees might best be answered by studying wild chimpanzees, but questions about whether chimpanzees are capable of learning symbolic communication are best answered in a captive setting.

One method for operationally defining animal **welfare** is to compare the behaviour of captive animals with that of their wild counterparts (**see: Measuring welfare**). Greater similarity between behaviour in the two settings indicates improved welfare of the captive animals. While this approach has limitations, it has proved useful. Some have proposed that captive environments could be evaluated by how well the animals therein would be expected to cope, were they to be released into the wild.

Recent trends in captive animal care place greater emphasis on the naturalization of captive environments. Enclosures are larger and more complex, with more natural materials, social groupings are more species-appropriate and enrichment procedures are offering challenges. In these ways, captive environments are becoming more 'wild'. On the other hand, habitat destruction and the rapid growth of the human population bring humans and wild animals into closer proximity. **Wildlife management** programmes removing or relocating animals, **ecotourism** programmes bringing humans directly to animals and **reintroductions** of captive animals to the wild are all examples of the increasing role of humans in the lives of wild animals. The combination of better modelling of the wild in captivity, and an increased direct influence of humans on animals in natural settings, is blurring some distinctions between captivity and the wild. (MAB)

See also: Domestication

Further reading

Carlstead, K. (1996) Effects of captivity on the behavior of wild mammals. In: Kleiman, D.G., Allen, M.E., Thompson, K.V., Lumpkin, S. and Harris, H. (eds) *Wild Mammals in Captivity: Principles and Techniques*. The University of Chicago Press, Chicago, Illinois, pp. 317–333.

Midgley, M. (1999) Should we let them go? In: Dolins, F.L. (ed.) *Attitudes To Animals*. Cambridge University Press, Cambridge, UK, pp. 152–163.

Veasey, J.S., Waran, N.K. and Young, R.J. (1996) On comparing the behaviour of zoo housed animals with wild conspecifics as a welfare indicator. *Animal Welfare* 5, 13–24.

Cardiac output

The term cardiac output (CO) refers to the amount of blood being pumped by the heart in a given period of time, usually per minute, and can be roughly calculated by using the following equation: CO (ml/min) = **heart rate** (beats/min) × stroke volume (ml/beat). (RM-F)

Carnivore

Carnivores eat the flesh of other animals. As adults, they typically have dentition that assists in the prehension of prey and allows cutting and chopping of muscular tissues (meat) and crunching of bones. Carnivores may be solitary or social species, with some being able to switch between the two as food availability and size dictate. (PCM)

Cartesianism

René Descartes' view of the world, as being divided into two major substances – the mental and the material – influenced a school of thought known as Cartesianism. Cartesians adopted this duality of mind and body. Taking this reasoning to its practical conclusion, Cartesians of the day believed that they were doing nothing wrong by nailing living dogs (which did not possess the relevant kind of consciousness) to boards and cutting them open without anaesthesia. Whatever 'painful' yelps that emanated from these research animals were dismissed as inconsequential.

Today, contemporary British philosopher Peter Carruthers, who speaks from a contractarian view, argues for a position consistent with the Cartesian line that animals do not possess conscious experiences in the morally relevant sense. Like Cartesians, Carruthers emphasizes the dissimilarities that exist between human beings and animals. Although accepting that animals have sensations and leaving it open that they have beliefs and desires (not founded upon propositional language), Carruthers contends that what matters from a moral point of view is the capacity to form beliefs about one's beliefs. Since animals' mental states are 'not able to be consciously thought about [by subsequent thought] … human beings are unique among members of the animal kingdom in possessing conscious mental states' (Carruthers, 1992).

Carruthers' argument, which rests on the premise that animals are unable to think about their thoughts, is a result of his commitment to a Higher Order Thoughts Theory of Consciousness. In this view, a mental state or experience is conscious just in case it is available to a thought that in turn is available to be thought about later. That is, to possess a conscious mental state requires something like 'access' consciousness and for one's mental states to be related to other mental states in the way described above. Carruthers contends that the inability of animals to reflexively access their thoughts rules out their capacities for 'representing to themselves different possible futures'. Furthermore, from a moral point of view, animals are not hardwired to reciprocate, because not only are they unable to adopt or agree upon social conventions and rules for moral conduct, but they also lack conscious experiences for guidance of behaviour. Carruthers holds that all animal consciousness can be likened to the unconscious event of 'absent-minded driving', where one is on autopilot.

A few concerns can be raised against Cartesians. First, it is unclear why exactly being able to retrieve one's thoughts or having higher-order beliefs is a necessary condition for being conscious in the morally relevant sense. Empirical observation contradicts the view that something like higher-order thoughts (i.e. rational abstraction or reflective consciousness) precludes animals from having phenomenal experiences like pleasure and **pain**. Furthermore, pain appears to be directly experienced and one's memory of pain does not appear to have the same qualitative feel as experiencing a painful event at the moment of occurrence.

Cognitive ethologists and animal welfare scientists like Donald Griffin, Marc Bekoff and Marian S. Dawkins, and many others, provide interesting insights into the cognitive powers and affective states of animals. Their research threatens the Cartesian contention that animals have no minds and are merely on autopilot. Furthermore, their research raises interesting philosophical questions about whether propositional language is necessary for **communication**. Perhaps animals have their own languages and communicate their thoughts in ways we cannot fathom. This latter insight raises questions about the propriety of judging the capacities of animals from a human-centric view. From an empirical point of view, these researchers also offer evidence that (reasonably) debunks some of the so-called 'uniquely human' abilities, like access to previous experiences, use of previous experiences to guide behaviour or general flexibility while interacting with one's world. (RA)

See also: Dualism, Cartesian

Reference

Carruthers, P. (1992) *The Animals Issue: Moral Theory in Practice*. Cambridge University Press, Cambridge, UK.

Caslicks

The Caslicks procedure is a surgery performed on the vulvar labia of the mare. This surgery is typically indicated in broodmares with a perineal conformation that might predispose the mare to uterine infections. Anatomical features that may be indications for the Caslicks procedure include poor contact or seal between the labia and a recessed anus, with cranial tipping of the vulva. These may be associated with poor body condition, age or parity. Mares with such conformational defects are often called 'windsuckers' because of the airflow permitted through the poor seal of the vulvar lips. The Caslicks procedure is often used on young female racehorses because of presumptive poor performance secondary to 'windsucking' and vaginitis.

This procedure is normally performed with the mare standing, and local anaesthetic infiltrated into the labia. A narrow strip of tissue crossing the mucocutaneous junction of the labia, from the dorsal commissure and extending to the approximate level of the pelvic floor, is excised and reapposed with sutures (Beard, 1991). With the surgical wound healed the mare has only a short ventral opening for the evacuation of **urine** (see Figs C.1. and C.2). The vulvar lips are surgically reopened prior to parturition; this is normally performed with only manual restraint of the mare, or with restraint and local anaesthetic. An episioplasty is less frequently performed along with the Caslicks: the vestibular mucosa cranial to the labia is dissected and the edges apposed to create a narrower vestibular lumen.

Another surgical procedure performed on the external genitalia of the mare is the clitoral sinusectomy or, more rarely, a clitoridectomy, as an aid in preventing the spread of transmissible venereal diseases. The greatest concern is *Taylorella equigenitalis*, the organism causing contagious equine metritis, which is commonly harboured in the clitoral sinuses.

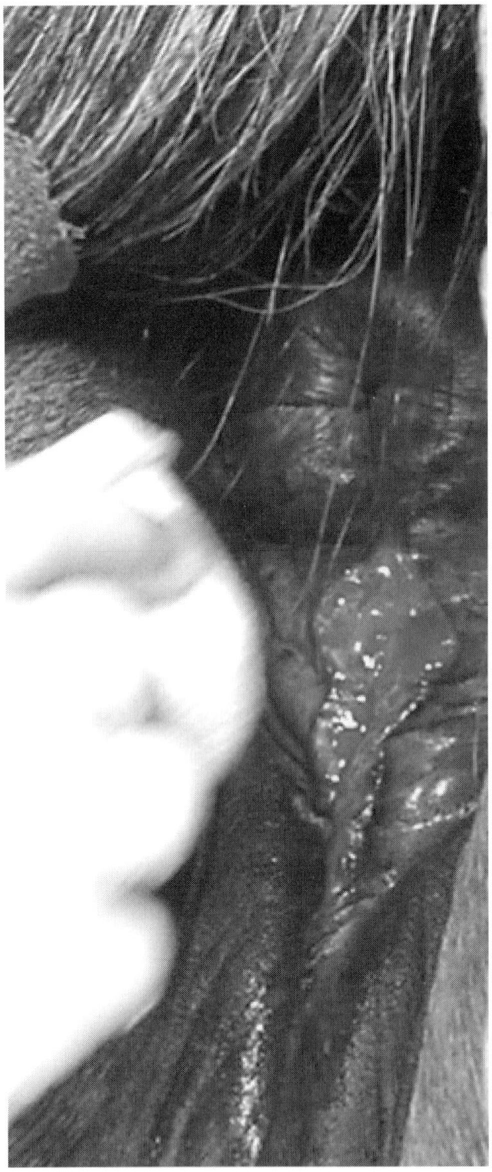

Fig. C.1. A strip of tissue at the mucocutaneous junction of the vulvar labia has been removed.

Cultural practices in certain human populations also incorporate surgical altering of the female external genitalia. Commonly called female circumcision or female genital mutilation, these surgeries may include clitoridectomy and excision of the labia minora and labia majora, with reapposition of the tissue edges similar to that performed in the Caslicks procedure. While numerous medical complications have been associated with this practice in humans, no systematic review of medical or behavioural consequences of the Caslicks, episioplasty, clitoral sinusectomy or clitoridectomy in the mare has been reported. (NKD)

See also: Circumcision

Reference

Beard, W. (1991) Standing urogenital surgery. *Veterinary Clinics of North America: Equine Practice* 7, 669–684.

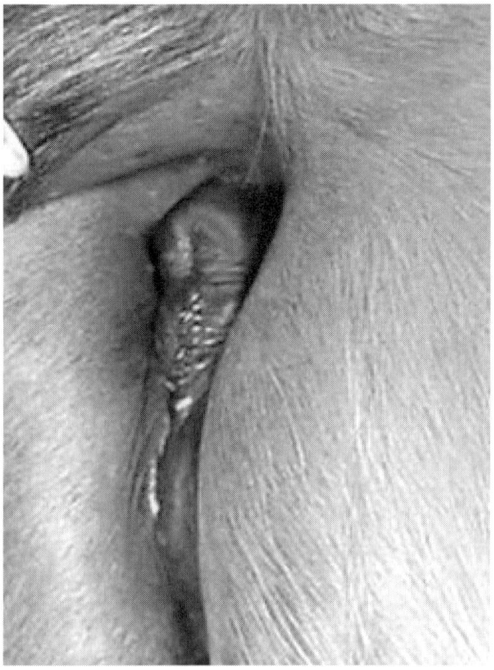

Fig. C.2. The Caslicks procedure completed, with the labia reapposed with sutures.

Castration

The term castration is sometimes used to refer to the neutering of either sex; however, this article will focus on its more common usage, i.e. the neutering of males. Not all male animals are required for breeding purposes, and they may have characteristics that would be undesirable in their offspring. Additionally, adult male animals can be difficult to handle and control because they fight with other adult males, are difficult to restrict or confine if there are oestrous females in the vicinity and may be aggressive towards handlers. Consequently, most young males are castrated, which involves removal or destruction of their testicles, resulting in dramatically reduced levels of circulating **testosterone** and the physical characteristics and behaviours that are largely controlled by this hormone. Thus, castrated males do not develop physical male secondary sexual characteristics, show fewer aggressive (**see: Aggression**) and **agonistic behaviours**, fewer **sexual behaviours** and tend to be easier to control and handle.

Castrated male livestock tend to have reduced liveweight gains, poorer feed conversion efficiencies and higher proportions of body fat than entire males at a given liveweight and, in some countries, castrated cattle (steers) are implanted with hormonal growth promotants to make liveweight gains and carcass composition more similar to entire males.

A number of devices and techniques are used to perform castration in livestock: (i) the bloodless castration clamp (e.g. burdizzo), which is applied for 5–10 s, two or more times to the scrotal neck in order to crush the spermatic cord without breaking the skin; (ii) the application of tightly fitting elastrator rubber rings or bands, which cut the blood supply to the testicles so that the scrotal and testicular tissues shrivel and eventually drop off; and (iii) surgical removal by cutting open the scrotal sac, removing the testes, cutting the *vas deferens* and breaking/cutting the blood vessels. These techniques are used with and without **anaesthesia/analgesia** and control for haemorrhaging by, for example, applying an emasculator or ligature, and cautery. Deaths from post-castration haemorrhage have been reported.

For livestock, the age at which castration can be conducted, whether it can be performed by the farmer or a veterinarian and the requirement for local anaesthetics are dependent upon the species and country in which it is performed. In general, rubber rings are used for young animals (a few days to a few weeks of age), and the clamp and surgery are used in animals of several weeks to several months of age. In recent years a variation of the rubber ring technique (the application to the neck of the scrotum of thick latex bands, which are then ratcheted to a high degree of tension using a special hand-held instrument) has been promoted for bulls of 12 months of age and older.

Extensive research has been conducted on castration, particularly in lambs and calves, although most of the lamb studies have involved a combination of castration and tail docking (**see: Docking – tail**). Castration studies have compared different techniques and the effects of using anaesthetics and analgesics, and the majority have used behavioural and plasma cortisol changes as indicators of stress and pain. Most have concentrated on the few hours post-castration and there can be little doubt that castration causes considerable acute (up to 12 h post-castration in calves) **pain** regardless of the technique if anaesthetics and analgesics are not used. There is some evidence that chronic pain can last for at least 42 days in calves castrated at ages ranging from 1 to 36 weeks. In calves, the use of a local anaesthetic (e.g. lidocaine hydrochloride/lignocaine) reduces plasma cortisol for up to about 1.5 h post-castration, whereas a systemic analgesic, such as ketoprofen, has been reported to reduce plasma cortisol to pre-castration levels. Reductions in feed intakes and average daily gain have also been reported in castrated calves for periods of up to 28 days post-castration. In addition, if techniques are used that result in open wounds, there can be complications at the site of castration, such as infections, abscesses and '**fly strike**' (fly larvae burrowing into and feeding on the tissue).

In cattle, castration at a young age (less than 3 months) is desirable, as there is some evidence that it results in less stress and pain. In lambs there appears to be no evidence that younger animals suffer less pain than older ones. Indeed, in one study, younger lambs appeared to experience more pain from surgical castration than did the older animals.

A number of studies have shown that surgical castration in calves and lambs results in greater stress in the hours following castration, as indicated by increases in plasma cortisol levels, compared with other methods. Although there are some contradictory results, it appears that the clamp has the least effect on growth rates, fewest complications at the castration site and lowest responses in terms of plasma cortisol and behavioural changes. However, it is not always effective, and a failure rate of 10% has been reported. The use of rubber rings for calf castration is reported to cause pain in the first 3 h after administration, chronic pain for up to 7 weeks post-application and abscesses. Castration by latex banding has been shown to cause necrotic wounds in 14-month old bulls, which persisted after the scrotum was lost (at about 56 days post-banding) and took several weeks to heal. Banded animals also had lower growth rates than surgically castrated animals for a 56-day period post-castration.

Severe lesions of the type found to result from latex banding in cattle have not been reported following the use of rubber rings in 1-week-old lambs, but have been seen in 3-week-old animals. Some work indicates that a combination of the application of a standard rubber ring followed immediately by application of the clamp distal to it causes less acute pain than rubber rings alone. However, the position and duration of application of the clamp appear to be critical in effecting a reduction in plasma cortisol levels. It has been suggested that the clamp disables innervation of the skin and tissue distal to the ring, and that the ring prevents inflammation in the dying tissues. Hence the first application destroys the blood supply and innervation and subsequent applications should, therefore, be made below the first. There is, also, evidence that application of the clamp promotes more rapid healing.

Less invasive methods of castration have been tried, for example injection of lactic or α-hydroxy-propionic acid into the testes, which causes sclerosis and atrophy. Several studies (on cattle ranging from about 3 to 9 months of age) have indicated that the method is not always effective in causing testicular atrophy and sterility. Furthermore, a proportion of animals (18–50%) can show development of secondary sexual characteristics. This method is reported to take about three times as long as surgical castration to conduct. There are conflicting reports as to whether chemical or surgical castration induces more acute stress and pain; one study reported higher plasma cortisol levels in surgically compared with chemically (α-hydroxy-propionic acid) castrated animals, while another stated that behavioural changes indicated greater pain in chemically (lactic acid) castrated than surgically castrated calves. The results from the latter study indicated that lactic acid caused scrotal necrosis in some animals, and wound healing times were about twice that of surgically castrated calves.

More recently, alternatives to castration have been explored, such as immunization against **gonadotropin-releasing hormone** (GnRH) and GnRH agonists. Immunization against GnRH is reported to reduce or eliminate the reduced growth rates and feed efficiency associated with castration, provide carcass quality comparable to steers and reduce aggressive and male–male sexual behaviours. (JCP)

See also: Aggression; Agonistic behaviour; Docking – tail; Fly strike; Free-range animals; Gonadotropin-releasing hormone (GnRH); Sexual behaviour; Testosterone

Cat

A carnivorous mammal of the taxonomic family Felidae, which is characterized by its sharp teeth and claws and also includes the big cats (e.g. lions) that can roar and the smaller wildcats (e.g. ocelots) that cannot. Most cats are relatively solitary, coming together only for breeding, the exception being lions, which live and hunt in a socially cooperative way. Domestic cats will live in loose social groups, which in the feral situation typically consists of female siblings or relatives, which may cooperatively rear young and show some affiliative behaviour.

The domestic cat – which is variously known as *Felis silvestris domesticus*, *F. domesticus* or *F. catus*, is a domesticated variant of the small wild cat (*F. silvestris*) – is believed to have diverged from its nearest wild relative (*F.s. lybica*) over 100,000 YBP, although the earliest evidence of close human association dates to only 9500 YBP, to a shared cat–human grave in Cyprus. This date is quite closely associated with the development of agriculture by man, and the domestic cat almost certainly arose as a result of a **commensalism** with humans who stored grain which thus provided concentrations of rodent prey for the cat. Recent work suggests that the domestic cat appears to show some behavioural changes typical of domestic **companion animals**, such as increased attachment and attention to human cues, though these features are not as strong as they are in the domestic **dog**, and no studies have compared domestic cats to their wild relatives. There are five distinct subspecies of the wild progenitor, which appear to have started to diverge about 230,000 YBP: *F.s. lybica* (Near Eastern wild cat), *F.s. ornata* (Central Asian wild cat), *F.s. cafra* (Southern African wild cat), *F.s. silvestris* (European wild cat) and *F.s. bieti* (Chinese wild cat).

Domestic cats are now one of the most popular companion animal mammals and are found in homes throughout the world, with over 76 million estimated to be living in the USA and 47 million in Europe (Pet Food Manufacturers Association). In industrialized countries they are increasingly kept indoors, sometimes because of local restrictions aimed at protecting indigenous wildlife (e.g. parts of Australia).

Domestic cats are well known for their predatory behaviour and, while they can kill large numbers of small mammals and birds, their impact on the ecology of areas where they are indigenous or their potential prey covers large geographical regions is more debatable. However, they undoubtedly have had an effect on smaller island wildlife populations where they are not indigenous (**see: Exotic species invasion**) and where local species have no similar natural predator, e.g. the almost flightless ka'kapo (New Zealand owl parrot) and, most famously, the Stephen's Island wren, whose entire population was wiped out as a result of the introduction of a single cat by the island's lighthouse keeper.

Although domestic cats are carnivores and are unable to synthesize taurine, dictating the need for a meat diet or one supplemented with this amino acid, they also eat invertebrates and some plant matter as a matter of routine. The maintenance of both large and small cats in **captivity** has given rise to welfare concerns. Large and small wild cats, despite their natural tendency to spend up to three-quarters of the day sleeping or resting (being **crepuscular** in their activity patterns), appear to be particularly prone to showing locomotor **stereotypies** when kept in restricted, barren or frustrating environments. This is not a common problem of domestic cats, although they will show stereotypic oral behaviours (**pica**: including fabric eating and self-grooming behaviours resulting in substantial hair loss; see Fig. C.3).

However, confinement within the home may give rise to other problems. It has been suggested that cats without access to the outside are at increased risk of a number of physical diseases, including interstitial cystitis (bladder inflammation), resorptive lesions of the teeth, obesity and hyperthyroidism. The relationship between these problems and **stress** is variable, with a physiological mechanism explicable in the case of cystitis; however, recent work suggests that the increased risk of hyperthyroidism may not relate to stress but rather to exposure to fire retardants used in some home furnishings.

None the less, keeping domestic cats indoors undoubtedly reduces the impact of their **predation** and the risk of injury or death in road traffic accidents, which are a major cause of **mortality** in this species. Homes can be enriched by exploiting the available three-dimensional space to make it more usable, since cats often prefer high-perch resting places (see Fig. C.4). Such environmental enrichment, together with appropriate provision and distribution of other key resources (feeding and drinking areas and latrines, together with safe passages to each) may also allow multiple cats to live together in a single home relatively peacefully, even if they do not form a social bond. The **personality** of domestic cats is an area that has attracted research and it appears that distinct types relating to their sociableness are consistently revealed. This has implications for the selection of cats to live together within multi-cat households, which are becoming more popular.

Dependence on solitary predatory behaviour, together with origins in an arid environment, are major influences on the nature of cats. Cats are generally quite heat tolerant, but may be subject to chilling when wet. They have largely binocular stereoscopic vision which allows them to judge distances well, and their hearing extends up to the ultrasonic frequency range (64 kHz), which helps with both the detection of auditory signals emitted by rodents and the localization of sound. Olfactory communication is perhaps the predominant medium between cats, with facial rubbing and urine spraying resulting in the deposition of chemical signals. Contrary to popular reports, the function of non-sexual **urine** spraying remains unclear, with many hypotheses being inconsistent either logically (e.g. a signal that helps indicate recent hunting activity success is unlikely to be a stable strategy) or from the available data (e.g their use as territorial marks). Cats may also use faeces in communication (middening) as well as scratching, both of which convey information visually and chemically. Taste appears less well developed, with cats lacking the ability to detect sweetness.

Cats are seasonally polyoestrous in their reproductive behaviour, being quite vocal and often urine spraying when in season. These characteristics, together with induced ovulation and superfecundity (use of several males to father a litter), help to maximize local genetic input during breeding. Typical litter sizes are between three and five kittens and may be reared collectively by feral cats in female colonies, usually of related individuals. The feeding of feral cats can result in very high local population densities, with an increased risk of disease and potentially public health problems as a result. The tendency to keep cats indoors and a failure to appreciate their nature are major reasons for the surrender of cats to **animal shelters** (**see also: Abandoned animals**), other common reasons relating to changes in living condition and allergic responses among family members, although some breeds (e.g. Devon rex) appear to be generally less allergenic.

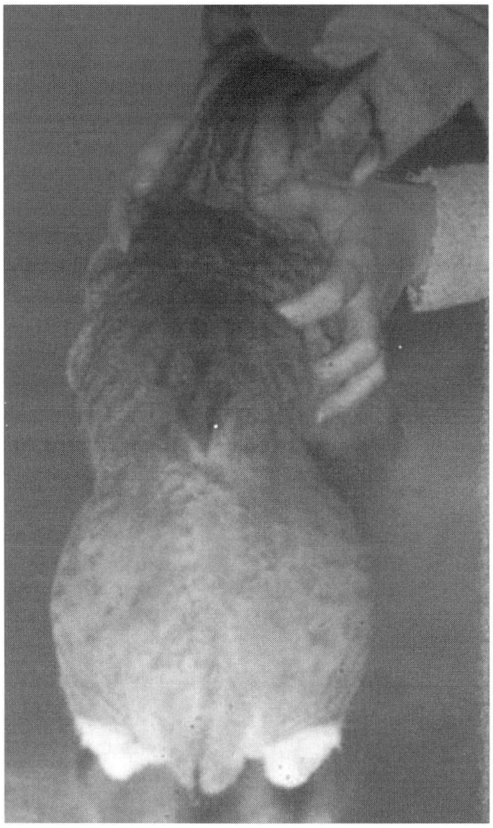

Fig. C.3. Domestic cat showing bilateral hair loss, which can arise as a result of stress-induced stereotypic hair pulling (image courtesy of S. Pipan).

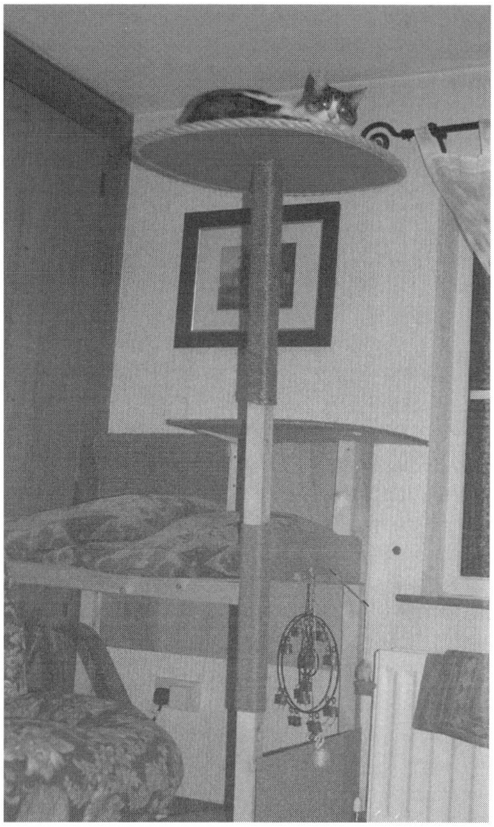

Fig. C.4. The vertical dimension of space is important to cats, and they often prefer to rest on higher platforms when given the choice (image courtesy of S.A. Mills).

Cat breeding has resulted in relatively little breed diversification on the original body form, compared with dogs, although specific mutations are now being preserved by some to create new breeding lines with distinct musculoskeletal or dermatological features (e.g. taillessness, dwarfism, hairlessness, folded ears, etc.). Some national associations refuse to recognize such animals as breeds, due to their associated health problems. (KT, DSM)

See also: Companion animal; Pet

Further reading

Beaver, B.V. (2003) *Feline Behavior: a Guide for Veterinarians*, 2nd edn. Elsevier Science, St Louis, Missouri.

Miller J. (1996) The domestic cat: a perspective on the nature and diversity of cats. *Journal of the American Veterinary Medical Association* 208, 498–502.

Rochlitz, I. (ed.) (2005) *The Welfare of Cats*. Springer, Dordrecht, The Netherlands.

Turner, D.C. and Bateson, P. (2000) *The Domestic Cat: the Biology of its Behaviour*, 2nd edn. Cambridge University Press, Cambridge, UK.

Catecholamine

Catecholamines fall within the amine class of neurotransmitters and hormones and include: **epinephrine** (adrenaline), **norepinephrine** (noradrenaline) and **dopamine**. These catecholamines are important in determining how the body responds to **stress**. All three are synthesized from tyrosine in sympathetic nerve cells, and derived sequentially from one another through the conversion of dopamine to norepinephrine, which in turn can be converted to epinephrine. (DCL)

Cathemeral

Cathemeral refers to a pattern of activity that is neither nocturnal nor diurnal, but more sporadic through both day and night. This activity pattern was first documented in lemurs by Tattersall (1979). (DSM)

Reference and further reading

Donati, G., Baldi, N., Morelli, V., Ganzhorn, J.U. and Borgognini-Tali, S.M. (2008) Proximate and ultimate determinants of cathemeral activity in brown lemurs. *Animal Behaviour* 77, 317–322.

Tattersall, I. (1979) Patterns of activity in the mayotte lemur, *Lemur fulvus mayottensis*. *Journal of Mammalogy* 60, 314–323.

Cattle

Types of cattle

Cattle were first domesticated about 9000 YBP and, since that time, they have become the major meat and milk producer in the world. They evolved from the aurochs, or *Bos primigenius* cattle, in two distinct subtypes, *Bos taurus* (European cattle) and *Bos indicus* (zebu cattle). Originally, it is believed, cattle were enticed into a symbiotic relationship with humans by offering salt and other food necessities. The last wild cattle from the *Bos primigenius* subtype were exterminated in Czechoslavakia in the 17th century. The two distinct subtypes, European cattle and zebu cattle, have been established as suitable domestic breeds for cool temperate and hot tropical conditions, respectively. The European-type cattle have been bred for high productivity because of the more suitable climate. Selected lines have been developed for milk and meat production, with the most popular of these for milk production being the Holstein–Friesian, which can have **lactation** yields of over 10,000 l, or several times a calf's requirements, and cows will lactate for approximately three-quarters of their life in the dairy herd. For meat production the **double-muscle** breeds, such as the Belgian blue, are increasingly popular, with growth rates exceeding 1.5 kg/day and a high muscling characteristic in the body, which can lead to dystocia problems in purebred calvings.

Major welfare concerns in cattle

The selection of cattle exclusively for the output traits in the milk and meat production breeds has given rise to many welfare problems, especially when they are kept in intensive conditions, for economic reasons (**see: Intensification of animal production**). The Holstein–Friesian cow tends increasingly to be kept year round in a concrete-floored barn system, supplied with feeds that have been conserved so that they can be stored and fed mechanically. The high productivity of intensively bred milk and meat production animals requires the feeding of energy- and protein-rich foods, so that the natural ability of cattle, as ruminant animals, to digest coarse plant fibres with the aid of the microorganisms in their rumen, is circumvented. The feeding of wet, acid silage (grass preserved in the acid byproducts of bacterial fermentation) leads to the production of low-dry matter faeces. Unless this is cleaned regularly from the floor, the enforced standing of dairy cows in a wet slurry of faeces and urine, on concrete, and with only narrow, concrete beds to lie on, gives rise to some serious **lameness** problems that are undoubtedly very painful for cows. The principal of these is **laminitis**, exacerbated by the feeding of energy-rich diets. These and other lameness problems caused by the environment that the cows are kept in can give rise to a prevalence of lameness in dairy herds of over 20%.

The prevalence of faeces in the cows' environment encourages farmers in intensive dairying regions to dock the tails of their cows, which prevents them being used to remove flies from the cows' body or as a signalling agent (**see: Docking – tail**).

The other major problem for the health of the dairy cow is **mastitis**, or intramammary infection, which is increasingly due to environmental pathogens invading the mammary glands via the teat canal, when the cows are lying on soiled bedding. Although many improvements have been made in hygienic management of dairy cows in the parlour and in the early detection and treatment of mastitis, the prevalence of the disease has not declined, largely due to the emergence of novel pathogens that colonize the mammary gland in her environment, rather than the milking parlour, which is now routinely disinfected between **milkings**.

Another concern for the high-yielding dairy cow is the limited amount of time that she has to eat sufficient food to supply the nutrients for production of perhaps 50–70 l/day of milk. It has been postulated that the high-yielding dairy cow in early lactation is permanently hungry, even though she is likely to have conserved feeds or pasture available for most of the day. Time available for lying down will be limited and so she may be permanently tired, as well as hungry. The average longevity of high-yielding dairy cows is usually no more than about three lactations so, given a 2-year rearing period, they live for approximately 5 years in total compared with their potential lifespan of about 25 years.

Apart from lameness and mastitis, one major reason for premature **culling** is reproductive failure, since cows that mobilize a lot of their own body's adipose reserves to support the high milk yields do not readily conceive. Conception rates in high-yielding dairy cows have declined significantly since the late 1980s, from about 60% down to about 40% in some herds. Reduced manpower to detect oestrus in cows is also partially responsible for poor reproduction rates. As well as the use of genetics to increase the potential milk yields in the dairy cow, an artificial **growth hormone**, bovine somatotropin, has been developed that will increase milk yields and have similar adverse effects on the cow's welfare to increasing productivity through genetic selection. The use of this artificial growth hormone is common in some intensive dairying regions, such as in the USA, but is prohibited in the European Union and Australia.

The development of separate lines in *Bos taurus* cattle for the production of meat and milk has led to some ethical and welfare problems for the offspring of dairy cows. Male calves are often of little value and may be killed soon after birth. Alternatively, they may be transported to a market and transferred to a cattle-rearing farm, but the quality of meat from calves that originated in the dairy herd is much below that of beef calves. Considerable attention is usually given to the rearing of the female replacements for the dairy herd, and their subsequent development into dairy heifers. However, the transition from a rearing unit to the dairy herd, including passing through a milking parlour two or three times a day, is inevitably a difficult one. The change in diets alone can cause significant lameness problems, but the mixing of young heifers with adult dairy cows can lead to a period of aggression and discomfort for both cows and heifers.

Intensively reared cattle suffer from inadequate space and metabolic diseases associated with the feeding of highly concentrated feeds, including acidosis and laminitis. Entire male cattle grow faster than castrated cattle, but they are more likely to have behaviour problems, such as riding each other and fighting. Excessive riding can cause broken penises and suffering on the part of the ridden cattle. Most cattle, however, reared on extensive pasture or rangelands, particularly in Australia and the USA, will utilize their ability to digest coarse fibre on grassland, rangeland or other land that cannot grow crops for human food. These land areas are not only less productive, often because of poor soil types, but they are also subject to very variable rainfall, and hence the unreliable quality and quantity of pasture grown provides a more variable feed supply than that offered to intensively kept farm animals, such as pigs and poultry, for example.

The ability of cattle to deposit adipose tissue when there is plenty of grass available, and then to mobilize it to sustain productivity when there is little feed available, has made them an extremely successful animal to utilize the extensive rangelands of the world today. However, some cattle farmers may be tempted to keep too many cattle in times of drought, because the loss of a breeding herd has a major impact on the farm's economic potential for future years. In times of drought farmers should sell cattle, provide them with supplementary food or rent lands for grazing. Nevertheless, inadequate supplies of food for cattle in times of drought do lead to significant cattle mortality.

In Australia, many cattle are reared in breeding herds on rangelands and they are then transported to **feedlots**, where they are kept at a high stocking density and fed conserved feeds, usually based on high-quality cereals. High temperatures will challenge such stock at high densities, and significant mortalities can result from a heatwave. The provision of shade is a simple but effective technique in ameliorating high temperatures (**see: Shade-seeking behaviour**).

Traditionally, the movements of cattle between farms or to markets or the abattoir was along droving routes, which provided an opportunity for animals to feed while they were moving. More recently, though, it has become normal for cattle to be transported in vehicles or by train. Many rangeland farms are a long distance from the destination of cattle, which then become tired, hungry and thirsty. Some cattle are transported overseas on ships, with journey times typically from 1 to 3 weeks. On the ships they are fed a pelleted ration and provided with water, but stocking densities are high and cattle may have difficulties adapting to conditions overseas (**see: Transport**).

Cattle are still used for draught, pulling ploughs, other machinery, carts or occasionally hauling timber. Because the work that the cattle have to do takes most of the day and they are often corralled into a small enclosure at night for safety, the biggest problem for these animals is inadequate food supply. As well as working for farmers they may also produce dung for fuel, which is a valuable commodity in subsistence economies. In India the heavy reliance on cattle for milk products, work and the production of dung has led them to be regarded as **sacred animals** by the Hindu population, so that they cannot be killed but are allowed to roam the streets. Because of restricted land availability many cattle have to scavenge on the streets and can eat plastic bags and other rubbish that is of no nutritive value. India has the largest cattle population in the world, and much remains to be done to improve the welfare of cattle in that country. In Muslim countries cattle are usually killed according to the principles of the Islamic faith, with a sharp knife cut to the neck to allow rapid exsanguination. No stunning is permitted, and hence the time to unconsciousness will be longer than in Western countries, and this has raised concern about their welfare (**see: Religious slaughter**).

There has been much criticism of the **welfare** of cattle in recent years, as well as their impact on the environment and on the health of the consumer. However, their ability to utilize poor-quality lands will ensure that they remain one of the foremost food producers in most countries of the world. They can be kept in both hot and cold climates and in diverse environments. Their endogenous heat production makes them particularly well suited to cold climates, but the heat stress resistance of *Bos indicus* cattle renders them particularly well placed to cope with tropical and subtropical conditions, as long as high productivity is not required. Being one of the most popular food production animals, there has been considerable research into the welfare of cattle, and major advantages to welfare could be gained from effective implementation of current knowledge. (CJCP)

See also: Herd health

Further reading

Phillips, C.J.C. (2002) *Cattle Behaviour and Welfare*. Blackwell Publishing, Oxford, UK.

Phillips, C.J.C. (2009) *Principles of Cattle Production*, 2nd edn. CAB International, Wallingford, UK.

Rushen, J., de Passille, A.M., von Keyserlingk, M.A.G. and Weary, D.M. (2008) *The Welfare of Cattle*. Springer, Dordrecht, The Netherlands.

Causal factor

How an animal behaves at any point in time depends upon what factors are motivating it at that time. The animal will be influenced by external factors, such as the environment or social situation, as well as by internal factors such as hunger or hormone levels. The **brain** monitors changes in all of these factors and interprets them, using the animals' previous experience of the situation, in order to make a decision as to what the animal will do. While many different factors may be influencing the animal, some of them will trigger particular behaviours, such as **stereotypies**, and these are referred to as causal factors.

In some cases the causal factor for a behaviour can be easy to distinguish; for example, hunger triggers foraging behaviour. However, in other cases it can be much more complex, such as with **feather pecking** in chickens, where one individual pecks at the plumage of another. Some studies suggest a lack of **dust-bathing** substrate is the causal factor; however, other studies suggest that the behaviour may be derived from foraging motivation. There has been no conclusive evidence that there is one specific causal factor, and it may be that a number of causal factors play a part in the expression of the behaviour.

When keeping animals intensively, **welfare** issues can result from internal causal factors motivating behaviours that cannot be performed in the environment in which the animals are being kept. A classic example of this is in pregnant sows; as they approach farrowing time changes in hormone levels cause them to become highly motivated to nest build. Since the causal factor is internal it occurs even when the environment is not suitable for nest building, e.g. when they are kept in a barren **farrowing crate**. The **frustration** that results from being unable to fulfil this behaviour can cause a prolonged birthing process and higher levels of stillbirth. (PoS)

Central nervous system

The central nervous system (CNS) consists of the spinal cord and the **brain**. The spinal cord lies in the vertebral canal, whereas the brain is located in the cranial cavity. The functions of the CNS are to receive, identify, integrate and interpret incoming sensory stimuli from the body's internal and external environments, then produce electrochemical impulses that are transmitted to the body's effector organs (muscles and glands) via the **peripheral nervous system**, including the **autonomic nervous system**, to initiate appropriative responses to the stimuli (adaptation). The responses aim to satisfy an organism's physiological requirements, such as survival, experience of positive emotions, avoidance of negative emotions and learning to improve performance. Disorders of the CNS may result in a permanent impairment of sensory, memory and motor functions, which affect animal well-being.

General structures of the central nervous system

Both the spinal cord and the brain are wrapped in three layers of connective tissues (meninges). They are, from inner to outer, the *pia mater*, the arachnoid, and the *dura mater*. The subarachnoid space lies between the *pia mater* and the arachnoid, filled with cerebrospinal fluid (CSF).

The cerebrospinal fluid is a clear, watery fluid that is produced by vascular plexuses (choroid plexus) in the ventricular system, and exits the CNS through the arachnoid villi. The CSF circulates within the ventricular system and the subarachnoid space. The ventricular system consists of a thin central canal of the spinal cord, the fourth ventricle of the medulla and pons, the third ventricle of the diencephalon and two lateral ventricles (representing the first and second ventricles) of the cerebrum. The essential function of the CSF is to prevent the spinal cord and the brain from both physical and physiological damage. The CSF provides physical supports for the brain and the spinal cord and maintains a relatively constant nutritional environment, including glucose, proteins and salts, for the **neuron(e)**s. Analysis of the CSF, such as for pressure and contents, is important in the diagnosis of many conditions affecting or damaging the brain and the spinal cord, such as infection, haemorrhage and mental retardation.

The brain receives arterial blood from a cerebral arterial circle fed by the internal carotid and the vertebral arteries. The venous blood is collected in cerebral veins (deep and superficial), passes to the sinuses and exits the brain for the systemic circulation through the internal jugular vein. The spinal cord receives arterial blood at different levels via the anterior and the posterior spinal arteries. The venous blood is collected in a venous plexus at the surface of the spinal cord, then passes to the epidural plexus at the surface of the dura and, finally, to the veins outside the vertebral canal. Damage to blood supply, such as from infarction, haemorrhage or atherosclerosis, may permanently impair brain and spinal functions.

There are three barriers within the central nervous system: (i) between blood and CSF – the blood–brain barrier (BBB); (ii) the CSF–brain barrier (CBB); and (iii) the blood–CSF barrier (BCF), intercalated between the blood, CSF and brain to protect the neurons from damage and to allow cross-exchanges of certain substances that are required by the central neurons for survival and action.

Neural structures of the central nervous system

Based on the colouring of tissue sections, the CNS can be divided into the white matter and the grey matter. The white matter contains axons, some of which form the peripheral nerves (the spinal cord) and cranial nerves (the brain). The grey matter contains nerve cell bodies (somas) and dendrites.

The CNS – both the brain and the spinal cord – consists of nerve cells (neurons), a component of the CNS for conducting electric impulses, and neuroglial cells (glia), an element of the CNS for supporting neurons. Neuroglia include astrocytes (fibrous astrocytes located in the white matter and protoplasmic astrocytes located in the grey matter), oligodendrocytes and microglia. The functions of astrocytes include: (i) structural support for neurons by providing a net frame; (ii) functional support for neurons by isolating and grouping of nerve fibres and synaptic terminals; (iii) metabolic support for neurons via regulating ions, transmitters and metabolites; and (iv) pathological support for neurons by repairing and isolating damaged nerve tissue. Oligodendrocytes form a myelin sheet around the long axons (nerve fibres), which protects and modifies the axons needed to send signals over long distances. Microglia are small phagocytes scattered throughout the CNS. Following injury or degeneration of the nerve system, these proliferate and transform into larger macrophages, which defend against inflammation and infection by phagocytizing and removing the debris.

The brain and the spinal cord consist of billions of interconnecting neurons (nerve cells). Bundles of their axons form nerve tracts or pathways within the CNS. Based on their functions, in general, there are three types of neurons: (i) the afferent neurons (sensory neurons), inputting information that they receive from the internal and external environments to the CNS; (ii) the efferent neurons (motor neurons), bringing instructions that fan out from the CNS to the body's effector organs (muscle and glands); and (iii) interneurons, receiving information from sensory neurons, processing that information, then sending the signals to the motor neurons by forming a complex, interconnecting electrical circuitry.

Neurons with the same or similar function integrate into a system (pathway), and there are four neuronal systems: (i) the sensory system, which acquires information about the external and internal environments; (ii) the integrative system, which processes the incoming information, evaluates this information within the context of past experience and makes decisions for action or inaction depending on the circumstances; (iii) the motor system, which converts decisions into commands for action (or inaction) by effector organs (muscles and glands); and (iv) the coordinating system, which organizes the patterns of commands to the effector organs to ensure operation in a smooth and efficient way. There are interactions among the different neuronal pathways within the CNS in response to stimuli.

Specific structures of the central nervous system

The spinal cord extends from the lower end of the brainstem down the vertebral canal. It consists of an H-shaped centre of grey matter and an outer zone of white matter. The grey matter contains neurons and their dendrites. Based on morphological and functional characteristics there are three major types of neuron: (i) afferent neurons (sensory neurons) located in the dorsal horn, the dorsal arm of the H, which receive input from skin and send their axons to the higher levels of the CNS; (ii) efferent neurons (motor neurons) located in the ventral horn, the ventral arm of the H, which send their axons out of the CNS to the skeletal muscles; and

(iii) interneurons located in both dorsal and ventral horns, which send their axons to other parts (segments) of the spinal cord. There is another group of neurons located in the intermediolateral cell column, belonging to the autonomic nervous system. This group of neurons send their axons out of the spinal cord to smooth muscles and glands through the ventral roots. The neurons located in the thoracic and upper lumbar segments of the spinal cord and in the sacral segment belong to the sympathetic and parasympathetic parts of the autonomic nervous system, respectively.

The white matter contains axons. Based on their functions, the axons are divided into funiculi or columns. There are three major types of axons conveying signals: (i) from the cord to higher levels of the CNS; (ii) from higher levels to the spinal cord; and (iii) between the segments of the spinal cord. Axons mediating communication between the CNS and other parts of the body make up the peripheral nerves (spinal nerves), which split into two parts, the dorsal root (sensory axons) and the ventral root (motor axons). These roots carry information to and from the spinal cord, respectively.

The spinal cord connects the peripheral nervous system to the brain and is involved in regulating sensory, motor and autonomic functions. Information from sensory receptors (nerve impulses) reaching the spinal cord through the dorsal roots is transmitted up into the brain. Signals arising in the motor areas of the brain travel back down the spinal cord and leave through the motor neurons via the ventral roots. Acting as a lower coordinating centre, the spinal cord is also responsible for some simple reflexes like the withdrawal reflex.

The brain is the major organ of the CNS. It controls vital functions, such as heartbeat, respiration, blood pressure and body temperature, and both sensory and motor functions. The brain also controls reactive functions, such as receiving and interpreting sensations from the nerves in response to stimuli, then initiating and coordinating those activities such as arousal, emotion and learning.

The brain consists of several specific structures that control physiological and psychological homeostasis.

The brainstem is the part of the brain that extends from the upper end of the spinal cord up the vertebral canal. It connects the cerebellum and cerebrum. It consists of the medulla (*medulla oblongata*), the pons and the mesencephalon (the midbrain), and the diencephalon.

The grey matter of the brainstem contains 10 of the 12 pairs of cranial nerves:

- oculomotor nerve (III);
- trochlear nerve (IV);
- trigeminal nerve (V);
- abducent nerve (VI);
- facial nerve (VII);
- auditory vestibular nerve (VIII);
- glossopharyngeal nerve (IX);
- vagus nerve (X);
- spinal accessory nerve (XI); and
- hypoglossal nerve (XII).

Many of the cranial nerves contain nerve fibres that conduct impulses out of the brainstem (efferent or motor fibres), while others bring impulses from sense organs (afferent or sensory fibres). Most cranial nerves are mixed, i.e. containing both motor and sensory fibres. There are four categories of cranial nerve fibres: (i) somatic efferent fibres, innervating skeletal muscles; (ii) visceral efferent fibres, supplying smooth muscle and glands; (iii) somatic afferent fibres, carrying sensory impulses from skin and mucous membranes; and (iv) visceral afferent fibres, transporting sensory impulses from the visceral organs. The white matter of the outer layer zone of the brainstem contains the major inflow and outflow nerve tracts from the brain and the spinal cord.

The brainstem plays an important regulatory role in basic functions, such as respiration, **blood pressure** and body temperature. The brainstem also acts as a highway for messages travelling between the other parts of the brain and the spinal cord. There are several major structures within the brainstem, including the reticular formation, the thalamus and the **hypothalamus**.

The reticular formation consists of groups of diffuse connections of scattered nerve cells throughout the core of the brainstem. It has functions in the central integration of coordination and in controlling the level of consciousness, which evoke preprogrammed or automatic reflex actions in response to stimulation. As an ascending activating system, it alerts the higher brain centres to the sensory stimuli, such as attention and consciousness, mental processes (mental state) and locomotory movement. The reticular formation also controls cardiovascular and respiratory functions, sleep, wakefulness and level of arousal.

The thalamus is located within the diencephalon, beneath the cortex and basal ganglia and above the hypothalamus. It consists of a massed group of neurons that are arranged into distinct clusters, named the dorsal thalamus, ventral thalamus, epithalamus and thalamic reticular nuclear group. It functions as a subcortical centre that influences visceral and somatic effectors, via the connections with the hypothalamus and the neostriatum. It also acts as a relay centre of sensory information to the cerebrum. All sensory input (except for olfaction) integrates in the thalamus, then forwards signals up to the somatosensory region of the cerebral cortex through the thalamocortical radiations. The signals from the somatosensory areas return to the thalamus on the way to the motor areas of the cerebral cortex.

The hypothalamus is located within the diencephalon. It consists of many smaller nuclei on each side of, or at, the midline. The hypothalamus is the principal integrating region for the entire autonomic nervous system. It produces hormones that affect other endogenous glands by which, it controls growth, sexual behaviour, metabolism, fluid balance and many other physiological variables. The hypothalamus is the source of nine tropic hormones:

- **antidiuretic hormone**;
- **corticotropin-releasing hormone (CRH)**;
- **gonadotropin-releasing hormone (GnRH)**;
- growth hormone-release inhibiting hormone (GH-RIH);
- growth hormone-releasing hormone (GH-RH);
- melanocyte-stimulating hormone-release inhibiting factor (MSH-RIF);
- melanocyte-stimulating hormone-releasing factor (MSH-RF);
- **oxytocin**;

- prolactin-release inhibiting factor (PIF);
- prolactin-releasing factor (PRF); and
- thyrotropin-releasing hormone (TRH).

The hypothalamus is the site that has close connections with the **pituitary gland** (the **hypothalamic–pituitary–adrenal (HPA) axis**) and serves to coordinate endocrine, autonomic and somatic motor functions to maintain body homeostasis in response to various stimuli, such as stress-induced changes of immunity, emotions and emotional reactions, mental functions and psychosomatic interrelations.

The cerebrum, the enlarged frontal structure of the brain, is divided into the cerebral hemispheres by the sagittal fissure. Each of the cerebral hemispheres contains a thick outer layer of grey matter (the cerebral cortex or *cortex cerebri*) and a central mass of white matter with some embedded nuclei, including the basal ganglia and the hippocampus. The surface of the cerebral cortex comprises gyri (folds) divided by sulci (fissures). Based on the borders formed by fissures and sulci and the main bones of the skull that overlie the cerebral hemisphere, the cortex is divided into four distinct lobes – the occipital lobe, parietal lobe, temporal lobe and frontal lobe, which can be further divided into approximately 50 cortical areas on a cytoarchitectonic basis.

The grey matter contains neurons that are arranged into six layers. The neurons receive impulses from and send impulses to lower parts of the CNS. The white matter contains nerve fibres that are organized into bundles (or nerve tracts). There are three main types of fibres: (i) association fibres, which link areas of the cortex within a single hemisphere; (ii) projection fibres, which link areas of the cortex to central brain structures and to the brainstem; and (iii) commissural fibres, including the *corpus callosum*, which link the two hemispheres.

The cerebrum functionally controls conscious thought, **decision making, perception, memory**, sensation, movement, speech and writing. The two hemispheres are lateralized with respect to function (hemispherical lateralization), i.e. in humans some higher activities such as speech and writing are controlled from one cerebral hemisphere, the dominant one (mostly the left hemisphere), whereas the non-dominant side (mostly the right hemisphere) is associated with visual/spatial orientation, artistic appreciation and creative thought, such as musical ability, facial recognition and the perception of social cues.

The cerebral cortex belongs to the grey matter and is rich in neurons. The neurons are arranged in six layers in a network manner, forming columnar sets among the vertically connected neurons. The cellular columns are the basic communication and information-processing units. These units exchange impulses with lower centres of the CNS. Anatomically and functionally the cortex is divided into different regions, such as the motor cortex (voluntary movement), somatosensory cortex (sensibility), visual cortex (vision), auditory cortex (hearing), olfactory cortex (smelling) and the limbic system, including the hippocampal gyrus and its adjoining cortical regions (learning and memory). The functions of the cerebral cortex are storage, analysis and integration of sensory information received from the internal and external environments to produce higher cortical function, such as abstract thought, memory, learning and **consciousness**; and initiation of somatic and vegetative responses, such as volitional movement and cardiovascular and gastrointestinal function. Damage to the cerebral cortex may cause permanent impairment of sensory, memory and motor functions, depending on the area(s) affected.

The basal ganglia are masses of subcortical nuclei, including the caudate nucleus, the putamen, the *globus pallidus* and the lentiform nucleus. Among them, the caudate nucleus and putamen are named the neostriatum. They are connected to the brainstem and the cerebellum through various nerve bundles, and are functionally involved in modifying motor output from the cerebral cortex, i.e. signalling and coordinating movements of skeletal muscles. Damage to the basal ganglia or the striatum itself causes movement disorders, such as Huntington's disease.

The limbic system (Broca's limbic lobe or the Papez circuit) consists of several groups of neurons, including the cingular gyrus, the hippocampal formation, the septal nuclei and the amygdala (some also include the mammillary nucleus and anterior thalamic nucleus). The limbic system receives inputs from various associated areas in the cerebral cortex and passes signals on to the *nucleus accumbens*. The functions of the limbic system are involved in long-term memory, learning and cognitive tasks (hippocampus), emotions, **fear** and **anxiety**, anger and **aggression** (amygdala).

The cerebellum ('little brain') is a massive, rounded structure located dorsal to the brainstem. It is linked to the brainstem by three thick nerve tracts (peduncles). The cerebellum consists of two hemispheres, which are further divided into lobes by the fissures and sulci. The cortex (the cerebellar cortex) of the hemisphere belongs with the grey matter and is rich in nerve cells, which are arranged in three main interconnected layers. In the white matter a central core consists of several nuclei, the central (deep) cerebellar nuclei.

The cerebellum functionally belongs to the motor system, which controls muscular coordination, balance and posture for the execution of movements. It receives information from organs such as muscle tendons and the balance organs in the inner ears, integrates the information, then sends the signals to the neurons of the nuclei and to the other regions of the CNS, resulting in smoothly coordinated movements and balance of the trunk and extremities. The cerebellum also regulates visceral function, such as heart rate and blood flow, during changes in posture. (H-WC)

Cephalopods

Cephalopods embrace the octopuses, squids, cuttlefish and chambered nautiluses. They are a class of mollusc. Cephalopod means 'head-foot', and the fact that cephalopods have their appendages, known as arms, branching from their heads distinguishes them from other molluscs. Cephalopods are an ancient group of animals that lived in the seas hundreds of millions of years ago. One extinct branch, the ammonites, is very well represented in fossil records. Ammonites had a coiled shell up to 1 m across – similar to present-day nautiluses. Their relatives, the belemnites, had an internal shell and they probably gave rise to most of the present-day cephalopods. Two factors probably drove the ammonites and belemnites extinct: competition with fish and the extinction event at the end of the Cretaceous period 65 million YBP, popularly ascribed to the collision of a meteor. No fossils of ammonites or belemnites have been found since that time, but a distant relative, the chambered nautilus, survives as a relict, a true living fossil.

Nautilus

Nautiluses have a coiled shell in which they live and to which they are attached. Like snails, they cannot be completely free of their shell. The shell is similar to a snail's shell, but serves a different purpose besides protection of the animal. A nautilus's shell is divided into chambers and it lives in the last outer chamber. The others are sealed, except for a thin thread of tissue that connects them and regulates the water level – and hence the air pressure – in each chamber. The animal thus regulates its buoyancy like a submarine. This is important, because nautiluses make daily vertical migrations on the reef face, upward at night and downward during the day, probably to avoid being seen by predators. Once they remove water from a chamber, they cannot readily replace it and, if they remove too much, they will float to the surface and be tossed around by wave action.

Today there are two genera of nautiluses extant, with only five or six species – one species has never been seen alive and is known just from shells washed ashore. This lack of diversity is one reason that nautiluses are considered a living fossil and that they may be dying out. They have a limited range in the tropical western Pacific, and several of the individual species are restricted to certain archipelagoes there. Also, thousands of these animals are being caught just for their shells, which are then sold in novelty shops to tourists and shell collectors. Because they are so limited in species and range, they are deserving of extra consideration, both in the wild and in captivity in aquariums or laboratories.

Other aspects of nautiluses' natural history, physiology and ecology also render them ripe for study. Their eggs have rarely been found in the wild and, although hundreds have been laid in captivity, few have hatched and they did not survive to maturity. Interestingly, hatchlings have several functional shell chambers capable of controlling buoyancy. Nautiluses appear to be the only cephalopod group that survives reproduction: all others are semelparous, i.e. they die after reproducing.

Preliminary indications indicate that nautiluses may live 10–15 years or more but, because they have not yet been reared to maturity, this is only a guess. If they live that long it would make them the longest-lived cephalopod, as others live only up to a few years. Although many aspects of nautiluses warrant further study, researchers must use prudence in working on them. They are difficult both to obtain live and maintain successfully. Their population status is unknown, mostly from lack of numbers and since they come from Third World countries. There is increasing evidence that they are being overfished and there is no regulation of their take.

Cuttlefish

Unlike nautiluses which are members of the subclass Nautiloidea, cuttlefishes and all the other cephalopods are coleoids. Cuttlefish have internal shells like their belemnite ancestors, although octopuses and squids have mostly lost theirs. The cuttlefish shell is thick and porous, providing structure and support to the body, as well as buoyancy. Cuttlebones sold in pet shops to supply calcium to psittacines are the shells of cuttlefish, obtained from animals used as human food or washed up on shores. Cuttlefish have eight arms and two long, prey-catching tentacles. They have two remarkable eyes replete with lens, cornea and retina. The eyes of cuttlefish and their kin have w-shaped pupils. The eyes are a good example of **convergent evolution**, as they are similar to vertebrate eyes. The eyes are used to find prey or mates, avoid predators and to help match their shallow water backgrounds in camouflage behaviour.

Most coleoids have the ability to change their colour using chromatophores in the skin, tiny cells that show their colour when stretched open by muscles under direct control of the brain. The chromatophores can show black, brown, red or yellow, and other cells show white or reflect blue or green, which gives them quite a diverse palette to work with. They can open chromatophores of all one colour, say red, or show patches of colour to match a mottled background. In addition, they have fine control of the other muscles in the skin and are able to vary the texture of the skin as well as the colour, from smooth to bumpy or even spiked.

Cuttlefish can change the colour and texture of their skin in a fraction of a second. Chameleons and flounders, other animals considered masters of camouflage, may take hours to change their colour. Cuttlefish can even change their colour in waves running across their bodies. These are dark bars passing over a light background on the skin. These 'passing clouds' wash over the whole body in less than a second. They are probably used as signals to other cuttlefish, although they have not been studied in detail. Passing clouds have been described in octopuses as possibly being used to startle a motionless crab into movement so that it could be seen.

Cuttlefish are relatively easy to maintain in captivity. Institutions such as the National Resource Center for Cephalopods, in Galveston, Texas, have kept many successive generations of them and have supplied them to aquariums and research laboratories much as white rats are. Although about ten species have been kept in captivity, we know little about their behaviour in the wild and almost nothing about the other species. Cuttlefish mate head to head, when the male passes a sperm packet into the female. She stores the sperm until she is ready to lay eggs, usually in coral crevices or other hard substrate. The eggs are attached individually in a cluster adjacent to each other. The eggs are roundish, typically about 1 cm across, with a distinctive point at the end away from the attachment. Cuttlefish frequently incorporate ink into the eggs, making them dark or black. Hatchlings are about 1 cm long after several months of direct development inside the egg; there are no larval stages as in other molluscs. They begin feeding immediately on live food, crustaceans and fish. They look like miniature adults and can replicate most of the behaviours of the adults, like inking, changing colour and even burying in the sand to hide from predators. For burying they sit on the bottom, blow sand back and forth with their water jet funnel, allowing it to settle down upon themselves, then wiggle from side to side to 'settle' the sand.

Cuttlefish and most other coleoids have an ink gland that produces a dark brown or black ink, which they shoot out their funnels to confuse predators. They can incorporate thick mucus into the ink to create a thick ink blob, resembling the animal itself, or they can shoot it into a cloud like a smokescreen to hide behind. Some cephalopods seem to be able to shape the ink blob just as smokers can blow a smoke ring. The ink may also have a noxious chemical in it, so it confuses both the sight and smell of a predator.

Only one cuttlefish species has been studied in detail, the common European cuttlefish *Sepia officinalis*, and its behaviours have not been studied well in the wild, although some of its habits were noted millennia ago by Aristotle. There are about 100 other species. Aquariums and research laboratories are just beginning to concentrate on the husbandry of some of these species.

Squid

Like cuttlefish, the 250 species of squid have eight arms and two long tentacles for catching prey, mostly fish. They have a fin on each side of the body used for slow swimming and manoeuvring, but their main propulsion is jetting by taking water into the body cavity and squeezing it forcefully out of the funnel in a water jet. The funnel is muscular and can be directed in all directions, so the squid can jet forward, backward and even sideways, although their fastest swimming direction is backward. When the fins are folded against the sides their torpedo-shaped bodies slide aqua-dynamically through the water, thus escaping from predators.

Squid also use schooling to avoid being eaten by predators, clustering together in groups of hundreds or thousands. Schools of shallow-water squid seem to disappear when a predator such as a bar jack or tarpon approaches – as a group, the squid turn a pale shade and jet away, frequently leaving ink blobs that resemble squid in their place to further confuse the predator. Since they school together, squid are the most social of the cephalopods but the behaviours have only been studied for a few species. Some squid get together in enormous groups to mate and lay eggs. Males grasp females and their arms turn bright red while they transfer a sperm packet to the female. Females lay white, finger-shaped egg cases containing about 200 eggs. They lay their eggs next to or attached to each other on the bottom, so that it is quickly covered with thousands of the clustered egg cases. Then both males and females die, so there is no parental care of the eggs or the hatchlings.

Squid are surprisingly difficult to care for. Only a few species have been kept in captivity and even fewer have been raised from an egg. Being open-water animals they are not accustomed to walls, and collide with them when startled. The practicalities of keeping this interesting group have to be worked out before we can begin to study their behaviours in the laboratory. Although there are some shallow-water squid, some members live far deeper and have related adaptation, such as large eyes and photophores (light-producing organs), possibly for luring prey, signalling to each other, counter-shading or perhaps startling potential predators. Some use ammonia in their tissues for flotation since this is more buoyant than water. We also know little about the giant squid living in the depths. These can reach 18 m in length and weigh more than 200 kg, making them the largest invertebrate. Giant squid are the main prey of sperm whales, which are adapted to dive deep to catch the squid, proving that there must be plenty of them in the depths, yet we have yet to see a living giant squid.

A side-branch of the squid family – the bobtail squid – is now providing us with an opportunity to study cephalopod behaviour. Several species live in relatively shallow water, but the majority of the 50 species live on the bottom in deeper waters. Although commonly called squid, these sepiolids bear little resemblance to the true squid. Most sepiolids live on the bottom, where they bury themselves during the daytime and come out at night to feed or look for mates. They are weak swimmers and have a rounded, stubby body ill-suited for fast swimming. They keep their tentacles in a pouch except when used for catching food, mostly small shrimp. They are easily kept in aquariums and many have been reared in captivity from their large eggs. One species, the endemic Hawaiian dumpling squid, is associated with a commensal, light-emitting bacterium. It uses this light-producing ability on its underside to eliminate its silhouette from below when swimming – a form of countershading.

Once buried, a stubby squid may exhibit another interesting behaviour: it may poke the arm tip of one of its second pair of arms up through the sand and wiggle it. This has been described as an angling behaviour. Squid, octopus and cuttlefish have all been described using this behaviour, so it may be conserved across the subclass.

Octopus

Octopus fossils have been dated to as far back as 65 million YBP. Octopuses have just eight arms and no tentacles, hence their name. There is some controversy over the plural of octopus, with octopi, octopods, octopodes and octopuses all used. Shallow-water octopuses have colour-changing ability and use it for the same reasons as cuttlefish and squid. Although octopuses are known for eight-armed crawling, they can also swim, using their funnels for water jet propulsion. However, they cannot swim very far, since they go into oxygen debt while swimming. Usually, they only swim far enough to elude a predator and camouflage against the bottom.

Octopuses mate in two ways, depending on the circumstances. The male mounts the female or extends a modified arm from a distance into her body, passing a spermatophore into her. She can store sperm for several months and then uses it when she lays her typically rice-sized eggs. Eggs of most species hatch into tiny octopuses that live in the plankton for 1–2 months before settling to the bottom. Some octopuses have larger eggs that hatch out into miniature octopuses that immediately take up a benthic existence. Very few of the small-egged species have been reared from an egg.

Sub-adult and adult octopuses are fairly easy to keep in aquariums or research laboratories, with some cautionary provisions. Octopuses are able to escape their enclosures – as primates can. Their escape abilities have achieved legendary status, as many public aquariums can attest to. The common legend goes that an octopus crawled out of its tank, crawled into another tank and ate its occupants, then crawled home, leaving its keeper befuddled on discovering the missing prey. This urban legend is actually based on fact, as an octopus kept at the Brighton (UK) aquarium in the late 19th century crawled out of its tank at night to consume lumpfish kept in an adjacent tank.

Octopuses take well to standard aquarium conditions and water quality, with the exception of exposure to heavy metals, which are toxic. They are extremely susceptible to trace amounts of copper, which affects their behaviour. One copper penny in a 60-l tank was enough to affect one researcher's octopus. Octopuses can learn in target training, run mazes and even open jars to obtain food, but much is yet unknown about how they use their intelligence. Their remarkable ability has resulted in them being afforded the same level of protection as vertebrates in some national legislation. (RCA)

Chemical communication

Chemical communication is ubiquitous among many invertebrates – particularly insects – nearly all vertebrates, including some birds, and bacteria. Chemical communication probably evolved as animals were able to use digestive exudates and secretions from their integument to provide **'honest' signals** about the location and features of the identity of the donor to its audience. Chemical communication differs from other forms of **communication** in that the donor does not have to be present for its message to be read, and chemical signals can persist in the environment.

As many terrestrial animals move about their home ranges they are surrounded by scent marks – some are their own and some are those of conspecifics. Animals investigating these scent marks can often determine many features about the individual that deposited them such as its sex, age, reproductive condition, diet, etc. These scent marks provide a signpost or bulletin board for the transfer of information between individuals. After gathering such information about the signaller, an individual may respond to these scents by seeking the signaller, avoiding it or altering its own behaviour in some fashion.

One type of olfactory communication is **scent marking**. Scent marking is ubiquitous among terrestrial animals. A current model of scent marking suggests that scent marking serves multiple functions; some functions are unique to each species, whereas other functions serve a general purpose across species, indicating one's presence in an area. This model provides several hypotheses. One hypothesis is that scent marking provides a chemical résumé of the signaller, which may include information about features of its identity. Another hypothesis is that scent marking provides a chemical résumé that accurately reflects current features of the signaller. A prediction of this hypothesis is that scent marking provides 'cheat-proof' signals. A third hypothesis is that animals match the scent marks with their signallers and use this information to distinguish between familiar and unfamiliar individuals, residents and intruders, dominants and subordinates, reproductively active and reproductively quiescent, opposite-sex **conspecifics**, adults and juveniles, related and non-related conspecifics, and other possible relationships. The last hypothesis of the model is that scent marking is a sexually selected trait used by individuals during same-sex competition and mate choice. A prediction of this hypothesis is that scent marking provides benefits to the signallers and their audience that increases their fitness relative to those individuals that did not scent mark or respond to scent marks. That is, the benefits of providing olfactory information to conspecifics outweigh the costs of being detected more easily.

Recent work has identified two specialized forms of scent marking, over-marking and self-grooming. Over-marking occurs when animals deposit their scent marks on top of or in close proximity to the scent marks that were deposited by a conspecific. The bulk of the data suggest that over-marking is akin to a modified bulletin board. The bulletin board allows the top- and bottom-scent donors to provide information to conspecifics, but the transfer of information to conspecifics generally favours the top-scent donor of the over-mark. That is, after investigating an over-mark, individuals often spend more time investigating the mark of the top-scent donor than that of the bottom-scent donor when the two marks presented separately. Studies have reported that animals have the capacity to distinguish between the top-scent mark and the bottom-scent mark, including areas in which the positions of the top- and bottom-scent donors change. These studies show that individuals behave as if the top- and bottom-scent marks of an over-mark are distinct, but that the top-scent mark has somewhat greater intrinsic value relative to the bottom-scent mark.

Such findings led to many hypotheses that surround the potential advantages to the top-scent donor of an over-mark that are not enjoyed by the bottom-scent donor. The first hypothesis is that the top-scent mark 'behaviourally' masks the presence of the bottom-scent mark. That is, individuals exploring an over-mark later respond selectively toward the mark of the top-scent donor relative to that of the bottom-scent donor. The second hypothesis is that over-marking provides an advantage to the top-scent donor but not to the bottom-scent donor in the transfer of olfactory information to animals investigating an over-mark. The third hypothesis is that an over-mark is an 'honest' signal, which is based on studies showing that dominant males over-mark the scent marks of subordinate males but subordinate males rarely over-mark the scent marks of dominant males.

Another specialized form of scent marking is self-grooming. Many terrestrial mammals self-groom when they encounter odours of opposite-sex conspecifics, suggesting that self-grooming is involved in olfactory communication between the sexes. The self-grooming hypothesis states that individuals can, by self-grooming, alter features of the odours that they produce. Predictions of this hypothesis are that by self-grooming, individuals produce or release odours that are more attractive than those produced by individuals that do not self-groom. Such odours may reduce agonism between the groomer and opposite-sex conspecifics, and/or increase the odour field that surrounds the groomer, making it more likely that it will be detected by nearby opposite-sex conspecifics.

Studies on species of voles suggest a role for self-grooming in reproduction. Sexually receptive voles are more likely to groom in response to odours of potential mates, and males are especially prone to grooming when encountering the odours of sexually receptive females. Sexually mature male voles also self-groom more than do sexually immature male voles when they are exposed to the odours of sexually receptive opposite-sex conspecifics. In addition, self-grooming patterns are also consistent with the mating system of voles. For example, prairie voles, which tend to be monogamous, self-groom more often in response to odours of their mates than they do in response to odours of unfamiliar opposite-sex individuals, probably conveying a message about their willingness to copulate with their mates. In contrast, meadow voles, which are promiscuous, self-groom more often in response to odours of unfamiliar opposite-sex conspecifics than they do in response to odours of a previous mate, conveying a message about their willingness to mate with the former. Interestingly, meadow voles and prairie voles do not spend much time self-grooming in response to odours of their opposite-sex siblings, suggesting further that self-grooming may be a form of mate attraction. Overall, meadow and prairie voles adjust the amount of time they self-groom in response to particular opposite-sex conspecifics, which would increase the likelihood of communicating with potential mates.

Chemical communication is an integral component of both reproduction and competition, as well as other aspects of behavioural ecology. Thus, it is not surprising that researchers are beginning to apply knowledge of scent communication to the management of captive animals, particularly in **captive breeding programmes**. Breeding success can be improved by providing species-appropriate opportunities for chemical communication or manipulating chemical signals to maximize mating opportunities. Because conspecifics and their odours can also induce stress, judicious management of the olfactory environment can also alleviate welfare problems. (MHF)

See also: Pheromones

Further reading

Gosling, L.M. and Roberts, S.C. (2001) Scent marking in male mammals: cheat-proof signals to competitors and mates. *Advances in the Study of Behaviour* 30, 169–217.

Chewing

Chewing (mastication) commences the breakdown of food, lubricates it for swallowing, adds buffers and, in some species, salivary amylase to the ingesta. Chewing of bones as a postprandial activity in dogs seems to have some additional functions that relate to dental care. Chewing may also have a function in dealing with arousal, e.g. mouthing of articles (including toys and bones) is seen as a greeting behaviour in some domestic dogs. Chewing and licking movements are seen in horses during training and have been putatively associated with stress reduction and even attempts to signal to the trainer. However, this a source of some debate since it could also be that these movements simply reflect dryness of the mouth, perhaps as a result of recent adrenaline activity.

Opportunities to chew are frequently inadequate in intensive husbandry systems since bulky, high-fibre foods are unpopular because they require more labour to handle and create more waste after digestion than concentrated rations. Furthermore, concentrated foods are thought to improve productivity or performance by providing readily digested energy sources that can be consumed more rapidly than less energy-dense (more natural) forages. Fundamentally, diets associated with low rates of chewing are implicated in the development of oral stereotypic behaviours and so it is appropriate to emphasize the importance of chewing as a significant element in the normal **time budgets** of domestic animals.

The amount of chewing per bolus of food, and therefore the time taken before deglutition (swallowing), depends on a variety of factors relating to the physical properties of the food, including its moisture content and size. For example, when fed to horses, hay requires four times as many chews as oats (per unit of weight) and therefore takes four times as long to consume. It has been estimated that a horse foraging exclusively on high-fibre substrates may chew 57,000 times per day, but that this can be almost halved under moderately intensive stable conditions, e.g. by a 500 kg horse on 5 kg of forage and 7.8 kg of concentrate. This reduction in daily chewing has a profound impact on salivation which, in horses, occurs only in association with mastication and not in anticipation of a meal. So, the chief consequence of feeding less fibre is that there is less chewing and less saliva generated. Since the buffering effects of bicarbonate are diminished under these circumstances, gastric acidity rises and the risk of gastric ulceration increases. The addition of chaff, a traditional means of increasing the time taken to consume concentrated feeds, works simply by increasing the total forage content of the ration. Horses and farm animals on low-fibre diets often supplement chewing opportunities by foraging on straw that has been provided as bedding.

The amount of chewing per day also depends on food availability. For example, free-ranging horses spend an average of 16–17 h/day grazing but, when forage is scarce, they increase their bite frequency and extend the grazing period, sometimes beyond 19 h. Rates of chewing can be increased by reducing the displacement of the lower jaw. That said, the extent to which animals can thrive on poor forage is clearly limited, because they require more energy for ingestion itself. The effect of exercise on chewing is of interest since exercised animals take fewer, but larger, bites than non-exercised controls when grazing.

In all mammalian species, restrictions on feeding behaviour, and especially the provision of discrete concentrated meals that involve minimal chewing, lead to digestive anomalies and behavioural frustration. For instance, free-ranging pigs devote significant portions of their time budget to rooting and chewing. In contrast, intensively managed pigs, fed concentrated foods and given limited access to fibre, can consume their daily ration in less than 1 h and have no normal substrates on which to chew. Various oral **stereotypies** have been described in tethered sows when not provided with straw. Examples include sham chewing, chain chewing, nipple playing, bar rubbing and **bar biting**. Inadequate **foraging behaviour** opportunities are also recognized as potential causes of snout rubbing when pigs rub their snouts on the flanks of other pigs, a behaviour that causes necrosis of target areas.

The timing of discrete meals has an impact on the timing of chewing activity. Nocturnal ingestion naturally punctuates periods of drowsing and sleep and, for this reason, stable managers seeking to reduce the behavioural changes inherent to stabling should leave horses with sufficient forage to last the whole night. In horses, periods without chewing are related to increases in gastric acidity, which predisposes the animal to gastric ulceration.

There is some evidence that chewing by mice may provide some help in coping with adverse stressors and, indeed, hippocampal development (Watanabe *et al.*, 2002). Chewing involves molar activity whereas gnawing involves the incisor teeth. Gnawing in laboratory mice can become stereotypic but is often labelled bar biting. It has been linked to escape attempts.

Rumination (the circulation of ingested material through the chambers of the stomach, interspersed by regurgitation and further chewing) brings with it a second opportunity to chew fibrous ingesta and can be accomplished in sternal recumbency, so it is an economical means of increasing the energy intake at pasture. Horses devote less time to chewing than ruminants but have to compensate for this by ingesting more food per unit of bodyweight per day. Inability to chew properly as a result of dental problems is one of the chief causes of weight loss in older herbivores and constitutes a **welfare** problem, especially if mastication becomes painful. Often referred to as broken mouth in aged sheep, this condition is sometimes treated by rasping the teeth, a procedure that is criticized since the evidence for its efficacy is inadequate and there is potential for postoperative chronic pain.

Chewing is sometimes presented as a problem in **companion animal** behavioural practice. Dogs, especially during puppyhood, readily chew organic substrates. Sometimes this is a natural response to teething and the solution is simply to provide an appropriate target for that behavioural need. Others chew items when left alone as part of a variety of responses to **separation anxiety**. For example, dogs may chew seat belts when left in cars, or may chew wooden doors and their frames when left inside a house. The latter case could be interpreted as an attempt to follow the owner, since the door targeted is often that through which the owner was seen to depart. Clearly, it would be unwise to treat symptomatically dogs that chew when left, without addressing the problem of their inability to cope in the absence of the primary caregiver(s).

Fur chewing and fur sucking are reported as unwelcome ingestive responses in cats (labelled **pica**). Most commonly seen in exotic breeds, notably Burmese and Siamese, fur chewing is poorly understood but it has been suggested that it may relate to generalized distress, low-fibre diets or even redirected suckling. (PDM)

See also: Feeding

Reference

Watanabe, K., Ozono, S., Nishiyama, K., Saito, S., Tonosaki, K., Fujita, M. and Onozuka, M. (2002) The molarless condition in aged SAMP8 mice attenuates hippocampal Fos induction linked to water maze performance. *Behaviour Brain Research* 128, 19–25.

Chicken

Domestic poultry of the species *Gallus gallus* are known by several popular names. ‘Chicken’ is a generalized term, similar in usage to ‘cattle’, that may refer to one or a group of adult birds of either sex and to egg layers or meat birds. More often the word chicken is associated with specially bred strains of birds that are termed broiler chicken(s) or just broilers and are used for meat production. The word hen refers specifically to mature females that lay eggs. Cockerel (or cock) frequently is used for mature males but occasionally can be used to describe young birds, for example an all-male flock of broilers. Young female birds are termed pullets when they are intended for egg production: the term is seldom used to describe female broilers. For the first few weeks after hatching young birds are known as chicks, and according to their destiny may additionally be described as either as layer chicks or broiler chicks.

All chicken are descended from the red junglefowl native to the tropical forests of South-east Asia. Junglefowl are omnivorous and spend much of their daylight hours in small groups of typically six to ten adults, seeking food, scratching and pecking in clearings, showing preference for insects and worms but also eating seeds and grasses. They are good flyers and at night roost high in trees to avoid predators. Females lay eggs in clutches of about 12 and then brood them, taking care to select a suitable nest site and to create a nest hollow with materials such as dried grass and a few feathers. Their characteristic pre- and post-laying behaviour routines are recognizable in current commercial hybrid hens, even in the absence of a nest site or suitable nesting materials. Despite hundreds of years of domestication and very intensive selection since the late 1950s, modern chicken retain many other instinctive behaviour patterns of their junglefowl ancestors, such as **dust bathing**, which is important for maintaining the feathers in good condition.

Apart from farmed fish, chicken are the most numerous, intensively bred and raised species of all farmed animals. Genetic diversity is low among the majority of birds that are commercially farmed, with only a few strains of egg layers or of broilers to be found in significant numbers worldwide. Relatively minute numbers of *G. gallus* remain genetically diverse as backyard and show stock. Commercial flocks are intensively selected hybrids with phenotypic uniformity. Broilers have been selected principally for rapid growth and high feed conversion efficiency, and to have white feathers and relatively large breasts. Layers have been selected to lay continuously rather than in batches and to lay large numbers of eggs, also with a high feed conversion efficiency. Broodiness has been selected against. According to consumer preference for eggshell colour, both white- and brown-feathered strains of hen are common and these lay white or brown eggs, respectively.

The sheer numbers of chicken needed to satisfy the growing demands of an expanding human population of approaching 7 billion is astonishing: over 50 billion broiler chickens are reared per year and some 54 billion t of eggs are produced worldwide. The world population of chickens is estimated by the Food and Agriculture Organization (FAO) to be about 16 billion at any one time, and this includes the large number of breeding birds needed to produce the broilers and layers.

Chicks

There is concern about the exposure of newly hatched chicks to the toxic sanitizing chemical formaldehyde routinely used in hatchers. Although chicks may be handled automatically on a series of conveyors, separators and counters, generally they are not physically damaged. The method used to kill unwanted male chicks of layer strains varies. In practice, maceration is humane and cost effective despite appearing unattractive. Operators routinely drop chicks into a bin filled with carbon dioxide, which is a heavy, acidic gas. This is a slower method of killing chicks that they find more aversive, but enables whole birds to be sold as feed for birds of prey and other markets. Chicks have yolk sac reserves to sustain them, but performance is reduced if they are transported for long periods or do not feed quickly on arrival at the farm. They need to be transported in environmentally controlled conditions as they do not have well-developed body temperature control systems.

Layers

Chicks are normally reared by specialist pullet farmers to 'point of lay' (POL) at 18 weeks of age, when they typically weigh 1.4 kg. They are then sold and transported to separate laying accommodation. Floor rearing tends to predominate, but chicks may be reared from 1 day old in cages. Pullets receive several vaccinations during rearing. Research has shown that it is very important that pullets experience features such as perches and nestboxes during rearing if they are to successfully use them in their laying house. Lighting is crucial as it influences age of maturity as well as egg size and numbers. Chicks have up to 24 h of light for the first few days to enable them to locate their food and drink easily and to feed frequently. Many lighting patterns are used but, in general, the number of hours of light per 24 h is first reduced and then increased from about week 16 to stimulate early maturity and ovulation (a step-down–step-up programme). Both during rearing and especially in lay, practical experience has found that low light intensity (< 5–10 lux) can reduce **feather-pecking** and **cannibalism**. More recent research has shown that poultry may prefer brighter light for behaviours such as feeding and that ultraviolet light is important for bird recognition and mate selection. Normal tungsten and fluorescent lights do not provide ultraviolet wavelengths.

From a **welfare** perspective there tends to be some trade-off between meeting the health and the behavioural needs of layers. The prevalent housing system is the wire battery cage, each holding three to five hens and arranged in several tiers. Compound feed such as pellets or mash is provided continuously in a trough outside the front of the cage, and water is accessed via nipple drinkers in the side of the cage. Abrasion of neck feathers from accessing feed is common. The battery cage system has been criticized for inadequate space allowance and for the inability of birds to perform behaviours that experiments have shown to be important to them, such as nesting and dust bathing (see Baxter, 1994). Most current strains of layers are prone to social stress leading to feather pecking and cannibalism. In an effort to control these problems, birds are routinely beak trimmed. Such problems can be worse in systems that allow birds more space and to be kept in larger groups than can establish a stable 'peck order', which is thought to be about 100 birds. Infestation with parasites such as coccidia and red mites is much more likely in non-cage systems that enable an extended behavioural repertoire.

By selecting for improved feed conversion efficiency, the size of hens has been reduced, with adults weighing on average 1.7–2.0 kg depending on strain. There has also been selection for high egg output, and these factors increase the risk of osteoporosis. In non-cage systems over half the population may sustain bone damage and fractures during lay (Wilkins *et al.*, 2004). At the end of lay, osteoporosis renders hens more susceptible to bone fractures, with possibly one in four from battery cages being affected during depopulation. Caged birds suffer disuse osteoporosis from insufficient weight-bearing and wing-flapping opportunities. Modified (or furnished) cages allow more space for exercise and provide a perch, which improves bone strength (**see: Housing**). Egg production peaks at 90% or more about 3 months after POL and slowly declines thereafter, although egg size tends to increase with age and can be manipulated by altering air temperature (and light). Most hens are slaughtered on economic grounds at 70–72 weeks of age because of the decline in production to about 50%. This compares with a lifespan averaging 7 years in backyard flocks. Some countries permit **forced moulting** by withdrawal of all feed (and sometimes water restriction), which stimulates egg production for a further 10 months once birds resume feeding. Feed withdrawal is a welfare issue, likely to be **aversive** and stressful to the hens.

Broilers

The overwhelming majority of broilers are reared from day-old chicks to **slaughter** in a single house on deep litter, which is based on locally available materials such as wood shavings, whole or chopped straw or rice hulls. Even in hot climates, extra heat is provided for young chicks and this is commonly via overhead gas brooders at several points in the house. Powerful single gas heaters can be used to heat the whole house if it is fully enclosed. Chicks usually develop feathers by 4 weeks of age, when controlled-environment house temperatures are then reduced and **free-range** birds given access to range. Water is usually provided automatically via lines of bell or nipple drinkers. Feeders are hoppers filled manually or via auger – or a trough that runs right around the house with a chain inside to distribute feed. A pelleted, compound ration that is generally based on wheat and soya is used with three formulations – chick starter crumbs, a grower ration that often contains a coccidiostat and a finisher ration for about the last 2 weeks to slaughter. Most broilers are slaughtered at 1.5–2.0 kg at 32–39 days of age. Lighter birds are preferred in a few countries and heavier birds up to 3 kg (7 weeks) are grown for catering outlets and for portioning. Male birds are heavier than females at the same age. Some producers rear single-sex flocks but many are mixed-sex ('as hatched').

The prevalent practice of *ad libitum* feeding of concentrated diets that are high in protein and energy reduces bird viability, and the principal welfare issues in broiler chickens arise from excessive growth rate. Intensive genetic selection has resulted in a proportion of birds within the population that are unable physiologically and metabolically to sustain such rapid weight gain. Examples of metabolic disorders include ascites and heart attacks (flip-overs). Most broilers suffer leg problems to a greater or lesser degree and these essentially relate to the inability of the immature musculoskeletal system to support their weight. With the withdrawal of several antibiotics that were previously in routine use, **lameness** due to infective agents is now more prevalent and may be the principal cause of lameness in some flocks. Overall, however musculoskeletal disorders predominate. There is evidence that many leg disorders are painful (see Mench, 2004). Many of the conditions have heritabilities between 0.1 and 0.4% and can be selected against.

Breeding companies have selectively reduced the incidence of tibial dyschondroplasia (TD), which is an abnormality of the growth plate in the long leg bone. They have been able to do this by using a lixiscope, which is a portable device that X-rays the legs of potential parent birds to show whether they themselves have the condition. Poor gait is highly correlated with body weight (Kestin *et al.*, 2001), yet it may be possible, by careful multi-trait selection, to improve leg health with little impact on growth rate.

The incidence of lameness is increased by prevalent husbandry practices such as long light periods and access to feed *ad libitum*, because these do not allow sufficient rest, discourage bouts of activity and encourage excessive eating and hence rapid growth. Manipulating the growth curve by restricting early rates of weight gain and allowing catch-up growth by slaughter age can have welfare and economic benefits. Studies have shown that increased food conversion efficiency resulted in a decreased incidence of lameness and growth-related diseases such as ascites and flip-overs. Lameness tends to increase with increasing stocking density, and this is thought to be due in part to restricted ease of movement and in part to deterioration in environmental conditions (e.g. air and litter quality).

The most widely used method of assessing lameness is gait scoring, which visually appraises the **gait** and assigns a score to the broiler's ability to walk. Lame birds spend less time walking and dust bathing and more time lying than sound birds. They also perform most of their **preening** and idling when lying and alter their feeding strategy to make fewer but longer visits to the feeder. Broilers are substantially less active than layer or relatively unselected strains of chicken. By selecting for increased food conversion and high growth weight efficiency, it is inevitable that 'lazier' animals are favoured.

Recent estimates of the incidence of substantial lameness (i.e. broilers with a gait score of 3 or higher) in broilers from European surveys range between 2.5 and 30.1%. There is no reason to suppose that these figures cannot be globally applied, which means that between 1.25 and 30 billion broilers suffer with lameness every year. There is a demonstrable need for the incidence of lameness, morbidity and mortality to be monitored in all countries that produce broilers, as economically sustainable losses do not equate with values acceptable for good welfare. Figures from the USA indicate that lameness costs both industry and birds, with leg problems accounting for 1.1% of deaths in production and 2.1% of downgrades and carcass condemnations (Morris, 1993, cited in Mench, 2004).

Growth rate may be reduced by husbandry manipulations such as feeding a less dense diet, meal feeding or reducing hours of light. On the whole a single, long dark period of 6–8 h confers more benefit than short, intermittent light and dark periods. Cannibalism is rare in broilers on a good diet and there seems no justification for maintaining dim lighting below 10 lux – over 20 lux is the recommended level.

Broiler breeders

There are five generations of selection and multiplication involved in the production of broilers. The parents of broilers need to produce a reasonable number of fertile eggs for hatching and thus to live for 1 year or more. Obesity dramatically reduces the fertility of both males and females, as well as reducing their lifespan. All generations of breeders have therefore to be severely feed restricted to prevent obesity, and this restriction leads to **frustration**, abnormal behaviour and other welfare issues (see Hocking, 2004).

Handling and transport

Catching and handling by humans is an aversive experience for chickens, which have minimal contact with humans during normal commercial rearing. All chicks are handled at the hatchery, where most handling is automated and they are then transported to rearing sites. Limited experimental evidence indicates that, provided **transport** is of short duration in thermally comfortable conditions, chicks do not experience much stress in transit. Pullets are transferred to the laying house. Not only are they valuable, but they also are metabolically very robust with good feather cover, and hence are resilient to any catching and transportation stressors. End-of-lay hens are metabolically exhausted, osteoporotic and hence susceptible to both cold stress and bone fractures. Broilers have the greatest heat output owing to their high growth rate and good feather cover. They are susceptible to heat stress during transit and **lairage**. Experimental evidence that birds find the whole transportation experience stressful and fearful is indicated by long periods in **tonic immobility** (TI) at the slaughter plant. For more details see Weeks (2007).

(CW)

References and further reading

Baxter, M.R. (1994) The welfare problems of laying hens in battery cages. *Veterinary Record* 134, 614–619.

Hocking, P.M. (2004) Measuring and auditing the welfare of broiler breeders. In: Weeks, C.A. and Butterworth, A. (eds) *Measuring and Auditing Broiler Welfare*. CAB International, Wallingford, UK, pp. 19–35.

Kestin, S.C., Gordon, S., Su, G. and Sorensen, P. (2001) Relationships in broiler chickens between lameness, liveweight, growth rate and age. *Veterinary Record* 148, 195–197.

Knowles, T.G. and Wilkins, L.J. (1998) The problem of broken bones during the handling of laying hens – a review. *Poultry Science* 77, 1798–1802.

Mench, J. (2004) Lameness. In: Weeks, C.A. and Butterworth, A. (eds) *Measuring and Auditing Broiler Welfare*. CAB International, Wallingford, UK, pp. 1–17.

Weeks, C.A. (2007) Poultry handling and transport. In: Grandin, T. (ed.) *Livestock Handling and Transport*, 3rd edn. CAB International, Wallingford, UK, pp. 295–311.

Wilkins, L.J., Brown, S.N., Zimmerman, P.H., Leeb, T. and Nicol, C.J. (2004) Prevalence of keel and furculum damage in laying hens: validation of a palpation method. *Veterinary Record* 155, 547–550.

Chimpanzee

The chimpanzee (*Pan troglodytes*) is one of the **great apes**, a category that also includes the **gorilla**, **orang-utan** and bonobo. There are more chimpanzees in captivity than any of the other great apes. Four subspecies of chimpanzee live in equatorial Africa in a variety of habitats, including subtropical forest, savannah and open grasslands. Captive chimpanzees live in **zoos**, biomedical research facilities, sanctuaries, and some are privately owned. The behaviour and **welfare** of chimpanzees have been fairly extensively studied. Wild chimpanzees are considered possibly endangered, but certainly vulnerable on the IUCN Red List, with major threats from habitat destruction and hunting for human consumption (the '**bushmeat**' trade). Captive populations are in better condition. For example, the chimpanzee population in the zoo community in the USA is considered self-sustaining.

Wild chimpanzees divide their time between terrestrial and arboreal life, generally moving quadrupedally on the ground, and brachiating and leaping through the tree canopy. They spend about 20% of their waking hours travelling, and generally build elevated nests nightly in the trees. Modern captive chimpanzee housing emphasizes spacious and complex outdoor enclosures, and extensive use of environmental enrichment techniques. Large yards with grass, trees and other foliage, and/or artificial structures encourage physical activity by providing opportunities to brachiate, swing, climb and locomote on the ground. Arboreal sleeping platforms and hammocks are favoured areas, particularly if the chimpanzees have materials for nest building. Chimpanzees should be given a variety of manipulable objects, and they will especially use those that are novel or destructible.

The social organization of wild chimpanzees comprises large multi-male, multi-female groups, with members of all ages interacting in a loosely structured community of 40 to 90 animals or more. Adult males are the social core of the group, and typically (although not always) females transfer out of their natal groups at maturity. Within the large community, there is a fission–fusion structure with smaller parties of chimpanzees, averaging four to eight members, travelling and feeding together, and later reuniting with other members of the larger community and forming new parties. Over the last three decades, more captive chimpanzees have been housed in large **social groups**, some with over 20 animals, which begin to mimic the species-typical group structure. Living in compatible social groups is the most critical feature in promoting the welfare of captive chimpanzees. Wild chimpanzees are aggressive to chimpanzees outside of their communities, and sometimes fight with them to the death. **Aggression** is a major issue in managing captive chimpanzees. In a very few cases they have killed one another, but less severe wounding is also of concern. Aggression is especially prevalent during the introduction of unfamiliar chimpanzees and among adult males, although affiliative behaviours typically predominate.

Chimpanzees are long lived, with some individuals living beyond 40 years in the wild and beyond 60 years in captivity. Females typically give birth to single infants after a gestation of about 8 months and have interbirth intervals of several years. The developmental period for young chimpanzees is long – infants nurse and are cared for by their mothers for several years, and are 10–15 years old or more before reaching physical and social maturity. Maternal competence is sometimes a problem, particularly among captive females that were not raised by their mothers and have not lived with infants.

Mother rearing or other forms of early **socialization** are especially important in facilitating normal behavioural development. It is preferable to allow mother rearing of infants for several years but, if this is not possible, infants can be raised by humans in a nursery. To normalize the development of nursery-reared infants, they should live with other young chimpanzees and they should be introduced to older chimpanzees later in their lives. If infants are raised without other chimpanzees, their behaviour can be disrupted throughout their long lives, including deficiencies in sexual behaviour, maternal behaviour and other social skills, as well as expressing abnormal and stereotypic behaviour. Individual housing of chimpanzees at any age usually leads to behavioural problems and should be avoided if at all possible.

Beyond social housing, many forms of environmental **enrichment** are effective in promoting the welfare of captive chimpanzees. Some enrichment techniques are designed to elicit behaviours characteristic of wild chimpanzees, and feeding enrichment is an example of this. The diet of wild chimpanzees includes a good deal of fruit, and they also eat a wide variety of leaves, pith, seeds, insects, eggs and meat. Foraging and eating account for the majority of their activity during daylight hours. Tool use is a hallmark feature of this species, and tools are often used in food gathering and processing. Attending to feeding strategies is important in maintaining the well-being of captive chimpanzees. Feeding routines should stimulate long foraging bouts, with multiple meals daily. A variety of foods should be fed that require processing and tool use. Other cognitive challenges can be presented since a variety of food puzzles and devices have been developed for chimpanzees.

Chimpanzees use different forms of sensory enrichment, including watching videotapes and television, looking into mirrors and listening to music. The highly developed cognitive abilities of chimpanzees are well documented, and these abilities should be stimulated by setting problems for chimpanzees to solve. One way to accomplish this is by offering computer tasks, which can be made more challenging over time. Some forms of enrichment give chimpanzees the ability to control their environment or to indicate choices. Choice and control are believed to be important in promoting psychological well-being of captive animals.

Captive chimpanzees' behaviour can be improved by positive interactions with humans. This is particularly important for chimpanzees living alone. Animal training techniques can be used to gain the voluntary cooperation of chimpanzees with routine husbandry activities (e.g. to move within their enclosures on cue), veterinary procedures (e.g. to examine body parts for wounds or evidence of ill health) and research procedures (e.g. to collect blood or urine samples). Training can address social problems among chimpanzees, such as reducing aggression. Training increases the ease and efficiency in working with chimpanzees, reduces animal distress and is cognitively stimulating. However, the presence and activity of humans can also be a negative influence on chimpanzees. For example, wounding may be higher when there is more human activity around the chimpanzees. (MAB)

Further reading

Brent, L. (2001) *The Care and Management of Captive Chimpanzees*. American Society of Primatologists, San Antonio, Texas.

IUCN (2004) *IUCN Red List of Threatened Species*. Available at: http://www.iucnredlist.org

Choice test

A choice test is an experiment in which a subject is required to choose between concurrently available resources or events. Choice tests are used to investigate an animal's **motivations** and **preferences**. Here, a 'resource' is defined as a commodity, or an opportunity to perform a behaviour. An 'event' is defined as either a stimulus or an experience such as a husbandry procedure. The term 'preference' denotes a difference between the strength of motivation to obtain or avoid one resource or event and the strength of motivation to obtain or avoid another, but also implies that the resources or events are alternatives.

To choose a resource or event, the animal is typically required to move into a chamber in which the resource is located or the event occurs. Since approach is an operant response, choice tests are really a subset of **operant tests**. Hence, the resources and events that support choice behaviour can be termed reinforcers, or **reinforcement**, defined broadly as outcomes that motivate animals to perform responses.

Research questions addressed by choice tests include:

(A) To ascertain whether an animal is motivated to obtain or avoid a resource or event, i.e. to ascertain whether it is a reinforcer. We might ask: 'Is bedding material wanted?' or 'Is a particular handling technique aversive?'.
(B) To ascertain whether an animal has a preference among alternative reinforcers. Alternatives satisfy the same positive motivation, or contribute to a common negative motivation. Hence, we might ask: 'Which bedding material is more comfortable to lie on?' or 'Which handling technique is less aversive?'.
(C) To measure the strength of an animal's motivation for a reinforcer, or the strength of its preference between two reinforcers, either in absolute terms or relative to the strength of its motivation for a comparator. When using a relative measure, it is possible to ascertain whether the strength of motivation or preference for the reinforcer of interest is high or low, provided that the comparator is a reinforcer: (i) whose consumption does not influence satiety for the reinforcer of interest, i.e. not a substitute or a complement (this applies primarily to positive reinforcers since it is often assumed that negative reinforcers do not interact); and (ii) that has a known value (e.g. food, electric shock). We might ask: 'How important is it that the animal has bedding?' or 'How strong is the preference for one bedding material over another?' or 'How aversive is a particular handling technique?'.

Choice tests designed to ascertain whether a motivation exists (A) operate by giving the subject a choice between the hypothesized reinforcer and a control (no reinforcer). They can distinguish positive reinforcers (in which case the reinforcer is chosen) from negative reinforcers (the control is chosen).

Tests designed to ascertain whether a preference exists (B) operate by giving the subject a choice between two or more alternative reinforcers. They work on the principle that an animal will interact more with a reinforcer it prefers. The choice may be between several positive reinforcers, several negative reinforcers or between positive and negative reinforcers. These tests cannot distinguish positive from negative reinforcers.

In practice, these two types of procedure (A and B) are frequently combined, giving the subject a choice between several reinforcers and a control. For positive reinforcers, the distinction between these procedures may be somewhat arbitrary, e.g. in a choice between a bare concrete floor and a floor with straw bedding, the bare concrete floor would be a control if 'bedding' was being evaluated, but an alternative reinforcer if 'floor substrates' were being evaluated. For negative reinforcers, the distinction is more robust.

Choice tests designed to measure the strength of motivation or preference (C) are based upon, or similar to, those designed to ascertain whether a preference exists. For positive reinforcers, one important difference is that the reinforcer of interest is compared with a reinforcer that the animal has an independent motivation to obtain or avoid. For negative reinforcers, there is no clear distinction between procedures B and C. Strength of preference (between the reinforcer of interest and an alternative) is assessed if the animal has access to an alternative during the test (in the same location as the comparator), whereas strength of motivation is assessed if the animal has no access to an alternative. These tests work on the principle that the stronger an animal's motivation to interact with one reinforcer compared with another: (i) the more it should interact with one compared with the other; or (ii) the more it should be prepared to give up or work for one compared with the other. Approach (i) is procedurally similar to tests designed to ascertain whether a preference exists (B); however, the validity of this approach is limited when comparing positive reinforcers that satisfy different motivations. Approach (ii) is more complex and has been developed to overcome these limitations.

Specific procedures

The procedures listed below are those that have been used by ethologists. The letters and numbers in parentheses indicate the research questions that have been, or could be, addressed by each procedure. The 'index' is the rule used to infer motivation from behaviour.

1. Discrete choice (A, B, Ci/ii). The subject is given one or a series of discrete choices between two or more locations (e.g. chambers, arms of a **T-maze** or Y-maze) containing either a reinforcer of interest or a comparator, or no reinforcer (control). *Index*: if the site of the hypothesized reinforcer is chosen more/fewer times than the control location, or by more/fewer subjects, then a motivation to obtain/avoid the reinforcer exists (A); if one location is chosen more/less often, or by more/fewer subjects than another, then motivation to obtain/avoid the reinforcer in the former is stronger (B, Ci). If subjects receive only one test trial each, or widely spaced trials are used (ensuring that only one option is chosen), then this procedure can ascertain whether subjects are prepared to 'spend' the comparator on the reinforcer of interest (Cii).
2. Discrete choice, varying the quantity of the comparator (Ci/ii). The subject is given a series of discrete choices between one location containing the reinforcer of interest + quantity X of a positively valued comparator (e.g. food) and another location containing no reinforcer of interest + quantity Y of the comparator (X or Y may be zero). The ratio X:Y may be varied between trials. *Index*: if the location containing the reinforcer of interest is chosen more/less often, then motivation to obtain/avoid this reinforcer is stronger than the motivation to obtain the quantity, Y–X or X–Y, of the comparator (Ci). If widely spaced trials are used and X:Y is varied between trials, this procedure estimates how much of the comparator the subject is prepared to spend on the reinforcer of interest (Cii).

Limitations: procedures C1 and C2. When comparing positive reinforcers that satisfy different motivations, the following limitations apply. (i) Non-exclusive choices (where subjects allocate some choices to reinforcer X and some to reinforcer Y) may be difficult to interpret when multiple choices are permitted in a short time period (massed trials), since motivational strength may be confounded with the number of bouts required to reach satiation. *Solution*: use widely spaced trials. (ii) Because sessions tend to be short, this procedure is particularly susceptible to bias caused by fluctuating internal/external variables. *Solution*: see General limitations, below.

3. Free choice: allocation of time (A, B, Ci). The subject is allowed to allocate a fixed amount of time between two or more chambers, containing either a reinforcer of interest or a comparator, or no reinforcer (control). *Index*: if more/less time is spent in the chamber containing the hypothesized reinforcer than in the control chamber, then the subject is motivated to obtain/avoid the reinforcer (A); if more/less time is spent in one chamber than another, then motivation to obtain/avoid the reinforcer if the former is stronger (B, C).

Limitation: procedure C3. When comparing positive reinforcers that satisfy different motivations, motivational strength is confounded with the time required to reach satiation. *Solution*: none.

4. Free choice: change in allocation of time, negative reinforcers (A, B/Ci). The subject is initially allowed to allocate a fixed amount of time between two chambers that may both contain no reinforcer, or that may both contain a standard positive reinforcer to encourage entry. The procedure is then repeated one or more times with a reinforcer of interest added to one chamber. *Index*: if the time spent in this chamber declines then the subject is motivated to avoid the hypothesized reinforcer (A); if the time spent in this chamber declines more for reinforcer X than for reinforcer Y, then reinforcer X is more aversive than reinforcer Y (B/C).

5. Free choice: allocation of time between a negative reinforcer and a positive comparator, both continuous properties of the environment, e.g. air quality versus temperature (Cii). The subject is allowed to allocate a fixed amount of time between one chamber containing the reinforcer of interest + the comparator and another chamber containing no reinforcer of interest + no comparator. *Index*: if less time is spent in the chamber containing the reinforcer of interest, then motivation to avoid this reinforcer is stronger than motivation to obtain the comparator.

6. Free choice: comparison of consumption, positive reinforcers (B). The subject is placed in a chamber containing two or more reinforcers (typically foodstuffs or fluids) for a session of fixed duration. *Index*: the reinforcer that is consumed in the greatest quantity is preferred.

Limitations: procedures B1–B6. Non-exclusive choices (where subjects allocate some time or choices to reinforcer X and some to reinforcer Y) may be difficult to interpret, particularly for positive reinforcers. A 90:10 choice might indicate that reinforcer X is preferred, if the 10% choice is merely due to the subject monitoring the alternative environment; and a 50:50 choice might indicate indifference between reinforcer X and reinforcer Y. However, two alternative explanations must be ruled out, as follows. (i) Fluctuation in an internal or external variable may cause preference to change during or between test sessions (see General limitations, below). (ii) Alternative positive reinforcers may have different uses, so the subject is in fact making a trade-off between two preferences, e.g. between a high-energy feed and a high-protein feed, or between a large cage that is preferred for exercise and a small cage that is preferred for nesting. A preference can only be established among reinforcers that differ along a single motivational dimension (Nicol, 1997). In order to test whether this is the case, the consequences of consuming/interacting with the alternatives should be compared, e.g. the range of behaviours exhibited, to see whether they have different uses. If they do, the resources should instead be compared using a measure of their substitutability, to test how well one substitutes for another with respect to a particular use.

7. Price elasticity of demand, or slope of the demand curve, cost imposed on access, positive reinforcers (B, Cii). The subject is placed in an apparatus consisting of two or more compartments, each containing either a reinforcer of interest, or a comparator, for a session of fixed duration. There are two costs associated with occupying one compartment: the reduced time that remains to be spent in other compartments and a cost imposed on access to the compartment (an operant response, e.g. a narrow gap that must be squeezed through, or a weighted push-door). The access cost is increased across trials and this constitutes an increase in the price of the reinforcer of interest (note that if access costs to all compartments are increased uniformly, this in fact constitutes a reduction in income, not a change in price – see: **Economics of behaviour**). The number of visits to the reinforcer of interest is taken as a measure of demand and plotted against price, using linear regression, to obtain a demand curve (the amount of time spent with the reinforcer, or the quantity consumed, is sometimes taken as a measure of demand instead, but then cost and demand may not co-vary). *Index*: if the absolute slope of this curve, or the absolute slope of a bi-logarithmic plot of the curve (i.e. the price elasticity of demand), is less for one reinforcer than for another, then motivation to obtain the former is stronger. Alternatively, if the slope of the curve for the reinforcer of interest is not significantly different from zero, or if its price elasticity of demand is less than 1 ('inelastic demand'), then motivation for this reinforcer is strong.

8. Price elasticity of demand, or slope of the demand curve, cost imposed on consumption, positive reinforcers (B, Cii). The subject is placed in an apparatus similar to procedure 7 for a session of fixed duration. The cost of spending time in one compartment is the reduced time that remains to be spent in other compartments. The cost of interacting with or consuming the reinforcer of interest may be increased in several ways: (i) by increasing procurement time of the reinforcer of interest (e.g. by burying it in substrate); (ii) by increasing the motivational importance of alternative reinforcers; or (iii) by subjecting the animal to an unpleasant stimulus (whose intensity may be increased across trials) while interacting with the reinforcer of interest. These manipulations constitute increases in the price of the reinforcer of interest. The time spent with this reinforcer ('expenditure') or the quantity consumed ('demand') is plotted against price, using linear regression, to obtain a demand curve (a plot of demand

against price) or an analogue of a demand curve (a plot of expenditure against price). *Index*: see C7 (a bi-logarithmic plot of expenditure against price is an analogue of the price elasticity of demand).

Limitations: procedures C7 and C8. (i) When income (the amount of time or energy the subject has to spend) is constant, price changes have income effects (Kirkden *et al.*, 2003). This problem can be ameliorated by ensuring that income is high. (ii) Empirical findings suggest that it is not valid to infer a strong motivation from inelastic demand, for two reasons. First, price elasticity of demand is influenced by procedural variables such as reward size and response type and, for a given reinforcer, it can vary widely between studies. Secondly, practically every reinforcer that has been evaluated in an operant test exhibits price inelastic demand. This is likely to be an artefact of the methodology of operant tests. Therefore, a comparator should be used to assess the relative, not absolute, strength of motivation.

9. Income elasticity of demand, or slope of the Engel curve (describing how demand changes as income varies), positive reinforcers (B, Cii). The subject is placed in an apparatus similar to procedure 7. The cost of interacting with all reinforcers is increased by reducing the session duration. The smaller the session duration, the more costly it is to spend a given amount of time interacting with one reinforcer, because less time remains in which to interact with the others. This manipulation constitutes a reduction in the subject's 'income'. The time spent with the reinforcer of interest ('expenditure') is plotted against income, using linear regression, to obtain an analogue of an Engel curve (which is a plot of demand against income). *Index*: see C7 (a bi-logarithmic plot of the curve is an analogue of the income elasticity of demand). *Limitation*: procedure C9. This procedure yields relative, not absolute, measures of motivation, since when income is limiting the sum of all elasticities must be zero.

Limitations: procedures C7 to C9. (i) Cost and reinforcer use may not co-vary (Mason *et al.*, 1998). For some reinforcers, the subject may be able to compensate for fewer visits (C7), or for reduced consumption/interaction time (C8, C9), by increasing consumption/interaction time on each visit (C7), or increasing the consumption/interaction rate (C8, C9). Motivational strength will then be confounded with ability to compensate. One solution that addresses the problem in C7 is to place a limit on consumption/interaction time, although this may devalue some reinforcers. A more general solution is to employ these procedures only with reinforcers and comparators for which little or no time (C7), or rate (C8, C9), compensation is exhibited. (ii) Motivational strength depends on deprivation level. When a single measure is extracted from a demand curve, this represents motivational strength at an average level of deprivation, which is of limited practical value (Kirkden *et al.*, 2003). There is no solution to this, although the consumer surplus could be used instead (**see: Operant test**). (iii) Motivational strength is confounded with the rate at which subjects satiate for a resource as they earn more (Kirkden *et al.*, 2003). Again there is no solution to this other than to consider using the consumer surplus instead. (iv) Motivational strength is confounded with the quantity consumed/interaction time when cost is low (Kirkden *et al.*, 2003). To solve this it is necessary to factor out this quantity/time in the analysis.

General limitations (all procedures)

1. Motivation/preference depends on context. It may be influenced by both external variables (e.g. ambient temperature, group size) and internal variables (e.g. reproductive status). An unchanging context may bias motivation/preference, while a fluctuating context may cause motivation/preference to vary during the experiment, possibly resulting in non-exclusive choices. It is therefore important to check for a correlation between preference and time or context. If this is an important factor, then it can be addressed by measuring motivation/preference separately at a range of different times/contexts.

2. Motivation/preference depends on experience. Animals tend, at least initially, to prefer environments that they are used to. Therefore subjects should have experience of all alternatives before testing, or else motivation/preference should be measured separately in subjects with a range of past experiences.

3. Motivation/preference for positive reinforcers depends on the availability of substitutes and complements, both during and between test sessions. Therefore, experimental reinforcers should not be available between sessions (a 'closed economy'), and the range of other reinforcers available in the test environment should be as similar as possible to the range available in the real-life environment, to ensure external validity.

4. Motivation/preference for positive reinforcers is influenced by the presence of cues emanating from/associated with test reinforcers at the location where choices are made (Warburton and Mason, 2003). Strength of motivation/preference measured in the test environment may therefore not reflect the strength of motivation/preference in the real-life environment. If the real-life environment lacks the reinforcer of interest, or the comparator, then cues should be reduced or eliminated and choice tests should be complemented by behavioural/physiological measures of frustration obtained in the absence of the reinforcer of interest.

5. Measures of motivation/preference do not reveal what is good for welfare in the long term, and so choice tests should be complemented by long-term welfare comparisons between environments that contain, or lack, the reinforcer of interest.

General advantages

1. Choice tests allow animals to express their own priorities, revealing what is important to them (Dawkins, 1990).

2. They are highly sensitive to differences between reinforcers (Rushen, 1990).

3. They can compare widely different reinforcers (Rushen, 1990).

4. They are better able to quantify motivational strength than measures of frustration, since it is unclear how much abnormal behaviour or physiological stress must be shown before motivation can be said to be strong (Dawkins, 1990).

(RDK)

See also: Aversion; Aversion learning; Avoidance behaviour; Behavioural elasticity; Conditioning – types of; Measuring behaviour; Operant test; Preference; Time budget; T-maze

References

Dawkins, M.S. (1990) From an animal's point of view: motivation, fitness, and animal welfare. *Behavioural and Brain Sciences* 13, 1–61.

Kirkden, R.D., Edwards, J.S.S. and Broom, D.M. (2003) A theoretical comparison of the consumer surplus and the elasticities of demand as measures of motivational strength. *Animal Behaviour* 65, 157–178.

Mason, G.J., McFarland, D. and Garner, J. (1998) A demanding task: using economic techniques to assess animal priorities. *Animal Behaviour* 55, 1071–1075.

Nicol, C.J. (1997) Environmental choices of farm animals. In: Forbes, J.M., Lawrence, T.L.J., Rodway, R.G. and Varley, M.A. (eds) *Animal Choices*. British Society for Animal Science, Penicuik, UK, pp. 35–43.

Rushen, J. (1990) Use of aversion-learning techniques to measure distress in sheep. *Applied Animal Behaviour Science* 28, 3–14.

Warburton, H. and Mason, G. (2003) Is out of sight, out of mind? The effects of resource cues on motivation in mink, *Mustela vison*. *Animal Behaviour* 65, 755–762.

Cholecystokinin

Cholecystokinin (CCK) is a hormone found within the **brain** and gastrointestinal system that has diverse roles in **sexual behaviour**, learning, **pain** and, most importantly, gastrointestinal (GI) function and feelings of satiety. Cholecystokinin originates as a preprohormone consisting of 115 amino acids, though following further processing can be found in many forms, including CCK-58, CCK-33 and CCK-8. The relative potency of each form varies with location and species (Reeve *et al.*, 1994). Typically, plasma CCK concentrations will rise within 15 min of a meal's initiation, peak after 30 min then fall to basal concentrations over 3–5 h.

Cholecystokinin is secreted by I-cells of the duodenum in response to the sensation of dietary fat and proteins. Positioning of I-cells allows for rapid detection of intestinal products, and subsequent CCK release into the bloodstream and surrounding tissues. Cholecystokinin affects GI function and secretory products through activation of enteric CCK receptors and stimulation of afferent vagal nerve fibres. The vagus nerve integrates a variety of other stimuli, i.e. other hormones and gut distension, and relays a signal of satiety to the hindbrain and forebrain (Moran and Schwartz, 1994). Cholecystokinin is believed to act synergistically with these other stimuli in generating feelings of satiety.

The effects of CCK on GI function include gallbladder contraction and subsequent release of bile, secretion of pancreatic enzymes and inhibition of gastric emptying.

Cholecystokinin will also have potent effects on feelings of satiety which, when strong enough, terminate feeding behaviour. Gibbs and Smith (1973) first demonstrated a reduction in the size and duration of a meal by exogenous CCK administration. Subsequent studies have confirmed this role, utilizing a variety of species and test procedures (Moran and Kinzig, 2004). Due to a relatively fast-acting nature and short duration of effect, CCK is believed to serve the primary role of ensuring the body does not overload itself nutritionally versus the induction of long-term satiation (Moran and Kinzig, 2004). (MT)

References

Gibbs, J. and Smith, G.P. (1973) Cholecystokinin decreases food intake in rats. *Journal of Comparative and Physiological Psychology* 84, 488–495.

Moran, T.H. and Kinzig, K.M. (2004) Gastrointestinal satiety signals II. Cholecystokinin. *American Journal of Physiology. Gastrointestinal and Liver Physiology* 286, G183–G188.

Moran, T.H. and Schwartz, G.J. (1994) Neurobiology of cholecystokinin. *Critical Reviews in Neurobiology* 9, 1–28.

Reeve, J.R., Jr, Eysselein, V.E., Ho, F.J., Chew, P., Vigna, S.R., Liddle, R.A. and Evans, C. (1994) Natural and synthetic CCK-58. Novel reagents for studying cholecystokinin physiology. *Annals of the New York Academy of Sciences* 713, 11–21.

Christianity

Christianity, rooted in **Judaism** and influenced by Hellenistic philosophy, has been the leading belief system in Europe since the early Middle Ages. As such, Christianity became, and still is, an integral and essential part of Western culture and moral thinking. A variety of perspectives regarding human–animal relations met within these three traditions, whereby no single view developed. Rather, within Western culture two lines of thought can be found: one stressing human superiority, the other pointing to human interdependence with creation. The Bible is open to both interpretations.

Humans being described as created in God's image, *Imago Dei*, is a central idea within the Judaeo–Christian tradition. 'Let us make man in our image, after our likeness' (Genesis 1:26). Even if there is no single interpretation of this bible quotation, its basic importance for a Christian belief system is undisputed. The most influential line of thought points at human superiority and rights to use animals for human purposes. This view is mostly taken for granted within Roman Catholic and Protestant Christianity today. Motivated by *Imago Dei*, animal experimentation, animal slaughter for human food and extensive breeding of farm and **companion animals** has become a widely accepted and common practice throughout Christian cultures. The medieval theologian **Thomas Aquinas** (1225–1274), the most influential Catholic scholar, motivated human superiority by theological and philosophical arguments. He divided humans from animals, arguing that human rationality showed his closeness to God. Irrational beings have no fellowship with human life, nor can they be friends of humans, since without rationality they cannot be wished good things. Care for animals, however, was still an issue for Aquinas. They should be used 'in conformity with the divine purpose, in order that they may not bear witness on us in the Day of Judgement'. For Aquinas, good care for animals was instrumental only to human eternal life, even if animals will share eternal life. However, according to Celia E. Deane-Drummond's recent Aquinas interpretation (2004), a good Christian becomes a responsible co-creator through treating animals in line with the cardinal virtues.

During the Renaissance human intellectual skills were idealized, and a growing interest in scientific knowledge during the Enlightenment led to increased stress on human rationality and their higher moral status. This was in line with theological interest, since the interpretation of *Imago Dei* in terms of human dominion was a means to avoid paganism and pantheism. In mainstream philosophical and theological writings the perspective described so far was the dominant line of thought.

However, this was not without criticism. One famous opponent was the German theologian and physician, Albert Schweitzer (1875–1965). In an acid remark he stated that philosophers and theologians had swept animals out of their theories as efficiently as a housewife kept her doorstep clean. According to this critical perspective, i.e. the interpretation stressing human interdependence with creation, humans have responsibilities towards the rest of creation exactly because they are an 'image' of God. Contemporary theological criticism of human dominion and its practice has a rich tradition of theological writings and arguments on which to build. For example, church fathers such as St John Chrysostom (4th century) emphasized human ability to feel compassion with animals thanks to a close relatedness between humans and animals: 'Surely we ought to show (other species) great kindness and gentleness for many reasons, but above all *because* they are of the same origin as ourselves' (Hume, 1962). Perhaps Christianity would have reacted with less indignation to Darwin's idea of fellowship between humans and other animals if Chrysostom's view had been predominant in theological writings.

St Basil of Caesarea (also 4th century) goes one step further by emphasizing human interdependence and interaction with the rest of creation: 'O God, enlarge within us the sense of fellowship with all living things, our brothers the animals to whom Thou gavest the earth as their home in common with us'. (Hume, 1962). In St Basil's liturgy Christians ask for forgiveness for being cruel towards animals – without a second thought for eternal life. Rather, his prayers are a very early expression of animals seen as valuable in themselves. St Basil's position has strongly influenced Eastern Christendom, and there is a direct connection to today's Orthodox liturgy. Expressing his holistic perspective blessing of creation and creatures is still part of a service. This is foreign to Western Christendom, in which animals and other parts of creation are largely excluded from the liturgical agenda. However, during the end of the 20th century, animal services came into fashion within both Catholic and Protestant Churches. Pet owners, farmers and zoo employees bring their beloved ones to a service where clergy bless the animals. This is often performed on 4 October, the feast day of St Francis. According to tradition, St Francis of Assisi (*c.*1182–1226) showed great compassion and had a humble relationship with animals.

A central issue in animal theology is the question of immortal soul. In traditional Christian interpretation eternal life is reserved for humans, but even here different positions flourish. Perhaps surprisingly for some, Martin Luther (*c.*1483–1546) considered the possibility that humans would meet animals in eternal life. There might be differences in the type of souls or immortality, but he did not deny the possibility, and rather regarded it as probable. More explicitly, the Spanish mystic St John of the Cross (1542–1591) argued in favour of animals' immortal souls and God blessing every animal. He pointed at the community of God the Creator with humans and animals, which are created and 'clothed in beauty' in the image of God. He thus extended the concept of *Imago Dei* to include also non-human animals. Other famous Christians arguing that animals have an immortal soul – and therefore should be treated well – were James Fox (1624–1691), James Butler (1692–1752) and John Wesley (1703–1791). Alexander Pope (1688–1744) took a contrary position when denying immortal souls in animals. He emphasized, however, that we ought to treat them with great compassion because they would not be compensated for the cruelty they had experienced in this life.

In contemporary discussions on animal theology the British theologian Andrew Linzey is well known. By relating to some of the above-mentioned theologians he refutes criticism from philosophical animal ethicists such as Peter Singer and Ursula Wolf that Christianity would support any use of animals. Rather, 'the uniqueness of humanity consists of its ability to become the servant species' (Linzey, 1995). Through his 'generosity paradigm' and from an ideal of following Christ, Linzey argues that all marginalized beings, including animals, are objects of direct moral concern. Moreover, due to their dependency and weakness they deserve greater respect than stronger beings, i.e. most humans. (HR)

References and further reading

Deane-Drummond, C.E. (2004) *The Ethics of Nature.* Blackwell Publishing, Oxford, UK.

Hume, C.W. (1962) *The Status of Animals in the Christian Religion.* Universities Federation for Animal Welfare (UFAW) Theological Bulletin, No. 2, UFAW, Wheathampstead, UK.

Linzey, A. (1995) *Animal Theology.* SCM Press, London.

Preece, R. (1999) *Animals and Nature. Cultural Myths, Cultural Realities.* UBC Press, Vancouver, Canada.

Preece, R. and Fraser, D. (2000) The status of animals in biblical and Christian thought: a study in colliding values. *Society and Animals* 8, 3.

Schweitzer, A. (1929) *Kulturphilosophie* II: *Kultur und Ethik.* Paul Haupt, Bern, Switzerland, 1923 [Cultural Philosophy II: Civilization and Ethics. Translated by John Naish]. A & C Black, London.

Schweitzer, A. (1932) *Kulturphilosophie* I: *Verfall und Wiederaufbau der Kultur.* Paul Haupt, Bern. Switzerland, 1923 [Cultural Philosophy I: The Decay and the Restoration of Civilization. Translated by C.T. Campion], 2nd edn. A & C Black, London.

Singer, P. (1990) *Animal Liberation*, 2nd edn. New York Review of Books, New York.

Wolf, U. (1990) *Das Tier in der Moral.* Klostermann, Frankfurt, Germany.

Circadian rhythm

Biological rhythms are probably of universal occurrence in animals, and it is usually easy to demonstrate adaptive explanations for them. They occur in periodicities that range from several seconds to several years but generally, however, there are four common types: circadian rhythms, **diurnal rhythms**, **ultradian rhythms** (over a shorter time period than a circadian rhythm) and infradian rhythms (over a longer time period than a circadian rhythm). Circadian rhythms have a period of approximately 24 h and are primarily endogenously regulated but may be entrained by exogenous stimuli, e.g. **photoperiod**. In humans, for example, the sleep–wake cycle has an endogenous periodicity but it normally becomes entrained by the light–dark (day–night) cycle. Similarly in domestic pigs, circadian rhythms or patterns in body temperature usually cycle around the onset of activity in the morning, but feeding and alterations in feeding times can be used to manipulate its periodicity (Ingram and Mount, 1973).

The area of the **brain** that is central in maintaining circadian rhythms is the suprachiasmatic nucleus, which is situated within the **hypothalamus**. It is located just dorsal to the optic chiasm and receives exogenous light signals through the retinohypothalamic pathway. Other signals originating in the thalamus, midbrain, hippocampus and pineal gland are also capable of exhibiting some regulatory functions. Likewise, there are many neurotransmitters involved in the coordination and expression of these rhythms including adrenaline (**epinephrine**), **serotonin** (5HT), **acetylcholine**, melatonin and **dopamine**. Disruption of the synthesis and/or release of these transmitters may lead to the interruption of circadian rhythms that can be expressed behaviourally, e.g. disturbance of normal sleep–wake cycles or patterns of activity. In diurnal animals, circadian rhythms are for the most part phased to optimize arousal during active and light portions of the day and to induce sleep (e.g. secretion of melatonin) at night. The opposite is obviously the case in nocturnal species. (RM-F)

Reference

Ingram, D.L. and Mount, L.E. (1973) The effects of food intake and fasting on 24-hourly variations in body temperature in the young pig. *Pflügers Archiv European Journal of Physiology* 339, 299–304.

Circumcision

Circumcision is a surgical procedure referring to removal of a section of the preputial skin of the penis. In the horse, the term circumcision typically refers to removal of a section of the internal preputial skin (see Fig. C.5). Surgical texts also name this as the reefing procedure or posthioplasty (Varner and Schumacher, 1991). Circumcision surgery in the male horse is medically indicated to remove a section of the internal preputial skin that has been injured or is affected by tumour. It is commonly recommended to remove as little of the internal preputial skin as necessary to avoid painful or inadequate erections and subsequent breeding unsoundness. The rate of return to breeding soundness has not been documented in horses undergoing this procedure.

As in the horse, circumcision in the bull is performed for medical reasons, most commonly to prevent recurrent preputial injuries. Related surgical procedures are also performed to create teaser bulls that are unable to successfully complete intromission. Circumcision has been reported as creating stallions as teasers in a pasture heat detection system (Silva *et al.*, 1994). However, in this case it is not the internal preputial skin but the epithelium of the shaft of the free portion of the penis that is removed, from just behind the glans penis to the attachment of the internal prepuce. Extensive removal of the epithelium, followed by surgical closure of the distal and proximal ends of the wounds, results in an effectively shortened penis and thus the inability to complete intromission. **Sexual behaviour** of the circumcised stallions measured up to 180 days postoperatively was similar to that of uncircumcised stallions and considered adequate to aiding in oestrus detection. Though reported in the literature, this procedure does not appear to be common in practice.

Use of the term 'circumcision' may effect comparisons with the surgical removal of the foreskin (prepuce) in human males. In the human there are gross and histological indications, as well as psychological assertions, that the foreskin plays an integral role in sexual gratification, in part due to its relative mobility even with the penis in the erect state. In the horse the internal prepuce does not appear to have the same degree of mobility during erection and thrusting, though detailed gross anatomical and histological studies have not been reported.

(NKD)

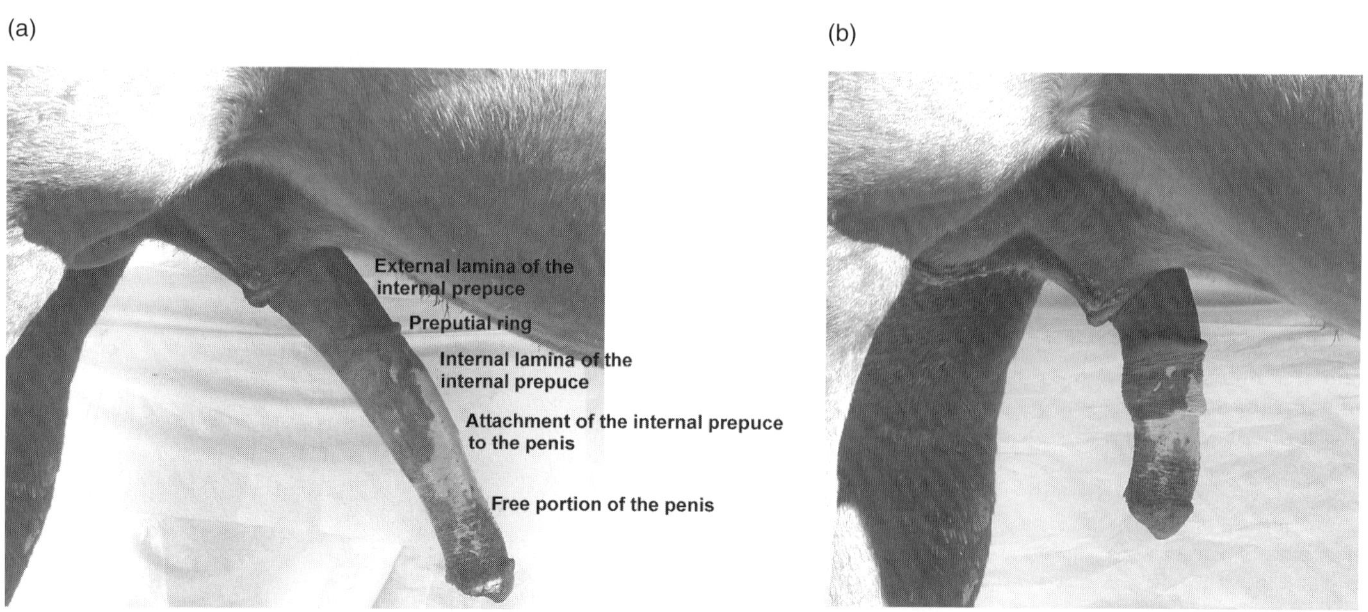

Fig. C.5. The erect (a) and non-erect (b) stallion penis. When non-erect, the free portion of the body of the penis telescopes into the looser internal and external lamina of the internal prepuce.

References

Silva, L.A.F., Fioravanti, M.C.S., Marques Junior, A.P. and Melo, M.I.V. (1994) Sexual behaviour of Mangalarga stallion teasers prepared by circumcision with shortening of penis. *Revista Brasileira de Reproduocao Animal* 18, 110–115.

Varner, D.D. and Schumacher, J. (1991) Diseases of the reproductive system: the stallion; preputial resection (reefing operation, circumcision, posthioplasty). In: Colahan, P.T., Mayhew, I.G., Merritt, A. and Moore, J.N. (eds) *Equine Medicine and Surgery*, 4th edn. American Veterinary Publications, Inc., Goleta, California.

Circus

A circus is an arena (usually round) in which humans and/or non-human animals perform for the entertainment of others. A circus may be a temporary, seasonal or permanent establishment, and may use wild or domestic species. Integral to their use is their training to perform entertaining activities. Very little scientific research has been carried out on the training or **welfare** of circus animals, possibly because of the relatively small size of the industry from a provision perspective (although consumer numbers are high). None the less, concerns can be divided into the following categories: (i) the accommodation of circus animals (**see: Confinement**), when both on the road and in winter quarters (**see: Housing, space**); (ii) the stress of **transport**; (iii) the training methods used (**see: Reinforcement – types of**); (iv) the stress of performance; and (v) the sourcing and dispersal of animals after their working life is ended. The mobile and transient nature of circuses can pose challenges to the enforcement of regulation of standards.

The use of wild animals is of growing concern, although international conventions such as CITES should regulate the sourcing of animals of conservation importance from the wild. However, these animals generally appear less adapted to confinement and may frequently show signs of stress when asked to perform in the arena, although many animals used in circuses may show no evidence of distress and some may appear to show signs of enjoyment. There has been an increased trend in using domestic species (especially **horses**, **cats** and **dogs**) or only human performers in many circuses in recent years as a result. Concerns are certainly justified when animals are asked to perform behaviours that may be biomechanically or physiologically very stressful, e.g. **elephants** balancing on their hindquarters or horses dancing on their hind legs, by which their backs may be hyperextended (see Fig. C.6).

Assessment of the welfare of such animals poses particular challenges. Reluctance to enter an area (including **transport** and accommodation facilities) and observational assessment of physical health and body condition are relatively simple measures of the individual that can be used at least as a preliminary measure, together with an assessment of the physical resources available and management regimes. In some countries, such assessment is a necessary condition for licensing of the circus, while in others little or no regulation exists. One study in the UK found that 70% of animals were not in peak condition, and that coprophagia and stereotypic behaviour were frequently shown by animals when confined (see Figs C.7 and C.8).

Fig. C.6. A circus horse being trained to rear on command. Prolonged or frequent hyperextension of the back can result in musculoskeletal and/or arthritic problems.

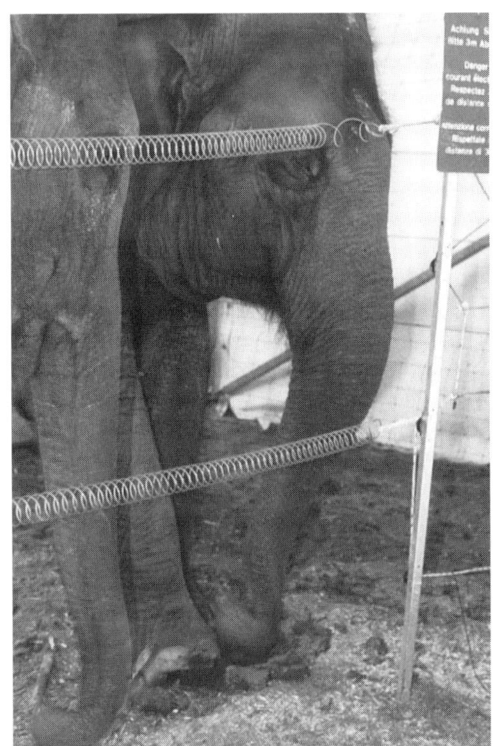

Fig. C.7. Restrained circus elephant manipulating faeces prior to ingestion.

Fig. C.8. Restrained circus elephant showing stereotypic weaving behaviour.

It has been suggested that this high prevalence may relate to a genetic influence in the founder population and high levels of **inbreeding** and, while this may offer an explanation, it does not invalidate the primary concern over the welfare of these animals.

It is generally recognized that circuses do not fulfil a meaningful role in animal conservation and their impact on education is minimal; however, their potential in this latter regard is probably underexploited as the science of animal training has developed to the point where they could be used to illustrate the importance of conditioning procedures and good communication in animal training.

Far from being a detrimental stressor, it has been argued that training, and potentially performance, may be enriching for the animal if conducted using appropriate methods, and the value of training as an environmental **enrichment** in **zoos** is increasingly recognized. However, it has been suggested that, unlike in **zoos**, where training is used to add environmental complexity and control, the training of circus animals is very repetitive and so potentially less enriching. For some, regardless of the methods used or their educational potential, the concept of an animal circus is rejected on ethical grounds (**see: Ethics**). (DSM)

Further reading

Circus Working Group (1998) *A Report into the Welfare of Circus Animals in England and Wales*. RSPCA, Horsham, UK.

Kiley-Worthington, M. (1990) *Animals in Circuses and Zoos. Chiron's World?* Little Eco Farms Publishing, Devon, UK.

Clever Hans

Hans was a Warmblood horse living in Berlin at the turn of the 20th century. His owner, Wilhelm von Osten, was a former schoolteacher. Von Osten, clearly confident of his own teaching skills as well as his horse's intellectual abilities, embarked on the ambitious task of educating Hans to complete cognitive tasks traditionally associated with scholastic exercises for humans. His efforts were rewarded when, after 4 years of training, Hans could demonstrate his seemingly remarkable intellectual abilities. Hans could perform mental arithmetic (add, subtract, divide and multiply numbers), could read and spell, and even answer questions on history, music, literature and geography, extraordinary feats for a non-human animal (though, prophetically, not for a human). Of course, Hans could not speak, but could give answers to specific questions by stomping his hoof. For example, when posed an arithmetic problem Hans would provide his solution by stomping his hoof the correct number of times.

Clever Hans and his owner quickly became famous, and attracted the attention of sceptical scientists, most notably Professor Carl Stumpf, director of the Berlin Psychological Institute. The obvious 'logical' explanation for Hans' abilities was, not surprisingly, that von Osten was giving discrete clues to his horse as to how to respond. These clues could have been given deliberately, in which case von Osten was a fraud, or else he could have been unintentionally providing reactions that reliably cued Hans to the correct answer. However, this theory was quickly shown to be wrong, because Hans still performed remarkably well when posed questions by someone other than von Osten, even when von Osten himself was absent.

Stumpf then enlisted a young German psychologist, Oskar Pfungst, in the task of investigating in detail Hans' remarkable abilities. Hans' answers were indisputably the product of mental calculations or acquired knowledge, and it was natural to attribute these to Hans himself. But, of course, whenever Hans was tested, it was in the presence of humans, and their ability to answer these questions was never in doubt. Was it still possible that the mental processes attributed to Hans were, in fact, being performed by other people present at the same time? Pfungst's insight was to recognize this possible confound and to devise a means of eliminating it. Pfungst's systematic approach has become a landmark in the history of experimental psychology. The critical experiment devised by Pfungst was to test Hans with the only people present being ignorant of the answers to the questions posed. To achieve this, Pfungst and his assistants prepared questions on cards, and these cards were then presented to Hans in a different room. The people present in the room with Hans, including the person holding the cards, could not see the questions written on the cards, and thus could not know the correct answer. Under these conditions, Hans performed badly. Clearly, then, it was the presence of people who knew the answer that determined Hans' performance. That is, people were somehow providing the answers to Hans. But how?

Pfungst and his collaborators were not about to reject one implausible conclusion – that Hans was capable of spelling and arithmetic – and replace it with an even less plausible one – that Hans was capable of mind-reading. Instead, Pfungst concluded that Hans was exquisitely sensitive to the behavioural reactions of people. That is, small involuntary movements were cues that governed Hans' responses. Pfungst then set about to test his theory and uncover which responses were responsible. Initially, Pfungst fitted Hans with blinkers, and then asked von Osten to stand beside Hans, out of view. Under these conditions Hans performed badly, indicating that visual cues were necessary to guide Hans' stomping. Pfungst then proceeded to make careful examination of von Osten's behaviour during each test. What Pfungst noted was that von Osten's posture and facial expression changed subtly across the course of each test. Specifically, von Osten would tilt his head forward, to look down at Hans' hoof, at the commencement of each test, but would lift his head back up when Hans' stomping reached the point corresponding to the correct answer. This latter response was accompanied by characteristic facial expressions – flaring the nostrils and arching the eyebrows. Importantly, these behaviours were not unique to von Osten. In subsequent laboratory studies conducted by Pfungst, 90% of people displayed similar behavioural reactions in anticipation of the correct answer to a question being tapped out by the experimenter. Such findings are a testament as to how uniform across individuals human behaviour can be.

Although these behaviours were clearly correlated with Hans' performance, that alone does not prove they controlled it. It is possible that both von Osten's and Hans' behaviours were determined by the answer to the question posed, or even that von Osten's reactions were in response to Hans' stomping, not vice versa. The clinching evidence came when Pfungst demonstrated that his own postural changes and facial expressions alone were sufficient to control Hans' responses. In the absence of any question posed of Hans, the horse would start stomping and then stop again solely in response to Pfungst tilting his head forward and then lifting it back again. Therefore, von Osten's subtle behavioural reactions were both necessary and sufficient to determine Hans' performance. Thus, it seems, Hans had not learned how to spell or do mental arithmetic, but instead had learned to start stomping his hoof in response to certain behavioural cues, and to stop stomping in response to other behavioural cues.

Pfungst went on to publish his investigations in 1909, in his book *Clever Hans (The Horse of Mr. von Osten). A Contribution to Experimental Animal and Human Psychology*. It should be noted that, before Pfungst's investigations, von Osten himself was unaware of these mechanisms – his own unconscious behaviours provided reliable information to his subject about how to respond.

The story of Clever Hans is an enduring one because it provides several important lessons. First, and most immediately, it highlights the pitfalls in our ability to draw conclusions about the psychological or mental states of other animals based on their behaviour. Hans is a classic illustration of the traps of anthropomorphizing – it is seductively easy to ascribe human psychological processes to explain the behaviour of other animals. A broader lesson concerns the appropriate use of scientific methodology to investigate behavioural processes. Pfungst devised the first test in which the experimenter was naive to the experimental condition (the question posed) and thus to the expected outcome, a procedure at the heart of 'double-blind' testing protocols that have become indispensable in many psychological and medical studies. The explicit lesson here is that animals, including humans, are exquisitely sensitive to picking up what is expected of them by an experimenter and behaving accordingly. (JH)

Further reading

Boakes, R.A. (1984) *From Darwin to Behaviourism: Psychology and the Minds of Animals*. Cambridge University Press, Cambridge, UK.

Clicker training

This refers to an application of the principles of positive **reinforcement** and the use of a secondary or conditioned reinforcer to train an animal to perform behaviours.

Clicker training derives primarily from learning theory, as developed by Skinner and Watson. These theories were applied by animal trainers such as the Bill and Marian Breland Bailey in the mid-1900s and entered first the marine mammal and later the dog and horse training fields, primarily through the writings of Karen Pryor in the 1960s. The name derives from the 'clicker', a small plastic box with a metal tongue, which, when depressed and released, makes a 'click' sound. The sound of the clicker is paired through classical conditioning with a primary reinforcer, usually food, until it acquires the status of a conditioned reinforcer. Once this association has been made, the sound is used to mark precisely the desired behaviour as it is displayed by the animal being trained. Delivery of the primary reinforcer then follows at a convenient interval. In this way the clicker may be referred to as a 'bridge'. As the conditioned reinforcer does not require direct interaction with the animal, it can be used accurately to mark the behaviour upon which the reinforcer is contingent, and so the accuracy of training can be improved and, in some cases, the speed of learning in practice can be increased. Once behaviours have been learned, a cue is then associated with the behaviour and the clicker is no longer needed as the learned behaviour can then be maintained through other forms of reinforcement.

Clicker trainers traditionally make use of other aspects of learning theory, for example shaping new behaviours through successive approximation, although luring of behaviours and capturing of complete actions also are used. Because the technique relies on communicating with the animal without the need for contact, it has been successfully applied to captive wild animals, where it is usually employed to improve husbandry and **welfare**, for example in training animals to cooperate with handling procedures instead of needing to make use of sedation. (HZ)

Further reading

Pryor, K. Website available at: http://www.clickertraining.com

Clinical behaviourist

A clinical behaviourist is a professionally trained individual, a veterinarian and/or a PhD, who assesses and treats behavioural problems of animals, usually domestic animals, but also exotic **pet** and **zoo** animals. The most rigorous training is that of veterinarians who, in many countries, have organized a specialty board. For example, in the USA, in order to be board certified, a veterinarian must have at least 1 year of clinical medical experience and then a 2–3-year residency, during which the major portion of time is devoted to treating animals with behavioural problems. In addition to the residency, the veterinarian must also have published a paper in a refereed journal, must submit three acceptable case reports and take 2 days of written examinations. There is also specialty recognition in Europe and Fellowship examination in Australia. The Animal Behaviour Society based in the USA also certifies applied behaviourists. In order to be certified the candidate must be a PhD (or have a Master's degree for certification as an associate), must have taken courses in behaviour, learning, etc. and submit case reports.

The sister organization outside North America has more recently developed a certification programme for graduates elsewhere as Certificated Companion Animal Behaviourists. It requires formal education in a range of topics, supervised experience and case histories, but no postgraduate qualification. Cases can be seen in a clinic setting or in clients' own homes. Most clinical behaviourists ask the owner to describe a detailed history of the problem and the owner's response to the misbehaviour, as well as the general management of the animals. Medical causes of behavioural problems must be eliminated, either by a referring veterinarian or at the time of the consultation. Board-certified veterinarians and clinical behaviourists use behaviour modification techniques and, in addition, the former can prescribe psychoactive or other drugs as required. The most common problems presented to clinical behaviourists are **aggression** in all species, **separation anxiety** and **fears**/phobias in dogs, **house soiling** in cats and **stereotypies** in horses. (KAH)

Clone

A clone is when two or more animals have the same genetic make-up, e.g. like identical twins, but the production is artificially induced (splitting the fertilized egg at a very early stage of fewer than eight cells, or transferring somatic cell nuclei to a fertilized egg that has been enucleated). The latter technique means that the clones are separated in time.

(DBM)

See also: Cloning

Cloning

Somatic cell nuclear transfer or somatic cell cloning is a way of producing another animal of the same genetic make-up as the donor animal. It was originally performed in frogs using intestinal cells as the source of DNA, but has now also been achieved from other somatic cells (mammary gland) and early embryonic cells (four- to eight-cell stage). In some species, e.g. cattle, cloning from somatic cells has been found to cause an exceptionally high incidence of abnormal fetuses (overgrown) but, in other species such as pigs, these problems have not been found. Offspring of the **clone** do not appear to be affected. Apart from frogs, cloning has been carried out with sheep, cats, cattle, dogs, mice and horses.

Reproductive cloning, when early-stage (e.g. four to eight cells or blastomeres) embryos are split by dividing the blastomeres and inserted into enucleated ova still with their *zonae pellucida*, has been successful, but few studies have been carried out on their **health** and **welfare**. (DBM)

Further reading

EFSA Report on Cloning (2008) EFSA website: http://www.efsa.europa.eu

Wilmut, I., Schnieke, A.E., McWhir, J., Kind, A.J. and Campbell, K.H.S. (1997) Viable offspring derived from fetal and adult mammalian cells. *Nature* 385, 810–813.

Cockfighting

Cockfighting is the pitting of two fighting cocks (roosters) against each other for **sport**. The fight takes place in a cockpit. The sport is closely associated with gambling, with bets made on the outcome of the match. The cocks are purpose-bred and trained for fighting. Cockfighting is banned in most Western countries, but continues to be a popular traditional blood sport in many Eastern countries. The sport has existed in many cultures, including those of India, China and the Middle East, since ancient times. It was introduced to Greece in the period 500–450 BC, was subsequently adopted by the Roman empire and spread to England, Wales, Scotland, Italy, Germany and Spain and their colonies. While opposed by the Christian church because of its association with gambling, it became a popular sport supported by the English monarchy from the time of Henry VIII. Cockfighting was popular in America from the time of early settlement until after the Civil War, with both Thomas Jefferson and George Washington being supporters of the sport. By the early 1800s the sport had become so well established in Britain that cockfights were regularly held in churchyards and schools.

A cockfighting match usually consists of several fights between birds. The birds are placed in a circular pit about 6 m in diameter, surrounded by the audience. Purpose-built arenas may have tiered seating around the pit. Depending on the tradition, a pair of birds is placed in the pit, or several are placed together (known as the 'battle royal'), until one bird emerges as the overall victor.

Despite its long-term popularity, the cruelty of the sport was recognized by Boswell, as recounted in his London Journal written in 1762:

> I then went to the cockpit, which is a circular room in the middle of which the cocks fight. It is seated round with rows gradually rising. The pit and the seats are all covered with mat. The cocks, nicely cut and dressed and armed with silver heels, are set down and fight with amazing bitterness and resolution. Some of them were quickly dispatched. One pair fought three quarters of an hour. The uproar and noise of betting is prodigious … I was sorry for the poor cocks. I looked around to see if any of the spectators pitied them when mangled and torn in a most cruel manner, but I could not observe the smallest relenting sign in any countenance.

As indicated above, birds sustain severe injuries during fights, including lacerations to the body and eyes, punctured lungs and broken bones. The end of the fight is usually signalled by the death of all bar one of the birds, although in some cases a time limit is set or birds may be withdrawn when severely injured. Birds may be fitted with extremely sharp metal spurs or blades, sometimes armed also with rowels (sharp discs on the end of the spur), designed to puncture or mutilate the opposing bird. The spurs are fastened to the bird's leg with a leather bracelet or similar and vary in length from around 3 to 8 cm.

Fighting birds are bred from game fowl, closely related to the Indian red junglefowl (*Gallus gallus*), from which domestic chickens originate. Breeding of gamecocks remains popular regardless of whether or not fighting is banned. The birds are highly colourful and with a distinctive erect stance and long spurs. Fighting birds are exercised when young to build up muscle mass. They are generally fed a high-protein diet and may also be massaged with an astringent solution to toughen the skin. Their feathers are clipped and the comb is trimmed to reduce the targets for opposing birds. Training may also include pitting birds against each other while wearing pads over their spurs. Birds are usually put to fight when they are between 1 and 3 years old.

Cockfighting behaviour stems from natural aggression between roosters. However, this is exaggerated in gamecocks through selective breeding and training for maximum aggression. Domestic roosters and their feral counterparts will challenge each other to establish dominance, territory and to gain access to hens, but such encounters are short lasting and rarely result in serious injury. Breeding game fowl has involved selecting and breeding from the most aggressive roosters to heighten fighting behaviour. This may be further supplemented by the use of pharmacological stimulants (such as caffeine, **epinephrine** or amphetamines) and the use of spurs to increase the injuries resulting from aggressive behaviours. Training to maximize aggression towards other roosters involves isolating the bird from conspecifics to reduce their experience of natural **social behaviours**. This lack of social experience means they have little ability to interpret signals used during male–male conflicts that would generally (in a more natural context) lead to a peaceful resolution of the encounter.

Cockfighting was first prohibited by law in Massachusetts in 1836, and in most other American states it is now expressly banned through specific legislation or is prohibited under general cruelty to animals legislation. The sport was prohibited in the UK in 1849 and, for more than a century, it has been illegal in many other countries, including Australia, New Zealand, Canada and most of Western Europe. In some countries the possession or sale of gamecocks and/or cockfighting spurs, keeping a cockpit, attending or promoting a cockfight or transporting birds for the purpose of fighting are listed as offences under animal welfare legislation. Despite these laws, the sport has continued underground in these countries, with illegal fights a regular occurrence. Gambling at these events is endemic and is based on wagers between individual spectators. Many thousands of dollars can change hands during a match, creating a tense and often violent environment. Cockfighting is often associated with other underground activities, such as illegal firearms and other weapons, illicit drugs and other blood sports such as dog-fighting. Prosecutions by organizations such as the **RSPCA** on cockfighting-related charges are a regular occurrence where cockfighting rings have been exposed or people have been found in possession of cockfighting equipment.

Cockfighting remains a popular legal sport in many places, including Latin America, Africa, South-east Asia, the Philippines and also in some parts of the USA (Lousiana, New Mexico). In some countries there is a close association between cockfighting and cultural or religious rituals. For example, in Balinese culture the ancient Tabuh Rah ritual to expel evil spirits involves three cockfights, each to spill blood on the ground as an offering. Cockfighting is a male-dominated activity and women may be specifically excluded from fights or related rituals. In South-east Asia the sport is so widespread that most men in rural villages will own at least one fighting cock and fights are part of regular village activity. (BJ)

Reference

Pottle, F.A. (ed.) (1951) *Boswell's London Journal, 1762–1763*. William Heinemann, London.

Codes of practice

Wherever humans use, house or manage animals, there is the potential to cause avoidable **suffering** if the animals' needs are either not understood or not awarded sufficiently high priority. National and international legislation relating to different areas of animal use often includes voluntary or mandatory codes of practice, with the intention of ensuring that animals are cared for appropriately and treated humanely. Codes of practice have been drawn up for many different types of human–animal interaction, including animal housing and care, **transport**, **killing** animals, the conduct of veterinary, husbandry or scientific procedures, breeding and sale. Further information on specific codes of practice can be obtained from national government departments that administer the relevant primary legislation, e.g. Agriculture and Agri-food Canada, the United States Department of Agriculture or the UK Department for Environment, Food and Rural Affairs (Defra).

There is a fundamental need for codes of practice, because they can protect animals from unacceptable treatment or levels of care, but it is important to note that the welfare of animals housed or treated according to a code of practice is not necessarily good. This is because codes of practice are drawn up not only to fulfil animals' best interests, but also to satisfy a range of different stakeholders. These usually include: (i) the regulatory bodies that will be monitoring the guidelines and ensuring compliance; (ii) relevant industries; (iii) animal behaviour and/or welfare societies; and (iv) individuals that will be regulated by and are required to implement the codes of practice. Balancing the different priorities held by each of these groups usually results in standards that represent a compromise position and have a number of functions, such as:

- minimizing or eliminating avoidable suffering;
- improving animal **welfare**;
- improving animal health;
- encouraging implementation of the spirit of a piece of legislation (e.g. by setting out appropriate environmental enrichment or techniques for humane killing);
- standardizing or harmonizing global practice;
- preventing malpractice that could present a risk to human health or wellbeing; and
- maximizing benefits to humans such as increased productivity, reduced economic costs or improved scientific validity.

Different codes of practice give different weightings to these criteria, depending on prevailing social, political and economic pressures and how these are brought to bear. There will always be conflicts between the needs of humans and animals; for example, optimal animal welfare often requires more expensive housing and higher staffing levels. The economic implications of this may be acceptable for **zoos**, yet very strongly resisted by sectors that aim to maximize animal productivity and minimize costs. Economic considerations can also result in very different standards within the same species. For example, husbandry guidelines for domestic fowl used in experiments include much better standards than most fowl experience in an agricultural setting, although they relate to the same species with the same requirements for basic wellbeing (see Table C.3).

Another possible reason for such discrepancies is that, historically, codes of practice were based almost entirely on perceived good practice within a particular sector at that time. In recent years there has been a rapid expansion of the literature on animal behaviour, **preferences**, **motivation** and physiological responses to different husbandry protocols and events such as **transport**. This is now routinely channelled into new and revised guidelines, but it can still prove difficult to change traditional perceptions on what is acceptable practice. Challenging the *status quo* is, of course, especially problematic for species where empirical evidence is still scarce or non-existent.

Whichever factors have influenced the development of a code of practice, in the majority of cases even recently updated guidelines will represent minimum standards, not best practice. As a consequence, codes of practice can actually prevent proactive welfare improvements if they are implemented to the letter without question. If an individual believes that the legislation regulating his/her activities is strict (which is likely, given that it restricts their actions), he/she will probably also believe that its codes of practice are equally rigorous and promote good welfare. It can be difficult for a regulator or inspector to change this attitude and encourage better practice, especially if he/she is operating in an advisory capacity and has no legal inducements. In some cases, especially for farmed and laboratory animals, there are voluntary, higher welfare standards that can be used to encourage a more animal-centred approach. One further advantage of voluntary guidelines is that they can be frequently updated to take account of new applications of technology to animals and new information about animal behaviour and welfare needs, whereas official codes are more likely to lag behind recent knowledge and public attitudes.

When considering and assessing a code of practice, it is often very useful to compare it with a range of voluntary standards, and also with equivalent codes of practice in other countries and for animals used in other contexts. Each of the recommended space allowances for laying hens in Table C.3 gives different priorities to animal welfare, but all claim to provide sufficient space for the birds' behavioural needs to be fulfilled. The standards from the two welfare bodies are significantly different, although both aim to set allowances that will improve significantly on current practice yet be attainable for the majority of producers. Evidently, both bodies have made different judgements as to what can feasibly be achieved as a first step towards improving the quality of life of the laying hen.

An ideal code of practice, free from such practical, political and economic constraints, should encompass the principle of a *duty of care* on the part of all those responsible for animal housing, husbandry, use and management. The **five freedoms** concept, as adopted by the UK Farm Animal Welfare Council (**FAWC**), provides a useful basis for defining a duty of care that reflects the basic needs of each species and strain. It was originally set out for farmed animals, but it can be partly or wholly applied in all contexts including **companion**, **laboratory** and **wild animals** in and out of captivity. At the time of writing, there is no such ideal code of practice that includes a duty of care, has an entirely objective basis and will ensure good animal welfare. It is therefore vitally important to approach and use all codes of practice critically, review the literature regularly and always implement better practice and/or encourage others to do so. (PH)

See also: Council of Europe; FAWC; Feedlot

Cognition

Broadly defined, cognition refers to 'the mechanisms by which animals acquire, process, store, and act on information from the environment' (Shettleworth, 1998). Given this definition, the study of cognition encompasses a wide range of topics, including sensory systems and **perception**, learning and **memory**, **decision making**, navigation, **communication** and **language**, and reasoning. Cognitive processes unquestionably play important roles in behaviour including, for example, the remarkable abilities of honeybees to communicate the location of a food source to their hivemates, crows to fashion tools or Clark's nutcrackers to recall the locations of thousands of cache sites. Our efforts to understand behaviours such as these can scarcely be complete without an exploration of cognition.

Table C.3. Comparison of floor space allowances[a] within some codes of practice for laying hens.

Organization	Space allowance per bird (cm^2)
United Egg Producers, USA[b]	393 (white Leghorn) 439 (brown egg-layers)
McDonald's	465
British Egg Industry Council	550
UK Farm Animal Welfare Council	600
RSPCA Freedom Food	855
UK Home Office, laboratory chickens 1800–2400 g (group housed)	1200
Draft revision of Council of Europe Convention ETS123 for the Protection of Vertebrate Animals used for Experimental and other Scientific Purposes, laboratory chickens 1800–2400 g (group housed)	1300

[a]This table lists space allowances only for simplicity, but it is also essential to provide a good-quality environment within the available space.

[b]These guidelines apply from 1 April 2005; producers will have had to increase space allowances to 432 cm^2 for white Leghorns and 490 cm^2 for brown egg-layers from 1 April 2008 to retain their 'Animal Care Certified™' status.

While 'cognition' is often assumed to imply complicated processes leading to declarative knowledge (rather than procedural, or route 'how to' knowledge), a broader definition of cognition is useful because it allows for the possibility that complex behaviours may be governed by relatively simple mechanisms. Such a definition does not, however, encompass issues of **consciousness**. The questions of whether animals are self-aware, understand emotions, have intentions or plan actions are ones that are especially difficult to answer because behaviours that might be motivated by conscious thought can usually be explained in terms that do not require reference to consciousness. While some may disagree, most scientists studying animal cognition use the information-processing definition of cognition and agree that cognition can and should be studied without reference to consciousness.

The study of cognition

Historically, the study of cognition has been pursued independently by psychologists and ethologists. Much of the early research in animal cognition focused on animal learning, and was primarily conducted by experimental psychologists. One dominant idea motivating much of the research in animal learning emphasizes that most learning can be explained by a few general processes. One of the most influential general process mechanisms is associative learning, the process by which animals learn cause and effect relationships and how to respond to them. Perhaps the best-known example of associative learning is Pavlovian or classical **conditioning**, where animals learn associations between a novel stimulus and an unconditioned stimulus.

In his famous experiment, **Ivan Pavlov** demonstrated that a dog can be trained to associate a sound (a conditioned stimulus, CS) with the presentation of food (unconditioned stimulus, US) and will salivate in anticipation of the food (conditioned response, CR) when it hears the sound (CS). Thus, an association is formed between the two stimuli, the sound (CS) and the food (US).

Another type of associative learning, instrumental conditioning (first described by Edward Thorndike) has also played an influential role in the development of associative theories. In instrumental conditioning, also known as **operant** conditioning, animals learn relationships between stimuli and responses based on the consequences (reward or punishment) of their actions. If the consequence is pleasant, the association between stimulus and response is strengthened and, if unpleasant, the relationship is weakened. For example, rats can be trained to press a lever if that action is followed by a reward such as the delivery of food.

From these basic concepts, associative theory has grown with the core assumption that the formation of associative relationships between two events (stimuli and responses) can explain most learning phenomena. The success of the associative paradigm can perhaps be attributed to a couple of factors. First, research in neurobiology on the formation of synaptic connections has provided a physiological explanation of associative mechanisms. Secondly, models of associative learning have succeeded in explaining many learning phenomena, thereby demonstrating how many seemingly complex behaviours can be explained without resorting to mentalistic explanations. However, associative explanations fail to account for all types of learning, and critics have cautioned against an overemphasis on associative accounts and general process views of learning.

The strong emphasis on general learning processes accounts in part for psychologists' overwhelming reliance on a few species – primarily rats, pigeons and monkeys – in their research. In contrast to experimental psychologists, biologists are guided by an evolutionary framework and recognize that species may differ in their cognitive abilities as a result of differing ecological pressures and evolutionary trajectories. Hence, biologists are interested in studying the cognitive differences in a wide variety of species and are mindful of the evolutionary and ecological factors that create such differences.

The field of **ethology**, the biological study of behaviour that flourished in the mid-20th century, was dominated by European zoologists whose primary concern was to study animal behaviour in relation to their natural environments. One of the founding fathers of ethology, **Konrad Lorenz**, described behaviour patterns as 'organs', specifically adapted to the functions they serve, and focused his studies on species-specific differences in patterns of behaviour. From this tradition, the field of **behavioural ecology** developed with a similar focus on asking functional, evolutionary questions about behaviour and studying a diversity of animals in their natural environments.

The separate evolutions of biology and **psychology** have led to disadvantages in both approaches to the study of cognition. In the psychological tradition, the clearest problem has been the lack of evolutionary, comparative considerations and awareness of the natural context of cognition. On the other hand, the ethological tradition, particularly behavioural ecology, has focused heavily on functional and evolutionary questions with little consideration of the proximate mechanisms of cognition. However, in the 21st century, the study of animal cognition has become increasingly integrated between the two disciplines. While traditional boundaries remain, both biologists and psychologists have called for a 'synthetic approach', as evidenced by the many labels used to describe the field that studies animal cognition (i.e. comparative cognition, cognitive ecology, **evolutionary psychology** and **cognitive ethology**); to truly understand cognition, we need to follow **Tinbergen's four questions** framework and integrate questions and answers about cognition from all levels (causation, ontogeny, function and evolution) into a coherent whole.

An evolutionary approach to cognition

An integrative understanding of cognition requires more than the sum of separate pieces and, as with all biological phenomena, behaviour and cognition cannot be fully understood without consideration of evolution. Although research focusing on proximate mechanisms is interesting in its own right, considerations of the function and evolution of cognition can inform and guide such work. From an evolutionary perspective, cognitive abilities are viewed as traits that are subject to natural selection; just as the beaks of Darwin's famous finches have evolved in shape and size in response to seed abundance and type, cognition has also evolved to solve ecological problems. This adaptive view of cognition has led several researchers to argue that the mechanisms of cognition are likely to be specialized for different information-processing problems, much as sensory systems have become specialized to gather different types of information from the environment. The belief that cognition is 'adaptively specialized' predicts that 'we should expect to find a collection of specialized computational devices designed for special purposes rather than a single general-purpose, problem-solving mechanism' (Shettleworth, 1998, p. 42). In contrast to the general processes view described earlier, this paradigm views associative learning as one specific 'computational device' or 'module' for learning about causal relationships between events. Other modules may exist for solving different types of problems, such as learning spatial relationships or social ones. The exact nature of these proposed modules, their neurological underpinnings and how they may function to produce cognition are not yet clear. They pose some of the most critical and exciting questions for cognition research today.

Animal behaviour and cognition

Great animal behaviourists like **Niko Tinbergen**, **Konrad Lorenz** and **Karl von Frisch** were also great naturalists who recognized the fundamental importance of observing behaviour in an animal's natural environment. While laboratory-based studies of behaviour are necessary, behaviour in artificial settings is often very different from behaviour in nature. Thus, the logical place to begin studying behaviour is to identify and observe it in the field, as do naturalists. Given that cognition plays a critical role in many aspects of an animal's life and how it behaves, the study of cognition should likewise include field-based observations and studies.

This need is perhaps best illustrated by cognitive research investigating how animals perceive their environments. As articulated by Jacob Von Uexkull's concept of '**umwelt**', animals inhabit unique sensory worlds. To understand how different animals process sensory information, it is necessary first to know what type of sensory information is being collected and how it is being used; such information comes from observations of the animal in its natural environment. For example, the star-nosed mole possesses a unique 'star' appendage that sprouts from the end of its nose. Its function as a highly specialized touch organ was deduced from observations of moles using the appendage to detect, identify and eat prey with remarkable speed. Subsequent research on this extraordinary organ has uncovered striking similarities in the somatosensory system of the star-nosed mole and the mammalian visual system, and shed light on general principles of how the brain organizes sensory information.

The naturalist perspective of animal behaviour has proved to be vital in other areas of research in animal cognition. One area that illustrates this well and also demonstrates how biological and psychological approaches are being integrated is research on spatial cognition in food-storing birds and central-place foraging insects. Like the Clark's nutcrackers mentioned before, several species of birds – primarily parids (chickadees and titmice), corvids (jays, crows and nutcrackers) and sittids (nuthatches) – hide food in a large number of locations at summer's end, and then recover the food as needed throughout the winter and spring. Using a combination of field observations and carefully controlled experiments in the laboratory, researchers have shown that many food-storing birds are using spatial memory to recall where they hid their seeds, with remarkable accuracy. The results suggest the possibility that the foraging ecology of food-storing birds has shaped specialized memory systems. If so, we might predict that certain species that rely more on food caches have a better-developed spatial memory.

One testing of this idea was a comparative study of the long-term spatial memory of four seed-caching corvids. Field observations of nutcrackers and pinyon jays show that they rely on cached seeds for a longer period of time than do Western scrub jays and Mexican jays, and thus would be predicted to have superior spatial abilities. In a laboratory setting, birds were allowed to cache and later given recovery sessions at varying time intervals. The results showed that there were differences in ability to locate hidden caches: nutcrackers and pinyon jays had better long-term spatial memory than the species that were less dependent on stored food, the Mexican and scrub jays. To test whether these results were specific to cache recovery tasks or were indicative of overall superior spatial memory abilities, comparisons of these species were also performed using different spatial tasks. Studies using radial mazes and operant tests, well-developed experimental paradigms in psychology, confirmed that the birds that relied more heavily on stored food, the nutcrackers and pinyon jays, performed better on spatial tasks than the species less reliant, the Mexican jays and scrub jays. Together, these studies provide strong evidence that spatial cognition in food-storing birds is shaped by its foraging ecology.

Foraging insects face different spatial challenges in nature from those of food-storing birds. For insects such as honeybees and ants that are central-place foragers, the task is to find their way to a goal, whether it is their nest or a good food source. Many such insects are known for their remarkable ability to accomplish this; honeybees, for example, will return repeatedly to a specific location of a rich nectar source several kilometres away. Much research in this area has focused on how animals learn and use landmarks and has used well-developed techniques from psychology, such as the radial-arm maze, to study spatial learning in the laboratory. These carefully controlled studies have revealed a great deal about how landmark information is encoded and used on a small scale.

However, studies of insects foraging in their natural environments have led to a richer understanding of spatial cognition. For example, field studies have revealed that *Catalgyphis* sp. ants use path integration to navigate in landmark-barren deserts, and honeybees use the sun as a directional compass. Using path integration and compass orientation probably involves a different set of computational problems and solutions than using landmark information does, and thus highlight the importance of studying spatial cognition in a variety of contexts. Moreover, extensive studies of honeybees and ants show that they use multiple strategies for navigation. In honeybees, for example, large-scale navigation involves the use of the sun as a compass and landscape features as landmark guides, while small-scale navigation near the goal involves the use of landmarks close to the goal, and odour cues. An integrative theory of spatial cognition requires an understanding of how various strategies are integrated to allow animals to find their way around. This can best be accomplished by combining insights from the laboratory with studies in the field, and by integrating insights from both biological and psychological perspectives.

Research in spatial cognition is but one of many areas that are successfully integrating psychological and biological traditions and developing the integrated answers about behaviour advocated by Tinbergen. Other areas, such as communication, are discussed in other entries of this encyclopedia. (CAW)

See also: Behaviourism; Deceit behaviour; Experimental analysis of behaviour; Imitation; Intelligence – comparative; Observational learning; Perception; Theory of mind

References and further reading

Balda, R.P., Pepperberg, I.M. and Kamil, A.C. (1998) *Animal Cognition in Nature*. Academic Press, New York.

Dukas, R. (1998) *Cognitive Ecology: the Evolutionary Ecology of Information Processing and Decision Making*. University of Chicago Press, Chicago, Illinois.

Kamil, A.C. (1987) A synthetic approach to the study of animal intelligence. *Nebraska Symposium on Motivation* 35, 257–308.

Shettleworth, S.J. (1998) *Cognition, Evolution, and Behaviour*. Oxford University Press, New York.

Wasserman, E.A. and Zentall, T.R. (2006) *Comparative Cognition: Experimental Explorations of Animal Intelligence*. Oxford University Press, New York.

Cognitive bias

Cognitive bias describes the altered perception of positive and negative events resulting from an individual's background emotional state. In human psychology, studies on individuals experiencing particular **emotions**, which can be either reported (e.g. **depression**) or experimentally induced, report a change or bias in information processing. These cognitive biases can be grouped into three types: attentional bias, memory bias and judgement bias (see Paul *et al.*, 2005 for a review).

In the field of animal welfare science, cognitive bias can be considered a subjective component of emotion, or affective state, providing an additional approach to physiological and behavioural assessments of **welfare**. Although the observation of cognitive bias may indicate a component of emotional state, the accompanying conscious experience of emotion in animals cannot be assumed (Burman *et al.*, 2008) (**see: Consciousness; Sentience**). In cognitive bias tests on animals, the subjects are taught to respond to positive and negative reference stimuli, and then presented with intermediate (ambiguous) probes, their response to which is taken as an indicator of their cognitive bias.

For example, Burman *et al.* (2008) trained rats in a spatial discrimination task to recognize two reference locations (rewarded and unrewarded). Once the rats had successfully discriminated between the two reference locations, three ambiguous probes, one midway between the two reference probes and the other two between the midway probe and each of the reference locations, were presented to the rats. In this example, the average group latency of two groups of rats was compared; the group of rats that had been kept in an enriched environment (assumed to have a relatively positive background emotional state) had a longer latency to approach the near-negative probe, indicating a more optimistic-type bias when compared with the group of un-enriched rats (assumed to have a relatively negative background emotional state), which had a higher latency to approach this probe, indicating a more pessimistic-type bias. (HFW)

Reference and further reading

Burman, O.H.P., Parker, R., Paul, E.S. and Mendl, M. (2008) A spatial judgement task to determine background emotional state in laboratory rats, *Rattus norvegicus*. *Animal Behaviour* 76, 801–809.

Harding, E.J., Paul, E.S. and Mendl, M. (2004) Cognitive bias and affective state. *Nature* 427, 312.

Paul, E.S., Harding, E.J. and Mendl, M. (2005) Measuring emotional processes in animals: the utility of a cognitive approach. *Neuroscience and Behavioural Reviews* 29, 469–491.

Cognitive dysfunction

Advances in the management and care of **companion animals** mean that an increasing number achieve their lifespan potential, and so all conditions associated with old age are being seen more frequently by veterinarians. **Ageing** brings both specific and general changes in organ structure and function, which can lead to behaviour problems that are often of welfare concern. Ageing and cognitive dysfunction are not the same, and the latter is not an inevitable consequence of the former. Unlike many age-related problems that reflect compensatory behaviour for the inevitable challenges and changes associated with the ageing body, e.g. increased irritability associated with the **pain** of arthritic change, cognitive dysfunction is a condition with a specific pathology, which is being defined increasingly well. A range of otherwise unrelated medical conditions or environmental stressors in the older animal may unmask cognitive decline. This does not, however, mean they are the cause of the problem, which is related to the ageing **brain**.

The ageing process is associated with two key variables, which give rise to the collection of problems labelled as age-related disorders; first, the increase in time the animal has been alive raises the cumulative risk for the animal to be afflicted with a random, spontaneous or time-dependent event. This includes an increased risk of spontaneous lesions such as those due to ischaemic damage or certain types of neoplasia. Secondly, there are physiological changes associated with ageing, which not only affect cells but also their non-cellular constituents after they have been produced.

Theories of ageing concern the cumulative effect of random mutations or cross-linkages in DNA and the effects of cumulative damage from free radicals (including reactive oxygen species) on tissues. In addition, the possibility of an internal clock in the hypothalamus controlling the process through the endocrine system has also been speculated upon. As a result of ageing, brain mass decreases, ventricle size increases and the meninges thicken. Some selective loss of nerve cells occurs but, more importantly, there is a loss of neural connections and the development of plaques in those cells that remain. In some these take the form of neurofibrillary amyloid deposits and granulovacuolar degeneration, which is associated with Alzheimer's disease in humans, which has analogues, known as cognitive dysfunction, a form of senile dementia in some domestic species, notably the dog and cat, which show some but not all of the changes seen in the early stages of human Alzheimer's disease.

Dementia has been defined as a 'global deterioration of intellectual and cognitive functions characterised by a defect of all five major mental functions: orientation, memory, intellect, judgement and affect, but with persistence of a clear consciousness' (Timiras, 1994). The acronym DISHA has been used clinically to describe the most common signs of cognitive decline associated with dysfunction in companion animal species: **D**isorientation; altered **I**nteractions with people or other pets; **S**leep–wake cycle alterations; **H**ouse soiling; and altered **A**ctivity level (Landsberg *et al.*, 2003).

However, these categories do not describe all of the signs that might be associated with cognitive decline, with its increase in anxiety, decrease in hygiene/self-grooming, altered appetite, decreased responsiveness to stimuli, and deficits in learning and memory (especially discrimination and reversal learning and spatial memory) also being recognized. A decline in learning and memory can be demonstrated in dogs, beginning as young as 7 years of age, using a variety of neuropsychological tests. However, clinical cases of cognitive dysfunction syndrome are seldom identified until the age of 11 years or older. Data on cats are less consistent and not as scientifically robust. Both species may be useful models of different aspects of the pathology of the condition in humans. (DSM, GML)

References and further reading

Cummings, B.J., Head, E., Ruehl, W., Milgram, N.W. and Cotman, C.W. (1996) The canine as an animal model of human aging and dementia. *Neurobiology of Aging* 17, 259–266.

Head, E., Milgram, N.W. and Cotman, C.W. (2001) Neurobiological models of aging in the dog and other vertebrate species. In: Hoff, P. and Mobbs, C. (eds) *Functional Neurobiology of Aging*. Academic Press, San Diego, California, pp. 457–468.

Landsberg, G.M. and Head, E. (2004) Aging and effects on behaviour. In: Hoskins, J.D. (ed.) *Geriatrics and Gerontology of the Dog and Cat*, 2nd edn. Saunders, St Louis, Missouri, pp. 29–42.

Landsberg, G.M., Hunthasuen, W. and Ackerman, L. (2003) The effects of aging on the behaviour of senior pets. In: Landsberg, G.M., Hunthausen, W. and Ackerman, L. (eds) *Handbook of Behaviour Problems of the Dog and Cat*, 2nd edn. Saunders, Edinburgh, UK, pp. 269–304.

Timiras, P.S. (1994) *Physiological Basis of Aging and Geriatrics*, 2nd edn. CRC Press, Boca Raton, Florida.

Cognitive ethology

The term cognitive ethology was coined by Donald Griffin, an opponent of radical **behaviourism**, in an attempt to encourage scientists to investigate mental states, such as **consciousness**, in non-human animals. It is a branch of **ethology** concerned with the influence of conscious awareness and intention on the behaviour of the animal. The evolution and context of the behaviour is important in the study of cognitive ethology, and animals are usually observed under fairly natural conditions and with biologically relevant tasks or stimuli. Cognitive ethological studies also emphasize broad taxonomic comparisons instead of focusing on a few representatives from a few taxa. The overall goal of cognitive ethology is to understand the phylogenetic distribution of mental processes and their significance for the evolution, development, causation and adaptiveness of behaviour (**see: Tinbergen's four questions**). (LMD)

See also: Cognition; Comparative psychology; Ethology; Evolutionary psychology

Further reading

Ristau, C.A. (ed.) (1991) *Cognitive Ethology*. Lawrence Erlbaum Associates, Trenton, New Jersey.

Cognitive map

Cognitive maps are a type of mental processing or **cognition** used to structure and store spatial knowledge into an internal representation of the environment. These 'mental maps' of the spatial environment can help enhance recall and learning of information by allowing visualization of images in order to reduce cognitive load. (LMD)

Cognitive model

Cognitive models are used for the comprehension and prediction of cognitive phenomena or processes. They usually focus on single cognitive phenomena or processes, on how two or more processes interact (e.g. **decision making**) or on how to make behavioural predictions for a specific task or tool. (LMD)

Cold stress

Cold stress refers to a fall in the core body temperature below the thermoneutral zone, and can result in clinical hypothermia. Animals are generally quite tolerant of low temperatures, but cold stress is more likely when these are accompanied by strong winds or draughts or damp conditions, which affect the insulation properties of the coat or pelage. The provision of shelter is therefore important in preventing this problem in these conditions. Animals adapted to living in cold climates often use fat deposits to provide insulation not affected by these conditions. When temperatures drop, an increase in basal metabolic rate is necessary to maintain activity, but food may be limited at these times and so the animal may enter a state of torpor or hibernation to minimize energy demands. (DSM)

See also: Critical temperature

Collar

Collars are bands of resilient, usually flexible, material placed around the neck or sometimes the head of animals to restrain and lead them, the premise being that where the head goes, the body will follow. They are used in dogs for restraint and leading, in cats to carry identity discs and in a variety of farm species for tethering. Apart from traditional 'flat collars' that work simply by applying pressure to one side of the neck, collars may be designed to constrict the neck. Examples include slip collars (that have a discrete section that can be tightened), choke chains (also known as check chains) and pinch collars or prong collars (with elements that protrude inwards when the collar is tightened).

Collars with prongs that face the dog's neck are controversial, since they can damage the soft tissues under the skin of the neck. That said, any collar that constricts the dog's neck can be used abusively and injudiciously to the extent that many trainers avoid using them or advocating their use. Good trainers can generally get the same or better results through heelwork (during which the dog walks close to the handler's leg), without the use of force.

The use of electric 'shock collars' is also highly controversial, especially in the hands of inexperienced pet owners or those seeking an alternative for consolidated shaping and positive **reinforcement** training. Their use has been outlawed in some countries. In contrast, 'head halters' or 'head collars' are popular among canine behaviour therapists since they are generally useful, especially in restraining and training large dogs or those that have undergone **habituation** to regular collar pressures. Pulling on the leash places pressure on the dog's muzzle and neck, via a ring under its jaw, and thus guides the head and consequently the body.

A further modification on the traditional collar is the 'spray collar', which emits an **aversive** but non-toxic atomized aliquot of fluid, such as a citronella-scented preparation. Some owners find bark-activated versions of these devices effective in controlling unwanted barking. The extent to which other unwelcome behaviours can be effectively eliminated by remote punishment mediated through such spray collars depends entirely on the operator's consistency and timing.

Modified collars may be used to prevent auto-grooming that is inappropriate (e.g. after surgery or in cases of self-mutilation); examples include so-called Elizabethan collars (in small animals) and neck cradles (in horses).

Because both **crib-biting** and **wind-sucking** are accompanied by characteristic flexion of the horse's neck, they can be discouraged by the application of a collar that punishes the flexion with discomfort. The typical (anti-)cribbing collar is a simple leather strap incorporating a galvanized, hinged arc that accommodates the trachea and allows normal breathing, even when tightened to the point where neck flexion and/or engulfing air into the oesophagus are not performed. It is possible that the presence of a constriction in this site not only makes crib-biting uncomfortable but may also make the behaviour less easy to consummate, since the terminal grunting in this oral-based stereotypy is known to involve distension of the cranial oesophagus. Modifications of the collar include the use of leather or metal spikes to heighten the discomfort associated with neck flexion, as well as designs made entirely of leather and often marketed as being distinctly 'humane' (even though they work along the same principles as above). Horses often adapt to the constriction of the collars, which are subsequently tightened, occasionally to the extent that skin trauma follows. Combinations of straps that can be tightened around different parts of the cranial neck have been advocated to reduce tissue damage at the poll.

Commercially available electronic training collars can be programmed to discharge when the neck is flexed or remotely controlled, so that the horse does not associate punishment with the presence of a human. The use of dummy collars before aversion therapy of this sort may avoid the horse associating punishment with the collar itself. It is reported that by applying an electric shock after the horse has been grasped, but before it has engulfed air, punishment and extinction of grasping may be achieved. However, follow-up studies in 60 cases of aversion therapy using electric shock collars indicated that only nine 'cures' were effected and, of these, three required further shocks after 9 months (Owen, 1982). One practical problem with such aversive techniques may be that the horse may not crib-bite for a long time after the first shock, so constant observation and considerable patience may be required for the trainer to maintain the contingency between crib-biting and punishment. A more profound ethical problem is that physical prevention or punishment may increase distress levels. It is possible that elimination of one stereotypic behaviour may precipitate unwelcome behavioural side effects, including the emergence of a modified or alternative stereotypy. (PDM)

Reference

Owen, R.R. (1982) Crib-biting and wind-sucking – that equine enigma. In: Hill, C.S.G. and Grunsell, F.W.G. (eds) *The Veterinary Annual*. Wright Scientific Publications, Bristol, UK, pp. 156–168.

Colostrum

Colostrum is the first fluid produced by the mammary glands for the initial few days following **parturition**. It is somewhat different from the milk of the main **lactation**, being high in carbohydrates, proteins, vitamins and minerals, and also containing antibodies that help to protect the offspring from diseases early in life. Animals that do not receive colostrum tend to be sickly and have an increased **mortality** rate. (LMD)

Further reading

Uruakpa, F. (2000) Colostrum and its benefits: a review. *Nutrition Reserarch* 22, 755–767.

Comfort behaviour

Comfort behaviour is a generic term given to a heterogeneous group of activities thought to be involved in body maintenance, which includes keeping clean and relieving skin irritation and muscle tension. For example, cattle swish their tails and stamp their legs to keep flies on the move; during **dust bathing**, chickens lower themselves into a sitting position, fluff their feathers and move their wings such that dust (or other material) moves through the wings and on to the back; pairs of horses stand nose to tail so that both animals benefit from the tail movements of the other; ducks dip their heads repeatedly in water as part of bathing; pigs can scratch the head and neck region by lifting a hind leg or scratch their hindquarters by sitting down and moving backwards and forwards.

Indeed, a variety of comfort behaviours have been described for birds, rabbits, pigs, cattle and other animals, including: wing flapping, feather raising, stretching (wing and leg), scratching, shaking (head and body), preening, beak wiping, yawning, sneezing, tail wagging, tail flicking, hoof stamping, dust bathing, water bathing, **grooming**, rubbing, ruffling, ground scratching and wriggling. Other terms for comfort behaviour include comfort movements and comfort activities. Some of these behaviours involve interaction with environmental resources (e.g. dust bathing), while other behaviours are self-directed (e.g. **preening**).

A variety of factors affect the performance of comfort behaviours, including spatial limitations, social factors and environmental stimuli. In some intensive housing systems space is restricted, and this may prevent the performance of comfort behaviours. For example, as cage height increases, hens are more likely to engage in head stretching, head scratching and body shaking. Similarly, when given increased cage area, hens are more likely to engage in head scratching, body shaking and feather raising (Nicol, 1987). These findings indicate that hens require adequate space in order to perform specific comfort behaviours.

In addition to spatial limitations, the presence or proximity of other animals may influence the performance of comfort behaviours. For example, the frequency of tail wagging, body shaking and preening increases when other hens are visible or in close proximity, while the frequency of other comfort behaviours, such as beak wiping, are not affected by visible contact with **conspecifics** (Nicol, 1989). Social grooming is another example of how performance of comfort behaviours can be influenced by the presence of other animals. For instance, when horses perform **allogrooming**, animals face one another and use their lips and teeth to groom the neck, withers and back of the other animal.

The performance of some comfort behaviours is closely linked to the availability of environmental stimuli or resources. For example, functional dust bathing requires a loose substrate for chickens to spread over their body and through their wings. In addition, comfort behaviours are often thought to relieve irritation from environmental factors such as parasites or weather. For example, male shelducks perform nearly twice as many body shakes in response to rain, compared with visual contact with another displaying male or no other stimuli (Sebbel *et al.*, 1998).

Examining the factors that affect the performance of comfort behaviours highlights the heterogeneity of this group of activities. Although it may be convenient to group behaviours such as yawning, preening and wing flapping into the same category, the function of these behaviours may differ. For example, preening is likely to be involved in feather maintenance, but birds may also preen during an aggressive social interaction (**displacement behaviour**). Similarly, wing flapping may be important to relieve muscle tension, but cocks may also flap their wings to communicate to other males (**see also: Communication**). Finally, yawning is thought to be important for forceful exchange of gases in the lungs, but it is also socially facilitated, indicating that it may have other functions.

Knowledge about the specific function of these activities is important when using comfort behaviour to assess animal welfare. Indeed, multiple functions for a single behaviour can complicate the interpretation of comfort behaviours when they are used as measures of animal **welfare**. In addition, there are other difficulties interpreting the importance of comfort behaviours when assessing animal welfare (**see: Welfare measurement**). For example, in the spatial limitation example described above, an increase in the frequency and the ability to perform comfort behaviours in larger cages is interpreted as good for the hens. However, an increase in the frequency of comfort behaviours may not always be a positive welfare indicator. For example, there may be clear differences in the frequency of comfort behaviours between animals housed in a parasite-ridden aviary compared with those housed in one well maintained. Hens may perform more scratching and preening in the parasite-ridden aviary, although this does not indicate that their welfare is better in this system. Like many other measures of welfare, comfort behaviours need to be interpreted in conjunction with other measures, such as parasite load.

Learning about the **motivation** to perform comfort behaviours is an area that has the potential for improving our understanding of the importance of these behaviours for animal welfare. The motivation to perform comfort behaviours can be assessed in a number of ways. Environmental costs, such as restricted **time budgets** or limited space, may be used to assess the importance of various comfort behaviours, particularly for self-directed comfort behaviours. In contrast, the importance of comfort behaviours that involve environmental resources can be assessed by manipulating the availability of these resources (**see also: Behavioural need**). Finally, motivation to perform comfort behaviours can also be assessed by depriving animals of the possibility to perform these behaviours and monitoring **'rebound' behaviour**. The physical benefits (e.g. feather condition, bone strength) of comfort behaviour can also be determined by preventing animals from performing them and assessing changes in condition. As with motivation, documenting the health benefits of comfort behaviours improves our understanding of the importance of these behaviours for animal welfare.

In addition to the group of comfort behaviours describe above, the term 'comfort' is used in other contexts, such as in relation to thermoregulatory behaviour (**see: Thermoregulation**) and differences in time budgets, e.g. **resting behaviour** in cattle. (CBT)

References and further reading

Nicol, C.J. (1987) Effect of cage height and area on the behaviour of hens housed in battery cages. *British Poultry Science* 28, 327–335.

Nicol, C.J. (1989) Social influences on the comfort behaviour of laying hens. *Applied Animal Behaviour Science* 22, 75–81.

Sebbel, P., Duttmann, H. and Groothuis, T. (1998) Influence of comfort and social stimuli on a comfort movement and a display derived from it. *Animal Behaviour* 55, 129–137.

Commensalism

Commensalism is a symbiotic relationship between two or more organisms from different species in which one organism benefits from the relationship while the other is unaffected by it. Commensal relations may form between and within plants, fungi and animals, although some argue that no relationship is completely neutral, and that therefore the relationship may be either mutual or parasitic. (LMD)

Communication

Communication is the transfer of information from a sender to a receiver, where both the sender and the receiver map a **signal** to a particular meaning (see Hauser, 1996, for several other definitions). Although some may consider information transfer as anything that reduces the receiver's uncertainty about future events, such a definition implies that communication is a one-way system and does not take into account manipulation, or even the intentional ploy of the sender, in some cases to enhance the receiver's uncertainty about future events. If the 'signal-to-meaning' mapping involves syntax (a formal structuring of the signals in relation to each other), this is termed **language**.

Signal-to-meaning mapping does not necessarily involve intention or conscious processing. Plants have also been found to communicate information with each other; for example, undamaged sugar maple seedlings and maize synthesize compounds that make them less palatable – or even slightly toxic – when damaged, conspecifics emit airborne signals when their leaves are damaged and chemical signals can even be emitted that attract predators of the likely perpetrators.

Signals may be distinguished from cues in that signals can be varied behaviourally by the sender (for example, an animal can call or not call, as well as possibly use many varieties of a call), and involve some cost for the signaller to produce. In contrast, cues are typically always present, such as the warning or 'aposomatic' coloration of some poisonous species, and there are no extra costs associated with each usage.

In the animal kingdom every possible sensory information channel is exploited for communication. Sound waves are used, for example in song, grunts and calls, such as: (i) the low-frequency calls made by elephants, which travel through the ground and can be picked up by other elephants via their trunks and feet; (ii) the visual signals that are used in the form of colours and decoration, body language, gestures, facial expressions, light signals, dance, display and even the way in which a nest is built and decorated, as by the hammerkop (*Scorpus ombretta*) to signal fitness to potential mates; (iii) chemical signals (**see: Chemical communication**), including **pheromones**, which play a role in partner selection, signalling alarm and the setting out of trails, aversive defensive chemicals, chemical territorial markings and social odours, such as the chemicals exchanged between lions (and domestic cats) when rubbing their heads against each other, enhancing bonding; (iv) the vibration used, for example, by the southern green stinkbug (*Nezara viridula*) to signal both gender and location through the vibration of leaves attached to plants or trees; (v) the electricity used by some aquatic species, for example, the frequency of the electric discharge of the whale-faced marcusenius (*Brienomyrus brachyistius*) is used by conspecifics to determine gender (**see: Electric fields**); and (vi) touch, for example, the licking used as a greeting by marmosets (*Callithrix jacchus*).

The specific channel used for communication depends on the signal content, sensory ability of the species concerned as well as the prevailing environment for communication; for example, pigs often use auditory signals to communicate rapidly socially, as their natural woodland habitat is a barrier to visual signals.

Communication is generally focused on the needs of the sender or the receiver, linked to the survival of the individual or the species: to find a mate, to defend, to threaten, to mark territory, to coordinate **hunting**, to determine or establish a social hierarchy, to know something about the state of the sender (e.g. whether prepared to mate), etc. The needs of the sender and receiver in an evolutionary context do not require that communication is always truthful: senders can benefit by using signals to mislead receivers. For example, roosters will sometimes peck at the ground, as if foraging, even when no food is present, in an apparent attempt to attract hens to their vicinity. 'Honest' signalling is more likely if the signals are expensive to produce. For example, crowing acts as an honest signal of social status, in part because subordinates are more likely to be pecked by flock-mates if they crow.

Communication is not limited to social animals, although communication among animals living in groups tends to be more varied (both in channel use and the repertoires employed) compared with those living solitary lives. For example, bonobos (*Pan paniscus*) use facial expressions, gestures, calls, grunts, touch, grooming and even sex in communication (de Waal and Lanting, 1997), while the solitary living giant kangaroo rat (*Dipodomys ingens*) communicates mostly by leaving scent trails in the sand and by foot drumming (Murdock and Randall, 2001).

The total repertoire of signal-to-meaning mappings is vast for human beings (the adult 'mental lexicon' contains 50,000–100,000 words, depending on the level of education) compared with the signal repertoires available to other animals. For example, vervet monkeys living in Amboseli National Park in Kenya have three different alarm calls to signal danger: (i) leopard alarm calls enticing vervets to run into the trees; (ii) eagle alarm calls causing vervets to look up; and (iii) snake alarm calls making vervets look down (Seyfarth *et al.*, 1980).

Attempts have been made artificially to increase the total set of signals that are naturally available to chimpanzees (Rumbaugh *et al.*, 2003; Hillix and Rumbaugh, 2004), dolphins (Herman *et al.*, 1984), parrots (Pepperberg, 2000) and dogs (Kaminski *et al.*, 2004; Rossi and Ades, 2008). For example, a border collie named Rico, who knew the names for at least 200 objects (as tested by retrieving these objects), was taught almost 40 names for new objects in single trials, and was shown to remember these new names after a period of 4 weeks, making his performance 'comparable to the performance of three-year-old toddlers' (Kaminski *et al.*, 2004).

Although the human-like ability of 'fast mapping' was claimed for the process of name acquisition by Rico, methodological problems in this and other animal studies have not refuted the suggestion that the size, structure, use and acquisition of the mental lexicon in humans is qualitatively different compared with other animals. The human mental lexicon encodes, among other things, complex relations (e.g. 'later' and 'conclusion'), abstract categories (e.g. 'planets' and 'recursive') and syntactic features (e.g. 'gender' and 'number'). With the (spontaneous) intention to convey a message, humans normally produce two to three words per second without repeating the same item and while making only one to two mistakes per 1000 words.

Humans learn signal to meaning mappings by using a **theory of mind** to understand word reference (e.g. to understand pointing, or to understand what the signaller attends to); humans do not have to be taught which items should be learned from among the total available input (e.g. 'this is not a car' does not contain a reference to a name ('not-a-car') as an entity), and humans learn words without reinforcement (Bloom, 2004; Markman and Abelev, 2004). Although research continues to investigate to what extent these and other features are part of the signal to meaning mapping in other species, it is quite unlikely that all the features found in humans will be found in any single other species, due to differences in the evolutionary histories of humans and non-humans, including our closest primate cousins.

Most signals have evolved for communication between members of the same species, but some signals are also used for interspecific communication. For example, prey species use a variety of signals in response to the presence of potential predators, such as 'alarm calls' by chipmunks or 'stotting' by gazelles (**see: Alarm reaction**). These signals may have evolved as a way of warning group mates, but they also function to inform predators that their presence is known and that the probability of success is therefore low. By attending to such signals the predator saves the costs associated with (unsuccessful) pursuit.

It is generally accepted that signals have evolved by natural selection out of behaviours that had non-signal functions (Hauser, 1996); for example, urinating has in some species developed into a signal for territory marking. Although historically emphasis has been given to the role of selection operating mostly on the sender to convey unambiguous information for the recipient, during competitive interactions selection also operates 'against individuals using displays that are highly predictive of their subsequent behaviour' (Hauser, 1996). While it is in the interest of the sender to deceive and to manipulate the truthfulness of a signal, it is in the interest of the receiver to decode a signal reliably, and to resist any manipulation successfully. In studying the evolution of communication, and in comparing the signal to meaning mappings between different species, it is important to focus on both senders and receivers and on both truthful and deceitful aspects of communication (**see: Deceit behaviour; Honest signals**).

For people working with animals, an understanding of their signals is especially important and, arguably, many of the animals that we have domesticated have become proficient at recognizing the signals we use (**see: Dog**). Although there appear to be some common properties to certain signals (for example, short, high-pitched rising tones tend to be used to communicate the need to speed up across taxa, and longer, low, descending tones tend to be used to inhibit behaviour), similar signals may also be used quite differently between taxa: for example, the baring of teeth by a human may represent a friendly smile, but a similar gesture by a dog can have a defensive/aggressive meaning. A failure to appreciate this is believed to be one factor contributing to the higher rate of dog bite injuries in children (**see also: Bite prevention programmes**). (EvdZ, DW)

References and further reading

Bloom, P. (2004) Can a dog learn a word? *Science* 304, 1605–1606.

De Waal, F. and Lanting, F. (1997) *Bonobo: the Forgotten Ape*. University of California Press, Berkeley and Los Angeles, California.

Hauser, M.D. (1996) *The Evolution of Communication*. MIT Press, Cambridge, Massachusetts.

Hauser, M.D. (2000) A primate dictionary? Decoding the function and meaning of another species' vocalizations. *Cognitive Science* 24, 445–475.

Herman, L.M., Richards, D.G. and Wolz, J.P. (1984) Comprehension of sentences by bottlenosed dolphins. *Cognition* 16, 129–219.

Hillix, W.A. and Rumbaugh, D.M. (2004) *Animal Bodies, Human Minds: Ape, Dolphin, and Parrot Language Skills*. Springer, New York.

Kaminski, J., Call, J. and Fischer, J. (2004) Word learning in a domestic dog: evidence for 'fast mapping'. *Science* 304, 1682–1683.

Markman, E.M. and Abelev, M. (2004) Word learning in dogs? *Trends in Cognitive Science* 8, 479–481.

Murdock, H.G. and Randall, J.A. (2001) Olfactory communication and neighbor recognition in giant kangaroo rats. *Ethology* 107, 149–160.

Pepperberg, I.M. (2000) *The Alex Studies: Cognitive and Communicative Abilities of Grey Parrots*. Harvard University Press, Cambridge, Massachusetts.

Rossi, A. and Ades, C. (2008) A dog at the keyboard: using arbitrary signs to communicate requests. *Animal Cognition* 11, 329–338.

Rumbaugh, D.M., Beran, M.J. and Savage-Rumbaugh, E.S. (2003) Language. In: Maestripieri, D. (ed.) *Primate Psychology*. Harvard University Press, Cambridge, Massachusetts, pp. 295–423.

Searcy, W.A. and Nowicki, S. (2005) *The Evolution of Animal Communication: Reliability and Deception in Signalling Systems*. Princeton University Press, Princeton, New Jersey.

Seyfarth, R.M., Cheney, D.L. and Marler, P. (1980) Monkey responses to three different alarm calls: evidence of predator classification and semantic communication. *Science* 210, 801–803.

Communitarianism

Communitarianism is an approach to ethical and social issues according to which elements of community (friendship, common purpose, tradition) possess intrinsic value and justify the imposition of rules and moral norms. This contrasts with theories according to which community matters only indirectly – for example, because it makes people well off (**utilitarianism**) or would be agreed to by reasonable people (**contractarianism**). The debate about communitarianism may be relevant to discussions on traditional forms of animal use. (PS)

Companion animal

Usually preferred in the scientific literature to '**pet**', this term usually refers to any animal kept primarily for companionship. It is therefore a functional definition, and strictly speaking could be applied to almost any non-human species. Companion animals may therefore belong to any of the groupings used by veterinarians (farm, small, equine and exotic), even though the companion animals that have received most research attention from ethologists (**dogs** and **cats**) fall into the 'small' category. However, being a member of one of these species does not automatically confer companion animal status; consider, for example feral cats, or dogs bred primarily for their meat or fur. The term 'companion animal' should therefore be applied to individual animals, not a breed or species. The current preference for 'companion animal' over 'pet' appears to stem from a desire to indicate that such animals have a status of their own, and are not simply the possessions of their owners.

Some authors restrict the use of 'companion animal' or 'pet' to domesticated species. However, this would technically exclude animals that were originally wild-caught and have been tamed (parrots, for example) and then develop a close relationship with an owner for many years. Moreover, it glosses over the differences in the extent of **domestication** that exist between and within the species traditionally regarded as 'companion'. For example, pedigree Persian cats fit the definition of domestication in that their breeding and nutrition are completely controlled by man but, strictly speaking, non-pedigree cats might not, because they generally select their own mates (from any cats nearby that have avoided neutering).

It has been suggested that companion animals should encompass all those shown special kindness and consideration by people, but this would include most wild animals in zoos, and also wild birds deliberately fed in gardens, neither of which would normally be thought of as being pets. The requirement for the animals to be under human control and kept primarily for companionship therefore seems to be essential. The definition is therefore focused on the owner's perception of the relationship; whether the animal perceives the owner as a 'companion' (member of the same affiliative social group) is a separate issue.

However, the assumption is usually made that in a successful companion animal–owner relationship, social benefits accrue to the animal as well as by the human partner. This is most obvious in dogs and also cats, which display species-typical affiliative behaviour towards familiar people. It is less apparent in some other animals that can be referred to as 'pets' – for example, fish in tanks (fish are probably the most numerous pets, in terms of total numbers of individuals). It is still less certain how to classify insects such as exotic spiders, or the crickets and beetles kept by children in Japan and referred to as 'mushi'.

Because 'companion animal' describes current status, it need not apply to an animal throughout its life. For example, working dogs, initially kept mainly for utilitarian purposes, may be 'retired' to their former keeper or elsewhere and thereby become companions. Logically speaking, a dog that is abandoned by one household and spends time at an **animal shelter** prior to being adopted by another should temporarily lose its 'companion' status, but this is probably one distinction too far. More problematic are animals that simultaneously fulfil companion and utilitarian roles. Examples of these are widespread both geographically and across species. Many of those who use dogs for guiding and searching functions form a strong emotional bond with them. Indian **elephants**, an undomesticated species that is easy to tame, are regarded as companions as well as working partners by their mahouts. Finally, many people would automatically exclude animals that are eventually destined to be eaten, even though some undoubtedly provide companionship for their owners; such relationships occur in a wide range of cultures, from the New Guinea village where pet **pigs** are slaughtered for ritual purposes to the modern hobby-breeder of rare types of domestic stock that are, for example, given names but are ultimately destined for the table.

Overall, it appears to be easy to classify some animals as companions – the family dog for example – but more difficult to define the boundaries of what should be excluded. This appears to reflect a continuum from those animals that are exclusively kept for companionship for their whole lives to those for which companionship is only a minor or temporary part of their relationship with man, and in which other, such as utilitarian or decorative, functions are predominant. Although it is incorrect to define a companion animal according to which species it is, it is self-evident that some species occupy such a role much more than others do, and also that there are qualitative differences between species in the type and extent of companionship experienced by their owners: otherwise, all species would be interchangeable. Some authors have wished to refine the definition of companion animal in order to address such questions as 'why do people keep pets?' or 'what benefits do people get from pet-keeping?' more effectively (**see: Anthrozoology**). Others have pointed out that, due to the differences between the species involved, such questions may be oversimplified, and recent research has indeed suggested that cat ownership and dog ownership (for example) may differ from one another both in motivation and effects. (JWSB)

Further reading

Eddy, T.J. (2003) What is a pet? (and commentaries). *Anthrozoös* 16, 98–134.

Serpell, J.A. (1996) *In the Company of Animals.* Cambridge University Press Cambridge, UK.

Comparative biology

Comparative biology concerns the study of homologous (**see: Homology**) and analogous (**analogy**) features in living systems and their relationship to each other. Comparative biology can extend into any of the life science disciplines from the molecular to the organismal or environmental, such as genetics, anatomy, psychology (**see: Comparative psychology**) and **ecology**. In some cases where the focus of the comparison is on the phylogenetic relationships between individuals, the discipline may be prefaced with the term 'evolutionary', as in **evolutionary psychology**. (DSM)

See also: Comparative method

Comparative method

The comparative method is a cross-species comparison that relates species differences in behaviour, anatomy or physiology to differences in species ecology. For example, variation in testicular size is closely associated with mating system in primates, bats, antelope, birds, amphibians, fish and insects. Essentially, males have relatively large testes in species whose mating system is characterized by female promiscuity – an adaptive response to the risk of sperm competition. In using the comparative method it is important to control for both confounding variables (especially body size) and the non-independence of species data. Because species within a genus tend to have similar characteristics due to shared common ancestry, analysis of species data reflects phylogeny rather than ecology. Thus, modern comparative methods attempt to distinguish independent evolutionary origins of character states from cases of identity by common descent. (PE)

Further reading

Harvey, P.H. and Pagel, M.D. (1991) *The Comparative Method in Evolutionary Biology*. Oxford University Press, Oxford, UK.

Comparative psychology

Comparative psychology is the study of the development and evolution of behaviour across species. Modern-day comparative psychologists study a wide range of animal behaviours. However, early comparative psychologists focused specifically on animal **cognition** and **intelligence**, basing their research on the belief that the origins of the human mind could be understood by studying the animal mind. **Charles Darwin** (1809–1882) is credited with being the father of modern comparative psychology because his theory of **evolution**, which revolutionized biological science, stimulated scientific exploration of the mental life of animals.

In his *On the Origin of Species*, Darwin suggested that natural selection occurred through biological principles. Darwin stated that the mental life and behaviour of organisms followed a continuum such that there were no qualitative differences in behaviour and mental functioning. In support of his idea of mental continuity between humans and animals, Darwin described: the cleverness of animals in making use of their natural environment (e.g. dam building in beavers); the adaptability of domestic or captive animals to unnatural environments, which he took as evidence of reasoning; their propensity to imitate humans or other animals, implying rational inference; their intercommunication of ideas and plans of action, suggesting cooperation; and their similar emotional reactions, suggesting that animals are capable of deceit, shame, a sense of justice and **altruism**. He did, however, concede that there may be quantitative differences across species such that animals could be divided between those that were capable of higher-level thought processes and those that were capable of only precursors for higher-level thought processes.

By stressing the continuity of species, Darwin made the study of animal behaviour crucial to the understanding of human behaviour. Although much of Darwin's theory of the continuity of mental functioning in humans and animals was based on anecdotal evidence, his theory of evolution served to generate a number of testable hypotheses and inspire new methods of scientific enquiry, which allowed scientists actively to pursue the collection of evidence for mind or intelligence in animals.

The focus of experimental research in comparative psychology following Darwin's theory of evolution took one of two approaches. Scientists either sought evidence to suggest that humans and animals were alike in mental functioning or that humans were fundamentally different from animals in mental functioning. One of the first scientists formally to study animal intelligence was **George John Romanes** (1848–1894). Romanes was directly influenced by Darwin's readings and sought to demonstrate the similarity between human and animal functioning by studying a wide spectrum of species, ranging in physical size and complexity from single-celled organisms to vertebrates. Romanes provided anecdotal evidence for animal reasoning and intelligence from watching the behaviour of animals in their natural environment.

Other researchers criticized his use of subjective evidence and sought to provide a more objective and experimental methodology for studying animal behaviour. For example, the theories of classical conditioning developed by **Ivan Pavlov** and **instrumental conditioning** developed by E.L. Thorndike together explained objectively a great deal of behaviour and seemingly cognitive behaviour observed in both humans and animals, strengthening the idea that humans and animals were not dissimilar in their learning abilities.

These two theories served as catalysts for the development of a new movement in psychology known as **behaviourism**, which dominated the field of psychology for several decades. Progenitors of this movement sought to remove all mentalistic terminology such as **consciousness**, mind, imagery and mental states from the interpretation of both animal and human behaviour. They believed, unlike Darwin, that the behaviour of organisms differed not in terms of the quality of responses to stimuli but in terms of quantity. That is, more complex organisms might be capable of emitting a greater variety of responses to a greater variety of stimuli, but the principles underlying those responses are no different to those governing the behaviour of less complex organisms. The widespread acceptance of the universality of behavioural and cognitive mechanisms led to a bias in searching for generalities in behaviour that still exists to some extent today.

In contrast to studies underscoring the generality of behaviour across species, some studies began to show that species may, in fact, differ both qualitatively and quantitatively in their abilities to perform similar tasks. Researchers studying the biological bases of animal behaviour in fields such as **ethology**, **behavioural ecology** and **sociobiology** criticized comparative psychologists for a number of factors, including focusing their research on behaviour observed in laboratory settings rather than in the field, studying a limited number of species and concerning themselves mostly with learning at the expense of ecologically relevant behaviour.

These fields caused comparative psychology to take a more evolutionary approach to their study of animal behaviour. It was argued that, for comparative psychology to survive and compete in the intellectual arena with ethology, behavioural ecology and sociobiology, it had to become more integrated with these fields by choosing species based on their phylogenetic origins rather than on convenience, by increasing fieldwork or research in semi-naturalistic settings, by adopting a more ecological approach and by incorporating the theory of evolution into the **anthropocentric** approach in their study of animal behaviour.

Comparative psychologists are now beginning to promote an ecological approach, as opposed to a strictly generalist approach, to learning and cognition. It has long been accepted by scientists that environmental pressures could create specializations in sensory, motor or physiological processes in animals through the process of natural selection. It is, therefore, not unreasonable to think that the same environmental pressures could create specializations in memory and brain function. The ecological approach takes into account the natural history of the organism by identifying and designing experiments to test learning abilities that may be important to the organism in nature, and determining how and why these capabilities may have evolved.

Thus the current field of comparative psychology seems positioned to bridge the gap between psychologists and ethologists by attempting to discover not only similarities, but also differences, both within and across species, based on their phylogeny and evolutionary niche. Modern comparative psychologists are interested in a wide range of behaviours, including orientation and locomotion, feeding and foraging behaviour, predator defence and prey–predator interactions, reproduction, **social behaviour** and **social cognition**, **communication** and **language**, learning, animal cognition or intelligence and **theory of mind**. They study animal behaviour in both the laboratory and the field, in basic and applied settings, using a variety of methodologies.

Comparative research on species differences in learning and memory has increased dramatically and is gaining in popularity in the scientific literature, and is important for those interested in applied animal behaviour and welfare. Understanding the similarities as well as the differences between species can be valuable for improving the management of animals in captivity, specifically in terms of feeding, providing social opportunities and designing **enrichment** that stimulates the animal mind in a manner that is appropriate for each individual and species. Knowledge of the foraging skills and learning capabilities of a species could also be helpful in designing reintroduction programmes. Finally, research from the field of comparative psychology provides insight into the intelligence of animals, which can be valuable for influencing public opinion and increasing the favourability and appreciation of a species. (LRT)

See also: Psychology

Further reading

Domjan, M. (1987) Comparative psychology and the study of animal learning. *Journal of Comparative Psychology* 101, 237–241.

Dore, F.Y. and Kirouac, G. (1987) What comparative psychology is about: back to the future. *Journal of Comparative Psychology* 101, 242–248.

Greenberg, G. (1987) Historical review of the use of captive animals in comparative psychology. In: Tobach, E. (ed.) *Historical Perspectives and the International Status of Comparative Psychology*. Lawrence Erlbaum Associates, Hillsdale, New Jersey, pp. 81–89.

Vauclair, J. (1996) *Animal Cognition: an Introduction to Modern Comparative Psychology*. Harvard University Press, Cambridge, Massachusetts.

Compulsive disorder

A compulsive disorder describes a repetitive or sustained behaviour that does not change much from one situation to another and which is not a normal part of the ethogram of the organism within the given context. Historically, within the animal science literature the terms compulsive behaviour/disorder (or obsessive–compulsive behaviour/disorder) have been used synonymously with the term stereotypy, although the two are clearly distinguished in the human literature. A compulsive behaviour is highly motivated towards achieving a specific goal and, while it can become ritualized and quite rigid in its execution, the behaviour remains goal-focused, while the term stereotypy refers to motor acts that are uncontrollably released and have no obvious goal. The importance of this distinction is increasingly becoming recognized in the clinical animal literature, as it may ultimately be of prognostic and treatment value.

Some compulsive behaviours, such as acral licking of the carpus in the dog, have been proposed as models of human obsessive–compulsive behaviour, and respond to similar treatments (serotonergic agents). However, the 'obsessive' dimension cannot be determined in non-human animals and, in humans, the obsession does not necessarily correlate with the compulsion and so this element of the terminology has been rejected by many. Compulsion can be determined by the high motivation to perform the behaviour even in the presence of physical and psychological barriers. Some stereotypic behaviours may be compulsive behaviours, e.g. weaving in the horse, but some compulsive behaviours do not fit the definition of stereotypic behaviour as they are not repetitive, e.g. light fixations in Border collie dogs, hence the preference for the term compulsion in the clinical literature. (DSM)

Further reading

Mason, G. and Rushen, J. (eds) (2006) *Stereotypic Animal Behaviour: Fundamentals and Applications to Welfare*, 2nd edn. CAB International, Wallingford, UK.

Concentrate feeding

Concentrated feeds, or concentrates, are usually based on cereals or other high-energy and protein feeds. They are generally made into a pellet, or compound, with the addition of a binding agent, most often sugarcane molasses. Concentrates are fed to pigs, poultry and horses, and to cattle in situations where high performance is required or forages are in short supply. The allocation of concentrates should take account of the animals' physiological state, lactating or non-lactating, pregnant or non-pregnant, etc. The majority of dairy farmers prefer to feed concentrates to those cows giving the most milk.

The risk of upsetting the digestion of animals that have evolved to eat large quantities of herbage, by feeding high-concentrate diets and causing problems such as low milk fat concentrations or acidosis, has led to several developments in concentrate feeding. First, if high levels of concentrate are fed, it is better if they are based on digestible fibre, such as that from beets, rather than starch from cereals. The starch in compound pellets is exposed to rapid degradation by gastrointestinal bacteria. In some cases fats may be preferable to starch-based rations, but in ruminants they are likely to need protection from rumen degradation. Secondly, the concentrates may be fed at a low level several times during the day, rather than just once or twice a day, e.g. when visiting the milking parlour. This will help both to reduce excessive bouts of acid production by bacteria that reduce the efficiency of digestive function and also, in the case of cattle, to improve the efficiency of ammonia capture by ensuring a more even degradation of the concentrate feed over time. The high cost of producing, delivering and feeding concentrates compared with forages ensures that they are used in limited quantities, but they are an essential part of the diet of high-producing or potentially undernourished livestock that cannot meet their energy demands from the available vegetation alone. (CJCP)

Concept formation

In 2002, Zentall *et al.* observed that 'there is, perhaps, no larger or more diverse literature within experimental psychology than that focused on categorization and concept learning'. Why, one may wonder, has so much scientific effort been invested in understanding how animals form categories and/or concepts. Why does it matter?

The fact is that animals often respond adaptively to objects or events they have never encountered before: a monkey meeting a strange, larger male will initially avoid it; a chicken moved to a new cage knows to drink from the new drinker and feed from the new feeder; and a pigeon in a **Skinner** box will peck at a new picture of a tree rather than at anything else if it had previously been rewarded for pecking at tree pictures.

Such appropriate and adaptive behaviours are expressions of an ability to categorize stimuli. When an animal responds adaptively to a new **stimulus** it is often because the new stimulus resembles others for which the appropriate response is already known: the monkey will have had previous encounters with larger males from which he has learned that larger males are likely to be aggressive, stronger and best avoided; the chicken will have had a drinker and feeder in its previous cage; and the pigeon will have learned to associate tree pictures rather than non-tree pictures with a food reward.

Generalizations across stimuli would be impossible without this ability to categorize. The animal would perceive every object or event as new and unique, and would have to respond to it as such. There is, therefore, clearly an advantage to be gained from being able to generalize across objects and events that are similar to each other, and from being able to sort them into categories on the basis of their resemblance. The crucial question with respect to concept formation is: which aspects of the stimuli do animals use to assess stimulus resemblance, i.e. category membership?

Adult humans often categorize objects and events with reference to abstract 'concepts'. The *Concise Oxford English Dictionary* defines a concept as an idea of a class of objects. However, in the terminology of this article, 'concepts' refer not just to classes of objects but also to classes of events and groups of any stimuli in general. Whichever stimuli (e.g. events, people, objects) we may refer to, the point is that, with concepts, category membership is on the basis of the abstract idea for which all members stand. The idea is 'abstract' in the sense that it is not necessarily dependent on any specific, concrete, physical features of the category members. Having the concept of 'growing', for example, is not based on what 'growing' things taste, smell, sound, feel or look like. Instead, it is knowing that all members of this category are increasing (in size, height, amount, degree and so on); examples would be a seedling tree, an infant or the human population.

Understanding whether and how animals form concepts is important because, if we care for animals and are concerned whether they are faring well under our care, we need to know how and to what extent they understand their environment; this includes whether they can, in fact, recognize an unfamiliar drinker as a new example of something similar learned in the past – and kind handlers as the ones in the blue overalls, and whether and how they recognize kin or discriminate between familiar and unfamiliar conspecifics (**see also: Kin selection**). Secondly, it may be important to know when, why and how the ability to form abstract concepts has arisen in the animal kingdom, and whether abstract concept formation is uniquely human or dependent on having **language**. A thorough review of all the theoretical and experimental work on the subject is clearly beyond the scope of this article. Instead, it aims to introduce the main relevant research approaches and, in doing so, draws on several thorough, but usually conflicting, reviews.

Much of our knowledge about category and concept formation in animals comes from laboratory studies with pigeons. Herrnstein and colleagues (1976), for example, investigated whether pigeons could learn to discriminate between images containing and not containing trees. In each training session, the pigeons were shown 80 different photographs, one at a time; 40 ± 5 of them were of scenes featuring trees ('tree-photos'). Note that the trees were of any type or description, and not necessarily the centre of attention in the pictures. 'Non-tree photos' were comparable to 'tree-photos' in every respect except the lack of a tree or part of a tree. Some images were presented more than once, some only once. If the pigeons pecked at the response key when a tree-photo was shown, they received a food reward; if they pecked for a non-tree photo, they received no food but a 10 s penalty delay before the next picture was presented. Over 120–131 such sessions, pigeons learned to discriminate between the two categories of photo: they pecked significantly more for tree-photos than for non-tree photos, and this discrimination applied even to pictures they had not seen before. Results such as these have been confirmed by many studies in different laboratories using different stimuli. Taken together, they show that pigeons can discriminate between different stimuli (photos in Herrnstein *et al.*'s case) on the basis of which category they belonged to ('tree' or 'non-tree').

Subsequent studies have demonstrated experimentally that many other vertebrate species ranging from chinchillas to carp are also capable of category formation, on the basis of the presence or absence of certain physical features. Not all of these studies have used visual stimuli. Johnston and Jernigan (1994), for example, showed that male hamsters can discriminate between familiar and unfamiliar females on the basis of their scent; carp, a fish species with particular sound sensitivity, can be trained to discriminate between blues and classical music (Chase, 2001).

The now standard experimental method in such categorization tasks is to train animals with a 'training set' of repeated exemplars (examples) of the two categories, then to test whether they can transfer any learned discrimination to novel exemplars. In Chase's study, for example, the carp were first trained with a training set of John Lee Hooker (guitar and vocals) and Bach (oboe concertos) exemplars by rewarding a button press with food when John Lee Hooker was playing, but not when Bach was. The carp were then tested for transfer with a variety of previously unheard recordings of blues and classical music (baroque and other) from several other artists and ensembles.

Animals may also form categories when the respective stimuli do not share physical features but lead to the same outcome, response or reaction. Consider the results of a study with **chimpanzees** by Savage-Rumbaugh *et al.* (1980). The chimpanzees were trained to sort a mixed bin (training set) of three food and three tool objects (stimuli) into two separate bins. In tests they were then presented with a mix of new food and tool items, and the chimpanzees sorted them as expected into 'food' and 'tools'. Food items in the tests clearly differed physically from each other and from the food items used in the training sets, but they had something else in common: they all made the chimps salivate (Pearce, 2007). The animals may thus have learned that anything that makes them salivate belongs to one category ('food'), while any stimulus that does not belongs to the other ('tool').

Such experimental evidence has since shown that animals can learn to group primary stimuli together using secondary stimuli, even when the former are completely arbitrary and unrelated (e.g. chairs and cars). Pigeons, for example, might be trained to lever-press when they are shown pictures of chairs and cars. When one of the stimuli (e.g. 'chair') is subsequently paired with a new response (e.g. button press), pigeons will tend to button-press also for the other stimulus (in this example, 'car'). This suggests that they had learned to treat chair and car pictures as members of the same category on the basis of their originally shared secondary stimulus (the lever-press) (see Urcuioli, 1990, for details of an elegant, if rather more complex, related experimental design). Such secondary **stimulus** generalization is, somewhat unhelpfully, also called 'mediated generalization', 'acquired stimulus equivalence', 'functional concept learning' or 'associative concept learning'.

Authors such as Thomas Zentall and Peter Urcuioli, who do refer to secondary stimulus generalization as associative concept learning, suggest it as one of several types of category formation by 'conceptualization' or, as Zentall and his colleagues put it, as one of 'three broad types of relations that appear to unite events into a category' (e.g. Urcuioli, 2001; Zentall *et al.*, 2002). In Zentall's terminology, the other two types of concept are 'perceptual' (as above in the examples of tree or music type) and 'relational' concepts (see below).

The important point is that secondary stimulus generalization too, however it is called, is a stimulus generalization process even if based on functional, acquired rather than perceptual stimulus equivalence and, as such, is unlikely to fulfil the 'abstract concept' criteria of others such as Chater and Heyes (1994) on the one hand, or Jeffrey Katz and colleagues on the other (e.g. Katz *et al.*, 2007). As mentioned above, categorization explicitly on the basis of abstract concepts as in humans has to be independent of any shared primary or secondary features of the category members. It has to be on the basis of the shared idea for which all stimuli in the category stand. Abstract concept learning thus cannot be accounted for by Herrnstein *et al.*'s kind of stimulus generalization, or Savage-Rumbaugh *et al.*'s, Urcuioli's or Zentall's kind of secondary generalization processes. It is hence considered a much more difficult cognitive feat, and one school of thought has argued strongly that human-type abstract concepts can only exist in conjunction with symbolic rather than absent/present labelling, that is they depend on language (e.g. Chater and Heyes, 1994).

However, a strong case has also been made for abstract concept learning without language based on judging the relationship between stimuli (see Katz *et al.*, 2007). This would be categorization via 'relational concepts' in Zentall *et al.* (2002). One much-studied example of such abstract stimulus properties is the 'same–different' relationship. In same–different discrimination tasks, animals are tested to see whether they can learn to distinguish pictures ('stimuli', 'exemplars') that show a group of identical symbols ('same') from pictures that contain various symbols ('different'). Animals are first trained on a training set of 'same' and 'different' pictures by conditioning one type of response for 'same' pictures and another response for 'different' pictures. Subsequent transfer of the discrimination to new 'same' and 'different' pictures is taken as evidence that the animals have learnt the abstract concept of 'difference' (or 'sameness').

While much methodological progress has been made in recent years in the study of 'relational concept' learning as abstract concept learning, a number of methodological challenges remain. These include avoiding the use of too few different exemplar pictures in the training set, to give animals a fair chance of acquiring the discrimination, or the use of more than two symbols in each picture, which might enable animals to discriminate based on 'global perceptual features' such as the levels of activity or entropy in the picture (Katz *et al.*, 2007).

Complementary evidence for animals using non-linguistic, yet abstract, concepts when they categorize stimuli may eventually be provided by the neurosciences. Not only have mirror **neuron(e)**s been identified but also – in the pre-motor cortex of macaque monkeys – neurons that are active during hand or mouth movements that involve manipulating or holding objects such as grasping food items, tearing, manipulating or holding them. Importantly, these neurons become active only when the effector movements (i.e. hand or mouth movements) actually result in a different end state from when the movement started; for example, a piece of fruit is *successfully* grasped, or an object is torn apart or flung away (see Gallese, 2003).

Of particular interest here is that some of these neurons are active whenever the monkey achieves a particular end state (e.g. successfully grasps a fruit), regardless of whether the grasping happened with the mouth, left, right or both hands, and that these neurons are located in the brain's *motor* system and might therefore have been expected to fire during effector actions regardless of end state. The fact that these neurons fire when the end state of an action is achieved independently of which effector was used is interpreted as evidence for an 'abstract kind of means–end representation' (Gallese, 2003) – and lies at the heart of Gallese and his colleagues' thesis that abstract concepts can and do arise in the sensory–motor system of non-human brains, and that human, language-dependent concept use 'exploits this pre-existing multi-modal character of the sensory–motor system' (Gallese and Lakoff, 2005). (SDEH)

References and further reading

Chase, A.R. (2001) Music discriminations by carp (*Cyprinus carpio*). *Animal Learning and Behaviour* 29, 336–353.

Chater, N. and Heyes, C. (1994) Animal concepts: content and discontent. *Mind and Language* 9, 209–246.

Gallese, V. (2003) A neuroscientific grasp of concepts: from control to representation. *Philosophical Transaction of the Royal Society of London B* 358, 1231–1240.

Gallese, V. and Lakoff, G. (2005) The brain's concepts: the role of the sensory–motor system in reason and language. *Cognitive Neuropsychology* 22, 455–479.

Herrnstein, R.J., Loveland, D.H. and Cable, C. (1976) Natural concepts in pigeons. *Journal of Experimental Psychology: Animal Behaviour Processes* 2, 285–302.

Johnston, R.E. and Jernigan, P. (1994) Golden hamsters recognize individuals, not just individual scents. *Animal Behaviour* 48, 129–136.

Katz, J.S., Wright, A.A. and Bodily, K.D. (2007) Issues in the comparative cognition of abstract-concept learning. *Comparative Cognition and Behaviour Reviews* 2, 79–92.

Pearce, J.M. (2007) *Animal Learning and Cognition*, 3rd edn. Psychology Press, Hove, UK.

Savage-Rumbaugh, E.S., Rumbaugh, D.M., Smith, S.T. and Lawson, J. (1980) Reference: the linguistic essential. *Science* 210, 922–925.

Urcuioli, P.J. (1990) Some relationships between outcome expectancies and sample stimuli in pigeons' delayed matching. *Animal Learning and Behaviour* 18, 302–314.

Urcuioli, P. (2001) Categorization and acquired equivalence. In: Cook, R.G. (ed.) *Avian Visual Cognition*. Available at: http://www.pigeon.psy.tufts.edu/avc/urcuioli

Zentall, T.R., Galizio, M. and Critchfield, T.S. (2002) Categorization, concept learning, and behaviour analysis: an introduction. *Journal of the Experimental Analysis of Behaviour* 78, 237–248.

Conditioning – types of

In the experimental study of animal behaviour, *conditioning* refers to the changes in an animal's behaviour that arise when it learns the association between two correlated events, which we will call Event 1 and Event 2. As the animal learns the Event 1–Event 2 association, the conditioned response emerges in its behavioural repertoire. This response is then observed to extinguish if the original association between events is broken (i.e. if Events 1 and 2 no longer occur together).

Conditioning has been divided into two main types – classical conditioning (also known as Pavlovian conditioning) and **instrumental conditioning** (also known as operant conditioning). The two types of conditioning differ at a basic operational level. In classical conditioning, the animal is exposed to correlations between external events. For example, the animal is presented with a neutral **stimulus** (a light or a noise) and this is followed by a biologically important stimulus (e.g. a noxious stimulus such as a shock, or a positive stimulus such as delivery of food). The initially neutral stimulus, known as the conditioned stimulus (CS), comes to provoke a response as a consequence of its being paired temporally with the intrinsically important unconditioned stimulus (US). The classic example of this is the preparation studied by **Pavlov** himself. Pavlov's dogs would hear a noise (CS) that was reliably followed by delivery of meat powder (US) into the mouth. After several such pairings of the CS and US, the dogs would come to salivate in response to hearing the noise. Salivation would then extinguish if the noise were repeatedly presented to the dogs without subsequent arrival of food.

Instrumental conditioning requires that the animal learns to produce a specified response as a result of its association with a positive or negative US (such as shock or food). For example, in the preparation developed by **Skinner**, hungry rats learn to press a bar (the instrumental or conditioned response) to receive a food pellet (the reward, the US). As in classical conditioning, the response then extinguishes if the US is withheld. In addition to learning to make a specified response (like pressing a bar) to obtain a positive reinforcer (such as food), animals can learn to make such responses to prevent the occurrence of a negative US such as a shock (i.e. a punisher).

In the early part of the 20th century, the separation of classical and instrumental conditioning had as much to do with geography as science. Pavlov and his disciples in Europe were devoted to the study of classical conditioning, while Thorndike and his disciples in North America were almost exclusively interested in instrumental conditioning. However, the two share more than just the term 'conditioning'. In both there must be some contingency between the CS or instrumental response and the US – that is, the two events must be correlated such that the likelihood of the US occurring is increased in the presence of the CS or instrumental response. Thus, conditioning is retarded or prevented if the contingency is reduced by the occurrence of either event without the other (i.e. the CS or instrumental response occurring without the US, or the US occurring without being preceded by the CS or instrumental response). Similarly, conditioning in both cases is sensitive to the contiguity between the events – conditioning is retarded if the temporal gap between the two events is increased. Both are subject to extinction if the US is withheld, and extinguished responding can be seen to recover spontaneously if the animal spends time removed from the experimental apparatus, or if the US is presented on its own (without the CS or instrumental response).

Several theorists have argued that classical and instrumental conditioning are functionally the same, differing only in the experimental procedures employed. The early behaviourists in North America believed that both types of conditioning involved the formation of stimulus–response (S–R) links (e.g. Hull, 1943). According to this view, conditioning involves a simple mechanism whereby, if an animal performs a particular response in the presence of a stimulus and this was followed by a positive outcome, then a connection between the stimulus and response would be formed such that the animal would now be more likely to produce the same behaviour in response to that stimulus. This S–R mechanism was believed to operate whether the response was instrumental or classically conditioned.

However, there is a considerable amount of empirical evidence indicating that any such explanation of classically conditioned responses cannot be correct. One line of evidence is based on the logic that, if these responses were learned because of some putative beneficial outcome, then preventing the animal from producing the response should prevent conditioning. Clearly, this is inevitable in the case of instrumental conditioning – preventing a rat from pressing a bar will prevent it from ever learning that the response produces a reward. However, this has been shown to be false in a variety of classical conditioning paradigms (Mackintosh, 1974). For example, injecting dogs with atropine can prevent them from salivating and therefore should, according to S–R analysis, prevent them from learning to salivate to a CS (e.g. a noise) that is paired with delivery of food. Several experiments have shown that dogs do, nevertheless, learn the CS–US association, and will immediately salivate to the CS if they are tested after the effects of atropine have worn off.

Many studies have since shown S–R theories to be incorrect in their description of most instances of instrumental conditioning, as well as classical conditioning. One critical source of evidence has been that both instrumental and classically conditioned responses are usually sensitive to changes in the value of the reward (Adams and Dickinson, 1981). For example, if a rat has learned to press a bar for food when hungry, then it will be much less inclined to perform this response when sated for food. This shows that instrumental responses are like an action produced to obtain a particular outcome, rather than being a reflexive response elicited by the sight of the bar (as argued by an S–R account).

Other experiments have identified a more fundamental distinction between instrumental and classically conditioned responses. This evidence relies on the impact of what is known as an 'omission schedule', where the US is delivered only if the animal does *not* produce the particular response in question (Mackintosh, 1974). Instrumental responses can be eliminated by introduction of an omission schedule. For example, rats that have learned to press a bar to obtain food can readily learn to stop pressing to obtain the same food (if the response–reward contingency is reversed). Classically conditioned responses, on the other hand, are not so flexible. For example, a dog cannot learn to stop salivating to a noise that signals the delivery of food. Similarly, pigeons cannot learn to suppress pecking at a small light that signals the delivery of grain. In each case, if the omission schedule is continued the classically conditioned response will eventually extinguish because the CS is no longer followed by the US. Such findings indicate that, whereas instrumental responses are more flexible and under voluntary control, classically conditioned responses are more reflex-like responses elicited automatically in anticipation of the US. (JH)

References

Adams, C.D. and Dickinson, A. (1981) Instrumental responding following reinforcer devaluation. *Quarterly Journal of Experimental Psychology B* 33, 109–121.

Hull, C.L. (1943) *Principles of Behaviour.* Appleton–Century–Crofts, New York.

Mackintosh, N.J. (1974) *The Psychology of Animal Learning.* Academic Press, London.

Confinement

The vast majority of animals kept by man do not stay where humans want them to be unless they are confined to some extent. Hence, the history of animal **domestication** and usage has involved varying degrees of confinement. If an individual animal learns that a period of confinement is associated with provision of food or other benefits, the confinement itself need not be aversive. However, enforced, prolonged, inescapable confinement will always be **aversive** and animals indicate this by escaping if possible. The ideal management of animals would involve encouraging them to stay where humans want them, rather than forcing them to do so. Although sheep or cattle on a hillside may be confined by walls, watercourses or cattle grids, and horses in a 20 ha paddock may be confined by fences, the most widely used meaning of confinement is the restriction of animals to areas where their normal activities are significantly curtailed, and so such situations will be discussed here.

Bears in the wild carry out many different behaviours and range over many square kilometres. Bears in a concrete-floored **zoo** cage measuring 20 × 20 m are severely restricted in what they can do, and many of their **needs** are not met, so their welfare is poor. If a group of these zoo bears are given an enclosure that includes a wooded hillside 400 × 400 m, climbing frames, observation platforms, a swimming pool and an indoor sleeping area, most of their needs are met and their welfare is much better than in the bare cage. However, in one zoo where such facilities were provided, the bears still spent a substantial amount of time walking around the perimeter of the enclosure, presumably in order to try to escape. They were still confined and not all of their needs were met. It seems that some animals never adapt fully to even the best zoo conditions, so it can be argued that they should not be kept in zoos. The welfare of rhinoceroses in zoos is often not good, and their breeding record in captivity is poor. However, if they are kept in 50 ha, semi-natural enclosures in their countries of origin, their breeding record is about eight times better, so such conditions, rather than confinement in zoos, should be used if conservation via repopulation of suitable, safe habitats is the objective.

Confinement of animals taken from the wild often results in very poor welfare associated with frantic escape attempts or suppression of all normal behaviour, often followed shortly by death. Yet wild animals are often collected for the **pet** trade, or for other purposes, then transported over long distances and kept in confinement (**see also: Transport**). Surveys of wild-caught animals transported by air show that mortality of birds, reptiles, amphibians, mammals and fish is often very high. The transport and keeping of wild-caught animals should be allowed only in exceptional circumstances and is never justifiable for pet keeping. If wild animals have to be caught and transported for a more important reason than pet keeping, they should be fully adapted to a captive condition of good quality before use and, for most species, before transport. Zoos should exhibit captive-bred rather than wild-caught animals, and these captive-bred animals should be fully adapted to both the confinement and the exposure to humans which they will experience in the **zoo**.

Animals kept for fur production, such as foxes or mink, have been bred continuously in captivity for a much shorter time than domestic animals: 40–80 years as compared with 10,000 years or more for dogs, cattle and sheep. Hence, they are changed very little in their characteristics except in respect of colour of fur. It is therefore relevant to know that foxes and mink range over many kilometres, spend time hiding underground, climb frequently, live solitarily for long periods and, in the case of mink, are adapted for swimming and spend time doing so during hunting, playing and exploring. Confinement in a cage less than twice the body length and often not high enough to rear on the hind legs results in poor welfare. The needs of these animals cannot be met in small, wire cages, and these species show **abnormal** behaviour and other signs of poor welfare in typical **fur farm** conditions. In good zoo conditions there are few signs of poor welfare, so it appears to be the close confinement per se that causes the greatest problem.

The most severe examples of close confinement, usually temporary, are those used to restrain almost all movements of animals so that they can be examined, treated for disease or moved without risk to human handlers. When mink are put in a metal frame that prevents their movements they show high adrenocortical and other responses, which indicate very poor welfare. Various animals that may be caught and held by humans are also severely affected (**see also: Handling**). Pets transported in small boxes may be substantially affected by the experience. However, the long-term consequences of short-term close confinement may be slight. Some individuals learn from such an experience to avoid or minimize the risk of experiencing it again.

Among the smallest cages used to confine farm animals are those in which geese and ducks are kept during the period of force feeding prior to slaughter for **foie gras** production. The wire cages allow the birds to sit or partially stand, with their head protruding through a hole in the top so that they can be force fed and can drink water from a trough in front. In large-scale animal production, one of the smallest space allowances used is that for laying hens in the USA. Many birds are in cages with four or five individuals allowed 350 cm^2 each. A hen with its feathers in a relaxed position occupies more space than this, so the birds can only fit into the cage if their feathers are compressed. These birds are able to turn around but cannot preen normally or stretch their wings, unless the space allowance is 550 cm^2 per bird in a five-bird cage. Birds must cooperate by crowding together in order that one of their number has room to show stretching or preening movements. Even with this larger space allowance, which is now required in the European Union (EU), no bird can flap their wings so the hens have insufficient exercise and develop brittle bones (**see also: Bone strength**).

Pens, crates and cages that do not allow individuals to turn around are used for calves, sows and some horses. Each of these animals, plus dairy cows, bulls and many other animals, are sometimes tethered. With the exception of possible adverse effects on the animals of the tether itself, the degree of confinement is often similar in these two circumstances. Animals are prevented from showing normal grooming behaviour if they are unable to reach their hindquarters. They are usually unable to show normal movements during lying down and are completely prevented from showing escape or locomotor behaviour. The use of larger pens for individual animals will, like the smallest ones, prevent animals from showing normal social interactions. Since most social interactions are friendly and some agonistic interactions serve a useful function, prevention of **social behaviour** for the members of a social species is a serious deprivation.

What are the consequences when animals have to live in individual pens or cages that do not provide for their needs? Confined mammals and birds have elaborate control systems that depend upon receiving inputs from environmental cues. If animals lack control because of an inadequate environment, they are likely to use coping methods and to be affected in some ways that we are not able to recognize and others which we can recognize. One consequence is an increase in abnormal behaviour such as **stereotypies**, reduced responsiveness or **self-mutilation**.

One form of stereotypy is route tracing. An animal that is strongly motivated to escape from confinement but which is not able to do so may repeat the locomotor or other pattern of behaviour which might, if not thwarted, have effected escape. Repeated walking over the same route with the same steps is a stereotypy that is readily recognized. Abbreviated movements such as drinker-spout circling in confined rodents, **bar biting** in confined sows and various head and body circling movements in confined mink are all stereotypies which might originally have had a motivation associated with potential escape routes. Other stereotypies shown by confined animals include **tongue-rolling** in calves kept in crates, **crib-biting** in horses kept in stables and drinker pressing in sows kept in individual stalls or tethers. The stereotypy is associated with lack of control over the environment and indicates poor **welfare** in the animal. In some cases the stereotypy might help the individual to cope with the environment, but it is likely to be solely pathological in most cases. However, whether or not the confined animal is helped to cope, the stereotypy shows that the individual is having difficulty coping with its environment and hence that its welfare is poor. The stereotypy usually shows that the present environment is lacking in some way, but may be a behaviour that continues after the animal is moved to a better environment. In the latter case the stereotypy is a scar of a previously difficult environment, but it still shows that the welfare of that animal is poor.

There are similar arguments for other abnormal behaviours indicating poor welfare in confined animals – for example, reduced responsiveness to normal stimuli in confined sows, self-mutilation in confined monkeys or parrots and injurious behaviour in confined hens and pigs. Confinement can also lead to substantial adrenal cortical responses, immuno-suppression, reduced reproductive success and early death.

Where confinement prevents normal levels of exercise which an animal is highly motivated to show, restoration of the possibility of carrying out the movements may lead to **rebound behaviour**, in which much of the previously prevented movement is shown as a high priority. Hens kept for a long period in cages that do not allow wing flapping and are then put into a more spacious environment will flap their wings for a long period. This rebound behaviour is shown with a higher priority than eating and drinking. Cattle, sheep and other animals that have been confined during winter periods will show rebound running and jumping when first given access to a field.

As mentioned above, confinement may result in reduced activity and responsiveness. The behavioural, physiological and brain changes associated with confinement have close parallels with depression in humans. Confinement may limit normal behaviour and normal ability to control interactions with the environment in the same way that social constraints can reduce activity and control. It is reasonable to consider many confined animals as being clinically depressed.

Companion animals may be closely confined or kept in small spaces in the ways described above for farm and zoo animals. They may also be free to move around significant usable spaces, such as a house or a garden, but unable to obtain resources or carry out behaviours which are of high value to the individual. A dog or cat that is motivated to hunt for food, social companions or a sexual partner may be prevented from doing this by confinement. Cats kept in houses are certainly deprived of aspects of life which are available to those cats that can leave the house by way of a cat flap.

The welfare of the cat that can leave the house is likely to be improved, although they are at greater risk of being killed on the roads or of being attacked by other cats or by predators. Some cats that are free to leave houses have a very substantial impact on the populations and welfare of wild animals. For example, cats have been found to bring as many as ten rodents and birds per day back to the house where they live. Some of these prey animals will have suffered greatly before being finally killed by the cat. A cat wearing a bell causes less harm. Dogs may also be confined to an extent that their welfare is poor because of lack of companionship from other dogs or from humans. Confinement with another dog as a companion usually causes much less of a problem then confinement in isolation.

Confinement to a particular area may be brought about by training the individual not to go outside a garden or other area. Dogs are sometimes confined by the wearing of an electric **collar**. This mechanism is so devised that the dog receives an electric pulse if it moves past a certain line. The pulse might be very severe, or it might be just sufficient to allow the dog to learn not to cross the line. The consequences of such a confinement system for animal welfare are very different according to the severity of the current used and the frequency of responses that incur pulses. Farm animals may also be confined by the stimulation they get when they touch an electric fence. Pigs learn rapidly not to touch an electric fence. Indeed, some individuals learn by watching others without ever themselves touching the fence. Such a confinement system can be a good one and have relatively little effect on welfare. However, certain individuals could suffer greatly because of its use. Electric fences have also been used to exclude wild animals as diverse as rabbits and elephants from agricultural areas.

Almost every individual experiences some degree of confinement during life, even if it is only restriction of activity by a parent or by potential danger. However, confinement of any kind makes control of life more difficult, and more extreme confinement leads to such difficulties that welfare is poor. Since confinement of animals used by man may be prolonged, some of the worst effects of humans on animal welfare are associated with confinement. In every case where animals are kept there should be careful consideration of their needs so that adequate resources are provided and harmful confinement is minimized. (DMB)

See also: Farmed animals

Further reading

Broom, D.M. and Johnson, K.G. (1993) *Stress and Animal Welfare*. Kluwer, Dordrecht, The Netherlands.

Fraser, A.F. and Broom, D.M. (1997) *Farm Animal Behaviour and Welfare*. CAB International, Wallingford, UK.

Mason G.J. and Rushen, J. (eds) (2006) *Stereotypic Animal Behaviour: Fundamentals and Applications to Welfare*, 2nd edn. CAB International, Wallingford, UK.

Conflict behaviour

When an animal is subjected to a **stressor**, its response may manifest as changes in its behaviour and the physiology of its **autonomic nervous system**, neuroendocrine system and/or **immune system**. Conflict behaviour describes one category of **stress**-induced behaviour changes that arise from conflicting motivations, especially when escape/avoidance responses are not consummated. It can appear as 'a set of responses of varying duration that are usually characterised by hyper-reactivity' (McGreevy *et al.*, 2005). When arising from a relatively brief exposure to a single stressor, these responses typically present as hyper-reactive behaviours such as increased postural tonus, **agonistic behaviours**, redirected **aggression** and displacement activities (Wiepkema, 1987). On the other hand, when conflict behaviours arise from chronic exposure to conflicting stimuli, responses can manifest as repetition and **ritualization** of original conflict behaviours, raised **corticosteroid** concentrations, physiological disturbances and gastric pathology. In addition, **stereotypies** may develop as well as injurious behaviours, such as **self-mutilation**.

Both innate and acquired responses increase an animal's ability to control and predict its **'umwelt'**, and these serve to avoid conflict. For example, a warning display such as the common mammalian behaviour of pinning the ears back as an aggressive threat is an efficient signal, since it lowers the need for active combat with **conspecifics** during agonistic encounters. Indeed, injury is uncommon during aggressive intraspecific interactions in most species because of the efficiency of such **threat postures**. Animals respond to such threats and are able to flee the stressful situation.

Because of the efficacy of these evolved mechanisms, Moberg (2000) observes that 'it is difficult to envisage a single chronic stressor in non-(captive) animals, with the possible exception of unrelenting pain or prolonged exposure to an extreme environmental condition such as severe cold'. Prolonged motivational conflicts are of great importance in any analysis of animal stress, especially in relation to animals in **captivity** that may be unable to resolve the stressful situation. Conflict behaviours manifest as apparent dysfunctions, in particular stimulus–response entities that reduce predictability and controllability of the environment. However, conflict behaviours may have evolved as adaptive processes to resolve stressful situations.

The term conflict behaviour is appropriate to both interspecific and intraspecific interactions. For example, in human–animal interactions where animals are captive, ridden, led or tethered, there may be no available solution to the imposed stressor. Horse riding is based on negative reinforcement, which comes about through varying degrees of discomfort imposed on the animal by the reins and rider's legs. If the rider incorrectly negatively reinforces undesired or inconsistent responses, or if the rider concurrently applies pressure via the reins and legs, confusion and conflict behaviours may arise as the horse is unable to resolve the stressful situation. In horse riding, conflict behaviours may manifest as increased aggressive and defensive behaviours, self-mutilation and hyper-reactivity, that ranges from increased muscle tone and body tension to bolting, rearing (standing on the hind legs), bucking (bounding from the hind legs to the forelegs where the hindquarters are kicked up), shying (sudden veering to one side for one or many strides), leaping, flipping over or rushing backwards. It is the responsibility of animal custodians and trainers to understand learning theory well enough to establish desirable learned responses in animals, without causing conflict behaviours. Evaluating problem behaviour in animals in relation to lost predictability and controllability – and the potential conflict that arises from this – is useful in explorations of mismanaged animal training. (AM)

See also: Emancipation; Motivation

References and further reading

McGreevy, P.D., McLean, A.N., Warren-Smith, A.K., Waran, N. and Goodwin, D. (eds) (2005) Defining the terms and processes associated with equitation. *Proceedings of the 1st International Equitation Science Symposium*, Broadford, Australia, pp. 10–43. Post-Graduate Foundation in Veterinary Science, Sydney, Australia.

Moberg, G.P. (2000) Biological response to stress: implications for animal welfare. In: Moberg, G.P. and Mench, J.A. (eds) *The Biology of Animal Stress: Basic Principles and Implications for Animal Welfare*. CAB International, Wallingford, UK, pp. 1–21.

Stolba, A., Baker, N. and Wood-Gush, D.G.M. (1983) The characterization of stereotyped behaviour in stalled sows by information redundancy. *Behaviour* 87, 157–182.

Wiepkema, P.R. (1987) Behavioural aspects of stress. In: Wiepkema, P.R. and van Adrichem, P.W.M. (eds) *Biology of Stress in Farm Animals: an Integrative Approach*. Martinus Nijhoff Publishers, Leiden, The Netherlands.

Consciousness

Consciousness is instantiated in a wide range of mental states found in humans and higher animals. These states include bodily feelings (e.g. **pain** and pleasure), perceptual sensations (e.g. of sounds, odours and colours) and thoughts, moods and emotions. Consciousness plays an important role in discussions of the **ethics** of animal treatment, because the ability to feel pain and discomfort is one reason why higher animals merit moral consideration.

In both psychology and philosophy, consciousness has proved hard to characterize and even harder to explain. The problem is that conscious mental states seem to be inherently subjective and non-physical. Modern science, in which physics is used to describe the world at the most fundamental level, is not well equipped to deal with such entities. Nor have philosophical attempts to identify conscious states with physical phenomena – such as behaviour patterns or states of the **central nervous system** – been successful. To date, moreover, no one has offered a convincing account about why consciousness evolved. Simple, causal sensitivity to information (of the sort found in a rudimentary form in ants and bees) would appear to leave mentally endowed creatures as well adapted to their environments as equivalent conscious sensitivity. Why, then, did consciousness develop? This question has encouraged some to regard consciousness as a causally inert phenomenon (**see: Epiphenomenon**).

Conscious states always involve *awareness* of something: I may have a wound on my arm that I am unaware of, but in normal circumstances it makes little sense to say that I have a feeling of pain but am not aware of it. On the other hand, consciousness is not the same thing as awareness, as I may be aware of the spacing of the stairs in my house as I descend them, but not consciously so. Indeed one way of conveying what is special about consciousness is by asking: What is the difference between the awareness in the stair example, which consists principally of an ability to adjust behaviour to the environment, and conscious awareness of the kind involved in noticing that the stair carpet is scarlet, or that one's knee is sore? Very many animal species display awareness of the former kind here. Fewer, presumably, experience the latter.

The problem of providing a scientific account of consciousness can be spelled out with four propositions, each of which seems plausible, but all of which cannot be true.

1. Bodily states and processes are objective and physical in nature, and therefore they have spatial coordinates and can be described and explained in biological, chemical and ultimately physical terms.
2. Conscious states are subjective, lack spatial coordinates and hence cannot be similarly described and explained.
3. Bodily states and conscious states interact, so that if I am pricked with a needle I feel pain and the feeling of pain makes me react.
4. Something with a subjective nature cannot interact with something with an objective nature.

At least one of 1–4 must be rejected, and attempts to solve the problem of consciousness can be classified according to which proposition is denied. Rejecting 4, **René Descartes** (1596–1650) famously embraced interactive **dualism** and even pointed to a specific organ in the **brain**, the pineal gland, where he took the interaction to occur. Rejecting 1, Bishop Berkeley (1685–1753) accepted a variety of idealism according to which, in reality, only conscious states exist. But today it would be eccentric, to say the least, to pursue either of these paths. That leaves 2 and 3. The denial of 3 is scarcely credible, although philosophers have certainly toyed with the idea that conscious and bodily states run in parallel, but do not interact, or alternatively with 'epiphenomenalism', i.e. the view that conscious states are epiphenomena brought about by bodily states but lacking causal influence on the physical world (like the sound of a whistle on a steam train, as it is sometimes said).

Contemporary philosophers and scientists are therefore strongly inclined to reject 2 and adopt one or other 'materialist' **theory of mind**, and the consciousness debate today is really a discussion between advocates of various forms of materialism and critics who doubt that the materialists have succeeded in describing and explaining consciousness. There are several materialist views, of which four shall here be mentioned:

1. **Behaviourism**: conscious states are nothing but dispositions to engage in certain kinds of behaviour. To say that an animal feels hunger is to say no more than that the animal is strongly disposed to eat (this recalls the awareness of the stairs mentioned above).
2. Reductive materialism: a conscious state is really nothing but the occurrence of a state in the central nervous system of the organism. For example, the feeling of a certain kind of pain may be identified with the firing of certain kinds of **neuron(e)**s in the prefrontal cortex. However, according to this view conscious states need not be identified with isolated neural events but may also reflect complex interactions between different parts of the brain.
3. Functionalism: conscious states are physical states of the organism with a functional role. Thus, to have a pain in the right hand is to be in a physical state that causes one to react appropriately, e.g. by avoiding the cause of the pain and doing things that alleviate it. Functionalism combines the insights of behaviourism and reductive materialism. Unlike reductive materialism, however, it allows that conscious states might be found in different kinds of system, not just in organisms with a central nervous system. According to the functionalist, it may be possible for machines to possess consciousness.
4. Strong AI (artificial intelligence): consciousness is related to the body in the same way that a computer program is related to the hardware on which that program runs. Just as one does not need to attribute subjective states to a computer in order to explain how it can run a program, one does not

need to appeal to conscious states to explain how a human or other animal will act.

According to critics, all forms of materialism suffer from the same problem: they simply ignore the key feature of these states, i.e. that they are subjectively experienced, that the holder of the states is aware of them. A prominent representative of these critics is the American philosopher Thomas Nagel, who argues in a paper entitled 'What is it like to be a bat?' (Nagel, 1974) that what is special about conscious states is that they are essentially tied to the subjective perspective: there is something it is like to experience the colour blue, rather than green, for example, and this is not captured in materialist theories of mind.

However, it is one thing to criticize materialist accounts of consciousness and another is to come up with a better approach. And here the critics have been less helpful. Another prominent critic, the American philosopher John Searle, claims that a solution to the problem cannot be given until science develops a new, and richer, way to describe biological phenomena that cuts across the present distinctions between objective and subjective states. Other philosophers claim that it is part of the human predicament that we are unable to solve the problem of consciousness.

The upshot is that applied ethologists, who are required to obtain science-based information about the conscious states of animals in connection with, for example, enquiries into animal welfare, are left in a very difficult situation. They cannot directly study the conscious states of animals – they must use objective parameters that they can interpret as signs of the conscious states they are investigating. (PS)

Reference and further reading

Avramides, A. (2001) *Other Minds*. Routledge, London.

Bennett, M.R. and Hacker, P.M.S. (2003) *Philosophical Foundations of Neuroscience*. Blackwell, Oxford, UK.

Chalmers, D. (1995) Facing up to the problem of consciousness. *Journal of Consciousness Studies* 2/3, 200–219.

Flanagan, O. (1992) *Consciousness Reconsidered*. MIT Press, Cambridge, Massachusetts.

Nagel, T. (1974) What is it like to be a bat? *Philosophical Review* 83, 435–456.

Searle, J.R. (1992) *The Rediscovery of the Mind*. MIT Press, Cambridge, Massachusetts.

Conservation

Conservation may be defined as either: (i) the act or process of conserving, the careful guarding of an asset (biodiversity, beauty, heritage, ecological processes, evolutionary processes); (ii) preservation or restoration from loss, damage, deterioration or neglect; or (iii) the improvement of our natural resources and environment. It implies the preservation and careful management of the environment and of natural resources in a given place (particular habitat, ecosystem, biome, biosphere as a whole). Recognizing the urgency of the global ecological crisis, conservation aims actively to improve by preventing loss, injury or other change of a population, community or ecosystem.

Social and ecological structures and processes are interrelated. Society has an impact on the environment, and therefore conservation aims to influence society's impact and the way that it interacts with nature and **biodiversity**, by promoting solutions through research, consultation, education and partnerships. Conservation is necessary because every living thing depends on other creatures. Together they form a community, a living web, in which all strands contribute to the stability of the whole. Conservation aims to prevent **extinction**, a process of population decline (i.e. endangerment) that reaches the end point of complete disappearance from Earth, causing losses of genes, adaptations and ecosystem processes even before the actual complete disappearance occurs. **Endangered species** exist in low population numbers and need intensive long-term management in order to survive. Endangerment is a broad issue, one that involves the habitats and environments where species live and interact with one another. Although some measures are being taken to help specific cases of endangerment, the universal problem cannot be solved until humans protect the natural environments where endangered species dwell.

Conservation biology, originally a branch of biology that deals with the effects of humans on the environment and biological diversity conservation (endangerment), now broadens its scope to the consequences of biodiversity loss to human and biosphere survival. Conservation biology priorities are: (i) to stop the degradation of the planet's natural environment and build a future in which humans live in harmony with nature; (ii) conserve the world's biological diversity; and (iii) ensure that the use of renewable natural resources is sustainable by promoting pollution reduction and reducing wasteful consumption.

Reasons for conservation

While the idea of conserving the environment has steadily gained political acceptance over the past few decades, people still misunderstand and ignore the goods and services that nature, biodiversity and ecosystems provide to humans. For example, 15,589 species of animals and plants are threatened with extinction and many ecosystems, including wetlands and forests, are being degraded and destroyed, even though these ecosystems provide humans with a large range of highly valuable services. The importance of well-functioning ecosystems in helping reduce poverty and improve livelihoods, societies and economies is increasingly clear to scientists and to society. This knowledge needs to be integrated into the decisions and actions of local, national and international policy makers in all sectors, as well as business leaders.

To achieve effective biodiversity conservation, one prime activity for scientists is to contribute to the understanding of what natural ecosystems provide to humans. But conservationists should also seek to ensure that this knowledge is used in practical ways by bringing together scientists, policy makers, business leaders and non-governmental organizations (NGOs) to impact the way the world values and uses nature. Only in this way will Earth's natural heritage be maintained for future generations, allowing them to thrive spiritually, culturally and economically.

A conservationist's mission is to conserve the Earth's living heritage and global biodiversity, and to demonstrate that human societies are able to live harmoniously with nature, saving wildlife and wild lands through careful science, international conservation and education.

The emergence of conservation biology

Conservation biology is the study and preservation of habitat for the purpose of conserving biodiversity. The term 'conservation biology' refers to the science, and is sometimes used also to encompass the application of this science. In simple terms, conservation biology is the scientific study of the phenomena that affect the maintenance, loss and restoration of biological diversity. Its research integrates biological with economic, anthropological, legal and other social science perspectives. Conservation biology has only recently begun to address the **ethics** of protecting biodiversity. Conservation biology provides scientific expertise to support the conservation and recovery of biological diversity in its natural state through applied research, education, planning and community service. Philosophically, the concern of this branch of biology is generally to help save the diversity of life on Earth through applied conservation research.

In the realm of research, biologists seek creative and effective ways to address a wide diversity of ecological problems, ranging from endangered species population decline to regional conservation planning for ecosystem conservation. This translates into developing better conservation tools, analyses and techniques. A practical aim of the whole field is to stop human-induced extinction. Conservation biology requires conservationists to implement conservation actions. The reasons for extinction are changing: in prehistoric times, natural disasters and competition with other species were the main causes; in historic times, overexploitation and **exotic species** introductions caused many extinctions. But today, the main problems facing wildlife are destruction of habitat and pollution. Conservation uses an array of tools to describe the phenomena that affect the maintenance, loss and restoration of biological diversity. As a consequence of proper conservation actions environmental health, economic vitality and informed land use decisions can be achieved. In practice, conservation actions are oriented at preserving the plants, animals and natural communities that represent the diversity of life on Earth, by protecting the lands and waters they need to survive.

A conservationist is someone who works to protect species from extinction, but little attention has been paid to the fact that populations are composed of individuals (see Fig. C.9). A population decline is somehow described as an ecological phenomenon caused by humans, but how this decline operates in individuals by affecting their behaviour, physiology and welfare is often neglected or unknown. Any process that affects species is indeed affecting the individuals of a given population. **Hunting**, habitat destruction, pollution or introduced species are significant animal **welfare** factors affecting individuals and triggering stress responses. Long-term **stress** and distress factors caused by conservation problems are likely to produce suffering, pain, behavioural problems or displacement.

Conservationists tend to focus on population trends (i.e. decline), and they have historically paid little attention to animal welfare. In fact, the decline of any given population is preceded by animal welfare problems and the link between both disciplines, although it should seem obvious, is not clearly understood by conservation biologists. Population decline is a mere symptom of an animal welfare problem. Wildlife harassment and habitat destruction (e.g. fire, land degradation, toxic waste, oil spills) are both suffering and distress factors affecting individuals, and direct mortality, injuries, diseases and behavioural perturbations are all clear signs of animal welfare consequences. An animal welfare approach to conservation biology underpins a more thorough and accurate understanding of the prime causes of population decline (i.e. conservation problems).

There appears to be a dichotomy between animal welfare and conservation biology. While conservation biology focuses on population decline in a given environment (reflecting the interdependence of all living things and prioritizing the population, community and ecosystem levels), animal welfare focuses on selected wild species' life history and fate as a consequence of human pressure (directly or indirectly). In fact, there are many rifts and disparities between these two groups, both in practice and at a theoretical level. Like conservation biology, the emerging field of animal welfare is increasingly becoming an area of immense interest and inquiry in the general discipline of wildlife management and environmental sciences.

The sciences of conservation biology and animal welfare have traditionally been conducted in two independent spheres. Conservation biology, a discipline focused on the causes of population decline and the study of extinction caused by human actions, has traditionally been the realm of ecologists and biologists. Animal welfare, on the other hand, is defined as the state of an individual animal or wild population in relation to its ability to cope with its environment, and has traditionally been the niche of physiologists, ethicists, behaviourists and veterinarians. Thus animal welfare scientists measure factors that may contribute to good or poor welfare, or measure physiological and behavioural states believed to reflect good/poor welfare in wild animals. Conservation biologists are concerned about the causes of population decline such as habitat destruction, human pressure and the effect of alien species. Priorities are set at different thresholds (see Fig. C.10). While conservationists pay attention to species that are already declining (a symptom), animal welfare scientists are focused on how different species react to human-induced stress (the cause). A meeting point for both disciplines is in endangered

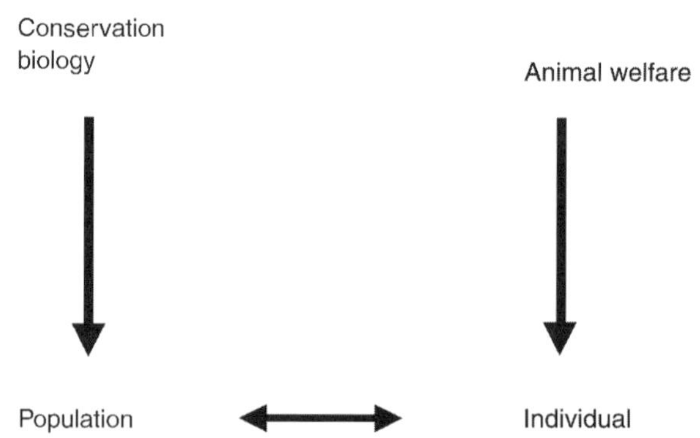

Fig. C.9. The mismatch between conservation and animal welfare approaches: an actual mismatch of priorities between animal welfare scientists and conservation biologists that can be solved by understanding that human-induced stress is a prime factor common to both disciplines.

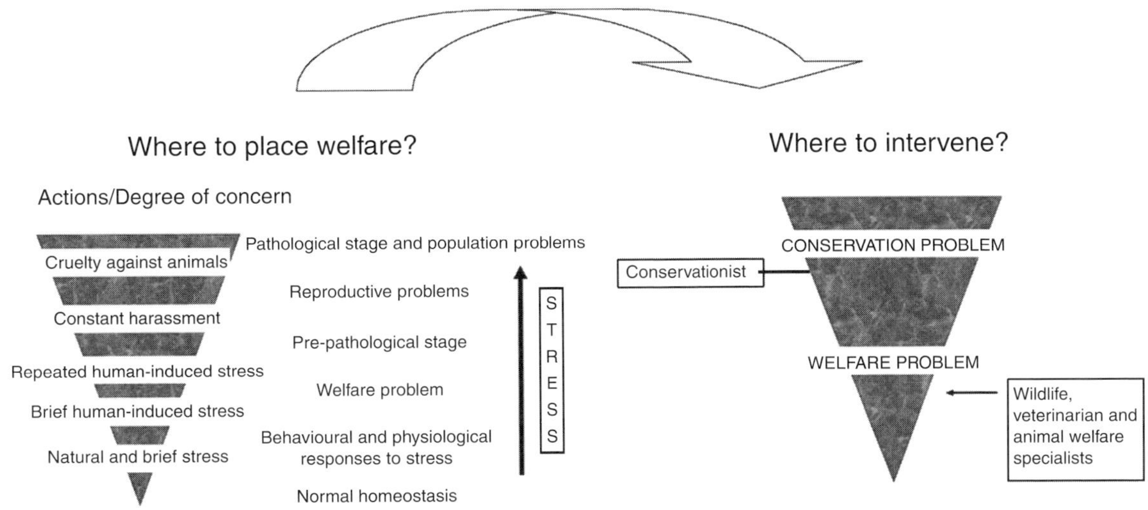

Fig. C.10. The critical point of intervention for a wild animal: an animal welfare perspective – a conservationist threshold that 'activates' their concern about a given species related to the diagnosis that population decline is occurring. An animal welfare scientist should intervene even before population decline occurs, by preventing the stress that causes welfare problems. Any natural stress or human-induced stress is not considered a conservation or animal welfare issue until it effectively becomes a repeated, long-lasting stress or distress that might drive the individual into the pre-pathological, reproductive or dead stage, when a population decline effect may be observed.

species, where only a few survive and rescue actions and management practices such as **translocation**, **captive breeding** or anti-poaching protection activities are prime tools for both disciplines.

On the other hand, animal welfarists and conservation biologists are in conflict when culling, hunting quotas or control techniques for invasive species are discussed as conservation tools. While conservationists' main priorities are to restrict ecological damage at any price, animal welfarists are concerned about the consequences of control, hunting, culling or sustainable use on the individual. This kind of issue gives rise to distrust and controversy and contributes little to the development of a common ground for both disciplines.

Although the approach to the species problem may seem different, ultimately both disciplines are focused on the direct or indirect effect of humans on wildlife. Moreover, both conservation biology and animal welfare are multidisciplinary, 'goal oriented' and based on the **precautionary principle**. It seems that the reasons for both disciplines to be separated are more related to lack of understanding of a common language than to differences in their ultimate goals.

Currently, society is asking for solutions to the biodiversity crisis, and animal welfare-oriented funding is contributing to wildlife conservation worldwide. It seems that the inclusion of animal welfare approaches is enriching the conservation biologist's tool kit for a better approach to conservation that integrates increasing societal demands. The dichotomy between individual- and population-level work can actually be detrimental to a full understanding of many life history processes affected by human disturbance.

Both conservation biology and animal welfare are based on ethical considerations of how humans interact with other non-human animals and the environment. The claim that we have ethical obligations to animals and their environment is valid and common to both disciplines, and it has been argued that animal welfare considerations should be incorporated into conservation policies (e.g. trapping standards, captivity recommendations, etc.). The fact that welfare is usually measured on an individual basis, whereas some wider goals of conservation are measured on populations, does not make them incompatible. Indeed, new assessment methods and more refined research approaches allow us to understand the ultimate causes of population decline (e.g. **biotelemetry**, **stress hormones** in faeces, video tracking of abnormal behaviours). Moreover, in many cases population decline can be explained by understanding behavioural and physiological changes that are the main realm of animal welfare science. Both disciplines are now crucial to understanding how to recover endangered species, conserve wildlife, control invasive species in a humane manner and solve conflicts between people and wildlife. The inclusion of animal welfare science into conservation biology is an opportunity to enhance research methods currently in use to understand population decline while meeting the ethical needs of our societies, demanding not only environmental protection but also proper care for wild species. (CB)

See also: Biodiversity; Conservation behaviour; Ecology; Endangered species; Reintroduction; Wildlife management

Conservation behaviour

Conservation behaviour is an emerging discipline, founded largely by behavioural ecologists (**see: Behavioural ecology**) who attempt to use behavioural research and theory to solve conservation problems facing diminishing animal populations in today's world. It differs from other areas of applied animal behaviour both because its roots are in the more theoretical approach adopted by behavioural ecologists and because its focus is predominantly on **conservation** in the wild. A series of influential books and papers since the end of the 20th century have spoken of the many implications that behaviour – particularly behavioural ecology – has for conservation.

That the behaviour of animals influences conservation is a truism, but the challenge for behavioural scientists has been to move from the *implications* phase of conservation behaviour to more active *applications* to solve real-world conservation problems. Past efforts have, perhaps, overemphasized function at the expense of proximate mechanisms. While a functional approach does afford testable hypotheses that can guide conservation behaviour research, it is the proximate mechanisms that can be manipulated in the service of conservation needs. Recent developments, however, have begun to synthesize the more mechanistic approach of applied ethologists and comparative psychologists (**see: Comparative psychology**) with the functional approach of behavioural ecologists, promising an increase in the conservation applications of the discipline.

Conservation behaviourists have identified several behavioural topics that are especially important for conservation application. In small populations – captive or wild – human intervention is becoming increasingly necessary to manipulate **social behaviour** and **reproductive behaviour**, to encourage breeding or to enhance the persistence of genetic diversity. Some conservation behaviour research strives to identify the right combination and density of individuals to get animals breeding. A great deal of research also addresses mating strategies, because some mating patterns can lead to loss of genetic diversity if individuals are over- or under-represented. For example, if one male is dominant or preferred by females, he can sire many offspring, while subordinate or less preferred males may leave behind no offspring. In small populations the loss of this genetic diversity can cause crippling **inbreeding** depression. Recent experiments have shown that it is possible to manipulate these outcomes and equalize breeding opportunities among competing males. Because animals use displays and signals to compete and court, studies of communication processes have been instrumental in solving this problem.

Several other topics are addressed by proponents of conservation behaviour. Foremost among these, perhaps, is dispersal and habitat selection. Understanding these behavioural processes is critical for designing reserves that contain the necessary resources and allow natural dispersal to take place. In most species the young of one sex disperses out of its natal area and settles far from its birthplace, thus avoiding inbreeding with its relatives. As habitat becomes increasingly fragmented, however, animals may not be willing to disperse across human-altered landscapes. Dispersal is a risky period in an animal's life, and dispersers must assess various aspects of habitat suitability and make a decision regarding where to settle. It is important to understand these behavioural processes to predict the potential effects that human activities – including conservation management – may have on animal populations.

Reintroduction programmes also rely – or should rely – heavily on behavioural research and may be the best-developed application of conservation behaviour research. Reintroduction involves the captive breeding and release of individuals into the wild or the translocation of wild animals from one location to another. These actions create major challenges for the welfare and survival of animals released into completely novel environments. While they are a worthwhile conservation action, these measures require comprehensive research programmes addressing many facets of behaviour to predict likely outcomes, devise optimal strategies and monitor individuals following release.

These are just a few of the activities pursued by conservation behaviourists. Others include human disturbance, human–animal conflict – such as crop raiding and livestock depredation – behavioural responses to invasive species and many more. Behavioural research is nowadays playing an increasing role in mitigating these emerging conservation crises. (RS)

See also: Dispersal and habitat selection; Intrasexual selection; Mate choice; Noise disturbance; Translocation

Further reading

Blumstein, D.T. and Fernández-Juricic, E. (2004) The emergence of conservation behaviour. *Conservation Biology* 18, 1175–1177.

Gosling, L.M. and Sutherland, W.J. (2000) *Behaviour and Conservation*. Cambridge University Press, Cambridge, UK.

Swaisgood, R.R. (2007) Current status and future directions of applied behavioural research for animal welfare and conservation. In: Swaisgood, R.R. (ed.) *Special Issue: Animal Behaviour, Conservation and Enrichment. Applied Animal Behaviour Science* 102, 139–162.

Conservation centre

There is no agreed-upon definition for what a conservation centre is or should be. The term 'centre' seems to confine the usage of this term to conservation programmes that are based primarily in a single geographic location. Thus, many major conservation organizations, such as The Nature Conservancy or World Wildlife Fund, are not considered conservation centres, although they may manage many such centres as part of their conservation effort.

Popular use of the term includes environmental education programmes or **ecotourism** vacation packages. Here, one might expect to see miles of hiking trails with an interpretive centre carrying a **conservation** message. Animal rehabilitation centres are another common form of conservation centre. In these, local wildlife is rescued, cared for and sometimes released back to the wild. Often, rehabilitation centres are more focused on individual animal welfare and may contribute little to the conservation of the species in the wild and, in at least a few cases, may compromise wild populations if releases are well-intentioned, but ill-conceived. For example, released animals may introduce disease, cause social disruption or reintroduce tame animals that will increase human–animal conflict. Many conservation centres are devoted to the conservation of a single species or a group of related species, such as **gorillas**, crocodiles or parrots.

Captive breeding programmes are another widespread form of conservation centre. Often run by government organizations, non-profit non-governmental organizations or **zoos**, these programmes usually maintain the goal of breeding

endangered species as a hedge against **extinction** in the wild and to provide candidates for release back to the wild. The new zoo ethic embraces conservation, and thus zoos are a fast-growing type of conservation centre. Many of the larger zoos – such as the Zoological Society of London and the Bronx (New York) Zoo's Wildlife Conservation Society – support programmes that integrate animal exhibition, captive breeding, field conservation and education. The San Diego (California) Zoo's Conservation Research Center, for example, espouses a mission to generate, share and apply scientific knowledge vital to the conservation of animals, plants and habitats. As part of a multidisciplinary effort, this centre targets several key areas in conservation, including sustainable populations, bioresource banking (the frozen zoo), wildlife health, habitat conservation, restoration biology and conservation education.

Conservation centres encompass a diversity of approaches but, as the label suggests, the term should be earned by making some meaningful contributions to conservation through a combination of activities such as protection of animals and landscapes, breeding programmes or research. (RS)

See also: Reintroduction

Further reading

San Diego Zoo Conservation Research. Website available at: http://zooconservation.org

Zoological Society of London. Website available at: http://www.zsl.org

Conspecific

A conspecific is a member of the same species. The provision of conspecific company is considered especially important when assessing the **welfare** of an individual member of a social species. However, the extent to which conspecifics are appreciated as such is dependent, not least, on **socialization** and filial **imprinting**. (PDM)

Conspecific attraction

Conspecific attraction describes the tendency of some animals to settle near individuals of their own species. It corresponds to an aggregation phenomenon which is beyond the density variation predicted by the quality of the environment. This behavioural phenomenon is particularly well known in colonial species such as seabirds, but recent studies have shown that conspecific attraction is widely distributed among a range of other, less social, species.

The conspecific attraction hypothesis suggests that, in many species, dispersing individuals looking for a place to settle and establish a new home will rely on the presence of **conspecifics** to assess habitat suitability. The general concept of conspecific attraction has been recognized outside the scientific community for generations. For example, hunters have long used duck decoys to attract ducks to a pond. Presumably the presence of these decoys falsely advertises that this is a safe place with good resources.

Judy Stamps of the University of California, Davis is largely responsible for developing and refining this concept in the scientific literature. In the 1980s she proposed that, even in relatively solitary, territorial species, habitat settlement was socially facilitated. Ecologists had long believed in the 'ideal free distribution' of animals in the landscape, wherein population density directly reflected habitat quality. Dispersing individuals were envisioned to settle in the most suitable habitat and to avoid settling in close proximity to conspecifics, which were thought of as competitors for resources. Territories should clump close together only where density of resources was high. Controlling for habitat quality, Stamps found that *Anolis* lizards actively preferred to settle adjacent to conspecifics and proposed that dispersers used the presence of conspecifics as cues to territory quality. Taking a more behavioural perspective than that of ecologists, Stamps suggested that dispersing individuals lack knowledge of habitat suitability, and could most easily assess habitat quality by its correlation with the presence of other conspecifics. Thus, dispersers may follow the rule of thumb that, if other conspecifics are present, then the area must be relatively safe and contain appropriate resources. Since this hypothesis was proposed, tests in a variety of vertebrate species have supported the notion that conspecific attraction is a widely distributed behavioural phenomenon, even in relatively asocial species.

This concept has profound implications for **conservation**. Ecologists regularly use animal density as an index of habitat preference, but the conspecific attraction hypothesis clearly suggests that patterns of animal density on the landscape may be unrelated to habitat suitability. This is increasingly true in human-altered landscapes. Animals may be pushed into marginal habitats or extirpated in areas of prime habitat suitability. Dispersers may choose to settle next to members of their own species in suboptimal habitat, and may even reject habitat that is more suitable for the species if conspecifics are absent. Thus, studies correlating the presence of a species with the habitat in which it is found may be misleading.

These observations pose a problem for conservation managers attempting to recruit animals into suitable habitat, such as newly created protected areas. If the species has been eradicated in the area, animals may not recolonize it through natural dispersal. If animals are reintroduced there they may not settle, instead choosing to disperse to another area where conspecifics can be found. This may be one explanation for why so many **reintroduction** programmes fail: released animals following simple behavioural rules of thumb for habitat selection and settlement may choose not to stay in a habitat that is perfectly suitable. It goes without saying that this kind of disruption, including risky and stressful long-distance dispersal, can also have significant **welfare** consequences. Thus, research to facilitate speedy settlement in suitable habitat has both welfare and conservation implications.

A promising strategy for addressing this problem is for conservation behaviourists to conduct research towards an understanding of the behavioural mechanisms underlying habitat settlement rules. Earlier studies have shown that the presence of actual individuals may not be necessary for conspecific attraction to operate. For example, playback of birdsong can attract territorial bird species to settle in unoccupied habitat. The key then is to understand which **signals** animals use to identify the presence of conspecifics and how they find ways of manipulating these signals to encourage settlement.

Recent research is beginning to tap into this realization, and the settlement patterns have been managed for conservation purposes in several species. Decoy models capitalizing on visual cues have been used to attract colonially nesting seabirds to safe places. In some cases simpler visual cues can be used. For example, griffon vultures can be attracted to new breeding sites by painting cliffs with white paint, imitating the accumulation of faeces. Grouse can be induced to establish new **lek** sites using acoustic playbacks of vocalizations. The role of olfactory signals in habitat settlement is less explored, as is the usefulness of this technique in manipulating the behaviour of mammals. Recent research, however, shows that translocated black rhino can be induced to settle next to 'virtual scent territories' by spreading conspecific dung in a way that mimics natural dung-marking patterns. For denning species, such as bears, olfactory cues may prove useful for attracting animals to safe denning sites in the same way that visual cues have worked to attract birds to nesting sites.

Behavioural researchers are only beginning to test the utility of applying such behavioural and ecological concepts to conservation management, but initial results suggest great promise for further study of conspecific attraction in a conservation context. (RS, SL)

See also: Affiliation; Bonding; Conservation behaviour; Dispersal and habitat selection

Further reading

Reed, J.M. and Dobson, A.P. (1993) Behavioural constraints and conservation biology: conspecific attraction and recruitment. *Trends in Ecology and Evolution* 8, 253–255.

Stamps, J.A. (1988) Conspecific attraction and aggregation in territorial species. *American Naturalist* 131, 329–347.

Ward, M.P. and Schlossberg, S. (2004) Conspecific attraction and conservation of territorial songbirds. *Conservation Biology* 18, 519–525.

Consummatory acts

Consummatory acts are characterized as being the relatively stereotyped actions occurring at the end of a behavioural sequence. Although ingestion of food and water may be consummatory acts, consummatory acts can include copulation, burrow digging and many other activities. Consummatory acts should not be confused with **modal action patterns** or fixed action patterns, **instinctive behaviour**, **innate** behaviour, **operants** or **reflexes**, although they may include such phenomena. The term 'consummatory act' was central to classical ethology, but became less so beginning in the 1970s as ethology was influenced by behavioural ecology and adaptationist thinking, the proponents of which were relatively uninterested in behavioural mechanisms. None the less, with the renewed interest in the control and organization of behaviour in animal cognition, behavioural neuroscience and applied areas such as conservation and captive animal welfare, the ideas underlying the concept are increasingly recognized as being both relevant and useful.

The term 'consummatory act' was first popularized, if not coined, by Wallace Craig in a classic paper published in 1918 (Craig, 1918). It was initially proposed as an attempt to systematize our thinking about behavioural organization, particularly the normally occurring instinctive sequences found in the behaviour of both human and non-human animals. The term nominally refers to a behaviour action, such as a bite or copulation, that follows **appetitive behaviour**, and thus cannot be effectively discussed separately from appetitive behaviour. In brief the sequence is as follows. A **motivational state** derived from a **need** or **drive** instigates the animal to engage in, typically, locomotor behaviour to locate important stimuli such as food, prey, a mate, or a nesting site. This *appetitive behaviour* is typically highly variable and may involve a variety of orientation processes as well as learning (using past experience) to locate such stimuli. When the stimulus is encountered, behaviour is elicited that leads to the performance of relatively stereotyped acts such as prey strikes and prey handling, copulation, burrowing, or even retrieving and splitting nesting material. The performance of these *consummatory acts* typically leads to the satisfying of biological needs, and are then followed by the third phase, *relative rest*. Aversions to dangerous stimuli involving predators or **conspecifics** would be handled similarly, involving flight or other anti-predator response such as freezing, crypsis, threatening, attacking, etc.

The actual sequential organization of behaviour is often more complex, of course, as more than one motivational system may be involved (Burghardt, 1973). Behaviour systems theory (Timberlake, 2001) is an updated version of the original conceptual framework developed by Craig and extended by other ethologists (e.g. **Lorenz**, 1981). In male stickleback fish reproductive behaviour, numerous consummatory acts have been described including chasing, biting, digging, gluing nest material, zigzag dancing, quivering and fanning eggs.

The usefulness of the separation of appetitive behaviour from consummatory acts was in the resolution of two apparently conflicting ways of looking at instinct. For zoologists, typically, instincts were highly stereotyped innate responses shaped by natural selection and largely impervious to any important role for individual learning. Psychologists, on the other hand, were more likely to view instincts as motivational factors leading to general biologically useful outcomes (e.g. **hunting**, eating, mating and fighting). The details of the instincts were thus quite open to learning and experience, especially in the mammals that psychologists were most inclined to study.

The sequential separation of appetitive behaviour and consummatory acts showed that both approaches were valid, but generally in different places in the sequence. Thus, almost all learning studies involve altering appetitive behaviour – mazes, problem boxes, operant conditioning, discrimination learning – even observational learning and imitation. What is modified is rarely the performance of the behaviours involved in responding to the attained 'rewards' such as eating, drinking or mating. Lesioning and other neural studies support the differentiation of consummatory and appetitive behaviour at both the neural and conceptual levels (Burghardt, 1973; Balthazart *et al.*, 1998). Arousal mechanisms, for example, have long been separated from copulatory mechanisms in studies of hormones and **sexual behaviour**.

One interesting aspect of consummatory acts is that animals often find the mere performance of the act, and the sensory stimulation involved, reinforcing, separate from **feedback** from the biological needs being satisfied. Mating is a clear example, but studies show that animals often stop eating or drinking long before the food or liquid has had its requisite physiological effect. This may be true of many other behaviours as well, and thus the opportunity to perform species-typical consummartory acts may be an important component of an effective incorporation of animal **welfare** concerns into captive management (Hughes and Duncan, 1988). It is also apparent that the quality of the stimulus responded to can either suppress or enhance performance of consummatory acts. This is a factor in feeding and weight regulation in many animals. More palatable food, regardless of nutritional value, is ingested more than less palatable items, and it has been suggested that this can lead to obesity in captive animal, pets and people, although some argue that the balance of nutrients is more important in this problem. In farm animals at least, it seems that palatability is more important in initial acceptibility, i.e. initiation of the consummatory act, than total intake, i.e. its persistence. The role of feedback thus needs to be evaluated in terms of both internal (physiological) and behavioural performance factors.

It is commonly noted that captive animals, especially without previous experience with normal stimuli, sometimes perform consummatory acts to an abnormal stimulus when deprived of species-typical stimuli. Such behaviour is often seen in predatory and mating behaviour, where stalking and mating attempts may occur towards alien species or even inanimate stimuli. Occurrence of such phenomena in captive animals need to be diligently recorded, as these may underlie many compulsively performed stereotypic behaviours found in captive animals, especially those kept in small or sterile enclosures.

Many studies show that experiences can alter some details of consummatory acts such as eating and mating behaviour, just as it takes some skill for people to learn how to use a spoon or chopsticks. However, the truly final end of instinctive sequences, such as swallowing food, sucking at the breast or thrusting in copulation is still largely stereotyped within a species, as compared with the appetitive phase.

Inexperienced animals, although highly motivated to perform, are often awkward and unsuccessful, but none the less they persist. **Play** behaviour may be a prime area in which consummatory acts are explored. For example, many animals manipulate objects using predatory movements (jumping, grabbing, biting, manipulating) but refrain from the final consummatory acts of ingestion. Similarly, in play fighting many movements are seen except for those that could truly injure or kill the opponent – or, more accurately, play partner. Play behaviour is often more often observed in well-cared for captive animals, and this may be due to the activation of one or more behaviour sequences in contexts not essential for survival. (GMB)

See also: Exploratory behaviour; Play

References

Balthazart, J., Absil, P., Gérard, M., Appeltants, D. and Ball, G.F. (1998) Appetitive and consummatory male sexual behaviour in Japanese quail are differentially regulated by subregions of the preoptic medial nucleus. *Journal of Neuroscience* 18, 6512–6527.

Burghardt, G.M. (1973) Instinct and innate behaviour: toward an ethological psychology. In: Nevin, J.A. (ed.) *The Study of Behaviour*. Scott Foresman, Glenview, Illinois, pp. 321–400.

Craig, W. (1918) Appetites and aversions as constituents of instincts. *Biological Bulletin* 34, 91–107.

Hughes, B.O. and Duncan, I.J.H. (1988) The notion of ethological 'need', models of motivation and animal welfare. *Animal Behaviour* 36, 1696–1707.

Lorenz, K. (1981) *The Foundations of Ethology*. Springer-Verlag, New York.

Timberlake, W. (2001) Motivational modes in behaviour systems. In: Mowrer, R.R. and Klein, S.B. (eds) *Handbook of Contemporary Learning Theories*. Erlbaum, San Francisco, California, pp. 155–210.

Contiguity

Contiguity describes the closeness of the relationship in either time (temporal contiguity) or space (spatial contiguity) between two events. In the original Laws of Association as applied to learning it is proposed that events that are highly contiguous are easily associated through learning, as are events that are more frequently associated or associated with an emotional consequence. However, more recent research has suggested that **contingency** (predictive value of one event relative to another) may be of more relevance than contiguity in predicting learning outcome. For example, events that are presented simultaneously are not necessarily easily associated, contrary to what the law of contiguity would suggest, and events that are not necessarily highly contiguous may be associated if one event reliably predicts another. (DSM)

See also: Conditioning – types of; Reinforcement – types of

Contingency

Contingency describes either the extent to which one event predicts another or to what extent the two events covary. The graphical representation of this relationship is referred to as a contingency space (see Fig. C.11). The contingency between events is important in predicting learning of the association.

Stimulus-contingent events occur only if a given stimulus is present (e.g. the delivery of food into the dog's food bowl when you call the dog for its meal), while response-contingent events only occur if a given response occurs (e.g. the dog is rewarded for coming when called only if it comes promptly). (DSM)

See also: Conditioning – types of; Contiguity; Reinforcement – types of

Contractarianism

According to the ethical position, contractarianism morality reflects a tacit agreement between *rational* agents who, motivated by self-interest, agree to rules or principles that indicate acceptable behaviour. This precludes animals, however. There is ongoing debate about whether dependency and reciprocity, instead of rationality or self-interest, should be the basis of an interspecific social contract. (RA)

Contra-freeloading

Contra-freeloading describes the phenomenon whereby an animal will work for a resource, in preference to accessing the same resource without work. The phenomenon has been reported as occurring in virtually all vertebrate species

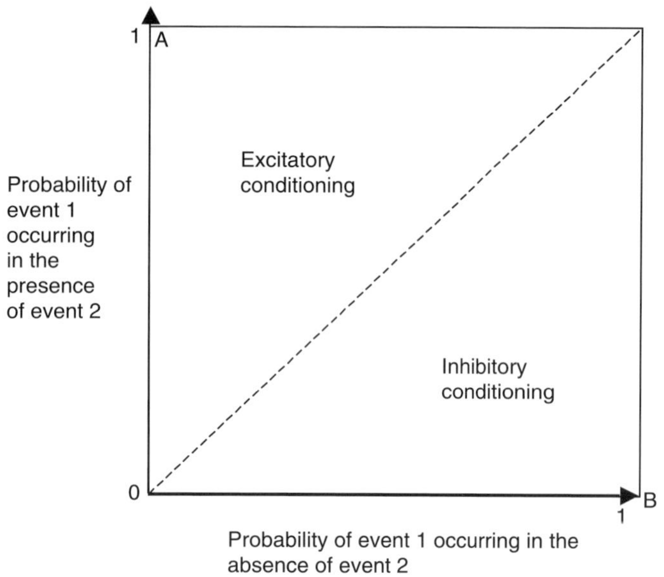

Fig. C.11. A contingency space of all possible probabilities for two events (1 and 2). At the extremes, if event 1 always occurs in the presence of event 2, and never in its absence (point A), then event 2 is a perfect predictor and so learning of this relationship (excitatory conditioning) can be expected. However, if event 1 never occurs in the presence of event 2, and always when event 2 is absent, then event 2 becomes a perfect predictor of the absence of event 1, and this relationship will be learned (inhibitory conditioning). In reality, most relationships fall somewhere between these extremes (i.e. sometimes together and sometimes event 1 alone), but it is the balance between the two probabilities that determines whether and what type of conditioning will occur to event 2. If the two probabilities are equal, then regardless of their absolute value (broken diagonal line), event 2 predicts neither the occurrence nor absence of the other event, and so no conditioning will occur.

investigated, with the possible exception of the domestic cat, although variation may occur between different breeds. Some studies that have claimed to investigate contra-freeloading using environmental **enrichment** devices have been criticized as a different food is offered in the enrichment device, which confounds the investigation.

The phenomenon is problematic to animal behaviour scientists, as it appears upon initial consideration to contradict traditional models of **motivation,** learning and **optimal foraging**. However it has been found that the tendency to contra-freeload declines with hunger and with an increased work effort, and that sensory reinforcement may be an important component of the process. Animals may choose to work for a resource in order to obtain additional information that is not available if only the free source is chosen (information gain hypothesis). This may have adaptive value as the animal learns about access to alternative food sources should the free source become unavailable. Thus it would seem erroneous to conclude from contra-freeloading studies that animals prefer to work for food, but rather that they will work to obtain information, and that food-based environmental enrichment devices do not necessarily improve welfare because they do not provide that information. (DSM)

Further reading

Inglis, I.R., Forkman, B. and Lazarus, J. (1997) Free food or earned food? A review and fuzzy model of contrafreeloading. *Animal Behaviour* 53(6), 1171–1191.

Osbourne, S.R. (1977) The free food (contrafreeloading) phenomenon: a review and analysis. *Animal Learning and Behaviour* 5, 221–235.

Young, R.J. (1997) The importance of food presentation for animal welfare and conservation. *Proceedings of the Nutrition Society* 56, 1095–1104.

Control – experimental

Research experiments are set up to investigate the effect of one variable at a time, and any measured effect is compared with animals that have not been so 'treated' – the controls. Controls may be given no treatment at all, or be given everything except what is under test, e.g. a new drug. A control may therefore be administered the same dosage regime but given the excipient (carrier solution or vehicle). A sham operation is another form of control where the same surgical procedure is used, except that an experimentally crucial part is omitted. On other occasions the control may be the best contemporary treatment, so that a new drug is compared with those drugs that are currently in use to treat a particular disease. Sometimes controls are used to make sure that the experimental test group is valid, e.g. a positive control might be run to ensure that a frozen–thawed microbiological culture was still virulent in a challenge test with a new batch of vaccine. (DBM)

Controllability

It has been suggested that the impact of an environmental challenge depends not only on its intensity and duration, but also on its predictability and controllability. In a classic experiment by Weiss (1972), pairs of rats were separated but had the same electric current applied to their tail at certain times. One of the pair had some control over the delivery of the shocks and some way of predicting and controlling the current, by pressing a lever when a light flashed, while for the other this cue and response had no value. Those without control or a way of predicting the shocks exhibited a much greater **stress** response. Similarly, a reduced predictability and controllability of positive events may also lead to a greater stress response.

Predictability is not necessarily linked to controllability; for example, food may be delivered according to a highly predictable schedule for animals in captivity, but it may have no control over the delivery of that food. It has been suggested that controllability may explain the phenomenon of **contra-freeloading**, although this explanation has been rejected by many. Environments with a high level of predictability and control are not necessarily beneficial to an animal's well-being because, although they may reflect certainty in the environment, they also suggest a lack of novelty. Given that higher animals have evolved in response to the pressures of a changing environment, it is suggested that their physiological systems may be adapted (both phylogenetically and ontogenetically) to function most efficiently within an environment with some level of unpredictability and uncontrollability, although there is no universal optimal value. (DSM)

See also: Allostasis

Reference and further reading

Bassett, L. and Buchanan-Smith, H.M. (2007) Effects of predictability on the welfare of captive animals. *Applied Animal Behaviour Science* 102, 223–245.

Weiss, J.M. (1972) Psychological factors in stress and disease. *Scientific American* 226, 104–113.

Wiepkema, P.R. and Koolhaas, J.M. (1993) Stress and animal welfare. *Animal Welfare* 2, 195–218.

Convergent evolution

This term describes the evolution of superficially similar (analogous) structures or behaviours in two or more distantly related lineages that have arisen through adaptation to similar ecological conditions rather than through shared, common ancestry. For example, the wings of a bat and a bird are similar, despite the fact that a bat is more closely related to a whale than it is to a bird. Thus, wings in these two groups have evolved independently (in parallel) yet serve similar functions. A classic example is the physical similarity between marsupial and placental mammals despite the fact that the two lineages have evolved independently on separate continents: the placental anteater and the marsupial anteater (numbat), the Tasmanian wolf and the placental wolf, the marsupial mole and the placental mole, and the flying squirrel (placental) and the flying phalanger (marsupial). (PE)

Cooperation

Cooperation is a voluntary act in which two or more individuals work together to bring about an end situation with greater benefits than they could have obtained by working individually. In cooperation, benefits to the individuals working together exceed the costs. African wild dogs (*Lycaon pictus*), travelling in a pack across the Serengeti, will spot a grazing gazelle and organize themselves around the prey. As the gazelle flees the approaching dogs, the pack will spread out around it so that a sudden turn or misdirection will give one dog a chance to intercept. While one dog grabs the gazelle by the nose, others attack from behind. Successful predations such as this will only occur through cooperative effort. In captive chimpanzees (*Pan troglodytes*), individuals often groom each other for long periods of time. If the chimpanzee group is fed later in the day, the groomer may solicit their friend for food. The latter will show greater tolerance than they would have without the previous **grooming**.

In a natural world in which every individual is considered most interested in its own survival and reproduction ('nature red in tooth and claw', as Alfred Tennyson put it), cooperation represents a puzzle. Why cooperation evolved is a central issue in evolutionary biology, which classifies it into two distinct types depending on the genetic relatedness among the cooperating parties. The first type, cooperation among related individuals, is probably the product of **kin selection**. The second, cooperation between unrelated individuals, may be a product of either **mutualism** or reciprocal **altruism** (see Table C.4).

Kin selection

William Hamilton's theory of **kin selection** is widely accepted as an explanation of cooperation among related individuals. Hamilton explained kin selection by noting that the greater the benefits of altruism (where the donor incurs the costs of action while the recipient benefits) between closely related individuals, the more likely cooperation will be worthwhile. This can also be explained in terms of an individual's inclusive **fitness**, defined as the combined effects on fitness of the individual's altruistic actions towards its own offspring and other kin. If a behaviour is detrimental to the actor yet greatly benefits its kin, there is still a net increase in inclusive fitness. Close relatives are more likely to carry similar genetic propensities for altruism; by cooperating with close kin, an individual is promoting the transmission of these altruistic 'genes' – if not in itself then at least in others. Kin selection explains why individuals are more likely to assist close kin than distant relatives.

When **Charles Darwin** observed social insects in which certain castes are sterile (such as worker ants or worker honeybees), he was baffled by the possibility of individuals working cooperatively while forgoing their own reproduction. This type of social organization, known as **eusociality**,

Table C.4. Cooperation delineated into different types based on category of genetic relationship (kin versus non-kin).

	Cooperation		
Subjects	Form	Description	Example
Related individuals	Kin selection	Cooperation dependent on the relatedness of the individuals involved; the closer their relatedness, the lower the effort needed to overcome the cost of cooperating	Eusociality in insects, where sisters cooperate to raise and defend the colony's young
Unrelated individuals	Mutualism	The benefits of cooperation are collected simultaneously by all parties	Cleaner fish eat ectoparasites off larger fish, which benefits both parties at the same time with, respectively, food and a reduced parasite load
	Reciprocal altruism	All the benefits of cooperation go to the partner, who reciprocates on a later occasion	Chimpanzee food for grooming exchanges in which the recipient of grooming later tolerates taking of food by the groomer

describes cooperative care of a colony's young by both reproductive and sterile castes. In many insect colonies, this is possible because the sterile worker insects are what Hamilton termed 'supersisters'. Honeybees, for example, are haplodiploid, where fertilized eggs mature only as females. All workers are female, and are related to one another by a factor of 3/4, a fraction that indicates the average proportion of one's own genes that are identical to the other's genes (**see also: Genetics of behaviour**). In sexually reproductive animals, including humans, a parent's offspring are related to the parent and to each other (if full siblings) by 1/2. An eusocial worker is thus more closely related to her sisters than she could possibly be to her own offspring, were she able to reproduce. This may explain why a sterile honeybee is willing to defend her sisters and the colony's younger generations by stinging an intruder, an act that is always suicidal.

Other animals, such as birds and mammals, also show kin-selected cooperation. Lions (*Panthera leo*) of the Serengeti in Africa live in prides, which typically consist of multiple adult males and females. Adult females breed synchronously, and spend much of their time together. A lioness will sometimes nurse young cubs that are not her own. She is much more likely to do so, however, when the cubs belong to close kin and the group's litter size is relatively small. In lions, it appears that communal nursing is highly influenced by the need of females to keep cubs together and protected (Pusey and Packer, 1994).

Florida scrub jays (*Aphelocoma coerulescens*) perform a particular form of cooperative breeding known as 'helping at the nest' that offers an important illustration of kin-selected behaviour. The Florida scrub jay is sexually monogamous and does not migrate. Helper birds assist more than 50% of monogamous pairs, and most helpers are previous offspring of the pairs they assist. Helpers often feed young (usually their siblings) and assist in defending the territory of the breeding pair (Woolfenden and Fitzpatrick, 1984). Predation is reduced due to the increased level of territory defence.

Scientists often study behaviour by attempting to determine how it benefits an individual's own fitness versus its inclusive fitness. Altruism lowers one's own fitness, but can be beneficial overall if the act increases the survivability of close relatives. Although cooperative breeding is detrimental to the helper's own breeding success at the time of cooperation, helping for a few years increases inclusive fitness because the cost of forgoing production of one's own offspring is outweighed by the benefits of helping to raise close kin. This behaviour can eventually lead to the creation of one's own territory and so positively affects one's own fitness. A young scrub jay that helps its parents increases the likelihood it will eventually take over their territory, or create another one nearby. After a helper disperses, the chances of forming one's own territory increase substantially. Cooperative breeding then both increases an individual's inclusive fitness and allows it to obtain valuable experience.

Mutualism and reciprocity

How an individual chooses a partner with whom to cooperate depends on two factors: (i) the proven effectiveness of partnering with a particular individual in cooperation; and (ii) the benefits received through their cooperative efforts. These return benefits may occur either immediately if partners share the pay-off of their cooperation or – in case all the benefits go to one individual – after a time interval. The first case is known as mutualistic cooperation. The second is reciprocal altruism, or delayed reciprocity. For a review of the literature on cooperation among unrelated individuals, see Lee Dugatkin (1997).

A typical example of mutualism is the aforementioned one of African wild dogs chasing a gazelle, in which all hunters may benefit at the moment the ungulate is killed. A pack of carnivores that cooperates to bring down prey will receive one large reward distributable across all of the hunters. Another good example of mutualism, in this case between species, is that of the bluestreak cleaner wrasse (*Labroides dimidiatus*), also known as the cleaner fish. Usually found on coral reefs in the Indo-Pacific, cleaner fish establish what are known as 'cleaning stations', usually in a cave or under a rock formation, and perform a particular 'dance' to attract larger fish. The larger fish then line up, waiting to be cleaned. The cleaner fish eat ectoparasites off the host's body, often entering the mouth and gill regions to complete the job. Although the larger fish may have the opportunity to eat these cleaner fish, they do not. The benefits of having ectoparasites removed now and on future occasions apparently outweigh the benefits of a quick meal. In exchange for reducing the parasitic load of a larger fish, the cleaners receive a nutritional benefit.

Reciprocal altruism is more complex than mutualism in that it requires the foregoing of benefits, at least for a brief while. In 'The evolution of reciprocal altruism', Robert Trivers (1971) postulated that such behaviour might evolve if one individual's altruism (defined as costly behaviour that benefits others) towards another is being repaid after a time interval by the latter's altruism towards the former. If for both exchanged acts the recipient's benefits (B) exceed the performer's costs (C), and both acts are equal in this regard, both parties will gain B – C.

Reciprocity thus boils down to 'I'll scratch your back, if you'll scratch mine'. How many animals show such tit for tat? In their quest for evidence of reciprocity, scientists initially focused on group-wide distributions of behaviour. If individual A often aids B, and B does the same for A, and a whole group follows this pattern, the result will be a positive correlation between given and received support. Monkeys and apes, for example, form coalitions within their groups, such as when two or more individuals gang up on a third. Coalitions show healthy correlations indicative of reciprocity. But these correlations do not tell us exactly how benefits find their way back to the original altruist. Do animals keep score? This is hard to imagine for small-brained animals. Possibly they just divide the world into 'buddies', which they prefer, and 'non-buddies', about which they care little. If such attitudes are mutual, relationships will be either mutually positive or mutually negative. Such symmetries can account for the simple reciprocity reported for fish, vampire bats, dolphins and many monkeys.

A widely cited example of reciprocal altruism involves blood sharing in vampire bats (*Desmodus rotundus*; Wilkinson, 1984). Vampire bat roosts consist of large groups of females that travel to collect blood meals. Individuals that have fed recently will return to the roost and sometimes regurgitate food to conspecifics that had been less lucky and remained unfed on that night. These regurgitations may save their lives: a vampire bat will starve to death if it does not receive a blood meal within approximately 2 days. In addition, vampire bats can distinguish between individuals within their colony, and have special partners with which most exchanges take place. Wilkinson's studies indicate that relatedness and previous exchanges play an integral role in the blood sharing among vampire bats, on a level that may be as simple as 'buddy' versus 'non-buddy' distinction.

Reciprocity becomes more complex if there is a contingency between given and received favours. If the behaviour of A towards B depends on what B did earlier to A, we are getting close to benefit exchanges based on **memory**. Two examples will be given of experiments in this domain: (i) cooperation between brown capuchin monkeys (*Cebus apella*), in which food is shared with a helper; and (ii) food-for-grooming exchanges among chimpanzees.

Captive capuchin monkeys can be positioned such that they can see each other but are separated by a mesh partition. In front of each monkey is a pull bar attached to a tray with two bowls (see Fig. C.12). Two test conditions can be compared, both requiring different amounts of 'pull' to bring the tray towards the monkeys. In the solo effort task, only one monkey has access to the pull bar and the bowl with food. The strength of one monkey is enough to pull in the tray. In the cooperation task, the tray is counterweighted such that the strength of both monkeys is required to pull in the tray. Both pull bars have to be utilized, but only one of the bowls has food.

In previous studies, it has been shown that these monkeys follow a pattern known as facilitated taking, where a food possessor will approach the mesh divider, dropping crumbs or whole pieces of food so that the partner can reach through the mesh and collect the food. This is an active form of sharing because the food possessor could just as easily stay away from the divider and hoard the food. When the two monkeys are involved in a cooperative pulling task, the food possessor is much more likely to participate in facilitated taking with the helper than in the solo effort trials, when there is no cooperation. In addition, the results indicate that the helper pulls more often in the cooperation trials when the preceding trial has been successful than if it had failed.

It appears that capuchin monkeys are able to recognize and reward other individuals that have helped them obtain food in the immediate past by allowing them more access to the rewards thus obtained. This may be a result of attitudinal reciprocity based on an 'If you're nice, I'll be nice' principle. The reason to talk of attitudinal reciprocity is that the interval between exchanges is very short, and it is unclear how much memory or record keeping is required to attain the payment-for-labour observed. When the interval between exchanged acts becomes longer reciprocity becomes more complex, as demonstrated in the chimpanzee. Food-for-grooming exchanges in chimpanzees are thus far the only well-documented case of so-called 'calculated reciprocity', defined by mental scorekeeping among individuals for given and received favours. Although chimpanzees have distinct social hierarchies where alpha males get the majority of the reproductive opportunities, food sharing is often peaceful and appears to be independent of social rank (see Fig. C.13).

In a study conducted by de Waal (1997), a captive group of chimpanzees was fed browse (i.e. large bundles of branches with leaves) twice a week during the spring and summer over a period of 3 years. Grooming behaviour was recorded for a period of 2 h prior to each feeding trial. Subsequent sharing during the food trials was then compared with data on grooming. This study indicated that chimpanzees were more likely to allow others to take food from them (by taking it from their hand or mouth) when the other had donated grooming services. In addition, chimpanzees that solicited food from individuals they had not groomed prior to being fed were more likely to be met with resistance and aggressive protests than individuals that had groomed the food possessor prior to the feeding. This indicates that chimpanzees not only remember having been groomed by a particular individual, but reciprocate the grooming with another positive behaviour, food sharing. This is in fact the only well-documented case of a memory-based contingency between given and received services.

Fig. C.12. The test chamber used for the cooperative pulling task in capuchin monkeys (drawing by S. Brosnan).

Fig. C.13. Chimpanzees sharing food.

Charles Darwin wrote 'Nothing is easier than to admit in words the truth of the universal struggle for life', and argued that this struggle would be quite evident between individuals competing for the same resources. Ironically, cooperation evolved precisely because it helped animals to outcompete less cooperative variants. In both intraspecific cooperation (as in chimpanzees and monkeys) and interspecific cooperation (between the cleaner fish and their 'client' fish), the benefits of working together outweigh those of working alone. (JP, FdW)

See also: Selection; Sociobiology

References

de Waal, F.B.M. (1997) The chimpanzee's service economy: food for grooming. *Evolution and Human Behaviour* 18, 375–386.

Dugatkin, L.A. (1997) *Cooperation among Animals: an Evolutionary Perspective*. Oxford University Press, New York.

Pusey, A.E. and Packer, C. (1994) Non-offspring nursing in social carnivores: minimizing the costs. *Behavioural Ecology* 5(4), 362–374.

Trivers, R.L. (1971) The evolution of reciprocal altruism. *Quarterly Review of Biology* 46, 35–57.

Wilkinson, G.S. (1984) Reciprocal food sharing in the vampire bat. *Nature* 308, 181–184.

Woolfenden, G.E. and Fitzpatrick, J.W. (1984) *The Florida Scrub Jay: Demography of a Cooperative-breeding Bird*. Princeton University Press, Princeton, New Jersey.

Coping

Individuals must continually cope with a certain level of **stress** imposed on them from their external environment. Aversive stimuli such as a change in temperature, light level, disease, restriction of movement and lack of food all impose a level of stress and, unless avoided or dealt with, may result in reduced **fitness**. The amount of stress that they place on an individual depends on their nature, intensity, duration and degree of novelty. Coping is the term given to the mechanism by which an individual deals with these stressors. Changes in physiology and behaviour are made by the individual and, if effective, result in a return to an optimal state, in a process referred to as **homeostasis**. The extent to which an individual can eliminate or negate the stress inflicted on it by making these changes reflects the extent to which it is coping.

Responses may include physiological changes in the body – for example, production of antibodies, an increase in heart rate or behavioural reactions such as vocalization or movement. To illustrate, if the air temperature within an animal housing system falls, thermoreceptors on the surface of an individual animal's skin will detect this change. The **brain** may stimulate the contraction of pilo-erector muscles, raising hairs on the surface of the skin, which helps to reduce heat loss in response; but, if the body temperature falls below a **critical temperature** threshold level, greater metabolic effort may be required to maintain the body temperature. The individual may also respond behaviourally and move to a warmer part of the housing system. If these responses are successful, the individual will soon return to optimal temperature and its fitness will not be compromised.

Levels of stress and coping

Individuals cope with low-level stress constantly, and behavioural responses may be unnecessary or minimal as they are generally, energetically speaking, quite expensive. More aversive stimuli may inflict a higher level of stress on the animal and a behavioural response may be elicited, for example by moving to a warmer part of their living quarters (level I). This level of stress is still low and, if the animal has the appropriate behavioural or physiological mechanism, it can cope with it relatively easily. More intense, prolonged, novel or noxious stimuli may impose a higher level of stress on the animal and it may have greater difficulty coping (level II). A greater intensity of response may be required, or a novel response that would not be utilized at lower levels – for example, shivering or huddling behaviour. The animal may be able to cope indefinitely with this level of stress but it is likely that its welfare will be compromised in time. Particularly noxious or prolonged stressors may exceed the animal's capacity to cope (level III). Fitness, as in potential for **reproduction** and survival, is compromised at this level and may even be affected at level II. Their tolerance of this level of stress is limited and, if it is not removed, the animal may collapse and die, for example from hypothermia.

Stimuli may have a compound effect in that a number of noxious stimuli may produce a level of stress greater than the sum of their parts. For example, an animal may be coping with a low level of stress (I), but the addition of other mild or more severe stressors raises the level to such an extent that the animal struggles (II) or even fails to cope (III), unless it is able to change to a different coping method (see Table C.5).

Coping behaviour is that behaviour which is performed with the aim of reducing the effects of an aversive stimulus on the fitness of an individual. The detection of noxious stimuli and the appropriate behavioural response to these is therefore an adaptation. Coping behaviour may involve ways of removing, reducing or avoiding the stimulus. In the example above, moving to a warmer part of their environment is an example of coping behaviour.

Consistent forms of coping behaviour are described as coping strategies. It has been suggested that the most common ones are 'escape', 'remove', 'search' and 'wait' (Wechsler, 1995). Avoidance of a noxious stimulus by physically escaping from it is often the first coping strategy used by an animal. Removal of the stimulus, for example, by attacking it, may be utilized, particularly if escape is not possible. 'Search' may be an effective strategy for some situations, including coping with hunger. 'Waiting' for the situation to change may be the strategy utilized when these other strategies are not possible or successful, and may explain apparently apathetic behaviour seen in caged animals (Wechsler, 1995).

Not all individuals or species will cope with an aversive situation in the same way. The kind of strategy an animal tends to utilize may depend on its individual tendency or 'coping style'. The idea of coping styles came from the observation that individuals may show consistent behavioural and physiological tendencies in a variety of environmental situations. These tendencies have been characterized on an ordinal scale from 'reactive' to 'proactive' or from 'passive' to 'active'. For example, Koolhaus and Bohus (1989) found that mice that were aggressive to an unfamiliar mouse in their cage tended to cope with the introduction of an electric shock prod into their cage by burying it with bedding material. They suggested that these mice had 'active' coping styles. Less

Table C.5. Levels of stress and coping ability (adapted from Broom and Johnson, 1993).

Stress level	Coping ability	Response	Tolerance	Fitness
III	Not coping	No response available, collapse	Not tolerable, lethal if continues	Reduced
II	Difficult	Large or extra responses needed	Indefinite, but barely tolerable	Reduced
I	Easy	Large or extra responses needed	Indefinite	No effect
Normal	Easy	Small homeostatic changes	Indefinite	No effect

aggressive mice tended to avoid the prod, which was defined as a more 'passive' coping style. In these situations, however, both styles were equally effective at avoiding the aversive stimulus (being shocked). The extent to which the concepts of active and passive coping styles can be applied to different species is a matter of some debate.

The **welfare** status of an individual or group of animals can be defined in relation to how well they are coping with the stresses imposed on them by their environment. For example, within intensive housing systems animals may have difficulty coping for a number of reasons and, by inference, their welfare status may be compromised. The **stressors** may be too numerous or too great to allow for effective coping, the animal may not have appropriate coping strategies in its behavioural repertoire because its species did not evolve in this environment and/or the animal may be prevented from utilizing its coping strategies effectively because of the restricted environment. For example, 'escape' or 'remove' may not be possible, and 'wait' may be not be effective in the long term since the situation may not change. It has been suggested that some **abnormal** behaviour patterns might be the result of unsuccessful coping strategies (Wechsler, 1995). (KT)

See also: Emancipation; General adaptation syndrome; Learned helplessness

References and further reading

Bolhuis, J.E. *et al.* (2006) Effects of rearing and housing environment on behaviour and performance of pigs with different coping characteristics. *Applied Animal Behaviour Science* 101, 68–85.

Broom, D.M. and Johnson, K.G. (1993) *Stress and Animal Welfare*. Chapman and Hall, London.

Koolhaus, J.M. and Bohus, B. (1989) Social control in relation to neuroendocrine and immunological responses. In: Steptoe, A. and Appels, A. (eds) *Personal Control and Health*. John Wiley & Sons, New York, pp. 295–304.

Wechsler, B. (1995) Coping and coping strategies: a behavioural view. *Applied Animal Behaviour Science* 43, 123–134.

Wiepkema, P.R. and van Adrichem, P.W.M. (1987) *Biology of Stress in Farm Animals: an Integrative Approach*. Martinus Nijhoff, Dordrecht, The Netherlands.

Coprophagy (coprophagia)

Coprophagia describes the ingestion of faeces. The faeces consumed by a coprophagic animal may be its own (e.g. **rabbits** consume their own faecal pellets, which then undergo a second wave of digestion), those of another species (e.g. **dogs** may consume **cat** faeces) or a **conspecific** (e.g. foals consume faeces of adult **horses**, especially from their dam).

Foals are the most closely studied of these groups. Often seen pawing at faecal material, foals may use coprophagia as one method of learning about the gustatory and olfactory features of plants consumed by the dam. The significance of this activity in learning is supported by the coincidental decline in the foal's interactions with toxic plant species during the transient phase of coprophagia. There may be an additional purpose, such as the acquisition of intestinal microbes and possibly deoxycholic acid, thought to play a role in the deposition of myelin. For these putative reasons, hand-reared foals should be given the opportunity to perform coprophagy, a practice that is highly unlikely to expose the foal to viable forms of endoparasites. Although coprophagia is rarely seen in adults, fibre-restricted rations, **frustration** (which probably results from the same deficit) and underfeeding in general can cause its appearance in adult horses.

When encountered in dogs, coprophagia is considered unwelcome by many owners. The reasons for its emergence are poorly understood, and therapies are often confined to reviewing the formulation of the diet, extinguishing attention seeking and using taste deterrents on target faeces (**see: Aversion**). The prevalence of this response among kennelled dogs suggests that environmental enrichment should be a part of more comprehensive therapeutic responses.

Rabbit accommodation should always include some solid flooring, because cages with wire floors compromise the occupant's ability to consume faeces, otherwise coprophagia, part of the animal's natural digestive cycle, may be jeopardized. (PDM)

Core area

The core area (also called centre of activity) is an intensively used area within the **home range**. Plotting an animal's locations on a map, it would be the place or places with a dense cluster of points. Alternately, viewing the data as bars of various heights representing frequency of use (the **utilization distribution**, UD), the core area would be the cluster of tall bars.

Sometimes, the core is defined as the smallest region encompassing 50% of the UD (the 50% isopleth). However, this value is arbitrary and often not particularly fitting for real-life situations: for example, suppose only 30% of the locations were clustered in a small area and the other 70% were rather evenly dispersed. Methods based on the actual dispersion of the data for each individual animal are more consistent with the concept of core area. However, no universally adopted method has emerged.

All methods assume that the UD is accurately measured. Problems arise if the sampling is insufficient (clusters of locations reflect periods when the data happened to be collected) or biased (certain areas appear to be heavily used only because the observer could more easily find the animal there).

Core areas may change seasonally or over a shorter time scale, with changing food, weather or biology of the animals (**breeding**, birthing, rearing of young, etc.). Moreover, multiple disjunct core areas may exist, corresponding with favoured sleeping, feeding or birthing sites (dens), or a portion of the home range not used by **conspecifics**.

It is generally presumed that core areas represent preferred habitat within the home range. As such, some **conservation** plans for threatened species call for the protection of core areas. This approach has drawn criticism because: (i) other large parts of home ranges may be just as important, but with less concentrated use (e.g. an expansive foraging area versus a core refuge area); and (ii) there is no general agreement on how to define core areas. (DLG)

See also: Territory

Further reading

Barg, J.J., Jones, J. and Robertson, R.J. (2005) Describing breeding territories of migratory passerines: suggestions for sampling, choice of estimator, and delineation of core areas. *Journal of Animal Ecology* 74, 139–149.

Bingham, B.B. and Noon, B.R. (1997) Mitigation of habitat 'take': application to habitat conservation planning. *Conservation Biology* 11, 127–139.

Kerr, G.D. and Bull, C.M. (2006) Exclusive core areas in overlapping ranges of the sleepy lizard, *Tiliqua rugosa*. *Behavioural Ecology* 17, 380–391.

Samuel, M.D., Pierce, D.J. and Garton, E.O. (1985) Identifying areas of concentrated use within the home range. *Journal of Animal Ecology* 54, 711–719.

Corticosteroids

Corticosteroids are **steroid hormones** produced by the cortex of the **adrenal gland**. These include: (i) the mineralocorticoids (e.g. aldosterone), which are vital for the regulation of sodium and potassium levels in the body; (ii) the precursors to the sex steroids (e.g. **androgens** and **oestrogens**), which contribute to establishing and maintaining secondary sexual characteristics; and (iii) the **glucocorticoids** (e.g. cortisol or corticosterone), which are critical in carbohydrate and protein metabolism. Corticosteroids exert their effects in the body by determining the nature and rate of gene transcription in target cells.

Mineralocorticoids

Mineralocorticoids, as their name suggests, have a role in mineral metabolism. Aldosterone is the principal mineralocorticoid involved in this process, and this hormone is involved predominantly in regulating the exchange of sodium and potassium ions in the distal tubule of the kidney. The resulting physiological effects are an increased resorption of sodium, increased resorption of water and increased renal excretion of potassium.

Aldosterone exerts its effects by stimulating transcription of the gene encoding sodium–potassium ATPase, leading to an increased expression of sodium–potassium pumps in the basolateral membranes of tubular epithelial cells. It also acts to stimulate expression of sodium channels in the tubular lumen, facilitating uptake of sodium in this region. The active resorption of sodium facilitates the passive flow of potassium into the tubular **urine**. The flow of potassium across the membrane is dependent on the rate of sodium delivery and the distal nephron flow; however, aldosterone allows potassium to flow across the membrane (in a restricted quantity) even when sodium levels are low. Aldosterone also exerts its effects on the colon, where it stimulates sodium resorption from this area, thereby enhancing potassium excretion in the faeces. It also acts on the sweat glands and salivary glands, using a similar mechanism to that described above.

Secretion of excess aldosterone also results in increased blood pressure. This is due, in part, to the retention of sodium, expansion of the extracellular fluid and an increase in cardiac output. In addition, the increase in sodium and water content of the arteriolar cells may increase, thereby causing the arteriolar lumen to narrow and increase resistance in the artery.

The three most significant factors affecting the regulation of aldosterone are: (i) the concentrations of potassium ions in the extracellular fluid (increased levels of potassium stimulate the production of aldosterone); (ii) the concentration of sodium ions in the extracellular fluid (decreased levels of sodium stimulate aldosterone production); and (iii) angiotensin II (a decrease in renal blood flow results in activation of the renin–angiotensin system and a concomitant secretion of aldosterone).

Precursors to the sex steroids

The precursors to the sex steroids (i.e. DHEA-S, DHEA and androstenedione) are relatively weak androgens secreted from the adrenal cortex. They are converted into the more potent androgen, **testosterone**, which acts to influence the formation and maintenance of secondary characteristics (e.g. pubic and axillary hair in humans) in both sexes. Testosterone has more influence on male than female sexual characteristics, as the production of this hormone from the testis in the male far exceeds that of the adrenal gland, rendering cortical secretion physiologically unimportant in this sex.

Glucocorticoids

The glucocorticoids are hormones involved in energy production. Secretion of cortisol (in primates, dogs, cats and most ungulates) or corticosterone (in rodents and birds) induces breakdown of protein to release glucose into the bloodstream, where it is used to provide energy to the organism. Cortisol is also involved in fat metabolism, muscle, renal and immune system functioning, maintenance of the skeleton and modulation of the activity of the **central nervous system**.

When an animal is exposed to a novel or disturbing situation it experiences **stress** to varying degrees. Any event, whether it is physiological or psychological, that disturbs the organism's **homeostasis** is able to cause a stress response and secretion of corticosteroids from the adrenal cortex. The difference between adaptational and maladaptational physiological responses is the duration and persistence of hormone secretion.

When exposed to acute conditions of stress, an animal prepares for an immediate response, typically in the form of the fight and flight response, and the **sympathetic nervous system** and **hypothalamic–pituitary–adrenal (HPA) axis** are activated. Activation of the HPA system causes corticotropin-releasing hormone (CRH) to be released from the **hypothalamus**, which appears to mediate a variety of anxiogenic and depressogenic behavioural effects associated with increased arousal, alertness and other changes characteristic of heightened alertness to the environment. CRH acts on the anterior lobe of the **pituitary gland** where, together with **vasopressin**, it stimulates the secretion of adrenocorticotropic hormone (**ACTH**). ACTH, in turn, exerts its effects at the cortex of the adrenal gland, causing the synthesis and secretion of glucocorticoids. Regulation of the activity of the HPA axis is performed, in part, by the glucocorticoids, which exert negative feedback effects at both the hypothalamus and pituitary gland to regulate the production of ACTH.

The receptors for cortisol exist in two forms, classified by the specificity of different steroids and neuroanatomical distribution: (i) the mineralocorticoid or type I receptors (MR), which bind both corticosterone and cortisol with high affinity; and (ii) the glucocorticoid or type II receptors (GR), which have approximately one-tenth the affinity of MRs but bind the glucocorticoid analogue dexamethasone.

The GR is most abundant in regions associated with the behavioural stress response (e.g. the hippocampus and hypothalamic paraventricular nucleus). MRs are mainly restricted to limbic brain regions, and the density is particularly high in the neurones of the hippocampus. MRs are thought to regulate the basal level of activity of the HPA system. The majority of the MRs are occupied when the animal is at rest and the corticosteroid levels are low; however, when the animal becomes stressed, and corticosteroid levels increase, the GRs become occupied. The relative binding of the hormone with either the GRs or MRs, therefore, seems to be an important determinant of limbic activity. This activation is influenced by factors such as affinity of the receptors, the concentration of corticosterone circulating in the bloodstream and regulation of receptor synthesis.

The predominant effect of cortisol is to facilitate the conversion of protein to glycogen. It does this by accelerating protein degradation and inhibiting protein synthesis, thereby augmenting mobilization of muscle protein for gluconeogenesis. The energy released by gluconeogenesis is subsequently utilized to facilitate behavioural activity as the organism responds to a stressful situation.

Even when an animal is 'unstressed' (i.e. undisturbed physically or psychologically), a daily rhythm in cortisol production can be observed in both diurnal and **nocturnal** animals. Maximal concentrations are measured concomitant with the onset of locomotor activity, prior to or at awakening. This occurs, therefore, in the morning in diurnal species and in the evening in nocturnal species. In diurnal species the levels of cortisol decrease gradually throughout the day to become a minimum just before the animal retires for the night. In nocturnal species the levels are maximal in the evening as the animal wakes. It is thought that high cortisol levels upon waking may serve to mobilize energy stores in preparation for activity.

When an animal experiences acute stress, e.g. by being placed in social isolation or by being exposed to certain novel stimuli, an unfamiliar conspecific or a predator or particular drugs (e.g. ether stress), it exhibits the 'fight or flight' response. Animals exhibiting this reaction will show an increase in locomotory and active behaviour, will orientate themselves towards the offending stimulus and will vocalize at an increased rate. It has been suggested that the rate, intensity and type of **vocalization**s that an animal emits when exposed to a disturbing situation can give an indication of the stress levels, and hence the psychological state of the animal at that time. The increase in these behaviours has, in some cases, been shown to correlate with increasing cortisol levels in response to various types of arousing stimuli. However, despite providing an indication of the psychological state of the animal, behavioural parameters alone do not always give a conclusive reflection of the amount of stress the animal is experiencing. Therefore, when aiming to assess the psychological state of an animal, it is useful to examine both behavioural and physiological parameters together to reach a more accurate conclusion.

The stress response can be induced in animals by exposing them to potentially disturbing situations. One traditional method of inducing the stress reaction in an animal is to expose it to **social stress**, which, in many cases, can be achieved by placing the animal in a group of unsuitable conspecifics. The degree of HPA axis activation (and hence the magnitude of glucocorticoid secretion) in response to social stress, is determined by two principal factors: (i) social context, defined by the position in the hierarchy, gender and the presence of conspecifics; and (ii) individuality, determined especially by early environmental experiences (although later conditioning can also affect this response) and genetic determinants, including parent–infant bonding. The social context and individuality may account, in part, for the varying results found when testing the same species under the same conditions.

Other factors which may affect HPA axis activity are group stability, size and composition (whether or not a **social group** is stable or whether there is competition for dominance). The magnitude of the adrenocortical response also varies between species, and indeed between breeds in domestic animals. This enhances the need for careful control of genetic diversity and use of large sample sizes when assessing corticosteroid response to a treatment.

Situations in which an animal experiences **pain** have been shown to be associated with increased glucocorticoid secretion. Studies have shown an increase in cortisol levels in response to procedures such as surgery, tail docking, **castration** and **mulesing** (the removal of skin folds in the sheep to prevent fly strike). In many of these studies the severity of the procedure correlates positively with the magnitude of the adrenocortical response. Adrenal activation in response to situations that are predictably painful can be reduced by the use of analgesics. However, it has also been shown that the adrenocortical response can be initiated even when anaesthetic is used, suggesting that the tissue damage itself, even if not perceived as painful by the CNS, results in corticosteroid secretion.

Chronic stress is characterized by a long-term elevation in glucocorticoid secretion. This results in a down-regulation of GR and MR mRNA and ligand binding in the hippocampus in response to prolonged receptor activation. Chronic elevation of glucocorticoids indicates a maladaptational adjustment to environmental conditions.

Chronic stress has also been associated with degenerative changes in the **brain**s of primates and rodents, mediated in part by the GRs. A sustained elevation of glucocorticoids results in neurotoxic effects and cell death, which potentially contributes to a loss of negative feedback control of the HPA axis. Chronic stress has also been shown to result in persistent inhibitions of granule cell production and structural changes in the hippocampus. Excess glucocorticoids interfere with nerve cell function in the hippocampus, interrupting processes in the synapses that are thought to be involved with memory and learning. They have also been shown to interfere with the uptake of glucose by hippocampal nerve cells, making them vulnerable to other insults, such as lack of oxygen occurring during strokes or epileptic seizures, which may eventually lead to the death of these neurones. If the feedback mechanisms fail various pathophysiologic consequences can arise, and hypersecretion of glucocorticoids at a chronically elevated concentration can have deleterious effects on the **immune system**, metabolism, **reproduction**, tissue repair and neurologic status.

When animals are exposed to chronic stress – for example, when they are housed in unnatural and confined spaces for long time periods – they may exhibit an alternative stress-induced reaction. Under conditions such as this, the fight or flight response can be gradually replaced with abnormal behaviour patterns such as stereotypic behaviour and apathetic behaviour. **Stereotypies** are repeated actions of a singular behaviour or pattern of behaviours that have no functional purpose and, in fact, may often be maladaptive. Some examples of stereotypies are the rocking of a primate reared in isolation (which possibly develops as a substitute for the proprioceptive and kinaesthetic stimulation that it would usually receive from its mother), cage circling of a confined and unstimulated marmoset or the pacing of a confined and unstimulated cat. Another less obvious stereotypy is the facial grooming of mice when they are exposed to a stressful or novel environment. In this case the behaviour is not affected by lesions that block sensory input from the facial area, suggesting that the performance of this behaviour is indicating unease, rather than a need to groom the facial area.

A number of techniques have been developed to measure cortisol levels. One commonly used method is to assay the cortisol levels in blood plasma. Assessment of cortisol levels using this method is useful to obtain an accurate indication of the levels of cortisol in the blood at any chosen point in time. However, care should be taken when using this technique, as the capture and restraint required during collection of the sample for assay, together with the adversiveness of venipuncture, may itself cause stress to the animal and confound the results obtained. Urinary and faecal cortisol analyses have also been used to examine daily fluctuations in cortisol and can be useful in examining cortisol levels in free-ranging animals or when samples are not required to be obtained at a high frequency. Saliva sampling is a useful alternative to these methods. It is non-invasive, is less stressful than collecting samples using blood-sampling techniques and can be used at frequent intervals in a majority of captive settings. It does, however, require that subjects are trained willingly to provide saliva samples in advance of the period of experimentation, to avoid stress being caused due to the procedure of sample collection. Cortisol concentrations are assessed using radioimmunoassay techniques. (NC)

See also: Abnormal; Adrenal gland; Apathy; Measuring welfare; Stereotypies; Steroid hormones; Stress; Temperament

Corticotropin-releasing hormone (CRH)

Corticotropin-releasing hormone is a peptide produced in the paraventricular nucleus of the **hypothalamus**, from where it is carried via the portal circulatory system to the anterior **pituitary gland**; it stimulates the production of adrenocorticotropic hormone (ACTH) and thus an increase in **corticosteroid** levels. CRH production is stimulated by the **cytokine, interleukin**-1β, which also plays an indirect role in sensitizing the **brain** to **pain** associated with the inflammatory response, although CRH itself may interact with receptors that stimulate positive effects. CRH is also produced in some peripheral tissues, including T-lymphocytes, but most notably the placenta, and an increase in CRH production towards the end of pregnancy results in the fetal surge in corticosteroid production that triggers **parturition**. (DSM)

Cost–benefit analysis

In utilitarian ethics, the consequences of any action are an important aspect of deciding what actions are right and wrong. In animal research, the cost–benefit (CB) approach is embodied in the UK 1986 Animals (Scientific Procedures) Act, and the Secretary of State for the Home Office is charged with weighing the likely adverse effects on the animals involved against the benefit (to humans, other animals or the environment) likely to accrue from the programme of work outlined in the project licence. In its simplest form, if the likely benefits outweigh the predicted harms, then an action can be said to have some justification. However, in practice, the CB analysis comes with a range of other questions. Is the objective worth pursuing? Are *avoidable harms* going to occur? How good are the predictions of the harms? Is there proportionality, that is are the **harms** commensurate with the anticipated benefits?

It is likely that higher levels of animal **suffering** will be permitted when the anticipated benefits are high (e.g. saving human lives, promoting quality of life for humans, developing an AIDS vaccine) than in a project where the benefits are less important, e.g. one that might make farm animals grow faster, or that results in meat being produced more cheaply or advancement in knowledge. The CB approach can be used to evaluate any area of animal use. For example, some consumers will only purchase free-range eggs as opposed to eggs from hens kept in battery cages for, in their view, the costs (for humans) of eating free-range eggs are outweighed by the perceived benefit to the birds in terms of their **welfare**. Using a similar CB argument, other persons will not eat eggs at all, because the harms imposed on hens kept in cages or free range are so great that they are not outweighed by the pleasure gained by eating eggs.

Several problems arise with this approach in research. First, the anticipated benefits may not occur, whereas the harms done to animals are more accurately predictable, e.g. postoperative pain. So the benefits have to be realistic. Some benefits, even though not achieved, may still provide useful information, resulting in increased knowledge. Secondly, in other areas of animal use, the benefits may be 'base' pleasures such as humans exploiting natural animal instincts as in dog fighting, or **baiting** of wild animals. Thirdly, costs and benefits are often in different 'currencies' for members of different species. Although it may be possible to propose a hierarchy of acceptability of benefits, with an animal experiencing cost above other animals of the same species who are above animals of a different species. So the type of pleasure or benefit has also to be considered. CB arguments can be used to justify, or otherwise, a range of issues such as the torture of human subjects, infringing the liberty of some for the greater good of the many. (DBM)

See also: Ethics; Utilitarianism

Council of Europe

The Council of Europe, established in 1949 by the Treaty of Rome, is based in Strasbourg, France. Following World War II, it was an attempt to reconcile and establish cooperation across Europe. It should not be confused with the European Community, and does not make laws, but encourages a wider membership to abide by conventions, protocols and treaties which set out principles on issues of common interest. Its main areas are democracy, rule of law, human rights and freedoms and, more recently, terrorism and crime. Its strength is in bringing together members for debate but, with little real power, could be seen as no more than a discussion forum. It is unelected, therefore potentially undemocratic, and the conventions are not legally binding, with nations able to choose whether to sign up to them, with no direct consequence should they not do so.

Sample European Conventions of relevance to animal behaviour and welfare include: (i) Convention for the Conservation of European Wildlife and natural Habitats (1974); (ii) Convention for the Protection of Animals during International Transport (1968, 1979); (iii) Convention for the Protection of Animals kept for Farming Purposes (1976, 1992); (iv) Convention for the Protection of Animals for Slaughter (1979); (v) Convention for the Protection of Vertebrate Animals Used for Experimental and Other Scientific Purposes (1986); and (vi) Convention for the Protection of Pet Animals (1987). (BG)

Counter-conditioning

Counter-conditioning, or response substitution, is a commonly used behavioural modification technique for the alleviation of behaviour problems in animals caused by inappropriate arousal to a given stimulus. The term literally means training an animal to do a behaviour that is counter to the one that a trainer wishes to eliminate, i.e. producing an incompatible response to a given stimulus that is expressed in preference to the undesired response. The technique is widely used in combination with systematic desensitization (**see: Desensitization – systematic**). By ensuring that the preferred behaviour is more rewarding, the animal learns to perform the new behaviour when exposed to the potentially problematic stimulus. There are two ways in which counter-conditioning is usually undertaken.

First, during exposure to the problem stimulus the animal can be presented with a stimulus that inherently arouses an alternative emotion (or respondent behaviour), which is counter to the emotion underlying the problem behaviour. For example, a toy may be produced to encourage playing or food to encourage eating when a thunderstorm starts; the responses elicited by these stimuli are incompatible with a fear response. If the animal begins to show signs of anxiety the intensity of the stimulus needs to be reduced before the counter-stimulus (food, toy, etc.) is withdrawn – for this reason recordings of the aversive stimulus may be used during the training process. Eventually the animal should learn that the problem stimulus is now a predictor of a positively reinforced (rather than aversive) event. As a result the problem behaviour is eliminated – this version of the process is also known as respondent counter-conditioning, or classical counter-conditioning.

Alternatively, the animal may be trained with reinforcers to undertake a specific action that is incompatible with the problem behaviour (e.g. a 'sit–stay'). This operant behaviour may initially be trained to a command or signal, which is then paired with the problem stimulus. It is obviously essential that the cue used to elicit the obedience response is more likely to be responded to than the aversive stimulus. This process is called operant counter-conditioning. Once the problem stimulus reliably predicts the counter stimulus and elicits an acceptable response, it is usually possible to fade out the cue if the behaviour is appropriately reinforced. (KT)

See also: Conditioning – types of; Fear

Further reading

Horwitz, D.F., Mills, D.S. and Heath, S. (2002) *BSAVA Manual of Canine and Feline Behavioural Medicine*. British Small Animal Veterinary Association, Quedgeley, UK.

Landsberg, G., Hunthausen, W. and Ackerman, L. (2003) *Handbook of Behaviour Problems of the Dog and Cat*. Saunders, Edinburgh, UK.

Courtship behaviour

Courtship behaviour precedes and accompanies the sexual union of male and female animals. The form, flamboyance and length of courtship vary considerably from species to species. Some species reproduce with very little interaction, while others have very complex courtship patterns that can last hours or even days. These often involve stereotypic movements and the display of bright colours, striking patterns and the production of loud **vocalizations**. The function of courtship behaviour is to create the circumstances in which successful mating is possible. Within this framework there are a number of aspects that should be considered.

The first step leading to a successful mating is locating a **conspecific** member of the opposite sex. For some species this is not an issue because males and females live together in permanent mixed-sex **groups**. However for solitary and sexually segregated species, attraction of a conspecific individual of the opposite sex, sometimes over a considerable distance, is an integral part of courtship behaviour.

The **signals** used by animals to attract mating partners are known as mate attractant signals. To be effective, mate attractant signals must provide information about the species, sex, receptivity and location of the signaller. In many species, only one sex gives the signal and only when it is receptive. This means that the signal need only be species specific and locatable, since sex and receptivity are implied by presence of the signal itself.

Different species use different kinds of signals to attract mates. Visual, auditory and chemical (odourant) signals are the most common, and the modality used depends to a large extent on the habitat in which the species lives and the senses that have evolved in association with other aspects of its life. For example, birds have a keen sense of sight and hearing but a relatively poor sense of smell, and so they most commonly attract their mates using visual signals, like brightly coloured feathers or vocal signals, like songs. Insects and mammals, on the other hand, have a well-developed sense of smell used to detect food and so they commonly use **pheromones** to attract mates. Habitat type also influences the kind of signal used. Birds that live in forests where visibility is restricted often use songs or other vocal signals to attract mates, while those that live in more open habitats mostly use visual displays. A good example of such a display is the tumbling flight used by lapwings and woodcocks to attract females over considerable distances.

The type of signal used to attract mates is also influenced by the sex of the signalling partner. Most commonly it is the males that produce mate attractant signals and the modality used is often auditory or visual. For example, in many birds, lizards, newts, fish, fiddler crabs and mammals, it is the males that are brightly coloured, that develop enlarged body structures and that perform elaborate displays. Females are seldom as brightly coloured as males and almost never produce structures or engage in elaborate displays like males do. Similarly, in animals that use vocal mate attractant signals, the signalling sex is almost always the male. This is true for frogs, insects and some birds although, in some bird species, both males and females produce long-range vocal mate attractant signals. There are also a few species in which females produce vocal signals to attract mates. Female elephants emit ultra-low-frequency sounds to attract males when they are in **oestrus**. Olfactory signals are sometimes used by males to attract females (e.g. some species of beetles), but are most commonly used by females. This is true of many species of mammals, snakes, spiders and moths.

Mate attractant signals must travel over long distances and be easy for the receiver to locate. Visual, vocal and olfactory signals are most effective for transmission over a distance, but are also used during close-range interactions between potential mating partners. Close-range auditory and visual signals are often reduced in amplitude and intensity but have a higher duty cycle ('on' time relative to 'off' time). For example, male frogs produce calls at a much faster rate when a female is close by. Chemical signals used at close range are low-volatility compounds (evaporate less easily than those used for long-range signals) and are detected by contact receptors (e.g. the antennae of insects and the vomeronasal organ of reptiles and mammals). Electric signals, used by some species of fish, are well suited to short-range exchanges since they only travel over distances of less than 1 m, and tactile signals are only used when potential mates are in close contact with one another.

All signals that form part of courtship behaviour, whether they are used for attracting mates at a distance or during close-range encounters, contain information about species identity. In order for offspring to survive and go on to reproduce, it is crucially important that animals mate with members of their own species. Hybrid matings seldom produce any offspring at all, and those that are produced are unlikely to produce viable offspring themselves. Courtship behaviour is a good way of facilitating mating between conspecific partners. For animals that use vocal signals, species identity can be encoded in either the temporal (time) or spectral (frequency) parameters of the signal, or both. For example, many frog species breed in large mixed-species assemblages. Males of each species attract conspecific females by producing unique, species-specific calls which differ from the calls of heterospecific males with respect to frequency, duration or repetition rate, or some combination of these parameters (see Fig. C.14). Females recognize calls by these differences and respond only to calls produced by conspecific males.

Species identity can also be encoded in visual signals. For example, firefly males produce flashes of light to which females are attracted. Different species of males produce signals that differ in the duration and temporal pattern of flashing. In fiddler crabs, males wave their major cheliped (an enlarged feeding claw) at passing females in an attempt to attract them to their burrows. Different species differ in the pattern of the waving display, as well as with respect to cheliped colour, shape and size (see Fig. C.15). All these may provide females with cues to species identity. For animals that use chemical signals, species identity is encoded in the chemical's size, shape and functional group. Pheromones generally consist of a carbon chain backbone with one or more double bonds in a specific configuration and one functional group at the primary end. In butterflies and moths, different species within a genus differ with respect to the functional group located on the end of the carbon chain.

Once a male and female are together, be it because they live in mixed-sex groups, through a chance encounter or by some form of mate attraction signal or behaviour, they must decide whether or not to mate. Often, one partner is more motivated to mate than the other. This is usually the male, because his energetic investment in gametes is lower than the female's. Males are therefore more likely to play an active role during courtship than females. Females are more cautious because they invest a significant amount of energy in the production of gametes, and so errors in species recognition or mate choice are more costly to females than males. Therefore, one important function of courtship behaviour is to induce sexual arousal in the partner, usually the female, which is less willing or motivated to mate. For example, in the smooth newt, males and females meet mainly by chance and initially the male is the active partner and the female is unresponsive. The courtship display by the male is therefore long and vigorous and is aimed at stimulating the female and eliciting a positive sexual response from her. Males posture to show off their bright coloration, and they also secrete a pheromone that the male wafts towards the female's snout by rapidly fanning his tail to create a current of water.

In addition to sexually stimulating the female, courtship behaviour functions to ensure copulation synchronization and/or synchronous gamete release. This often involves communication of receptive stages by both male and female. Copulation synchronization signals are usually visual or tactile and are common in many species of mammals and birds where fertilization is internal. In rodents, for example, the female adopts a stereotypical posture (**lordosis**), which signals the male to mount. Intromission by the male cannot be achieved if the female does not adopt this posture. In smooth newts

Fig. C.14. Vocalizing male painted reed frog, *Hyperolius marmoratus*. Male frogs attract females from a distance using species-specific vocal signals. Male *H. marmoratus* use the same signal at close range, although they increase their rate of signal production (calling rate) as the female approaches.

Fig. C.15. *Uca vocans* fiddler crabs, male and female. The male attempts to coerce the female into his burrow using his rear ambulatory legs, in order to mate with her.

fertilization is internal, but the male does not have a penis. Instead, he releases a spermatophore, or sperm packet, on to the substrate, which the female then picks up with her cloaca (genital opening). After a complex courtship display by the male, the female signals her receptivity by touching the base of the male's tail. Once this occurs, he releases a spermatophore. Males will not deposit a spermatophore until the female has signalled that she will pick it up.

Synchronous gamete release is very important for both sexes in fish and amphibians, where fertilization is external and eggs and sperm are released directly into water. In these animals, the male must release his sperm immediately after the female releases her eggs. In the three-spined stickleback, males defend territories containing a nest and display their territorial status through bright body coloration. When a receptive female (her receptivity is indicated by her swollen, egg-filled belly and a 'head up' display) enters the male's territory, he performs a 'zig-zag' dance, which involves alternate broadside and face-on postures. If the female responds to this display, the male swims towards his nest and the female follows. The male then stimulates the female with a series of prodding movements at the base of her tail, which induce her to release her eggs. The male then enters the nest and releases his sperm. Like sticklebacks, courtship in many other animals consists of a chain of interactions between the male and female. Neither partner will commence with the next movement in the sequence until the other partner has responded to the preceding movement. This ensures that both partners reach the stage of gamete release at the correct time.

Courtship signals not only ensure that mating occurs between males and females of the same species, but they also facilitate choice of mating partner from among *individuals* of the same species. The role of courtship signals in conspecific mate choice is especially apparent in those species that mate in groups. Where a number of males and females come together to mate (e.g. on mating grounds or '**leks**'), there is great potential for rivalry between members of the same sex for access to members of the opposite sex. The effect of this rivalry on courtship behaviour depends to a large extent on differences between males and females in reproductive potential. Female reproductive potential is much more limited than male. Because the production of eggs is a major energetic investment for females and because, in most species, females are the ones that care for the young, the number of eggs a female can produce in a lifetime is limited. Males, on the other hand, invest very little in each sperm and in many species play little or no role in the care of the young. The number of offspring a male can produce over a given period of time is therefore potentially much higher than a female.

This difference in reproductive potential leads to the evolution of different reproductive tactics. To ensure the survival of their limited eggs, females need to ensure that they mate with males of the highest quality. This leads females to be choosy about the males they mate with, selecting and copulating with those males that are likely to offer them benefits that will maximize their limited reproductive potential. Males, on the other hand, will attempt to maximize their reproductive potential by mating with as many females as possible. Among animals that live in groups, this leads to fierce competition among males for access to females. Competition can take the form of intense fighting and aggression between rival males and/or the evolution of elaborate sexual displays, which not only attract the attention of females, but do so more effectively than the displays of other males.

There is now ample evidence in the literature that females of many species do not mate randomly, but selectively mate with males whose displays generate the strongest stimuli (e.g. are the most colourful, loudest, largest or most complex). A number of hypotheses have been proposed to explain female choice of mating partners. One hypothesis (the 'good genes' model) suggests that females choose males that show particular displays/traits because they are costly and therefore provide females with reliable cues of male genetic 'quality', which is inherited by the female's offspring. According to this hypothesis, only the highest-quality males are able to support the cost of producing extreme forms of the display (e.g. longest tails, brightest colours, loudest calls). Low-quality males either pay a higher survivorship cost compared with high-quality males for a given amount of energetic investment in the display/trait (for example, take large antlers in deer: low-quality males are unable to support the cost of growing large antlers and are more likely to die than are high-quality males) or they produce an inferior display (e.g. in some birds low-quality males produce smaller, less ornate feathers than do high-quality males, or their feathers are in worse condition).

Traits used by females to assess male genetic quality are therefore known as indicator traits. An example of an indicator trait is the red plumage seen in many bird and fish species. The size and intensity of red patches are often indicative of the levels of carotenoid pigments in feathers and scales. Carotenoid levels are often associated with immunological competency; sick males often exhibit a reduction in red coloration compared with healthy males, and only males with good immune system function (and hence disease resistance) can invest in intense colour production and remain viable.

A second hypothesis is that some male signals do not indicate genetic quality but, instead, provide females with information about how males can directly affect the survival of offspring. For example, a high calling rate by some male birds may indicate to the female that he can defend a high-quality nest site, and this may directly affect the survival of the female's chicks.

A third hypothesis is that some male displays do not provide females with any benefits at all (either genetic or direct), but have evolved to exploit female sensory predisposition towards some stimuli. Females of many animals have pre-existing biases towards specific stimuli (such as feature detectors for prey items or food plants) and are attracted to male courtship displays that incorporate or exaggerate such stimuli. For example, in the fiddler crab, *Uca musica*, some males build mud 'hoods' at the entrance to their burrow. Females are attracted to those males whose burrows have hoods because such objects are used by females to hide from bird predators.

There is evidence that all three types of female choice occur and that they are not mutually exclusive but may operate simultaneously. For example, in house finches females that choose males with larger, redder plumage patches obtain direct benefits because these males offer better paternal care for nestlings. Male redness is also heritable, so it is possible that females gain genetic benefits for offspring as well, e.g. higher growth rate.

Finally, it should be noted that females are not the only sex to exercise choice. There are a few species where male investment in a reproductive event exceeds that of females, which results in the sex roles being reversed. When males are a limiting resource for females (e.g. species where males care for the offspring of a single female or where there is a significant cost associated with sperm production), females compete for males and may develop ornaments and courtship displays. Males, on the other hand, are selective about their choice of mate. Complete sex role reversal occurs in several fish species (e.g. seahorses and pipefish), dendrobatid frogs and several bird species (e.g. spotted sandpiper and jacana).

To summarize, the biological function of courtship behaviour is to ensure successful mating occurs between conspecific members of the opposite sex. Within this framework location of potential mates, sexual arousal in partners, synchronization of mating and choice of mating partner are all integral parts of courtship behaviour. (MD)

See also: Breeding; Communication; Emancipation; Intersexual selection; Intrasexual selection; Mate choice; Mating behaviour; Sexual behaviour

Further reading

Bradbury, J.W. and Vehrencamp, S.L. (1998) *Principles of Animal Communication.* Sinauer Associates, Inc., Sunderland, Massachusetts.

Gould, J.L. and Gould, C.G. (1996) *Sexual Selection: Mate Choice and Courtship in Nature.* Scientific American Library, USA.

Crepuscular activity

Behaviour that is most common during the periods of dawn and dusk is described as being crepuscular. This is in contrast to **diurnal** (daytime), **nocturnal** (night) and **cathemeral** (at periods during both day and night) periodicities of behaviours. As such the timing of crepuscular behaviours is probably controlled endogenously, but these are linked to exogenous factors such as temperature changes or, most commonly, changing light levels. Additional factors such as risk of predation or foraging efficiency at dawn and dusk may also influence the crepuscularity of behaviours. In the murid rodent *Arvicanthis ansorgei*, its crepuscular behaviour is correlated with distinct peaks of the adrenal **steroid hormone** corticosterone, indicating a hormonal basis for its increased activity at these times of day.

There are several obvious examples of crepuscular activity. The first is the dawn chorus, when usually passerine birds in broad-leafed woodlands at temperate latitudes concentrate most of their singing during dawn and dusk. There are two adaptive reasons for this crepuscular activity. First, the acoustic properties of the air, particularly at dawn, are suitable for the effective transmission of bird songs at the commonest frequencies, because air turbulence is minimized and temperature gradients are ideal. For example, the recorded songs of the white-throated sparrow, *Zonotrichia albicollis*, and the swamp sparrow, *Melospiza georgiana*, were transmitted through open grassland and closed forest. It was found that the signal transmission quality of the songs was more consistent at dawn than at midday. The signal transmission quality is probably essential for the identity function of a bird's song. However, the swamp sparrow's song was more consistent over increasing distances up to 100 m compared with that of the white-throated sparrow. The dawn chorus also appears to have an important territorial function compared with diurnal singing, which aims to attract females (**see: Courtship behaviour**). Temporary removal of rival males surrounding the territory of a male chipping sparrow, *Spizella passerina*, led to a cessation or reduced singing at the dawn chorus, but did not affect diurnal singing. The return of a single rival male led to a resumption of normal singing at dawn.

Secondly, this is a period of the day when the colour vision of birds is less effective, so they are less able to forage effectively. Insect prey may also be inactive owing to low temperatures and hence would be difficult to locate. However, this may depend on the abundance of food. In experiments with great tits, *Parus major*, it was found that when food was abundant and easy to find, great tits would be less likely to sing or chase away rival birds trespassing on their territories. Therefore, they concentrate their singing at the time of the day when its transmission is most effective and foraging is least efficient.

Mammalian predators such as lions, *Panthera leo*, often hunt crepuscularly. In the Serengeti activity increases dramatically at dusk and is maintained throughout the night, when it declines equally dramatically at dawn. Most of their hunting occurred just after dark, between 19.00 and 20.00 hours and in the early morning, between 02.00 and 04.00. At dusk mammalian vision is changing over from the daytime, cone-mediated type to the nocturnal, rod-mediated type. Neither is working at its greatest efficiency, so that prey species often have great difficulty in seeing predators. Therefore, hunting efficiency is often greater at dawn and dusk. The behaviour of potential prey animals is similarly influenced by the crepuscular and nocturnal behaviour of predators. For example, African buffalos, *Syncerus caffer*, all stand and become much more vigilant at dusk and can only relax again at dawn.

A less obvious example of crepuscular activity is animals utilizing microorganisms to digest their food. They gain an advantage by extending their feeding activity into the crepuscular period to maintain substrate at consistent levels compared to just diurnal feeding.

Bats emerge from their roosts at around dusk, but their actual time of emergence will depend on their flight ability and feeding ecology. For example, noctules, *Nyctalus noctula*, emerge on average 5 min after dusk, but the long-eared bat, *Plecotus auritus*, only takes flight after almost 1 h (54 min on average). Emergence times are a compromise between emerging early enough to get enough food to survive and emerging too early and being at risk of being preyed upon. Noctules depend on the dusk peak of flies, but are fast and high flying, so that their risk of **predation** is lower than that of slow-flying species such as the long-eared bat, which gleans moths and non-flying prey from tree vegetation. These prey can be found throughout the night, and so the long-eared bat plays it safe and emerges much later to minimize predation.

Nightjars and related birds also hunt at dawn/dusk and through the night. Field studies of the standard-winged nightjar, *Macrodipteryx longipennis*, and long-tailed nightjar, *Caprimulgus climacurus*, in the bush savannah of West Africa have shown that their foraging activity is highly crepuscular; peak foraging occurs at dusk, there is less at dawn and least activity through the night. This pattern of foraging activity correlated with the activity of the nightjars' insect prey. Dusk foraging intensified in periods during the new moon, when lack of moonlight precluded nocturnal foraging.

Swallows, martins and bats often fly over fresh water at dusk in pursuit of insect prey. Many aquatic insects are polarotactic, i.e. they detect aquatic habitats by seeing how they reflect polarized light. At dawn and dusk the sun is low in the sky, and this provides the maximimum ability to detect water surfaces polarotactically. However, it is only at sunset that air temperatures are sufficiently high to allow the swarming flights of aquatic insects that attract aerial crepuscular predators.

Visual communication may also be influenced by crepuscular activity. Another nightjar species, the common nighthawk, *Chordeiles minor*, has white patches on its wings, tail and throat, which are important in sexual and agonistic communication. The white patches on the wings and tails of males increase with age, supporting their role, combined with flight displays, in visual communication at low light levels at dusk.

Like humans the great apes are diurnal species, which must limit their nocturnal behaviour, because their eyes are adapted for colour diurnal vision. Because they must maximize their diurnal foraging, an important crepuscular behaviour of chimpanzees, *Pan* spp., and gorillas, *Gorilla* spp., is the construction of nests to rest in during the night. Rather than returning to a particular site, chimps construct their nests close to where they have been feeding. Chimpanzees construct a new nest each night by selecting a stable base in a tree 9–12 m above the ground, then bend in larger branches to which are added smaller branches and leafy vegetation to create a green mattress. This crepuscular behaviour allows chimpanzees' maximum flexibility in their diurnal foraging. (ACK)

Crib-biting (cribbing)

Crib-biting and **wind-sucking** are the most common equine oral **stereotypies** (others include lip-licking, licking the environment and self-biting). The distinction between crib-biting and wind-sucking is often blurred because there are different definitions in several English-speaking countries. For example, in the UK, wind-suckers are distinguished from crib-biters because they do not hold on to fixed objects while engulfing air, whereas in Australia wind-sucking includes grasping and engulfing air. The critical unifying feature of these responses is the characteristic arching of the neck, a posture that allows **horses** to draw and transiently hold (i.e. engulf) air in the cranial oesophagus. Although horses prefer to grasp wood rather than metal (perhaps because it cushions their teeth), they will crib-bite on metal if no other substrate is available. Often confused with crib-biting by novice observers, wood-chewing is not a stereotypic behaviour since it is not invariant and is seen in free-ranging horses, and so is likely to have a function in digestion or nutrition. That said, it has been suggested that the function of crib-biting may also relate to digestion by generating saliva that may buffer gastric acid accumulation resulting from intermittent feeding of **concentrates**.

The prevalence of wind-sucking/crib-biting in UK Thoroughbred populations has been estimated at 4.2%, and in several countries has been shown to be greater in Thoroughbreds than in other breeds. However, it remains difficult to separate the effects of management and breed because, although they are among the most reactive of breeds, Thoroughbreds are generally raced and therefore managed very intensively.

Like other stereotypic behaviours, crib-biting is not recognized in free-living feral horses but is not purely a product of **domestication**, since it is reported in captive examples of wild equidae such as the onager (*Equus hemionus onager*), mountain zebra (*E. zebra*) and Przewalski horses

(*E. przewalskii*). Management practices that depart from the trickle feeding typically found in free-ranging equidae, such as feeding discrete concentrated meals that require little mastication, have a greater effect than housing practices on the incidence of crib-biting, but a single causative factor is rarely identified in its aetiology. It is not clear why some horses start crib-biting and others do not, despite being managed identically. If a horse exhibits one oral stereotypic behaviour, it has a greater chance of having a second oral stereotypic behaviour, when compared with normal horses. Some ethologists maintain that, far from being undesirable, horses that go on to perform stereotypic behaviours are simply fortunate since they may have a response, albeit an imperfect one, to the insults of intensive management.

There has been much debate about the function of stereotypic behaviours (Mason, 1991). One influential school of thought argues that they enable animals to cope with **stress**. For example, transient decreases in heart rates have been demonstrated in association with bouts of crib-biting, and measures to prevent crib-biting are linked to increases in some physiological stress parameters. However, in contrast, it has also been suggested that crib-biters may have higher basal sympathetic activity because they have a higher overall mean heart rate (Minero *et al.*, 1999). Experimental studies to assess the validity of the stress-coping hypothesis in several species have produced equivocal results.

Stereotypic behaviours may help horses cope with suboptimal environments or provide direct and immediate rewards that bring their own intrinsic gratification, possibly through **endorphin** release, which has been implicated as a possible source of reinforcement for crib-biting. **Opioid** antagonists can reduce crib-biting significantly, which suggests that at least one of the perceived benefits of crib-biting (from the horse's perspective) is mediated by opioids. However, resting behaviour in crib-biters also significantly changes with opioid antagonists, so the reduction in crib-biting may occur because of generalized sedation. Furthermore, there is recent evidence that stereotypic horses persist in non-rewarding activities longer than control horses (Hausberger *et al.*, 2007; Hemmings *et al.*, 2007), possibly reflecting alteration of basal ganglionic activity (Hemmings *et al.*, 2007).

It is also possible that a given stereotypic behaviour may retain a function within the motivational system from which it is derived. So, crib-biting may provide a route to normal feeding and digestive activity within environments that limit normal forage intake (e.g. an intensive training programme characterized by the provision of diets with a high concentrate:roughage ratio). Support for this possibility has come from a prospective study of a population of foals, which showed a peak in the emergence of crib-biting at 20 weeks of age (Waters *et al.*, 2002). This coincided with a significant increase in the feeding of concentrates. Foals and adults have been shown to crib-bite less when fed antacid supplements.

Horse lore emphasizes the importance of mimicry rather than environmental deficits, believing that exposure to a stereotypic neighbour may increase the likelihood of the behaviour's development or performance. Such social influences, known in bank voles, may also affect stereotypic behaviour levels in horses, despite there being no current indication that horses can learn by observation. Even if **observational learning** is not involved, the possibility that having a stereotypic neighbour may arouse observing horses and so predispose them to developing stereotypic behaviour cannot be ruled out. Since crib-biting is popularly regarded as being transmissible by mimicry and is associated with health problems, horses exhibiting it are often isolated. Wind-sucking (but not crib-biting, due to its possible confusion with wood-chewing) must be declared at some auctions in the UK and tends to lower the value of affected animals.

The deleterious effects of crib-biting vary, with some owners reporting no ill effects while others regard diminished performance, digestive disorders, incisor erosion and failure to thrive as important sequelae. Some veterinarians maintain that crib-biting leads to chronic colic, flatulent colic and a distended abdomen. This is supported by epidemiological evidence that epiploic entrapment is associated with a history of cribbing/wind-sucking (Archer *et al.*, 2004). However, the extent to which these signs can be linked directly to aerophagia has been challenged by radiographic studies indicating that negligible gas is ingested during crib-biting (McGreevy *et al.*, 1995). It is possible that crib-biters have concurrent gastrointesitnal disturbances that cause these signs.

The need to prevent stereotypies for aesthetic and occasional health reasons has prompted searches for permanent cures (**see: Collar**). Physical or surgical approaches that are commonly attempted to prevent crib-biting vary considerably in success. Surgery is falling from favour as the causes of oral stereotypic behaviours – and especially the role of gastric acidity in its ontogeny – are becoming recognized, but a humane remedy that is effective for every crib-biter remains elusive. It seems that the continual reinforcement of stereotypic behaviours contributes to their resistance to therapy. Data demonstrating increasing prevalence with age suggest crib-biting can become emancipated from its initiating causes. So, while stereotypies may arise in response to adverse management, they may persist in more enriched environments. That said, it is better to implement environmental **enrichment** programmes prophylactically rather than therapeutically. Notwithstanding their effects in maintaining normal behaviour in youngstock, they are also likely to increase learning ability while, at the same time, promoting musculoskeletal health.

(PDM)

References and further reading

Archer, D.C., Proudman, C.J., Pinchbeck, G., Smith, J.E., French, N.P. and Edwards, G.B. (2004) Entrapment of the small intestine in the epiploic foramen in horses: a retrospective analysis of 71 cases recorded between 1991 and 2001. *Veterinary Record* 155, 793–797.

Cooper, J.J. and Nicol, C.J. (1994) Neighbour effects on the development of locomotor stereotypies in bank voles *Clethrionomys glareolus*. *Animal Behaviour* 47, 214–216.

Hausberger, M., Muller, C., Gautier, E. and Jego, P. (2007) Lower learning abilities in stereotypic horses. *Applied Animal Behaviour Science* 107(3–4), 299–306.

Hemmings, A., McBride, S.D. and Hale, C.E. (2007) Perseverative responding and the aetiology of equine oral stereotypy. *Applied Animal Behaviour Science* 104, 143–150.

Mason, G.J. (1991) Stereotypies: a critical review. *Animal Behaviour* 41, 1015–1037.

McGreevy, P.D., Richardson, J.D., Nicol, C.J. and Lane, J.G. (1995) A radiographic and endoscopic study of horses performing an oral stereotypy. *Equine Veterinary Journal* 27, 92–95.

Minero, M., Canali, E., Ferrante, V., Verga, M. and Ödberg, F.O. (1999) Heart rate and behavioural responses of crib-biting horses to two acute stressors. *Veterinary Record* 145, 430–433.

Waters, A.J., Nicol, C.J. and French, N.P. (2002) Factors influencing the development of stereotypic and redirected behaviours in young horses: findings of a four-year prospective epidemiological study. *Equine Veterinary Journal* 34, 572–579.

Critical period

In general, a critical period is a limited time in which an event can occur that will result in a lasting change. From an animal behaviour point of view it refers to the early stages of an animal's life in which the animal is particularly sensitive to certain types of experience or stimuli. Exposure to the correct stimuli during these early life stages will result in appropriate, species-typical behavioural changes; while animals that were not exposed to the stimuli or were exposed to inappropriate stimuli during the critical periods may not perform certain behaviour patterns or may display them inappropriately later in life. There are criteria that must be met for an early stage of development to be called a critical period: (i) they must have set beginning and end points; (ii) there must be an intrinsic effect – something in the animal must change so that experience that formerly had an impact on behaviour can no longer affect the behaviour; and (iii) the extrinsic stimuli or experiences must be identifiable.

Critical periods are especially important during **imprinting**. One of the best-known examples comes from **Konrad Lorenz**, who demonstrated that greylag geese would imprint on the first suitable moving stimulus they saw between 13 and 16 h after hatching. Lorenz was able to imprint the geese on himself and is often pictured with a line of goslings following him. The term critical period is now generally replaced by **sensitive phase**. Critical periods are characterized by abrupt beginnings and ends and, if the appropriate experience is not achieved during this time period, the animal will not be able to learn the correct behaviour; in contrast, sensitive phases have gradual beginnings and ends and, while they are the period of maximum sensitivity in which the appropriate experience may occur, the correct behaviour may still be potentially learned at a later stage.

Ensuring an animal receives the appropriate early experiences is important for the development of later behaviour patterns. However, with the **intensification of animal production**, many animals are not kept in natural surroundings and do not experience the correct stimuli during these critical periods. As a result, their behaviour may also be altered and different from their wild or less intensively raised counterparts. In some cases this can result in suboptimal behaviour, but the differences are not necessarily of welfare significance. However, from a conservation perspective the failure of filial and sexual imprinting during critical periods can have catastrophic effects on attempts at successful and sustainable **reintroduction**. (LMD)

Further reading

Bateson, P.P.G. (1979) How do sensitive periods arise and what are they for? *Animal Behaviour* 27, 470–486.

Bateson, P.P.G. and Hinde, R.A. (1999) Developmental changes in sensitivity to experience. In: Bolhuis, J.J. and Hogan, J.A. (eds) *The Development of Animal Behaviour*. Blackwell Publishing, Oxford, UK.

Critical temperature

Animals perceive thermal information in the form of extreme ambient temperatures, relative humidity or wind speed (or a combination of these), which are detected by thermoreceptors, skin dryness (particularly in the throat and nasal passages) and mechanoreceptors, respectively. The optimum temperature is usually referred to as the comfort or thermoneutral zone, above and below which (the critical temperatures) animals have to work physiologically to sustain their core body temperature, through increasing either heat loss or production. They learn the importance of being in an optimum temperature, and young animals are more likely than older animals to use areas where the ambient temperature is below the comfort zone; older animals learn to use favourable microclimates and exploit exposed areas at times when ambient temperatures are increased.

Cold temperatures are generally less of a problem than excessively hot ones. For example, the lower critical temperature of adult cows is –23°C and, at sub-zero temperatures, the availability of fodder is more of a problem than the temperature. However, they are prone to heat stress, especially if they have a high metabolic rate because of high productivity and/or consume a high-fibre diet that produces considerable heat. Initial responses to high temperatures include increased respiration rate, which occurs as low as 21°C, but further increases in temperature (above 25°C) reduce food intake and hence the heat of digestion.

Perception of temperature, as with most other stimuli, is relative rather than absolute. Response to localized thermal loading to one part of the body is reduced if the ambient temperature is low. There are genetic differences in the susceptibility of animals to thermal load, which may derive from differences in perception, but are more likely to derive from differences in endogenous heat production and heat dissipation. For example, Holstein–Friesian cattle are particularly prone to heat stress, compared with *Bos indicus* cattle, but mixed groups may follow the less resistant cattle in reducing grazing time. Pigs and poultry have less endogenous heat production than ruminant animals and are therefore more susceptible to cold temperatures. (CJCP)

See also: Heat stress; Thermoregulation

Further reading

Silanikove, N. (2000) Effects of heat stress on the welfare of extensively managed domestic ruminants. *Livestock Production Science* 67, 1–18.

Critically endangered

A species of animal or plant is considered critically endangered when, generally speaking, its prospects for survival in the wild are so poor that special measures must be taken to prevent its **extinction**. A more formal definition is provided by the International Union for the Conservation of Nature (IUCN). There are several criteria and sub-criteria, but these include a reduction of the population by more than 90% in the preceding 10 years, a geographic range of less than 100 km^2 and a population of fewer than 250 mature individuals. (RS)

See also: Endangered species

Further reading

IUCN Redlist. Available at: http://www.iucnredlist.org

Cropping of ears

Cropping of ears involves the surgical removal of a distal portion of both of the ears of a dog, usually specific breeds such as boxers, Great Danes, Dobermann pinschers and schnauzers. The historical reasons for ear cropping are similar to those for tail **docking**. The aim was to reduce injury to the ears and to make it harder for such dogs to be caught by the ears. The practice was common in dogs bred for guarding, fighting and **hunting** of small animals. Although few dogs are used for certain of these purposes now, some breed fanciers argue that cropped ears are an inherent part of the historical breed standard. However, others regard the practice of ear cropping as an unnecessary and painful **mutilation**. In many countries, including Australia, Belgium, Denmark, Finland, Germany, the Netherlands, Norway, Sweden and the UK, the practice is banned under prevention of cruelty to animals' legislation. (BJ)

Cross-suckling

In 'nature', unweaned animals obtain their milk by suckling from their mothers. Occasionally, however, animals may suckle from animals other than their mother. For example in **cattle**, this can involve an unweaned calf suckling from a lactating cow other than its biological mother, a calf sucking on another calf or weaned calves and even adult cattle suckling another lactating cow (also called 'milk stealing' or 'inter-sucking'). However, the cases that have received most attention from researchers involve unweaned calves sucking at other calves (see Fig. C.16).

Unfortunately, there is inconsistent use of the terms 'cross-sucking', 'cross-suckling' and 'inter-sucking', which are often used interchangeably. Terminologically, 'suckling' is usually used to describe the suite of behaviours that the neonate performs during **nursing**, and tends to imply milk transfer. Although actual sucking is the most obvious of these behaviours, and is the consummatory behaviour that is the most essential to ensure milk transfer from the mother to the young, other appetitive behaviours besides sucking (such as butting the udder, switching teats, stripping teats, etc.) do occur. Most often, the sucking between unweaned animals, where no milk can be transferred, is referred to as cross-sucking, a usage that we follow here. Inter-sucking tends to be used to describe sucking involving older including adult animals.

When kept with their mothers, the young of many species almost exclusively suckle from them. However, when the mother is absent (e.g. as a result of **death** or removal), they may attempt to suckle from other lactating animals. In cattle, usually, these attempts are unsuccessful, since cows are capable of recognizing their own offspring and reject suckling attempts by other calves. Occasionally, such attempts are successful, particularly if the calf attempts to suckle from behind rather than from the side, which is the more normal position. An approach from the rear may make it more difficult for the mother to detect by smell that the calf is not her own. In some cases, however, cows will adopt another calf and allow it to suckle freely in the normal position. Use of such 'nurse cows' is a common feature of some types of cattle rearing (e.g. dual-purpose cattle rearing in many tropical countries). Interestingly, such cases of suckling do not have the same physiological effects on the cow (e.g. those resulting in prolonged lactation or anoestrus) as when the cow is suckled by its own calf. This suggests that it is the psychological component of the mother–offspring bond which is important for these physiological effects rather than the suckling and milk transfer itself.

Although beef cattle usually suckle from their mother, most dairy and veal calves (at least in most industrialized countries) are separated from their mothers soon after birth and are fed

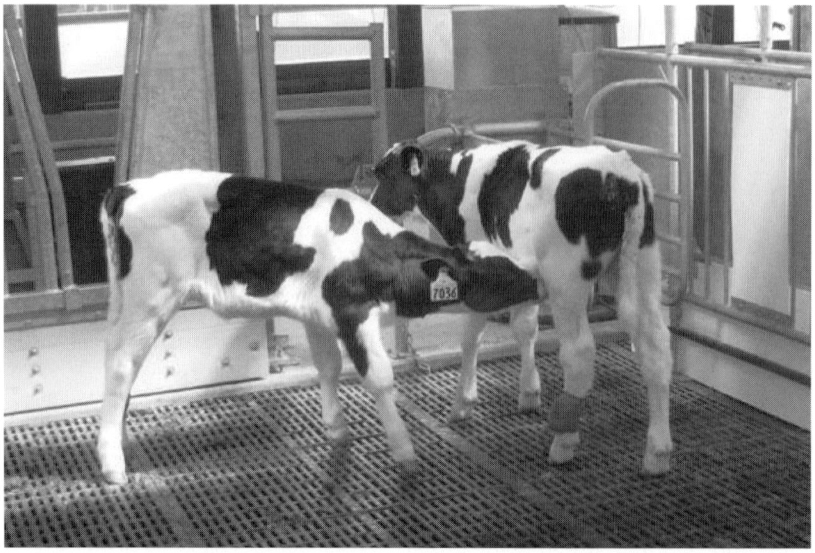

Fig. C.16. An example of cross-sucking between two unweaned dairy calves.

milk either from a teat feeder or from a bucket. In the case of bucket feeding, they have little opportunity to actually suck in order to obtain their milk. How much of a threat to their **welfare** does this constitute? Calves that are housed individually and fed milk from a bucket suck avidly at the bucket, bars of the pen, chains, bits of rubber, ears and mouths or other calves – or indeed almost anything that they can get into their mouths. When milk-fed calves are housed in groups, calves will often suck at the belly, inguinal area or the mouth and ears of another calf (**see: Vacuum behaviour**). Similar sucking behaviours have also been documented in other species when they are early weaned and/or hand reared, e.g. cats and horses, but it is in cattle that most research into this problem has been conducted.

The occurrence of such behaviour may have implications for the welfare of animals in a number of ways. First, many researchers have suggested that young 'need' to suck and that providing milk from a bucket or other source is not sufficient to ensure good welfare, even if the animals' nutritional needs are met. Cross-sucking therefore may be a sign of behavioural deprivation (**see: Behavioural need**). Secondly, cross-sucking may lead to injuries (including teat damage, when the behaviour is directed at other animals) and drinking **urine** directly from another calf, in the case of cattle. Finally, some of the ways in which this behaviour is controlled may themselves have detrimental effects on animal welfare. For example, many dairy farmers believe that cross-sucking by milk-fed calves will lead to 'milk stealing' from lactating cows when these calves grow up.

For these reasons, the occurrence of cross-sucking discourages dairy producers from keeping milk-fed calves in groups, despite the animal welfare (and labour) advantages of this form of rearing. A number of physical devices have been developed and used to prevent cross-suckling by older animals, especially to prevent milk stealing in cattle. These are generally attached to the muzzle of the initiator or (less commonly) to the udders of the recipients of the behaviours, and are designed to physically prevent the animal sucking. In some cases surgical intervention has been used, e.g. slicing the tongue of the animals. There have been few systematic studies of the effectiveness of such techniques, which in all cases require some labour and expense, and will cause distress to the animals. Surgical interventions, in particular, are ethically questionable and are illegal in a number of countries (**see: Mutilation**). It would be far better to be able to limit the incidence of cross-suckling through understanding the causes of the behaviours and developing feeding and housing systems that prevent it occurring.

Considerable research has now been carried out to try to understand the causes of cross-sucking between unweaned calves in particular, and to find means of preventing it. To be able fully to control cross-sucking (and to decide on the importance of sucking deprivation for animal welfare), we need to understand better the causes and consequences of sucking behaviour (**see: Motivation**). Why do calves suck and what satisfies their sucking motivation? At a simple level, the motivation to suck would seem to result from the calves' hunger and, under normal circumstances, the performance of sucking, leading to the ingestion of milk, would reduce this hunger.

However, research results have recently demonstrated that this model is over-simplistic, or at least does not apply to cross-sucking. For example, most cross-sucking in group-housed, milk-fed calves occurs around the time that the animals drink milk. Nevertheless, in contrast to what might be expected, the occurrence of cross-sucking is highest immediately after the meal rather than before the meal. This is the opposite of what we would expect if hunger was the main motivation. Furthermore, cross-sucking is reduced when the calves can suck a dry teat after the meal, even if this does not alter the amount of milk drunk by the calf. Finally, cross-sucking can be greatly reduced (and will almost disappear) when the calves are weaned off milk. Clearly, young calves do not suck simply because they are hungry, and drinking milk is not the only factor that reduces their sucking motivation.

If it is not hunger, what is it that motivates calves to suck and what can be used to reduce this motivation? When milk-fed calves are given milk to drink, they will often suck a dry teat once they have finished drinking. This non-nutritive sucking is thought to be similar to cross-sucking (except that it is directed to a dry teat rather than to another calf).

Research using this behaviour has allowed us to answer some of the questions about sucking motivation. First, sucking motivation seems to be stimulated (rather than reduced) by the ingestion of milk, at least in the short term. When calves drink milk from a bucket they are then highly motivated to suck on something. Experiments that have varied the concentration of the components of milk show that it is primarily the taste of lactose (or sugars) that stimulates sucking in this way. The quantity of milk drunk does not seem to be very important. This sucking motivation is reduced if the calf can suck at something, even if no milk is ingested. Thus, it is the performance of the sucking behaviour itself, rather than the ingestion of milk, which reduces sucking motivation. If the calf cannot suck at anything, the sucking motivation gradually declines over the following 10 min. Together, these facts explain why cross-sucking occurs most often soon after the milk meals, why it is high if calves drink their milk without being able to suck and why allowing the calves to suck a dry rubber teat can reduce cross-sucking.

Using this information, we can develop feeding management systems for calves that allow us to control and limit the amount of cross-sucking. The most important aspect is to allow the calves sufficient opportunity to suck for long enough to satisfy their sucking motivation, either during or after the ingestion of milk. For example, use of automated milk feeders or other teat-feeding systems (such as a bucket fitted with a teat) that allow the calves to suck their milk reduces the incidence of cross-sucking. Even providing a dry teat after the meal can substantially reduce cross-sucking. However, the use of teat-feeding systems is not invariably associated with reduced cross-sucking, suggesting that the details of feeding management are important.

Automated milk feeding systems can be designed to allow the calves to suck the teat for a sufficient time after the meal to satisfy their sucking motivation. For example, cross-sucking is increased if teat-fed calves are not able to continue to suck the teat after milk ingestion, but a swing door can prevent calves from being displaced from the teat by other calves, which allows them to perform longer bouts of non-nutritive sucking on the teat and results in a lower incidence of cross-sucking. Displacements of calves from the teat feeder can also be reduced by feeding the calves a larger quantity of milk. Slower rates of milk delivery can reduce the incidence of cross-sucking. Simply restraining calves so that they cannot suck at other calves for 15 min or so after a milk meal can help reduce the incidence of the behaviour. However, this leads to calves remaining highly motivated, but unable to suck, which is not ideal for their welfare (**see: Frustration**). In any case, techniques to prevent cross-sucking are most effective the earlier they are implemented: once calves have learned to suck at other calves, the habit is much harder to break.

It is important to remember that sucking is a vital behaviour for the survival of young mammals, so it is not surprising to find that they are highly motivated to suck. Simply providing an adequate quantity of milk, but without letting the animals actually suck at something, is not sufficient. Research has shown that calves that can suck to obtain their milk lie down sooner in more relaxed positions than calves that drink their milk from a bucket. Furthermore, the secretion of hormones associated with digestion and metabolism, such as **insulin** and **cholecystokinin**, has also shown to be increased when calves can suck a teat.

Milk stealing is most often a problem when weaned animals, including adult cows, suckle from lactating cows that are kept for milking. In such cases, the animals have been reported to drink many litres of milk per day, making this a problem with considerable economic potential. Milk stealing has been reported to be associated with more frequent teat injuries and may increase the chance of **mastitis** from certain bacteria (e.g. *Corynebacterium pyogenes*). The incidence of this behaviour within herds of lactating animals varies greatly. Surveys in Sweden and Switzerland reported that the majority of farms had some problems with milk stealing, but often only a few cows were involved. The incidence is most often reported to involve somewhere between 0 and 10% of the animals, although on some farms over 50% may be involved. The few reports on this behaviour that are available suggest that certain animals are far more often involved (both as recipient and initiator) than others, and that certain pairs of animals appear to be involved more often than others. There has been little systematic research done on this behaviour, perhaps because the relatively low rate of occurrence makes it difficult to study. Epidemiological surveys have so far produced no consistent findings as to the effect of housing or feeding management on the occurrence of the behaviour. Surveys have reported that milk stealing by cows tends to be more common on farms that have a higher incidence of cross-sucking between calves and between older heifers, but the causal link between these behaviours is unclear. A low feed allowance at weaning appears to be implicated. Apart from the suggested link to teat injuries and mastitis, we have little information as to whether milk stealing is important for animal welfare, despite its obvious economic importance. (JRu, AMdP)

See also: Urine drinking

Further reading

de Passille, A.M. (2001) Sucking motivation and related problems in calves. *Applied Animal Behaviour Science* 72, 175–187.

Jensen, M.B. (2003) The effects of feeding method, milk allowance and social factors on milk feeding behaviour and cross-sucking in group housed dairy calves. *Applied Animal Behaviour Science* 80, 191–206.

Keil, N.M. and Langhans, W. (2001) The development of intersucking in dairy calves around weaning. *Applied Animal Behaviour Science* 72, 295–308.

Lidfors, L. and Isberg, L. (2003) Intersucking in dairy cattle – review and questionnaire. *Applied Animal Behaviour Science* 80, 207–231.

Crypsis

All traits that reduce an animal's risk of becoming detected when it is perceivable to an observer represent crypsis. This definition distinguishes crypsis from hiding and masquerade (mimicry), the primary function of which is to prevent recognition. Visual crypsis can encompass background matching, whereby an animal matches the colour, contrast and texture of the background, self-shadow concealment (countershading) and disruptive coloration, in which markings create false edges that hinder detection of the true outline of the animal.

In order to match the background some animals attach substrate to their bodies (for example, caddisfly larvae) or even other animals and plants (for example, decorator crabs attach anemones, coral polyps and seaweeds to their exoskeletons). Some animals are able to change colour and texture incredibly quickly. Cuttlefish have groups of coloured chromatophores (cells containing pigment granules enclosed within an elastic sac) that can change in size via muscular contractions, enabling the animal to change colour very rapidly. Crypsis may also be non-visual. Many animals (especially insects) use a chemical disguise to prevent detection. For example, the caterpillar of the moth *Biston robostum* avoids predation by ants by having a cuticular chemical signature that resembles the host plant upon which it feeds. However, the fact that the chemical signature resembles the host plant makes it difficult to distinguish between crypsis and mimicry. (PE)

Further reading

Stevens, M. and Merilaita, S. (2009) Animal camouflage: current issues and new perspectives. *Philosophical Transactions of the Royal Society, Series B Biological Sciences* 364, 423–427.

Culling

Culling is the selective **slaughter** of a number of animals. This has been a traditional practice in the human relationship with animals throughout history. The deliberate removal of unwanted stock has lead to the development of the breeds of domesticated animals that exist today, through the culling of undesirable animals and their removal from the gene pool. Current practical **breeding** programmes for desired traits, involving minimization of phenotypic standard deviation (through the use of best linear unbiased prediction (BLUP)), necessarily involve a large proportion of each generation being discarded, and often culled.

Culling is a normal husbandry practice to remove from herds or flocks animals that are no longer productive, or to remove undesirable animals from the breeding stock. Commonly, 20–25% of stock are culled from groups of animals annually. Poor fertility is the most common cause for culling in dairy herds in the UK (Esslemont and Kossaibati,

1997), sheep (although **mastitis** is also a significant cause) and sows. Other reasons for culling livestock include: (i) poor levels of production (growth, egg laying, litter numbers, milk yield); (ii) poor mothering ability; (iii) difficulties during **parturition**; (iv) poor performance of offspring; (v) **disease** (acquired and inherited); (vi) **lameness**; (vii) other physical abnormalities; (viii) poor sight; (ix) weakness; (x) poor condition; (xi) old age (including reducing generation length to maximize genetic gain); and (xii) being the wrong sex. In sheep, additional reasons include undesirable characteristics of the fleece and 'broken mouth' (inability to graze efficiently). Culling on a large scale is an epidemiological defence against disease threatening livestock (examples include Newcastle disease and avian flu in poultry, swine fever in pigs, foot and mouth in cattle and sheep and infectious pancreatic necrosis in salmon). In addition, culling provides protection against diseases threatening other animals, including humans (for example, the culling of dogs and wild canids during outbreaks of rabies).

Welfare concerns may also lead to culling of stock. Disease, parasitism and inherited or acquired bone or muscular deformities may cause a permanent state of **suffering** in an animal. Examples might include bone deformities, joint fracture or trauma causing pain in broiler chickens, scrapie and **footrot** in sheep and mastitis and lameness in dairy cattle. Indeed, in this regard a high rate of culling may reflect good welfare management – demonstrating a keen awareness by the **stockperson** of the suffering of their stock and efforts to eliminate, or at least not prolong, that suffering. Culling animals that are genetically predisposed to excitability or nervousness during **transport** and handling can reduce the **stress** experienced by future generations of animals during these procedures. Therefore culling per se is not necessarily an indicator of poor welfare, especially considering that a dead animal is not a suffering animal. However, culling diseased animals is a costly and unnecessary exercise if the animal is otherwise productive and the problem is treatable. Replacement animals will need to be paid for and these come with no disease-free guarantee. The culling of animals simply because they have an infection should be carefully considered and the specific agent identified; for example, mastitis infection by *Streptococcus agalacticae* is relatively treatable compared with *Staphylococcus aureus*. Animals may be also culled for demonstrating undesirable behaviours, such as inter- or intraspecies **aggression**.

Culling is also used as a tool in the management and **conservation** of populations of animals in the wild or in animal collections. Animals are removed from captive populations by culling when they are surplus to the holding capacity of a **zoo** or animal park. It is a measure used to ensure balanced social dynamics in respect of age and sex, to prevent the spread of disease – including the spread to domesticated animals; for example, the culling of badgers (*Meles meles*) in the UK is a perceived attempt to reduce the risk of bovine tuberculosis to domestic cattle. Wild animals may also be culled to reduce the risk of genetic contamination, the inadvertent mating of domestic animals by the wild type (such as wild boar with domesticated pig breeds).

Culling is also used to reduce the environmental pressures concomitant with overpopulation of groups of animals in the wild. Indeed, the practice of culling animals in the wild is further defended as a means of supporting the stronger members of a group by removing the weaker. Non-native species may be culled to reduce environmental pressures on preferred native species, examples being the red (*Sciurus vulgaris*) and grey (*S. carolinensis*) squirrels in the UK; or to prevent hybridization of pure native species with non-native species (the ruddy duck, *Oxyura jamaicencis*, and the white-headed duck, *O. leucocephala*, in Europe).

Animals are also culled to protect plants of particular value – either cultivated crops, grassland (in particular rabbits, *Oryctolagus cuniculus*) or botanical specimens of special interest (which includes the culling of various species of deer in Europe and the flying fox, *Pteropus poliocephalus*, in Australia). Wild animals are also culled to prevent damage to fencing designed to keep out/in predators/livestock (elephants in African reserves and camels damaging Australian dog fencing). Culling of wild animals can also provide a source of income for local human populations, through the attraction of shooting or safaris for tourists, or the value of the carcasses, including the harvesting of ivory from elephants.

Culling of wild carnivores is also practised throughout the world to prevent predation of livestock (including the culling of seals to protect wild fish stocks) and the dangers some animals pose to humans. Stray animals, particularly dogs, are culled, as are verminous animals, a group that can include rats, mice, moles, rabbits, stoats, mink, pigeons and innumerable invertebrate species, depending on an individual's definition as much as national legislation. For example, vermin are not defined in UK law.

Unwanted **draught animals**, racing dogs, other working animals and **pets** are routinely culled, including those that have outgrown their usefulness and those deemed not to have 'made the grade' at an early age. **Laboratory animals** are commonly culled following completion of scientific procedures. (DA)

Further reading

Esslemont, R.J. and Kossaibati, M.A. (1997) Culling in 50 dairy herds in England. *Veterinary Record* 140, 36–39.

Culture

Culture is broadly defined as behavioural traits, shared by a population or subpopulation, that are acquired through **social learning** (i.e. learned from others through observation or interaction) and transmitted within or between generations (i.e. horizontal versus vertical transmission). Several reports have been made of cultural behaviours among various animals such as primates, cetaceans, birds or fish (e.g. tool use, foraging behaviour, vocal dialects). One of the best-known examples is probably the sweet potato washing first observed 50 years ago in the Japanese **macaque** population of Koshima, that has not been seen in other populations of macaques.

However, the question of animal cultures is still the subject of debate mainly based on the definition of the term 'culture'. Although the broad definition above has been adopted by many researchers working with non-humans, some psychologists restrict it to transmission only through **imitation** and teaching, and thus name it 'tradition' in animals rather than 'culture', while others prefer to use the term 'proculture'.

According to the definition of psychologists, in order to speak of culture in animals, one should demonstrate that the transmitted behaviour is not a result of any social learning process such as **local enhancement**, for instance, but stems only from an imitation process. If there is increasing experimental evidence that some animals have good imitation skills, no conclusive proof of such social learning could be drawn from observations in the wild populations demonstrating so-called cultural behaviours.

Ethnographic methods are an alternative approach that, in this case, tries to demonstrate that a behaviour is cultural by excluding all alternative explanations such as individual rather than social learning, ecology or genetics. This approach favours a categorical perspective, whereas behavioural variations can probably result from a combination of several factors. Alternatively, **translocation** of individuals between populations or of populations of animals between sites has been carried out successfully in fish to demonstrate culture.

(SL)

Further reading

Laland, K.N. (2008) Animal cultures. *Current Biology* 18, R366–R370. Available at: http://www.cell.com/current-biology/fulltext/S0960-9822(08)00235-2 (accessed 25 September 2009).

Laland, K.N. and Janik, V.M. (2006) The animal culture debate. *Trends in Ecology and Evolution* 21, 542–547.

Cynomolgus monkey

Macaca fascicularis (also known as long-tailed **macaque**, crab-eating macaque, Java monkey) is a diurnal, arboreal macaque, native to forest areas of South-east Asia and introduced into Mauritius over 300 years ago. It is primarily used in drug development and testing in the USA, Western Europe and Japan. *M. fascicularis* is now almost as widely used as the rhesus macaque, *M. mulatta*, in some areas of academic research, e.g. neuroscience. *M. cynomolgus*, *M. irus* and *M. philippensis* are names that were formerly used for this species.

(DBM)

Cytokines

Cells of the **immune system** communicate with each other, and with other organs and tissues, using cell signal proteins termed cytokines. Cytokines induce and orchestrate the specific response necessary to combat the threat (be it a pathogen or stressor) at hand. Cytokines are predominantly produced by immune cells such as macrophages, dendritic cells and T-cells of the lymphocyte population. Cells of the immune system and other tissues contain receptors for these cytokines that lead to further cell signals and cellular activation. These receptors may be on the cell surface of the recipient cell or the receptors may be soluble. Some receptors are functional and activate the recipient cell, while others are decoys and play an important role in regulation of the immune response. Cytokines can be grouped by function, including acute phase cytokines, immune response cytokines and haematopoeitic growth factors. Another very important group of cell signals are the chemokines, with the primary functions of cell recruitment and trafficking.

Acute phase cytokines are produced when a pathogen or injury is detected. The first group are inflammatory mediators and include **interleukins** (IL)-1α and -1β, IL-6 and IL-8; tumour necrosis factors (TNF)-α and -β; and macrophage chemotactic proteins (MCP)-I and -II. IL-1 has receptors on many tissues, including the **brain** and liver. Between IL-1 and TNF-α, sickness behaviours such as anorexia, fever and somnolence are induced. Interestingly, IL-1 production in response to stressors is species- and stressor-specific (restraint stress and isolation stress induce IL-1 differently). IL-6 and IL-1 stimulate production of acute phase proteins by the liver, which act to bring the animal back to **homeostasis**. Acute phase cytokines are enhanced with some stressors such as **transport**, **weaning** and food deprivation (**see: Feed restriction**).

The second group contains immune response cytokines. Cytokines involved in activation of Type I or Type II immune responses include IL-2, IL-4, IL-10 (which acts as a suppressor), IL-12 and IL-15. Supporting the production of immunoglobulins are interleukins 2, 4 and 8 by promotion of differentiation of lymphocytes and proliferation, both of which are necessary for a polyclonal antibody response. IL-2, which is produced by T-cells, is also important for B-cell maturation, supporting adaptive immune responses.

Many of the inflammatory cytokines can be destructive if not controlled rapidly. Therefore, several cytokines have a suppressive effect; among them are IL-10, which is instrumental in down-regulating macrophages, and IL-1 receptor antagonist (IL-1RA), which, along with decoy receptors, keeps the IL-1 response from running rampant. Macrophage inhibitory factor, produced by T-cells and pituitary cells, inhibits macrophage migration, stimulates macrophage activation and induces steroid resistance, thereby allowing for lymphocyte differentiation and proliferation and antibody production.

Haematopoietic growth factors constitute another group of cytokines responsible for growth and differentiation of cells. IL-6 is a member of this group, as are IL-11, granulocyte colony-stimulating factor (G-CSF), granulocyte macrophage colony-stimulating factor (GM-CSF), IL-3 and stem cell factor. These cytokines play various roles in maturation and differentiation of immune cells from the bone marrow and in the blood.

Interferons are a group of cytokines with anti-viral properties – IFN-α, IFN-β, IFN-ω and one immune interferon (IFN-γ). Interferons can arrest viral production and even remove them from the infected cells without killing the infected cell. Their role is unique and has proved crucial in therapeutic control of viral diseases in many species.

An additional group of proteins, termed chemokines, is responsible for recruitment of cells to sites of infection or inflammation. Some chemokines recruit and stimulate maturation of dendritic cells (a phagocyte responsible for initiating adaptive immunity). This is a critical process, since Langerhans and dendritic cells are an entirely different phenotype before their recruitment to immune tissues and maturation. Immature dendritic cells can phagocytize foreign particles, but are not major cytokine producers. After stimulation with chemokines, they lose their ability to phagocytize particles, but become extraordinary antigen presentation cells, stimulating the T-cells with cytokines during presentation, thus engaging the adaptive immune response.

An elaborate control system exists for many of these cytokines. Left unchecked, they would destroy healthy tissues and deplete the animal of available metabolic resources. Some receptors are produced in a soluble form to provide additional control, particularly of inflammatory cytokines such as IL-1 and TNF-α. For instance, IL-1 has redundant control. Besides having soluble receptors and an antagonist, a dummy receptor is produced with responses similar to IL-1RA.

Epithelial cells, fat cells, hepatocytes, fibroblasts, osteoblasts, chondrocytes and bone marrow cells are some of the non-immune cells that produce cytokines. Generally, they do not produce cytokines in great abundance – unlike the immune cells – but help to control the immune activity in the microenvironment. For instance, the lungs constitutively express higher IL-1RA than IL-1. Lung tissue is very fragile, so induction of cells is not beneficial under healthy conditions. The **hypothalamus** has receptors for IL-1 and TNF, but also expresses IL-1 and IL-1RA when necessary. Many immune cells contain receptors for cortisol. Cortisol at low concentrations can stimulate some immune functions, but at high concentrations (acute or chronic) it can be suppressive, resulting in increased susceptibility of the animal to disease. This has been demonstrated by reduced cytokine production under these circumstances.

In summary, cytokines are produced to communicate between immune cells and tissues and with non-immune tissues to elicit an immune response appropriate to the challenge. An intricate maze of controls and counter-controls is in place to drive the animal back to homeostasis. These mechanisms are intertwined with neurophysiological systems, metabolic processes and with the HPA axis (**hypothalamic–pituitary–adrenal axis**) through cytokine communication. Cytokines and chemokines continue to be discovered, so the complexity of the immune communication network is still not fully known. (SE)

Further reading

Janeway, C.A., Travers, P., Walport, M. and Schlomchik, M. (2001) *Immunobiology: the Immune System in Health and Disease*, 5th edn. Garland Publishing, New York.

Matarese, G. and La Cava, A. (2004) The intricate interface between immune system and metabolism. *Trends in Immunology* 25, 193–200.

Pastoret, P., Griebel, P., Bazin, H. and Govaerts, A. (1998) *Handbook of Vertebrate Immunology*. Academic Press, San Diego, California.

D

Dark cutting

Dark cutting is a meat quality defect caused by a lack of the normal acidification of meat during *rigor* development post-slaughter. Between **slaughter** and chilling of the carcass, glycogenolysis occurs in the muscle tissue, which converts glycogen, the muscle energy store, to lactic acid, thus decreasing the pH of the meat. Dark cutting is related to low muscle glycogen at the time of slaughter, which can be indicative of animals having experienced **stress** during the period prior to slaughter. (JCP)

See also: DFD meat; Lairage; Slaughter

Darwin, Charles Robert (1809–1882)

The Society for the Prevention of Cruelty to Animals was founded in London when Charles Darwin was 15 years old. Darwin was actively involved in alleviating pain and suffering among animals, which grew out of his conviction of an evolutionary continuity in emotional states between animals and people. In his autobiography, Darwin reflected on the problems of **vivisection**:

> I have all my life been a strong advocate for humanity to animals ... [but] ... know that physiology cannot possibly progress except by means of experiments on live animals [which have been anesthetized, when possible] ... it is justifiable for real investigations on physiology; but not for mere damnable and detestable curiosity.

Charles Darwin was born on 12 February 1809 near the town of Shrewsbury, England. His father, Dr Robert Darwin, was a physician, while his mother, Susannah Wedgwood, was a daughter of the founder of the famous pottery firm. As a young boy, he thrived on wandering in the country, collecting insects and plants, and one of his strongest passions was for birds. He would take notes on their activities, as well as shoot them, which was a common scientific practice at the time. In fact, Darwin was quite a marksman when it came to shooting birds, a hobby that served him well while on the voyage of the HMS *Beagle*.

Darwin's initial career trajectory took him to Edinburgh, Scotland to study medicine, but after a couple of years he changed his mind. He moved to Cambridge, England to study for the clergy but, again, decided against that career. With the support of his uncle, he dropped out of school and became a crew member of the HMS *Beagle*. This vessel sailed from England on 27 December 1831 and re-docked on 2 October 1836. Upon return from his 5-year adventure, Darwin married his cousin, Emma Wedgwood, and settled in London.

While living in London, Darwin took regular trips to the Zoological Gardens, where he made copious notes about animals and quizzed the keepers about their knowledge of animals. One of his main interests at the time was trying to evaluate the **cognition** abilities of apes in an effort to gain insights into the origins of the mental powers of people. An orang-utan named 'Jenny' had arrived at the Zoological Gardens in 1837, and she became a subject of his work. Darwin recorded her responses to music, mirrors and different food items. But Darwin was not comfortable living in the city, so in 1842 he moved with his family to the countryside in Kent, about 22 km south-east of London. Their residence was called Down House, and he remained there until his death on 19 April 1882.

Charles Darwin's primary contribution to biology was not in inventing the idea of evolution, but in providing a mechanism to account for the diversity of life on Earth, which he called 'natural selection'. His logical arguments were based upon three premises. First, variation within a species is a fundamental aspect of biology. Such reasoning ran contrary to religious fundamentalists, who considered variation as deviations from an ideal model organism. Secondly, differences among individuals in trait expression can be inherited. Although the mechanism of inheritance was not discovered until about 100 years after Darwin's proposal, the transmission of traits from generation to generation was a requirement of natural selection. Thirdly, individuals differ in their reproductive output, either by producing more offspring than others (differential fertility) or by dying sooner than others (differential mortality). The latter concept formed the basis of the idea that 'nature red in tooth and claw' drove evolution.

In essence, Darwin's idea was that changes over time were an outcome of the inheritance of traits that differed across individuals and resulted from differential reproduction. He was so convinced that his model was correct that he proposed a testable hypothesis that, if true, would completely destroy his idea: 'If it could be proved that *any part* of the structure of *any one species* had been formed for the *exclusive good of another species*, it would *annihilate my theory*, for such could not have been produced through natural selection [italics added]'.

While best known for proposing a mechanism of biological change over time, Darwin was a voracious reader, prolific writer and incessant correspondent whose insights into natural history resonate today in terms of applied animal behaviour and welfare. One of his early perceptions concerned reproductive processes in captivity. He noted that: '[m]any facts clearly show how eminently susceptible the reproductive system is to very slight changes in the surrounding conditions. Nothing is more easy than to tame an animal, and few things more difficult than to get it to breed freely under confinement, even when the male and female unite.' He puzzled over why 'weak and sickly' animals could breed in the wild, while 'long-lived and healthy' animals had problems in captivity. His astute observations remain an enigma today: what is the mechanism whereby reproductive success is hindered in captivity among apparently well-fed and healthy animals?

Darwin's concepts about evolutionary processes enabled him to provide a possible reason why animals in captivity might have reproductive failures. He wrote: 'It is by no means rare to find certain males and females which will not breed together, though both are known to be perfectly fertile with other males and females … the cause lies in an innate sexual incompatibility of the pair'. Mate choice was a pillar of sexual selection that could have profound implications, not only for variation in reproductive output among wild animals but for reproductive success in captive animals.

In the absence of a clear understanding of genetics, Darwin nevertheless thought that inheritance of traits and genetic relatedness influenced reproduction. One topic he addressed still haunts zoological institutions, i.e. kinship and mating success. He concluded that:

> A cross between individuals of the same species, which differ to a certain extent, gives vigour and fertility to the offspring; and that close interbreeding continued during several generations between the nearest relations, if these be kept under the same conditions of life, almost always leads to decreased size, weakness, or sterility. Devising optimal breeding programmes for captive animals that incorporates genetic information, as well as an understanding of partner compatibility and social adeptness, remains a challenge.

Perhaps Darwin's major foray into setting the framework for animal welfare issues was his perspective on animal **emotions**. For example, he wrote: 'Animals which live in society often call to each other when separated, and evidently feel much joy at meeting', later commenting that 'Young chimpanzees make a kind of barking noise, when pleased by the return of any one to whom they are attached'. About a century after he wrote these words, field studies of the natural behaviour of wild chimpanzees revealed that reunions are accompanied by a cacophony of calls, gestures, expressions and physical contact. The extent to which our shared ancestry with chimpanzees indicates continuity in emotional states remains a contentious issue that reverberates today within the context of **animal rights** and **ethics**.

Charles Darwin was a scientific navigator mapping novel insights into natural history. Although best known for developing ideas about mechanisms of evolutionary change, his prescience was truly amazing: he sought to alleviate pain and suffering in animals, but recognized that animal experimentation was critical for scientific progress; he systematically recorded animal behaviour among both domestic and captive exotic animals in order to obtain insights into biological principles transcending taxa; he recognized that reproductive deficiencies in captivity could arise from physiological as well as behavioural problems; he understood that **inbreeding** could have negative consequences for reproductive success and that mongrels were often healthier than pure-breeds; he struggled with the dilemma of whether the evolutionary continuity between people and apes was based on quantitative or qualitative differences; and he was a brilliant scientist who floated a plethora of ideas, while recognizing that:

> False facts are highly injurious to the progress of science … but false views, if supported by some evidence, do little harm, for every one takes a salutary pleasure in proving their falseness; and when this is done, one path towards error is closed and the road to truth is often at the same time opened.

On 26 April 1882, Charles Robert Darwin was laid to rest in Westminster Abbey, London. (FBB)

See also: Evolution

References and further reading

Browne, J. (1995) *Charles Darwin: Voyaging*. Princeton University Press, Princeton, New Jersey.

Browne, J. (2002) *Charles Darwin: the Power of Place*. Princeton University Press, Princeton, New Jersey.

Darwin, C. (1859) *On the Origin of Species*. John Murray, London [republished by The Modern Library, New York, 1958].

Darwin, C. (1871) *The Descent of Man and Selection in Relation to Sex*. John Murray, London [republished by The Modern Library, New York, 1958].

Keynes, R. (2001) *Annie's Box*. The Fourth Estate, London.

Death

Death is the extinction or cessation of life, and may be used to describe the state of an individual or group of related individuals, e.g 'death of the dodo'. In the latter sense it is synonymous with **extinction**.

The determination of the point of death in animals is confounded by difficulty in defining the living state, especially in simpler organisms. Within a medical setting death is usually defined in relation to the absence of electrical activity within the **brain**, although **consciousness** may be lost before this, and brain activity can be artificially suppressed temporarily with drugs or as a consequence of hypoxia or hypothermia. At a practical level, the assessment of brain activity is rarely made to determine death in animals, and so the absence of detectable cardiac and respiratory activity is used to determine the point of death. Destruction of the cardiorespiratory centres is often seen as a priority for a rapid death.

For many, death is not a **welfare** issue, but rather the manner of death is. Thus during the **killing** of an animal, e.g. during **slaughter**, welfare may be protected by rendering the animal insensible as rapidly as possible, for example by stunning. None the less, some argue that welfare relates to both the quantity and quality of life, positing that an animal whose life expectancy is artificially reduced has less welfare within it and thus a reduced level of welfare.

Ageing ultimately leads to death, but the evolution of this process and the advantage to fitness conferred by programmed cell death are not fully understood, as it is not a universal process in the animal kingdom (e.g. hydra). (DSM)

Deceit behaviour

Deceit behaviour is of interest in applied animal behaviour and animal welfare science, mainly because of what it may tell us about the animal's mind: does it show that animals can have intentions? Do deceiving animals intentionally manipulate the *behaviour* of others (first-order intentional deception)? Or do they even understand, and knowingly manipulate, the *beliefs and thoughts* of others as do deceiving humans (second-order intentional deception)? Or is deceit behaviour deceptive merely in its effect, not in its intention?

An animal acts 'deceitfully' or 'deceptively' if its behaviour leads another into responding against its interest. The actor benefits from its behaviour, the deceived recipient pays a price. This is demonstrated by the deceptive flashing behaviour of some female fireflies. Female fireflies generally flash to attract conspecific male mates. Females of the genus *Photuris*, however, use two different flashing patterns: one attracts males of their own species, which they will then mate with; the other mimics the flashing pattern of *Photinus* females and hence attracts misguided *Photinus* males, which the *Photuris* females will then eat.

Deceptive behaviour occurs in many kinds of competitive interactions between or within species. Many animal displays are classic models of such interactive behaviours. They have provided much information about the conditions under which dishonest displays, as examples of deceptive behaviour, may evolve. In an exchange of aggressive male threat displays, for example, both participants would benefit from misleading their opponent about their fighting potential by inflating their displays. However, they would also gain from being able to call their opponent's bluff, that is from being able to accurately assess their opponent's actual fighting potential.

Selection should thus favour individuals that respond only to honest displays. Honest displays are those that can only be produced by an animal that is displaying the truth. For example, when the ability to display is directly related to **fitness**, displays are by necessity honest and deception difficult. A case in point is the roaring display of red deer (*Cervus elaphus*). Roaring uses the same muscles of the thorax as are used in fighting. Roaring is thus an honest display of fighting ability as it cannot be faked by a stag with inferior muscle power. Because the responses of the receivers determine how displays evolve, and receivers are selected for their ability to respond only to honest displays, most displays in competitive signalling systems are honest. None the less, deceptive displays can occur even in evolutionarily stable systems where animals make assessment errors and the deceptions are infrequent enough for receivers to encounter honest displays, at least on average (Johnstone, 1997).

The idea that deceit behaviour may be intentional comes from behavioural observations in the vertebrate taxa, where it tends to be more complex, variable and flexibly deployed than in invertebrates like *Photuris*. A well-studied example is the injury-feigning behaviour of ground-nesting plovers (*Charadrius* spp.). When a fox approaches the nest or young, the plover will start to move away from the fox awkwardly, arching its wings and dragging one or both behind as if broken. The fox, distracted from the nest, may follow the plover, which will maintain the display until the fox is at a safe distance from the nest. The bird will then suddenly take off and fly back to the nest. Field tests have shown that plovers display more vigorously towards 'dangerous' intruders than towards 'non-threatening' intruders; they will also adjust the direction of their movement depending on the relative positions of the nest, the intruder and themselves so that the intruders are never led closer to the nest.

Some argue that such behavioural flexibility is possible only if the animal has a goal or purpose in mind, which continuously guides its behaviour. The plover thus *intends* to lead the fox away from its nest (Ristau, 1991). However, a problem with interpreting deceptive behaviour as intentional purely on the basis of behavioural flexibility is the mounting evidence that animals can perform highly flexible behaviours on the basis of complex sets of previously learned contingencies.

The most complex and flexible patterns of deceptive animal behaviour have been reported in primates (Byrne and Whiten, 1988). Many primates live in highly structured and long-lived social groups, in which great potential benefits can be derived from manipulating and exploiting group members. This has been proposed as a major factor driving the evolution of intelligence in primates (Humphrey, 1976). Primate deceit behaviour covers a wide range of strategies, such as signalling the presence of non-existent predators (e.g. vervet monkeys, *Cercopithecus aethiops*), using **conspecifics** as 'social tools', thereby misleading a third party (e.g. savannah baboons, *Papio cynocephalus*), and misinforming dominants about the location of desirable food items when foraging in groups (e.g. mangabey monkeys, *Cercocebus torquatus*; chimpanzees, *Pan troglodytes*).

Careful analysis of the data on primate deception shows that, in almost all cases, the possibility cannot be excluded that the primates might have learned to manipulate the behaviour of others to their own ends, without intentionally manipulating the beliefs of others or understanding the mechanism of deception. However, some cases of primate deceit behaviour are hard to explain as learned without understanding, and a strong case has been made for apes intentionally deceiving by manipulating the beliefs of others (Byrne, 1995). Evidence of such second-order intentional deception would show that apes could attribute mental states to others, and had some **theory of mind** ability. Because possessing a theory of mind has been suggested as one of the hallmarks of human-like consciousness, evidence of intentional deception in a given species would lead us to fundamentally reconsider our attitudes towards it.

Research on the mechanisms by which deceptive strategies might arise has recently been expanded to laboratory studies using farm animals and corvids, where learning explanations can be more easily controlled than in field experiments of primates. Results suggest that, similarly to mangabey monkeys and chimpanzees, ravens (*Corvus corax*) and domestic pigs (*Sus scrofa*) develop potentially deceptive behavioural strategies to avoid exploitation by dominants when foraging together (Bugnyar and Kotrschal, 2002; Held *et al.*, 2002). (SDEH)

References and further reading

Bugnyar, T. and Kotrschal, K. (2002) Observational learning and the raiding of food caches in ravens, *Corvus corax*: is it 'tactical' deception? *Animal Behaviour* 64, 185–195.

Byrne, R.W. (1995) *The Thinking Ape*. Oxford University Press, Oxford, UK.

Byrne, R.W. and Whiten, A. (eds) (1988) *Machiavellian Intelligence: Social Expertise and the Evolution of Intellect in Monkeys, Apes, and Humans*. Oxford Science Publications, Oxford, UK.

Held, S., Mendl, M., Devereux, C. and Byrne, R.W. (2002) Foraging pigs alter their behaviour in response to exploitation. *Animal Behaviour* 64, 157–166.

Humphrey, N.K. (1976) The social function of intellect. In: Bateson, P.P.G. and Hinde, R.A. (eds) *Growing Points in Ethology*. Cambridge University Press, Cambridge, UK, pp. 303–317.

Johnstone, R.A. (1997) The evolution of animal signals. In: Krebs, J.R. and Davies, N.B. (eds) *Behavioural Ecology: an Evolutionary Approach*, 4th edn. Blackwell Science, Oxford, UK, pp. 155–178.

Ristau, C.A. (1991) Aspects of the cognitive ethology of an injury-feigning bird, the piping plover. In: Ristau, C.A. (ed.) *Cognitive Ethology*. Lawrence Erlbaum Associates, Hillsdale, New Jersey, pp. 91–126.

Decision making

All animals that live in variable environments need to make decisions about what behaviours to perform, where and when. Normally, animals are able to pursue only one behaviour at a time, even if they have competing strong tendencies to do more than one thing. For example, after a cold winter's night a horse may be motivated to feed, drink and sleep, but how does it decide which to do first? Decision making is the process by which all of the physiological and sensory information available is integrated with an assessment of current environmental threats and opportunities, to permit the animal to choose its next course of action.

The biggest theoretical advance in understanding animal decision making was the view taken by McFarland (1993) that animals evaluate every aspect of their state (although not necessarily by conscious processing of this information) and assess the costs of each course of action using a common cost-function or 'currency'. If the horse in our example chose first to eat hay, it would benefit from reduced hunger, but would incur the cost of increased thirst. Only by evaluating these costs and benefits on the same scale can the best decision be made. McFarland (1993) argued that evolution has produced animals that make decisions that tend to maximize their **fitness** (which could, for example, be measured in terms of lifetime reproductive success).

Houston and McNamara (1999) develop these arguments, demonstrating that fitness can be treated as a single unifying currency that encompasses all other biologically relevant currencies. From a practical perspective, however, immediately measurable currencies such as energy or time are evaluated (**see: Economics of behaviour; Motivation**), and usually suffice as a surrogate measure of fitness. By using fitness as a single unifying currency, the value of seemingly disparate activities can be compared. In this way, if, at a given moment in time, an animal has a choice between grazing, drinking or engaging in social activity, we can identify which of these behaviours should be conducted by comparing the changes each of them would bring about in the animal's fitness.

If animals really are evaluating all possible courses of action on the basis of a common currency, then their decisions should be self-consistent and rational. Thus, if an animal prefers A to B, and B to C it should also prefer A to C. There is growing interest in evaluating these features of animal choices, as violations can highlight the currencies that are actually used by animals to evaluate alternative options (Bateson, 2004; Houston *et al.*, 2007).

Of course, decision making involves the **brain**, and the economic perspective taken above does not involve itself with the actual neural processes involved. However, recent work examining human patients with specific types of damage to the prefrontal cortex is providing some new insights into the neuroscience of decision making. These patients make very poor short- and long-term decisions that are not generally in their own best interests, although their **intelligence** is unaffected. This work has resulted in a growing appreciation of the role of brain areas such as the amygdala and the ventromedial prefrontal cortex in decision making, in how decisions are linked with emotional processes and how costs and benefits are encoded by neural systems (Bechara and Damasio, 2005).

Barnard (2007) argued forcefully that the decision-making propensities of animals were of crucial importance in animal welfare. He thought that observing the choices that animals made was the only way to make sense of other welfare indicators that on their own 'provide only equivocal evidence about welfare'. Interest in how animal decisions might be related to animal welfare can be traced back to early studies examining the environmental choices of farm animals (**see: Aversion; Preference**) and to more recent work examining demand functions. Finding that decision making is a strongly emotional process only increases its animal welfare relevance.

(CJN)

References

Barnard, C. (2007) Ethical regulation and animal science: why animal behaviour is special. *Animal Behaviour* 74, 5–13.

Bateson, M. (2004) Mechanisms of decision making and the interpretation of choice tests. In: *UFAW International Symposium on Science in the Service of Animal Welfare*, 2–4 April 2003, Edinburgh, UK, pp. S115–S120.

Bechara, A. and Damasio, A.R. (2005) The somatic marker hypothesis: a neural theory of economic decision. *Games and Economic Behaviour* 52, 336–372.

Houston, A.I. and McNamara, J.M. (1999) *Models of Adaptive Behaviour: an Approach Based on State*. Cambridge University Press, Cambridge, UK.

Houston, A.I., McNamara, J.M. and Steer, M.D. (2007) Do we expect natural selection to produce rational behaviour? *Philosophical Transactions of the Royal Society of London B* 362, 1531–1543.

McFarland, D. (1993) *Animal Behaviour*. Longman, Harlow, UK, Chapters 24 and 25.

Deer

A ruminant, ungulate and member of the cervid family. The males of most deer species characteristically lose their antlers and grow them anew each year. Historically the term 'deer' has included other wild animals outside the cervid family, so some care is needed when considering old documents. Terminology for maturity and sex of animals is not always used in the literature consistently. Mature males (of 5 years), usually red deer, are termed 'stags', mature females (over 3 years), usually red deer, are called 'hinds', while males of smaller deer species are usually 'bucks' and the females 'does'; newborns are 'calves'.

The deer has evolved as a forest animal, and has since prehistory been an animal preyed upon by humans. The deer is, however, not a domesticated animal, or at least its domestication has been so recent as to be unlikely to have affected its behavioural and other needs for good welfare, in comparison with traditional domesticated species. Its long flight distance, agility, speed and social behaviour have made it unsuitable for **domestication**. It is an animal that is captive, however: in farms, parkland, **zoos** and research institutes (see Fig. D.1).

In Europe, and indeed worldwide, farmed deer are most frequently red deer (*Cervus elaphus*), but there are also significant numbers of farmed reindeer (*Rangifer tarandus*) and fallow deer (*Dama dama*), and smaller numbers of others: sika (*Cervus nippon*), shital (*Axis axis*), wapiti (*Cervus canadensis*) and white-tailed deer (*Odocoileus virginansus*). Perhaps as a result of this, scientific research has tended to be most commonly reported on the red deer. Most deer species are social animals, forming single-sex groups. In the male groups, levels of **testosterone** and aggression increase prior to the rutting season, in the early autumn. These aggressive sparring contests prepare the males for (in the case of red deer) the defence of territories – these being areas occupied by the hinds. Fights between stags are common. A resident stag defends his own territory and harem against invaders, but also frequently reasserts the border with his neighbours and tries to increase the numbers in his harem. This is expensive energetically, and the males can lose up to 30% of their bodyweight during the rut, and up to 80% of the accumulated reserves of fat. This loss is through a combination of the increased energy expenditure and a reduction in the time available to spend feeding.

Calving is usually in the late spring, and immediately prior to this the females will seek a hidden, isolated site. Following **parturition** the calf will normally stand within 30 min, and start to suckle within 40 min. Within a week the calf will be able to run with the herd. Calves are totally reliant on milk as a food source for the first 30 days of life. Suckling usually continues until the onset of winter, and may continue until the following summer in non-gestating females, but the frequency and duration of suckling bouts decreases over the lactation period. Common diseases of deer include tuberculosis, malignant catarrhal fever and yersiniosis; the most important parasite, both clinically and because of its effect on production parameters, is the lungworm *Dictyocaulus viviparus*.

Farming of deer is usually extensive, involving grazing with supplementary feed as conditions require, though intensive

Fig. D.1. Captive deer in a research facility (image courtesy of M. Ots).

systems for young stock are becoming more frequent. Their meat is a high-value product, with lower levels of fat and a higher meat:bone ratio than many other commercial meats, and their production is perceived by the public to be good in terms of animal welfare. There may be overwintering of animals, especially in cold, wet climates to protect the animals, to preserve the quality of the pasture and to better control management. This can lead to problems of **pica** and aggression, leading to increased skin damage, though the first can be ameliorated by providing browse and the second by grouping according to size. Calves are usually housed over the winter, as they have less fat stored to enable them to survive than the adults.

It is common farm management practice to remove the antlers before the rutting season. The antlers begin to develop at puberty, and each year in the spring. They have a covering, the velvet, that is vasculated and innervated. Harvesting of this velvet, for the preparation of oriental medicines, takes place in some systems (but is illegal in some countries) and is presumably painful while in this innervated state. Anaesthetics are recommended for this procedure, although the choice of short-acting drugs can leave some animals unprotected. By the end of September in the northern hemisphere, growth of the antlers ceases, the blood vessels constrict and the velvet peels away. The deer assist this process by rubbing the antlers against rough bark and browsing. The exposed hard antlers are retained during the rut, but drop off naturally in the early spring. Antler growth can be prevented by disbudding or **castration**. Reproduction is managed on-farm by introducing single stags to groups of hinds, and it is quite common for farms to operate with single stags. Timing of mating is arranged to ensure calving is in spring, as this gives the calves more time to put on weight during the time of highest forage availability, and it also reduces the likelihood of bacterial infection. Early calving can be achieved by improving the nutrition of the females; heavier females have higher conception rates and reach puberty earlier. Overfeeding should, however, be avoided.

Deer are a popular **game animal**, mainly through stalking, tracking and shooting, but also by **hunting** with dogs. It has been argued that the last method was the most stress-free method of culling red deer (to prevent populations exceeding the carrying capacity of a habitat), but this has been disproved, and there is good evidence that the best method of culling or slaughter from a welfare perspective is stalking. Hunting with bows is a popular recreation activity in some countries (including the USA), though is illegal in others (including the UK). The **welfare** problems caused by inaccurate shooting by bow, leaving the deer to escape wounded, have been reported, including pain, ultimately lethal wounding, increased likelihood of predation and septicaemia. Such cases may be as high as nearly 50% of those shot, greater than those reported for rifle shooting. This may be a particular problem for less well-regarded trophy animals such as females, or those with less spectacular antlers, which may not be thought worth following and despatching. Shot animals may survive for days.

On-farm **slaughter** of animals, whether by shooting or in a mobile slaughter facility, is much preferred to their transport to abattoirs that might be some distance away, as facilities suitable for their slaughter are infrequent and not necessarily near to the site of production. **Transport** of deer is a welfare problem, as they are not as used to human contact as more domesticated species. **Stress** is induced not just from the journey but also through handling, loading and unloading. Capture myopathy is a common outcome during the transport of deer and during poor handling, indicating damage and presumably pain to the animal, but also results in an impairment in the quality of the meat. There is evidence that this can be ameliorated by the injection of drugs, though administration is itself a problem as a stress-inducing event.

The stress resulting from handling can be reduced by handling in low-light conditions, although this might compromise the safety of the animals and their handlers. It has been suggested that portable handling facilities, designed specifically for the purpose of deer handling, would improve the efficiency of handling procedures and the welfare of both the deer and the handlers. Deer can become aggressive towards handlers, particularly stags during the rutting season, hinds during calving and animals that have been singled out from their group. Young animals and those unfamiliar with either the handler or the handling procedure are more likely to panic. Signs of aggression include flattened-back ears, grinding of teeth, a fixed stare and snorting.

Deer commonly feed on mixed grazed/browsed fodder, and their digestion is similar to other ruminants. However, their basal metabolic requirement is high and they need more energy proportional to their body weight than, for example, sheep. However, a ration high in concentrates, particularly processed grains, can cause acidosis, while high levels of fibrous material can cause ruminal and omasal impaction. Dominant animals are particularly at risk of the former because of their successful competition for access to feed. The feeding pattern of deer is seasonal, with increased voluntary feed intakes leading to greater growth and extensive fat deposition during the spring and summer, supplying them with resources to survive winter periods when feed is in short supply.

Deer can compensate for poor-quality forage by increasing intakes. The stocking rate can have a significant effect on intakes from grazing, with high densities in areas of limited good pasture being associated with reduced body weights compromising their **health** and welfare through the winter. This is commonly used as an argument for the **culling** of wild deer populations, as well as more managed populations in parkland. It is recommended to supply supplementary feed during the winter, with wet winters being particularly deleterious in terms of body weight loss.

It is common in Europe and North America for wild deer to be offered supplementary feed in the winter, to ensure a viable population for hunting and to reduce damage to the environment through overbrowsing and grazing of the available forage. There are several problems with this, not least of which is that the deer can become reliant on the supplementary feeding, reduce their normal foraging behaviour (on occasion actually to zero) and therefore actually lose weight and condition. Other problems include unequal access to feed, with dominant stags benefiting most and subordinate animals least, and increases in the risk of infections and parasite load from abnormal concentrations of animals in a small space at the feeding stations.

The consideration of the welfare of deer is not restricted only to that arising from farming systems or hunting. Wild deer have been shown to flee from areas used for leisure purposes, such as orienteering, and the energetic costs of this flight might be considerable. The disturbance by the public during visits to parks housing deer is thought to be problematic, and deer show little habituation to such disturbance, although there is conflicting evidence regarding this. Increasing time spent in vigilance behaviour has been found to be at the expense of feeding in the behavioural budget of deer, so disturbance may have an impact on production parameters. Occasions of high mortality of both red and fallow deer in two parks in the UK have been attributed to cumulative stress in the population from visitors (which includes not only the effects of humans alone but also, and more significantly, those of humans with their dogs).

In addition to stress there are direct effects on injury and mortality from disturbance, running into fences, damage from attack by dogs and accidents with motor vehicles. Visitor disturbance is thought to be the reason behind high rates of breeding problems observed in deer. This may result partly from the disturbance of stags, affecting their sexual activity, but is particularly reflected in high rates of calf mortality. This is thought to be due to the abandonment of calves by their mothers following disturbance causing a reduction in suckling behaviour and, perhaps more importantly, reducing their opportunity to form a bond with their calves. This is thought to be at least partly because of the difficulty in finding suitable sites in the restricted areas of parks to conceal calves in the first day of life, which allows bonding and suckling to take place uninterrupted and in security.

Pacing along boundary fences by deer at calving time has been observed to increase, and is thought to be either an expression of effort to find a suitable site or frustration behaviour at not being able to do so, or even to relieve the discomfort of approaching parturition (although overall activity does increase at this time, this is surpassed by pacing itself). This pacing behaviour is more intense when humans are visible and in close proximity. Advantageous ground cover that could offer hinds the choice of such a site, isolated from **conspecifics**, includes areas of tall plants and heather, though hinds did not choose woodland when offered the opportunity in one study. It was presumed that this was because the deer preferred cover that gave them a view of the surrounding territory. Such a consideration is also, of course, appropriate in other deer-keeping contexts, such as under farming, hunting or other captive management systems. In deer farms for example, the deer typically graze grassland without the provision of such cover.

Assessment of the health and welfare of deer in the wild in particular, but also in managed populations, should be undertaken as non-invasively as possible. Monitoring methods, such as observation of normal and abnormal behaviour, and the analysis of faeces, hair and saliva remaining on grazed pasture rather than sampling of blood, for example, are preferable. Wild non-native deer, such as the sika, muntjac (*Muntiacus reevesi*) and Chinese water deer (*Hydropotes inermis*) in the UK, may have their own welfare problems, including the large, and increasing, numbers involved in traffic accidents. These deer populations are the result of escapes and deliberate releases from animal collections, mostly from the last century (**see: Exotic species invasion**). Despite their apparent success in an alien environment, they may face environmental stresses, including exposure to infections, to which they are not well adapted. They may also be treated with less sympathy by the human population than native deer, particularly when causing damage while trespassing in gardens. They are considered responsible for the destruction of plants of conservation value and cultivated crops, particularly the tops of root crops and saplings, through grazing and trampling.

(DA)

Further reading

Ewbank, R., Kim-Madslien, F. and Hart, C.B. (eds) (1999) *Management and Welfare of Farm Animals*, 4th edn. UFAW, Amersham, UK.

The DEFRA Code of Recommendations for the Welfare of Livestock – Deer. Available at: http://www.defra.gov.uk/foodfarm/farmanimal/other/index.htm (accessed 25 September 2009).

Defecation

Defecation is the act of eliminating solid or semi-solid faecal waste through the anus. Material that has been eaten but cannot be digested is stored temporarily in the rectum. The need to eliminate this material occurs when stretch receptors in the wall of the rectum are stimulated as a consequence of an increase in the volume of material. Relaxation of the anal sphincters allows the material to exit the body. This usually occurs daily and may be under voluntary or involuntary control. Loss of control may occur as a consequence of psychological (e.g. fear response) or physiological stress (e.g. disease). Defecatory behaviour and the quality of faeces can be used in the assessment of animal **welfare**. (KT)

See also: Elimination behaviour

Dehorning

Livestock are dehorned to minimize injury to other stock and to stockpersons, and to reduce carcass and hide damage. Dehorning or disbudding (the term usually used when the procedure is conducted on very young animals) is mainly conducted on cattle, although it may also be performed on goats. If conducted correctly, dehorning means that the animal will not develop horns. Dehorning is different from 'horn tipping' or amputation of part of the hard horn, a procedure that is sometimes conducted on adult animals. Horn tipping is much less of a welfare issue than dehorning because it does not normally lead to damage of innervated tissue, although animals must still be restrained for the procedure to be conducted safely (**see: Handling**).

Mature male deer may have their antlers removed for ease of handling prior to transportation or to harvest velvet, but the procedure is different from dehorning in that the antlers will normally grow again the following year. Velvet antler removal is the surgical amputation (by a saw or embryotomy wire) of living, vascularized and innervated tissue, and is banned in some countries. Antler growth can be prevented by the application of a rubber elastrator ring to the developing pedicle, although the technique appears more reliable in fallow than red deer. Behavioural changes indicate that the procedure may cause some discomfort and/or pain. Antler growth can also be prevented by surgical removal of the pedicle with a disbudding iron, as used on cattle, although there are reports that it is not always effective.

Horns grow from a specialized area of the skin (the corium or 'horn bud') that unites with the cornual process of the skull. The corium produces keratin that forms the outer shell of the horn. The age at which the corium and cornual process join varies with breed/genotype, with reports of it occurring at 7 days of age and at 2 months of age. Evidently, destruction of the corium when it is still 'free-floating' is much simpler, results in a smaller wound and is, therefore, likely to cause less **pain** and **stress** to the animal (see Fig. D.2). Once the corium attaches to the skull, the horn core becomes a bony extension of the skull and the hollow centre of the core opens into the frontal sinuses of the skull. Thus, dehorning at this stage can result in open sinuses, making animals vulnerable to sinus infection and **fly strike** (flies laying eggs in the wound, which leads to larvae living in and feeding on the tissue).

The age at which dehorning/disbudding can be conducted, whether it can be performed by the farmer or a veterinarian and the requirement for **anaesthesia/analgesia** are all legally dependent upon the species and country in which it is performed. Given the way in which the horn develops, dehorning should be conducted at as young an age as possible. Not only does this minimize the extent of tissue damage and, hence, lead to less blood loss and more rapid healing, but also the animals are more easily handled and restrained, which results in reduced stress in them. Animals should be dehorned at a time of year that avoids wet weather and high fly/insect numbers, as this reduces irritation and potential for infection. If flies are a problem then repellents can be applied around the wound, but not on it. High standards of hygiene are also important in avoiding **sepsis**.

There are several dehorning methods that are commonly used, and which one is used tends to depend upon the extent

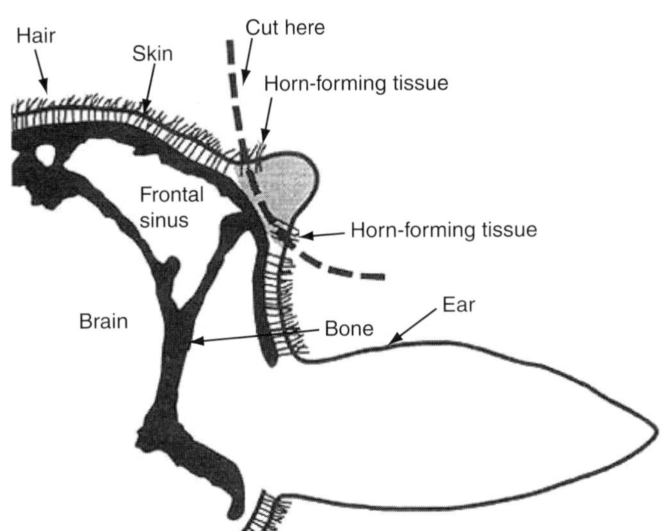

Fig. D.2. Diagram showing horn bud in relation to the skull and frontal sinuses, and the requirement to remove horn-forming tissue to prevent horn regrowth (reproduced with the permission of New South Wales Department of Primary Industries, Australia).

to which the horn is developed. All methods of dehorning involve the firm restraint of the animals and produce varying degrees of tissue damage. At a very young age (less than 7 days) caustic chemicals applied to the horn bud of calves will generally destroy it, but there is the risk that the chemicals can get into the eyes, especially if applied during wet weather. Hot iron dehorning/disbudding (also called cautery or thermal dehorning) is best suited for calves up to about 8 weeks of age (see Fig. D.3). A circular iron is heated by fire, liquid petroleum gas, butane or electricity until it is red hot and applied firmly over the horn bud. The iron is rolled over the horn bud several times to ensure that all of the horn-forming tissue around the bud is destroyed. In due course the horn bud drops off. This and the caustic chemical method have the advantages that there is no blood loss and no wound, and so can be performed at any time of the year.

The other commonly used techniques involve amputation of the horn or horn bud and, in all cases, to prevent regrowth a ring of skin at least 1 cm wide must be removed with the horn bud. For young calves (up to about 4 months of age) a curved dehorning knife can be used to slice off the horn level with the skull. For older calves (up to 12 months of age) the most commonly used devices are cup (see Fig. D.4) or scoop dehorners (see Fig. D.5). Both devices have cutting blades that amputate the horn bud adjacent to the skin and also remove some of the underlying tissue and bone.

Excessive bleeding is a major concern when older, larger cattle are dehorned and there are reports of deaths from exsanguination. Research on cattle, investigating behavioural and plasma cortisol changes, indicates that cautery dehorning and dehorning by amputation are painful and stressful procedures, and that the pain and stress can be alleviated by the use of local anaesthetics and/or systemic analgesics. Dehorning 5–6-month-old calves by amputation without anaesthesia or analgesia causes plasma cortisol levels to peak at 0.5 μg/dl at 1 h post-amputation, then to decline rapidly to plateau values maintained between 1.5 and 3 h post-amputation, returning to pre-amputation levels by about 6 h post-dehorning.

In cattle, horn development is typically classified into three phenotypic groups: (i) horned animals, with fully developed horns attached to the frontal bones, although horn size and shape may vary considerably; (ii) polled or hornless animals, in which the corneus appendages are totally absent; and (iii) scurred animals in which there are incomplete horns that are not attached to the frontal bones.

Thus, there is an alternative to dehorning; the selective breeding of polled animals. Although the exact gene or genes for horned, scurred and polled status have not been identified, it has been determined that the polled gene(s) is dominant/epistatic to that for horned. A third locus, the African horn

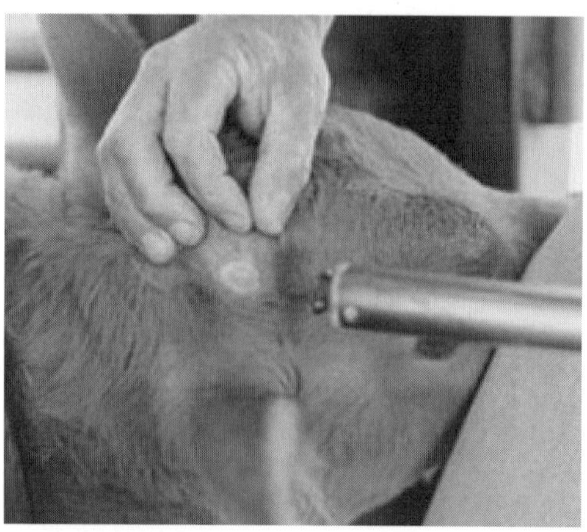

Fig. D.3. Hot iron dehorning (reproduced with the permission of New South Wales Department of Primary Industries, Australia).

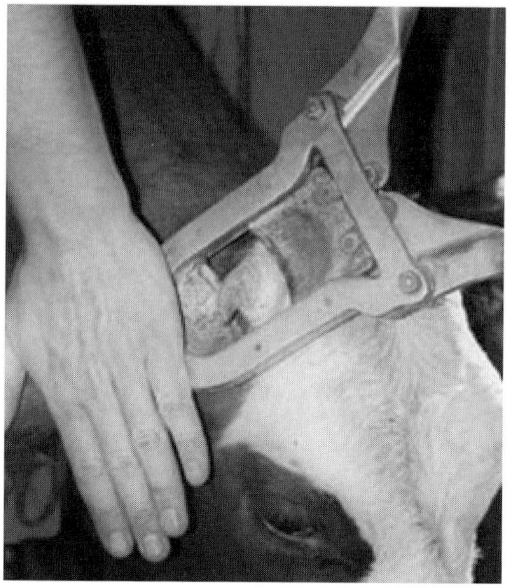

Fig. D.4. Dehorning using cup dehorners (reproduced with the permission of New South Wales Department of Primary Industries, Australia).

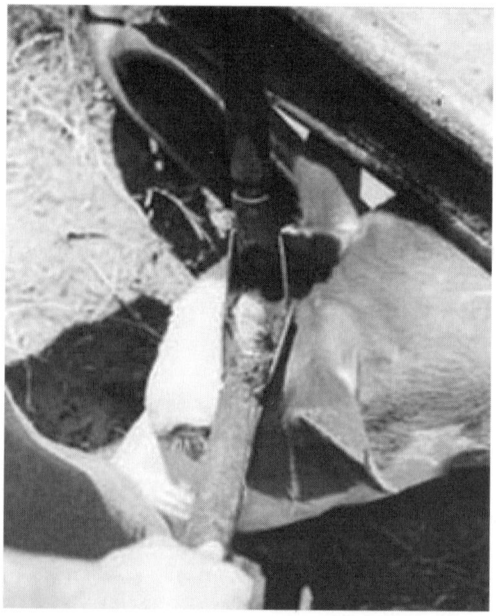

Fig. D.5. Dehorning using scoop dehorners (reproduced with the permission of New South Wales Department of Primary Industries, Australia).

locus, has been postulated, which is epistatic to the polled locus. It is possible that the scurred and African horn factors are different alleles at the same locus. Linked genetic markers for the poll gene(s) have been identified, tested and shown to be accurate (about 95%) in some *Bos taurus* breeds. However, the inheritance of the genes is still poorly understood in *Bos indicus* cattle, and research is being conducted to discover genetic markers related to horn status. There is evidence indicating that there are no differences in production traits between polled and horned cattle in a number of breeds and crossbreeds. (JCP)

See also: Fly strike; Free-range animals; Handling

Denerving

Denerving or local **anaesthesia** is applied for a variety of reasons, some more supportive of **welfare** than others. The primary reason for local anaesthesia is to reduce the pain of minor surgical procedures, such as **dehorning** or wound suturing, or to produce regional blocks, for example prior to a rumenotomy. Another use is in diagnosis. Blocking a nerve such as the posterior digital nerve in a lame horse will allow the veterinarian to determine the location of the lesion that is producing **lameness**. Similarly temporary blocking of the infraorbital nerve with lidocaine can be used to determine whether neurectomy could possibly be useful in equine **head shaking**. Head shaking is a relatively common problem of horses and can render the horse unrideable. It is usually a response to pain or discomfort somewhere in the head or upper cervical region. Another trigger can be bright sunshine, i.e. a photic reaction. The antihistaminic cyproheptadine has been more successful than neurectomy as a treatment of head shakers.

Chemical denerving of the caudal nerves, usually with alcohol, is used to keep the tail limp for Western pleasure classes. In such classes the ideal horse is one that can be used for recreation and not working cattle, and therefore should not shy and should be very obedient. When cued to jog (trot) or to lope (canter) it should move very slowly, but it should stop abruptly on command. It should keep its head below the level of the withers. The horse is judged on quietness, and if it lashes (swishes) its tail it will be penalized. Fortunately, electromyography can be used to determine whether the tail is innervated or not. Chemical restraint has also been used to produce both a lowered head and an inactive tail. Blood samples to be tested for drugs are taken from horses in major shows, and nowadays even long-acting phenothiazine tranquillizers, such as fluphenazine, can be detected (**see: Neurectomy** for surgical denervation). (KAH)

Deontology

Most ethical theories are said to fall into either consequential or deontological standpoints, although in practice a combination of both is usually operative. Utilitarian theories are consequentialist insofar as good actions are assessed according to the beneficial consequences that accrue from them. In general, utilitarian theories are primarily concerned with raising the quality of lives – a goal that is shared by every ethically informed animal owner, veterinarian, doctor, nurse and biomedical researcher. Yet while considerations of utility provide important guidelines in practical ethics, many philosophers and ethicists have argued that **utilitarianism** restricts attention to the maximization of benefit and cannot provide a complete account of moral philosophy.

One of the key alternatives to utilitarianism is deontology, which is derived from the Greek term *deon* meaning 'absolute duty'. Some deontologists speak of an absolute duty to follow divine commands, whereas others attribute the source of duty to natural law. Deontologists speak of duties to be fulfilled regardless of the consequences, and these will include telling the truth, keeping promises, respecting the autonomy of others, never treating people as means to others' goals and treating others as you would want to be treated yourself. Problems in medical practice are said to occur when a conflict of duty arises. Consider the conflicting imperatives to: (i) do everything possible to combat illness and preserve life; and (ii) show respect for a patient's right to self-determination, which may include an express wish to forgo certain forms of life-sustaining therapy. If both duties are to be accepted as absolute then the conflict cannot be resolved.

Deontology is largely associated with the rule formalism of **Immanuel Kant** (1724–1804), who maintained that appeals to utility cannot account for some of our most strongly held moral beliefs. Whereas the utilitarians, **Jeremy Bentham** (1748–1832) and John Stuart Mill (1806–1873), adopted a consequentialist theory, seeking the maximization of happiness (which they equated with 'good'), Kant insisted that consequences do not make an action right or wrong. Rather, it is the principle upon which the agent acts that is the morally significant factor. His emphasis on the principles that underpin moral actions has some obvious advantages. Very often we carry out an action with the best intentions, only to see the outcome frustrated by unintended circumstances because, once initiated, the outcome of an action may be influenced by a network of causes and effects beyond our control. A veterinarian may administer a drug with the best intentions and yet, because of some unforeseen circumstances – e.g. an undiscovered side effect – the animal may suffer greatly or die. The consequences are bad, but unintended. Do we now conclude that the veterinarian acted immorally?

Consider a case from the other side: a careless and irresponsible veterinarian misdiagnoses an illness or prescribes the wrong medication, but the ill effects are abrogated by a series of accidents; the animal expels the medication and manages to recover. The result is beneficial, but the veterinarian's action is morally questionable. It is unjust to limit the moral assessment of certain actions to the consequences; common sense and justice inform us that the agent's motives and intentions should be included in the assessment.

According to Kant's deontology an act is moral only if it springs from a 'good will': that is, an act arising out of a rationally based sense of duty, as opposed to acting out of inclination. Duty, maintains Kant, is governed by respect for the moral law; an act is good not because we enjoy it, or because it brings happiness or pleasure or relieves suffering, but because it is performed out of respect for the moral law. This is expressed in Kant's famous categorical imperative: 'I ought never to act except in such a way that I can also will that my maxim becomes a universal law'. This formal statement is predicated on the moral equality of all, and expresses the view that the measure of right action is that it can be universalized without violating the equality of all human beings. It implies treating others as ends, not as means, and respecting the autonomy of others.

Although criticized for excessive formality, Kant's deontology has practical application. Consider a researcher who intends to carry out experiments with a certain drug on companion animals. The predicted outcome of the research is that it would bring considerable benefit, but there is a risk of adverse side effects for the animals in the experiment. Were the owners of these animals informed of the risks they would probably withhold consent. The researcher lies, saying that no risk is involved. The researcher can be said to be acting on the principle that 'whenever your research is imperilled by the reluctance of subjects to participate in it, it is permissible to lie'. This maxim cannot be willed as a universal law; if it were it would amount to universally endorsing the making of false promises in the interests of scientific progress, and this would indicate that the very meaning of making promises (truth telling) would be jeopardized. The maxim could not be universalized without self-defeat. It should be stressed that Kant's deontology does not appeal to the consequences of dishonesty, but is designed to show that some principles cannot be universalized without involving the author in a contradiction.

Kant's rule deontology has been criticized for excessive formality and the difficulties it faces when providing a guide for conduct when facing a conflict of duties. Nevertheless, his insistence that to be moral we should require that individuals are treated as 'ends', not as 'means', is the basis of our respect for autonomy, the doctrine of informed consent and a resistance to paternalistic exercise of authority in professional conduct. (DL)

See also: Animal rights; Ethics; Regan, Tom

Further reading

Kant, I. (1964) *Groundwork to the Metaphysics of Morals* [translated by Paton, H.]. Harper and Row, New York.

Depression

In relation to humans, depression describes a condition of persistent negative mood, and as a clinical disorder is the most commonly diagnosed psychiatric illness in modern Western society. It encompasses a number of more specific conditions including major depressive illness, dysthymia, bipolar disorder and seasonal affective disorder. In humans, to a greater or lesser extent, each of these diseases is characterized by feelings of **apathy**, hopelessness, helplessness, inability to experience pleasure and worthlessness and, in severe cases, can result in suicide. Signs and symptoms of depression include sleep disturbances, increase in fatigue, changes in appetite, social withdrawal and a reduced ability to concentrate. Various theories regarding the development of depression have been proposed, but no single explanation regarding its underlying aetiology is accepted. There is also still debate regarding the distinctiveness between depression and other negative affective disorders such as **anxiety**, due to the overlap in many of the diagnostic criteria and presenting symptoms between the different conditions. Given the subjective nature of the properties of depression when applied to humans, analogy with the term when applied to animals should be with caution.

Ethological laboratory models of depression for the screening of antidepressant drugs are often based on the induction of suppressed behaviour through manipulation of social **stressors**, e.g. through social isolation and changes in social hierarchy or the use of physical stressors that may be chronic or acute, such as those based on the induction of **learned helplessness**, the tail suspension test (the animal is suspended by its tail with sticky tape for 6 min) and the forced swim test (the animal is forced to swim repeatedly in a bath for 15 min and the amount of time spent immobile recorded). The effect of these drugs on spontaneous behaviour, such as **exploratory behaviour** in either an **open-field test** or **elevated plus-maze**, may also be used as part of the screening process, since these drugs are expected to increase such activity.

In animals, the term depression is usually applied to situations where the normal behaviour of the individual is altered such that they appear less active and exhibit reduced interaction with their environment, **conspecifics** and, especially in the case of **companion animals**, possibly with human members of their social group. This behavioural depression has a physiological basis in situations of illness through the production of **cytokines**, and has been studied in conjunction with other effects of these mediators, such as the suppression of appetite.

Behavioural depression as a result of negative affect is more difficult to determine in animals, as mood cannot be assessed through the articulation of feelings. However, certain theories regarding the evolution and development of depression in humans – neuroanatomical correlates combined with data on neurotransmitter levels (especially **serotonin**) associated with depressive conditions – make speculation in this area of interest to those concerned with animal **welfare**. One of the theories regarding the evolutionary value of depression speculates that a depressed mood may be linked to an increased withdrawal from situations of conflict and a yielding of status in a dyad (**see: Hierarchy**), which helps to reduce the risk of exclusion from a **social group** following conflict.

Elevated mood is hypothesized to be associated with the ability to hold a resource, depressed mood with the surrendering of the resource, within conscious individuals (**see: Consciousness**). Both could potentially occur as acute or chronic events, and the function of depression for the individual would thus be the ability to remove oneself from a ritualized conflict without sustaining life-threatening injury. As social groups depend on **cooperation**, individuals that yield in these situations and inhibit their behaviour are less likely to provoke further agonistic interaction (**see: Agonistic behaviour**), and so could remain within the population without further compromising their **fitness**.

The neurotransmitter alterations that appear to correlate with the unsuccessful outcome of a conflict situation seem to bear some similarities to the alterations described in individuals experiencing depression, which adds some support to this theory. Evidence for a genetic component to depressive illness in people may add further support for this trait being part of the normal range of human emotional processes. However, it may be that modern circumstances provoke an extreme or abnormal level of response that results in a less adaptive clinical condition. Postulated proximate triggers (**see: Tinbergen's four questions**) for the condition include loss of status, abilities or interpersonal conflict, and the impact of these may be exacerbated when they have a perceived lack of **controllability**.

Arguments based on evolutionary continuity may provide a theoretical basis to suggest that depression might have parallels in animal species with highly evolved **social behaviour** and systems. This may be reinforced by new experimental approaches to the assessment of the subjective states of animals, such as **cognitive bias** tests and, although we cannot know the subjective experience of another, the **precautionary principle** may be applied to protect their welfare as we understand more about the biology of the condition. (HZ)

See also: Emotion

Further reading

Fuchs, E. and Fliugge, G. (2006) Experimental animal models for the simulation of depression and anxiety. *Dialogues in Clinical Neuroscience* 8, 323–333.

Gross, R. (2005) *Psychology: the Science of Mind and Behaviour*, 5th edn. Hodder Arnold, London.

McArthur, R. and Borsini, F. (2006) Animal models of depression in drug discovery: a historical perspective. *Pharmacology Biochemistry and Behaviour* 84, 436–452.

McGuire, M.T., Troisi, A. and Raleigh, M.M. (1999) Depression in evolutionary context. In: Baron Cohen, S. (ed.) *The Maladapted Mind: Classic Readings in Evolutionary Psychopathology*. Psychology Press, London.

Price, J., Sloman, L., Gardner, R., Gilbert, P. and Rohde, P. (1999) The social competition hypothesis of depression. In: Baron-Cohen, S. (ed.) *The Maladapted Mind: Classic Readings in Evolutionary Psychopathology*. Psychology Press, London.

Descartes, René

René Descartes (1596–1650), the professed 'father of modern philosophy and modern natural science', was a philosopher and mathematician who was also a leading exponent of the mechanistic view that emerged after the Middle Ages. As is commonly imputed to Descartes by many scholars, he argued that human beings alone were members of the moral community because they alone were 'thinking beings,' i.e. beings who possessed the mental faculty of reason, had propositional language, who could reflect abstractly and who possessed the 'faculty of judgment'. Animals were conceived as nothing more than nature's complex automata. He considered them biological robots, no different in kind from a sophisticated clock. Under this interpretation, Descartes held that animals were incapable of conscious mental states. They 'do not perform any action which indicates that they can think', he argued. Here, since animals are but unconscious 'wind-up' material objects, treating them as mere objects of use should raise no ethical concerns whatsoever.

Descartes' view on animals received dissent from a notable contemporary, Pierre Gassendi, a priest and logician, who insisted that we ought to focus on certain other mental capacities of animals besides their lack of rationality. Gassendi called attention to animals' perceptive abilities and their ability to suffer. He also argued that our ability to reason makes it incumbent upon us to be humane and to exercise compassion towards animals. Gassendi argued that 'There is no pretence in saying that any right has been granted to us by [the moral law] to kill any of those animals which are not destructive or pernicious to the human race'.

Interestingly, Descartes did not deny that animals were capable of all 'sensation'. They possessed **memory**, perception, fear, joy and hunger. However, he denied that these experiences are conscious in the relevant sense, i.e. that there is something it feels like for an animal to experience these sensations. He held that animals were limited in their hardwiring. For Descartes, animals 'sense' their environments just like a thermometer 'apprehends' the temperature and registers it. By analogy, since a thermometer is not conscious, attributing intentionality to it, i.e. to suggest that it can 'feel or experience cold or hot' is **anthropomorphism**, bordering on preposterous. (RA)

See also: Cartesianism; Dualism, Cartesian

Reference

Descartes, R. (1988) Discourse on method. In: Cottingham, J. *et al.* (eds) *Descartes: Selected Philosophical Writings*. Cambridge University Press, Cambridge, UK.

Desensitization – systematic

Systematic desensitization is a commonly used behavioural modification technique for the alleviation of behaviour problems in animals caused by inappropriate **arousal**. Typical patients are animals that become inappropriately aroused when in the presence of a particular stimulus or situation – for example, fear-related behaviour in the presence of fireworks, aggression towards other dogs and overexcitement towards visitors.

The process in animals is an adaptation of a psychotherapy technique originally developed by Wolpe (see Wolpe and Lazarus, 1969), and is based on the concepts of **habituation** or conditioning relaxation in response to the problem situation. The animal is exposed in a controlled situation to low levels of the arousing stimulus according to an increasing gradient, and rewarded when it remains relaxed or shows an appropriate response. An increase in the level of the stimulus is not made until the animal reliably fails to react to the previous level. In this way the technique aims to raise the threshold for a response to this particular stimulus. The process is usually speeded up by rewarding the animal for remaining relaxed (respondent **counter-conditioning**) or training it to perform a different response in the presence of the arousing stimulus (operant counter-conditioning).

The first stage is to identify the arousing situations. These are then broken down into the relevant component stimuli that might contribute to the animal's response. These might include the sight of the object, its smell, its sound or the location (taste or touch may also be involved). It is particularly important to do this if the arousing situation is complex, such as a visit to the veterinary clinic. Then the therapist and pet owner should work out a schedule for exposing the animal to very low levels of these stimuli. Intensity is controlled in a number of ways, e.g. through controlling distance, volume, etc., depending on the stimulus (see Table D.1). Single stimuli may be presented to the animal and, once a reliable level of desensitization to these has occurred, multiple stimuli may be presented together (usually at an initially lower intensity than the individual stimuli) until the animal is desensitized to the

Table D.1. Commonly used methods to control the intensity of a stimulus during desensitization programmes (stimulus gradient).

Stimulus quality		Possible scenario useful for:
Distance	Exposure to other, controlled objects at decreasing proximity	Over-arousal towards animate and inanimate objects, e.g. other people, animals, vehicles
Volume	Play a recording of the noise at increasing levels	Fear-related behaviour to loud noises, e.g. fireworks, hair clippers
Visual similarity	Exposure to similar situations involving similar objects	Over-arousal in specific situations (may involve animate or inanimate objects), e.g. visiting the veterinarian, car journeys

whole scenario. The type of stimulus and the levels by which it is introduced make up the stimulus gradient. The stimulus gradient and rate of progress depend on the situation that is provoking the arousal.

A combination of methods for controlling intensity may be used within a given desensitization programme; for example, desensitization to car journeys may begin with approaching the car (distance) and sitting in another, novel, car (similarity).

Once the stimuli have been ordered from the least arousing to most arousing, the desensitization programme can begin. Figure D.6 lists the basic process.

If the animal begins to show signs of inappropriate arousal to the stimulus, the exposure should cease and the animal should then be distracted and ignored until it calms down. The stimulus should then be reintroduced at lower levels than the previous exposure. It often helps if sessions are relatively short and frequent and cease when the animal is relaxed in the presence of the stimulus. As desensitization is a learning process the subject is prone to forgetting, and so new sessions should start with the stimulus presented a lower level than that at which the previous session finished and, once the goal has been achieved, this should be regularly reinforced by exposure to the original arousing stimulus on an intermittent basis. It is also useful occasionally to expose the animal to lower levels of the stimulus to which it reliably fails to react during the training process, so exposure is less predictable; however, exposure to the natural stimulus during training should be avoided if possible, to present disruption to the learning process. In an ideal situation both the stimulus and the animal should be under careful control during the rehabilitation process. (KT)

See also: Conditioning – types of; Fear

Further reading

Horwitz, D.F., Mills, D.S. and Heath, S. (2002) *BSAVA Manual of Canine and Feline Behavioural Medicine*. British Small Animal Veterinary Association, Quedgeley, UK.

Landsberg, G., Hunthausen, W. and Ackerman, L. (2003) *Handbook of Behaviour Problems of the Dog and Cat*. Saunders, Edinburgh, UK.

Wolpe, J. and Lazarus, A.A. (1969) *The Practice of Behaviour Therapy*. Pergamon, New York.

Despotic species

Despotic species have a steep dominance hierarchy, in which only one or a few individuals display **dominance** while all others show **submissiveness**. Displays of **aggression** are rare but fierce when they do occur, and resource access is biased towards high-ranking individuals. This affects the social **grouping** of the animals and a clear spatial structure emerges, with high-ranking individuals in the central area and subordinates at the periphery. With this type of dominance hierarchy females can attain higher dominance rankings than males, and females tend to select only high-ranking males as mates. (LMD)

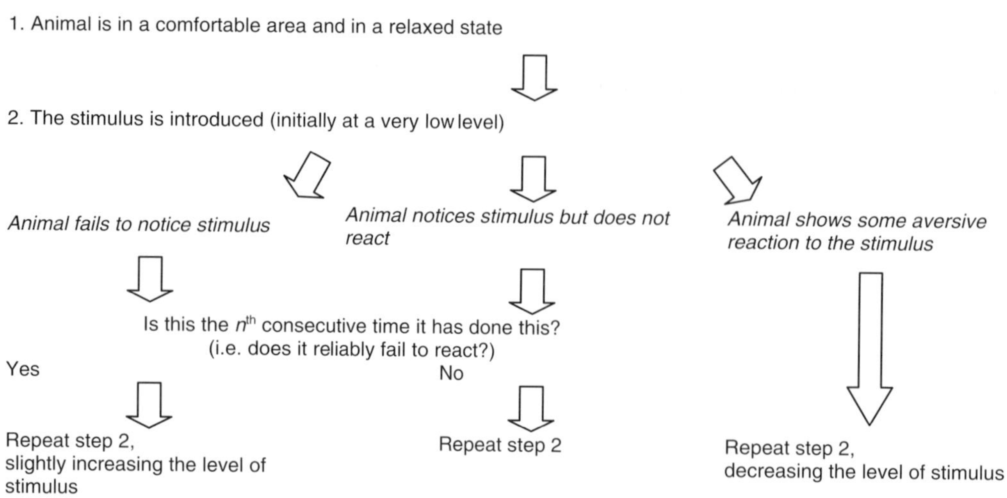

Fig. D.6. Simple desensitization programme.

Development of behaviour

The development of behaviour can be considered in both a proximate and an ultimate sense. In the former it relates to **ontogeny**, and in the latter to **phylogeny**. In either case, variability in the process that facilitates adaptation is an important feature.

This listing focuses on the processes influencing development within the lifetime of the individual (ontogenetic development), the other process being considered the **evolution** of behaviour.

Historically, the nature–nurture debate has featured prominently in the investigation and description of behavioural development, with advocates of the former placing an emphasis on **'instinctive' behaviour** processes and the importance of genetics, to the point of arguing for biological determinism (or preformationism) in the development of some behaviours. While the opposing perspective, epitomized by radical versions of Watson's **behaviourism**, has suggested that the environment is all-important to development and has placed a greater emphasis on the lability of the process (although it is important to recognize that flexibility does not just come from learning). Needless to say the reality is somewhere in between, with both genetic and environmental factors being essential for normal development and able to effect change.

However, there may be differences between the two in the proportion of the variability seen that can be ascribed to genetic versus environmental factors, and this can be quantified by careful determination of the **heritability** of the trait under consideration. Within the context of behavioural development, such calculations are complicated by the dynamic relationship that exists between the two. The concept of **epigenesis** proposes that the genotype interacts throughout life with the environment to produce a changing phenotype. Both environmental and genetic factors may, however, be important in limiting the variability of responses seen during development; for example, exposure to certain endogenous or exogenous stimuli may start to canalize the response so that it becomes more stimulus-specific. This is sometimes referred to as a shift from a more open to a more closed programme.

Although development is a continuous process throughout life, it is not an even process, with genetic and environmental factors influencing the current phenotype throughout life to varying degrees at different times, producing discontinuities in development. For example, **sensitive phases/critical periods** are times of particularly rapid development, when the animal may be prone to learning particular types of association and during which the first exposure to certain environmental stimuli can have a rapid and long-lasting effect (e.g. resulting in **imprinting** and **bonding**).

Even though it is tempting to think of these phases of development as being periods when environmental influences are most important (see the nature–nurture perspective mentioned above), it is important to appreciate that the timing of these developmental phases is regulated by both genetic factors and experience. For example, although general timings may be offered for the socialization phase of a dog as occurring between 3 and 12 weeks of age (Scott and Fuller, 1965), the onset of this and many other developmental processes may be delayed in the absence of appropriate stimulation, and curtailed sooner after a certain level of stimulation. Thus 3–12 weeks may not represent the time span throughout which an individual dog is particularly sensitive to learning about other individuals, but rather the average range for dogs in general (some starting earlier and some later). Therefore, it may be that the timing for a given individual is much shorter, depending on the timing, quality and quantity of stimulation received.

Just as some developmental processes appear to emphasize the role of the environment, others emphasize endogenous factors – for example, maturational changes (e.g. onset of **sexual behaviour**), which depend on the development of the body to a certain degree before they can be expressed functionally. The timing of these, too, depends on a balance between genetic events and environmental influences, e.g. plain of nutrition, presence of members of the opposite sex.

These examples serve to emphasize the important difference between chronological time (days, weeks, months, etc.) and individual developmental time (i.e. stage of development) in the onset of specific developmental processes, and the danger of simple generalizations. Related to the concept of individuals developing according to their own history is the importance of recognizing that development does not represent a process during which an animal progresses towards being a fully functional being, but rather that an animal has different functional stages to its life, each of which is adaptive to normal circumstances, and development represents the process – including transition – from one stage to another. In a given circumstance a given stage of development may be most appropriate. Thus the **neonatal period** maximizes **fitness** by exploiting the maternal resource available at minimal cost.

Given the dual influence of environmental and genetic factors on the developmental process, and the adaptive value of different life stages, there is the potential to select for and exploit differences in the process. Thus broilers have been selected for and managed in a way that exploits the rapid growth phase associated with earlier stages of development. In dogs it has been suggested that they differ from wolves as a result of selection for the retention of juvenile characteristics into adulthood (paedomorphosis), and that different breeds have evolved as a result of selection heterochrony (changes in the relative timing of developmental processes), especially in relation to the development of the hunting sequence, which allows dogs to fulfil different roles more easily (see Coppinger and Coppinger, 2001).

Thus traditional rescue breeds (St Bernard, Pyrenean Mountain dog, etc.) have been selected (either directly or indirectly) for high levels of competence in the location of heterospecifics (an early part of the sequence to appear in puppies, which coincidentally is associated with a high growth rate and so larger adult size), whereas herding sheepdogs have been selected for stalking and chasing without the bite. Since these behaviours develop later, they are not associated with a phase in life with such high growth rate, and so these breeds tend to be smaller and have more adult 'lupine' facial features (see Fig. D.7).

The scientific investigation of the factors that might facilitate, initiate or maintain the development of a given behaviour is methodologically very challenging, not least because the same end point may be reached by different processes in different individuals (equifinality), and interactions

Fig. D.7. Comparison of Border collie (front) and Pyrenean Mountain dog (back) head profiles. Note the more domed head and more rounded features of the larger Pyrenean Mountain dog, which is generally more paedomorphic (puppy-like) in appearance.

between different experiences may play an important role in development (e.g. in the balance that occurs between response **generalization** and **discrimination**). Cross-sectional epidemiological studies may be used to reveal factors of potential significance, although association does not indicate causality. For example, it has been found in several cross-sectional studies that horses that develop **weaving** behaviour are often also those with reduced social contact (increased risk odds ratio).

However, three plausible explanations are that: (i) horses that weave are generally kept away from other horses because of (unfounded) concern that the unaffected horses may copy the behaviour (i.e. weaving results in a lack of social contact); (ii) lack of social contact is stressful for horses, and this predisposes them to weaving (i.e. lack of social contact results in weaving); or (iii) the sort of horses that weave are also the sort of horses to have reduced social contact due to some common factor, e.g. breed, use in racing, etc. (i.e. that lack of social contact and weaving both result from some other factor). The latter can be assessed by examining interrelationships with other factors, but this depends on the ability to identify this common factor first. It is possible for all three statements to be true simultaneously because one does not necessarily preclude the other. In an attempt to focus on the role of specific factors, **isolation** experiments have been proposed, in which the effect of depriving the animal of specific stimuli during development is assessed. There are numerous methodological challenges with such work.

First, the level of isolation used is often unrealistic and results in an unnatural level of consistency in the environment, which may have its own effects on behaviour, beyond that which is being studied; it is also important in any study of behavioural development to distinguish between effects that impact on a specific behaviour and more general effects on behaviour (isolation has general effects on emotionality).

Secondly, it may be difficult to determine exactly what the subject is being isolated from, as the perceptual world of other species is different from our own and it may be that certain stimuli can substitute for each other in certain circumstances (whether or not this occurs in normal circumstances remains unknown as a result of the experiment) – thus a negative result does not mean that the given factor is not important. Thirdly, in some experiments it is not possible to distinguish between the post-isolation effect and the treatment effect, i.e. the result is due to the animal's response to release from isolation rather than to the isolation itself.

Finally, there are also problems with interpreting results due to an interaction between the age of subjects, the timing of the treatment, testing and time elapsed between treatment and testing (see Martin and Bateson, 2007). Related to this is the timing of assessment and importance of distinguishing between short-term and long-term, or transient and more permanent, effects. Many animals can compensate for a period of deprivation during development, and may return to normal should circumstances allow at a later date; however, it may also be that short-term disruption can result in long-term effects indirectly: for example, in the absence of exposure to people a dog may be fearful upon first encounter, and the response of the person to this initial encounter could change the long-term developmental pathway. Someone who gives the dog time to investigate and assess the situation may help the dog to overcome its wariness of people and ultimately allow it to become socialized, whereas the individual who insists on handling it like a normally socialized dog is likely to provoke an extreme fear response that reinforces anxiety towards people in future.

These criticisms and others (relating to the analysis and interpretation of results) apply to one of the best-known series of studies into behavioural development, namely the work of Scott and Fuller (1965) on dogs, the conclusions of which have now become almost lore among those working with dogs – despite their questionable validity – being heavily biased towards the deterministic view of behaviour that dominated thinking at that time. Many of the difficulties outlined above are equally applicable to other experimental methods of investigating behavioural development, and serve to emphasize the need for multiple testing and methodologies before firm conclusions about the influence of different factors on development can be made. (DSM)

See also: Play

References and further reading

Albrecht, P., Gould, S.J., Oster, G.F. and Wake, D.B. (1979) Size and shape in ontogeny and phylogeny. *Palaeobiology* 5, 296–317.

Coppinger, R.P. and Coppinger, L. (2001) *Dogs: a New Understanding of Canine Origin, Behaviour, and Evolution*. University of Chicago Press, Chicago, Illinois.

Martin, P. and Bateson, P. (2007) *Measuring Behaviour – an Introductory Guide*, 3rd edn. Cambridge University Press, Cambridge, UK.

Scott, J.P. and Fuller, J.L. (1965) *Genetics and the Social Behaviour of the Dog*. University of Chicago Press. Chicago, Illinois.

DFD meat

DFD stands for dark, firm, dry, and describes the appearance of meat. DFD meat is also referred to as '**dark cutting**' and is a meat quality defect resulting from a lack of the normal acidification of meat post-slaughter. Between **slaughter** of the animal and chilling of the carcass, glycogenolysis occurs in the muscle and, in this process, glycogen is converted to lactic acid. Glycogenolysis decreases the pH of the meat. Good-quality lamb and beef has a final or ultimate pH value close to 5.5, and above a pH of 5.8 the colour of the meat darkens progressively with increasing pH; meat with an ultimate pH of greater than 5.9 is generally classed as DFD meat. The term 'black beef' has been used to describe meat of a neutral pH value (around 7.0), which is the value generally found in the live animal.

Cut DFD meat does not bloom or brighten when exposed to the air, and appears dark coloured because less light is reflected, the presence of deoxymyoglobin and the reducing sugar content affecting colour development. The dry appearance of the meat is thought to be a result of an unusually high water-holding capacity (both raw and cooked), causing the muscle fibres to swell with tightly held water. As a consequence of this water-holding capacity, DFD meat loses little water during storage. The appearance of DFD meat does not appeal to consumers. DFD meat is variably reported as being either more or less tender and juicy than good-quality meat (differences may depend upon the ultimate pH). However, it has a reduced flavour or a noticeable 'off' flavour. The meat has a reduced shelf life and, in particular, does not survive well in vacuum-packaged anaerobic storage due to the absence of glucose as a substrate for bacteria. Instead, the amino acids are used by the bacteria, producing spoilage odours much sooner than would happen with good-quality meat.

This category of meat is important from an animal welfare perspective, as it can be indicative that animals have experienced stress pre-slaughter. In healthy, well-fed cattle, the concentration of muscle glycogen generally ranges between 60 and 120 μmol/g, although can be as high as 200 μmol/g in grain-fed cattle (**see: Feedlot**). Rapid glycogen breakdown in living muscle may be triggered by an adrenergic mechanism, a contractile mechanism or a combination of both. These two rapid mechanisms are activated by the **sympathetic** and somatomotor **nervous systems**, respectively, and act by increasing the concentrations of cyclic AMP and calcium ions (**see: Stress**). Additionally, low muscle glycogen can arise through slow depletion, as a result of low energy intake over an extended period.

Cattle DFD meat has been found to be correlated with a number of factors. The incidence is higher in the following situations: (i) in bulls rather than in steers or heifers; (ii) when groups of unfamiliar animals have been mixed; (iii) in oestrous females; (iv) when animals have been food-deprived or had poor-quality food over an extended period; (v) after long-haul **transport**; and (vi) after cold exposure. Some reduction in glycogen levels pre-slaughter can be accommodated and have no adverse effects on meat quality, but if levels fall below 40–57 μmol/g the ultimate pH will increase. Thus, a combination of factors is more likely to predispose animals to glycogen depletion than a single factor.

Mixing unfamiliar animals can lead to high levels of aggressive interactions, particularly between bulls, and mounting behaviour, which is also seen between oestrous females. Physical activity per se does not necessarily result in glycogen depletion, but rather it is the intensity of activity that determines glycogen levels. Furthermore, mixing induces emotional stress because social relationships and the social hierarchy are disrupted (**see: Hierarchy; Social stress**). Mixing cattle pre-slaughter probably has the most detrimental effect on glycogen loss and should be avoided.

Muscle glycogen levels will, also, evidently be influenced by initial glycogen levels, which may be determined by nutritional history. Feedlot cattle, for example, are less predisposed to producing DFD meat, which can, partially, be attributed to their superior nutritional status and concomitant high muscle glycogen levels. Furthermore, such animals may experience less stress pre-slaughter than cattle coming from pasture, because of increased exposure to and experience of handling and humans.

Glycogenolysis varies between muscle and fibre types; muscles along the back and hind legs appear to be most susceptible; **epinephrine** has a marked glycogenolytic effect on slow-twitch fibres but not on fast-twitch, but the opposite is seen when muscle is given low-frequency stimulation. These differences reveal why mounting activity is closely correlated with DFD meat and also suggests that in situations that are not physically demanding it is the emotional state of an animal that is critical in relation to glycogen loss.

There is some evidence that the character or temperament of cattle affects meat quality and the incidence of DFD meat. The mechanism has not been fully determined, but it may be that animals of different characters also differ in the proportions of muscle fibre types (fast and slow twitch), or that animals with poor temperaments do not cope well with the stressors associated with transportation and slaughter and, therefore, glycogen loss is rapid, or it could be a combination of such factors.

Cold exposure may deplete glycogen through the development of hypothermia, shivering and non-shivering thermogenesis (**see: Cold stress**). Cold exposure can occur during transportation or when animals are washed/hosed prior to slaughter.

Transportation over distances of less than about 400 km probably has little effect on glycogen depletion and ultimate pH, but over much greater distances there may be slight effects. Attempts have been made to attenuate transportation stress using mixtures of sugars and electrolytes provided to the animals on-farm immediately prior to transportation, or while in lairage. The results on the effectiveness of such treatments in reducing the incidence of DFD meat are equivocal.

Rest post-transportation potentially allows glyconeogenesis to occur, raising muscle glycogen levels. However, glycogen repletion rates are slow in ruminants (compared with monogastrics), ranging between 0.1 and 1.0 μmol/g/h, depending on the extent of glycogen depletion and inanition, and access to feed. To avoid DFD meat, cattle with subcritical levels of muscle glycogen would require in excess of a 48 h rest and recovery period (**see: Lairage**). For bulls there is a clear reduction in DFD meat with short rest periods, and the same may apply to cattle that have been minimally stressed pre-slaughter. There is an indication that short rest periods (less than 12 h) may reduce DFD meat. (JCP)

Diagnosis

A diagnosis is a decision about the cause of the disease that is affecting an animal. In UK law only a veterinarian can formally make a diagnosis in an animal. (DBM)

Diarrhoea

Usually refers to loose faeces, but can also be used to describe a very frequent passage of normal stools. 'Looseness' can be anywhere on a spectrum from being a less solid stool to a frank watery excretion due to transudation of fluid from the gut wall into the gut lumen. In some species diarrhoea may even contain mucus due to hyper-stimulation of the mucous glands in the intestinal wall. (DBM)

Diet selection

Diet selection at its simplest level considers the attainment of nutrients from available foods in order to meet physiological requirements for maintenance, growth and **reproduction**. The study of the concept of diet selection and associated feeding behaviour has interesting applications for the applied animal sciences. Understanding the causal mechanisms underlying diet selection has practical implications for the development of diets or feeding systems that promote animal health, behaviour, production and environmental sustainability. In a natural setting nutrients are obtained from a variable environment, in which sensorily distinct foods vary spatially and temporally in their nutrient quantity and quality. Equally variable is the animal's physiological requirement for nutrients, which change according to life stage, reproductive stage and within the range of normal activity patterns, including the fight or flight response.

The challenge faced by an animal is to select a diet from this varied environment that meets its needs. It is worth noting that diet selection may be less complex for captive and domesticated species, as often a 'balanced' diet proportionate to their requirements is readily provided. However, the animal will still be faced with the same problem of deciding whether to eat, what to eat (if choices are available) and how much to eat. Whether an animal can select a nutritional diet, when given the opportunity, has been the subject of considerable research and debate, given that the wisdom of the body is often called into question when animals and humans fail to select appropriate diets either by the consumption of toxic foodstuffs or the long-term selection of foods that lead to obesity or metabolic conditions.

Models of **optimality** have been used to understand the rules governing diet selection and the adaptive significance of **foraging behaviour** patterns. **Optimal foraging** theory uses a univariate approach to the understanding of feeding behaviour, by predicting that the decisions that animals use to select a diet will maximize their net rate of energy intake (Stephens and Krebs, 1986; Hughes, 1993). Alternative models have examined foraging decisions in relation to the quantities of the three main energy sources within foods, the macronutrients: protein, fat and carbohydrate. This area of research has grown from the premise that the relative proportions of different macronutrients within individual foods influence selection and consumption patterns.

One model, the geometric framework of nutrient balancing, developed by Raubenheimer and Simpson (1997), is based on the idea that the proportions of different foods that are ingested is dependent upon the degree that a particular nutrient intake target can be reached. The intake target is the optimal combination of nutrients that, if ingested, will maximize an individual's performance and minimize post-ingestive processing costs. This latter model also takes into account the fact that animals may not always act to maximize nutrient intake as, in certain situations, trade-offs develop between macronutrients and other dietary requirements, such as sodium.

Selection between different foods is often based on the familiarity of the organoleptic characteristics, i.e. those relating to the flavour, texture and temperature of the food item, as identified by olfaction and taste perception. Many species are neophobic and will either avoid novel foods completely or ingest them in small quantities. Sensory characteristics do not act in isolation to drive food choice. Signals are received from the gut, the liver, circulating factors and adipose tissue. The **central nervous system** integrates these signals with those from the external environment, which, coupled with prior, learned experience, modify and regulate the animal's **feeding** behaviour.

Post-ingestive **feedback** from the gastrointestinal tract acts to modulate dietary choice. Signals that relate to gastrointestinal tract distension, nutrient feedback and malaise act in association with organoleptic properties to influence the degree that ingestion is initiated, maintained or terminated. Thus the sensory signals generated prior to and following ingestion are the cues that the animal can use to associate with the consequences of eating particular foods (Booth and Thibault, 2000). If food properties remain stable, the organoleptic characteristics will reliably predict the positive or negative consequences of ingesting particular foods. This association has been demonstrated experimentally by pairing a flavour with a nutrient infusion that was administered intragastrically. Favourable post-ingestive consequences increase the acceptance and the ingestion of a food. For example, Villalba and Provenza (2000) investigated whether lambs' intake of wheat straw was guided by organoleptic properties or from feedback from the gastrointestinal tract. Lambs were fed the same source of wheat straw and the simulated post-ingestive consequences were manipulated by either a water or starch intraruminal infusion. Wheat straw was eaten sparingly or avoided when presented to lambs in 5 min tests with or without an intraruminal infusion of water. However, ingestion of wheat straw was enhanced, as measured by an increase in intake rate, when ingestion of straw was paired with an intraruminal infusion of starch. The strength of the association ensured that the effect persisted, since Villalba and Provenza (2000) found that the association persisted for 9 days after the starch infusion was removed.

The decision made by the animal to start or terminate a feeding bout over and above other behavioural patterns is dependent upon a multitude of factors. Broadly, these can be classed into external and internal factors. External factors include the aspects within the immediate location such as the presence or absence of appropriate food items, social conspecifics, natural predators and environmental conditions. Internal factors, on the other hand, include influences from physiological processes that act to maintain the internal **homeostasis**. For example, food intake slows as the meal progresses, which may reflect feedback from the gastrointestinal tract.

Internal and external factors are not mutually exclusive – they interact; for instance, free-ranging horses have been observed to cease grazing and seek shelter during the inclement weather conditions of the winter months. This strategy may allow them to conserve energy while the nutrient content of the available forage is low. However, feral horses with low body condition scores have been observed to maintain grazing patterns, even during adverse weather conditions, presumably due to a greater need to replenish body reserves. (SR)

References

Booth, D.A. and Thibault, L. (2000) Macronutrient-specific hungers and satieties and their neural bases, learnt from pre- and post-ingestional effects of eating particular foodstuffs. In: Berthound, H.-R. and Seel, R.J. (eds) *Neural and Metabolic Control of Macronutrient Intake*. CRC Press, Boca Raton, Florida, pp. 61–91.

Hughes, R.N. (1993) *Diet Selection. An Interdisciplinary Approach to Foraging Behaviour*. Blackwell Scientific Publications, Oxford, UK.

Raubenheimer, D. and Simpson, S.J. (1997) Integrative models of nutrient balancing: application to insects and vertebrates. *Nutrition Research Reviews* 10, 151–179.

Stephens, D.W. and Krebs, J.R. (1986) *Foraging Theory*. Princeton University Press, Princeton, New Jersey.

Villalba, J.J. and Provenza, F.D. (2000) Postingestive feedback from starch influences the ingestive behaviour of sheep consuming wheat straw. *Applied Animal Behaviour Science* 66, 49–63.

Dietary fibre

Dietary fibre describes feed components that are less easily digested. Dietary fibre is digested by microbial enzymes in the rumen of ruminant animals and the hindgut of monogastric animals, but not by enzymes produced by the animals themselves. It is a normal component of the diet of herbivores. Most usually it is obtained from plant-derived material, such as cellulose, hemicellulose, fructans, gums, pectins and lignin, but can also include such other undigestible components as oyster shells and synthetic polymers. Usual sources of dietary fibre added to animal diets include: straw, hay (particularly from mature grass), ensiled forages, grain husks, legume seed coats, sugarbeet pulp and some by-products. Different sources of fibre will contribute to effective digestion to varying extents and will usually have both chemical and physical properties that are important for adequate digestion. However, some sources – for example, highly comminuted fibre – will have only the chemical properties.

A diet high in fibre is the normal diet of the grazing ruminant. Intensive production systems that incorporate energy-rich concentrates at the expense of dietary fibre risk physical damage to the ruminal epithelium, but are fed because they can satisfy the energy requirements of animals that have been bred for high levels of production. In acidosis caused by inadequate fibre intake, there is accelerated fermentation in the rumen leading to reduced ruminal pH. This disrupts the normal range of microbial flora in the rumen (and therefore the products of fermentation – increasing propionic and butyric acid production while reducing that of acetic acid) but also damages the ruminal wall itself, causing lesions, abscesses and malformation of the ruminal papillae. This acidotic state is presumed to be uncomfortable for the animal, and certainly leads to reduced intakes and the multitude of welfare and production impacts arising from a poor state of nutrition. Ruminal acidosis has also been associated with laminitis and displaced abomasum.

The behavioural act of **rumination** is the process of repeated chewing and addition of saliva in order to increase the surface area of dietary fibre particles and release soluble nutrients from the severed ends of fibre particles. Rumination bouts can last up to 30 min. It is a normal behaviour, and should be encouraged, in addition to which it is performed when the animal is at ease (**see: Comfort behaviour**), so is a good indicator of **welfare**. Feed that is low in dietary fibre will not be regurgitated, and in this case the animal will spend much less of its time ruminating, which can result in behavioural disorders. Dietary fibre, spending longer in the rumen than more rapidly fermentable dietary components, ensures that the rumen remains active for longer. This is an advantage to ruminants in colder environments, as the fermenting rumen is a considerable source of heat to the animal.

Dietary fibre is also important for the monogastric animal, feeding the natural microbial population of the hindgut, increasing the transit time of digesta through the gut and increasing the bulk of faeces. The hindgut microflora are able to digest the fibre and supply volatile fatty acids as an energy source, thereby lowering the pH and discouraging the proliferation of other harmful microorganisms (particularly coliforms) that can cause inflammation of the gut wall. The increase in transit time should result in greater digestive efficiency, as the digesta spend a longer period of time in each sector of the gut. Reduced time spent chewing on low-fibre diets can impair dental and jaw health in the rabbit. Additionally, the provision of dietary fibre (straw) in the bedding of the pig provides the animal with the opportunity to perform rooting behaviour. Pigs choose to spend about 20% of their time rooting in straw when given the opportunity to do so, and will preferentially select straw bedding over concrete housing.

Fractions in the diet that require more complex ingestion and digestion engage the animal in natural activities that can offset **boredom** and the expression of **stereotypies** deriving from inactivity. Improvements in the welfare of captive animals by the addition of dietary fibre into the diet, through encouraging the expression of normal behaviours and decreasing the incidence of **abnormal** behaviours, have been observed in production systems (cattle, pigs, laying hens), in **zoos** (giraffe, elephant and possibly gorilla) and in other captive animals (horse and rabbit).

The horse is a grazing animal that can spend 12 or more hours a day feeding at pasture. If their diet is low in dietary fibre the feeding time is much reduced, as the easily ingested and digested concentrates pass quickly through the GI system. This can lead to behavioural problems such as **pica** and stereotypic behaviours and health problems, principally colic.

Similarly, an animal without a great deal to occupy its time, and that in the wild would have spent that time seeking food (estimated to be 75% of the time in the elephant), may feel a strong **motivation** to feed. The provision of dietary fibre, which might satisfy such a need without concomitant over-nutrition, has been recommended for pigs. Perhaps related to this satisfaction of the motivational drive to feed, and therefore not leaving an animal in a state of **frustration**, increasing dietary fibre for pigs has been shown to reduce incidents of **tail biting** and other expressions of **aggression** and for rabbits has reduced the incidence of fur chewing, bar biting and aggression.

The addition of dietary fibre to the diets of dogs, cats and rabbits has been used to assist in the treatment of diabetes, colitis, constipation and obesity. Dietary fibre in poultry also speeds the passage of digesta and, as with the monogastric mammal, improves the environment for beneficial gut microbes at the expense of harmful bacteria. Intakes are increased, and this may reduce **cannibalism** in laying hens by increasing the time spent feeding. Herbivorous reptiles, of which there are few, such as the iguana and the tortoise, digest fibre in much the same way as monogastric herbivorous mammals. Digestive problems have been reported for the tortoise fed insufficient dietary fibre.

Perceived problems of adding fibre to animal diets include the generation of greater amounts of faeces, the welfare problems arising from larger amounts of slurry in bedding and the difficulties of their removal. Against this, however, amounts of nitrogenous material in the animal waste are reduced, in both pigs and poultry. Ammonia emissions and other odorous compounds have thus been found to be reduced in the manure and, subsequently, bedding of laying hens, broilers and pigs, thereby improving the environmental conditions for these animals and positively affecting their **health**, welfare and levels of production. (DA)

Discrimination

Discrimination (cf. **generalization**) arises when examining whether an animal's behaviour is controlled specifically by one stimulus rather than another. As an example, consider the case in which an animal has learned to produce a particular action in the presence of a specific cue. While generalization refers to the animal's tendency to perform the action in response to other cues, discrimination is the extent to which the animal does not perform that action in response to other cues. A *generalization gradient* is the rate at which the tendency to perform the action changes with test cues that are more distinct from the trained cue. For example, if the animal has been trained to respond to a 1000 Hz tone, it will also respond (though less vigorously) to a 500 Hz tone or a 2000 Hz tone, and may also respond, though even less vigorously, to a 100 Hz tone or 10,000 Hz tone. Importantly, the generalization gradient can be altered by explicit discrimination training.

Any form of training necessarily consists of trials in which the target cue (e.g. the 1000 Hz tone) is presented and the animal is rewarded for producing the specific action. During explicit discrimination training, these trials are intermixed with trials in which other cues (such as 500 and 2000 Hz tones) are presented and the animal is not rewarded if it performs the action. Not surprisingly, the animal learns to suppress responses to these non-rewarded cues, and thus successfully discriminates between them. Interestingly, while explicit training of this sort is the most effective way to establish discrimination, an improvement in the ability to discriminate between cues (i.e. a decrease in the tendency to generalize) can be achieved simply by exposing the animal to these different cues prior to training (Mackintosh and Bennett, 1998). (JH)

Reference

Mackintosh, N.J. and Bennett, C.H. (1998) Perceptual learning in animals and humans. In: Sabourin, M., Craik, F. *et al.* (eds) *Advances in Psychological Science*, Vol. 2: *Biological and Cognitive Aspects*. Psychology Press/Erlbaum, Hove, UK and Taylor & Francis, London, pp. 317–333.

Disease

Clinical disease may have profound influences on animal behaviour. In the later stages of infection in many serious diseases (for instance, BSE in cows), the behaviour of animals is affected to such an extent that it is obvious even to the untrained eye. This may include changes in the posture, **vocalization** and sequences of behaviour. A sick animal might be reluctant to stand, eat or drink and may show general signs of malaise. Some of these behavioural changes may also provide direct clues as to which body part or system is affected. For example, an animal kicking at its belly probably has an abdominal pain that is caused by colitis or other digestive system disturbances. Cats and dogs often tilt their heads when they suffer from ear infections. Excessive scratching that leads to skin damage and bleeding may be indicative of the presence of skin parasites or other skin infections. **Lameness** results in characteristic changes in gait. Animals with fever adopt a crouching position and may show shivering. Many diseases lead to characteristic changes in behaviour, which are an important aid to selecting the most appropriate intervention by the animal keeper and veterinarian and are often described in veterinary handbooks.

Much more challenging is the detection of the, frequently, more subtle changes in behaviour associated with **subclinical** disease and with the early stages of diseases that have the potential to develop clinical signs. In both cases, early detection of problems has important benefits, both for the **health** and **welfare** of animals and for economic reasons. In order to recognize such behavioural changes, a sound understanding of the 'normal' animal behaviour routine is required. Knowledge of the social dynamics within a group of animals and changes in them may also reveal the presence of disease. Dominant animals that usually feed first when fresh food is presented to them or enter the milking parlour first may start to hang back during the course of subclinical illness. Experienced personnel that know individual animals can be excellent at detecting such changes. However, the decrease in contact between stockperson and stock in most production systems means that this is frequently no longer the case. Systems that automatically record and analyse (aspects of) animal behaviour will, therefore, increasingly be relied upon for the detection of such changes.

In most cases of subclinical disease the clearest behavioural modification seen is a change in the animal's **feeding** and **drinking behaviour**. Infectious challenges that lead to subclinical disease generally result in a reduction of the daily food intake of at least 10–20%. The occurrence, extent and duration of this so-called 'pathogen-induced anorexia' will depend on the type of challenge and the ability of the animal to eventually overcome it. In bacterial infections the reduction could occur within a number of hours post-challenge and last for only a few days. In infections with gastrointestinal parasites there is a substantial delay in the onset of 'anorexia', which may last for a number of weeks (see Fig. D.8).

Automatic recording of water and food intake at the group or individual level may thus facilitate early detection of disease problems. For some systems, automatic detection of behavioural changes caused by subclinical disease is most appropriately carried out at the group level. In the UK, for example, pig producers are encouraged to install water meters that measure daily water consumption per housing unit, and to use changes in water consumption as a problem identification aid. A slowing in the drinking rate from electronic milk feeders is advocated as an early indicator of health problems in systems for dairy and beef calves.

Other systems may be based on the capture of (feeding and drinking) behaviour of individuals, especially for larger animals such as cattle and sows that may be equipped with electronic tags. This allows a much more subtle approach to health challenges individual animals may face. For instance, locomotory problems (often accompanied by lameness) may not result in changes in daily food intake, but can be recognized from profound changes in short-term feeding behaviour such as eating rate. Such characteristic changes are likely to be due to hungry animals that attempt to reduce the time spent feeding because standing at the feeder is painful. Automatic analysis of animal feeding and drinking behaviour may be linked with other automated recording systems (such as data obtained in the milking parlour for dairy cows) to facilitate early detection of disease from subtle changes in behaviour.

The final category of changes in behaviour caused by health challenges includes modifications that are undetectable even by analysis in terms of number and duration of bouts of behaviour. Such changes appear during the early stages of health challenges and affect mainly the complexity of behaviour, as indicated by fractal analysis. For instance, a reduction in the complexity of the pattern of vigilance and feeding behaviours has been observed in grazing animals that are infected by parasites. Standard measurements of behaviour cannot reveal such changes, since in the above case both parasitized and non-parasitized animals spent the same amount of time feeding and performed the same number of head lifts. However, the temporal patterns of these behaviours, i.e. their complexity, were significantly affected by parasitism.

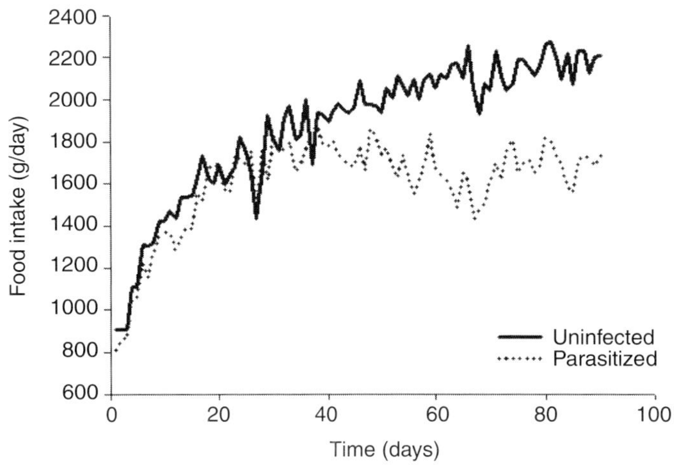

Fig. D.8. The daily food intake of sheep infected with a gastrointestinal parasite and their non-infected counterparts. Sheep were infected with the parasites on day 0. The reduction in their food intake appeared 5 weeks post-infection and was evident 100 days post-infection, while sheep still harboured parasites in their gastrointestinal tract.

Good stockmanship is based on a profound knowledge of normal animal behavioural patterns, and the manner in which they are affected by health challenges can be an important aid in the early detection of disease. However, as interactions between livestock and farm personnel decrease, more sophisticated systems to detect changes in animal behaviour caused by subclinical disease will be required. Detection of subtle changes in behaviour is consuming in terms of both time and resources. In order for these changes to be used as a means of early detection of health and welfare problems, the following system needs to be in place: (i) one that automatically captures these behaviours; (ii) one that analyses the behaviours and compares them with 'normal' behaviours; and (iii) one that delivers warnings that arise from those modifications of behaviour, but also minimizes the number of false warnings. Automation in the capture of behaviours, such as feeding, is now becoming increasingly available, even in conventional systems of production. The challenge lies in incorporating these into an automated early health detection system.

(IK, BT)

Further reading

Gonzáles, L.A., Tolkamp, B.J., Coffey, M.P., Ferret, A. and Kyriazakis, I. (2008) Changes in feeding behaviour as possible indicators for the automatic monitoring of health disorders in dairy cows. *Journal of Dairy Science* 91, 1017–1028.

Dishabituation

Dishabituation refers to the reinstatement of a reflex response to a stimulus to which the animal had previously become habituated (reduced responding as a result of learning about the non-significance of the event). Dishabituation arises as a result of 'forgetting' about the previous learning experience, rather than due to a new association between the stimulus and its consequences (this would be a new form of learning). It has been proposed that dishabituation may arise in some circumstances as a result of non-specific stress in animals (stress-induced dishabituation), and so the reappearance of a response may be an indicator of non-specific **welfare** problems in the management of the animal. (DSM)

See also: Habituation

Disinhibition

The term disinhibition is used in a variety of contexts within applied animal behaviour. Within the context of behavioural control, it is used to describe less restrained behaviour. Some drugs (notably **benzodiazepines**) are reported to result in the disinhibition of behaviour (disinhibition of **aggression** is of particular concern with these drugs), i.e. as a result of their tranquillization effect the animal is less restrained in its behaviour and so may be more willing to engage in an aggressive encounter, where previously it would not.

Within learning theory the term disinhibition is used particularly in relation to classical conditioning (**see: Conditioning – types of**) to refer to the spontaneous appearance of a previously extinguished response in association with a novel stimulus. (DSM)

Disorders of behaviour

The term behavioural disorder is often used to describe a syndrome of one or more **abnormal** behaviours exhibited by an individual, and so the term is often used interchangeably with '**problem behaviour**'. However, strictly speaking, the term should refer only to conditions incorporating malfunctional responses, since the disordering of behaviour reduces its function. Disordering produces a variant of a normal behaviour pattern. Malfunctional behaviour will typically be **maladaptive**, but the reverse is not necessarily the case, i.e. a maladaptive behaviour is not necessarily malfunctional, because it is possible for normal adaptive behaviour patterns to be maladaptive if they are expressed in an inappropriate context, e.g. escape attempts when escape is not possible. (KT)

See also: Ethopathy

Further reading

Mills, D.S. (2003) Medical paradigms for the study of problem behaviour: a critical review. *Applied Animal Behaviour Science* 81, 265–277.

Dispersal and habitat selection

Animals that leave an original site, **home range** or **territory** and eventually settle in a new location are called dispersers. Natal dispersal occurs when an individual leaves the area in which it was born and settles in a new area where it will spend biologically significant amounts of time. Many animals make additional movements after natal dispersal, but typically the distances travelled during natal dispersal are longer than those traversed later in life. Hence, for many animals, natal dispersal is the period during which individuals are most likely to move long distances through unfamiliar terrain before settling in a new location.

Within a given species, there is often considerable variation in dispersal tendencies or distances. For instance, in some squirrels, most females settle close to their natal home range, but a few females leave their natal neighborhood and make long-distance movements before settling in a new home range. Sex differences in dispersal tendencies and distances are also commonly observed. In mammals, males usually disperse farther than females, whereas the reverse is typically true for birds. Because the behavioural processes affecting the behaviour of dispersers may vary as a function of dispersal distance (see below), it is sometimes convenient to categorize individuals in terms of their dispersal distances, e.g.: (i) 'non-dispersers' or 'philopatric' individuals, which remain in or very close to their natal site; (ii) 'local dispersers', which settle in new sites within the same neighbourhood; or (iii) 'long-distance dispersers', which travel long distances before settling in a new location.

Within a given population, some individuals are usually more likely to disperse than others. Typically, natal dispersal occurs at a particular age, life stage and/or season. In addition, individual differences in the tendency to disperse or in behaviour during dispersal may be related to the condition of individuals at the time/age or stage of dispersal, conditions in the habitat of origin at the time of dispersal or interactions between these factors.

While it is clear that dispersers often differ from non-dispersers within the same population, we currently lack a general framework to predict which individuals are most likely to become dispersers in any given species. Once in a while, there are clear morphological differences between dispersers and non-dispersers. Thus, insects such as aphids have both winged and wingless forms, and it is reasonably easy to predict that the former would be likely to disperse over longer distances than the latter. However, diagnostic morphological features that can be used to distinguish dispersers from non-dispersers are less often encountered in vertebrates. In naked mole rats (a highly specialized fossorial mammal), dispersers have higher body mass than non-dispersers of the same age in the same cohort, and larger body masses for dispersers have been reported for a scattering of other species. However, there are many other vertebrates in which there are no discernable differences in mass between dispersers and non-dispersers, or in which dispersers are lighter than non-dispersers.

Because dispersal is a behavioural process, behavioural factors are as likely to generate variation in dispersal behaviour as are inter-individual differences in physiology or morphology. A variety of behavioural differences between dispersers and non-dispersers has been reported in insects, and recent studies of great tits, an avian species, indicate that dispersers and non-dispersers differ with respect to a **personality** type (fast–slow), which is also related to a host of other behavioural traits (including aggressive and exploratory behaviour) in this species. These and similar studies suggest that it might eventually be practical to identify potential dispersers before they leave by scoring particular diagnostic behaviours or other traits.

One reason why applied biologists should be concerned about correctly identifying potential dispersers is that these are the individuals most likely to be willing to accept and settle in an unfamiliar habitat. Hence, if the goal is to move animals from a familiar to a novel location in captivity, to release captive-raised animals into the field or to transfer free-living animals from one area to another, success is more likely if the translocated animals are at the age, stage and condition in which they would naturally be most likely to move and settle in a novel habitat.

From the perspective of the animals that are doing the dispersing, it is important to remember that the goal is not simply to travel away from the point of origin, but also (and more importantly) to locate a new high-quality area in which to live. Hence, dispersal is intimately related to habitat selection, where habitat selection is the behavioural process that governs a disperser's decision about when and where to settle after leaving its habitat of origin.

Traditionally, most theoretical and empirical studies of habitat selection have been based on the assumption that animals can freely choose between different types of habitats, and choose the habitat in which they are likely to have the highest fitness after settlement. This assumption is embedded within the most influential general model of habitat selection (the **ideal free distribution**), which assumes that every disperser is free to settle in the habitat in which it will enjoy the highest fitness. Since the ideal free distribution model was introduced in the 1970s, theoreticians have devised many variants of the original model, including models which consider the effects of assessment errors, conspecific or heterospecific competitors, or cooperative conspecifics on the choices animals make when selecting new habitats.

However, until recently, most habitat selection theory has focused on situations in which habitat selection is low cost or cost-free, in the sense that animals are assumed to have an unlimited amount of time (or energy) available to them to visit, assess and select habitats that are available to them. In the context of habitat selection by dispersers, this assumption may be reasonable when individuals disperse over relatively short distances, when travel costs are low and when animals are able to sample many different new habitats prior to choosing a new one. However, this assumption is less likely to be true for long-distance dispersers that have a limited amount of time or energy available in which to search for a new habitat.

Recently, field workers and modellers have become interested in factors that might affect habitat selection for time- or energy-limited dispersers, particularly those that encounter relatively few suitable habitats after leaving their habitat of origin. In this situation, theory suggests that habitat selection is unlikely to be an 'all or nothing' phenomenon, and that individuals should often be more selective (more choosy) at the beginning of the search period than at the end of it. Data supporting this prediction are available for the larval dispersers of many types of marine invertebrates, where declines in selectivity during search are explained by the 'desperate larva' hypothesis. This colourful phrase reflects the main reason why one would expect disperser selectivity to decline over the course of the search period: it may be better to be non-choosy (willing to settle in less preferred types of habitat) than to be dead (as a result of continuing to 'hold out' for a highly preferred type of habitat). Experimental data indicating that selectivity declines during search have also been obtained for insects searching for new habitats, hosts or oviposition sites. However, thus far we lack evidence that this phenomenon occurs in vertebrates.

Another traditional assumption of habitat selection theory is that all of the members of a given population prefer the same type of habitat. That is – with appropriate allowances for errors in habitat classification or assessment – most current studies assume that every individual in a population should rank alternative potential habitats in the same way. However, there is another body of literature that suggests that this may not always be the case. For instance, the concept of 'habitat **imprinting**' suggests that exposure to particular cues in a natal habitat may increase a disperser's preference for new habitats that contain those same cues.

Similarly, entomologists have found that early experience with a particular type of host, food or oviposition site may increase a disperser's preference for the same type of item, a phenomenon they have called 'preference induction'. With respect to habitat selection by dispersers, these and related terms fall under the 'umbrella concept' of natal habitat preference induction (NHPI), which includes a variety of processes by which experience in a natal habitat increases a disperser's level of preference for new habitats that contain cues comparable to the cues in its natal habitat. Thus far, NHPI has been documented in a wide range of taxa, including social and non-social insects and a scattering of vertebrates.

The concept of NHPI has a number of interesting implications for problems in applied biology. Indeed, many of the best empirical studies of NHPI to date have been conducted by entomologists whose goal was to use parasitoid wasps as biological control agents. These scientists have identified a number of ways in which early experience with a given host and with habitat features associated with that host can affect the habitat preferences of newly emerged parasitoids. One impetus for this line of research was a series of failed attempts at biological control, when it was found that parasitoids raised in captivity on convenient artificial hosts and habitats were ineffective at controlling their target host after being released into natural landscapes.

In addition, NHPI has interesting and potentially important implications for problems in conservation biology, particularly those involving the transfer of animals born and raised in one habitat to a new, unfamiliar habitat. Success rates for captive release and translocation studies are often low because the newly released animals refuse to settle at the new location, and instead emigrate away from the release site into surrounding areas. Problems inducing animals to settle in unfamiliar habitats occur even when other evidence indicates that these habitats are highly suitable for the members of that species. An NHPI approach argues that a mismatch between the experience and cues in the habitat of origin and conditions and cues in the new habitat may contribute to an individual's reluctance to settle in habitats that are perfectly suitable for the members of their species. Hence, a perspective on habitat selection that assumes that habitat preferences are not fixed, but are instead affected by an individual's experience in its previous habitat, may be helpful for increasing the success rates for management programmes that have the goal of inducing animals to settle in habitats that are new to them. (JAS)

See also: Conspecific attraction; Reintroduction; Translocation

Further reading

Bullock, J.M., Kenward, R.E. and Hails, R.S. (2002) *Dispersal Ecology*. Blackwell Publications, Malden, Massachusetts.

Clobert, J., Danchin, E., Dhondt, A.A. and Nichols, J.D. (2001) *Dispersal*. Oxford University Press, Oxford, UK.

Stamps, J.A. and Swaisgood, R.R. (2006) Some place like home: experience, habitat selection and conservation biology. *Applied Animal Behaviour Science* 102, 392–409.

Displacement behaviour

A displacement behaviour usually arises when there is motivational conflict or **frustration**, and is a normal behaviour that is performed in a different, and apparently inappropriate, situation from the one in which it is usually performed. A typical displacement behaviour in cats is a grooming lick between the shoulder blades, and it might occur if a cat suddenly appears in an unexpectedly aversive environment (e.g. comes downstairs and sees there is a dog in the house). It is often a simple behaviour that is of no relevance to the current situation. (KT)

See also: Emancipation

Further reading

Rowell, C.H.F. (1961) Displacement grooming in the chaffinch. *Animal Behaviour* 9, 38–63.

Sherwin, C.M. and Nicol, C.J. (1993) A descriptive account of the pre-laying behaviour of hens housed individually in modified cages with nests. *Applied Animal Behaviour Science* 38, 49–60.

Tinbergen, N. (1951) *The Study of Instinct*. Oxford University Press, Oxford, UK.

Display behaviour

Display behaviour is a specially evolved pattern of behaviour that functions to convey information between individuals. The 'specially evolved' caveat relates to those aspects of an animal's behaviour that are highly exaggerated, conspicuous and ritualized (**see also: Ritualization**), which allow the distinction between display and **communication** to be made. Displays are usually associated with either mating rituals (**see: Mating behaviour and systems**) or contests over resources (**see: Agonistic behaviour**), and generally function to aid information transfer between individuals, with the vigour or magnitude of the display supposedly providing honest and reliable information about the signaller.

For example, during animal contests, displays tend to enable competitors to resolve conflicts without resorting to costly physical combat, because the display provides reliable information about the fighting ability of the contestants. However, displays do not always convey reliable information as they may manipulate the receiver into behaving in a particular way that is beneficial to the sender and costly to the receiver. For example, during contests over cavities in rocks and coral, mantis shrimps (Stomatopoda) use the 'meral spread' threat display (the spreading of the raptorial and walking appendages) as an indicator of motivation and aggressiveness during an encounter. This is usually a reliable indicator of fighting ability, although it is known that some newly moulted individuals produce this display despite the fact that they have soft exoskeletons and would thus be poor competitors during a contest. With manipulation comes counter-selection on the part of the receiver not to be manipulated by the deception (**see: Deceit behaviour**). Thus, receivers should (in theory) not respond to displays (or elements of the display) that convey dishonest information. In effect, receivers select for honest signallers, although the selective pressure to manipulate or deceive others will ensure that displays are not 100% honest 100% of the time. (PE)

Dissection

The cutting up of a body to reveal its parts. Normally as an educational exercise in courses on anatomy, principally for medics, professions allied to medicine, veterinarians and zoologists. (DBM)

Distress

Derived from the Greek word *dus* and having a connotation of poor function, or malfunction. Applied to **stress** it infers that an animal is not coping with its situation in some way, leading to a stress response involving stimulation of a sequence of glands, i.e. the **hypothalamus** releases hormones that stimulate the **pituitary gland** to release hormones that stimulate the cortex of the **adrenal gland** to release **corticosteroids** that affect various body systems in various ways. For distress to occur, it has to persist to a point where it becomes pathological in some way, i.e. outside the normal range of an adaptive stress response. Some use the term distress to describe any situation where there are raised blood corticosteroid levels. However, this is too blunt an approach, as corticosteroids can also rise during exercise and during pleasurable events. Furthermore, distress can be used to describe short-term experiences, such as transient pain, as well as the long-term effects of not coping, such as fear, frustration and so on. (DBM)

Diurnal pattern or rhythm

A diurnal pattern is a biological pattern occurring with a pattern of about 24 h. It should not to be used synonymously with **circadian rhythm**, even though they both have approximate 24 h rhythmicity. Strictly speaking, a true circadian pattern is endogenously generated – i.e. exists without external cues. Diurnal rhythms are generated in response to external cues such as the day–night cycle. Both free-living animals and domestic animals housed in a variety of systems show strong diurnal patterns in behavioural activity. For free-living animals, the major factor influencing behaviour is the need to acquire nutrition. The amount of time spent each day engaged in feeding-related activities will vary depending on whether the animal is a herbivore, omnivore or carnivore but, in all cases, will still show diurnal variation.

Herbivores may show variation in activity related to when it is coolest, warmest or safest to forage. Many free-living ungulates show peaks in activity around sunrise and towards sunset, when there are thermoregulatory benefits. Not surprisingly, carnivores will match their activity patterns to their prey species – for example, African wild dogs make most of their kills in the 2 h after sunrise and in the 1 h after sunset. As with foraging, other behaviours will also show diurnal patterns and synchronicity within the herd/group. For example, free-ranging pigs may intersperse periods of foraging-related activity with resting, social interaction, wallowing, etc.

Within the applied animal behaviour and welfare field, diurnal patterns have been most studied with regard to behaviour. As well as general activity, specific behaviours studied include **stereotypies**, **feeding**, **dust bathing**, drinking, standing, walking and feather pecking. With pigs, it has been shown that **bar biting** of confined gestating sows shows a diurnal pattern, related to feeding time, with a peak in occurrence over the 2 h immediately following **feeding**. Pigs show diurnal patterns in lying behaviour and in posture, showing an increase in lateral lying and lying in contact with pen-mates during the night, and these amounts can be influenced by fluctuations in the external temperature. Drinking by newly weaned piglets shows a diurnal pattern, with maximum intake around late afternoon and minimum intake before dawn. It has been speculated that this pattern is so stable in healthy pigs that changes in pattern could be used automatically to detect disease.

Poultry have diurnal patterns in feeding behaviour and, unsurprisingly, **perching** behaviour that can be influenced by day length. Free-range broilers show maximal outdoor use at sunrise and sunset, and those birds subject to natural light as opposed to artificial light show earlier perching, presumably as light gradually begins to fade. For broiler-breeders that are fed restricted diets, diurnal patterns exist in resting, preening, standing, walking and stereotyped pecking. Poultry also show diurnal patterns in dust bathing and **feather pecking**.

For grazing ruminants such as dairy cattle and captive red deer, there are again diurnal patterns of activity, but superimposed upon these are **ultradian rhythms**, most probably associated with rumination. Horses also show ultradian patterns in grazing superimposed upon diurnal patterns. Again, for grazers, patterns can be influenced by management, and diurnal patterns of dairy cows have been shown to differ between a continuous stocking system and a strip-grazing system.

Diurnal patterns have also been investigated physiologically, but there is the difficulty in discerning whether patterns are circadian (i.e. endogenous) or truly diurnal, and also the confusing use of the terms diurnal and circadian as if they were interchangeable. For example, researchers have investigated heart rate over the course of 24 h and found it to be higher during daytime than during night-time. However, activity levels are also higher during daytime and, although it has been reported that a circadian pattern exists for heart rate, without evidence of rhythmicity in the absence of light cycle and activity patterns, the pattern should be described as diurnal rather than circadian.

In general, investigations of diurnal patterns have not been fully exploited in animal **welfare** research, and many more opportunities exist for their use as a welfare assessment tool.

(RM-F)

Further reading

Berger, A., Scheibe, K.M., Eichorn, K., Schiebe, A. and Streich, J. (1999) Diurnal and ultradian rhythms of behaviour in a mare group of Przewalski horse (*Equus ferus przewalskii*), measured through one year under semi-reserve conditions. *Applied Animal Behaviour Science* 64, 1–17.

Kjaer, J.B. (2000) Diurnal rhythm of feather pecking behaviour and condition of integument in four strains of loose-housed laying hens. *Applied Animal Behaviour Science* 65, 331–347.

Madsen, T.N. and Kristensen, A.R. (2005) A model for monitoring the condition of young pigs by their drinking behaviour. *Computers and Electronics in Agriculture* 48, 138–154.

DNA analysis

Deoxyribonucleic acid (DNA), the 'building blocks of life', consists of a double helix of two strands of alternating sugar (deoxyribose) and phosphate groups linked to nitrogenous bases (the four bases are adenine, cytosine, guanine and thymine: thymine always pairs with adenine, and cytosine with guanine). These bases are arranged in specific sequences that code for the production of proteins. The study of genes and their sequences is known as genomics, and the translation of the genes into proteins as proteomics.

There are thought to be around 30,000 genes in the mammalian nucleus, many of them shared between the various species. DNA analysis aims to sort out the sequences within the genome and to identify the sequence of each gene – as well as the flanking sequences of non-gene-coding DNA – and their functions. Some of these flanking sequences repeat at intervals (micro- and mini-satellites) that enable individuals to be identified, forming the basis of DNA fingerprinting. It is so discriminating that it is believed that every human being on this planet can be identified, apart from identical twins. The genetic basis for behaviour is a growing subject of research at the molecular level. (DBM)

Docking – tail

Tail docking is performed on a variety of species: **pigs**, **sheep**, **horses**, **dogs** and, most recently, **cattle**. The reasons for tail docking are to prevent injury to the tail, to prevent spread of **disease**, to improve appearance or to facilitate harnessing. The welfare issues concerning tail docking fall into two categories: (i) the **pain** caused by the initial removal; and (ii) the consequences of loss of the tail to the animal's long-term **welfare**, behaviour and comfort.

Pain

The method of tail docking differs from species to species and country to country. In general, puppy, piglet and horse tails are cut off, but lambs and cattle may be docked using either knives or bands. When the two methods are compared, the initial pain is greater with cutting, but the ischaemic pain of the bands lasts longer. The signs of pain following docking also vary with the method. Puppies shrieked a mean of 25 times when their tails were amputated (Noonan *et al.*, 1996). Cattle abduct their tails. Lambs show abnormal postures for days after banding (Wood and Molony, 1992). Piglets vocalize at the time of docking, but less so when in response to an injection of penicillin. There are many other behavioural responses as well, but these remain to be quantified.

Consequences of docking

The function of the tail elucidates the welfare problems that might arise from its loss. Tails are used in **communication** in almost all species and for insect control in ungulates. Finally, in carnivores, and possibly in other classes of animals as well, the tail is used for balance. Tail position is an especially important signal in dogs. The stiff upright tail indicates arousal and **aggression**. The wagging tail indicates anticipation, usually pleasurable. The lashing tail in the horse, cow and cat is a sign of frustration. Raised tails function as a **play** signal in calves and can be a sign of oestrus in many species.

The depressed tail, held close to the body, indicates **fear** in almost all species. Both calves and lambs wag their tails while suckling (**see: Nursing**). In conventional pigs, the curl of the tail is an indicator of well-being in that a flaccid tail connotes fatigue or discomfort. In piglets, the tightly curled, tense tail indicates that the sow is letting down milk, i.e. the piglet is swallowing milk. Whether pigs use tail position and curliness as signals has not yet been ascertained. Some breeds, notably Vietnamese pot-bellied pigs, do not curl their tails.

The balance function of the canine tail is most apparent when the dog has developed hind limb lameness or pain. At that time, the dog will deflect the tail in an exaggerated manner, presumably to help keep its balance while minimizing weight bearing by the painful limb. No one has compared the number of falls suffered by docked dogs such as Rottweilers and Australian shepherds with those of undocked dogs of the same breed. These dogs may have learned to compensate for the loss of their tails.

Tail docking of draught horses, hackney ponies and other horses that wear cruppers (a belt attached to the saddle and looped around the tail head to prevent the saddle slipping forward) renders them vulnerable to insect bites. The long-haired tail of the horse functions as a more efficient fly swatter than the tufted tail of cattle. The number of flies settling on docked horses or the frequency of self-grooming and rubbing movements have not been quantified. Because the tail functions in communication, one would expect aggression to escalate more often because the conspecfic was not aware of the threats implied by a lashing tail. The docked horse would have to lift its hind limb or actually kick rather than simply lash its tail.

The history of tail docking of cattle is recent and, for that reason, this **mutilation** may be the easiest to legislate against. Originally, tail docking was performed to reduce the spread of leptospirosis from cow to person or cow to cow in New Zealand. The practice spread to the USA, where it was alleged to reduce the risk of **mastitis** because the cows would not become so dirty. In fact, in the free-stall situation, docked cows have neither a cleaner coat nor a lower incidence of mastitis than intact cows (Stull *et al.*, 2002).

The real reason for the popularity of bovine tail docking is that it prevents the milker from being hit in the face with a manure-laden tail. One alternative is simply to clamp the tail out of the way during milking; another is a system in which the cow is above the milker so that the human's face is out of tail range. Finally, **robotic milking** could solve the problem entirely.

Pigs are tail docked at birth or within a few days of birth. The reason for docking porcine tails is the problem of **tail biting**, a sporadic behaviour that injures the victim's tail and can cause ascending infections that will lead to condemnation of the most valuable part of the carcass, the hams. For that reason, some **slaughterhouses** will not accept pigs with intact tails. Farms that do not have tail-biting problems are, therefore, forced to dock tails. Tail biting apparently arises when pigs are foraging and encounter a tail that is similar in size and shape to the worms and snakes they might eat in a free-ranging situation. Once the tail is bitten and the pig tastes blood, it may bite more frequently because it apparently prefers iron-rich substrates such as blood. Iron-rich supplementation might help reduce levels of tail biting. Bedding in which the pigs can forage would probably also reduce the problem.

(KAH)

See also: Castration

References

Noonan, G.J., Rand, J.S., Blackshaw, J.K. and Priest, J. (1996) Behavioural observations of puppies undergoing tail docking. *Applied Animal Behaviour Science* 49, 335–342.

Stull, C.L., Payne, M.A., Berry, S.L. and Hullinger, P.J. (2002) Evaluation of the scientific justification for tail docking in dairy cattle. *Journal of the American Veterinary Medical Association* 220, 1298–1303.

Wood, G. and Molony, V. (1992) Welfare aspects of castration and tail docking of lambs. *In Practice* 14, 2–7.

Dog

The canid or dog family consists of true foxes (vulpini, which consist of species of the genera *Vulpes* and *Urocyon*), basal canids (or basal dogs, which include the raccoon dog and bat-eared fox) and the more diverse true dogs (canini). The latter include the genus *Canis*, which consists of coyotoes, the Ethiopian wolf and various genetic lines of the grey wolf (*Canis lupus*), including dingos, domestic dogs, Arabian wolf, Indian wolf, red wolf and the New Guinea singing dog, some of which are considered distinct species in some texts. Also included among the true dogs are a range of other genera whose members often include the terms 'dog', 'fox' or 'wolf' in their title, such as the African wild/hunting dog (genus *Lycaon*), the Asian wild dog (genus *Cuon*, also known as the Dhole – it is worth noting that several other genera also include the name Dhole), bush dog (genus *Speothos*), crab-eating and Falkland Island fox (genus *Cerdocyon* and *Dusicyon*, respectively) and maned wolf (genus *Chrsocyon*).

Domestic dogs are believed to have diverged from the ancestral wolf in excess of 15,000 years ago, possibly up to 50,000 years ago, although some estimates place an upper limit potentially in excess of 100,000 years ago. Since this time the two have evolved in their own niches and so comparisons between the domestic dog and modern grey wolf (especially the North American timber wolf, which is so often pictured as the 'proto-dog' in popular texts) are of limited value, and the latter is not representative of the progenitor of the former; rather, they are perhaps best considered more like 'cousins'. It is generally accepted that the success of dogs is inextricably linked to the success of humans and the close relationship that has existed between them; indeed, the two may have co-evolved and have certain adaptations to aid their intra-specific communication (see below). One feature of the ancestral wolf that may have given rise to this potential is its natural range around the glacial edges, which would have favoured the conservation of genetic diversity given the relatively unstable environment.

Wolf–dog analogies are common, but there are several important limitations that need to be considered when such comparisons are made. First, as already mentioned, most comparisons appear to be with certain subspecies of North American grey wolf (of which there may be around 20) rather than with some of the potentially more relevant Asiatic subspecies. There are around a dozen subspecies of Asiatic lupine wolf, and genetic studies suggest that only certain lines within some of these may have contributed significantly to the domestic dog population, emphasizing the danger of generalizations concerning the relationship between dog and wolf.

A second problem concerning such analogies focuses on comparative behavioural studies. Much of the information gleaned on wolf behaviour is derived from studies of captive wolf populations. In this context competition for resources and **social stress** may be greater, as **space** is more limited and feeding more controlled externally by managers. As a result there may be an overemphasis on behaviours related to **resource-holding potential** and competition, especially **aggression** and **agonistic behaviour**, and the importance of **hierarchy**, at the expense of recognition of the role of **affiliation**, **cooperation** and **bonding** in its **social behaviour**. Some have even suggested that, since wolves are not adapted to captivity through **domestication**, using the behaviour of wolves in captivity to understand dogs is like using the behaviour of dysfunctional people and those with mental health problems to understand the normal behaviour of humans.

A third potential problem with wolf–dog analogies relates to the changes that have occurred in the dog as a result of domestication. It seems that this may have resulted in both physical and behavioural paedomorphosis compared with the wild type, the young being more playful and cognitively flexible. In which case, analogies should be between the behaviour of immature rather than adult wolves (in whom hierarchy is much less rigidly defined) and adult dogs.

On top of this are the changes associated with selection for diverse functions in specific **breeds** of dogs, which are also likely to affect their behavioural style. The diversity in form and behaviour of dog breeds may reflect both the diversity of the ancestral population as well as pleiotropic effects associated with the processes of domestication and the selection for tameness (which appears to result in additional phenotypic changes), as well as specific selection for certain traits during the process of breed diversification.

In addition to obvious differences in physical form, it appears that the structure and functionality of the eye may also vary with conformation (McGreevy *et al.*, 2004), with long-nosed breeds (doligocephalic) possessing a high acuity area within a visual streak, and short-nosed breeds (brachcephalic) more a focal spot (similar to primates). The former may be better suited to scanning the horizon and the latter for focusing on closer objects – and especially faces. This may explain in part why the ancient lapdogs tend to be shorter nosed and the sight hounds long nosed.

Although breeds of dogs are classified by Kennel Clubs into various functional groups (working, terrier, pastoral, etc.), this is not a phylogenetic classification, and distantly related breeds may appear in the same group. Genetic analysis suggests four broad groups (Parker *et al.*, 2004): (i) an ancient, largely Asiatic, group, containing the Shibu Inu, chow-chow and Akita, but also the basenji and malamute; (ii) a mastiff group, including boxers, Rottweilers and bulldog types; (iii) a herding group, consisting of many sheepdogs and the borzoi; and (iv) a more recent working and hunting group, including many terriers, spaniels, setters, etc., as well as the Border collie.

Originally, breed was linked to function but, with the advent of breed societies and showing, the form of the animal is now more often linked to a more arbitrary breed standard, often with an emphasis on physical appearance rather than function. This, together with the closure of breed stud books, has resulted in very restricted gene pools and, combined with a very high level of **inbreeding**, has resulted in a very high prevalence of genetic diseases in dogs. While some of these may be used as models of human conditions, their widespread distribution in the general population is a serious cause for concern.

Dogs appear to show an inherent sensitivity to human stimulus enhancement, although this may vary between breeds. Hand-reared and pet dogs also show greater attachment and attention to humans compared with hand-reared wolves, and this may explain in part some of the difference in learning performance between the two species, with the latter being more independent and less likely to seek assistance from a human handler during problem-solving tasks. The tendency to human attachment may also partly explain some of the problems seen in pet dogs, such as **separation anxiety**. Other adaptations that may have historically aided the development of the human–dog relationship include the playfulness of dogs and their social cognitive ability, especially in tasks relating to interspecific **communication**. This includes their capacity for specialist processing of information within human faces (Guo *et al.*, 2008) and the use of barking, which, intuitively, humans seem to be able to relate consistently to specific contexts (Pongracz *et al.*, 2005).

It has been hypothesized that dogs became domesticated as a result of scavenging from waste sites following the development of human settlements, with the advantage of obtaining substantial protein sources – possibly cooked and more readily digestible – with relatively little effort, providing strong selective pressure against **neophobia** and towards attraction to humans. Thus dogs may be considered omnivorous scavengers rather than obligate carnivores. Although dogs seem to be highly gregarious in the feral state, they do not necessarily form packs, especially when food resources – such as those that can be provided in association with people – are abundant.

Dogs exist within society in diverse contexts, ranging from the wild and feral through the free-ranging and cared for but unowned, to the owned and potentially highly restricted **pet**, working and sport animal, **laboratory animal** or **farmed animal**. The **welfare** issues obviously vary with context (see Stafford, 2007, for further details). Free-roaming dogs may directly and indirectly impinge on the welfare of others, as they are a major cause of road traffic accidents, zoonotic infections and damage to livestock, although they may also be predated upon by large carnivores, including other canids.

Because bitches typically cycle twice a year, overpopulation can rapidly become a significant problem and, because dogs are a common vector for human infection by rabies, many regions take measures to control **feral** animals. **Abandoned animals** are also a significant problem, with more than 1% of the global population euthanased each year as a result. Neutering is encouraged to reduce unwanted and stray animals, but its effectiveness is often limited in practice. Neutering does, however, also have beneficial effects for the individual (reduced risk of mammary tumours and elimination of risk of infections of the womb in the female, reduced risk of prostatic diseases in the male, reduced roaming and other male typical behaviour), but may also increase reactivity and the risk of developing certain fears.

Problem behaviour is the most common reason for the relinquishment of dogs to **animal shelters** and is also a major reason for the **euthanasia** of pets. Problem elimination and lack of training appear to be particular behavioural risk factors for relinquishment, along with other non-behavioural factors such as not being neutered, being young, of mixed breed, obtained at little cost and unrealistic owner expectation. Aggression is a common reason for relinquishment, but is also quite prevalent in non-relinquished dogs and the most common reason for owners to seek professional behavioural advice. Other common reasons include separation problems and specific fears, which are also a cause for welfare concern.

Training is integral to the utility of the dog kept in most forms of captivity, whether it be a pet or used in work or sport. The methods used are increasingly attracting attention from both a welfare and ethical perspective (**see: Ethics**). There is growing advocacy for reward-based methods, but it is important to recognize that no training can be purely reward-based, since the absence of reward is a form of negative punishment (**see: Reinforcement – types of**), and failure to acknowledge this may result in the development of inappropriate or less effective training schedules and psychological distress. The use of strict schedules of positive reinforcement may also reduce the dog's ability to both problem solve independently and cope in psychologically stressful situations, and therefore potentially increase the risk of emotional problems. The training of dogs is therefore a complex but important welfare issue deserving further attention.

The dog was the first species to become domesticated and is one of the most ubiquitous species on the planet. It exhibits great variability, in both form and function, which arises as a result of latent potential both phylogenetically and ontogenetically. Their success has been supported by their close association with humans, and has resulted in them showing some remarkably human-like abilities. However, a rise in interest and acceptability of the dog as a topic for behavioural research suggests that this may often be the result of their great perceptual abilities and accommodation of people, and not due to similar processes at a mechanistic level. As a result the continued study of the dog may offer new insights into a range of important biological processes. (DSM)

References and further reading

Cooper, J.J., Ashton, C., Bishop, S., Ferguson, A., McKinley, S., West, R., Mills, D.S. and Young, R.J. (2003) Clever hounds: social cognition in the domestic dog (*Canis familiaris*). *Applied Animal Behaviour Science* 81, 229–244.

Guo, K., Meints, K., Hall, C., Hall, S. and Mills, D. (2008) Left gaze bias in human infants, rhesus monkeys and domestic dogs. *Animal Cognition* (DOI 10.1007/s10071-008-0199-3).

McGreevy, P.D., Grassi, T.D. and Harman, A.M. (2004) A strong correlation exists between the distribution of retinal ganglion cells and nose length in the dog. *Brain, Behaviour and Evolution* 63, 13–22.

Miklosi, A. (2007) *Dog Behaviour, Evolution and Cognition*. Oxford University Press, Oxford, UK.

Parker, H.G., Kin, L.V., Sutter, N.B., Carlson, S., Lorentzen, T.D., Malek, T.B., Johnson, G.S., DeFrance, H.B., Ostrander, E.A. and Kruglyak, L. (2004) Genetic structure of the purebred domestic dog. *Science* 304, 1160–1164.

Pongracz, P., Miklosi, A., Molnar, C. and Csanyi, V. (2005) Human listeners are able to classify dog barks recorded in different situations. *Journal of Comparative Psychology* 119, 136–144.

Stafford, K. (2007) *The Welfare of Dogs*. Springer, Dordrecht, The Netherlands.

Domestication

Domestication is the process of enfolding a species into human society. It is distinct from both taming and the husbandry of wild animals in **captivity**. Only a very few animal species have proved capable of being domesticated, though any mammal or bird, if taken into husbandry at a very young age, will imprint on its human carers, becoming tamed as a result and accepting the carer as a socially dominant parent. However, on reaching adulthood, non-domesticated species exhibit their natural adult behaviours and can prove unmanageable. In contrast, domesticated species continue to accept the social **dominance** of humans into adulthood. The species that have proved amenable to domestication all have natural patterns of social behaviour that can be moulded in this way as part of the domestication process, typically, including dominance hierarchy and lack of territoriality.

In several species, notably the **dog**, many adult behaviours are in fact juvenile behaviours retained past the age when, in the natural state (i.e. in the progenitor species, the grey wolf) they would be superseded.

Presumably domestication, which for animals was first achieved about 10,000 years ago (though perhaps much earlier for the dog), involves genetic changes, though what these were remains unknown. Taming of young animals taken from the wild would have been the first step, and individuals that exhibited a greater response to the taming process would have been selected. Likewise, a tendency to retention of juvenile behaviours would be favoured.

Species differ in the completeness of the enfolding process. Fish and invertebrates probably do not have cognitive abilities of the kind that permit this enfolding. Considering mammals, Clutton-Brock (1987) visualized a spectrum, from 'man-made animals' (dog, **sheep**, **goat**, **pig**, **cattle**, equines) to 'exploited captives' (**cat**, **elephant**, **camelids**, reindeer, water **buffalo**, yak, Bali cattle and mithan). Small mammals (**rabbit**, ferret, **rodents** and fur-bearing carnivores) have been bred in captivity for much shorter periods of time, so should be regarded as being in the early stages of domestication. A similar spectrum can be envisaged for birds, with the **chicken**, duck and **turkey** as the most completely domesticated, and guinea fowl and **ostriches** less so. (SJGH)

See also: Buffalo; Farmed animals; Flight zone; Free-range animals; Imprinting; ***individual species entries***

Reference

Clutton-Brock, J. (1987) *A Natural History of Domesticated Mammals*. Cambridge University Press/British Museum (Natural History), Cambridge and London, UK.

Dominance

The term dominance can be used in both a genetic context and a **social behaviour** context. This article focuses on its use within animal behaviour. Animals that live or come together in groups develop various kinds of relationships with one another (**see also: Grouping**). One form of relationship is of a dominant–subordinate nature, where the dominant animal(s) has a higher status, recognized by influence or control over another or others in the group. Dominance is not an attribute or trait of an individual animal, but is an aspect of a dynamic relationship between two or more individuals, which may be of the same or different species.

For dominance to exist there must be a disparity or difference between the individuals, and this disparity can occur in any one (or more) of the factors or attributes that affect the outcome of an interaction. The disparity must also be detected by the participants and the appropriate responses made. The relationships between individuals can be: (i) learned, based on individual recognition and/or recognition of the disparities and their consequences; (ii) already established, based on cues that are correlated with disparities; or (iii) both mechanisms can operate.

Dominance will usually be most evident in situations in which two or more animals are crowded together or come into close proximity; the dominant individual apparently moves wherever it wishes while the subordinate(s) gives ground and moves away. Thus, the behaviour of the subordinate animal(s) is changed and/or inhibited by the presence of the more dominant animal. Indeed, this principle provides the basis of numerous techniques that have been developed for assessing or measuring dominance. Methods include displacement from a particular resource, such as food or water, or just displacement in space. Agonistic interactions (involving offensive and defensive behaviours such as threats, attacks, submissions and flight; **see also: Agonistic behaviour**) are also recorded because these can reveal 'winners' and 'losers', or whether interactions and relationships are unresolved.

There are many factors that influence the dominance potential of an individual and, thus, the outcome of an interaction. A major influence is reported to be the 'aggressiveness' of the individual, which can be considered to be an animal's propensity for displaying offensive or attack behaviours that have the potential to cause physical injury to another individual (**see also: Aggression**). Aggressiveness is influenced by several factors that are, also, likely to interact and influence each other. Factors affecting aggressiveness include:

1. Hormone levels (particularly the **androgens**). As a result of hormonal influences and concomitant morphology, such as large body size, broad shoulders and large horns or antlers, it is often the case that males are dominant to females. Hormonal changes, such as occur seasonally and according to reproductive status, will influence aggressiveness and dominance, for example during the breeding season and after the birth of offspring (**see also: Maternal behaviour**). Furthermore, hormones affect the expression of secondary sexual characteristics (e.g. antlers), and these characteristics are often non-arbitrary signals or cues of an animal's dominance potential, e.g. antler size is positively correlated with dominance status. In ruminants, the presence of horns affects dominance, with animals possessing horns generally being dominant over those that do not.
2. Genetic predisposition, which means that individuals inherit an aggressive character, or that certain breeds and genotypes tend to be more aggressive/dominant over others. For example, it has been reported that Hereford bulls are generally dominant to Brahman bulls. Further, it has been suggested that physical fighting skills, such as agility and speed, are likely to be partially inherited.
3. Social and other experiences. Experience can take two forms: (i) being involved in an interaction and emerging the winner or loser; and/or (ii) learning social skills from others in the group. Research has shown that, in comparison with socially experienced heifers, naive ones in their first encounter spent longer fighting, were less submissive and took longer to establish their first social relationship, which was less stable. These findings indicate that some social skills, at least, have to be learned. Animals that have a history of winning encounters are more likely to be dominant than those with a history of losing. Previous winners tend to increase their readiness to attack, while consistent losers appear to generalize to all encounters and perform submissive behaviours that result in termination of the interaction. As a consequence they are likely to retain a low rank or status. Additionally, dominance potential can be influenced by the dominance status of the individual's parents. This may be a consequence of group members avoiding the young animal when it is close to its mother or father, and then this avoidance being carried over into situations when the young animal is alone. The young animal may also learn aggressive and social skills from its parents. It is probable that all of these factors are involved.
4. Other factors that affect dominance potential in many species, including sheep and goats, are body size and body weight, with larger and/or heavier animals being dominant to smaller/lighter ones. These factors appear less important in cattle, where dominance is correlated best with age and/or length of time spent in the herd. In most species, older individuals are dominant to younger ones until such time that age or injury impairs the physical ability of older animals to fight.

There is also a so-called 'home court advantage', which is said to influence dominance potential, with individuals that are on home ground (**see also: Home range**) or holding/defending a territory (**see also: Territoriality**) having an advantage over intruders during social interactions. It has been suggested that an unfamiliar environment heightens fear and increased defensiveness and, hence, an intruder's intrinsic level of aggression and offensive behaviours will be reduced in contrast to an animal in a familiar environment.

Being the dominant individual is said to have advantages, such as having access to the highest-quality resources or, if resources are restricted and defendable, access to the resources at the expense of other, less dominant, animals. If such a system operates, then it is probable that the **welfare** of the dominant animal will be enhanced. The corollary is that the welfare of subordinate animals could be poor. However, being dominant also has associated 'costs' for the individual in some situations, as dominance may require constant vigilance and defence of resources, which may reduce the animal's welfare and **fitness**. It is probable that there are degrees of subordination, with some individuals that will frequently challenge others in order to elevate their status, while others will rarely or never do so. In some circumstances the welfare and fitness of these latter animals will be better than any other animal in the group, but rarely would the welfare and fitness of the former be good.

Whether dominance influences access to resources appears to be species- and situation-specific; for example, in one study subordinate goats were found to be heavily parasitized as a consequence of having access to food prevented by dominant animals until the food had become spread out, trampled and contaminated. In contrast, in a study of beef cattle with severe restriction on feeding space, but more or less *ad libitum* availability of food, dominant animals did not prevent subordinates from accessing the feeding stall. Dominant cattle have, however, been shown to select the 'best' resting sites, which could adversely affect subordinates if they were forced to rest in adverse microclimates, such as unsheltered sites or areas with poor air quality (e.g. high levels of noxious gases, high temperature, high humidity).

There are very evident adverse consequences for welfare, such as physical exhaustion and injuries, in situations where there is frequent aggression and fighting, as a result of disruptions to social relationships. For these reasons, mixing of unfamiliar animals should be kept to a minimum. Social tension may not always be manifested overtly, e.g. through physical interactions, but this does not mean that individuals are not experiencing **social stress**. Social stress, like other kinds of stress, if sufficiently severe, can have profound effects on welfare, such as increased susceptibility to **disease** and illness as a consequence of impaired immune function. (JCP)

See also: Hierarchy

Dopamine

Dopamine (DA, 3,4-dihydroxyphenylethylamine) is the immediate metabolic precursor of **norepinephrine** and **epinephrine**. DA itself is a neurotransmitter activating dopamine receptors, and a neurohormone inhibiting **prolactin** release from the **pituitary gland**. DA cannot cross the blood–brain barrier. Within the **central nervous system** (CNS), DA is synthesized from the amino acid L-tyrosine, derived from dietary proteins and phenylalanine metabolism in the liver, which is taken up into dopaminergic **neuron(e)s** from plasma. Tyrosine is hydroxylated by the enzyme tyrosine hydroxylase (TH, rate-limiting enzyme) to 3,4-dihydroxyphenylalanine (L-DOPA). L-DOPA is decarboxylated by L-aromatic amino acid decarboxylase (DOPA decarboxylase) to form DA.

Following synthesis, DA is transported to axon terminals and stored in synaptic vesicles. It is released into the synaptic cleft in response to depolarization of the axon terminal (action potentials). There are several metabolic pathways of the released DA, e.g. coupled to dopaminergic receptors, retaken by axon terminals and astrocytes, or metabolized by monoamine oxidase (MAO). Homovanillic acid (HVA) is a DA metabolite. The content of central DA is maintained via: (i) regulation of TH activity; (ii) regulation of MAO activity; and (iii) feedback inhibition by catecholamine compounds.

In the CNS, DA mediates a wide range of brain functions involved in motor activities, motivational and sensory information, domestic behaviour, **reproduction** and **coping** strategies to stressful conditions. The actions of DA depend on multiple factors, including the characteristics of its receptors. Two basic receptors, D1 and D2, have been identified based on the characteristics of functions in pharmacological and biochemical mechanisms of signal transduction. All D1 receptors are located on postsynaptic structures, whereas D2 receptors are found on both pre- and postsynaptic structures, and all DA auto-receptors (receptors activated by DA for regulating its self-release from presynaptic terminals) located on somatodendritic sites or on presynaptic terminals are of the D2 type.

Other subtypes of DA receptors have been isolated and categorized as D1-like or D2-like according to their nucleotide sequences and their pharmacological expressions. The D1-like receptors include the D1 and the D5 receptors, whereas the D2-like receptors include D2, D3 and D4. The D1-like and D2-like receptors display differing affinities for the second messenger system in transferring signals. The D1-like receptors activate adenyl cyclase, whereas D2-like receptors inhibit adenyl cyclase activity and suppress Ca^{2+} current but activate K^+ current.

In the CNS, as a neurotransmitter, DA mainly presents in three nerve systems: (i) the dopaminergic nigro–neostrial pathway, in which dopaminergic **neuron(e)** bodies are located in the *substantia nigra* of the midbrain and send axon terminals to the caudate nucleus–putamen complex (the *neostriatum*); (ii) the dopaminergic midbrain–mesolimbic forebrain system, where the nerve cell bodies are located in the ventral tegmentum of the midbrain and send axons to the head of the caudate nucleus, the *nucleus accumbens*, the *tuberculum olfactorium*, the amygdaloid nuclei and the cortex; and (iii) the dopaminergic tubero-infundibular system, where the neuron bodies lie in the region of the arcuate nucleus of the hypothalamus and the axons terminate in the median eminence.

Each dopaminergic system has distinctive functions in regulating physiological **homeostasis**. The dopaminergic nigro–neostrial pathway is involved in the integration of incoming sensory stimuli and regulation of movement (controlling muscle tone and movement). The degeneration of dopaminergic neurons of the nigro–neostrial pathway is the hallmark of Parkinson's disease, and of several other neurological and psychiatric disorders. The dopaminergic midbrain–mesolimbic forebrain system regulates cognitive, reward, mood, arousal and sensorimotor integration.

Experimental studies show that selective lesions of the dopaminergic neurons of the midbrain mesolimbic forebrain system cause cognitive deficits in rats and primates, especially when the mesocorticolimbic component of the DA is altered. The dopaminergic tubero-infundibular system is involved in neural regulation of the functions of the **hypothalamic–pituitary–adrenal (HPA)** system in response to internal and external stimuli. The endogenous DA secreted in the hypothalamus also exhibits catecholamines' tonic inhibition of luteinizing hormone-releasing hormone (LHRH) release. Administration of exogenous DA also causes a reduction in the LHRH level from the **hypothalamus** and less **luteinizing hormone** (LH) secretion from the pituitary. These effects of DA could have been verified by administration of the DA receptor agonist, 2-Br-alpha-ergocryptine, or blocked by use of DA receptor blocker, haloperidol.

Although each of the dopaminergic sites has distinctive functions in the regulation of physiological homeostasis, they act in a complicit pattern in the **brain**. The effect of DA on one brain site influences and may even act in opposition to that on another region. It has been indicated that DA activity in the prefrontal cortex can inhibit DA activity in the *nucleus accumbens*. However, acting at the basal ganglia, DA appears to strengthen the capacity of the lower stress-response control centre. Thus, stimulation-induced DA actions are brain region- and their connection-dependent.

Pharmacological studies also indicate that the effects of administrated chemical compounds, such as amphetamine, on physiological and psychological homeostasis are dependent on the brain regions where the interactions between the compounds and DA are conducted. Amphetamine is a reversible inhibitor of MAO, and causes displacement of DA from the storage vesicles into the cytoplasm. Amphetamine acts in promoting activity, such as exploration, at the *nucleus accumbens*, but in the *ventrostriatum* it acts in increasing the tendency to perform stereotypic behaviours.

In particular, as a key neuroregulator, DA affects behavioural exhibition and contributes to behavioural adaptation in response to various stimulations. Disturbance of DA properties has been linked to psychiatric illness such as mood disorders (unipolar and bipolar), social anxiety disorder and **depression**; and dysfunctional behaviours including aggressiveness, stereotypic and compulsive behaviours, as well as changes of cognitive function. Greater concentrations of DA have been found in the brains of Japanese quails with aggressive behaviour.

Alterations in DA concentrations are also linked to stereotyped cage pecking and **feather pecking** in birds. Feather pecking can be reduced by subcutaneously injected haloperidol (a dopamine blocker) and D2 receptor antagonist. Similarly, in the bidirectionally selected Roman high (RHA/Verh)- and low (RLA/Verh)-avoidance rats, increased DA activity occurred in the RHA/Verh rats, associated with increased locomotor activity, a marker of anxiety and stressful status. In addition, experimental stress-induced aggression in mice is associated with heightened levels of DA in the brain. The aggressive reaction can be decreased – facilitated either by DA administered centrally or L-DOPA (a precursor of DA) administered peripherally. Aggressiveness can also be facilitated by administration of apomorphine, a D2 receptor agonist, but was suppressed by spiperone, a D2 receptor antagonist, and haloperidol and pimozid, both DA blockers.

Dopamine is a potential substrate for synaptic plasticity and memory mechanism in animals. DA-dependent synaptic plasticity plays important roles in an animal's ability to learn to adapt to its environment. In particular, the dopaminergic system and D2 receptors in the medial prefrontal cortex and limbic system are involved in reward learning and appetitive conditioning under both stressful, emotionally unpleasant stimuli and rewarding, emotionally pleasant stimuli. These DA networks are involved in planning, decision making and finalized movement, which may directly or indirectly affect an animal's capability of coping with its environment. Associations between an increase in DA activity and stress-induced unpleasant learning have been found in animals, such as levels of DA correlating with the tendency to develop **stereotypies**.

In contrast, other studies have shown that stress-related learning capacity is preserved after DA depletion in the CNS by cutting the medial forebrain bundle, or following 6-hydroxydopamine treatment. These studies suggest that the role of DA in learning effects is not uniform, which may be related to differences in types of stimuli and duration and frequency of stimulus presentation, such that controllable foot shock but not uncontrollable shock is associated with DA release in the *nucleus accumbens* in rodents. In addition, the differences in dopaminergic and behavioural responsiveness are consistent with the interpretation that coping strategies of animals are based on the inheritance pattern and phenotypic correlations of behavioural, physiological and neuroendocrine variables. (H-WC)

Double muscling

Double muscling refers to the exceptional muscle development seen particularly in some livestock breeds, most notably Belgian blue and Piedmontese **cattle**, but also in other species such as the Texel **sheep**. It arises as a result of variation in the gene regulating the development of myostatin, which regulates muscle development. Extreme development of the muscles may be desirable from a productivity perspective (increased carcass meat), and it is suggested that the meat from these animals is also more lean, but in some cases double muscling of the rump can be so extreme as to inhibit natural **parturition**. The prevalence of the condition is therefore a cause for concern and, in the case of the Belgian blue, it is estimated that elective Caesarean surgery is used in 90% of cases, due to the frequency of the allele in the population. (DSM)

Draize eye and skin irritancy test

Based on a scientific paper by John Draize and colleagues (Draize *et al*., 1944), this test was used, almost without question, until the 1980s to test the irritancy of new chemicals and their formulations. The test consisted of placing increasingly diluted solutions on to the skin or into the eyes of **rabbits** until a dilution was obtained that caused no irritancy. At the same time, the ability of the eye to recover from such an **injury** was determined, a chemical being labelled as 'corrosive' if repair did not occur. Largely due to protests from those in the antivivisection movement, scientists developed a hierarchy of tests that much reduced the animal suffering that undoubtedly occurred, particularly in the eye test. For example, a chemical that showed skin irritancy was assumed to be an eye irritant; chemicals with extremes of pH – being very alkaline or very acid – were not tested but assumed to be irritant and corrosive; and chemicals were tested at high dilutions before more concentrated solutions were applied. (DSM)

Reference and further reading

Draize, J.H., Woodard, G. and Calvery, H.O. (1944) Methods for the study of irritation and toxicity of substances applied topically to the skin and mucous membranes. *Journal of Pharmacology and Experimental Therapeutics* 82, 377–390.

Wilhelmus, K.R. (2001) The Draize eye test. *Survey of Ophthalmology* 45, 493–515.

Draught animals

Horses, **cattle**, donkeys, water **buffalo** and **camelids** are the best known draught animal species, but **goats**, **dogs**, reindeer,

elephant, yaks and a few other species have also been important, together with the **mule** (an interspecific hybrid – the progeny of a male donkey and a female horse). Today, about 400 million animals are used for draught worldwide. Most of these are cattle and water buffalo, whose numbers are increasing while those of working equines have not changed greatly, worldwide. Any agricultural system than can produce more food than is needed to support the human population has the opportunity to keep animals to reduce the physical work people need to do – and to enable tasks to be completed rapidly to take advantage of seasonal weather patterns. A horse eating 4 kg of oats a day can do as much work as ten strong men who, together, would consume nearly 7 kg of grain daily (Smil, 2000).

In history, draught animals have provided large amounts of energy to drive machinery, as well as to work on the land or transport produce. While many of these uses are things of the past, many new functions have emerged. In much of the developing world, growth in transport services is being met by draught animals, especially **equids**, in both city and country, and the contribution of these animals to assuring national food production and local food distribution is as deserving of attention as that of the animals that work in the field.

Different species have different merits as work animals. Oxen are slow, but at the end of their working life they have a much greater meat value than horses. However, the speed with which horses work is a big advantage, especially in crop cultivation. Technologies have been developed to allow the power of the draught animal to be applied effectively – some technologies are more suitable than others. Poorly fitting harnesses and pack saddles are major causes of suffering, while some otherwise excellent designs for implements and accessories are far too expensive for the developing world. Attitudes towards animals can vary enormously between (and within) human societies, and this can condition the priority placed upon the welfare of draught animals. Within the context of Third World development, it must be remembered that the poorest members of society are too poor to own or to have the use of draught animals. Also, a spread of the use of draught animals into cities could mean that the people who were then responsible for animals may have had no prior experience or family connection with them – and this could prejudice welfare.

Draught animals have an extremely important role to play in world food security. Today's human population is about 6 billion; it is expected to level off at 9 billion by 2050. At present 800 million are undernourished. The major cause of undernourishment is poverty, and improving access to food (food security) depends on a reduction in poverty. Food security is now emphasized more than food production in social policy. It is defined as having access at all times to enough food for an active, healthy life. For country dwellers, food security requires food to be produced locally because of high transport costs and uncertainty of supply. Draught animals clearly have an especially important part to play here. Household members are often working for wages in cities, and this means labour shortages are frequently a real problem for farmers. In principle, draught animals also have some unique benefits – notably provision of manure. However, in practice, it is now clear that inorganic fertilizer is necessary for large-scale crop production, and considerable labour is needed to spread manure on fields. Draught animals can utilize crop residues, though for adequate work performance a high-energy supplement may need to be fed. For each 1 kg of grain produced there is at least 1 kg of straw, and Smil (1999) calculated that the world's annual harvest of 2750 million t of dry matter was accompanied by production of 3750 million t of crop residues.

The present status of draught animals and their future prospects vary greatly between countries. Essentially, they are a component in the portfolio of strategies available to farmers. Fuel and motor vehicles have to be bought at world prices, and their price and availability will influence the future status of draught animals. So too will the availability of fodder and grazing land. As cropping regimes are intensified and more land is cultivated, less grass and forage will be available, but there will be an increase in crop residues and by-products of food processing. In most countries a range of draught animals is available, and this gives flexibility. Much valuable research has been conducted on such matters as energy nutrition of draught animals, with cattle as the main focus. However, work should also be directed towards neglected species, primarily the donkey, as this species has a definite role in achieving poverty reduction and improving the status of women in some regions.

In principle, there is scope to make draught animals more suitable for their purpose by selective breeding. Many draught horses in particular are defective in conformation, and it would be preferable not to breed from such animals. It might seem an attractive option to bring in animals from elsewhere to improve the breed, but the risk then exists of losing locally adapted genotypes.

Animal **welfare** in developing countries is often poor, and many charities based in the developed world seek to improve conditions for Third World draught animals. Projects are usually designed so that the benefits of improved welfare are tangible and of direct benefit to the owners. Such work often focuses on farriery, the treatment of injuries and veterinary care generally, and the design of affordable and effective harnessing and vehicles. (SJGH)

References and further reading

Hall, S.J.G. (2005) The horse in human society. In: Mills, D.S. and McDonnell, S.M. (eds) *The Domestic Horse: the Origins, Development and Management of its Behaviour.* Cambridge University Press, Cambridge, UK, pp. 23–32.

Smil, V. (1999) Crop residues: agriculture's largest harvest. *Bioscience* 49, 299–308.

Smil, V. (2000) Horse power. *Nature* 405, 125.

Dr Hadwen Trust

Dr Hadwen (1854–1932) was a general practitioner in Gloucester, UK, who was concerned by the suffering of animals killed for meat and so became a vegetarian at the age of 21 or 22 – originally taking on a bet from a fellow student that he could not live 6 months without meat. He also claimed that he was in much better health than his carnivorous friends, writing: 'For my part I am quite satisfied with my trial of Vegetarianism, and it would take more than mortal power to persuade me once again to make my stomach a graveyard for the purpose of burying dead bodies in'. In 1932 at the age of 78, and still active in general practice, he died from a severe heart attack. The Dr Hadwen Trust, founded in 1970, was the brainchild of Sidney Hicks, then General Secretary of the BUAV (British Union for the Abolition of **Vivisection**), and the Trust is dedicated to funding exclusively non-animal techniques to replace animal experiments, thus benefiting humans and other animals. (DBM)

Drinking behaviour

Drinking behaviour is stimulated by the **motivation** provided by thirst, although there may be social reasons for drinking – e.g. it is often observed that grazing cattle move to water at the same time. The reason for this coordination may relate to safety or, in the case of young animals in particular, lack of knowledge concerning the site of water availability. This happens when calves are weaned into yards, and have not previously seen water troughs. Similar scenarios can exist with adult animals in unfamiliar environments. **Psychogenic** polydipsia describes a condition resulting in excessive drinking as some form of **stress** response; and close confinement in a barren environment, e.g. a **farrowing crate**, can result in manipulation of the drinker and excessive water intake. In both of these conditions medullary washout (loss of ability to concentrate **urine** in the kidney) can occur, with serious clinical consequences. Understanding drinking behaviour is of particular importance in the dairy industry, given the high rate of fluid production required by animals, and therefore much research has been undertaken in cattle, but the principles may apply to other species.

The water requirements of animals are met from three major sources: (i) free drinking water; (ii) water contained in feed; and (iii) metabolic water produced by oxidation of organic nutrients. Water in animals is lost through urine, faeces and inevitable evaporation from the body surface and the respiratory tract. Many species experiencing **heat stress** can lose a significant amount of water through sweating, panting and drooling.

It is difficult to quantify total water requirements, which will vary with live weight, health status, physiological state (growth, stage of lactation or pregnancy), level of activity (rate of growth, level of milk production), body condition, coat colour, diet and ration composition, dry matter intake, climatic conditions (ambient temperature, relative humidity, solar radiation and wind speed), management system (e.g. **feedlot** or **grazing**), access to shade and water quality and temperature.

Many animals do not have to drink each day. For example, under mild climatic conditions cattle can go without drinking water for 2 or more days, especially where pasture is lush. However, where water is freely available grazing cattle tend to drink just after sunrise and again late in the afternoon. Generally, more water is consumed in the afternoon than in the morning. Animals usually drink just prior to and after eating, primarily to maintain osmalarity in the contents of their gastrointestinal tract.

The amount of water consumed per day increases as ambient temperature increases. Heat stress stimulates both energy and water metabolism simultaneously. For example, the amount of water consumed by cattle during periods of heat stress is usually greater than the amount needed for metabolism and evaporation. Water intake may be increased two- to threefold during heat stress, and especially when the air temperature is above 30°C. There is a strong relationship between ambient temperature, dry matter intake and water requirements. Feeding concentrate therefore has an important impact on drinking behaviour, dependent on temperature. For example, the water requirements for cattle under various thermal environments are: 2–4 l/kg dry matter intake for temperatures below 25°C; 4–10 l/kg dry matter intake for a temperature range of 25–35°C; and 8–15 l/kg dry matter intake when the temperature is greater than 35°C. Young farm animals and high-yielding producers (lactating females) also frequently exhibit higher intakes of dry matter per kilogram of body weight, as a result of **concentrate feeding**. They also have a higher water:liveweight requirement, because of increased requirements for growth or lactation. For example, in the 15–25°C temperature range young cattle and lactating cows may require up to 50% more water per day than non-lactating and mature cattle exposed to similar climatic conditions.

Output of water in milk can significantly increase water requirements, and a dairy cow producing 40 l of milk/day may drink as much as 135 l/day under thermoneutral conditions. As with dry matter intake, the relationship with yield is also temperature dependent, so the differential water intake between high and low yielders will be even greater when they are exposed to heat stress. The high-producing animal also has faster dehydration rates due to higher rates of water turnover. Higher water intake is also useful in dairy animals, because the high specific heat of water allows cattle to store heat during the day and dissipate this heat at night.

In hot environments, a major physiological reaction by animals is to increase water intake. This can result in an increase in body water content. As water is lost during a heat challenge a temporary deficit occurs, and increased body fluid mineral concentration stimulates the hypothalamic thirst centre to increase water consumption. However, there may be occasions when animals will not seek water under these conditions, e.g. they may be reluctant to move from shade when there is intense solar radiation. This may result in a reduction in productivity, since a lack of water intake is also associated with reduced feed consumption.

Post-mortem data from cattle that have died from heat stress suggest that dehydration is a major cause of death, even where the cattle have had adequate access to water. Reported lower water intakes for *Bos indicus* than *Bos taurus* cattle may be a function of their smaller body size, dry matter intake and reduced sweating rate. However, there is limited evidence that *B. indicus* cattle can go for longer periods than *B. taurus* without drinking.

Water consumption can be difficult to quantify under field conditions. Total water consumption may be confused with total water usage, and extraneous factors may impact on drinking behaviour to skew consumption. Many species will use water for cooling. For example, horses and cows will dip their heads in water troughs and splash water out of the trough; horses may dunk hay as a matter of routine. Under intensive conditions dominant animals may stand at the water trough during periods of hot weather and block access to lower-ranking animals. Also, in species where drinking behaviour is socially facilitated, inadequate access to water at the same time as others may result in reduced water intake.

The effects of drinking water temperature on water intake are variable. For example, in a hot environment *B. taurus* cattle consume more feed and gain more weight when given access to cool water (18°C) compared with non-cooled water (32°C). Furthermore, the intake of cool water is less than of hot water, as cool water produces cooling of the body via contact with the digestive tract. Hot drinking water may reduce the ability to lose heat, and therefore the intake of water may decrease. Provision of shade over drinking stations and placing of pipes below ground will be beneficial in reducing the temperature of drinking water in hot climates. However, if drinking water is too cold intakes may also fall. Under very cold conditions (−10°C or lower), cold drinking water may impair the animals' ability to maintain heat balance and intake will fall. In addition, frozen water will have a major impact on water intake. To overcome this, heated water troughs or bowls may be used for animals in captivity.

Drinking water flow rates into water troughs and bowls can influence the water intake, behaviour and performance of animals. Restricting water flow may result in troughs not filling fast enough to give all animals adequate access to the water. Under these circumstances dominant animals may spend considerable time at the trough. Others may not have adequate intakes and, as such, are prone to greater production losses. Water quality is also an issue. Water high in salts and other compounds may lead to a reduction in intake. Water with a high soil content and water that looks and smells different from the water the animals are used to may also reduce intake. This could have serious consequences if animals are exposed to heat stress but of limited concern during periods of cold weather. For example, there have been recorded instances where cattle have died from dehydration during periods of hot weather rather than drink muddy river water. High levels of algae in water troughs may also reduce intake.

The type of diet fed and its physical form also affect water consumption. For example, lush green pastures, silage and supplements – such as brewers' grain in the case of herbivores or wet food diets in the case of cats and dogs – will provide considerable amounts of water and therefore result in less water intake. Diets high in protein, sodium and potassium and high in roughage tend to result in a higher consumption of water than those with lower protein and/or lower roughage. It is also suggested that, in some species, diets that are high in silica or dissolved salts will also result in an increased need for water. However, if animals are offered only salty water and salt intake (from feed and water) is high, water consumption may drop, even during periods of hot weather, as the need to balance electrolytes outweighs the need to drink.

In conclusion, drinking behaviour is influenced by a complex interaction of climate, nutrition and the animal. Water systems should be designed to ensure that the animals have access to water as required and that the systems will be able to cope with peak demands. (JG)

Further reading

Houldcroft, E., Smith, C., Mrowicki, R., Headland, L., Grieveson, S., Jones, T.A. and Dawkins, M.S. (2008) Welfare implications of nipple drinkers for broiler chickens. *Animal Welfare* 17, 1–10.

Kadzere, C.T., Murphy, M.R., Silanikove, N. and Maltz, E. (2002) Heat stress in lactating dairy cows: a review. *Livestock Production Science* 77, 59–91.

Drive (drive theory)

Drive, or 'motivational energy', is a concept introduced into motivational and learning theory that is based on the premise that animals are born with innate **needs** that help them to main a state of **homeostasis**, and that these needs are met as a result of an internal drive that energizes behaviour towards relevant goals. Relevant stimuli may then guide the behaviour towards these goals but are not considered part of the underlying drive, unless their value is learned through reinforcement. Some make a distinction between a drive, which is defined as the latent energy for a behaviour, and an 'urge', which is a physical expression of this through the ensuing behaviour. Thus the drive builds up and, when the animal acts on this, it has an urge to behave in a particular way. The concept of drive is sometimes invoked to explain the observation that a behaviour may be performed with more vigour when the opportunity arises, following prolonged deprivation. A similar concept in Freudian psychology and some fields of **ethology** is referred to as an 'instinct'.

A key premise of drive theory is that it is the deprivation of the need that motivates behaviour, although it does not necessarily exclude the potential role of stimulus attraction in motivation. As a result the emphasis is on the alleviation of negative emotional states, potentially at the expense of a more balanced consideration of motivation that includes the importance of positive hedonic factors in the regulation and expression of behaviour (**see: Emotion; Feelings**). Individuals may still eat even if their gut is already full or they are not hungry. In addition, critics of drive theory point out that many behaviours, e.g. resting and sleeping in the absence of other priorities, do not appear to be driven as such.

Drive denotes the internal energy that provides the force, often represented as a mental process, needed to orientate the animal towards the goal that will satisfy its needs. However, further criticism is attracted by this unconventional, confusing and potentially misleading use of the concept of energy as a causal agent rather than as a property of a physical entity.

Both specific drives associated with specific needs and a non-specific drive associated with general behavioural activity have been postulated in the regulation of behaviour. However, it is pointed out by critics that this concept would then seem to demand an almost limitless number of drives to account for all functional behaviours.

Drive may be an inferred state that is not directly observable, but it is suggested that as time passes without the need being satisfied, the drive increases until the goal is achieved through some physical process such as behaviour and the drive/urge reduced. For some, most notably advocates of the psychologists Clark Hull and Kenneth Spence, it is the reduction in drive that provides **reinforcement** for the behaviour and results in learning. However, critics of drive reduction theory argue that many reinforcers neither reduce physiological needs nor do they have a history of association with any such events (i.e. they cannot be considered secondary reinforcers); for example, in the case of babies the focusing of an image may be used as a behavioural reinforcer.

Although advocates of drive theory have proposed a number of amendments to the concept to accommodate some of the criticisms levied at it, critics also argue that the concept is now too vague and imprecise to be useful.

This term has now fallen out of favour as a scientific context and has been replaced by other motivational concepts (**see: Motivation** and **Decision making**), although it is still widely used in popular writing to describe behavioural tendencies (e.g. the chase drive of a sheepdog, bite drive of a guard dog, etc). (LMD, DSM)

Dualism, Cartesian

The term 'dualism' refers to philosophical theories concerning the relationship between the mind and the body. Dualists have maintained that the terms 'mind' and 'body' refer to two distinct entities. Arguments supporting dualism stem from the 17th-century mechanistic concept of science: that facts about the objective world, as revealed by the physical sciences, are not facts about how they appear from any particular point of view, whereas facts about subjective experience are relative to the point of view of an individual conscious subject. The most famous defender of dualism was **René Descartes** (1596–1650), who held that human beings were composed of two substances – material and non-material – but that their essential nature was non-material.

The dualist assumption that the mind is distinct from the body is apparent in various widely used expressions. Thinking about doing something is different from actually doing it. Thinking about mowing the lawn is different from actually mowing the lawn. The former is said to refer to a mental event while the latter refers to a physical action. For the dualist this involves two distinct substances, mind and body, which raises the classical problem as to how the two interact.

Dualism would appear to be intuitively obvious. Whereas the body is a physical structure, and easily identified, the mind is more problematic. Many scientists involved in animal welfare refer to knowledge regarding the animal's physical structure but express scepticism with regard to their knowledge of the animal's subjective awareness of experience, as well as its motivation. Physical structures can be measured, weighed and are subject to the influence of gravity. Thoughts cannot be measured or weighed, and are free from gravitational influence. This would suggest that the mind occupies an entirely different dimension from physical structures. On the other hand, a physicalist explanation of mental phenomena also has intuitive appeal. Changes in the chemical structure of the **brain**, caused by the intake of hallucinogenic drugs, will influence the kind of thoughts one has. Damage to certain parts of the brain will impair certain mental functions. Instead of thinking of mind and matter as two distinct entities, it is possible to adopt a monistic view and claim that there is just one thing, the brain, that is a physical entity.

Some living beings are said to possess minds, while others apparently do not. Human beings possess minds but the existence of a mental life in many animals is a matter of dispute. Usually, however, the mind is associated with activities involving the body. A being with a mind can engage in conversation, laugh, cry, remember, empathize, deceive, persuade, design, build, solve problems and find a better way of doing things. Mindless things like rocks, trees, dummies and corpses cannot do these things.

Having a mind is bound up with the capacity for intentional activity: rocks and trees may fall, but they cannot jump. Making decisions and carrying them out is a form of intentional activity. Although inanimate objects cannot make decisions there is a sense in which various mammals can be said to choose or make decisions. A dog or a horse may choose not to eat or not to jump over a hurdle. The extent to which this provides evidence of a mental life in these animals is a matter of dispute. There is also a close association between certain forms of intentional activity and moral responsibility. A moral agent is capable of carrying out actions that are influenced by a moral principle. Various animals and inanimate objects may move and cause serious injury, but they are not believed to be conforming to a moral principle and cannot be held morally responsible, and they cannot be said to have a duty to refrain from causing harm.

A major ethical issue is therefore bound up with the question of whether or not non-human animals have some or all of the characteristics associated with minds. In this respect reflections on dualism, on the relationship between mind and body, focus on the age-old philosophical question concerning our relationship to nature: what kind of duties do we have to other beings? For it is clear that we have a different set of duties to beings with minds than to those that do not possess a mental life. We may have a duty not to harm or destroy a plant or an ecosystem, but we have no duty to refrain from insulting it or lying to it. A being with the capacity to reflect is entitled to an apology should we treat it badly, but this obligation does not extend to all living creatures, even though we have a duty to care for them.

Dualists have not, however, provided a satisfactory account of the relationship between mind and body, and behaviourists and materialists have challenged their assumption that mind and body are two separate things. Advocates of **behaviourism** deny the existence of mental phenomena, insisting that it is simply a question of how bodily and mental characteristics are related. The behaviourist seeks to avoid commitment to two kinds of substance by *reducing* talk about emotions, thoughts, beliefs, desires etc. into complex talk about behaviour. Central-state materialists attempt to identify talk about mental states with processes in the brain. For the behaviourists and materialists there is no such thing called a mind, hence there is no possibility of dualism and the question of how the mind relates to the body is out of place.

Although the debate between dualists and materialists is centuries old it is far from settled. Questions about minds and the world transcend scientific inquiry involving both metaphysics and ethics; metaphysics seeks the best opinion of what kind of beings we are and what kind of world exists, which raises questions regarding the kind of life we should lead and the scope of our duties towards other beings. Reflections on the nature of mind have a direct bearing on our considerations of the welfare of other sentient creatures. The importance of dualism lies in its acknowledgement of consciousness – and hence a moral conscience – in a modern scientific materialist world view.

In 1641 the French philosopher and mathematician René Descartes published his *Meditations on the First Philosophy*, which discusses the foundations of knowledge, the existence of God and the distinction between mind and body. Descartes established the philosophical foundation for the modern scientific world view, bringing the rigour and certitude of mathematics to philosophical and scientific inquiry. His mechanistic view of the universe depicted all kinds of matter as being of the same type, and subject to the same mechanical principles. Cartesian dualism, or 'two-substance dualism', represents Descartes' belief that mind and body are two distinct substances. According to Descartes, knowledge of the mind obtained through introspection carried greater certainty than the sensory-based claims regarding knowledge about the physical world. This belief is a development of his mechanistic standpoint and his sceptical method, which entailed the demand that acceptable scientific truths should emerge from a barrage of doubts and counter-argument. The scientist, held Descartes, should only accept truths that cannot admit the possibility of doubt. Descartes was concerned that a description of the world that relied upon sensory observations was unsatisfactory – as sensations can be misleading – and that a physical description of the world should be based on the certainties associated with mathematics.

In *Meditations*, Descartes employed a variety of strategies to undermine the alleged certainty of sensory-based truth claims. The method of doubt was designed to reject all beliefs whose falsity was conceivable. Descartes invoked two sceptical hypotheses: first, the dream hypothesis – there are occasions when we have thought ourselves to be awake when we were actually dreaming, so how can we be certain, asked Descartes, that we are not dreaming at any particular time, and the objects of sensory experience are illusory? He insisted that it was unwise to trust anything that had deceived him in the past and relegated the claims of sensory observation to a secondary role. This is not to say that Descartes saw no role for empirical studies; his scepticism was directed at philosophical claims that sensory observations could provide the foundations of scientific knowledge. The objection to the dreaming hypothesis is fairly obvious. As an argument designed to demonstrate the inadequacy of sensorily derived knowledge it does not work; we can only discover that the senses have deceived us by invoking those situations where the senses are reliable. The awareness that we had mistakenly believed ourselves to be dreaming is merely derived from sensations we regard as reliable.

His second hypothesis was much stronger, casting doubt on the certainties of mathematical reasoning. It involved the supposition of a 'malignant demon' of the 'utmost strength and cunning' bent on deceiving him every way, with regard to the perception of colours, shapes and all forms of sensory input. Even the principles of geometry and mathematics could have been planted by the demon such that they would be perceived falsely. Thus, having eliminated all possible grounds for certainty, Descartes arrived at one unshakeable foundational statement of truth: the meditator's self-verifying certainty of his own existence, for whatever is doubted the meditator must exist in order to perform the act of doubting. This 'I' is still thinking and existing even when it doubts whether it is thinking or existing. Hence the Cartesian principle, *'Cogito ergo sum'* – 'I think therefore I am'.

The certainty of the *cogito* principle applies only to a conscious being. It marks the foundation of Cartesian dualism; the self-standing truth of the mind's existence has epistemic priority over sensory-based information. Cartesian dualism was to have profound consequences for the status of non-human animals, where the possession of **consciousness** was denied. From the Cartesian standpoint all living beings were analogous to machines. Humans, however, possessed souls and were conscious beings (Descartes did not draw a distinction between the mind and the soul). In contrast Descartes maintained that animals lacked consciousness, could not reason, reflect, participate in discourse and, consequently, like machines, lacked any moral significance in their own right.

Descartes' reflections supported the emerging 17th-century scientific consensus which held that humans, who have the capacity to reason, use **language** and construct mathematical schemes, reveal a mental life independent of their physical nature, whereas other animals are wholly lacking in mind are akin to mechanical automata. Animals were, like objects, morally significant because they were valuable to people. The Cartesian tradition of treating a vast range of living phenomena as machines, incapable of any awareness of **suffering**, was scientifically attractive and marked a departure from the medieval view that animals were natural servants of mankind. Many of Descartes' followers employed this doctrine to justify what is now perceived as the infliction of enormous suffering on animals in the name of science.

Critics of Cartesian dualism have pointed out that the postulation of mind and body as two incompatible substances prohibits an explanation as to how they allegedly interact. Descartes speculated that the point of interaction was the pineal gland, located in the brain. But as this is a material substance it cannot represent the point at which the non-physical mind and the physical body interact. Problems also arise with regard to mental attributes, such as sensory awareness. Most animals are aware of hunger, thirst, pain and so on, which, based on a purely Cartesian model, would have to be denied to most animals.

Cartesian dualism is not the dominant theory in contemporary philosophy of mind where physicalist alternatives predominate. Two-substance dualism is frequently dismissed as the doctrine of the 'Ghost in a Machine'. It is also maintained that the possession of mental functions is a matter of degree, rather than an 'all or nothing matter'. But the Cartesian legacy of separating mental and physical functions, with the latter wholly explicable in mechanical terms, laid the foundations for the modern scientific approach to medicine and animal research. (DL)

See also: Cartesianism; Ethics

Further reading

Campbell, K. (1984) *Body and Mind*, 2nd edn. University of Notre Dame Press, Notre Dame, Indiana.

Descartes, R. (1954) *Philosophical Writings*. Thomas Nelson and Sons, Ltd, Edinburgh, UK [translated by Anscombe, E. and Geach, P.T.].

Popper, K.R. and Eccles, J.C. (1977) *The Self and the Brain*. Springer, New York.

Dust bathing

Dust bathing is an important maintenance behaviour that consists of tossing and rubbing dust into the **feathers** or fur to maximize contact between dust and feathers or skin. It is found in a number of species of bird, particularly the Galliformes, and is observed in a number of species of mammals, such as **horses**, baboons and desert **rodents**.

In birds, the behaviour begins with the appetitive phase, during which the individual scratches and pecks at the ground, often forming a shallow hollow. Following this is the consummatory phase, in which the bird lies on its breast, fluffing up its feathers, a process referred to as tossing. This consists of sequences of leg scratching, bill raking, head rubbing and vertical wing shaking, which cause the dust-bathing substrate particles to penetrate the feathers and come into contact with the skin. This behaviour is followed by rubbing, in which the bird lies on its side, feathers are flattened and the wings are pressed against the body, accompanied by side and head rubbing, which result in an increase in contact between the substrate and the downy part of the feathers and the skin. In the final stage of dust-bathing behaviour the bird may stand and perform body and wing shaking, which functions in removing the substrate from the feathers.

The majority of studies of dust bathing have focused on laying hens or quail. Adult hens, when provided with a suitable substrate, tend to dust bathe approximately every other day, with the average bout lasting approximately 30 min. It is a behaviour that shows a **circadian rhythm** and tends to occur mostly during the middle of the day or when conditions are otherwise warm and bright. It has been suggested that it is a behaviour that may be socially facilitated, as hens in a group often synchronize their dust bathing. This may, however, be due to the shared presence of external factors, such as heat, light and a suitable substrate, that initiate dust-bathing behaviour. Among mammals, the frequency of dust bathing varies according to species and the function of the behaviour.

Studies have demonstrated a variety of functions of dust bathing among different species. Examples include **communication** and marking, **courtship behaviour** and **thermoregulation**. However, in birds, dust bathing appears mainly to function in maintaining good feather and skin condition, through the removal of stale and excess lipids that have been applied to the feathers from the uropygial gland. This removal of stale lipids functions in improvement of feather quality, necessary for a number of reasons such as thermoregulation, flight performance, camouflage or attractiveness to potential mates. The barbs of the downy part of the feather stick together with the build-up of stale lipids. Dust bathing in a suitable substrate reduces this and causes the downy part of the feathers to appear fluffier. Because fluffy down is a good insulator, dust bathing may save energy by preventing heat loss caused by poor feather condition. Different substrates are thought to vary in their effectiveness at removing excess lipids. For example, it has been shown that, following a period of litter deprivation, hens on wood shavings spent more time dust bathing than those on sand, but the removal of lipids was the same, suggesting that wood shavings may be less effective at lipid removal and that birds must dust bathe for longer in order to achieve the same effect. This is likely to be a result of litter particles of differing sizes reaching different areas of the feathers.

One additional possible function of dust-bathing behaviour is the removal of **parasites**. It is possible that the dust-bathing substrate acts to suffocate or desiccate ectoparasites, or even physically remove them from the skin and plumage. In addition to this, a number of ectoparasites are known to be attracted to lipids, and so the removal of excess lipids may help in controlling such parasites. There is, however, little experimental evidence to support this theory as a function of dust-bathing behaviour.

There does not appear to be a particular '**sensitive phase**' during which preference for a dust-bathing substrate is established. Provided adult hens have previously encountered a suitable substrate for a brief period at any point during rearing, they will readily accept it for dust bathing as adults. Naive chicks initially show no preference for any of a variety of substrates, and can be trained to dust bathe on apparently suboptimal substrates such as straw or feathers. Birds will, however, show clear preferences for certain substrates over others. Studies have shown that, when given a choice, chicks show a preference for peat or sand over woodshavings, feathers or straw. Such preferences may well reflect the relative effectiveness of different substrates at removing stale lipids.

Dust-bathing behaviour appears to be strongly internally motivated, and birds that have been prevented from dust bathing show a marked **rebound behaviour**, displaying increased dust-bathing activity. Studies have shown that hens have a strong motivation to gain access to a litter substrate, and will work to gain access to a suitable substrate for dust bathing. It is likely that litter deprivation in birds may result in reduced welfare through the prevention of dust bathing, as well as other litter-related behaviours such as **pecking behaviour** and scratching. Motivation to perform dust bathing continues to exist, even when the behaviour is apparently functionless.

Caged hens, housed on wire flooring lacking a suitable loose substrate, will readily go through the sequence of behavioural elements of dust bathing on the wire floor. This has been described as a **vacuum behaviour**, although this is not entirely correct, as birds tend to perform this behaviour near to the feed, flicking the feed on to themselves as a substrate. This behaviour is better termed 'sham' dust bathing, and does not appear to affect motivation to perform real dust bathing if given the opportunity. Similarly, featherless hens have also been shown to dust bathe, and also show an increase in dust bathing following deprivation.

Dust bathing is a highly important behaviour to many Galliforme species of bird, and one that is shown to be performed by captive birds even in the absence if its functional consequences (removal of stale lipids). It is likely that, in conventional battery cages or furnished cages that lack a suitable dust-bathing facility, the **welfare** of laying hens may be severely compromised. (RJNM)

See also: Laying hen housing

Dynorphin

Dynorphins are a class of naturally occurring **opioids** that are found throughout the **central nervous system**, but in highest concentrations in the **hypothalamus**, medulla, *pons*, midbrain and spinal cord. They are produced from prodynorphin, which is cleaved by proprotein convertase 2 (PC2) to give dynorphin A, dynorphin B and α/β-neo-endorphin. Sometimes this processing is incomplete, resulting in the release of a conjoined dynorphin A and B molecule known as 'big dynorphin'. Dynorphins act primarily through kappa (κ)-receptors, but also have some affinity for mu (μ)-, delta (δ)- and bradykinin receptors.

Dynorphins are produced in a number of different areas of the CNS, and their function is somewhat dependent on the site of production. As with other endogenous opioids, dynorphin is implicated in the **pain** response, being analgesic and being six to ten times more potent than morphine when applied to spinal cord κ-receptors. However, it may also have a pain-inducing effect via bradykinin receptors.

With relevance to the animal **welfare** field, dynorphin has been implicated in **stress**, **exploratory behaviour** and **depression**. Imposed stressors in rodents, such as forced swim tests, foot shocks and immobilization, have been shown to elevate dynorphin A and B in the hippocampus. Exposure of rats to novelty also induces an increase in hippocampal dynorphin B, and the concentration of dynorphin B is negatively correlated with exploratory behaviour. Increased dynorphin within the *nucleus accumbens* is also implicated with **learned helplessness** – an animal model of **depression**. In pigs, decreased levels of κ-receptors in the frontal cortex have been associated with high levels of oral **stereotypic behaviour**, potentially as a result of down-regulation in response to high dynorphin levels in the brain. Sows housed long term in close confinement have been shown to have higher levels of dynorphin in the frontal cortex compared with group-housed sows, with correspondingly higher levels of stereotypic behaviour and inactivity, which has been suggested as being akin to learned helplessness seen in rodents.

Dynorphin has also been shown to play a role in: (i) **appetitive behaviour**, with high levels increasing appetite; (ii) **thermoregulation**, with high levels shifting the body towards hypothermia; (iii) addiction, with high levels in the *striatum* and *substantia nigra* in response to repeated exposure to cocaine; and (iv) **circadian rhythms**, with high levels in the **pituitary gland** and low levels in the hypothalamus during the day, and the reverse during the night. (JNM-F)

E

Echolocation

Echolocation (also known as biosonar) is the use of the reflection of pulses of high-frequency sound waves by an individual to help it orient itself as it moves through the environment. The source direction of higher-frequency sounds is more readily detectable, and the use of very short pulses allows precise calculation of the distance of physical objects that reflect the sound. Thus individuals can potentially gain good resolution of physical objects in their environment (including prey) that they may not be able to see. Echolocation was first identified in bats, but is also used by some species of cave-dwelling birds (e.g. the cave swiftlet), nocturnal birds (e.g. oilbirds), terrestrial mammals (e.g. shrews and tenrecs – an example of **convergent evolution**) and marine mammals (e.g. toothed whales, including the dolphin and porpoise). (DSM)

Ecological niche

The meaning of the term 'niche' has a long and somewhat winding history. Grinnell originated the concept in an obscure paper in 1914, and then developed it during the next decade with particular application to birds and mammals. Wholly without reference to Grinnell, Elton in 1927 devoted an entire section to niches in his influential book *Animal Ecology*. Consistent with its then prevalent figurative usage as a 'recess', both authors considered the niche as a place or role in the ecological community and sought to understand the existence of ecological equivalents – species that have similar ecological characteristics in different communities. Both authors included species' relations to food resources as well as to predators. Both included spatial aspects, mainly at the finer scales, to describe such relations; these scales extended to coarser values for Grinnell. Elton sometimes included a set of species in a single niche. Mostly, however, the concepts presented by Grinnell and Elton were very similar.

Hutchinson revolutionized the meaning of ecological niche in 1957 by focusing on the occupant rather than the recess or role, thereby allowing the niche to change as a function of the other species in the community, especially as affected by resource competition. Hutchinson phrased his concept quantitatively as a set of resource ranges along multiple dimensions, forming a hypervolume within which a species population could survive (Hutchinson niche). Such a hypervolume would typically be difficult to ascertain empirically and, perhaps for that reason, MacArthur and Levins in 1967 defined the niche as a utilization distribution giving the fraction of resource use for a species population – e.g. the percentage of foods of different sizes eaten by the population as a whole, as determined by sampling the diets of representative individuals.

The utilization distribution extends to as many resource dimensions as are useful; Fig. E.1 illustrates a one- and two-dimensional resource utilization distribution. Utilization distributions were much easier to determine empirically than the data on population growth needed for Hutchinson's concept and, indeed, ecologists had been gathering such data for years in studies of resource partitioning (see below). Finally, something of a revival of Hutchinson's niche concept was developed in the 1990s by Chase and Leibold, culminating in a recent (2003) book. These authors used population-dynamical theory to delineate those values of resource density taken jointly (each resource is plotted on a separate axis, as in Fig. E.2a) for which a species population has a zero or positive population growth rate, those values being defined as its niche. Unlike Hutchinson's concept, axes were not limited to resource kinds, but also included predators as well as physical stresses (see Fig. E.2b, c, d). While exciting as a theoretical concept, the Chase–Leibold niche has the same disadvantage as Hutchinson's in that it could be difficult to determine in nature, especially for large organisms having many food species; it is probably most practically applied to microcosms – communities of necessarily small species in experimentally contrived containers.

The utilization–distribution concept is perhaps the most useful of the three for those working with vertebrates, and ecologists have focused on three of its properties: niche axes; niche width/breadth; and niche overlap.

First there are the *niche axes*. The possible axes or dimensions of an ecological niche can be classified into three broad types. First, habitat (or place) dimensions specify where a species is found, and can range from fine-scale (called microhabitat), such as feeding height in a tree or perch diameter, to broad-scale (called macrohabitat), such as vegetation zone. A species' foraging behaviour results in a utilization distribution over a set of habitat categories and, in turn, implies a characteristic set of prey eaten from each. Secondly, food-type dimensions specify characteristics of the food items themselves, e.g. size or hardness. Figure E.3 gives an example of the food-size niche for five tern species. Thirdly, time dimensions specify when individuals of a species population are active; time can be daily or seasonal. The former especially reflects behaviour, e.g. in some species individuals favour early and late times while other species have individuals active at midday. Frequency histograms of percentage activity by hour or by month can be used to represent the daily or seasonal time niche, respectively.

The second property, *niche width/breadth*, describes the degree to which a species' utilization is widely spread over a set of resource categories (generalized) versus concentrated on a few (specialized). Typically, niche width is measured as the

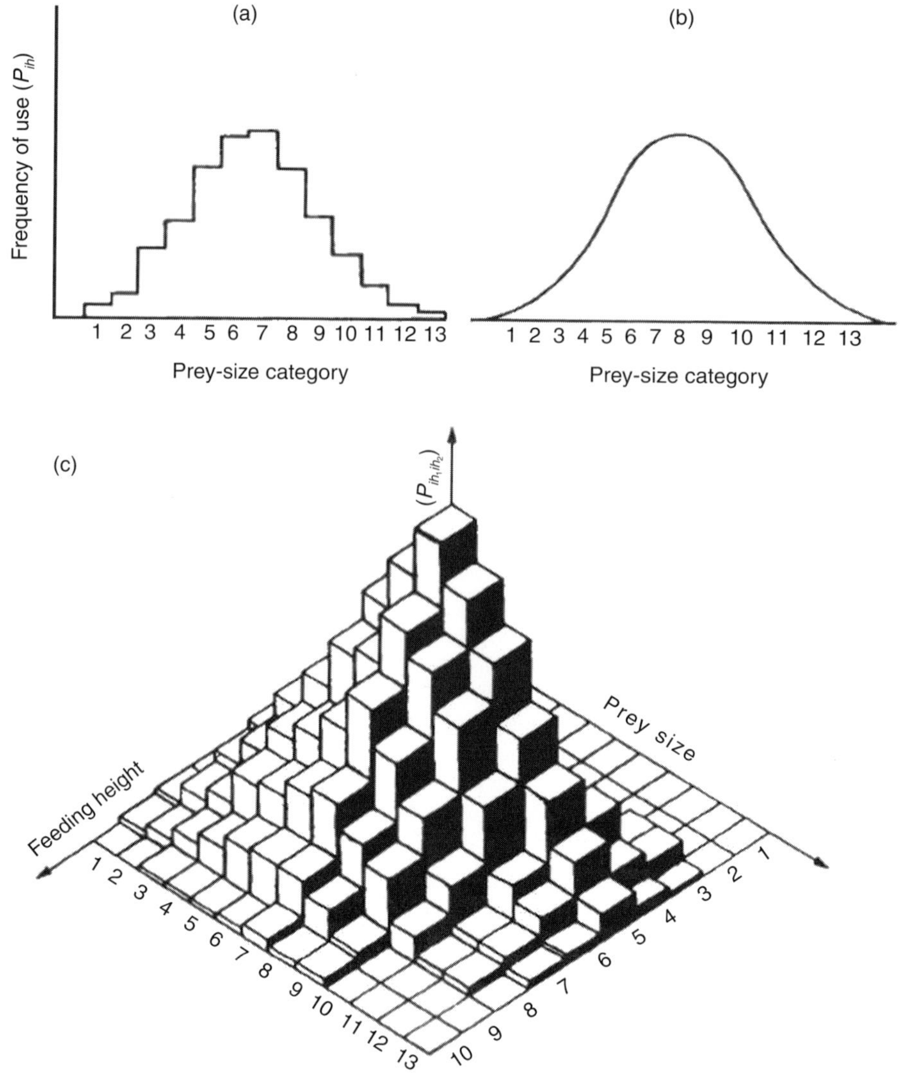

Fig. E.1. The ecological niche as a utilization distribution. (a) A one-dimensional niche, where the dimension is prey size. Numbers refer to prey-size categories, indexed by b. (b) The same utilization smoothed; (c) a utilization for two resource dimensions, prey size and feeding height (after Schoener, 1989).

standard deviation of the utilization distribution. In Fig. E.3, the species of tern at the top has the greatest niche width, while that at the bottom has the smallest niche width. The greater the niche width, the more generalized the species, while the smaller the niche width, the more specialized it is. Ecologists also measure niche position as the average or mean of the utilization; in Fig. E.3 the more robust the beak of a species, the more to the right is its niche position (i.e. the greater is its mean food size). Often there is a positive correlation between niche position and niche width for arithmetic prey categories; were log prey size to be used instead, the positive relation would be substantially reduced.

The niche axis approach is limited to types of niche properties that can be ordered along an axis, i.e. those that are quantitative. However, qualitative categories, such as prey taxa, can also be used; then niche width is measured as the diversity of fractional use over a set of these categories, e.g. as:

$$H' = \sum_{i=1}^{S} \rho_i \log \rho_i$$

where ρ_i is the fractional use of resource category í. With such a measure, a species eating relatively evenly over a set of prey taxa would have a greater niche width than one concentrating on only a few taxa.

The third property, *niche overlap*, results from resource partitioning studies, those documenting differences in the utilization niche between species occurring in the same general area. Species have been found to differ commonly in habitat, somewhat less commonly in prey type and least commonly in time (perhaps because the latter does not result in as great a difference in the resource populations being utilized). But how different must the niches of species be in order for them to

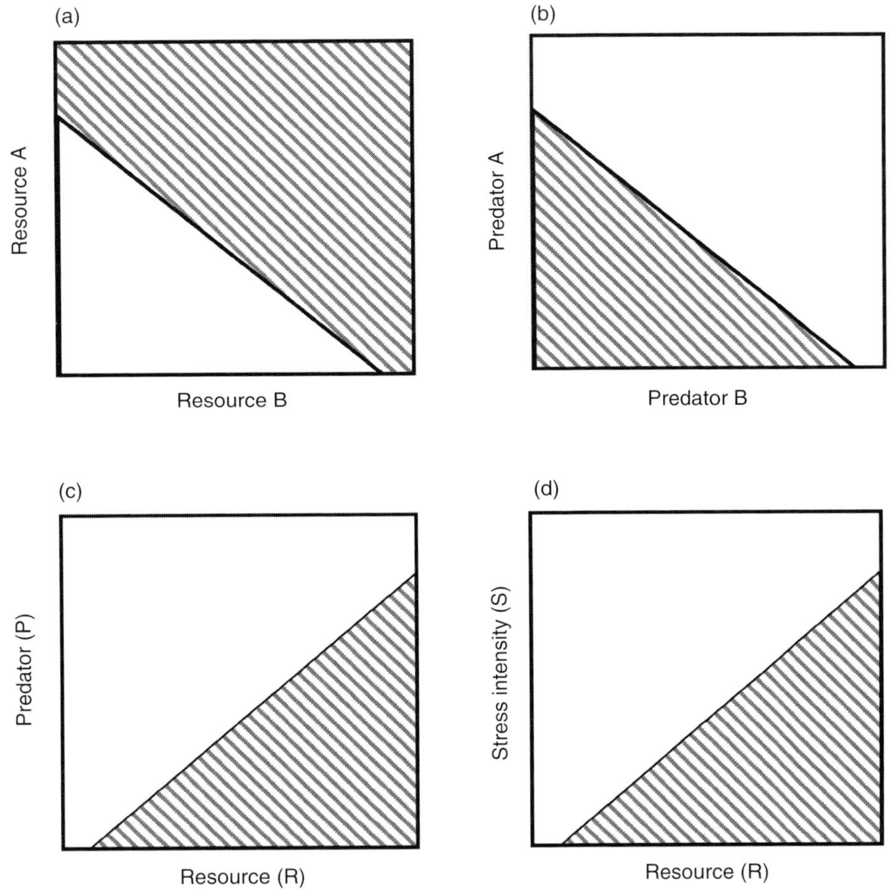

Fig. E.2. Examples from four modules of the requirement component of the niche for a species interacting with two limiting factors. The zero net growth isocline (ZNGI), where dN/dt = 0 (thick solid line) denotes the boundary of the species' niche (birth rate = death rate). Shaded areas are where birth rates are greater than death rates, and are included within the species niche. (a) A species with two substitutable resources, A and B; (b) a species with two predators, A and B; (c) a species with a predator (P) and a resource (R); (d) a species with stress (S) and a resource (R) (modified from Chase and Leibold, 2003).

coexist, i.e. for one species not to be eliminated by competitive exclusion? This can be studied using niche overlap – the degree to which two utilization distributions overlap. Under certain assumptions about the existence and nature of competition, the greater the niche overlap the less likely it is that the species can coexist. That degree of niche overlap just allowing coexistence is called the limiting similarity; any greater overlap would eventually result in competitive exclusion. Limiting similarity is measured in units of d/w, where d is the distance between the niche positions of two species and w is the standard deviation (d is divided by w because, for the same d, the larger the w, the larger the overlap).

Some species, such as the particle-feeding gastropod *Hydrobia*, show a substantial difference in utilization distributions of food sizes when occurring together, but take about the same food sizes when occurring in a solitary state. Differences in utilizations (niche differences) are reflected in differences in phenotypic distributions, in this case body size. The appropriate phenotypic trait is called an indicator for the niche; another example is beak size for food size in the terns of Fig. E.3. Vertebrates sometimes show quite regular differences in phenotypic indicators, even where there is a rather large number of coexisting species. Thus felids and mustelids show extraordinarily regular phenotypic spacing when the diameter of the canine is the indicator. (TWS)

References and further reading

Chase, J.M. and Leibold, M.A. (2003) *Ecological Niches*. University of Chicago Press, Chicago, Illinois.

Dayan, T.D., Simberloff, D., Tchernov, E. and Yom-tov, Y. (1990) Feline canines: community-wide character displacement among the small cats of Israel. *The American Naturalist* 136, 39–60.

Elton, C. (1927) *Animal Ecology*. Sidgwick & Jackson, London.

Grinnell, J. (1914) An account of the mammals and birds of the Lower Colorado Valley with especial reference to the distributional problems presented. *University of California Publications in Zoology* 12, 51–294.

Hutchinson, G.E. (1957) Concluding remarks. *Cold Spring Harbor Symposia on Quantitative Biology* 22, 415–427.

Keller, E.F. and Lloyd, E.A. (1992) *Keywords in Evolutionary Biology*. Harvard University Press, Cambridge, Massachusetts.

MacArthur, R.H. and Levins, R. (1967) The limiting similarity, convergence and divergence of coexisting species. *American Naturalist* 101, 377–385.

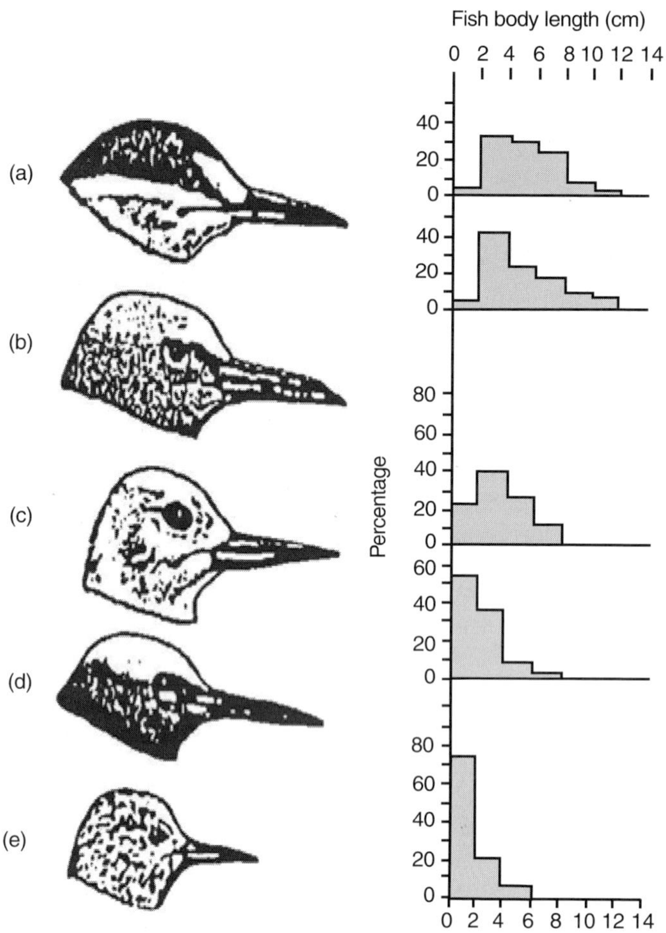

Fig. E.3. Utilizations for five species of tern on Christmas Island, as studied by Ashmole. Niche axis is food (fish) length. (a) *Sterna fuscata* (based on 803 items); (b) *Anous stolidus* (139 items); (c) *Gygis alba* (224 items); (d) *Anous tenuirostris* (1911 items); (e) *Procelsterna cerulea* (702 items) (after Schoener, 1986).

Schoener, T.W. (1986) Resource partitioning. In: Kikkawa, J. and Anderson, D.J. (eds) *Community Ecology: Pattern and Process.* Blackwell Scientific Publications, Oxford, UK.

Schoener, T.W. (1989) The ecological niche. In: Cherrett, J.M. (ed.) *Ecological Concepts (75th Anniversary Symposium of the British Ecological Society).* Blackwell Scientific Publications, Oxford, UK.

Ecology

Ecology is the scientific study of the distribution and abundance of organisms, and how those organisms interact with each other and with their environment, which includes both living (e.g. competitors for food) and non-living (e.g. atmospheric nitrogen) components. Ecology relies upon and builds upon natural history, the careful observation and description of natural phenomenon but, as a science, ecology seeks to explain those observations in a way that allows extrapolation to novel situations.

Ecological questions often cannot be addressed in the same way that scientists would address other types of questions. In hypothetico-deductive scientific enquiry, carefully designed experiments are conducted so that empirical data can be used to reject alternative hypotheses until only one likely explanation is still plausible. However, in ecological research there are often too many uncontrollable variables for rigorous control and experimentation, so variables may interact in ways that cannot be detected through a reductionist approach. In addition to methodological concerns, the ecological question of interest may in fact not be *whether* factor A plays a role in producing observation B; the important question may ask *how much* influence factor A has on observation B, and whether this influence is the same across all levels of factors C, D, and E. One example of an ecological question not easily amenable to traditional hypothesis testing is the question: 'Under current governmental regulations, how will polar bear reproduction be affected by noise associated with petroleum extraction?' However, this type of question can be addressed with mathematical models built upon the answers to questions such as: 'What is the hearing range of female polar bears?'

Ecology is a historical science; current patterns and processes have been shaped by **evolution**, past stochastic events and ongoing responses to current conditions. For example, the distribution of the cattle egret has changed dramatically since its colonization of North America several decades ago, but that colonization would not have been successful without traits the egret possessed due to past natural selection. As a consequence of egret colonization, current survival rates of some North American grasshoppers may now depend on whether past predation by other bird species has shaped grasshopper anti-predator behaviour that coincidentally thwarts hungry egrets.

Ecologists study questions at various levels of organization and complexity, from broad enquiries spanning the entire biosphere (e.g. the water cycle) to enquiries that focus on individual organisms (e.g. the **seasonality** of **reproduction**). Each level of organization exhibits properties not apparent at lower levels of organization, but dependent on characteristics of those smaller units (e.g. genetic differences among neighbouring populations can be influenced by dispersal of individuals). Ecologists sometimes consider the operation and impact of small-scale factors on individuals, but ecologists typically focus on questions regarding larger scales – individuals, populations, communities and ecosystems. A population is the group of potentially interbreeding individuals of the same species. Population ecologists seek to understand the factors determining the rate at which individuals are born, reproduce and die, and how changes in those rates alter the size of the population. An ecological community is comprised of the populations of different species that interact with each other in a particular habitat. Community ecologists seek to understand interactions between species, species diversity and persistence and community productivity. Landscape ecologists strive to understand the spatial and temporal heterogeneity of communities, while ecosystem ecologists investigate the interactions of communities of organisms and their abiotic environments.

An ecologist might specialize in using certain methodologies to address questions (e.g. molecular ecologists) or focus on specific types of each level of organization (e.g. freshwater ecologists). Although many subdisciplines of ecology have relevance for applied animal behaviour and welfare, several subdisciplines are particularly relevant. Agroecologists apply ecological knowledge in the development of sustainable agricultural practices, including animal husbandry. Conservation ecologists strive to restore and maintain long-term functional levels of **biodiversity**. Disease ecologists attempt to understand and control the factors contributing to the spread of diseases. Ecotoxicologists endeavour to identify and minimize the impact of chemical and biological pollutants at various ecological levels, including the physiology of individuals. Sensory ecologists investigate the way in which animals receive and respond to information about their environment, which may guide human efforts to maintain species-typical psychological, physiological and reproductive processes in a captive or domestic setting. Urban ecologists seek to identify how human cities alter ecological processes: the preservation and restoration of undeveloped habitats; the degree to which those habitats are connected; and the levels of environmental pollutants that determine which species persist in urban areas. Wildlife managers traditionally focused on the management of populations of **free-ranging animals** for sustainable human consumption, but now utilize population and habitat management for the broader conservation of biodiversity.

As an example of the application of ecological principles, consider the alteration of carrying capacity in the conservation of eastern bluebird populations in North America. The carrying capacity for an organism is the population size that can be sustained in a given patch of habitat. The ecological factor(s) limiting an animal's carrying capacity may be food, water, shelter, refuges from predators or breeding sites. Modifying the factor limiting the carrying capacity can be one way to alter a population's size, for better or worse. Populations of eastern bluebirds increased dramatically as native forests were converted into fields suitable for bluebird foraging adjacent to wooden fences providing nesting cavities. Through the early 1900s, however, competition with introduced competitors for nesting cavities, replacement of wooden fences with metal fences and pesticide use reduced the carrying capacity and populations of eastern bluebirds. Subsequent widespread installation of thousands of birdhouses designed specifically for eastern bluebirds increased carrying capacity throughout large areas, which resulted in rebounding bluebird populations.

From an applied standpoint there is increasing recognition that, although ecology developed from a descriptive field into an explanatory field, ecologists must now also predict how organisms will fare under conditions that ecologists have never seen and organisms have never experienced. Human activities such as **hunting**, fishing and farming were made possible by the types of observations that underlie ecology; the ongoing challenge for ecologists is to understand and address the ecological ramifications of increasing human resource use.

(RCVH)

See also: Behavioural ecology; Biodiversity; Conservation; Dispersal and habitat selection; Ecological niche; Home range; Territoriality; Translocation; Wildlife management

Further reading

Begon, M., Townsend, C.R. and Harper, J.L. (2006) *Ecology: from Individuals to Ecosystems*, 4th edn. Blackwell Publishing, Oxford, UK.

Real, L.A. and Brown, J.H. (1991) *Foundations of Ecology: Classic Papers with Commentaries*. University of Chicago Press, Chicago, Illinois.

Economics of behaviour

Economic theory can provide a beneficial tool for the study of animal preferences and welfare. By applying economic models and techniques to animal behaviour, we can infer the value that animals place on particular behaviours or resources, which is a fundamental key in assessing animal welfare. Methods based on economic models have been used to assess the value that various species place on a number of different resources. For example, experiments have investigated the value of **dustbathing** substrates, or nestboxes for hens, social contact for pigs and various **enrichments** for laboratory mice and mink.

Common assumptions

Although there are some differences between human economics and animal behaviour, the similarities are sufficient to allow us to draw comparisons in a range of behavioural disciplines such as behavioural ecology, operant psychology and animal welfare. In all cases we are interested in the choices being made by a subject, whether a human in a supermarket, a plover in a field or a rat in a **Skinner** box. Although the decision-making processes of humans and other animals may be quite different from one another, the assumptions made in the models show enough areas of agreement to justify the drawing of parallels between economics and animal behaviour.

In both cases it is assumed that the consumer is rational. 'Economic man' is assumed to choose the best possible combination of resources under all circumstances, his primary motivational principle being the attainment of an optimal state (Lea, 1978). The assumption of rationality among consumers means that each consumer approaches each buying decision with a blank, unbiased mind, and then rationally considers all the available options. One of the basic tenets of behavioural ecology is that natural selection will shape animals' behaviour so as to maximize their inclusive **fitness**. Natural selection should therefore design animals' decision-making processes so as to allow them to make the most cost-effective choices between different options, which will minimize the cost:benefit ratio of the behaviour.

Some other logical assumptions, which were originally proposed for human models of **preference**, also apply to animal systems. The first is known by economists as two-term consistency. This means that if resource A is sometimes chosen over resource B, but resource B is at other times chosen above resource A, the consumer is indifferent in choosing between resources A and B (Varian, 1997). This supposition is taken a step further with the assumption of three-term consistency, or transitivity of preference. If preferences are transitive, as models of choice behaviour assume, then if the consumer prefers resource A to resource B, and resource B to resource C, then resource A should be preferred to resource C.

So the assumptions of human economics and animal behaviour can be seen as analogous. In human economics we assume that the consumer makes rational decisions based on perfect knowledge of the market; and in animal behaviour we assume that natural selection has shaped animals' decision-making processes such that they make optimal choices.

Economic models of choice

Simple economic models are descriptive, and describe the different choices that a consumer can make within the constraints of a particular budget; more complicated models can be used to compare demand between very different goods or resources to determine consumer preferences between the different choices available.

Budgets

The budget constraint on a consumer describes the total amount available to spend. For two resources, x and y, there are various combinations ('bundles') of amounts of the two resources that a consumer can buy. A budget line can be drawn by joining all the bundles that the consumer can afford, as shown in Fig. E.4.

In Fig. E.4 $m/p1$ is the amount of good 1 that a consumer could buy if all his money was spent on good 1, and $m/p2$ is the amount of good 2 that could be purchased if all the budget was spent on good 2. If the income of the consumer or the price of either of the goods changes, then the budget set will also change.

Budget sets and constraints can also apply to animal behaviour. In this case, time and energy are usually assumed to be the currencies with which the animals are working. In some species, or at certain periods during an animal's life, one or the other of these currencies may become dominant in determining behaviour. Short-lived species such as butterflies that have a limited time to mate before dying are likely to be more restricted by their time budget than by energy constraints. Foraging bees, though, are limited by the energy constraints of flight. If time and energy are considered as currencies in an animal's budgets, then it is easy to see how their budgets could be changed in ways similar to those of humans. An increase in the day length of a diurnal species could have the effect of increasing income (and therefore budget), as could an increase in the energy value of the animal's food source. Changes in the price of goods can be likened to changes in the ease of executing a particular behaviour pattern, such as finding and exploiting a food source, which could affect the time and energy remaining available to the animal.

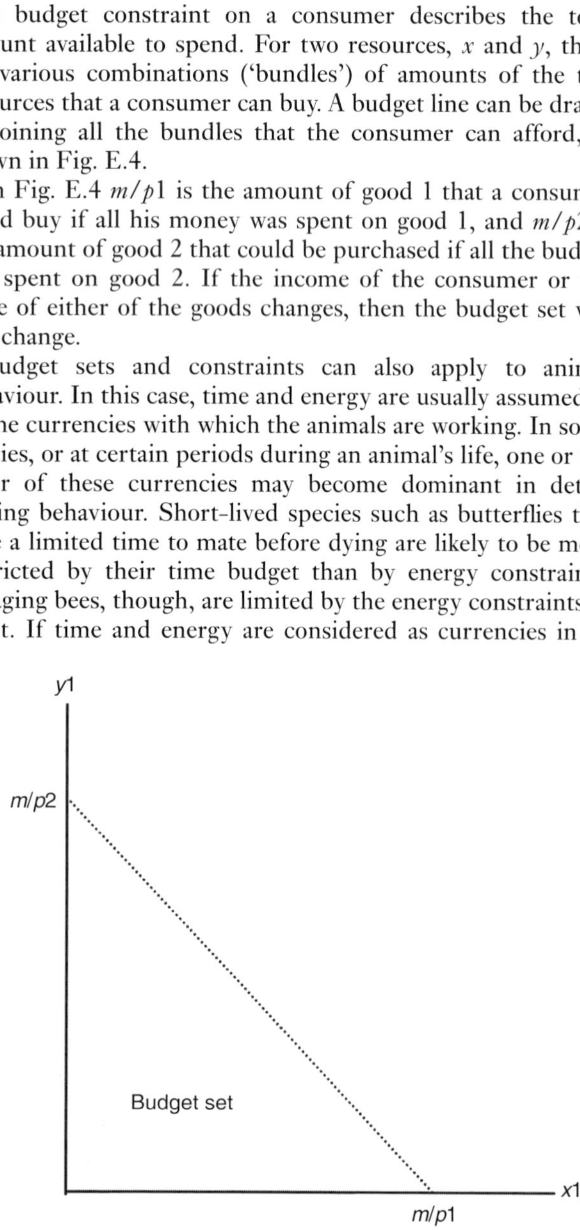

Fig. E.4. The budget set of a consumer. *m*/*p*1 is the amount of good 1 that a consumer could buy if they spent all their money on good 1. Likewise, *m*/*p*2 is the maximum amount of good 2 that could be bought with this budget.

Budget sets can therefore be constructed to describe animal behaviour options, and provide a useful basis from which to start investigating the influences on an animal's behaviour. The mechanism by which the consumer actually decides upon one of these options is not described by this simple model, but can be analysed economically by creating more elaborate models of consumer choice.

Utility

The relative value of a particular chosen bundle of goods can be labelled by assigning it a utility value. The better bundles in a budget set are assigned higher utility values than less ideal bundles. The magnitude of the difference is of no significance, as the relative utility is deduced simply by the position of the bundle in the list. In animal behaviour, utility is commonly used by behavioural ecologists, who use its analogue, fitness, to describe the relative advantage of behaviours or resources to the animal. For domestic animals, fed and protected by humans and often killed when young, the evolutionary consequences of behaviour have become divorced from the proximate need to carry them out (Dawkins, 1990). This means that animal welfare researchers place more emphasis on what the animals want to do than on the ultimate fitness consequences of their choices. Attention has therefore rested on determining the preference for bundles rather than speculating about the evolutionary advantage of choosing them.

Preference and indifference

The preferences that a consumer shows for certain goods or bundles of goods are most simply described by presenting them graphically. For any particular bundle (e.g. $x1$, $y1$) on a graph, we can identify all the bundles that the consumer prefers: all the bundles on the boundary of this set are the bundles between which the consumer is simply indifferent ($x1$, $y1$) (Fig. E.5; Varian, 1997). This boundary line is the indifference curve for that bundle. Different curves are, by definition, different levels of preference for the bundles of goods, so bundles that offer more of the goods will be higher and to the right of less preferred bundles.

Optimal choice and consumer demand

Indifference curves give us information about the way in which the goods on offer are perceived by the consumer, but still do not help us determine which bundle the consumer will regard as the optimal choice. In order to develop a model of a consumer's optimal bundle choices, we need to integrate the budget set of the individual with the indifference curves for the resources in question. Because consumers will always choose the most preferred bundle they can afford, we can determine the optimal choice of a consumer if we know the consumer's budget and the indifference curves for the goods. The optimal

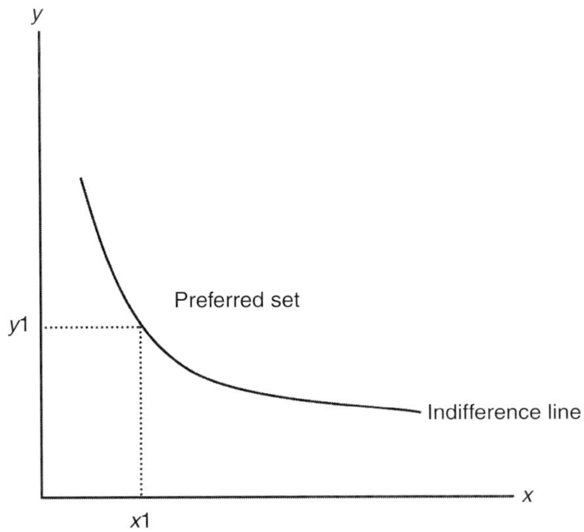

Fig. E.5. The indifference curve for the bundle $x1,y1$.

bundle is usually the point at which the indifference curve is at a tangent to the budget line, as described in Fig. E.6.

All the points on line A would be preferred to any on the other two lines, but they are beyond the consumer's budget, represented by the dotted line. The points on B are preferred to any on C because B is the higher indifference curve, but only one point on B intersects the budget line, so this is the optimal choice for the consumer – the best possible bundle the consumer can afford. When prices or income change, however, the optimal choice for the consumer will also change. The consumer's demand function is the mathematical function that relates optimal choice to different prices and incomes. The function $x1(p1, p2, m)$ describes the relationship between the quantity of the good demanded ($x1$) and all the factors that will influence the demand for the good, namely its price ($p1$), the price of all other goods ($p2$) and the consumer's budget or income (m). It can be derived either by holding income constant and altering the good's price or by holding price constant and altering the consumer's income, and plotted graphically as in Fig. E.7.

Demand curves typically have negative slopes because, as the price of a normal good increases, the demand for it will decrease (Varian, 1997). They give information about the responsiveness of the consumer to changes in price or income but, to make meaningful comparisons between goods that are not on offer in the same bundle, we need to subtract measures such as elasticity, consumer surplus and reservation price from the curve.

Price elasticity

Elasticity is expressed as the ratio of the price to the quantity bought, multiplied by the slope of the demand function. The price elasticity of demand is defined as:

$$\textit{price elasticity} = \frac{\textit{change in quantity demanded (\%)}}{\textit{change in price (\%)}}$$

By using percentages or ratios, the units of price and quantity are eliminated, so comparisons can be made between different goods. The sign of elasticity of demand is usually negative, because for normal goods the amount demanded will

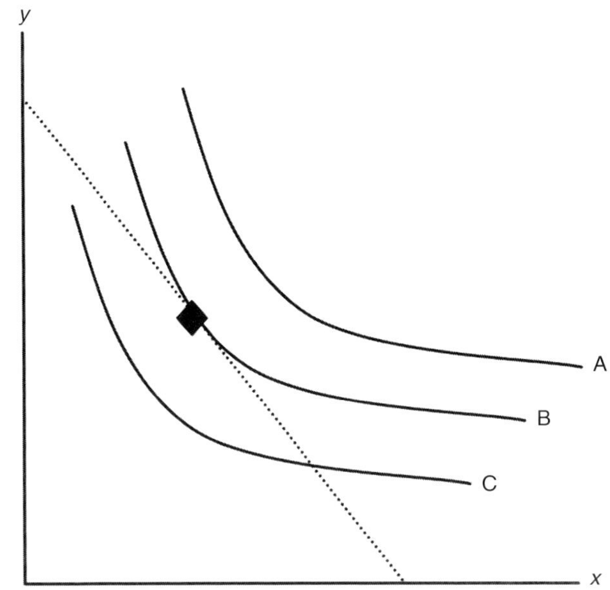

Fig. E.6. The optimal bundle that the consumer can afford: where the budget line (broken) is at a tangent to an indifference curve.

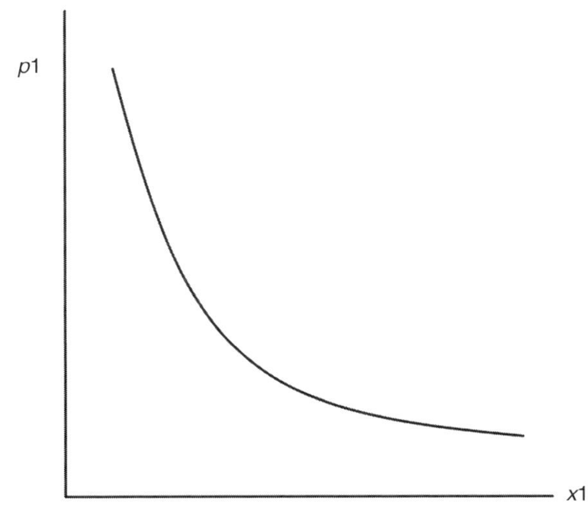

Fig. E.7. The demand curve obtained by plotting the change in price against the change in amount of x bought.

decrease as the price increases, although conventionally the minus sign is often dropped in verbal discussion of demand.

Income elasticity

Income elasticity is very similar to price elasticity, but describes changes in demand with changes in income or budget rather than prices of individual resources. It is defined as:

$$\textit{income elasticity} = \frac{\textit{change in quantity demanded (\%)}}{\textit{change in income (\%)}}$$

For normal goods, the amount bought will increase as income increases, so they have positive income elasticities.

When a demand curve is plotted on log–log coordinates the curve becomes a line, and the resulting function of this line takes the form of Y = *a*X + *b*, where Y is the log of quantity purchased and X is the log of the price. On such lines, the slope, a, is the elasticity coefficient (Varian, 1997). The more a resource is influenced by price or income, the greater the elasticity.

Consumer surplus
The area bounded by a demand curve is called the consumer surplus (Ng, 1990; Varian, 1997) and is used by economists as a measure of the utility of a good or service when assessing human welfare. It has been suggested (e.g. Ng, 1990) that this would be a more valid way of measuring the welfare value that animals derive from resources than using elasticity of demand. Resources for which larger consumer surpluses are displayed are viewed as more important than those with a smaller area beneath the curve.

Maximum price
The maximum price that a consumer is willing to pay for a resource or service can be used to assess the value of the first-obtained unit of a particular good. This measure is sometimes referred to in economics as willingness to pay, or reservation price. A higher maximum price would indicate that the consumer places a higher value on obtaining a unit of this resource. This, too, could therefore draw useful information from demand curves about the welfare benefits of a good (Mason *et al.*, 1998).

Economic analogies in animal behaviour and welfare
The assumptions of economics and the evolution of behaviour discussed above are similar enough to justify the adoption of economic techniques for use in animal behaviour studies. Economists may be interested in what particular goods a person will buy, and how these decisions will be affected by a change in the price of the good or the amount of money the person has to spend. In a similar way an ethologist may be interested in what behaviour an animal will carry out given a number of options, and how this decision will be affected by time or energy constraints. However, the aim of animal welfare studies is to extrapolate further than this, so it is assumed that, if one behaviour is chosen above another, this is not just better for the animal in terms of its inclusive fitness, but also that its importance carries some more proximate benefit to the animal. For such studies, the utility or value of different behaviours or animals' motivation to perform them is determined from measures of preference or quantified using consumer demand measures, and these are used to infer the degree of welfare benefit of providing these different opportunities.

Animals' budgets can be experimentally manipulated, e.g. by altering the energy available from their food. In animal behaviour resilience, which is the extent to which particular behaviours are performed when animals' **time budgets** are reduced, is a type of income elasticity. The price of individual resources or of performing particular behaviours can be altered by imposing a cost, e.g. pressing levers or pushing weights, before the animal can access them. The number of times a cost is paid or the time spent paying the cost can be taken as a measure of consumption, and demand curves can therefore be constructed from which the various measures of resource value can be taken.

Dawkins (e.g. Dawkins, 1990) first introduced this method of assessing the importance that animals place on resources or behaviours. By reducing hens' time budget, their demand for litter in which to dust bathe was compared with demand for food, a resource of known importance. Since then, the slope of the curve has been the most frequently used measure of value (e.g. in pig, hen, calf and mouse studies). Some authors (e.g. Ng, 1990) have challenged this, advocating consumer surplus and maximum price instead, and some researchers have adopted these techniques. There have also been considerable differences in the way in which these experiments have been designed (e.g. whether or not the subjects are allowed to control their own bout lengths after paying for a behaviour). However, commentaries on the design of such experiments (Ng, 1990; Mason *et al.*, 1998) have helped to clarify which methods will provide the most valid results. Provided that the animals' behaviour conforms to the economic assumptions, and that the experiments are carefully designed (Mason *et al.*, 1998), such techniques have considerable potential for evaluating animal welfare. (HJW)

See also: Choice test; Laying hen housing; Motivation; Preference

References

Dawkins, M.S. (1990) From an animal's point of view: motivation, fitness and animal welfare. *Behavioural and Brain Sciences* 13, 1–61.

Lea, S.E.G. (1978) The psychology and economics of demand. *Psychological Bulletin* 85, 441–466.

Mason, G., McFarland, D. and Garner, J. (1998) A demanding task: using economic techniques to assess animal priorities. *Animal Behaviour* 55, 1071–1075.

Ng, Y.-K. (1990) The case for and difficulties in using 'demand areas' to measure changes in well-being. *Behavioural and Brain Sciences* 13, 30–31.

Varian, H.R. (1997) *Intermediate Microeconomics: a Modern Approach.* W.W. Norton, London.

Ecotourism

Ecotourism is tourism in which the main attraction is the natural environment and which tries to teach some **ecology**, minimize environmental impacts and contribute to **conservation**. Wildlife is a major component of many ecotourism products. Ecotourism can be both a tool and a threat in wildlife conservation. Economic, social and political returns from ecotourism can help to protect wildlife from poaching and wildlife habitat from clearance. Ecotourism can also modify wildlife behaviour quite significantly, however, through either disturbance or **habituation**. There are also many tourism products that are marketed as ecotourism even though they make no attempt to meet its goals, and some of these involve chained or caged wildlife and other abuses of animal welfare. Ecotourism products are structured and packaged differently in different continents and countries. Very broadly, tropical rainforest areas tend to promote **biodiversity**, with the potential to see many different species. Tours that promote individual icon species or animal families are the exception: examples include mountain **gorilla** in Uganda, **orang-utan** in Borneo, **giant panda** in western China, quetzal in Guatemala, tiger in southern Nepal, lemur in Madagascar and possum in north Queensland, Australia.

In more open landscapes, individual icon species are a major feature in ecotourism marketing. Examples include: the so-called Big Five game (lion, **elephant**, **rhinoceros**, **buffalo**, leopard) and associated charismatic megafauna of eastern and southern Africa; wolves, bald eagle and grizzly **bear** in Alaska; polar bear in the circumpolar High Arctic; condor in the Andes and puma in Patagonia; and kangaroo and koala in Australia. Similarly, some marine ecotours promote individual species such as whale, walrus, manta ray or elephant seal, whereas others promote diverse ecosystems such as coral reefs. Many ecotourism products also feature plants, scenery, cultural attractions and adventure activities, but watchable wildlife is commonly a key selling point.

Ecotourism enterprises use both public and private land and water under various tenures, but **wildlife management** issues are commonly most contentious in public protected areas such as World Heritage Areas, Biosphere Reserves and National Parks. Many such reserves were established for recreation as well as for conservation, and many are currently experiencing an increasing influx of visitors, principally for independent recreation but also on commercial tours. While such visitors may help in providing political support and, in some cases, funding for conservation of wildlife habitat, they may also have significant impacts on animal behaviour and population dynamics. There are even instances where tourism has threatened the survival of entire species. Equally, of course, there are instances where tourism has been instrumental in species conservation.

While the types of impacts that wildlife watchers and ecotourists can produce on birds, terrestrial vertebrates and marine mammals are broadly understood, the detailed mechanisms and quantitative effects differ greatly between species and depend heavily on activities and equipment, visitor numbers and behaviour, season and so on. Scientific research to date is quite inadequate to predict such impacts in sufficient detail for informed wildlife, habitat and visitor management.

In general, impacts that are direct, immediate and easy to observe are more heavily studied than those that are indirect, diffuse, delayed and not obvious to the naked eye. The distances to which walkers, joggers, dogs or vehicles can approach various bird species either feeding, roosting or nesting, for example, have been recorded on numerous occasions, but particularly for shorebirds in north-eastern USA (Buckley, 2004). There are far fewer cases where the energetic consequences for individual birds, the population consequences of reduced reproductive success, or the implications of less obvious impacts such as habitat modifications or the introduction of avian diseases have been assessed. There is at least one known instance, however, where a single light aircraft overflight caused the immediate death of 88% of the eggs and chicks of a colony of white pelicans in Canada.

Severe reductions in breeding success have been recorded for various northern hemisphere waterbird species subject even to very low levels of disturbance. Nesting albatross on sub-Antarctic islands exhibit symptoms of severe stress when approached by tourists, even if the birds remain on the nest. Remote telemetry has demonstrated that nesting penguins approached by tourists in the Antarctic experience increased pulse rate and temperature, physiological indicators of **stress**. Concerns have been expressed that they may be exposed to avian diseases through contact with tour boats. Many colony-nesting birds suffer greatly increased predation when normally vigilant nesting adults are distracted momentarily by bird-watching tourists or other humans.

A variety of similar behavioural and physiological responses to disturbance has also been found for terrestrial and marine mammals. Avoidance of tracks and trails, increased vigilance and decreased feeding, elevated pulse rates and energetic costs, introduction of diseases and interference with intra- and inter-species interactions have all been reported on occasion (Buckley, 2004). Disease introduced by tourists, for example, led to the death of six lowland gorillas in the Democratic Republic of Congo in 1988 (Buckley, 2003). A classic study by MacArthur *et al.* (1982), using remote telemetry, found that bighorn sheep in North America suffer physiological and energetic stress when approached by hikers or dogs, even if the sheep do not run away. Sperm whales subject to overflights by helicopter tours at Kaikoura, New Zealand apparently spend less time on the surface replenishing oxygen supplies. Trails made by hikers through dense forest vegetation may increase predation by feral foxes and cats on native Australian mammals.

Most of the interactions between tourists and wildlife are inadvertent. Most ecotourists prefer to watch undisturbed wildlife engaging in natural behaviours, even if in practice the tourists' ignorance of animal behaviour, or enthusiasm for a close encounter, does lead to disturbance. There are some situations, however, where interactions between tourists and wildlife are deliberate, and there is controversy as to whether or not these may constitute ecotourism. The two best-known examples are: first, deliberate habituation through feeding; and, secondly, so-called consumptive wildlife tourism such as **hunting** or fishing.

Most protected area management agencies specifically ban feeding of birds and animals. Reasons commonly given are that: (i) feeding leads to **abnormal** animal behaviour, which is not what visitors should see; (ii) some animal species become aggressive when fed and may injure visitors; (iii) feeding and/or touching of wild animals may lead to the spread of diseases in local animal populations; and (iv) by boosting local populations of more aggressive species, feeding may indirectly reduce populations of more timid species, which may include rare or threatened species.

Feeding of wildlife by commercial tour operators, however, is extremely widespread. It may occur openly where it is permitted in public parks or private land, or covertly where it is not. Examples of surreptitious feeding include wallabies at the Gordon River landing in Tasmania and polar bears in Churchill on Canada's Hudson Bay. Examples where animals are fed by tourists as part of the tourist product include parrots at O'Reilly's Rainforest Guesthouse and dolphins at Tangalooma Resort, both in Queensland, Australia. However, many of the world's best-known wildlife tourism operations, both private and public, are particularly careful not to associate tourists and food. Examples include open-vehicle safaris in the private game parks of southern Africa and the management of camping permits at Denali National Park in Alaska (see Buckley, 2003 for further case studies).

Controversy over recreational hunting and fishing, especially through commercial outfitters, is even more intense than for wildlife feeding. Some people are opposed to hunting on both ethical and philosophical grounds (**see: Ethics; Morality**); others are not opposed to hunting, but do not think it should be considered as ecotourism; others again are prepared to countenance hunting as a form of ecotourism if it contributes to maintaining populations of rare or endangered species. Similar considerations apply for fishing.

The issues are commonly complex. For example, managing a hunting reserve for specific target species may reduce populations of other species that are of greater conservation significance. For example, gamekeepers in the UK may shoot golden eagles on grouse moors, whether legally or not. Conflicts may also occur between photographic and hunting operations. In areas where hunting is permitted, animals rapidly become wary, spoiling the opportunities for wildlife watching. Where animals that are habituated to wildlife tourists, however, move into areas subject to hunting, they are unafraid of humans and hence become easy prey, as in some notorious East African examples (Buckley, 2003). Advocates of hunting tourism argue that hunting can generate a great deal more revenue than photographic safaris, with far less infrastructure investment. In practice, however, it seems that, with the exception of countries with stable and effective legal and wildlife management systems, the issue of high-value hunting licences can soon become subject to abuse, with major detrimental effects on wildlife populations. This has apparently occurred, for example, in a number of African nations. Where strong legal and wildlife management frameworks do exist, however, parks agencies both public and private have been able to raise significant funds for conservation by breeding icon wildlife species and selling them either to game reserves or to hunting preserves.

In conclusion, therefore, it appears that: (i) watchable wildlife forms a major component of many ecotourism products and hence makes a very significant contribution to ecotourism revenues; (ii) while the best ecotourism products go out of their way not to influence animal behaviour, the majority do indeed produce behavioural impacts; and (iii) ecotourism may produce either positive or negative effects on wildlife conservation and animal welfare, depending on the particular circumstances. (ECB)

References and further reading

Buckley, R.C. (2003) *Case Studies in Ecotourism*. CAB International, Wallingford, UK.

Buckley, R.C. (ed.) (2004) *Environmental Impacts of Ecotourism*. CAB International, Wallingford, UK.

MacArthur, R.A., Giest, V. and Johnston, R.H. (1982) Cardiac and behavioural responses of mountain sheep to human disturbance. *Journal of Wildlife Management* 46, 351–358.

Weaver, D. (2001) *The Encyclopedia of Ecotourism*. CAB International, Wallingford, UK.

Egg production

Worldwide there are approximately five thousand million hens that lay eggs for human consumption. In addition there are breeding birds whose eggs are fertilized and hatched to produce more layers and meat birds. One consequence of strains of hens bred specifically for egg laying is the redundancy of the male chicks. These ought to be humanely disposed of at hatching, using maceration rather than the prevalent practice of slow suffocation in carbon dioxide, which is an irritant gas (see HSA, 2001).

Original strains of hens laid eggs in batches or clutches and then became 'broody' regardless of whether the eggs were removed. Such behaviour has been bred out of modern laying hybrids. Table E.1 indicates how the productivity of hens has been increased through genetic selection and altered diets and housing. It is now approaching the biological maximum of 1 egg/day, particularly early on in the laying cycle.

As the ovum, or egg, passes down the oviduct the shell is added, the whole process taking 24–25 h. Among other factors, shell thickness depends on the quality of the diet. It used to be thought that hens drew on the reservoir of calcium in the skeleton to help calcify the eggs and that this accounted for the osteoporosis commonly seen in birds at or before the end of lay (**see: Bone strength**). However, current evidence demonstrates that a major influence is the high levels of **oestrogen** associated with egg laying that preclude the formation of bone, along with genetic and other factors (Whitehead, 2004). Most hens in intensive laying systems develop osteoporosis and have a high risk of fracturing bones. To improve economic efficiency, hens have also been selected to be smaller and thus have a lower maintenance requirement for feed. This compounds such welfare risks and reduces the longevity of the birds.

Laying hens in commercial production systems are usually culled after 50–70 weeks of age, whereas those in backyard flocks generally live for several years. A few producers, notably in the USA, use the practice of **'forced' moulting** where, after about 1 year of production when egg output has declined, feed and water are withheld from the hens for several days. This causes them to shed feathers (moult), which then regrow, and it also has the effect of rejuvenating the hens' egg production. There are substantial **welfare** issues associated with withholding feed – and especially water – from hens, and the practice is therefore not permitted in most countries, including the European Community.

From the mid-20th century, scientific experiments revealed that egg size and production could be manipulated by altering temperature and lighting patterns. An increased day length is now routinely used to photo-stimulate pullets to advance their sexual maturity so that they come into lay by about 18 weeks of age (see Lewis and Morris, 2006, for more details). A typical egg production curve for intensively managed layers is given in

Table E.1. UK housed hen egg production, 1936–2006 (eggs per bird per year) (from FAWC, 1997).

Year	Production
1936	149
1946	121
1956	170
1966	202
1976	245
1986	279
1996	310
2006	311

Fig. E.8, which shows that younger birds lay more eggs per week. Egg mass output is relatively constant at a given age, so birds can produce several smaller eggs or fewer larger eggs.

A number of factors influence egg size. Larger eggs tend to be laid by larger birds and egg size increases with the age of the hen, mainly due to her growth and increased bodyweight. If larger eggs are required at the point of lay, pullets are allowed to grow bigger before reaching sexual maturity. In general, warm environmental temperatures result in larger eggs as the hen has to partition less feed to maintain body temperature; however, heat-stressed birds (above about 26°C) lay fewer, smaller eggs. The number and size of eggs can be manipulated by lighting patterns, with longer day lengths leading to increased feed intake and higher egg weight.

Eggshell colour is related to the genetics of the bird, so white-feathered hybrids (e.g. those derived from Leghorns) lay white eggs and brown-feathered birds (e.g. those derived from Rhode Island reds) brown eggs. Unusually coloured eggs, such as blue-green laid by Araucana hens, have a novelty market.

Eggshell quality has been used as a non-invasive indicator of hen welfare (Dawkins *et al.*, 2004). This and other work shows that shell thickness may decline when hens are stressed (particularly **heat stress**), eat insufficient feed or when dietary components are inadequate. Blood spots on the shell surface may indicate the presence of red mites. Blood streaks are indicative of vent pecking, which may be associated with either young birds laying relatively large eggs or an early onset of lay. (CW)

See also: Chicken; Laying hen housing

References

Dawkins, M.S., Edmond, A., Lord, A., Solomon, S. and Bain, M. (2004) Time course of changes in egg-shell quality, faecal corticosteroids and behaviour as welfare measures in laying hens. *Animal Welfare* 13, 321–327.

FAWC (1997) Report on the welfare of laying hens. MAFF PB 3221, UK.

HSA (2001) *Code of Practice for the Disposal of Chicks in Hatcheries*, 2nd edn [reprinted 2004 with minor amendments]. Humane Slaughter Association, ISBN 1 871561 06 X.

Lewis, P. and Morris, T. (2006) *Poultry Lighting, the Theory and Practice.* Northcot, Andover, UK (see also http://www.poultrylighting.co.uk).

Whitehead, C.C. (2004) Skeletal disorders in laying hens: the problem of osteoporosis and bone fractures. In: Perry G.C. (ed.) *Welfare of the Laying Hen;* Poultry Science Symposium Series 27. CAB International, Wallingford, UK.

Electric field

The ability to detect and produce electric fields is confined to aquatic species because, unlike air, water conducts electricity, and sea water is more conductive than fresh water because it has a greater concentration of salts. The diversity of aquatic species exhibiting **electroreception** and electrogenesis is great, including the sharks and rays, teleostean fishes, lungfishes, coelacanths, salamanders, frogs and a marsupial mammal, the platypus (see Fig. E.9).

These species use the electrical sense for a variety of purposes. They detect their prey or mates by the electrical fields associated with the movement of salt water, a conductor, through gill cavity and muscle contractions, and migrate between habitats by orienting to the seafloor's local pattern of magnetization or Earth's dipolar field. They actively produce weak electric fields around themselves, and detect the proximity of objects of different conductivity by the distortion of this electric field. Finally, they immobilize their prey by generating strong electric fields. It is likely that electro-reception was an early feature of vertebrate evolution, because the capability is possessed by the lamprey, the sole survivor of the large ancestral group of jawless fishes, the Agnathans, which were present 600 million years ago. However, this ability was lost once in their evolution, and was later evolved independently by the knifefishes, elephant fishes and catfishes.

Electroreception

In order to understand electroreception, it is useful to review the basics of a simple electrical circuit (see Fig. E.10). The simplest circuit consists of a battery, a resistor and a switch. The battery contains many electrons, and can be compared to

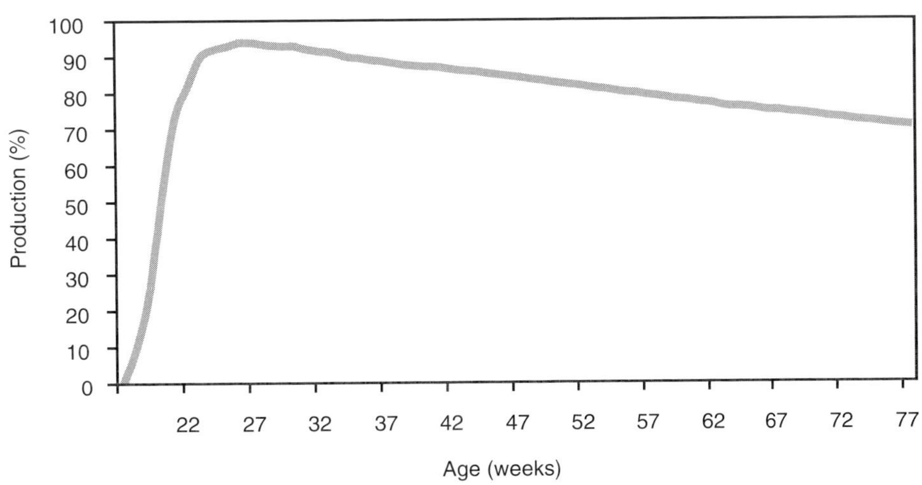

Fig. E.8. A typical egg production curve.

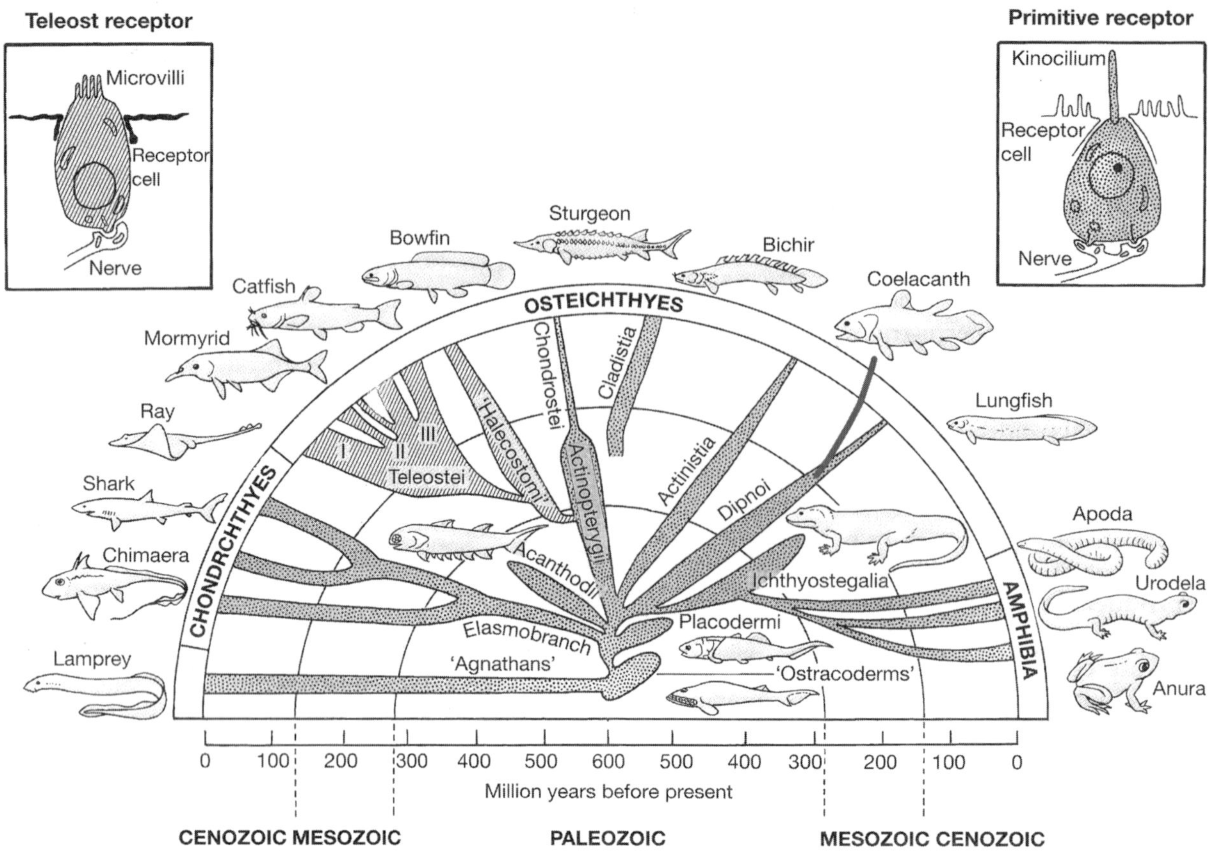

Fig. E.9. Phylogenetic distribution of electrosensitivity. Stippling and cross-hatching indicate two classes of electroreceptors, as shown at upper left and right.

a reservoir of water; once a switch is thrown, thus closing the circuit, electrons will flow through the circuit, just as water would pass through an opening in the dam into a riverbed. However, the rate of electron flow is modulated by the resistance of the conductor, just as water flow is impeded by the width of the river bed and the number of rocks in its bed and lining its sides. Ohm's law demonstrates that the current, or electron flow through the circuit, is equal to the voltage, the density of electrons in a battery divided by the resistance of the conductor. In Fig. E.10 the battery is symbolized by vertical lines alternating in height, a resistor by the saw-shaped lines and the switch by the two small circles, one of which has a line extending from it.

The electric field shown is a dipole, consisting of a positive and a negative charge separated by a given distance, and this can be described first by equi-current lines, curved lines leading between the charges that indicate the density of electrons flowing through the medium – more flowing in a straight line between the poles, fewer at a greater distance to either side of this line. The field can be described secondly by concentric oval rings around each charge, indicating the tendency for current to flow between the two poles at greater distances. The asymmetrical concentration of electrons in the parts of any aquatic animal produces a dipolar electric field with a voltage. The magnitude of this voltage will change in time as, for example, a fish opens and shuts it mouth, drawing water though its gill cavity and passing it by its gills to respire, or contracts its muscles as it propels itself forward by beating its tail. The alternating electric fields, such as those produced as a small fish buried in the sand breathes, are detectable at night by a shark or ray with an organ sensitive to these fields.

Indeed, the sharks and rays, which are dominant predators in the marine environment, are very sensitive to electrical gradients and can perceive an equipotential gradient of as little as 1 nanovolt, or one billionth of a volt, over a distance of 1 cm. Most living animals produce electrical potentials that are much greater, often in the magnitude of thousandths of volts, over a distance of 1 cm. The cartilaginous fishes, including the sharks, rays and chimaeras, detect these fields with an organ termed the ampullae of Lorenzini (see Fig. E.11).

This consists of a minute circular opening, or pore, in the epidermis, with a tubular canal containing a non-conductive wall filled with an electrically conductive gel that leads to a lumen, or cavity, lined with voltage-sensitive cells, each having a single kinocillium projecting into the lumen. Because the canal runs for some distance beneath the epidermis, the receptor cell detects the relatively large difference in voltage, or potential electron flow, between the charged body fluids within its tissues and the electrolytic solution comprising the surrounding environment. The nerve at the base of each receptor cell fires at a constant rate, and its rate of firing is depressed in the presence of a voltage whose polarity is

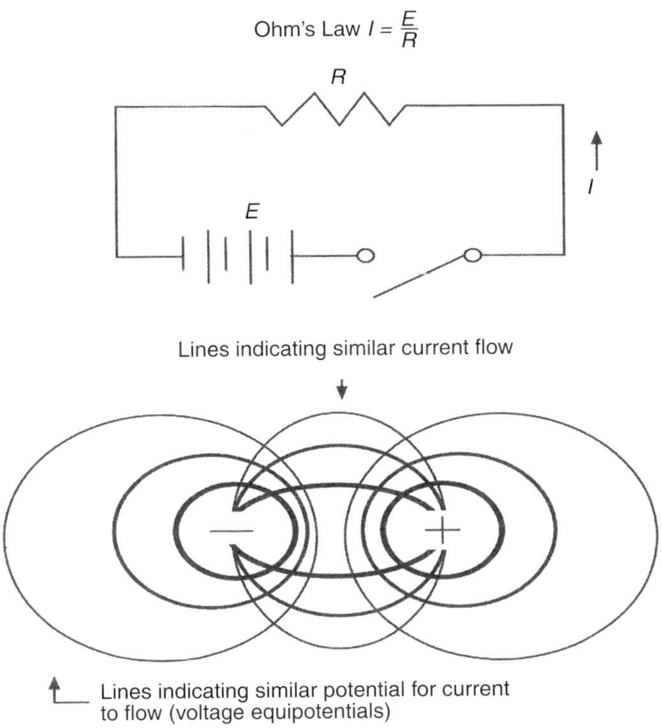

Fig. E.10. Simple electrical circuit for illustrating Ohm's law and dipole shown with equi-voltage (or equipotential) and equi-current lines. In Ohm's Law, I indicates the current or electron flow, E denotes the voltage or potential flow and R expresses the resistance in the circuit to this flow of electrons.

positive with respect to the internal body of the shark or ray. These canals are distributed over the head on sharks, with the highest density surrounding the mouth. They are not only scattered over the heads of rays, with the highest density near the mouth, but also extend outward on the underside of their laterally enlarged pectoral fins.

The sharks and rays use this organ to detect prey or mates. A shark can locate a fish, such as a flatfish or cigar-shaped wrasse that is buried in the sand out of sight, by detecting its electromagnetic discharges. All muscle activity generates electric potentials, or voltage changes, such as short-duration pulses of electron flow caused by the release of neurotransmitters by motor **neuron(e)**s, or longer-duration pulses of electron flow associated with muscle contractions. Furthermore, the movement of water, a conductor, through the gills during respiration also induces an electron flow.

Juvenile scalloped hammerhead sharks, which live in shallow bays, swim at night 20 cm above the bottom, searching for buried prey. The laterally elongated rostrum, or hammerhead, may be an adaptation to increase the scanning surface and effectiveness of the receptor system – the pores of the ampullae of Lorenzini extend all of the way to the distal ends of the laterally expanded head near the eyes. In a sense, the selective pressure on this species of shark is consistent with the impetus for an engineer to design a coin detector with a wide disc. Even the parrotfish, which escapes predation from an olfactory-sensitive predator – the moray eel – by secreting a mucilaginous cocoon around itself to contain its excreted metabolites, is detected by the shark or ray using its electroreceptors. Rays use the electroreceptors that are distributed over their enlarged pectoral fins to detect their mates. Male

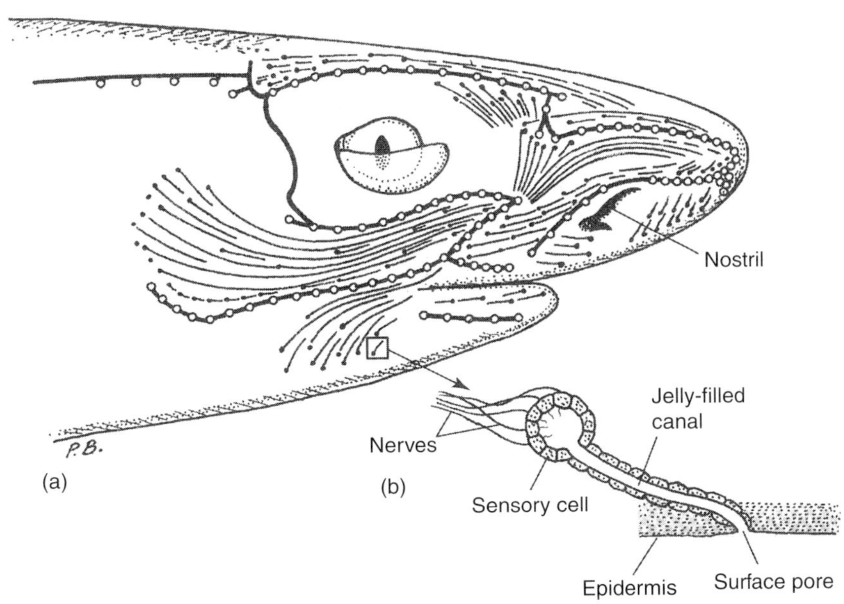

Fig. E.11. Ampullae of Lorenzini. (a) Distribution of ampullae on head of spiny dogfish. Open circles represent openings to the canals. (b) Ampullary organ, consisting of surface pore and canal filled with conductive jelly leading to lumen lined by sensory cells.

round stingrays swim into shallow sandy waters in the Gulf of California, where they locate females from a distance, change their course and move directly to them to inspect them, orienting to the equipotentials of their pulsing electrical fields (see Fig. E.12).

Finally, elasmobranches use their electrical sensitivity to navigate. The shark's body can be imaged as a conductor, and its movement through the Earth's dipolar field produces a perceptible electrical voltage. The shark can maintain a constant heading by swimming in such a way as to keep constant the voltages perceived by its ampullae of Lorenzini. Indeed, many species of sharks – such as blue, scalloped hammerhead and tiger sharks – swim in a highly directional manner for long distances at night, well below the sea surface and well above the seafloor, thus far away from fixed visual references such as the sun, moon and bottom topography that could alternatively guide their movements. There are also patterns of magnetization in the seafloor that may be perceptible to sharks. The seafloor consists of alternating bands, leading north and south, of strongly and weakly magnetized crust, and this pin-striped pattern is broken by strong dipolar fields of small islands and underwater mountains (seamounts). These two patterns are created by the cooling of magma extruded over long periods of time from the Earth's core, and the infrequent reversal of the polarity of the Earth's field gives these tiny magnetic particles a parallel or antiparallel orientation to the current polarity of the Earth's field. The night-time foraging excursions of hammerhead sharks, 20 km to and from seamount and islands, have been shown to follow magnetic ridges and valleys from the seamount associated with lava flows.

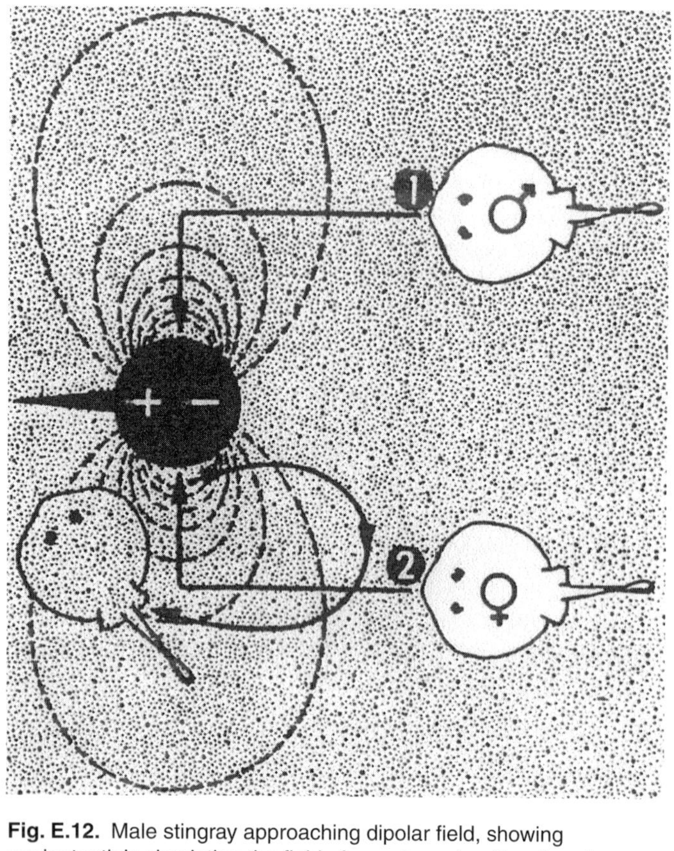

Fig. E.12. Male stingray approaching dipolar field, showing equipotentials simulating the field of prey, to mate with a female stingray.

Electrogenesis

Fishes also produce electrical fields, which can be both weak and strong. The knifefish (Gymnotidae) and elephant fish (Mormyridae) produce weak fields around themselves, and can move around in turbid waters by avoiding rocks and other obstacles to movement by perceiving the distortion to their field caused by these objects, which differ in conductivity from fresh water (see Fig. E.13). The signals of males and females of these species differ in the waveform of their signals, and it is thought that they identify their mates on the basis of discharge pattern. The freshwater electric eel and catfish and the marine teleost, the stargazer and marine elasmobranch – the torpedo ray – produce electrical discharges strong enough to stun their prey.

These electrical discharges are produced by modified muscles, termed electrocytes, that have lost the ability to contract, but have gained the ability to generate ionic current flow. Each electrocyte resembles a small comb standing on its side, with a smooth side, which is innervated, opposite a rough or multi-fingered side. The inside of the membrane on the smooth side of the electrocyte is negatively charged with respect to the surrounding extracellular fluids, due to the exclusion of sodium ions by the cell membrane pump. When the cell is stimulated by the nerve discharge, the sodium ions pass rapidly through the smooth side to produce a positive voltage, while the ions stay in place on the rough surface, which is not innervated. This results in a voltage potential across the entire cell, including the smooth and rough surfaces, of 151 mV. The electrocytes are arranged in stacks, and a single stack may contain as many as 10,000 layers. Their voltages sum like those of multiple batteries arranged in series, each with its positive lead next to the negative lead of the adjacent battery, to generate potentials as high as 600 V, as recorded from the electric eel in the Amazon River of South America.

The knifefish of South America and elephant fish of Africa, which produce weak electric fields, live in freshwater rivers and lakes. They produce discharges that are either continuous waveforms or series of pulses, yet they discharge often in contrast to those fishes that produce strong voltages. This is because they are nocturnal and live in highly turbid waters, where vision is of limited use, and they use these fields to navigate in their local environment. These fields can extend outward as far as 50 cm from the body, and the field is distorted by objects that are either less or more conductive than fresh water. Rocks are highly conductive, while other fishes, invertebrates and even plants are more conductive. These distortions of the field produce a change in the distribution of the equipotential lines across the fish's body. Because of the energetic cost of producing this field, the majority of the weakly electric fishes produce pulsed electrical fields, varying in frequency between 50 and 300 cycles/s.

The skin of these teleosts has a specialized electric receptor, the tuberous organ, which probably evolved much later than the ampullae of Lorenzini at the beginning of the Mesozoic age, 280 million YBP. These receptors are derived from the

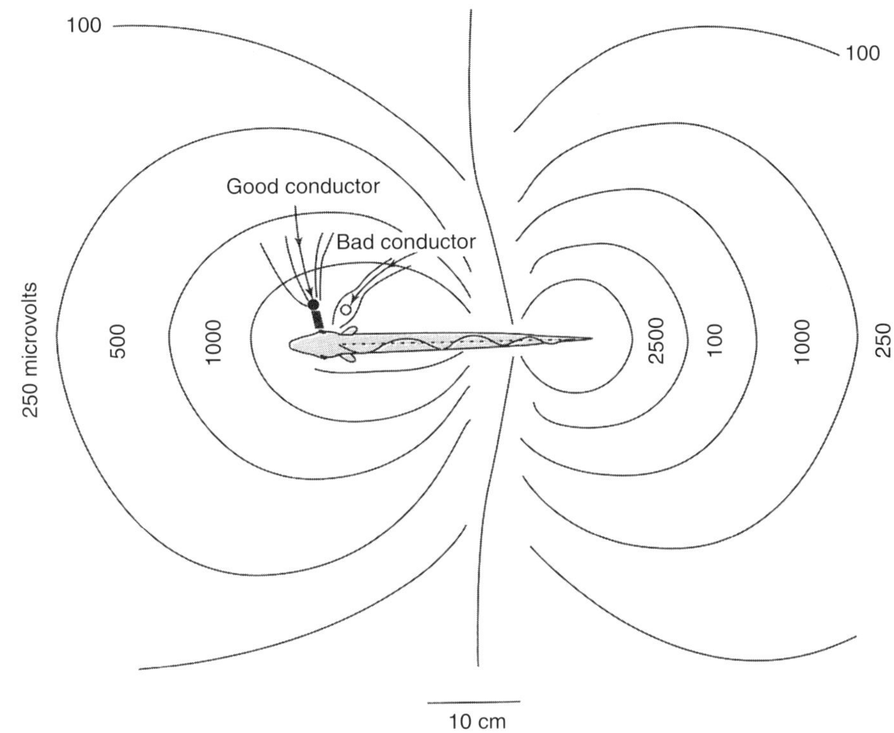

Fig. E.13. Electroreception by fishes. Distortion of lines of equipotential by two objects, one with low conductivity and the other with high conductivity.

neuromast receptors of the lateral line, and differ from the ampullae both anatomically – in that they have many small microvilli at their surface – and physiologically – in that they respond to an external voltage that is positive with respect to that of the cell, by increasing their rate of discharge. During each organ discharge, an inhibitory command is issued to the electroreceptors so that they are insensitive to their own discharge, yet between the discharges they turn on and report the distortion of the electrical field by the presence of a foreign object. The nature of their discharges varies with their sociability and habitat. For instance, social species that live in shallow and narrow streams have pulses of short duration in order to avoid interference with the discharges of their neighbours. In contrast, solitary territorial species have longer electric discharges. The waveforms of males and females differ, and these are used during courtship by sexually active males to identify receptive females. Yet this social advertisement has a cost – a predator such as the electric eel can use its electrosensitive organs to locate these fish by orienting to their **courtship behaviour** signals.

The primary function for the generation of weakly electric fields is navigation, and hence they are relatively continuous. Therefore, these species are largely confined to rivers and lakes with fresh water of low conductivity. It would be difficult to maintain this ability in the marine environment – the energetic cost of maintaining fields at any distance from themselves in the highly conductive marine waters would be prohibitive. The fish that produce strong fields need to do so infrequently to stun their prey. These fish exist in fresh water, such as the electric catfish in the Nile River and the electric eel in the Amazon River. However, there are also species in the marine environment with the same capability. These species must be close to their prey to stun them, due to the high conductivity of salt water. For that reason, the stargazer remains motionless on the bottom, looking upward for above-swimming prey, and produces a voltage sufficient to stun it as it passes closely over. The torpedo ray drifts slowly through the water at night and alters its path ever so slightly so that it can move close to small fish, and then discharges its electrical organs while moving its pectoral disc into a concave shape around the prey, so that its body is as close as possible to the prey before discharging its electrocytes.

In conclusion, the evolution of the electrical sense has played a major role in adapting species to aquatic and marine environments. It has enabled these species to find their way in an environment where vision is of less value than in the terrestrial environment. It is also effectively used to detect prey and find mates and, hence, the sense has evolved at least twice during the evolutionary history of animals on our planet.

(APK)

Further reading

Hopkins, C.D. (1981) On the diversity of electric signals in a community of mormyrid electrical fish. *Science* 21, 211–222.

Kalmijn, A.J. (1981) Electric and magnetic field detection in elasmobranch fishes. *Science* 218, 916–918.

Westby, G.W.M. (1988) The ecology, discharge diversity, and predatory behaviour of gymnotiform electric fish in the coastal streams of French Guiana. *Behavioural Ecology and Sociobiology* 22, 341–354.

Electric/electronic training aid

Numerous electrically activated training devices are available for use on animals worldwide. Their use is a contentious and emotional issue and is often justified by the suggestion that they are a last resort and hence an alternative to **euthanasia**. They are often seen (and marketed) as a quick-fix answer to problems that may have been ongoing for years. The use of these training aids is legal in some countries and illegal in others.

Electric training aids have been used to confine animals, to deter unwanted behaviours such as barking and to attempt to train new or desirable behaviours. They are available in several versions: (i) operator controlled (i.e. the operator decides when to trigger the device); (ii) environmentally controlled (e.g. where the location/distance of the device relative to the animal within the environment triggers the device); and (iii) animal controlled (i.e. the animal's behaviour triggers the response, e.g. volume of bark).

Devices can include **collars** (used on dogs and horses) or electric mats (cats) that may be paired with a noise (e.g. hissing from compressed air), a smell (e.g. citronella collars) or a discharge of electricity between two metal probes in contact with the skin of the neck (e.g. the electronic stimulus/shock collar). They operate on the basis of positive punishment, whereby the devices are activated to produce an aversive stimulus at the time the animal exhibits the unwanted behaviour.

For these electric training aids to be effective on their own they must, by definition, provide an aversive stimulus and so have the potential to inflict **pain** and **suffering** on the animal. The correct use of punishment is critical if it is to be effective. That said, training aids that inflict punishment may actually impede learning, especially in anxious animals. In some cases, they are used at a low level to interrupt behaviour and facilitate positive reinforcement of the correct behaviour.

Concerns about the possible consequences of the use of some of these devices for animal welfare relate to the extent to which they inflict pain. The type of pain may be acute at the time the stimulus is applied or may be chronic, affecting behaviour subsequently (as shown by several studies). The extent to which individual animals suffer pain varies considerably both between animals and within the same animal under different environmental circumstances.

Research has shown that, in some cases, the use of electronic devices may cause **distress** and potentially harmful responses, both physical (e.g. skin damage) as well as psychological (e.g. increased **anxiety** or **fear**) in some animals. They may also exacerbate the unwanted responses. For example, they may increase reactivity in an aroused or aggressive animal. Additionally, since a dog that is adversely affected by the collar cannot remove it, distress and possibly even **learned helplessness** may occur. None the less, they can be effective, leading to a debate concerning the **ethics** of their use.

Some of the behaviours that electric training aids are employed to change are normal behaviours. Attempting to stop normal behaviours may well be detrimental to the animal's welfare, if the behaviours are highly motivated. An alternative approach to stopping the undesirable behaviour is to consider its **motivation** and consider allowing another acceptable outlet for the behaviour, or to eliminate the stimulus for the behaviour.

For example, when attempting to modify a dog's barking or contain a dog in a yard it is important to first establish the reasons for the barking or attempts to escape. Treating the problem symptomatically, e.g. only by stopping the barking or confining the dog, may be ineffective as well as harmful. Failure to address the underlying causes of the behaviour often causes the problem to continue, and may also lead to further anxiety or possible pain.

Operator-controlled collars are difficult to use consistently within the guiding principles of learning theory criteria. For example, to be effective, punishment needs to be administered at the time of the behaviour. Even experienced trainers find that this is not always possible. The punishment also needs to be applied every time the behaviour occurs but, unless the operator is present 24 hours a day, 7 days a week, this criterion cannot be satisfied.

Electronic boundary collars are often used to confine dogs and other animals to specific areas of a property, for example when traditional fencing is impractical or prohibitively expensive. They may also be used to restrict access to certain areas (e.g. flower beds). If they succeed in keeping the dog inside a defined area, hence preventing roaming and thus reducing the chances of a road traffic accident, then they may have positive welfare implications for the dog, the owner and the community. However, if their use results in the dog being attacked by another dog entering its property, it would appear that the welfare of the resident d og can also be compromised.

The use of electric training aids is problematic. Even if these devices are used only by people with specialized training, under very controlled, exceptional or specific circumstances, they can still be misused and abused, and can cause pain and physical, as well as psychological, damage to the animal. With increasing knowledge about behaviour modification it is clear that alternative methods are usually available.

The fundamental problem with such devices is that they attempt to control signs (for example, barking or escaping) without addressing the possible underlying causes for the behaviour. They may exacerbate the problem and may also have significant welfare implications for some animals.

As veterinary behaviour therapy has developed since the 1990s, the use of 'punishment' in behaviour modification has fallen out of favour and has been largely displaced by the use of reinforcement (any approach that encourages the animal to exhibit the desired behaviour). (KS)

Further reading

Juarbe-Diaz, S. and Houpt, K.A. (1996) Comparison of two antibarking collars for treatment of nuisance barking. *Journal of the American Animal Hospital Association* 32, 231–235.

Polsky, R. (1994) Electronic shock collars: are they worth the risks? *Journal of the American Animal Hospital Association* 30, 463–468.

Polsky, R. (2000) Can aggression in dogs be elicited through the use of electronic pet containment systems? *Journal of Applied Animal Welfare* 3, 345–357.

Tortora, D.F. (1983) Safety training: the elimination of avoidance motivated aggression in dogs. *Journal of Experimental Analysis of Psychology: General* 112, 176–214.

Electrocardiogram

The transmission of electrical activity throughout the heart gives rise to an electrical field that can be continuously

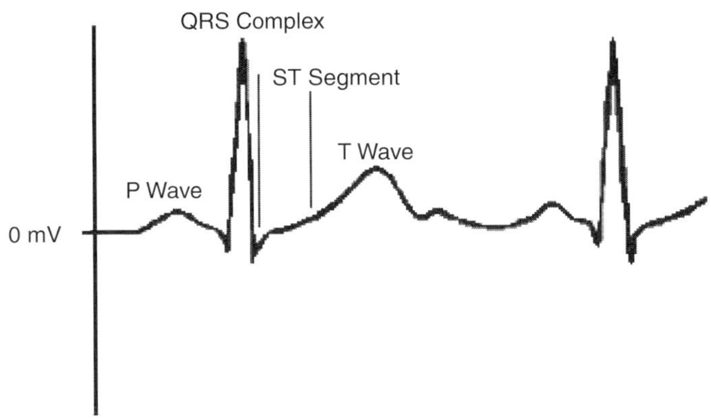

Fig. E.14. Electrocardiogram (ECG) showing the location of the P, S and T waves and the QRS complex.

detected through appropriately placed sensory electrodes on the body surface. The electrodes register voltage differences produced by the heart that can be plotted on to paper, or displayed online, to form an electrocardiogram (ECG). Generally, in humans 12 to 15 leads are attached to electrodes and placed in a bipolar arrangement on the body. In animals, however, fewer leads (e.g. three to five) or implantable biotelemetry transmitters may be used for easier and longer-term bipolar or unipolar ambulatory monitoring.

A normal ECG has three principal components that represent different facets of cardiac activity (see Fig. E.14): the P wave, the QRS complex and the T wave. The small, rounded P wave represents atrial depolarization, and results in the contraction of the atria. Following on from this is the PR (PQ) segment, which represents the movement of depolarization through the atrioventricular node. The magnitude of this is too weak to be detected, so it is characterized by a 0 mV segment. Next comes the QRS complex, which corresponds to the movement of the depolarization from the atrioventricular node to the interventricular septum. This stage is also referred to as ventricular depolarization. Once again, voltage returns to 0 mV before ventricular repolarization occurs, forming the T wave.

Repolarization of the atria is not evident on an ECG, as it occurs during ventricular depolarization. In medical environments the ECG is frequently used for the observance of cardiac defects, arrhythmias, heart disease and so on. In animal behaviour research, the ECG has been successfully used as a tool to evaluate and compare different **welfare** states, including response to stressful events and painful procedures.

(RM-F)

Electroception/electroreception

Electroception is the ability to perceive electric impulses. Many primitive fish (e.g. sharks) can detect electric fields generated by other animals and thus use passive electroception. Among mammals, only the monotremes (e.g. platypus) have provided convincing evidence of being electroreceptive. Other animals, such as electric fish, use active electroception, as they can generate electric fields: they are electrogenic. Electroception can be used for electrolocation (i.e. location of objects) in the context of prey location or navigation. It can also be used for electrocommunication among animals that can modulate the electrical waveform they produce. (SL)

See also: Communication; Electric field

Electroejaculation

Electroejaculation involves the use of a rectal probe to electrically stimulate erection and ejaculation in male animals; it is a long-established technique, with the first report of its use being in 1936. It is routinely used for captive species where semen collection is required for assisted-breeding programmes. For domestic species where regular semen collections are required, males are generally trained to ejaculate into an artificial vagina or in response to manual manipulation, but electroejaculation is still used for untrained or recalcitrant males where semen samples are required for breeding-soundness examinations.

Use of electroejaculation has a number of advantages: (i) it is independent of the male libido status, and so does not require the presence of teaser oestrous females or dummy mounts; (ii) it does not require training, which can be a time-consuming process that cannot be justified for a one-off collection; (iii) it can be used on anaesthetized or sedated animals where there are serious safety or handling considerations; and (iv) it has a high success rate in inducing ejaculation.

Emission, transfer of semen from storage in the *ampullae* and *vasa deferentia* into the pelvic urethra, erection and ejaculation are all achieved by a combination of smooth muscle contraction and vascular mechanisms initiated by the **parasympathetic** and **sympathetic nervous systems**. Stimulation is via a cylindrical probe with a series of longitudinal ventral electrodes; older-type probes had ring electrodes around the probe, which resulted in additional dorsal stimulation of motor nerves, causing muscular contractions in the back and hind limbs that could be sufficiently severe to cause soreness and bruising. Ventrally oriented electrodes focus the electrical stimulation in the required area, reducing unwanted motor stimulation.

The majority of modern systems use a sine wave pulse with a frequency of 20–30 cycles/s and a variable output up to 16 V (< 900 mA) that is applied intermittently in series stimulations lasting a few seconds, with rest periods in between; voltage may be increased with each successive stimulation until ejaculation is achieved. More advanced systems allow a set stimulation sequence to be pre-programmed into the probe. In large domestic species stimulation may be preceded by transrectal massage of the accessory glands to sexually stimulate the animal; in some cases massage alone may induce ejaculation, but success rates are much more variable than those achieved using electrical stimulation.

The major problems with the use of electroejaculation are the **welfare** considerations consequent on the degree of **pain** or discomfort associated with the procedure. The extent to which the procedure is painful is not clear; in human cases where electroejaculation has been used to overcome ejaculatory failure, the procedure is known to be painful and only performed under anaesthetic. However, with domestic livestock, general **anaesthesia** carries its own risks and a number of studies have investigated the acceptability of electroejaculation on welfare grounds. Within the UK, welfare **codes of**

practice stipulate that the process should only be performed by a veterinarian, and electroejaculation without anaesthesia is discouraged. Some European countries have gone further, banning the process completely and even banning the import of semen collected without anaesthesia. Electroejaculation has been shown to be an aversive procedure, but only to an extent comparable to other routine stock procedures requiring handling and a degree of **restraint**, such as shearing of sheep.

Levels of pain associated with the procedure may be mitigated by good operator procedures and the use of up-to-date probe designs, with protocols involving minimal stimulation. While aspects of electroejaculation usage may be controversial, it remains a routine and widely used procedure in domestic livestock production, and has a growing role in assisted-reproduction protocols for endangered and exotic species. However, electroejaculated semen tends to be of lower quality than a natural ejaculate. (MRC)

Electroencephalogram

An electroencephalogram (EEG) is a non-invasive and painless test that detects electrical impulses (activity) in the **brain**. It consists of different kinds of waves traced on graph paper, representing brain activities.

When a group of **neuron(e)s** is excited simultaneously (synchronously), the action potentials of each neuron sum to generate one large surface signal. The signal can be recorded by placing small metal electrodes on the external surface of the head, and is shown as EEG waves on either graph paper or a computer screen. Based on frequency range, the various brain waves have been categorized and named:

- alpha (α) waves – relatively slow, with a frequency of 8–12 Hz;
- beta (β) waves – with higher frequency (13–30 Hz) and lower amplitude;
- theta (θ) waves – with a frequency of 4–7 Hz;
- gamma (γ) waves – with a frequency > 30 Hz;
- delta (δ) waves – slow, with a frequency < 4 Hz but large in amplitude.

The EEG waves vary dramatically and often correlate with particular states of behaviour, such as level of attentiveness, sleeping and waking, and pathological changes, such as seizures or coma. In general, high-frequency, low-amplitude rhythms are associated with alertness and waking or the dreaming stage of sleep; low-frequency, high-amplitude rhythms are associated with non-dreaming sleep states and with the pathological state of coma. When the EEG changes from α to β waves, from synchronization to desynchronization (activation) of the EEG, an arousal is produced.

In particular, through examining the characteristics of wave patterns, the EEG provides us with a convenient window on the structure and functional states of the brain in both healthy individuals and those with psychiatric illness, such as **anxiety**, mood disorders and post-traumatic stress disorder. When an animal is sick, hurt or stressed, electrical messages to and from the brain can be changed or blocked, resulting in changes in EEG waves. The EEG test can pinpoint the location of these neural changes, and has been used to help in the diagnosis and find the causes of certain neurological and psychiatric conditions, especially seizure disorders, strokes, brain tumours, head trauma, metabolic dysfunction, vascular disease and infections of the nervous system, as well as certain mental responses, such as those associated with **social stress**, **emotion** and **cognition**, all of which affect brain function. In addition, a high-resolution EEG structure analysis has been used to examine stress-induced states and their correlations with the resonance activation of the brain.

However, EEG has certain limitations in practice. One limitation is that most rhythms have no direct linkage with the brain function. Instead, they may be just by-products of the interconnections of various forms of excitatory feedback of the brain circuits, regardless of stimulators. For example, a similar EEG could be recorded from the instability or oscillation of the brain actions induced by either an audio amplifier or a human stadium ('Mexican') wave. In addition, EEG is difficult to interpret and subject to some errors. Recently developed, quantitative EEG gives a much more accurate and representative view of the location within the brain of the alterations of EEG activity, through the qualitative assessment or overall appearance of the rhythms and amplitudes and their changes. (H-WC)

Elephant

Elephants are the largest living land mammals, with adult males standing nearly 4 m tall and weighing up to 7000 kg. Besides their immense size, elephants are known for their dexterous trunk, large tusks and matriarchal family structure. There are two genera of living elephants (Proboscidea: Elephantidae) – the Asian elephant (*Elephas*) and the African elephant (*Loxodonta*). The former may have three subspecies, with Indian (*E. maximus indicus*), Sri Lankan (*E. m. maximus*) and Sumatran (*E. m. sumatranus*) forms, although some recent evidence suggests that the Indian and Sri Lankan forms are not distinct, but a Borneo subspecies may be reinstated.

Currently, the African Elephant Specialist Group (AfESG) of the International Union for the Conservation of Nature (IUCN) recognizes only a single African species with two subspecies, the savannah elephant (*Loxodonta africana africana*) and the forest elephant (*L. a. cyclotis*). African elephants are tallest at the shoulder, with large, Africa-shaped ears, two 'fingers' on their trunk tip and lozenge-shaped loops on their teeth. The forest subspecies is considerably smaller than the savannah elephant. Asian elephants are tallest at the head, with India-shaped ears, a single trunk tip 'finger' and compressed loops on the teeth. Female Asian elephants have reduced or no tusks.

Elephants are endangered in the wild, with population estimates of approximately 660,000 for all African elephants and estimates ranging from 30,000 to 50,000 for Asian elephants. While fewer than 1000 African elephants are maintained in captivity, primarily outside of range states, there may be as many as 15,000 Asian elephants in captivity, mostly in range states. Humans have a long history of using Asian elephants for labour, especially in lumbering. Habitat loss and degradation are a continual problem for the maintenance of elephant populations, with direct loss through **hunting** for ivory an important contributing factor. Human–elephant conflict involves competition for space and food, resulting in crop loss and sometimes death of humans as well as harassment, injury or death of elephants. Management of elephant populations, all of which are confined to some degree, is controversial yet critical to the long-term coexistence of humans and elephants.

Elephant society centres on calves that are cared for by a matrilineal assemblage of related females and their older offspring. The youngest elephants are most susceptible to predation and, when alarmed, elephants perform numerous alert behaviours, including formation of a phalanx around the calves. The eldest female is generally the matriarch, with a linear dominance **hierarchy** among females established on the bases of age and size. Group size is larger with older matriarchs, who are a valuable source of information to facilitate survival. The family is the core social unit, although forest elephants may reside more as mother–offspring groupings. Related families may travel together in kin groups, and these may unite into larger social groups, especially in savannah African elephants. Although likely to be a rare occurrence, unrelated females may join a group or coalesce into a functional herd, especially after alterations in normal herd composition such as through heavy hunting.

Captive elephants outside of range states usually reside in unrelated groups, often with no or few calves. In the wild, males remain in the natal group until after 10 years of age, although they commonly range away from the herd core more than same-aged females before this age. Males may travel in bachelor groups, especially when younger, or more commonly alone as adults. Males reach adult female size by their late teenage years but, in general, they are not viable mates until at least their mid- to late 20s.

Elephants breed year round, although seasonality is evident in areas of pronounced wet and dry seasons. The oestrous cycle lasts 14–16 weeks, with 3–5 days of receptivity. Females usually mate for the first time between 10 and 16 years of age. Unique to elephants, there are two **luteinizing hormone** (LH) peaks, with an anovulatory surge 3 weeks before the ovulatory surge. Female Asian elephants release an oestrous **pheromone**, (*Z*)-7-dodecen-1-yl acetate, in their **urine** that increases in concentration from the first to the second LH surge. Female African elephants also appear to release a chemical signal by which males can detect forthcoming ovulation. Adult males enter a rut-like condition known as must (musth), during which they demonstrate increased **testosterone** levels, chemical signal release through the temporal glands and urine, **aggression**, competitive ability, and association with females leading to offspring production. Auditory and postural cues also signal **oestrus** and must. The gestation period is 22 months, with a single offspring – or rarely twins – undergoing an extended developmental period following **parturition**.

Elephants occupy a wide variety of habitats, from forests to deserts. Asian elephants tend to graze more, while African elephants tend to favour browsing, but all elephants have an expansive herbivorous diet in which they use their dexterous trunk, resilient tongue, strong tusks and huge body to feed on small plants to large trees. The sexual dimorphism of elephants is reflected in slightly different foraging patterns, with males opting more for quantity while females select for quality of vegetation.

Elephants also show dimorphism in their communication patterns. Females have a larger repertoire of **vocalizations** and may use infrasonic vocalizations not only for distant communication but also for regular, nearby interchanges. Some elephants have been recorded mimicking trucks, other elephants and even a human handler. Both sexes use chemical signals extensively, with females showing greater interest in body regions that may release signals, such as the genitals, than in excretory products, whereas males regularly inspect urine and faeces. Auditory and olfactory stimuli may be useful enrichment tools in captivity but, without additional **reinforcement** or variation in presentation, elephants are likely to habituate quickly.

Elephants live in complex social groups and can use their trunks to fashion tools that can be used to remove ectoparasites or to obtain food. Elephants can be trained for a variety of tasks, which has made them valuable in labour and entertainment. In the wild, they demonstrate caring for injured or recently deceased relatives. Recent increases in elephant aggression towards humans and their machinations have raised the question of chronic **stress** in elephants driving them to unusually erratic behaviour. Further research on elephant behaviour and **conservation** is needed to maintain healthy elephant populations in a sea of humanity. (BAS)

Further reading

Fowler, M.E. and Mikota, S.K. (eds) (2006) *The Biology, Medicine and Surgery of Elephants*. Blackwell Publishing, Ames, Iowa.

Sukumar, R. (2003) *The Living Elephants: Evolutionary Ecology, Behaviour and Conservation*. Oxford University Press, New York.

Elevated plus-maze

The elevated plus-maze is an apparatus used in behavioural tests of **fear** or **anxiety** in **rodents**. It is shaped like a plus sign with four arms, two of which have vertical sides and an end wall, whereas the other two arms have no sides or wall. It is usually raised approximately 30 cm from the floor. Rodents exhibit **thigmotaxis**, i.e. they prefer to stay close to vertical surfaces, so when they are placed in the elevated plus-maze, less fearful animals make more frequent and longer visits to the open arms. This test is used widely in the development of anxiolytic (fear-reducing) drugs for humans. (CS)

Elimination behaviour

Elimination behaviour includes defecation and urination. Young mammals change their diet as they undergo the transition from neonate to juvenile and there is a consequent shift in their water balance and digestive function, as reflected by a decreased frequency of urination and an increased frequency of defecation. It is worth noting that **urine marking** and **scent marking** are not considered forms of functional elimination.

As adults, some species (such as horses and dogs) are prompted to defecate by the sight of faeces or the action of a conspecific defecating. Drinking and eating are also thought to stimulate defecation, possibly via a gastro-colic reflex. This has important implications for those seeking to house-train domestic animals, since it is necessary to take the animal to the appropriate area for elimination immediately after feeding it. Other critical precursors to elimination in young animals include waking and **play**.

While ruminants often eliminate indiscriminately, other species (e.g. horses) show considerable care in the selection of defecation sites, since they return repeatedly to areas not used for grazing. Preferred substrates for defecation can have a critical influence on the success of house-training strategies in **pet** species. Cats' preference for deep, sandy litter is a good example, since the absence of key properties in the litter offered can precipitate **house soiling**.

Although coprophobia is a common feature of domestic animal behaviour, faecal material is very useful in communication between **conspecifics**. For example, stallions indicate their presence by piling their faeces, a strategy that avoids confrontation between stallions. Middening by cats and defecation against trees by dogs are similar examples of faecal marking.

Arousal in the form of both **fear** and excitement can also stimulate animals to defecate. Indeed, increased frequency of both defecation and urination has been used as an indicator of social distress in isolated horses and laboratory **rodents**. As with defecation, some species – and most often the males – use **urine** as a marker. Stallions generally visit latrine areas to urinate, but also use urine to mask the **odours** of conspecifics.

Urination in females often becomes elaborate when they are in **oestrus**, as it seems to provide a visual stimulus to mating partners and also provides chemical signals. An example is the clitoral winking that occurs in mares after urination, but which is far more frequent when they are in oestrus. In oestrus, mares pass less urine at each urination. Urination often occurs as a spontaneous solicitation and in response to a **teaser**.

The number of urinations per day reflects the balance between water intake and water loss, e.g. through sweating, and so is negatively correlated with exercise. Although it may reflect the presence of cystitis or urinary calculi, polyuria (excessive urination) is usually a consequence of **polydipsia** (excessive drinking) that can be somatic or, far more uncommonly, **psychogenic** in origin. Accurate determination of water intake is an important step in defining the extent of the problem, while direct observation of the animal's drinking and eating behaviour can help to identify contributing behavioural anomalies, such as excessive use of salt licks. (PDM)

Emancipation

Emancipation refers to the process by which behaviour becomes independent from the originally eliciting stimuli and motivational (proximate) states. Emancipation is used for analogous processes on evolutionary and ontogenetic timescales (**see: Evolution; Ontogeny**).

In evolutionary terms, emancipation relates to the evolution of signals and displays, and goes together with **ritualization**. Many signals and displays are derived activities, originating from behavioural responses to motivational conflict (e.g. **displacement behaviour**, intention movements). Activities derived from motivational conflict have frequently evolved in the context of agonistic encounters (threat, **appeasement**) and **courtship behaviour**. Through emancipation, the behaviour became increasingly independent of motivational conflict. For example, ritualized displacement activities are among the most frequent courtship displays in many birds.

Ontogenetic emancipation refers to an analogous process in behaviour development, for example in stereotypic behaviour (**see: Stereotypies**). Stereotypies also originate mainly from behavioural responses to conflict and thwarting (e.g. adjunctive behaviour, **appetitive behaviour**, displacement behaviour, intention movements and **redirected behaviour**). In the course of development, these source behaviours become increasingly independent from the original situation of conflict or thwarting, and elicited by a wider range of stimuli and motivational states. Eventually, stereotypies may persist even under conditions where they would not normally develop. For example, environmental **enrichment** introduced after 3 months fully abolished stereotypies in bank voles, while the behaviours were only partially abolished if enrichment was introduced after 6 months and after 9 months it had no effect on stereotypic behaviour.

This change in functional context goes together with changes in form and performance of the stereotypy that are analogous to ritualization. In the evolution of **signals** and displays, ritualization renders the behaviour less ambiguous and more conspicuous through exaggeration (behavioural and morphological), rhythmic repetition and rigidity in form. Similarly, stereotypy performance increases in frequency and duration, while becoming increasingly repetitive and invariant. Two different ideas were proposed to explain emancipation of stereotypies. First, stereotypies might be learned responses to cope with adversity. Emancipation may thus reflect generalization of the **coping** response from the original situation of conflict or thwarting to other adverse situations. Second, stereotypies might reflect acquired **brain** dysfunction. According to this view, emancipation might either reflect progressive sensitization of the repeatedly activated neural pathways mediating the behaviour or, alternatively, progressive disruption of inhibitory mechanisms of behavioural control. (HW)

Embryo

From the stage that an egg is fertilized, the resulting zygote divides to become an embryo, i.e. from the two-cell stage onwards. (DBM)

See also: Fetus (foetus)

Embryo transfer

Embryo transfer is the artificial transference of an **embryo** from one female reproductive tract to another, normally carried out in the first few days after fertilization, with the aim of it developing into a newborn animal in a surrogate mother that is not its dam. The embryo may be genetically manipulated *in vitro* before being transferred (**see: Transgenics**). It is important that there is synchrony between the age of the embryo and the recipient uterus or oviduct in terms of the number of days following ovulation, otherwise the embryo will not implant. (DBM)

Further reading

FAWC Report on the Welfare of Dairy Cattle – Embryo Transfer. Available at: http://www.fawc.org.uk/reports/dairycow/dcowrtoc.htm (accessed 24 September 2009).

Emotion

The terms 'emotion' and 'affect' are often used synonymously, and there is no universally accepted definition of either term. Affect may be considered as referring to the range of emotional processes that can bias **cognition** and the expression of behaviour, and includes emotional reactions, mood and aspects of **temperament** and their conscious representations as **feelings**. Emotion is not a unitary entity, but it arises, in one form or another, as a result of appraisal of the circumstances by the individual, and highlights the personal significance of stimuli. It may or may not involve self-perception. Emotional influences can lead to differences in response to stimuli that may arise as a result of generalization from previous experiences. The concept of emotion is therefore inextricably linked to the processes of **reinforcement** and **motivation**.

The subjective nature of the phenomenon has posed difficulties for its scientific investigation, although it has historically been of great interest, with four elements of emotional processes being emphasized in its investigation: its role in appraisal; the effects of emotional arousal; the function of its expression; and its role in preparing the animal for future action. **Charles Darwin** (1872) famously addressed the subject in his text *The Expression of Emotions in Man and Animals*, in which he documented the consistency of expression of analogous emotional states in different races, and also proposed gen eral principles associated with their expression across the animal kingdom, most notably the 'principle of antithesis'. This states that signals with opposite meanings are often conveyed using opposite expressions. Thus a frown is in many ways the physical opposite of a smile, and confidence is often expressed through extraversion and expansion, while **fear** is conveyed by introversion and withdrawal.

Darwin concluded that many of these expressions have an innate basis, although they may be influenced by learning and culture. He suggested that certain stimuli give rise to a central emotional response that, as a result of selection, has been directed towards evoking specific bodily expressions that communicate this. However, William James (James, 1884) – another early pioneer in the field – took the opposite and, perhaps what might initially appear counterintuitive, view, by suggesting that the bodily changes arose first and it was awareness of these that gave rise to the feelings; for example, by feeling our heart rate increase and running away in response to the sight of a bear, we become scared. In so doing he emphasized the role of the emotional response in arousal, appraisal and preparation for action. He argued that, if it were otherwise, then there would be no function for the emotion, i.e. we could run away from a bear without having to feel scared of it.

However, a fundamental problem with the model proposed by James related to which emotion we should feel in any given situation, as it was argued that the autonomic response is similar even for opposite feelings, e.g. happiness and fear both evoke a strong response from the **sympathetic nervous system** (see: **General adaptation syndrome**). Walter Cannon (Cannon, 1927) therefore proposed that the emotional and physiological responses arose independently and simultaneously following appraisal of the situation, the latter preparing the body for action and the former giving it a qualitative dimension. These approaches have dominated our thinking about emotion, and debate continues as to what is an emotion. Each of the four elements of the emotional process mentioned above is briefly considered below.

Defining the features of emotional processes

If emotional processing involves personal appraisal, it is important to determine the criteria that may be used in this process. It has been proposed that there are two stages to this process. Primary appraisal involves evaluating the threat or benefit posed by the current situation to the individual. Secondary appraisal is then used to evaluate the options available to manage the situation and whether these are likely to be successful. Both may influence the final assessment. Thus the presence of an **aversive** stimulus (primary appraisal) that the animal obviously cannot tolerate, escape from or adequately control (secondary appraisal) may lead to sadness, whereas the same stimulus given in a context that the animal does feel it can cope with may give rise to hope and a very different behavioural strategy. Both types of appraisal may be affected by previous experience. From a welfare perspective it would be useful to determine upon which dimensions appraisal is made, as these will largely determine the individual's well-being at any given time. Secondary appraisal may be particularly important if the importance of **allostasis** as a mechanism for safeguarding welfare is recognized.

Emotional processes are undoubtedly associated with changes in arousal, with the balance between the **parasympathetic** and sympathetic **nervous systems** being an integral part of this change. However, it is now recognized that it is possible to distinguish the physiological response associated with some emotional states (McNaughton, 1989). No single measure may distinguish one emotion from another but, by measuring **heart rate** and diastolic and systolic **blood pressure**, it is possible to distinguish between **fear**, anger, sadness, mirth and sexual arousal in humans (see Table E.2), and this presumably reflects different neurochemical cascades within the body as well.

Such physiological distinction means that it is not necessary for emotions to be conscious in order to be qualitatively different. Thus, it is suggested that the term 'feeling' be used to describe the conscious representation of an emotional process, in order to avoid confusion and distraction related to the issue of **consciousness** when trying to scientifically evaluate the subject of emotions in animals. A change in arousal may not only be a consequence of emotional processing, but it may also influence processing. For example, if animals are administered the **beta-blocker** propranolol, which inhibits the effect of the sympathetic nervous system, **aversion learning** may be disrupted. These drugs are also used clinically in man and other animals to assist in the management of fears and phobias, since by reducing arousal they potentially reduce the magnitude of the perceived threat. Thus arousal does not seem to determine the emotion, but it may influence it.

Table E.2. Relative change in three cardiovascular parameters associated with different emotional states, showing how they may be distinguished physiologically (original data from various sources and summarized in McNaughton, 1989).

Emotional state	Heart rate	Systolic blood pressure	Diastolic blood pressure
Fear	R	R	R
Anger	RC	R	RC
Sadness	L	R	R
Mirth	R	U	U
Sexual arousal	U	R	R

R, raised; RC, raised considerably; L, lowered; U, unchanged.

The expression of emotion may play a similar role to arousal in providing **feedback** to the individual that influences the emotional response (biofeedback). In humans it has been found that by subconsciously manipulating the facial expressions of people (by asking them to hold a pen horizontally using either just their lips at the front of their mouth or across their back teeth, to create more of a frown and smile, respectively), their judgement of ambiguous stimuli may be biased in line with the facial expression associated with a given emotion. This phenomenon is also exploited clinically within the field of animal behaviour modification, by encouraging animals to adopt relaxed postures during **systematic desensitization** to minimize arousal to the problematic stimulus. The expression of emotion also has communicative value to both **conspecifics** and heterospecifics; for example, dogs appear to process differently novel faces portraying different emotional expressions (**see: Facial recognition**) and respond differentially to commands with different emotional quality spoken for the first time by a trainer. However, the role of learning in these phenomena remains uncertain.

The influence of emotions on preparing the body for action is diverse. The emotional response may not only bias the type of action taken, and so certain emotional states – such as sexual arousal, intricately linked to the specific motivational states, and the tendency to engage in **reproductive behaviour** (**see: Motivation**) – but it may also prepare the body for learning about certain types of association, e.g. conditioned avoidance, which affect its future state of preparation. Conditioned emotional responses may play an important role in increasing behavioural flexibility and optimizing **fitness**, through the preparation of the animal for future likely experiences. The importance of emotion on learning varies with the task involved, and emotional influences may be minimal in certain forms of procedural learning, such as that involved in the acquisition of specific motor skills.

Ultimately, emotions help animals to cope with and minimize the impact of **stress** by preparing for it in both the short and longer term. Given that different types of **stressor** have different relevance in different species, it is likely that they will have different emotional processes and potentially associated experiences. For example, socially related emotional processes such as embarrassment or jealousy would not be relevant to an animal with little social complexity in its interactions, such as a cat. It might be that there is a relatively small number of primary emotional processes associated with key biological functions, e.g. acquisition of key reinforcers and avoidance of specific dangers, with secondary emotional processes arising as a result of the interaction of these or more specific contextual circumstances; alternatively, a negative emotion associated with danger may give rise to fear, but negative emotion associated with the absence of others may give rise to loneliness. However, a consensus on the primary emotional reactions of even man has yet to be reached, and so agreement in non-human animals is an even more distant prospect. None the less, a neuroscientific approach in which specific circuits are identified may be more fruitful (Panksepp, 1998).

Emotional reactions, mood and temperament

Emotional processes are not just immediate reactions, but have a more pervasive effect on behaviour, and they may be distinguished on the basis of their temporal stability. From an evolutionary perspective, the advantage of biasing attention, **perception**, **cognition**, **arousal** and behaviour according to prevailing conditions in the immediate, medium and longer term is quite clear. An emotional reaction may be considered the immediate affective response to a new situation, while a mood may be considered a particular emotional predisposition to behave in a certain way over a longer time span, and **temperament** may represent a trait that describes the affective style of an individual and is the stable emotional process. Although this distinction between these three emotional processes may be considered somewhat arbitrary, it may, none the less, be conceptually useful (see Table E.3).

Moods are less clearly defined than specific emotional reactions, and some suggest there are basically only two moods (a positive, optimistic, one and a negative, pessimistic, one). Moods generally last longer than specific emotional reactions and are less closely associated with antecedent events or such intense arousal and expression, but they still play an important role in biasing cognition, shifting information-processing priorities, to bias perception and potentially the accessibility of information and emotional reactions that are relevant, or not, to the underlying mood state. Thus, within a negative mood state, the negative attributes of a situation may be accentuated: there may be a bias to behave either negatively or cautiously, and access to positive memories may be reduced. Experimental

Table E.3. Distinction between emotional reactions, mood and temperament (adapted from Goldsmith, in Ekman and Davidson (1994)).

Aspects	Emotional reaction	Mood	Temperament
Temporal and other associations	Short-lived after removal of eliciting stimulus	Episodic and often related to a series of events with a particular quality	A general disposition or trait of the individual, activated by particular contexts
Cognitive components	Stimulus appraisal and election of a particular behavioural response	Management of emotions and stimulus attribution	Interacts with a wide range of cognitive processes during development to produce personality
Antecedents	Species-typical or learned stimuli of personal significance	Cumulative effect of emotional reactions and endogenous biochemical states, such as diet or chronobiological effects	Genotypic and early experiential interaction
Universality versus individuality	Universal species-typical phenomenon	Universal states, individualized with respect to management of emotions	Contributes to the definition of individual differences
Development	Specific reactions develop at different early life stages	Unknown	Organization and stability emerge, especially duration maturation

methods used to assess mood have tended to focus on biases in perception, the retrieval of information and cognitive flexibility, e.g. **cognitive bias** tests.

Temperament includes a particular emotional predisposition within the individual, and there has been much work on the identification of temperament traits in animals and their physical correlates – for example, with laterality and the distribution of hair whorls. Although the temperament of individuals may be classified using a range of terms relating to their emotional response in different situations, many of these can be traced to differences in the sensitivity of three interconnected emotional systems that are linked to environmental reinforcement (Gray, 1991).

These go by various names, but basically involve a system that: (i) responds to signals of reward or the termination of punishment (Behavioural Approach System, Behavioural Activation System, Behavioural Facilitation System, Foraging/Expectancy System or Seeking System); (ii) responds to positive punishment and non-reward (Fight–Flight System or Behavioural Aversive System); or (iii) responds to signals of punishment, signals of non-reward, innate species-typical fear stimuli and novelty (Behavioural Inhibition System). This system functions as a comparator that stops ongoing behaviour when there is a discrepancy between what is predicted and what is detected in the environment.

All three systems have a neuroanatomical basis with predictable emotional outcomes from their activation. The Behavioural Approach System consists of dopaminergic projections from the midbrain to the basal ganglia and both their thalamic nuclear and neocortical associations, which make up the caudate motor system and *accumbens* motor system. The system is usually tonically active and may create the emotions that underpin hope, happiness and optimism. The Fight–Flight System consists of the *amygdala*, medial **hypothalamus** and central grey area. Activity within this circuit produces the emotions associated with defence, which may be felt as panic, rage or pessimism. The Behavioural Inhibition System lies within the septohippocampal system. Differences in the operating level of this system may be responsible for variations in trait anxiety, while specific activation may result in a more specific anxiety response.

It is suggested that other emotions may arise as a result of the blending of differential activation of these systems in specific contexts, in much the same way that the blending of activity in different retinal pigments results in the sensation of colour. However, some authors, e.g. Panksepp, propose a wider range of emotional circuits derived from evaluation of social situations, which may be particularly relevant to cognitively complex gregarious species including primates and non-primate species such as the rat and the dog. (DSM)

References and further reading

Bates, J.E. and Wachs, T.D. (eds) (1994) *Temperament – Individual Differences at the Interface of Biology and Behaviour*. American Psychological Association, Washington, DC.

Cannon, W.B. (1927) The James–Lange theory of emotions: a critical examination and an alternative theory. *American Journal of Psychology* 39, 106–124.

Ekman, P. and Davidson, J.J. (eds) (1994) *The Nature of Emotion – Fundamental Questions*. Oxford University Press, Oxford, UK.

Gray, J.A. (1991) The neuropsychology of temperament. In: Strelau, J. and Angleitner, A. (eds) *Explorations in Temperament: International Perspectives on Theory and Measurement*. Plenum, New York, pp. 105–122.

James, W. (1884) What is an emotion? *Mind* 9, 188–205.

McNaughton, N. (1989) *Biology and Emotion*. Cambridge University Press, Cambridge, UK.

Panksepp, J. (1998) *Affective Neuroscience*. Oxford University Press, Oxford, UK.

Rolls, E.T. (1999) *Brain and Emotion*. Oxford University Press, Oxford, UK.

Empathy

An attempt to understand what it feels like to be another being in that situation, e.g. to be castrated without an anaesthetic, to be hungry or to be in **pain**. (DBM)

See also: Anthropomorphism

Enculturation

Culture may be defined as a set of behaviours that is not species-specific, but only displayed by some animals in a **group**, one **social group** or a population. This set of behaviours is transmitted through some form of **social learning**, and the term enculturation describes the process by which an individual acquires these behaviours. Traditions are considered precursors of cultural differences and refer to behaviours that are transmitted through some form of **imitation**. Culture and tradition in animals have become topics of intense investigation in recent decades, since scientists realized that large differences between individuals or populations of the same species may exist in their behavioural repertoire (**see: Individual differences**).

One of the most popular and early systematic observations on the development of a cultural behaviour in animals is potato washing in Japanese **macaques** (*Macaca fuscata*). Japanese researchers were giving sweet potatoes to a troop of Japanese macaques on the beaches of Koshima Island, off the coast of Japan, to habituate the monkeys to their presence and start behavioural observations of them. They observed a 1.5-year-old female (later named Imo, which means 'potato' in Japanese) bringing a potato to the nearby river, then dipping it into the water to remove the sand from the potato before eating it. Imo's mother and sibling soon started to display this behaviour after having observed Imo and, over time, the whole troop mastered potato washing. Koshima macaques have also learned to wash the potatoes in sea water to add salt to them, or to wash wheat in water to remove the sand more easily (as the wheat floats up to the top, it can be picked out without sand).

Culture and tradition are currently intensively studied, particularly in primates, to understand the social, environmental and cognitive factors that affect their occurrence and transmission. A growing body of data is showing that behaviours that can be defined as culture or tradition are common in several animal species. Animal dialects (e.g. in bird songs), fidelity to migration routes, the capacity to incorporate a new skill to extract a given food item, or differences in the behavioural repertoire used to attract a mating partner or to catch a prey are all examples of behaviours that can be defined using the term 'culture'.

The development of some of these cultural differences or traditions probably does not require complex cognitive abilities. However, at least with respect to **tool use**, animals may need to possess some capacity to causally link a behavioural sequence with an outcome. This may be the case, for example, for a great tit (*Parus major*) opening a milk bottle with a stick or for a **chimpanzee** (*Pan troglodytes*) using a stick to catch termites (i.e. termite fishing). This sequence of behaviour may be reinforced through a trial-and-error process that does not require any conscious capacity to plan an action ahead. Recent findings, however, indicate that chimpanzees are very selective in their choice of tools to be used for a given task (e.g. termite fishing or nut cracking) and, when found, they can bring a tool with them for hundreds of metres or several minutes before reaching the place where they are going to use that tool. These observations suggest that, at least in some species, advanced planning, mental mapping and object manipulations may occur and that they are not skills that are exclusive to human beings.

During the enculturation process, several factors have to be considered. The diffusion rate of a particular behaviour (i.e. how quickly a new behaviour spreads in a group) may depend on how often animals encounter each other and how close they are during these encounters, or on the size of the social group and its cohesion, as all of these factors affect the occurrence of imitation and social learning. The spread of some cultural behaviours, such as a new extractive method to obtain food, may also be affected by the number of 'scroungers' in a group – that is, the number of animals obtaining food using the work of others. The identity of the animal that first displays a new behaviour may also affect the diffusion rate. For example, the diffusion rate may be faster if a new behaviour is displayed by a dominant individual, as the process will be at the centre of the group and thus there will be more chances for imitation.

Finally, the number of relatives an 'inventor' or 'expert' has may affect the diffusion rate, as imitation is favoured among relatives. Contrasting results are currently available on which animals are more likely to develop new behaviours. Some studies suggest that juvenile and/or subordinate animals should be more likely to become 'inventors' of a new behaviour, particularly in relation to food extraction, as these animals are usually prevented by older and/or more dominant animals from accessing rich and energetically plentiful food patches, or they may develop new social skills to obtain some benefit. Other studies indicate that the reverse may be true.

An elegant experiment on enculturation is the study conducted by Victoria Horner and colleagues on chimpanzees (Whiten *et al.*, 2005). They used three groups of captive chimpanzees and one apparatus ('pan pipes') from which the animals could gain food through the use of a stick and using two distinct techniques (i.e. 'poke' and 'lift' techniques). Horner and colleagues trained the dominant female in the first group to gain food from the apparatus using the 'poke' technique and the dominant female in the second group to gain food using the 'lift' technique. These two females were isolated from their group during training so that no other animal could observe them gaining food from the task. The two females were released back into their group when training was completed. No animal was trained in the third group. Horner and colleagues found that all the animals in the group where an 'expert' was available quickly learned how to gain food from the apparatus and consistently followed the technique used by the 'expert', even though some animals managed to discover by chance the other technique. No animal in the untrained group discovered how to gain food from the apparatus. These results indicate that social learning is essential for the diffusion of such cultural behaviours. Moreover, they show that chimpanzees tend to conform to cultural norms similarly to humans. (BM)

References and further reading

Avital, E. and Jablonka, E. (2008) *Animal Traditions: Behavioural Inheritance in Evolution.* Cambridge University Press, Cambridge, UK.

Laland, K.N. and Janik, V.M. (2006) The animal cultures debate. *Trends in Ecology and Evolution* 21, 542–547.

Whiten, A., Horner, V. and de Waal, F.B.M. (2005) Conformity to cultural norms of tool use in chimpanzees. *Nature* 437, 737–740.

Endangered species

For many, the typical 'endangered species' are animals such as **giant pandas**, struggling to exist in their shrinking habitat, **rhinoceros**, decimated by poaching, or the bald eagle, threatened by pesticides. Most endangered species, however, are not so charismatic, although they all contribute to the rich diversity of life on Earth.

Most people will recognize that the label 'endangered species' denotes that a species of animal or plant is so few in numbers that it is at risk of extinction, but the term also has a more formal, often legal, definition. The International Union for the Conservation of Nature (IUCN) has developed a system for designating varying levels of vulnerability to extinction, as have many countries, provinces and states around the world. The IUCN enlists the support of more than 10,000 scientists worldwide to review data and determine whether species are at risk. The results of these evaluations are published in the *IUCN Red List*, which in 2007 listed more than 16,000 species of animals and plants threatened with extinction.

The IUCN recognizes three levels of endangerment: **critically endangered**; endangered; and vulnerable. A designation of 'endangered' indicates that the species is at very high risk of extinction in the wild. The criteria for listing a species are strict and systematic and include, for example, evidence that the species numbers have declined by more than 70% in the previous 10 years and that this trend has been stopped *or* that the numbers have declined by more than 50% but the cause of reduction has not been removed, and the species is continuing to decline. Participating biologists also use several other criteria, such as geographic range and absolute population size, to determine species status.

The Endangered Species Act of 1973 in the USA – and similar laws in other countries – has played a pivotal role in the legal protection of endangered species. Using its own set of criteria not too dissimilar from those of the IUCN, the US Fish and Wildlife Service (USFWS) charges its biologists with listing endangered species and developing and implementing plans to recover listed species. This form of legal protection prevents Federal agencies from jeopardizing endangered species and restricts 'take' (killing) and trafficking (as in the pet trade). The Act also requires that USFWS authorities take action to establish **conservation** programmes, with the goal of stabilizing populations to the point that the species will no longer require legal protection. A group of experts has convened in a 'recovery team' to develop and implement a conservation strategy or 'recovery plan'.

For species facing more immediate risk of extinction, recovery plans sometimes call for establishment of **captive breeding** and **reintroduction programmes** (e.g. the California condor), but more often they aim to protect the existing population in the wild, as well as the habitat on which it depends. In recent years recovery plans have focused less on single species conservation, which have met with limited success, and have been replaced with multiple-species habitat conservation plans. The goal of these plans is to preserve entire landscapes and ecosystems that contain multiple endangered species, rather than piecemeal approaches sometimes associated with single-species recovery plans. (RS)

See also: Conservation centre

Further reading

IUCN Red List. Available at: http://www.iucnredlist.org

Nicholopoulos, J. (1999) The endangered species listing program. *Endangered Species Bulletin* 24, 6–9.

Endocrinology

Endocrinology is the branch of biological science that deals with the endocrine system of glands in the body and their secretions (hormones). The endocrine system functions as a signalling system, using hormones as signalling molecules that induce a response in target cells. Hormones are mainly transported from the secreting gland to the target organ via the bloodstream. Major specialized glands of the endocrine system include: thymus, **adrenal glands**, **pituitary gland**, **hypothalamus**, pineal gland, thyroid gland, parathyroid gland, ovaries and testes. Other organs of the body also have an endocrine function, including the liver, pancreas, stomach, kidneys, skin, heart, adipose tissue and striated muscle.

Hormones have a wide variety of functions and modes of action, and are important in regulating **growth rate**, metabolism, tissue function, puberty, development and mood. Hormones are classified into three types: **steroid hormones**; protein hormones (including peptide hormones); and amine hormones. Steroids are derived from cholesterol and include such hormones as cortisol, corticosterone, **oestrogen** and **progesterone**. Protein and peptide hormones are made up of amino acids (ranging from three to over 200 residues) and include such hormones as **growth hormone**, **oxytocin** and **ACTH**. Amine hormones are derived from single amino acids and include such hormones as **dopamine**, **epinephrine** (adrenaline) and **norepinephrine** (noradrenaline). Hormones act by binding to specific receptors located on the surface of the target cells and inducing a response within that cell. Within the animal behaviour and welfare field, hormones of all three types have been subject to widespread research, with perhaps the most researched parts of the endocrine system being the **hypothalamic–pituitary–adrenal (HPA) axis** (CRH → ACTH → Cortisol) and the hormones of the adrenal medulla (epinephrine and norepinephrine), both of which are implicated in an animal's response to **stress**.

Some of the major challenges involved in using endocrine measures in animal welfare research revolve around the sampling method. Direct measures of circulating hormones invariably involve taking blood samples. Depending on the method, this may involve restraining the animal and inserting a needle into a superficial blood vessel to collect a sample – all of which can induce a stress response from the animal. The animal can be catheterized under **anaesthesia**, which may allow serial samples to be drawn without as much restraint, although this may restrict the ability to keep the animal housed in its original group as the catheter may be subject to unwanted attention from pen-mates. Alternatively, hormones and hormone metabolites may be quantified in saliva, **urine** and faeces. (JNM-F)

Endorphin

Endorphins – α-endorphin, β-endorphin and γ-endorphin – are endogenous, morphine-like substances with a similar chemical structure and clinical function in analgesic effects as morphine (endorphin is a short term for endogenous morphine).

The endogenous endorphins do not cross the blood–brain barrier. In the **central nervous system** (CNS), endorphins and their immediate precursor, β-lipotropic hormone (β-LPH), distribute in the **neuron(e)**s of the arcuate nucleus of the **hypothalamus** and the limbic system. Their fibres project to different nuclei within the hypothalamus, mainly to the periventricular region, and to several regions within the CNS, including the mesencephalic periaqueductal grey area, the thalamus, the septal area, the amygdala complex, the *locus coeruleus*, the raphe nucleus and the spinal cord (**see: Brain**).

Endorphins are the basis of a diverse system (i.e. the **opioid** system) associated with physiological functions and pathological disorders. The multiple functions of endorphins are dependent upon binding to their specific receptors, which are widespread within the CNS. Three well-characterized types of **opiate** receptors have been identified by both pharmacological bioassay techniques and radiolabelled opioid binding studies, and originally designated as μ- (mu), δ- (delta) and κ- (kappa) receptors, as well as several subtypes within each class. Each of them has a distinct selectivity profile and a unique pharmacological profile. The three major classes of opioid receptors have recently been renamed OP1, OP2 and OP3 receptors, respectively, by an International Union of Pharmacology (IUPHAR) nomenclature committee. Another σ- (sigma) receptor may also be present in the brain.

Through binding and activation of these receptors, endorphins play important roles in the interaction of an organism with different **stress** factors, providing anti-stress and pain relief effects, which affect behavioural and physiological **homeostasis**, such as eating, drinking, **thermoregulation** and energy metabolism, immunity, mental illness and mood, **learning** and **memory**, **reproduction**, general activity and **locomotion**, and other behavioural and neurological disorders, most of which are associated with the functions of the hypothalamus and the limbic nervous system.

The distribution of endogenous endorphins and opiate receptors corresponds to the CNS areas where electrical stimulation can relieve pain. As components of an intrinsic pain suppression system, endorphins are involved in regulating the pathogenesis of **pain**. An activated endorphin system triggers an endogenous **analgesia** cascade by: (i) modulating nociceptive signals within local circuits, such as in the dorsal horn of the spinal cord (the gate control theory of pain), resulting in reduction or inhibition of nociceptive messengers forwarded to the somatosensory system; and (ii) by increasing activity of the supraspinal cord pain inhibitory systems, such as those in the periaqueductal grey matter and the hypothalamus.

These functions of endorphins suppress pain sensation and, hence, increase analgesia by leading to an increase in the pain threshold and **hyperalgesia** (intensity of pain), as well as **allodynia** (pain elicited via activation of normally non-painful Aβ-fibres by a non-noxious stimulus such as touch). Analgesia and its correlations with the activation of the endogenous opioid system have been found in experimental animals that suffer from pain caused by various stimuli. Analgesia also can be reached following an intraventricular administration of endorphin, which directly activates opioid receptors on the neurons of the somatosensory system.

Endorphin, similar to other endogenous opioid peptides, is sensitive to different stress factors. Significant endorphin activation by increasing the synthesis of peptides in opioid-containing neurons and the transformation of the inactive form into an active one has been observed under various intense stimulations. Provoked activation of the opioid system is part of an organism's physiological or pathophysiological adaptive processes in response to stress. Several cellular mechanisms for endorphins in the regulation of the stress response have been identified through activity: (i) on the functions of the **autonomic nervous system**, affecting cardiovascular response to stress, such as **heart rate**, respiratory rate, **blood pressure** and temperature; (ii) on the functions of the hypothalamic–pituitary–adrenal axis, affecting synthesis and release of stress proteins – including ACTH and cortisol; (iii) as neuroregulators, exerting their effect on the functions of other neurotransmitter systems, such as **dopamine** and **serotonin**; and (iv) on the functions of the other brain regions such as the thalamus, the septal area and the amygdala complex, which mediate a variety of behaviours such as **depression** and **aggression**. For example, brain endorphins and **enkephalins** reduce aggression in animals, as indicated by endorphin levels being negatively correlated with aggression scores, and highly positively correlated with attack latency.

In addition, endorphins modulate animals' acute response to social conflict with an aggressive, dominant counterpart. Following injections, endorphins ameliorate stress-induced behavioural changes. These results indicate that stress-induced enhanced activations of endorphins serve to counterbalance the effects of stress hormones. These processes diminish harmful consequences of frequent overstimulation by chronic stress, increase organism resistance to stressors and lead to physiological function normalization. In humans and animals, disturbances in the regulation of endorphins might lead to stress-related physical and mental illness.

Endogenous endorphin also acts within the brain to mediate reproductive function. Endorphins inhibit the release of **gonadotropin** secretion, such as **luteinizing hormone** (LH) release in the **pituitary gland**. There are several neuroendocrine mechanisms of endorphins in modulating reproduction: (i) acting directly on the pituitary *pars distalis*; (ii) influencing gonadal functions via the regulation of hypothalamic LH-releasing hormone (LHRH) release; and (iii) regulating other neurotransmitters such as dopamine and serotonin, both of which have tonic, inhibitory effects on **sexual behaviour** and reproduction. These results indicate that endorphins have an integrative role between hormonal and neuronal functions. Under stimulation, endorphin levels are usually increased, mediating the inhibitory influence of both acute and prolonged stresses on reproductive functions, and exhibiting a long-lasting influence on brain serotonin content and sexual behaviour. Following beta-endorphin injection, brain serotonin levels are significantly decreased, while sexual activity (**see: Lordosis quotient; Meyerson index**) is increased in experimental animals.

Endorphins play an important role in the regulation of feeding behaviour and maintenance of energy **homeostasis**. Opioid neurons in the hypothalamus are direct targets of leptin, an adipostatic hormone, in regulating energy homeostasis via integrating peripheral and central information. Endorphins stimulate food intake, increasing the incentive value of food as a primary reinforcer. Endorphins modulate eating and energy metabolism through regulation of the function of melanocortin, since both endorphin and melanocortin are processed from **pro-opio-melanocortin** in the opioid neurons and putatively co-released at axon terminals.

Finally, endogenous endorphins as modulators are also involved in the cross-talk between the nervous and immune systems, directly or indirectly through the HPA axis. Endorphins increase natural killer (NK) cell cytotoxicity and immune responses at the sites of **inflammation**. In stressful stimulations, activation of endorphins has inhibitory effects on immune cells. (H-WC)

Engram

An engram is the hypothetical 'footprint' of a **memory** in neural tissue – i.e. the physical and/or biochemical alteration that is thought to occur in the **brain** in response to external stimuli during memory formation and which persists during memory storage. (JNM-F)

Enkephalin

The endogenous enkephalins – met-enkephalin and leu-enkephalin – are a group of small peptides found within the nervous system. In the **central nervous system** (CNS), enkephalins are synthesized from proenkephalin. Proenkephalin- and enkephalin-containing **neuron(e)s** are widespread in the CNS, including the hippocampus, the amygdala, the cingulate cortex, the entorhinal cortex, the septum, the *substantia nigra*, the **hypothalamus** and the spinal cord.

Enkephalins have multiple roles in the regulation of physiological and pathological processes, including **pain** processes. Enkephalins display analgesic effects by: (i) directly modulating the processing of nociceptive information in the local circuits, inhibiting the release of pain-related neuropeptides such as substance P and calcitonin gene-related peptide in the dorsal horn of the spinal cord and the trigeminal nucleus of the brainstem, where the integration of nociceptive signals is located; and (ii) indirectly blunting the distressing affective components of pain by reducing psychological and physical **stress**-associated emotional and affective states.

The widespread distribution of enkephalins and their receptors throughout the limbic system and the hypothalamus indicates that the enkephalin system is a major regulating system in modulating stress and non-stress states. Through the limbic system, enkephalins affect **memory**, **emotion** and **motivation** via the regulation of neural transmission, such as inhibiting gamma-aminobutyric acid (GABA) release from inhibitory interneurons, resulting in increased excitability of neurons of the hippocampal and dentate gyrus. These changes may facilitate memory-associated long-term potential and its related **neuronal plasticity**.

In addition, enkephalins acting in the modulation of **learning** and memory may also occur through alteration of peripheral autonomic function. These autonomic afferents communicate with the memory trace in the CNS through central regular pathways. Enkephalins, as neuromodulators, co-locate with **dopamine** and regulate behaviour via the regulation of the striatal output pathway, and also regulate secretion of stress hormones via the **hypothalamic–pituitary–adrenal (HPA) axis**. These modes of action indicate that changes of the enkephalin system, including enkephalin and its receptors, may cause damage to the adaptive cellular mechanism underlying the response of the CNS to stressors.

Enkephalins such as met-enkephalin also function as immune modulators, linking both the nervous system and the **immune system**. Met-enkephalin is involved in humoral- and cell-mediated immune reactions. Experimental studies have shown that enkephalins promote the proliferation of mouse splenocyte and the production of IL-2 and IL-6 in a dose-dependent manner. Thus, met-enkephalin appears to be an important immunomodulatory signalling molecule exerting regulatory actions concerning the expression of pre-inflammatory cytokines. (H-WC)

Enrichment

Enrichment is a broad concept, which encapsulates any change to an animal's environment that leads to positive ramifications, and thus improved or enhanced **welfare**. Enrichment is frequently considered to stimulate captive animals, both mentally and physically.

The concept of enhancing the environments, and thus the lives, of captive animals was recognized and suggested as being beneficial by Heini Hediger in the 1950s (Hediger, 1955). However, efforts were not really focused until the work of Markowitz, who coined the term 'behavioural engineering', later known as behavioural enrichment. Markowitz suggested that captive animals were motivated to work for rewards, so he and his colleagues set up situations in which the expression of a desirable behaviour would be rewarded (Markowitz, 1982). This followed simple **operant** conditioning rules, whereby any behaviour that was positively rewarded would be more likely to occur in the future, and thus its expression would increase (**see: Conditioning**). For example, to achieve the aim of increasing activity levels in four Diana monkeys, they were provided with a food reward after completing a complex series of movements around their enclosure. The monkeys quickly learned that if they performed this series of behaviours they would gain food, and so they became more active and obtained more food rewards. Much behavioural engineering was implemented to elicit species-specific behaviours, increase activity levels, offer cognitive challenges and provide an easy method of monitoring health.

Other researchers considered the assertion that captive animals were motivated to work for rewards as being **anthropomorphism**. It is now generally accepted that this phenomenon, termed **contra-freeloading**, is frequently observed, and captive animals are seen to actively work for a resource in their enclosure, even when it is also freely available. Behavioural engineering was more accurately criticized for creating **abnormal** behaviours; behavioural expression was not always linked to environmental context and sometimes occurred at abnormally high levels. Most of the arguments against behavioural enrichment were offered on the basis that an alternative, environmental enrichment, approach was superior.

Environmental enrichment aimed to provide captive animals with opportunities in their environment that would stimulate 'natural' behavioural expression. For example, to increase activity levels in the same group of Diana monkeys, the environmental enrichment approach would provide a large and complex enclosure. This would be intended to stimulate the monkeys to investigate and interact with their environment, and thus result in the desired elevated levels of activity. Naturalistic enclosures were thus integral, though not mandatory, to the evolution of environmental enrichment.

In many contemporary enrichment programmes both behavioural and environmental enrichment techniques are employed; for example, puzzle feeders are provided in a large, dynamic and complex enclosure. The two approaches are similar in their underlying mechanisms, but very different in their function. In contrast to behavioural enrichment, animals' responses to environmental enrichment can be variable in form and timing.

Enrichment aims

Much enrichment aims to promote in captive animals specific behaviours that are observed in their wild **conspecifics**, to include reduction of **stereotypies**, increase in general activity, improvement in physical fitness and enhancement of social interactions. In addition, it is hoped that enrichment will create situations that confer welfare benefits, for example by providing choice or opportunities to gain control over the environment.

Behavioural expression is determined by neural activity, and thus if enrichment can modify behaviour it is not surprising that **brain** morphology, development and function and associated neural networks are also affected. The cerebral cortex is associated with 'higher' cognitive processing – for example, problem solving or complex movement – and is more receptive to enrichment than other areas of the brain. Neurogenesis, the creation of new neuronal cells, can be promoted through the use of enrichment. This increases brain plasticity enabling greater adaptation to environmental change through improved **learning** and **memory**. Indeed, learning deficits in mice, caused by lead exposure, can be reversed when given enrichment.

Enrichment can also affect an animal's interactions with its environment; for example, fearfulness has been shown to decline in animals given enrichment. The physical exercise associated with enrichment has also been shown to aid recovery from brain damage, illness and also the deleterious effects associated with **ageing**. To date, many studies identifying the greater benefits of enrichment have been restricted to laboratory and some domesticated species. It is likely that, as we continue to study the impact of enrichment, we will discover that its benefits go well beyond what we initially expect or see.

Types of enrichment

Enrichment comes in many shapes and forms. Categorizing enrichment into meaningful groups can help with its practical implementation and can also give an indication of its function. It has been suggested that enrichment can be subdivided into five elements, which are not mutually exclusive: (i) food-based – anything involving food; (ii) physical – any part of the animal's environment, be it permanent or temporary structures and objects within the enclosure; (iii) sensory – stimulating any of the animal's senses, including visual, olfactory, tactile and auditory; (iv) social – whether this be provided by animals or humans; and (v) cognitive – which requires problem solving to differing degrees of complexity.

Food-based enrichment

Many species have evolved to spend a large proportion of their day searching to process and eat food (**see: Foraging behaviour**). Captive environments are frequently restricted and offer few opportunities for foraging or **feeding**, and the food provided frequently requires little processing. So it is probably a fair assumption that many captive animals spend less time on, and express a lower diversity of, feeding and foraging behaviours than their wild counterparts. To this end, many feeding enrichments aim to prolong the feeding experience by making the acquisition of food more difficult, whether food is hidden, hard to obtain in a puzzle or, once found, is low in calories, requiring more food to be eaten. As animals are highly motivated by food, it is not surprising that these types of enrichments are among the more successful.

Foraging can be stimulated by simply hiding food in **flooring** substrates, log piles, cardboard boxes or paper sacks. Alternatively, processing the food can be made more demanding by putting food in puzzle feeders or wire cages. These types of enrichment have been shown to increase enclosure use, activity levels and behavioural diversity, while decreasing incidences of **aggression** and **abnormal** behaviours.

Food can also be presented in such a way that species-specific behaviours are required to obtain it. For example, researchers provided food for black and white ruffed lemurs on the roof of their enclosure, and also using suspended feeding baskets. These methods were successful in encouraging the lemurs to use the vertical space in their enclosure and gain food as they would in the wild, using suspensory feeding postures. Enrichment can also be used to provide animals with the opportunity to use tools to gain access to their food; **tool use** has been observed in some primate and bird species.

As much enrichment is food based, care needs to be taken that the provision of enrichment does not indirectly lead to more food being provided than necessary, with resultant obesity. As a rule, enrichment should use food from the animal's daily ration; alternatively, food low in calories can be used so that long periods of feeding can be achieved.

Physical enrichment

Good enclosure design incorporates enrichment through the inclusion of beams, platforms, ponds and anchors for fixing perching and other objects. Many of these features are permanent, but many of them can be modified with some imagination or moved to provide animals with a different environment. For example, a pond can be filled with water or different substrates, and small bits of food can be provisioned in either the water or the substrates. Also perching, the use of ropes, beams and/or branches, can be rearranged or renewed within an enclosure, essentially 'rejuvenating' the exhibit. Many species show a high level of **motivation** for privacy or hiding places within their enclosure. Leopard cats were observed to lie in hiding places within their enclosure when their urinary cortisol levels were high, which was considered as showing that, when stressed, the cats sought out privacy. When provided with hiding places, as well as perching, the cats displayed increased exploratory behaviour and reduced levels of stereotypical behaviours, relative to that seen in previously unenriched enclosures.

Another method used to provide privacy or protective areas within the group is the provision of vertical panels within an enclosure. The effectiveness of vertical barriers depends greatly on the species' natural history. In primates the ability to escape visually from one another reduces aggressive encounters in stump-tailed macaques, but increases aggression in pig-tailed macaques. In one study, both of these primates were maintained as single-male, multi-female troops, but the source of aggression (and thus social tension) was different. In the stump-tailed macaque group the males interrupted female–female aggression and, in the pig-tailed macaque group, the male initiated male–female aggression. Accordingly, the visual barriers reduced social tension and aggression instigated by the male, as the females were able to hide from him. However, an increase in aggression was observed when the male was prevented from dispersing aggression between females, as he was unable to see them all.

The variety of objects used as enrichment is vast, from hammocks for laboratory-housed mice to straw bales for broiler chickens and balls for common seals. The degree to which objects can stimulate interest and the longevity of this interest are both greatly affected by many factors, including individual and species differences, group composition and size, and enclosure features, including which objects are currently present. For example, a group of 28 chimpanzees was provided with an uprooted tree as an enrichment item and spent 41.9% of the first day interacting with it, but thereafter interaction declined to 3.5%, whereas simple objects, like cardboard tubes and coloured key rings, led to significant and long-term reductions in stereotypic wire gnawing (reduced by 40%) and aggressive head pecking and mortality (from 1.06 to 0.57%) in laboratory-housed mice and laying hens, respectively.

Sensory enrichment
Frequently we forget, as 'highly visual' animals ourselves, that animals gain information about their environment using many different modalities (**see: Signal**). In captivity we can easily manipulate the visual, olfactory, auditory and tactile information that animals' receive. With some imagination it is also possible to manipulate thermal, humidity and electromagnetic gradients and the tastes available to captive animals.

Despite the fact that many people seem to use sensory enrichment, there are few empirical studies of its effectiveness or benefits in the literature. It is not surprising that the senses we are familiar with are used most frequently, so reports of using visual, auditory and olfactory sources of information are common. Television and radio have been used with many different species and found to have a variety of effects, depending on the programme shown or the type of music played! Researchers provided chimpanzees with a variety of videotapes showing different scenes (chimpanzees, other animals and humans). It was unsurprising that single-housed chimpanzees watched the television more than those socially housed, though there was a lot of individual variation.

When using sensory information care needs to be taken. It is important that the information being conveyed is understood. In particular, when playing back conspecific **vocalizations**, what messages are you sending? It is essential that the context is known when the vocalizations are recorded. Equally, when providing olfactory information in the form of faeces, it is necessary to ensure that they are from known healthy animals and do not provide a route for the transmission of disease.

Social enrichment
Sources of social enrichment can be provided through cage mates, whether of the same or a different species, and people, whether keepers or visitors – although interspecific contact may not substitute for intraspecific enrichment and vice versa. All social enrichment provides dynamic and unpredictable sources of stimulation, which may be one of the most enduring and effective ways of delivering enrichment to captive animals. Within the laboratory industry there is a lot of evidence demonstrating the importance of housing primates in social groups. As such it is accepted that the welfare of social animals will be enhanced in captivity if they are housed with conspecifics. Social housing has also been shown to enhance the impact of other enrichment methods. Social groups are inherently difficult to manage, however, as they provide a myriad of pressures that can result in stress or physical injury (**see: Social behaviour**).

Whether social interactions between different species are enriching or not again depends on many variables. Mixed-species exhibits in **zoos** have been successful with some combinations of animals but not with others, so care needs to be taken when putting different species together. Choosing species with different ecological niches will reduce the likelihood of competition, and thus aggression. In some circumstances the presence of humans will be enriching to captive animals, but again this depends on many variables, from the attitude and personality of the human to the past experiences of the animals.

Cognitive enrichment
Enrichment can provide captive animals with a mental challenge, usually a complicated route to food. Much has been learned about the cognitive abilities of a wide array of species by providing them with enrichment. Enrichment devices are seldom the only source of food available; generally, other sources of food are available concurrently in the enclosure, or regular meals will ensure the animal does not 'need' to use the enrichment. So why do animals take up the challenge to use these cognitively demanding tasks? It may be that the food in the device is particularly favoured, or maybe undertaking the task is stimulating. It has also been suggested that animals are strongly motivated to seek out information from their environment, which can be provided by learning from enrichment (**see: Exploratory behaviour**).

Summary

Enrichment methods are highly variable, and their success cannot necessarily be generalized between species, or sometimes between individuals. Modifications to general housing and husbandry routine can result in enrichment; however, changes can be made and no visible change in behaviour observed, or stress may be increased. It is imperative, therefore, that any change made to animals' environment that aims to enrich is appropriately monitored. In some instances, enrichment may provide a more dynamic environment in which animals can display more choice within and control over their environment, which is hard to quantify and may not lead to obvious behavioural changes.

There is good scientific evidence that enrichment is extremely beneficial on many levels. Through its incorporation into a wide variety of housing and husbandry regimes, in farms, laboratories, zoos and for companion animals, the application of this knowledge improves animal welfare. (VM)

References and further reading

Carlstead, K., Brown, J.L. and Seidensticker, J. (1993) Behavioural and adrenocortical responses to environmental changes in leopard cats (*Felis bengalensis*). *Zoo Biology* 12, 321–331.

Hediger, H. (1955) *Psychology of Animals in Zoos and Circuses*. Butterworths Scientific Publications, London.

Markowitz, H. (1982) *Behavioural Enrichment in the Zoo*. Van Nostrand Reinhold Company, Inc., New York.

Mason, G.J. and Rushen, J. (eds) (2006) *Stereotypic Animal Behaviour: Fundamentals and Applications to Welfare*, 2nd edn. CAB International, Wallingford, UK.

Shepherdson, D.J., Mellen, J.D. and Hutchins, M. (1998) *Second Nature: Environmental Enrichment for Captive Animals*. Smithsonian Institution Press, Washington, DC.

Wurbel, H., Chapman, R. and Rutland, C. (1998) Effect of feed and environmental enrichment on development of stereotypic wire-gnawing in laboratory mice. *Applied Animal Behaviour Science* 60, 69–81.

Young, R.J. (2003) *Environmental Enrichment for Captive Animals*. Blackwell Science, Oxford, UK.

Environmental ethics

Environmental ethics undertakes to examine and articulate, from philosophical and ethical perspectives, our attitudes towards animals and nature, paying particular attention to how human beings should relate to 'Mother Earth' and to certain non-human beings with whom they share the planet. Questions related to the kind of moral relationship we ought to have with wild and domesticated animals, plants, ecosystems, inanimate natural objects and the planet, respectively, are posed and debated. Topical concerns given serious consideration include climate change, sustainability, consumption habits, obligations to future generations, pollution, species **extinction**, deforestation, species and wilderness management, unbridled agricultural expansion and thoughtless land development, human overpopulation and world poverty, and the respective local and global policy frameworks and institutions best suited to address these concerns.

Environmental ethicists come in many forms. Animal liberationists and protectionists typify *individualism*. Individualism is the view that only individuals (like sentient creatures) should be the proper focus of direct moral consideration. Holists, in contrast, contend that moral concern should be directed at the level of species or ecological systems rather than at individuals. A healthy debate continues between environmental ethicists and proponents of animal welfare and **animal rights** on whether the interests of individuals should remain subordinate to concerns related to the preservation of biotic communities. Notable scholars embroiled in these issues include Baird Callicott, Mark Sagoff and Bryan Norton.

Another way in which debates in environmental ethics are framed is in terms of human-centred perspectives (or **anthropocentricism**) and bio- or ecocentricism. Biocentrists contend that human beings are one among many interdependent members of Earth's community of life and that human beings have no greater inherent worth relative to other organisms. Another variant of this view, 'enlightened' anthropocentricism, contends that by spotlighting obligations to future generations of human beings we might help to curtail unrestrained and thoughtless use and misuse of Earth's resources.

Other notable environmental ethics include the land ethic and its contemporary offspring, deep ecology, coined by Aldo Leopold and Arne Naess, respectively. Leopold argued that our desire to control and conquer nature might be moderated if we identified with nature and habituated the norm that 'a thing is right when it tends to preserve the integrity, stability and beauty of the biotic community. It is wrong when it tends otherwise.' A parallel critique of the tendency to dominate nature is also made by ecofeminists like Karen Warren and Val Plumwood, who argue that there is a correlation between the former and the domination of women and other vulnerable individuals or groups in society. (RA)

Further reading

Callicott, B. (1980) Animal liberation: a triangular affair. *Environmental Ethics* 2, 311–328.

Norton, B. (1991) *Toward Unity among Environmentalists.* Oxford University Press, New York.

Palmer, C. (2003) An overview of environmental ethics. In: Light, A. and Rolston III, H. (eds) *Environmental Ethics: an Anthology.* Blackwell, Oxford, UK, pp. 15–37.

Sagoff, M. (1995) Animal liberation and environmental ethics: bad marriage, quick divorce. In: Sterba, J.P. (ed.) *Earth Ethics: Environmental Ethics, Animal Rights and Practical Applications.* Prentice Hall, Englewood Cliffs, New Jersey, pp. 166–177.

Epidemiology

Epidemiology is the study of the factors that influence health and disease at a population level. Epidemiology combines biology, clinical medicine and statistics to elucidate the **aetiology** of **diseases** or illness. Epidemiological techniques have not only been applied to examine illness and disease in animal populations, but also more broadly in the field of animal welfare, for example to study risk factors for behavioural problems. In many cases, **health** and **welfare** are intrinsically linked, and thus the gaining of a clearer understanding of the factors that cause illness or disease in animals also elucidates factors that impact the animal's welfare.

To carry out an epidemiological investigation, a number of different methods may be employed: surveillance and descriptive studies are used to study distribution; and analytical studies are used to study determinants (causes, risk factors). Surveillance can be defined as the systematic collection (usually involving survey), orderly consolidation and analysis of data, with prompt dissemination and feedback of the results to those who need to know, particularly those who are in a position to take *action*. Descriptive studies may take the form of case reports, which may document characteristics of individuals, and ecological studies, which may compare characteristics of populations. Surveillance and descriptive studies can be used to generate hypotheses about the illness (or welfare parameter) of interest and suspected causal factors, which can then be tested using analytical studies, such as observational studies (including cross-sectional, case–control and cohort studies) and experimental studies.

Epidemiological techniques have been increasingly used in farm animal welfare in particular, where relatively large populations of genetically similar animals may be kept in diverse housing systems. Surveys of farmers have yielded useful information about risk factors associated with such measures as lameness, abnormal behaviours and morbidity and mortality. Epidemiological techniques are also being used in the development of on-farm welfare assessment systems (**see also: Farm assurance scheme**). (JNM-F)

Epigenesis

In modern evolutionary theory the term epigenesis refers to the phenotypic development of an organism or structure as a result of the interaction between the **genotype** and the environment. It contrasts with the theory of preformation, which states that the development of an organism is predetermined and not affected by environmental variables. (BM)

Epinephrine

Epinephrine (adrenaline) is classified as a **catecholamine** and acts as both a neurotransmitter and a neurohormone. Upon stimulation of the **sympathetic nervous system**, epinephrine is synthesized and released from a small number of nerve cells in the **central nervous system** and a much larger number of cells in the medulla of the **adrenal gland**, which release epinephrine into the vascular circulation. Activation of sympathetic **neuron(e)s** in the adrenal medulla causes the release of **acetylcholine** from nerves that terminate on chromaffin cells in the adrenal medulla. Receptor activation causes chromaffin cells to produce and release epinephrine and some **norepinephrine** into the blood. Epinephrine acts as an agonist for α_1-, α_2-, β_1-, β_2- and β_3-adrenergic receptors. However, the potency by which epinephrine can activate a receptor is less for β-receptors compared with α-receptors. Once epinephrine is released into the blood it can reach and affect all tissues possessing adrenergic receptors.

Epinephrine causes a multitude of effects that are both tissue- and receptor-specific. Through circulation in the blood, epinephrine can cause sympathetic effects at tissues receiving little or no direct sympathetic innervation, such as adipose tissue and blood cells. Almost all cells in the body have adrenergic receptors and thus the body is altered in a variety of ways when epinephrine is released into the circulation. Sympathetic activation occurs in response to stimuli associated with **fear**, **pain**, **anxiety** and **stress**. When released, epinephrine causes the following effects:

- increased heart rate;
- increased stroke volume;
- relaxation of vascular smooth muscle to increase blood flow to muscle and **brain**;
- constriction of vascular smooth muscle to decrease blood flow to the skin and intestine;
- mobilization of glucose from liver and fatty acids from adipose tissue; and
- stimulation of the conversion of glycogen to glucose in the liver.

Epinephrine has been measured and used in research to assess the **welfare** of animals. The basis for this measure is that an epinephrine release can be reliably produced when animals are subjected to **aversive** stimuli such as electric shock, **restraint** and **injury**. Furthermore, research with humans has shown that, when challenged with a mental test as compared with a physical challenge such as cold, epinephrine is released at a greater rate during mental challenges and norepinephrine is released at a greater rate during physical challenges. Thus, researchers have a tool that can indicate to what degree an animal is physically and mentally challenged by a potential stressor by measuring epinephrine and norepinephrine. However, reliable measurement of epinephrine has proved difficult; because epinephrine is released in response to nerve activation, its increase in the bloodstream is almost instantaneous.

Obtaining a blood sample from a research animal without the actual blood sampling procedure increasing circulating epinephrine is therefore challenging. Simply sticking a needle in the subject will increase the amount of epinephrine in the blood, thereby masking any effect of the situation that we are attempting to quantify. To avoid this problem, researchers have established protocols that fit animals with indwelling catheters to allow blood collection without venipuncture. This alone is not enough, however, because the simple act of entering the animal's pen can cause epinephrine to spike. Gentle **handling** and training the animal to the procedure prior to blood collection, in combination with establishment of a jugular catheter, have been successful in allowing base concentrations of epinephrine to be collected.

Epinephrine has proved useful as one measure in assessing animal welfare; however, epinephrine is maintained at an elevated level in the body for only a very short time (measured in minutes). Assessment of chronic stress in research and production animals has relied more heavily on measurement of **glucocorticoids**. (DCL)

Epiphenomenon

An epiphenomenon is something that is caused by something else but lacks causal powers of its own. It is unclear whether anything could meet this dual requirement, and in fact the term is often used informally to refer to an inconsequential by-product of a process. Epiphenomenalism is a rarely advocated theory in the philosophy of mind that treats mental states as causally inert products of physical activity. (PAR)

See also: Consciousness

Epistemology

In philosophy the term 'epistemology' refers to the study of **knowledge (see also: Procedural knowledge)** – what it is, how we acquire it and what its limits are. Philosophers have found knowledge hard to characterize in general terms, and various theories of knowledge continue to have supporters. Most theories start from the insight that knowledge differs importantly from mere true belief: by guessing I might come to believe truly that it is exactly 4.29 pm, but I would not thereby know that it is 4.29 pm. The key question is then: what is required, in addition to true belief, if a person is to know that something is the case?

After this there is little consensus. Until well into the 20th century it was commonly assumed that a true belief becomes knowledge when the subject can *justify* holding it. However, in the mid-1960s it was shown that a person can be justified in holding a true belief and still lack knowledge. Two obvious options then opened up: one was to make the conditions of justification more stringent; the other was to look for something other than justification as the special feature of knowledge (e.g. to say that the true belief and the fact must covary in some way). Further problems ensued: stringent conditions of justification might imply that we know very little at all and, when the justificatory element is left behind, it can easily turn out that we know rather more than we take ourselves to know. Moreover, any condition will be hard to spell out in a way that is generally satisfactory.

Scepticism is the view that we suffer from a radical lack of knowledge, usually in a particular domain. Some domains of knowledge do seem to be more 'distant' than others: a belief that you now have an itch may seem more secure than a belief about the distant past, other people's minds and subatomic particles. The sceptic challenges us – often by pointing to the limits of what can be observed – to show that knowledge extends into these domains. (PAR)

Equid

Equidae include Prewalski's **horses** (the Mongolian wild horse), domestic horses, African wild asses (and domestic donkeys), Asian asses, Grevy's zebras, common zebras and mountain zebras. They all belong to the order Perissodactyla (odd-toed ungulates), the suborder Hippomorpha and the superfamily Equoidea. It is believed that long-term geographic isolation of equid populations in the late Pleistocene led to the distinct species that exist today, with true horses (*Equus* *caballus*) emerging from the Eurasian lowlands north of the great mountain ranges, while the asses occupied the arid zones of Asia. (PDM)

Ethics

Broadly speaking, ethics is the systematic study of values, virtues, moral concepts, principles and/or practices as a way of helping us to determine what we ought to do. Ethics is not about who can scream the loudest during a verbal dispute, nor is it simply a matter of feelings, tastes or preferences. Instead, ethics deals with justifying or thinking through one's point of view in a principled way.

Ethics may be divided into the private/individual and public/social spheres or domains. While they are distinct, they certainly overlap. The private/individual sphere involves questions directed at the individual moral agent, e.g. 'What should *I* do about the treatment of animals in my laboratory?' Here, individuals may be assigned responsibility for performing harmful actions and receive praise for performing beneficial ones.

Ethics in the social sphere involves questions related to the nature of *our* public policies, laws and regulations, i.e. how the aforementioned should be established and promoted as the basis of a good society. Here, the attention is to social structures and their consequences and typically involves the distribution of goods by various state, civic or commercial institutions. Questions such as 'What ought I to do to bring about just or fair regulations?' and 'How should I challenge morally unjust social policies or regulations?' reflect the continuity between the personal and social domains of morality.

According to one influential view, ethical principles are *action-guiding* in nature and take the form of 'should'. For example, 'One should take proper care when experimenting on canines'. Scientific principles are *informative* in nature and take the form of 'can'. For example, 'When subjected to high levels of ammonia, pigs can experience irritated throats and eyes and suffer from lung damage'.

Ethics is not coextensive with or identical to the law or religion, but can certainly inform both these domains of human life. That is to say, we can separate legal questions from ethical ones, i.e. legal: 'What *does* the law say about experimentation using primates?'; ethical: 'What *should* the law or regulations say about experimentation using primates?'

Likewise, we can make the distinction between religious and ethical questions. For example, religious: 'How do the respective faith tenets in **Hinduism**, **Islam**, **Christianity**, **Buddhism** or **Judaism** say we must treat livestock?'; ethical: 'What ought members of a pluralistic society say about the moral status of livestock and how they should be treated?'

With the foregoing discussion in mind, we can construe animal ethics as follows: it is the systematic study of how we ought to relate to non-human animals (including domesticated and wild species). More specifically, it is the careful endeavour to determine our responsibilities or duties towards non-human animals (which need not have a legal or religious basis) and to establish principles that serve to guide our moral interactions with them (from the spheres of both private and social ethics). When inquiring into the nature of our obligations towards animals we should pay particular attention to the following: (i) What are the reasons given in support of someone's (and our own) view that animals have moral standing or that they do not? (ii) Are there moral principles (or action-guiding rules) that are implied by this reason or set of reasons?; (iii) What are the attractive features associated with this view?; (iv) What are some short-comings?; and (v) How does our treatment of animals reflect good or exemplary moral character?

Animal welfare is unlike other areas of science in that the primary motivation for this field of enquiry is often based on the view that we have direct moral duties to animals rather than being a result of scientific curiosity. Thus, our notions about the proper treatment of animals are influenced not only by the empirical findings about animals but also by our ethical values. Our ethical values have evolved over time to include a growing concern for animals and what matters to them. Animal science has contributed immensely to ethical enquiry into the nature of our obligations to animals as well. Knowledge about anatomical similarity, biological functioning and a deeper appreciation of the mental lives of animals have all contributed to a better understanding of how animals can be benefited and harmed as a result of our actions and omissions. Careful ethical thinking and continued interweaving between these two fields of enquiry can also better shed light on other dimensions that influence animal **welfare**. This includes exposing the interplay between technological systems (such as housing and management practices), economics, regulation, ethical values, different conceptions and measurements of animal welfare, and knowledge about the animals themselves. One of the central debates in animal ethics concerns the moral status of animals. Conceptually, two broad views can be distinguished: extension approaches to moral status and community-oriented approaches to moral status.

Extension approaches

Proponents of extension approaches assume that the moral status of animals (and thus duties owed to them) hinges on whether they possess some essential characteristic or set of characteristics. These characteristics include the capacity for **pain** or pleasure, self-awareness and the capacity to have future-oriented projects. Apart from the noticeable disagreement among different camps regarding which characteristic should count as definitive for affording animals moral status, extension approaches also tend to offer 'a one size fits all' answer as to how we ought to treat animals, thus providing little genuine guidance in particular cases. The main accounts associated with the extension approaches include exclusive views and inclusive views.

Exclusive views

Exclusive views on the moral status of animals deny that animals deserve moral consideration in their own right. Two broad views include: (i) human-centrism – the main proponents being **René Descartes** and **Immanuel Kant**; and (ii) **contractarianism**, a leading proponent being Jan Narveson.

HUMAN-CENTRISM Proponents of human-centrism are committed to the view that human beings have dominion over the shape of animals' lives and, in particular, only those beings that have **consciousness**, reason and/or **language** should be afforded moral status and legal protections. As the argument runs, since only human beings possess consciousness, language and reason, they alone should be afforded moral status and/or protected by legal rights. It also follows that it does not matter to the animals themselves how they are treated.

One consequence of this view is that animals have value only as a means to human ends. From an animal welfare standpoint, injuring animals, **hunting**, rodeos and confining animals for exhibitions should raise no moral eyebrows. The treatment of animals should be judged as being either moral or immoral only by virtue of its connection to human interests.

A general worry about human-centrism is that, if we exclude certain animals from the moral community because they lack language or reason, then we must also (as a matter of consistency) exclude some categories of human beings who do not possess either language or the capacity for reason or abstract thought. Thus, human infants and certain non-paradigm human beings will also be excluded from being the recipients of moral respect in their own right. This certainly challenges our deeply held moral intuitions about who counts and who does not, from the moral point of view.

CONTRACTARIANISM A close cousin of human-centrism is contractarianism, which has an affinity to egoism. Jan Narveson, a contemporary Canadian thinker, is a main proponent of this view. According to Narveson, morality is a function of a contract between members (contractors) of comparable abilities or political–ethical 'powers'. Animals do not possess moral status or rights because they, unlike human beings, cannot be parties to a social contract that upholds certain reciprocal obligations. Once again, differences between animals and ourselves are highlighted as the basis of keeping animals out of the moral arena. On this view, it is presumed that only human beings can cooperate with and act in kind towards each other.

However, this view does not preclude having duties *about* animals, even if we do not have direct duties to them. These indirect duties are predicated on acting in ways that promote *our* own benefit. For example, we are required to refrain from injuring a cat, not because we think that it will harm the cat itself if we mistreat it. It is a moral affront to harm the cat because it results in harming the moral agent who is invested emotionally (for example) in the animal. We may also be worried that this co-contractor could take some legal action against us for injuring 'their property'.

Contractarianism suffers from the same concerns levelled against human-centrism. The capacities that are appealed to in denying that animals have moral status will also leave certain categories of human beings who are not contractors, like children and mentally incompetent human beings, as being undeserving of respect or protections in their own right. There is little incentive to regard these human beings as mattering for their own sake. Treating them with respect may matter only as a side condition to observing one's contractual obligations to other contractors to whom they belong.

Inclusive views

In contrast to exclusive views, inclusive views on the moral status of animals ascribe moral status to animals in their own right based on the animals' possession of certain moral status-making characteristics. Two broad views include **utilitarianism**, a key proponent of which is **Peter Singer**, and the rights-based approach, a key proponent being **Tom Regan**.

UTILITARIANISM Utilitarians consider the rightness or wrongness of an act to be determined solely by the consequences that will impact the well-being of the affected individuals. Singer contends that we should include the interests of *sentient* animals in our utilitarian calculations. To discriminate against them simply on the basis of species membership is tantamount to **speciesism**. He concludes that practices such as agriculture and biomedical research that employ animals should cease, since the suffering they inflict on animals is not outweighed by the benefits produced for all the affected parties.

The primacy of **sentience** as *the* morally relevant criterion is not uncontroversial. Some argue that Singer's criterion of sentience may be too simplistic as a moral category, since it fails to reflect our phenomenal experiences of animals by ignoring differences in quality and richness of life across species. There *are* morally relevant species differences between animals and human beings, including the ability to take responsibility for their actions, to reciprocate in kind and to desire self-respect.

Some utilitarians are considered reformists. Sensitive to the upward struggle to establish strong or equal moral status for animals relative to human beings, these utilitarians call for piecemeal changes within current systems where animals are used. For example, utilitarian reformists might call for more attention to disease modelling and enriched housing systems and for greater scrutiny of the appropriateness of techniques, endpoints and experimental design. This perspective is sometimes mirrored in work pioneered by **Russell and Burch** (1959) on the **three Rs principle**, and often serves to underpin how many Institutional Animal Care and Use Committees function.

RIGHTS-BASED Taking a rights-based approach on the issue of inclusion, Tom Regan contends that, in contrast to utilitarianism, certain actions ought not to be done to morally considerable subjects or rights-holders, regardless of the goodness of their consequences. He argues that it is unethical to sacrifice the interests of morally considerable individuals in order to benefit others. The basis of Regan's argument is his notion of subjects-of-a-life. Subjects-of-a-life (including some animals) have inherent worth and deserve moral respect/have rights. They should not be considered or used as mere resources. Regan concludes that we should not use animals even if we do so humanely or if suffering is minimized – or despite a great benefit to a large number of people. He takes an abolitionist stance with respect to most forms of animal use.

One major criticism of this view is that the rights-based approach does not offer clear guidance on what we ought to do in cases of conflicts of rights. Some have criticized Regan's definition of subjects-of-a-life for its lack of specificity. If 'psychological identity over time' and 'sense of one's own future' are construed broadly they encompass almost any animal, including organisms like *Planaria*, in which case it is not a useful moral category. However, if they narrowly denote 'reflective self-consciousness' then they may be misapplied to most animals that Regan counts as subjects-of-a-life (Russow, 1988).

Another criticism of the strong **animal rights** view is that it lumps animals into one amorphous group with the same moral identity. By emphasizing only species similarities and egalitarianism, the vital interests that moral agents or human beings uniquely have are ignored. Also, by de-emphasizing the 'humanness of animals', we might be better able to appreciate the natures of animals and devote our energies to promoting responsibilities based on care for the kinds of creatures animals are in and of themselves (Anthony, 2003).

Community-oriented approaches

In contrast to the extension models of the moral status of animals, community-oriented approaches contend that moral concepts (including the question of moral status) are a function of common enquiry and consensus negotiation, and should be applied on the basis of some form of community. Our circles of shared and mutual dependencies ground the basis of specific obligations.

This view is eclectic in scope in that it incorporates elements of interspecific equity and human virtues such as restraint and diligence, and promotes an intimate understanding of animals' points of view. It rejects the 'one-size-fits-all' application of a single abstract moral principle to the animals issue and encourages context or case-by-case specific analysis.

Some key shortcomings associated with this approach include worries about whether all the relevant parties will be represented in consensus dialogue or debate. Furthermore, since it must be determined antecedently who counts as a member of a community and why (in order to have a seat at the table, so to speak), it presupposes that we have already worked out the (quite controversial) basis of the moral status of animals beforehand. There are two categories of this approach: (i) ethics of care; and (ii) **communitarianism**. The main proponents of the former include Mary Midgley and Nel Noddings (**see: Relational ethics of care**), and of the latter **Bernard Rollin**.

Ethics of care

On the ethics of care front, moral obligations are a function of our relationships with others. Our sense of community or connectedness with others (including animals) is the basis of our ethical obligations to animals. What is important is the discharge of responsibilities based on virtues of care, compassion and empathy, and not safeguarding of rights per se.

Worries about this view include the fate of animals that we regard as being within a social community – for example, certain wild animals or animals considered as 'pests'. Also, the notion of community is fluid and precarious, and hence the protections afforded to animals that may have once been considered members of the moral community – due to their proximity and high visibility – may wane in light of their physical remoteness and our emotional attachments to them (as in the case of terrestrial and aquatic animals raised for food).

Communitarianism

Bernard Rollin is a proponent of a social or consensus ethic, where rights and responsibilities are a function of an implicit social contract among human beings that stipulates mutual expectations for conduct (including the treatment of animals). Under Rollin's view, animals have a distinctive **telos** or 'nature', or central defining properties that capture the good of the being in question and form the basis for moral respect. In the case of food production and use of animals in research, Rollin argues that they are permissible so long as we minimize or abolish production systems that go against these animals' telos.

A general worry about Rollin's view of telos is that it is unclear whether or not animals' 'natures' are, in fact, metaphysically real properties, as he seems to suggest. The identity of distinctive animal natures remains controversial (especially given evolutionary theory that species (and their traits) are always in flux). Furthermore, Rollin's emphasis on respecting the telos of animals, and its conservative streak, may not go far enough in opposing traditional uses of animals or conditions that are not really good for their welfare. It also does not address very well the acceptability of changing the natures of animals through those genetic means that may bypass welfare concerns but that pose an affront to conceptions of animal integrity and dignity. (RA)

Further reading and references

Anthony, R. (2003) Ethical implications of the human–animal bond on the farm. *Animal Welfare* 12, 505–512.

Armstrong, S. and Boltzer, R. (2003) *The Animal Ethics Reader*. Routledge, London.

Rachels, J. (1983) *The Elements of Moral Philosophy*. McGraw-Hill, New York.

Russell, W.M.S. and Burch, R.L. (1959) *The Principles of Humane Experimental Technique*. Methuen, London.

Russow, L.M. (1988) Regan on inherent value. *Between the Species* 4, 46–54.

Sandoe, P., Crisp, R. and Holtug, N. (1997) Ethics. In: Appleby, M. and Hughes, B.O. (eds) *Animal Welfare*. CAB International, Wallingford, UK, pp. 3–18.

Ethogram

An ethogram was originally considered to be a list of the complete behavioural repertoire of a species. However, there are inevitable difficulties in achieving such a list for most species, since behaviour varies depending on environment and individual experience. In addition, the chances of reaching agreement on definitions of much of the complex behaviour exhibited by many species are slim, and so this concept of an ethogram is not particularly useful or practical. The current, more common, usage of the term is that an ethogram is a list of the types of behaviour performed by particular individuals in a particular environment (and usually derived for a particular purpose – see below). In addition to a list of behaviour types, an ethogram should include precise descriptions of each behaviour (see Table E.4).

It is absolutely fundamental to any study of animal behaviour to define what behaviour types are being observed and recorded, and therefore production of an ethogram is always the first step in any animal behaviour research. To produce an ethogram it is necessary to observe the animals for some time in order to become familiar with all the types of behaviours they perform and what purpose each behaviour appears to have. Sensible decisions can then be made as to how broad or narrow the behaviour categories need to be. Often ethograms include broad categories of behaviour that are subdivided into more precise forms of that behaviour, e.g. a broad category might be **locomotion**, which could be divided into walking, running, climbing and jumping (see Table E.5). It is usually necessary, for the purpose of analysing data, to make each category of behaviour in an ethogram mutually exclusive, so the definitions must be written with care, e.g. for a ruminant the behaviour 'lying at rest' may be performed at the same time as 'ruminating', so needs to be defined more precisely as 'lying at rest, not ruminating'.

For behavioural research purposes it is important to classify each behaviour in an ethogram as a behavioural state or a behavioural event, as these require different recording methods to obtain an accurate estimate of how often or for how long they are performed. A state behaviour generally has a long (measurable) duration and includes behaviours such as rest, walk and feed. An event behaviour has a very short duration such as yawn, scratch or roar. Because state and event behaviours are usually recorded by different means and used in separate analyses, it is acceptable in these cases to include non-mutually exclusive states and events in an ethogram – e.g. a baboon may perform lying rest at the same time as yawning.

Table E.4. An ethogram developed to determine the broad activity budget of hamadryas baboons in captivity.

Behaviour	Description
Locomotion	Fast or slow, horizontal or vertical movement around the enclosure not including movements involved in another behaviour (e.g. play, aggression)
Immobile	Stationary standing, sitting or lying and not performing another behaviour (e.g. grooming)
Foraging and feeding	Searching for (e.g. in bark substrate), picking up, placing in mouth and chewing of food items, including chewing of wooden structures in the enclosure
Playing	Apparently functionless manipulation of objects or rapid movements such as running, somersaulting, etc. alone or with others
Grooming others	Grooming another individual
Being groomed	Being groomed by another individual
Autogrooming	Grooming self
Aggression	Any aggressive behaviour to another individual, including visual and vocal threats, chasing and fighting
Mounting	Mating behaviour – male mounts female with back feet on female's ankles

Table E.5. An ethogram developed to investigate changes in behaviour of captive blue duiker (*Philantomba monticola*) when moved to new enclosures.

Main category	Subcategory	Description
Rest	Standing rest	Standing inactive for long periods, not ruminating
	Lying rest	Lying inactive, eyes open, not ruminating
	Sleep	Lying inactive, eyes closed, not ruminating
Maintenance	Eat provided food	Intake of pellets/grain/fruit
	Eat provided browse	Intake of browse provided by keepers
	Eat enclosure browse	Intake of browse growing in enclosures
	Eat soil/insects	Licking/intake of unidentified items on ground
	Drink	Intake of water
	Urinate	Excretion of urine
	Defecate	Elimination of faeces
Ruminate	Standing rumination	Rumination while standing
	Lying rumination	Rumination while lying
Travel	Walk	Slow movement not accompanied by investigation of surroundings
	Run	Rapid movement not accompanied by investigation of surroundings
	Jump	Movement over an obstacle
Investigation	Explore	Slow movement accompanied by frequent pauses to sniff/look at objects and surroundings
	Pause alert	Short pause and observation of surroundings during exploration
	Sniff	Stationary sniffing at air or objects in enclosure
Groom	Standing autogroom	Scratch, with foot or teeth, or lick self while standing
	Lying autogroom	Scratch, with foot or teeth, or lick self while lying
Interact	Groom other	Scratch or lick other animal
	Groomed by other	Scratched or licked by other animal
	Reciprocal groom	Simultaneous groom other and groomed by other
	Rub scent glands	Pressing together of pre-orbital scent glands with other animal (blue duikers only)
	Scent mark object	Marking objects in enclosure with pre-orbital gland
	Genital sniff	Male sniffing at rear of female
	Chase	Fast movement after another animal
	Flee	Fast movement away from another animal

Ethograms for the same species may differ depending on the individuals, environment and, most of all, the purpose of the research. For instance, an ethogram for wild lions will probably include **hunting** behaviour, possibly broken down into more precise stages of hunting such as stalking and chasing, whereas, an ethogram for lions in a **zoo** is unlikely to include hunting since captive lions are generally not provided with live prey. Similarly, an ethogram for juvenile macaques will probably include **play**, but may not for adult macaques. Even in the same environment for the same individuals ethograms will differ depending on the research question. For instance, if the research question is 'Does group size affect aggression in mandrills?', then the ethogram will include many types of **aggressive behaviour** such as threat, display, spatial displacement and fight. However, if the question is 'Does group size affect activity levels of mandrills?', then the ethogram may only include one broad category, '**aggression**', or may even group aggression into a wider category of '**social behaviour**'. (AP)

See also: Ethology; Time budget

Further reading

Lehner, P.N. (1998) *Handbook of Ethological Methods.* Cambridge University Press, Cambridge, UK.

Martin, P. and Bateson, P. (1993) *Measuring Behaviour: an Introductory Guide*, 2nd edn. Cambridge University Press, Cambridge, UK.

Schleidt, W.M., Yakalis, G., Donnelly, M. and McGarry, J. (1984) A proposal for a standard ethogram, exemplified by an ethogram of the bluebreasted quail (*Coturnix chinensis*). *Zeitschrift für Tierpsychologie* 64, 193–220.

Ethological relevance

Ethological relevance refers to stimuli that are relevant to an animal and how it behaves. That is, animals have evolved to respond in certain ways to certain stimuli and ignore other stimuli. When conducting animal behaviour experiments, it is important only to ask the animals to perform ethologically relevant tasks or to interact with ethologically relevant stimuli, as lack of response may not indicate inability to learn or a **cognitive dysfunction**. (LMD)

Ethology

Ethology was defined by **Niko Tinbergen**, one of the founding fathers of the field, as 'the biological study of behaviour'. It is a word particularly applied to the body of theory built up by **Konrad Lorenz** and by Tinbergen in Europe during a period of about 40 years in the middle of the 20th century, culminating in their award, with **Karl von Frisch**, of the Nobel Prize for Physiology or Medicine in 1973. While much of that theory has been superseded, the approaches that they pioneered have had a lasting impact on studies of animal behaviour throughout the world.

At that time, most others who studied behaviour did so from a psychological viewpoint. The closest this came to ethology was the field labelled '**comparative psychology**' in the USA, but even this was limited in its perspective compared with ethology. Most psychologists then were particularly interested in mechanisms, especially of learning, and their primary aim was to shed light on human behaviour. To the extent that they studied animals, this was usually rats or pigeons in carefully controlled laboratory conditions, the aim being to discover fundamental laws of behaviour that could be generalized to human beings. The ethological approach was in stark contrast to this, with studies on a wide variety of species, often birds, fish or invertebrates, usually in as near to natural conditions as possible. Tinbergen's observations on the courtship behaviour of sticklebacks, and Lorenz's on imprinting in young geese, are well-known cases in point.

As biologists, the early ethologists were impressed with the diversity of nature and the wonderful ways in which the behaviour of animals is adapted to their particular ways of life. While they did develop general theories – for example, Lorenz's ideas about **motivation** – the evolutionary basis of behaviour and how it had led to differences between species was never far from their minds. The key role of genetics in leading to such differences led them to stress biological constraints on behaviour, often referring to the behaviour patterns they studied as 'innate' or driven by 'instinct'. Nothing could have contrasted more with the approach of psychologists, with their prime interest in learning, and this led to some sharp disputes, particularly in the 1950s.

In one of his most important papers, Tinbergen outlined the subject matter of ethology. He pointed out that, if one wanted to understand a particular aspect of behaviour, one had to seek answers to four different types of questions, those concerned with causation, function, development and evolution. To know the causes of behaviour one had to look at the mechanisms underlying it. What external stimuli elicit it? What neural circuits are involved? How is it influenced by hormones and other aspects of the animal's internal state? Function is quite a different matter, for here one is interested in what use the behaviour pattern is to the animal that shows it. What is its selective advantage? How does it enhance the animal's **fitness**? Questions about cause and about function are both important, but they are very different and also demand totally different answers. A male fish may court because he has **testosterone** in his bloodstream (cause) *and* because this behaviour attracts females to mate with him (function). We need to answer both sorts of questions and also keep them separate in our minds, for light cannot be shed on questions of one sort by answers of another.

Beyond these two primary questions, sometimes referred to as ultimate questions (function), as opposed to proximate questions (causation), are the two others: How does the behaviour develop during the lifetime of the individual? From what was it derived during evolution? The developmental question concerns when the behaviour appears and what particular experiences, if any, shape it. For example, does the animal copy other individuals? The evolutionary question looks at the derivation of the behaviour during the course of evolution. If a male shows a particular display, what normal everyday activities might it be derived from and why might females respond positively to these so that selection favoured their evolution into more striking and stylized displays?

At the time when Tinbergen outlined these questions, all four of them were active subjects of study. While Lorenz occasionally referred to development as a subject for embryologists, his own studies of **imprinting** were a counter-example. Both he and Tinbergen generated models of motivation to try to understand why animals sometimes do one thing and sometimes another, an issue central to understanding causation. Lorenz's comparative study of the displays of ducks, geese and swans was a fine example of how evolution had shaped the behaviour of animals, and the experiments of Tinbergen on eggshell removal by gulls are classic examples of how the scientific method can be applied to answer functional questions.

In his book *Sociobiology* in 1975, **E.O. Wilson** predicted that 'ethology and comparative psychology ... are destined to be cannibalized by neurophysiology and sensory physiology from one end and **sociobiology** and **behavioural ecology** from the other'. In many ways this prediction has come true. Studies of causation at the behavioural level now receive much less study than they did in the 1970s, and this is partly because of the rise of '**neuroethology**', work aimed at understanding mechanisms of behaviour at the level of the nervous system. On the other hand, behavioural ecology, the main focus of which is understanding the adaptive significance of behaviour, has become the major focus of studies at the behavioural level.

While these changes have somewhat eclipsed the rounded approach outlined by Tinbergen, there is recent evidence of a swing back, with the increasing realization that the adaptiveness of behaviour is limited by causal and developmental constraints. Thus, while it is true that the detailed theories put forward by the early ethologists receive rather little attention today, their general approach to studying the biological basis of behaviour has left a rich legacy. The bald dichotomy between what is learned and what is instinctive has been left behind. Studies of animal behaviour today, by both biologists and psychologists, are concerned with a wide range of species in the field and in the laboratory, and have been greatly enriched by the perspective that the pioneers of ethology provided. (PJBS)

See also: Evolution; Experimental analysis of behaviour; Motivation; Phylogeny; Tinbergen's four questions

Further reading

Kruuk, H. (2003) *Niko's Nature. The Life of Niko Tinbergen and his Science of Animal Behaviour*. Oxford University Press, Oxford, UK.

Slater, P.J.B. (1999) *Essentials of Animal Behaviour*. Cambridge University Press, Cambridge, UK.

Thorpe, W.H. (1979) *The Origins and Rise of Ethology*. Heinemann, London.

Tinbergen, N. (1963) On aims and methods of ethology. *Zeitschrift für Tierpsychologie* 20, 410–433.

Ethopathy

An ethopathy is a behaviour not found within the **ethogram** of free-ranging members of the species. As such, an ethopathy may be considered a form of **abnormal** behaviour. An ethopathy may be a behavioural problem, but behavioural problems are not necessarily ethopathies. It has been suggested that the occurrence of ethopathy is a cause for concern deserving of scientific attention to evaluate its **welfare** significance. (PDM)

Eugenics

A distinction is usually drawn between positive eugenics – the deliberate selection of desirable physical and behavioural traits – and negative eugenics – the attempt to 'breed out' undesirable characteristics through prevention of reproduction by means of **castration** or sterilization. During the 19th and early 20th centuries, eugenics was widely influential in policies aimed at human improvement, but was discredited by Nazi abuse, although it still persists in racist propaganda and behavioural genetics.

The scientific development of eugenics has its roots in the 19th century. It involved improvement of the inherited characteristics of a species by deliberately multiplying desirable qualities and eliminating undesirable ones through selective breeding. It was largely assisted by the emergence of the science of statistics, as formulated by the German mathematician Carl Friedrich Gauss, and applied to a variety of empirical observations of plant and animal propagation. (DL)

See also: Breeding

Euphagia

Euphagia, or the exhibition of **nutritional wisdom**, involves selection of suitable diets through the detection of nutritional or anti-nutritional properties of feeds and comparison with body status of specific nutrients or toxins. Thus, when an animal becomes deficient in a particular nutrient, such as sodium, it is suggested that it will develop a hunger for that nutrient that will attempt to cause the animal to search for it and replenish body levels. Such an ability could easily be inherited, either genetically or from exposure to post-ingestional signals *in utero*; there is evidence of both mechanisms. However, the neurological requirement to recognize and correct deficiencies of all the major nutrients would be very great and is unlikely to exist. Indeed, avoidance of toxic elements in feeds is notoriously imprecise and often results in poisoning of stock when suitable alternatives are not available. To date, an ability of some animals to select feeds high in energy, protein, sodium, calcium and iron has been detected in a variety of species, mostly domestic, and avoidance of the toxic elements lead and cadmium in feeds demonstrated in domestic ruminants. (CJCP)

European regulation of animal welfare

Regulation of animal **welfare** across Europe operates both through agreement to conventions produced by the **Council of Europe** and other international bodies such as the Organization for Economic Cooperation and Development (OECD) (**see: International regulation of animal welfare**) and the enforcement of instruments introduced through the European Community (EC) in the form of primary legislation, such as treaties, and secondary legislation, such as regulations, directives and decisions. European law has established 'four freedoms' relating to cooperation between states; this includes the ability of members to market freely across the EC.

From an animal welfare point of view, this meant there could be no restriction or veto on meat from countries where animal welfare was not as high as in the nation importing the meat, although the concept of higher welfare standards producing a different and higher-value product that can be protected by law is being pursued by some. The Treaty of

Rome in 1957 brought in the Common Agricultural Policy, which was seen as a 'cash cow' for overproduction. Reform means that links between production and subsidies have lessened, with more available for animal welfare and environmental protection.

For a considerable time, animals were considered in European law as merely articles of trade. The Treaty of Amsterdam, in force from 1999 and brought in by Protocol, introduced the idea of animals as 'sentient beings', with member states required to have 'full regard to animals' welfare'. However, this was weakened by the proviso of exceptions for national customs and religious and cultural rites, thus still allowing, for instance, **bullfighting**. (BG)

Eusociality

Eusociality is the most complex form of sociality and is characterized by the cooperative rearing of young, an overlap of generations and specialization between reproductive and one or more sterile castes that specialize in foraging or parental care (workers) or defence/attack (soldiers). Eusociality is most commonly found in the Hymenoptera (ants, bees and wasps), but is also evident in some termites, aphids, thrips, shrimps (*Synalpheus*), a species of beetle (*Austroplatypus incompertus*) and two mammalian species of mole rats, although inclusion of the latter depends on how strictly eusociality is defined: in the narrow-sense definition, the irreversible development of non-reproductive castes would exclude the mole rats. The evolution of sterile worker castes has been used to challenge the concept of Darwinian selection (**see: Darwin, Charles; Evolution**).

The most widely held view is that eusociality evolved via **kin selection**. Essentially, by helping close relatives raise additional offspring, the sterile castes can increase the frequency of their genes in the next generation through the production of non-descendent kin. This may be their only 'reproductive' option, as their ability to produce descendent kin is limited by ecological constraints or reproductive suppression by the queen. In some respects, the non-reproductive workers are analogous to the somatic cells of an organism: the cells themselves do not pass on copies of their genes, rather they function together to transmit their genes by proxy, via the germ line cells. (PE)

Eustress

Eustress, literally meaning good **stress**, is a positive form of stress or a stress that may be perceived as pleasurable. Eustress can have similar physiological effects on the body as **distress**, such as increased **heart rate** and elevated **glucocorticoids**, which in themselves are not necessarily harmful to animals. Examples of eustress include copulation or running during a **play** bout. (LMD)

Euthanasia

Literally meaning a good death (from the Greek *eu*, having a connotation of goodness, and *thanatos*, death). When applied to animals it is usually applied to the means by which an animal is killed, rather than the reason for it being killed (the human sense). Acceptable methods should normally cause a rapid loss of **consciousness**, be painless, not cause discomfort or **distress**, e.g. through being **aversive** or unpleasant in some way, be aesthetic to the person carrying the procedure out, be easy to perform, be repeatable without loss of efficiency and be minimally invasive. Methods may be divided into inhalant methods, non-inhalant pharmaceutical agents and physical methods. In some cases, one method may be preferred for social reasons, e.g. the increasingly common use of lethal injection to induce euthanasia in horses in preference to shooting, as this is less distressing for the owner. There are reasons why some acts of euthanasia, regardless of the method, may be unacceptable (the human sense), such as **breeding** more animals than one needs for research resulting in a surplus that has to be killed, and killing animals in ways that cause avoidable **suffering** for ritual reasons. (DBM)

See also: Religious slaughter; Sacrifice; Slaughter

Further reading

Crawford, R.L. and Adams, K.M. (2007) *Animal Euthanasia*. Special Reference Brief Series, SRB 2007-01, USDA National Agricultural Library, Animal Welfare Information Center, Washington, DC. Available at: http://www.nal.usda.gov/awic/pubs/Euthanasia07/animal_euthanasia.shtml (accessed 24 September 2009).

Evolution

> Nothing in biology makes sense except in the light of evolution.
>
> T. Dobzhansky

In the early 1800s, species were viewed as immutable entities that did not change from the time they were created until they became extinct. Variation among individuals within a species was considered to be unimportant – imperfections around an ideal form. This view is known as essentialism. Essentialism can be traced back to Plato and was thoroughly incorporated into Christian theology. Fossils, under this view, were assumed to be species that had gone extinct. New species were thought to be created *de novo* (from scratch).

Jean-Babtiste Lamarck (1744–1829) was one of the few in his time to suggest that species evolved. Lamarck viewed fossils not as extinct species but instead as ancestral forms of modern species. He hypothesized that 'lower' forms arise through spontaneous generation and gradually evolve into 'higher' forms because of an intrinsic drive towards increased complexity. The great diversity of life forms was a consequence, he argued, of organisms adjusting to environmental conditions within their lifetimes and transmitting these modifications to their offspring (inheritance of acquired characteristics). Thus, Lamarck's theory could account for the overall trend towards increasing complexity in the fossil record and the observation that organisms are usually well adapted to the environments in which they are found. All that would be required would be vast amounts of time.

Geologists in those days had already concluded that the earth was millions of years old (the modern estimate is 4550 million years), and they knew that fossil species were found in a succession of sedimentary layers corresponding to different geological periods. For example, fish first occurred in the Silurian, reptiles in the Carboniferous, mammals in the Triassic, placental mammals in the Cretaceous, and so on. These patterns were repeated around the globe. Nevertheless, geologists were quite adept at reconciling their findings with the prevailing essentialist view. Lamarck's theory was rejected, but largely for the wrong reasons. The geologists correctly pointed out that simpler forms often replace more complex forms in the fossil record, which contradicted Lamarck's notion that complexity increases over time within lineages. We also know now that characteristics acquired during an individual's lifetime generally are not passed on to offspring (with some important exceptions, such as culturally inherited traits). Nevertheless, Lamarck deserves credit for proposing the first coherent theory of biological evolution.

Charles Darwin himself was a geologist, so he knew a great deal about fossils, but this is not what turned him into an evolutionist. In 1831, at the age of 22, he took a 5-year voyage as the naturalist on the HMS *Beagle*. The crucial observation he made on that voyage was that species varied so much from place to place that it was impossible to draw clear distinctions between species and varieties within species. For example, mockingbirds in the Galapagos differed so much from island to island that the ornithologist to whom Darwin shipped specimens was compelled to call them separate species. Intraspecific (within species) geographic variation clearly conflicted with the conventional wisdom that species were distinct and immutable, with only random variation around ideal forms. Darwin rejected the idea that species were unchangeable, and began to suspect that one species could evolve into another. That sent him in search of more evidence and a mechanism.

Some of the kinds of evidence that Darwin gathered are listed below. Not all of these specific examples were available in Darwin's time. Absent from this list are categories of evidence discovered after Darwin's time, such as modern molecular genetics.

Convergent forms

Different kinds of animals occupy similar ecological roles on different continents. The comparison of Australian marsupials to placental mammals elsewhere in the world is a classic example (marsupials are mammals in which the young are born at a very early stage of development and carried in pouches). In their external features, marsupial moles closely resemble placental moles, marsupial mice resemble placental mice, marsupial flying phalangers bear a striking resemblance to flying squirrels, and so on. Despite these outward similarities, however, all marsupials are more similar to each other skeletally, reproductively and genetically than they are to any of the placentals. This kind of pattern, known as convergence, looks more like an accident of history than the divine scheme of a creator. It would be hard to argue that marsupials are better designed for living in Australia than are placentals. Marsupials tend to become extinct when they come in contact with their placental 'counterparts' (i.e. their ecological competitors). Marsupial Tasmanian wolves, for example, were driven to extinction in part because of competition with dingos – placental dogs introduced to Australia by the Aborigines.

Embryology

The phrase '**ontogeny** [development] recapitulates **phylogeny** [evolutionary history]' describes the idea, promoted by Ernst Haeckel, that 'higher' organisms such as humans reveal their evolutionary history by developing through the adult stages of 'lower' organisms such as fish. This idea is false and has been discredited, but it brings to mind certain aspects of development that are true. First, early-stage embryos across the vertebrate taxa (that is, animals with backbones) are practically indistinguishable. General features shared by all vertebrates, such as gill arches and spinal cords, develop first. Class-level characteristics develop from these generalized features; characteristics of the particular order then develop from these class-level features, and so on, until finally the characteristics of the species emerge. The embryonic structure that develops into the forelegs of lizards develops into the wings of birds and the arms of humans. In fish, gill arches become bony structures supporting the gills. In humans, one of the gill arches is converted into the lower jaw, another forms the eustachian tube between the ears and mouth, and others are reduced to tiny bones in our ears. Some structures, such as the tail, are lost completely. Human embryos never go through a stage that resembles an adult fish; instead, fish and human embryos start off with the same structures and then diverge from each other as they develop. These patterns of development are exactly what we would expect if fish, reptiles, birds and mammals all descended from a common ancestor and had subsequently been modified in different ways over time.

Vestigial structures and behaviours

Examples of vestigial structures and behaviours include: non-functional eyes on cave fish; wings on flightless birds (e.g. ostriches) and beetles; remnants of pelvic and hind limb bones in some snakes and whales; teeth in the embryos of toothless baleen whales; ear-wiggling muscles, wisdom teeth, tail bones and appendices in humans. Some excellent behavioural examples are provided by parthenogenic vertebrates. These are female-only species that reproduce asexually, i.e. the eggs develop without being fertilized by a male. Nevertheless, in some parthenogenic species, **reproduction** is enhanced by male courtship. In the Amazon molly (*Poecilia formosa*), courtship is performed by the unwitting males of a closely related sexual species. In the parthenogenic whiptail lizard *Cnemidophorus uniparens*, male-like courtship is provided by other females. The existence of vestigial structures and behaviours is very difficult to explain without invoking descent from an ancestor that made good use of them.

Normally unexpressed traits

Examples of normally unexpressed traits include: humans born with monkey tails; lizard scales produced by mouse skin when exposed to chicken tissue; rapid reversion of feral domesticated pigs to wildtype; fish teeth induced to develop in frogs (frogs ordinarily lack teeth). It takes a creative mind indeed to explain such phenomena without invoking evolutionary descent.

Jury-rigged adaptations and design flaws

Close examination of biological structures often reveals idiosyncrasies that suggest historical constraints. Evolution has no way of backing up and doing it over again the right way; it can only build on what it already has to work with. The vertebrate eye, for example, is superficially a wonderfully well-designed structure. It has been used over and over again as evidence that biological organs were designed by a creator – the classic argument for design. In truth, however, the vertebrate eye is wired backwards. Light needlessly passes through layers of cells in order to reach the photoreceptive pigments. The eyes of cephalopods (squid, octopuses, cuttlefish) are wired 'correctly' (convergently evolved).

The fossil record

For obvious reasons, fossils provide the hardest evidence for evolution. In Darwin's time, there were very few examples of clear intermediates, the missing links between major groups of organisms. Fossilization is an extremely rare event, so the fossil record will always be patchy, but no self-respecting palaeontologist would maintain that we lack intermediates today. For example, *Archaeopteryx* is a clear intermediate between reptiles and birds. This creature had bird-like feathers and wings, but it also had teeth instead of a beak, a long bony tail and claws on the leading edge of the wing. Modern birds do not have teeth, long bony tails or claws on their forelimbs – these are reptilian traits. Equally clear fossil intermediates have been found between **reptiles** and mammals, between land mammals and whales, between **chimpanzees** and humans, and between many other living (and extinct) taxonomic groups.

Ample evidence for evolution was available by the early 1800s, but each piece of evidence had been gathered separately and interpreted in light of the prevailing theory (the Creation). Embryologists, for example, believed they were working out the divine developmental plan. The reason Darwin's book *On the Origin of Species* had such a tremendous impact was that he brought all of the evidence together and showed that it was far more consistent with a particular model of evolution (in which all organisms descended from one or a few common ancestors) than it was with special Creation (in which all species were created *de novo*). Darwin's greatest contribution – though it was not recognized immediately – was to provide a better mechanism for evolution than Lamarck did. That mechanism, of course, was natural selection (**see also: Selection**).

Biological evolution can be defined as: (i) any change in the characteristics of organisms over many generations (classic definition); or (ii) any change in the genetic constitution (gene frequencies) of a population of organisms over time (population genetic definition). The first version is closer to the common sense meaning of evolution, and is the definition used in practice by palaeontologists. The second definition is less restrictive. Anything that qualified under definition (i) would qualify under definition (ii), but not vice versa. Changes in the frequency of genes that have no effect on the phenotype are still considered to be evolution.

Artificial selection (**see also: Breeding**) produces a change in gene frequencies by separating the parents into two groups, selected and discarded, that differ in gene frequencies. Natural selection is subtler, and generally has much weaker effects; it merely favours genes that have positive effects on survival and reproduction. Over modest amounts of geological time, however, even relatively weak selection can have huge effects.

If certain basic conditions hold, evolution by natural selection is not only possible, it is inevitable. Those conditions are: (i) variation among individuals in heritable traits that affect survival or **reproduction**; (ii) variation is universal; (iii) most variable traits are at least somewhat heritable (offspring tend to resemble their parents); (iv) virtually all traits have some effect, however slight, on survival or reproduction. Thus, evolution by natural selection must be rampant.

The term microevolution refers to evolution within a single lineage. Macroevolution is the term used for speciation (the formation of new species) and the evolution of higher taxonomic groups.

Common misconceptions about evolution

1. Evolution is random. Natural selection, the primary driving mechanism of evolution, is anything but random – no more random than selective breeding.
2. Evolution occurs within individuals, or should somehow be directly observable. Evolution occurs between generations, not within individual animals. Major changes take millions of years.
3. Evolution violates the laws of thermodynamics. This is one of the most common and sophisticated-sounding arguments against evolution, because it seems to be based on the laws of physics. In fact, it is based on a misconception. The first law of thermodynamics basically states that you cannot get something from nothing. More formally, the total amount of energy in the universe is conserved – it never increases or decreases. This means, for example, that you cannot build a machine that produces more energy than it consumes. The second law states that, in a closed system, the net effect of energy transfer is towards increased entropy (disorder). Closed systems can never become more organized – overall, order always decreases. But evolution is an order-increasing process. Organized living systems have evolved from disorganized non-living matter, and simple organisms have evolved into more complex ones. Creationists have used this to argue that evolution is impossible, based on the second law of thermodynamics. But living organisms are not closed systems. The earth itself is an open system – energy pours in constantly from the sun. The only system that is truly closed is the universe itself. Order arises from disorder all the time. Complex trees grow from simple seeds. Snowflakes form from water vapour.
4. Natural selection is untestable. One Creationist argument goes like this: natural selection is untestable; to be scientific, a theory must be testable; therefore, natural selection is not a valid scientific theory. This sounds perfectly logical. The only problem is that the starting premise is false: natural selection is testable.
5. Evolution leads to progress. Modern evolutionists agree that evolution has no goal. Two hundred and fifty million years ago, there would have been no way to predict that mammals would evolve from reptiles. If it happened all over again, we can be certain of only one thing: the outcome would be quite different. Certain broad evolutionary trends, such as that towards increasing complexity or size, are undeniable. But this is only true on the average (e.g. virus evolution, mammalian size reduction over the last few million years). Given that life started from a single cell, there are few places to go but 'up'. Amoebas have just as long an evolutionary history as do humans. Bacteria are much more numerous and ecologically successful than insects; insects are much more numerous and ecologically successful than vertebrates. There is no justification for thinking that humans are somehow the most highly evolved organisms; we just have huge brains.

Processes other than selection that cause evolution

1. Genetic **mutation**. **Mutations** cause changes in gene frequencies, hence evolution. More importantly, mutation provides the raw material for natural selection.
2. Gene flow. Movement of individuals between populations can cause changes in population gene frequencies, hence evolution.
3. Genetic drift. Gene frequencies can change purely by chance (random sampling error), especially in small populations. Specific mechanisms include:

- Founder events. When a small population becomes isolated from a larger one (for example, by dispersing to an island), rare genetic variants (alleles) are likely to be lost and gene frequencies, in general, are likely to change.
- Random sampling during mating. Parents do not contribute all of their genes to each offspring, just a random half. This can result in chance variation in gene frequencies. Rare alleles, say the allele for purple eyes (if there was such a thing), are likely to get 'lost', i.e. disappear from the population, just by chance.
- Bottlenecks. A bottleneck is a population crash, followed by recovery – for example, because of overhunting, Northern elephant seals (*Mirounga angustirostris*) were reduced to 20 individuals by 1890, when hunting was banned. By 1993, the population had increased to about 30,000, but genetic variation remained extremely low. The Southern elephant seal (*Mirounga leonina*) went through a less severe bottleneck and shows correspondingly higher levels of genetic variability.

(GFG)

Further reading

Futuyma, D.J. (1998) *Evolutionary Biology*, 3rd edn. Sinauer, Sunderland, Massachusetts

Ridley, M. (2003) *Evolution*, 3rd edn. Blackwell, Oxford, UK.

Evolutionary psychology

Evolutionary psychology is the study of the way in which a number of psychological traits or higher mental functions may have formed as either adaptations or functional productions of natural **selection**. Evolutionary psychologists argue that many behaviour patterns, especially human behaviour patterns, have evolved to solve problems in past environments. Such adaptations may include the ability to acquire **language**, maternal and paternal **motivations**, and the ability to infer the emotions of others. Behaviours and **emotions** that are found worldwide in man are thought to be more likely to reflect evolved traits. These traits then interact with cultural or environmental stimuli to produce specific behaviours – for example, the specific language learned. (LMD)

See also: Comparative psychology

Exaptation

Exaptation refers to the process whereby a (physical or psychological) characteristic acquires a function separate from that which led to its original development through natural **selection**. For example, **feathers** probably evolved as an adaptation of dermal structures to aid **thermoregulation**, and only after this did their use in flight become a possibility, and so the development of feathers for use in flight may be considered to have arisen as a result of exaptation. Exapted traits may therefore have features that appear to be redundant, since these features relate to their original function. However these apparently redundant traits can themselves have latent potential to provide the basis for new developments. Exaptation may also help to explain why some traits may be suboptimal or not fully functional. The development of **ritualization** and certain cognitive abilities can often be explained more accurately in terms of exaptation rather than adaptation. The term exaptation is preferable to the alternative 'preadaptation' that is often used, since the latter suggests that evolution is a teleological process. (DSM)

Further reading

Gould, S.J. (1991) Exaptation: a crucial tool for evolutionary psychology. *Journal of Social Issues* 47, 43–65.

Gould, S.J. and Vrba, E.S. (1982) Exaptation – a missing term in the science of form. *Paleobiology* 8, 4–15.

Exotic companion animal

Exotic companion animals can include both indigenous and non-indigenous species, and both wild and captive-bred individuals. The principal distinction of exotic species used here is that they are non-domesticated species (**see: Domestication**). The range of exotic species kept as **companion animals** is vast. It includes invertebrates, such as spiders, scorpions, stick insects and mantises; and mammals, such as spiny mice (*Acomys sp.*), chipmunks (*Tamias* spp.), wallabies (*Macropus* spp.) and common marmosets (*Callithrix jacchus*), in addition to many birds, especially parrots and finches, **reptiles**, **amphibians** and the majority of tropical aquarium fish.

With the exception of tropical aquarium fish, reptiles and amphibians represent the majority of exotic companion animals. Reptiles are becoming increasingly desirable pets. As an indication of the popularity of exotic pets in 2007, it was estimated that 13.4 million reptiles were kept in 4.8 million households in the USA alone. A significant proportion of these are CITES (Convention on International Trade in Endangered Species)-listed, with 300,000 iguanas being legally imported into the USA in 2005. The vast majority of exotic animals for sale as companion animals are wild caught, and a number are endangered. Their trade is regulated by CITES. Additionally, national legislation often governs the import of wild-caught individuals of many species.

Within the UK the Department of the Environment, Transport and the Regions (DETR) is the primary body concerned with exotic pets. For example, it regulates trade in Mediterranean tortoises (spur-thighed, *Testudo graeca*, Herman's, *Testudo hermanni*, and marginated, *Testudo marginata*), and a DETR certificate is required to sell, buy, exchange or exhibit these species. In the UK, the keeping of large and/or potentially dangerous exotic species is also regulated under the Dangerous Wild Animals Act (1976).

As well as **conservation** issues, the exotic pet trade gives rise to serious **welfare** concerns. Many of the animals involved are injured or die during their capture. **Mortality** rates vary by the type of animal, the country of origin, the techniques used during capture and **transport**, the species' ability to withstand the associated trauma and its ability to adapt to **captivity**. Mortality before export is typically high; for example, up to 50% of Senegalese finches and waxbills may die before leaving their country of origin. These mortality rates are commercially acceptable, as long as some individuals survive to the point of sale to the exporter, as wild capture costs very little for those involved.

The welfare of most exotic pets may also be compromised during shipment due to overcrowding, poor temperature control, **stress** and exposure to infection. Once in captivity, poor husbandry such as a lack of space, inappropriate temperature and/or feeding regimes and dietary deficiencies can result in a very poor prognosis, with an estimated 90% of pet reptiles dying within their first year in captivity. Poor welfare in reptiles and amphibians is a particular problem as it is far more difficult to detect than it is in mammals, for the simple reason that they lack the facial expressions and **vocalizations** that draw attention to their **pain** and **distress**. Consequently, many suffer in silence. Their **suffering** is also often more prolonged than in mammals due to their slower metabolic rate, meaning that they take longer to heal or die.

Exotic companion animals present a number of challenges to owners, some of which are specific to exotics and some of which are more general. Providing suitably sized accommodation for large or active species such as primates, crocodilians, large lizards (monitors and iguanas) and snakes (pythons and boas) can be difficult. There may also be temperament considerations, as some can be aggressive – for example, male iguanas and some snakes. Perhaps the most universal challenge when keeping exotic pets is their nutrition. Many species require a specialized diet, not easily provided in captivity, such as live invertebrates (flies, crickets, etc.) or dead vertebrates (baby mice, rats, piglets, etc.). Furthermore, the nutritional requirements of many exotic species, not just of reptiles and amphibians, are not yet fully understood.

Malnutrition is often a key component of compromised welfare in exotic animals. It has been estimated that 15% of pet reptiles are malnourished. Critical factors include the quality and quantity of food and water, and in many reptiles the calcium:phosphorus ratio in their diet. The slow metabolism of reptiles also makes them particularly prone to obesity, which in turn is linked to fatty liver syndrome, arteriosclerosis and kidney calcification. The correct provision of water is just as essential as food, for inadequate drinking water or misting can cause gout and bladder stones (calculi).

In addition to food and water some reptiles have a requirement for ultra-violet B (UVB) radiation, as its action enables the synthesis of vitamin D, which is required for the adsorption of dietary calcium. An unbalanced calcium: phosphorus ratio leads to the development of potentially fatal metabolic bone disease (MBD). Animals suffering from MBD are susceptible to bone fractures. Although **bone strength** can be restored through the correct use of UVB, calcium supplements and a proper diet, skeletal deformities, such as curvature of the spine and a shortened jaw, will be permanent.

Many exotic pets have other special requirements. For example, particular care must be paid to temperature and humidity, and how these are provided. Ideally, the cage should have a temperature gradient to allow the animal to choose its preferred temperature, as incorrect temperatures can have serious metabolic consequences. In snakes and chelonians, if the temperature is too high, food may decompose before it can be digested; if it is too low, digestion is halted and food may begin to rot in the intestines. A correct humidity level is also critical in preventing a number of health problems, including respiratory and fungal infections and difficulties during sloughing. The substrate used in cages is also important. Some animals, such as chinchillas (*Chinchilla lanigera*), require a soil- or sand-based substrate suitable for burrowing or **dust bathing**, while others may require water for swimming or bathing, or branches and ledges for climbing. However, not all substrates are suitable for all species; for example, sand can cause intestinal impaction in some species. Ticks and mites can also be picked up from contaminated substrate materials and these can harbour diseases and be a source of irritation.

(AS)

See also: Animal enclosure; Conservation; Endangered species

Exotic species invasion

Exotic species (also called 'introduced', 'alien', 'non-native' or 'non-indigenous' species) are species that are not native to a particular habitat but have been accidentally or intentionally introduced there. If a species is able to colonize new areas beyond the initial site of introduction, and in the process to cause significant ecological or economic damage, then that species is considered 'invasive'. It is estimated that only about 1% of all species that are introduced to new habitats become problematic but, due to the increasing globalization of the world's economy, more and more species are being transported around the globe and the rate of invasion is increasing worldwide. The ecological impact of invasive species can be profound, and is one of the biggest threats facing imperilled animal species, second only to habitat loss and degradation (**see also: Extinction**). The economic costs of invasive species are also increasing, and are estimated to be over US$120 billion per year in the USA alone.

What makes one exotic species a successful invader while another fails to establish itself? There are two main reasons. First, the new habitat may lack the limiting factors that would keep the invading species' population in check. These limiting factors may be predators, **parasites** or competing species that are present in the invader's native habitat but missing in the new habitat. Without limiting factors, the population of the invading species increases and may overwhelm the native plants or animals in its new home. The second reason why some exotic species become successful invaders is that they possess behavioural traits that allow them to exploit the resources in their new habitat with remarkable efficiency. Invasive species are often generalists in terms of the food resources they can eat, and they may also possess a high level of **behavioural elasticity** that allows them to take advantage of other novel resources they encounter in their new habitat.

One of the most devastating examples of the impact of an invasive animal species is the ecological havoc wreaked on the island of Guam by the brown tree snake (*Boiga irregularis*). The brown tree snake is native to Australia, Indonesia, Papua New

Guinea and the Solomon Islands, and was accidentally introduced to Guam some time in the early 1950s. By 1968 the snakes had dispersed throughout the 541 km^2 island and, by the 1990s, had reached peak densities of up to 100 snakes/ha. Brown tree snakes are generalist predators and will feed on lizards, birds and small mammals, as well as bird and **reptile** eggs. It wasn't until 30 years after the initial introduction that researchers realized this snake was largely responsible for the mysterious decline of many native species. The brown tree snake has caused the local extinction of nine of the island's 13 native forest-dwelling bird species (including the global extinction of the Guam flycatcher, *Myiagra freycineti*), as well as over half of the native lizard and bat species. The native species are unable to protect themselves against this exotic predator, and today only isolated pockets of native populations remain.

Part of the reason that researchers did not make the connection between the disappearance of the birds and the presence of the exotic snake was behavioural; brown tree snakes are arboreal and **nocturnal**, so they are hidden during the day and extremely hard to locate at night. The impact of the brown tree snake on Guam also includes significant economic damages, as these climbing snakes will often short-circuit power lines, causing many power outages. Local poultry businesses have also been devastated by the loss of eggs to the predatory snake. Efforts to control the brown tree snake are ongoing, as are efforts to prevent the transport of the snake to new, vulnerable habitats.

The role of behaviour in determining whether an animal becomes a successful invader is well documented. It is also essential, however, to understand the role of behaviour in native species' ability to respond to invasive animals and plants. Invasive species can impact native species directly through **predation** or competition, but also indirectly by altering the ecosystem upon which native species depend. Often these changes occur very rapidly, and native species may not have time to evolve adaptations to the new conditions. On the other hand, if native species possess enough behavioural elasticity, they may be afforded more time to adjust to the changes that are occurring – but their evolutionary trajectory will be forever altered.

In some cases native animals may, to their detriment, seek out an invasive species if they mistakenly see the invader as a resource. For example, females of several native *Pieris* butterfly species will lay their eggs on invasive plants (e.g. *Thlaspi arvense* and *Allaria petiolata*) that are toxic or even lethal to their caterpillar offspring. There is some evidence that these females may prefer to lay their eggs on the invasive plant rather than on a more appropriate native plant. In that case, the invasive plant becomes an 'ecological trap' because the normal cues the females use to find egg-laying sites are now maladaptive and lead to reduced survival of their offspring.

In other cases, however, the native species may learn and/or evolve behaviour that helps them to compensate for the presence of an invader. For example, red-legged frog tadpoles (*Rana aurora*) from areas that have been invaded by an exotic predatory bullfrog (*Rana catesbeiana*) will reduce their activity and hide when they detect chemical cues from the bullfrog. Tadpoles from other areas that have not been invaded by bullfrogs do not show any change in behaviour when presented with the same chemical cues. Whether the anti-predator behaviour in tadpoles that live in areas with bullfrogs is learned or has evolved over several generations is unclear. In either case, the change in behaviour is crucial to the survival of these frogs, as it is very difficult and expensive to control established bullfrog populations.

Controlling any invasive species is difficult and expensive, and may even be impossible if an invasion has gone undetected for a long period of time. The survival of native species may depend on whether, in the short term, they are able to adjust their behaviour to compensate for the presence of the invasive species and whether, in the long term, they can evolve adaptations that allow them to coexist with the exotic invader. (JCN)

See also: Conservation; Evolution; Predation; Wildlife management

Further reading

Kiesecker, J.M. and Blaustein, A.R. (1997) Population differences in responses of red-legged frogs (*Rana aurora*) to introduced bullfrogs. *Ecology* 78, 1752–1760.

Mooney, H.A. and Cleland, E.E. (2001) The evolutionary impact of invasive species. *Proceedings of the National Academy of Sciences* 98, 5446–5451.

Pimentel, D., Zuniga, R. and Morrison, D. (2005) Update on the environmental and economic costs associated with alien-invasive species in the United States. *Ecological Economics* 52, 273–288.

Rodda, G.H., Frittes, T.H. and Chiszar, D. (1997) The disappearance of Guam's wildlife. *Bioscience* 47, 565–574.

Schlaepfer, M.A., Sherman, P.W., Blossey, B. and Runge, M.C. (2005) Introduced species as evolutionary traps. *Ecology Letters* 8, 241–246.

Experimental analysis of behaviour

The experimental analysis of behaviour is a scientific discipline that arose from the philosophy of **behaviourism**. Behaviourism is a natural science approach to **psychology** that emphasizes the study of environmental influences on observable behaviour. The philosophy is associated with John B. Watson, an American psychologist who advocated that psychologists define psychology as the study of observable behaviour rather than as the study of the **mind** or **consciousness**. The behaviourist movement was meant to counter Sigmund Freud's psychoanalysis, in which the unconscious mind was believed to play a large role in human behaviour. By avoiding the terms consciousness and mind and focusing on observable behaviour, psychologists could not only study human behaviour scientifically, but they could also study animal behaviour and compare it to human behaviour.

The behaviourist movement was popularized by **B.F. Skinner**, an influential psychologist who developed a theory of operant conditioning to explain how much behaviour developed. Skinner believed that the processes of learning and natural selection are very similar to each other. Whereas Skinner acknowledged that heredity may play a role in behaviour, he also believed that much of our behaviour and that of other animals results from having learned what will happen after we perform the behaviour. For example, behaviours that lead to favourable outcomes are more likely to be repeated and increase in the future. These favourable outcomes are often referred to as reinforcers. Behaviours that lead to poor outcomes are less likely to be repeated and decrease in the future. Poor outcomes are often referred to as punishers.

Behaviours that are influenced by their consequences are known as operant behaviours. Therefore, **operant conditioning** is a type of learning in which the future probability of a behaviour is affected by its consequences. Operant behaviour is more voluntary and flexible than stimulus-elicited or reflexive behaviour. Viewing human and animal behaviour in this way implies that behaviour can be influenced by controlling the environmental consequences for a particular behaviour or class of behaviours.

Principles derived from studies conducted in the field of the experimental analysis of behaviour have been very influential in guiding animal management and husbandry in laboratories and **zoos**. For example, the principles of operant conditioning have been incorporated into animal training programmes in which positive **reinforcement** (rewarding stimulus) is provided after a particular response to increase the likelihood that the response will occur again. In this way, new behaviours or patterns of behaviour can be shaped over time and the frequency of desirable behaviours can be increased.

Common reinforcement provided in captivity includes food, the opportunity to engage in social activity and manipulable non-food items designed to sustain interest and activity levels. Operant conditioning can also be used to decrease **abnormal** or stereotypic behaviours that can appear in captive animals, presumably as an artefact of the captive environment. The frequency of unwanted behaviour can be decreased by either punishing the behaviour or by removing the reinforcement that is sustaining the behaviour. Punishment involves the weakening of a behaviour through either application of an **aversive** stimulus (positive punishment) or the removal of a reinforcing stimulus.

From an animal **welfare** point of view, **punishment** is not the method of choice for deterring behaviour. Early experiments by Skinner and others led them to conclude that punishment was not a very effective method for controlling behaviour, and that it often had only temporary effects at best. Research has shown that there are risks associated with punishment if it is not provided correctly. First, punishment does nothing to increase appropriate behaviour – it only decreases the unwanted behaviour. Secondly, punishment can lead to escape and **avoidance behaviour**. Animals in zoos and laboratories will often hide or escape from the source of punishment. Thirdly, punishment can sometimes lead to either an aggressive response or a strong emotional response. Finally, punishment can be physically or mentally harmful if not applied appropriately, leading to high levels of **stress**. Discovering and eliminating the reinforcement for a behaviour, combined with positive reinforcement for desirable behaviour, is generally a preferred means of eliminating unwanted behaviour as opposed to punishment, when considering the welfare impact of the processes involved.

In conclusion, the experimental analysis of behaviour provides a framework for systematically and scientifically evaluating the variables that contribute to the presence or absence of behaviour, patterns of behaviour, frequency of behaviour and strength of behaviour. Because of this, it is a powerful field for understanding animals and their interactions with the environment. Most importantly, the experimental analysis of behaviour has led to more humane and effective techniques for managing the behaviour of captive animals. Specifically, the principles of operant conditioning can be used to increase the quality of captive environments, improving both the physiological and psychological well-being of animals in captivity. (LRT)

Further reading

Tarou, L.R. and Bashaw, M.J. (2007) Maximizing the effectiveness of environmental enrichment: suggestions from the experimental analysis of behaviour. *Applied Animal Behaviour Science* 102, 189–204.

Experimental playback

Experimental playback allows auditory **communication** to be experimentally manipulated. Essentially, animal **vocalizations** are recorded and then played back through loudspeakers and the response of the receiver(s) monitored in order to identify the function of the vocalization. For example, vervet monkeys (*Cercopithecus aethiops*) are known to give different alarm calls in response to different predators: a loud barking alarm is given in response to leopards, a short double-syllable cough signals the presence of an aerial predator (eagle), while a 'chutter' call is given in response to the presence of a snake. That the monkeys responded differently to the different alarm calls can be demonstrated by playing back the calls through loudspeakers: when the bark was played, vervet monkeys ran to and climbed nearby trees, the double-syllable cough caused the monkeys to look up and run to the nearest bush and the 'chutter' resulted in vervets standing on their hind legs while scanning the surrounding grass.

Experimental playback allows the observer control over potentially confounding variables. For example, cuckoo chicks elicit a higher parental feeding rate than the host bird's own chicks. This could be due to the greater mouth gape of the cuckoo chick, but it could also be due to the different begging call of the cuckoo chick. By playing back a cuckoo chick begging call at the nest when no cuckoo chick was present, it was revealed that the call alone was capable of stimulating parent birds to bring more food back to the nest. Experimental playback can also reveal costs of chick begging: eggs in artificial tree swallow nests were more likely to be predated when the taped sounds of begging chicks were played nearby. (PE)

Exploratory behaviour

It is difficult to define exactly what is meant by exploratory behaviour; indeed, one of the early researchers on this topic sadly observed that it consisted of a large range of behavioural acts that 'have nothing very tangible to hold them together apart from our failure to recognize a specific biological function that can be associated with them'. Nevertheless, it was quickly recognized that exploratory behaviour is an important subject because a large part of an animal's behaviour is characterized not so much by activities that obviously serve its biological needs but rather by pronounced tendencies to explore, investigate and seek out new forms of stimulation. In its broadest sense exploratory behaviour refers to all activities concerned with gathering information about the environment.

Types of exploratory behaviour

During the early period of research into exploratory behaviour a number of important distinctions between different types of exploration were made. One such distinction is between *extrinsic* exploration, which has a clear goal – namely to gather information about a conventional reinforcer (e.g. a hungry animal searching for food), and *intrinsic* exploration, which has no such goal in that it is directed at stimuli of little obvious biological consequence. From an **adaptive** viewpoint it is doubtful whether this distinction has much validity, because changes in the animal's environment not related to conventional reinforcers may still potentially have repercussions for survival and hence should be investigated. Nevertheless, the extrinsic/intrinsic distinction has been important in the development of motivational theory. While the **motivation** for extrinsic exploration can at least partly be explained on the basis of the **reinforcement** obtained as a result of the information gathered, explaining the motivation for intrinsic exploration is far more difficult, as here the stimuli involved appear to be rewarding in themselves.

Another distinction is between *specific* exploration, defined as gathering information about a specific event or object, and *diversive* exploration, which is the gathering of information from any source. For example, the former is the type of behaviour we perform when we are 'looking for' something in particular, while the latter is encompassed by such terms as 'recreation' and 'entertainment'. Specific and diversive exploration are thought to be caused by different motivational states – specific exploration being linked to 'curiosity' about a particular stimulus change, and diversive exploration to '**boredom**' created by a general lack of **stimulus** change.

The final major distinction is between *inspective* exploration (also called *forced* exploration) – which is responding to stimulus change – and *inquisitive* exploration (also called *free* exploration) – which is responding for stimulus change. Inspective exploration involves the animal adjusting to new stimuli that are forced upon it, as, for example, when a rat is placed into a novel maze. In contrast, inquisitive exploration involves the animal freely seeking stimulus change, as, for example, when a rat learns to bar press solely for exposure to novel stimuli. The adaptive significance of responding to a change in the environment is clear, but there has been much debate on why animals should also actively seek change.

Determinants of exploratory behaviour

The strength and direction of exploratory behaviour are determined by a large number of factors, but experimental studies have revealed a number of stimulus characteristics that stand out as being the most influential. These include properties such as novelty, surprisingness, ambiguity, incongruity and complexity. The term 'collative variables', i.e. variables brought together for comparison, was coined to cover these characteristics because, apart from complexity, they are properties of an animal–environment interaction rather than of the environment itself. The collative variables imply an expectancy by, and induce uncertainty in, the animal.

Methods used to study exploratory behaviour

Research on exploratory behaviour usually involves exposing the animal to environmental stimuli that it has either never before experienced or not experienced for some time – conventionally termed the novel stimuli or novel environment. Many experiments on inspective exploration have involved briefly putting an animal into a relatively homogeneous novel environment, such as a maze or open field, that is devoid of food, water, objects and other animals, and from which it cannot escape. In such studies the degree of movement shown (e.g. by marking out the floor of the open field in squares and counting the number crossed) is often taken as a measure of 'locomotive exploration'. There are, however, several problems with this interpretation; for example, locomotory scores are very likely to be contaminated by changes in activity levels and attempts to escape from the apparatus. Furthermore, the usefulness of this general approach is questionable because in the wild it will be very rare that an animal suddenly finds itself in a novel environment from which it cannot escape.

Experimental protocols that more closely resemble the conditions under which animals encounter novelty in the wild involve either: (i) exposing the animal to a number of discrete novel stimuli within a familiar area; or (ii) allowing it to move freely between a familiar area and a novel environment. In these tests, measures such as the number and duration of contacts with the various novel objects, and the latency to enter and time spent in the novel environment are used as indices of exploration. In many studies the novel stimuli do not impinge upon the animal until after it has made a specific response. Thus the resulting behaviour is inquisitive exploration because the animal responds for, rather than to, the environmental change. Typically the animal is placed in a Skinner box and an environmental change, such as a brief exposure to a novel visual stimulus, is made contingent on a lever press or similar operant. Experiments of this kind have shown that response-contingent novelty acts in much the same way as a conventional reinforcer like food or water and produces learning – i.e. a relatively permanent rather than a temporary change in behaviour.

Some studies have manipulated the dimension of stimulus complexity to see how an animal's response depends not just on the **novelty** of an environmental change but also upon the nature of that change. In such experiments stimulus complexity is manipulated by varying the number of distinguishable elements within the stimulus and the degree of difference between them. There is good evidence that complex stimuli are preferred in **choice tests**, elicit more exploration and make stronger sensory reinforcers than simple ones. However, there is also evidence that highly complex stimuli are actively avoided and, as discussed below, this finding has important theoretical implications.

Theories of exploratory behaviour

For the most part theories of exploratory behaviour have failed to take account of the adaptive significance of exploration and, instead, have concentrated upon its immediate proximate causes. Initially, exploration was thought to be a secondary or acquired **drive** serving one or other of the classic drives, e.g. searching for food. However, the results of latent learning experiments have shown that animals would explore and learn about their environment even when their primary **needs** had been satiated. A secondary drive status for exploration was therefore dropped. Instead, it was proposed that this behaviour had its own drive because deprivation and satiation effects were found within exploratory behaviour similar to those found within classic need states. For example, lengthening the interval between exposures of novel stimuli increased exploratory behaviour and it was also found that exploration decreased within the session following such an interval.

Two main modifications of **drive theory** were proposed. One argued that novel stimuli motivated exploration, i.e. a curiosity drive, while the other proposed that it was the lack of novelty that energized this behaviour, i.e. a boredom drive. Unfortunately, these explanations were in large part circular because, unlike the other primary drives, there were no independent physiological correlates for curiosity and boredom, and so these drives were both inferred from and used to explain exploratory behaviour (e.g. an animal was said to be bored only because it subsequently explored).

Difficulties arose with both formulations. It was hard to explain inquisitive exploration using the concept of a curiosity drive because here the animal is seeking novel stimuli and yet a curiosity drive is not present until after the animal has come into contact with such stimuli. While a curiosity drive could theoretically explain inspective exploration, because novel stimuli are already present to both elicit and then reduce the curiosity drive, it could not predict the empirical fact that animals would avoid 'very' novel stimuli. The boredom concept could account for inquisitive exploration as an attempt to reduce boredom by a self-produced exposure to new stimuli. Furthermore, it could also account for the direction of inspective exploration in that the most novel object is the one least resembling the boring situation and thereby the most favourable. However, like the curiosity concept, it had difficulty in explaining why extreme novelty should be **aversive**. In addition, boredom theory predicted that the longer the period an animal spent in a monotonous environment the greater would be its urge to explore, and this was found not to be the case. Under prolonged sensory restriction exploratory behaviour declined and was replaced by a preference for little or no change in environmental stimulation.

Since explanations of exploratory behaviour using either curiosity or boredom concepts alone were not successful, the remaining attempts to use drive theory involved both components. It was argued that all incoming sensory stimulation contribute to a general drive, or **arousal** level, and that for a given task there was an optimal level of arousal that the animal strived to maintain. Theories differed about the form of the relationship between sensory input and arousal, but all concluded that there is an optimal level of sensory input that the animal prefers to maintain. Thus, if the level of environmental stimulation falls below the optimum the animal will seek novelty, while if it rises above the optimum the animal will strive to withdraw from that environment. Unfortunately there was no generally accepted mechanism whereby the arousal state of an animal could be measured, let alone its discrepancy from a vaguely defined optimal level. In addition it was proposed that arousal was determined not just by the novelty of a stimulus but also by such factors as its intensity and meaning and by internal stimuli of the animal associated with primary drive states. As one researcher remarked, the arousal concept provided 'the basis for explaining everything and predicting nothing' and, as a result, it had relatively little impact on animal studies.

Some theories of exploratory behaviour have recognized its adaptive significance. The ability to adapt to unfamiliar and novel situations is a fundamental requirement for animals living in complex and stochastic environments, and this must involve a comparator process in which the current stimuli are compared to past experience. '**Cognitive model**' theories of exploratory behaviour argue that the animal builds up a neuronal representation of spatial and/or temporal relationships within its environment (i.e. a cognitive model) that it then uses to find its way around and to anticipate events. Neurophysiological studies have implicated the hippocampus in such a function. The advantage of having a model of your environment is indicated by experiments showing that familiarity with the locality greatly improves, for example, an animal's ability to escape predators. The goal of exploratory behaviour was thought to be the elimination of any discrepancy between incoming stimuli and the stimuli predicted by the cognitive model. It was proposed that the size of the discrepancy determined the form of exploratory behaviour: a small discrepancy triggered inspective exploration, while a large discrepancy resulted in withdrawal from the novel stimuli. However, it was difficult to account for inquisitive exploration on this basis because inquisitive exploration necessarily involves moving away from familiar stimuli towards novel ones that cause larger discrepancies.

The concept of a cognitive model is also a central plank of the more recent 'information primacy' theory of exploratory behaviour. Here, the term information refers to the reduction of uncertainty: thus when an animal explores a novel environment it reduces its uncertainty about the stimuli experienced and gains information, but when it moves through a completely familiar environment there is no uncertainty to be reduced and no information is gained. It is argued that information-gathering behaviour is continuously present, forming a substratum upon which other, more obviously goal-directed, behaviours occur. We have therefore moved from a position where exploration simply served the reduction of primary needs to one in which uncertainty reduction is given pride of place and need states have to be sufficiently intense to be able to break into and direct the information-gathering behaviour.

The proposal that animals have an inbuilt tendency to continually reduce environmental uncertainties is supported by neurophysiological studies of the extended lateral hypothalamic corridor. This is thought to form a 'seeking system' that: (i) spontaneously learns about uncertain environmental events that may predict resources; (ii) translates correlations between such events into perceptions of causality; and (iii) seeks information to confirm predictions about causality. Electrical stimulation in this brain area triggers very vigorous exploratory behaviour in animals, and in humans gives rise to feelings of 'intense interest'.

One prediction from the information primacy theory is that if an animal is given the choice between feeding most efficiently in a familiar situation or less efficiently on the same food in an unfamiliar situation it will choose the latter, but that this preference will decline with increasing hunger. Empirical support for this prediction comes from **contra-freeloading** experiments in which animals are given the choice between feeding freely from a dish of food or working (e.g. searching, solving puzzles, bar pressing) for the same food. When no uncertainty is associated with either food source the animals prefer to eat freely from the most profitable source, i.e. the dish. However, if uncertainty is associated with the food for which they have to work (e.g. it is camouflaged), they choose this less profitable food source rather than the dish, and this preference decreases with increasing hunger. One reason why it has generally been reported that animals forage where it is easiest to obtain the largest amount of food may be because the experimental protocols commonly used in optimal foraging experiments permit only the minimal expression of information-gathering behaviour. The subjects are deprived to a degree that ensures near maximum foraging efficiency, and the short trials are terminated before the animals have begun to vary their behaviour as a result of satiation.

Information primacy theory argues that animals do not have an optimal level of environmental stimulation that they strive to maintain, as proposed by arousal theories, but rather an optimal rate at which discrepant sensory input is processed, i.e. a preferred rate of information processing. Experiments on human attention indicate there are capacity limitations on the amount of such processing that may be conducted at any given time, and the information primacy theory proposes that the processing capacity currently available determines the preferred level of stimulus change. When this capacity is inadequate for the efficient processing of the uncertain stimuli **anxiety** accrues and the animal leaves the possibly dangerous situation. Thus inspective exploration is explained on the basis of the need to reduce the uncertainty associated with the novel stimuli present, and the avoidance of very novel stimuli by there being insufficient capacity to process the discrepant inputs. When the capacity is greater than that required, another negative state, boredom, occurs. Boredom ensures that the available processing capacity is effectively used to improve the reliability of the cognitive model by triggering inquisitive behaviour to seek, and then gather information about, unfamiliar and novel stimuli. A positive affective state exists when the level of environmental change approximates the capacity available to process the discrepant stimuli.

The previous levels of discrepant sensory input experienced in the environment are thought to determine the available processing capacity – i.e. capacity increases with time spent in a complex and stochastically varying environment and decreases in a constant simple environment. A prolonged period under sensorily restricted conditions does not, therefore, lead to a high level of exploratory behaviour, because the lack of discrepant input in that environment has resulted in a greatly reduced capacity to process novel stimuli. Thus the **welfare** of animals kept under such conditions is compromised because the processing capacity becomes so small that any environmental changes create high levels of anxiety and **stress**. (IRI)

Further reading

Archer, J. and Birke, L. (1983) *Exploration in Animals and Humans*. Van Nostrand Rheinhold, Wokingham, UK.

Berlyne, D.E. (1960) *Conflict, Arousal and Curiosity*. McGraw-Hill, New York.

Fowler, H. (1965) *Curiosity and Exploratory Behaviour*. Macmillan, New York.

Inglis, I.R. (2000) The central role of uncertainty reduction in determining behaviour. *Behaviour* 137, 1567–1599.

Extinction

The term extinction is used in several ways in biology, referring to the elimination of a species or a behavioural response in certain circumstances.

The high rate of extinction of species in geologically recent times (the last 100,000 years) has been linked to man's spread and activity, although the evolutionary record points to mass extinctions occurring periodically throughout life's evolutionary history. The current rate of extinction appears to be the highest it has been since the end of the Mesozoic era about 65 million years ago, when the Earth was subject to a number of natural catastrophes including major volcanic eruptions and meteorite impacts. Within **conservation**, the term extinction may be used to refer to the elimination of animals within a given geographic region, rather than to the Earth as a whole. The term 'extinct in the wild' is used by the International Union for Conservation of Nature and Natural Resources (IUCN) to refer to species that are only maintained in captivity, some of which it may not be possible to reintroduce due to the loss of natural habitat.

Extinction can also refer to the disappearance of a conditioned behavioural response to a conditioned stimulus (**see: Conditioning – types of**) as a result of elimination of its **reinforcement**. Extinction can be successfully employed as a behaviour modification technique (e.g. to stop a dog jumping up for attention), but often the behaviour will intensify before it starts to decline, as part of a phenomenon known as an 'extinction burst' due to **frustration** of the conditioned response. The range of conditioned behaviours treatable in this way is therefore practically quite limited. (KT)

Eye white responses

Emotional experiences, at a basic level, enable animals to react to and interact with the surrounding environment. The amount of eye white (or sclera) shown by an animal is considered to be closely linked to its current emotional status. More eye white is visible when the upper eyelid is raised, and this is thought to be orchestrated by the **sympathetic nervous system**. In general, negative **emotions** such as **fear**, **frustration** and surprise (i.e. response to unpredictability) cause the eye to widen, revealing a greater proportion of the sclera. In contrast, less eye white is visible when animals are in a more positive emotional state, performing behaviours such as **grooming** and resting.

The benefits to the animal showing increased sclera are not completely understood, but it does communicate clearly to other animals that a heightened awareness of the environment is required. This response is deemed to be innate in most animals, and learned and automated very early on in human emotional development. The brain of the observer responds to increased sclera before any conscious awareness takes place, preparing the observer for prompt evasive action. The amygdala of the **brain** is involved with the reflex response of an observer, and humans who have sustained damage to the amygdala have difficulty in recognizing a fearful expression or are unable to detect differences between the gradients of fearfulness. In studies on animals, cattle that were either startled or thwarted from performing a desirable behaviour and cows separated from their calves all showed a significant increase in the proportion of visible eye white.

The increase in visible sclera under negative emotional conditions and the decrease under positive emotional states indicates that the eye white response may be a useful and reliable indicator for the assessment of welfare. (LW)

See also: Measuring welfare

F

Facial perception and recognition

Faces are probably the most important visual stimuli in the social environment of visually dominant species, such as humans. They not only provide visual information about an individual's gender, age and familiarity, but also convey significant cues to affective state (e.g. anger, happiness), cognitive activity (e.g. concentration, **boredom**), **temperament** and **personality** (e.g. hostility, shyness). It is not surprising that the ability to recognize these cues and to respond appropriately plays an important role in the social life of higher primates – and possibly of other species as well.

Humans have extraordinary sensitivity and capability for detecting, remembering and recognizing faces. Soon after birth, infants show a preferential response to faces. Our ability to process facial information develops quickly through an experience-dependent process of gradual specialization that gives us a marked advantage in recognizing **conspecific** (or even own race) faces. Adults can differentiate the expression and identity of briefly and peripherally presented faces even when focal attention is not fully available (Grossmann and Johnson, 2007). It has been further argued that face perception and **recognition** (at least in humans) involves a face-specific cognitive and neural mechanism based on empirical findings such as:

1. Face and non-face object processing depends on different **brain** areas. A few distinct neuroanatomical regions in the cerebral cortex (i.e. fusiform gyrus) have been predominantly associated with face processing.
2. Face and non-face object processing can each function independently without the other. The selective impairments of face and object recognition (prosopagnosia and visual agnosia, respectively) have been observed in neurological patients.
3. Face and non-face object recognition process information differently. Whereas reliable object recognition can depend on individual parts (featural processing), face recognition is more holistic and relies heavily on the overall configuration in faces (configural processing). The hallmark commonly used to differentiate these two processes is the face inversion effect, in which detrimental recognition performance is more evident for inverted faces than for inverted non-face objects.

The critical role of faces in social communication is not restricted just to humans. In the **phylogeny** of primates, there is an increasing trend towards larger and more complex **social groups** in which individuals rely more on facial signals for **communication**. Like humans, non-human primates (e.g. chimpanzees and rhesus monkeys) can display, perceive and respond to sophisticated facial expressions, a crucial component for successful interpersonal relations. They are also able to perceive differences between individuals of their own species based on facial cues alone. For instance, monkeys at an early age can respond appropriately to the expressions of other individuals and recognize their faces. They can even discriminate the faces of unfamiliar individuals after only a short exposure to sets of their images. While inspecting faces, they exhibit a similar eye-scanning pattern to that of humans, allocating a disproportionate share of fixations to facial regions that provide diagnostic information for face processing (e.g. eyes for detecting identity and gaze direction). Once the faces are turned upside down, they also suffer from the face inversion effect, at least towards conspecific faces. Studies such as these provide compelling evidence that non-human primates can perceive facial identity and facial expression based on purely visual cues, even when presented as two-dimensional black-and-white face images.

The sensitivity of detecting facial signals and the capacity for differentiating individuals based on faces are spread beyond primates. With experimental protocols such as match-to-sample or visual paired comparison tasks, many behaviour studies have observed that mammals such as cattle, sheep, pandas, dogs, birds such as budgerigars, and insects such as wasps and bees are able to detect various facial cues and recognize **conspecific** faces based on visual cues alone.

For some species, their remarkable facial perception and recognition capability may be extended beyond faces of their own species. The faces of other species they can process, however, are subject to the animal's exposure and differential experience towards these species and their behavioural significance. For instance, some captive non-human primates (i.e. chimpanzees and monkeys) can respond to human facial cues such as gaze direction, discriminate individual human faces and demonstrate the human face inversion effect. Domestic dogs also exhibit this phenomenon, despite their greater genetic distance from humans compared with that from other primates. It has been argued that the ability to extract information from human faces and respond appropriately may have had a selective advantage during the process of **domestication**. Consequently, domestic dogs may have evolved some human-like strategies for face processing. This hypothesis has been supported by empirical observations, such as the finding that domestic dogs can use internal representation of their owner's face to correlate facial and vocal signals; they can use faces to differentiate human individuals, and their sensitivity to human directional communicative cues exceeds that of non-human primates in certain tasks, such as following human gaze direction to find hidden food (see Hare and Tomasello, 2005).

Overall, sensitivity to faces can be found in many species. This sensitivity seems to be inherited or developed at an early age, and could be shaped by the animal's experience according to its adaptive behavioural significance. It is unclear, however, whether a similar cognitive or neural mechanism in face processing is shared by these diverse species. (KG)

References and further reading

Farah, M.J., Wilson, K.D., Drain, M. and Tanaka, J.N. (1998) What is 'special' about face perception? *Psychological Review* 105, 482–498.

Ghazanfar, A.A. and Santos, L.R. (2004) Primate brains in the wild: the sensory bases for social interactions. *Nature Reviews Neuroscience* 5, 603–616.

Grossmann, T. and Johnson, M.H. (2007) The development of the social brain in infancy. *European Journal of Neuroscience* 25, 909–919.

Hare, B. and Tomasello, M. (2005) Human-like social skills in dogs. *Trends in Cognitive Sciences* 9, 439–444.

Kanwisher, N. (2000) Domain specificity in face perception. *Nature Neuroscience* 3, 759–763.

McKone, E., Kanwisher, N. and Duchaine, B.C. (2006) Can generic expertise explain special processing for faces? *Trends in Cognitive Sciences* 11, 8–15.

Tarr, M.J. and Cheng, Y.D. (2003) Learning to see faces and objects. *Trends in Cognitive Sciences* 7, 23–30.

Valentine, T. (1988) Upside-down faces: a review of the effects of inversion upon face recognition. *British Journal of Psychology* 79, 471–491.

FAP/FMP (Fixed action pattern/Fixed motor pattern)

Fixed action patterns (FAPs) are unlearned (**innate**), repetitive movements produced by a neural network known as the **innate releasing mechanism (IRM)** in response to external stimuli (sign stimuli/releasing stimuli) that are relatively unaffected by **feedback** from the stimuli and are species-specific. Their relative lack of dependence on environmental feedback means that such behaviours, which often relate directly to survival or **fitness**, can be performed effectively the first time the animal attempts them. Some well-known examples of FAPs include egg retrieval in the greylag goose, **courtship** displays in male sticklebacks and **grooming** sequences in mice.

The term 'fixed action pattern' has been criticized for its implied rigidity of behaviour patterns, leaving little room for individual variation or even variation in different bouts of the behaviour. While some behaviour patterns do have this quality, others show more variability in their performance, and thus the term **modal action pattern (MAP)** was devised to account for less rigid but still predictable behaviour patterns.

Many consummatory behaviour patterns can also be classified as FAPs, such as eating, drinking or copulatory behaviour. When an animal has **motivation** to perform a behaviour pattern, levels of **appetitive behaviour** are increased to attempt to achieve the consummatory behaviour. However, with the **intensification of animal production**, a number of consummatory behaviours are not possible or their performance is restricted. Thus, it has been theorized that some forms of **stereotypies** derive from the inability to perform these FAPs and satisfy their underlying motivations. (KT, LMD)

See also: Instinctive behaviour

Further reading

Dixon, L.M., Duncan, I.J.H. and Mason, G.J. (2008) What's in a peck? Using 'fixed action pattern' morphology to investigate the motivational basis of feather pecking in laying hens. *Animal Behaviour* 76, 1035–1042.

Lorenz, K. and Tinbergen, N. (1938) Taxis and instinctive behaviour pattern in egg-rolling by the greylag goose. In: Lorenz, K. (ed.) *Studies in Animal and Human Behaviour*, Vol.1. Harvard University Press, Cambridge, Massachusetts, pp. 316–350 [translated by R. Martin].

Tinbergen, N. (1951) *The Study of Instinct*. Oxford University Press, Oxford, UK.

Farm Animal Welfare Council

The Farm Animal Welfare Council (FAWC), created by an Act of Parliament in the UK in 1979, is an independent committee that gives advice to the government and produces reports available for the public, including the agriculture industry. Its terms of reference are to keep under review the **welfare** of farm animals on agricultural land, at market, in transit and at the place of **slaughter**, and to advise the government of any legislative or other changes that may be necessary. FAWC statements on attitudes and practicable welfare policy have had influence in many countries. (DMB)

Farm assurance scheme

How does the scheme work?

Quality assurance can be provided for any product that has been certified to meet defined standards. The term 'farm assurance' may be applied to the quality assurance of most farm products, but in this article will refer specifically to the assurance of products with a farm animal origin, from the birth of the animals, through rearing and **transport**, up until the point of **slaughter**.

Farm assurance provides guarantees for consumers on a range of areas such as food safety, product traceability, environmental impact and animal **welfare**. This is provided through possession of a defined set of standards, owned by a farm assurance scheme, which are formally checked (audited) on a farm by a farm assurance assessor, who is, ideally, independent of the scheme. The various schemes may assure to the minimum level of legislation compliance or have standards that go far beyond this. Schemes may also have different ethical emphases: for example, organic standards provide specific assurances about environmental issues and, in the UK, the **RSPCA**'s Freedom Food Scheme has an animal welfare focus.

Assurance schemes for farms were developed from the 1980s onwards as a voluntary private-industry initiative to differentiate members' products within the marketplace. As retailers have come under increasing pressure to demonstrate food safety through traceability, and subsequently other forms of product provenance such as animal welfare, farm assurance has grown.

The first critical component of a useful farm assurance scheme is a set of meaningful, well-defined standards with which auditors are able to assess members' compliance. Animal welfare standards have typically been based upon some of the 'inputs' that affect an animal's welfare, such as the **housing**, feed and **stocking density**; however, other inputs such as genetic make-up and **stockperson** ability have been less easy when defining standards. In order to better evaluate the effect of these inputs on the welfare of farm assured animals there have been recommendations from organizations, including the Farm Animal Welfare Council (**FAWC**) and Compassion in World Farming, to include some 'outcome' measures of animal welfare (direct assessments of animal welfare; **see also: Welfare measurement**) such as physical **health** and behaviour. This is being investigated by a number of farm assurance schemes and, in a separate initiative, the European Union (EU) has funded a large pan-European project to develop welfare outcome measures for cattle, pigs and poultry, termed 'Welfare Quality'.

There has been little research directly considering whether belonging to a farm assurance scheme results in an overall improvement in animal welfare; many farm assurance schemes aim to be almost universally available to farmers by setting welfare standards at, or marginally above, the legal minima with which they must, in any case, comply, and are therefore not likely to produce significantly improved welfare. One study to evaluate the effect of belonging to a high-welfare, but input-based, farm assurance scheme on the welfare of dairy cows showed an equivocal effect, with some welfare outcome measures, such as **mastitis**, being better on the assured farms whereas others, such as **lameness**, were worse. However, a review of the welfare of animals on organic farms concluded that, as long as animal health can be maintained, **organic farming** systems are in a better position to provide good animal welfare compared with conventional systems. With the increasing use by farm assurance schemes of appropriate standards relating to animal welfare to accurately assess and monitor improvements in animal welfare, there will come the potential to improve animal welfare over time for members.

The second critical component of a useful farm assurance scheme is accurate auditing of welfare standards by farm assurance assessors. A thorough routine inspection may be able to identify farms that comply with standards on any given inspection day, but a better approach is to provide assurance that there are systems in place on a farm to ensure continuing compliance with the standards beyond the inspection day. Whether welfare inputs or outcomes are being assessed, there are several important auditing considerations. The assessors must be standardized in the interpretation of, and judgement of compliance with, the standards. This process should be ongoing beyond the initial training, for example by having multiple assessors at an assessment, or assessing photographic/video footage of a farm. The timing of the assessment and the interval between assessments need to be considered, and may vary depending on species. For example, to understand the range of conditions on a farm and welfare the animals may experience, it may be useful to visit a farm at various parts of its cycle. For a beef farm this might be at different times of the year, but for a broiler chicken unit this would be over the rearing period of a few weeks. If benchmarking is desirable then all farms would need to be visited at a comparable time.

Who wants farm assurance?

Farm assurance may be desirable or required by some or all parts of the food market, from processors to retailers and consumers. For processors and retailers farm assurance can provide evidence, when required, of due diligence in ensuring that appropriate product standards, particularly relating to food safety, have been met. The decision about whether and how to pass this information on to the general consumer depends upon the market advantage of doing so. There may be no visible evidence to a consumer, but information is available on request or general claims may be made, such as 'all our meat is farm assured'. More specifically, farm assurance logos may accompany products and, at times, this may be accompanied by more detailed information such as 'from sheep that graze grass all year round'. In many countries the market has responded to an increasing desire by consumers for knowledge of the welfare provenance of animal products, and consequently for higher-welfare products themselves.

The World Trade Organization (WTO) does not currently allow trade restrictions on products solely on the basis of animal welfare (**see also: International regulation of animal welfare**). If the desire by consumers for welfare-differentiated products continues to increase, it may be that the WTO accepts that appropriately certified minimum welfare standards could be a condition attached to international trade to benefit consumers rather than to protect producers. An alternative scenario if the current WTO rules continue is that higher-welfare products are denied the opportunity to differentiate themselves on the global market, thus reducing the incentive for their production. Some international trade in farm assured products exists at present, where a specific customer requires compliance with farm assurance standards. For example, large EU retailers buying chicken from Thailand or Brazil may insist upon compliance with equivalent farm assurance standards by their national suppliers. (SM)

Further reading

FAWC (2005) *Report on the Welfare Implications of Farm Assurance Schemes*. Farm Animal Welfare Council, London.

Lymbery, P. (2002) *Farm Assurance Schemes and Animal Welfare – Can we Trust Them?* Compassion in World Farming Trust, Godalming, UK.

Main, D.C.J., Whay, H.R., Green, L.E. and Webster, A.J.F. (2003) Effect of the RSPCA Freedom Food scheme on the welfare of dairy cattle. *Veterinary Record* 153(8), 227–231.

Spoolder, H.A.M. (2005) Animal welfare in organic farming systems. In: *Organic Farming Food and Health Congress of the Soil Association*. John Wiley & Sons, Ltd, Newcastle, UK, pp. 2741–2746.

Farmed animals

Farmed animals are those animals whose production is used to benefit humans who, in return, claim ownership, provide protection, decide on matings and take responsibility for food supply. A spectrum exists of human–animal relationships of the farming kind. Animals that are kept in **confinement** or enclosure throughout their lives and whose **reproduction** is controlled are very obviously farmed. Examples are familiar farm livestock species, bees and silkworms. Animals that are free ranging and only gathered periodically for individuals or products to be harvested are also farmed, but in a less intensive manner (**see: Free-range animals**). A term like 'ranching' would be more appropriate for this activity. **Zoo animals**, being undomesticated, are not regarded as being farmed.

People breeding animals for sale as actual or potential breeding stock would generally describe themselves as breeders rather than farmers.

Many attempts have been made to add to the number of species that are farmed. **Wild animals** can be farmed, and this process involves the confinement of animals that have not been selected for their tolerance of husbandry. This kind of farming can pose serious **welfare** issues, but the high value of many of the products (such as **bear** bile, musk and gourmet meats) provides a strong financial inducement.

Small animals such as **rabbits** and **guinea pigs** have been popular with rural development specialists seeking to establish new income-generating activities for rural people in the developing world. In the developed world, **game animal** farming (mainly of red **deer** and fallow deer) is a well-established alternative farm enterprise. Farming of fish and shellfish (molluscs and crustaceans) is locally very important. The longest-established aquaculture systems are probably those of China, where ponds are stocked with complementary species of carp, and the system is integrated with the farming of crops, **pigs** and ducks. More than 220 aquatic species are now farmed and, in 1997, 29 million t of fish supplies came from this source, while the catch of wild fish was 95 million t (Naylor *et al.*, 2000).

The use of farm animals has changed radically over the last 200 years as industrial societies have developed. In the developed world today, farm animals are primarily producers of commodities (meat, milk, eggs and, to a much lesser extent, wool). As such, emphasis is on the production of young animals for **slaughter** or for an unnaturally brief life as an egg or milk producer. Previously, the main value of livestock was in the form of the adult animal, yielding products and services such as milk, eggs, wool, manure and labour during as long a life as possible. Also, herds and flocks were smaller, there were more intense and individual relationships between the animals and their carers, and farm animals in general were much more integrated into human society.

These changes are particularly clear in the case of **chickens**. Before the industrialization of farming, these were essentially farmyard scavengers, though certain aspects of their biology have proved fundamental to the development of modern poultry industries (Hall, 2004). In a sense, they have undergone a second, technology-led, form of **domestication**. They are, at least physically, very adaptable to novel feeding and housing conditions and incubation can be mechanized, while their reproductive attributes suit them for intense selective breeding. They have high fecundity, a short generation interval, eggs that remain fertile in storage and sex-linked plumage characteristics, all of which promote rapid response to selection. Sexual maturation and egg laying respond to artificial light, so the **seasonality** of **egg production** can be abolished.

On a global scale, farm animal production is changing in complex ways (Owen *et al.*, 2004). In the developed world, production is growing only slowly, and many markets are becoming much more concerned with issues of product quality, especially in relation to animal welfare and food safety. In the developing world, as incomes rise demand is increasing. It is predicted that in 2020 China will consume one-third and India one-fifth of global milk production. Thus we can expect to see a great intensification of animal production in the developing world. At the same time, the problem of rural poverty in the developing world demands solutions. Farm animals would be expected to play a big part here, but local production can easily be threatened by the cheap imports made possible by a globalized market system. Certainly, farm animals are vital for draught power in the developing world, and they have much to contribute to rural development, provided the benefits reach the poor. World changes in livestock numbers are illustrated in Table F.1.

Assuring consistency between censuses is never easy, but some comparisons can be made (Hall, 2004). The most dramatic changes have been in the avian species. Increases in **goats** have mainly been seen in China and India and, as global trade in goats is small, the increases have been in production for local consumption. Pig numbers have increased, while international trade has increased at a greater rate. The rise in **cattle** numbers reflects increasing demand for milk, at least in the developing world, while in developed countries the increase in milk supplies has come from increased yields. **Sheep** generally use land that could support crops, and their decline presumably reflects competition for this resource, though patterns of change differ greatly between countries. **Horse** numbers have increased in richer countries while declining in the less wealthy. Some developing-world countries are particularly dependent economically on livestock exports, and these might be especially vulnerable to globalization (see Table F.2).

Some wealthy countries are big exporters of live animals, such as France (US\$1.6 × 10^9 in 1999) and Canada (US\$1.0 × 10^9) (CAB International, 2002). All these patterns of trade can change rapidly in response to exchange rate fluctuations, disease outbreaks and market developments generally.

Livestock systems are often closely integrated with the rest of agriculture. Before artificial fertilizers became widely used, manuring by livestock was vital for adequate crop yields, and crops and animals were farmed in rotation systems of various degrees of sophistication. In much of the world this no longer

Table F.1. World livestock populations, 1992–2001 (× 10^6) (from CAB International, 2002).

Species	Population (1992)	Population (2001)	Percentage change
Donkey	43.7	42.0	−3.89
Water buffalo	153.1	167.6	+9.50
Cattle	1,302.8	1,360.1	−4.41
Goat	597.6	693.5	+16.00
Horse	60.3	58.3	−3.40
Pig	869.6	927.7	+6.68
Sheep	1,161.3	1,059.1	−9.80
Camel	18.2	19.1	+5.17
Other camelids	5.41	6.20	+14.50
Mule	15.1	13.5	−10.40
Rabbit	519.7	481.6	−7.33
Rodents	22.9	14.9	−34.90
Chicken	11,500.0	14,700.0	+27.30
Duck	652.1	915.4	+40.40
Turkey	223.4	241.4	+8.04
Geese	163.5	240.6	+47.20
Beehives	57.9	58.9	+1.68

Table F.2. Countries with the highest dependence on live animal exports, based on FAO data from 1999 (from CAB International, 2002).

Country	Total merchandise exports (US$ × 10^6)	Live animal exports (US$ × 10^6)	Live animal exports/total merchandise exports (%)
Somalia	150	62	41
Chad	211	33	16
Sudan	596	87	15
Mali	665	95	14
Niger	276	29	11
Burkina Faso	255	20	8
Mauritania	448	33	7
Namibia	1246	83	7

applies, and now crop and animal systems are physically separated, though animal industries are often vertically integrated so production of feedstuffs is contractually linked with production of animals for slaughter. However, crop husbandry leads to large quantities of crop residues (world annual crop production has been estimated at about 2750 × 10^6 t, and crop residues 3750 × 10^6 t; Smil, 1999), much of it ideal for feeding to ruminant livestock. Technical innovations to ensure this material is fully exploited are highly desirable.

Farming of animals in high density always poses environmental challenges, whether it be the engendering of new diseases, the accumulation of veterinary drug residues in people or the environment, the disposal of large quantities of dung and other wastes or the destruction of natural habitats to provide land for production of animal feeds. But this is not new. In many smallholder systems, livestock are so intimately associated with human habitation that diseases arising in one species can easily pass to another; this is the source of novel influenza viruses.

Generally, livestock production is the responsibility of ministries of agriculture, with environment and public health being overseen by other ministries, while non-governmental organizations may be the principal ambassadors for animal welfare (Wickens, 2001) and for the conservation of farm animal genetic resources (Hall, 2004). Whether this system of governance, which is often adversarial, is in the best interests of society as a whole is debatable. Similarly, agriculture is not usually held accountable for the environmental damage it does, nor is it often recompensed for the environmental protection and **conservation** it frequently provides. In the developed world these issues are beginning to be addressed. Recent rates of subsidy for agricultural activities (US$82,500/km^2/year in the European Union (EU); James *et al.*, 1999) are unlikely to remain so high for much longer (in the USA the corresponding figure is US$16,100). Also, with recent developments in the EU agricultural regime, the relevance and emphasis of extensively husbanded livestock is being shifted towards maintenance of landscapes, and this will lead to extensification. However, the bulk of livestock production will come from intensive farming, though niche markets for more traditional products will develop further.

The diversity of livestock **breeds** provides the genetic resources necessary for sustainable development of livestock husbandry, and it is very important that this biodiversity, which is recognized by the 1993 Convention on Biological Diversity, should be conserved and characterized.

Intensive livestock production is often, perhaps almost always, inimical to animal welfare as it is conceived today, especially when this is assessed in terms of the need of the animal to practise normal behaviours that are important to it. While good welfare can add to the value of animal products in some markets, for the vast majority of consumers and opinion leaders the welfare of farm animals is of secondary importance to economic considerations. Meat quality, for example, can be demonstrably better in animals kept and transported in good conditions, but the growing importance of food processing means that the economic incentive for enhanced animal welfare is diminishing.

Classical animal breeding, based on performance recording and **selection** (e.g. Simm, 1998; Hall, 2004) can be extremely rapid in achieving goals, and gene technology is opening up new possibilities (**see: Transgenics**). Societies are more willing to accept novel technologies when the result is new medical treatments or products, and are generally less so when the outcome is increased profit from greater yields of milk, meat, eggs, etc. There is much scope for bioethicists to lead debate on innovations in animal breeding (Mepham, 2005).

The farming and keeping of animals under domestication has led to a new kind of interspecies relationship, to add to the familiar catalogue of **predation**, **parasitism**, symbiosis, etc. Artificial **selection** has replaced **sexual selection** and, to a great extent, natural selection as a determinant of the **genotype** of animals. Although under domestication there have been very extensive genetic changes in the skeleton, patterns of fat and muscle deposition, organ size, reproductive characteristics and certain aspects of behaviour, some aspects of the phenotype have remained relatively unchanged. For example, some behaviours such as nest building in sows and the motivation of **mink** to swim remain prominent in farmed animals, in spite of long histories of husbandry.

The science of applied **ethology** (Jensen, 2002) was developed in order to understand the behaviour of farm and other captive animals. In parallel with insights from veterinary medicine it has enabled welfare to be assessed objectively and production systems improved. **Anthrozoology**, the study of human–animal interactions, is a growing area that also contributes to farm animal welfare, as well as to the understanding of the requirements of **companion animals**. (SJGH)

References

CAB International (2002) *Animal Health and Production Compendium.* CAB International, Wallingford, UK.

Hall, S.J.G. (2004) *Livestock Biodiversity.* Blackwell Publishing, Oxford, UK.

James, A.N., Gaston, K.J. and Balmford, A. (1999) Balancing the Earth's accounts. *Nature* 401, 323–324.

Jensen, P. (ed.) (2002) *The Ethology of Domestic Animals: an Introductory Text.* CAB International, Wallingford, UK.

Mepham, B. (2005) *Bioethics: an Introduction for the Biosciences.* Oxford University Press, Oxford, UK.

Naylor, R.L., Goldburg, R.J., Primavera, J.H., Kautsky, N., Beveridge, M.C.M., Clay, J., Folke, C., Lubchenco, J., Mooney, H. and Troell, M. (2000) Effect of aquaculture on world fish supplies. *Nature* 405, 1017–1024.

Owen, E., Smith, T., Steele, M.A., Anderson, S., Duncan, A.J., Herrero, M., Leaver, J.D., Reynolds, C.K., Richards, J.I. and Ku-Vera, J.C. (eds) (2004) *Responding to the Livestock Revolution. The Role of Globalisation and Implications for Poverty Alleviation.* BSAS Publication No. 33, Nottingham University Press/British Society for Animal Science, Nottingham, UK.

Simm, G. (1998) *Genetic Improvement of Cattle and Sheep.* Farming Press, Ipswich, UK.

Smil, V. (1999) Crop residues: agriculture's largest harvest. *Bioscience* 49, 299–308.

Wickens, S.M. (ed.) (2001) *Science in the Service of Animal Welfare. A Chronicle of Seventy-Five Years of UFAW.* Universities Federation for Animal Welfare, Wheathampstead, UK.

Farming of animals

Animal **domestication** can be viewed as a process by which captive animals adapt to humans and the environment that they provide for the animals. Since domestication implies change, it is expected that the **phenotype** (observed characteristics of the individual as they have developed under the combined influence of genes and experiences) of the domesticated animal will differ from the phenotype of its wild counterparts. Adaptation to the captive environment is achieved through genetic changes occurring over generations, and environmental stimulation and experiences during an animal's lifetime. Thus domestication can be viewed as both an evolutionary process and a developmental phenomenon.

The environments of modern farms may differ markedly from the natural habitats of the ancestors of today's farm animals. Furthermore, today's farming systems are highly variable. The least restrictive of these is rangeland systems, in which animals are generally confined only by fences and have limited human contact, whereas the most restrictive systems house animals individually or in dense groups indoors, at times with bouts of intense human contact. Within all systems, there can be a range of conditions provided that markedly impact on the behaviour, productivity, **health** and **welfare** of the animals, ranging from the simple provision of water and food for rangeland animals to highly sophisticated environmentally controlled sheds where temperature, humidity and light may be tightly controlled, as well as food and the social environment.

The desire by consumers and governments in Western society in the post-World War II period for cheap and safe food has driven improvements in housing and production methods to increase animal productivity, improve the quality of food and lower the cost of food. However, there appears to be growing public concern about livestock production in terms of the effects on human health and safety, environmental effects, **intensification of animal production**, welfare of livestock in general and genetically modified foods and feeds for livestock. It is likely that these strong and, at times opposing, economic, health, environmental and ethical forces will continue to drive changes in livestock production. The need to produce cheap and safe animal products will remain, but a range of social issues will shape livestock production.

History of farm animal domestication

An important behavioural trait that influenced the domestication of ungulate and galliform species was their social organization: these species could live in relatively large groups without marked year-round **territoriality**. The formation of the social **hierarchy** or social organization in many of these species is considered to be an important mechanism that functions in controlling **aggression** and so allowing these animals to be kept in large groups with fewer chances of injuring one another or expending large amounts of energy competing for resources, such as mates or feed, that may be limited. This trait has been further emphasized through artificial **selection** during domestication. Other traits that have favoured domestication – and, in particular, livestock production – include: (i) promiscuous **mating behaviour**, in which males can mate with large numbers of females; (ii) early **bonding** between dam and offspring, which allows the offspring to remain in the **social group**; (iii) **precocial** young, which allows the young to join the social group early; and (iv) high **adaptability**, which allows these animals to adapt to a range of environments, and thus they do not require highly specialized requirements.

Wild **sheep** (*Ovis orientalis*) and **goats** (*Capra agagrus*) appear to have been domesticated at about the same time in the Near East – around 11,000 YBP – while remains of domesticated **cattle** (*Bos primigenius*) and **pigs** (*Sus scrofa*) are first seen in archaelogical records around 8000–9000 YBP at various sites in Asia. **Horses** (*Equus feus*), asses (*Equus africanus*), **camelids** (*Camelus* spp.), water **buffalo** (*Bubalus bubalis*) and **chickens** (*Gallus gallus*) all appear to have been domesticated in different parts of Asia and North Africa about 5000–7000 YBP. In the New World, llamas and alpacas (*Lama* spp.), **guinea pigs** (*Cavia* spp.) and **turkeys** (*Meleagris gallopavo*) underwent a similar process at various times and locations.

Some authors have suggested that it is unlikely that Paleolithic and Neolithic people consciously domesticated animals for specific economic or practical purposes. Domestication in the early stages was therefore perhaps largely an unconscious process on the part of humans, in which tame or semi-tame wild animals were gradually brought under increasing levels of human control. Furthermore, different species may have been domesticated in different ways. For example, species that could be herded, such as the wild ancestors of domestic sheep, goats and llamas, would have been followed and hunted by these nomadic human groups long before people began to play an active role in guiding the movements of the animals, protecting them from predators or interfering selectively in their **reproduction**. Large and potentially dangerous animals, such as wild cattle, may have been coaxed gradually into semi-dependent relationships with humans by the provision of salt licks close to villages. Other species such as **dogs**, pigs and poultry may have been captured or collected when young and then kept primarily as **pets**.

The relationship between humans and domesticated animals is sometimes viewed as a successful '**adaptive** strategy' in an evolutionary sense, a kind of symbiosis in which both humans and the animals have benefited. The notion is often used to justify the use of domestic animals by humans, and is supported by the fact that most domestic animals are more numerous nowadays than their wild ancestors, some of which are extinct. While the argument may be most applicable for **companion animal** species (such as the dog and **cat**), the notion of symbiosis is less persuasive in relation to farmed food animals. Nevertheless, the relationship between humans and animals in modern agriculture has components of symbiosis in that, in return for the animal products or results that they provide for humans, the animals are maintained and cared for by humans, although the appropriate standards of welfare that should be afforded to these animals is contentious.

Domestication process

Some authors have noted that domestication is a process rather than an event, and that it is difficult to define the point at which a tame or captive wild animal can be classed as domesticated. In general, truly domestic animals exhibit some obvious genetic divergence from the ancestral 'wild type', due to the effects of artificial selection over many generations.

The main genetic changes influencing the development of the domestic **phenotype** result from artificial selection, natural selection in captivity and relaxed selection. Artificial selection has generally focused on productivity characteristics, and there are numerous examples of remarkable gains in many production traits in farm animals in recent decades. Growth rate of broiler chickens and pigs and milk yield of dairy cows have all dramatically increased over recent decades through artificial selection. Several specific behavioural characteristics have also been selected. Selection of silver foxes for non-aggressive behaviour towards man commenced in the late 1940s in Siberia. As a result of this selection, foxes can now be bred that show few defensive responses to people, presumably because they are less fearful. They consequently resemble domestic dogs in their behaviour, with the unselected control population continuing to exhibit wild-type behaviour, including strong defensive responses towards humans.

All of the selection imposed on captive populations that cannot be ascribed to artificial selection must be 'natural'. Animals selected to be parents of the next generation may not always reproduce and, if they do, they do not always provide the theoretically expected number of offspring. Reproductive failure may result from physiological or psychological **stress** caused by such factors as sensory and locomotor deprivation, social incompatibility, dietary deficiencies and **parasitism**. Thus natural selection has also operated on farm animals.

In many cases, relaxed selection can be expected to accompany the transition from the wild to **captivity**. Certain behaviours important for survival in nature (e.g. food finding and **predator avoidance**) may lose much of their **adaptive** significance in captivity. As a result, both genetic and phenotypic variability for such traits are likely to increase. In comparison with beef cows, dairy cows that typically have their calves removed at birth exhibit less intense **maternal behaviour** – less licking of newly born calves and less stationary behaviour during teat seeking by the calf are apparent in dairy cows. There are some very obvious examples in farm animals where domestication has resulted in the deterioration in some **fitness** characteristics. Artificial selection for large breast size in domestic turkeys has nearly eliminated the ability of males to copulate naturally, and females require **artificial insemination**. Artificial selection of laying hens for non-broody behaviour has resulted in strains of chickens that normally do not incubate eggs or brood chicks.

Although there are many reported differences between wild and domestic stocks, there is little evidence that domestication has resulted in the loss of behaviours from the species repertoire or that the basic structure of the motor patterns for such behaviours has been changed. In nearly all cases, behavioural differences between wild and domestic stocks are quantitative in character and best explained by differences in response thresholds. These comparisons are difficult because of problems in both determining an appropriate wild population and interpreting differences between wild and domestic populations under one environment, in nature or in captivity. However, studies of farmed and wild Atlantic salmon (*Salmo salar*), both reared in either captive or wild environments, indicate that farmed salmon show fewer predator responses, while also showing increased growth, increased disease resistance and decreased stress responses.

The absence of certain 'key' stimuli in the physical environment of farm animals can influence adult behaviour. For example, social experience during rearing markedly affects the development of the **sexual behaviour** of boars, while heterosexual sexual experience influences the sexual behaviour of rams.

As discussed earlier, an important behavioural trait that influenced the domestication of farm animals was their social organization. The formation of social hierarchies or social organization in many of these species allows these animals to be kept in large groups with fewer chances of injuring one another or expending large amounts of energy competing for resources, such as mates or feed, that may be limited. This trait has also been further accentuated through artificial selection.

However, research on the effects of domestication using, for example, domestic **rodents** as a model, suggests that while the response thresholds for **agonistic behaviour** may be readily modified by early experience, response thresholds for defensive agonistic behaviour have been increased during domestication. It was a common belief for many years that the aggressive behaviour of the Norway rat (*Rattus norvegicus*) had been so reduced by domestication that what fighting behaviour remained was relatively immature and seldom resulted in physical harm to opponents. However, under certain rearing and stimulus conditions, the attack and other offensive aggressive behaviours of domestic Norway rats can be just as intense as those of their wild counterparts. It is the defensive behaviours of Norway rats that have become attenuated by the domestication process, perhaps as a correlated effect of selection for ease of handling.

Unless wild animals have been habituated to the presence of humans, they tend to avoid close contact with humans. In captivity, the capacity to adapt to the presence of people and frequent handling is an important fitness-determining factor.

The two types of learning, conditioned **aversion** responses and **habituation** to humans, occurring both early and subsequently in life, are probably the most influential factors affecting the behavioural responses of farm animals to humans. Through **conditioning**, the **fear** responses of farm animals to humans are regulated by the nature of the experiences occurring around the time of interactions with humans. Thus farm animals in situations in which they frequently interact with humans may, through conditioning, associate humans with rewarding and punishing events that occur at the time of these interactions, and thus conditioned fear responses to humans may develop. Reduced fear of humans may also be achieved by positive associative conditioning, in which the animal's fear of humans is reduced by the latter's role as a secondary reinforcer. As providers of such necessities as food, water, shelter and **grooming**, humans become secondarily associated with such positive stimuli and the threshold for avoidance behaviours is raised. Over time, young domesticated animals that may have had limited experience with humans may habituate to the presence of humans and thus may perceive humans as part of the environment without any particular significance. Habituation will occur over time as the animal's fear of humans is gradually reduced by repeated exposure to humans in a neutral context; that is, the human's presence has neither rewarding nor punishing elements. Even wild strains of rats and **deer** that are highly fearful of humans will habituate to humans over time.

The presence or absence of certain stimuli, intraspecific aggressiveness and interactions with humans can have profound effects on the development of specific traits in farm animals. It appears that these factors, together with genetic changes accompanying domestication, can also contribute to more general traits in farm animals. For example, it has been argued by some that the single most important effect of domestication on behaviour is reduced responsiveness (i.e. sensitivity) to environmental change. This characteristic is observed in most populations of farm animals and encompasses a wide variety of behavioural traits such as the response to novel stimuli, intraspecific interactions and response to the presence of humans. Reduced responsiveness to environmental change can be seen as an adaptation to living in a biologically 'safe' environment in which there are limited opportunities for perceptual and locomotor stimulation, frequent invasions of personal space, with little opportunity to escape from dominant **conspecifics**, and frequent association with humans, who are prone to cull individuals that are difficult to inspect and handle.

Failure to adapt

The process of farm animal domestication implies a transition from nature to captivity. However, 'nature' and 'captivity' are only extremes on a continuum. Free-living wild populations are found in environments ranging from wilderness areas to semi-wild preserves where animals are protected or managed and supported to some degree. Captive populations are found in environments ranging from small laboratory cages to complex **zoo** enclosures to large breeding parks or corrals.

Farm animals are sometimes provided with an environment that is physically similar to the habitat of their wild ancestors. Behavioural and physiological adaptations to such an environment will be readily achieved. Very often, however, the captive environment does not match the ancestral environment and **adaptability** is challenged. Welfare problems and indeed productivity problems may arise, although behavioural and physiological responses are earlier and more sensitive indicators of adaptation to the environment than productivity.

Impact of domestication

Several authors have emphasized that domestic animals are not inferior or degenerate animals in comparison with wild animals – they are simply adapted to a different environment than that of their wild counterparts. Behaviours that may be considered to have been lost through domestication are usually expressed when the right key stimuli or experiences are provided. In most cases, behavioural differences between wild and domestic stocks are quantitative in character and best explained by response thresholds. Once a species has become adapted to its captive environment, reproductive and growth rates are typically enhanced through improved nutrition, health and climate, reduced energy expenditure and, in some cases, reduced stress in captivity. (PHH)

See also: Organic farming

Further reading

Price, E.O. (2002) *Animal Domestication and Behavior*. CAB International, Wallingford, UK.

Serpell, J. (1986) *In the Company of Animals: a Study of Human–Animal Relationships*. Blackwell, Oxford, UK.

Stricklin, W.R. and Mench, J.A. (1987) Social organisation. In: Price, E.O. (ed.) *Farm Animal Behavior*. Maple-Vail Book Manufacturing Group, York, Pennsylvania, pp. 307–322.

Farrowing crate

For hundreds of years, farrowing accommodation for domestic sows remained much the same: a bedded indoor or outdoor pen. However, with the intensification of animal production in the 1940s, the concept of the farrowing crate was developed, its aim being to reduce the amount of space within which the sow and her litter could be kept, while reducing piglet deaths from overlying (crushing), a major source of pre-weaning mortality. In the 1940s and 1950s the sow occupied the crate for a few days before and after **parturition**. By the 1960s crates had increased in popularity, and were used to house the sow and litter for the entire lactation. Initially, farrowing crates were constructed of wood, which wore badly and was difficult to clean. Subsequently a variety of materials was tried, including bricks and concrete, before the advent of the metal crate.

Currently, a typical farrowing crate is a rectangular metal structure, a little larger than an 'average' sow's body size. Thus, the crate may be too short or small for large, old sows (which continue to grow during their reproductive life), while too roomy for gilts and smaller sows, which may be able to turn around inside. Each crate is situated within a pen, and farrowing rooms will commonly contain multiple pens. Piglets can move freely through both the sow's crate area and the remainder of the pen (which will sometimes contain a heated and/or bedded 'creep' area), while the sow is fixed in one place. The pen floor may be slatted or perforated, partly slatted or solid, and will sometimes contain bedding such as straw. A feed trough and usually a separate drinker will be incorporated for the sow. Sows are generally moved to the crate from **gestation** accommodation a few days before they are due to farrow, and remain there until the piglets are weaned, often at about 3–4 weeks of age, although crates may be used for a shorter period (e.g. until only a few days after parturition). Piglets are most vulnerable to overlying and other causes of death during their first 2–3 days of life.

There are many variants on this 'typical' crate design: its sides may be straight, sloped or skirted, its top may be flat or peaked and its bottom bars may be straight or 'fingered'. Hydraulics (which allow the crate sides to move outwards as the sow lies down, providing more space) and floating bars may be included, as well as additions such as the 'blowaway', which emits a puff of air over the piglets as the sow lies down, in theory to encourage them to move out of the danger area beneath their mother's body. In practice, few of these design variants are based on research, but have arisen due to manufacturers' and farmers' guesswork or hunches.

The farrowing crate is undoubtedly a helpful management tool, especially in larger **pig** units. It confines the sow in one place, allowing piglets to be easily handled without risk of worker injury by a protective sow. In low-labour systems, crates are often combined with slatted floors and no bedding, reducing the need for cleaning.

Crates do appear to have some beneficial effects on piglet mortality. In the UK, pre-weaning mortality levels averaged 25% in 1966, but had dropped to around 10–15% by the mid-1990s, largely due to the widespread use of farrowing crates, although a corresponding decrease in weaning age over this time period must also be taken into account. However, movement restriction and lack of exercise may impede the progress of parturition, leading to increased numbers of stillbirths. It has also been suggested that crate use changes the nature of piglet crushing, with fewer piglets becoming trapped under their mother's rear end but more beneath her chest.

The effects of farrowing crates on sows' behaviour and welfare are less positive. Crate use violates at least one of the **five freedoms**, the freedom to express normal behaviour, for although the occupant can stand up, lie down and roll over, it provides insufficient space to allow turning around, a behaviour commonly exhibited by unrestrained sows around the time of birth. Two more of the 'freedoms' (freedom from discomfort and freedom from **fear** and **distress**) are also threatened, as detailed below. An early version of the five freedoms, recommended in 1965 in the **Brambell Report**, stated that 'an animal should at least have sufficient freedom of movement to be able without difficulty to turn around, groom itself, get up, lie down and stretch its limbs'. Standing up and lying down are often more difficult in farrowing crates: the lack of space means that sows are unable to perform normal sequences of these posture changes, and may result in an abrupt final drop of the sow's hindquarters during lying down – which can endanger piglets – rather than a gentle and controlled lowering. The lack of space (and usually absence of substrate) prevents the sow from performing normal pre-farrowing nesting behaviour (**see also: Sow behaviour**), which may lead to distress. Sows' responses to confinement in the farrowing crate may depend on their gestation environment: several authors have reported that freedom during gestation is likely to lead to increased agitation upon being placed in a farrowing crate.

Several alternatives to the farrowing crate exist, the simplest being the indoor bedded pen, which may incorporate rails or other piglet protective devices. Such pens can be very successful, particularly if they are sufficiently large, well designed and attended by skilful caretakers, but at least in the UK their use tends to be confined to smaller, less intensive units. Outdoor pig production thrives in many countries, and farmers achieve acceptable levels of pre-weaning mortality by accommodating farrowing and lactating sows in bedded arcs or huts. Variants on the conventional farrowing crate, such as the ellipsoid crate, Freedom crate and Ottawa crate, have been designed, with the aim of protecting piglets but allowing sows more freedom of movement. These designs have seen limited success, probably because of the extra space and labour required. Similarly, a voluntary crate, which the sow leaves for food and exercise, has so far gained little popularity. (MH)

Further reading

Fraser, D. (1990) Behavioural perspectives on piglet survival. *Journal of Reproduction and Fertility* 40(Suppl. 1), 35–43.

Fear

Fear is an emotional reaction, induced by the perception of stimuli associated with danger, which leads to protective defensive reactions. Fear-inducing stimuli may include **novelty**, threats to which animals have been exposed over evolutionary time, intense stimuli, social **signals** and stimuli that have acquired fear-inducing characteristics through learning (Gray, 1987). Because fear is an **emotion**, it can be viewed as an adaptive system of coordinated changes in behaviour, physiology and **cognition**. In defining and measuring fear, it is important to take into account the types of stimuli to which an animal is reacting, the quality of an animal's earlier experiences with these stimuli, the type of behavioural and physiological responses the animal shows and the consequences of these reactions at both the proximate and ultimate levels. In **applied ethology**, emphasis has often been put on fearfulness, a **personality** trait reflecting the general tendency to show fear responses to different threats (**see also: Temperament**). However, different fear responses may not be correlated and the reaction shown by the animal may depend on specific characteristics of the eliciting **stimulus**.

Several cognitive processes are central to the concept of fear. Perception of threatening stimuli leads to a focusing of **attention**, **arousal** and comparison of the stimuli with memories of earlier experiences, followed by decision making and the coordination of behavioural responses. Fear is expressed behaviourally as suppression of other behavioural systems and attempts to avoid, terminate or escape exposure to threatening stimuli. In some cases, fear may also be expressed as immobility or defensive attack, the specific response depending on situational variables and species as well as on the animal's predisposition to a reactive or proactive **coping** style.

Immobility may be observed either when an animal has the possibility of hiding from, and thus escaping detection by, a predator or when it is held by a predator and has no possibility of escaping. Defensive attack may be shown by some species in the close proximity of a predator or **conspecific** when the chances of escaping are low (**see also: Aggression**).

Fear is also associated with a range of physiological and endocrine responses characteristic of activation of the sympathetic–adrenal–medullary and **hypothalamic–pituitary–andrenocortical (HPA)** axes. These include both neuroendocrine (adrenaline, noradrenaline, corticotropin-releasing factor, adrenocorticotropic hormone (**ACTH**), **glucocorticoids**, beta-endorphin) and physiological (**blood pressure**, changes in **heart rate** and heart rate variability, stress-induced hyperthermia) changes generally associated with **stress**.

The central nucleus of the amygdala is among the most important centres of the **brain** mediating fear responses. It has connections to many other areas of the brain, including the **hypothalamus**, dorsal motor nucleus of the vagus, parabranchial nucleus, *locus coeruleus* and paraventricular nucleus. Neural systems implicated in fear include noradrenergic, dopaminergic, serotonergic and GABAergic systems. Pharmacological treatments aimed at alleviating fear typically target GABAergic systems using **benzodiazepines** such as diazepam and chlordiazepoxide, or the serotonergic system using drugs such as buspirone.

The diversity of empirical correlates of fear, as outlined above, underpins the fact that fear may be ascribed to animals on the basis of well-defined behavioural and physiological criteria without implying that the animal must be undergoing the same kinds of conscious experiences that a person might have. However, although the animal's subjective experience is by definition inaccessible to an outside observer, it is normally assumed that fear is experienced as uncomfortable and therefore detrimental to **welfare**. Under conditions to which an animal is well adapted, fear responses protect the individual from exposure to damaging stimuli. On the other hand fear is typically associated with the activation of systems mediating physiological stress, and prolonged exposure to fear-eliciting stimuli may therefore function as a chronic **stressor**.

Fear versus anxiety

It is useful to distinguish between fear and **anxiety**. Whereas the former is a response to stimuli associated with real and immediate danger, the latter is a response to stimuli associated with potential or more ambiguous danger. Although it may not always be easy to stringently separate the two, empirical justification for the distinction between fear and anxiety can be found in the work of Blanchard *et al.* (1990), who studied rats living in visible burrow systems. Rats exposed to the actual presence of a predator – and thus a real danger – initially flee and become inactive. In the absence of an escape route, rats may freeze or exhibit defensive threat or attack, the latter behaviour becoming more predominant with the increasing proximity of the predator. This type of behaviour, given the context, may be considered as indicating a state of fear, which functions to ensure survival in the face of immediate threat.

On the other hand, rats exposed to an area in which they have previously encountered a predator – a situation clearly indicative of potential danger – engage in cautious approach and investigative risk assessment behaviour. The perception of mild threat thus causes a general increase in anxiety that helps to locate the source and type of danger and to plan possible ways of dealing with it (Marks and Nesse, 1994). In addition to justifying the conceptual distinction between fear and anxiety, the above example partly illustrates that **fitness** may suffer if fear and anxiety reactions are deficient, excessive or inappropriate in form (Marks and Nesse, 1994).

Fearfulness

Fearfulness is a component of personality and can be defined as the propensity to perceive and react in a similar way to a variety of threatening or potentially threatening stimuli (Boissy, 1995). Measurement of fearfulness presupposes that the measured variables provide information about stable characteristics of an individual that can be generalized to other times and other similar situations. This is different from measures of fear that may sometimes be very specific to a given situation and may not show stability over time, but change according to situation-specific variables. As an example, avoidance of a novel and conspicuous object will lead to a reduction in fear over time, but the animal's general fearfulness will remain stable and influence behaviour in the same way the next time it is exposed to a similar situation.

Fearfulness is influenced by both genetic and environmental factors. The effect of genetics is documented by a large number of **selection** experiments. Fearfulness is also one of the characteristics that have been altered through **domestication** and selection for production characteristics, as illustrated by the fact that thresholds for the elicitation of different fear responses are normally found to be lower in ancestral animal breeds than in modern production breeds. On the other hand, a basal level of fearfulness has been maintained in all domestic populations because of the importance of this personality trait for survival. Fearfulness is also continually modulated by pre- and postnatal developmental processes, as well as by learning. Prenatal stress increases fearfulness and early postnatal stress may reduce fearfulness. These effects may be very dramatic. There are therefore two major ways in which fearfulness may be reduced in efforts to improve welfare in domestic animals, through either genetic selection or manipulation of the prenatal or postnatal environment.

Cost and benefit of fear

Under extensive production conditions, **predation** by carnivores may be considerable, and normal fear reactions may play a role in determining which individuals survive. Breeds showing more adaptive defensive reactions may be better suited than others. Even under intensive production conditions, many common stockmanship procedures, such as herding, often presuppose that animals avoid humans. A **stockperson** may sometimes treat animals that do not avoid him or her abusively, thus reinforcing learned avoidance reactions. On the other hand, research within applied animal behaviour has focused largely on the negative implications of intense and chronic states of fear and anxiety in farm animals. Because farm animals are widely protected from predation, fear and anxiety elicited by contact with humans and other environmental stimuli are generally considered unnecessary and maladaptive, reducing production and welfare under normal production conditions. Hypotheses predicting fitness-reducing effects of fear and anxiety may be formulated in terms of exaggerated and prolonged activation of stress-mediating endocrine systems, as interfering with cognitive functioning or as redirecting attention away from a task at hand towards stimuli eliciting fear or anxiety.

Measuring fear

Methods used for measuring fear and anxiety may involve the quantification of both conditioned and unconditioned behaviour. Conditioned responses involving conflict include the fear-induced inhibition of food-reinforced lever pressing, water-reinforced licking and food-reinforced key pecking. Unconditioned responses involving conflict that have been used as indicators of anxiety include behaviours in the **approach test**, **social interaction test**, **elevated plus-maze**, **light–dark test**, **open-field test** and **novel object test**, as well as a diversity of other specific tests. Aversively stimulated behaviours that have been used for studying fear include fear-potentiated startle, **vocalization** (separation distress, vocalization in response to pain, startle, attack and drug withdrawal), defensive burying, stress-induced hyperthermia and stretched approach posture, in addition to a range of other species-specific responses.

Several of the tests mentioned above are based on the observation of **conflict behaviour**. Animals have competing motivations to avoid and approach dangerous stimuli. The acquisition of information about the degree and type of danger facilitated by approach and investigation is crucial to survival because it allows the animal to show adaptive stimulus-specific responses. Animals forced into close proximity to dangerous stimuli primarily show immobility, desperate escape attempts or aggressive behaviour. Tests using forced exposure to **aversive** stimuli can be used to assess the form of fear responses, but may have limited utility in assessing the level of stimulation at which fear is elicited in different individuals. The threshold of stimulation at which fear reactions are elicited can more easily be assessed by the voluntary approach and avoidance behaviour expressed by a freely moving animal.

The observation that freely moving animals withdraw from threatening stimuli when the motivation to avoid becomes stronger than the motivation to approach is formalized by Miller's model of **approach–avoidance conflict** (reviewed by Gray, 1987). The model implies that tendencies to approach dangerous or potentially dangerous stimuli will be inhibited by fear or anxiety at a distance from stimuli at which the motivation to avoid becomes stronger than the motivation to approach. The strength of both approach and avoidance increases with increasing proximity to the object, but the strength of avoidance increases more rapidly than the strength of approach. Miller's model thus provides a framework within which to evaluate the relative levels of stimulation at which fear and anxiety responses are elicited based on the approach and avoidance behaviour of freely moving animals responding to dangerous or potentially dangerous stimuli.

Validity

Many of the tests commonly used to quantify fear responses are subject to very little validation prior to their use (**see also: Validity**). This is regrettable in that recorded measures may be very ambiguous and give little information about fear itself. In the following, fear of humans will be used as an example of how measures can be validated using several techniques. Tests of the internal validity of putative measures of fear of humans involve testing for predicted effects of a treatment thought to affect fear of humans. Aversive treatment by humans or medication with anxiogenic drugs prior to testing would be predicted to increase avoidance of humans in freely moving animals, as well increasing the severity of physiological stress responses shown during forced exposure. Pleasant treatment by humans or medication with anxiolytics, on the other hand, would be predicted to reduce avoidance in freely moving animals, and reduce physiological indicators of stress during forced exposure. Confirmation of these effects on the response variable would entail a partial internal validation of the methods used for measuring fear. If measures are not affected by the treatment in the predicted direction, either the sensitivity or the internal validity of the putative measures of fear may be considered to be low. To test for external validity, recorded measures of fear of humans may be used to predict zootechnical performance (e.g. milk production, egg production, growth, immune function, maternal ability) or other aspects of animal behaviour that are thought to be affected by variability in fear.

If the effects of several different treatments are tested, preferably in different studies, the results can be used to evaluate the specificity (convergent and discriminant validity) of putative measures of fear of humans. Testing for convergent validity entails testing the hypothesis that alternative measures of fear are empirically related. This may involve testing for predicted correlations between alternative measures of fear. It could also involve testing the hypothesis that several different aversive treatments by humans increase fear, whereas several rewarding treatments by humans reduce fear. To evaluate discriminant validity, it would be relevant to test the prediction that measures thought to reflect different behavioural systems – such as fear of humans and aggressiveness towards conspecifics – are not correlated. This would establish that conceptually unrelated constructs are also empirically unrelated. These methods for validating indicators of fear should be used more frequently in future studies.

Challenges for future research

The state of knowledge regarding fear in farm animals has a number of weaknesses that could be ameliorated through future studies. First, there is a general lack of validation of tests used to measure fear. As part of future research, tests could be properly validated following the guidelines suggested above. There are also a limited number of studies testing for predicted associations between existing variation in fear and anxiety and production of fitness-related behavioural parameters in a number of species. Existing studies have primarily been descriptive, and have focused on documenting associations between the way in which animals are handled and their growth and reproduction. Documented associations between handling and productivity may therefore be an effect of handling on stress itself and need not be mediated by fear.

There has been a primary focus on the personality trait of fearfulness as a general response pattern, and its relevance for predicting general response thresholds and adaptability in relation to a range of challenges. Furthermore, responses to fear- and anxiety-eliciting stimuli are rarely considered as representing distinct states and are rarely studied as adaptive mechanisms. In order to study fear responses as adaptive mechanisms it is necessary both to distinguish between fear and anxiety and to focus on differences between fear-eliciting stimuli and corresponding behavioural responses that confer an adaptive advantage.

Fear–welfare relationship

Short-lasting and adaptive fear responses are likely to safeguard an individual's welfare by leading to responses that protect it from physical damage. Chronic and exaggerated fear responses to stimuli that an animal cannot avoid, on the other hand, are likely to be associated with reduced welfare. This can be illustrated for **chickens** living in loose-housed production systems. Moderate and adaptive fear response lead to moderate avoidance of a stockperson and keep birds from being trampled under the feet of the stockperson. On the other hand, hysterical responses where birds flee and pile up at one end of the house may cause significant damage by scratching, and mortality by crushing and suffocation. (AMJ)

See also: Handling

References

Blanchard, R.J., Blanchard, D.C., Rodgers, J. and Weiss, S.M. (1990) The characterization and modelling of anti-predator defensive behaviour. *Neuroscience and Biobehavioral Reviews* 14, 463–472.

Boissy, A. (1995) Fear and fearfulness in animals. *Quarterly Reviw of Biology* 70, 165–191.

Gray, J.A. (1987) *The Psychology of Fear and Stress*. Cambridge University Press, Cambridge, UK.

Marks, I.M. and Nesse, R.M. (1994) Fear and fitness: an evolutionary analysis of anxiety disorders. *Ethology and Sociology* 15, 247–261.

Feather

Feathers are specialized, avian, epidermal structures of which the two main types are the externally visible contour feathers and the insulating down feathers. The vane of contour feathers consists of numerous barbs, branching from the main shaft at 45°, connected to larger numbers of barbules that possess microscopic hooks that allow the barbules to adhere closely to one another. Down feathers appear fluffy because they lack these hooks and hence do not form a closed vane. The embedded feather quill occupies the follicle, to which are attached feather muscles, capable of elevating or lowering the feather. The walls of the follicle, the dermal papilla that extends into the end of the quill and the feather muscle are all well supplied with general somatic afferent nerve fibres and nociceptors.

There is behavioural and electrophysiological evidence that feather removal by pulling is painful in the domestic **chicken**. Initial feather removal results in escape behaviour and subsequent feather removal results in a passive, crouching immobility behaviour typical of chickens' response to **pain**. **Electroencephalogram** recordings indicate a progressive increase in the high-amplitude, low-frequency activity following successive feather removals. This type of activity is similar to that seen in other species at times of **stress**. Chickens also show temporarily increased **blood pressure** following feather removal.

The plumage of the domestic chicken is important for insulation and prevention of **injury**, in addition to the limited flight of which it is capable. The lipids secreted by the uropygial or preen gland serve to reduce feather wear and waterproof the feathers. The secretions of the gland are spread over the plumage as part of **preening**, which tends to have a **diurnal rhythm** with peaks in mid-morning and afternoon. Both preening and **dust bathing** serve to maintain plumage quality. There is some evidence that the function of dust bathing is to remove stale lipids from feathers, increasing the 'fluffiness' of down feathers and improving their insulating properties. The feed consumption of layer hens increases as more feathers are lost, since more body heat is lost. This is of economic concern to farmers, who want to maximize the conversion of feed into eggs.

Feather loss or damage can be caused by abrasion, more common in caged hen systems than in other forms of housing. It is more frequently caused by **feather pecking** in all systems, in which one chicken pulls the feathers or pecks at the plumage of another bird. A validated and extensively used method of plumage scoring in laying hens is described by Tauson *et al.* (1984), in which the body is divided into five individually scored parts. A total score of 20 represents an undamaged plumage, whereas the lowest possible score of 5 indicates a heavily damaged plumage with no or very small areas of the body covered in feathers. The level of severe feather pecking present is known to adversely affect the plumage score, but other factors such as cage design can also have a marked effect. In most laying hen flocks, bird plumage score decreases throughout the period of lay.

Feather pecking is a widespread problem in laying hens and can also occur in caged laying hens and other poultry species. It is a **welfare** problem because of the pain associated with feather removal, the possible relationship with **cannibalism** and the widespread use of **beak trimming** as prophylaxis against the behaviour. It is thought that this **abnormal** behaviour is related to management conditions on farms. Various risk factors have been identified, including large flock size, high stocking density, pelleted food rather than mash, nutritional deficiencies, poor litter quality, little access to perches during rearing, and high-intensity lighting. It is possible that by exposing skin, feather damage and loss may lead to cannibalism, when a bird will peck at the flesh of another, causing injury and sometimes death. However, the relationship between feather pecking and cannibalism is not yet clear.

It is thought that feather-eating behaviour in laying hens is a relatively common occurrence. This is a form of **pica** and can lead to crop impaction; it has been observed in hens suddenly placed in unfamiliar flocks. High levels of feather eating in reared pullets can be associated with outbreaks of feather pecking and cannibalism. However, it remains unknown whether feather eating promotes feather pecking, or vice versa.

Feather-plucking or -chewing psittacine birds, such as the grey parrot (*Psittacus erithacus*), are commonly presented in veterinary practice. This should be differentiated from feather pecking in poultry, in that the psittacine is removing its own feathers rather than those of other birds. However, this may simply reflect the fact that psittacines are more frequently individually housed than group-housed. Feather plucking syndrome can have a variety of medical and psychological causes, which in themselves may represent a welfare problem. Wild psittacines generally live and forage in flocks. Hand rearing of young birds in isolation can result in **anxiety** and poor acquisition of the learned components of preening, which may lead to feather plucking. Other psychological causes, such as **boredom**, anxiety and phobias, may result from the conditions in which the bird is kept.

Some consider feather plucking to be similar to self-injurious behaviour in monkeys, in which rearing and **weaning** in isolation results in a large proportion of primates developing locomotor **stereotypies**. In these circumstances, retraining the bird or environmental **enrichment** may resolve the symptoms. Environmental enrichments that increase foraging opportunities for parrots have been shown to increase plumage scores, in a similar way to chickens. Feather plucking may be seen following the practice of wing clipping, which prevents the bird from flying. Concern has been expressed by some avian veterinarians that the practice of clipping one wing can result in the bird injuring itself while attempting to fly. Other methods of control, involving training the bird, have been advocated. (GD)

See also: Self-mutilation

Reference and further reading

Chitty, J. (2003) Feather plucking in psittacine birds. *In Practice* 25(8), 484–493.

Gentle, M.J. and Hunter, L.N. (1990) Physiological and behavioural responses associated with feather removal in *Gallus gallus* var. *domesticus*. *Research in Veterinary Science* 50, 95–101.

Green, L.E., Lewis, K., Kimpton, A. and Nicol, C.J. (2000) Cross-sectional study of the prevalence of feather pecking in laying hens in alternative systems and its associations with management and disease. *Veterinary Record* 147, 233–238.

Savory, C.J. (1995) Feather pecking and cannibalism. *World's Poultry Science Journal* 51, 215–219.

Tauson, R., Ambrosen, T. and Elwinger, K. (1984) Evaluation of procedures for scoring the integument of laying hens – independent scoring of plumage condition. *Acta Agriculturae Scandinavica* 34, 400–408.

Feather pecking (feather plucking)

Feather pecking is an **abnormal** behaviour found mainly in barren, frustrating and stressful intensive production environments in which the performance of a number of normal behaviour patterns is not possible. It is defined as the pecking at and removal of feathers of one bird by another, although in singly kept animals, such as parrots, self-plucking may occur. It can be further subdivided into gentle feather pecking, where the feathers are gently pecked at and nibbled but not removed, and severe feather pecking, where the feathers are vigorously pecked at and removed. It is severe feather pecking that is a **welfare** concern, as removal of feathers can cause bleeding and stimulate **cannibalism**.

There is growing evidence that feather pecking in domestic poultry (especially severe feather pecking) is related to the **motivation** for **foraging behaviour** and gut function. For example, birds kept on a substrate at which they can direct foraging behaviour feather peck less than birds kept on wire floors. High-fibre diets also result in lower feather pecking levels compared with low-fibre diets. **Chicken** feathers are considered non-nutritive, but are often consumed by the feather pecker after removal. Feather consumption has been shown to affect digestion in a manner similar to insoluble fibres, which are generally at quite low levels in a standard chicken feed. An analysis of the fixed action patterns involved in feather pecking, foraging and a number of other types of pecks found that the movement patterns involved in feather pecking were similar to those in foraging for all measures, also indicating similar underlying motivations behind those pecks. Thus, feather pecking may be a way for birds to satisfy foraging motivation and increase insoluble fibre in the diet, thus improving gut function.

However, there are many other factors that affect feather pecking. For example, birds provided with opportunities for **dust bathing** or other types of environmental **enrichment** also show decreased levels of feather pecking. Chickens genetically selected for high levels of feather pecking were found to have lower **serotonin** levels than those selected for low feather-pecking rates. Low levels of serotonin have been implicated in other behavioural disorders, such as **depression** and **compulsive disorder**. Thus, high levels of feather pecking may be indicative of a mental abnormality in birds, and this is the primary hypothesis relating to its expression in captive **pet** birds, such as parrots. Early experience also affects feather pecking. Chicks reared with a hen (a practice never seen on commercial farms) displayed less feather pecking later in life than chicks reared by a standard method. Feather pecking can be considered a stereotypic behaviour, and thus it is possible that some birds continue to feather peck even if they have been moved to enriched conditions due to perseverance of the behaviour, adding to the difficulty of solving the problem of feather pecking.

Feather pecking is a welfare concern because the removal of feathers has been found to be painful, and blood from feather removal may stimulate cannibalism. However, current methods in attempting to control feather pecking on the farm also have welfare concerns and thus are not acceptable as long-term solutions. Currently, feather pecking is largely controlled by **beak trimming** the birds and/or putting them in battery cages; in the case of parrots, a collar may be used to restrict the head movement necessary for plucking. Beak trimming, the removal of the last third of the beak, does not stop feather pecking but makes feather removal more difficult, decreasing the related cannibalism. Beak trimming has been associated with acute and chronic **pain** and, as the beak continues to grow, trimming may have to be performed a few times throughout the life of the bird. Battery cages or small wire cages limit contact and even movement, which decreases the amount of feather pecking possible, but these will be banned in a number of countries, including those within the European Union, in 2012. (LMD)

See also: Laying hen housing

Further reading

Dixon, L.M. (2008) Feather pecking behaviour and associated welfare issues in laying hens. *Avian Biology Research* 1, 73–87.

Dixon, L.M., Duncan, I.J.H. and Mason, G.J. (2000) What's in a peck? Using fixed action pattern morphology to identify the motivational basis of abnormal feather pecking behaviour. *Animal Behaviour* 76, 1035–1042.

Harlander-Matauschek, A., Piepho, H.P. and Bessei, W. (2006) The effect of feather eating on feed passage in laying hens. *Poultry Science* 85, 21–25.

Riber, A.B., Wichman, A., Braastad, B.O. and Forkman, B. (2007) Effects of broody hens on perch use, ground pecking, feather pecking and cannibalism in domestic fowl (*Gallus gallus domesticus*). *Applied Animal Behaviour Science* 106, 39–51.

Feedback – positive and negative

Feedback systems are regulatory circuits that help to maintain some controlled parameters around an optimal value. This control is operated through a **signal** (a feedback loop) that is referred back to the system when some parameters change. A simple example of a feedback system at the behavioural level is the maintenance of optimal body temperature in **reptiles**. Reptiles are ectothermic animals that expose their bodies to sunlight to warm up. Body temperature needs to be maintained within a narrow range in order to assure the proper functioning of the organism. Therefore, a feedback system operates so that a reptile exposes its body to direct sunlight when body temperature is too low or, conversely, it moves to the shade to cool its body temperature down.

Feedback systems are observed at every biological level, from genes and cells to organisms and ecosystems. Two types of feedback may be distinguished: (i) positive feedback, where the signal tends to accelerate or increase the occurrence of a particular process; and (ii) negative feedback, where the signal operates by reducing or slowing down a process. Some degree of oscillation is a recurrent characteristic of feedback systems, resulting from either the imperfectly tuned inputs of positive and negative feedback or the fact that conditions change too rapidly or too often for the system to remain stable through feedback. (BM)

Feeding

Feeding is perhaps the most basic, primitive and persistent of all behaviours, and makes a central contribution to most animal functions. It is due to the consequences of the action of feeding that the animal is able to survive, grow and eventually reproduce. Like most behaviours, it allows animals to achieve their ultimate goal, that of maximization of **adaptive** fitness. While the contribution of feeding behaviour to **fitness** is accepted unquestionably for wild and unselected, extensively kept animal populations, its function and purpose for most domestic animals has been the subject of debate. This is because uninhibited feeding behaviour may lead to 'unfit' adults that are unable to reproduce. The greatly reduced reproductive ability of broiler breeders – the parents of broiler chickens – that have been fed a high-quality food *ad libitum* throughout their lives is a good case in point.

The intensive selection of domestic animals for certain traits, spectacular as its effects have been, has had only a minor effect on behaviour, including feeding behaviour, in an evolutionary context. It is artificial selection for 'unusual' traits, such as excessively high rates of milk and meat production, that has led directly to reduced fitness and not changes in feeding behaviour that have contributed to it. Nevertheless, in view of the unusual extent to which traits in domestic animals have been selected for, it is useful if the idea of fitness as a goal is broadened to account for the outcomes of feeding behaviour in such animals. It is now accepted that domestic animals have current output goals that they seek to achieve through their feeding behaviour. Such output goals include their rates of growth, milk, wool and egg production and the level of fatness. The consequences of this suggestion are, in most cases, identical to those of having fitness as a goal.

The components that constitute feeding behaviour relate to the following questions: (i) when to eat; (ii) where to eat; (iii) what to eat; and (iv) how much to eat during a particular period of time. The relevance of these questions to the feeding behaviour of the various kinds of domestic animals will depend on their system of rearing. Extreme examples are animals offered a limited amount of a particular food once daily, such as in the rearing of broiler breeders, on the one hand and ruminants in extensive grazing systems on the other; in this latter case all four questions apply. The scientific issue is how these four components contribute to the 'desire' of animals to achieve their output goals and what are the factors that modify them.

When to eat

Animals do not distribute their feeding behaviour randomly in time, but cluster it into discrete bouts. The smallest unit of feeding behaviour is an event, such as a visit to a feeding trough, during which the animal exhibits uninterrupted feeding. For foraging animals, this can consist of a number of consecutive bites or pecks taken from forage. Such feeding events are separated by time intervals of variable duration, during which feeding does not occur (see Fig. F.1 for cows). Short intervals will usually be the consequence of random events that lead to the disruption of feeding behaviour; such disruption may arise from competition or other interaction with **conspecifics**, sudden noise, etc. Feeding can also be facilitated by conspecifics, especially in young and growing animals, so that members of the group tend to feed together. Long intervals are considered to be a consequence of the

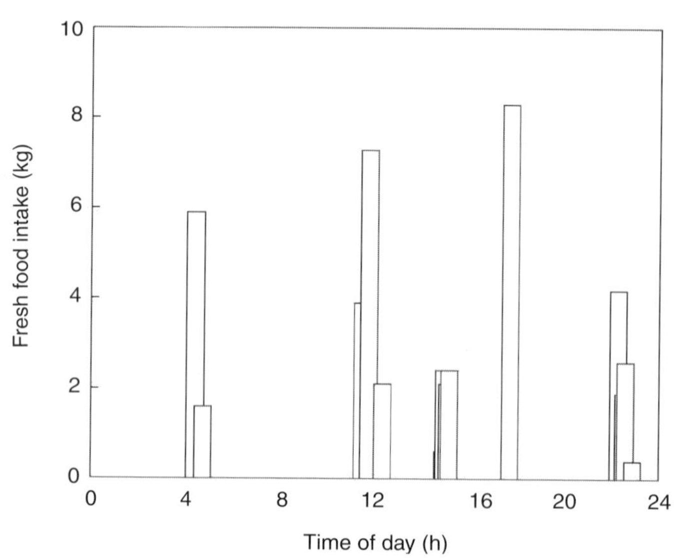

Fig. F.1. Temporal distribution of the feeding behaviour of a cow; each bar represents a feeding event and events are clustered into bouts.

underlying physiology of the animal and its **habituation** to its environment; they are used to define a feeding bout or *meal*. A meal may be expected to start when the animal is motivated to eat – i.e. it is hungry – and to stop when the animal is satiated.

In the extreme case, the duration and size of a meal will be highly correlated with the duration of its preceding or succeeding interval: these are the widely used pre- and postprandial correlations. However, the distribution of meals is mainly dictated by a **diurnal pattern** that in turn is determined by events that are associated with the animal's management. Farm animals show increased intensity of feeding behaviour immediately after fresh food has been offered to them, around the time of **milking** and at the onset of natural or artificial lighting when the first meal of the day is eaten. Figure F.2 shows an example taken from data on pigs. The natural history of the species will dictate this pattern, as laboratory **rodents** still take their meals during darkness, and raise the possibility that feeding behaviour may have a **circadian rhythm**, i.e. it is the result of an internal clock.

While the distribution of feeding behaviour into meals seems to reflect the 'preferred' behaviour, animals are also characterized by great flexibility with regard to when to eat. For example, animals kept in hot environments modify their feeding behaviour so that most of their meals occur mainly during the cooler parts of the evening. The frequency of the meals can be affected by competition, due to the limited number of animals that can be accommodated at any one time at the feeder. Increases in animal pressure per feeder may thus increase meal frequency while decreasing meal size. In simple terms, the distribution and size of meals can be seen as the means the animals have at their disposal to achieve the goals of their feeding behaviour.

Where to eat

This option is available only to animals that are reared extensively. The question is not independent of what to eat (see below), the difference being only a matter of scale. With other things being equal, animals are expected to forage at sites that offer the highest nutrient and energy intakes per unit of effort. For example, grazing ruminants prefer tall over short swards, and swards of higher nutrient content. At the same time, animals would be expected to prefer foraging sites that minimize the risk of exposure to predators, pathogens and weather elements. The fact that both cattle and sheep avoid areas contaminated with faeces, where the concentration of pathogens is at its highest, is consistent with this view. On the other hand, site selection by sheep seems to be unaffected by exposure to cold, presumably because they have a high capacity to cope with such conditions.

In relatively unmanaged and natural environments animals are faced with the complex problem of choosing between sites that simultaneously offer benefits, such as high rate of nutrient intake, and costs, such as exposure to **parasites**. In these cases a reasonable expectation would be that site selection is the outcome of a trade-off between these benefits and costs, and that animals prefer sites that offer the highest net benefit. This theoretical prediction has been confirmed by experiments performed on a small scale.

What to eat

This is an issue relevant only to animals that are given free and continuous access to two or more foods as a choice. The most complex choices are faced by grazing animals that are required to choose not only what plant to eat, but what specific part of it as well, e.g. stem or leaf. Less complex choices are offered to

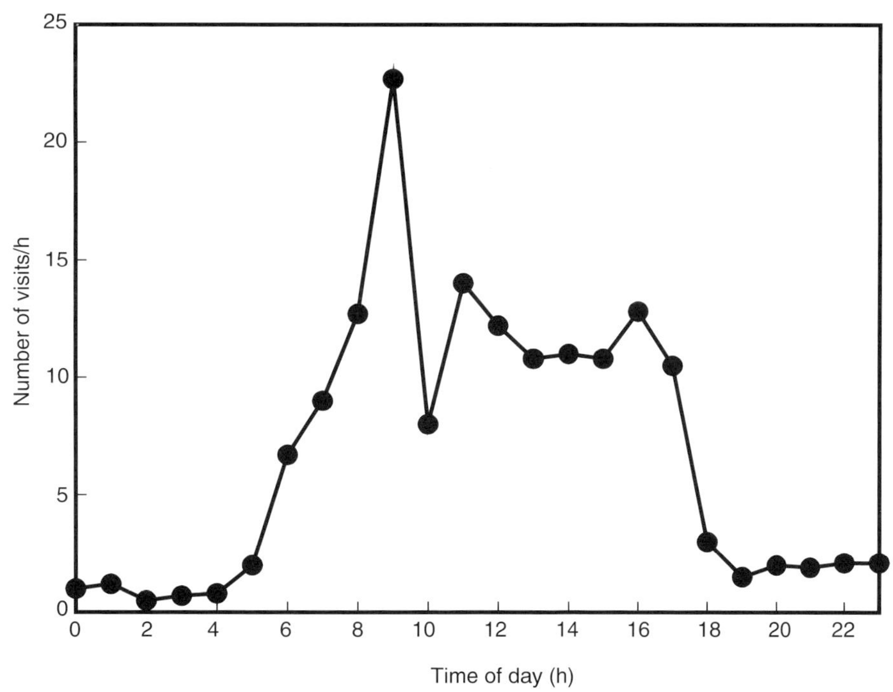

Fig. F.2. The number of visits to a feeder per hour for a group of pigs. Most visits take place during the time that artificial lights are on (06:00–18:00 hours).

cattle given a heterogeneous mixture of silage and concentrates, and to poultry given access to a heterogeneous mixture of cereal and a protein concentrate. In such cases the choices made by the animals are neither random nor purposeless, but appear to follow some specific rules.

Some of these rules appear to be the consequence of hard-wired behaviour and have been pre-shaped by **evolution**. Animals still exhibit innate **preferences** for sweet-tasting foods (usually sources of energy), even when these do not have an energetic value, and avoidance of bitter foods (in nature usually associated with toxins), even when the foods do not contain any toxins. Most of the rules of diet selection, however, are the consequence of learned behaviour. Such rules include the selection of diets that: (i) maximize nutrient and energy intake rate; (ii) meet the animal's requirements for energy and nutrients; (iii) minimize intake of toxins and exposure to pathogens; and (iv) reduce the likelihood of digestive disturbances. There are examples in the literature to support the evidence of behaviours consistent with these rules, provided that the animal has had the opportunity to learn about the properties of the foods offered and their consequences, and is offered choices that allow such behaviours to be exhibited. Whether these rules are followed in the long or short term will depend on the time scale of the consequences of diet selection. While meeting nutrient requirements might not matter in the short term (see below), consistent exclusion of toxins from the diet would be of significance both in the long and short terms.

In a manner similar to the choices offered during site selection, animals are frequently faced with complex food choices. These choices may include instances in which animals are given access to two or more foods as a choice, a combination of which cannot be an appropriate diet – i.e. a diet that meets all above criteria. In these cases the principles of trade-off behaviour will apply, so that the selected diet is the outcome of trading between the benefits and the costs associated with each food. These benefits and costs, however, will depend on the internal state of the animal. For example, an animal already immune to parasites may select for tall swards that are associated with the presence of faeces and hence parasites. This would be because consumption of parasites by an immune animal is associated with very small costs. Similarly, the perception of costs and benefits may be affected by the animal's motivation for food. A highly food-motivated animal would be prepared to include in its diet foods that under normal circumstances it would have either avoided or preferred to a lesser extent.

How much to eat

This is perhaps the most frequently asked question in relation to feeding behaviour, as it relates directly to the applied implications of the control of food intake and animal productivity. Given the proposition that animals will seek to achieve their current output goals through feeding behaviour, then the expectation is that they will eat as much as is necessary to achieve this and no more. When animals are given access to a balanced food, the amount of food eaten will be dictated by their output goals: fat **genotypes** will be expected to eat more than genotypes that have been selected against fatness; and animals of high-yield capacity, such as the high-yielding dairy cow, will eat more than their unimproved counterparts. This suggestion precludes the possibility that animals may be eating simply in order to derive pleasure, i.e. reinforcement from the action of feeding per se.

When an animal is given access to an unbalanced food, a reasonable expectation is that it will modify its rate of food intake in a manner that brings it closer to achieving its output goals. Therefore, when an animal is given access to a food that is low in a food resource (energy or nutrient), it would be expected to increase its food intake in such a way that the appropriate or required intake of that nutrient would be achieved. The consequence would be that animals would eat all other nutrients in excess. Pigs and chickens given access to a food low in protein have a higher food intake, and are considerably fatter, than animals on a high-protein food, presumably because they attempt to eat for protein.

It is also possible that, by attempting to eat for the first limiting food resource, the animal reaches it capacity to deal with the excess. Such capacities are defined by the animal's physiology and anatomy, and may include the capacity of the gastrointestinal tract and the maximum ability to deal with excess nutrients, such as protein, and toxins (detoxification rates). The environment also influences other capacities in which the animal is kept, such as the maximum capacity to dissipate excess energy as heat to the environment. When these capacities are met the animal will not be able to increase its food intake any further. These suggestions are well demonstrated by the changes in the intake of animals whose food intake is progressively diluted with an indigestible material (see Fig. F.3). Initially, animals increase their food intake so that their energy intake is constant. On reaching their capacity for bulk, food and energy intakes start to decline, since intake is now set to constant bulk.

The question is whether the regulation of food intake, both in terms of what and how much to eat, is achieved over the short term – on a meal basis for example – or in the longer term. When the spontaneous nature of feeding behaviour is maintained, there is very little evidence that either the amount or the composition of the food eaten is regulated in the short term in a tight manner. In farm animals there is considerable evidence for large variation in the values of the variables used to describe short-term feeding behaviour, such as number of meals and amount of food eaten per meal. Animals with similar average daily intakes can have twice the number of meals (with half the average intake per meal) compared with other animals that consume similar amounts in total. In addition, there is much evidence that changes in environmental conditions that have no effect on daily food intake can be accompanied by very large changes in short-term feeding behaviour. An example is that either an increase in the number of animals in a group with a constant number of feeders, or a decrease in the number of feeders for a given group size, almost invariably leads to considerable changes in feeding behaviour, while having little effect on daily or longer-term food intake. Short-term feeding behaviour may, therefore, be seen simply as the strategy that is used by the animals to achieve their food intake and that ensures achievement of their output goals.

Given the basic and unremitting nature of the feeding behaviour, it is hardly surprising that animals exhibit specific behavioural problems when they are deprived of its control, e.g. given restricted amounts of food. The occurrence of this is

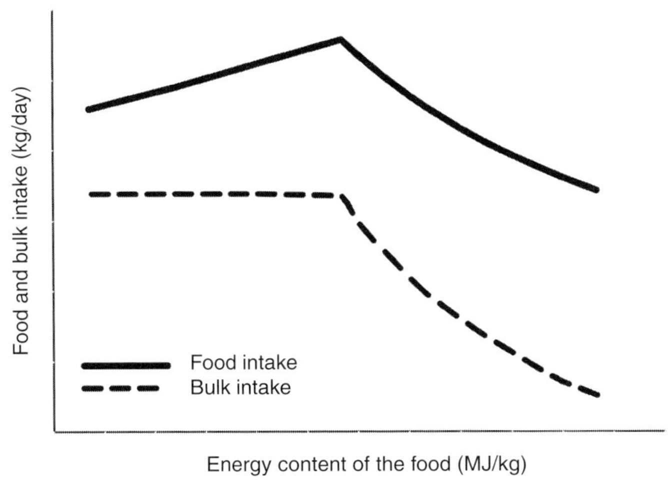

Fig. F.3. The food and bulk intake of an animal whose food is progressively diluted with an indigestible material (bulk). Dilution of the food with bulk reduces its energy content (from right to left).

more frequent than not, as even **pet** animals are given a defined amount of food a couple of times a day. Such behavioural problems include food-related **stereotypies**, **tail biting**, etc. and are considered to be a reflection of **redirected** (feeding) **behaviour** towards the most appropriate stimuli. In some instances feeding behaviour may be directed towards suitable foraging substitutes (such as straw bedding), but in many others the physical constraints of the environment are such that it is directed towards pens, cages or even towards conspecifics. Such behaviours are widely considered '**abnormal**' and by many as indicative of reduced animal **welfare**, and are of public concern. (IK)

See also: Foraging behaviour

Further reading

Kyriazakis, I. (1997) The nutritional choices of farm animals: to eat or what to eat? In: Forbes, J.M., Lawrence, T.L.J., Rodway, R.G. and Varley, M.A. (eds) *Animal Choices*. Occasional publication of the British Society of Animal Science No. 20, Edinburgh, UK, pp. 55–65.

Feeding systems

A feeding system is a term often used in association with farming practices. In general, the feeding system describes the type of **feeding** regime adopted by the **stockperson** or owner and often encompasses the method by which the food is delivered.

Access to natural or formulated feeds varies with the feeding system adopted. For example, dairy cows may be given access to pasture alongside concentrate feeds (cereal-based feeds) during the grass-growing seasons when herbage is plentiful. When conserved forage or formulated diets are provided, they are either presented on an *ad libitum* basis or they are fed as a restricted or controlled ration. During *ad libitum* feeding, appropriate food is freely available for self-selection by the animal throughout the day, whereas restricted feeding ensures that food provision is more controlled.

Reasons for restricted feeding vary by species. For example, the amount of concentrates fed to dairy cows may be limited due to the risk of gastrointestinal problems, and limiting the food intake of pigs can achieve a better food conversion ratio and a leaner carcass. Restriction is achieved either by reducing the amount of time the animal has to access the feed or by reducing the total amount offered at one or multiple feeding intervals. It is important to understand the feeding behaviour of the species when designing or proposing feeding systems, since this may not only affect the utility of the system, but also impact on the welfare of the animals within it. For example, if feeding is normally socially facilitated (**see: Social facilitation**) but provision is not made for such activity, intake may be suboptimal and the risk of **aggression** increased.

The method of feed delivery differs between feeding systems. The requirements of *ad libitum* systems are that all animals should have equal access to a continual supply of food and that food waste is minimized by ensuring that food stays in the feeding device or feeding area. In this type of system problems may arise due to aggression or hoarding, and thus food should be dispersed so that all individuals can gain access to the feed and no single individual or group monopolizes the food source.

Feed delivery to group-housed animals on a restricted feeding regime can be more complex, as the stockperson or owner needs to be sure that each animal is receiving no more than its allocated ration. This can be achieved by providing separate feeding stations or fitting animals with a transponder on a collar. The transponder identifies the individual when it enters the feeding area, allowing a pre-allocated amount of food to be dispensed (this may be a full daily ration or part ration depending, in part, on the natural feeding patterns of the species). The system can be programmed to dispense the food after predetermined time slots (interval schedules), which prevents animals from regularly blocking entry to the feeding system and ensures that the animal does not eat the entire ration at once. It is important to make sure that an animal cannot be displaced from the system once the food is dispensed (e.g. food is dispensed only once the animal is within a secure station and not as it enters), and that the animal can safely consume the food and leave the system (usually through a separate exit) without being injured or threatened by other members of their group.

The type of feeding system adopted is dependent upon the role or use of the animal. For example, the broiler chicken has been genetically selected for an increased rate of growth that has resulted in an increased appetite. To maintain rapid growth the birds that are destined for **slaughter** are fed an *ad libitum* ration high in protein from day one to finishing. However, breeder parents of broiler parent stock need to have their feed intake controlled, often limited to one meal per day, to prevent reduction in reproductive performance and health problems associated with rapid and excess weight gain. The consequences of this type of feeding regime for the chicken include chronic **hunger**, increased aggression between individuals and performance of **abnormal** behaviours that are indicative of poor **welfare**. This highlights the point that, in some circumstances, feeding systems can be in conflict with the natural feeding behaviour of the animals in question, since, in the case of production animals, the feeding system chosen may be based on maximizing economic return through optimizing measures of **health** and production. (SR)

Further reading

Ewbank, R., Kim-Madslien, F. and Hart, C.B. (1999) *Management and Welfare of Farm Animals. The UFAW Farm Handbook*, 4th edn. Universities Federation for Animal Welfare, Potters Bar, UK.

Feedlot

A feedlot is an enclosed area or building where ruminants are, generally, fattened rapidly for **slaughter**. However, animals may be put into a feedlot at other times of their life when pasture is not growing, such as during extended dry periods or when the ground is covered in snow and/or frozen. In these situations the aim of keeping animals in a feedlot is to provide them with sufficient feed to maintain, rather than increase, body weight and body condition.

The animals that enter feedlots have usually been sourced from extensive, grazing-based systems (**see: Free-range animals**) and, consequently, experience dramatic changes in their environment and management on transition to the feedlot. Additionally, animals from different properties may be mixed to comprise a pen-group, which results in disruption of social relationships (**see: Hierarchy**). Animals may have also been subjected to periods of transportation prior to feedlot entry (**see: Transport**). All of these factors contribute towards **stress** in the animals, which may reduce their immunocompetence (**see: Immune system**). This makes the animals susceptible to **disease** and it is, therefore, very important that the animals are closely monitored during the early phase of lot-feeding for signs of ill health. One major advantage of close confinement in a feedlot is that it is relatively easy for all animals to be checked for injury or signs of ill health, and this checking tends to be conducted at least daily. Thus animals can be treated promptly, reducing the duration of any **pain** and **suffering**.

The usual aim of lot-feeding animals is to improve their nutrition, which increases growth rate and the rate of fattening, and thus reduces the age at which animals can be slaughtered. The slaughter of young animals has consequences for aspects of meat quality, such as a whiter fat colour and a greater proportion of tender meat cuts.

Grain feeding is the usual nutritional regime adopted in lot-feeding, and this has the potential to create a number of **health** and **welfare** problems because the rumen is designed for a forage-based diet. Normally, the ruminal environment supports a population of protozoa and mainly Gram-negative bacteria, and these microbes convert carbohydrates to volatile fatty acids, carbon dioxide and methane. With readily fermentable carbohydrates, such as grains, there is a change to the fermentation pattern and microbe populations and total volatile fatty acids increase, resulting in conditions such as acidosis, ruminal distension (**bloat**), **laminitis** (**lameness**), systemic dehydration, rumenitis, liver abscesses and, in extreme cases, sudden **death**.

Rigorous nutritional management is required to avoid the deleterious conditions associated with feeding high-grain rations. A gradual introduction to grain will allow the ruminal microbes to adapt, and sudden changes in diet composition should always be avoided. Rations should start at about 60–70% roughage, with a gradual decrease in roughage and increase in grain over a 3–4-week period until the finishing ration, containing about 70–80% grain, is achieved. Most rations also contain antibiotics that act as ruminal modifiers and, depending upon the level of inclusion in the diet, will act to increase the efficiency of feed use and improve liveweight gain (growth promotants) or, at therapeutic levels, will have anti-acidosis and anti-bloat effects in order to minimize the problems associated with high-grain diets.

The size and dimensions of a feedlot yard or pen will depend upon the species, the number of animals in the group and the stocking rate (average space allocated to each animal (**see: Stocking density**)). Pen shape may affect access to feed; generally, pens are rectangular with feed available from a trough or bunk on one long side, which maximizes feeding space and limits the distance that animals have to move to reach feed.

Many countries have standards or **codes of practice** that stipulate minimum space allowances for animals, not only for general space, but also for feeding and drinking space. Stocking rate is critical, as it impacts on the health, welfare and productivity of the animals. High stocking rates may result in social tension (**see: Social stress**) because individuals are unable to avoid or escape from each other and thereby terminate aggressive interactions (**see: Aggression**). It has been suggested by some workers that the '**buller-steer syndrome**' that occurs in feedlot cattle, in which a particular animal or animals is/are persistently mounted and ridden by group-mates, is associated with the establishment and maintenance of the social hierarchy. Other researchers believe that the syndrome may result from multiple stressors, one of which may be social dynamics associated with large group sizes (rather than space allocation per se). Hormonal status (natural variability and from hormonal growth promotant implants) has also been implicated in the syndrome.

Space restrictions may make it difficult for some animals to access resources such as a resting site, food and water. High stocking rates may also mean that contagious diseases are spread easily and rapidly through the group. Stocking rate can also affect animal health and welfare by its impact on the moisture content of the 'pad', or soil/excreta substrate. Pad moisture content determines the production of noxious gases (e.g. ammonia), odours and dust, which can lead to or exacerbate respiratory problems. Wet, boggy conditions can lead to **footrot**, foot lesions and lameness.

One disadvantage with feedlots is that they provide a relatively unstimulating environment for the animals, and oral **stereotypies** may be observed in animals that are in feedlots for extended periods (**see: Boredom**). These stereotypic behaviours are probably a result of frustration of the motivations associated with grazing, combined with the consequences for the animals' **time budgets** of the low levels of roughage in the diet (**see: Tongue-rolling**). Some workers have also suggested that the buller-steer syndrome may be related to an unstimulating environment.

Feedlots need to provide protection for the animals from climatic extremes. In cold climates it is essential to provide protection from winds and a dry area for the animals to rest. In tropical and subtropical climates there can be high mortalities through excessively high thermal loads (generally a combination of high air temperature and high relative humidity) on the animals. The provision of shade is recommended, but the positioning and levels need careful consideration to avoid exacerbating **heat stress**. Animals may group tightly together under shade and actually reduce the opportunity for heat loss; and, following wet weather, shade delays the drying of the pad, which can raise relative humidity locally (thus adding to the heat load) and also create problems for foot health. (JCP)

See also: DFD meat; Flooring

Feed restriction

Feed restriction is a type of management procedure in which the amount of feed the animal would consume to eliminate **hunger** sensations is limited by one of multiple mechanisms. Feed restriction is related to other types of management scenarios that induce hunger, i.e. qualitative dietary restriction, in which the animal consumes a full amount of a diet that is nutritionally inadequate due to an absent nutrient or other required component. Thus qualitative restriction provides for *ad libitum* consumption of an inadequate diet, whereas feed or *quantitative* dietary restriction provides a complete diet, though not enough to satiate the animal.

Feed restriction can be imposed either indirectly or directly in relation to the caretakers' provisions. In cases where feed is restricted indirectly, some aspect of the environment prevents the animal from consuming an adequate amount. For instance, animals on pasture, where considerable distance exists between foraging opportunities, are limited by the travel time between foraging sites. Even though unrestricted in their movement, animals are unable to meet their nutritional needs and may enter a state of negative energy balance.

In another example of indirect restriction, animals can be restricted from feeding by **conspecifics**. Livestock typically maintain similar daily routines for eating, drinking and sleeping activities. However, when access to the feed source is obstructed by herd-mates at the feeder, animals must adjust their feeding times accordingly. This will become a concern when low-ranking animals are not able to gain access to the feeder for long enough periods of time to ingest what would be their sufficient daily ration.

A variety of environmental conditions can induce changes in **feeding** behaviour, of which increased ambient temperature is the most severe. The process of food digestion releases heat as food particles are broken down into their basic components. Homeothermic animals, or animals that maintain body temperature by balancing heat loss and production, will normally dissipate heat produced from digestion to the surrounding environment. As ambient temperature rises, heat is less able to exit the body. If temperatures continue to rise, the animal acts as a sink and pulls in heat from the environment. In such situations, animals decrease intake to reduce heat produced from digestion and thereby maintain required body temperature. If temperatures are severe enough that these responses are ineffective, animals will exhibit decreases in weight gain – or even show weight loss – and eventually lower their overall metabolic rate.

Animals may experience feed restriction imposed directly by caretakers, typically for some production benefit. In this classification, animals are fed some amount of feed that is less than the amount consumed if allowed to feed *ad libitum*. Direct quantitative restriction is a common production regimen used in several livestock scenarios.

Restricted feeding for sows/gilts

Restricted feeding of gestating sows and gilts is a popular method to restrict weight gain and prevent obesity, improve feed efficiency, increase **lactation** yields and reduce feed costs. Modern swine production utilizes an animal that has comparatively reduced fat reserves and appetite compared with earlier generations of animals. Combined with an increased litter size, metabolic requirements have grown in spite of reduced endogenous energy reserves. These conditions typically force the sow/gilt into a catabolic state in which she is unable to satisfy her lactation demands and, as a consequence, will lose significant amounts of body fat and other essential tissues. Increasing intake during **gestation** to allow the sow to establish reserves during lactation is one method of correcting for such deficiencies; however, a restricted diet during gestation has been shown to be far more efficient and effective.

Salmon-Legagneur and Rerat (1962) reviewed theoretical work suggesting that increasing feed levels during lactation versus gestation was a more efficient means to deliver nutrients for lactation production versus endogenous reserves. These theoretical predictions were validated by weight gain and 16% greater milk production in sows that were fed restricted levels during gestation and overfed in lactation compared with a second treatment group receiving the reverse. Thus, increasing lactation intake should be the producers' primary goal. Interestingly, restricted feeding during gestation can serve to increase lactation intake. Sows receiving a restricted intake during gestation eat significantly larger quantities during lactation, the converse of which has proved true as well. Further advocating the benefit of such a regimen, animals receiving low feed amounts in gestation and high amounts in lactation consume less feed overall, suggesting a greater feed efficiency.

Interval feeding

Interval feeding is a technique used primarily in swine production to reduce labour requirements where animals have to be fed manually, although some improvements to feed efficiency have been found. In such systems, sows are fed a bulk amount once per interval where the interval may range up to 3 days. Animals receive the same amount of feed over the course of the interval: only its distribution within the interval is altered. Interestingly, Douglas *et al.* (1998) reported that sows on an interval feed schedule exhibited reduced behavioural signs of hunger, although such conclusions need further investigation before application.

Forced moulting

The technique of **forced moulting** used by the poultry industry provides a third specific example of feed restriction for production benefit. Forced or induced moulting is practised by poultry producers to rejuvenate layers in their operation. The practice typically involves all or a combination of: feed removal for 4–14 days; water removal for 1–3 days; and a decrease in light exposure to 6–10 h/day. Adoption of these practices is believed by some to mimic a natural biological process of poultry that occurs as a result of seasonal decreases in food availability and daylight exposure. Such conditions will reduce or eliminate gonadotropic support of the hen's ovary via decreased release of **gonadotropin-releasing hormone (GnRH)**. Forced moulting has been shown to increase egg production by approximately 10% over non-moulted birds. Egg quality has also been shown to improve as well.

Broiler breeders

Genetic selection for rapid growth rates and high feed intake has resulted in broiler chickens that reach **slaughter** weight at 7 weeks of age. The rapid growth rate results in **welfare** issues associated with **bone strength** and organ growth. It also has consequences for the breeding flock, where egg laying does not start until around 24–26 weeks of age. In order for broiler breeders to reach reproductive age, feed restriction has to be practised to prevent obesity and to control **growth rate**.

Overall, the practice of feed restriction, though necessary in production systems that need to control weight gain of breeding animals and birds, may result in chronic hunger of individuals. Feed restriction therefore may have large negative welfare consequences for affected individuals. (MY)

See also: Malnutrition

References and further reading

Close, W.H. and Cole, D.J.A. (2000) *Nutrition of Sows and Boars.* Nottingham University Press, Nottingham, UK.

Douglas, M.W., Cunnick, J.E., Pekas, J.C., Zimmerman, D.R. and von Borell, E.H. (1998) Impact of feeding regimen on behavioral and physiological indicators for feeding motivation and satiety, immune function and performance of gestating sows. *Journal of Animal Science* 76, 2589–2595.

Kyrizakis, I. and Savory, C.J. (1997) Hunger and thirst. In: Appleby, M.C. and Hughes, B.O. (eds) *Animal Welfare.* CAB International, Wallingford, UK, pp. 49–62.

Salmon-Legagneur, E. and Rerat, E. (1962) Nutrition of the sow during pregnancy. In: Morgan, J.T. and Lewis, D. (eds) *Nutrition of Pigs and Poultry.* Butterworths, London, pp. 207–223.

Feelings

Feelings, such as love and happiness, are defined as conscious emotional states generated by physiological and/or psychological processes that may in turn be triggered by sensory perceptions (e.g. visual or tactile) or memories. Scientists have traditionally stressed the importance of **consciousness** when trying to distinguish between feelings and other emotional expressions. This has resulted in the frequent association between **emotions** and instinctive responses on the one side and feelings, a **theory of mind** and abstract thinking on the other. Recent advances in animal **cognition** seem to suggest that some non-anthropoid species may be able to experience some sort of human-like feelings. Moreover, some scientists disagree with the view that feelings have to have a conscious basis and prefer a 'loose' definition of the term to include unconscious emotional states. These considerations have led some scientists to propose that the use of anthropoid primates, and possibly of other species too, for animal experimentation and their housing in **zoos** and laboratories should be banned. (RV)

Further reading

Panksepp, J. (2005) Affective consciousness: core emotional feelings in animals and humans. *Consciousness and Cognition* 14, 30–80.

Feminist ethics

Feminist ethics takes as a starting point a critique of the dominant traditions within ethical theory for having a sexist, male bias. Feminists are concerned with creating gender-equal moral principles, policies and practices, and supplementing masculine ways of moral reasoning by giving prominence to women's experiences. Varieties include maternal and feminine **ethics**. In the case of animal ethics, one popular strand is the ethics of care. Mary Midgley, for example, contends that our responsibilities towards animals are a function of the relationships or community we form with them. The greater our sense of community with certain animals, e.g. **pets** and farm creatures, the greater our sense of responsibility. (RA)

Fence-line behaviour

Fences serve as physical, visual and/or psychological barriers to keep animals within or outside a particular area. Factors that keep animals from penetrating a fence-like barrier are rooted in the physical properties of the fence (i.e. height, porosity, horizontal or vertical spacing within the fence, sharp barbs, voltage, etc.) and the cognitive realization by the animal that it cannot pass through, go over or go under without consequences (i.e. by being physically stuck, receiving an electric shock, jabbing by sharp barbs, etc.). Many farm species learn, with relatively few trials, and often only one trial, to recognize the properties of a fence that allow them to penetrate or be repelled by a particular fence and they are able to associate and apply that knowledge to other locations where they find similar barriers with similar visual cues. However, future invisible or virtual fences may operate by guiding animals via auditory cues or vibrations from their **collars** or ear tags to avoid delivery of an electric shock (Butler *et al.*, 2006).

Motivational changes, such as **hunger**, **fear** or **isolation** will cause animals to try to breach fences that they would normally respect, which explains why fences in more intensive systems or near handling facilities are more formidable in construction. Ruminants also appear to have some awareness of fences acting as a barrier or as protection, separating them from individuals on the opposite side. For example, the normal flight zone that a ruminant would display between a predator, a human or another more dominant **conspecific** is physically reduced when a fence separates the two individuals. Likewise, separating a dam and her suckling offspring, on opposite sides of a fence, will elicit overt signs of **distress** (i.e. increased **vocalizations** and pacing) despite the fact the individuals may only be separated by a few centimetres (see Fig. F.4). However, fence-line **weaning** offers a better alternative, causing fewer overall signs of distress, compared with weaning by abrupt and remote separation of mother and offspring (Price *et al.*, 2003). (JMS)

References

Butler, Z., Corke, P., Peterson, R. and Rus, D. (2006) From robots to animals: virtual fences for controlling cattle. *International Journal of Robotic Research* 25, 485–508.

Price, E.O., Harris, J.E., Borgwardt, R.E., Sween, M.L. and Connor, J.M. (2003) Fenceline contact of beef calves with their dams at weaning reduces the negative effects of separation on behaviour and growth rate. *Journal of Animal Science* 81, 116–121.

Fig. F.4. Photograph of a cow and her calf showing fence-line contact following separation at weaning (image courtesy of Derek Haley).

Feral

An individual or population of domesticated animals that is no longer living in **captivity** may be refereed to as being feral. Such animals have reverted to a wild or semi-wild state, with little or no dependence on humans. Feralization describes the process by which a domestic animal becomes feral, and may involve selection within this non-captive environment. Changes that occur include an increase in their flight distance from humans and possibly the formation of social bonds with other feral **conspecifics**. Feral species may have an important impact on the **ecology** of the areas in which they live, e.g. feral cats can eliminate local wildlife where such a predator does not naturally exist. (KT)

Ferret

A ferret is a small carnivorous mammal of the weasel family. It is similar to a weasel but has longer fur (known and sold as fitch). Ferrets are domesticated strains of the polecat developed for **hunting game animals**, particularly **rabbits**. Polecats grow to nearly 60 cm long, including the tail. The fur of polecats is dark brown above, with yellow patches on the ears and face, whereas the ventral surfaces are nearly black. Their scent glands, located either side of the anus, emit a powerful fetid odour. It has been used for research into influenza in humans, as it is one of the few animals that develop the **disease**. (DBM)

Fetus (foetus)

A stage in prenatal development when the organs become recognizable. At this point the developing **embryo** is known as a fetus, and the remainder of the **gestation** involves mainly fetal growth rather than organ development. (DBM)

Field study

Any study carried out in the 'field', which in its broadest sense is a location where the animals under study can move about at will and potentially interact with members of their own and/or other species. A classic field study does not involve any manipulation of the situation. This does not mean, however, that hypotheses cannot be tested; for example, behaviour could be compared between different existing situations or populations. It is also possible to undertake experimental field studies, in which the classic experimental method of variable manipulation is applied. (KT)

Further reading

Martin, P. and Bateson, P. (1993) *Measuring Behaviour: an Introductory Guide*, 2nd edn. Cambridge University Press, Cambridge, UK.

McFarland, D. (1981) *The Oxford Companion to Animal Behaviour*. Oxford University Press, Oxford, UK.

Fish welfare

Although some of the earliest animal welfare legislation, the United Kingdom Protection of Animals Act (1911), specifically covered fish as well as birds and other domestic animals, public concern and scientific interest in the welfare of fish has notably lagged behind that of the terrestrial animals affected by man. This was probably primarily because of a lack of empathy and understanding of a species adapted to a natural environment so very different from our own. However, there is a recent, growing interest in potential **welfare** issues, as research brings to light the true complexity of fish and the mechanisms with which they are equipped to adapt physiologically and behaviourally to survive.

The term 'fish' covers a wide range of animals. Most scientific work has been carried out on teleost fish (bony, ray-finned fish) and, in terms of the number of species and biomass, teleosts are the typical vertebrates and also those most used by man. This article is primarily concerned with the welfare of teleost fish. An exhaustive discussion of the concept of fish welfare and the issues surrounding fish welfare can be found in Huntingford *et al.* (2006) and others (see References and further reading).

The concept of 'fish welfare' is perhaps only relevant in terms of the consequences to the fish of actions by man. Human activities that affect fish welfare include our effect on the fish environment and, although this is rarely noted, it is our effect on the environment as a whole, through pollution and other disturbances to the ecosystem, including loss and modification of habitat, that may degrade the welfare of the greatest number of individual fish.

Commercial fishing and sport fishing present many potential welfare problems. The process of capture from the wild is likely to be stressful and, in many cases, involves a prolonged **death**, as fish may be subject to injury, dramatic change of pressure, exhaustion, oxygen deficit, crowding and dragging in nets. Once landed there is generally an urgency to process a catch. Fish flesh deteriorates much more rapidly than that of warm-blooded animals, as the enzymes of cold-blooded animals continue to function efficiently after death. The enzymes of warm-blooded animals work less efficiently as the body changes temperature, cooling post-mortem. The requirement to keep fish very fresh up to the point of consumption often leads to a compromise of their welfare, with fish being transported live and even served prepared, but alive and conscious, as may be experienced in some sushi restaurants. In the case of commercial sea fishing, the fish are not humanely killed as is required in most land-based **slaughterhouses** processing domesticated animals. While alive and conscious, fish may be eviscerated, packed in ice and otherwise processed. Demersal fish (bottom feeders), in particular, have been shown to take up to 40 min or more to die following evisceration and packing in ice (Robb and Kestin, 2002).

Besides commercial fishing, the scale of sport fishing should not be underestimated. For example, the 2006 National Survey of Fishing, Hunting and Wildlife-Associated Recreation (US Census Bureau, 2006) shows that in the USA there are 30 million anglers (approximately 13% of the population) spending US$42.2 billion on their sport each year. In part due to depletion in fish stocks from overfishing, there has been a dramatic increase in aquaculture, producing fish for the table. Within Europe, aquaculture has been the fastest growing animal production sector, with an increase in both the number of species farmed and the total amount of fish produced. Farmed fish in Europe alone now include such diverse species as carp, salmon, trout, tilapia, catfish, sea bream, sea bass, eel, turbot, halibut, barramundi and cod. The farming of each new species requires the development of specialized knowledge and associated techniques if potential welfare problems are to be avoided. For example, mistakes in water quality, temperature or feeding when rearing from eggs through larval stages can lead to deformity and **abnormality** of almost all the fish in batches containing hundreds of thousands of animals. Fish may be reared at high stocking densities under highly constrained conditions, affecting their ability to carry out normal behaviours such as feeding, shoaling and **predation/parasite** avoidance. Factors such as these are thought to be the reason why the majority of farmed salmon and trout suffer high levels of fin erosion and why injuries, especially of the eyes, are more prevalent than in wild populations.

Other potential welfare problems in aquaculture include increased exposure to disease, poor water quality and the effects of husbandry procedures such as handling, grading, **transport** and harvest. The majority of fish are not stunned after harvest and death can be prolonged. Many fish are allowed to asphyxiate, often on ice or in an ice slurry. Salmon are frequently exsanguinated by gill cutting, a method which leads to death only after several minutes. A full review of **slaughter** methods is given in Robb and Kestin (2002), in which the majority of methods are found to be unsatisfactory. Electric stun/kill of fish at harvest has been shown to be effective, and methods are beginning to be developed for a range of farmed species and to be adopted by industry, often with support and encouragement from large buyers such as the supermarkets (see, for example, Lines *et al.*, 2003).

Fish are used in scientific research and for toxicity testing of products and of, for example, water supplies. Of the 3.2 million scientific procedures licensed to be carried out within the UK, 10% were carried out on fish (Home Office, 2008). Fish in research may be subjected to deliberately imposed treatments that may have a variety of adverse effects on growth, behaviour and physiology. Fish are also kept for private and public display. Ornamental fish are often captured from the wild using sublethal doses of poison and may undergo extensive transport. In captivity, as in aquaculture, their care depends on owners providing an appropriate environment. It is not uncommon to find predator and prey species kept within the same enclosure, a practice generally frowned upon in, for example, a terrestrial **zoo**.

Potential welfare problems exist only if fish have the ability to suffer. At present the evidence either way is not entirely satisfactory. Rose (2002) puts forward a strong argument that fish are unable to 'experience' **pain** and are unable to suffer, stating 'the reactions of fishes to noxious stimuli are nociceptive and without conscious awareness of pain'. His argument is in the main part based upon a comparison between the neural structures present in fish and those present in mammals, with particular reference to man. An alternative view is summarized in Huntingford *et al.* (2006), which they relate only to adult fish. They draw on the current literature showing that, of those investigated so far, several fish species are capable of learning and integrating information and that this requires processes more complex than simple associative learning. Based on this evidence and supposition they state that 'where there is evidence of fish species with sophisticated cognitive and behavioural processes, the experience of suffering may be a real possibility'. A strong piece of work in support of Huntingford *et al.* (2006) by Sneddon *et al.* (2003) found that when rainbow trout were administered bee sting venom around the mouth they suspended feeding, showed a greatly increased rate of opercular beat and moved to the bottom of the tank to rock gently from side to side. Current research in this area is turning to functional and structural magnetic resonance imaging (MRI) techniques to help investigate **brain** functioning and remodelling during **aversive** episodes. Whatever the eventual outcome of the argument, for the time being it is suggested that the '**precautionary principle**' is adopted. (TK)

References and further reading

Home Office (2008) *Statistics of Scientific Procedures on Living Animals*. The Stationery Office, London. Available at: http://www.homeoffice.gov.uk/rds/pdfs08/spanimals07.pdf (accessed 25 September 2009).

Huntingford, F.A., Adams, C., Braithwaite, V.A., Kadri, S., Pottinger, T.G., Sandoe, P. and Turnbull, J.F. (2006) Current issues in fish welfare. *Journal of Fish Biology* 68, 332–372.

Lines, J.A., Robb, D.H., Kestin, S.C., Crook, S.C. and Benson, T. (2003) Electric stunning: a humane slaughter method for trout. *Aquacultural Engineering* 28, 141–154.

Robb, D.H.F. and Kestin, S.C. (2002) Methods used to kill fish: field observations and literature reviewed. *Animal Welfare* 11, 269–282.

Rose, J.D. (2002) The neurobehavioral nature of fishes and the question of awareness and pain. *Reviews in Fisheries Science* 10, 1–38.

Sneddon, L.U., Braithwaite, V.A. and Gentle, M.J. (2003) Do fishes have nociceptors? Evidence for the evolution of a vertebrate sensory system. *Proceedings of the Royal Society of London Series B* 270, 1115–1121.

St-Hilaire, S., Ellis, T., Cooke, A., North, B.P., Turnbull, J.F., Knowles, T.G. and Kestin, S.C. (2006) Fin erosion on rainbow trout on commercial trout farms in the United Kingdom. *Veterinary Record* 159, 446–450.

US Census Bureau (2006) *National Survey of Fishing, Hunting and Wildlife-Associated Recreation. Quick Facts*. Available at: http://www.census.gov/prod/2008pubs/fhw06-qkfact.pdf (accessed 20 November 2009).

Fission–fusion social system

This is a type of social **grouping** in which all individuals belong to a parent group that can fracture (fission) into smaller stable subgroups or individuals to seek out dispersed resources, such as forages or mates, but afterwards return to the larger parent group (fusion), often using contact calls to facilitate this reunion. There can be fluid movement of individuals between subgroups and even between different parent groups, such that group composition and size are frequently changing. In order for a **social group** to be spatially coherent, its members need to synchronize their activities. However, synchronization can be costly to individuals if it causes them to put off an activity that would benefit them more than what the group is doing. This will occur more in groups comprised of different ages, sexes and sizes, since optimal time allocation is likely to differ between these individuals. Thus, division into smaller subgroups can facilitate **synchronization of behaviour** in individuals with similar requirements. Species that have a fission–fusion social system tend to have very large **brains** relative to their body size, suggesting that this type of system may be cognitively demanding (**see: Intelligence – comparative**), perhaps relating to the difficulty of managing social relationships with individuals which are out of contact for long periods of time. (LMD)

Fitness

In **evolution**, success is defined by the persistence of genes in a population. There are three ways to do so and, together, they define an individual's inclusive fitness. First, survival and doing things that help an individual survive maximize personal fitness. Secondly, **reproduction** and doing things that help an individual reproduce maximize direct fitness. Thirdly, helping relatives survive and reproduce, since kin share common genes (**see: Kinship**), maximizes indirect fitness. (DTB)

See also: Dominance; Individual fitness; Sexual behaviour

Five freedoms

The basis for the five freedoms was first documented in the **Brambell Report** in 1965, but not formalized as such until later. They comprise:

- Freedom from thirst, **hunger** and **malnutrition** (by ready access to fresh water and a diet to maintain full **health** and vigour).
- Freedom from physical and thermal discomfort (by providing an appropriate environment including shelter and a comfortable resting area).
- Freedom from **pain, injury** and **disease** (by prevention or rapid diagnosis and treatment).
- Freedom to express most patterns of normal behaviour (by providing sufficient space, proper facilities and company of the animal's own kind).
- Freedom from **fear** and **distress** (by ensuring conditions and treatments that avoid mental **suffering**). (DBM)

See also: FAWC; Laying hen housing; Measuring welfare

Further reading

Brambell Report (1965) *Report of the Technical Committee to Enquire into the Welfare of Animals Kept under Intensive Livestock Husbandry Systems*. Chairman: Professor F.W. Rogers Brambell. Cmnd. 2836, December 1965. Her Majesty's Stationery Office, London.

Farm Animal Welfare Council (1993) *Second Report on 'Priorities for Research and Development in Farm Animal Welfare'*. Ministry of Agriculture Fisheries and Food, Tolworth, London.

Flehmen

The term flehmen was introduced by Schneider in 1930 and refers to a conspicuous behaviour performed by ungulates and members of many other taxa in response to **pheromones** from other animals, usually **conspecifics**, but also other species. Pheromone is the term used for a chemical that transmits messages between animals. The flehmen response involves a characteristic **posture**: the animal stands erect, extends its neck, raises its head, flares the nostrils, opens its mouth slightly and curls the upper lip, often exposing the upper gum (performs a 'grimace'; see Fig. F.5).

Flehmen is generally performed in response to the smelling and/or licking of **urine** and/or the anogenital region of another animal. It is often performed repeatedly to the same stimulus and may be maintained for a minute or more. The behaviour is thought to facilitate the transport of relatively non-volatile compounds to the **vomeronasal** (or Jacobson's) **organ**, which is a chemoreceptor organ located in the vomer bone between the nasal passages and the mouth. Thus the vomeronasal organ appears to be used for the perception and analysis of pheromones, allowing the animal performing flehmen to gain information about the animal emitting the pheromones.

Generally, flehmen is performed by a male to the urine of a female, is often seen in a sexual context and, thus, one function of flehmen may be to allow a male to determine the reproductive status of a female. However, this cannot be the only function, as it occurs interspecifically, and in some species it is performed during female–female and male–male encounters. There are also reports that, in some species, females perform flehmen to neonates and amniotic fluid. Thus it may be that

Fig. F.5. Sable antelope (*Hippotragus niger*) showing flehmen to urine (image courtesy of, and reproduced with the permission of, Kaci Thompson, University of Maryland, College Park, Maryland).

flehmen is used for individual **recognition**, as in offspring recognition by mothers, and recognition of certain attributes of an individual, such as social status, which may be revealed through hormonal levels. (JCP)

See also: Sexual behaviour

Flight behaviour

To help grazing animals avoid predators, they have wide-angle or 360° panoramic vision. This enables them to continually scan for predators while they are **grazing**. The principles of flight behaviours are similar in **horses**, **cattle**, **sheep**, **deer**, antelope and other grazing animals. To facilitate scanning for danger during grazing, the retina has a horizontal band with a high density of sensitive cells called the visual streak. Grazing animals are dichromats, which are partially colour blind, and their retinas are most sensitive to the colours yellowish-green and bluish-purple. Dichromatic vision may provide better vision at night and make detection of motion easier.

Sudden movement is very stimulating to the nervous system. Research by Joseph LeDoux indicates that sudden movement stimulates the amygdala, which is that part of the **brain** that controls **fear** responses. This results in the grazing animal taking action against the danger. Small, fine-boned grazing animals, such as antelope, avoid predators by fleeing, but animals such as cattle, with heavier bones and bodies, will also turn and fight a predator instead of running. It has been found that those cattle most likely to become agitated in an auction ring and ram the fence were those that were also most likely to flinch and jump in response to the waving of the auctioneer's hand when he took a bid. The animals with the flighty **temperament** were the most sensitive to rapid movement (Lanier and Grandin, 2000).

A novel sudden movement is more likely to elicit flight behaviour compared with a familiar one. For example, Nyala antelope in a **zoo** have little fear of children running by the fence outside their exhibit, but the novelty of people moving around on the roof of a building within the enclosure can provoke an intense flight reaction and the animals may run into the fence. In the life of a grazing animal, new things are both **aversive** and attractive. This is the paradoxical effect of novelty. Cattle may baulk and refuse to move if they are driven by people towards a novel strange flag, but that same flag will attract the cattle and cause them to approach if placed in the middle of their pasture. Novelty is aversive when it is suddenly introduced and attractive when the animal can voluntarily approach it (**see: Approach test**).

Flight zone size and principle

The size of the **flight zone** is the distance an animal will allow a person to approach before it moves away. The artificial removal of the flight zone is the process of **taming** (Hedigar, 1968). Tame animals will allow themselves to be touched and handled. Other factors affecting the size of the flight zone include genetics and whether the animals have become agitated. Herds of cattle that have been handled quietly will have a smaller flight zone compared with herds that have been handled roughly by people who yell, scream and hit them. The size of an object will also affect the size of the flight zone: when a person approaches full face, the flight zone will often be larger than when approaching with a smaller sideways profile.

The design of handling facilities can reduce the size of the flight zone. The size of the flight zone diminishes if there is a solid fence that blocks vision between a person and the animal. Solid sides help to prevent dangerous flight behaviour when wild, untamed **grazing** animals are handled. Blindfolding an animal with completely opaque material can also help to maintain calm.

The use of solid sides is most important when animals with a large flight zone are handled (**see also: Restraint**). Solid sides should be installed on restraint devices, single-file races, truck loading ramps and the crowd pen that leads up to the single-file race. When completely tame animals are handled, such as a tame draught water **buffalo** that is trained to lead or a completely tame dairy cow, solid sides are not needed. However, a water buffalo that has become **feral** and allowed to run wild will have a large flight zone and solid sides will be a necessity. Grazing animals are controlled by vision. A flimsy solid fence made of thin opaque plastic or cloth may hold them in because they will not run through a solid barrier. Groups of animals can be moved by people holding a long strip of plastic. A solid top on a single-file race or a **restraint** device will inhibit rearing. Wild horses, bison and other animals will often rear up when a person gets too close. A piece of cardboard or other solid material placed over the top of the race will often stop rearing.

Handling principles

Cattle can be moved most efficiently if the handler works on the edge of the flight zone. The animals move away when the flight zone is penetrated and stop when the handler retreats. The edge of the flight zone is not distinct or set, and approaching an animal quickly will enlarge the flight zone. Excited cattle have a larger flight zone, and eye contact with the animal will also enlarge the flight zone.

To make an animal go forward, the handler should stand in the shaded area marked A and B in Fig. F.6. The handler must stay out of the blind spot right behind the animal's rear. At close quarters, a handler is likely to get kicked if he/she enters the blind spot. In pastures, entering the blind spot will cause the animal to turn and look at the handler; the animal does this to determine where the handler is. To make an animal move forward, the handler must remain behind the point of balance at the shoulder (see Fig. F.6). To make the animal move backwards, the handler should be positioned in front of the point of balance. Handlers should avoid standing in front of the point of balance when they want to make an animal move forward.

Use of the movement patterns shown in Fig. F.7 will induce cattle and other grazing animals to move forward in a race. The animal will move forward when the handler walks past the point of balance in the opposite direction of desired movement. The handler can turn cattle in a small pen by using a flag on a stick to block the animal's vision on one side. Moving the flag towards the right side of the animal's head will make it turn left. When an animal is faced head on it will turn right if the handler moves left and vice versa.

Survival behaviour pattern

Grazing animals have two basic survival behaviour patterns relating to flight. They are: (i) turn and face a potentially

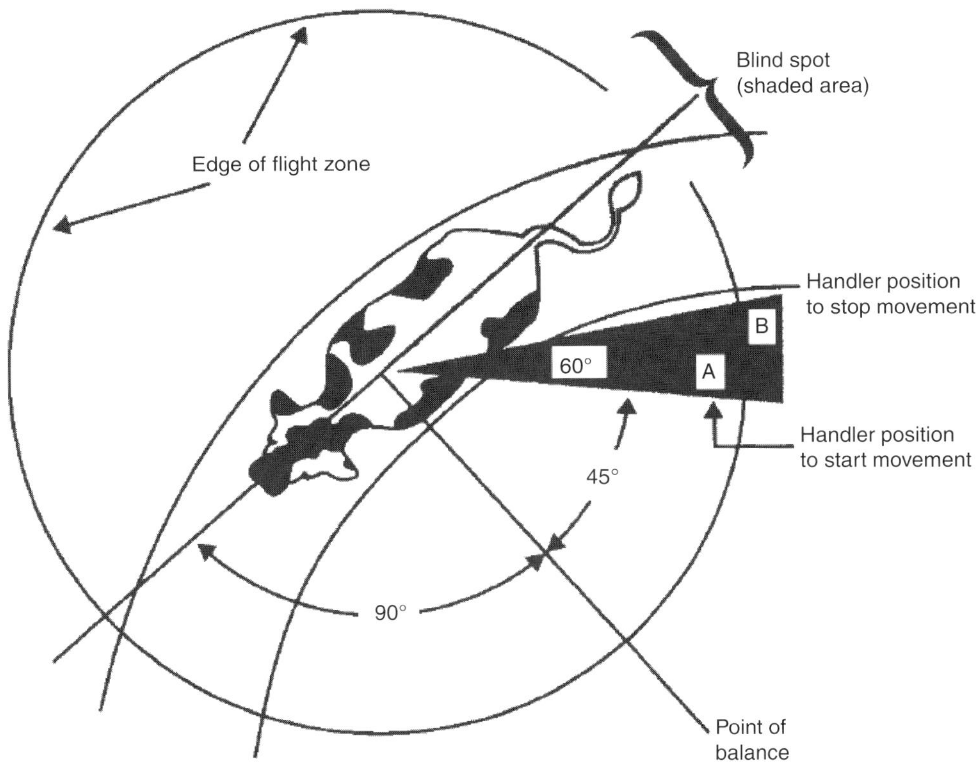

Fig. F.6. Flight zone diagram showing the most effective handler positions for moving an animal forward.

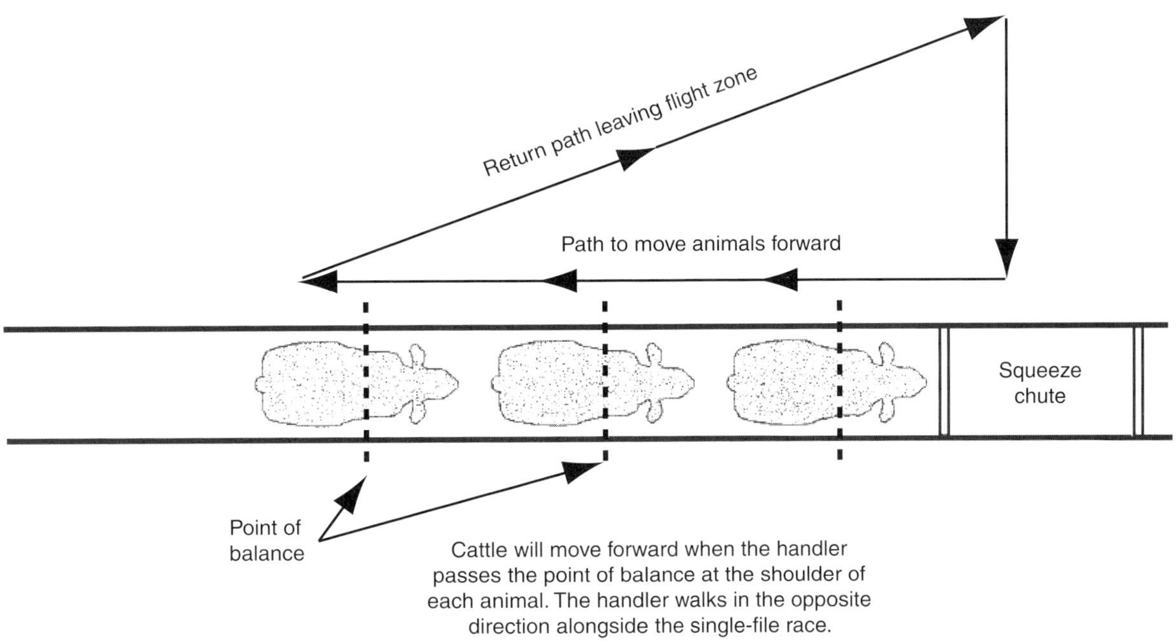

Fig. F.7. Handler movements to induce cattle to move forward in a race.

threatening stimulus but keep a safe distance (flight zone); and (ii) use the point of balance (Grandin, 2000). Turning and looking at potential danger enables the animal to know where a potential predator is. A distant predator that can be seen is not going to eat them. Grazing animals appear to detect whether or not the predator is looking at them. A lion that is looking at them is a greater threat to them than a lion passing through. Point of balance helps a grazing animal get away from a chasing predator. This appears to be a hard-wired behaviour that helps the animal evade capture.

Depending upon the species of grazing animal, the strategy for protection from predators varies from pure flight to turn and fight. Antelopes and horses are more flighty and survive by fleeing. Cattle use both fleeing and turning on a predator and attacking, and the cape buffalo attacks. Other grazing animals stay together in larger herds to provide safety in numbers. When the herd is threatened, the animals bunch together. (TG)

See also: Handling; Predator avoidance

References and further reading

Grandin, T. (1987) Animal handling. *Veterinary Clinics of North America: Food Animal Practice* 3, 323–338.

Grandin, T. (2000) Behavioural principles of handling cattle and other grazing animals under extensive conditions. In: Grandin, T. (ed.) *Livestock Handling and Transport*. CAB International, Wallingford, UK, pp. 63–85.

Hedigar, H. (1968) *The Psychology of Animals in Zoos and Circuses*. Dover Publications, New York.

Lanier, J.L. and Grandin, T. (2000) The relationship between reaction to sudden intermittent movement and sounds to temperament. *Journal of Animal Science* 78, 1467–1474.

LeDoux, J.E. (1996) *The Emotional Brain*. Simon and Schuster, New York.

Flight zone

The flight zone is the area around an individual that it actively attempts to keep free from interference from other animals or objects. Intrusion within the flight zone elicits an escape response. The distance from an animal's head to the edge of the flight zone is known as the flight (initiation) distance. The size of the flight zone is thought to be determined by an animal's genetics, sex, age, body size, **health**, reproductive condition, **temperament** and its perception of the threat. The amount of contact with other animals and the quality of that contact will also affect the length of the flight distance, with **wild animals** typically having a much larger flight zone than their domesticated counterparts (**see: Domestication**). Habitat type and the distance of travel required to reach a refuge can influence flight distance, as can the speed, agility and behaviour of the encroaching animal and the direction from which it is advancing. **Cattle**, for example, will have a larger flight zone when approached from the front than from the side or back. For arboreal species, the horizontal, vertical and diagonal position of the approaching animal can influence the length of their flight distance.

Understanding the flight zone is important when **handling** animals. Handlers working on the edge of a flight zone are able to control an animal's movement by approaching and withdrawing from the animal. However, penetrating the flight zone too deeply may cause the animals to flee. Conversely, a handler standing outside an animal's flight zone may have difficulty getting that animal to move. Sufficient handling and human socialization can decrease an animal's flight zone to the point where the flight zone may disappear. Animals with no flight zone can be difficult to handle and, on occasions, aggressive towards humans. (MKP)

See also: Restraint

Flocking behaviour

Flocking behaviour is shown by many gregarious species, but particularly refers to when a group of **sheep** or birds cluster together. Many sheep breeds exhibit this behaviour, which originated from a response to the presence of a perceived predator. It arises from the advantage of having a greater number of eyes looking out for predators, which increases the chances of their detection (**see: Vigilance**). **Grazing** can occur in relative safety when many individuals are keeping watch. As a consequence dietary intake is increased as more time can be spent grazing rather than on the lookout for predators. In areas where no natural predators exist for the species, such as Iceland for sheep, the local breeds of sheep show little flocking behaviour.

Flocking behaviour is a useful trait for many farmers, as a large, compact group of stock is much easier to handle and move than a fragmented set of individuals. The sheep exhibit this behaviour because they perceive the humans and any working **dogs** to be threats, and thus cluster together. However, the behaviour can give rise to problems in situations where feed is limited. Overgrazing of highly palatable areas will occur readily, which leads to environmental damage and **welfare** problems of **malnutrition** and, in extreme cases, **starvation**.

Connected to flocking behaviour is hefting – where sheep naturally graze around the areas in which they were born and grazed as lambs within the flock. This behaviour is highly beneficial for shepherds on open mountain land, as it negates the need for expensive walls or fences (often on land on which they may be difficult to construct). In combination with flocking behaviour, these two interlocking behavioural traits give the shepherd an advantage when managing the stock. (MM, HO)

Flooding

When an animal experiences a **stimulus** that evokes an emotional response such as **fear**, a behavioural response is elicited, which may be associated causally or coincidentally with the removal of the initial stimulus. As a consequence, the response will be reinforced and thus it is likely to be repeated at future exposures. Also, if a non-significant but arousing stimulus is removed before **habituation** occurs, the converse process, **sensitization**, may result. Such responses may form the basis of problem behaviours, as they can be difficult to eliminate if the animal is never given an opportunity to experience the stimulus and learn to tolerate or exhibit an alternative, more desirable response. One of the behaviour modification techniques that can be employed to change such responses, if they are inconsistent with the level of threat posed by the stimulus, is flooding.

Flooding (also referred to as response prevention) involves exposing the animal to the stimulus, while preventing it from escaping or avoiding the stimulus, until the negative response that is exhibited ceases. The aim of the procedure is to enable the animal to either habituate or learn an alternative appropriate response to the stimulus, through the removal of reinforcement for the behaviour (**extinction** of the learned response).

The stimulus can be presented at full strength from the earliest exposure or, alternatively, graded exposures to ever-increasing strengths can be presented. For elimination of the response to occur it is important that the stimulus is not withdrawn before cessation of the response occurs, because if the animal still remains in a state of heightened **arousal** and the stimulus is withdrawn, the response may be negatively reinforced (**see: Reinforcement – types of**) and hence strengthened, or sensitization may occur as a result of the heightened arousal experienced. Such incorrect implementation of flooding is one of the significant risks of the procedure.

Alternatively, an inadequate time lapse between exposures that does not allow complete physiological and behavioural abolition of arousal before the next exposure may lead to a similar outcome. Another cause for concern is that, if the animal is being restrained and exposed to uncontrollable **aversion**, **learned helplessness** may result. In this case the animal will be passive, and may superficially appear to have improved, but its **welfare** is seriously compromised. Even if the animal is not experiencing **distress** due to the stimulus (e.g. it is being used by an animal that gets overexcited), it will be subjected to **frustration** if the response has been previously reinforced, and this in itself can be highly aversive and seriously compromise the animal's welfare. Therefore, flooding should only be used as a clinical tool for behaviour modification with caution, after the rejection of other alternative methods of eliminating the behaviour (e.g. **systematic desensitization** and **counter-conditioning**), and only following a careful assessment of all aspects of the welfare and safety for all people and animals involved in the procedure.

(HZ)

Further reading

Lieberman, D.A. (2000) *Learning, Behaviour and Cognition*, 3rd edn. Wadsworth Publishing Co., Belmont, California.

Mills, D.S. (2002) Learning, training and behaviour modification techniques. In: Horwitz, D., Mills, D.S. and Heath, S. (eds) *BSAVA Manual of Canine and Feline Behavioural Medicine*. BSAVA, Quedgeley, UK.

Flooring

The floor is the point of intimate contact between animals and their **housing**. The type of flooring used in the housing affects the **welfare** of the animals in a number of ways. Use of hard or uncomfortable flooring surfaces in the areas where animals rest or **sleep** can substantially reduce the amount of time that the animals lie down. This is particularly true for large animals such as adult cattle or pigs. The precise effect of this on animal welfare is hard to determine, but adequate sleep and rest are clearly essential for the welfare of most animals (**see: Resting behaviour**). Furthermore, research on dairy cows suggests that a reduction in lying time, if sufficiently large, can increase the incidence of **lameness**.

Most research on resting surfaces has used lactating dairy cattle, and one firm conclusion is that cubicles for dairy cows should not have bare concrete floors, which have been found to reduce resting time by up to 20%. This is associated with an increased incidence of swollen (front) knee joints that comes about because cows initially put their weight on their front knees when both lying down and standing up. The swelling in the knees makes both standing up and lying down difficult, with the result that, when cows must lie down on hard floors, they spend more time standing and lie down less often.

Preference tests have shown that, in general, cattle, horses, sheep and pigs prefer to lie on well-bedded surfaces, e.g. deep straw. Straw bedding can provide the means for pigs to express their rooting behaviour. Growing pigs raised on straw show less nibbling and rooting at pen fixtures and pen-mates than those raised on bare floors. However, concern about bacterial or other contamination (for example, leading to increased **mastitis** in lactating cows or foot-pad dermatitis in broiler chickens) has led farmers to reduce their use of organic bedding. Sometimes this results in animals having to lie on bare concrete floors, although in the dairy industry, particularly, there is much interest in alternatives such as sand, mats or mattresses (or even waterbeds). It is essential for the welfare of dairy cows that floors be sufficiently soft.

The type of flooring used in lying areas can also be important for **thermoregulation**. Thermoregulatory requirements of animals will vary with the age of the animal, the type of production and the climatic area in which the animals are kept. Evidence for the importance of thermoregulatory function in floors is that shorn sheep prefer deep straw bedding over bare floors, but this preference is less obvious in unshorn sheep. Growing pigs prefer deep straw bedding at colder temperatures, but prefer bare concrete floors at high temperatures. Young calves, lambs and piglets are most likely to suffer from hypothermia, and so require warm, well-insulated lying surfaces. In contrast, high-producing dairy cows in peak **lactation** and older milk-fed veal calves generate considerable metabolic heat and tend to suffer more from **heat stress** than **cold stress**. Their prime thermoregulatory requirements are, therefore, for losing body heat. Heat stress can be a particular problem for beef cattle in many of the hotter parts of the world. For example, heatwaves are a common occurrence in the south-western states of the USA, and thousands of cattle have been known to die as a result. Lying down on a cool, highly conductive surface can aid the animals in losing heat. Wetting the ground in beef **feedlots** has been shown to help the animals to cope with high temperatures.

The type of flooring in areas where the animals walk or stand is important for animal welfare, since poor flooring choices can lead animals to slip (leading to injury) or can directly injure the animals' hooves or feet. To avoid slipping, floors need to have sufficient traction. Floors become slippery when wet or covered with **urine** and manure, and smooth surfaces – such as smooth concrete – should be avoided. Surface friction can be increased by adding some texture to the surface, for example by grooving the concrete. This is preferable to smooth concrete but there are still problems with the hardness of concrete. Softer floors provide more traction than harder floors, since the hooves sink into the floor. Cattle prefer to stand on softer surfaces than concrete, and the increased forces on the hoof when cattle walk on concrete increase the risk of physical **injury** to the hoof. The use of concrete floors in cattle houses has been implicated as a risk factor for increased lameness in dairy cows. Some dairy farmers cover the areas where cattle walk with special rubber mats, which are soft but have high surface friction. This increases the speed at which the cows walk and reduces the chance of slipping. Using softer flooring where cows walk and stand can reduce the amount of time the cows spend standing in their stalls, while softer floors in front of feed bunks may increase the time spent feeding.

Good drainage is essential in ensuring that floors are dry. Wet floors increase the chance of slipping but also result in the hooves of cattle, sheep and pigs becoming softer. The hooves of cattle absorb water rapidly and become soft as a result. Increased softness increases the chance of hoof lesions. Wet floors, especially floors covered with slurry, harbour bacteria, thereby increasing the risk of transmission of infectious diseases. Wet floors have been implicated as a risk factor in dermatitis in dairy cows and, in beef feedlots or **lairage** facilities, have been found to have large bacterial counts.

Slatted floors, in which solid slats are separated by gaps or 'slots', are commonly found in indoor housing for pigs, beef and veal cattle, and in some indoor dairy barns. Generally, these are made of concrete, but wooden slats are also used, especially in veal production. These have some advantages in that they allow urine to flow through so that floors remain drier. Cows housed on slatted floors tend to have drier, harder feet. Slaughter pigs kept on partly slatted floors show less evidence of *Salmonella* infection than those on solid floors. The extent to which faeces is also removed depends on the relative size of the slats and slots and on the amount of animal traffic. Often, the performance of slatted floors in removing faeces is disappointing. Furthermore, poorly designed slatted floors can interfere with normal locomotion of the animals and can increase the risk of physical injury to the hoof, especially if slots are too wide or the slats have rough edges. Slatted floors are a risk factor for **tail tip necrosis** in indoor-housed beef cattle. Another major problem with slatted floors is poor air quality in barns, due to high **stocking density** and inadequate ventilation. (JRu)

See also: Laminitis

Further reading

Mills, D.S., Eckley, E. and Cooper, J.J. (2000) Thoroughbred bedding preferences, associated behaviour differences and their implications for equine welfare. *Animal Science* 70, 95–106.

Rushen, J., de Passillé, A.M., von Keyserlingk, M.A.G. and Weary, D.M. (2008) *The Welfare of Cattle*. Springer, Dordrecht, The Netherlands, 310 pp.

Telezhenko, E. and Bergsten, C. (2005) Influence of floor type on the locomotion of dairy cows. *Applied Animal Behaviour Science* 93, 183–197.

Tuyttens, F.A.M. (2005) The importance of straw for pig and cattle welfare: a review. *Applied Animal Behaviour Science* 92, 261–282.

Fluctuating asymmetry

Symmetry in body form, particularly bilateral symmetry, is a widespread trait among animals. However, perfect symmetry rarely, if ever, exists. Typically, individuals with bilateral symmetry exhibit some small differences from an exact mirror image on both sides. These random, small deviations from perfect symmetry are termed fluctuating asymmetry.

By definition, calculations of fluctuating asymmetry are obtained by measuring the same morphological trait on each symmetrical side (e.g. the size of the left and right ears on an animal). This yields a signed value which, when measured over many individuals, should have a unimodal distribution with a mean of zero. In fact, the distribution of deviations is used as a defining characteristic to distinguish fluctuating asymmetry from other types of asymmetry such as directional asymmetry or antisymmetry.

While the exact mechanism generating these small random deviations is largely unknown, fluctuating asymmetry is thought to be a measure of developmental instability. The phenotypic expression of a trait is determined by both internal and external factors. Given one's **genotype** and environmental conditions, the left and right characters in an organism with bilateral symmetry should develop identically. However, the developmental pathway is subject to small, random errors that can create disturbances to cellular growth. This developmental noise can produce small, random variation from perfect symmetry. The likelihood of developmental noise resulting in fluctuating asymmetry depends on an organism's capacity to normalize and control variance in cellular growth despite perturbations. Developmental stability thus acts to reduce morphological differences, minimizing fluctuating asymmetry. Conversely, developmental instability reflects an individual's sensitivity to developmental noise, evidenced by the phenotypic expression of these stochastic processes that create fluctuating asymmetry.

The degree of developmental instability of an organism is influenced by the interaction between its genetic composition and environment. Environmental and genetic stressors, such as unfavourable environmental conditions, food limitations, toxins, **parasites**, **inbreeding**, **mutations** and hybridization, can tend to increase developmental instability. Because of the genetic basis of developmental instability, fluctuating asymmetry may play a role in **individual fitness**. In general, individuals with greater asymmetry have lower reproductive success and/or reduced survivorship. This relationship may be due to the allocation of energy needed during development. Animals developing under suboptimal conditions may be more susceptible to developmental noise because less energy is available for the metabolic mechanisms that control developmental stability. Furthermore, animals in suboptimal conditions need to allocate more energy for basal metabolism than for growth or **reproduction**. This in turn results in reduced **fitness**.

For example, western fence lizards (*Sceloporus occidentalis*) were found to have a higher degree of fluctuating asymmetry at sites that were designated for off-road vehicle use; while this observational study was only able to demonstrate a correlation between sites with off-road vehicle use and fluctuating asymmetry, it may indicate a potential effect of added environmental **stress**. Similarly, parasites can place a considerable drain on their host's energy budget, which can negatively affect the development of their host. In fact, parasitism has been shown to be positively correlated with fluctuating asymmetry. More asymmetrical individuals also may be more at risk to **predation** or less able to compete. In one study observing the interactions of gemsbok (*Oryx gazella*) at waterholes, females with asymmetrical horns tended to be subdominant. Morphological asymmetry, particularly in traits that affect mobility, may make it more difficult for individuals to escape predation and therefore become easier prey items. For instance, a comparison of fluctuating asymmetry among prey items consumed showed that birds consumed domestic flies (*Musca domestica*) with asymmetrical wings and tibia more frequently than those that were more symmetrical.

Fluctuating asymmetry may also be an important factor in **sexual selection**, in terms of competition for mates or **mate**

choice. Significant negative relationships between fluctuating asymmetry and mating success have been found. For example, barn swallows (*Hirundo rustica*) were found to have lower fecundity when their tail feathers were experimentally manipulated to be more asymmetrical. This negative relationship may be more profound in males because, in many species, females tend to be choosier regarding mate selection and males tend to engage in more contest competition for access to mates. Due to the force of sexual selection, the negative effect of fluctuating asymmetry on mating success is also stronger when it involves secondary sexual characteristics. While potentially increasing mate attractiveness or competitive ability, elaborate sexual ornaments can be energetically costly and more susceptible to environmental variation. Thus fluctuating asymmetry in secondary sexual traits may reflect the genetic quality of the individual as well as the degree of environmental stress that the individual experienced during development. Both mechanisms can produce a negative relationship between size of sexual traits and fluctuating asymmetry.

Genetic factors may also influence the prevalence of fluctuating asymmetry. Using both wild and inbred lines of fruit flies (*Drosophila* spp.), the distribution of fluctuating asymmetry was shown to be directly related to the genetic differences between individuals and their level of developmental stability. The breakdown of genetic co-adaptations, as seen in hybridization studies on fruit flies and ninespine stickleback fish (*Pungitius* spp.), can lead to increased developmental instability. In addition, mutations or chromosomal abnormalities can create morphological asymmetries.

However, the relationship between fluctuating asymmetry and fitness is complicated by many confounding variables. For instance, the **heritability** of fluctuating asymmetry appears to be somewhat low. Furthermore, increasing environmental variance of phenotypic traits can also affect their heritability. Patterns between fluctuating asymmetry and fitness may be obscured in high-quality environments and only detected under environmental stress. Measurement error can also be a significant problem, since fluctuating asymmetry is typically less than 1–2% of mean trait size. Measurements of fluctuating asymmetry over several traits are needed to more accurately estimate developmental instability, since single-trait fluctuating asymmetry tends to underestimate the relationship for developmental instability. An added complication is the fact that the expression of fluctuating asymmetry differs among morphological traits, fitness components and stress factors.

Despite these complications, fluctuating asymmetry may be an important indicator of fitness and environmental stress. This relationship may be due to direct or indirect effects. As our knowledge of the mechanisms of developmental instability increases, the application of fluctuating asymmetry as a tool to understand these patterns will become more useful. (LAN)

See also: Intrasexual selection; Mate choice; Selection

Further reading

Dongen, S.V. (2006) Fluctuating asymmetry and developmental instability in evolutionary biology: past, present and future. *Journal of Evolutionary Biology* 19, 1727–1743.

Møller, A.P. (1997) Developmental stability and fitness: a review. *The American Naturalist* 149, 916–932.

Fly strike

Fly strike (also known as fly-blown) is a cutaneous myiasis caused by parasitic dipterous fly larvae feeding on the host's necrotic or living tissue. This disease is common in **sheep**. The female flies lay their eggs on the animal in damp areas, mainly in the urogenital or anal areas. The eggs hatch within a few hours and the larvae begin feeding on the animal's tissue, rapidly leading to bacterial infection and possibly **death** if left untreated. Fly strike in sheep is of particularly significant economic and **welfare** importance. Prevention can include general good **health**, good hygiene, regular monitoring of the animal, **mulesing** and tail docking (**see: Docking – tail**). (SL)

Foie gras

Foie gras (or 'fat liver' in English) is a French delicacy in which the fattened liver of a duck or goose is made into a paté, the taste of which is described as rich and buttery. Migratory birds, such as ducks and geese, are used for foie gras as they have a high capacity for weight gain, especially in the liver, and in the wild may double their weight when preparing for migration. Birds in foie gras production are fed forages for the first few months of life, then the gavage or 'completion of fattening' phase begins in which birds are force-fed through a funnel fitted with a long tube (gavage) that forces the feed into the animal's oesophagus. This feed is usually comprised of maize boiled with fat, and is fed for 12–15 days twice a day for ducks, and 15–18 days up to four times a day for geese, with the birds receiving approximately one-third of their body weight in feed each day. The end result is a liver with fatty deposits approximately six to ten times the size of an ordinary liver. Around 75% of the world's production and consumption of foie gras is in France, with roughly 24 million ducks and half a million geese slaughtered each year, with the USA and China also having large markets.

Production of foie gras is controversial, and its production and sale have been banned in a number of European Union countries. Criticisms of force-feeding include the potential for oesophageal damage and increased risk of infection from the gavage, and impaired liver functioning from the excess fat deposited. As the liver increases in size, the abdomen becomes distended and the liver puts pressure on other organs, making breathing and walking difficult. Additionally, during the force-feeding phase birds are usually kept in small cages that do not allow for much movement or performance of many normal behaviour patterns, and the birds may undergo **beak trimming** at a young age to decrease **feather pecking** and **cannibalism**. In addition to the animal **welfare** concerns associated with foie gras production, recent research has found a possible link between foie gras and amyloidosis-related disorders, such as Creutzfeldt–Jakob disease or bovine spongiform encephalopathy, and Type II diabetes. However, more research is needed to establish this link more convincingly. (LMD)

Further reading

Scientific Committee on Animal Health and Animal Welfare (1998) *Welfare Aspects of the Production of Foie Gras in Ducks and Geese*. Available at: http://ec.europa.eu/food/animal/welfare/international/out17_en.pdf (accessed 5 December 2008).

Solomon, A., Richey, T., Murphy, C.L., Weiss, D.T., Wall, J.S., Westermark, G.T. and Westermark, P. (2007) Amyloidogenic potential of foie gras. *Proceedings of the National Academy of Sciences* 104, 10998–11001.

Follicle-stimulating hormone (FSH)

Follicle-stimulating hormone is a glycoprotein hormone produced in the anterior **pituitary gland** that stimulates the maturation of the germ cells in the **gonads** of both sexes, i.e. in the female it causes the growth of the follicles in the ovary (hence the name) and in the male it stimulates the sertoli cells of the testes, which control the maturation of the spermatozoa. The release of FSH is controlled by the release of the peptide hormone **gonadotropin-releasing hormone (GnRH)** from the **hypothalamus**. FSH production – and thus fertility – can be suppressed as a result of chronic **stress**, which (together with an increase in **endorphin** and **enkephalin** levels) suppresses the release of GnRH, although the effect does not appear to be reliable, and FSH levels are not widely used in this context. (DSM)

Follower species

Among the **precocial** ungulate species, those in which the young immediately start moving around with their mother and keep in close contact with her are designated as follower species. They are generally considered to be the opposite of the **hiders**. Either strategy relies on the early establishment of a strong mother–young bond (**see: Bonding – parent–offspring**). The mobility of the young follower, combined with birth synchrony, which leads to a large number of young animals being present at the same time, is part of an anti-predator strategy based on flight and protection gained from the group in open habitat. (SL)

Footrot

Footrot is the term given to a painful condition in which bacteria infect the hoof of **sheep** or, less commonly, **cattle**, **goats** and **deer**. A major problem in many temperate sheep production systems, footrot is caused by two different bacteria – *Fusobacterium necrophorum* and *Bacteroides nodosus* – but the latter invades the skin only if it has already been damaged. *F. necrophorum* can survive for considerable amounts of time in many weather conditions and situations, though infection rates increase dramatically with wet conditions.

The condition causes **lameness**, as the animal finds it painful to bear pressure on the affected foot. In the worst cases the animal is unable to walk. Other symptoms range from a distinctive nod when walking (mild) through hobbling, stopping to hold a hoof off the ground and, in some of the most severe cases, not being able to use either front leg. One distinctive sign is highly discoloured knees.

Footrot has a detrimental impact on the **welfare** and productivity of affected animals. Feed intake is reduced, especially if the flock is fed concentrate, as affected animals cannot compete equally at the feeding trough. Painful feet also hinder effective mating, especially in countries like the UK where the autumn mating season corresponds to the warm, moist weather conditions that favour severe outbreaks.

Housed animals are particularly at risk from footrot, and thus the regular provision of clean bedding is essential. Treatments include the use of antibiotics and foot trimming (cutting away the overgrown hoof). This stops the hoof from rubbing and damaging the soft flesh and, more importantly, allows air to reach the **inflammation**. Neither of the bacteria responsible can survive in the presence of oxygen (Henderson, 1990). Spraying with an antibiotic aerosol is typically used after trimming.

For control of footrot, farmers may use footbaths (containing formalin or zinc sulfate) in conjunction with foot trimming. Contamination with mud and dung decreases the effectiveness, and often the time each animal spends in the footbath is inadequate. In conjunction with these treatments, eradication is possible by selection of resistant ewes, culling of the worst cases and utilization of clean pasture, though difficult in practice. (MM, HO)

Reference

Henderson, D.C. (1990) *The Veterinary Book for Sheep Farmers*. Farming Press, Tonbridge, UK.

Foraging behaviour

Foraging behaviour is the set of actions an animal takes to acquire energy and nutrients through **feeding**. The foraging behaviours used by an animal have evolved in concert with the physiology and morphology of the animal and the environment in which it exists. For example, **grazing** animals are herbivores found in vegetated areas, having specialized digestive systems that allow them to break down plant cellulose for energy. The set of behaviours used may be determined either by the individual (e.g. a grazing moose) and/or a group of individuals (e.g. a pack of wolves capturing a deer). For the moose, the foraging behaviour is one of a solitary animal moving through its environment locating specific types of vegetation on which to browse. For the pack of wolves, the foraging behaviour is one of a group of animals working together to bring down the deer so that all members of the group can feed.

The selection of food items for consumption by a foraging animal is dictated by many factors such as taste (e.g. clover versus grass for a horse), size limits (e.g. food items smaller than the gape limit of a fish's mouth), ease of capture (e.g. fast-moving minnow versus slow-moving snail), visual markings (e.g. avoidance of toxic insects by birds based on markings on the insect), abundance (e.g. more abundant food items may be easier to obtain in quantity) and energy content (e.g. a low-caloric-density insect versus a high-caloric-density fish).

Food availability and the condition of the animal affect the degree of **selection** an animal uses in deciding whether or not to consume specific food items. If a preferred food item, based on the selection criteria used by the animal, is present in adequate amounts an animal will tend to feed primarily on that food item. If the preferred food item is not available or is scarce, an animal will tend to feed on less preferred items. For example, a fish may prefer to feed on other fish but feeds on insects because no other fish of a size within the gape limit of its mouth are available, while insects are available. These behaviours are further affected by the recent foraging history of the animal. If an animal has been feeding successfully and has good energy stores it will tend to be more selective in what it chooses to consume, even choosing not to consume readily available, less preferred food items while searching for more preferred food items. If an animal has not recently been successful during foraging efforts and the state of its energy stores has been declining, it will tend to be much less selective in the food it chooses to consume, feeding on any food available.

Other factors that influence the degree of behavioural selection employed include the presence of intra- and inter-specific competitors and predators. The more effective an individual animal is as a competitor (e.g. faster, stronger, more observant, etc.), the more selective it can be in the presence of other competitors. Less effective competitors can not be as selective in choosing what to consume and must feed on less preferred available food items in order to survive. The presence of potential predators alters which food items are selected, based on the perceived degree of threat that a predator presents when feeding on the various food items available in different habitats. For example, a small fish may prefer a particular aquatic insect found in open water but feeds on a less preferred insect that is found in shallow, inshore waters because feeding on the latter results in less exposure to predation from larger fish.

The selective aspect of an animal's foraging behaviour is further influenced by prevailing environmental conditions (e.g. for fish, oxygen levels, temperature, ammonia concentrations, presence of toxins, etc.) in which their food is found. When preferred food items are found in conditions that are marginal for survival and may even be life threatening (e.g. below critical oxygen levels necessary for survival), an animal may choose a less preferred food item in a less threatening environment and/or spend only limited time foraging in the marginal and/or life-threatening environment. For fish, this may mean that they enter waters in which oxygen levels are below the critical level necessary for long-term survival for only very short periods in order obtain a preferred food item.

The type of food item being consumed and the habitat in which it is found help to determine the foraging behaviour employed by an animal. For instance, the sea otter uses its ability to swim in order to successfully feed on fish and molluscs found in the ocean. The leopard uses its ability to stalk and rapidly pursue its prey in order to feed on antelope on the open plains.

Foraging behaviours can be either passive or active and may be different for different species feeding on the same food item. The classic passive foraging behaviour is demonstrated by the sit-and-wait predator (e.g. antlion and many species of fish) that camouflages itself, waits for its prey to come to it and then swiftly attacks to capture the prey. Active foraging behaviour encompasses a wide range of behaviours used in the search and/or pursuit and capture of food items (e.g. grazing by sheep, a hawk feeding on rabbits, lions feeding on zebras). An example of how different foraging behaviours can be employed for the same food item is one in which one species of fish captures an insect on a tree branch overhanging the water by leaping from the water to grab the insect, while another species of fish grabs the insect once the fish has knocked it into the water through the use of water as a projectile.

Finally, the foraging behaviours employed by an animal are determined by all of the factors previously described (i.e. food item, availability, prevailing environmental conditions, presence of competitors and predators, etc.). These complexities associated with the act of foraging with which an animal must contend are the reasons why there is such a wide range of foraging behaviours employed among and within species – and even within an individual animal. It is this continuously evolving repertoire of foraging behaviours available to animals that allows them to be successful in the complex environments in which they exist. (MLW)

See also: Appetitive behaviour; Consummatory acts

Further reading

Alcock, J. (2005) *Animal Behaviour: an Evolutionary Approach*, 8th edn. Sinauer Associates, Sunderland, Massachusetts.

Forced mating

Within a free-range environment, animals will display **courtship behaviour** prior to copulation. In doing so, the state of receptiveness of the female is manifested, and both the male and female are physiologically prepared for impending copulation.

This contrasts with the commercial breeding of some domestic species, in which it is considered normal for the female to be restrained for mating. Examples arise in **dog** breeding, where bitches are muzzled and prevented by manual restraint from crouching and spinning, and in **pig** breeding, where boars may be given access to sows that are tethered by the neck and thus prevented from escaping copulation when non-receptive.

In domestic **horse** breeding, in-hand mating, rather than paddock mating in which the mare and stallion are at liberty to court and copulate without human interference, is common. For in-hand mating, mares are commonly restrained, sometimes chemically and often physically (with hobbles and/or a **twitch**). In-hand mating is favoured for many reasons. These include: (i) its convenience (being conducted during the working day); (ii) its commercial advantages that optimize the use of a valuable stallion and avoid him servicing only specifically favoured mares; and (iii) its perceived safety, as the mare's behaviour can be managed. This has merit but, if the female is ready for mating, **restraint** should not be required. Forced matings have been likened to rape by some observers.

There are significant differences between forced matings and those that occur in a natural situation. Neither the male nor the female is able to exhibit normal courtship behaviour, and even intromission by the male may be different, often being more forceful and curtailed to the minimum duration of contact necessary for ejaculation. Consequently, the **welfare** of animals is compromised during this process. It is not surprising, therefore, that the conception rate from forced matings is often reported to be significantly lower than for natural situations. (AW-S)

Forced moulting

Moulting refers to the replacement of old **feathers** by new ones. Natural moulting is stimulated by the decreasing day lengths preceding winter. Poultry moult naturally about once a year to maintain good plumage and, during this moult, stop laying so that energy can be focused on feather growth. The birds will halt **egg production** for around 2 weeks, and this allows the reproductive tract to rest and repair itself. After the moult, egg production will peak at a slightly lower level than before, but the eggs will be of a higher quality.

Forced moulting is performed after the hens have had a year of egg production and when the numbers of eggs laid have dropped, egg quality is lower, the bird's skeleton is depleted of calcium and the hen is overweight. If hens were switched to shorter day lengths at this point, they would gradually stop egg production and would moult naturally. However, this would be a slow process and hens would start and stop moulting at different times. Forced moulting shortens the non-productive period to about 8 weeks.

Forced moulting involves artificially stimulating a whole flock of hens to moult simultaneously. This is usually achieved by shortening the hours of light provision and withholding feed from the birds for 5–21 days at a time or until they have lost 25–35% of their initial body weight. Mortalities of 1.25% are expected during this feed withdrawal. After the moult, egg production and egg quality increase. Not all producers withdraw feed altogether, while others combine feed withdrawal with a period of water withdrawal. An alternative technique involves feeding the birds a low-quality food during the moult so that, although the birds will not be starved, they will still lose weight due to poor nutrition.

The most obvious **welfare** concern involved with forced moulting is the **starvation** of flocks of birds. In fact, this practice has been banned in most European countries and Canada because it is considered cruel and violates the **five freedoms**. The increased **mortality** rate that occurs can also be considered a crude indicator of the decreased welfare during starvation. **Chickens** have evolved to be highly motivated to forage and consume food continuously throughout the day. Thus, deprivation of food acts as both a psychological and a physiological **stressor**. Frustration of feeding can also lead to increased **aggression** and stereotypic object pecking, which are signs of decreased welfare.

A further concern involved with forced moulting is the increased shedding of *Salmonella enteritidis* by hens subjected to this procedure. The USDA Farm Animal Well-Being Task Group estimated that human salmonella infections from eggs could be reduced by 2.1% if forced moulting was eliminated and alternative practices adopted. (LMD)

Further reading

Duncan, I.J.H. and Mench, J.A. (2000) Does hunger hurt? *Poultry Science* 79, 934.

Webster, A.B. (1995) Immediate and subsequent effects of a short fast on the behavior of laying hens. *Applied Animal Behaviour Science* 45, 255–266.

Webster, A.B. (2000) Behavior of White Leghorn laying hens after withdrawl of feed. *Poultry Science* 79, 192–200.

Fostering

In general terms, 'fostering' refers to bringing up and nurturing; specifically, it can be defined as the provision of parental care and nurturance to unrelated offspring. Spontaneous adoption and provision of care to others' young occur widely throughout free-living animal communities in response to a variety of circumstances. In domesticated species, artificial fostering is common in pig production, but more difficult and less frequently used in other species.

Fostering is used in commercial pig production to equalize litter sizes. Selection for larger litters has meant that domestic sows may sometimes produce 18 or more live piglets, yet they generally only have 12 or 14 viable teats, and can therefore only nurse this number of piglets. Producers also use fostering to produce 'litters' of equal body weight, and sometimes single-sex litters. Fostering in pigs is generally quite straightforward, and simply consists of removing one or more piglets from a sow and transferring them to a foster-mother. This procedure is best done as early as possible after farrowing, ideally within 24–48 h. By about 3 days, unused teats will have dried up and it will be difficult for an additional, newly resident piglet to obtain milk. For several days after giving birth sows appear unable to distinguish their own piglets from aliens and will not usually reject foster-piglets. The reason that sows do not recognize their own litters immediately is thought to be evolutionary: in a natural environment, sows will distance themselves from group members before farrowing and give birth in secluded nests, remaining alone with their litter for a week or more. Therefore there is no need for a dam to recognize her offspring, since the only piglets present in her nest will be her own. Recognition between sow and piglets develops gradually over several days (**see also: Sow behaviour**).

Sows are likely to show **aggression** if they receive foster-piglets 4 days or more after they have farrowed. Piglets fostered after 7 days do not gain weight at a normal rate. However, older piglets can successfully be fostered on to a sow that has recently farrowed. As an alternative to weaning, older 'runts' may occasionally be fostered on to a newly weaned sow that still has lots of milk, although these sows are very likely to recognize fostered piglets as aliens, which can lead to problems with **rejection**.

If carelessly carried out, fostering may stimulate aggression (from sow to piglets and between piglets) disturbance and impair milk intake and growth, negatively affecting the **welfare** of piglets and sow. The nursing/suckling sequence is a complex series of interdependent **vocalizations** and behaviours, susceptible to interruption by noise or other types of disturbance. Disturbance at the time of nursing can inhibit the sow's milk letdown, leading to incomplete **nursings**. This is more likely to occur once piglets have established a **teat order** within the first few hours to days after birth. Once the teat order has been formed, piglets return to their 'own' teat at each suckling and, if alien piglets are introduced, there can be competition between resident and alien piglets for control of the same teat. Immediately to several days after fostering, fostered piglets may be absent from successful milk letdowns. While repeated fosterings are stressful and likely to reduce piglets' body weight at **weaning**, judicious fostering can reduce pre-weaning **mortality** and increase growth and weaning weights of piglets that would otherwise have been disadvantaged.

While the frequency of fostering will depend on unit policy, manager preferences and the reasons that fostering is used, and so may vary with time, on an average pig farm a significant proportion of piglets and sows are likely to be involved in fostering. The procedure is easier in an intensive situation, so likely to be more common and involve higher proportions of animals on indoor, **farrowing crate** units than those using other farrowing systems. Piglet fostering is easier in larger units, where several litters of the same age will be available for exchange. On most pig farms, breeding is synchronized so that farrowing occurs ideally over 1 or 2 days each week. Some units organize farrowing in batches every 3 weeks – a labour-saving measure, which also maximizes the number of same-age litters and so facilitates easy fostering.

In cross-fostering, two or more litters are completely mixed then reallocated to foster mothers. Large piglets may be put on to one sow and small ones on to another, or the procedure may be used to form single-sex litters. In pig farming, cross-fostering is sometimes used as a synonym for fostering. Cross-fostering also refers to a research method whereby infants are transferred from their birth mother to be reared by an unrelated female (possibly even of a different species) in order to separately examine the effects on their development of genetics and environment. Shunt-fostering in pigs is a more complex procedure involving several steps and litters: the oldest litter is weaned early, while a litter about 1 week younger is moved to the first sow. A litter younger still is moved to the second sow, and the step can be repeated again, resulting in a litter-free, newly farrowed sow that is then free to act as a foster-mother for an entire litter.

Sows' universal acceptance of offspring in their nests contrasts with other domestic species such as cattle and sheep, in which dams develop recognition of their own offspring very soon after birth and will reject aliens. Near-immediate bonding is necessary in these animals because they may give birth in groups, or mothers may leave their offspring soon after birth to forage. In these species, fostering is more difficult, as the dam is likely to recognize and reject the alien offspring, and is therefore performed only in exceptional circumstances, e.g. the **death** of a cow or ewe. Giving the alien infant the appearance and/or smell of the dam's own newborn (for example, by covering it with the skin of the dead infant or a jacket soaked in birth fluids), using a masking odour, tranquillizing the dam or restraining her while enforcing contact with the foster infant may facilitate acceptance. In cows, presenting the foster-calf immediately after the cow has given birth, while she is in the **sensitive phase** of maternal awareness, increases the chance of it being accepted as her own. (MH)

Fowl hysteria

Fowl hysteria (or avian hysteria) is a condition in which otherwise normal **chickens** seem to react with **fear** and become highly excited and out of control, running and flying without provocation. Bouts of hysteria are cyclic in nature, with periods of activity from 1–3 min long that occur every 5–30 min. Affected birds are very susceptible to any unusual movement, noise or light and these can also initiate bouts of hysteria. Most cases of fowl hysteria have been observed in growing birds from 7 to 17 weeks of age and, once an outbreak of hysteria has occurred in a flock, it generally continues for weeks or months. The appearance of fowl hysteria is thought to be linked to the **intensification of animal production**, as it is usually found in barren, stressful and crowded conditions. In addition to poor housing conditions, fowl hysteria is also an animal **welfare** concern due to the damage the birds can do to each other during bouts of hysteria. When hysteric, birds can trample each other, fly into walls and windows, pile up and possibly suffocate each other. Study of fowl behaviour has led to the use of partitions in poultry houses, which has been successful in controlling fowl hysteria by separating eating and resting areas and allowing birds to escape from each other, reducing **social stress**. (LMD)

Further reading

McBride, G. (1968) Behavioural measurement of social stress. In: Hafez, E.S.E. (ed.) *Adaptation of Domestic Animals*. Lea and Febiger, Philadelphia, Pennsylvania, pp. 360.

Free-range animals

All non-human animals are influenced by humans to greater or lesser degrees, but for free-range animals this influence is reduced compared with animals that are confined to a small area, cage or container in a building, or in a pen or yard, i.e. those that are kept/housed 'intensively' (**see: Intensification of animal production**). Free-range implies that the animals are, at least, partially independent of humans and are able to move unrestricted over relatively large areas of land, in contrast to intensively kept animals that are almost entirely dependent upon humans for their care and management.

The extent of human influence and control places animals into certain categories:

- Domesticated: humans exert the greatest control over domesticated livestock, **laboratory** and **companion animals** and, particularly, those confined in small areas.
- Captive: captive animals in **zoos** and wildlife parks are influenced by humans in a way similar to domesticated animals, but there is less control of their **breeding**, particularly in relation to selection for specific characteristics and traits.
- Semi-wild: animals in nature reserves and parklands generally experience little direct human control, except when populations exceed the carrying capacity of the areas to which they are restricted (artificially or naturally), and then individuals may be culled.
- Wild: wild native and **feral** animals live independently of humans and there is no human control or deliberate interference, except for the **slaughter** of individuals for consumption and/or population control, or where there is conflict between human and animal interests, as for animals regarded as 'pests', which usually results in **culling** of the animals (**see: Pest control – ethics of; Wildlife management**).

Free-range animals are represented in all of these categories, but within the domesticated and captive categories there is also considerable variation in the extent to which the animals are independent of humans and free to move. For example, among domesticated livestock, the term 'free-range' is used to describe **chickens** and **pigs** that are confined to relatively small areas of land and are given the majority of their food and water by humans. In contrast, free-range **cattle** and **sheep** can roam over hundreds of square kilometres and have to find their own food and water, although food and water sources may be supplemented in some instances.

There is a general perception among people in most developed countries that the low level of human management (or interference) experienced by free-range animals actually results in superior **welfare** compared with their intensively kept counterparts. This perception appears to be based on the belief that situations allowing animals to range free and providing greater freedom for them to express their normal behaviour patterns are good for the animals, while those that confine animals and restrict their behaviours are bad (**see: Confinement**). Following this line of argument leads to the conclusion that wild animals have the highest standards of welfare and domesticated ones have the poorest. This is patently not the case when one considers the consequences of disease and **parasites**, **predation**, climatic extremes, food shortages and so forth. The welfare issues for the animals are clearly illustrated even if only free-range livestock, all of which are cared for by humans, are considered. In addition to being provided with food and water, free-range chickens and pigs are generally provided with man-made shelters, protection from predators, regularly treated for parasites and disease and promptly treated for injuries. In comparison, free-range cattle and sheep have to find their food, water and shelter, are subject to predation, treatment for parasites and disease may be limited and injuries may not be detected to allow prompt treatment.

Certainly, depending upon the level of human intervention, the ability of free-range animals to perform normal behaviour patterns can be critical for their survival. One consequence of reduced human management is that the environment of free-range animals tends to be much more variable and extreme than the environment of intensively kept animals. This means that the animals in these variable environments must be flexible if they are to survive, and their behaviour, together with their physiology, gives animals the ability to adjust and adapt to their environments (**see: Adaptive**). **Wild animals** rely more or less completely on the flexibility of their behavioural and physiological mechanisms to survive, while, at the other end of the spectrum, domesticated species are heavily reliant on humans to provide a stable environment. The extent to which the 'natural' capacities of domestic animals have been changed presumably depends upon the intensity of **selection**, during **domestication**, for particular characteristics.

Both natural and artificial selection have contributed to adaptation and survival, and their relative contributions will depend upon the level of intervention by humans. For wild animals, selection will be largely natural, although selective harvesting/culling will lead to artificially induced changes in the population. Artificial selection plays a much greater role in domesticated species, as humans often determine which animals survive and then select which animals will reproduce. However, in some free-range animals, e.g. cattle and sheep, natural selection may be expected to play a greater role, because humans have less control over the survival of individuals, even if they control partners for reproduction.

Adaptation is a key factor in reducing welfare problems. In any given environment, some populations, **breeds** and individuals will be less prone to parasitism and diseases, and less susceptible to thermal **stressors**, nutrient shortages and so forth. For example, some breeds of cattle are more resistant to certain endo- and ectoparasites and show greater heat tolerance and more efficient utilization of poor-quality pasture. In harsh tropical environments, these breeds have improved welfare and productivity compared with breeds that do not have these adaptations. However, placing these tropically adapted cattle into a cold, temperate environment would severely compromise their welfare because they are not adapted to that environment.

Some of the factors that affect the behaviour and welfare of free-range animals include the following.

Environmental hazards

Environmental hazards include flooding and fires, having the potential to cause **injury** and **death**. Animals that have a good knowledge of the home range or territory may be able to escape the hazard, or seek and find refuges. Animals that are restricted in their movement by fences or natural barriers may be unable to escape the dangers.

Shortages of food and water

Shortages may be a result of natural fluctuations or could be caused by animals being confined, artificially or naturally, to an area too small to support the population. Animals that normally migrate seasonally to avoid food shortages may be prevented from doing so, and this could result in food shortages. To prevent starvation in such situations, animals in the care of humans require their natural food source to be supplemented.

Animals may also suffer through the shortage or absence of specific nutrients in their diet. Again, those animals managed by humans require supplementation, while wild animals may be able to locate natural sources of the nutrient in their home range, provided any restrictions to their movements do not prevent this.

For **grazing** free-range animals there can be an interaction between the location of water and feed in some environments. For example, in arid areas the position of the water source may restrict the area that is used for grazing because the animals will need to return to drink on a regular basis. This can result in overgrazing of the land around the watering point and, potentially, animals could starve even though other parts of the paddock or pasture might contain ample food.

The animals' abilities in relation to foraging and diet selection will also affect welfare. Foraging and selection strategies will be influenced by the time it takes to search for and consume a specific food type, the relative nutritional qualities of the food, its ease of digestion, whether it contains toxic substances and so forth. There are situations in which free-range animals have suffered and died from consuming toxic substances, for example through consumption of poisonous plants. Such situations generally arise when grazing animals are placed on pastures containing plants with which they are unfamiliar, or when they are suffering from food shortages as a result of overgrazing.

Some individuals may experience food shortages or have access only to poor-quality food as a result of their rank within the social **hierarchy**.

For free-range animals there may well be a trade-off between feeding and exposure to risks, such as climatic extremes and predation, and this trade-off is likely to be greater in situations where there is reduced human management.

Predation

Predation causes considerable suffering through injuries inflicted on the prey animal. Young animals and those that are sick or injured are most vulnerable to predation. Group members remain vigilant for predators when grazing and, should an attack take place, mothers will defend their offspring by attacking the predator. In some species, in the face of attack from predators, the adults – and males in particular – will form a protective cordon around the young.

Humans are obliged to protect from predation, as far as possible, animals in their care. However, for some free-range animals, such as grazing species that are allowed to roam large areas, it is extremely difficult to provide this protection. In such situations, the method usually adopted is to control predator numbers to reduce the probability of predator attacks, but methods of **predator control** can, in themselves, cause welfare problems.

Climatic extremes

Examples such as high heat and humidity, or cold temperatures in combination with strong winds and/or precipitation, are a threat to welfare if animals are unable to shelter, and animals may suffer from hypo- or hyperthermia. To avoid **heat stress**, animals will seek out and stand in shade or even in cool water (**see: Shade-seeking behaviour**). When resting, they may dig a scrape in a shaded place to expose a cooler surface and lie fully or semi-recumbent to maximize contact with the cooler area. To avoid **cold stress**, animals will huddle together against walls, rocky outcrops or vegetation, or in depressions in the ground, to get shelter from the wind. An inability to shelter generally results from animals being confined to areas in which there is no shelter (natural or artificial), or there is insufficient shelter, resulting in competition for it. Crowding beneath shade can exacerbate heat stress because air movement around the animals is reduced. Animals low in the social hierarchy are the ones most likely to be unable to gain access to shelter.

Diseases and parasites

Diseases and parasites will jeopardize the welfare of animals that do not have some natural resistance to them. Animals will have some resistance to endemic diseases and parasites, but may experience considerable suffering from exotic diseases and parasites until they develop resistance. This is illustrated in the use of disease to control populations of pest animals, for example myxomatosis to control rabbits. Many diseases can be prevented, or their effects attenuated, through vaccination.

Grazing animals will attempt to alleviate the effects of ectoparasites, such as biting flies, by grouping together, swishing tails and stamping hooves or walking under low vegetation. Rubbing against vegetation and rolling on the ground may help to dislodge flies and ticks. For animals in the care of humans, further protection from ectoparasites can be provided by the administration of pesticides in sprays, rubs and dips.

If animals become ill there may be limited opportunity for timely treatment because the illness may not be detected promptly. Even if detection is rapid, there can still be considerable practical problems in capturing and restraining animals for treatment. Indeed, in some instances capture and restraint may exacerbate the welfare problem and **euthanasia** may be the best solution.

Injury

Many instances of injury will threaten the welfare of free-range animals because, as with illness, in many situations there will be limited opportunity for prompt human intervention to alleviate pain and suffering. Many free-range animals in the care of humans may roam over considerable areas and, as a consequence, it is impossible to closely monitor and check every animal frequently. If an injury that requires treatment is detected then the animal has to be restrained or moved to restraint facilities, and these actions could compromise welfare further. In some situations euthanasia is the most appropriate action.

Culling

Culling involves the killing of particular individuals or capturing them for relocation. Killing animals is not a welfare issue, provided it is done quickly and humanely, with minimal **suffering** caused to the animals. The main risk with killing free-range animals is that unsuccessful attempts will result in a slow, painful death and/or injury.

In addition to the adverse impacts on welfare as a result of possible injuries, culling may cause welfare issues in any animals that are dependent upon a culled animal, such as offspring that may die from starvation. If large numbers of animals are culled in a particular area then there may be adverse impacts on the welfare of other species in that area, such as the predators and prey of culled species. Culling can also lead to the breaking down and disruption of social relationships.

Specific management practices

Culling may, in some instances, jeopardize the welfare of free-range animals more than the same practices conducted on animals that are more intensively kept, as a consequence of the size of the areas over which the free-range animals move and the relative infrequency with which the animals are closely handled. Free-range animals may be more fearful of humans because of their limited exposure to and experience of them, and this greater **fear** has the potential to compromise welfare in any situations in which the animals are in close proximity to or are handled by humans.

The extensive nature of keeping free-range livestock means that animals are gathered from large areas and this can result in heat stress, dehydration and exhaustion. Costs are high in gathering animals from large areas and so they tend to be gathered, for the conduct on them of various procedures, a small number of times per year. This can mean that invasive husbandry procedures, such as **castration**, **dehorning** and tail docking (**see: Docking – tail**) are not performed at optimum times. **Restraint** and **handling** stress are likely to be greater, wounds larger and healing times slower when animals are older and larger. (JCP)

Frustration

The term 'frustration' is used in many different ways: (i) the motivational and/or (ii) affective state that results from being thwarted or blocked from a goal; (iii) the emotional state elicited by non-reinforcement of a response that previously was reinforced; and (iv) the procedure of blocking or thwarting of behaviour (which may or may not lead to an emotional state of frustration). Thus, like the term '**stress**', the term frustration has many different usages that can become confused, and this confusion is confounded further by the frequent use of the term in a more general way outside of a scientific context.

Since it difficult to know the emotional state felt by an animal, even though such outcomes are of **welfare** importance, it is perhaps more useful to use the term in a scientific context to refer to those of the animal's responses to thwarting or blocking that can be objectively observed. This definition of frustration also distinguishes the term from the emotional consequences of deprivation, which might otherwise also be referred to as frustration.

For example, laying hens in battery cages without a suitable substrate to perform **dust-bathing** behaviour are deprived, but not necessarily frustrated in this latter sense, whereas hungry cows that want to get to the feeding station but whose way is being blocked by dominant **conspecifics** are being frustrated. However, this distinction between deprivation and frustration may be false since, if the performance of a specific behaviour is rewarding in itself (**see: Behavioural need**), it might be argued that deprivation of this activity results in the same process of non-reward as might occur when an animal is blocked from some external stimulus. Thus, whether a behaviour is self-rewarding or enables an animal to achieve a certain goal (e.g. food, a nest, access to a mate), then preventing the occurrence of the behaviour is a form of frustration that may lead to a state of emotional frustration, suffering and reduced welfare.

Behavioural expressions of frustration include: **ambivalent behaviours**, **displacement behaviours** and **redirected behaviours**, which may appear unrelated to the original context of the frustration. For example, many animals that are physically prevented from reaching food when they expect to feed show pacing, **aggression** or **grooming**, behaviours all indicative of frustration in this situation. Some animals also seem to express their frustration vocally (e.g. the gakel-call in laying hens). Abnormal injurious behaviours (e.g. **feather pecking** in laying hens, **tail biting** in pigs) and stereotypic behaviours (e.g. sham chewing in pigs, **tongue-rolling** in giraffes) are often observed when animals are confined and restricted in the performance of certain behaviour patterns. Stereotypic behaviours may result from the blocking of **foraging behaviour**. In addition, the specific form seems to be predicted from the type of natural foraging activity (e.g. Clubb and Mason, 2003).

Behavioural signs of frustration, however, do not tell us exactly how motivated or frustrated an animal is. The strength of the **motivation** to perform the behaviour determines the intensity of frustration and, indirectly, the effect on the welfare of the animal. This may be deduced from a range of physiological and/or behavioural measures. Behavioural demand tests can give an indication of the intensity of the motivation or frustration (**see: Economics of behaviour**). Typically in these tests, an animal has to pay a 'price' for access to a condition where it can interact with an object/conspecific or where it can perform a certain behaviour. 'Price' in this case is defined as the amount of effort an animal is prepared to put in to gain access to a preferred commodity or condition, usually by performing an **operant response**, such as pressing a lever, pecking a key or pushing through a weighted door. When an animal has to work harder to gain access to its most preferred environment but still persists, it suggests which environments or commodities are important to them.

The emotional expression of frustration may vary between individuals or circumstances. It may present as an intensification of the frustrated behaviour (anger), aggression (rage) or **depression**, and while the ability to measure such states directly is not possible, an awareness of their potential occurrence may be important. (PHZ)

See also: Laying hen housing; Vocalization

Reference and further reading

Clubb, R. and Mason, G. (2003) Animal welfare: captivity effects on wide-ranging carnivores. *Nature* 425, 473–474.

Dawkins, M.S. (1988) Behavioural deprivation: a central problem in animal welfare. *Applied Animal Behaviour Science* 20, 209–225.

Jensen, P. and Toates, F.M. (1993) Who needs 'behavioural needs'? Motivational aspects of the needs of animals. *Applied Animal Behaviour Science* 37, 161–181.

Petherick, J.C. and Rushen, J. (1997) Behavioural restriction. In: Appleby, M.C. and Hughes, B.O. (eds) *Animal Welfare*. CAB International, Wallingford, UK, pp. 89–105.

Wood-Gush, D.G.M. (1963) The control of nesting behaviour of the domestic hen. I. The role of the oviduct. *Animal Behaviour* 11, 293–299.

Fur farming

Animals whose pelts are in demand are often farmed intensively; examples include **mink**, Arctic **fox**, blue fox, **rabbits** (angora) and sometimes **ferrets**. Because of the small cages involved there is normally quite a high level of stereotypic behaviour (**see: Stereotypies**) around the time of feeding, e.g. pacing, shown by the animals, which has given rise to concern and debate over the **welfare** of these animals. (DBM)

G

GABA

Gamma amino butyric acid (GABA) is the main inhibitory transmitter in the **central nervous system** of mammals. Drugs such as the **benzodiazepines**, which act through stimulation of GABA receptors, are thought to have an anxiolytic (anti-anxiety effect) by modulating ascending noradrenergic and serotonergic input to the hippocampus, which forms part of the behavioural inhibition system (**see: Emotion**) and is involved in assessing any discrepancy between perceived and predicted events. Discrepancies, as might occur in a novel environment, result in behavioural inhibition and a heightened state of arousal, together with **anxiety**.

GABA can also have excitatory effects, especially in young animals and in insects where it has a largely excitatory effect on muscles. These opposing effects are possible because GABA receptors regulate ion flow across the membrane and so the orientation of the flow determines the effect. During development, as **neuron(e)**s mature, the polarity of the receptors may reverse, e.g. in the hippocampus and neocortex, so that GABA changes from having a stimulating to an inhibitory effect. Defects in this reversal may contribute to the onset of some forms of epilepsy. Since GABA also stimulates neurogenesis in the young, it may also play a role in the differential effect of novelty during development, having an enriching effect and increasing brain complexity in the young, but having an anxiogenic effect in the mature individual. (DSM)

Gait

Gait is a coordinated and rhythmic pattern of **locomotion** resulting in progressive movement (Barrey, 1999). Quadrupedal locomotion has three distinct gait patterns – the walk, the trot and the gallop – and these can be further classified as symmetrical and asymmetrical (Howell, 1944). In symmetrical gaits the gait pattern of a limb pair is repeated by the subsequent limb pair, half a stride later (Alexander, 2003). The walk is a symmetrical gait, with two, three or four legs supporting the animal during the stride and typically the order of limb sequence is left front, right hind, right front, left hind. The trot is also a symmetrical gait in which diagonally opposite legs move in phase with each other (Alexander, 2003).

During walking and trotting, there is an equal interval of time between footfalls. Asymmetrical gaits, like the gallop, have no symmetry to the motion cycle – the time between footfalls is not equal (Alexander, 1992), and the gait pattern of a limb pair in the first part of the stride is not repeated by the following limb pair (Dagg, 1977). During the gallop, the two feet of a limb pair strike the ground in rapid succession, followed by a longer period before the next limb pair is set down (Alexander, 1992).

Terrestrial locomotion in birds is limited to four gait patterns – the walk, the run and the hop, with legs in-phase or out-of-phase. During walking and running the left and right legs alternate, half a stride out of phase with each other (Alexander, 2003). Birds may not use every gait pattern; for example, chickens use the walk and the run but rarely the hop.

Until recently, gait in farm animals had not been investigated thoroughly, but this is changing due to increased concern about animal welfare and the economic costs associated with **lameness**. Currently, two approaches are used to analyse gait: subjective assessments and objective biomechanical measures, such as those from force platforms, kinematics, electromyography and accelerometers. (FF)

References

Alexander, R.M. (1992) *Exploring Biomechanics: Animals in Motion*. Freeman, New York.

Alexander, R.M. (2003) *Principles of Animal Locomotion*. Princeton University Press, Princeton, New Jersey.

Barrey, E. (1999) Methods, applications and limitations of gait analysis in horses. *Veterinary Journal* 157(1), 7–22.

Dagg, A. (1977) *Running, Walking and Jumping*. Wykeham Publications London, Ltd, London and Basingstoke, UK.

Howell, A. (1944) *Speed in Animals*. Chicago University Press, Chicago, Illinois.

Game animal

Game animals are defined simply as any free-ranging animals hunted for subsistence, spiritual or recreational purposes, but lurking beneath this simple definition are complex issues involving population **ecology**, animal behaviour, genetics and **evolution**, **conservation** policy, land management, human rights and **ethics**. It is perhaps unfortunate that the English language also employs the word 'game' in reference to human endeavours specifically oriented towards competition, caprice and manipulation (although the definition of 'game' used above is independent of the word's other meanings, and indeed may pre-date them). As the discipline of wildlife management has broadened to adopt an ecosystem-centred view in Western cultures, 'wildlife' has increasingly been adopted by management agencies in preference to 'game'. Central to the concept in any case is that wild animals are deliberately killed to meet human needs or desires.

Hunting as a human endeavour pre-dates historical documentation; meat is an accepted part of human diets in most cultures and leather remains a popular and useful material. Since the era of domesticated animals, however, most humans have faced choices: consume animals cultivated specifically for human use or consume animals that have evolved principally from the forces of natural selection. Although most meat and leather used by humans is now undoubtedly of domestic origin, hunting has not disappeared from many cultures, nor is there any indication of it doing so soon.

Population biologists have a fundamental understanding of the patterns expressed by species commonly hunted, but have rarely developed specific models of carrying capacity- and population-level responses to hunting on geographic and temporal scales of interest to human societies. This may seem surprising, given the ubiquity of hunting, but is understandable when considering the complexity of forces affecting any given animal population. Because hunting as a human activity long pre-dated modern science, most management and control of hunting developed from taboos, traditions or rules of thumb rather than from experimentally derived data. Still, it is difficult to deny the success many modern societies have had in maintaining game populations in the face of increasing human density and modern weaponry. In short, most vertebrate populations equilibrate in the presence of moderate, human-caused off-take because the environmental conditions faced by survivors are improved by the reduction in the density of **conspecifics**. Where management systems are mature, achieving sustainability of terrestrial species subject to hunting is now commonplace.

Deer (*Odocoileus* spp.), despite being brought close to extinction by settlers' overhunting in the 18th and 19th centuries, are probably more abundant in North America today than prior to European arrival. Wapiti (*Cervus elaphus*), pronghorn (*Antilocapra americana*) and American black **bears** (*Ursus americanus*) have also made dramatic (if incomplete) recoveries in the presence of hunting. Despite extensive legal hunting, native doves, grouse, quail and introduced pheasants (i.e. *Phasianus colchicus*) have avoided the unfortunate fate of the passenger pigeon (*Ectopistes migratorius*) in North America (which was hunted to extinction – both for commercial purposes and to reduce tree damage – before effective conservation measures were adopted). Underlying these recoveries has been the transformation of hunting from an uncontrolled, open-access activity to an egalitarian, state-sanctioned one in which hunters willingly accept various restrictions in order to provide similar hunting opportunities to their neighbours and descendants. More importantly, interest in game animals has provided strong motivation for habitat protection; hunters have encouraged governments to restrict agriculture, forestry and industrial developments in deference to maintaining some wild habitats. Early hunter–conservationists in North America supported not only hunting, but also complete protection for national parks. In more densely populated Europe, hunted populations of red deer (*Capreolus elaphus*), roe deer (*Capreolus capreolus*) and various Galliform birds have also persisted (albeit under more restricted access to hunting privileges). When societies decide to prioritize conservation, hunting per se is clearly not an impediment to population persistence.

Nevertheless, even when demographically sustainable, human **predation** exerts quantitatively and qualitatively different effects on game animals than do non-human mortality sources (e.g. **parasites**, **diseases**, predators). Hunting usually changes behavioural patterns of game animals (e.g. choices of food and cover), at least during time periods when they are pursued, often to the detriment of the survivors. By altering existing social patterns, hunting can also produce cascading ecological effects, such as changing the magnitude of inter-deme gene flow, or the intensity of intra-species infanticide.

Perhaps more pernicious is that by selectively removing certain age classes (usually adult males), intensive hunting has the potential to alter the genotypes of populations produced by natural selection, resulting in species that evolve (at least partly) in response to artificial selection. In Europe, such management has often taken the form of culling animals with 'undesirable' antler forms, risking unknown pleiotropic effects (wherein a gene influences numerous phenotypic characteristics). In contrast, hunts that too intensively remove males with the largest (i.e. most desired by hunters) secondary sexual characteristics may inadvertently produce selection for males that grow slowly (as well as other, unknown genetic effects). Such artificial selection can be minimized by keeping human kill rates sufficiently moderate and/or by management regimes that produce age-/sex-specific mortality patterns similar to those with which the target species evolved.

Subsistence hunting is rarely questioned on moral grounds, as by definition the alternative is human **starvation**. Still, as human populations increase and formerly subsistence societies become increasingly integrated with market forces, subsistence hunting has often been found to be unsustainable. Solutions to overhunting for **bushmeat** usually require development of more functional social feedback systems that prioritize long-term benefits flowing to a satisfactorily defined social group over short-term, individual benefits.

Hunters in modern Western culture have developed numerous codes of ethics, but these all presuppose that hunting itself is ethical. The morality of deliberately killing animals in the modern world is a legitimate question. Defences of hunting that call upon notions of hunters' self-restraint or their spirited defence of wildlife habitats fail to answer this ethical question directly. Ultimately, the acceptance of hunting is an individual decision, and will depend on what sort of hunting is envisaged and how one conceives of oneself in relation to the natural world. Among those to whom hunting is optional, most agree with their opponents that hunting is essentially atavistic, primitive and uncivilized. They disagree with anti-hunters about whether these are negative or positive values. In accepting their role as predators, thoughtful modern hunters embrace the responsibility towards wildlife that killing requires. In this moral stance, consumption of wild meat becomes not merely communion with nature but akin to Communion, a physical acknowledgement of mankind's destructive behaviours towards the natural world, and thereby a recommitment towards minimizing them. (RBH)

See also: Farmed animals

Further reading

Caughley, G. (1985) Harvesting of wildlife: past, present, and future. In: Beasom, S.L. and Roberson, S.F. (eds) *Game Harvest Management*. Caesar Kleberg Wildlife Research Institute, Kingsville, Texas, pp. 3–14.

Harris, R.B., Wall, W.A. and Allendorf, F.W. (2002) Genetic consequences of hunting: what do we know and what should we do? *Wildlife Society Bulletin* 30, 634–643.

Milner-Gulland, E.J., Bennett, E.L. and SCB 2002 Annual Meeting Wild Meat Group (2003) Wild meat: the bigger picture. *Trends in Ecology and Evolution* 18, 351–357.

Petersen, D. (ed.) (1996) *A Hunter's Heart: Honest Essays on a Blood Sport.* Henry Holt and Company, New York.

Reiger, J.F. (2001) *American Sportsmen and the Origins of Conservation.* Oregon State University Press, Corvallis, Oregon.

Game theory

Why animals behave the way they do is one of the central questions in animal behaviour research. While some behaviours are easy to understand, others are much harder to predict. Game theory is one tool that can be used to help determine what, when and why behavioural strategies are employed by individuals. For example, will a male elephant seal opt to fight or flee when challenged over access to mates? Should an individual stickleback fish inspect to see whether a predator is present or let others in a group inspect instead? Game theory is instrumental in helping to answer these questions and many other animal behaviours.

John von Neumann first developed the mathematical basis for game theory in 1928, but it was not until von Neumann and Oskar Morgenstern's book *Theory of Games and Economic Behaviour* in 1944 that game theory became recognized as a mathematical means to analyse different strategies. Games can be used to assess which strategy results in the highest pay-off, given a set of strategies and choices made by other players in the game. There are three main elements for a game: (i) two or more players; (ii) a set of strategies; and (iii) a corresponding set of pay-offs for each of the strategies. One of the goals of game theory is to determine the strategy set where each player has chosen the best response and cannot acquire additional benefits by changing their strategy given the other players' strategies. This strategy set solution is called the Nash equilibrium after John Nash, who developed the concept.

While initially used in the field of economics, game theory has since been applied to biology to model the behaviour of individuals given the strategies and behaviour of others in the population. John Maynard Smith significantly contributed to game theory in biology by introducing the concept of an evolutionary stable strategy (ESS). Equivalent to the Nash equilibrium, an ESS is a strategy that, through the process of natural selection, can become established and cannot be eliminated by any other strategy introduced into the population. In this case, the pay-offs correspond to the effects each strategy has on an individual's relative **fitness**. For a new strategy to become established and evolutionarily stable, it must yield a higher fitness than other strategies in the population so that its frequency will increase and resist invasion by other strategies with lower fitness pay-offs. Evolutionary games can be used to model the emergence and stability of behaviours in a population based not only on the benefits and costs of each strategy, but on the interactions between individuals.

One of the first evolutionary games applied to animal behaviour was Maynard Smith and Price's hawk–dove game. While simple, this game provides insight into animal conflicts over resources. Two different behavioural strategies are considered: hawk and dove. Individuals using the hawk strategy always fight for the resource and do not withdraw until either they win the resource or they are injured and lose the contest. Doves, on the other hand, will only display as a threat to fight, but will back down and relinquish the resource if the other player escalates the fight. Considering only pair-wise contests between individuals, a hawk may encounter another hawk or a dove, and likewise a dove may encounter another dove or a hawk. Each of these contests has fitness consequences, or pay-offs, as a result of either winning or losing the contest. In hawk–hawk interactions, half the time the individual will win the contest and half the time it will be injured and lose. If the contest is between hawk and dove, the individual with the hawk strategy will win the resource and the dove will lose, but not be injured. Contests between two doves result in each splitting the resource. Based on these pair-wise interactions, a pay-off matrix can be constructed using the value of the resource (V) and the cost associated with fighting (C) (see Fig. G.1).

These assumptions provide the rules of interactions and reward structure for the game. Using this pay-off matrix, we can analyse the model and determine the ESS. In this case, the ESS depends on the value of the resource compared with the cost of fighting. If $V > C$, then the hawk strategy is the only ESS because the pay-offs for hawk–hawk $((V–C)/2)$ would be greater than for dove–hawk (0), and the pay-offs for dove–dove ($V/2$) would be less than hawk–dove (V). However, neither the hawk nor dove strategy becomes an ESS if $C > V$. Instead, a mixture of hawk and dove strategies is maintained in the population at equilibrial frequencies. This mixed ESS can occur from individuals either adopting mixed strategies, where they play the hawk strategy for a proportion of the time and the dove strategy the other proportion of the time, or from individuals that use pure strategies (either always dove or always hawk) occurring at equilibrial proportions in the population. This game can be made even more complex to reflect more closely behaviour observed in animal populations by changing the assumptions of the model, adjusting the pay-off matrix or **resource-holding potential** of individuals, allowing repeated conditional interactions or adding additional strategies, such as bourgeois or anti-bourgeois, to account for different behaviour by **territory** owners and intruders.

Another classic game is the prisoner's dilemma, which has been used to model the evolution of **cooperation**. In this game, the optimum is achieved when both prisoners cooperate with each other and remain silent when interrogated by the police. However, this optimum is not a stable equilibrium point if the pay-off to betray and testify against the other prisoner is greater than the pay-off to remain silent. The only ESS is the betray strategy, even though the pay-off is lower when both prisoners betray than if both were to cooperate – hence the dilemma. If repeated interactions and learning are incorporated into the model, a tit for tat strategy, where one player

	Hawk	Dove
Hawk	$\frac{(V-C)}{2}$	V
Dove	0	$\frac{V}{2}$

Fig. G.1. Pay-off matrix constructed using the value of the resource (V) and the cost associated with fighting (C).

matches the strategy used by the other in previous interactions, emerges as a potential solution if the probability of interactions with the same player is high. Thus, game theory shows how reciprocal **altruism**, as well as other behaviours, may become established in a population.

While games are merely simplified mathematical models of real behavioural systems, they can be used to help generate new hypotheses and test predictions generated from the model. Game theory helps to define the conditions needed for a strategy to be conserved in a population. This in turn helps animal behaviourists better understand and investigate the behavioural strategies of individuals based on their interactions with others. (LAN)

Further reading

Dugatkin, L.A. and Reeve, H.K. (eds) (1998) *Game Theory and Animal Behaviour*. Oxford University Press, New York.

Maynard Smith, J. (1982) *Evolution and the Theory of Games*. Cambridge University Press, Cambridge, UK.

General adaptation syndrome

The general adaptation syndrome (GAS) is a term coined by **Hans Selye** to describe what was believed to be the body's common response to **stress**. There are three stages to the GAS: an initial alarm phase, followed by a period of resistance leading ultimately to adaptation or exhaustion. The alarm phase, which lasts a matter of seconds, is characterized by activation of both the sympathetico-adrenal medullary system (**see also: Sympathetic nervous system**), which results in the release of **epinephrine** and **norepinephrine**, and the **hypothalamic–pituitary–adrenal (HPA) axis**, which results in the release of **corticosteroids**. These serve to mobilize energy reserves and prepare the animal for action, with an increase in heart rate and blood pressure and a diversion of blood away from the gut and towards skeletal muscle. Also at this time the levels of certain anabolic hormones like **insulin** fall as resources are diverted to maximize a fight or flight response. During the resistance phase the level of adrenal hormones remains high but the level of anabolic hormones returns to normal. The success of this phase depends on the magnitude of the alarm response and previous experience, which may indicate success or failure. In the event of a prolonged period of resistance due to a failure to adapt, the animal may suffer from a range of stress-related illnesses and ultimately die. The initial focus of exhaustion is related to the consequences of prolonged immunosuppression as a result of persistently high levels of corticosteroids.

The GAS in broad terms is a non-specific response that occurs in response to many psychological and physiological stressors, even those that are believed to be pleasurable, such as sexual arousal, and so it is not a sufficient indicator of compromised welfare alone. Nor is it a necessary condition for reduced welfare, because there is a range of **stressors**, e.g. hyperthermia (**see: Critical temperature**) and dehydration, that do not provoke an increase in costicosteroids (because such a response would actually exacerbate the impact of this type of stressor).

While initially useful in helping to describe how the body copes physiologically with stressors, the concept of the GAS is criticized for being oversimplistic, with more recent models emphasizing the context-specific nature of stress responses (**see: Emotion**) and need for a wider variety of measures to evaluate their impact on **welfare** (**see also: Measuring welfare**). (DSM)

Further reading

Broom, D.M. and Johnston, K.G. (1993) *Stress and Animal Welfare*. Kluwer Academic Publishers, Dordrecht, The Netherlands.

Moberg, G.P. and Mench, J.A. (2000) *The Biology of Animal Stress – Basic Principles and Implications for Animal Welfare*. CAB International, Wallingford, UK.

Generalization

The term generalization is used in relation to two different learning phenomena. Most often it refers to the extension of a response to stimuli that are similar but not identical to those with which it was originally associated (stimulus generalization). As such the concept of generalization is the opposite of **discrimination**. Thus a dog that is taught to sit on command by a specific trainer may respond appropriately to the command 'sit' spoken by another individual as a result of stimulus generalization. While many species generalize the stimuli to which they learn a response, the response is not necessarily identical and there may be a decrement in response initially, when any of a wide range of features is changed, for example when the animal is asked to sit outdoors when it has been trained indoors. From a practical perspective, it is important to recognize and accommodate this decline so as to encourage the animal to generalize as desired.

Response generalization refers to a change in the form of the response that occurs to a particular stimulus, i.e. a wider range of behaviour occurs to a given trigger. In high-performance work this may not be desirable (e.g. obedience competition work), and differential reinforcement of only a narrow range of high level responses (**see: Operant response**) can help to minimize the tendency for response generalization. (DSM)

Genetic bottleneck

When the number of breeding organisms is restricted, the genetic variation within the population is reduced because some genes, found in only a small number of individuals, will disappear by mere chance. Subsequent generations are genetically very different and less genetically diverse than the ancestral population as a result. Typically, rare alleles will be lost but in some cases by chance they will become greatly increased in frequency. Restrictions of this kind are termed genetic bottlenecks. Studies of fossil DNA show, for example, that **domestications** of wild species were genetic bottlenecks as the variation in the wild progenitor (such as the aurochs) was greater than that in the domesticated form (cattle). During **breed** formation, when small numbers of particularly esteemed animals were used to found breeds, further genetic bottlenecks resulted. Thus the Thoroughbred horse owes much of its ancestry to three stallions (Cunningham *et al.*, 2001) while the famous C57BL/6 and C57BL/10 strains of laboratory mice are all descended from female mouse no. 57 on the Massachusetts farm of pioneer mouse breeder and schoolteacher Abbie Lathrop (Silver, 1995).

When small numbers of males inseminate many females, this is effectively a genetic bottleneck and a rare allele can become frequent as a result. The gene for bovine leucocyte adhesion deficiency (BLAD) became widespread in the North American Holstein breed because of the heavy use of a single bull that carried the recessive gene (Hall, 2004).

Genetic bottlenecks do not appear to have as large an effect on quantitative genetic variation (broadly speaking, inherited characteristics whose values follow a normal distribution in a population, such as body weight) as might be expected. Thus, one would not expect bottlenecks to have a sizeable effect on behaviour phenotypes, though this has not been explicitly studied. (SJGH)

See also: Captive breeding programme; Inbreeding

References

Cunningham, E.P., Dooley, J.J., Splan, R.K. and Bradley, D.G. (2001) Microsatellite diversity, pedigree relatedness and the contributions of founder lineages to thoroughbred horses. *Animal Genetics* 32, 360–364.

Hall, S.J.G. (2004) *Livestock Biodiversity. Genetic Resources for the Farming of the Future*. Blackwell Publishing, Oxford, UK.

Silver, L.M. (1995) *Mouse Genetics. Concepts and Applications*. Oxford University Press, Oxford, UK.

Genetic engineering

Genetic engineering is the process of actively and directly manipulating the genetic material of an organism, with the objective of giving this organism new characteristics for which there is an agricultural, biomedical or other application. This engineering can be either *somatic* (in which the germline cells are not affected, so that the change only affects the phenotype of the engineered individual) or *germline* (in which all cells including eggs or sperm cells are affected and the change is passed on to offspring). (AO)

Genetics of behaviour

Animal species usually have characteristic patterns of behaviour, and certain species are especially suited to **domestication** because of these patterns. Like all phenotypic characteristics, behaviours are expressions of interactions between the environment and the animals' genotypes. Genes code for enzymes, and behaviour is controlled by the operation of the neuroendocrine system and, to varying extents depending on the complexity of the animal, by the **brain**. Mutations in relevant genes may lead to altered forms of enzymes in these systems, leading to changes in the ways in which the systems operate.

That differences in behaviour, whether between species or within species, have a genetic basis can be tested by raising experimental animals in identical environments. This is frequently done in species such as the fruit fly and laboratory mice. Genetics of behaviour differences are sometimes very simple, even being controlled by a single gene. In some cases, a behavioural change has been observed in wild animals and the underlying genetic cause has been deduced. For example, the blackcap, a small migratory bird, has evolved a new pattern of winter migration from its breeding grounds in Germany and this clearly has a genetic basis (Berthold, 1991), while experimental populations of mink have been selectively bred to exhibit different responses ('fearful' or 'confident') to a novel stimulus (Malmkvist and Hansen, 2001).

Under domestication, certain behaviours became fixed, such as neotenization in dogs. The hypothesis would be that there had been genetic variation for this trait in the earliest domesticated dogs, which was rapidly reduced by human selection so all subsequent dogs have been uniform in this respect. But some behaviours of our domesticated animals have not changed (Hall, 2004). Our pigs are descended from wild boar, and both the domesticate and her wild **conspecific** build a nest before giving birth, though under intensive husbandry conditions this may be impossible for the farmed sow. It seems likely that the welfare consequences of preventing an animal from performing a behaviour that has a strong genetic component may be more serious than preventing the performance of one that is more dependent on environmental factors (**see: Behavioural need**).

There are several examples of the demands of modern animal husbandry having led to heritable changes in behaviour. Wild boar and wild junglefowl use, respectively, feeding strategies that are more energetically costly than those of their domesticated counterparts. When breeds differ in behaviour, this can be an indicator of a possible genetic basis, though proper experimental design is essential.

Modern molecular genetics is leading to an understanding of how changes in genes can be linked with the appearance of new behaviours. Compared with the fruit fly, the honeybee has more complex genetic systems for **vision**, **olfaction**, **learning** and **memory** – consistent with its evolution of a social lifestyle (Honeybee Genome Sequencing Consortium, 2006).

Study of the genetics of behaviour has several important practical consequences. If **zoo animals** are to be reintroduced to the wild, care must be taken not to operate a mating system that favours traits adapting the animals to captivity – in other words, domestication of zoo animals must be avoided. With **companion animals**, while behavioural differences between breeds can be very evident (most notably with dogs), differences between individuals within breeds can also be important. For example, in Japan it costs about US$23,000 to train a guide dog and, among Labrador retrievers, only about 30% of those trained prove suitable for use. It is claimed (Fuyuno, 2007) that variation in a gene that codes for an enzyme that, in humans, is involved in mood regulation can indicate the ability of dogs to concentrate. If so, this could lead to a valuable aptitude test. With farm animals, understanding of the genetics of behaviour could lead to development of breeds of livestock that are adapted to particular husbandry conditions. Though superficially attractive, this might be seen as ethically suspect in that, arguably, conditions should be improved to suit the animals rather than the animals being changed to suit the conditions (Hall, 2004; **see also: Ethics**). (SJGH)

References

Berthold, P. (1991) Genetic control of migratory behaviour in birds. *Trends in Ecology and Evolution* 6, 254–257.

Fuyuno, I. (2007) Biobank provides leads for selecting guide dogs. *Nature* 446, 119.

Hall, S.J.G. (2004) *Livestock Biodiversity. Genetic Resources for the Farming of the Future*. Blackwell Publishing, Oxford, UK.

Honeybee Genome Sequencing Consortium (2006) Insights into social insects from the genome of the honeybee *Apis mellifera*. *Nature* 443, 931–949.

Malmkvist, J. and Hansen, S.W. (2001) The welfare of farmed mink (*Mustela vison*) in relation to behavioural selection: a review. *Animal Welfare* 10, 41–52.

Genome

A genome comprises all of the genetic information or hereditary material in one set of chromosomes in an organism. Humans and higher animals have two genomes – a nuclear chromosomal genome and a mitochondrial genome. Whereas the nuclear material arises from contributions from both parents as a result of fertilization of the egg to produce a **zygote**, the mitochondrial genome is derived from the egg alone and so the maternal line. If the rate of mutation within DNA is known, differences in mitochondrial DNA can be used to provide an evolutionary clock for comparing species or breeds. (LMD)

See also: Genetics of behaviour; Genotype

Genotype

A genotype is the genetic make-up of an individual, encoded in the DNA, which is the 'blueprint' for building and maintaining that individual. The genotype underpins **phenotype**; however, the expression of traits can be changed through environmental factors or the influence of other genes. A genotype includes not only the dominant or expressed genes but also the recessive or masked genes. (LMD)

See also: Genetics of behaviour

Gentling

Although sometimes used in other species to describe **habituation** programmes, gentling is commonly used in **horse** handling. It is now becoming accepted as a better term than 'breaking-in' for the earliest steps in the foundation training of horses. Gentling emphasizes that fear and therefore flight responses (both in the short and long term) are controlled most effectively by **instrumental conditioning** and habituation rather than by force. Traditional breaking techniques that use roping and other forms of restraint to train a horse that attempts to escape are not to be recommended. Concern for the welfare of horses trained in this way relates to the likelihood of their developing **learned helplessness**. Similarly, this approach to training may have the disadvantage of compromising the horse–human bond and, as such, is losing favour with riders who regard their horses as **companion animals**.

Gentling techniques that expedite the formation of a bond between horse and handler are becoming increasingly popular. Operant conditioning largely based on negative reinforcement – but also some positive reinforcement – forms the basis of all such methods. Primary reinforcers can take the form of physical contact (e.g. rubbing the neck can be analogous to **allogrooming** by a **conspecific**). Mild punishment may also have a role, so the horse may learn that moving away from the handler spontaneously is associated with **aversive** outcomes such as being chased (the application of so-called pressure), an outcome that continues until the horse turns to approach the handler (negative reinforcement). As with all negative reinforcement regimes, it is critical that handlers use only sufficient pressure to achieve the desired effect and remove the pressure as soon as the correct response is made.

Opposition to gentling comes particularly from those who argue that it is time-consuming compared with traditional breaking. However, successful horse whisperers have been instrumental in countering this assertion since they can often reach training milestones faster than those using traditional restraint and force. (PDM)

Geophagia

The deliberate eating of earth or soil. A wide range of species has been observed to exhibit geophagia, examples including species from the following classes: mammals, birds and reptiles, and indeed including humans and many invertebrate species. The advantage of geophagic behaviour was presumed to improve the intake of minerals that may otherwise be deficient, in particular phosphorus, calcium, copper and iron. However, it has recently been proposed that other advantages may be derived, involving **self-medication**. One is detoxification. Herbivorous animals exhibit geophagia more often than omnivores or carnivores, and it is thought that such animals are at a greater risk of poisoning from plant-derived alkaloid compounds. It may be that sodium is of particular importance in metabolizing these toxins into harmless products, and sodium-rich soils have been found to be particularly sought after by animals. Aside from the mineral provision, clay (which may actually have a low mineral content) binds toxins and removes them from the body in expelled faeces. Mycotoxins and bacteria, as well as plant-derived toxins, can all be bound by clay in this way. Clay also has an antacid effect and is thought to line the intestinal wall, thus protecting it from mechanical damage by plant toxins and from bacterial and parasitic infestation. Rats have been shown to eat more kaolin (china clay) than controls when experimentally infected with pathogenic bacteria. Indeed, kaolin is used as a human medicine for gastrointestinal problems. Despite this, geophagia may also be an **abnormal** behaviour – rats have been shown to eat clay when stressed. Geophagia carries a risk of infection from soil-dwelling bacteria, nematodes and other pathogens and **parasites**. (DA)

Further reading

Krishnamani, R. and Mahaney, W.C. (2000) Geophagy among primates: adaptive significance and ecological consequences. *Animal Behaviour* 59, 899–915.

Gestalt perception

Gestalt (meaning 'shape' in German) psychology developed in Berlin at the beginning of the 20th century. Max Wertheimer is credited as the founder of the movement, although its origins go back to the work of various philosophers, including **Immanuel Kant**. The main theoretical principle of the Gestalt psychology is the view that biological phenomena have to be studied in their wholeness. In other words, Gestalt psychologists used a (w)holistic approach to study human behaviour. This approach contrasts with the traditional physical reductionist scientific methodology, which divided the object of investigation into parts that could be analysed separately in order to reduce the complexity of the object. The Gestalt movement was particularly active and influential in the area of visual perception. According to this view, we perceive and recognize objects and figures as whole forms and not just

Fig. G.2. Gestalt perception predicts that the dog in this picture would be perceived without the observer first focusing on its parts.

as a sum or collection of simple shapes, lines and curves. An example of a Gestalt perception would be the figure of a Dalmatian dog in the shade of the trees (see Fig. G.2): according to Gestalt psychologists we would perceive the dog as a whole all at once, instead of first recognizing its parts (e.g. its nose and tail) and then inferring the dog from them.

Gestalt psychologists described a number of principles and laws that control our perception. The two most important principles are the principle of totality and the principle of psycho-physical isomorphism. The principle of totality states that perception – or any conscious process – needs to be analysed by simultaneously considering the physical and mental processes involved. The principle of psycho-physical isomorphism links conscious experience to cerebral activity. (RV)

Further reading

Palmer, S.E. (1990) Modern theories of Gestalt perception. *Mind and Language* 5, 289–323.

Gestation

Gestation, also known as pregnancy, is the period of time in which a mammalian female supports the life of a **fetus** or fetuses from fertilization to **parturition**. Gestation begins when an ovum is fertilized by a sperm in the oviduct to form an embryo. The embryo then migrates from the oviduct to the uterus, where implantation (or attachment) and recognition of pregnancy are initiated. Typically, after the embryo reaches the uterus, the mother produces an excess of **progesterone** (from the *corpus luteum*) to maintain the pregnancy, and progesterone concentrations remain high through the duration of the pregnancy. Once the embryo has migrated from the oviduct into the uterus, morphogenesis of the placenta occurs. The placenta functions not only to separate the maternal and fetal organisms, but it also serves as the fetal gastrointestinal tract, lung, kidney, liver and endocrine glands. Another important function of the placenta is to keep maternal and fetal blood separate while still allowing nutrient transfer.

As gestation continues, the uterus expands with the growing fetus. The uterus expands in three stages, although the myometrium layer remains tight to prevent premature expulsion. At the end of gestation, there is a chain of events that leads up to parturition. **Oestrogen** concentrations begin to increase, causing the pelvic ligaments to gradually become more relaxed as gestation continues, and the hormone relaxin also aids in this process. At the end of gestation, parturition is stimulated by a complex interaction of endocrine, neural and mechanical factors. The length of gestation is determined by a multitude of factors including genetic (breed and fetal **genotype**), maternal (age), fetal (litter size, sex and endocrine function) and environmental (nutrition and temperature) factors. (ELS)

Giant panda

The giant panda (*Ailuropoda melanoleuca*) is a **bear**-like **carnivore** inhabiting bamboo forests of mountainous regions in China. Molecular studies confirm pandas as one of eight extant bears, but evolving separately from them since about 20 million years ago. Fossilized dental remains indicate a commitment to bamboo as its principal food source by at least 3 million YBP. Adult males may reach 1.8 m in length and over 100 kg in weight. Females are usually smaller, but size dimorphism is not rigidly evident. Pandas readily ascend trees. They usually feed in a sitting posture and can easily stand bipedally. An unusual characteristic is an enlarged wrist bone that functions somewhat like a human thumb and provides dexterity in handling food items.

As much as 90–98% of the diet consists of the leaves, shoots and stems of bamboo. Despite adaptations in the forepaws, teeth and jaws for bamboo consumption, the panda retains the digestive system of its carnivore ancestry, and is therefore unable to digest cellulose, a main constituent of bamboo. Pandas solve this problem by rapidly passing prodigious quantities of bamboo through the digestive tract on a daily basis. Mating and birth are strictly seasonal, separated by a period of about 5–7 months. A female has a single **oestrus** during the spring, lasting from 1 to 3 days. Like all bears, giant pandas undergo a delay in embryonic implantation of about 2–3 months. The period of fetal growth lasts about 50–60 days, and results in birth weights of about 110–115 g or 1/800th of the dam's weight. Twinning occurs in nearly 50% of all litters, but typically only one cub is raised. Captive pandas may live beyond 30 years in captivity, but their lifespan in the wild is estimated at about 20 years.

Consistent with giant pandas' largely solitary nature is a strong reliance on **olfaction** for **communication**. A large scent gland surrounding the anus is used to deposit olfactory messages at strategic locations within a panda community. Tree trunks, rocks or clumps of grass are used on which a mark is left that can be monitored by others. Scent is known to convey information on identity, sex and possibly social status for marking males. Males use scent to identify the areas where they live, whereas females' primary use is in signalling oestrus. Males locate oestrous females by tracking their scent. At close range auditory and visual modalities are used in courting. From one to five males attend a female in heat, engaging in often severe competition for mating.

Pregnant dams seek parturition sites in hollow bases of large trees or in limestone caves. A fetal life of only 50–60 days results in a neonate that is highly **altricial**, depending on the dam for warmth, nourishment, positioning at the breast and stimulation of voiding. Denning is mandated for the first 100–120 days of life, soon after which cubs ascend trees for resting. Dams cuddle and position infants on their ventrum nearly continuously for the first 2–3 weeks of life. The maternal investment is such that they are typically able to care for only one infant in the case of twins, leaving the other to die. Eyes begin to open at roughly 45 days, and **locomotion** is seen at 75–80 days. By 14 months the dentition is sufficiently developed for bamboo consumption; weaning occurs at 18–24 months.

Giant panda populations, both wild and captive, are subject to an array of **stressors**. Few studies have been conducted on wild-living pandas, yet known circumstances imply a considerable human impact on their well-being. It is known, for instance that the species cannot survive outside the bamboo forests, but satellite recordings of a sharp decline in habitat availability in recent decades suggest that competition for resources is an ongoing process. The reproductive cycle of bamboo flowering, followed by die-off and 5–10 years for regeneration, has resulted in starvation for some populations in recent times. Bamboo cycles vary according to species and, in locales where only one or two survive, the primary food source for pandas is periodically in short supply. Because much of their natural range is shared with farmers, pandas are subject to diseases carried by domestic livestock. In some areas pandas and humans compete directly for bamboo shoots and possibly other resources. Atmospheric and water pollution have the potential to adversely affect panda well-being, as do trace minerals for populations in proximity to roads and mining operations.

Giant pandas readily develop **stereotypies** when confined in static, albeit naturalistic, captive environments. These typically include pacing, pirouetting, head tossing, exaggerated pushing off from structures, tongue flicking and body swaying. Providing enrichment is essential in ameliorating the constraints imposed by captivity. Scientific investigations have revealed that pandas respond to a range of objects, with benefit derived from their varied properties in stimulating a diversity of behavioural responses. Based on both behavioural and corticoid measures, providing individuals free choice in access to different parts of the habitat reduces their **stress** load.

Fossils from northern Myanmar and Vietnam, and across China from western mountains to eastern shores, indicate a widespread distribution of giant pandas in Asia during the early Pleistocene. Today, they occur only in bamboo forests in Sichuan, Shaanxi and Gansu Provinces, ranging over a total area of about 13,000 km^2. Census data acquired in the mid- to late 1980s placed the wild population at about 1000 individuals. New surveys concluded in 2002 placed its numbers at about 1600. A recent population assessment using DNA extracted from droppings suggests an even larger population. Approximately 200 pandas are maintained in **captivity**, primarily in China. This population has achieved remarkable growth within the last decade through improved breeding.

China regards the giant panda as a national treasure, and commonly refers to it as belonging to the world. Substantial progress in **conservation** has occurred in recent years as a result of collaborations with non-governmental organizations such as the World Wide Fund for Nature and several American **zoos**. Prior eras of gifting and short-term commercial loans to zoos internationally have given way to lending agreements that, in some cases, generate funds for conservation of the wild population. Cooperative international arrangements have been implemented to provide training in reserve management, community-based programmes and captive breeding. Added to these efforts, the Government of China is investing billions of dollars in replacing marginal farmland with forest and in upgrading programmes for protected areas. The number of panda reserves had been expanded from 14 in the early 1990s to a total of 56 by 2006. (DGL)

See also: Endangered species; Wildlife management

Further reading

Lindburg, D. and Baragona, K. (eds) (2004) *Giant Pandas: Biology and Conservation*. University of California Press, Berkeley, California.

Swaisgood, R.R., White, A.M., Zhou, X., Zhang, H., Zhang, G., Wei, R., Hare, V.J., Tepper, E.M. and Lindburg, D.G. (2001) A quantitative assessment of the efficacy of an environmental enrichment programme for giant pandas. *Animal Behaviour* 61, 447–457.

Wildt, D.E., Zhang, A.J., Zhang, H., Janssen, D. and Ellis, S. (eds) (2006) *Giant Pandas: Biology, Veterinary Medicine and Management*. Cambridge University Press, Cambridge, UK.

Giraffe

Giraffe (*Giraffa camelopardalis*) are the smallest megaherbivores; adult males weigh an average of 1200 kg, while adult females average 800 kg. Giraffe live in sub-Saharan Africa, where they are usually found in savannah and riverine habitats, although in Namibia they survive in desert-like conditions. They are listed as conservation-dependent by the International Union for Conservation of Nature (IUCN) and are common wherever adequate habitat remains to support them, but decreasing in number over the past decade (except in South Africa). They are readily distinguishable from other animals by their height (as adults, males can be 6 m tall, while females are usually 4–5 m tall) and pelage. Spot patterns can be used to distinguish individual animals. Giraffe are ruminants in the order Artiodactyla, and share the family Giraffidae with okapi. Until recently, all giraffe have been considered members of a single species, but genetic work published in 2007 suggests that there may be up to six distinct species.

Giraffe reproduce throughout the year, and females become pregnant while still nursing their previous offspring. Among ungulates, species can be grouped into two types according to their maternal behaviour: **hiders** (in which young remain at a distance from their mothers, often concealed in vegetation) and **followers** (in which young remain close to their mothers at all times). Giraffe fall into the former category, with mothers spending time away from calves in order to: increase calves' crypsis; reduce the cost of protection for the mother; conserve calf energy and water; decrease calf heat loads; and enable the adult female to negotiate terrain that would be difficult for the calf. Initially, females leave calves lying out singly, but older calves often rest together in groups. Bonds between mothers and offspring are persistent (as long as 22 months), at least until the next parturition. Females are observed in closer proximity to their mothers than to males in the second year of life, and some recent reports suggest that mothers and daughters may maintain bonds into maturity.

Feeding and ruminating account for about 80% of the daytime activity budgets of wild giraffe. They mainly eat *Acacia* and *Grewia* species that require them to use their tongue to obtain leaves while avoiding thorns and stinging ants. Giraffe feed for longer and eat different plant species during the dry season than during the wet season. Oestrous females reduce feeding time while engaged in mating activities, and pregnant giraffe eat higher-protein and higher-fat food resources than do lactating females.

Field studies have concluded that giraffe form only loose social bonds, and frequent changes in social partnership are typical (**see: Social flux**). However, defining a group of giraffe has been problematic because of the potential disparity between giraffe and human perceptual fields. Giraffe may use their height and infrasonic (low-frequency) **vocalizations** to maintain contact over long distances, as has been reported for **elephants**. Therefore, group membership may be more consistent than previously reported. The population of females within a study area appears to be fairly consistent, while there is significant fluctuation in the male population. Social groups are more stable and larger when resources are plentiful, particularly in the wet season.

As in many ungulates, sociality differs by sex. Females with young calves live in nursery groups that are fairly stable in composition, and may form the basis for lying-out groups of calves. Bulls are likely to be found in social groups until they are fully mature, when they are typically solitary. Younger bulls frequently spar with each other, standing parallel and swinging their heads towards each other. Adult bulls occasionally spar, and rarely engage in intense fighting (which does not develop from sparring). Age and size are fairly good predictors of dominance among males, and larger, older bulls have a higher success rate in inducing females to urinate. By tasting and smelling a female's urine (**see: Flehman**), males can identify oestrous females and will perform **mating behaviour** with only those females in **oestrus**.

From a welfare perspective, the prevalence of **stereotypies** in captive giraffe may indicate a need for improvement in captive environments. **Zoo**-housed giraffe and okapi perform both locomotor and oral stereotypic behaviour; a survey found that 79% of the individuals performed at least one type of stereotypic behaviour. Licking of non-food objects was the most common form of stereotypic behaviour, reported in 72% of the animals in the survey. Over 29% of the giraffe and okapi paced and 3% performed other forms of stereotypic behaviour, which included **self-mutilation**, head tossing and tongue playing. Feeding motivation appears to be related to oral stereotypies in giraffe and okapi, while aspects of the physical environment predict locomotor stereotypies. These results corroborate research on domestic hoofstock.

In the zoo, giraffe are often fed hay and manufactured grain feed, with occasional browse. None of these foods is difficult to process and they are normally provided in open-topped feeders. At Zoo Atlanta (USA) and the Paignton Zoo Environmental Park (UK), experimental changing of feeding techniques caused repetitive licking to decrease dramatically, supporting the idea that motivation for feeding and/or **rumination** is related to oral stereotypy in giraffe. More research is needed to reduce the prevalence of stereotypic behaviour in the giraffe. (MB)

Further reading

Bashaw, M.J., Tarou, L.R., Maki, T.S. and Maple, T.L. (2001) A survey assessment of variables related to stereotypy in giraffe and okapi. *Applied Animal Behaviour Science* 73, 233–245.

Baxter, E. and Plowman, A.B. (2001) The effect of increasing dietary fibre on feeding, rumination and oral stereotypies in captive giraffes (*Giraffa camelopardalis*). *Animal Welfare* 10, 281–290.

Ciofolo, I. and le Pendu, Y. (2002) The feeding behaviour of giraffe in Niger. *Mammalogy* 66, 183–194.

Dagg, A.I. and Foster, J.B. (1976) *The Giraffe: Its Biology, Behaviour, and Ecology*. Van Nostrand and Reinhold, New York.

Pratt, D.M. and Anderson, V.H. (1985) Giraffe social behaviour. *Journal of Natural History* 19, 771–781.

Glucagon-like peptide-1

Glucagon-like peptide-1 (GLP-1) is important in the regulation of blood sugar, appetite and food intake. Nutrient balance is maintained by inputs arising from gastrointestinal distension, intestinal hormones, nutrients present within the gut and in the blood following absorption. Nutrients in the gut trigger the release of GLP-1 together with other gastrointestinal peptides such as **cholecystokinin** and ghrelin. There are two forms of GLP-1 produced but they appear to be equipotent in their effects, which include: stimulation of **insulin** and inhibition of glucagon release; inhibition of gastric emptying; reduction in gastric acid production; and suppression of food intake by stimulating a sense of satiety. (DSM)

Glucocorticoid

Glucocorticoids are one of two types of **corticosteroid** hormones produced from the cortex of the **adrenal gland**, the other type being mineralocorticoids. Whereas the latter have a primary effect on salt/water balance, the former – as their name suggests – affect glucose metabolism, causing blood glucose to rise. These hormones are particularly associated with the **stress** response (**see also: General adaptation syndrome**). The most important glucocorticoids are cortisone and corticosterone, with one or the other being the dominant glucocorticoid, dependent on the species. Other effects of the glucocorticoids include anti-inflammatory and immunosuppressive effects, effects on **cognition** and a role in the maturation of the lung.

There are many synthetic glucocorticoid drugs (often referred to as **steroid hormones**) that are often used to control inflammation or immune-related conditions in animals, and these drugs (e.g. prednisolone, dexamethasone and betamethasone) are much more potent than the natural hormones. Side effects of these drugs include increased appetite and thirst, and their excessive use can result in iatrogenic (drug-induced) Cushing's disease. The cognitive effects of these drugs in animals that might affect welfare remain largely unknown. (DSM)

Goad

The term 'goad' describes both an instrument, such as a flexible pole, stick or club, used to move animals on and the act involving its use. The principle behind the use of a goad is to apply negative reinforcement (**see: Reinforcement – types of**) to encourage the animal to move away from the goad in order to guide its movement; however, it is often used inappropriately by being applied without a clear **contingency** related to the animal's behaviour or with an inappropriate intensity that may cause the animal to panic. This gives rise to **welfare** concerns about its use. Electric goads (sometimes known as cattle prods or 'hotshots') are available that discharge an electric current when they come in contact with the skin of the animal. The use of the goad, together with other negative interactions with a **stockperson**, increases the risk of high lactate and cortisol levels and **PSE meat** in pigs, together with a more general avoidance of people. (DSM)

Goal-directed behaviour

Goal-directed behaviour may refer to any behaviour or sequence of behaviours that is performed in order to achieve a perceivable goal. Goal-directed behaviour can be inferred by its reliable termination when the individual perceives specific stimuli that, by their presence, imply that the goal has been achieved. For example, **foraging behaviour** might cease when the individual perceives signals of satiety, **courtship behaviour** might cease when eggs are observed in the nest and **grooming** behaviour might cease when the irritation disappears. Frequently, observable events are used to infer the goal of a behaviour (and this may be subject to bias), although experimental behavioural studies may be used to assess the **motivation** towards certain goals more objectively. (KT)

See also: Economics of behaviour

Further reading

Manning, A. and Stamp-Dawkins, M. (1998) *An Introduction to Animal Behaviour*, 5th edn. Cambridge University Press, Cambridge, UK.

Goal emulation

The phrase 'goal emulation' was first coined by Whiten and Ham (1992). They used it to refer to a type of **social learning** (i.e. a mechanism in which the observation of other individuals or the products of their activities facilitates or enables the acquisition of novel behaviour) (**see also: Observational learning**). Whiten and Ham intended the term goal emulation to refer to the same mechanism that was first described and labelled as plain 'emulation' by Wood (1989). Wood argued that children will not only attempt to imitate the form of others' actions, but also 'try to emulate them by achieving similar ends or objectives'. 'More specifically, emulation includes "instances where children achieve common goals to those modelled, but do so by using idiosyncratic means that were never observed"' (Whiten *et al.*, 2004). Whiten and Ham prefixed 'emulation' with the word 'goal' in order to distinguish it more clearly from **imitation** (i.e. learning to do an act by seeing it done), because 'dictionaries tend to equate emulation with imitation'.

Tomasello (1990) was the first person to discuss emulation with respect to non-human (animal) behaviour. When he first discussed the term, it is clear that he intended it to refer to (goal) emulation, as previously described by Wood. He stated that: 'A focus on the demonstrator's goal may lead the observer to be attracted to and seek to attain the goal. The observer may then attempt to "emulate" the demonstrator's behaviour, that is, to reproduce the completed goal by whatever means it may devise.' Hence, in contrast to imitation, when emulating, an observer learns about the goal of a demonstrator's behaviour, not the topographical details of the actions used. In other words, an emulator learns about the ends not the means by which others achieve goals.

For example, an observer chimpanzee sees a **conspecific** use a delicate twig to winkle termites out of their mound. The demonstrator slowly inserts the twig into a hole in the mound, gently agitates it from side to side for a few seconds and then slowly extracts it, revealing several guard termites that have attached themselves to the twig's surface. In this case, the goal is ingestion of termites. In contrast to imitation, if the observer were learning via emulation, it would learn nothing about the precise behavioural strategy employed by the demonstrator. Instead, the observer would be attracted to and seek to attain the goal (i.e. extracted termites) by whatever means it may devise. The observer may eventually end up re-inventing, through its own independent problem solving, the same or a very similar fishing technique (especially if it is exposed to discarded twigs near the mound), but it might also hit upon a completely different strategy such as smashing down the mound through use of brute force. If it emulates rather than imitates, the observer learns nothing from watching the demonstrator about the strategy used (i.e. the fishing technique); instead it learns about, and its subsequent behaviour is directed by, the goal (i.e. attainable termites).

As Whiten and Ham (1992) acknowledge, the use of the word 'goal' in relation to emulation is cognitively loaded. It could be taken to mean that, when emulating, an observer attributes goals to the demonstrator. If characterized in these terms, emulation would constitute evidence of mental state attribution. Tomasello (1998) categorically refutes that he ever intended emulation to refer to a mechanism by which observers perceived a demonstrator's actions as goal-directed. An alternative, and much less cognitively complex, characterization is that observers can attend to results, and may consequently set goals for themselves that function to attain similar results. Whiten *et al.* (2004) argue that both characterizations of the term are valid. Hence, they distinguish result and goal emulation from one another and both are placed within the superordinate category of 'end-state emulation' (i.e. copying only the end or outcome of an action sequence).

Tomasello (1998) expanded the original meaning of emulation to incorporate observational learning about the properties of objects and the potential relations among them. As he explains:

> It so happens that when I observe others manipulate objects I learn a lot about those objects and their affordances for my own actions – much more than when I observe the objects sitting idle. Thus, if we give naïve chimpanzees a rock and a nut, they may not discover on their own how to crack the nut open. However, if they see another chimpanzee do it, they might learn from this observation that nuts can be opened, which creates a new possibility for them, and they might even learn something about the rock's role in the process.

Emulation thus characterized involves learning not about the behaviour used to manipulate objects, but about such things as an object's causal relationship with other objects or substances, that it might contain other objects or substances within it, its length, solidity, rigidity, function and so forth.

The ability to learn about the affordances of objects by observation is likely to involve highly complex cognitive operations. It may be equally, if not in some cases more, cognitively complex than imitative learning. Learning about the properties or affordances of objects seems a quite different process from learning about the results or goals of another's actions. Hence, Whiten *et al.* (2004) use the label 'affordance learning' (of properties, relations and functions) and distinguish this from end-state emulation.

In summary, emulation is a term that has come to mean different things to different people. Recently, many of the contradictions and sources of confusion have been clarified. It seems most helpful to follow Whiten *et al.*'s (2004) lead and distinguish end-state emulation from affordance learning. It also makes sense to distinguish between result and goal emulation. Result emulation is a relatively cognitively simple process in which observers attend to results and, consequently, set goals for themselves that function to attain similar results. Goal emulation is a process by which an observer 'regards the model (i.e. demonstrator) as an organism with goals and seeks to reproduce what it infers to be the model's goal on a particular occasion' (Whiten *et al.*, 2004). There does not seem to exist, as yet, any unequivocal empirical evidence of goal emulation in non-human animals. (DC)

References

Tomasello, M. (1990) Cultural transmission in the tool use and communicatory signalling of chimpanzees? In: Parker, S. and Gibson, K. (eds) *Language and Intelligence in Monkeys and Apes: Comparative Developmental Perspectives.* Cambridge University Press, Cambridge, UK, pp. 274–311.

Tomasello, M. (1998) Emulation learning and cultural learning. *Behavioural and Brain Sciences* 21, 703–704.

Whiten, A. and Ham, R. (1992) On the nature and evolution of imitation in the animal kingdom: reappraisal of a century of research. In: Slater, P.J.B., Rosenblatt, J.S., Beer, C. and Milinski, M. (eds) *Advances in the Study of Behaviour.* Academic Press, San Diego, California, pp. 239–283.

Whiten, A., Horner, V., Litchfield, C.A. and Marshall-Pescini, S. (2004) How do apes ape? *Learning and Behaviour* 32, 36–52.

Wood, D. (1989) Social interaction as tutoring. In: Borstein, M.H. and Bruner, J.S. (eds) *Interaction in Human Development.* Erlbaum, Hillsdale, New Jersey, pp. 59–80.

Goat

Goats are ruminant animals, similar in size to **sheep**. Roughly 95% of the 765 million goats in the world are found in the developing countries of Asia, Africa and South America.

They have a well-founded reputation for eating a wide variety of feedstuffs. Similarly to other ruminants, they have a complex stomach comprising the rumen, reticulum, omasum and abomasum. The bacteria and protozoa found in the rumen and reticulum break down the foodstuffs, including cellulose and hemi-cellulose, which are fermented to shorter-chain fatty acids that can be absorbed from the rumen.

As do other ruminants, goats regurgitate and re-chew their foodstuffs, which increases the surface area for the microorganisms to work upon. This, in combination with the high concentration of microorganisms in the ruminal liquor, allows goats to survive on very low-quality, high-fibre feeds. However, for high levels of productivity and health, good-quality nutrition is essential.

Goats are agile animals, and indeed they often are found in trees. This allows them access to some of the most inaccessible feed. In combination with their ability to utilize low-quality feeds and relative drought tolerance, they can survive in conditions where sheep and cattle cannot.

Goat production

Goats are kept throughout the world for a variety of purposes. The commodities they produce include meat, milk, fibre, skins, horns and dung. Dung can be collected and used either for fuel or to fertilize household vegetable gardens. Goats are often more suitable for smallholder production (in both temperate and tropical situations) than cows, as their products are in more manageable quantities. Linked to this is another advantage – because more individuals can be kept (roughly five goats graze the same area as one cow), the risk to the farmer from losses due to disease, stealing, etc. is reduced. They are usually inexpensive to purchase and their fast reproductive rate means that herd numbers can build up rapidly (to a level determined by the availability of feed). This can be particularly important after events such as natural disasters and war.

Goat meat is unlike pork and beef in so far as it is consumed throughout the world, and there are no religious taboos associated with its consumption. It is fairly lean and does not have the acrid, fatty smell produced when cooking lamb, and is thus more acceptable than lamb in many parts of South-east Asia, especially the Philippines.

Carcass yield is typically between 35 and 50%, similar to that for sheep (Steele, 1996). In young kids the yield is less, as the head, rumen and internal organs make up a greater proportion of the live weight than those of an adult goat. However, kid meat is highly valued, especially from very young (8–12 weeks) kids in the Caribbean and Latin America, where it is known as *cabrito*.

Carcass fat content, especially subcutaneous and intramuscular, is typically less than lamb. Abdominal fat is greater, however, though with correct butchering this can be tailored to suit the consumer's requirements. Many breeds of goat are used for meat production, often as a by-product of a milk enterprise, though the Boer **breed** (originating from South Africa) is a developed meat breed with good carcass conformation and growth rates. These large animals have been exported throughout the world. Males weigh typically 130 kg and females 80 kg (Steele, 1996).

Goats' milk is also popular throughout the world. Its main advantage over cows' milk is that it has a higher percentage of small fat globules, which in turn means it forms a fine curd in the stomach. This allows it to be more easily digested by humans, notably those who are allergic to cows' milk. Chemical composition is very similar to cows' milk, however, though protein and total solids concentrations are slightly higher. This means that it is very suitable for use in cheese making. A number of breeds have been developed specifically for milk production, the most important being the Saanen, Toggenberg, Alpine and Anglo-Nubian (females from these breeds are 50–65 kg in weight). With these breeds productivity levels are greatly improved, notably daily milk yield and the length of lactation. Unimproved tropical breeds typically produce 0.5–1.0 l/day for 125 days, depending on conditions and breeds, whereas specialist dairy goats in temperate intensive conditions can yield up to 5 l/day (Steele, 1996). In non-pregnant goats **lactation** can continue for up to 2 years, though production levels do decrease.

Products from goats' milk include fresh milk, cheeses, butter and ghee, and yoghurts. Cheese is particularly important in Western Europe (especially in France), where high-quality, niche products are highly valued by consumers. As a result, large, intensive flocks are found, complete with automated **milking** machines similar to those used to milk cattle.

Goat fibres are also highly valued, specifically Mohair and Cashmere. The former is a single coat of coarse, non-hairy fibre that is found only on the Angora breed. Its uses include furnishings, rugs and blankets, but the fine fibre from Angora kids is used in clothing. Cashmere is the fine-fibred undercoat found on a goat. Most breeds of goat have this, but not in sufficient quantities to make it economically useful. However, some breeds, such as the Kashmiri, have been developed over many centuries so that they have large quantities of this underwool. It can be spun in the same way as a sheep's fleece and is used for the production of high-value clothes.

Dietary requirements

Goats differ from sheep in one major way – they are browsers rather than grazers. Thus they exhibit a behavioural preference for foliage from trees and bushes in addition to grasses and plants. They have a well-founded reputation for eating almost anything, which can range from plant material through plastics to electricity cables. This means that care must be taken when looking after goats: fields must be constantly checked for objects that may harm the animals if ingested, particularly if electric fences are used to control grazing. However, the positive element to this is that they can be used to clear overgrown land of weeds. As long as the area is fenced adequately, stocking rates of 10 goats/ha can be maintained on land to control and eliminate weeds (Steele, 1996). This works most effectively when the goats are moved in rotation around pastures ahead of cattle or sheep. The practice has been used in areas throughout the world, particularly in New Zealand, South Africa and the USA. Castrated males have proved to be the most effective in this respect, as they do not have the nutritional demands of pregnant or lactating females.

Goats' milk has a reputation in some quarters for having a tainted taste. In reality, this arises from a number of factors. First is the influence of diet, especially if a goat eats a shrub or weed such as ivy. In addition, certain feeds such as kale, turnip, swede, garlic and cow parsley can taint the milk, and should be avoided. Therefore, for milk production care must also be taken to prevent this from occurring and thus tainting the milk. Other factors that increase milk taint include genetic inherited traits, vitamin B_{12} deficiency, **mastitis** and poor hygiene and milking technique (Hetherington and Matthews, 1992).

Typical feed dry matter requirements are 3% of the animal's body weight per day for meat production, 4–5% for dairy production in less developed areas and around 8% in intensive dairy systems. The grazing behaviour exhibited by goats is highly selective. Being agile increases their range of feeding options. Initially, when put on to fresh pasture they are unselective, but as their intake increases so too does their selectiveness. They graze individually rather than in flocks, often seen standing on their hind legs in order to reach browse from overhanging trees, and in some instances even climbing on to low branches.

As mentioned previously, goats are especially difficult to keep enclosed. They will easily escape from the majority of fences used to restrict cattle and sheep. Some will jump over, so fences of at least 1.2 m are necessary, though Steele (1996) suggests 1.5 m as a minimum height. Some individuals can wriggle and burrow under wires, so consideration must also be given to this aspect. Electric fences can be used, though if the goats are horned it is strongly advised not to use the net type of fence due to the risk of entanglement and subsequent death of the animal. Goats have the ability to learn how to operate latches and bolts, and so these must always be mounted in sites inaccessible to the animals.

Breeding

As with sheep and cattle, the practice of flushing (enhanced feeding and nutrition before mating) is highly effective at increasing conception rates and litter sizes.

Conception can be through either natural mating or **artificial insemination** (AI), the latter depending on availability within an area as this is not as widespread as the use of AI in cattle, for instance. Goats are in **oestrus** throughout the year in the tropics but are seasonal in temperate areas, though less so than sheep. In many dairy goat herds, artificial lighting is used in spring in order to manipulate day length and thus have year-round milk production.

Gestation is typically 145 days, though kidding can be anywhere in the region of 140–150 days depending on various factors, including litter size – European breeds tend to be nearer 150 days; in Saanen it can even be a few days longer than this. The udder starts to produce and fill with milk prior to kidding (**parturition**), and in some goats a small amount may leak just prior to the birth.

Average litter sizes vary from 1.5 in breeds such as the Boer and Alpine to 1.9 in the Saanen, though these increase slightly with increased body size of the nanny and increasing age.

If vaccinations are administered 2–4 weeks before kidding, the protection will be passed on to the kid via high levels of immunoglobulins in the colostrum.

Health

In general goats are hardy animals, certainly more so than sheep. However, they do need veterinary care and consideration. Although somewhat resistant to illness, once ill their condition deteriorates rapidly. They are also particularly susceptible to pneumonia, mainly as a result of insufficient or poor-quality housing.

As a result of their inquisitive nature and willingness to eat anything, choking can also be a problem. Likewise, plant poisoning can be a problem under poor management conditions. Control of weeds such as ragwort is essential, though most evergreen shrubs should be considered poisonous unless known to be otherwise, including rhododendrons, azaleas and laurels. Goats can browse members of the *Prunus* family (cherry, plum, etc.) but, although the fresh leaves produce no problems, the dry and wilted leaves are poisonous.

Their browsing behaviour also means that goats are more susceptible to internal **parasites** than sheep or cattle. Reasons for this again go back to their willingness to eat anything and browsing of feed that other livestock cannot reach. The four groups of internal parasites that infect goats are cestodes (tapeworms), nematodes (roundworms), trematodes (flatworms) and protozoa (unicellular organisms). The nematode species *Haemonchus contortus* is considered to be responsible for the vast majority of cases; it attaches to the wall of the abomasum and feeds off the blood of the animal. This loss of blood renders the affected animal anaemic and sluggish. Control of intestinal parasites in goats is very similar to that for other ruminants, and includes grazing management (resting pasture to decrease numbers of infective eggs and larvae), avoiding the grazing of wet and swampy areas, drenching with anthelmintics, and numbers within the digestive tract can be limited by the feeding of fodder and browse with a high tannin content.

Other conditions for consideration include foot-and-mouth disease, footrot, mastitis and enterotoxaemia. This latter **disease**, also known as pulpy kidney, is an acute, often fatal clostridial disease (*Clostridium perfringens* type D) arising from bacteria normally found in the digestive system and the soil. However, if there is a sudden change in diet, particularly if large quantities of cereals are fed, these bacteria multiply rapidly and produce the toxin that causes the condition. Thus it is vital that all changes in a goat's diet are phased in gradually wherever possible.

Vaccinations are available to lessen the risk and are highly recommended. Most enterotoxaemia vaccines also contain an anti-tetanus dose (*Clostridium tetani*), which protects the animals against the possible fatal infection from the numerous scratches and pricks that result from their browsing behaviour.

A good indication of an animal's general health is the nature of its faeces, and the goat is no exception. The dung should consist of small pellets similar to that of rabbits and sheep, and certainly not liquid in any way.

Housing

While in general goats are fairly healthy animals, they do need year-round **housing**. This need is exacerbated by cold and wet conditions, when goats become particularly susceptible to pneumonia. Good ventilation is essential to prevent this condition, though draughts at animal level must be avoided. Slatted floors are often used to prevent a build-up of dung and associated health problems.

Social behaviour

Goats are social animals, and it is a general consensus that they should not be grazed individually. Traditionally, however, many farms kept a single goat tethered away from the rest of the livestock. The advice now is to always keep a goat with other animals, either other goats, sheep, cattle, horses or even donkeys. (MM, HO)

References and further reading

DEFRA (2006) *Codes of Recommendations for the Welfare of Livestock: Goats.* Available at: http://www.defra.gov.uk/animalh/welfare/farmed/othersps/goats (accessed 11 September 2006).

Hetherington, L. and Matthews, J.G. (1992) *All About Goats.* Farming Press, Ipswich, UK.

Peacock, C., Devendra, C., Ahuya, C., Roets, M., Hossain, M. and Osafo, E. (2005) Goats. In: Owen, E., Kitalyi, A., Jayasuriya, N. and Smith, T. (eds) *Livestock and Wealth Creation.* Nottingham University Press, Nottingham, UK, Chapter 19.

Steele, M. (1996) *The Tropical Agriculturalist: Goats.* Macmillan Education Ltd, London and Oxford, UK.

Gonad

The gonads comprise the testes in the male animal, for which the term is most commonly used, but the term also encompasses the female homologue, the ovaries. In both sexes the principal function of the gonads is gamete production – spermatozoa by the testes and oocytes by the ovaries – but in both cases there is an important secondary endocrine function. Gonadal hormones have regulatory roles in the processes of gamete production, but are also important for a wide range of other parameters linked to reproductive success, including cyclical changes within other parts of the reproductive tract and in the initiation and control of reproductive and **courtship behaviour** patterns.

Although there are significant differences between the testis and the ovary both in terms of morphology and function, there are also many underlying similarities, which is unsurprising as male and female gonads develop from a single undetermined bipotential primordium consisting of a mixture of germ cells and somatic cells. This primordium within the embryo is poised to activate one of two alternative developmental programmes in response to a sex determination signal early in the course of fetal development. The primordial germ cells can be identified at an early stage of development (day 7 in the mouse) originating at the base of the allantois at the posterior end of the primitive streak, where a small population of cells (approx. 45 in the mouse) is allocated to the germ cell line; these cells migrate to the genital ridge where they proliferate and aggregate together within the non-differentiated developing gonad. These primordial cells will become either, in the male, prospermatogonia in a state of mitotic arrest or, in the female, oogonia arrested in meiosis I.

In addition to the germ cells there are somatic support cells partitioned with the germ cells within the seminiferous tubules in the testis or within follicles in the ovary. These are the Sertoli cells in the male, the first cell population to undergo determination, and the follicular cells or granulosa cells in the female. Thirdly, there are endocrine somatic cells, the testicular Leydig cells or ovarian thecal cells, which are steroidogenic cells located in the stromal tissue of the testis and ovary. Each of these three types of cells in the early embryo has bipotential and can form either male or female cell types within the morphological architecture of the testis or the ovary.

In mammals the choice of developmental pathway to testis or ovary is dependent on genetic factors in the form of specific genes present on the Y chromosome, although in other vertebrates other trigger systems may operate, with many **reptiles** for example utilizing environmental cues and allocating sex according to egg incubation temperature. Ovarian formation is considered to be the default condition that occurs in the absence of activation of testis-determining Y chromosome genes in mammals; however, the completion of the ovarian pathway must also involve its own specific genetic factors.

In both sexes, the gonads have an important role in fetal development. In 1947 Alfred Jost demonstrated that rabbit fetuses castrated *in utero* all developed female genitalia irrespective of whether they were genetically XX or XY, thereby demonstrating that expression of the male reproductive phenotype is dependent on the presence of the fetal testes. In both sexes steroid hormones from the fetal gonads supplement the endocrine function of the maternal–placental unit. Some species, for example the horse, show a period of pronounced gonadal hypertrophy during development, so that the gonads are for a time the most significant structures within the fetal abdomen, before subsequently regressing to the much smaller size seen in the neonate.

The male phenotype is dependent on the secretion of two testicular hormones: first, anti-Müllerian hormone (AMH), produced by the Sertoli cells and responsible for degeneration of the Müllerian ducts, the precursors for the female tract; and, secondly, the steroid **testosterone**. Testosterone, secreted by fetal Leydig cells, initiates differentiation of the Wolffian duct into the epididymis, vas deferens and seminal vesicles, and causes the urogenital swellings to develop within the scrotum and penis. In females, **oestrogen** secreted from the fetal ovaries appears sufficient to induce the differentiation of the Müllerian duct into its various components – the oviducts, uterus and cervix – while in the absence of androgen support the Wolffian duct regresses.

This developmental role for the testes *in utero* is concluded at **parturition**, and endocrine activity of the gonads largely ceases with steroid hormone production maintained at very low levels until puberty is reached. This juvenile hiatus may not be evident in small, short-lived mammals that tend to reach puberty early but, in larger mammals, the delay to puberty may ensure that important growth and maturation processes are completed before reproductive function begins. At puberty gametogenic and steroidogenic activity in the gonads is stimulated by rising levels of hypothalamic and pituitary hormones, particularly resulting from release of **gonadotropin-releasing hormone** (GnRH) from the **hypothalamus**. The trigger for resumption of GnRH production remains to be fully elucidated, but it is clear that its timing can be influenced by environmental factors such as **photoperiod** and by aspects of nutritional status. GnRH stimulates the release of the gonadotropins **follicle-stimulating hormone** (FSH) and **luteinizing hormone** (LH) from the **pituitary gland**. These hormones are identical in males and females, and the steroidogenic cells within the gonadal stroma (Leydig cells in the testis and thecal cells in the ovary) are receptive to LH while the supporter cells within the germ cell compartment (Sertoli cells in the testis and granulosa cells in the ovary) are sensitive to FSH. In response to raised gonadotropin levels the gonads increase steroid production, which has effects on both the activity of the gonads themselves and a series of secondary systemic effects establishing the secondary sex characteristics.

Gonadal morphology does differ between the sexes. The adult testis has a connective tissue capsule, the *tunica albuginea*, and a parenchyma packed with finely coiled seminiferous tubules lined by the germinal epithelium with a mixture of germ cells and Sertoli cells. Spermatogenesis occurs within the germinal epithelium, with developing spermatozoa moving progressively from the tubule wall to be released into the lumen as they complete their maturation. From the seminiferous tubules newly created sperm are collected in the *rete testis* and transported from the testis into the epididymis for further maturation and storage. The endocrine Leydig cells are situated in the interstitial tissue between the seminiferous tubules. The cells of the tubule walls, together with a ring of Sertoli cell–cell junctions, combine to form the blood–testis barrier, which makes the tubules an immunologically privileged site, shielding haploid spermatozoa from autoimmune attack. Seasonal breeders may show changes in the size of the testes between the breeding and non-breeding seasons, these changes being directly related to the level of sperm production and, in many livestock species, measurements of testicular size have traditionally been used as an indicator for fertility assessment.

The adult ovary has a fibrous core surrounding a parenchyma containing numerous primordial follicles, each containing a single oocyte within a layer of granulosa cells. The number of these primordial follicles is effectively fixed during fetal development, and the pool of follicles is progressively depleted throughout the animal's breeding life. During the breeding season cohorts of follicles are recruited to begin a process of growth and maturation that culminates, depending on species, in the production of a single or small group of ovulatory follicles ready to release their oocytes into the oviduct of the female tract for fertilization. Interstitial tissue surrounding the follicles contains the endocrine thecal cells, which act in association with the granulosa cells to produce oestrogen in the follicular phase of the cycle and subsequently **progesterone** in the post-ovulatory luteal phase.

Initiation of gametogenesis may be regarded as 'physiological puberty', and the presence of spermatozoa in the ejaculate – or follicular growth and ovulation – means that these animals are technically fertile; however, in many cases it is not until the secondary sex characteristics – gender-related characteristics not directly affecting the reproductive tract – are fully established that animals exhibit their complete reproductive behavioural repertoire. For example, males may need anabolic steroids to increase in size and strength or to develop other endocrine-dependent structures in order to attract females or compete with other males. With some large mammals this may mean that successful breeding may occur only some time, even years, after gonadal competence is achieved.

In mammals, gender identity and related gonadal morphology are fixed in the course of fetal development and, while some genetic conditions or other developmental problems can result in ambiguous or otherwise non-functional gonads, these conditions are generally permanent and not alterable in later life. However, this is not the case in all vertebrates, so, for example, some species of fish will demonstrate serial hermaphroditism in which reproductively functional female animals may, in response to environmental or social cues, change their sex and become functioning fertile males. Hermaphroditism is common in invertebrate species but, despite the dual potential of the vertebrate primordial gonad, does not appear to be a common evolutionary choice in these animals.

The single bipotential primordium responsible for both testis and ovary means that both originate at the same fetal site adjacent to the developing kidneys. However, while the ovaries remain abdominal throughout life, the testes in the large group of Boreoeutherian mammals (a clade of placental mammals) undergo migration to become situated in the scrotum, a pouch of skin in the inguinal region outside the main body wall. Migration occurs in two phases: (i) a transabdominal phase usually occurring approximately halfway through the gestation period, although precise timings vary with species; followed by (ii) an inguinal–scrotal phase where the testes pass through the inguinal canals, bilateral openings in the abdominal wall, into the scrotum. The fetal testes are attached to the abdominal wall by a ligamentous structure, the gubernaculum, which, as the fetus grows, remains the same length, having the effect of drawing the testes down away from their origin in the mesonephric region. The reason for the scrotal position of the testes in this group of mammals has long been the subject of debate. Migration is a complex and highly regulated, androgen-dependent process that is generally completed by, or soon after, birth. In cases where migration fails and the testes remain within the abdomen, termed cryptorchidism, a not uncommon condition in **companion animals**, pigs, horses and humans, although rarer in cattle and sheep, spermatogenesis is compromised and the animals are infertile.

The scrotal position means that the testes are maintained at some 3–4°C lower than core body temperature, and the importance of this temperature difference is emphasized by a number of other thermoregulatory adaptations. The scrotal wall has little insulation associated with it either in terms of hair, fur or fat layers, and a slip of muscle within the spermatic cord, the cremaster muscle, allows the scrotum to be held close to the body wall under cold conditions or to be lowered away from the abdomen when environmental temperatures rise. In addition, the main testicular artery supplying the scrotum is highly coiled and is in close association with a network of testicular veins to form an efficient counter-current heat exchange system in the so-called pampiniform plexus, which cools arterial blood as it leaves the abdomen and simultaneously re-warms returning venous blood. While this small temperature difference appears necessary for spermatogenesis in this group of mammals, a number of mammalian species and all birds retain abdominal testes with normal fertility, while in some other species there is a temporary migration of the testes into the scrotum during the breeding season followed by withdrawal back into the abdominal cavity in the non-breeding season.

From the earliest recorded times the gonads have been know to affect aspects of behaviour. **Castration**, or removal of the gonads, most commonly the testes of the male, has been extensively used as a method in livestock husbandry for reducing male–male aggression and territorial behaviour, making male animals generally more tractable and eliminating or reducing unwanted displays of sexual behaviour. Castration is also used in **production animal**s as a method of increasing growth rates and also to avoid undesirable flavouring in meat such as boar taint in pork. Regulatory effects on behaviour are endocrine-based and must be considered as unsurprising, as the function of the gonads is to coordinate appropriate and timely behavioural responses in order to facilitate the production, delivery and combination of competent gametes and thereby ensure reproductive success. (MRC)

Gonadotropin

Gonadotropins (also called gonadotrophins) are hormones that are so called because their first identifiable effects were on stimulating the **gonads**. The gonadotropins include **luteinizing hormone** (LH) and **follicle-stimulating hormone (FSH)**, both produced by the anterior **pituitary gland**, and chorionic gonadotropin, produced by the placenta during pregnancy. **Stress** has a differential effect on pituitary gonadotropin production, depending on the nature of the stressor; for example, acute cold or **heat stress** causes an increases in LH but does not cause a rise in FSH, but, if prolonged, the level of both hormones may fall as the pituitary gland biases activity towards the production of adrenocorticotropic hormones over gonadotropins, the process and consequences of activation of the former resulting in suppression of the latter. The differential production and release of gonadotropins by gonadotroph cells in the pituitary arises as a result of the actions of a range of hormones, including activin, inhibin, follistatin and the sex steroids (**see: Oestrogen; Testosterone**).

Gonadotropins can be used artificially to induce ovulation or superovulation in females for embryo transfer, but there is not necessarily cross-species compatibility. For example, mares do not respond to the gonadotropins of other species, and even equine FSH can be unreliable in its effects. (DSM)

Further reading

Moberg, G.P. and Mench, J.A. (2000) *The Biology of Animal Stress – Basic Principles and Implications for Animal Welfare*. CAB International, Wallingford, UK.

Gonadotropin-releasing hormone (GnRH)

GnRH (also known as luteinizing hormone-releasing hormone (LHRH) or luteinizing hormone-releasing factor (LHRF)) is released in a pulsatile pattern from cells in the **hypothalamus** into the portal blood stream, where it has its primary effect on the gonadotropic cells of the anterior **pituitary gland**, which produce **follicle-stimulating hormone (FSH)** and **luteinizing hormone** (LH) – both forms of **gonadotropin** – as a result. In females, the frequency of the pulses varies with the **oestrus** cycle, whereas in males it is more consistent. The amount of GnRH produced varies with the plane of nutrition and the maturity of the individual, being very low in the undernourished and prior to sexual maturity, resulting in the suppression of sexual activity and infertility at these times. Immunization against GnRH can be used as a form of non-surgical **castration** (chemical castration) in some species such as pigs. (DSM)

Gorilla

Gorillas are the largest living primates. Males may reach 275 kg in body weight in the wild and stand up to 1.75 m tall; females are considerably smaller, reaching on average 100 kg body weight. Until recently there was thought to be only one species of gorilla, *Gorilla gorilla*, with three subspecies ranging from the mountains of eastern Africa through central Africa to Nigeria. However, recent molecular and morphological studies have confirmed that there are two gorilla species, each with two subspecies: (i) the western lowland gorilla, *G. gorilla*, from much of western and central Africa (including the critically endangered Cross River gorilla, *G. g. diehli* from Nigeria and Cameroon); and, 1000 km away, (ii) the eastern gorilla, *Gorilla beringei*, with two subspecies, the mountain gorilla (*beringei*) of the Virungas of Rwanda, Uganda and Democratic Republic of Congo (DRC) and the Impenetrable Forest in south-western Uganda, and the eastern lowland gorilla (*graueri*) in east–central DRC.

Western and eastern lowland gorillas inhabit lowland tropical rainforest, but Cross River and mountain gorillas are found in montane (highland) rainforest and bamboo forest up to an altitude of 3500 m.

Gorillas are mostly terrestrial foragers, especially the montane subspecies, but females and juveniles may be arboreal, using suspensory **locomotion**. Silverback male western lowland gorillas may also climb large trees. Locomotion on the ground is usually quadrupedal, and typically involves knuckle walking, but gorillas may stand and even take a few steps bipedally, especially when foraging on aquatic plants in bais (swamps). Being diurnal, gorillas sleep at night. They do not have a **core area** where they rest (although their **home ranges** may have a core area that is used more intensively), but make nests of whatever vegetation is available where they have been foraging. Nests are usually constructed in trees, particularly for the mountain gorilla; for some western lowland gorillas nests may mostly be built on the ground.

Up to 22 **vocalizations** have been recorded from gorillas, including long-distance hoots that can be heard up to 1 km away, aggressive roars and growls, and barks and grunts to encourage coordination of the social group. The most dramatic form of **communication** is the chest-beating display, particularly of adult males, which occurs when strange gorillas or even humans approach too near. The display begins with some hoots, then the gorilla stands bipedally and beats its chest alternately up to 20 times with slightly cupped hands, and finally the gorilla runs about on all fours, dramatically breaking vegetation.

Gorillas have highly stable social groups comprising usually an adult, silver-backed male, which dominates several females with their dependent young. Other black-backed younger males associate often temporarily with the group before moving off to live alone for a few years, before returning when mature to try to take over or establish a new group. Group sizes range typically from two to 34 individuals in all subspecies of both species. In some mountain gorilla groups there may be more than one mature male that may associate for up to 10 years. These multi-male groups usually contain closely related males such as fathers and sons or brothers and half-brothers. Even so, these males have weak social relationships with each other and are competitors for mates. Non-breeding males may also form groups until such time as they take over or establish their own groups, and here the relationships are more amicable.

Owing to the high stability of a gorilla social group, all young females leave, usually voluntarily, to join a neighbouring group or to establish a new group with a young silverback thus avoiding inbreeding. Some females move between several groups, but in other groups females may cooperate to prevent incoming females joining that group. In mountain gorillas any young transferring with a female are almost invariably victims of infanticide, whereas this has not been observed in eastern lowland gorillas. Some multi-male groups split to found new single-male groups. Although most gorilla social groups are as described above, in some parts of western Africa fission–fusion **social systems** have been observed, which may reflect local availability of sparse food resources spread widely over large home ranges. When two gorilla groups meet they may: ignore or avoid each other; interact amicably for a while; or there may be open hostility, particularly between adult males.

Gorillas breed throughout the year. Females have a menstrual cycle like humans and other great apes, which lasts 27–28 days, with peak fertility lasting 1–3 days in mid-cycle. Although it is usually the dominant silverback that mates with all the breeding females within a group, females do select mates when opportunities arise, such as when two groups meet or in those groups where there is more than one adult male. Gestation ranges from 251 to 295 days (mean 258 days), and the single young weighs about 2 kg at birth. Twins have been recorded, but are very rare. The young are weaned at 3–4 years, so that a female may only give birth every 3.5–4.5 years. In accordance with the predictions of the **parental investment** theory, female mountain gorillas appear to invest more in the rearing of male young, as shown by higher and slightly increasing levels of **grooming** throughout infancy and generally later weaning.

Western lowland gorilla group home ranges range from 7 to 23 km^2, with a high degree of overlap between groups, whereas mountain gorillas' ranges are only 4–11 km^2, also overlapping but rarely at the same time. Groups travel 148–5500 m/day while **foraging** and their diets can comprise up to 100 plant species, as has been recorded for eastern lowland gorillas. Mountain gorillas feed on more than 50 plant species, of which about three-quarters are vines, herbaceous plants and parts of trees, including leaves, bark, stems and roots. Western lowland gorillas have equally diverse diets, but these tend to include more fruits (e.g. 40%). Gorillas have a keen sense of smell and are able, for example, to reject overripe fermenting fruits. They also develop complex manipulative skills for dealing with this wide range of foods, but do not use tools. (ACK)

See also: Great ape

Further reading

Robbins, M.M., Sicotte, P. and Stewart, K.J. (eds) (2001) *Mountain Gorillas*. Cambridge University Press, Cambridge, UK.

Taylor, A.B. and Goldsmith, M.L. (eds) (2003) *Gorilla Biology, a Multidisciplinary Approach*. Cambridge University Press, Cambridge, UK.

Granivore

Granivores are animals that eat seeds as the main part of their diet, and can be found among both vertebrate and invertebrate species, especially mammals, birds and insects. Granivory can be divided into pre-dispersal seed consumption and post-dispersal seed consumption, and these different types of consumption require different **foraging behaviour** strategies. Some plants have evolved defences, such as toxic tannins, to defend themselves against granivores, while some granivores have co-evolved adaptations to these plant defences. (LMD)

Grazing

Grazing is a method of **feeding** and, in particular, relates to the cropping and ingestion of growing grasses. The grazers are predominantly the mammals, and common examples include **rabbits**, kangaroos, **horses**, donkeys, **cattle** and **sheep**. Grass is also eaten by snails, slugs, grasshoppers, tortoises and geese. Natural grasslands often develop as a consequence of insufficient rainfall to support forest and are largely made up of perennials that propagate by lateral stems that spread beneath the soil surface and produce a mat of vegetative shoots. Regular grazing can act to preserve and improve the grassland, as cropped leaves continue to grow and new grass growth is higher in nutrient content.

The structure of the mouth and jaw influences how the herbage is cropped. Horses gather plants into their mouth using the sensitive prehensile lips and cut through the tough, fibrous stems using their upper and lower incisors, whereas cattle and sheep have no upper incisors, but have instead a horny pad that acts as a cutting board for the lower teeth. For example, cattle graze herbage by collecting it into the mouth and compressing it against the upper palate with the tongue and lower incisors. The herbage is then severed from the plants by jerking the head upwards. This is repeated several times a minute, typically 30–70, and the animal moves its head from side to side as it walks. Associated with grazing bites are occasional chewing bites or manipulative movements of the tongue or lips to manoeuvre the herbage in the mouth. These are more common when the herbage is long and fibrous. Sheep and **goats** feed more selectively than cattle due to their smaller mouthparts and are able to nibble at the shortest of grass due to their split upper lip. Cattle therefore cannot manipulate individual plant items to the same degree as sheep and goats.

To effectively digest the useable nutrients locked up within the rigid plant cell walls, grazers must first (using the molar teeth) grind down the plant matter into smaller fragments. Ruminants regurgitate food into the mouth, where it is further ground down by the molars ('chewing the cud'). This increases the available surface area for enzymatic digestion and microbial fermentation. Proteins, lipids and soluble carbohydrates are broken down by enzymatic digestion. No mammalian enzymes are capable of breaking down the structural components of plant cell walls. Cellulose is one such structural component, and many herbivores depend upon it as a major energy source. Instead, symbiotic microorganisms that live within the digestive tract are able to break down the cellulose and other structural components to organic acids through the process of fermentation. Microbial fermentation occurs in specialized and enlarged chambers that are located either in or near to the foregut or the hindgut. Ruminants are foregut fermenters, whereas rabbits and horses are examples of hindgut fermenters.

To cope with a nutritionally variable environment, grazers adopt a patch-feeding strategy whereby preferred plant communities are regularly visited and sampled. Thus selection of grazing sites is not uniform across the whole pasture due to both positive selection of desirable areas and avoidance of undesirable areas. For example, selection occurs across the surface of the pasture (horizontal selectivity); within any pasture there are tall areas of herbage close to faecal deposits that are left ungrazed. The rejection around each faecal deposit will be greater in undergrazed swards because animals have the choice of other, clean areas to graze. Herbage around dung deposits is rejected initially because of the smell of the faeces, and later because of its maturity. Typically it is lighter and browner than 'grazed' green herbage, and this indicates that it has a low nutritive value. Whereas darker green pastures are usually preferred because this indicates higher nitrogen content, dense pastures are preferred because they give a greater bite weight.

Selection also occurs within the depth of the forage canopy (vertical selectivity). When determining the height above ground level at which to defoliate the sward, grazing animals probably balance a high leaf:pseudostem ratio (which would maximize nutrient content of the herbage) with achieving an adequate total intake. According to optimal foraging theory, the composition and volume of the plant eaten and the energetic cost of grazing are optimized by the animal to supply its nutrient requirements at least cost.

The grazing action of cattle allows them to select a greater proportion of leaf material than the sward contains as a whole, thereby increasing the digestibility of their diet. This is primarily because they graze the upper strata of the sward (i.e. above the pseudostems, normally about 2 cm above the soil surface), and this layer contains more leaf material. However, cattle do not exclusively graze tall swards and leave short areas untouched. When offered a choice of 3- or 6-cm grass swards, cattle spend only just over one-half of their grazing time on the tall sward. Horses have also been noted to reside significantly longer on patches of long grass that had been mown to a height of 15 cm. Sheep, on the other hand, select individual tillers – shoots that sprout from the base of a grass – and sever them much closer to the ground than cattle, providing less frequent but more severe defoliation. This has its effect on the sward itself: because of the reduced frequency of defoliation, more tillers escape grazing altogether and become rejected due to their maturity, especially if the stocking rate of the sheep is too low. With cattle this usually occurs only in herbage around dung deposits, and even this is limited if the stocking rate is high. This ability of cattle to keep a pasture relatively free from mature herbage has led to their sometimes being kept principally to defoliate mature pasture on hill sheep farms. There are even greater advantages if cattle and sheep actually graze together, as the sheep do not show an aversion to cattle dung deposits and will prevent herbage near it from becoming mature and wasted. The sweeping side-to-side grazing action of each individual animal combines with the cohesive grazing action of a group of grazing animals to result in a sward that is all, or nearly all, defoliated to a common height, with the exception of areas around dung deposits. Herbage height largely determines the bite size or mass, with a tall sward giving the greatest ease of prehension and therefore minimum foraging time.

In a mechanistic model of the grazing process, the feed intake can be determined from the grazing time multiplied by rate of intake. The latter can be further subdivided into bite mass and rate of biting, and the bite mass determined from the herbage density and bite size or volume. Bite volume is the product of the bite area and bite depth. Bite area is a function of palate breadth and the distance between the palate and the tongue when the mouth touches the sward. Bite depth is the main factor regulating the other behavioural factors, and it is largely determined by sward height. Hence, on a 10-cm lush spring pasture, intake rate of cattle is about 25–30 g dry matter (DM)/min, whereas a 5-cm autumn pasture will only be eaten at a rate of 15–20 g DM/min.

The useable energy gained from plants is far less than from animal protein, and thus large grazing mammals have to spend much of their time feeding in order to acquire sufficient daily nutrients and energy. For cattle, the maximum grazing times and biting rates normally occur at about 10–12 h/day and 65–70 bites/min, respectively, although longer grazing times of 13 h/day have been recorded on sparsely vegetated rangeland. This gives a normal maximum of 50,000 bites/day, although individual cattle achieve more. High-yielding cows may have to balance the requirement for long grazing times with the requirement for adequate lying and ruminating times. This emphasizes the need to minimize the time that high-yielding cows are kept off the pasture for milking. Grazing lactating dairy cows typically have about five meals per day, each lasting on average 110 min. Cattle with lower intake requirements in relation to their weight (e.g. dry cows, mature bullocks) have fewer and shorter meals. Normally, the first meal begins shortly after dawn, followed by two to three meals between morning and afternoon milking, and the longest, most intensive meal in the evening ending around dusk. This is to provide sufficient food to digest during the night period. There is often a short meal averaging 30 min at about 01.00, after which the rest of the night is spent ruminating and resting.

The horse, as a hindgut fermenter, has a slightly different digestive strategy from the ruminant and compensates overall by having a higher voluntary intake and rate of gastrointestinal passage. Free-ranging horses spend 13–18 h each day grazing and browsing, and similar total foraging times of between 15 and 18 h have been recorded with pastured horses. In general, horses take regular grazing bouts throughout each 24 h period, although time of day affects bout length, as longer feeding bouts have been recorded at dawn and in the late afternoon.

Grazing is a very important commensal system ecologically between plant and animal, allowing both to thrive while maintaining good utilization of land and other resources.

(CJCP)

Great ape

The great apes or anthropoid apes belong to the superfamily Hominoidea, whose living representatives include the **orang-utans** (*Pongo* spp.), the **gorillas** (*Gorilla* spp.), the **chimpanzees** (*Pan* spp.) and humans (*Homo sapiens*). Until recently humans were regarded as being sufficiently different from the other great apes that they were assigned to separate families, the Hominidae for humans and the Pongidae for the great apes. However, recent molecular studies have confirmed the very close relationships between all the great apes and, today, all are regarded as being hominids, or with only the orang-utans remaining in the Pongidae.

Being phylogenetically so close to humans, there has been a vast amount of behavioural research aimed at all great apes, both in the wild and in captivity. This research often helps us to directly or indirectly gain some insight into the behaviour and **ecology** of our now extinct ancestors. The great apes show a spectrum of closeness of relationship with humans, from the distantly related orang-utans to the most closely related chimpanzees and bonobos. Comparison of behaviours, cognitive abilities, **social systems** and ecology provide us with interesting insights into how australopithecine and early *Homo* may have evolved and which factors (and in what sequence) were most important in the evolution of our own species.

Non-human great apes all inhabit tropical forests, including rainforests, montane (highland) forests and woodland savannah. Unlike us, the other great apes are more arboreally adapted, ranging from the orang-utans, which move mainly by suspensory climbing, to the more terrestrial gorillas, which typically move quadrupedally by 'knuckle walking'. Intermediate between the gorillas and orang-utans, chimpanzees and bonobos move around and forage both on the ground and in the trees. Orang-utans are mainly fruit eaters, but gorillas feed mainly on leaves, pith, shoots and other herbaceous vegetation, although western gorillas feed more on fruit than eastern gorillas. Chimpanzees and bonobos feed mostly on fruits, pith and nuts, but also consume leaves, invertebrates, birds' eggs and small vertebrates. However, adult males may hunt cooperatively to kill larger prey such as **monkeys** and duikers, which are shared among the group members.

Social systems are very different between the non-human great apes. Adult orang-utans are normally solitary, the only consistent social groupings being when male and female come together for mating and mothers with their dependent young. This social system is also found in nocturnal prosimians and is termed a **noyau social organization**. Despite this apparent solitary social system, orang-utans have overlapping **home ranges** and are able to communicate and maintain a wide but dispersed social network. Fierce battles may occur between adult males, but younger males forage with females for weeks or months at a time. Although both males and females may become sexually mature by 7 or 8 years old, it may be many years before adult males become fully mature and are able to mate with females. Dominant males may suppress the social and physical maturation of subordinate males, who only develop fleshy cheek pads at full maturity.

Gorillas are the largest living apes and, like orang-utans, display considerable sexual dimorphism, with males weighing up to three times as much as females. Gorillas live in social groups of about ten individuals, led by a dominant silverback male with several females and their dependent young. The group may also contain one or more younger black-backed males. Mountain (*Gorilla beringei beringei*) and some western gorilla (*Gorilla gorilla* sspp.) groups remain very cohesive, although with overlapping home ranges with those of other groups, but some western gorilla groups have been observed with a **fission–fusion social system**, whereby small groups split off and rejoin the main group when foraging, perhaps reflecting their typically larger home ranges with more dispersed foods. Gorilla society is unusual among primates in that males and females may move between social groups and the relationships are relatively weak between unrelated adult females, but are strong with the dominant silverback.

This pattern contrasts with the complex and dynamic social groupings of chimpanzees (*Pan troglodytes*), which have a fission–fusion social system. Interactions between group members are frequently noisy, boisterous and aggressive. Within a common shared home range, individuals may forage alone and come together temporarily to interact socially. Adult males are usually gregarious, and their **dominance** within a hierarchy may depend on their social skills and the presence of brothers or close allies, as well as their age or physical attributes. They often patrol their home ranges, and hunt and forage together. Among females the strength of relationships may vary geographically; in eastern Africa females forage more solitarily and have weak social bonds and interactions with other females, whereas in western Africa females are more social and often forage together. Although undoubtedly all great apes develop local traditions or cultures relating to aspects of their behaviours and ecology, chimpanzees show a very wide range of local influences and traditions on almost every aspect of their social lives, including **communication**, social dynamics, tool use, etc.

Mating systems also vary from consortship (i.e. a male and female going off together), through mate guarding within the group to promiscuity. What particularly marks out chimpanzees from other great apes is their use of tools, which are used to process tough foods (e.g. nuts) and reach inaccessible resources (e.g. water or termites). All great apes build nests from vegetation at hand in order to construct a bed for sleeping at night. Nests are usually in the trees, but mountain gorillas rarely climb trees and build theirs on the ground. Chimpanzee home ranges are aggressively defended by males, which tend to be closely related to each other, thereby ensuring long-term and stable communities. In contrast, females move between groups to avoid inbreeding. Bonobos (*Pan paniscus*), on the other hand, are far less aggressive, and even though their home ranges overlap with those of neighbours, inter-group relations seem to be mostly peaceful. Like chimpanzees, bonobos have a fission–fusion social system, but males and females form mixed-sex foraging groups. Relationships between females are also much stronger than those between female chimpanzees, but give way to stronger male–female affiliations in larger foraging groups. Amicable interactions between bonobos often involve sexual behaviours, including copulations between males and females, and genito–genital rubbing between females. Bonobos use larger food patches annually, which appears to be related to the greater social cohesiveness of the sexes.

Being so closely related to humans, the great apes have been subject to a vast amount of behavioural research, often in laboratories or captivity, since the second decade of the 20th century. The earliest experiments involved simple **discrimination** tests, which tested the ability of chimpanzees, in particular, to discriminate between shapes and colours of objects, and problem solving, such as how to use objects to reach inaccessible food items. A whole array of more or less complex experiments have been devised to elucidate our understanding of the **memory** and cognitive skills (**see: Cognition**) of all great apes, and much publicity has been generated in those few individual chimpanzees and gorillas that have been taught various forms of non-vocal **language** in order to communicate with humans and other apes. (ACK)

Grooming

Grooming is a natural behaviour, and its function is thought primarily to be cleaning of the outer surface of the body, including removal of both dirt and **parasites**. Grooming may, however, be equally important as a **display behaviour** and a **social behaviour**. Many species of animals spend a large part of their time self-grooming. Grooming involves cleaning the surface, also known as the integument, of the body, including hair, skin, feathers or scales. As the integument may function as a temperature regulator, a sense organ, a protective device, a moisture barrier and a signalling device, its care is critical in body maintenance.

Animals differ in their methods of grooming. Mammals tend to lick themselves, use their teeth to nibble and/or use their forepaws or, in the case of primates, their dextrous hands, to rub and manipulate their hair. Companion animals may spend a large amount of time in self-grooming. Cats are known for their fastidious attention to self-grooming. Birds use their bills to realign and dress feathers with oil from a gland at the base of the tail, a behaviour known as **preening**. Invertebrate animals also spend time in grooming rituals. Insects rub their antennae, while crustaceans have appendages with setae (hairs) specialized for grooming. Like the insects, they keep their antennae and gills clean.

Licking the body surface may help to keep animals cool, and behaviours such as **dust bathing**, commonly observed in birds, may help with this process, too, as well as with the removal of parasites. In birds with a deformed beak or bill, the resulting parasite infestation can reduce the insulating properties of the plumage.

Grooming may also help to heal wounds. Ants have metapleural glands, located on the thorax (the region between the head and body of arthropods), which secrete antibiotic substances that are smeared over their bodies to keep bacteria and fungi from flourishing in their nest. Animal saliva may produce proteins that help in wound healing.

Grooming can also serve as a social signal. Small male rodents groom not only to keep their fur clean but also to attract females. The longer a male grooms himself the more interest a female shows, possibly due to the odours that the males are emitting. In some birds a ritualized display of grooming is often part of a male's courtship display, a bright and healthy coat attracting potential mates. In the chimpanzee males groom one another more when females are present. Social insects also use grooming as a communication tool. Worker honeybees groom the queen, picking up a pheromone that is then transferred through the nest.

The grooming of one individual by another is known as **allogrooming**, and will often be reciprocated by **conspecifics** in bouts of mutual grooming. Allogrooming is commonly observed between mother and young, and is necessary for the cleaning of juvenile animals, which may not yet be able to toilet or self-groom.

Mutual or social grooming is readily demonstrated in adults of many animal species. Horses regularly groom one another, a response that has been shown to reduce heart rate. Primates, particularly the apes, help to remove parasites and dirt from one another's coats. Cats that are familiar with one another or related are more likely to groom one another and to spend time in proximity to one another. Birds, too, may groom one another, especially when forming pair bonds. Primate relatives, such as baboons and chimpanzees, spend more time grooming near-relatives than more distantly related individuals. This mutual grooming, therefore, serves not only to clean areas of the body that are difficult to reach, but also reinforces the bond between kin.

Studying this intraspecfic mutual behaviour can help scientists to understand the social systems of animal species. **Dominance** hierarchies (**see: Hierarchy**), for instance, may be determined by observing grooming interactions. Some scientists have taken this one step further and hypothesized that social grooming, as a **communication** tool, may even be linked to human **language** development. Where animal groups, such as some human societies, are large, it is difficult to maintain effective communication among all members by grooming, and language may have developed as a similarly effective way of communicating. In both grooming and speech, endorphins are released.

Grooming may be a less pleasurable activity when it is demonstrated excessively. This is a common behaviour problem in some cats and dogs that lick themselves excessively, resulting in hair loss, damaged skin or infection. Proximal causes of such over-grooming include itchiness from parasites, such as fleas or other foreign bodies on the coat, skin allergy or disease. Boredom and frustration have also been postulated as possible causes of this becoming a stereotypic behaviour. Excessive grooming may also be a product of anxiety, and the presence of an unknown feline in the vicinity may be enough to set off an excessive grooming episode in some cats.

Cases of over-grooming may require treatment in the form of anti-inflammatory or anti-anxiety medication, as well as rigorous attention to flea treatments, environmental change and behaviour modification. When licking is excessive, companion animals can be discouraged with the use of an Elizabethan collar, but this sort of intervention does not address the cause and can lead to further frustration.

Wild animals in captivity may also display excessive grooming behaviour, possibly as a result of boredom or frustration resulting in displacement behaviour, replacing more appropriate behaviour that may have been performed in the wild. Stereotypic or compulsive and repetitive grooming behaviour has been observed in many species, including giant pandas, parrots and chimpanzees. Behavioural enrichment programmes, now included in the management programmes of many zoos and wildlife parks, are helping to reduce this unwanted behaviour.

In addition to reducing the parasite load and keeping the animal's coat healthy, grooming, such as stroking or petting the animal for pleasure, is a mutually beneficial exercise and an important part of caring for animals, and therefore has a significant role to play in enhancing the human–pet bond. (JR)

See also: Compulsive disorder; Intraspecific behaviour; Self-mutilation

Further reading

Curtis, T.M., Knowles, R.J. and Crowell-Davis, S.L. (2003) Influence of familiarity and relatedness on proximity and allogrooming in domestic cats. *American Journal of Veterinary Research* 64, 1151–1154.

Dunbar, R. (1998) *Grooming, Gossip and the Evolution of Language.* Harvard University Press, Cambridge, Massachusetts.

Stewart, D. (1998) The importance of getting clean (grooming behaviour in animals). *International Wildlife* March/April. Available at: http://www.nwf.org/nationalwildlife/

Group

Animals are often found in close associations or groups. The study of why animals group together is a fundamental question in **sociobiology** and **behavioural ecology**. Animals may be brought together in groups by some localized resource or attraction, such as moths around a light or animals surrounding a water source in the desert. These groupings are not necessarily due to the animals being social in nature. If an animal's habitat and resources are locally distributed, they may overlap with other animals and these animals may then be found in clumps or groups. Additionally, if the habitats and resources of animals are patchy in distribution then the animals may also appear to be living in **social groups**, even though they do not engage in much **social behaviour**. Non-social animals tend to have random patterns of distribution and thus do come in contact with each other from time to time. Social animals, however, tend to have even or clumped distributions, which keep animals at the correct distances from each other, and they maintain those positions during specified time periods.

Animals may group together to gain advantages over living a solitary life. The strongest selective pressure to live in groups is thought to be protection from predators. Predators may hesitate to attack a large group of animals and, even if they do attack, the risk to any one individual of the group is greatly decreased. Groups of animals also have better access to resources, as a larger number of individuals are available to search for the resources; they potentially have access to mates without prolonged searching and they may have help in finding and/or building shelter. (LMD)

See also: Grouping

Grouping

Animal associations are observed when groups are more beneficial for individual members than solitary action or existence. Predator mobbing is an example of a group-related benefit. Grouping may be distinguished into aggregations (e.g. an association of animals without any clear organization over an abundant source of food) and **social groups**, characterized by some degree of cohesion and coordination of movements among group members. (BM)

See also: Relationship; Social behaviour

Growth hormone (GH)

Growth hormone (also called somatotrophin or somatotropin) is a polypeptide anabolic hormone produced in a pulsatile pattern by the anterior **pituitary gland** affecting cell growth and division, particularly within the musculoskeletal and immune systems. Overproduction results in accelerated growth, while underproduction during development can result in pituitary dwarfism.

From a production point of view, growth hormone is of interest because it increases food conversion ratio, lean muscle mass, bone mineralization and milk production. However, its use in animal production is controversial. Transgenic species can be produced that mature much more quickly and efficiently (e.g. growth rate in salmon may increase 11-fold), but there are concerns about the **welfare**, ecological and potential food safety implications of this. The exogenous administration of growth hormone is advocated by some and opposed by others, with the administration of bovine somatotropin in dairy cows being licensed for use in **cattle** in the USA, but banned within the European Community on welfare grounds following a report from the EU Scientific Committee on Animal Health and Animal Welfare in 1999, which concluded that its use was associated with an increased risk of foot disorders, **mastitis** and reproductive problems. Advocates for its use have more recently suggested that, by increasing efficiency, its use can help to reduce the environmental impact of the industry used to support the feeding of cattle (e.g. production of fertilizers and feedstuffs), as well as reducing methane and carbon dioxide emissions from cattle, which are a significant source of greenhouse gases and global warming.

Growth hormone production follows a diurnal pattern, with surges occurring during the day and especially during **sleep**. The production of growth hormone is affected by the balance of specific stimulatory and inhibitory release factors (e.g. **growth hormone-releasing hormone (GH-RH)** and **growth hormone-inhibiting factor (GH-IF)**), which are produced alternately in the **hypothalamus** to magnify the pulsatile pattern of GH release, less specific factors (e.g. ghrelin, which has a stimulatory effect, and **oestrogen** and **corticosteroids**, which have inhibitory effects) as well as other physiological factors, such as levels of blood sugar, dietary protein, sleep and exercise. Negative feedback on production also occurs in relation to circulating levels of growth hormone and growth hormone-like factors such as **insulin-like growth factor**. **Stress** has a variable effect on growth hormone levels, with both a decrease (in response to **social stress** associated with overcrowding in rats) – but more typically an increase – reported. This rise blocks the effect of **insulin** and helps to maintain higher blood glucose levels in preparation for action (**see: General adaptation syndrome**). (DSM)

Further reading

Moberg, G.P. and Mench J.A. (2000) *The Biology of Animal Stress – Basic Principles and Implications for Animal Welfare*. CAB International, Wallingford, UK.

Growth hormone-inhibiting factor (GH-IF)

GH-IF, also known as growth hormone-inhibiting hormone (GH-IH), somatotropin release-inhibiting factor (SR-IF) or somatostatin, is a peptide hormone produced in a pulsatile pattern within the **hypothalamus** that is secreted into the portal blood supply, from where it is carried to the anterior **pituitary gland**, and stimulates the production of **growth hormone (GH)**. GH-IF appears to have wider direct physiological effects, such as suppressing digestive function, and its production may be suppressed during suckling (**see: Nursing**), although these effects are less consistent. (DSM)

Growth hormone-releasing hormone (GH-RH)

GH-RH, also known as growth hormone-releasing factor (GRF or GH-RF) or somatocrinin, is a peptide hormone produced in a pulsatile pattern within the **hypothalamus** that is secreted into the portal blood supply from where it is carried to the anterior **pituitary gland** and stimulates the production of **growth hormone** (GH). Production in the hypothalamus alternates with growth hormone-inhibiting factor, which has an antagonistic effect on growth hormone production, resulting in the latter being produced in more marked surges and troughs. (DSM)

Growth rate

Growth rate is most often measured as weight (or mass) gain over time. A common production measure, growth rate allows producers to assess flock/herd productivity. Growth rate curves, in conjunction with additional flock/herd information such as feed efficiency, can be used to determine the ideal time to market an animal. Growth rates are also used to compare strains or breeds and to compare production between years or seasons. Although there is a heritable component to growth rate, there is also a large environmental component. Growth rate is affected by numerous factors including diet, **stress**, immunocompetence and genetics. Growth rate is an ideal measure for assessing the production efficiency of new diets and genetic lines. With environment held constant in experimental paradigms, growth rate is often used as a measure of stress or stress-coping ability between animals of different strains, breeds or treatments, but should not be used as a stand-alone determinant of an animal's welfare. Animal production research has accelerated the growth rate of production animals over the past half century through improved breeding, nutrition, veterinary care, production practices and the use of pharmaceuticals.

Although growth rate varies greatly between and within species, it is most often a logarithmic function. Animals experience the most rapid growth in the time immediately following birth or hatch, and experience a gradual deceleration of growth rate. A plateau in growth rate is usually achieved around or shortly following puberty. Although animals may continue to gain weight following puberty, there is generally a slow rate of growth. In many cases animals that have passed their sexual maturity exhibit a reduction in weight or negative growth rate. Negative growth rates are also associated with injured or diseased animals. Egg-laying-type hens also exhibit a negative growth rate during the induced moulting period. (RD)

Further reading

Sibly, R.M. and Hone, J. (2002) Population growth rate and its determinants: an overview. *Philosophical Transactions of the Royal Society B* 357, 1153–1170.

Guinea pig

The guinea pig, *Cavia porcellus*, comes from South America and does not originate from Guinea or Guiana. The 'pig' part is thought to derive from its resemblance to a suckling pig, and 'guinea' as a result of the countries visited by the trading vessels in the South Americas and Africa. In the 16th century the Dutch introduced it into Europe where it failed to gain acceptance as a food, but it did as a **pet**. From there it was used as a research animal and to diagnose tuberculosis in humans. (DBM)

H

Habit

A behaviour or sequence of behaviours that has been performed with such regularity in response to a certain stimulus, or at certain times, that it has become a fixed, semi-automatic response. The performance of the behaviour may occur with some regularity and may appear to be independent of its original reward. Habitual behaviour becomes entrenched within the behavioural predispositions of the individual following repetition, and this developmental element distinguishes it from fixed action patterns. Habits are often referred to in a negative context, although they may not necessarily be detrimental to the individual's welfare or fitness. For example, habits might include successful ritualized behaviour sequences, as occur, for example, in courting displays, as well as behaviours with more negative connotations such as stereotypical pacing within confinement. (KT)

See also: Compulsive disorder; Repetitive behaviour; Stereotypies

Further reading

Dickinson, A. (1985) Actions and habits: the development of behavioural autonomy. *Philosophical Transactions of the Royal Society of London, Series B* 308, 67–78.

Habituation

Habituation is a form of learning found widely in animals, including many invertebrate phyla. It is said to have occurred when the presentation of a previously arousing **stimulus** no longer elicits a response. Thus, it can be defined as learning *not to respond* to a stimulus. Behaviourally it may be very similar across phyla, but clearly different processes are involved in different types of habituation; for example, habituating to a noise is not the same as habituating to a toe pinch (which may be organized at spinal level). This emphasizes an important behavioural principle – similar phenomena may operate through very different mechanisms in different contexts. Habituation can improve biological fitness, since to respond to every novel or alarming stimulus, regardless of its significance, may waste energy and thus be maladaptive. The process and rate of habituation are dependent on three features of the causal stimulus: its nature, frequency and regularity. For successful habituation, a stimulus must be presented well beyond the point of habituation in order to minimize the risk of **spontaneous recovery**. Spontaneous recovery describes the rebound in response strength when a stimulus has undergone extinction or habituation.

Both domestic species and captive wild animals habituate to the presence and activities of humans and other species, as well as to their surroundings. This is also evident in behaviour development, when a puppy, kitten or foal quickly adapts to a novel environment. Livestock that live beside a railway line soon become accustomed to the visual and auditory stimuli of passing trains. Furthermore, ridden animals, such as **horses**, donkeys, camels and elephants, show remarkable habituation to riding paraphernalia, such as saddlery. The early stage of exposure to the girth strap is sometimes difficult in horses because of their strong innate flight response to touch in the regions associated with the girth. However, habituation occurs when their innate predator-removal mechanisms, such as bucking and rearing, do not result in the removal of the girth pressure and so are not negatively reinforced. Forced exposure to a harmless stimulus is sometimes used in behaviour therapy, where it is often referred to as '**flooding**' or 'stimulus flooding'. Given the dangers of severe expressions of the flight response and its resistance to extinction, flooding should be used with care and progressively at the lowest threshold of fear. Habituation differs from the **extinction** of a learned response, as it is an instinctive response that is being eliminated, rather than one that has been previously conditioned.

Preventing the animal's legs from moving by means of hobbles during exposure to aversive stimuli has been employed for centuries by horse trainers. In this process, the animal is restrained with one or more legs trussed, and with the animal sometimes thrown to the ground. The horse is then exposed to repeated maximal presentations of aversive stimuli until it no longer responds. The mechanism relies on an emerging stimulus–response relationship that trains the redundancy of leg movement during presentation of an aversive stimulus (i.e. the animal acquiesces in the presence of that stimulus). There may be some evidence of stimulus **generalization** here, too, in that trainers often report a diminution of general aversive reactions as an outcome of such procedures, but such exposure may simply be inducing a state of **learned helplessness**. In the past, police horses have been hobbled to train them to habituate to various aversive stimuli that they might encounter in their work. The folklore of **horse training** (and some modern 'horse whisperers') often claims that the process is one of hierarchical intervention, that the animal quietens because it accepts the dominance of the trainer and subsequently learns to 'trust' the trainer, but this is a misrepresentation of its scientific basis.

In other habituation procedures, the flight response may still be exhibited, yet complete escape is thwarted, for example in the 'round pen' training technique championed by Monty Roberts and many others interested in 'New Age' horse training. Here the animal is generally presented with an aversive stimulus in the form of human aggression (e.g. the trainer may be at the centre of the round pen flailing a rope at the animal while it runs around in flight). Because escape is impossible, the animal's flight response is never reinforced and its motivation to flee may diminish, i.e. the response becomes habituated. Then, as the horse slows down, the human may become passive and relatively non-threatening, providing negative reinforcement of the habituated response. The trainer then shapes the horse's response so that the animal will gradually approach. Adherents to these methods claim great success in reforming previously traumatized horses. However, there are safety and **welfare** concerns surrounding this procedure, despite claims that it is more 'welfare friendly'.

Gradual habituation is a more common form of desensitization procedure, where the aversive stimulus is generally maintained at a low threshold until the animal habituates. For example, thunder-phobic dogs may be gradually exposed to mild recordings of thunderclaps. The intensity of the thunderclaps is gradually raised until the dog becomes habituated to thunderclaps of typical intensity.

During ontogeny, captive animals should be progressively habituated to the presence of humans and other potentially aversive stimuli. (AM)

See also: Desensitization – systematic; Memory

Further reading

Leussis, M.P. and Bolivar, V.J. (2006) Habituation in rodents: a review of behavior, neurobiology, and genetics. *Neuroscience and Biobehavioral Reviews* 30, 1045–1064.

Hamster

Hamsters are mainly nocturnal and live naturally in burrows in dry, rocky or brushy, steppe-type ground. There are several species of hamster: the Golden or Syrian hamster (*Mesocricetus auratus*), the European hamster (*Cricetus cricetus*), the Chinese hamster (*Cricetulus griseus*) and the Djungarian hamster (*Phodopus sungorus*). They are used in the laboratory and as pets. The most common laboratory hamsters are derived from *M. auratus*, which has a grey belly and a reddish brown coat. This species is thought to have been introduced to the Western world from a few pairs caught in Syria in 1930. (DBM)

Handling

The handling of animals can be defined as the manual or mechanical carrying, moving and/or transportation of fauna. The correct handling of animals is important for their **welfare**, as incorrect handling may result in **stress** and **injury** to the animal. It may also have health and safety implications for the handlers involved.

Animals differ in their acceptance of handling by humans. Exposure to humans from an early age (**socialization**) generally makes animals more amenable to handling. Exposing and bonding animals to humans from an early age, otherwise known as **imprinting**, was first demonstrated in goslings and is important in a number of animal species, particularly birds.

There is often a specific time period during the early part of an animal's life in which acceptance of humans occurs. Such **sensitive phases** have been identified in **companion animals**. During this time period or age range, which reflects the sensory and motor development of the animal, particular events are likely to have long-term effects on the individual's development. In puppies, the sensitive period for interspecific encounters is generally considered to be between 2 and 13 weeks of age, with a peak between 6 and 8 weeks. In kittens it is between 2 and 14 weeks, with a peak between 2 and 7 weeks. During this time positive experiences with humans will tend to ensure a long-lasting and pleasant association with handling. **Aversive** experiences will be remembered and handling avoided. Similarly in **horses**, intensive handling of young foals supposedly results in more tractable horses.

Animals with very little or no exposure to human contact during sensitive periods of socialization will be difficult, and often impossible, to handle. Such individuals are often referred to as **feral** animals.

Animals that are handled from an early age are often more confident than those not exposed to regular human contact. In some cases, overly confident animals will challenge humans, which makes them more difficult to handle, especially if the challenging behaviour is reinforced. Knowledgeable and positive methods of handling should ensure that this does not become an issue.

Animals benefit from correct human handling throughout their lives. **Farmed animals** may be stressed by human contact when unaccustomed to it, leading to decreases in productivity, reduced reproductive performance and reduced immune function. Stressed animals may be more flighty, excitable, fearful or generally more difficult to handle and may also have lower milk production or poorer meat quality. High levels of **vocalization** in many species may be indicative of incorrect handling.

Handling that is inappropriate, inconsistent, clumsy or excessively firm may result in both acute and chronic stress responses, reflected in elevated plasma cortisol concentrations. Gentle handling decreases physiological stress responses and keeps animals calm and easier to handle.

Being unaware of the correct way to approach and physically handle animals can often compromise human safety and enjoyment of companion animals. Children must be shown how to handle both companion and farm animals. Young children must be supervised when interacting with animals, even small companion animals or pets, as either party may be injured or stressed as a result of a negative encounter. An animal, when lifted, should be supported by two hands. A child should be taught not to restrict an animal with strong holds, i.e. hugs, which are unnatural for the animal, especially when the animal is unknown to the handler. Handling with the use of species-appropriate **collars**, leads and harnesses should be encouraged.

Understanding animal behaviour and the behavioural ecology of the species can help to reduce stress levels in animals and make them easier to handle. Caution should be taken, however, as injured or frightened animals may become stressed, agitated and cause injury. Some animals may be less frightened in the dark, so covering them with a blanket may help with ease of capture and holding, especially when the animal is injured. Animals such as **cattle** may be frightened of dark or shaded environments, so provision of adequate, evenly distributed lighting may reduce **anxiety** in these animals. Approaching animals such as horses or **dogs** from an angle where they cannot see the advance may cause them to startle. Animals such as **cats** and dogs may respond aggressively to unseen approaches. Loud noises or some **odours** may also cause a fear response in some animals.

An awareness of the influence of rank on group movements may facilitate handling in production contexts. Herd animals, such as cattle, have a leader that they will follow and this can be used to the handler's benefit. By focusing on moving the herd leader, the whole pack can be encouraged to follow with relative ease. In contrast, when solitary, livestock may be difficult or even dangerous to handle and move. Some breeds of animals may be more reactive to human handling than others. More flighty or excitable breeds, strains or individuals may need more frequent handling by humans to get them acclimated to human touch and movement. When the animal is unused to human contact, entering its **flight zone**, the distance within which it will attempt to escape, may cause it to move away. Tame animals normally have a shorter flight distance (perhaps zero distance) than wild animals.

Animals that are to encounter humans rarely, such as livestock on large rural properties, should ideally still be exposed to human contact if they are to benefit at a later stage, when being loaded on to trucks, for instance, or when receiving veterinary treatment. Bruising and injuries lower productivity, and about half of all bruises are thought to be caused by rough, careless handling.

Aversive handling techniques are generally discouraged nowadays. Use of electric prods (**see: Goad**) for livestock and punishing collars such as check chains in dogs is decreasing in favour of more positive methods of handling. Understanding the animal's own behavioural repertoire and using motivators such as food are now considered superior to harsh punishment in handling animals.

Offering individuals a choice is often advocated as a method of enhancing an animal's welfare. For instance, when offered a choice of negative human contact (shouting and hitting) or positive contact (pail feeding), cattle chose the positive option. They also chose humans talking in a gentle voice as opposed to shouting. Positive interactions between stockpeople and livestock can influence their welfare and productivity.

The welfare of farmed animals and companion animals is of major concern to animal welfare and **animal rights** groups. The correct handling of animals can ensure their welfare, so it is important for all animal professionals to have an understanding of the correct techniques of animal handling. (JR)

See also: Branding; Dehorning; Fear; Free-range animals; Lairage; Stockperson; Transport

Further reading

Grandin, T. (2007) *Livestock Handling and Transport*, 3rd edn. CAB International, Wallingford, UK.

Hand rearing

Hand rearing is primarily required for orphaned neonates, but may also arise as an integral part of dairying and some **fostering** programmes (e.g. foals may be removed from their dams so that more valuable orphaned foals can be fostered). Capturing and hand rearing neonates can result in dangerous or flighty **wild animals**, and is common within the illegal trade in exotic animals (e.g. for pets and as street performers).

Hand rearing requires knowledge of each species' natural history, experience in the practicalities of hand rearing and a considerable amount of equipment (e.g. heat lamps and caging), which cannot easily be provided on an ad hoc basis. It is best performed by experienced specialist carers, and the first few hours of an orphan's time away from its parents are critical. Ill-advised feeding by inexperienced carers can prove catastrophic. Many countries have legislation that prohibits unlicensed personnel from hand rearing non-domestic orphans, a move that reduces the transfer of wild animals into the **pet** trade.

It is critical in neonates to avoid dehydration and to feed correctly. Hand-reared mammals usually need some form of immunoglobulin supplement to replace what would have been supplied by **colostrum**; colostrum hand stripped from a **conspecific** is usually fed. In the absence of colostrum from the same species, interspecific colostrum is better than none (e.g. goat colostrum can be given to lambs).

Birds are not fed specific immunological boosters, but their rations must be carefully mixed to reflect the sort of diet parent birds typically regurgitate for their young. Clearly, this depends on whether the parents are **carnivores**, **insectivores**, **granivores** or **piscivores** (e.g. raptor orphans need pulverized whole-body feeds because meat alone can rapidly cause an imbalance in calcium metabolism). Errors in diet formulation can lead to **starvation**, **malnutrition** and osmotic diarrhoea.

Experienced personnel are skilled in presenting food in ways that encourage the young to eat. Latex delivery teats are available in multiple sizes from veterinary suppliers. The frequency of feeding should mimic the **ethogram** as closely as is practical, which requires a serious commitment from carers, especially in providing night-time feeds for neonates.

Handling and **gentling** neonates may facilitate feeding and health checks but may not be entirely appropriate. Loss of **fear**, often interpreted erroneously as a loss of respect, may make animals dangerous to handle as juveniles. Furthermore, mal-**imprinting** may reduce the biological fitness of hand-reared animals when they are reintroduced to the wild. For example, filial imprinting may reduce an animal's fear of humans, increasing its chances of subsequent capture, while sexual imprinting on human carers may misdirect the courtship repertoire and reduce sexual **fitness**.

Ideally, hand-reared animals should occupy a physical environment analogous to that of a free-ranging conspecific (e.g. orphan joey kangaroos are generally kept in a bag like the marsupial pouch). An appropriate rearing environment can greatly increase success rates when young animals are reintroduced to the wild after weaning. Inappropriate dependence on shelter, for instance, may fail to prepare animals for normal weather patterns in the native habitat. The company of conspecifics, especially for neonates, can be of critical long-term importance in social species, such as equidae.

Hand-reared animals should be weaned gradually at the time appropriate for the species. Their release into a captive social group of conspecifics or into the wild should be planned to maximize their chances of acceptance and minimize the risk of injurious **aggression** or **predation**. (PDM)

See also: Development – behavioural; Reintroduction; Sensitive phase

Harms

Some systems of animal husbandry, animal **housing** and animal use cause animals to suffer in some way by imposing an environment or a manipulation procedure on animals that can cause them **pain**, **distress** or other unpleasant feelings. Harms can be divided into two types: (i) those that are incurred to achieve an objective, end or a goal – i.e. the intended use of an animal, such as for egg or meat production – and may be the minimum needed to achieve those objectives (*necessary* harms); or (ii) more harm than the minimum necessary may be caused, even though that extra harm is not needed to obtain the accepted objective. In this case, where more harms than necessary are caused, they can be described as *avoidable* harms, and we have an ethical obligation not to cause them. For example, if **castration** of an animal is necessary for the best interests of that animal (e.g. testicular cancer) and if it is done without an anaesthetic or postoperative **analgesia**, then more pain and distress is caused than is necessary, i.e. that suffering was avoidable and so avoidable harms were caused. (DBM)

See also: Suffering

Harrison, Ruth

Born 24 June 1920 and a graduate in English, Ruth Harrison developed great expertise in the methods used for the keeping and managing of farm animals. The publication of her book *Animal Machines* in 1964 changed the attitudes of many people and led directly to the Act of Parliament that protects farm animals in the UK. She commended efforts to increase food production but questioned the morality of keeping or treating animals in ways that were contrary to their biological functioning and gave them little opportunity to adapt, so that poor **welfare** resulted. She pointed out that the welfare of calves is poor if they are kept in small crates in which they could not turn around or groom themselves in a normal way, could not interact with other calves and received a diet lacking in iron and fibre – so the practice is cruel. She criticized the keeping of hens in battery cages and other systems for the intensification of animal production (**see: Intensification of animal production**). Indeed, she visited farms and spoke with experts on husbandry, biology and veterinary medicine until she was herself an expert. She kept up to date, presenting her knowledge to governments and others from the 1960s until a few months before her death in 2000.

Ruth Harrison read and understood scientific papers on farm animal welfare and sat on advisory committees such as **FAWC** and the Council of Europe Standing Committee of the Convention on the Protection of Animals kept for Farming Purposes. She would not allow inaccurate statements to go unchallenged, and politely reminded committee members of the necessity to focus on the welfare of the animals. She encouraged the scientific study of animal welfare and, as Chair of the Farm Animal Care Trust, promoted many new developments in animal welfare assessment and housing system design. (DMB)

Further reading

Harrison, R. (1964) *Animal Machines*. Stuart Publishing, London.

Head shaking

Head shaking is a condition involving recurrent, intermittent, sudden and apparently involuntary bouts of head tossing that may be so extreme as to throw the animal off balance. The term is most often used in relation to **horses**, and the condition may be so severe as to unseat the rider. Sneezing and snorting are frequently present, accompanied by attempts by the horse to rub its nose on the ground, a foreleg or nearby objects. Although the condition can occur at rest, more cases are obvious during exercise, especially at trot. Head shaking is not a diagnosis but a presenting complaint, and more than 60 disorders that may present as head shaking are now recognized, although establishing a diagnosis either ante- or post-mortem is often very difficult.

While a seasonal pattern of signs has been interpreted by some as a suggestion of an allergic rhinitis or condition linked to **photoperiod**, many other factors also correlate with the sunnier months of the year. These include an increased tendency to ride out, higher aerial dust burden, an increase in ozone, an increase in plant volatiles – such as those that are produced by linseed – and an increase in some moulds, any or all of which may result in irritation leading to head shaking.

The occurrence of certain specific behavioural features may suggest irritation of a specific area; for example, horses that shake their heads horizontally, either in addition to or instead of the classical vertical movement, or rub their ears, may be more likely to have a pain focus around the ears, such as might occur with ear mite infestation. 'Flipping' the nose, sneezing, snorting, 'clamping' the nostrils as if to close them and attempts by the horse to rub its face on a foreleg are all reported to be responses to naso-facial irritation or **pain**.

Since many cases remain without a definitive diagnosis or effective treatment, they may be labelled as a form of **stereotypy**, although a physical cause is suspected in the majority of cases, with increased attention focusing on the potential for referred neuralgic pain in the trigeminal nerve, which results in discomfort within the nose. This may be possible because many branches of this nerve are quite superficial and located close to riding gear (e.g. bit and bridle) that could cause damage. Because head shaking has a heterogeneous aetiology and a diagnosis is rarely made, treatment also remains problematic despite recent advances. (DSM)

Headstarting

Headstarting is a **conservation** strategy in which juvenile animals, either bred in **captivity** or collected from the wild, are reared in a safe captive environment for varying lengths of time prior to release into natural habitats. The rationale behind headstarting is based on the observation that larger juveniles usually have a much higher probability of surviving the neonatal period than do smaller ones. If young animals can be reared in a captive environment free from predators and other threats of life in the wild, theoretically a greater proportion will reach sexual maturity and be recruited into the adult breeding population. Headstarting was initially applied to the recovery of declining sea turtle populations in the 1970s, and has since been adopted as part of integrated recovery plans for a number of other **reptiles**, including freshwater turtles, tortoises and iguanas.

Concerns have been raised that headstarting is not appropriate for all species and should not be used indiscriminately as a species conservation tool. In sea turtles, headstarting fails to address adult mortality, the major cause of population decline, and deprives the ocean environment of maturing turtles, which probably play an important ecological role in the marine food chain. For long-lived species with protracted developmental periods, it may not be feasible or cost-effective to headstart enough juveniles to offset reductions in survival to adulthood. In contrast, headstarting appears suitable and effective as an interim conservation measure when applied to species for which severely reduced juvenile recruitment due to **predation** by invasive species is the major threat to wild populations.

Even when headstarting is feasible from a demographic perspective, animals reared in captivity for eventual release must possess the behavioural competency to survive and reproduce in the wild if **welfare** concerns are to be addressed and conservation benefits are to be realized. Minimally, release candidates must be able to locate refuges and avoid predators, acquire and process suitable food resources, locomote and navigate effectively on natural terrain, and interact socially in an appropriate manner with **conspecifics**. It should not be assumed that captive-reared animals, even those with apparently innate behavioural repertoires and without parental care, will be competent in a wild environment. However, it may be possible to enhance the competency of headstarted animals by combining species-specific training with a well-designed captive environment based on a solid understanding of the **behavioural ecology** and **ontogeny** of the species in question.

Some of the most successful examples of headstarting include programmes focused on iguanas in the Caribbean and the Galapagos Islands. The primary threat to these species is excessive predation on juvenile age classes by a variety of invasive species, including **feral** cats, mongooses, dogs, rats and pigs. Estimated remaining wild populations for several taxa are so low that emergency headstarting programmes have been initiated in an attempt to bolster recruitment while effective predator control efforts can be instituted. These programmes are successfully returning animals to the wild, and initial data on survival rates are promising.

Iguanas appear to be good candidates for successful headstart-release programmes for several reasons. They retain their natural wariness when carefully managed in captivity, are generalist herbivores that do not require extensive training in order to obtain and process natural foods and are capable of rapidly integrating into wild breeding populations. Nevertheless, it is likely that success might be further enhanced by providing experiences in captivity that familiarize release candidates with key elements of the specific natural environment into which they will be released. In particular, controlled experimental studies comparing post-release foraging abilities and growth rates of headstarted animals reared on artificial and natural diets would help elucidate the degree to which foraging competency can be improved through experience. Experimental studies focusing on the impact of exposure to simulated predators or predator cues on post-release survival would be useful in determining whether pre-release training programmes are likely to be cost effective. Comparative studies of post-release behavioural competency in animals headstarted in structurally simple and complex environments with and without exposure to conspecifics would also aid in the design of optimal captive paradigms. An especially promising area for future research is the use of artificial retreats with which animals are familiarized prior to release in order to encourage post-release settlement (**see also: Dispersal and habitat selection**).

Selection of a suitable release site and identification of appropriate release candidates are also important variables influencing restoration success that would benefit from further study. Comparative studies of headstarted animals released near to and far from known nest sites, into high- and low-density populations and at sites experiencing greater and lesser degrees of habitat disturbance and human visitation would be valuable in teasing out the potentially interacting effects of these variables on long-term survival under a range of environmental conditions. Experimental studies to determine the optimal age and size at which to release headstarted individuals into the wild are another important component in developing a successful release strategy.

For several reasons, including considerations of behavioural competency, in-country headstarting programmes are preferable to those for which the captive phase is far removed from the release site. Locally administered programmes reduce the potential for transmission of exotic diseases between captive and wild animals, and allow headstarted animals to develop immunity to local pathogens prior to release. In-country programmes eliminate the stress of long-distance transport and allow release candidates to be reared in an environment where they can more easily be exposed to local environmental conditions, including predator and conspecific cues, natural food resources and local climatic conditions. Finally, on-site programmes can maximize opportunities for outreach, education and active involvement of local communities in restoration efforts. (ACA)

See also: Reintroduction

Further reading

Alberts, A.C., Lemm, J.M., Grant, T.D. and Jackintell, L.A. (2004) Testing the utility of headstarting as a conservation strategy for West Indian iguanas. In: Alberts, A.C., Carter, R.L., Hayes, W.K. and Martins, E.P. (eds) *Iguanas: Biology and Conservation*. University of California Press, Berkeley, California, pp. 210–219.

Cayot, L.J., Snell, H.L., Llerena, W. and Snell, H.M. (1994) Conservation biology of Galápagos reptiles: twenty-five years of successful research and management. In: Murphy, J.B., Adler, K. and Collins, J.T. (eds) *Captive Management and Conservation of Amphibians and Reptiles*. Society for the Study of Amphibians and Reptiles, Ithaca, New York, pp. 297–305.

Frazer, N.B. (1992) Sea turtle conservation and halfway technology. *Conservation Biology* 6, 179–184.

Heppell, S.S., Crowder, L.B. and Crouse, D.T. (1996) Models to evaluate headstarting as a management tool for long-lived turtles. *Ecological Applications* 6, 556–565.

Pritchard, P.C.H. (1979) 'Headstarting' and other conservation techniques for marine turtles. *International Zoo Yearbook* 19, 38–42.

Health

Health is a term that is open to multiple interpretations and definitions and, while many individuals have a specific concept of health, an all-encompassing definition is difficult to establish. Hence various approaches to defining health should be considered, including their limitations. Health could be defined as 'the absence of **disease**'. An animal considered free of disease could be considered healthy but, conversely, if no disease is identified that does not guarantee a healthy animal.

To illustrate: faecal egg count levels can indicate a helminth infection of the gastrointestinal tract, but the animal may show no clinical signs of disease and could therefore be considered healthy. However, in the absence of a faecal egg count, it is also possible to have a high level of helminth infection with clinical signs. The difficulty lies in determining the point at which, in this case, the naturally occurring **parasite** infestation becomes clinically significant and thereby negatively affect the animal's health status.

In that context, disease can be defined as a 'sub-optimal performance of one or more body systems', with varying degrees of clinically apparent signs as a result of disease pressures and multiple system involvement. These body systems can include the gastrointestinal, reproductive, nervous, cardiovascular, immune, lymphatic, endocrine, musculoskeletal, integumentary and respiratory systems, separately or in combinations. Conversely, this definition can be used to define health as 'the status of performance of body systems and their interactions whereby poorer performance is usually associated with increased likelihood of clinical signs'. Based on this definition, the poor performance of body systems can result in a great variety of signs including those that are expressed through behavioural changes. Realistically, behavioural changes indicating decreasing health status are likely to be caused by one or multiple body systems functioning sub-optimally or out of balance. A behavioural sign is not a diagnosis, as a given sign may arise from different causes (**see: Aetiology**). These competing explanations for the presenting sign form the basis of differential diagnoses. To illustrate, apparent avoidance behaviour in a cow could be due either to the neuropathy of BSE or the discomfort of trauma-induced pain. A depressed isolated animal could be suffering from a neurological disorder, toxic fever or could simply be ready for parturition. (FLR)

Hearing

Selective pressures have resulted in different animal species having varying abilities to perceive sound. There are three main selective pressures determining hearing range for each species. First, the animal must be able to detect a sound. This ability allows an animal to determine the presence of sound-producing objects in its environment. Secondly, the animal must have the ability to localize the source of the sound, allowing it to react behaviourally and to either approach or avoid the sound source. Finally, the animal must be able to identify the sound and respond accordingly.

Assessment of the absolute hearing sensitivity of an animal (i.e. its ability to hear sounds of varying intensity) can be assessed using training and reward techniques. One procedure used to assess hearing sensitivity involves training a thirsty animal to touch a plate with its nose when it hears a sound. Correct responses are rewarded by the animal being given water. If an animal responds incorrectly (i.e. when no tone is present) the testing is halted for 5–15 s. Once an animal has been trained to respond reliably to hearing a sound, then sound intensity testing can follow. This is assessed by reducing the sound intensity until the animal does not respond. The ability of an animal to hear sounds of different frequencies can be assessed using a similar method. Testing can be performed until the animal has received sufficient water.

An animal's hearing ability is related to its specific requirements to survive. Birds have a relatively poor sensitivity and ability to perceive sounds of different frequencies. However, exceptions to this general rule include the pigeon, which is able to hear at low frequencies, presumably to assist navigation, and the barn owl, which has excellent sensitivity that assists it in locating prey during nocturnal **hunting**.

Mammals have a higher degree of variability in their ability to hear sounds. Unlike birds, they are generally able to hear sounds above 10 kHz, and some mammals – for example bats and porpoises – are able to hear sounds above 100 kHz. Hoofed mammals have hearing abilities similar to that of humans, but are also able to hear ultrasonic sounds. **Carnivores** tend to be able to hear higher frequencies than hoofed mammals, with cats possessing the ability to hear sounds of approximately 79 kHz. The orders Rodentia and Lagomorpha are able to hear still higher frequencies, and the house mouse has been shown to hear frequencies of up to 92 kHz. Their lower frequency limit, however, is higher than that of other mammals.

The ability of mammalian vertebrates to hear ultrasonic sounds is due to the three auditory ossicles in the middle ear, which transmit sound from the eardrum to the cochlea. Non-mammalian vertebrates possess a single middle ear bone that is less effective in transmitting high-frequency sounds. Mammals also possess a pinna (the external ear). This structure acts as a filter and amplifies sound originating from in front of the focal animal while decreasing the amplitude of the sound originating from behind the ear. It also has the function of making the ear canal asymmetrical, i.e. the animal is able to distinguish between sounds originating in front of its body and sounds originating from behind.

The determination of the ability of an animal to localize sound is more difficult to assess, as localization ability is dependent on the specific sound being localized and the location of the sound source. The ability of an animal to accurately localize a sound source is increased in the horizontal plane compared with the vertical plane. Localization acuity is also increased when the sound source is located directly in front of the animal, and decreases at locations to the side of the animal. An additional factor affecting the ease of determination of localization ability is spectral composition. Broadband noise with an abrupt onset is easier to localize than narrowband noise. Some animals vocalize using pure tones, which are very difficult to localize, to make it more difficult for predators to establish the position of the individual.

There are two basic methods for determining an animal's localization acuity. The first is to assess an animal's 'two-point sensitivity' by training it to discriminate between the same sound presented from two loudspeakers and then reducing the angle between the speakers until the animal can no longer discriminate between the two. The second method examines the degree of error as an animal naturally orients its head towards the source when it hears a sound. Songbirds have been shown to have a threshold for separation of between 20 and 30°, hens have a separation of about 4°, while studies have shown that the barn owl can orient to a frontally placed sound with an average error of approximately 2°. The barn owl is thought to possess this degree of orientation accuracy to assist it in locating prey during darkness. The ability of mammals to accurately localize a sound source varies between species.

Elephants and humans have been shown to have a sound localization ability of approximately 1°, while cattle have been shown to have localization ability of only 25°. Some species are unable to localize brief sounds.

Upon hearing a sound most animals orient their head towards the source of the sound. It has been shown that the animal will direct the fovea of the eyes towards the sound to maximize the visual acuity in this area (**see: Vision**). Most mammals have a broad area of maximal visual acuity covering almost the entire horizon of the eye; this is known as the visual streak. These species, therefore, require less precise auditory information to locate the sound source. (NC)

Heart rate

The heart rate (HR) is defined as the number of heart beats per unit time. Usually the unit time is 1 min, and thus HR is usefully quantified in terms of beats per minute (bpm). Activity within the cardiovascular system depends on input from the **autonomic nervous system** (ANS) as well as through local control mechanisms: cellular, metabolic, hormonal, receptor/reflex, etc. For the most part, HR is determined by the electrical discharge rate of the cardiac pacemaker (sinus node) and electrical conduction in other parts of the heart tissue. The behaviour of the sinus node is influenced by both divisions of the nervous system that regulate automaticity and rhythmicity. Of most interest here is the extrinsic control exhibited by the ANS.

The sinus, or sino-atrial (SA), node is an anatomical region of the heart that is located in the epicardial groove between the right atrium and the superior vena cava. The node contains two types of cells, elongated and round. Of particular interest are the round, or pacemaker, cells that are capable of spontaneous depolarization that initiates electrical activation within the heart itself. In the absence of autonomic innervation or during complete autonomic blockage, the discharge rate of the pacemaker cells represents intrinsic HR. Initiation of a cardiac cycle primarily begins with the SA node, though the atrial pacemaker may also perform this role depending on local neurohormonal and neural conditions. Spontaneous depolarization within the pacemaker cells rapidly spreads across the right and left atria. Discharge rates within the pacemaker cells depend on ionic gradients and membrane permeability. Cardiac action potential in ventricular cells begins with a rapid increase in resting membrane potential (depolarization, phase 0), followed by a fall (phase 1) to a plateau (phase 2), and then a rapid return (repolarization, phase 3) to resting levels (phase 4). Sino-atrial and atrioventricular (AV) cells, however, differ during phases 4 and 2, giving rise to the pacemaker potential that is characterized by a progressive depolarization during phase 3 and absence of the plateau in phase 2.

Cardiac muscle is innervated through both the **sympathetic** (SNS) and **parasympathetic** (PNS) divisions of the ANS, through the cardiac plexus, with both divisions continually transmitting impulses: that is, they are tonically active. Intrinsic HR is, for the most part, slightly higher than that seen when PNS activity dominates, but varies according to the distribution of pacemaker cells with the node and the nature of transmembrane potentials. The ANS regulates cardiac activity by altering the discharge rate of the SA node or by changing conduction velocity in other areas of the heart. Action potentials are generated by both shifts in the concentration gradients of sodium (Na^+), potassium (K^+) and calcium (Ca^{2+}) across the cell membrane and ion channel permeability and the resultant cell electrical charges relative to its immediate environment. Ion channel permeability, which directly affects cell charge, is influenced by autonomic neurotransmitter signals. Parasympathetic nerve terminals release muscarinic acetylcholine, which alters ionic currents across cell membranes by increasing K^+ conductance. Administration of a muscarinic receptor antagonist, such as atropine, attenuates parasympathetic influences and results in an increase in heart rate as the ratio of sympathetic influence increases by default.

Parasympathetic nerves are located on the epicardial surface and innervate the heart wall themselves. Parasympathetic vagal nerves (left and right) innervate the SA (right vagus) and AV (left vagus) conducting pathways as well as the atrial muscle. Activity within the right vagus decreases discharge rate within SA cells, which slows HR, a negative chronotropic effect, and reduces conduction velocity by decreasing the force of cardiac contractions. Activity in the left vagus slows AV conduction of impulses from the atria to the ventricles. Too much parasympathetic control may cause several degrees (first to third) of AV block. Excessive AV blockade is harmful and is characterized on an electrocardiogram (ECG) by autonomously occurring P waves and QRS complexes. Extreme levels of PNS regulatory dominance can decrease HR in humans to as low as 20–30 bpm.

Parasympathetic nerve terminals release **acetylcholine**, which is rapidly decomposed by the enzyme cholinesterase that is present in large quantities in both the SA and AV nodes. The effects of alterations in parasympathetic regulation are both immediate and relatively short lived, enabling the PNS to exercise beat-by-beat regulation of cardiac activity. A further mode of influence is wielded by parasympathetic nerves that terminate on sympathetic nerves and impact on cardiac activity by inhibiting the release of sympathetic neurotransmitters. Left and right sympathetic (cardio-accelerator) nerves are distributed throughout the heart and form an extensive cardial plexus. Left and right nerves have different influences on cardiac activity: left nerves are more concerned with inotropic (muscular contractility) conditions, whereas right nerves have more of a chronotropic impact on heart rate. Sympathetic nerve terminals release the catecholamine **norepinephrine** (noradrenaline) that binds with β-adrenergic receptors throughout the heart, including the nodes, and has an excitatory influence on cardiac cell contractility and HR. Sympathetic regulation is somewhat slower than the near-instantaneous parasympathetic effects. Exocytotic release of norepinephrine from the nerve terminal is thought to occur relatively slowly. Also, unlike parasympathetic acetylcholine mechanisms, a secondary messenger (adenyl cyclase) is involved, which slows the process down considerably. The result of this is a much more gradual alteration in cardiac activity, but it is one of extended duration relative to its parasympathetic counterpart.

Sensory regions, termed baroreceptors and chemoreceptors, located within some blood vessels (e.g. carotid sinus region, aortic arch) may also influence HR, based on the feedback of information concerning blood flow within certain vessels and its nutrient and gas content. However, higher cortical regions of the **brain**, for example the thalamus, **hypothalamus** and forebrain – which are implicated in the expression of **emotions** – may override the more localized effects of these receptors, resulting in corresponding changes in blood pressure.

Heart rate has been widely used as a welfare indicator with the assumption that HR is increased when the animal is encountering stressful conditions. However, the difficulty in interpretation comes with the assignment of emotional state. In humans, it is known that HR can increase in pleasurable, exciting situations as well as during negative emotional situations. It is also influenced by purely physical factors such as **posture** and **locomotion**. For these reasons, there has recently been movement towards measures of cardiovascular function that may contain detailed information as to the relative inputs of the ANS (**heart rate variability**), and thus perhaps elucidate the emotional state of the animal. (RM-F)

Further reading

Koeppen, B.M. and Stanton, B.A. (2008) *Berne & Levy Principles of Physiology*, 6th edn. Mosby/Elsevier Health Sciences, Philadelphia, Pennsylvania.

Papillo, J.F. and Shaprio, D. (1990) The cardiovascular system. In: Cacioppo, J.T. (ed.) *Principles of Psychophysiology*. Cambridge University Press, Cambridge, UK, pp. 456–512.

Heart rate variability

Heart rate variability (HRV) is a measure of the beat-to-beat variations in **heart rate**. The time interval between successive heartbeats (interbeat interval, IBI) in healthy animals is not of fixed duration. Beat-to-beat variation occurs, reflecting the continuous interplay between ionic membrane currents responsible for sinus node automaticity, the **central nervous system** (CNS) and autonomic activity. Heart rate variability has attracted an increasing amount of attention over the last few years, as animal welfare scientists seek methods to obtain more information from cardiovascular data.

All physiological signals can be examined in both the time and frequency domains to separate the contributory frequencies that form the absolute signal. In cardiac research, both time and frequency analysis of cardiac activity are generically referred to as heart rate variability analysis. While the clinical importance of HRV was first recognized in the mid-1960s, the first research into these rhythmic variations was not published until nearly a decade later. This work emphasizes the inexactness of drawing inferences about the behaviour of the **autonomic nervous system** (ANS) from mean HR parameters. Different levels of activation in the discrete branches could potentially have the same consequence on cardiac activity, i.e. identical mean HR. As the popularity of HRV research has increased, so too have the number and types of equipment capable of automatic and precise detection of IBI. It is now possible to obtain continuous **electrocardiogram** (ECG) information over a long period of time, such as 24 h, using ambulatory monitors.

Measurement of HRV

Numerous methods have been used to extract information from a time series of IBIs. There are a number of ways to analyse IBI data to gain information on autonomic activity: time domain methods, frequency domain methods and non-linear methods. Simply, time domain methods measure the parameters of HRV versus their time, frequency domain methods measure parameters versus their frequency and non-linear methods look at the complexity in the structure of HRV.

Time domain methods

There are several statistical approaches that may be used to evaluate heart rate variability in the time domain. These methods are essentially based on descriptive statistics and are perhaps the most straightforward way to manipulate IBI data. They can broadly be split into two categories: (i) measures derived directly from IBIs; and (ii) measures based on the difference between successive cycles. The simplest time domain measures to calculate are mean HR and mean IBI variables. Units for these variables are beats per minute (bpm) and milliseconds (ms), respectively. Additional methods derived directly from IBIs include standard deviation of normal-to-normal intervals (SDNN, also termed standard deviation of IBIs) and variance. The variance is simply the square of the SD and both are considered to be a good estimate of all components influencing overall heart rate variability. The mean of the standard deviation of all 5-min epochs within a 24 h recording (SDNNIDX) also reflects total variability, as does a coefficient of variance calculated by dividing SDNN by the mean R–R (as measured by ECG) interval length.

Analysis of measures derived from differences between successive cycle lengths is more involved than methods based on beat-to-beat intervals, but provides additional information about autonomic regulatory control. Several measures in this category provide good estimates of **parasympathetic nervous system** (PNS) regulation, including the root mean square of successive differences (RMSSD). This is calculated by finding the square root of the mean of the sum of the squares of differences between neighbouring beats. Parasympathetic activity can also be estimated by counting the number of interval differences of successive intervals that differ by a value greater than 50 ms (NN50). This measure, when divided by the total number of intervals, yields pNN50. All of these measures are highly correlated, not only with each other, but also with the amount of power attributable to the high-frequency variation on a power spectral density (described below), which in turn represents parasympathetic activity. Also, very high correlations have been shown to exist within measures by repeated monitoring of cardiac activity in individuals. Time domain indices are more easily calculated than frequency indices. They have been used extensively to assess cardiac function following myocardial infarction and have proved useful for assisting detection of those at risk of further life-threatening cardiac events and sudden cardiac death. However, their interpretation is restricted to inference about total variability and parasympathetic regulatory input only and they do not provide a measure of modulations in the sympathetic branch.

Frequency domain methods

The most commonly reported method of HRV analysis is that of frequency domain analysis. While time domain parameters are a good estimator of overall variability and variability attributable to activity within the parasympathetic branch of the ANS, they are less powerful than frequency domain approaches and do not reliably quantify sympathetic control. There are two main approaches presently used to calculate the components of power spectral densities (PSDs): (i) non-parametric, Fast Fourier Transformations (FFT); and (ii) parametric, autoregression modelling (ARM). Both approaches have their relative merits, with both describing the dynamic properties of the heart period signal similar to the way a prism decomposes light into its spectrum.

A spectrum is essentially a histogram that expresses the rhythmic components of the signal that it is decomposing. The FFT approach is extensively used in the assessment of second-order rhythms in heart rate variability. FFT breaks any signal down into its constituent frequency components by partitioning the signal into different categories of variances that describe the different contributory frequencies that are contained in the signal.

The following two assumptions are made about data used in FFT analysis: (i) the data are discrete, i.e. the data series was created from a continuous sample that has not been spliced together from more than one source; and (ii) the data are stationary, i.e. their statistical properties (mean, variance, etc.) do not alter over time. Data generated from FFT can be visually presented using a PSD, which plots the amplitude (or power) of the component signals as a function of their frequencies. Spectral power for any frequency range is derived from the PSD by quantifying the area under the spectral curve at any given frequency band, for example 0.0–0.5 Hz. Typically, the Fourier approach is more appropriate for the analysis of shorter-term HRV (256 or 512 beats), as it is quite unusual for any biological signal to remain statistically stationary for any length of time. Consequently, longer segments of data usually violate the stationarity requirements needed to undergo FFT analysis.

While shorter windows of data exhibit greater stationarity, the resolution of their spectral estimates will be reduced. Although better resolution is achieved with longer segments of data, they are far less likely to meet the stationarity requirements to undergo FFT analysis. An alternative approach to FFT is ARM, which does not have the stringent stationarity laws of FFT analysis and so can be applied to longer signals. It can be used to analyse unequally spaced time series and has greater frequency resolution over FFT, making it more accurate for determining slow signals in short data series. For instance, to estimate low-frequency signals, FFT requires that the data contain at least one full cycle of the frequency of interest. ARM, on the other hand, requires just one-quarter of a cycle from which it can construct the full cycle. A further advantage of ARM over FFT is that it is less susceptible to spectral leakage. This occurs when power from one frequency component is not sufficiently separated from other frequency components, and is a common criticism of FFT. To reduce this leakage and further improve frequency resolution it is necessary to first 'window' data destined for FFT analysis. FFT, nevertheless, offers a distinct advantage over ARM – the relative ease of the actual computations involved – which, in part, explains its popularity among clinical scientists. As ARM fits the model to the data, prior information regarding the time series of the data is required in order to select an appropriate model/order. ARM, on the other hand, is a better analytical tool for decomposing extremely short segments of data and data where the statistical properties do not meet the assumptions required by FFT.

Interpretation of power spectral density

Early investigations on the PSD of cardiac IBI data, gathered during a variety of pharmacological, surgical and physical manipulations of the ANS, all highlight the presence of three principal spectral bands in humans, and later work has reported on their presence in other mammals. The precise location of these bands varies from species to species and may be predicted using respiratory rate and/or pharmacological interruption of the separate branches of the nervous system. The three bands are: (i) a high-frequency (HF) component; (ii) a low-frequency (LF) component; and (iii) a very low-frequency (VLF) component. In humans, these bands are centred around 0.25, 0.10 and 0.04 Hz, respectively. A fourth component, ultra-low frequency (ULF, <0.0033 Hz), is sometimes detected in spectra derived from longer-term recordings. Spectral components can be expressed as absolute values or as normalized units, where the discrete power bands are presented as a proportion of the total power in the spectrum. Normalizing the data gives a clear picture of the interplay in PNS/**sympathetic nervous system** (SNS) regulatory control, particularly during recording when the total power is expected to fluctuate greatly. Normalized units, however, should always be presented alongside their corresponding absolute values to provide a complete picture of the autonomic activity under investigation.

An example of a pig-derived PSD, generated using FFT and showing the three principal components found in short recordings (512 beats), is presented in Fig. H.1. In humans, the VLF component is thought to represent peripheral

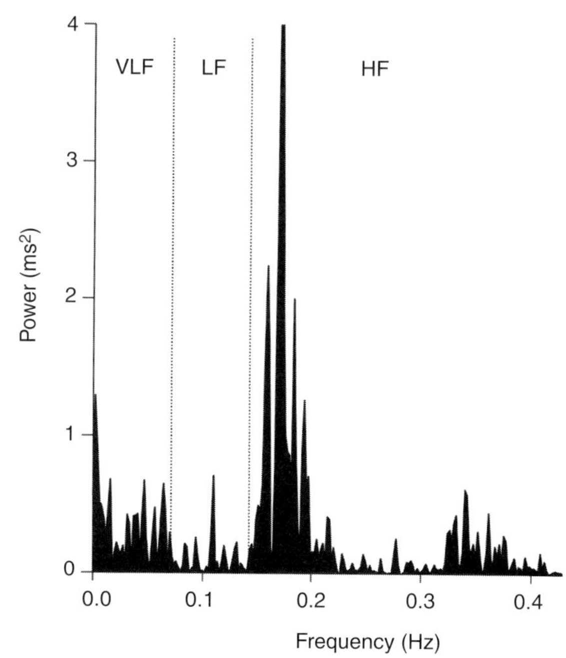

Fig. H.1. Power spectral density of pig IBI data showing the location of the VLF, LF and HF power spectral bands and a prominent respiratory peak within the HF band. IBI, interbeat interval; VLF, very low frequency; LF, low frequency; HF, high frequency.

vascular mechanisms and thermoregulation, as well as activity within the renin–angiotensin system. The VLF component cannot be detected reliably in short-term signals by FFT analysis as they represent the much slower modulations in cardiac activity, and a single cycle typically takes longer to complete than the time available in a short signal (usually 5 min or 512 beats). Likewise, estimates of ULF are only statistically stable when obtained from long-term signals. In humans, this frequency band represents much of the power in 24 h spectra. It is thought to reflect neuroendocrine circadian periodicity, but the origins of the power in this band are presently not well defined.

Of particular interest to animal welfare scientists are the LF and HF components that are known to reflect autonomic activity in humans, dogs, rats, pigs, horses and cattle. Power in the HF component is generated by respiratory modulation of parasympathetic (vagal) activity. The LF spectral component is mediated jointly by both sympathetic and parasympathetic regulatory activity. This has been determined experimentally in a number of studies using selective blockade or enhancement of one or both branches of the nervous system. For instance, atropine, a competitive antagonist of parasympathetic muscarinic cholinergic receptors that inhibits parasympathetic influences, diminishes all power in the HF spectral band while simultaneously reducing power in the LF band. Delivery of propranolol, a non-selective β-adrenergic (sympathetic) receptor blocking agent, causes a significant decrease in power in the LF band but power in the HF band remains unchanged. HF power has been found to exhibit a dose-dependent decline in magnitude following step-wise increments of atropine dosage. Such dose-dependent responses substantiate the sensitivity and reliability of using frequency domain parameters as estimators of autonomic regulatory activity. Complete blockade of both branches produces spectra with relatively no power, analogous to those seen in humans and animals following transplantation, due to an absence of innervation in the transplanted heart.

While estimates of sympathetic activity contributions to spectral power are more difficult to separate than parasympathetic input, a quantitative measure can nevertheless be obtained when specific information concerning HF power is available. As some PNS modulations emerge in the LF band, direct interpretation of LF power as SNS activity would in fact produce an overinflated estimate of the power attributable to sympathetic modulations. The LF:HF power ratio, however, can be used to quantify both SNS activity and sympathovagal balance. High values indicate a predominance of sympathetic regulatory control, whereas low values represent more PNS control. For instance, below 1 the ratio reflects parasympathetic dominance and, above 1, sympathetic dominance. A number of experimental manipulations have demonstrated how the LF:HF ratio responds to specific ANS manipulations. For instance, when posture is shifted from lying to sitting and then to standing, manoeuvres known to increase SNS and decrease PNS regulatory activity, the magnitude of the ratio has been found to increase correspondingly. So, when PNS activity is known, estimates of SNS activity can be achieved using the LF:HF ratio, which may also be used to evaluate whether any change in balance occurs in either a reciprocal or non-reciprocal fashion.

Pharmacological or surgical interception of PNS activity abolishes HF power and causes a reduction in the LF spectral component, not only in humans but also in horses, dogs and rats. Total, LF and HF spectral components also change chronically with ongoing illnesses, cardiac disease, age and other biological occurrences that effect autonomic activity and reactivity. For example, adult humans, when supine, show a progressive decline in total power with age while the LF:HF ratio remains fairly constant. The LF:HF ratio during postural changes increases as a function of age, indicating a reduction in autonomic reactivity in older subjects.

Horses have a characteristic pattern in their power spectra, with the HF band occurring between 0.07 and 0.60 Hz and the LF band at 0.01–0.07 Hz when resting. The spectral analysis of HRV in pigs closely resembles that in humans. The HF peak occurs between 0.1 and 0.5 Hz and the LF below 0.1 Hz. This is not surprising given the physiological similarities (e.g. mean HR and respiration rate) between adult humans and adult pigs. In adult cows, LF and HF reportedly occur between 0.0–0.25 Hz and 0.25–0.58 Hz, and in calves between 0.0–0.15 (LF) and 0.15–1.00 Hz (HF). The central location of either of these frequency peaks may shift depending on the physiological status of the individual; nevertheless, the width of the bands is sufficiently broad to accommodate any such repositioning. Without controlled pharmacological manipulations (i.e. response curves) of ANS activity, it is difficult to pinpoint the precise location of these frequency peaks or predict how they will respond to **stress**.

HF power can also be expressed as a function of total spectral power (HF/TOTAL = PNSI), and LF power as a function of HF power (LF/HF = SNSI), to yield parasympathetic and sympathetic activity indicators, respectively. These normalized units of HF and LF power have been consistently reported to respond appropriately to repeated orthostatic challenges known to either increase SNS or decrease PNS regulatory control. Moreover, normalization reduces the effects due to alterations in total power and underscores the reciprocal or non-reciprocal relationship that reflects the different patterning of PNS and SNS regulatory control of the heart. (RM-F)

Further reading

Task Force of the European Society of Cardiology and the North American Society of Pacing and Electrophysiology (1996) Heart rate variability. Standards of measurement, physiological interpretation and clinical use. *Circulation* 93, 1043–1065.

Von Borrell, E., Langbein, J., Despres, G., Hansen, S., Leterrier, C., Marchant-Forde, J.N., Marchant-Forde, R.M., Minero, M., Mohr, E., Prunier, A., Valance, D. and Veissier, I. (2007) Heart rate variability as a measure of autonomic regulation of cardiac activity for assessing stress and welfare in farm animals – a review. *Physiology and Behavior* 92, 293–316.

Heat stress

When the body is challenged with an elevated external temperature that begins to raise body temperature beyond the physiological homeostatic body temperature range, the animal is considered as being exposed to thermal stress. Thermal stress is a serious consideration for animal **welfare** because the inability to cope with this **stress** can rapidly lead to

suffering and **death**. Animals have a physiological thermal zone, termed the thermal comfort zone, which is a range of temperatures within which the body is able to effectively operate without invoking extraordinary measures. If the body temperature begins to exceed the upper critical temperature, then physiological and behavioural mechanisms are initiated that help to dissipate heat and bring the body back into its thermal neutral zone (**see also: Thermoregulation**).

To increase temperature loss, the animal's body will alter its vasculature to increase vasodilation to peripheral areas, such as the skin, to allow heat loss through convection, radiation and conduction. In addition, the animal will increase respiration and perspiration, if possible, to dissipate heat through evaporation. Increased cardiac output, including increased blood volume, will occur to increase heat transfer from the body core to the periphery. The animal will also invoke behavioural responses that aid in decreasing core body temperature. Many animals will seek shade and water to help cool the body. Swine, unable to sweat sufficiently, will enter water or a mud wallow to cool the body using evaporation. If the animal is in a windy environment, it may orient its body broadside to the wind to increase the surface area that is cooled. In all species appetite is typically depressed, which prevents further creation of heat from digestion. In modern livestock production, producers can incorporate artificial shade, high-output fans and/or misters in their facilities to help keep animals cool. Livestock rapidly learn to use shade and misters during periods of elevated temperature.

Livestock are often able to withstand relatively hot conditions if given the chance to become accustomed to the heat slowly. Death losses of livestock in summer months typically occur from sudden spikes in environmental temperatures rather than from gradually increasing temperatures. Swine are especially susceptible to heat stress due to their typically high body fat content and their inability to sweat sufficiently. All livestock and poultry are at risk to heat stress when being transported. **Transport** of these species typically involves moving a large number of animals into a limited amount of space for the trip. In an environment that is not overly hot, these livestock can still succumb to heat stress and even death without proper ventilation. When environmental temperatures start to rise by even a few degrees, the temperature inside the transportation vehicles can reach critical values at an alarmingly fast rate. Therefore, during raised temperatures, all livestock and poultry should be transported during night-time. (DCL)

See also: Branding; DFD meat; Feedlot; Free-range animals

Hedyphagia

Hedyphagia is the selection of food items according to the amount of pleasure derived from them, or their hedonic value. Natural selection has evolved mechanisms for animals to be satiated by foods that are good for them, but not by foods that do them harm or are of no benefit. The primary mechanism for discrimination between foods is by their taste, and four primary taste sensations have evolved to assist in this process: (i) sweetness for foods with high concentrations of simple carbohydrates; (ii) sourness for foods with a high acid:base ratio; (iii) saltiness for foods with high concentrations of sodium and, to a lesser extent, potassium; and (iv) bitterness for foods with potentially toxic organic compounds, e.g. many plant alkaloids. However, there is limited evidence that animals can recognize the inherent value of foods from their taste. They may adapt to potentially toxic or bitter compounds, however, and this could be due to increased hunger in the absence of alternative, safe, foods. They can also develop aversions to sweet foods. In relying on sensory feedback, the hypothesis of hedyphagic selection does not often have sufficient regard for the post-ingestional consequences of the feed. (CJCP)

Herd health

The health of herds of farm animals, especially **cattle** and **pigs**, is an interaction between the animals, their environment and **disease**-causing organisms. As they are gregarious animals, cattle and pig herds should be considered not only as a group, but also as individuals, if the health of one animal differs markedly from that of the group. Bringing cattle or pigs indoors presents one of the most severe challenges to their health, as the contact between animals is increased and disease organisms are better able to survive in the more constant and generally benign environment compared with that outdoors. However, in the case of cattle, the nutrition of high-yielding cows may be better controlled, and their ability to rid themselves of pathogens may be better than if they were permanently at pasture, as the immune system functions best in a well-nourished animal.

The advent of widespread use of vaccines and antibiotics in the latter half of the 20th century heralded a low incidence of infectious diseases for most pigs and cattle, especially dairy cows, which are often treated annually at the end of **lactation**, as well as routinely when there is any evidence of bacterial infection. In the long term antibiotics will have limited value, as the ability of pathogenic bacteria to mutate and escape control is undoubtedly greater than our ability to find new antibiotics. Farmers must therefore be prepared to use more prophylactic and vaccination measures, such as reducing the stocking density of cattle and keeping them cleaner to prevent **disease** transfer.

Since the late 1950s, farm size has increased considerably in most cattle production systems in industrialized countries, as enterprises expand and small family farms are amalgamated into larger units (**see: Intensification of animal production**). Cattle are also more likely to be housed, often to the detriment of their welfare. The eradication of many disease organisms was achieved in the past through careful husbandry, but this was relaxed in the latter part of the 20th century with the perceived prospect of effective cures and better prevention strategies for most of the infectious diseases. The good husbandry that used to be essential for disease eradication is exemplified by the measures taken to prevent the spread of rinderpest more than 150 years ago in the UK:

- The byres of infected animals had to be washed and left empty for 2 months.
- Persons attending sick animals were prohibited from going near healthy stock.
- The sale of sick cattle was prohibited.
- Sick cattle were slaughtered and buried.
- All cases of the disease had to be reported.

These simple measures restricted the transmission of the disease and eventually led to its eradication in the UK. Nowadays, there is considerable movement of cattle and this reduces the effective immunity that develops for a range of pathogens in a specific area.

An alternative method of controlling disease is to reduce the stocking density of housed cattle and hence contact between animals. However, the reduction in output per unit area is only likely to be practical if product prices are increased or farmers are directly compensated for economic losses. Some reduction in cattle stocking density can be achieved through mixed farming systems, for example with **sheep**, because many pathogens are species-specific. However, the increased mechanization of farms has necessitated the adoption of simple farming systems, usually relying on the farming of single species of animals and crops (monoculture). It is a technological challenge for the future to design mixed farming systems that minimize disease risks but are still efficient users of land, labour and capital.

There are many challenges to the health of cattle, some of which have been long known, such as foot-and-mouth disease, **mastitis** and **lameness**, and some of which are new, such as *Escherichia coli* O157 and bovine spongiform encephalopathy (BSE). The type of **housing** is the dominant influence on the incidence of many of the common disorders of dairy cattle, especially lameness and mastitis.

Lameness

In relation to lameness, a scourge of modern housing systems, concrete wears the hooves more than earth, but the constant wetting of the foot in cubicle passageways covered in deep slurry can erode the soft heel bulb. This predisposes the cow to **laminitis**, and sometimes leads to the toes losing contact with the ground and unchecked growth (slipper foot). Excessive walking on concrete stretches the white line and wears the sole, thus weakening the bond between the wall and the sole. In straw yards the abrasion on the hoof is minimal, leading to increases in toe length until the cows are out at pasture. Interdigital infection is more common than in cubicle houses, as straw may be pushed up between the claws, causing a lesion that is open to infection, especially with *Phlegmona interdigitalis*. Excessive growth may close up the interdigital space, trapping dirt and causing infection. Long digits are common in modern cattle units, due partly to excessive growth as a result of the high nutrient density of their diet and partly to inadequate wear as a result of low levels of activity and a smooth concrete floor or soft strawed yard to walk on. Trimming reduces the load bearing of the heel, improves **locomotion** in cattle and reduces the hardness of the sole and abaxial wall. It is normally done by a method developed by Dutch cattlemen.

The condition of cow tracks can influence the prevalence of lameness in a herd. Stony, muddy tracks provide an uneven surface that can stress the sole and lead to lameness and poor welfare. The ideal surface is absorbent and provides a firm surface for walking on – bark chippings are recommended, but need constant care. They should be laid over a porous membrane and aggregate for good drainage. Concrete is very durable and is better than stony tracks, but not as comfortable underfoot as bark chippings. However, a concrete track can be used by both cows and vehicles, whereas if bark chippings are used, separate tracks are required because a vehicle's wheels will create ruts. Cows should never be hurried down a track by a herdsperson using a dog or motorcycle.

Mastitis

Mastitis is an opportunistic infection of the mammary gland, the severity of which is dependent mainly on the delay in inflammatory response to the infection. It is not one disease, but a wide range of possible infections by up to 100 possible pathogens, making it a multifactorial disorder.

Mastitis causes a major reduction in the welfare of the cow, due to the fever that it induces, the localized pain in the mammary gland and the possible **stress** of **isolation** and treatment. The cost to the farmer is considerable, with losses arising for a number of reasons, including: (i) penalties imposed by the dairy purchasing the milk because it has a high somatic cell count (SCC); (ii) reduction in milk yield from the infected gland, which may persist into the subsequent lactation; (iii) withholding of the milk from sale after antibiotics have been applied; (iv) drug and veterinary costs; (v) reduced value of the affected cow; (vi) replacement costs; (vii) increased labour for managing the sick cow; and (viii) costs of associated diseases, in particular reproductive failure.

Mastitis also presents risks to human health, as milk containing antibiotics that is drunk by humans may induce bacterial resistance, or the consumption of unpasteurized milk contaminated with bacteria could lead to the consumer acquiring **zoonotic disease**. There are penalties in most countries for dairy farmers supplying milk contaminated with antibiotics, and insurance against this possibility is now available in some countries. Not only is milk contaminated with antibiotics unfit for human consumption, it is also unsuitable for cheese or yoghurt production, as the antibiotics impede the fermentation process. Dairy cows that contract mastitis before pregnancy is established have a delay in the interval to first service, on average about 3 weeks, and an average of one extra service is required for each conception.

Control measures for mastitis are based mainly on cleanliness and the regular use of antimicrobial chemicals and antibiotic drugs. They have been well established for at least 25 years and, if correctly implemented, should enable a dairy herd to have between 100,000 and 150,000 somatic cells/ml of milk in the bulk tank, fewer than 20 cases of mastitis/100 cows/305-day lactation, less than 12% of the herd infected at any one time and a recurrence rate of less than 6%. Fewer than five doses of antibiotics should be used for each case of clinical mastitis. These herd health targets, although attainable, may not in future be adequate for milk purchasers, who may devise species-specific penalties/incentive payments to regulate individual bacteria.

The continuing successful treatment of mastitis will require researchers to develop better control measures, as the bacteria will mutate and develop resistance to the current range of antibiotics. So far resistance has only been observed in response to *Staphylococcus aureus* infections when these are treated with penicillin or when cloxacillin is used for dry cow therapy. *S. aureus* infections during lactation are best controlled by clavulanate/amoxycillin, and for dry cow infections cloxacillin is still the best antibiotic to use. Most other infections can be controlled with penicillin, although clavulanate/amoxycillin is best for *E. coli* infections. When an antibiotic is chosen, careful consideration needs to be given to the milk withdrawal time (from the time of insertion of the last tube). Some products are a combination of antibiotics. When using antibiotic therapy on dry cows, the persistence should be noted, this ranging from 7 days for ampicillin to 42 days for cloxacillin.

The level of mastitis has not declined significantly since the late 1980s, despite widespread adoption of control measures, and in future there will probably be a return to an emphasis on maintaining cows in very clean conditions. This will be particularly relevant in controlling environmental pathogens, which have become more common as infectious pathogens have been largely brought under control by antibiotics. Slurry disposal will need to be carefully planned and attention paid to fly control and ventilation of buildings.

Heifers are particularly vulnerable to mastitis as their defences are not well developed, and they can exhibit twice the incidence of mastitis found in older cows. Infection chains from older cows to heifers are common, where bacterial infections are transmitted in the milking parlour.

Part of the reason why mastitis incidence has not declined relates to the increase in milk yields of cows in most areas. High-yielding cows tend to have wide teat canals, so the ingress of bacteria into the teat cistern is facilitated. Increasing milk yields through the injection of the growth hormone analogue (bovine somatotropin or BST) also increases mastitis. BST can stimulate the immune system in the mammary gland, but the extra mobilization of body tissues to support increased milk production may generally depress the immune function, leading to an increased mastitis frequency.

Herd health assurance schemes

There is increasing concern among the general public that farm conditions are not always satisfactory for the health and welfare of cattle. The other major concerns that the public may have are for food quality and environmental quality. Some of these concerned members of the public are prepared to pay more for cattle products that have been produced to an assured health standard (**see also: Farm assurance schemes**). Standards may be set by animal welfare charities, veterinary associations or large traders, such as the major supermarket chains, and by the government for specific diseases, e.g. enzootic bovine leucosis. Membership of such schemes may be required by the major supermarkets. Such standards will usually focus on the health of the herd and individuals within it, but also on related issues such as hygiene on the farm, the quality of housing, plant and equipment, feedstuff and water storage facilities, stockmanship and the ability of a farm to manage an emergency. Monitoring is on a regular basis and may be performed by veterinarians or, more usually, by staff specifically trained for the task, who have a checklist to examine different parts of the farm to assess their adequacy.

The underlying principles of most herd health assurance schemes are to ensure that the cattle have: (i) adequate space, in particular to allow for sufficient exercise indoors; (ii) freedom from aggression by other cattle, e.g. by providing adequate feeding and drinking facilities; (iii) adequate floors to walk on; (iv) a comfortable and clean bedded area; (v) regular veterinary care; (vi) competent supervision by stockpeople; and (vii) adequate transportation away from the farm where necessary. Herd health assurance schemes are useful not only to monitor disease frequency – to enable comparisons to be made with acceptable standards – but also to determine risk factors contributing to disease and to implement control measures to improve performance.

At the end of the 20th century, the desire for further improvement in the herd health status of cattle and limitation of the spread of disease, particularly zoonoses, led to the introduction of cattle-tracing systems, which became mandatory in the European Union member states from 1999. Under the system operating in the UK, cattle are required to have 'passports' that contain details of: (i) each animal's breed and sex; (ii) its date of birth and, eventually, its death; (iii) its dam's number and any movements that the animal has made throughout its life; and (iv) any government financial support that has been received for the animal. The scheme enables government authorities to trace cattle easily if there is a disease outbreak and to assure members of the public that the authorities have control of cattle movements.

Never before have cattle diseases received so much attention from the public, veterinarians and farmers as in the years since the late 1980s. The farming industry has been repeatedly criticized for failing to provide a healthy product for consumption, and the government has been accused of not safeguarding human health and of not acting on scientific advice. From the difficulties that have been encountered, there is now emerging a system of dairy cattle monitoring and herd health care that should ensure a significant reduction in the risk to consumers, provided that the very considerable economic cost to the industry does not limit farmers' ability to implement the new measures. The systems put in place, such as the movement scheme and farm assurance schemes, should serve as a model for anyone wishing to reassure their consumers that beef and dairy products are safe foods to eat and that they have been produced in a high-welfare system, in relation to their physical needs, with attention paid to the environmental impact. (CJCP)

Heritability

Some characteristics (**phenotypes**) are determined only by an organism's genes, while others are influenced by the environment as well as by genes. Eye colour in people is purely genetic, while milk yield of cows is partly genetic and partly a result of the action of the environment.

Heritability is a measure of the importance of genes in determining the value a characteristic may have. Eye colour is, effectively, 100% genetically determined, while milk yield in cows is much less genetically determined (it has a typical heritability of 25%). Heritabilities vary widely among different characters. For example, in beef cattle, survival from birth to weaning is only 5% heritable, while in sheep, greasy fleece weight is 60% heritable. The heritability of a particular character may differ at different stages of an animal's life, reflecting how the role of genes in determining the phenotype may change during the animal's development. Thus, in principle, an animal could be selected for breeding on the basis of its phenotype at a young age, but there might turn out to be no response to this selection because the role of genes in determining the phenotype at that age is only minor.

When selection pressures on a character are strong, the gene for that character that has the greatest survival value will be selected. This means that genetic variation is reduced, as the poorer-performing genes are selected out and only the optimum genes remain in the population. Thus, beef cattle have the best available genes for pre-weaning survival, and there is almost no genetic variation for this character, indicated by low values for heritability. The situation is different in wool sheep – although it is usually a commercial advantage to have a heavy fleece, it would make survival different under certain circumstances. So, genes for heavy fleece weight will sometimes be advantageous and at other times a disadvantage – and the genetic variation that exists for this character reflects the overall balance of advantage and disadvantage. It would not really be practicable to attempt selective breeding for a character that has a low heritability, as there is only a small amount of genetic variation to work with.

Heritability can be measured by calculating the correlation between the values parents have for a character and the values for their offspring, because this measures the importance of genes in determining the latter. Generally, rather few measures have been made of traits relevant to welfare. Certain aspects of the docility of Limousin cattle have a heritability of about 22%, and this is enough for a breeding programme to be justified with a view to improving their temperament. **Tonic immobility** (TI) in chickens, which is thought to reflect **fear**, is highly heritable. However, selection for a short duration of TI would not necessarily improve welfare, because the resulting animals might be just as afraid but less able to express their fear behaviourally through tonic immobility.

(SJGH)

Further reading

Hall, S.J.G. (2004) *Livestock Biodiversity. Genetic Resources for the Farming of the Future*. Blackwell Publishing, Oxford, UK.

Nicholas, F.W. (1996) *Introduction to Veterinary Genetics*. Oxford University Press, Oxford, UK.

Heterochrony

Heterochrony refers to phylogenetic changes in the relative timing or rate of developmental processes between individuals (**see also: Development – behavioural**). Two forms of heterochrony are recognized: (i) paedomorphosis, in which juvenile characteristics are retained into adulthood (i.e. certain developmental traits are lost or delayed); and (ii) peramorphosis, in which additional development occurs during the **ontogeny** of an individual. The exaggeration of certain traits as might occur as a result of **sexual selection**, such as the extensive antlers of some species of deer, is an example of peramorphic change.

Many species – and especially domestic variants – appear to show a mix of paedomorphic and peramorphic change. For example, some giant **dog** breeds such as the St Bernard appear to have many paedomorphic characteristics (a more domed skull and shortened muzzle more typical of the ancestral neonate, as well as a more juvenile behavioural repertoire), but their large size may be explained, at least theoretically, in terms of either a paedomorphic or peramorphic process or a combination of these. This is possible because there is a range of processes that can give rise to heterochrony that are neither exclusive nor necessarily easy to distinguish.

In the example above, it might be argued that a delay in the onset of maturity (post-displacement) – and the associated slower adult growth pattern as a consequence – results in a prolonged period of accelerated growth typical of a juvenile (a paedomorphic trait). However, it might also be argued that it is primarily a delay in the ending of the juvenile phase of development (hypermorphosis) that results in an extended period of increased growth (a peramorphic trait). While the later than average age of sexual maturity of this breed might be viewed by some as tending to favour the paedomorphic explanation, it should be noted that this breed still matures much earlier than its wolf ancestor, which might suggest that more extensive developmental recombination has occurred during the development of the breed, resulting in a mixture of paedomorphosis and peramorphosis both within and between different traits. Sweeping generalizations concerning the heterochronic changes associated with **domestication** and breed diversification are therefore best avoided.

There are three basic mechanisms that might result in each form of heterochrony: (i) a change in the offset timing (ending) of development; (ii) a change in the onset timing (start) of development; and (iii) a change in the rate of development. The consequences of each are given in Table H.1.

While simple generalizations about paedomorphosis and peramorphosis can lead to false conclusions, it is reasonable to suggest that different environmental circumstances tend to favour the occurrence of different heterochronic processes; for example, in stable environments (in which k-selected species will tend to flourish), there are often advantages associated with having a prolonged developmental period (i.e. hypermorphosis and neoteny are more likely to occur) in order to facilitate greater competitive ability as an adult, such as through the increased experience gained from **play**, **exploratory behaviour** and possibly opportunities for **social learning**. Conversely, in less stable environments (which tend to favour r-selected species), more rapid development (acceleration) with an earlier completion (progenesis) may be favoured in order to maximize reproductive potential by minimizing generation time.

(DSM)

Further reading

Klingenberg, C.P. (1998) Heterochrony and allometry: the analysis of evolutionary change in ontogeny. *Biological Reviews* 73, 79–123.

Table H.1. Types of change in development seen as a result of different heterochronic mechanisms.

	Heterochronic change	
Developmental feature changed	Paedomorphosis	Peramorphosis
Offset time	Earlier – progenesis	Later – hypermorphosis
Onset time	Later – post-displacement	Earlier – predisplacement
Rate	Slower – neoteny	Faster – acceleration

Smith, K.K. (2001) Heterochrony revisited: the evolution of developmental sequences. *Biological Journal of the Linnaean Society* 73, 169–186.

Hiders

Among the **precocial** ungulate species, those in which the young remain concealed near the birth site for some time following parturition (usually several days), while the mother can be away for long periods of time, are designated hiders. The young of those species avoid predators largely through **crypsis**. Hiders are generally considered to be the opposite of the **followers**. However, the distinction between those two categories of species might not be so rigid, as some studies suggest that the response pattern of the young within a species may be flexible depending on ecological factors. (SL)

Hierarchy

When animals come together in a **group**, generally in order to cohabit, they establish social relationships between group members (**see: Social behaviour**), and one outcome of these relationships is the establishment of a so-called **dominance** hierarchy or social hierarchy. The term 'hierarchy' suggests that the animals are arranged in some kind of static graded or ranked order, but in most animal groups this is an oversimplification, as relationships between individuals are often dynamic and situation-specific and may involve coalitions formed with other group members.

In 1922 Schjelderup-Ebbe, studying the social organization of flocks of domestic hens, used overt **aggression** (pecking each other) as a measure of dominance and described a 'peck order' (hierarchy) among the birds. The simplest form of hierarchy is one in which one individual dominates all other group members, with no rank distinctions among the subordinates; this is a 'despotic' hierarchy. More commonly, hierarchies have multiple ranks in which individual A dominates all others, individual B dominates all but A, individual C dominates all but B and A, and so on down to the most subordinate individual that dominates no other; such a hierarchy is termed a 'linear' hierarchy. Many hierarchies are more complex, and contain animals of equal status, triangles or other circular elements – e.g. B may dominate C, which dominates D and E, which are of equal status and both dominate B. The hierarchical relationships are said to be established and stable if the pattern of the previous outcomes allows an accurate prediction of the next outcome.

The form of the hierarchy may be dependent upon the way in which the hierarchy develops, and the method of development appears to differ with species, depending upon the factors that influence the dominance potential in a particular species, and also whether relationships are established instantaneously or develop over time. For cattle, length of time in the herd appears to be a crucial contributor to dominance. Thus, if new members are added to a group in succession, then the hierarchy is more likely to approximate linearity than if several new members are added at the same time.

The function of the hierarchy is debatable, although its role in reducing aggression receives much support. This reduction in aggression appears to arise from individuals being able to assess the dominance potential of others by individual recognition, the recognition of attributes or cues associated with those attributes, and during encounters. The dominant animal may assert its dominance through aggression most frequently and vigorously when the relationship with another is least stable, such as when the relationship is newly established or has recently been reversed. This suggests an inverse relationship between frequency and intensity of aggression and the degree to which relationships are established and stable. The mixing of unfamiliar animals is reported to invariably result in an increase in levels of **agonistic behaviour** interactions. This indicates that at least some group members need to interact to physically determine social relationships and do not rely on cues and traits that are indicative of dominance. Evidently, situations such as mixing unfamiliar animals that result in high levels of aggression can cause excessive physical exertion, exhaustion and **injury**, and adversely affect welfare.

The social hierarchy is often regarded as having an all-pervasive influence on the life of animals, governing not only social and aggressive interactions but also influencing access to all resources. Thus, the hierarchy is seen as determining: when and where individuals move and rest; where, when and what they eat; with whom mating occurs and how successfully they reproduce. If this is the case, then there are obvious consequences for the **welfare** of group members depending upon their position within the social hierarchy. If dominant individuals have to be extra-vigilant, and spend time and energy frequently defending resources and their position in the hierarchy, then their welfare may be poor. In the case of species in the care of humans, some of the effects on welfare may be attenuated because resources may never be limited to the extent that particular individuals are totally deprived. However, the welfare of some individuals could be compromised because the resources they obtain may be of inferior quality; for example, there are reports that individuals at the bottom of the hierarchy access poor-quality food.

It is reported that large groupings of livestock animals can lead to poor welfare in certain individuals, because high levels of aggression arise as a result of difficulties in establishing a stable hierarchy with large numbers of individuals. If correct, this explanation implies that individual **recognition** is used in the development and maintenance of the hierarchy, and aggression occurs because it is difficult for individual recognition to occur in very large groups. In contrast, if hierarchies were to be established and maintained by individuals **learning** that certain traits or cues are associated with dominance, then after a few initial encounters individuals would recognize the cues and know the social relationship without further aggressive encounters occurring.

It is often noted that there are more aggressive interactions among individuals in the middle of the hierarchy than among those at the top. This is probably a consequence of the disparity in attributes that contribute to dominance potential being less clear-cut among the middle-ranking individuals. Individuals at the bottom of the hierarchy are often reported as having considerable aggression directed towards them, and may be constantly 'bullied' or harassed. Such harassment is more likely to occur when space is restricted and the subordinate individual is forced into close proximity to others, or is unable to escape from and/or show unambiguous submissive behaviours to individuals dominant to it. There would be obvious adverse welfare consequences for such subordinate animals: they risk injury, may have limited access to resources or access only to resources of poor quality, and may experience severe **stress** resulting from social tension (**see: Social stress**). (JCP)

Reference

Schjelderup-Ebbe, T. (1922) Soziale verhältnisse bei vögeln. *Zeitschrift für Psychologie* 90, 106–107.

Hill sheep

Hill sheep are hardy animals that have usually been developed to survive the tough conditions found on uplands and mountains. They do not merely survive but are productive on the poor-quality **grazing** typically found in these areas. Typically they have a thick fleece, which helps them deal with the cold and exposure they face, especially at lambing time. Hill breeds of **sheep** characteristically have a good mothering ability, give lots of milk and keep a careful watch on their offspring. Indeed, ewes become particularly disturbed if their lambs are taken away from them and will often jump out of a pen in order to search for them. Many breeds have a relatively small head, which means lambing is easier and needs minimal assistance.

Although the fleece is thick, in many breeds it has a high content of coarse and black fibres, which decreases its quality and thus its value. However, traditional carpet manufacture has utilized this hard-wearing wool, and centres of carpet manufacture are often found close to upland areas.

Hill sheep breeds tend to be smaller in stature than lowland breeds: a Welsh Mountain ewe weighs only approximately 45 kg. The conformation of the animal is often poor, especially the rear quarters, which show little or no rounded muscling. Combined with the lighter weights, this means prices paid for carcasses of hill sheep breeds are significantly lower than for those from lowland breeds. Hill sheep brought on to improved pasture or given supplementary feed for finishing can easily become over-fat, as they put on a greater proportion of fat to lean muscle compared with lowland breeds.

Hill sheep breeds are strongly associated with the areas from which they originate, and have typically been refined to suit the conditions found there. Examples from the British Isles include Welsh Mountain, Scottish Blackface, Swaledale, Lonk, Herdwick and Loaghtan. The Charmoise breed, imported from France, is also sometimes found on farms with slightly more favourable conditions, as it has better conformation than many British breeds. (MM, HO)

Hinduism

The origin of Hinduism is in the multitude of different belief systems with roots dating back to 3000 BC around the valley of the Indus in India. There have always been, and still are, a multitude of Hindu belief systems, but in 1966 India's highest court stated some common core elements: tolerance of different life views; a cyclical world course; reincarnation; and Vedic (the precursor to Hinduism) scriptures being regarded as central. Originally in Vedic times, animal sacrifice was common. In the *Ishopanishad* (scripture), the spiritual way to salvation is stressed, which, in combination with the aim of escaping rebirth through good *karma* (effect of deeds), often leads to a more humble relation to other life forms.

Thanks to influences from **Buddhism** and **Jainism**, the ideal of *ahimsa*, non-violence, became influential in Hinduism. **Vegetarianism** is regarded as a practical conclusion of these elements and is usually highly regarded, or an ideal for many people.

A great advocate for non-violence and practising vegetarian was Mahatma K. Gandhi (1869–1948). According to him, strict vegetarianism 'is for building of the spirit and not of the body. Man is more than meat. It is the spirit in man for which we are concerned' (Gandhi, 2002 (1959)). In fact, Gandhi's moral standpoint was what we would call veganism today. He also regarded it morally wrong to use milk products because they, like meat, 'bring with them the defects of the animal from which they are derived' (Gandhi, 2002 (1959)). Thus Gandhi argued for care for sentient animals with the focus on human spiritual salvation, i.e. striving towards relief from rebirth through reverence for life. However, some Hindu teaching not only accepts but also prescribes meat eating, and in fact only a small proportion of Hindus are vegetarian. This is due to both different interpretations of the principle of *ahimsa* and a gap between religious ideology and practice. The difficulty in translating ideology to practice becomes explicit when thinking of the cow. She is regarded as sacred, but often suffers from **hunger**, **stress** or **injury**. This is due not only to an interpretation of sacred as prohibiting the killing of cows, but there is also large-scale industrial milk and meat production in India. (HR)

References and further reading

Chapple, C.K. (1993) *Nonviolence to Animals, Earth and Self in Asian Traditions*. State University of New York Press, Albany, New York.

Gandhi, M.K. (2002 (1959)) *The Moral Basis of Vegetarianism*. Navajivan Publishing House, Ahmedabad, India.

Preece, R. (1999) *Animals and Nature. Cultural Myths, Cultural Realities*. UBC Press, Vancouver, Canada.

Preece, R. (2001) *Awe for the Tiger, Love for the Lamb. A Chronicle of Sensibility to Animals*. UBC Press, Vancouver, Canada.

Smart, N. (1998) *The World's Religions*. Cambridge University Press, Cambridge, UK.

Homeostasis

A fundamental characteristic of living organisms and open systems, homeostasis is the property to regulate the internal environment and to maintain its stability through dynamic adjustments. An example of homeostasis is the capacity of endothermic animals to maintain a stable body temperature despite what happens in the external environment. Planet Earth is another example of a homeostatic system. (BM)

See also: Feedback – positive and negative

Home range

The concept of a home range seems implicitly understood, but is difficult to define and measure precisely. It is often defined as the area that an animal routinely uses. Alternatively, the home range may be defined as the area that an animal is familiar with (which would be larger than the area that it normally uses) or the area that it uses a large proportion of the time (hence excluding areas used only occasionally).

Most animals do indeed have home ranges – they do not wander aimlessly across the landscape. It was once thought that species like polar bears and African wild dogs did not restrict their movements to a defined range, but it was later learned that they just have very large ranges.

Why do animals have home ranges? Familiarity with an area enables animals to better utilize the resources. It is believed that many, maybe most, animals have a sort of **cognitive map**, enabling them to return to good feeding spots, water holes, shaded rest areas and places that offer safety from predators or shelter from storms. If they did not reuse areas and gain familiarity with the distribution of resources, they would be less efficient in obtaining these resources and also subject to greater risks. The trade-off, though, is that by remaining in one area they may deplete their resources. Moreover, resource availability may change radically from season to season or year to year, prompting some animals to shift their home range or make seasonal excursions or migrations. If an animal consistently migrates to the same area, that too should be considered part of its home range. But if it makes occasional sallies to new areas on occasion, our perception of its home range becomes more muddled.

To an individual animal, the distinction between what is within versus outside its home range is probably not black and white, unless it is strictly territorial. A **territory** is a home range (or part thereof) exclusively used by an individual or group, often abutting another distinct area used by another individual or group. The borders may be marked by scent, sign or a worn pathway from frequent patrols. Non-territorial home range boundaries are far more vague. Thus, human observers often have a hard time discerning the boundaries and measuring the area.

Researchers studying home ranges usually obtain a series of point locations, either from visual observations of recognizable individuals or, more commonly, from individuals that are tracked by radio telemetry over some period of time (a year or more, if the intention is to estimate an annual home range). Home range is then often delineated as the area enclosed by connection of the outermost points. In the early days of development of this technique, it was quickly recognized that different observers perceived the outermost points differently. To standardize the process, a definition was imposed that all outside angles formed by the perimeter must be convex – hence yielding what is called a minimum convex polygon (MCP).

While simple and intuitively appealing, the MCP technique has problems: namely, the encompassed area may include places that the animal never (or very infrequently) uses. Imagine that an animal uses areas on both sides of a large swamp, but not the swamp itself. Its MCP range would nevertheless include the swamp. Such areas of non-habitat can be excised from the calculated area if they are obvious and well delimited (lakes, glaciers, agricultural fields, towns, etc.). But, in most cases, there is a gradient among habitats in their degree of use. Suppose open grasslands are traversed only at night, to get from one forest patch to another; should they be excised out or be included in the home range?

This dilemma led researchers to develop methods that considered the *amount* of use in various portions of the utilized area (the so-called **utilization distribution**, or UD). Graphically the UD is portrayed not as dots on a map, but rather as bars of varying heights representing the number of times, or amount of time, that each spot was visited. Mathematical procedures can then be applied to assess the area of use from the UD. The result depends on the procedure chosen (there are at least half a dozen, with kernel estimators currently favoured) and on how well the data reflect the true home range and fit the assumptions specific to the procedure.

These UD-based procedures measure not the full area of use, but the area that the animal uses some proportion of the time (usually 90 or 95%); inner contours (isopleths) can be drawn to indicate increasing degrees (probabilities) of use. Travelways connecting commonly used areas are usually not included, due to their infrequent use, resulting in home ranges that often look like disjunct blobs rather than a contiguous area. This makes sense if the aim is to delineate areas of common usage, but is not very useful if the goal is to ascertain how large an area to protect (e.g. as a park or reserve) to contain the home ranges of a certain number of animals; in that case, the spatial separation of all the little home range blobs is also relevant. No currently developed procedure is available for doing this. In that regard, the traditional, albeit flawed, MCP has value in outlining the total physical area occupied, even if not all of it is used. In terms of ecological and behavioural questions, however, the UD is more instructive for assessing patterns of use within the home range, relative to food sources, habitats, **conspecifics**, competitors, predators or other threats.

Home range size is largely a function of the size of the animal, its sex, **social system** (including group size) and availability of food. Larger animals, being both more mobile and requiring more food, have larger home ranges. Males typically have larger home ranges than females, not only because they are larger, but because in many species males seek to breed with multiple females.

High food availability enables animals to persist on smaller home ranges. Thus, within taxa, home range size may be a metric of habitat quality; hence, home range size tends to increase with latitude. Even across taxa, home range size reflects biomass of available food: for example, giant pandas may have 100,000 times the biomass of food per unit area of a polar bear (the former subsisting on bamboo, the latter on seals); as a consequence, polar bears require areas at least 20,000 times larger. It follows that carnivores, in general, require larger home ranges than equal-sized herbivores.

However, like everything in ecology, these sorts of relationships have numerous confounding variables. High food availability supports high animal densities, either squeezing them into small home ranges or causing so much intraspecific overlap, and hence resource competition, that home ranges need to be bigger. It has been suggested that highly fluctuating food resources lead to expanded home ranges (in preparation for times of food failure) with more home range overlap (because in most years there is a surplus of food within each range).

(DLG)

See also: Core area; Dominance; Territoriality

Further reading

Damuth, J. (1981) Home range, home range overlap, and species energy use among herbivorous mammals. *Biological Journal of the Linnean Society* 15, 185–193.

Gomper, M.E. and Gittleman, J.L. (1991) Home range scaling: intraspecific and comparative trends. *Oecologia* 87, 343–348.

Kelt, D.A. and Van Vuren, D. (1999) Energetic constraints and the relationship between body size and home range area in mammals. *Ecology* 80, 337–340.

Kernohan, B.J., Gitzen, R.A. and Millspaugh, J.J. (2001) Analysis of animal space use and movements. In: Millspaugh, J.J. and Marzluff, J.M. (eds) *Radio Tracking and Animal Populations*. Academic Press, San Diego, California, pp. 125–166.

Powell, R.A. (2000) Animal home ranges and territories and home range estimators. In: Boitani, L. and Fuller, T.K. (eds) *Research Techniques in Animal Ecology. Controversies and Consequences*. Columbia University Press, New York, pp. 65–110.

Hominoidea

This super-family of species, the so-called '**great apes**', comprising **gorillas**, **chimpanzees** and **orang-utans**, are the closest relatives of humans. They are the largest living primates, have no tail, they have a barrel-shaped chest and their forelimbs are longer than their hindlimbs, which are used to ambulate. (DBM)

Homology

Two structures are homologous when they share the same phylogenetic origin regardless of their current function. An example of homology is the wings of bats and the forelimbs of terrestrial mammals, as these two structures evolved from the same ancestor, although they currently have differing functions. (BM)

See also: Analogy

Homosexual behaviour

Homosexual behaviour describes within-sex **sexual behaviour**, although there is some disagreement over the sorts of behaviours that should be assigned to this. It is usually taken to include: **courtship behaviour**, mounting, genital contact, genital manipulation, pair bonding, affection and **parental behaviour**. Historically there has been some recognition of this behaviour among animals, particularly among bird species. But homosexual behaviour in animals was perhaps most famously, and at the time shockingly, reported in the 1950s with fish (sticklebacks) and **rats** – the first probably as a result of a lack of females, the second thought to be the result of **stress** through overcrowding. Observations have, more recently, been reported in a wide range of species across classes, including birds, **reptiles**, **amphibians**, crustaceans, octopi, spiders and insects, as well as many mammalian species. It perhaps therefore should not be seen as abnormal, but as part of an animal's complex sexual repertoire.

This is a subject that is fraught with the possibility of bogus analogies with human behaviour and, perhaps as a result of this, there has been very little research in animals and it is little understood. Homosexual behaviour is more often observed in younger animals, by members of both sexes, within the context of **play**. Male homosexual behaviour has been identified with the expression of **aggression**, and it has been surmised that it is therefore more common in species with social hierarchies. In females a similar aetiology is proposed, and it is most often observed during, or around, **oestrus**. **Macaque** females have demonstrated kin avoidance in their homosexual partner selection. Recent work, including work on the guppy and primates, seems to indicate that the social environment is important, with manipulation of the sex ratios of captive groups varying the degree of expression of male–male homosexual behaviour. This has implications for the husbandry of captive animals, and particularly those kept for **captive breeding programmes** for **conservation** purposes, although engaging in homosexual behaviour does not necessarily preclude normal reproductive behaviour and outcomes with other mates.

It has been reported variously that expressions of this behaviour are more commonly seen in captivity than in the wild. On the other hand, there is contrary evidence, at least for primates, with observations of homosexual behaviour being more common in groups in free-range environments. Indeed, accurate judgements regarding the natural expression of this behaviour in the wild are likely to be compromised, as captive animals are observed more frequently and intensively than those in the wild. Homosexual behaviour shown by larger, more striking, animals also features in the scientific and lay literature more often than for other species, presumably because they are more obviously noticed than other animals.

Suggested mechanisms determining the expression of homosexual behaviour have included neural and hormonal factors, the latter either during the animal's lifetime, perhaps during puberty or while *in utero*. Rats subjected to prenatal maternal stress are reported to have expressed altered sexual behaviour in maturity, but the evidence is equivocal. There is evidence from a range of animal classes, including insects as well as vertebrates, that homosexual behaviour has a genetic basis. Evolutionary pressure in favour of the retention of this behaviour in a population would at first sight seem to be unlikely, as it is a sexual behaviour that can result in no offspring. However, it may be that, as a demonstration of social **dominance**, it might allow individuals to enhance their position in the dominance **hierarchy** and thus improve their successful choice of sexual partners. Among females, mounters rather than mountees have been found to be the dominant individuals in the main. It has been speculated that repeated mounting of co-females may reduce the frequency of their solicitations to males, thus leaving the field freer for the dominant mounters. There might be a further evolutionary advantage to the group rather than to the individual, by promotion of social cohesion and the support and protection this provides. The individual may not pass on their own genes, but by sacrificing themselves to the nurture of close kin, they will improve the possibility of some of their genetic endowment being passed on to the subsequent generation.

The act of engaging in homosexual behaviour may help group cohesion too; bonobos (*Pan paniscus*) have been observed exhibiting homosexual behaviour after periods of aggression in the group, which seems to help in defusing the conflict. It should also be noted that among animals homosexual behaviour is most often not an exclusive contact; female individuals may show homosexual behaviour, yet also mate successfully with males, allowing any genetic component to be retained in the population. Indeed, practising homosexual behaviour may maintain reproductive readiness for whenever a suitable mate should become available.

Domesticated animals are thought to engage in homosexual behaviour more often than wild animals, though there are exceptions. It has been said that the normal social construction of mountain **goats** and **sheep** is basically homosexual, with

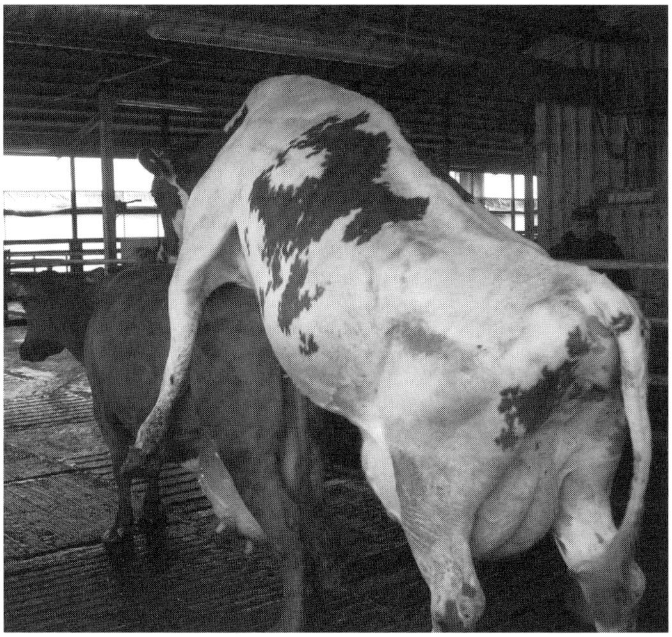

Fig. H.2. Homosexual behaviour: a dairy cow being mounted by another.

dominant males treating, and responding to, subordinate males and oestrous females in the same way. It is believed that domesticated animals may have been selectively bred to perform homosexual behaviours. Those obviously showing sexual behaviours might have been more easily observed and selected for mating. Current methods of oestrus detection and semen collection of some species of domestic animals could scarcely function without the ready expression of this behaviour. For example, most methods of oestrus detection of dairy cows rely on detecting female-on-female mounting behaviour (see Fig. H.2), while semen collection relies on the ready mounting of bullocks by the donating bull.

From an animal welfare science perspective, incidents of homosexual behaviour have been suggested as indicators of stress. However, reports of homosexual behaviour as an **abnormal** behaviour and therefore an indicator of stress must, in the light of the above, be questioned, unless other evidence is provided. Until more is known regarding the mechanisms and norms of this behaviour for each species, with respect to age, stage of sexual cycle, social dominance, social structure, presence of other-sex individuals and degree of **domestication**, conclusions from observations of homosexual behaviour in animals and their welfare should be treated with caution. (DA)

Further reading

Sommer, V. and Vasey, P.L. (2006) *Homosexual Behaviour in Animals: an Evolutionary Perspective*. Cambridge University Press, Cambridge, UK.

Honest signals

Signals are the means by which animals communicate with each other. However, defining **signals** and **communication** within a framework of animal behaviour is fraught with difficulties. Animal signals are unlikely to be 100% honest 100% of the time. When the interests of the signaller and receiver are the same, signals tend to be sufficiently conspicuous to be detected, but remain largely unexaggerated. An **analogy** might be the brake lights on the back of a car – it is mutually beneficial for the driver that is braking (the signaller) to signal to the driver behind (the receiver) that he/she is stopping; thus both parties avoid a costly accident. In this case the signal is honest as there is no benefit to the signaller in giving out false information about the intention to stop.

However, in some cases, the interests of the signaller and receiver do not coincide. When this occurs, signallers (and thus signals) can manipulate the response of receivers. For example, consider two dogs in conflict over a piece of food. The weaker dog that gives a threat display that results in the stronger rival backing down would have successfully manipulated it. Thus there is selection on the threat display to become more exaggerated. However, there is now selection on the receiver not to be manipulated by exaggerated signals (i.e. to detect and respond to true information conveyed by the signaller). Thus the co-evolution of manipulation and counter-manipulation may ensue, potentially resulting in exaggerated signals (the song of the nightingale or the feathers of a peacock) that essentially convey honest information about the qualities of the signaller. These exaggerations may further ensure a degree of honesty, as only individuals in good condition can afford the cost of producing such elaborate displays. There are also clear cases of dishonest signals – e.g. Batesian mimics signal that they are poisonous or dangerous by exploiting conventional signals used by other species, yet carry no such defences against predators. (PE)

Further reading

Stamp-Dawkins, M. (1995) *Unravelling Animal Behaviour*, 2nd edn. Longman, London.

Horse

Most equidae are plains feeders. They generally do not defend territories in the wild but rather occupy **home ranges** and prioritize membership of a cohesive **social group**. Like other social herbivores, they capitalize on **conspecific** companionship for added safety, mutual comfort and enhanced detection of food. Apart from dispersal as juveniles, they avoid becoming isolated. With long noses that allow them to graze while maintaining surveillance above the sward, horses rely largely on caution, speed and agility as their chief means of self-preservation. Contrary to some reports, they do not have a ramped retina but accommodate using ciliary muscles in the same way as most other mammals.

Horses have a large caecum, and so are able to digest material that ruminants generally overlook. Carrying such a voluminous fermentation vat obliges them to have tremendous muscular power to shift their considerable bulk from rest to top speed at times of danger. The same muscles that propel horses from danger provide power to their hindlegs, which can be extended explosively to kick with tremendous accuracy and effect. Because of its small stomach, the horse is a so-called 'trickle feeder' and must forage frequently rather than eating and then ruminating, so periods of fasting are more profound insults to equidae than to members of many other herbivore species.

Unfortunately, **housing** and stable management rarely take full account of the horse's need to socialize and to forage frequently. Accommodation generally imposes confinement and social limitations on them, while feeding regimens often constitute radical shifts from the horse's natural **time budgets**. These departures are critical in the **ontogeny** of common unwelcome behaviours in intensively managed horses. For example, social isolation has been identified as an important factor in the emergence of **weaving**, while discrete meals that allow gastric acidity to build up have been implicated in the development of **crib-biting** (and **wind-sucking**).

Horses have almost 350° vision, but their caudal blind spot accounts for some of their tendency to startle readily when stimuli appear from the rear. With dichromatic colour vision, they can be likened to colour-blind humans who see pale blues and yellows but cannot easily distinguish these from reds. Although their sense of taste is used to discriminate between safe and toxic plants, its accuracy is variable. Horses have highly developed accessory **olfaction** and an ability to hear within and beyond the range of human hearing. From a rider's perspective it is important to note that horses have predictable zones of very sensitive cutaneous sensation.

Because riders provide horses with cues in the form of pressure (e.g. from their legs to the horse's sides) and remove the pressure when the horses have made the desired response, equitation relies on negative reinforcement. However, successful innovations that accentuate the positive are being developed. For example, **clicker training** offers a simple method for shaping appropriate responses and therefore provides a basic framework for retraining many unwelcome responses, especially in the unridden horse. Horses are remarkably adaptable and tolerant, and their ability to learn facilitates their behavioural flexibility in novel environments. That said, it does not appear easy for horses to copy novel behaviour patterns, which suggests that imitation of **stereotypies** is unlikely. Although **habituation** is the pivotal technique for overcoming flight responses, most **horse training** involves **operant** conditioning and shaping that relies on reserving reinforcement until an improved response appears. Consistency and timing are the hallmarks of good training and are prerequisites if classically conditioned cues are ever to replace operant relationships.

In the free-ranging state, horses form social hierarchies that are complex and rarely linear. They seldom have the equivalent of an alpha individual, because the social roles of leadership and defence are more critical than that of simple domination. Social hierarchies determine the order in which members of a group access resources, but also increase stability in the band and decrease **aggression**, **injury** and **distress**. Social rank is not determined by weight, height or sex so much as by age and length of residency in the group. In addition, **matrilineal** dynasties can be observed in many bands of horses, while the rank of stallions is very much context-dependent.

Individual attachment, most notably in the form of pair bonds, is the fabric of social groups. While stallions tend to form weaker pair bonds than mares, juveniles or bachelors, this does not justify isolating them from conspecifics within domestic contexts. Domesticated horses have similar social organization to their free-ranging counterparts but often show more aggression because of various impositions, including **social flux**. Housed horses benefit from appropriate social contact with **conspecifics**.

Because horses have excellent vision, **communication** among them can involve very subtle body language. Ear position and head posture seem to be the most important variables in non-vocal communication between individuals, while the position of the tail can help to coordinate the movements of a group. To signal their readiness to mate, mares use olfactory cues in urine and visual cues (e.g. the characteristic posture of an oestrous mare during urination). Stallions avoid aggressive interactions by means of dung piles and the rituals attached to them.

The importance of strength and athleticism in horses for those who work with them and use them in competition means that locomotory health is often of prime concern. Movements *in utero* are of tremendous significance in priming a foal for **precocial** existence, and it seems that all horses have a **behavioural need** to stretch and locomote. **Gaits** have evolved for economical energy use at any given speed, and it is clear that whips, conditioning and warming up can influence **locomotion** in ridden horses. Kinematic studies are useful for studying **lameness** and the within-gait differences required in dressage, but there is room for considerably more scientific investigation of the cause-and-effect relationships between riders and horses, especially since the prevalence of wastage for behavioural reasons is unacceptably high.

In the free-ranging state, horses may graze and forage for up to 19 h a day. Food selection allows them to adjust their intake of nutrients to suit their current situation, and indulge their innate dietary preferences for sweet-tasting foods and short, young grasses high in fibre and carbohydrate, while avoiding poisonous alternatives. There is also individual variation in **preferences** and **aversions**, and horses may learn which foods to select from both conspecifics and personal experience.

Concentrated feeds encountered in domestic settings are associated with reduced saliva production and increased gastric acidity. In addition, periods without food precipitate yet more gastric acidity and the risk of gastric ulceration. This is compounded by a lack of forage that, in cross-sectional epidemiological studies, has been identified as the most important management factor linked to the development of stereotypic behaviours. Crib-biting and wind-sucking have also been associated with concurrent disorders of the digestive system (e.g. there is a significant association between stomach ulceration and crib-biting in foals). Therefore, treatments and management styles that reduce the incidence and severity of oral stereotypies by normalizing ingestive behaviour offer humane avenues for the improvement of the welfare of all stabled horses.

Grazing behaviour at pasture can be affected by the distribution of faeces and urine. Faecal material makes the grass less appealing, even though the grass itself is readily consumed if presented to horses without contamination. This prompts the development of latrine areas and so-called lawns, the main body of the pasture that, as a result, carries fewer **parasites**. Urination and defecation are profoundly influenced by age and sociosexual factors. For example, breeding stallions may use urine to mask olfactory stimuli left by other horses.

Self-grooming can take many forms, including rolling, shaking, rubbing, scratching, nibbling and licking. Body care also involves movements that facilitate pest avoidance and behavioural **thermoregulation**. Humans often spend a great deal of time **grooming** their horses and tending to their cosmetic appearance. While usually well intentioned, this can occasionally be inappropriate and even abusive. Inappropriate grooming includes excessive removal of sebum – which denies horses some of their natural weatherproofing – and brushing and pulling hairs to the extent that the horse is irritated and the horse–human bond is compromised. The removal of vibrissae at the end of the nose is inappropriate, as these are thought to be the chief means by which horses feel objects that are out of their line of vision. On the other hand, mutual grooming is of tremendous importance to horses because it affirms social bonds and, by lowering heart rates, is thought to have a role in **stress** reduction. So, grooming by humans can even affirm the bond between horses and their human carers.

Using a specialized apparatus in their lower limbs that relies on a system of reciprocal tension, equidae can rest with minimal muscular exertion. This conserves energy and allows them to rest while standing, thus reducing the need to enter sternal or lateral recumbency. In domestic contexts, the quantity and quality of **sleep** can be influenced by age, rank, group size, diet and environmental stimuli, including housing.

Stabling can compromise a horse's ability to safely perform many normal maintenance responses, such as rolling. It also affects the horse's ability to undertake locomotory and interactive **play**, behaviours that are recognized as being important for physical and psychological well-being. Play among foals and between foals and adult horses also has a significant role in the development of social behaviour, including learned inhibition of agonistic responses. Prior to 1 month of age, play is more likely to be solitary or with the mother, whereas after 1 month of age play is more likely to be with peers.

The behaviour of stallions has been broadly misrepresented in popular culture as being focused on domination rather than on benign harem maintenance, and this may have given rise to the notion that they are predominantly aggressive animals. In fact, the leadership role of stallions within a family band is largely confined to keeping foals safe from predators and mares away from other stallions. In the free-ranging state, a harem stallion interacts with his mares almost continuously and so can avoid errors in the timing of copulation. Free-ranging matings are generally more productive than those orchestrated by human managers, who seek to use courtship primarily in teasing and the development of an erection, and limit the number of copulations per stallion in a putative bid to maintain sperm reserves. Because of the poor timing of managed matings and the fact that they are not fully receptive, mares are regularly restrained during service so that they cannot harm the stallion if they attempt to kick.

Seasonality has little effect on a stallion's **libido**, and this has allowed the emergence of so-called shuttle stallions that travel between the northern and southern hemispheres serving mares throughout the year. Factors that do affect the sexual responses of stallions include individual preferences, visual stimuli, olfactory stimuli and learning. The latter includes learning to avoid mating because of discomfort associated with peri-coital stimuli, including forced dismounts, unreceptive mares and penile and musculoskeletal lesions. It is now accepted that stallions can be allowed to mount without an erection and that, if it becomes necessary to force a stallion to dismount, this should be accomplished without aversive stimuli that can have unpleasant associations later. Allowing the mare to facilitate the stallion's dismount by walking forward reduces the risks of unwelcome associations with copulation.

Among the human interventions that are frequently employed to modify libido, pharmaceuticals and **castration** are perhaps the most important. The social rank of a stallion is surprisingly static after gelding, but **testosterone** concentrations of stallions have been found to drop when males are housed together in what amounts to an analogue of bachelor bands found in the free-ranging state. So, if safety can be maintained, increased contact between mares and stallions can improve herd fertility and reduce the prevalence of unwelcome behaviours in stallions. Behaviour problems in the stallion, including failure in reproductive behaviour, **self-mutilation**, aggression and other handling problems, generally reflect the failure of management regimes to meet the behavioural and physiological needs of stallions and, indeed, mares.

Masturbation, previously a source of concern to many stallion owners, who linked it with a variety of outcomes ranging from low sperm counts to social disgrace, is part of the normal **ethogram** of male horses. However, stallion rings (constrictive circumferential bands that prevent vascular engorgement of the penis) are still sometimes used and bring with them the inherent risk of abrasions and scarring. Teaser males may be used for oestrus detection, but are usually prevented from breeding by barriers, because they are of insufficient height or have been vasectomized. The welfare of such stallions is regularly questioned, especially where surgical procedures (such as **circumcision**) are used to render a stallion unable to consummate sexual behaviours.

Mares in the free-ranging state can be categorized as either social dispersers (so-called mavericks that leave the natal band spontaneously), loyal to a single stallion or part of a multi-stallion band. They tend to form strong alliances with their daughters and other resident females and usually take a leadership role within a natal band, perhaps because they tend to be resident longer than any male and therefore have the most mature bank of information about the band's home range.

Being seasonally polyoestrous, mares show a cyclical active oestrus (7.1 ± 4.2 days) and quiescent dioestrus (16.3 ± 2.9 days) throughout the breeding season (152 ± 50 days). Oestrus in the mare is characterized by **courtship behaviours** such as abrupt halts during locomotion, raising of the tail (before being mounted or after being mounted), clitoral winking (especially during teasing), frequent urinating and pre-copulatory behaviour such as sniffing and nibbling. The role of mares in courtship, especially in approaching and following stallions, has frequently been underestimated, but it is now recognized that mares have a significant influence on the success of matings.

Some mares may exhibit oestrous behaviour during pregnancy because of high **oestrogen** levels from accessory *corpora lutea*. More commonly, a pregnant mare's behaviour does not change markedly during **gestation** until immediately prior to **parturition**, when obvious changes include increases in both walking and recumbency. Foalings tend to occur at night and regularly involve some form of social isolation, perhaps to avoid **mis-mothering**. Social attachments, such as the bonds that develop between mare and foal, hold equine social units together. In the free-ranging state it normally takes 2 weeks of close proximity and frequent interactions for the foal and mare to consolidate their bond. The spatial relationship within the dyad can be characterized by the number of 'approaches' and 'partings' made by either member. Although initially the mare's bond with the foal is much stronger than vice versa, foals do more work to maintain contact with their dams, but this tendency declines as they approach **weaning** age.

In free-ranging horses, weaning is very gradual and mares may permit **suckling** by yearlings even when they have a new foal at foot. Weaning in domestic contexts, on the other hand, is often marked by abrupt changes in the foal's social environment and nutrition. To reduce weaning stress in a herd situation, mares can be periodically removed so that foals are left together in familiar surroundings. Although creep feeding of foals prior to weaning may reduce some forms of distress and optimize the post-weaning growth rate of the foal, it has been linked with the emergence of oral stereotypies. Antacid supplements appear to have merit in reducing such undesirable sequelae.

Horse training and handling have undergone a recent renaissance, with **'horse whispering'** (horse handlers who challenge traditional horse-handling dogma and advocate novel and often gentle methods of training) showing how effective subtle handling can be in the foundation training of young horses and the retraining of older horses that have developed unwelcome responses. Central to their message is the importance of paying attention to the body language of horses whenever working with them. Whether they realize it or not, good handlers of horses are students of equine behaviour. They recognize that horses rarely forget aversive procedures. By keeping horses as calm as possible, they ensure that the type of restraint required for current and future handling can be kept to a minimum.

Behavioural wastage in the horse industry is unacceptably high, largely because effective training of horses is compromised by widespread ignorance of learning theory among riders. Furthermore, equine education is clouded by the implication that horses have some moral role in their own training and by the use of opaque terminology and **anthropomorphisms**. Contemporary horse-training dogma provides numerous obstacles to effective equine learning, such as training too many responses simultaneously and one signal overshadowing (a process whereby the presence of a second stimulus interferes with learning an association between the target stimulus and some important event) another. If novice riders were coached in learning theory and recognized the principles of operant conditioning before learning to develop an independent seat on a horse, the industry and many riding school horses would benefit.

Horse training and management techniques are notoriously resistant to change. However, the scientific data supportive of the case for bitless bridles, not least as a means of showcasing outstanding horsemanship, and for podiatry that avoids the use of metal shoes (because they may have deleterious orthopaedic consequences) are emerging.

To resolve learned responses, equine behaviour therapy relies on: (i) consideration of the ethological relevance of unwelcome behaviours; (ii) elimination of pain and discomfort as proximate causes; and (iii) application of learning theory. The reasons behavioural modification fails include inconsistent application of learning theory and a lack of reconciliation of the task required of the horse and its physical ability to comply. (PDM)

Further reading

McGreevy, P.D. (2004) *Equine Behavior: a Guide for Veterinarians and Equine Scientists*. W.B. Saunders, London.

Mills, D.S. and McDonnell, S.M. (2005) *The Domestic Horse: the Evolution, Development and Management of its Behaviour*. Cambridge University Press, Cambridge, UK.

Horse training

Despite their now-diminished role in agriculture and warfare, **horses** have a new and highly popular role in equestrianism for **sport** and leisure in the developed world. For example, horse racing codes include races for Thoroughbreds, Standardbreds, Arabians, quarter horses, ponies and endurance horses. Other equestrian pursuits include polo, polocrosse, dressage, eventing, showjumping, equitation, **hunting**, many show-riding codes (turn-out, riding classes, pony and hack classes, side-saddle, etc.), Western riding, Australian stock horse, games (barrel racing, etc.), rodeos, cattle cutting, gaited-breed riding, pony club, many carriage-driving codes, myriad show breed classes and leisure-riding activities.

Ever since humans first domesticated the horse, their interactions have involved training schemes that engaged the animal's perceptive and motor abilities for human purposes, such as the keen olfactory sense of canids and the locomotory abilities of equids. Although correct and successful horse training methodologies comply with the general principles of equine learning, horse training has traditionally been understood from an anthropomorphic viewpoint in which the horse has some voluntary involvement and is motivated by benevolence or malevolence. One of the problems with **anthropomorphism** is that it embraces the possibility that horses can be held responsible for their behaviour. One of the problems contributing to such notions of culpability is that there is no consistent descriptive terminology among horse enthusiasts. Furthermore, an appreciation of the learning theory that forms the broad base of contemporary training in many other species such as guide dogs and animals in **zoos**, **circuses** and oceanaria is largely absent in contemporary systems of equitation, training, coaching and manuals of horsemanship.

Correct horse training involves **habituation**, operant conditioning and classical **conditioning**. During the earliest phases of horse handling, which may take place soon after birth (**see: Imprint training**) or at **weaning** time, horses habituate to their surroundings, to humans and their movements as well as to training paraphernalia (e.g. head **collars** and lead reins). Because much of the apparatus used to facilitate training in horses operates through various levels of pressure, negative **reinforcement** is the chief operant learning process. Thus the trainer targets a particular response, and maintains or increases the pressure until the onset of that targeted response. For example, the horse is negatively reinforced to move forward when anterior lead pressure occurs, because the lead pressure is removed immediately after the animal steps forward. The optimal moment to release the lead pressure is after the horse's forelegs have moved forward. Similarly, the horse learns to decelerate its forward motion from lead pressure in the posterior direction.

The rules of negative reinforcement also apply 'under saddle' when the horse moves forward from the pressure applied by the rider's legs. The forward movement is then classically conditioned to occur in response to lighter versions of the pressure signals from the rider's legs, or perhaps seat movements or voice. The two reins are used negatively to reinforce decelerations when a release of pressure occurs as soon as the horse decelerates. Decelerations are classically conditioned to be elicited from light rein signals and bracing movements of the rider's seat or from vocal commands such as 'whoa'. The horse learns to turn its forequarters through the pressure and release of the single rein, so that tension in the right rein results in the right turn, and in the left rein results in the left turn. Some trainers use the rein contralateral to the turn direction to turn the horse in what is known as a neck-rein or an indirect rein, where the rein is moved toward the midline of the horse and pressed against the lateral aspect of the neck. In some equestrian codes horses are also trained to move their hindquarters laterally. This response is elicited by the pressure of only one of the rider's legs and, again, the movement is negatively reinforced by the removal of that pressure. Eventually, the sideways movement is classically conditioned to be elicited from the mild pressure of the rider's single lower leg.

Harness horses learn to go forward through negative reinforcement of the driving whip on their rump; they are then classically conditioned to move forward after a verbal command. In addition, they must habituate to the vehicle they pull. Some horses habituate poorly to visual stimuli to their rear, so harness trainers frequently use 'blinkers', eye-covers that allow only forward vision.

Most horse trainers and coaches fail to recognize the negative reinforcement that is inherent in the early training of leading signals. Instead, they cite compliance and responsiveness as evidence of the horse's willingness to please. In fact, horses that have learned via classical conditioning to respond to light signals have, nevertheless, initially learned through negative reinforcement.

Positive reinforcement can assist in shaping and maintaining stable responses. The methods of positive reinforcement may involve primary reinforcers, such as food, or caressing the horse at the base of the withers (an area where rubbing is known to lower heart rate), or conditioned secondary reinforcers such as verbal praise. Secondary positive reinforcement occurs when an inherently neutral (and therefore unrewarding) stimulus such as the trainer's voice is paired with a primary reinforcer.

Stimulus **generalization** also occurs in horse training where the animal is able to discriminate less context-specific signals from the rider, driver or handler and respond appropriately. Horse people observe that this tends to occur largely in older, more experienced animals, so generalization appears to be an ontogenetic process.

Because of the invasive nature of negative reinforcement, horses are susceptible to **stress** and **conflict behaviours** when the pressures are disproportionate, uncomfortable or distressing, or when they are not released at all, not released at the correct moment or released inconsistently. The solution lies in the correct application of negative reinforcement, both at the outset of training and in the repair of the unwelcome stimulus–response relationships.

Because the incorrect application of negative reinforcement causes stress in animals, it would be beneficial if trainers learned the correct application of negative reinforcement. Many people are becoming attracted to the use of positive and, particularly, secondary positive reinforcement as a viable alternative in horse training; this includes techniques such as '**clicker training**'. Although there is little evidence that the use of secondary reinforcements can rehabilitate horses suffering from negative reinforcement training errors, it could be more effective than regular positive reinforcement, because a secondary reinforcer may be delivered at the precise moment of the correct behaviour, whereas primary reinforcers can often be delivered only after some delay.

Many horse riders and trainers appear to assume that subjective mental states account for the horse's responses. For reasons yet to be established, equestrian technique has been more influenced by the science of **ethology** than the science of **psychology**. In her review of animal cognition, Shettleworth (2001) points out that the diversity of approaches between ethology and psychology, in understanding the animal mind, present a dyadic problem in the analysis of behaviour. Whereas ethology arose mostly in Europe within the context of zoology, with its strong emphasis on evolution, psychology emerged in North America without the influence of any evolutionary perspective that characterizes zoology. Meanwhile, despite its relevance to behaviour modification in humans and animals, the demise of behaviourism in the latter part of the 20th century meant that horse trainers were shielded from learning about it. So it is perhaps not surprising that the New Age horse trainers were more influenced by the *prima facie* appropriateness and relatively simplistic interpretation provided by ethology in understanding horses. In an appealingly simple traditional construction, ethology reveals the existence of various **innate** processes, such as the hierarchy, social and herding instincts, and others, but makes little reference to the more proximate mechanisms, such as reinforcement, that establish such behaviours. The pitfall of the singular ethological approach is that it fails to explain the proximate mechanisms of ethological phenomena, and this in turn promulgates inappropriate expectations (via subjective mental interpretations) to be placed upon the animal. Many New Age horse trainers have been influenced by the ethological approach, and place a great deal of emphasis on the equine 'peck order', emphasizing that: the horse should be made to see the human 'as a leader'; the human should be *dominant* over the horse and, conversely, the horse should be *submissive* to humans; and the horse should be taught to 'respect' humans and not to invade the human space. The common terms that unite the New Age methods include 'respect', 'willingness', 'submission', 'leadership' and 'attitude'. In human–horse interactions, these concepts may have negative **welfare** implications for horses and can lead to punitive measures that bear little relation to the original symptomatic behaviour.

The horse wastage statistics for the horse industry are high, and significant wastage of performance horses is for behavioural reasons (Ödberg and Bouissou, 1999). It is also significant that the statistics on wastage came from countries with well-established equestrian traditions, so there is no reason to believe that figures would be significantly different in other countries.

In addition, there is a considerable degree of danger posed by horse riding in modern training contexts. For example, in the USA the national horse-related death rate is 219 per annum, with most deaths occurring in younger females and middle-aged males (Bixby-Hammett, 1992). About 20 Australians are killed as a result of horse-riding activities each year and a further 3000 hospitalized. So the annual death rate from horse riding is about one per million head of population. It is widely reported that most deaths are of young female amateur riders (which correlates with a greater proportional representation in amateur equestrian pursuits by women), followed by older (age range 30–50 years) professional male riders. In terms of deaths and injuries, horse riding is more dangerous than motorcycle sports or car racing, and as dangerous as rugby.

The danger inherent in human–horse interactions reflects the fact that a horse weighs about 600 kg, can gallop at 65 km/h and has a powerful and dangerous kick. While most researchers rightly suggest that reductions in death and injury rates in horse-related activities would be achieved with increased supervision of children and the use of safety equipment, such as riding helmets and vests, few point to horse training or behaviour modification as a potential alleviating factor. Cripps (2000), however, sees the situation differently: 'It's important also to learn about how horses behave'. Clearly, the wastage statistics and the human injury statistics suggest that coaching in learning theory for horse people and, particularly, for trainers in all disciplines, is an essential step in the development of modern equestrianism.

(AM)

See also: Instinctive behaviour

References

Bixby-Hammett, D.M. (1992) Pediatric equestrian injuries. *Pediatrics* 89, 1173–1176.

Cripps, R.A. (2000) Horse-related injury in Australia. *Australian Injury Prevention Bulletin* 24, Flinders University, Adelaide, South Australia.

Ödberg, F.O. and Bouissou, M.-F. (1999) The development of equestrianism from the baroque period to the present day and its consequences for the welfare of horses. The role of the horse in Europe. *Equine Veterinary Journal Supplement* 28, 26–30.

Shettleworth, S.J. (2001) Animal cognition and animal behaviour. *Animal Behaviour* 61, 277–286.

Horse whispering

Traditionally, force has been used in **horse training** in the belief that human **dominance** and equine submission are critical to success and the development of so-called 'respect' on the horse's part. As a result, many horses learned unwelcome evasions, resistances and flight responses. Historically, horse whisperers might then be called upon to remediate such horses. They locked themselves in stables to weave their magic and emerged with impressive results but no intention of divulging the techniques used. Their deftness and the absence of any apparent **vocalization** as part of their practice gave rise to the beguiling notion of their needing only to whisper to horses to get the desired results. It was to their advantage to let the uninitiated believe that whispering, rather than training, was involved.

Recently, the use of applied learning theory has begun to expose simple, humane and effective mechanisms in horse training. As we appreciate how to communicate with horses sensitively and consistently, misunderstanding and misinterpretation by both parties will diminish. These advances are being matched by greater sharing of knowledge among practitioners. As an example, in the latest wave of 'natural horsemanship', trainers use techniques similar to the whisperers of old but do so in public. Natural horsemanship trainers are often talented observers of horse behaviour who respond consistently and swiftly to subtle cues during training. They apply **aversive** stimuli to prompt a flight response and then, when the horse slows down, move towards them, or offer space-reducing affiliative signals; they remove the pressure and thus negatively reinforce the desired responses. Learning theory can be used to explain almost all of the behavioural modification that goes on in such contexts.

Numerous horsemen have demonstrated their ability to exploit the marketplace and, largely without reference to learning theory, teach their methods for commercial reward. Unfortunately, not all followers of these methods are as effective as the good trainers. Poor technique can lead to disappointing results and eventually frustration in practitioners, which may result in abuse, confusion and **conflict behaviour** in horses. Scientific assessment of how novel techniques operate and questioning of how the language relating to horse training and riding relates to what is known through **psychology**, **ethology** and veterinary science, has helped to deepen our understanding of the benefits and limitations of these techniques. Burgeoning demand for best practice that is underpinned by empirical data rather than belief systems has prompted the emergence of equitation science as a scientific discipline in its own right. (PDM)

Further reading

International Society for Equitation Science website: http://www.equitationscience.com

House-soiling

House-soiling, or urination and **defecation** in inappropriate places, is one of the most common problems reported by people who live with **companion animals**. Feline **urine** spraying is also commonly stated as a house-soiling problem, but is functionally distinct.

Urination is the natural process by which the bladder is emptied, while urine spraying, also a natural behaviour, is a form of scent deposition that helps to demarcate **territory** and signal social or reproductive status. In **cats**, spraying is generally performed from a standing or squatting position, releasing a small amount of urine against a vertical surface. Urinating involves squatting and releasing a large quantity of urine on to a horizontal surface.

Dogs also commonly urine mark, especially males, who will typically cock their legs on to vertical surfaces and release a quantity of urine. Subsequently, dogs passing the marked spots obtain information by sniffing the odorous deposits that are conveniently located at nose height. Leg lifting, a behaviour normally observed in sexually mature male dogs, emerges in juveniles with increasing **testosterone** production. Some females are also observed to take up this leg-cocking practice while others will squat and, as males do, over-mark urine scents left by others.

House-soiling has a number of triggers, including physiological and behavioural factors. Medical problems causing house-soiling include kidney and urinary problems, and all house-soiling problems should be medically investigated before, or at the same time as, behavioural causes are considered. Feline lower urinary tract disease, for instance, may be extremely serious, even life-threatening.

Behavioural causes of house-soiling include dislikes or preferences for certain substrates, such as particular litter types or grass as opposed to concrete. **Anxiety**-related causes, such as the presence of other animals, may also result in inappropriate **elimination behaviours**.

Treatment includes implementing a behavioural treatment programme through:

- Appropriate areas and/or times for elimination, e.g. cats may require more litter trays.
- Different positioning or different litter.
- Dogs may need to be taken out more regularly on to suitable substrates.
- Appropriate cleaning methods to avoid reoccurrence by animals that tend to return to odourous spots.
- Neutering of cats and dogs to reduce marking behaviour.
- Reducing anxiety by keeping routines constant or, in the case of **separation anxiety**, gradually building separation tolerance and using distractions such as toys or food treats when the owner leaves. Where anxiety is the cause of house-soiling problems, medication such as antidepressants may help, in conjunction with a behaviour treatment programme. (JR)

See also: Scent marking; Urine marking

Housing

When animals are kept for farming, research, teaching, entertainment or as **companion animals**, they usually require housing. There is immense variation in the design and construction of housing according to the requirements of the animals, their keepers and the reasons for the animals being housed.

Farm animal housing

Farms are businesses, and therefore economics are often the governing factor in decision making. A major cost of many farming operations is the housing and therefore it follows that, to minimize farming costs, housing is often as basic as possible. This relates to not only the quantity of space, but also the quality of space. As a consequence, the vast majority of farm animals are housed in small, barren, minimalist environments (**see: Intensification of animal production**). In contrast, some farm animals are reared with access to large amounts of open, varied space (extensive systems), with relatively little housing because, under some circumstances, this is the most economical method, or because this can attract premium prices when the animals or their products are sold at the market place.

Intensive housing systems

Globally, the animal species farmed in greatest numbers are meat hens (broilers), layer hens (**see: Chicken**) and **pigs**. Being the most widely farmed species, the greatest economic pressures have been brought to bear on the housing of these, resulting in the widespread adoption of intensive systems for them. This drive for economy of animal production has been so great that, in the case of **laying hen housing**, it is legal in some countries to house birds in cages that provide each animal with floor space less than a sheet of A4 paper. The hen will be housed in a cage this size for its entire life, which might be several years. The cages usually have a wire mesh floor and sides, provide food and water (except when these might be withdrawn for several days during **forced moulting** to reinvigorate production) and little else to interact with other than cage-mates.

While such highly intensive systems offer some benefits to the animals (protection from **predation**, environmental variables usually maintained within comfortable limits, generally a clean environment, receipt of appropriate medication), they prevent the performance of many highly motivated behaviours, which leads to **frustration** and the performance of **abnormal** behaviour such as **feather pecking**, **cannibalism** and **stereotypies**.

Intensive housing systems sometimes involve highly restrictive environments. Modern domestic sows can be very large and are often highly fecund. However, they are not always particularly good mothers. Piglets are sometimes cannibalized or injured by their mother, or crushed because she has trouble rising and lying due to her great weight. Parturient sows can also be highly aggressive towards humans. As a consequence of these factors, many preparturient sows are placed in **farrowing crates**. These enclose individual sows in metal frames that allow them to stand and lie, take one or two steps forward or back, but they prevent her from turning around or being able to attack **stockpersons**. The sow is placed in the farrowing crate a day or two prior to the expected birth, and remains in there until several days after. There have been attempts to design less restrictive farrowing crates: some of these are circular, have a rail around the edge to reduce the possibility of the sow crushing her piglets when she lies down, and a creep area into which the piglets can escape but which the sow cannot enter.

Intensive housing systems can be very large and contain many thousands of animals. In the UK, a single house of caged layer hens might contain over 100,000 birds, and a flock of 20,000 broiler chickens might be reared in a single house. Due to scales of economy, there are often several such houses on a farm, meaning that hundreds of thousands of animals could be on a single farm, often under the charge of only a small number of stockpersons.

Semi-intensive housing systems

There have been recent attempts to develop housing that offers the benefits of both intensive and extensive housing systems for farm animals. The most notable of these is the development of ‘furnished’ cages for layer hens. They are designed to incorporate key **welfare** features such as increased space allowance, a nesting box, perch, dust bath and scratching areas. Research has shown that hens are highly motivated to perform behaviours associated with these key features, and it is believed that providing them produces a housing system that offers all the benefits of intensive cage systems and some of the benefits of extensive systems. In Europe, legislation has been passed stating that conventional cages for laying hens should be replaced by furnished cages by 2012. These systems are designed to house colonies of various sizes, some as small as four or five birds per cage, whereas others have larger colonies, e.g. 60 hens. Concerns remain about the welfare of hens in some of these systems, especially with regard to feather pecking, particularly for larger groups.

Other semi-intensive housing systems have been developed such as the Alex Stolba family pen for pigs, which aims to keep pigs together as a family unit throughout their production life. These systems require a high standard and effort of husbandry and management, which is possibly why they have not been adopted to any great extent.

Extensive housing systems

Some animal housing systems have always reared animals in large areas of natural space, e.g. range **sheep** or **cattle** in the USA or Australia. Essentially, the animal lives a free-roaming life and might be handled only once or twice a year. These systems are still operated because the minimal amount of housing required means the overall farming method remains viable. In more recent years, there has been a trend to reverse the confines of intensive farming, particularly for pigs and poultry, and give animals access to considerably greater amounts of space, although not the amount that would be given to range animals. This has advantages for welfare, indicated by an increased behavioural repertoire and more species-specific behaviours; however, the animals incur an increased risk of other welfare compromises such as disease or predation.

Laboratory animal housing

Economic concerns have also had a considerable influence on the design of housing for **laboratory animals**. For humans, it is essential that laboratory cages are inexpensive, easily cleaned, practical and have great longevity. In addition, from the scientist's point of view, it is essential that the animals' environment remains unchanged during studies, or at least is highly controlled. Very often, this requirement extends between studies. The easiest way of achieving all these requirements is to make the housing as simple as possible. As a consequence, laboratory animal housing is generally small, barren and minimalistic. For example, laboratory mice are typically housed in plastic cages that might be approximately 15 × 30 cm, food and water are supplied through the wire-mesh cage roof, and the floor might be covered with 1–2 cm of sawdust. Several mice will be kept in a cage of this size, and they will receive no stimulation other than that which they receive from the scientist as the procedure for which they are being kept.

This type of intensive housing clearly prevents many species-specific behaviours, some of which are demonstrably highly motivated, e.g. for mice, tunnelling, **nesting behaviour**, extensive **locomotion**. Other laboratory species, perhaps most notably those usually considered as companion species under other circumstances, are likely to be housed in less minimalistic conditions in the laboratory. **Cats**, **dogs**, **rabbits**, **horses**, etc. might be offered more playthings, a greater variety in their diet, a more complex environment, perhaps even access to the outdoors; however, the majority of laboratory animals will still be housed for a great proportion of their time in small enclosures that are only a few of their body lengths in any one direction, and containing little else other than food, water and **conspecifics**.

The small and minimalistic nature of laboratory housing has notable effects on the behaviour and physiology of the animals they contain, indicating that animal welfare is compromised. This type of housing can also create a secondary concern in that the biological functioning of the animals can be so abnormal that the validity of the science is reduced, thus negating the main purpose for housing the animals in the first instance.

Home-cage behaviour

Mice in standard laboratory cages self-administer greater quantities of anxiolytic (a drug that reduces **anxiety**) than mice in larger, enriched cages, indicating that underlying anxiety might be greater in animals in standard laboratory housing. When kept in standard housing, bouts of spontaneous nocturnal locomotion by mice are less intense and distinct than seen in mice in larger cages, and both cage shape and size can influence the mating behaviour of male **rats**. Standard laboratory cages are either white or transparent; however, it has been shown that the home-cage colour can influence food intake and body weight of mice.

External tests

Standard laboratory housing can influence a range of responses outside the home cage. In tests of emotionality (e.g. elevated plus-maze, open-field, shuttle box), **rodents** from standard housing often behave in a manner indicative of being more anxious or fearful than animals from enriched or larger cages. Jumping spiders from smaller cages progress less distance in a searching test and respond less vigorously to videotaped prey. Spiders from non-enriched standard cages are less active in the open-field test and only respond to videotaped prey from shorter distances.

Effects of laboratory housing on sensory development and neurogenesis

It is usually assumed that the sensory development of animals in standard laboratory housing is normal, and therefore any responses dependent on these senses will also be normal. However, mice raised from birth in larger, enriched housing have 18% better visual acuity than mice from standard housing. The effects of standard housing on behaviour can occur due to influences on neural development and plasticity. Enriched cages can enhance neurogenesis or, stated conversely, standard housing can limit neurogenesis. The implications of this effect can be profound because the links between neural development, behaviour and physiology are still not greatly understood. Certainly, there are many reports that enriched cages can improve recovery from brain trauma compared with standard, barren housing.

In transgenic mice that develop Huntington's disease, the decline in performance in the RotaRod test and the loss of peristriatal cerebral volume in the **brain** is considerably quicker for animals in standard housing compared with enriched. In wild-type mice that do not develop the disease, enriched housing considerably enhances performance in the RotaRod test, although there is no improvement of grip strength, indicating that the enhanced performance is not due to increased muscular strength. It has been suggested that enriched housing alters gene expression in the normal mouse brain, which modulates the course of the disease.

Effects of laboratory housing on physiology
Several studies show that characteristics of standard laboratory housing can influence physiology. Compared with animals in smaller, standard housing, daily energy expenditure and food intake of mice in larger cages can be greater or lesser, depending on the species. The mean baseline temperature of **hamsters** is significantly higher in animals housed in small cages compared with larger, and the reaction to fever-inducing lipopolysaccharide is highest in animals housed in large cages and smallest in animals in small cages. In studies of rats housed in cages with grid floors, plastic floors or bedding, there are significant differences in systolic and diastolic **blood pressure**, **heart rate** and body temperature.

Effects of the social environment
Standard laboratory housing often entails housing gregarious animals in isolation or solitary animals in groups. Housing animals contrarily to their normal social behaviour can result in responses that almost certainly reduce their welfare and the validity of the data gained from them. Singly housed hamsters (normally a solitary species) recover more quickly from mercury poisoning than group-housed hamsters, which show a reduced body weight gain, possibly due to high **social stress** elevating their susceptibility to intoxication. Plasma triglyceride levels and food intake are lower in isolated rats than in grouped rats, although there is no difference in body weight. Singly housed mice have higher heart rates and tend to have a reduced body temperature, although there is no difference in activity. Compared with group-housed mice, singly housed animals are smaller, have less soft lean tissue and bone mineral content, and lower bone mineral density. In terms of behaviour, singly housed rats are more active under several conditions of environmental novelty, but have an increased latency in an emergence test, indicating greater fearfulness. In addition, it appears that singly housed rats are less able to learn a task reversal than group-reared rats.

Designing more appropriate animal housing

Concerns about the effects of animal housing, whether these are for animal welfare or scientific validity reasons, can be avoided by redesigning housing from less anthropocentric and anthropomorphic viewpoints.

Understanding the animal's senses
To design and build more appropriate animal housing, we should try to reduce our anthropocentricity and consider how the animal perceives the environment with its own senses (**see: Anthropomorphism**). Sometimes this requires fundamental research on which sensory modalities the animal possesses, how well developed these are and their importance to the animal. This knowledge can then be introduced into applied studies of how the housing might affect these senses and possible methods of rectification if they raise concerns.

One example of developing this knowledge and its application is bird vision. Birds have different vision from humans: they can see in the ultraviolet range whereas most normal adult humans cannot, and birds have a much higher critical flicker fusion rate (the rate at which a flashing light is perceived to be a steady light). Birds are often housed in rooms illuminated by fluorescent lamps. These emit very little ultraviolet (UV) light and flicker on and off at 50–200 Hz. This means that the bird's vision is potentially distorted by the absence of ultraviolet wavelengths, and the birds are probably perceiving fluorescent lamps as flashing on and off. Behavioural studies have shown, however, that even if birds are perceiving fluorescent lamps in this way, it does not appear to be particularly **aversive** to them and therefore is unlikely to be a great welfare compromise. Fluorescent lamps that emit UV have recently been developed for birds and are now commercially available.

Understanding the animal's natural behaviour
Studies on the behaviour of animals in their natural environment can reveal that the housing we provide is often inappropriate. Many natural behaviours are thwarted by standard housing, e.g. burrowing by laboratory mice, nest building by farmed pigs, dust bathing by farmed hens (**see: Frustration**). It is not until we study their behaviour in a more natural environment that we realize that, despite years of domestication and hundreds of generations of housing under artificial conditions, the animals have retained the motivation to perform many of these behaviours.

Understanding the animal's motivation
If certain characteristics of housing are desirable to an animal, it will choose these in a preference test, or avoid them if they are undesirable. Similarly, we can measure the strength of **motivation** that the animal has to gain access to, or avoid characteristics of, the housing. This method is based on the premise that animals (usually) behave in ways that maximize their welfare, and that the more strongly motivated an animal is for some aspect of its environment, the more likely it is to suffer if we do not provide it. In this way, we can assess whether we should be providing animals with larger housing, grid or solid floors, nestboxes, etc. (CS)

See also: Dust bathing

Further reading

Littin, K., Acevedo, A., Browne, W., Edgar, J., Mendl, M., Owen, D., Sherwin, C., Wurbel, H. and Nicol, C. (2008) Towards humane end points: behavioural changes precede clinical signs of disease in a Huntington's disease model. *Proceedings of the Royal Society B – Biological Sciences* 275, 1865–1874.

Richter, S.H., Garner, J.P. and Wurbel, H. (2009) Environmental standardization: cure or cause of poor reproducibility in animal experiment. *Nature Methods* 6(4), 257–261.

Sherwin, C.M. (2004) The influences of standard laboratory cages on rodents and the validity of research data. *Animal Welfare* 13(Suppl.), 9–15.

Young, R.J. (2003) *Environmental Enrichment for Captive Animals*. Blackwell Publishing, Oxford, UK.

Human–animal bond

The human–animal bond describes the dynamic social relationship that exists between people and non-human animals and, while this can exist with any species, most attention has been paid to the relationship with **companion animals**, as this is where there are most clearly mutually beneficial outcomes from a **bonding** relationship. Humans have the potential, at least in theory, to develop a bond with any species of vertebrate or invertebrate with which they have contact, but the nature of the resultant bond will vary depending, at least in part, on the ability of the species concerned to reciprocate within the relationship and the degree of **empathy** and understanding demonstrated by the human partner. Companion animals are often kept primarily for social or emotional reasons rather than for economic purposes (Laurent, 2000). Relationships with non-human animals differ significantly from **conspecific** human relationships in that there tend to be fewer social and potentially psychological barriers applied by the human, since non-human animals appear non-judgemental and their attentiveness to their caregivers is in most cases largely unconditional.

The study of the human–animal bond comes within the wider realm of human–animal interactions, which is covered within the discipline of **anthrozoology**, and much of our knowledge in this field has focused on the benefits to people of developing affectional bonds with companion animals (**see also: Affiliation**). These benefits can be psychological and/or physical, and investigations into beneficial physiological responses, which reflect the interface between these effects, have been the focus of a numerous studies. Physical benefits may be derived through many routes; for example, as a result of an increase in exercise encouraged through **play** or shared exercise (e.g. walking a dog or horse riding), or as a result of decreases in **blood pressure** associated with the calming effect of tactility and/or the emotional support provided by contact with an animal (De Mello, 1999; Friedmann *et al.*, 2007). Psychological benefits include encouragement of social interactions, provision of social support, reducing loneliness as well as the direct provision of perceived friendship and unconditional love.

For a human–animal bond to be considered successful, the benefits of such a relationship must outweigh any potentially detrimental effects for both of the individuals concerned. Explanations of the balancing of both the positive and negative aspects of the human–animal bond incorporate concepts relating to **biophilia**, **economics of behaviour** and **attachment** theories. Animals are generally provided with food, shelter, companionship, protection and medical prophylaxis/treatment in return for the fulfilment of the emotional, social or utilitarian needs of their owner or caregiver. The financial, emotional and social costs of caring for the animal should not exceed the perceived benefits, because this may jeopardize the quality and/or longevity of the resultant bond; however, one major challenge is identifying a common currency for this equation.

Not all consequences of the bond are positive. The relationship is dynamic and influenced by internal as well as external factors; for example: (i) the **personality** and experience of the individuals concerned; (ii) the **health** of the partners, including their physical and mental well-being at any given time; (iii) environmental circumstances, including household membership in relation to people and other animals; and (iv) relocation or periods of **stress** and bereavement. Where owner expectation is not met, either because the animal does not fulfil their requirements or the work involved in caring for the animal exceeds that anticipated by or available from the owner, the risk of the relationship breaking down is increased, since it is ultimately the human that controls the continuity of the relationship.

When the owner fails to understand or provide for the husbandry, behavioural and welfare **needs** of the animal, the risk of that animal experiencing physical and/or mental stress is increased. This may manifest itself in either physical illness or behavioural disturbance, which may strain the bond with the owner(s); rarely will a socially developed species like a **dog** leave its owner, but this may not be true for a species like a **cat**, with access to a free environment potentially including other humans.

Various scales have been developed in an attempt to measure the human–animal bond (e.g. Pet Attitude Inventory (PAI), Lexington Attachment to Pets Scale (LAPS), Pet Relationship Scale (PRS), Monash Dog Owner Relationship Scale (MDORS) and Companion Animal Bonding Scale (CABS)). These tend to evaluate the positive perceptions in the relationship, but there are difficulties or unanswered questions relating to their validity. These tend to relate to: species variation and the resultant differences in the type and quality of bond possible; asymmetry of the bond and the assessment of the perspective of both the human and other animals within the relationship; and **reliability**, construct **validity** and feasibility for practical application.

The effects of problem behaviour on the human–animal bond are of increasing interest to those working within the veterinary behavioural field. The specific influence of the development, management and overall impact of different behaviour problems on both the stability and success of the human–animal bond has been neglected historically and may provide useful insights into the difference between enduring versus failing human–animal bonds. Evaluation of animal behavioural problems may provide a useful guide to the evaluation of the bond from the animal's perspective, as these can develop in physiologically normal animals as a result of their inability to cope with the demands placed upon them by both their social and physical environments.

There is growing recognition within the veterinary profession that, in order to fulfil its role in providing care for the many species of companion animals, it must recognize and support the human–animal bond. This support should include provision of advice on animal husbandry, behaviour and welfare so that problems can be prevented as well as treated. Attention should be paid to the provision of emotional support when animals become ill, require surgery or hospitalization and, particularly, before and after the decision to euthanize an animal. Thus, veterinary clinics should be designed to provide facilities for all members of the family to visit, especially when animals are hospitalized, without disrupting their essential function. (CLC)

See also: Relationship

References

De Mello, L.R. (1999) The effect of the presence of a companion animal on physiological changes following the termination of cognitive stressors. *Psychology and Health* 14, 859–868.

Friedmann, E., Thomas, S.A. and Cook, L.K. (2007) A friendly dog as potential moderator of cardiovascular responses to speech in older hypertensives. *Anthrozoos* 20, 51–63.

Laurent, E.L. (2000) Children, 'insects' and play in Japan. In: Podberscek, A.L., Paul, E.S. and Serpell, J.A. (eds) *Companion Animals and Us: Exploring the Relationships between People and Pets*. Cambridge University Press, Cambridge, UK, pp. 61–89.

Humane societies

A humane society is an organization that aims to stop and prevent animal (and/or human) **suffering**. Animal-centred humane societies include such large organizations as the UK's Royal Society for the Prevention of Cruelty to Animals (**RSPCA**), the **Humane Society of the United States (HSUS)** and the World Society for the Protection of Animals. There are also a great many local humane societies that are not affiliated to any of the large organizations and that have similar aims but perhaps cover a much smaller geographical area or a target species. (JNM-F)

Humane Society of the United States (HSUS)

The Humane Society of the United States (HSUS) was founded in 1954 in order to promote the protection of all animals. Headquartered in Washington, DC, and with over 10 million members, the HSUS is the largest animal protection organization in the world, working primarily in the USA but also internationally to reduce animal **suffering** and create meaningful social change for animals by advocating public policy, investigating cruelty, working to enforce existing laws and educating the public. (JNM-F)

See also: Humane societies

Further reading

HSUS website: http://www.hsus.org/

Humanism

Humanism as a philosophy of life stresses the importance of living ethically through reason and conscientious compassion. Most forms of humanism are civic minded and place a high premium on active participation in social justice issues. Some variants are committed to improving the lives of animals and our relationship with nature. (RA)

See also: Ethics

Hunger

'Hunger' can be used to describe a craving, most often applied to the sensation of needing food following a period of non-feeding in order to achieve or maintain **homeostasis** and ideal body mass (total body fat stores plus circulating concentrations of energy substrates). Hunger was traditionally described as a 'homeostatic drive', a behaviour (under negative feedback control) that initiated feeding as a direct result of changes detected within the animal's physiology. The modern view now regards hunger as a '**motivation**' (in accordance with the idea of those specific drives once thought responsible for different aspects of behaviour being superseded by the notion of 'motivational state'), whereby behaviour is determined by the **brain** as a result of both internal motivation and the animal's perception of environmental stimuli (external causal factors such as smell and food proximity).

It is misleading to think of hunger as a unitary variable, measurable along a single scale (e.g. 'low'- or 'high'-level hunger), as the relationship between hunger and behaviour is not direct. There are many different aspects of hunger that may alter an animal's **feeding** behaviour – e.g. specific hunger for a deficient compound or anticipation of physiological changes demonstrated by the existence of habitual feeding times in animals provided with food *ad libitum*. The development of 'state–space representations of motivational systems' has proved advantageous in the portrayal of hunger as a multidirectional vector, with axes representing fat, protein, carbohydrate, etc.

The sensation of hunger, and its associated physiological state, can normally be temporarily alleviated by food consumption. The feeling of satisfaction and sensation of fullness following a meal, together with a lack of desire to eat, is termed 'satiety', the opposite of hunger. 'Appetite' is often used synonymously with hunger, yet may be considered a more hedonic term, encompassing a desire for food (perhaps certain food items) and the enjoyment of eating, and as such is widely influenced by psychological factors.

The physiological system controlling food intake (via hunger and satiety) is extremely complex, involving numerous internal and external factors, and comprising the means for both short-term (daily) and long-term (monthly) energy management. The currently accepted model is of a central feeding system whereby two major peripheral systems, a short-term satiety system (transmitting meal-related signals) and a long-term system (concerned with adipose tissue and overall energy balance of the body), convey information to the **hypothalamus** (Jensen, 2001).

The hypothalamus functions as an integrating relay centre in the control of all eating behaviour in both mammalian and non-mammalian vertebrates. It receives nerve impulses from other parts of the brain and from a variety of sensory receptors, including chemoreceptors, olfactory and taste receptors, and stretch receptors in the gastrointestinal (GI) tract. A cluster of **neuron(e)s** within the lateral portion of the hypothalamus has been termed the 'hunger centre', because stimulation of this area causes voracious feeding (even when an animal is full). A further cluster of neurons within the ventromedial hypothalamus has been termed the 'satiety centre', because stimulation of this region causes animals to stop eating, even if they have undergone starvation for several days. The 'hunger centre' appears to be constantly active but is subject to inhibition by the satiety centre. When food is freely available feeding initiation is likely to arise following a reduction in satiety signals and subsequent removal of feeding centre inhibition.

Short-term control of ingestion appears to be primarily regulated by neural signalling pathways emanating from the GI tract. Activation of stretch receptors (mechanoreceptors) as a result of gastric distension and chemoreceptors signals the presence of food within the GI tract and initiates a reflex that activates the satiety centre and depresses the feeding centre.

Many animals (excluding cats) increase their food intake when the rate of glucose utilization in the brain falls. The glucostatic theory of hunger predicts that the rise in plasma glucose concentration (and the rate of cellular glucose utilization) that occurs during eating (as ingested carbohydrate is digested and absorbed) is detected by glucose receptors in the brain and leads to cessation of hunger. Conversely, in the post-absorptive period, plasma glucose concentration and glucose utilization decrease, removing the signal to the brain glucose receptors (reducing satiety centre activity and removing feeding centre inhibition), and stimulating the individual to eat. A detailed review of the role of glucose in meal initiation is provided by Campfield and Smith (2003).

Several peptide hormones have been identified as playing a role in food consumption. **Cholecystokinin** (released from the stomach and duodenum during eating and responsible for stimulating fat and protein digestion), glucagon (released by the pancreas when blood glucose is low), **insulin** (synthesized in the pancreas and responsible for initiating glucose uptake and storage) and leptin (released from adipose tissue) all act to suppress appetite. Ghrelin, conversely, demonstrates orexigenic (appetite-stimulating) properties, and as such is termed the 'hunger hormone'. When administered experimentally it has been observed to increase food intake in both humans and many other vertebrate species (although interestingly not in sheep). Ghrelin appears to be involved in short-term energy balance and meal initiation, because plasma concentrations peak before a regular meal and rise during fasting. Levels fall during feeding, then progressively increase to another peak prior to the next meal. Ghrelin also seems to play a role in long-term regulation of energy balance, as chronic administration of ghrelin in rodents leads to hyperphagia (excessive hunger) and weight gain. It also appears to stimulate gastric acid secretion and regulate GI tract mobility. It is possible that hunger pangs (stomach rumbling) experienced during short-term fasting are due to increased gastric motility induced by ghrelin levels. Comprehensive reviews of this hormone and the many other afferent signals that regulate food intake are provided by Chaudri *et al.* (2006) and Higgins *et al.* (2007).

The mechanical activities of chewing and swallowing are known to influence the hunger centre. In dogs both oropharyngeal (tasting, swallowing) and gastric stimuli are necessary to induce satiety. The horse, on the other hand, appears to depend more on pregastric oropharyngeal signals as an indication of satiety than other species. Discussion of ingestive behaviour in the main domestic mammal species can be found in Houpt (1998).

Knowledge in this area of physiology has increased significantly over recent years, yet due to the complexity of the feeding mechanism it is highly likely that many signals, receptors (both location and nature) and the means of interaction and integration of the many component inputs remain, as yet, unidentified.

Although food intake is primarily controlled by a combination of receptors (glucose, mechano-, etc.), it is also strongly influenced by the reinforcement, both positive and negative, of such things as smell, taste and texture (food palatability). The smell of a favourite food can activate hunger even when, physiologically, we do not 'need' to eat. Animals fed to satiety on one kind of food will often continue feeding if given access to a wider range of foods. Wirtshafter and Davis (1977) demonstrated that rats given a normal 'rat food' diet and then provided with a variety of sweet and palatable foods (canteen diet) will eat a great deal more than usual and often became obese. Similarly, pigs show a marked preference for sweet substances and will eat more and gain weight when their food palatability is enhanced.

Many examples of nutritionally driven diet selection can be found in the literature (Forbes, 1995). Aversion to diets lacking sodium, calcium, thiamine, riboflavin and specific amino acids have all been demonstrated in preference to an alternative that contains the missing component. Salt hunger is a particularly well-documented phenomenon in both humans and herbivores subjected to a sodium-deficient diet.

Reproductive state has a noticeable effect upon food consumption in female animals. **Oestrus** (and associated oestrogen levels) correlates with a reduction in food intake in several species (e.g. pigs, dogs, cattle), whereas increased food consumption occurs during pregnancy/**lactation**. Food intake is influenced thermostatically by the regulation of body temperature in response to environmental conditions, a process mediated through the anterior hypothalamus. Animals generally eat more in cold weather because more energy is required to maintain body temperature and thus prevent hyperthermia. Psychological elements such as stress or depression, disease, etc. can also effect food consumption. When the body temperature rises to pathological levels (as occurs during fever) food intake decreases, while cancer sufferers often demonstrate a debilitating lack of hunger (cachexia).

Hunger is an easy concept to understand but a difficult sensation to quantify in animals. Operant conditioning tests have been successfully employed as a direct measure of hunger but have the drawback of being impractical for use in large-scale studies or on production lines. Increased incidence of **stereotypies** in sows (e.g. biting, sham chewing and nosing/licking the feeder/floor when food is not present) have been found to be closely associated with restricted feeding levels, thus supporting the use of this method as an indirect indicator of hunger and feeding motivation (see Meunier-Salaun *et al.*, 2001). Feed-restricted broiler breeders also demonstrate behavioural abnormalities characteristic of hunger, including high motivation to feed at all times, hyperactivity, stereotyped pacing before feeding time and stereotyped over-drinking and pecking at non-food objects after feeding.

Hunger appears to increase locomotor activity in many animals, including laboratory rats, and it is theorized that in the wild such a response may increase the animal's chance of finding food. Offspring of many animal species (including birds, mammals and insects) signal their nutritional needs (hunger level) to the parents by conspicuous vocal, visual or tactile begging displays. Cold and hungry piglets removed from the sow and placed in isolation vocalize more than control piglets, while calves vocalize significantly more in response to milk deprivation. Increased gakel-calling (a vocalization widely believed to express **frustration**) from hens occurs with an increasing hunger state (in addition to increased pacing and displacement preening). (GC)

See also: Malnutrition

References and further reading

Campfield, L.A. and Smith, F.J. (2003) Blood glucose dynamics and control of meal initiation: a pattern detection and recognition theory. *Physiological Reviews* 83, 25–58.

Chaudri, O., Small, C. and Bloom, S. (2006) Gastrointestinal hormones regulating appetite. *Philosophical Transactions of the Royal Society B* 361, 1187–1209.

Forbes, J.M. (1995) *Voluntary Food Intake and Diet Selection in Farm Animals*. CAB International, Wallingford, UK.

Higgins, S.C., Gueorguiev, M. and Korbonits, M. (2007) Ghrelin, the peripheral hunger hormone. *Annals of Medicine* 39, 116–136.

Houpt, K.A. (1998) *Domestic Animal Behavior for Veterinarians and Animal Scientists*, 3rd edn. Iowa State University Press, Ames, Iowa.

Jensen, J. (2001) Regulatory peptides and control of food intake in non-mammalian vertebrates. *Comparative Biochemistry and Physiology (A)* 128, 471–479.

McFarland, D. (1999) *Animal Behaviour: Psychobiology, Ethology and Evolution*, 3rd edn. Longman Scientific and Technical, London.

Meunier-Salaun, M.C., Edwards, S.A. and Robert, S. (2001) Effect of dietary fibre on the behaviour and health of the restricted fed sow. *Animal Feed Science Technology* 90, 53–69.

Wirtshafter, D. and Davis, J.D. (1977) Set points, settling points, and the control of body weight. *Physiology and Behaviour* 19, 75–78.

Hunting

Hunting is the chasing of an animal, usually with the intention of killing it. The reasons are various but include hunting an animal for its pelt (e.g. seal), for food (e.g. wild birds), population control (e.g. **rabbits**), to conserve a flock or herd (e.g. **deer** control) or other industries (e.g. seal hunts to protect fishing, rabbits for crop protection), for prestige (e.g. when killing the 'big five' **game animals** (African lion, **elephant**, **buffalo**, **rhinoceros** and leopard) and for 'sport'. Hunting may be carried out with **dogs** (e.g. hunting foxes, game birds where the dogs according to breed may point, set up, retrieve), by trapping (e.g. for pelts) or stalking (deer, the 'big five'). Killing the hunted animal usually involves the use of guns, but may also involve crossbows or other physical methods, or with some prey the dogs may kill the hunted animal. (DBM)

See also: Sport – use of animals in

Hyperalgesia

Hyperalgesia refers to a state of increased sensitivity to **pain** and may arise as a result of either local neural changes (e.g. damage to sensory nerves) or systemic factors (e.g. following **cytokine** production in response to disease). Hyperalgesia also refers to an increased area of pain, and is caused by a phenomenon of 'wind-up' of **neuron(e)s** in the dorsal horn of the spinal cord. Hyperalgesia differs from **allodynia** in so much as a painful **stimulus** is required to determine the former, unlike the latter, which may be triggered by non-painful touch. (DSM)

Hypothalamic–pituitary–adrenal (HPA) axis

The hypothalamic–pituitary–adrenal (HPA) axis refers to an organ system comprising the **hypothalamus**, the **pituitary gland** and the **adrenal gland**. The hypothalamus is an organ formed of many separate nuclei and lies in the **central nervous system** below the thalamus. The pituitary gland lies below the hypothalamus and the two are connected by a tract of **neuron(e)s** called the hypophyseal stalk. The adrenal glands are a pair of endocrine glands that are located beside the kidneys; they are comprised of a medulla and a cortex. The term 'axis' refers to the fact that these three organs are able to control one another by stimulating production of hormones, as well as by terminating the production of hormones.

The HPA axis has been the main focus of much research on the **stress** response; however, it is important to remember that the HPA axis is always active, and thus quantifiable measurement of any of its components does not indicate a definitive stress response. The function of the HPA axis is to help maintain **homeostasis** by regulating glucose availability. When an animal is challenged with a **stressor**, activation of the HPA axis is heightened and thus its products can serve as a physiological indicator of stress in animals. When a stressor is perceived, sensory input from the thalamus activates neurons in the paraventricular nuclei of the hypothalamus to release **corticotropin-releasing hormone** (CRH).

The axons of these nuclei terminate in the hypophyseal portal vessels, which bathe both the hypothalamus and the pituitary gland. Thus hormones entering this portal system allow communication between the two organs. Once in the portal blood, CRH binds to receptors on the plasma membrane of corticotroph cells in the anterior pituitary gland. Activation of this receptor causes the corticotroph cells to secrete adrenocorticotropic hormone (**ACTH**) and β-lipotropin into the peripheral vasculature. Binding of CRH to its receptor also causes the corticotroph cell to produce proopiomelanocortin (POMC), the precursor to several hormones including ACTH and β-lipotropin. Once in the blood, ACTH reaching the cortex of the adrenal gland binds to its plasma membrane receptor on cells in the *zona fasciculata* and the *zona reticularis*, two of three cell layers in the adrenal cortex. Activation of its receptor causes the production and release of **glucocorticoids** into the peripheral circulation.

In livestock, cortisol is the glucocorticoid that is produced in greater quantities, whereas in rodents and birds corticosterone is produced in greater quantities. Glucocorticoids in the blood are then bound to protein carriers which help to transport them and decrease their degradation. **Corticosteroid**-binding globulin binds cortisol and, to a smaller extent, cortisol is also bound by albumin. A certain percentage of the hormone remains unbound and it is this portion that is able to bind to its receptors in target tissue to cause an increase in glucose and free fatty acids. Free circulating glucocorticoid is able to feed back negatively on both the hypothalamus and the pituitary gland to decrease further production of cortisol or corticosterone. By binding to receptors in the paraventricular nucleus, cortisol decreases the production of CRH and, by binding directly to the corticotrophs in the anterior pituitary gland, ACTH production is decreased, thus decreasing the production and secretion of glucocorticoids.

One characteristic of the HPA axis that has made it useful in the determination of chronic stress in animals is that under chronic stress the cells of the *zona fasciculata* and *zona reticularis* in the adrenal gland will undergo hyperplasia and hypertrophy. This occurs in response to the heightened and constant stimulation of the adrenal gland by ACTH from the pituitary gland. In the cadaver, researchers can collect the adrenal gland

and look for size, weight and morphological differences to determine whether they have been subjected to chronic stress. Similarly, in the live animal, researchers can administer ACTH to the subject and measure the resultant concentrations of glucocorticoid to determine whether the adrenal gland is more capable of producing cortisol or corticosterone. If so, this would indicate that the animal has been exposed to a significant amount of stress to cause its adrenal glands to enlarge, although in some circumstances the gland may become hypo-responsive and produce a below-normal response.

Challenges to using circulating levels of glucocorticoids in the measurement of stress deal with hormone dynamics and varied causes for its elevated concentrations in the body fluids. Dependent upon the assay system used, glucocorticoids can be measured as total, bound or free concentrations. It is only the free concentrations that are available to the animal's body to use; however, the bound portion represents a pool of readily available hormone. Thus, measuring total hormone versus free hormone will present a different view of the welfare status of the animal. Another challenge is that glucocorticoids can be elevated in situations that are not thought to be aversively stressful to an animal. For example, glucocorticoids have been shown to be elevated in animals during copulation, **play** and exercise. Objective measurement of stress in animals requires a multidisciplinary approach to increase the confidence in asserting when an animal is stressed and its **welfare** diminished. (DCL)

Hypothalamus

The hypothalamus is the part of the **brain** that is located below the thalamus and above the brain stem, forming the ventral part of the diencephalon (**see: Central nervous system**). Among other functions it provides a link between the nervous system and the endocrine system via the **pituitary gland**. The hypothalamus influences activity within the **autonomic nervous system** and is responsible for synthesizing and secreting neurohormones (hypothalamic-releasing hormones – **corticotropin-releasing hormone (CRH)**, **gonadotropin-releasing hormone (GnRH)**, **growth hormone-releasing hormone (GH-RH)**, growth hormone-inhibiting hormone (GH-IH), prolactin-releasing hormone (PRH), prolactin-releasing inhibiting hormone (PRIH), thyroid-releasing hormone (TRH)), which either inhibit or stimulate the secretion from the pituitary gland of other hormones. The hypothalamus also plays an important role in regulating body temperature (**see: Thermoregulation**), **hunger**, thirst (**see: Drinking behaviour**) and **circadian rhythms**. (LMD)

I

IACUC

IACUC stands for the Institutional Animal Care and Use Committee and is a requirement under US federal law for all institutions that use **laboratory animals**. The committee's role is to oversee and evaluate the institution's animal care policy and use of animals. (DSM)

Ideal free distribution (IFD)

The ideal free distribution (IFD) is a model of how animals distribute themselves between patches/habitats in relation to resource availability. The model assumes that animals behave in a way that optimizes some aspect of their day-to-day behaviour (usually foraging). The model also assumes that animals have ideal knowledge about resource availability and are free to move between patches (e.g. they are not prevented from accessing a resource by despotic competitors). A key prediction of the IFD is the resource-matching rule: essentially, animals will distribute themselves in proportion to resource availability. For example, should patch A contain twice the amount of resource of patch B, twice as many animals will settle in patch A than in patch B. (PE)

Imitation

In everyday language, the word imitation is used to describe any instance of copying the behaviour of others. However, within the field of animal behaviour, imitation refers to a very specific method of learning. Imitation takes place when one individual observes another (typically a **conspecific**) and reproduces a relatively faithful copy of their behaviour, including both their actions and the results caused by those actions. Imitation is one of several methods of learning collectively known as **social learning**. Social learning incorporates all learning that is influenced by the behaviour of others, and is distinct from individual learning (such as trial and error), which is not affected by others. Learning from others can be advantageous because it allows new behaviours to be learned faster, and with less potential risk than trial and error. For example, a young bird may learn which trees contain edible – rather than poisonous – berries by learning where other birds feed. The bird therefore saves energy and reduces risk because it does not have to fly to every tree and sample every berry.

Social learning includes all methods of learning that result from either interacting directly with another, from observing the behaviour of another or even from interacting with the left-over products of another individual's behaviour. For instance, black rats (*Rattus rattus*) can learn to efficiently extract seeds from pine cones if they are simply given pine cones that have already been opened by another rat. Put simply, all methods of social learning are influenced in some way by the behaviour of others.

Historically, imitation was viewed as a relatively simple form of learning that was widespread among animal species. However, recent studies suggest that imitation is a cognitively complex process (see below) that is not as prevalent as once thought. In humans, imitation is thought to play a fundamental role in language acquisition, learning about objects, acquiring cultural behaviours as well as identifying with, and understanding, the minds of others (**theory of mind**). For this reason many researchers have focused on studying which non-human species have the capacity to imitate.

In order to identify imitation, it must first be distinguished from a variety of other methods of social learning, such as **goal emulation** or **local enhancement**. Each method differs in how faithfully an observer reproduces the behaviour of another (the model). However, imitation is the only form of social learning whereby an observer learns to reproduce the model's actions in addition to the results of those actions. Hence imitation results in the highest fidelity match to the model's original behaviour, and can be identified by copying actions and results.

Imitation can be investigated using a 'two-action' paradigm. Two-action paradigms involve a task that can be solved by using one of two different actions (A or B). One example is an 'artificial fruit' that has been used to show that **chimpanzees** (*Pan troglodytes*) can imitate (see Fig. I.1). This artificial fruit is a box containing food. The lid of the box is held closed by two bolts that must be removed before the lid can be opened. The

Fig. I.1. Photograph of a chimpanzee pulling the bolts out of the 'artificial fruit' in order to open the lid and retrieve a food reward. Photograph reproduced with kind permission from A. Whiten.

bolts can be removed by either pushing them (action A), or pulling them (action B) out of their holders (see Fig. I.2). To test for imitation, chimpanzees are divided into two groups. One group observes a model pull the bolts out to open the lid, while the second group observes a model push the bolts out. Observers from each group are then given an opportunity to open the box on their own. If each group uses the same action as their respective model (push or pull) significantly more than the alternative to achieve the same result (open box), this indicates imitation. Two-action paradigms provide a particularly useful tool for studying imitation because they can be adapted for use with a wide range of species.

There are, however, disagreements about the details of how two-action tests are conducted. Some argue that simply copying the same action (pull or push) is not enough – observers must use the same body part as the model to perform the action. For example, marmosets (*Callithrix jacchus*) that observed a demonstrator use their mouth to remove the lid of a food container also used their mouth when opening the container themselves. Marmosets that observed a model remove the lid with their hands used their hands instead. There are further debates about how novel the two alternative actions must be, and about whether or not an observer must understand the intentions of the model. Nevertheless, despite these questions, the two-action paradigm is generally accepted as the standard way to test for imitation.

Imitation is typically studied in captivity because controlled experimental conditions are necessary for two-action tests (**see: Laboratory study**). Such studies have revealed that imitation can be affected by the relationship between the model and the observer. Observers are less likely to imitate models that are aggressive towards them than those that are tolerant of being observed. Hence imitation, and social learning in general, may be more prevalent in tolerant than despotic animal societies. Similarly, evidence for imitation may be affected by the difficulty of the task. If a problem is too easy, an individual may not need to imitate because they can solve the task on their own. Hence it has been suggested that the strongest evidence for imitation comes from reproduction of arbitrary actions, rather than from solving puzzles. Chimpanzees and **orang-utans** (*Pongo pygmaeus*) taught to copy human actions on command (like the game 'Simon says') were able to imitate completely novel gestures.

Imitation is believed to be cognitively complex because an observed behaviour (visual input) is converted into action (motor output). Several theoretical frameworks have been proposed to explain this process, the best known being the 'active intermodal mapping' hypothesis, which suggests that

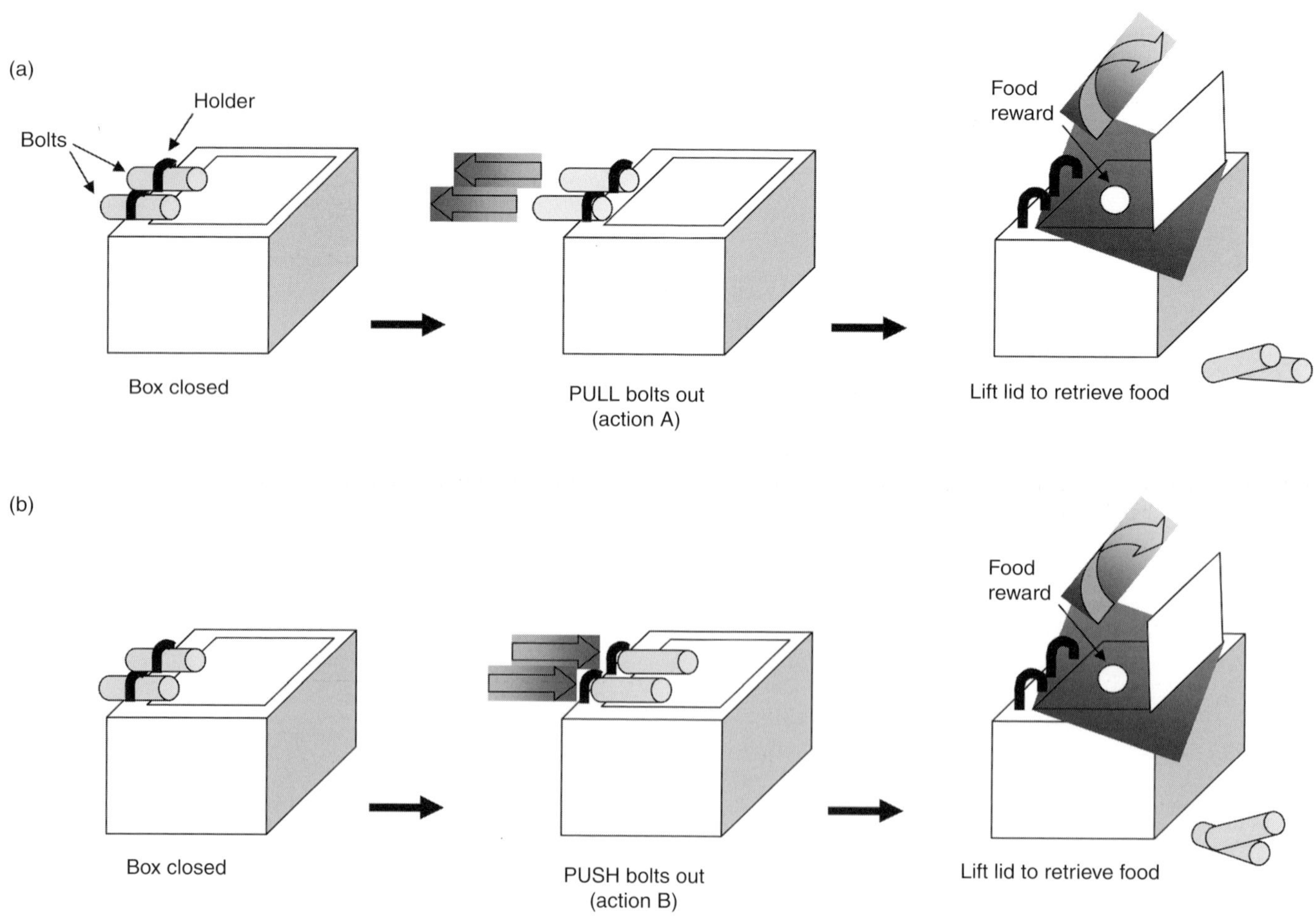

Fig. I.2. Schematic representation of the two alternative actions used to open the 'artificial fruit' two-action paradigm used to demonstrate imitation in chimpanzees: (a) pull bolts out to open lid and retrieve food; or (b) push bolts out.

imitation is possible because an individual can determine when its own body movements match observed actions via a perceptual feedback system. Such hypotheses are supported by the recent discovery of specialized **neuron(e)s** in the inferior frontal lobe of primates, known as mirror neurons, which appear to be specialized for behaviour matching. These neurons are thought to be involved in imitation because they have both sensory and motor functions, and are activated when an individual performs a behaviour (such as reaching for an object) as well as when observing the same behaviour performed by another. Further studies will be needed to determine whether similar neurons exist in non-primate species.

Vocal imitation has typically been treated separately from action imitation because it is thought to involve different cognitive pathways that match auditory and vocal domains rather than visual and motor domains. The debate over the relationship between vocal and action imitation remains unresolved. (VH)

See also: Cognition; Comparative psychology; Social behaviour

Further reading

Heyes, C.M. (1994) Social learning in animals: categories and mechanisms. *Biological Review of the Cambridge Philosophical Society* 6, 207–231.

Hurley, S. and Chater, N. (eds) (2005) *Perspectives on Imitation: from Cognitive Neuroscience to Social Science*. MIT Press, Cambridge, Massachusetts.

Whiten, A., Custance, D.M., Gomez, J., Teixidor, P. and Bard, K.A. (1996) Imitative learning of artificial fruit processing in children (*Homo sapiens*) and chimpanzees (*Pan troglodytes*). *Journal of Comparative Psychology* 110, 3–14.

Whiten, A., Horner, V., Litchfield, C.A. and Marshall-Pescini, S. (2004) How do apes ape? *Learning and Behavior* 32, 36–52.

Zentall, T. (2001) Imitation in animals: evidence, function, and mechanism. *Cybernetics and Systems* 32, 53–96.

Immune system

The primary functions of the immune system are, first, to prevent disease by preventing pathogen entry and colonization and, secondly, to clear ongoing infections. The immune system is not a static system. It is now believed that a healthy immune system is one that has been exposed to antigens throughout development, thereby developing an adequate education of self and non-self and immunological memory. Some aspects of the immune system are well conserved among species, but occasional pronounced differences can be observed. This article will focus on general immunology, with some reference to species-specific differences.

The six common classes of internal disease-causing organisms are: viruses, bacteria, fungi, protozoa, nematodes and helminths (the latter two are varieties of worm). The latter three constitute the discipline of parasitology and the first three are within the discipline of microbiology. The efficiency of the immune response is relative to the disease with which it is being challenged, and the immune system is arbitrarily divided into two systems, innate immunity and adaptive immunity. Some disease challenges are controlled best by an innate immune response, while others are controlled best by adaptive immunity. Adaptive immunity requires prior exposure to antigen, while innate does not. However, these two systems are closely connected, and are derived from the same cell source.

The cells of the immune system originate from the bone marrow. Myeloid progenitor cells are the precursors for monocytes, dendritic cells, mast cells and granulocytes. Lymphocytes (derived from lymphoid progenitor cells) mature in the bone marrow or the thymus (see Table I.1).

Monocytes in the blood mature into macrophages as they migrate into the tissues. Some become specific and are named to reflect the tissue in which they reside; macrophages in the liver are Kupffer cells, while those residing in the brain are termed microglial cells, and those residing in the kidney are intraglomerular mesangial phagocytes. Those residing in the lungs are simply termed alveolar macrophages. Macrophages serve multiple roles. They are crucial in both innate and adaptive immunity.

Dendritic cells are highly specialized in their role in recognizing and presenting antigen to lymphocytes in the lymph tissues. Immature dendritic cells travel from the blood into tissues, where they continually sample the environment. Once they encounter a pathogen, they quickly mature and migrate to lymph nodes.

Mast cells are bloodborne, but differentiate as they move into tissue. They primarily reside near small blood vessels and, when activated (through immunoglobulin E cross-linking), secrete substances that cause vascular permeability. They are renowned for this role during allergic responses.

The granulocyte population is also known as polymorphonuclear (PMN) leucocytes, having a distinctive, multi-segmented nucleus. These cells are full of densely staining, granule-containing lysosomes, which are used for bactericidal activity. Neutrophils, eosinophils and basophils are members of the granulocyte population. Neutrophils are possibly the most important cells of innate immunity. Hereditary deficiencies of neutrophils lead to fatal bacterial infections. They are the third phagocytic cell of the immune system (besides macrophages and dendritic cells), but function only in innate immunity. In birds, analogous cells are called heterophils. Eosinophils have dual roles; they are believed to be important during parasitic infections because of observed increasing numbers. However, they also play a role in allergic responses. Basophils function similarly to mast cells and eosinophils, with their primary role in allergic responses.

Lymphocytes originate from bone marrow. There are two major types of lymphocytes: B-lymphocytes (B-cells), which differentiate into antibody-producing plasma cells when they are activated, and T-lymphocytes (T-cells), which can be

Table I.1. Cells of the immune system.

Lymphoid progenitor cells	Myeloid progenitor cells
B-cells	Monocytes/macrophages
T-cells ($\alpha\beta$ and $\gamma\delta$)	Dendritic cells
NK-cells	Mast cells
	Granulocytes
	Neutrophils (polymorphonuclear leucocytes, PMN)
	Eosinophils
	Basophils

divided into two classes – one is the cytotoxic T-cell and the other is the helper T-cell, which activates other cells such as the B-cell and macrophages. Lymphocytes mature in the bone marrow or the thymus.

Primary (or central) lymphoid organs, where lymphocytes are generated, are the bone marrow and thymus. B- and T-lymphocytes originate from the bone marrow, but only the former mature there. T-lymphocytes migrate to the thymus, where they mature and develop the ability to discriminate self from non-self. The secondary lymphoid organs, which are widely distributed throughout the body, are where adaptive immune responses are initiated and where lymphocytes are maintained. It is to the secondary lymphoid organs that dendritic cells and macrophages carry their antigens to present to B- and T-cells to start an immune response. The secondary lymphoid organs include lymph nodes, the spleen and gut-associated lymphoid tissues (GALT). Included in GALT are the tonsils, adenoids, the appendix and specialized structures in the small intestine called Peyer's patches.

Leucocytes move among the tissues and organs and through the lymphatic and circulatory systems. Chemotaxis to a site of inflammation is initiated by chemokines (**see also: Cytokines**) and by peptides produced by neutrophils. Adhesion molecules on the cell and integrins on the endothelial surface become more tightly bound as the concentration of the chemoattractant increases. This leads to what is termed rolling and, eventually, diapedesis (movement of the cell through the blood vessel by squeezing between endothelial cells).

Innate immune response

The immune system begins with physical barriers. Skin and mucosal surfaces serve to keep pathogens at bay. The respiratory tract, gastrointestinal tract and urogenital system comprise the mucosal immune interface. The mucosal immune system is where the host most frequently encounters pathogens. Mucosal immunity must not only recognize pathogens, but it must also be capable of discriminating commensal microbial populations. Within the mucosa are immune cells that interact immediately with pathogens. Multi-fenestrated cells (M-cells) 'sample' the lumen of the gastrointestinal tract and interact with subepithelial lymphocytes and dendritic cells. The dendritic cells take the antigen to lymph tissues. Gamma-delta (γδ) T-cells are a subset of lymphocytes that are prolific in the gut mucosa. This subset of cells varies in phenotypic markers, predominant location within the body, abundance and function among species and age of animal within species, but generally has an innate immune function, requiring no previous experience with the antigen.

Cells of the innate immune system recognize foreign antigen without prior exposure and can trigger an appropriate immune response. Subsets of the innate lymphocyte population include the γδ T-cell and natural killer (NK-) cells. Both the NK-cells and γδ cells are phenotypically and functionally different among species. For instance, NK-cells are a prominent and important innate immune cell in humans and pigs, but appear to be inefficient at lysing tumour cells unless they have been activated by cytokine in cattle. However, they still appear to be efficient at lysing virally infected cells. Additionally, the bovine NK-cells are phenotypically distinct from NK-cells of other species. γδ cells comprise a large proportion of circulating lymphocytes, up to 30% of peripheral blood mononuclear cells (PBMC) and afferent lymph cell populations, and there is typically a greater percentage of PBMC in young calves than in mature cattle. In contrast, young piglets have only 3% γδ cell population in the blood, but greater percentages in the GALT. NK-cell activity is developed postnatally in pigs as opposed to prenatally in many other species, so the neutrophil is a very important innate immune cell in the neonatal piglet.

Neutrophils are phagocytic cells that ingest and kill foreign antigen without prior exposure (see Fig. I.3). Particles that have become opsonized (coated with antigen-specific antibody) are ingested much more efficiently. Neutrophils are a predominant source of antimicrobial peptides. These are peptides produced by neutrophils that use several mechanisms, to either destroy microbes or inhibit their replication through molecular intervention.

Convergence of innate and adaptive immune systems

Macrophages and dendritic cells are two important phagocytic and antigen-presenting cells. They interact with T-cells to initiate an immune response that will lead to cytotoxic T-cell activity or to antigen production. This interaction occurs at the T-cell receptor. A 'self' signal, termed major histocompatability complex (MHC), is generated by the macro-

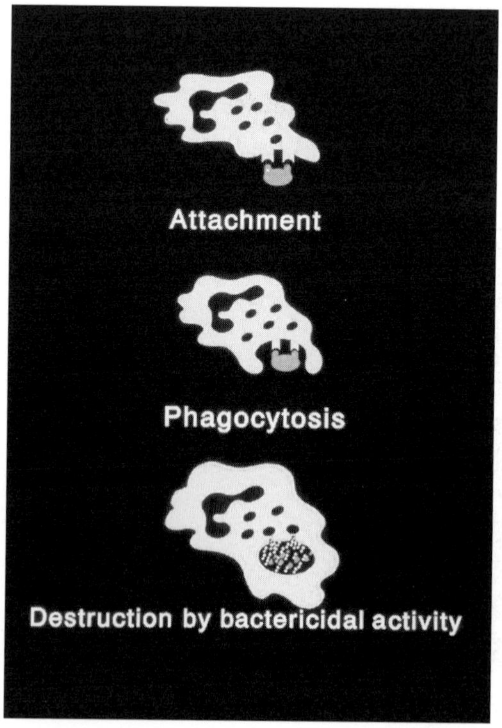

Fig. I.3. The phagocytosis process by neutrophils begins with attachment of the neutrophil, sometimes mediated by binding of the fixed constant (Fc) region of the antibody to the Fc receptor on the neutrophil. Phagocytosis ensues, with the phagocyte membrane surrounding the pathogen. The vesicle of the bound pathogen is called a phagosome. Lysosomes (vesicles containing enzymes, proteins, peptides and oxygen and nitrite intermediaries) fuse with the phagosome, creating a phagolysosome where the pathogen is destroyed by an oxidative burst.

phage and presented on the cell surface with the antigenic peptide and is required for this signal to be complete (see Fig. I.4).

Macrophages are frequently the first cells to encounter pathogens. Macrophages carry their cargo to lymph nodes for presentation to T-cells. They release a number of cytokines and chemokines that initiate an appropriate immune response and recruitment of accessory cells. The receptors that signal the presence of pathogens also upregulate co-stimulatory molecules on both macrophages and dendritic cells, leading to the adaptive immune response. Macrophage interaction with T-cells leads to cell signals and cytokine production (Interferon (IFN)-γ) that activates the macrophage to become bactericidal through nitric oxide, oxygen radical and proteases. As could be expected, this process can be destructive to surrounding tissue, and is therefore tightly controlled by the encoding of mRNA for IFN-γ with a sequence that reduces IFN-γ's half-life.

Macrophages and dendritic cells differ in initial location, in trafficking throughout the body and by the type of cytokine that they produce. Langerhans cells are dendritic cell precursors found in the skin. Immature dendritic cells have a different appearance than those that are activated and have begun translocation to the secondary lymph tissues. The immature dendritic cell has low MHC protein expression and lacks co-stimulatory molecules necessary to interact with T-cells. In transit to lymph tissues, they stop phagocytizing antigen and their morphology changes. The lysosomal proteins and MHC class II molecules become distinctly separate. In the lymph nodes, dendritic cells express high levels of MHC–peptide complexes and co-stimulatory molecules. It is here that the innate and adaptive immune systems converge.

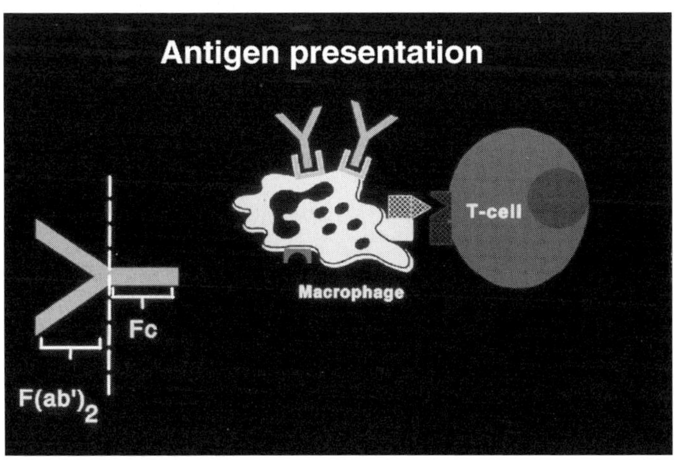

Fig. I.4. Antibody structure and antigen presentation by macrophage to a T-cell via the T-cell receptor and the major histocompatability complex (MHC). The antibody depicted on the left has a fixed constant region (Fc) and a variable region ($F(ab')_2$), which is the antigen recognition and binding component. Macrophages contain Fc receptors to facilitate binding and engulfing of opsonized pathogens. After processing within the macrophage, selected peptides of the antigen are expressed on the surface of the macrophage along with the 'self' antigen (MHC). This complex is recognized by the T-cell receptor, leading to activation of the T-cell and macrophage and cytokine production.

Dendritic cells reside in T-cell areas and are potent stimulators of naive T-cells. Both dendritic cells and macrophages express toll-like receptors (TLRs) (pattern recognition molecules) that discriminate pathogen-associated molecular patterns (PAMPs) of various pathogens. TLRs are expressed on the cell surface. Each receptor or heterodimer recognizes discrete PAMPs (see Fig. I.5).

Although the exact agonist of TLR7 and TLR8 has not been identified, these receptors have been shown to mediate activation of antiviral immunity. Following stimulation of the TLRs, a cascade of signals begins leading to the production of inflammatory cytokines, such as TNF-α and IL-6, and upregulation of the co-stimulatory molecules CD80 and CD86 on dendritic cells. Several accessory molecules have been identified that are required for cell activation – MD2 and CD14, which associate with TLR4, and the cytoplasmic factor MyD88, which mediates binding to intracellular domains. A second co-factor, TIRAP (also called MAL), is essential for MyD88 to mediate induction of an important cell signal.

TLRs play a vital role in the adaptive immune response; they are necessary for priming TH1 and TH2. Signalling through TLR4 has been demonstrated to induce both TH1 and TH2 responses, depending on the dose of LPS used. The lower expression of TLR1 was correlated with low antibody responders compared with normal responders following vaccination with *Borrelia burgdorferi* (the causative agent of Lyme disease).

Once macrophages have been activated by TLRs, the pathogen is internalized and the peptides are expressed on the cell surface in conjunction with MHC I or II, depending on the type of antigen that is being presented. Virus particles are presented with MHC class I molecules and bacterial particles are presented with MHC class II. T-cells express receptors specific for the MHC and antigen. Cytotoxic lymphocytes (CD8+) interact with MHC class I-expressing macrophages. These cells produce cytokines and enzymes for efficient cell-killing activity. Conversely, T-cells with a receptor specific for the antigen and the MHC class II antigen secrete cytokines that enable antibody production and clearance of the organism by those routes. These include both the TH1 and the TH2 subsets. It is at this point that the type of cytokines present in the milieu can affect the type of effector T-cell that is generated.

In the lymph node, dendritic cells that have encountered viruses and some bacteria induce NK-cells to produce IFN-γ by IL-12 production. This results in a CD4+ T-cell that secretes IL-1, TNF-β and IFN-γ. With other stimuli such as parasites, the NK-cells produce IL-4 and the dendritic cells are not stimulated to produce IL-12. This scenario results in a CD4+ cell that produces IL-4, IL-5 and IL-13. Each of these subsets produces cytokines that act negatively on the opposite subset. To complicate the issue, the differentiation of CD4+ subsets is influenced by the amount, affinity and avidity of the peptide. CD4 T-cells presented with low levels of a poorly binding ligand will differentiate preferentially into TH2 cells that produce IL-4 and Il-5. This makes them most effective in stimulating naive B-cells to differentiate into plasma cells and making antibody. In contrast, T-cells presented with high levels of ligand with strong binding to the T-cell receptor secrete IL-2, TNF-β and IFN-γ and are effective in activating macrophages.

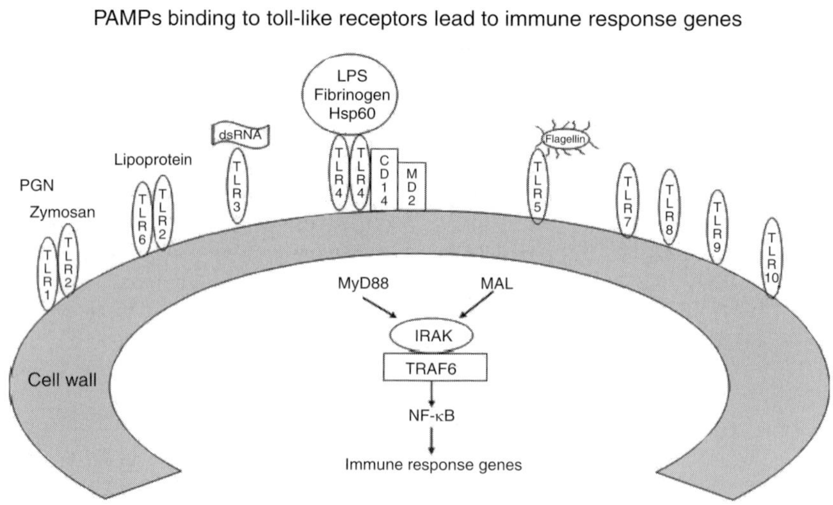

Fig. I.5. Toll-like receptors 1 to 10 are expressed on the cell surface. They recognize specific pathogen-associated molecular patterns (PAMPs, such as: (i) peptidoglycan or lipoproteins from Gram-positive bacteria; (ii) zymosan from yeast cell walls; (iii) double-stranded (ds) RNA; (iv) lipopolysaccharide (LPS) from Gram-negative bacteria; (v) fibrinogen (an acute phase protein) and heat shock proteins (Hsp); (vi) flagellin from bacteria; and (vii) CpG, a component of bacterial DNA). The PAMPs of TLR7 and 8 have not been defined but are believed to be involved in viral recognition. Binding of these receptors leads to activation of cytoplasmic proteins involved in signal transduction, leading to production of appropriate cytokines (immune response genes).

Adaptive immunity

The antibody is composed of a fixed region that is recognized by phagocytes (see Fig. I.4.) and a variable region that is flexible and conforms to hold the antigen peptide. Five isotypes of antibody are recognized, including immunoglobulin D (IgD), IgA, IgG, IgM and IgE. Subclasses of IgG vary by species, but four are recognized in humans. Additionally, while the major isotype classes are consistent among species for function, the subclasses are not. Very little immunoglobulin D is produced at any time. B-cells express both IgD and IgM on their surface. Immunoglobulin A and M form polymers that are unique to their functions. The repetitive binding site of IgM increases its avidity (ability to remain bound to the antigen), possibly because of its weaker affinity (strength of the binding) for each of those sites. IgE is in very low concentrations, but is responsible for tight binding to mast cells that can trigger the release of chemicals that cause sneezing, coughing and vomiting. IgA is the predominant form of antibody contained in colostrum and therefore passive immunity; however, IgG is also transported and plays a vital role. Polymerization of IgA is necessary for its transport through epithelia. IgG is the predominant circulating isotype that is involved in the presentation of antigen to phagocytes.

Passive immunity is the transfer of antibody from a maternal source to either a **fetus** or neonate. Antibody transfer in humans is through the placenta; however, many farm animals receive maternal antibodies through colostrum. Colostrum is the initial lactation secretion produced after **parturition**. It is high in energy, nutrients and antibodies compared with milk secretions that ensue within the first week of lactation. The neonate has a small window of opportunity for optimum antibody absorption: the ability to absorb the antibodies begins to decline as early as 12 h after birth. After 24 h the colostrum can continue to have a limited protective effect by interactions with pathogens within the lumen of the gastrointestinal tract.

There are three functions of antibodies. The first is neutralization. Many bacteria injure cells by secreting toxins. The neutralization function of antibodies is to recognize and keep the toxin from binding to cells. The second function is action by opsonization. This is the coating of an antigen with antibodies or complement. The Fc portion of the antibody is then bound by phagocytes, such as neutrophils, and the antigen ingested and killed. The third function of antibodies is through activation of plasma proteins known as complement.

Complement is a unique system that not only functions as an effector arm for antibodies, but can also act independently with activation early in infection. It comprises a number of plasma proteins that work in a triggered-cascade system. There are three distinct pathways for complement activation. Complement works to provide its protective effects through: (i) covalent binding to pathogens, allowing them to be engulfed by phagocytes; (ii) some of the smaller, cleaved components from the cascade acting as chemoattractants for phagocytes and inducing activation; and (iii) the terminal complement component damaging the cell walls of bacterial membranes by creating pores.

The immune system is a complex system, but with a relatively small number of types of cells involved. It begins with basic innate immune responses, these eventually leading to interaction with cells of the adaptive immune system. Through these interactions of cells, cytokines and proteins, immunity is developed against specific antigens. Finally, memory cells are left that can mount a rapid antibody response if the antigen is encountered again. (SE)

Further reading

Imler, J. and Zheng, L. (2004) Biology of toll receptors: lessons from insects and mammals. *Journal of Leukocyte Biology* 75, 18–26.

Janeway, C.A., Travers, P., Walport, M. and Schlomchik, M. (2001) *Immunobiology: the Immune System in Health and Disease*, 5th edn. Garland Publishing, New York.

Kabesch, M. and Lauener, R.P. (2004) Why Old McDonald had a farm but no allergies: genes, environments, and the hygiene hypothesis. *Journal of Leukocyte Biology* 75, 383–387.

Pastoret, P., Griebel, P., Bazin, H. and Govaerts, A. (1998) *Handbook of Vertebrate Immunology*. Academic Press, San Diego, California.

Rook, G.A.W. and Stanford, J.L. (1998) Give us this day our daily germs. *Immunology Today* 19, 113–116.

Imprint training

While **imprinting** refers to the learning process whereby a neonate may follow the first moving object it sees during a **critical period** of development, its relevance to **horses** is less clear. In the past 20 years or so, the term imprinting or imprint training has been extended to represent an equine training protocol conceived by Robert Miller. Miller (1991) claimed initially that the neonatal foal was subject to a similar critical period and that this facilitated a range of related training exercises. He subsequently acknowledged that the training process is not related to the imprinting process, and recent work suggests that the foal may learn from the response of the mother to people during handling (Henry *et al.*, 2003).

As a **precocial** neonate, the newborn foal must get to its feet as rapidly as possible and be capable of fleeing from predators and learning all the concomitant behaviours necessary for escape and the features of their world that are not threatening. So, it seems reasonable to train the newborn foal to react appropriately and at a time in its life when there is a lower chance of injury to both horse and human as a result of a **flight behaviour** response.

Miller's programme involves serial **habituation** of the foal, followed by training distinct cues that elicit specific responses. The foal's body is manipulated by the handler, a process Miller describes as 'stimulations'. He prescribes multiple (30 to 50) stimulations of each body area. Imprinting begins with the handler drying the newborn foal. Handling starts at the foal's head and continues in a preset pattern all over the body, moving from one area to the next when the foal relaxes (habituates). All areas of the body are covered, including ear canal, nostrils, mouth, perineum and external genitalia.

The next phase involves introducing various devices, such as clippers and a rectal thermometer. The imprinting programme continues with the introduction of head collars, and a desensitization process in which the foal is accustomed to crackling plastic (again, until it shows no fear), gunshots, loud music and many other potentially aversive stimuli. On the following day the foal, adjacent to its dam, is re-exposed to the desensitization process, but this time for no more than 15 min. The next stage involves restraint, where the handler's arms envelop the foal's thorax, squeezing rhythmically until habituation occurs.

The merits of these exhaustive interventions have been questioned by recent studies of early equine handling. The existence of a discrete early critical period in horses has been challenged (Williams *et al.*, 2003). Equine ethologists have suggested that a training window may exist around **weaning** time, which is usually at approximately 6 months of age, when the animal is in a state of continual heightened **arousal**. This state offers the potential for rapid learning (Bateson, 1979).

There are potential risks in handling neonatal foals, since the mare–foal relationship could be damaged, the handler could be injured by an aggressive or protective mare, foals may be injured if they struggle and, later in life, the 'imprinted' animals might present a danger to humans because they become targets for play or **sexual behaviour**. Imprinting of horses remains an area in which more research is needed. While it seems clear that Miller was wrong in his initial claim that his regime was a form of imprinting, there may be some justification for some limited early intervention. (AM)

References

Bateson, P. (1979) How do sensitive periods arise and what are they for? *Animal Behaviour* 27, 470–486.

Henry, S., Hemery, D., Richard, M. and Hausberger, M. (2003) Human–mare relationships and behaviour of foals toward humans. *Applied Animal Behaviour Science* 93, 341–362.

Miller, R.M. (1991) *Imprint Training of the Newborn Foal*. Western Horseman Inc., Colorado Springs, Colorado.

Williams, J.L., Friend, T.H., Collins, M.N., Toscano, M.J., Sisto-Burt, A. and Nevill, C.H. (2003) Effects of imprint training procedure at birth on the reactions of foals at age six months. *Equine Veterinary Journal* 35, 127–132.

Imprinting

Imprinting refers to the ontological learning process whereby an animal identifies an object (**see: Bonding – parent–offspring**) and subsequently follows and responds to it in preference to other stimuli. In **precocial** birds, such as chickens, ducks and geese, imprinting is largely a response to visual stimuli, so the first moving object seen during the so-called '**critical period**' (between 26 and 32 h of age) is imprinted and followed throughout the juvenile period. Furthermore, when the imprinted bird is sexually mature, it will attempt to mate with objects similar to the one it imprinted upon (Immelmann, 1972). This learning, first identified by **Konrad Lorenz**, is virtually irreversible so, clearly, imprinting has evolved to promote survival of newborn animals and to shape their future **reproductive behaviours**. Sound and **olfaction** are also involved in imprinting, with the importance of each sense depending on the species.

Knowledge gained through imprinting is retained for life and difficult to reverse. It establishes an individual animal's preference for a certain species. An inappropriately imprinted neonate will prefer to follow the learned stimulus (such as a human or even a balloon or a stick) rather than a member of its own species.

Not all behaviours are affected by imprinting. In sexually dimorphic species, imprinting of mate choice is sex specific. A male chicken will search for attributes in a mate that are similar to those of his mother, whereas a female will not. Interestingly, sexual responses in human-imprinted falcons are evoked only with a combination of human and avian stimuli.

Imprinting seems more important in precocial species, in which the offspring are less dependent on their mothers for food and warmth, than in **altricial** species, which keep their vulnerable, and often nude, young in nests or burrows. Altricial neonates are unlikely or unable to stray from their refuges in the first few days of life, so imprinting is unnecessary. Instead, they learn similar lessons later in life during what are known as '**socialization periods**'. Once the animal's sensory, motor and thermoregulatory systems are functional, they learn to move away from their mother and interact with **conspecifics** and other species. The socialization period varies according to the species. In dogs it is from 3 to 14 weeks and in cats 2 to 7 weeks, while in primates it is usually 6 to 12 months. Stimuli that the youngsters of each species are exposed to during these periods will be accepted as 'normal'.

Aversive experiences during imprinting create a more robust imprinting response than usual. When obstacles are placed in the runway between a duckling and the imprinted object, the following response is more energetic and determined. It is likely that this feature of imprinting is an adaptive anti-predator response. (AM)

Reference

Immelmann, K. (1972) Sexual and other long-term aspects of imprinting in birds and other species. *Advances in the Study of Behavior* 4, 147–174.

Inbreeding

Inbreeding is often used to describe the mating of closely related organisms. The result is that offspring tend to be homozygous, i.e. to have inherited copies of the same gene from each parent. Thus, it has been used by animal breeders to fix desirable inherited characteristics in breeds, i.e. for making a breed of domesticated animals uniform in appearance or performance. An inbred animal is one that has inherited from both parents copies of a piece of DNA belonging to an ancestor that is in the pedigree of both parents. The inbreeding coefficient is a measure of how likely it is that a particular pair of alleles is identical to that of an ancestor – put another way, it is the proportion of the gene pairs of an individual that are identical.

It is usually a disadvantage for organisms to have inherited identical copies in this way, in other words to be homozygous – if the gene is defective the organism will be at a disadvantage. If the organism has received a different copy from each parent, it will be heterozygous and, if one of the gene pair is defective, this will often be masked by the other, functional, one.

Homozygosity means that an organism is genetically less well equipped to deal with varying circumstances. This need not be a problem with **farmed animals**, where the art of husbandry is to remove environmental constraints on growth and production. But in the wild, homozygous organisms tend to be less viable than non-inbred (outbred) individuals and, as would be predicted, many species have evolved mechanisms to restrict inbreeding. To do this, animals must be able to recognize their relatives. The mechanism that is currently the strongest candidate as an indicator of relatedness is based on the set of genes known as the major histocompatibility complex (MHC). Laboratory mice clearly prefer to mate with individuals whose MHC type differs from their own (Brennan and Zufall, 2006), and there is some evidence (based on tests of preferred body odours) that humans do likewise. Female fur seals also move across a breeding colony to locate males that are more distantly related (Hoffman *et al.*, 2007).

If populations are inbred for many generations, the probability of extinction is high (Falconer and Mackay, 1996), but populations that do manage to survive may, as a result, have been purged of harmful recessive alleles (Visscher *et al.*, 2001). However, enough breeds have been inbred over time and still retain harmful recessive genes (this is particularly evident in dogs) to mean this is not a valid general strategy.

Inbreeding therefore poses a particular problem if there are many defective genes in a population. If the genes are only slightly defective then the result may be a relatively minor decrease in performance; the same is true if many genes code for a trait, as only a proportion of them will be present as identical gene pairs. For example, detailed milk records of thousands of cows whose pedigrees are known show that a 1% increase in inbreeding of dairy cows leads to a decrease of 6 days in total productive life, a reduced lifetime production of 177 kg of milk and an increase of 0.6 days in age at first calving. This is not life threatening but, if there are genes present in the population that would definitely be harmful when homozygous, inbreeding can lead to serious health and welfare problems, or even to mortality.

With all populations some degree of inbreeding is inevitable, but domesticated animals are in general not adversely affected provided it is not too intense, and occasional outbreeding (use of animals from a different breed or from a segment of the breed that has been kept relatively separate) will usually rectify harmful effects that are coming apparent.

As inbreeding leads to less adaptable animals, there is a welfare argument for restricting it. Whether this can be done depends on whether **breed societies** and individual breeders feel it is important to do so. The welfare of domesticated animals would be promoted by avoidance of inbreeding, and matings should always be planned with this in mind. Modern computer techniques enable selection programmes to be designed that improve performance without increasing inbreeding excessively.

The relationship between inbreeding and behaviour is, essentially, that there are behavioural mechanisms in wild animals for restricting inbreeding. (SJGH)

See also: Captive breeding programmes; Domestication; Genetic bottleneck; Kinship

References and further reading

Brennan, P.A. and Zufall, F. (2006) Pheromonal communication in vertebrates. *Nature* 444, 308–315.

Falconer, D.S. and Mackay, T.F.C. (1996) *Introduction to Quantitative Genetics*, 4th edn. Longman, Harlow, UK.

Hall, S.J.G. (2004) *Livestock Biodiversity. Genetic Resources for the Farming of the Future*. Blackwell Publishing, Oxford, UK.

Hoffman, J.I., Forcada, J., Trathan, P.N. and Amos, W. (2007) Female fur seals show active choice for males that are heterozygous and unrelated. *Nature* 445, 912–914.

Visscher, P.M., Smith, D., Hall, S.J.G. and Williams, J.L. (2001) A viable herd of genetically uniform cattle. *Nature* 409, 303.

Incentive motivation

Incentives are external stimuli that affect **motivation**. These incentives might, for example, be the smell of a certain food for a rat, or the sight of a hen for a newly hatched chick. The value, or motivational strength, of a given incentive may be affected by learning. (BAF)

Inclusive fitness

Inclusive fitness is a mechanism to calculate the spread or otherwise of a gene or genes through a population, that takes into account the effect that bearers of that **genotype** might have on different sorts of relatives. It is closely tied to the concept of **kin selection** and the evolution of **altruism**. The success (or otherwise) of the gene spreading through the population is determined from Hamilton's rule: $rB - C > 0$, in which r is the coefficient of relatedness between the giver and receiver of aid, B is the benefit to the recipient due to the act of the altruist (often measured in terms of *additional* offspring gained due to the help given by the altruist) and C is the cost to the altruist (often measured in terms of offspring *lost* due to the diversion of resources from personal reproduction to assisting a relative).

In reality, the cost is often difficult to measure as it would require knowing how many offspring an individual gave up in order to help a relative raise more offspring. Thus inclusive fitness is a tool to help biologists compare the relative success of two or more behavioural strategies rather than an absolute measure of an individual's genetic contribution to the gene pool. (PE)

See also: Fitness

Individual differences

Individual differences are literally any variables that may be used to differentiate between individuals with some degree of reliability. Differences may be categorical (giving rise to distinct types of individual – typology) or continuous (giving rise to scores along a scale to differentiate individuals – the trait-based approach); thus height, sex, age, coat colour or behavioural or psychological factors may all be used as the basis for distinguishing between individuals. The issue of whether individual differences exist in relation to a given behavioural variable can be addressed by answering the question of whether certain types of individuals are more similar to themselves than others over time and contexts. This can be assessed by looking for significant differences in specific behaviours or by considering a range of responses using multivariate methods such as factor analysis, principal components analysis and cluster analysis. Certain types of consistent individual differences in behavioural tendency in animals are often referred to as constituting the animal's **personality** or **temperament**.

Although individual differences in behaviour may be qualitative (types) or quantitative (traits), personality is composed of a collection of differences in behavioural style, and so this is described most completely by using a trait-based approach. However, certain groupings within a trait, e.g the extremes, may be labelled as a particular type. For example, in response to a challenge, many animals are classified as being either 'active' or 'passive' copers, with the former tending to show escape behaviour and the latter tending to freeze. This classification represents their predominant behavioural style. It does not imply that a given individual will not show the other style on occasion, and it is also possible for many individuals to fall between these extremes; for example, the term 'active coper' could be defined as the upper ten rather than the top fifty percentiles of animals responding most actively to a restraint test.

Whereas most traits would be expected to have a normal distribution, this may not apply if there is a positive or negative bias in the advantage conferred by any aspect of the trait, or if artificial selection has occurred to shift the norm in a particular direction. For example, Sheppard and Mills (2002) found that sensitivity to punishment was reported to be negatively skewed, but sensitivity to reinforcement was positively biased in a population of pet dogs, which would be consistent with the selection associated with domestication (towards neophilia and away from **neophobia**).

Historically, individual differences have been seen by some as an inconvenience to experimental behaviour science since, if adequate measures are not taken to reduce them, there may be a high level of variance in a study, reducing the power to detect differences due to any treatment or other experimental variable, necessitating the use of larger sample sizes and increased resources as a result. For this reason, subjects may be matched or balanced within an experiment (matched pairs design) in order to control for certain variables and, while this is often done in relation to physical characteristics, it is less often done for behavioural or psychological characteristics, although the use of certain strains or lines may help to reduce such variation.

There are several problems with balancing: first, it may not be known upon which parameters subjects need to be matched (or important factors may be accidentally overlooked, especially when the experimenter focuses on physical parameters); secondly, it may be difficult to find subjects that are closely matched. One way round this is to use subjects as their own control (within-subjects design, cross-over study or repeated-measures design), but this is not always feasible (especially if the study involves invasive sampling or the death of the animal to collect samples) and there may be an order effect (effect of exposure to the treatment on subsequent response to repeat exposure) or effect of the interaction between the treatments and order (cross-over effect) and, while these can be assessed statistically, if they arise, they greatly complicate the interpretation of the results.

Another alternative approach to the balancing of subjects between groups is to randomize subjects into the groups (randomized study), with partial (semi-randomized study) or no (fully randomized study) balancing between the groups and to use larger sample sizes to minimize the impact of uncontrolled variables. It might be argued that the fully randomized study is not in accordance with the **three Rs principle** that is adopted by many researchers interested in protecting animal welfare, although others may argue that it is the most scientifically pure type of study and so intellectually superior.

The importance of individual differences has been emphasized by several biological fields (including psychology). Individual differences are 'the raw material upon which natural selection acts' (Slater, 1981), and arise as a result of ecological variation in the distribution of essential resources. **Game theory** predicts that it will arise as a result of **evolution**, when no single trait (including any behaviour) is optimal in all circumstances. As such, individual differences may also reflect the variability and unpredictability of the environment inhabited by a species that would prevent total adaptation. This is illustrated by the development of ever more elaborate and precise displays by many species living in relatively stable tropical environments, and the conservation of variation in animals living in more variable environments. Individual differences may also arise as a result of the advantages associated with **kin recognition** and the communication of individual identity (e.g. to prevent unnecessary **agonistic behaviour**).

The importance of individual differences in animal management, **conservation** and **welfare** is increasingly recognized as increased emphasis is focused on the importance of the individual's welfare and its value as a genetic resource. Different individuals may respond to a similar disease challenge or load from a **stressor** differently as a result of individual differences (personal risk factors), and so the physical and psychological **health** of an individual cannot be understood fully without an appreciation of the individual that is affected.

Recognition and appreciation of this is one of the essential skills of a good clinician. One of the challenges of medicine and the development of good husbandry practices is the identification of significant risk factors of clinical relevance relating to individual differences, i.e. those factors that are of real importance, and this will vary with the condition under consideration. For example, in the case of a potentially fatal problem, small risk factors relevant to small individual differences are important but, in the case of less serious conditions, such variation may be of less relevance.

To date, most work has focused on physical differences, e.g. skin colour and the risk of squamous cell carcinoma in cattle and cats, with less attention generally paid to behavioural and psychological differences, despite their potential importance in susceptibility to even infectious disease. The notable exception to this is the study of individual differences in behaviour as a risk of behavioural problems, e.g. **aggression** and the risk of **cannibalism** in chickens, attachment behaviour and the risk of **separation anxiety** in dogs, etc.

An appreciation of the importance of individual differences in behaviour is therefore important to many aspects of applied animal behaviour and welfare. (DSM)

References and further reading

Bates, J.E. and Wachs, T.D. (eds) (1994) *Temperament: Individual Differences at the Interface of Biology and Behavior*. American Psychological Association, Washington, DC.

Cooper, C. (1998) *Individual Differences*. Academic Press, London.

Manteca, X. and Deag, J.M. (1994) Individual variation in response to stressors in farm animals: a review of methodology. *Animal Welfare* 3, 213–218.

Sheppard, G. and Mills, D.S. (2002) The development of a psychometric scale for the evaluation of the emotional predispositions of pet dogs. *International Journal of Comparative Psychology* 15, 201–222.

Slater, P.J.B. (1981) Individual differences in animal behaviour. In: Bateson, P.P.G. and Klopfer, P.H. (eds) *Perspectives in Animal Behavior*. Plenum, New York, pp. 35–49.

Individual fitness

Evolution by natural selection is a potent force of adaptive change. Simply put, those individuals whose genes spread through the population have greater **fitness** than those whose genes do not. An individual's genes can spread directly and/or indirectly, and this distinction is important. **Charles Darwin** recognized possible ways that his hypothesis of natural selection could be challenged if individuals did things that directly benefited others at some cost to themselves. For instance, we often see animals helping feed or care for others, or emitting potentially costly alarm calls. How could these potentially costly behaviours be adaptive to individuals? The solution is that individuals share genes with relatives (they share more genes with close relatives and fewer with more distance relatives (**see: Kinship**) and, by helping relatives survive and reproduce, individuals are helping their own genes survive and reproduce.

The British geneticist J.B.S. Haldane once said that he'd be willing to sacrifice his life for two brothers or eight first cousins. Working though the maths, we see that saving two brothers will generate 1 fitness unit (2 brothers × 0.5, the coefficient of relationship between an individual and a full sibling), as will saving eight first cousins (8 × 0.125, the coefficient of relationship between an individual and his first cousin). Thus, Haldane's back of the envelope calculation was correct.

There are a number of common confusions about the term fitness. Darwinian fitness is often mistakenly confused with physical condition. Evolution by natural selection does not inevitably select those in the best condition to reproduce; rather, those that reproduce have the highest Darwinian fitness. Envision an elephant seal. We typically think that large-bodied males are better able to defend harems and therefore have higher reproductive success. In most years this is certainly so, and we see the largest males obtain the most copulations and sire the most young in a colony. However, what happens when insufficient food is available? We know that, in such conditions, females of some species may save energy by not breeding. Because large-bodied individuals require absolutely more energy than smaller-bodied ones, large males may be the most vulnerable individuals when food is drastically restricted. In this case, small-bodied individuals might survive and be alive to reproduce the following year. Thus, the largest animals, in apparently the best physical condition, might not have the highest fitness.

Elephant seals also illustrate another strategy. Large males exert a lot of energy defending their harems and fight viciously with other large males. Smaller male elephant seals look more like females. Because they have no chance of winning a fight with a large male, they do not try. Rather, on occasion, smaller males will try to mate with a female in a large male's harem. These males are referred to as 'sneakers', and this illustrates a condition-dependent strategy whereby large males fight for reproductive success and smaller males try to steal it. Such condition-dependent strategies are common and illustrate another way that 'being in the best physical condition' may not necessarily translate to having the highest fitness.

Fitness is a relative concept. Fitness will not lead to the best conceivable individuals or traits. Rather, heritable traits will evolve so that populations consist of individuals with those beneficial traits. When conditions change, what determines the best trait might change, and thus we will have selection for another trait. Moreover, traits do not evolve to some '**optimal**

fitness'; rather, traits evolve such that individuals with them do better than individuals without them.

Fitness is constrained in many ways. The fluctuating selection described above, whereby large animals might be favoured in years of plenty and smaller animals might be favoured in sparse years, will maintain heritable variation and prevent fixation of a particular trait. Additionally, if there are genetic correlations among traits, selection for a particular trait may be constrained by its correlated trait. Since we expect traits to be correlated, and since linkage is common, 'optimal fitness' will not be seen.

Individuals act in many other ways to maximize their individual fitness. Any time we see animals acting in ways that favour their personal survival, we see evidence of animals behaving in ways that suggest that individual fitness is important. Consider a fight between two male bull elk (a large deer). These opponents have a formidable set of weapons – large, multi-pronged antlers. When two opponents meet they engage in a complex assessment ritual. They bugle, because the lowest frequency of an individual's **vocalization** is directly related to body size (larger animals produce lower-frequency vocalizations). This is an unbluffable display, and many potential competitors simply avoid larger opponents. They engage in parallel walks whereby opponents 'size each other up'; they lock antlers and push a bit. Only if very well matched will individuals initiate full-scale combat, because combat is risky. Animals could break a leg by falling, or be skewered with an antler. If an individual is small, and therefore likely to lose, it is much better to run away and come back and fight another day.

Fatal fighting is thus rare, and the conditions under which we see it are revealing. Fatal fighting is seen when the magnitude of direct benefit is large, and there is considerable skew in who obtains it (meaning that not all individuals will be able to obtain a needed resource) and when individuals have only a limited period of time to obtain it. In a mating context, we expect to see fatal fighting among animals with little residual reproductive value (i.e. among older ones that will die soon). We also expect to see more fatal fighting in species that have extremely short lives with a very circumscribed mating period. (DTB)

Infanticide

Infanticide is 'a behaviour that makes a direct and significant contribution to the immediate death of an embryo or newly hatched (born) member of the performer's own species' (Mock, 1984). Infanticide by males and non-relatives occurs in many species, generally serving to increase the fitness of perpetrators (e.g. male lions have been reported to kill nursing cubs after a pride take-over in order to trigger females' return to receptivity). The phenomenon of mothers killing their own infants has been documented in diverse mammalian species including humans, **elephants**, silver foxes, **rodents** and domestic **dogs** and **cats**. Mothers of other species, including those commonly kept on farms, as companions, in laboratories or otherwise in association with humans, occasionally display violent behaviour towards their newborn.

Domestic ruminant mothers such as cows and ewes are rarely overtly violent, but neglect or abandonment will lead to the death of a minority of infants. Perhaps because it is so unpleasant, infanticide has frequently been described as an **abnormal** behaviour, but is probably more accurately characterized as a normal response to a particular set of circumstances. Sarah Blaffer Hrdy (1979) suggested that mothers might increase their **inclusive fitness** by killing their infants because of their low social status, in response to lack of resources, due to disturbance or because infants were abnormal or too numerous. All of these factors decrease the probability of offspring surviving to sexual maturity, and killing infants reduces the mother's investment where the probability of successful rearing is low.

The maternal repertoire of the domestic **pig** includes some of the most violent acts of any seen in commonly farmed species. The phenomenon was referred to in the 1920s as 'parturient psychoneurosis', and as 'hysteria' as recently as the 1980s. In pigs, 'savaging' starts during or soon after farrowing. The sow, which may or may not be generally restless, threatens and may attack piglets that are close by. The attacks can result in injury or death of one or more piglets, or occasionally of a whole litter. Savaging episodes can be very brief, consisting of a single or a few snaps, or continue for several hours in the absence of intervention. **Aggression** may spontaneously cease at any time, or may stop once piglets start to suckle. Savaging is not generally followed by **cannibalism** of dead piglets.

Reports of the frequency of savaging have varied widely: levels may change with time and it has been reported to be a 'contagious' behaviour, occurring in sporadic outbreaks. A large-scale Canadian survey of commercial hybrid pigs reported that 3.4% of gilts (first-parity sows) and 1.2% of multiparous sows killed one or more piglets. Counting dead piglets alone will certainly underestimate the magnitude of aggressive behaviour: studies involving videotaping of farrowing in farmed wild boar and domestic gilts observed that 30% displayed some aggression towards their newborn, although few attacks resulted in **injury** or fatality. In common with many species where **mis-mothering** is more common in young females, savaging occurs more frequently in sows having their first litters, and there is some evidence that sows which savage their first litters are more likely to be infanticidal with subsequent litters. Gilts that savage at their first parity may not be kept by farmers for a second farrowing.

Traditionally, pig farmers might treat the savaging sow with a diverse range of home remedies, which included giving her beer, perhaps due to the now discredited belief that the behaviour arose due to extreme thirst, smearing her face and the backs of the piglets with a kipper, smacking her with the body of a dead piglet or the application of a rubber boot top, secured with string, around her snout. Modern treatments include the administration of sedatives such as azaperone and/or temporarily removing the litter until the sow has calmed down.

Wild boar farrow in secluded nests under dense cover; hence, the degree to which they or free-living sows giving birth in semi-natural conditions are infanticidal (if at all) is unknown. While it is unlikely that savaging is a **captivity**-induced behaviour, intensive production might increase its occurrence. The effect of the farrowing environment is unclear – it has recently been reported both that the behaviour is more frequent in open pens than in **farrowing crates** and that there is a tendency for more savaging when gilts farrow in crates.

While the **stress** of confinement and thwarted nesting motivation in a farrowing crate may increase a farrowing sow's tendency to behave aggressively towards her piglets, it is easier for piglets to escape from attacks when their mother is restrained in a crate, so the consequences of piglet-directed aggression may be less severe. Regardless of whether farrowing crates are used, the farrowing environment in large, modern pig units can be noisy, busy and disturbing. Sows are unable to isolate themselves prior to giving birth, as they would choose to do if free living (**see also: Sow behaviour**), and these factors could contribute to savaging.

With few scientific studies addressing the causes of infanticide in pigs, savaging remains a poorly understood behaviour. Possible evolutionary correlates have been investigated, but there is little evidence to link piglet-directed aggression with factors such as dominance status or litter size. While other species (e.g. dogs, cats, mice) will preferentially kill small or sick young, this is probably not the case with pigs. Especially in farrowing crates, the first piglet that walks past an infanticidal sow's face is likely to be attacked – this may be an early-born or vigorous piglet. Savagers may be more nervous prenatally than non-savagers. Prepartum postural behaviour has been reported to differ in sows that go on to savage. This suggests a role of **pain** or discomfort in triggering the behaviour, as does the reported link between savaging and prolonged parturition.

Modern pig production means that gilts (unlike their free-living counterparts) are unlikely to have encountered a piglet until they give birth. At the time of birth, they are often unable, due to restraint in a farrowing crate, to turn around and establish recognition through nose-to-nose contact. These factors suggest that a simple fear of novelty could contribute to some piglet-directed aggression, which is supported by a finding that 24-hours-a-day light around farrowing (which could have helped gilts to see and recognize piglets more easily) substantially reduced savaging deaths. (MH)

References and further reading

Blaffer Hrdy, S. (1979) Infanticide among animals: a review, classification, and examination of the implications for the reproductive strategies of females. *Ethology and Sociobiology* 1, 13–40.

Blaffer Hrdy, S. (1999) *Mother Nature: a History of Mothers, Infants, and Natural Selection*. Pantheon Books, New York, pp. 288–317.

Mock, D.W. (1984) Infanticide, siblicide and avian nestling mortality. In: Hausfater, G. and Blaffer Hrdy, S. (eds) *Infanticide: Comparative and Evolutionary Perspectives*. Aldine Publishing Company, New York, pp. 3–30.

Inflammation

Inflammation in essence is an immunological response that is usually initiated as a result of trauma or infection. As an immunological response it encompasses a complex of specific reactions that together create the 'typical' symptomatic appearance of inflammation. These include reddening, heat, pain and swelling. In addition to these sensed or visible symptoms, inflammation also includes a cellular immune response involving the attraction of granulocytes to the site of the infection/trauma (**see: Immune system**). Inflammation also causes increased vaso-permeability, allowing the immune cells to enter tissue. These granulocytes are responsible for the non-specific removal of invading microorganisms.

To illustrate the complexity of the inflammatory process, it is worth pointing out that the chemotactic **cytokine** released at the moment of trauma or infection is one of many cytokines that are responsible for the onset of the general signs such as the reddening, heat, pain and swelling. Inflammation can also occur as an overreaction against innocuous invaders of the host's own tissue, resulting in an allergic or autoimmune reaction. This is referred to as hypersensitivity, of which there are four distinct types: immediate (allergic, Type 1, e.g. anaphylactic shock); cytotoxic (Type 2, e.g. autoimmune haemolytic anaemia); immune complex related (Type 3, e.g. vaccine reactions); and delayed (cell-mediated, Type 4, e.g. response to tuberculin testing). (FLR)

Inherited disease

An inherited disease is a disease that is primarily transmitted genetically. There may also be an environmental trigger or other component (epigenetic) making an animal predisposed to a specific disease, e.g. type of exercise when young and hip dysplasia in certain breeds of dog. An individual inherits half of its genes from its dam and the other half from its sire, and each gene is comprised of two alleles, one from each parent. If they both code for a disease then that animal will be homozygous for that disease. If the alleles are different and one allele is dominant to the other, the degree to which that animal may have that disease may vary (depending on the disease), or it may simply be an all-or-none effect, and so it may not have the disease at all. If the disease is influenced by more than one gene (i.e. it is polygenic, polymorphic), then the expression of that disease may vary according to various factors determined by the expression of other genes and the environment (penetrance). It is hoped that the sequencing of the genome of various species may help us to better understand such diseases and the factors that influence their expression. (DBM)

See also: Breed societies; Breeding

Further reading

Sewell, A.C., Haskins, M.E. and Giger, U. (2007) Inherited metabolic disease in companion animals: searching for nature's mistakes. *The Veterinary Journal* 174, 252–259.

Inherited disorder

It is suggested that virtually all animals carry some deleterious genes. Because of natural selection, deleterious genes tend to occur at a low frequency, and hence the incidence of any particular defect is usually so low as to not cause major concern. Additionally, in **free-range animals**, the chances of these deleterious recessive genes expressing themselves as disorders are reduced by juvenile dispersal strategies that reduce **inbreeding**. In contrast, the mating of relatives (inbreeding) increases the frequency of homozygotes, thus permitting the expression of disease. By creating artificial groups, the establishment of breeds increases the chances of individual mating with related **conspecifics**. Closed stud books are especially likely to compound this problem because they prevent the introduction of 'fresh' genes.

Inherited disorders are now recognized in all established breeds of dog and cat, but are also reported in horses, production species and a growing number of captive exotic species. **Dogs** provide the best example of the problem. As a species, they show unique morphological diversity between breeds and have been intensely bred within the closed stud book system since Victorian times.

Pedigree dog breeders compete to produce phenotypes that conform to a written standard that may include traits that compromise the quality of life. Many of the morphological and behavioural traits valued by early dog domesticators had direct utility and functionality, and were incorporated into the first breed standards when dogs left the working arena and entered the world of dog shows. Unfortunately, these days some breed standards place more importance on appearance than on functionality.

That said, many inherited disorders cannot be blamed on any breed standard; rather they emerge by chance as a result of the inbreeding that is an inherent risk of breeding within the current framework that pursues so-called purity (being related) rather than fitness for purpose. Strategies to identify harmful genes and avoid breeding from carriers are, at first glance, an adequate response, but they effectively reduce the number of individuals available for breeding within a breed and thus paradoxically can increase inbreeding. The prevalence of inherited disorders will continue to rise unless the rules of breeding are changed and traits compatible with quality of life are selected for as a priority. (PDM)

See also: Breed societies; Breeding

Further reading

Sewell, A.C., Haskins, M.E. and Giger, U. (2007) Inherited metabolic disease in companion animals: searching for nature's mistakes. *The Veterinary Journal* 174, 252–259.

Injury

An injury is any damage to body tissues that may be either invasive, and involve breaking the skin and exposing deeper layers of the body, or non-invasive, but cause deeper damage such as bruising (bleeding), or stunning, as with a **captive bolt**. Some diseases make body tissues more susceptible to injury, e.g. lupus erythematosis. (DBM)

See also: Inflammation

Innate

Innate refers to a characteristic that is largely affected by the **genotype** of the animal. Innate characters, including behaviours, result from the evolution of the species rather than from environmental factors or those acquired during an individual's lifetime. Innate behaviours are performed effectively the first time they are expressed. Examples of innate characters include eye colour, begging behaviour and yawning. (KT)

See also: FAP/FMP; Innate releasing mechanism; Instinctive behaviour

Innate releasing mechanism (IRM)

An innate releasing mechanism (IRM) is a hypothetical, internal, neurally and/or hormonally controlled process that is responsible for releasing a fixed-action behaviour pattern (**FAP/FMP**) following sensory input from a specific stimulus (the **releasing factor**). The mechanism prevents the performance of the behaviour until the appropriate stimulus is perceived. For example, an IRM releases begging behaviour in herring gull chicks when the chick perceives, among other stimuli, the visual stimulus of the red dot on the parent's beak. (KT)

Further reading

McFarland, D. (1993) *Animal Behaviour*, 2nd edn. Longman Scientific and Technical, Harlow, UK.

Tinbergen, N. (1951) *The Study of Instinct*. Oxford University Press, New York.

Tinbergen, N. and Perdeck, A.C. (1950) On the stimulus situations releasing the begging response in the newly hatched herring gull (*Larus argentatus* Pont.). *Behaviour* 3, 1–39.

Insectivore

Strictly speaking, the term insectivore refers to a mammalian species whose diet is largely made up of insects, and which belonged as a result to the formerly recognized order insectivora. Birds and other species, including some plants, that have a similar diet may be referred to as having an insectivorous diet, but are not generally referred to as insectivores. (DSM)

Insight and relational mechanisms

Relational mechanisms lie at the core of human cognitive organization, expressed in everyday language by terms such as 'bigger than' and 'smaller than' and in formal symbolic systems by logical arguments such as 'A > B'. Ordered as a series, e.g. A > B > C, such relationally based mechanisms support transitive inferences of the sort that, given A > B and B > C, A must be bigger than C – an important deductive form of reasoning believed to be the cornerstone of rational choice behaviour. Relations are also central to the human ability to classify objects and events using rules of group membership. Reflected in linguistic acts of reference such as 'that is a canary', our further ability to denote the canary as (also) a bird reflects hierarchical principles that crucially underwrite our ability to combine and recombine words into sentences and communicate propositions about the world.

Historically, it was held by many that such relational competences were unavailable to 'the brute mind', a view offered by the father of American psychology, William James, and reinforced by the work of his student Thorndike, whose classic studies of problem-solving behaviour with cats at Columbia University based on 'blind' trial-and-error mechanisms led to a characterization of non-human **intelligence**, retaining a powerful influence over many comparative psychologists to this day. Designed ostensibly to evaluate the extent to which cats locked in a box would immediately connect a successful response – such as pressing on the door panel – with escape and inducing in turn a sort of 'aha' response, Thorndike found instead that cats only gradually acquired the successful response, selecting it with progressive frequency from a range of competing responses within their repertoire. Thorndike concluded that the connection between response and outcome was forged as the result of blind trial-and-error processes rather than the outcome of strategic planning and a perception of the causal relationship between action and success.

For Kohler, on the other hand, such trial-and-error characterizations were the result of bad experimenter practice that disabled the subject from seeing the problem 'in the round', rather than a reflection of normative mechanisms. In the early part of the 20th century, Kohler presented apes with problems that were meaningful 'at first sight' and without having to make any overt response at all. For example, confronted with alluring bananas out of reach, they could recruit sticks to augment their reach or stack boxes to gain height and recover the prize. In contrast to Thorndike, Kohler found that apes would often make sudden solutions without requiring trial and error when they were in a position to perceive the putative relationships between the different components of a problem, instead of seeing the task piecemeal as in the case of Thorndike's cats.

This disparity between Thorndike's and Kohler's characterizations of animal intelligence provoked the philosopher Bertrand Russell to comment dryly that American rats learn by trial and error, while European rats learn by insight! For Kohler, however, 'insight' was a pale reflection of what humans might consider as thinking, regarding it as the result of a sudden (involuntary) perceptual reorganization. Since then, augmented by a new and burgeoning interest in comparative and evolutionary precursors to thought and **language**, a range of new paradigms has been evolved, often based on intensive learning procedures to parallel the life history and deep cultural immersion of humans. Critically too, some of these have been designed to eschew manipulation constraints that have long embargoed the deployment of free classification and other object-based tests so effective in charting human cognitive growth. As a consequence, contemporary comparative research reveals that non-human primates (at the very least) share some core conceptual representations, arguably the precursors of human propositional thought, reasoning transitively, classifying hierarchically and exhibiting flexible executive skills in serial ordering tasks. Of great significance for the issues of evolutionary origins and continuity, these ordering competences, moreover, also based on temporal relations, suggest significant evolutionary precursors for the highest forms of human achievements including language.

Some of these key building blocks of evolutionary cognitive growth, based on relational competence, are illustrated in the sections below.

Systematic relational competences

For Kohler, 'insight' in apes is driven by relationally based perceptual laws of organization derived (as he put it) from 'silent physiological operations' not accessible to conscious scrutiny. Yet humans, at least, are well aware of object relationships per se (and not merely the objects themselves). In language, for example, simple declaratives such as 'Jane is bigger than Henry' reflect an explicit grasp of the material (size) relationship in question. Humans also compute the inverse relation 'smaller than' from the declarative, as in 'Henry is smaller than Jane'. While contemporary evidence for relational learning by non-humans is strong, it is not systematic in this sense, one of the major problems being that the choice of relational tests used with animals has been somewhat *à la carte*, based on the isolation of a single relation such as 'brighter than', using choice methods that conventionally require the subject to choose an object either as 'brighter than' or 'darker than' its neighbour.

Figure I.6 shows an important evolution of the conventional procedure. Here, the subject learns a 'language' of conditional signs where each sign signals which relational rule (within a set of five) must be chosen. Thus, if all objects varying in size are black, for example, the subject must select the larger/largest object (irrespective of its absolute size); if all white, by contrast, then the subject must choose the smaller/smallest object. As illustrated in Fig. I.6, this method has revealed that squirrel **monkeys** can acquire five related size rules, operating them systematically in ways very similar to that of 4-year-old children (McGonigle and Chalmers, 2002). In showing that rules are both systematic and available to non-verbal animals (at least primates), and that the type of relation itself can be linked with an arbitrary sign, we can conclude that core relational competences are reflected in language but not determined directly by language per se.

Relational competences and reasoning

Humans operate relations at a number of different 'levels'. First, they can detect these perceptually, as in the examples given above. However, a significant evolutionary advance centres on the human ability to reason symbolically as in situations where no direct perceptual information is present. Given several connected logical arguments, such as 'Jane is bigger than Henry' and 'Henry is bigger than John', human adults can order these items at a conceptual level and make the transitive inference that 'Jane is (therefore) bigger than John'. A key issue is the extent to which more abstract procedures of this sort are available to non-humans.

While the long-standing lack of positive evidence might suggest that they are not, one of the key difficulties here has been to present such tasks in a meaningful way to non-verbal subjects. With procedures that eschew language, we illustrate ways this can be done (see Fig. I.7). In an adaptation of a reasoning paradigm designed for use with very young children (McGonigle and Chalmers, 1977), squirrel monkeys were trained on four connected pairs of tins of varying colour and weight; each pair was either heavy or light, but (crucially) the weight relation could not be perceived directly. The outcome showed that all six novel pairs (of the ten pairs that derive permutatively from the term series as illustrated in Fig. I.7) were discriminated on the basis of a transitivity rule without further training. This was the first demonstration of 'reasoning' by non-humans under the most stringent conditions that developmental and experimental psychologists have been able to devise, now replicated extensively with a variety of species. However, significant variation in the procedures used since the first demonstration may support simpler solutions in certain cases.

'In the mind's eye'

Of course, no single source of evidence is definitive in this most difficult area. Instead, evidence is needed from tests assessing cognate cognitive competences. In linear transitive reasoning with human adult subjects, for example, subjects

Squirrel monkeys (left) being trained on a series of five conditional rules of relation such that, e.g. 'if all blocks are red' then choose the smallest one, 'if all blocks are green', choose the second smallest, etc. until all five rules can be operated concurrently and in random order. The array changes randomly from trial to trial so that the rewarded item cannot be identified by a unique location. Performance on five-rule training as depicted followed training on smaller numbers of rules. Five-year-old children were given similar training and showed highly similar acquisition profiles to monkeys during the final single random alternation phase, as the graph illustrates (from McGonigle and Chalmers, 2002).

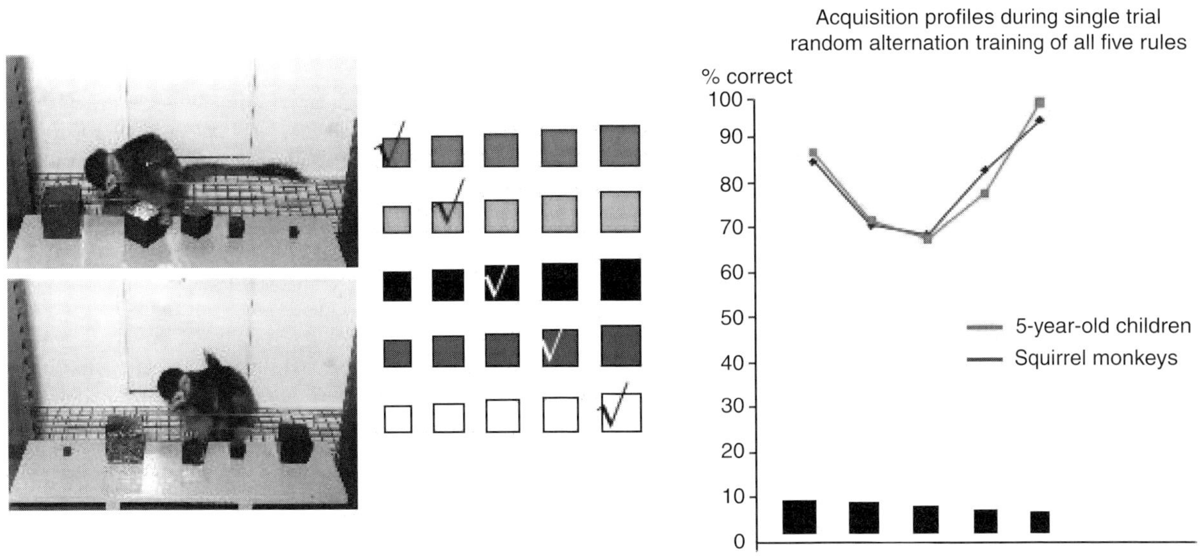

Fig. I.6. Systematic relational learning: the acquisition of five size rules.

While children are told that, e.g. white (A) is bigger than red (B); red (B) is bigger than green (C), etc. and then asked about the difference between white (A) and green (C) – a transitive relationship – monkeys are trained to choose red over white, green over red and then given the novel non-adjacent pairings such as white and green. Results show that monkeys make strongly transitive choices for all novel pairings just like 6-year-old children (McGonigle and Chalmers, 1977) – even for the pair BD where choice of either item has been equally often rewarded during training.

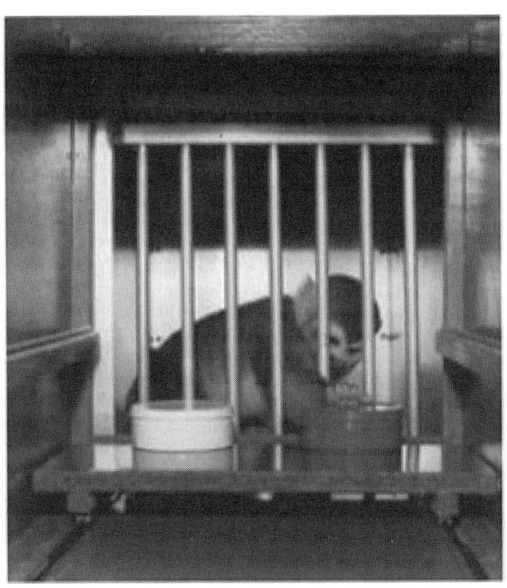

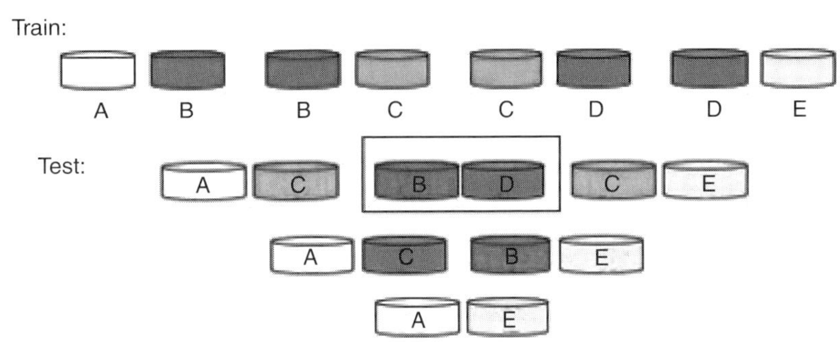

Pair	Six-year-old children (*n* = 10) B	C	D	E	Pair	Squirrel monkeys (*n* = 8) B	C	D	E
A	(98)	70	57	80	**A**	(98)	100	100	100
B		(98)	**70**	88	**B**		(93)	**90**	76
C			(98)	78	**C**			(89)	87
D				(98)	**D**				(97)

Fig. I.7. Transitivity: rational choice without direct perceptual support.

often report the use of a mental image where test items are aligned along an imaginary spatial vector (usually a vertical one) with, e.g. Jane at the top, Henry in the middle and John at the bottom of it. Coinciding with subjects' reports of this imaginal process, it has also been found that decision times are often faster for transitive inference tests. That is, it takes longer to retrieve the pairwise information as given than to generate the inferences that derive from it, and this is known as the symbolic distance effect (SDE).

Explained conventionally as being due to the fact that inference items are more remote on a spatial representation and less likely to be confused in a spatially based read-off, this phenomenon is also found in other tests of the mental representation of orders by humans when asked to compare a range of familiar objects conveyed as pictures or words 'in the mind's eye'. Once again, decision times are inversely related to ordinal distance and the mode of presentation is also important, with lexical representations of objects providing the slowest (absolute) decision times overall – although the psychophysical functions are similar. While we cannot hope to extract 'think aloud' data from non-verbal subjects, we can assess their decision time profiles obtained under similar conditions of test. As Fig. I.8 shows, squirrel monkeys show a significant SDE for both perceptual and 'symbolic' modes of test, recording also, as with humans, slower decision times overall with the 'symbolic' mode – although the psychophysical functions are similar. This pattern of results based on decision time data points further in the direction of a 'symbolic' competence in non-humans.

Serial ordering and executive control

Humans exhibit impressive feats of ordering as a core adaptive feature, controlling actions and words in extended productions that exhibit flexibility, capable of substantive recombinations. During human cognitive development, moreover, both the length of the child's utterance and the scope and extent of their ordering – as measured by the classic size seriation task – are a significant index of their cognitive growth. In the case of non-humans, by contrast, ordering mechanisms must be inferred from data conventionally based on binary choice paradigms, thus leaving a huge disparity between human and non-human behaviour in this domain.

Based on touch screens, new technology, however, has enabled investigators to assess the serial ordering competences of non-humans without requiring the high levels of manipulation required to place objects in a neat row, or to sort them into collections and classes. With touch screen procedures, the response requirements for each item are low-cost – a simple touch registered by the computer. Figure I.9 illustrates this contrast, first showing sequences achieved by

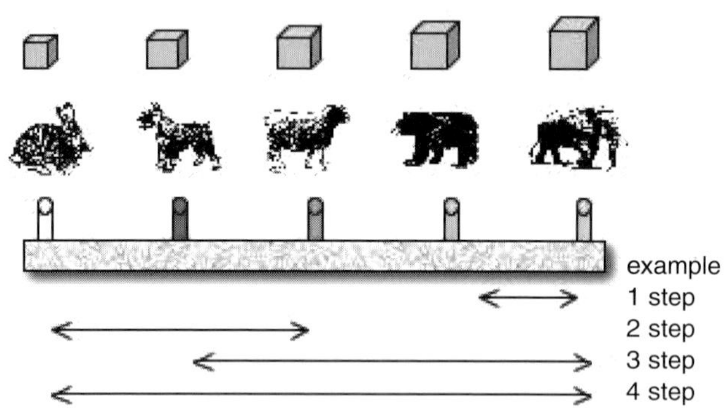

The schema on the left shows different sorts of stimuli – actual objects, pictures representing objects scaled to look the same size and items from transitivity testing whose size relations are taught verbally, e.g. 'white is shorter than red'. These items can be ranked in terms of their ordinal distance from one another (in 'steps') along a dimension such as size. The schematic graph below left shows the inverse linear reaction time function usually obtained with humans when asked e.g. 'which is larger?' of a pair, whether they are real or merely pictorially or verbally represented. The remaining two graphs below show the actual functions obtained for four individual monkeys (McGonigle and Chalmers, 2002) using real size differences (left) and items from transitivity tests (right) whose relations were taught using colour and weight differences.

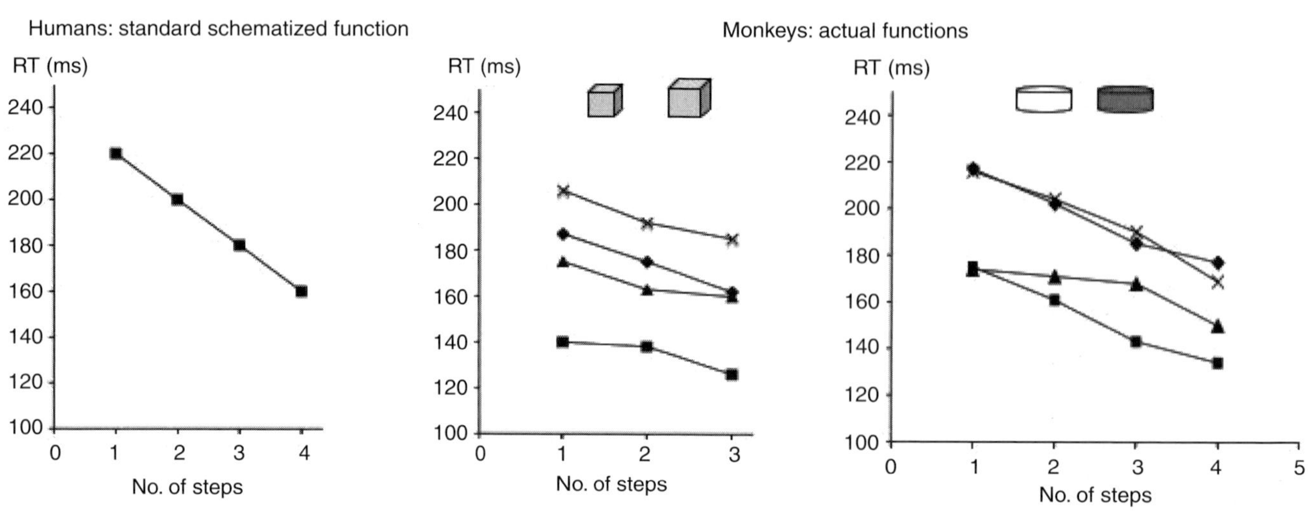

Fig. I.8. The symbolic distance effect: reaction-time measures of represented linear orders.

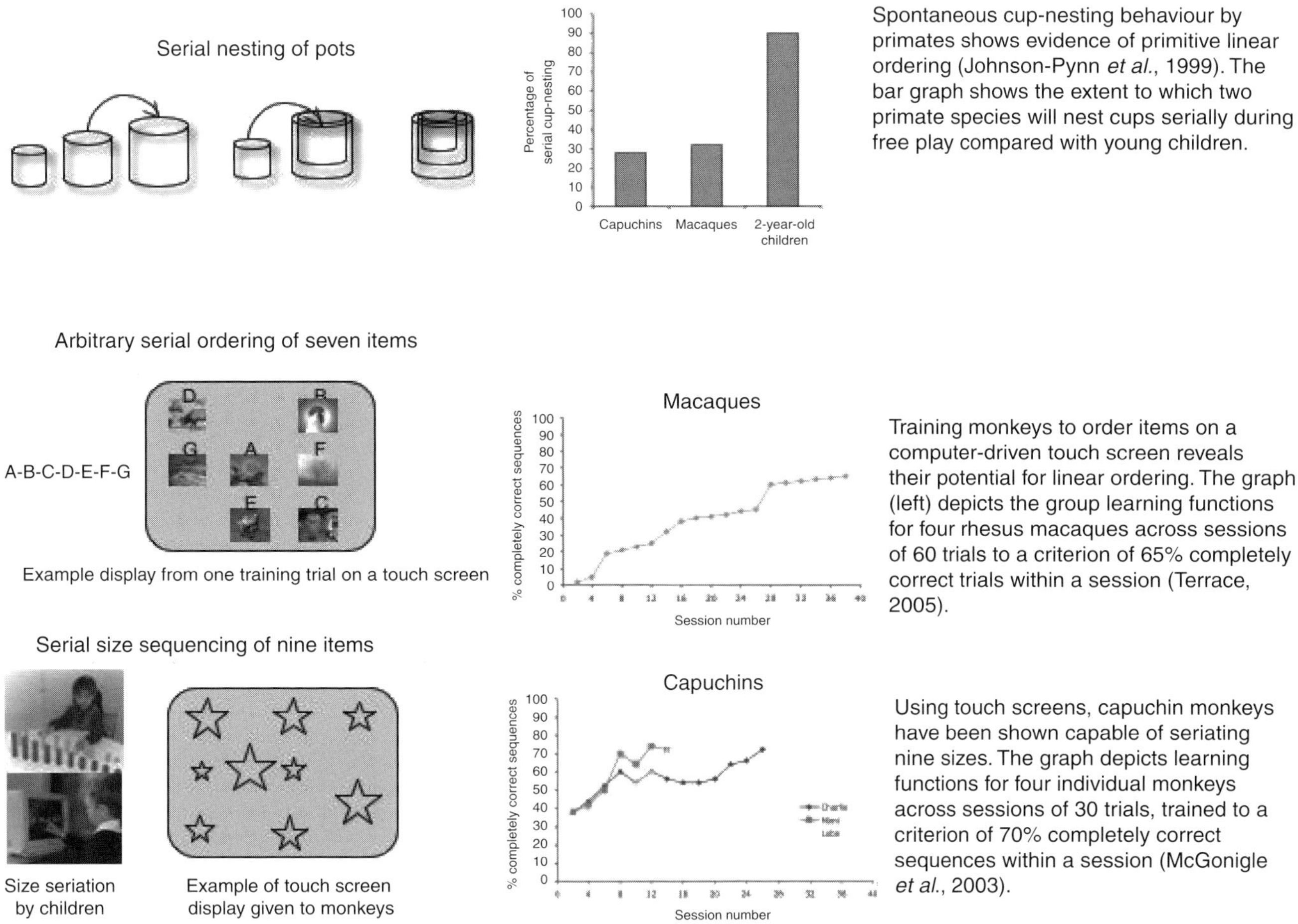

Fig. I.9. Linear ordering: serial nesting of real objects, arbitrary list learning and ordering of elements according to a (size) relational rule.

primates required to nest cups in a manipulative test of principled seriation (Johnson-Pynn *et al.*, 1999).

As can be seen, the number of objects seriated is small. In contrast, touch-screen-based seriation is limited more by the size of the screen than by the primates' competence to seriate. This is illustrated both by the sequencing of a fixed arbitrary list by rhesus **macaque**s where the elements have no material connection one to the other (Terrace, 2005), and also size-based seriation by capuchin monkeys (McGonigle *et al.*, 2003) using an iterative size rule (choose bigger). The monkeys' size seriation performance converges with that of 6-year-old children, both in terms of the length of the production and in its style (to count as successful, each production must be without error).

Categories and hierarchical organization

Humans deploy both linear and hierarchical structures in language and thought. The latter are revealed in the 'all/some' relationship and in the asymmetry between the use of the superordinate such as 'animal' and subordinate such as 'monkey'. Without such structures, **Darwin**'s taxonomy would have been impossible. Based on a complex of equivalence and difference relations, the only option available to assess such high-level, adaptive competences in non-verbal subjects is to require them to sort objects into groups and collections (see Fig. I.10). However, the high manipulatory demands entailed when using such conventional methods may be responsible for their meagre returns – a persistent result is that only one class can be identified from a group of objects, the rest left scattered and ungrouped. Given, however, that (even) non-human primates are relatively poor at object manipulation, conventional tests may have underspecified their competences in this regard. Figure I.10 shows, by contrast, how sequencing on a touch screen can reveal substantial classificatory skills in monkeys, where capuchin monkeys (*Cebus apella*) classify nine test icons – presented in a randomized array on each trial – into three separate collections based on shape (McGonigle *et al.*, 2003).

In the first phase of tests, the exemplars within each category are identical; the monkey must interrogate all the exemplars from the category that must be sequenced first, then all the exemplars from the second category, then the final one, before the production is complete. In the second phase, the exemplars from each category vary in terms of their relative size. In this condition, the monkey has to order each exemplar as well as each category. As there are only three sizes of exemplar common to all categories, the monkey can only seriate accurately by coordinating the ordinal position of each exemplar with the ordinal position of the category from which

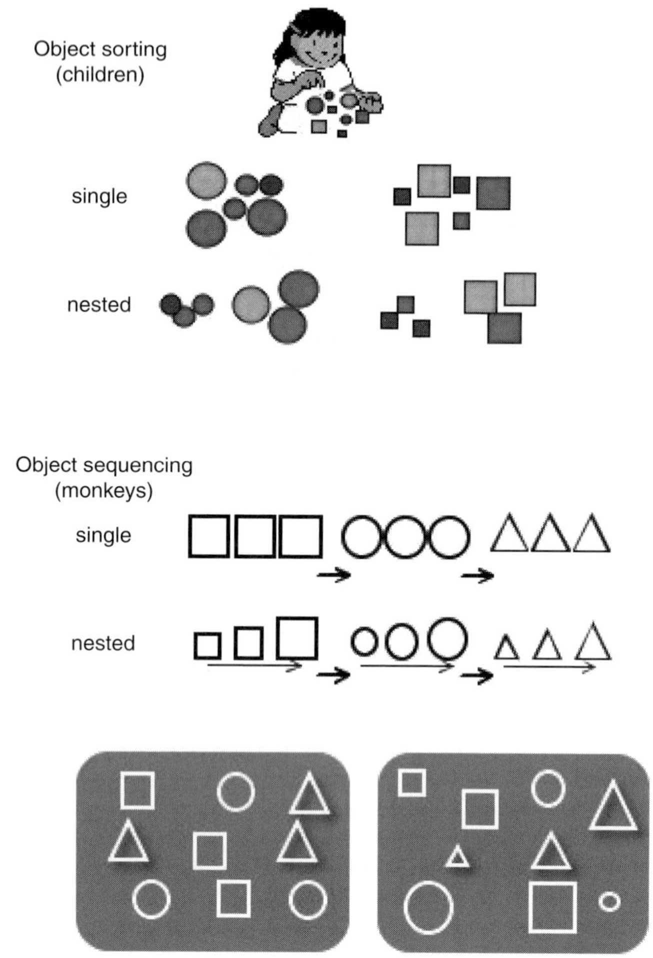

Children's early classification of objects is assessed during tasks that require them to put similar items together in spatial collections. The first signs of a true classificatory principle during such sorting show the use of a single feature (e.g. circles and squares) and only later show hierarchical nestings of, e.g. shape by size or colour.

In tasks that require items to be sequenced on a computer touch screen, capuchin monkeys have been found capable of both single-dimensional ordering as well as ordering that requires nesting up to three shapes by three sizes according to a hierarchical principle, 'select shape, then size'. This is shown by training the monkeys first to order the items by shape alone. In a second phase both the shape categories as well and their exemplars, now varying in size, must be sequenced. Capuchin monkeys can maintain an 80% level of completely correct hierarchically organized sequences of this sort (McGonigle *et al.*, 2003).

Sample displays for training monkeys on simple (left) and nested (right) sequencing of nine items. The positions of the items vary randomly from trial to trial. Monkeys are rewarded with a peanut following completely correct trials.

Fig. I.10. Classification: hierarchical classification capabilities revealed in non-human primates.

it is derived, and this requires a hierarchical form of control. These data show that monkeys can organize information hierarchically, now regarded as a vital precondition for the evolution of language itself.

Conclusions

From the classic work of Kohler deriving from the early part of the 20th century, insight and relational competences have evolved as new paradigms and new technology empower the investigator to ask more focused question and eschew many of the linguistic and manipulative constraints on the expression of animal **cognition** that have traditionally left it underspecified. Now, insight and the 'suddenness' of a solution, as emphasized by Kohler, are less important than the unearthing of mechanisms for a principled solution that can underwrite planning and goal-directed behaviour in a non-reactive, intelligent way. Based on studies of ordering and executive functioning in particular, comparative research is beginning to reveal important evolutionary precursors to human cognition and language alike. (BMcG, MMcG)

See also: Comparative psychology

References

Johnson-Pynn, J., Fragaszy, D.M., Hirsh, E.M., Brakke, K.E. and Greenfield, P.M. (1999) Strategies used to combine seriated cups by chimpanzees (*Pan troglodytes*), bonobos (*Pan paniscus*), and capuchins (*Cebus apella*). *Journal of Comparative Psychology* 113, 137–148.

McGonigle, B. and Chalmers, M. (1977) Are monkeys logical? *Nature* 267, 694–697.

McGonigle, B. and Chalmers, M. (2002) The growth of cognitive structure in monkeys and men. In: Fountain, S.B., Bunsey, M.D., Danks, J.H. and McBeath, M.K. (eds) *Animal Cognition and Sequential Behavior: Behavioral, Biological and Computational Perspectives*. Kluwer Academic, Boston, Massachusetts, pp. 269–314.

McGonigle, B., Chalmers, M. and Dickinson, A. (2003) Concurrent disjoint and reciprocal classification by *Cebus apella* in serial ordering tasks: evidence for hierarchical organization. *Animal Cognition* 6, 185–197.

Terrace, H.S. (2005) The simultaneous: a new approach to serial learning. *Trends in Cognitive Sciences* 9, 202–210.

Instinctive behaviour

Traditionally, a behaviour was considered as instinctive when it was displayed in a fully functional form from its first appearance, and it showed a very limited degree of variation during the course of **ontogeny** and across individuals of the same species. This definition was based on the assumption that instinctive behaviour is under exclusive genetic control in contrast to learned behaviour, which may show greater variability among individuals and also in the same animal at different developmental stages. This view draws a clear dichotomy between instinctive–fixed and learned–plastic behaviour. Moreover, this dichotomy has led to the misleading view, which has attracted the attention of some scientists and of the popular public for decades, that instinctive behaviour is prevalent in 'lower' animals whereas learning characterizes 'higher' animals and humans. The current view rejects this dichotomy on the basis of two considerations. First, a clear distinction between learned and instinctive behaviours is basically impossible, as every behaviour is, to some extent, under genetic control and even apparently fixed, while instinctive behaviour shows some degree of variability between and within individuals.

Imprinting is a classic example of a process that is traditionally considered instinctive, but in which learning is now known to play an important part. Imprinting is defined as a **sensitive phase**, usually in the early stages of an animal's life, during which attachment to a particular object/animal takes place. Imprinting is partially fixed, as during the sensitive period animals are genetically predisposed to react to an object or animal to which they are exposed; but it also is a type of learning, as imprinted animals learn to react to objects/animals and to improve their capacity to discriminate between them over the sensitive period. Secondly, the view that the importance of instinct versus learning is linearly related to **phylogeny** is over-simplistic and does not take into account the importance of the **ecological niche** of each species. Indeed, regardless, at least to some extent, of phylogenetic relations, species facing relatively fixed ecological conditions over their lifetime may efficiently 'rely' largely on instinctive behaviour whereas species living under highly variable ecological conditions have a stronger need to modify, through learning, their response to the environment. (RV, BM)

See also: Epigenesis; FAP/FMP; Innate

Instrumental conditioning

The term instrumental conditioning (also called operant conditioning) refers to a form of associative learning/**conditioning** whereby the consequences of a behaviour influence its subsequent occurrence. For instrumental conditioning to occur, an action must be associated with a biologically relevant event (**reinforcement** or **punishment**) that changes either the probability of the behaviour recurring or some aspect of its form. (DSM)

See also: Conditioning – types of

Insulin

Insulin regulates plasma glucose concentrations, production of alternative energy sources and the ability of glucose to enter cells of the body. The **brain**, an obligate glucose user, requires a constant supply of glucose. In periods where plasma glucose is reduced, i.e. fasting, an organism will maintain a steady supply of glucose to the brain by stimulating the production and release of glucose stored as glycogen in the liver, inducing non-neural tissues to use endogenous energy reserves (i.e. free fatty acids) and restricting entry of glucose into neural tissue only. In contrast, when glucose concentrations are high, these conditions are reversed. Insulin is the principal regulatory agent of these processes, where its presence allows for the systemic uptake and storage of glucose during periods of high plasma glucose concentrations.

Stored and released from pancreatic β-cells, insulin is composed of two straight peptide chains referred to as A and B chains, with 21 and 30 amino acids, respectively. The two chains are connected by two disulfide bridges; a third intra-chain bridge connects two portions of the A chain. Insulin originates as a precursor, preproinsulin, composed of four consecutive peptides: the N-terminal, the B chain, a connecting or C-peptide and the A chain. During synthesis of preproinsulin, the N-peptide is cleaved, whereupon the remaining molecule is transported to the Golgi apparatus for final packaging. In this packaging process, the C-peptide is enzymatically cleaved and separated from the actual insulin molecule, though both are stored and released from granules in the pancreatic β-cells. The position of the pancreas on the hepatic vein 'upstream' from the liver allows that organ to quickly respond to alterations in plasma glucose concentrations as a result of pancreatic secretions. **Neuron(e)s** within the brain may also possess the ability to produce and secrete insulin, although these results are weakly supported, particularly for the adult brain. The mechanisms described in this definition will not consider the brain's release of insulin.

The major releasing cue for insulin is increased plasma glucose concentrations, sensed by a negative feedback mechanism in the β-cells. Insulin release is controlled by an enzyme in β-cells – glucokinase – which serves as the first and rate-limiting step of insulin release. Glucokinase is active when plasma glucose concentrations are within an appropriate physiological range, initiating a G-protein system that causes granular fusion with the cell membrane and insulin release. The duration of this entire process is less than 60 s once heightened glucose is sensed. Conversely, a drop in plasma glucose concentrations will inactivate glucokinase, halting the G-protein system and subsequent insulin release.

Although plasma glucose concentration is the principal factor affecting insulin release, other factors can be potent as well. A variety of hormones (i.e. **cholecystokinin** and **glucagon-like peptide-1**) released by the gut in response to the ingestion of food enhance the glucose-dependent release of insulin. Products of digestion, such as amino acids, also enhance insulin release. Even the sight and smell of food preceding ingestion, referred to as the cephalic phase of digestion, induce insulin secretion, ensuring maximal uptake of glucose from the blood supply before being lost through excretion. Other factors inhibit insulin secretion, such as leptin, a hormone released from adipose tissue that decreases energy intake over both short- and long-term periods. Insulin will inhibit its own secretion as well.

Upon release into the bloodstream, insulin comes into contact with target cells in a number of tissues, principally the liver, adipose tissue and skeletal muscle. Once contact is made with the target cell, insulin binds with the extracellular α portion of a glycoprotein embedded in the cell wall that has an intracellular β portion extending into the intracellular space. Binding of insulin to the extracellular portion initiates intracellular tyrosine kinase activity and allows for phosphorylation of insulin receptor substrates (IRS-1 and IRS-2). These substrates serve as docking sites and allow for the activation of other protein kinases, ion channels and cellular processes that results in the transcription or repression of genes that induce plasma glucose-regulating processes, i.e. activation of GLUT-4 that facilitates transport of glucose into the cell.

The specific processes that insulin induces are extremely broad and only briefly described here. During periods of high plasma glucose concentrations, insulin promotes the uptake and storage of plasma glucose into glycogen, thus reversing plasma glucose concentrations. These effects are seen in the liver and skeletal muscle. Principally in the liver, insulin also inhibits hepatic glycogenolysis, the breakdown of glucose from glycogen, and gluconeogenesis, the production of glucose from precursor amino acids. The liver will also store glucose in the form of free fatty acids in response to insulin.

Skeletal muscle, in addition to storing glucose, will oxidize a percentage for energy production and inhibit the uptake of free fatty acids, an alternative energy source, as a result of insulin binding. During periods of high plasma glucose, adipose tissue responds to insulin by promoting glucose transport into the adipose cell, where it is stored as triglycerides. Hormone-sensitive adipose tissue lipase, which causes the breakdown of triglycerides, is inhibited by insulin. Insulin promotes the transport of protein and amino acid into all tissues as well as inhibiting proteolysis, the breakdown of stored proteins.

The importance of insulin for normal functioning becomes apparent when this regulating system is dysfunctional, as is the case for type I and type II diabetes. Type I diabetes, or insulin-dependent diabetes mellitus, is caused by a genetic abnormality that eliminates or greatly reduces insulin production and secretion. Without insulin, glucose use by insulin-dependent tissues is greatly reduced and hepatic gluconeogenesis is continuous, resulting in extremely high levels of plasma glucose. Excessive plasma glucose will be excreted in the urine, pulling water and salts with it as a result of changes in osmotic potential, and induce **hunger**, causing the organisms to eat more, which worsens the situation. Additionally, products of lipolysis and ketogenesis can cause a drop in blood pH and become fatal if prolonged. These and other symptoms are hallmarks of Type I diabetes. Type II diabetes, or non-insulin-dependent diabetes mellitus, has become a considerable problem in contemporary times as obesity and sedentary lifestyles have become commonplace. In this condition the body's cells become resistant to insulin, resulting in abnormally high plasma glucose concentrations. (MT)

Further reading

Berne, R.M., Levy, M.N., Koeppen, B.M. and Stanton, B.A. (1998) *Physiology*. Mosby, New York.

Insulin-like growth factor (IGF)

Insulin-like growth factor (IGF) is a growth-hormone-like substance responsible for the development of a variety of tissues. In response to **stress**, growth hormone levels may rise, but IGF levels will generally fall. IGF is also of interest in relation to breed differentiations; for example, variations in the size of different breeds of dog have been traced to early variation (**mutation**) in the genes regulating the production of IGF. (DSM)

Intelligence – comparative

An interesting aspect of human **cognition** is our seeming desire to classify others according to their level of intelligence (higher cognitive capacities). Ranking humans according to their intelligence has been divisive and controversial, mostly because measurement of intelligence is by no means straightforward. The most widely used psychometric test, the intelligence quotient (IQ) test, has been criticized because it is unable to measure intellect free from the influences of culture and previous education. It is also widely agreed that human intelligence is more than the sum of the parts that can be measured. Despite substantial criticisms, similar tests continue to be applied to human populations, e.g. cognitive ability tests are commonly given to 11-year-old UK school children, so that their subsequent academic progress can be related to this measure of their 'baseline' reasoning ability.

If assessing differences in human intelligence is so difficult, it might seem that comparing different animal species would be a hopeless task. And yet newspapers and websites are full of articles that ask such things as 'are dogs smarter than cats?'. Many scientists, too, seem quite happy to make certain judgements, e.g. 'animals differ in intelligence and humans are usually considered to be by far the most intelligent'.

Flowing from these assumptions, a large body of work has sought to understand what aspects of **brain** structure and function might underpin intelligence. Brain size is an obvious candidate but, as the brain size of elephants and whales exceeds that of humans, it is often dismissed. The most commonly cited correlate of intelligence is therefore *relative* brain size (i.e. the ratio of brain size to body mass). At first sight this seems appropriate; human brains comprise approximately 2% of body mass, a relatively high proportion for a large mammal. However, the relationship between relative brain size and intelligence is less clear when smaller mammals are considered. The brains of shrews, for example, comprise 10% of their body mass, but few would argue for shrews to be considered as being more intelligent than humans. These difficulties have not stopped the search. Other candidate features that have been examined as possible correlates of intelligence include: (i) the encephalization quotient (deviation of brain size of a species from expected brain sized for a standard species of the same taxon); (ii) the absolute or relative size of the cerebral cortex; (iii) the size of the prefrontal cortex; (iv) the number of cortical **neuron(e)**s; or (v) the conduction velocity of the cortical fibres.

It could be convincingly argued that this whole approach is at best a 'fishing' exercise. Searching at increasingly greater resolution is almost guaranteed to uncover some neuro-anatomical feature of the human brain that is larger, denser or faster than that of any other species. But this does not prove that the feature in question is responsible for human intelligence, or that we could in any way assume that animals with these larger or faster brain structures were more intelligent than their smaller- or slower-brained cousins. Rather than making assumptions about the intelligence levels of different animals and then searching for neural correlates, others have argued that we should devise tests of intelligence that can be applied across different species.

Experimental attempts to compare the intelligence of different species have focused on differences in performance on cognitive tasks such as learning set formation. Such tests involve comparing an animal's later performance with its own initial performance, thereby measuring the extent to which the animal 'learns to learn'. The advantages of this approach are that many confounding factors, such as individual or species differences in motivation to achieve a food reward, can be eliminated. Early results suggested that rhesus monkeys became progressively faster at learning successive discriminations, showing a grasp of the general principles involved and applying these to new exemplars, whereas rats and squirrels showed little tendency to improve their learning speed over successive tasks. However, subsequent work shows that the exact nature of the cues used can significantly affect the ability to form a learning set. Dolphins and horses have both found it difficult to form a learning set when visual stimuli are used, but have been able to form one if auditory or spatial cues (respectively) are employed. If the specific nature of the cues provided can influence apparent intelligence within a species, then the difficulties of drawing conclusions about comparative intelligence between species should be very apparent. Species unarguably differ in their perceptual capacities and will adapt differentially to any testing environment. It may be that we can never fully overcome these confounding factors to provide a measure of pure intelligence.

There are two contrasting views about the nature of animal intelligence. The 'ecological' view of learning emphasizes the role of evolution in shaping cognitive mechanisms shaped for particular needs within specialised niches. This view predicts that there will be biological constraints on learning – for example, that animals will be able to associate novel tastes with subsequent poisoning but not novel sounds, or that chickens could be trained to work for a litter substrate by treading a panel but not by pecking a key – but it also raises doubts as to the existence of any unitary process of intelligence. Some adherents of this 'ecological' view believe that evolutionary pressures have forced solutions to specific problems that are dealt with by specialized psychological modules that may map on to particular brain regions or sub-structures. The logical outcome of this view is that all species are equally intelligent in their own ways.

The alternative, 'general process', view regards intelligence as a set of learning mechanisms, similar across all vertebrate species, that can be brought to bear flexibly to solve a range of problems. Proponents of this view argue that they have been unable to find any convincing examples of qualitative differences in cognitive processes between vertebrate species. They argue that associative formation is the critical problem-solving process available to animals and that the same mechanisms are relevant in all ecological niches. This view also regards all animals as equally intelligent – MacPhail's null hypothesis – although for different reasons. The truth may lie somewhere between the two claims, and it is perhaps most likely that cognitive function relies on general mechanisms complemented by a few specializations, such as the role of the hippocampus in spatial memory in food-storing birds, or the role of the Broca's area in human **language** capacity.

Scientists may not know whether dogs are smarter than cats, but at least they are starting to ask smarter questions.

(CJN)

See also: Comparative psychology; Conditioning

Further reading

Healy, S.D. and Rowe, C. (2007) A critique of comparative studies of brain size. *Proceedings of the Royal Society B* 274, 453–464.

MacPhail, E.M. and Bolhuis, J.J. (2001) The evolution of intelligence. *Biological Reviews* 71, 341–364.

Neisser, U. *et al.* (1996) Intelligence: knowns and unknowns. *American Psychologist* 51, 77–101.

Pearce, J.M. (1997) *Animal Learning and Cognition*. Psychology Press, London.

Roth, G. and Dicke, U. (2005) Evolution of the brain and intelligence. *Trends in Cognitive Sciences* 9, 250–257.

Shettleworth, S.J. (1998) *Cognition, Evolution and Behaviour*. Oxford University Press, New York.

Intensification of animal production

One of the greatest achievements of the 20th century was the dramatic increase in the efficiency of the production of food from plants and animals. Animal production scientists studying genetics, nutrition, animal management and animal disease have made possible a substantial increase in animal protein availability per person, despite a rapidly expanding human population. For example, **chicken** meat has changed since the late 1950s from a luxury, which few in any country could afford, to a widespread source of protein. However, some of the changes in animals and animal production methods that have made this revolution possible have had significant costs for the animals.

Genetic selection of animals in the course of animal breeding has changed the animals rapidly as knowledge of genetics has advanced. The rate of change is likely to be accelerated further if transgenesis and other genetic modification procedures are permitted (**see: Transgenics**). Selection has promoted fast growth, efficient conversion of feed into animal product and use of nutrients for growth and production of meat, milk, eggs, etc. In parallel with changes in the genetics of the animals, nutrient requirements and the means of supplying them have become much better understood. The magnitude of the effects is emphasized by the meat (broiler) chicken, which took 65 days to reach a weight of 2 kg in 1970 but only 35 days in 2000. Most of the rapid growth in animals kept for meat production, such as chickens, **turkeys**, **pigs** and beef **cattle**, is concentrated in the muscles and the gut, which is processing the food. Limb growth and the growth of the cardiovascular system are much slower; a consequence of which is a potential for leg and cardiovascular problems. Another is that, while meat-producing animals are fed *ad libitum* and grow very fast, breeding animals have to be starved in order to prevent obesity problems.

Other problems related to intensification of animal production are those associated with the reduction of space allowance in living conditions (**see: Confinement**) and reduction in labour costs. Buildings cost money so, if more animals can be put in each building, costs can be lowered. In some systems, animals are housed individually so that feeding a known ration is easier and management does not necessitate having to deal with **aggression**. However, individual housing of social animals has been shown to cause poor **welfare** in all species that have been investigated. The use of small pens can cause lack of both exercise and much-needed resources. Minimizing space allowance can disrupt normal biological functioning. For example, pigs defecate in identifiable living areas but, if stocking densities are too high, they have insufficient space for differentiation of sleeping, lying and feeding areas. This is disturbing for pigs, and an animal that is normally relatively clean and careful not to injure group members may become dirty and perform injurious behaviours such as **tail biting**. In order to facilitate development in intensive systems, **mutilations** of animals such as **castration**, **tail docking**, **beak trimming** and **dehorning** may be used. These farm operations may be painful or prevent normal activities.

Dairy cows have been affected by intensification of farming, in that selection and feeding for high milk yield lead to increased levels of leg and foot disorders, **mastitis** and reproductive disorders, all of which involve poor welfare. High density of animals in housing conditions increases disease, while being kept tied for long periods also results in poor welfare. Calves kept for veal production are adversely affected by individual housing, lack of exercise, inadequate space for normal behaviour and lack of fibre and iron in the diet. Beef cattle have difficulties in coping with high stocking densities and may have leg disorders if growth is too fast.

Breeding sows were first put into stalls or tethered so that more could be kept in a building, individual feeding was easier and fighting could not occur. When the welfare of these animals has been compared with sows in well-managed group-housing systems, the welfare is clearly shown to be worse in stalls and tethers. The confined sows had high levels of **abnormal** behaviour, such as **stereotypies**, unresponsiveness and high levels of aggression, together with abnormal body and brain physiology. Farrowing sows have problems trying to cope with confinement in **farrowing crates** but, in indoor systems, piglet mortality is too high in various kinds of farrowing pens. Young pigs kept in solid or slatted-floored pens at a high stocking density, and with no material to root in or manipulate, show abnormal behaviour such as tail biting and excessive aggression. High density and lack of opportunity to show high-priority behaviour result in poor welfare.

The effects of intensification that result in the poorest welfare in chickens kept for meat production (broilers) are disorders of the legs, which result in difficulty in walking or inability to walk. Clinical diseases include tibial dyschondroplasia, femoral head necrosis, valgus–varus deformity, rickets and various forms of dermatitis resulting from birds with weak legs resting on alkaline litter. In addition to leg weakness, fast-growing broiler chickens are susceptible to ascites. These common conditions, some of which also occur in turkeys and ducks, are a consequence of genetic selection, high stocking density and poor maintenance of litter quality.

Laying hen housing is quite different from that for meat chickens. Laying hens may suffer from broken bones because of lack of activity in cages, poor-quality handling and calcium metabolism problems associated with production of large numbers of eggs. If kept in battery cages, exercise is always insufficient and birds are usually not provided with a perch, a nesting place or litter or sand for manipulation and **dust bathing**. High stocking density and high metabolic demand cause problems that are also associated with the housing system used and how well it provides for birds' needs. In some countries, laying hens may also be forced to moult so as to enter a second laying cycle, usually by depriving the birds of feed for a 2-week period.

High stocking density also leads to problems in farmed fish. Trout and salmon can have insufficient oxygen in some circumstances when space allowance is low. They may also have insufficient access to food, may injure one another and are more susceptible to some diseases. Animals kept for fur production, which are not domesticated in the same way as common farm animals, have poor welfare when kept in small, barren cages that do not fulfil their needs (**see: Fur farming**).

Intensification of animal production has led to poor welfare in many farmed animals, but welfare can be improved at relatively little cost if appropriate laws and codes of practice are enforced. (DMB)

See also: Free-range animals

Further reading

EU Scientific Veterinary Committee (2001) *The Welfare of Intensively Kept Pigs*. Reports of the Scientific Veterinary Committee (Animal Health and Animal Welfare Sections). Office for the Official Publications of the European Communities, Luxembourg, pp. 179–437.

Fraser, A.F. and Broom, D.M. (1997) *Farm Animal Behaviour and Welfare*. CAB International, Wallingford, UK.

Inter-animal distance

This term was coined by Hediger (1941), describing it as the maximum species-specific distance at which **conspecifics** approach each other. Importantly, this description not only assumes that inter-animal distance is based on an animal's ability to represent distance, but also on its ability to represent itself in a distance matrix, and its ability to distinguish conspecifics from other entities. The exact distance at which an animal tolerates a conspecific or another animal is highly context-dependent, e.g. puppies tend to accept a smaller inter-animal distance than older dogs, social species tend to accept a smaller inter-animal distance than solitary species, confined compared with open spaces enhance tolerance for smaller inter-animal distance and being sexually active or enjoying a lower ambient temperature enhances tolerance for smaller inter-animal distance (e.g. Bogusch, 2002).

In his later work, Hediger (1955) refers to 'critical distance' as the minimum distance at which an animal attacks and to 'flight distance' as the minimum distance at which an animal flees (**see: Flight zone**), while 'personal distance' refers to the maximum distance at which conspecifics feel comfortable, and 'social distance' refers to the maximum distance at which within-species groups feel comfortable. All of these notions are highly context-dependent, as described above. Since Hediger's (1955) publication in English, some of his notions have been incorporated into Anglo-Saxon-based anthropology and

psychology, and have been referred to as 'proxemics' (Hall, 1966) or personal space (e.g. Sommer, 1959). Recently, Graziano and Cook (2006) have speculated that the body-centred receptive fields in the ventral intraparietal area and in the pre-central gyrus form the neural basis for the concepts of 'personal space' and 'flight zone' in macaques. (EvdZ)

References

Bogusch, C. (2002) Zum Sozialverhalten rudellebender Wildcaniden in Gefangenschaft. Vergleichende Untersuchungen an *Canis lupus*, *Cuon alpinus*, *Lycaon pictus* und *Speothos venaticus*. Unpublished Doctoral Thesis, Philipps-Universität, Marburg, Germany.

Graziano, M.S.A. and Cook, D.F. (2006) Parieto-frontal interactions, personal space, and defensive behavior. *Neuropsychologia* 44, 845–859.

Hall, E.T. (1966) *The Hidden Dimension*. Anchor Books, Peterborough, UK.

Hediger, H. (1941) *Biologische Gesetzmaessigkeiten im Verhalten von Wirbeltieren*. Mitteilungen der Naturforschung Gesellschaft, Bern, Switzerland, pp. 37–55.

Hediger, H. (1955) *Studies of the Psychology and Behaviour of Captive Animals in Zoos and Circuses*. Butterworth Scientific Publications, Guildford, UK.

Sommer, R. (1959) Studies in personal space. *Sociometry* 22, 247–260.

Interleukin

An interleukin is a form of **cytokine** that was originally identified as being produced by white blood cells and that has a wide range of stimulatory effects on other white blood cells. More than 30 different types of interleukins are now recognized, and many are not produced exclusively by white blood cells, although the term is still used. (DSM)

See also: Enkephalin; Immune system

International regulation of animal welfare

Since there is no global legislature, international regulation of animal welfare in a global economy is a potential cause for concern and tends to be done by convention, agreement to which by national governments is often voluntary. Following the publication of a convention, agreements may be developed dealing with specific issues arising from the principles laid out in the convention. For example, ASCOBANS is an agreement developed out of the Convention on the Conservation of Migratory Species of Wild Animals, which deals specifically with the conservation of small cetaceans in the Baltic, North-east Atlantic, Irish and North Seas.

There are a number of conventions that impact on animal welfare and especially animal conservation, perhaps the best known of which is CITES (the Convention on International Trade in Endangered Species of Wild Fauna and Flora, or the Washington Convention), which aims to ensure that international trade does not threaten the survival of plants and animals (**see also: Endangered species**).

The Convention on the Conservation of Migratory Species of Wild Animals (also known as CMS, or the Bonn Convention 1979), mentioned above, is concerned with the conservation of certain migratory species over land, sea and air. The Bern Convention prohibits snares, certain devices and modes of trapping as well as poisoning to kill wild animals and, more recently, a Convention for the Protection of Animals Kept for Farming Purposes has been published, and this is now enshrined in certain countries through their own national legislation, such as the Animal Welfare Act 2006 in the UK.

Different bodies are involved in convening the meetings that may lead to the publication of a convention, dependent on the special interest being discussed. Thus CITES developed from a meeting of the World Conservation Union (formerly IUCN), while the Convention on 'Good Laboratory Practice', which allows the recognition of **laboratory animal** test results internationally (and so reduces the number of **animal experiments** conducted globally), was produced by the Organisation for Economic Co-operation and Development (OECD). (DSM)

Further reading

Convention on International Trade in Endangered Species of Wild Fauna and Flora. Available at: http://www.cites.org

Harrop, S. (2000) The international regulation of animal welfare and conservation issues through standards dealing with the trapping of wild mammals. *Journal of Environmental Law* 12, 333–360.

Inter-observer reliability

Inter-observer reliability is the degree to which different observers give consistent results when observing the same phenomenon. There is the potential for an observer's past experience and sensory and perceptual abilities to bias what is attended to in the phenomenon being studied. Thus, a measure of the validity and objectivity of the data is the extent to which the observers agree about what happened. (LMD)

Intersexual selection

Intersexual selection is one of two forms of **sexual selection** (**see also: Intrasexual selection**). The theory of sexual selection was developed by **Charles Darwin** (Darwin, 1871). He explained that competition for mates between individuals within a population results in differences in their reproductive success. The characteristics expressed by the most successful individuals are favoured by sexual selection as a result of their genetic material spreading through the population.

Intersexual selection occurs when an individual chooses one mate over another based on their particular morphological or behavioural characteristics. Preferences for mates can lead to the evolution of elaborate ornaments or secondary sexual traits – for example, the colourful tail of the peacock (*Pavo cristatus*). This is because competitors with the largest, most elaborate or conspicuous traits gain access to higher-quality or more numerous mates than unsuccessful ones, and so experience a higher reproductive rate.

Although secondary sexual traits can enhance reproductive success, they may also impose large costs. These may include impeding foraging ability, energetic costs or being more conspicuous to predators (**see: Predation**). The expression of secondary sexual characters will vary according to the cost of a trait balanced against potential reproductive benefits. The balance between the costs and benefits exerted by ornaments involved in intersexual competition frequently differs between the sexes, and can lead to sexual dimorphism (morphological differences between the sexes) within a species. In general, the less particular of the two sexes tends to exhibit the greatest degree of ornamentation.

Males are frequently less particular and the more elaborate of the sexes due to the differing roles of males and females in **reproduction**. During reproduction females produce large gametes in comparison with those of males. Due to the asymmetry in gamete size, females tend to invest more than males per reproductive attempt. Females gain the most from being selective and lose the most from making mistakes, and so are typically the more particular sex. In many species males are polygamous. Male reproductive rate is constrained by the number of females that they can fertilize. Some may mate with many, others none, and so it is this sex that most frequently competes for mates. Males with traits that attract the most females will have the highest reproductive rates, resulting in a strong selective pressure for those traits. Monogamous species of males also compete for females, often for different reasons. Preferred males tend to be chosen first by females. Males that gain access to females before other males will gain a selective advantage, as those females tend to be in better condition and have higher fecundity. Breeding earlier in a breeding season is also an advantage, as resources are in greater abundance and a second brood can sometimes be fitted into one season.

For female preference to drive the evolution of exaggerated male ornaments there are two requirements. First, the preference should be directional. This usually means that the strength of a preference for an ornament is proportional to its size. Secondly, there needs to be sufficient variation between males in the size or quality of their ornaments for females to discriminate between them.

Several theories have been proposed that attempt to explain how mate preferences are maintained. Discriminating between potential mates may impose energetic costs on an individual or expose them to predators. Mate preferences will be selected against unless the benefits of being selective outweigh the costs. The two main theories, direct and indirect selection, consider how mate preferences may benefit individuals.

Direct selection refers to direct benefits such as increasing female fecundity and/or survival. For example, females gain direct reproductive benefits from choosing males with high-quality resources. If the quality of the male's resources correlates with his secondary sexual traits, females can identify which males have the best-quality resources. Territory size, parental care and sperm quality may all be advertised by male secondary sexual traits. For example, Howard (1978) found that male American bullfrogs (*Rana catesbeiana*) defend breeding sites of variable quality. Larger males obtain the better-quality sites and females prefer these larger males.

Indirect selection occurs when preferred males provide genetic material that produces offspring with high reproductive value. This may be offspring with enhanced mating success or survival. For example, many male secondary sexual traits are heritable and correlate with viability. Female preferences for male traits that indicate genetic quality are selected for as their offspring have a high survivability (good genes theory). For example, in the barn swallow (*Hirundo rustica*), female preference is correlated with male tail length (Möller, 1988). Males with longer tails sire offspring that experience fewer parasites in the nest and, as a result, have increased survivability (Möller, 1989, 1990). Parasite resistance is genetically passed down from these high-quality males to the offspring, so that mates of preferred males benefit from higher fecundity. Traits that advertise male quality act as a 'handicap' to the bearer such that low-quality males experience a greater cost to maintain them than they gain from the reproductive benefits (Zahavi's Handicap Principle). This ensures that advertising is honest. For example, barn swallow males with experimentally lengthened tails experience reduced foraging ability, and poor growth the following moult.

Heritable secondary sexual traits that do not correlate with viability may also be of indirect benefit to a female. The Fisherian runaway (or sexy sons) hypothesis predicts that once a preference for a trait has originated, a male trait and the female preference for that trait become genetically linked. Females with this preference gain from having male offspring that are attractive to other females, and this enhances their overall reproductive success. The preference evolves as a correlated response to the evolution of the male trait. If female preference is directional and there is sufficient variation in the male trait, males with the trait become sexually selected such that the female preference and the male trait evolve in a self-reinforcing or 'runaway' fashion.

Debate on intersexual selection has, up until recently, focused on the evolutionary forces maintaining secondary sexual traits and the preferences for them. Current interest is now focusing on how traits and their preferences might have originated, and how intersexual selection may contribute to speciation. (TL)

References and further reading

Andersson, M. (1994) *Sexual Selection*. Princeton University Press, Princeton, New Jersey.

Darwin, C. (1871) *The Descent of Man, and Selection in Relation to Sex*. Murray, London.

Howard, R.D. (1978) The evolution of mating strategies in bullfrogs, *Rana catesbeiana*. *Evolution* 32, 850–871.

Möller, A.P. (1988) Female choice selects for male sexual tail ornaments in the monogamous swallow. *Nature* 332, 640–642.

Möller, A.P. (1989) Viability costs of male tail ornaments in a swallow. *Nature* 339, 132–135.

Möller, A.P. (1990) Effects of a haematophagus mite on the barn swallow *Hirundo rustica*: a test of the Hamilton Zuk hypothesis. *Evolution* 44, 771–784.

Interspecific behaviour

Interspecific interactions can take a variety of forms. On a purely ecological level, one can look at so-called assemblage rules, i.e. how species' communities build up and which species promote or hinder coexistence. On the behavioural level, many interspecific interactions can be classified as competitive. One can distinguish competition due to similar diets, competition because the species are in a predator–prey relationship (**see: Predator avoidance**) and competition because the species are in a host–**parasite** relationship. Competition due to similar diets might often be difficult to observe, as species may have found 'evolutionary' ways to reduce behavioural competition. The smaller species might simply avoid the presence of the more dominant species and thereby avoid **aggression**.

Predator–prey and host–parasite interactions are different, as hosts and prey cannot solve the conflict by avoiding their enemies. These relationships are best described as an 'evolutionary arms race' that may lead to co-evolution: both sides try to stay ahead in their struggle and any successful **mutation** selects for a successful counter-mutation. Co-evolution in host–parasite interactions is much better documented than for predator–prey relationships. The problem with predator–prey systems might be that hunting techniques and escape strategies are often much specialized. Take the example of an antelope and a pursuit hunter: faster-running predators may select for faster-running prey that selects for faster-running predators, etc., but to show that running speed has really co-evolved over time may seem like an impossible task.

A promising way to look for co-evolution between predator and prey is to look for systems in which: (i) hunting strategies and escape behaviour are quite flexible; and (ii) the outcome of the 'arms race' might depend on variation in ecological conditions. These assumptions are met in the 'arms race' between chimpanzees and their main prey, red colobus monkeys. A comparison between the Taï forest in the Ivory Coast and Gombe in Tanzania revealed that two parameters mainly influence the outcome of the 'arms race', namely relative size differences and the structure of the canopy. In Taï, size differentials between red colobus and chimpanzees are large (about 1:6) and the canopy is high and closed. As a consequence, red colobus males do not often defend their group members (they risk being killed as well) and all monkeys rely on **crypsis** at early stages of an encounter and escape routes during an actual attack. In response, chimpanzees actively search for their prey and then try to stalk their prey and hunt in a highly coordinated way to block prey escape routes. In response to stalking, red colobus associate with Diana monkeys that are particularly able to detect any sort of ground predators. Chimpanzees therefore avoid associated red colobus groups and mainly hunt in solitary groups. In Gombe, in contrast, red colobus males are large relative to the chimpanzees (about 1:3) and the canopy structure is broken. As a consequence, stalking by chimpanzees is not possible and the red colobus do not need another species as watchmen. As the monkeys cannot hide in the open canopy, chimpanzees do not have to search for their prey. Due to the limitation in horizontal escape routes, solitary attacks by chimpanzees can be successful and the main counter-defence is male red colobus aggression. Thus, co-evolution can be observed with respect to several components of predator and prey behaviour. A major unsolved question, then, is whether these adaptations and counter-adaptations are due to evolution on the genetic level or whether we are dealing with cultural evolution based on learning and tradition.

Mixed-species associations are a widespread phenomenon that occurs in a variety of primates and other mammals, birds and fish. Some of these associations may occur mainly due to chance encounters of individuals/groups with overlapping home ranges. In many cases, however, associations are actively formed and yield benefits to the partners. Benefits can either accrue due to increased foraging success or due to reduced **predation** risks. Often, benefits due to one primary advantage might create additional secondary advantages. For example, a reduced predation risk due to the association with another species may lead to reduced vigilance, which frees time that can be allocated to foraging. The relative importance of foraging and anti-predation benefits may vary according to the system under investigation. In the case of African forest monkeys, the anti-predation benefits appear to be the main selective force for mixed-species associations. In coral reef fish, however, commensalistic foraging associations are very common: several species that dig in the sand in search of prey are followed by a variety of species looking for flushed prey items. There is also apparent interspecific cooperative hunting occurring between groupers and moray eels in the Red Sea. On the other hand, many mixed-species shoals of fish have probably evolved to reduce predation pressure.

The frequent occurrence of mixed-species associations under natural conditions may be of animal welfare relevance, for example in **zoos**. Where space and intraspecific competition may often limit the number of individuals in a cage/enclosure, addition of allospecific individuals could augment the group without increasing conflicts. Group augmentation could reduce stress factors, such as being stared at by visitors, through dilution effects. An efficient way to use the occurrence of interspecific associations under natural conditions would be to merge adjacent cages/enclosures of species that tolerate/seek each other's presence. Under these conditions, individuals would have more space and larger group sizes while nothing has changed for the zoo in terms of animals per square metre. The only caution would be that species with different diets would probably fit together better under captive conditions than species with similar diets as, in the latter case, there is potential for strong competition between the species.

Also, visitors might find mixed groups more interesting to watch: mixed-species groups should be more active than monospecific groups due to perceived increased safety, and more social interactions will occur. (RB)

Further reading

Bshary, R. and Noë, R. (1997) Anti-predation behaviour of red colobus monkeys in the presence of chimpanzees. *Behavioural Ecology and Sociobiology* 41, 321–333.

Futuyma, D.J. (1986) *Evolutionary Biology*. Sinauer Associates, Sunderland, Massachusetts.

Noë, R. and Bshary, R. (1997) The formation of red colobus–Diana monkey associations under predation pressure from chimpanzees. *Proceedings of the Royal Society London Series B* 264, 253–259.

Intra-observer reliability

Intra-observer reliability indicates how stable an individual's responses are over different time points. A large difference in responses demonstrates low intra-observer reliability and a small difference demonstrates high intra-observer reliability. (LMD)

Intrasexual selection

Intrasexual selection is one of two forms of **sexual selection** (**see also: Intersexual selection**). Intrasexual selection occurs in the form of competition between members of the same sex (usually males) for access to mates. Similarly to intersexual selection, intrasexual selection may lead to the evolution of conspicuous ornaments, known as secondary sexual traits. These are involved in contests between members of the same sex and would otherwise be detrimental to survival according to natural selection. For example, display, song, coloration, size and weaponry may all be used during intrasexual competition. These ornaments may be totally absent in one sex, or present in a more elaborate form in one sex than the other, known as sexual dimorphism. Males are generally the more elaborate of the sexes in terms of secondary sexual traits, due to the different roles of males and females in **reproduction**.

While secondary sexual traits have no direct role in reproduction, they may increase reproductive success during competition. This competition may occur in the form of pushing, wrestling or other physical conflict. However, competitors often refrain from actual combat in order to avoid **injury** and, instead, engage in indirect assessment of strength or status via their secondary sexual traits. For example, male common toads (*Bufo bufo*) assess body size and thus the fighting capability of competitors through the pitch of their croak, so that males with deeper croaks are avoided (Davies and Halliday, 1978). The fighting ability of an individual can be also assessed through a series of ritualized displays.

In another example, red deer stags (*Cervus elaphus*) compete for females during each rut to obtain the largest harem, and therefore the greatest reproductive success. To avoid potentially fatal injuries, males will withdraw from contests with males that they are unlikely to beat (Clutton-Brock and Albon, 1979). Male fighting ability is assessed first by the rate of roaring, as an indicator of physical condition. Stags that are well matched in their roars will walk parallel to each other, possibly to provide a closer assessment of their physical prowess. Only if the stags are still well matched will a battle commence, involving interlocking of antlers and pushing. As both winner and loser may be injured during these fights, it is of benefit to both competitors to settle a contest at an early stage, and escalation into actual fights is rare.

Secondary sexual traits used to assess other males are not always directly related to their combative strength. These types of secondary sexual traits may act as indirect advertisements of male fighting ability and are often termed 'status-signalling badges'. However, they may still act as honest indicators of male quality. For example, only the highest-quality males can maintain conspicuous secondary sexual traits while still avoiding predators. This indicates that a male bird with more colourful or brighter plumage would be more likely to win a contest than a more cryptic male. Badges of status are supported by dominant behaviour, and this also prevents cheating. For example, among male Harris sparrows (*Zonotrichia queula*), those with the darkest plumage are dominant over paler ones. Experimental darkening of plumage of subordinate males does not raise their status, as their behaviour does not alter (Rohwer and Rohwer, 1978). Dominant males are more aggressive towards other males, and it is this behaviour that backs up the visual signal of dark plumage.

Although intrasexual selection generally favours more dominant males, under certain conditions less dominant males can also experience high reproductive rates. In some situations these males may even experience greater success than dominant males. The differences in the costs and benefits of being a dominant versus a subordinate male may lead to alternative mating strategies. For example, in the coho salmon (*Oncorhynchus kisutch*), there are two genetic morphs of male with two distinct life history pathways (Gross, 1985). All the young fry start off their life in streams where they remain for 1 year, and then all migrate to the oceans. Females return to the streams to spawn when they reach maturity, at 3 years of age. Hooknose males also mature at 3 years, and also return to the spawning grounds. These males have conspicuous red coloration and enlarged jaws and teeth. A second type of male, the jacks, mature a year earlier and return to the streams after 2 years. They are relatively small and mimic females, as they lack any secondary sexual traits. On reaching the spawning grounds, females excavate nests where they deposit their eggs. Hooknoses fight between each other for access to the eggs of the spawning females, using their hooked jaws and teeth. The jacks avoid conflict and, instead, use their small size and inconspicuousness to sneak close to the spawning females. This enables the jacks to surreptitiously fertilize the eggs. As the jacks can get close to the females' nests, they have a good chance of fertilizing the eggs. However, the larger size of hooknoses means that they produce more sperm with each ejaculation.

Intrasexual selection can occur in even more subtle forms, such as via sperm competition. In some species females copulate with a large number of males and store their sperm in the body to be used for fertilization later. Male competition occurs by preventing other males' sperm from fertilizing the female's eggs. Some males have a modified penis, which can remove other males' sperm before depositing their own sperm. Other males deposit a sealant into the female's genital opening after copulating to prevent other males from fertilizing her; this is referred to as a copulatory plug.

The boundaries between intersexual and intrasexual selection are not always clear. For example, during sperm competition females may express a choice about which male's sperm is utilized for fertilization. Females may also 'eavesdrop' on male contests when making a choice about which male to mate with. Therefore, the evolution of a particular secondary sexual trait may be influenced by both inter- and intrasexual selection simultaneously. (TL)

References

Clutton-Brock, T.H. and Albon, S.D. (1979) The roaring advertising of red deer and the evolution of honest advertisement. *Behaviour* 69, 145–170.

Davies, N.B. and Halliday, T.R. (1978) Deep croaks and fighting assessment in toads (*Bufo bufo*). *Nature* 274, 683–685.

Gross, M.R. (1985) Disruptive selection for alternative life histories in salmon. *Nature* 313, 47–48.

Rohwer, S. and Rohwer, F.C. (1978) Status signalling in Harris sparrows: experimental deceptions achieved. *Animal Behaviour* 26, 1012–1022.

Intraspecific behaviour

Intraspecific behaviour refers to all interactions that occur between members of the same species. Over their lifetimes, animals may engage in a wide variety of cooperative and competitive behaviours (both within and among the sexes) that are associated with **communication**, **territoriality**, **breeding** behaviour and **group** living. In the wild, even animals that are primarily solitary during their adult lives almost always interact with siblings or other **conspecifics** when they are young, and with potential mates once they become sexually mature.

An appreciation of intraspecific behaviour is vital to: the establishment of successful **captive** rearing and **breeding programmes**; the maintenance of healthy **zoo** populations; the improvement of the welfare of domestic animals that are housed together; and the management of threatened populations in the wild. Accounting for behavioural interactions among conspecifics goes beyond separating aggressive individuals or bringing together potential mates; it must also including a firm grasp of the underlying **social behaviour** and breeding behaviour of the species one is trying to manage.

Signalling, communication and territoriality

Animals communicate with each other in a wide variety of ways. Visual, olfactory and auditory 'signals' evolve over many generations in response to the relationship between the producer of a signal, the recipient and the environment inhabited. For example, male tungara frogs (*Physalaemus pustulosus*) produce different frequency **vocalizations** that not only provide information on their location but also vary in relation to their body size. In several bird species found in noisy environments (such as near moving water), tail flicking and body bobbing have evolved as methods of communication, presumably because visual displays of territoriality are more efficient than vocalizations.

Animals can passively, as well as actively, defend their territories against encroachment by conspecifics. Passive territoriality includes methods of communicating without direct contact between individuals, such as through the use of scent. Giant pandas, for example, use scent glands located beneath their tails to advertise site-specific dominance. They may mark objects by squatting, backing up, cocking a leg or even performing a handstand to emphasize their stature. Many other species engage in scent-marking behaviours, not only through the use of scent glands but also through urine or faeces. In contrast, active territoriality often includes chasing or attacking intruders. In birds, however, active territoriality also includes singing, which is used by males not only to attract mates but also to advertise their claim to a particular area so that conspecifics can identify the 'owner' of that area.

How and when animals defend territories is an important consideration in many applied contexts. Forcing normally territorial individuals to live too closely together might lead to aggressive displays, chronic **stress** or weight loss in less competitive individuals and, in extreme cases, **injury** or even **death**. Conservation biologists and wildlife managers must also take territory and/or **home range** size into account when designing reserves, since reserves that are too small or fragmented may not be able to accommodate viable populations.

Breeding behaviour and alternative reproductive tactics

In order to maximize their reproductive success, individuals must be able to recognize and attract members of the opposite sex and select the best mates. Over evolutionary time, natural selection has led to the development and maintenance of an incredible diversity of mating and social systems. These include alternative strategies that exploit the energy of individuals other than the genetic parents, in order to assure offspring success.

Depending on the resources available, parents might pursue a number of alternative reproductive tactics. Female wood ducks (*Aix sponsa*), for example, can choose between establishing a nest in which they lay their eggs and incubate them, or laying them 'parasitically' in the nests of other females and providing no parental care. Younger, less experienced, females more often lay eggs parasitically; younger females frequently arrive later at the breeding area and thus have fewer nesting options than older females that have arrived earlier.

Males also pursue alternative reproductive tactics. For example, the dominant male in a group of savannah baboons (*Papio cynocephalus*) tends to follow the female around during oestrus, attempting to monopolize mating opportunities. However, subordinate males may work together in coalitions to competitively exclude the dominant male and gain access to the female. In species such as salmon, alternative tactics culminate in the presence of multiple male forms, or 'morphs', that compete for access to females. Salmon morphs consist of larger and competitively dominant 'hooknoses' and smaller, furtively mating 'jacks'. Males of the two morphs follow different developmental trajectories, are strikingly different in size and display completely different behaviours when attempting to mate (**see: Intrasexual selection**).

Group living

Many animals live in groups for part or all of their lifetimes. What are the advantages of living in a group, and what are some of the costs? Living in a group has many benefits, including increased **predation** vigilance, more effective avoidance of attacking predators (due to a dilution effect), reduced exposure to the elements and improved ability to find and defend certain food resources. There are costs to group living as well: (i) moving in a group may increase the chances of being detected by predators; (ii) disease can spread more quickly; and (iii) food sources can run out more rapidly. Thus, individuals should only live together when the benefits of group living consistently outweigh the costs.

In many cases, members of a social group are close kin (**see: Altruism**). This high degree of genetic relatedness may partially explain why some species breed cooperatively. Cooperative breeding is the social system in which adult individuals, in addition to the genetic parents, aid in the rearing of young. This system is rare in vertebrates but occurs in a few mammals (e.g. Latin American marmosets and tamarins, lions and African wild dogs) and in about 3% of bird species, including the white-throated magpie-jay of Central America. Among cooperatively breeding species, some individuals (frequently the offspring from previous years) act as babysitters, helping to feed and protect offspring that may not be their own. Individuals may opt to help rather than establish their own nests if territories are unavailable, food is extremely limited or there are intrinsic benefits to remaining in the group (such as opportunities to learn parenting or foraging skills, or increased chances of breeding directly within the group). One fascinating example of a cooperative breeder is the naked mole-rat (*Heterocephalus glaber*), a furless South African rodent that spends its entire life in large underground colonies. Similar to honeybees, each mole-rat colony has a single reproductive 'queen' and many non-reproductive workers that assist in colony maintenance and rearing of young. (ECB, RWVB)

See also: Breeding; Sociobiology

Further reading

Alcock, J. (2005) *Animal Behavior: an Evolutionary Approach*. Sinauer Associates, Sunderland, Massachusetts.

Caro, T. (1998) *Behavioral Ecology and Conservation Biology*. Oxford University Press, New York.

Koenig, W.D. and Dickinson, J.L. (2004) *Ecology and Evolution of Cooperative Breeding in Birds*. Cambridge University Press, Cambridge, UK.

Krebs, J.R. and Davies, N.B. (1993) *An Introduction to Behavioural Ecology*. Blackwell Scientific Publications, Oxford, UK.

Introspection

Introspection is the process of 'looking' into one's own mind in order to 'see' its content – e.g. what one thinks, feels or desires. Few psychologists or philosophers today would conceive of introspection on the model of sensory perception, however. Current orthodoxy is that there is no 'inner eye'. Rather, one discovers what one thinks and feels by wondering (or thinking about) what one would say. (PAR)

Iron provision

Iron (Fe) is required for the formation of haemoglobin, the body's major oxygen carrier, and for oxidizing catalysts. It is also present in blood as ferritin and transferrin and in muscles as myoglobin. Milk has a particularly low concentration of iron in relation to other minerals, but in the suckling period there is often a store of iron at birth in the neonatal liver. The restriction of iron availability to bacteria limits their growth in the mammary gland and thus helps to prevent **mastitis**. Lactoferrin binds iron and thus controls its release in milk. It is particularly present in the milk residues in non-lactating mammary glands and in mastitic milk, suggesting that it has a protective role in limiting iron availability for bacterial growth.

Iron deficiency may be found in milk-fed calves and may be deliberately induced to produce pale meat, which consumers associate with tenderness. The deficiency is typified by anaemia, reduced appetite, poor growth rates and inability to cope with exercise. Iron also plays a part in the formation of immunoglobulins, so that iron-deficiency anaemia is associated with reduced immunocompetence. Iron is very effectively recycled within the body and there is little urinary excretion. Iron absorption is enhanced in animals that are iron deficient, but the milk-fed calf will still develop anaemia within 8–12 weeks unless iron supplements are provided. Veal calf producers aim to create low muscle myoglobin contents without adversely affecting appetite or growth rate. This requires approximately 25–50 mg Fe/kg of dietary DM, although iron availability is variable. Milk contains only 5 mg Fe/l, and the stores provided to the calf at birth are therefore vital. Calves that suckle their dams at pasture invariably start consuming a few leaves of grass before the anaemia becomes established, and therefore are not at risk. The iron content of leaves is much greater than that of seeds or milk, and iron deficiency does not occur in cattle fed forage-based diets. Indeed, an iron supplement should not be added to the diet of adult cattle at risk of hypocupraemia, because it will inhibit the absorption of copper. Piglets reared indoors, in intensive systems away from natural sources of iron that they may encounter through foraging, may require additional iron supplementation to prevent anaemia.

Iron status can be assessed from the haemoglobin content of blood. There is a critical level at which the muscle myoglobin content will be reduced, but not growth rate or the animal's ability to exercise. Legally, in the European Union (EU), all calves must be fed sufficient iron to ensure that this minimum blood haemoglobin level (4.5 mmol/l) is met. If less than this, then a supplement of between 25 and 50 mg Fe/kg should be provided. Providing solid feed as required by EU legislation will also help to prevent low blood haemoglobin levels. There is a need for consumers to be more aware that red veal meat is of equally good quality as white meat and will indicate that the calves have not suffered impaired welfare as a result of hypoferraemia. (CJCP)

ISAE

The International Society for Applied Ethology (ISAE) was created in 1966 as the Society for Veterinary Ethology, with a primary membership of UK-based veterinarians. It quickly expanded to encompass clinicians and non-clinical researchers all over the world working in all areas of applied animal behaviour. It was renamed in 1991 to reflect this growth and it continues to be the principal learned society for workers in the field of applied **ethology**. (JNM-F)

Further information

Homepage of the ISAE: http://www.applied-ethology.org/

Islam

Islam shares the same biblical tradition as **Judaism** and **Christianity**, and thus builds on an ambivalent attitude towards non-human beings. The Islamic scriptures, the Koran and the Hadith, describe the ideal human–animal relationship. An animal-friendly perspective comes to the fore through several themes. One is the reward for those who are gentle towards animals: 'There is a meritorious reward for every act of charity and kindness to every living being' (Hadith Awn 7 222, 2533). A related theme is the reason for this – God loving all his creatures: 'All creatures are like a family of God; and He loves the most those who are the most beneficent to His family' (Hadith Mishkat 3, 1392). Thus, from this perspective of equality, 'A good deed done to an animal is as meritorious as a good deed done to a human being, while an act of cruelty to an animal is as bad as an act of cruelty to a human being' (Hadith Mishkat Bk 6, Ch. 7, 8, 178).

Even if the authenticity of these passages is disputed, i.e. whether they can be traced back to Mohammed or not, it shows that respect for animals was expected from a faithful Muslim in his time. From the 16th to the 18th centuries, Europe was greatly influenced by Islamic thinking, including with regard to human–animal relationships. Complaining about European brutality towards animals, the philosopher Francis Bacon and authors like Alexander Pope and Richard Steele made references to Muslims, appreciating their care for animals.

Halal slaughter takes place in the name of Allah, during recitation of special verses of gratitude. As described in the Koran, calm treatment of the animal and a quick but smooth cut of the animal's throat is important. Blood is the sign of life and so traditional halal **slaughter** proceeds without stunning in order to bleed the animal. Freedom of religious expressions comes into conflict with many countries' legal framework for animal protection. However, negotiation on this issue has led some Muslim groups to accept stunning, for instance with electricity, which is the main method used in New Zealand and other countries. (HR)

See also: Religious slaughter

Further reading

Preece, R. (2001) *Awe for the Tiger, Love for the Lamb. A Chronicle of Sensibility to Animals.* UBC Press, Vancouver, Canada.

Isolation

Isolation is a word used in varied contexts by people concerned with **ethology** and **welfare**. The term may be used, for example, in relation to issues of disease transfer, naturally occurring isolation of free-ranging groups of animals by geographical boundaries or the artificially occurring isolation imposed during animal management.

In relation to issues of disease transfer, the term 'isolation' can be applied to techniques used to generate animals free of particular disease-causing agents (such as the existing specific pathogen-free lines of rodents). In such cases, the stocks of animals are generally re-derived using an isolator to eliminate specified infective agents (such as viruses, bacteria and helminths) before being subsequently barrier-maintained to prevent reinfection. Elimination of disease agents is especially important in animals with immune deficiencies, such as nude mice, that are very vulnerable to infection. There is sound evidence that mild or 'latent' diseases can influence behaviour and physiology; these influences can alter the results of laboratory tests. In a related fashion, 'isolation' can be used as a synonym for quarantine, a technique used to reduce the chances of disease spreading to stock from newly introduced animals. New stock (in laboratories, on the farm, in **zoos** or in the country) is frequently held in relative isolation before being allowed to interact with existing residents. Individual laboratory, companion and zoo animals may also be isolated after surgery (to reduce the chances of **conspecifics** damaging sutures, etc. in anaesthetized subjects), sometimes with behavioural repercussions (see below).

Geographical isolation on 'islands' (areas cut off by geographical barriers – not always by the sea, as valleys and mountains may also restrict movement between locations – from the rest of the environment) provides opportunities for speciation (the derivation of new species from organisms colonizing first or already present). This is because these isolated locations provide niche opportunities, since these defined areas lack competition from specialist competitors, as found elsewhere. In addition, the pressures exerted by particular predators may be lacking. As a consequence, problems (the new species can become endangered or even rendered extinct) are frequently encountered when competitors or predators are later introduced (either deliberately or accidentally). Human (anthropogenic) effects are increasingly common, especially with the ease of modern transport systems. Good examples are the endemic bird species found on Mauritius and the Hawaiian islands that have been subsequently endangered by the rat, cat, mongoose, goat and monkey species variously introduced by humans. The lemurs of Madagascar provide an example in which endangered primates are the focus.

In terms of management, isolation can take many forms. Isolation (as in the hatching of bird eggs in an incubator or the use of **hand rearing**) is one of the deprivation techniques used to establish whether a behaviour is largely genetically programmed. The assumption is made that, if the behaviour is expressed by animals that have not had the opportunity to copy the activities of peers or adult counterparts, it must have a strong genetic component. Isolation experiments may also be used to determine whether key experiences in so-called '**sensitive phases**' of development are essential for the genesis of specific behavioural elements, e.g. the 'dialect' component of bird song in the chaffinch (Thorpe, 1958) or aspects of vision in kittens. In both cases, the developing animal is restricted in terms of the communications it can receive from members of its own species. Such restrictions are generally limited to key stages of early development. The general strategy used in such experiments is to attempt to show that a particular experience has a less potent and/or not so lasting influence when the animal is exposed to it before and after the specified time, but the analysis of the results of such experiments is often quite complex due to potential interactions between time of treatment, duration of treatment, time of assessment and impact of assessment technique on the response obtained.

There has recently been renewed interest in the fact that isolating neonate rat and mouse pups restricts the development of their cerebral cortices (it decreases their size and weight), reduces the complexity of the neuronal connections (as studied histologically) and impairs the animal's learning capacity as expressed in a range of simple tests. Some of these effects can be ameliorated by the use of inclusions (toys) or '**gentling**' (handling). These studies generally confirm that social deprivation in early life has profound and lasting effects in a range of animals. The studies seem to be currently limited to mammals, but we know also, for example, that birds show '**imprinting**' in response to early influences. Hand rearing of birds may also result in the subjects fixating on inappropriate parents, siblings or mates. It is very difficult to reverse these effects. This means that such animals cannot be subsequently released in **reintroduction** programmes and have to be maintained in captivity. They may not only be unable to thrive following release, but can damage wild populations.

In his classical studies on mother–infant interactions, Harry Harlow showed that isolating baby female primates from their mothers resulted in their showing markedly impaired parenting skills (i.e. they were generally neglectful and inefficient mothers) when they became reproductively active in later life (e.g. Harlow and Suomi, 1971). This may also be of concern in **captive breeding programmes**, as it suggests that hand rearing (sometimes the only viable option for deserted or maltreated young) may well lead to the trans-generational passage of such deficits. A consequence is that the offspring of these animals will need to be 'helped' in this manner for generations.

The term isolation is most commonly used in the area of animal welfare, however, as a synonym for individual housing – for example, the housing of **laboratory** or **companion animals** in single cages, without direct physical contact with their conspecifics. One should note that such animals are rarely truly isolated as they generally have visual, auditory and/or olfactory contact with animals in neighbouring cages, tanks or enclosures. The same communication possibilities are certainly true of the housing conditions inherent in some intensive farming systems designed to reduce potentially damaging **aggression** by physically separating the subjects (**see: Intensification of animal production**).

Most early studies suggesting that isolation is stressful employed female mice, but this conclusion was based on the increased adrenal weights of such animals that do not accurately reflect **'stress'** (individual housing increases **oestrogen** production in females and this hormone increases adrenal size in mice). Individually housed male mice generally show an increased propensity to fight. Valzelli (1973) claimed that the increased aggressiveness of such isolates was a consequence of the stress of 'social deprivation' (**see: Social stress**). From this viewpoint, the animals are considered to 'need' contact with their peers and, without such exposure, to show increased irritability. This increased irritability is then thought to reduce 'tolerance' and to increase the likelihood of the animal attacking a 'stranger'. This phenomenon was used to test anxiolytic drugs, as it was argued that a drug that reduced fighting must have ameliorated the social deprivation effect. Other workers have operated from the alternative viewpoint – that the isolated rodent is the animal least subjected to the social stress of crowding (in a standard cage it is at the lowest density). These viewpoints are actually diametrically opposite, with the former group assuming the captive animal needs social stimulation, whereas the latter maintains that they crave solitude.

Brain and Benton (1983) advanced a third alternative. They reviewed the available studies on a variety of behavioural and physiological attributes in isolates and group-housed male mice. Besides the well-documented augmentation of aggressiveness, individually housed mice show increased reactivity in a novel arena, higher motor activity in response to novelty, reduced turnover of a range of neurotransmitters, reduced adrenal weight, lower basal plasma **glucocorticoid** titre, heavier sex accessory glands, greater secretory activity of **pheromones** from the preputial gland, high gonadal **androgen** secretion, better coat condition, improved response to helminth **parasites** and better immunological responses to a range of (soluble and particulate) antigens than most group-housed counterparts.

It is self-evident that many of these changes are inter-linked. For example, androgens increase the size of the preputial gland as well as its production of sebum (a major pheromone in this species). Androgens improve coat condition by also increasing sebum production from the sebaceous glands associated with the hairs and may well increase body weight by their anabolic effects. Some workers have linked the elevated androgen levels in isolated male mice to their increased social aggression. In a similar fashion, the glucocorticoids from the adrenal cortex may be immunosuppressive and can change neurotransmitter levels in the brain.

As noted earlier, 'isolated' mice rarely completely lack social communication with their conspecifics. Investigations with mice have shown that negating olfactory communication reduces, whereas increasing these cues intensifies, the subsequent aggression characteristic of 'isolates'. Specific touch communication via the mystacial vibrissae (whiskers) seems only to have a minor effect on the later propensity to fight in mice. Other forms of **communication** have also been investigated. In the case of vision, opaque contact lenses have been used and do have a minor impact on behaviour. Congenitally blind mice are, however, perfectly capable of showing complex social behaviours such as fighting and mating. Anosmic (lacking the sense of smell) male mice, however, will neither fight nor mate. The impact of audible (to the human ear) sound and ultrasound (frequencies higher in pitch than those we can detect) do not appear to have been investigated, which is surprising as sound is used in conflict in a range of rodent and other species. Another study that attempted to totally isolate male laboratory mice from sensory communication with all conspecifics (in special ventilated boxes with activated charcoal filters) found that this procedure actually markedly reduced aggressiveness in these animals. These subjects appeared hyperreactive but largely fearful when paired with docile opponents. The studies in combination suggest that some forms of sensory communication (especially odour in mice, as vision is not very important in these basically nocturnal animals) actually modulate the effects of isolation on aggression, but that the main factor responsible for the associated behavioural change is lack of direct physical contact.

Strong parallels between isolates and socially dominant male mice were obtained using the same behavioural and physiological measures employed in the isolated versus group-housed mouse comparisons. There is evidence, using genetic markers in mice, that only the socially dominant animals are reproductively active. The changes in isolated mice might also prepare them for breeding. The fact that many of the behavioural and physiological changes in the isolated mouse seem beneficial and that they parallel changes in the social dominant make it appear unlikely that one can simply dismiss isolation-induced aggressiveness as a consequence of social deprivation. Indeed, they suggest that the changes reflect a propensity to assume a form of **territoriality** in males of this species. If this is the case, and mice are not always a social species, group housing may not always be appropriate in these animals even if the intention of such housing is environmental **enrichment**.

A review of the literature comparing isolated and group-housed laboratory rats has been undertaken. It was initially anticipated that rats (because wild rats are generally colonial rather than individually territorial) might respond very differently to such housing than mice. It was noted that the comparisons were, however, often confounded by the imposition of stress on the animals during, for example, blood sampling. Studies in the rats also showed that a very wide range of hormones were changed by very minor disturbances such as opening the laboratory door or switching on the lights. There was also some evidence that the normal mechanisms of behavioural inhibition were impaired in isolated rats. They appeared slow to abandon previously learned behavioural strategies when the conditions were changed (so that the outcome was no longer rewarding) and were generally slow to habituate (**see: Habituation**). This suggests that rats become less behaviourally flexible as a result of the isolation. Somewhat surprisingly, in general, the results with rats were often similar to those with mice (social organization did not actually appear to have much of an effect). It was concluded, however, that isolation versus group housing comparisons actually look at differential housing rather than 'social deprivation' or crowding. The choice of the baseline in such comparisons is really arbitrary. Having only two conditions also makes interpretation of the results difficult.

As one might expect, the age at which male rats and mice are differentially housed alters the amount of aggressive behaviour that they show following individual housing. The highest level of aggressive behaviour following a fixed period of individual housing in mice was seen in intermediately aged animals. Isolating mature group-housed subjects, however, may result in relatively high levels of aggressive behaviour in some lines of mice. Conversely, male mice cohabiting with other males generally show low incidences and intensities of fighting. This may be a consequence of their habituating to the **odours** of other mice (there may be a colony smell), developing a learned inhibition by the establishment of a social hierarchy or gonadal hormone secretion in subordinates being suppressed (this would reduce the production of the pheromones that normally stimulate attack in males of this species). It is likely that only despotic dominants from each group rapidly evidence marked fighting behaviour. It may take longer periods of isolation to produce similar changes in subordinate animals.

Mature, outbred, albino mice of the Swiss line show a very rapid (within 24 h) increase in aggressiveness following isolation. This ability is retained over a substantial period of adult life, suggesting that this line shows a strong tendency to assume a territorial habit. Age (rather than time in isolation) seems the key to determining the response to individual housing of Swiss mice. It must, however, be noted that mice from different genetic lines show very varied responses to individual housing. In some lines of mice, pairing a male with a female for breeding activity also augments his aggressiveness when he subsequently encounters other males. Reproductive activity also augments testicular androgen production, and breeding males lack direct physical contact with other males for the duration of the breeding attempt (they are effectively 'isolated' from them). It can be impossible subsequently to socially house such animals, which can create husbandry problems (individual caging takes up a great deal of space and other resources).

Having said that, the responses of laboratory rats and mice to isolation are remarkably similar in spite of their very different social organizations, and one should not get the false impression that all animals show identical behavioural and physiological responses to such housing. In golden hamsters, the female rather than the male is the more aggressive sex and her behaviour is largely influenced by **progesterone** (rather than androgen). Androgens increase the size of the adrenal cortex in the golden hamster. The details of responding to isolation in this species are likely to be different from more 'traditional' rodents.

There is an urgent need to fully understand the effects of isolation on animals as imposed at different times in their lives. For example, isolating animals in adulthood generally makes it difficult to subsequently group house them. If the studies on rats and mice are representative, the particular aspects of behaviour and physiology changed are likely to be wide ranging and complex. Studies on a much wider range of species and contexts are needed for laboratory, farm, zoo and companion animals. (PFB)

References and further reading

Brain, P.F. and Benton, D. (1983) Conditions of housing, hormones, and aggressive behavior. In: Svare, B. (ed.) *Hormones and Aggressive Behavior*. Plenum, New York, pp. 351–372.

Harlow, H.F. and Suomi, S.J. (1971) Social recovery by isolation-reared monkeys. *Proceedings of the National Academy of Science* 68, 1534–1538.

Scott, J.P. and Fuller, J.L. (1965) *Genetics and the Social Behavior of the Dog*. University of Chicago Press, Chicago, Illinois.

Thorpe, W.H. (1958) Further studies on the process of song learning in the chaffinch (*Fringilla coelebs gengleri*). *Nature* 182, 554–557.

Valzelli, L. (1973) The 'isolation syndrome' in mice. *Psychopharmacologia* 31, 305–320.

Istwert

The term istwert is used in applied animal behaviour to refer to the perceived reality of an animal at a given time. In this context it is recognized that animals respond to the world as they perceive it, rather than necessarily to the physical reality. The concept embraces the notion that different species (and different individuals) live in different perceptual worlds (umwelts) and it is this that influences their behaviour and welfare. Thus, an approaching individual may be perceived as a potential threat by some, but a source of social support by others. The physical reality at the time is the same for the two groups, but they have markedly different experiences and impacts on welfare. (DSM)

See also: Merkwelt; Sollwert; Umwelt; Wirkwelt

J

Jainism

Jainism has its origins in India, and was founded in the 6th century BC by Mahavira. *Ahimsa*, non-violence, is its central principle. 'With the three means of punishment, word, thoughts, and deeds, you shall not injure any living being' – Jaina Sutra (Preece, 2001). It is applied not only through vegetarianism, for Jains also actively work for animal welfare, for instance by running animal hospitals. (HR)

See also: Buddhism; Hinduism

Reference and further reading

Chapple, C.K. (1993) *Nonviolence to Animals, Earth and Self in Asian Traditions.* State University of New York Press, Albany, New York.

Preece, R. (1999) *Animals and Nature. Cultural Myths, Cultural Realities.* UBC Press, Vancouver, Canada.

Preece, R. (2001) *Awe for the Tiger, Love for the Lamb. A Chronicle of Sensibility to Animals.* UBC Press, Vancouver, Canada.

Smart, N. (1998) *The World's Religions.* Cambridge University Press, Cambridge, UK.

Japanese quail

Japanese quail (*Coturnix japonica*) are small terrestrial birds of the pheasant family, which are farmed for food and used in **animal experiments**. Important quail behaviours include **foraging**, **dust bathing** and social interactions, so welfare is poor in barren cages. Good housing should provide cover, e.g. shrubs, and nesting boxes for laying hens. Thoughtful husbandry and early **habituation** to humans are essential, since rapid, upward escape flights can cause head injuries. (PH)

Journey times

Generally, longer journeys will be more stressful to animals than shorter ones under the same conditions. **Transport** time can therefore be a significant factor influencing animal welfare. This is particularly likely for animals transported under more austere conditions, such as those sent to **slaughter**, where there are commercial constraints on space allowances and levels of comfort. The method of transport is often related to journey times. Transport by air is relatively rapid, but that by road – and especially that by sea – is likely to be longer. Air travel is usually restricted to especially valuable animals, for example **pets**, breeding stock, racehorses and exotic and **zoo animals**. Individual journeys for slaughter animals are often part of a more complex process in which they pass through live auction markets. In this case the total travelling time, when all stages of the overall journey are added together, may be very prolonged.

Journeys within national boundaries

In North America calves and piglets are widely transported from breeding to fattening units, the process taking up to 24 h. In Australia animals can spend several days travelling to slaughter. Even in a small country such as the UK, journey times from farm to slaughter of up to 16 h have been recorded, although average times are much shorter (2–5 h). Published average figures from surveys carried out in the UK are 2.7 h for broiler chickens, 2.2–4.5 h for turkeys, 2.3 h for pigs and 4.7 h for sheep. Because of the trend for animals to be killed in fewer, larger slaughter plants, these times may tend to increase in the future. There is a particular problem with spent hens (birds culled after their economically worthwhile egg-laying life) in the UK, which may experience very long journeys simply because of the very small number of poultry processing plants that will accept them. Against this tendency is the realization by some more enlightened sectors of the slaughter industry that long transport is potentially detrimental to both animal welfare and the quality of the final product, this therefore leading towards pressure for shorter journeys and better-integrated delivery systems.

Extremely large numbers of broiler and layer chicks and turkey poults are transported from hatcheries to growing units worldwide. For example, worldwide in 2002 about 40 billion broiler chicks were sent to rearing units. Because they are newly hatched, and still reliant on their yolk sac for sustenance, the journey times are necessarily limited; however some chicks undoubtedly travel for up to 24 h. European legislation prohibits longer times without the provision of food and water.

Journeys across national boundaries

Exported meat animals can travel very long distances. Within Europe about 10% of commercial animal transport consists of journeys longer than 8 h. Large numbers of sheep have been transported from the UK to Spain, Italy and other countries in southern Europe, journeys taking several days, cattle are exported from Ireland to southern Europe and pigs are exported from The Netherlands to Spain and Italy – journeys of up to 1500 km sometimes lasting up to 40 h. The EU imports more than 2 million animals a year and exports about 300,000 cattle to North Africa; calves are sent from Poland to Italy and horses from central and eastern European countries, especially Poland, are imported into southern Italy, journeys taking perhaps up to 90 h. Outside Europe, Australia is one of the largest live animal exporters, sending large numbers of sheep by sea to the Middle East. This trade can often be of particular concern from the point of view of animal welfare, because of the length of the journey and the potentially poor conditions at sea.

In the USA, transport of livestock by rail or on board ships is limited to a maximum of 28 h without feed and water. However, this rule does not appear to apply to road transport. Long journeys in Europe must by law be interrupted at intervals by rest periods at designated staging points. However, a concern is that there is sometimes poor enforcement of these requirements, as there is of maximum permitted journey times, by the relevant national government authorities. Many animals, particularly those that have a relatively low individual economic value, such as animals exported for eventual slaughter for meat, may therefore travel longer than is allowed. Controlling this trade is made all the more difficult because it crosses national boundaries. An integrated and strict enforcement of legislation would undoubtedly improve the welfare of a significant number of animals.

The welfare implications of long journeys

Transported animals are often confined in an unfamiliar environment and subjected to noise and sometimes vibration. They may be mixed with unfamiliar animals and exposed to environmental temperatures and humidity outside their normal experience. They may suffer long periods of food and water deprivation. It is likely that many animals will experience **fear** and **distress** and become fatigued by the physical exertion associated with the journey. Some may become bruised or otherwise injured through falling. The longer the journey the greater will be the effects. In extreme cases individuals will die. Where large numbers of animals are transported confined under often poor conditions, mortality rates are often high. For example, the mortality of broiler chickens transported to slaughter averages about 0.2–0.3% and can, under particularly adverse conditions, rise to above 10%. Even at the average figure, because the total numbers transported are so high, the overall numbers of birds dying are considerable. Based on the number of broilers slaughtered worldwide, if the average transport mortality were 0.2%, this would equate to at least 80 million birds every year. Mortality rates in exported exotic birds and reptiles are often very high because of the difficulty of maintaining suitably stable conditions of temperature in journeys that may last days, even by air.

Moving animals provides a mechanism for the spread of disease, both between individuals transported together and geographically. The latter is especially a problem with animals unloaded and rested at staging points, where the potential for cross-infection is high. The **stress** associated with transport may also make animals more susceptible to infection by suppressing their immune responses. Disease has obvious **welfare** implications as well as the economic costs. An example is the outbreak of foot-and-mouth disease in the UK in 2001, the initial rapid spread of which was at least partly attributable to the extensive movement of animals throughout the country.

(PDW)

Judaism

Within Jewish scriptural tradition the relationship between Yahweh and his beloved Creation is thoroughly considered. The Torah principle 'do not destroy' has explicit reference to treating animals well in Talmud. In order to be kosher, **slaughter** has to be performed without **stunning**, with a sharp, long knife, accompanied by prayer and calmness. However, in industrial kosher slaughter, animal **welfare** is much compromised.

(HR)

See also: Islam; Religious slaughter

Further reading

Preece, R. (2001) *Awe for the Tiger, Love for the Lamb. A Chronicle of Sensibility to Animals.* UBC Press, Vancouver, Canada.

Smart, N. (1998) *The World's Religions.* Cambridge University Press, Cambridge, UK.

Juvenile behaviour

It is common in long-lived species for individuals to have a prolonged period of immaturity prior to reaching adulthood. Early in their lives, young birds and mammals are dependent on one or both parents for nourishment, but even after the point at which they are fledged or weaned there is often an extended period of close association with their parents. During this time, the behaviours they commonly exhibit may differ markedly from those seen in adults.

There have emerged two schools of thought regarding juvenile behaviour. On one hand, behavioural development can be thought of as a process by which young animals add behavioural components to their repertoire in a building-block fashion, continually refining those behaviours until they reach adult form. Under this view, a juvenile is incapable of expressing fully competent 'adult behaviour', and repetition of the juvenile (incompetent) form of the behaviour allows the juvenile to acquire the competency necessary in adulthood. A second perspective has more recently emerged, which views the behaviour of immature individuals as being adapted (as a result of natural selection) for a particular life stage. Natural selection operates at all stages of the life cycle, thus the behaviour of juveniles may differ from that of adults because they experience different social and physical environments than adults. Within this context, the distinctive behaviours that juveniles exhibit may enhance their ability to survive to adulthood.

These two views are not mutually exclusive; some motor patterns in **altricial** species require a degree of physical development in order to be expressed in their adult form, and some others (for example, bird song in passerines) require practice to reach their adult forms. On the other hand, there is considerable evidence that many of the behaviours that characterize the juvenile period are *not* rudimentary forms of adult behaviour.

This is most clearly seen in the **play** behaviour of species that produce **precocial** young. Social play and fighting share many similar behavioural components, and it has been suggested that play is a way of practising these specific motor patterns prior to their use in 'serious' contexts. However, play and fighting reliably differ in the sequence of motor patterns, the presence/absence of **vocalizations** and other fundamental features. Moreover, infants of precocial species are often capable of engaging in fighting behaviour in its adult form at the same age at which they are also engaging in social play. In the case of juvenile play it is clear that it is not solely a precursor to adult fighting, and it has recently been suggested that juvenile play has a broad impact on **brain** growth and development.

One large class of behaviours seen primarily in juveniles is care-soliciting behaviour. These behaviours may include vocalizations (e.g. the begging vocalizations of nestling birds and infant primates), specific postures (e.g. necks extended upwards with open mouths in nestling birds) and tactile behaviours (e.g. licking of the mother's mouth in infant carnivores and ungulates) that elicit care from other individuals. These behaviours are typically conspicuous and repeated. Rather than being accurate indicators, they appear to exaggerate the juvenile's need for parental care.

Exaggerated demands for parental care are somewhat paradoxical, since one might predict that the interests of parent and offspring would be in agreement, enabling each offspring to indicate its needs to the parent honestly and subtly. In reality, the interests of parent and offspring are not in perfect agreement because they are not genetically identical. Offspring share only half their genes with their mothers (because half their genome is inherited from their fathers) and roughly half of their genes with each sibling (fewer if they have different fathers). From the perspective of the mother, her available parental resources are best distributed evenly among her present and future offspring. Any given offspring, however, benefits from skewing that investment towards itself and away from its siblings. This sets the stage for parent–offspring conflict over the allocation of parental resources (**see: Parental behaviour**).

Evolutionary biologists have proposed several different interpretations of the exaggerated care-soliciting behaviours. First, offspring may be increasing their conspicuousness to predators, in effect blackmailing their parents into providing more care in exchange for silence. Secondly, if parents are more likely to feed the offspring with the most obvious solicitations, the exaggerated behaviours may enable one offspring within a brood to out-compete its siblings as they scramble for a share of parental resources. Thirdly, these care-soliciting behaviours may be honest demonstrations of offspring vigour that allow parents to preferentially provide care to those offspring most likely to survive in times of limited resources. While each of these proposed explanations has some empirical support, no single hypothesis adequately explains exaggerated care-soliciting behaviour in all species. This early period of dependency is critical for future survival and reproduction, making it likely that multiple factors have shaped the nuances of the parent–offspring relationship.

The process of **domestication** has favoured the persistence of a variety of juvenile characteristics into adulthood, a phenomenon known as neotenization. This is most readily seen in domesticated dogs, which exhibit puppy-like behaviours throughout their lives. These include a tendency to show submissive behaviour (in deference to their human companions), a general lack of **aggression** compared with their wild relatives and the tendency to solicit care-giving behaviour (for example, by licking their owners). Another neotenous feature, an extended **socialization** period, is seen in a variety of domesticated carnivores. Most species of wild cat, for example, have only a brief period during kittenhood when they tolerate the presence of other individuals before they become more solitary adults. Domesticated cats, however, more readily coexist with other cats and with humans. While it is unclear whether these juvenile behavioural characteristics were traits that were directly selected for by humans, the persistence of juvenile behaviours has made it easier to maintain these species in close association with humans, contributing to the success of these domestication events. (KVT)

See also: Kinship; Parental investment

Further reading

Lewkowicz, D.J. and Lickliter, R. (2002) *Concepts of Development: Lessons from the Laboratory*. Psychology Press, New York.

Wells, J.C.K. (2003) Parent–offspring conflict, signaling of need, and weight gain in early life. *Quarterly Review of Biology* 78, 169–202.

Juvenile period

The period between infancy and adulthood, which is characterized by significant anatomical, physiological and behavioural changes associated with adaptation to adult life, is referred to as the juvenile period. The length of the juvenile period differs among species and it is usually longer in species with a longer lifespan. In mammals, the juvenile period starts after **weaning**. (BM)

K

K and r selection

K and r selection is a concept used to explain life history variation. Essentially, r-selected species tend to have small adult size, early maturity and produce many small offspring (e.g. many insects). In contrast, K-selected species tend to be large, have delayed and iteroparous reproduction (successive batches of offspring) and produce relatively few, large offspring with greater levels of parental care (e.g. many mammals). This is a continuum, and thus many strategies fall between the two extremes. r-selected populations typically occupy short-lived (ephemeral) environments that are characterized by low levels of resource competition (r-selecting environments). Thus, r-selected species tend to be colonizing species. K-selected populations occupy constant or seasonally predictable environments that are characterized by crowded conditions and intense resource competition. Under such K-selecting environments there are few opportunities for offspring to become established, and thus the best strategy is to invest in offspring quality rather than in offspring quantity. (PE)

See also: Fitness; Inclusive fitness

Kant, Immanuel

The German philosopher Immanuel Kant (1724–1804) is an iconic figure not only for his brand of moral philosophy, *deontological ethics* (from the Greek *deontos*, meaning duty) (**see: Deontology**); his influence on Western thought and civilization is extensive. Kant's *Critique of Pure Reason* (1781) is one of the seminal works in philosophy. In it, Kant lays out his trademark metaphysical view of nature and epistemological accounts. Kant also wrote on aesthetics, natural science, mathematics, religion and, of course, moral and political philosophy.

With respect to his moral philosophy, Kant disliked the kind of consequentialist thinking later to develop within **utilitarianism**. In contrast to utilitarianism, Kant claimed that an action is right not because of its good consequences but because of the intrinsic character of the action itself. Furthermore, moral excellence does not depend on what we do, but on our motives for doing it.

Kant grounds **ethics** in reason alone and believed that reason prescribes/proscribes our moral obligations. He propounded the *Categorical Imperative*, a core feature of which is the *universalizability test*. The central feature of the test is to ask: 'What if everyone does as I do?' If like behaviour cannot be imagined without contradiction and without exception, then the act is prohibitive. Kant's test is often likened to another deontological account, the *Golden Rule*: 'Do unto others as you would be done by.'

The most famous version of Kant's Categorical Imperative is the *Formulation of Respect for Humanity*, which grounds rights-based ethics: 'Act such that you treat humanity, whether in your own person or in the person of any other, never simply as a means, but always at the same time as an end.' Persons deserve respect because they can chart and bring about their own life plans through exercise of their reason, and they can be held morally culpable. Their moral significance precedes their instrumental value to others.

For Kant, animals do not have direct moral significance. Instead, Kant held that we can have duties *about* animals or indirect duties *towards* animals. By his thinking, we should regard animals as 'friends' of sorts. If we mistreat them, then we are essentially prescribing to mistreat our friends, which is a contradiction in our will. (RA)

Further reading

Kant, I. (1963) *Lectures on Ethics* [translated by Infield, L.]. Harper & Row, New York.

Kant, I. (1964) *The Groundwork of the Metaphysics of Morals* [translated by Paton, H.J.] Harper & Row, New York.

Killing

It has always been widely accepted that there are circumstances in which human purposes legitimize the killing of animals. Today, few people would object to the veterinary **euthanasia** of severely ill animals. Acceptance of even large-scale killing of animals (e.g. in **culling** and **pest control**) seems to be nearly universal. In most countries, sports involving the shooting of animals are legal and cause no public outcry, although significant numbers of people object to the practices involved. Most people appear to accept that many animals used in biomedical research are eventually killed. Last, but not least, billions of young and healthy animals are, of course, **slaughter**ed in farm animal production. However, attitudes to the killing of animals are developing all the time and differ between persons and cultures. The practice of killing animals to meet human needs is questioned in some Eastern religions (**see: Buddhism, Hinduism**) and has certainly provoked public debate in the West, especially over recent decades.

Growing concerns about animal welfare in industrial farming practices have led to the promotion of **vegetarianism** and **veganism** and to questions about our need, and right, to eat animals at all. Some people seem to believe that the life of an animal has no value in itself, though they would accept that people can value animals, just as they can value (care about) teddy bears and mountains. The main ethical theories invoked in discussions of animal **ethics** seem to agree that living, sentient animals have inherent value. However, they disagree over whether an individual animal, such as a deer, is replaceable. Utilitarians believe that animals are replaceable and that the killing of animals can be justified. Adherents of animal rights, on the other hand, claim that the killing of a healthy animal is a violation of its right to life. (PS)

See also: Pest control – ethics of; Slaughter

Further reading

Sandøe, P. and Christiansen, S.B. (2007) The value of animal life: how should we balance quality against quantity? *Animal Welfare* 16, 109–115.

Kin recognition

Social recognition (i.e. recognition of individuals or various categories of **conspecifics**) is a necessary basis of living in stable social groups in which individuals communicate and interact. As part of this, kin recognition – that is, recognition of genetically related invidiuals – has been widely studied, particularly since the work of Hamilton (1964) and his concept of **inclusive fitness**.

The concept of inclusive fitness states that the global fitness of an individual consists of his/her own reproductive success, together with the improvement in **individual fitness** that the individual can provide to related animals. Hamilton's theory, namely **kin selection**, thus suggests that helping one's relatives to reproduce benefits the helper by improving his/her inclusive fitness (proportionally to the genetic proximity of the receiver). This implies that related individuals have to be recognized and therefore makes kin recognition a prerequisite of the kin selection theory.

A large number of behaviours indicative of nepotism (i.e. preferentially helping related individuals), and supporting Hamilton's theory, have been observed across species. Parental investment, alarm calls, greater tolerance during competition, support during social interactions or **grooming**, for instance, preferentially directed towards kin, are discriminative interactions that suggest that these related individuals are recognized.

Nepotism and optimal **mate choice** are usually the two main accepted benefits of kin recognition. With regard to kin selection, being able to recognize one's relatives is clearly adaptive as it facilitates the expression of nepotism by identifying the individuals to whom this behaviour should be directed. Kin recognition can also improve direct individual fitness and be of welfare importance. For instance, recognizing one's own mother increases survival, especially among species in which only the mother takes care of her young. In the context of mate choice, Bateson's (1983) theory of optimal outbreeding hypothesized that kin recognition would prevent **inbreeding** and favour the choice of the optimal mate (i.e. neither genetically too close nor too far).

Mainly since the 1980s have the mechanisms and sensory basis underlying kin recognition been widely studied. The term 'recognition' relates to the cognitive mechanisms (i.e. neural processes) enabling animals to classify conspecifics in different categories, whereas **discrimination** corresponds to the expression of different behaviours towards different individuals. As recognition is not visible, discrimination is therefore usually used to infer it, but it should be noted that the absence of discrimination does not necessarily imply the absence of recognition. In experimental situations, the existence of kin recognition can be assessed through tests based, for instance, on the simultaneous or successive presentation of kin versus non-kin individuals. In the simultaneous choice test paradigm, it is not only the existence of recognition that is evaluated but also the existence of a preference for kin. In any case, kin discrimination is inferred through either the observation of affiliative behaviours (e.g. physial proximity, grooming, **play**) or the absence of **agonistic behaviours**.

Four main mechanisms of kin recognition were first proposed: spatial recognition, recognition genes, direct association and phenotype matching.

Spatial recognition is based on the geographic localization of individuals (e.g. being in the same nest). However, as animals then behave discriminatively towards kin individuals based on environmental cues and not on their individual characteristics, it has been suggested more recently that spatial recognition should no longer be considered a mechanism of kin recognition.

The existence of recognition genes was suggested by Hamilton (1964), who hypothesized that a super-gene or group of genes would code for kin recognition. This theory has been named the 'green beard effect' by Dawkins (1982), and supposes the existence of an innate reference template in the brain. The existence of such genes has been widely discussed, and candidate markers relate to histocompatibility regions of the chromosome, e.g. the major histocompatibility complex (MHC) and other genes resulting in chemical discrimination.

Direct association and phenotype matching were initially considered as two dichotomous mechanisms until the late 1980s, when Richard Porter suggested the use of the terms 'direct familiarization' and 'indirect familiarization' corresponding to a single mechanism, familiarization. Recognition through familiarization is based on experience and appears to be the fundamental mechanism of kin recognition in many species. During direct familiarization, individuals become familiar as a result of direct contact (i.e. individual characteristics are learned and memorized) and are subsequently recognized.

This mechanism is thought to be the most widely found and has been demonstrated in numerous vertebrates using, for instance, cross-fostering or embryo transfer experiments. Non-kin young raised together then behave towards each other as if they were genetic siblings. In the context of indirect familiarization, individuals that have not been previously encountered are discriminated against based upon their phenotypic resemblance to the subject animal (a process called self-matching) or to one of their relatives with whom the subject is familiar (i.e. phenotype matching). In this case it is not individual but family characteristics (which are reliably correlated to the **genotype**) that are used. It has, for instance, been demonstrated that lambs display evidence of recognizing their twins from whom they had been separated at birth, as well as the twins of familiar partners. It should be noted that the recognition mechanisms cited above are not mutually exclusive.

The neural mechanisms on which kin recognition is based are still not exactly known. It is supposed that the animal builds templates that are memorized and against which a comparison of the characteristics of the individuals encountered can be made. But how these templates are built, where they are stored and how the matching is evaluated remain uncertain, even though some decision-making models have been suggested (Tang-Martinez, 2001).

Kin recognition depends on learning the phenotypic characteristics of kin individuals. This implies that those individuals have to: (i) wear labels giving information about their individual identity or the family group to which they belong; (ii) be able to perceive their conspecifics' labels and form an internal representation based upon them (i.e. a template comprising the traits of kin individuals) and to which the subsequently perceived labels will be compared; and (iii) evaluate the kinship of the individuals encountered thanks to a decision rule.

Complex phenotypical traits that are, to a certain extent, individually distinctive and will be used for recognition are called 'signatures'. The sensory basis used for kin recognition tends to depend on those that are predominant for communication in the species concerned (e.g. **olfaction** in rodents). Recognition can depend mainly on a single sensory basis or on several. In rodents, for instance, anosmic animals (i.e. those deprived of olfaction) are no longer able to discriminate their brothers from strangers. By contrast, in sheep, several studies have demonstrated that visual as well as acoustic cues and **odours** play a role in twin recognition and that the suppression of only one of those sensory cues does not disrupt recognition.

It has been shown that phenotypic signatures have a genetic basis and can thus indeed be used for kin recognition. However, such signatures, especially olfactory cues, can also be modulated by environmental factors such as the diet or maternal olfactory labelling (e.g. licking in the **cat**), which can then influence recognition. The development of kin recognition, its efficiency and the mechanisms used can also vary depending on several factors such as the learning period (e.g. pre- and postnatal learning), the kin individual's relationship (e.g. mother versus siblings), the **ecology** of the species and whether it shows temporal variation. (SL)

References and further reading

Bateson, P. (1983) Optimal outbreeding. In: Bateson, P. (ed.) *Mate Choice*. Cambridge University Press, Cambridge, UK, pp. 257–277.

Dawkins, R. (1982) *The Extended Phenotype*. Oxford University Press, Oxford, UK.

Hamilton, W.D. (1964) The genetical evolution of social behaviour, I & II. *Journal of Theoretical Biology* 7, 1–16; 17–52.

Hepper, P.G. (ed.) (1991) *Kin Recognition*. Cambridge University Press, Cambridge, UK.

Tang-Martinez, Z. (2001) The mechanisms of kin discrimination and the evolution of kin recognition in vertebrates: a critical re-evaluation. *Behavioural Processes* 53, 21–40.

Kin selection

The term 'kin selection' is often used synonymously with the term 'indirect selection', but its original definition embraced the evolutionary effects of both parental care and descendent offspring and **altruism** directed at non-descendent kin (i.e. relatives other than offspring). Indirect selection encompasses only the latter. This definition recognizes that the evolutionary consequences of a parent helping a descendent offspring are in fact genetically identical to the 'parent' helping a full sibling. Thus kin selection is a process in which genes may increase in frequency when bearers help close relatives produce additional offspring, so long as the benefits of giving aid to relatives outweigh the genetic costs of giving up the chance to produce descendent kin themselves (i.e. Hamilton's rule is obeyed; **see: Inclusive fitness**). (PE)

Kinship

Animals may benefit from behaving preferentially towards their kin, and this simple insight has revolutionized the study of animal behaviour. One solution to the apparent paradox of **altruism** is that animals benefit not only by obtaining direct **fitness**, but also by helping relatives obtain fitness (**see: Kin selection**). Because relatives share genes, helping kin can help spread an individual's genes. This insight, solidified by the evolutionary biologist W.D. Hamilton, created the important aim of quantifying kinship.

Kinship, the probability that genes are identical by common ancestry, can be calculated from detailed genealogies or from molecular techniques that focus on estimating those identical alleles shared by descent. In diploid organisms, the sperm and egg each contain a single strand of DNA that are combined during fertilization and first expressed in the **zygote**. Because each individual in a litter or clutch obtains half its DNA from its mother and half from its father, parents and offspring have a coefficient of relationship (often abbreviated r) of 0.5. Within a litter, the exact genes passed on to an individual are likely to be different, but the coefficient of relationship between full siblings is 0.5. Grandparents and grandchildren similarly share a 0.25 coefficient of relationship, while first cousins have a 0.125 coefficient of relationship.

A general method of calculating the coefficient of relationship is to draw a genealogy and then sum the pathways by which two individuals could share genes. Each generational link is counted, and r = the number of pathways × 0.5^L, where L is the number of generational links. (DTB)

Further reading

Krebs, J.R. and Davies, N.B. (1993) *An Introduction to Behavioural Ecology*, 3rd edn. Blackwell Scientific Publications, Oxford, UK.

Knowledge

Knowledge refers to that which is known; but the basis on which we know, as opposed to simply believe in, something is a source of considerable debate among philosophers. The philosophical study of knowledge is called **epistemology**. (DSM)

Kwashiorkor

Kwashiorkor is a type of **malnutrition** thought to be caused by insufficient protein consumption. It usually affects younger animals, especially when switching from a high-protein milk diet when nursing to a high-carbohydrate and -starch diet after weaning, and is most prevalent after famine or when the majority of the diet is uncharacteristically made up of starchy vegetables. Kwashiorkor is characterized by a swollen abdomen, alternating bands of light and dark hair and weight loss. (LMD)

L

Laboratory animal

A variety of animals are used in research and, when confined in a laboratory for research purposes, are termed by the species or generic name 'laboratory mouse/rat/dog', etc. In the UK, where good statistical data are collected, approximately 85% of laboratory animals used in research in 2004 were **rodents** (mice (67%), **rats** (18%), **guinea pigs** and other rodents (2%)); 6% fish; 4% birds; 2% other mammals (e.g. **rabbits**, **dogs**, **cats**, primates, **ferrets**, **pigs**, **equids** (**horses** and related species) and ruminants (**sheep**, **goats**, **cattle**, **deer**)); and 1% **reptiles** and **amphibians**. (DBM)

See also: Animal experiments

Further reading

Khamis, H.J. (1997) Statistics and the issue of animal numbers in research. *Contemporary Topics in Laboratory Animal Science* 36, 54–59.

Laboratory study

A laboratory study is a research project carried out on a **laboratory animal**. This may involve an animal specifically bred for the purpose or the importing of an animal into the laboratory environment. Studies using wild-caught animals may be conducted in the laboratory instead of in the field for scientific or practical reasons; for example, there may be a requirement for close monitoring or a controlled environment where single variables can be changed, in order to allow specific research questions to be answered. Laboratory studies and **field studies** have different implications for the health and welfare of the animal, all of which should be taken into account when deciding whether individual projects are justified. Special welfare considerations are necessary when considering the use of wild-caught animals in a laboratory study (see Table L.1) especially in relation to identifying areas where trapping, housing or techniques will need to be refined. (PH)

See also: Animal experiments; Biotelemetry; Confinement; Ethology; Field study; Three Rs principle

Further reading

Canadian Council for Animal Care (CCAC) (2003) *Guidelines on the Care and Use of Wildlife*. CCAC, Ottawa. Available at: http://www.ccac.ca/en/CCAC_Programs/Guidelines_Policies/GDLINES/Wildlife/Wildlife.pdf (accessed 4 November 2009).

Table L.1. Welfare issues associated with the use of wild-caught animals in the laboratory versus the field.

Laboratory studies	Field research
Individuals are temporarily or permanently removed from the field – could impact on other animals of same or different species	Presence of researchers or study itself could have adverse effects on other individual animals or ecosystem
Animals will need to be transported to the laboratory – transport is a highly potent stressor, even for purpose-bred animals	Can be difficult to balance minimizing animal holding time with ensuring that animals are fit for release, not distressed and/or drug residues are at acceptable levels
Laboratory environment is unnatural, impoverished and extremely stressful for wild-caught animals – severely reduced opportunities for performing natural behaviours, e.g. foraging, exercise, socialization	Animals are in their own environment, which may also speed recovery from capture stress and use in procedures
Veterinarian should be on site or on call, to treat animals and advise users and care staff	Veterinary resources could be remote
More opportunities for regulators to supervise researchers	Reduced control and supervision of researchers
Risk to the welfare of other animals in the unit if pathogens introduced; quarantine periods add to captivity time	Compromised health, welfare and fitness may be hard to predict and detect, including long-term effects, e.g. impacts of marking and external instrumentation still not fully understood
In less commonly used species, may be little or no expertise on housing, care, analgesia, etc.	Acceptable standards of anaesthesia, analgesia and asepsis are harder to maintain during field surgery
May require fewer animals since conditions are controlled, experimental animals cannot be lost	If recapture is necessary, animals will experience additional stress
Animals can rapidly lose condition and/or learn inappropriate behaviours, e.g. tameness; may have to be rehomed or killed rather than released	Animals should retain fitness, post-study release more feasible

Sherwin, C.M., Christiansen, S.B., Duncan, I.J., Erhard, H.W., Lay, D.C., Mench, J.A., O'Connor, C.E. and Petherick, J.C. (2003) Guidelines for the ethical use of animals in applied ethology studies. *Applied Animal Behaviour Science* 81, 291–305.

Lactation

Lactation, the secretion of milk from specialized mammary glands, is one of the major distinguishing characteristics of the taxonomic class mammals. The production of nutrient milk is thought to have evolved from secretions of apocrine-like glands associated with hair follicles, which may have provided moisture and anti-microbial agents to parchment-shelled eggs of pre-mammalian species.

In all cases the mammary glands are located ventrally in pairs lateral to the midline. However, there is considerable variability both in the number of glands and in their location along the midline. Glands may be located in any, or all, of the thoracic, abdominal or inguinal regions; species with small numbers of offspring such as ruminants and horses have between two and four glands in the inguinal region, whereas others such as primates have a single pair of glands in the thoracic region.

Species with multiple litter sizes – e.g. many carnivores, rodents or pigs – have between six and 16 glands arranged in pairs spread across all three regions. For most species both males and females have mammary glands but in males they are rudimentary in structure, lacking the post-pubertal development seen in females and are non-functional. In some species males may lack external teat structures and, in other cases, glandular tissue may be completely absent. There are, however, many anecdotal accounts of lactating males and it seems clear that, given appropriate endocrine support, males of a number of species can achieve a functional lactation state. There is at least one well-described exception to this rule, the Dayak fruit bat (*Dyacopterus spadiceus*), where both sexes lactate and share feeding of the offspring.

While there is a large degree of variability in the number, position and external appearance of the mammary glands, the internal structure of mammary tissue does not vary much between mammals. The functional unit of the gland is the alveolus, a small, sack-like structure with a single layer of secretory epithelial cells surrounded by a layer of contractile myoepithelial cells and a connective tissue basement membrane. Milk is synthesized and secreted by the epithelial cells into the lumen of the alveolus before being squeezed into a duct system by contraction of the myoepithelial cells under the influence of **oxytocin**. Alveoli are arranged into lobules consisting of 100–200 alveoli contained within a connective tissue sheath. A contractile duct system collects milk from individual alveloli and lobules and conveys it to the external teat. In mammals producing a large volume of milk, ducts may lead into large, open cisterns allowing for greater milk storage.

Milk production is not a continuous process in mammals, its onset being timed to coincide with **parturition**, and it continues for as long as nutritional support is required by the offspring. The attempted termination of suckling by the mother versus the persistence of the young can be viewed as the product of within-species evolutionary conflict. Lactation is energetically expensive, and the mother aims not to invest more than is necessary to maximize her **fitness** (i.e. the best strategy is not to give more milk than is necessary to ensure the survival and optimal development of the young). By contrast, suckling is a very efficient way of obtaining nutrients for the young, and so the offspring maximizes its fitness by being dependent upon milk and persisting with suckling for as long as possible. The demands of the two are therefore in conflict, and an optimal **weaning** age for a given species emerges as a result. None the less, the process of weaning may see quite aggressive behaviour being exhibited by both the mother and her offspring towards each other. It has been suggested that this is also an important time of emotional development in the young, as they learn to cope with **frustration**.

Mammals will undergo a series of reproductive cycles, each accompanied by a period of lactation, and therefore mammary tissue may undergo a number of repeated cycles of growth and differentiation, followed by active milk production and then involution and regression. The maintenance of lactation once established is termed galactopoeisis and is regulated by two interrelated factors, the presence of galactopoietic hormones and the continued removal of accumulated milk. The two most important regulatory hormones for lactation are the pituitary hormones oxytocin and **prolactin**, or leuteotropic hormone (LTH). Inhibition of galactopoietic hormones will depress milk production, depending on the species and lactation stage; however, milk removal is the major factor in maintaining secretion. As milk components accumulate in the alveolar lumen, a feedback inhibitor of lactation is secreted from the alveolar epithelial cells and blocks further milk secretion from the epithelial cells. Removal of milk by suckling (or by **milking**) removes the inhibitor factor and allows for continued milk secretion. Galactopoiesis continues up to weaning when, with the suspension of suckling (**see: Nursing**), the mammary gland involutes and the cycle will start again only with a new pregnancy. The relationship of lactation to the overall reproductive cycle varies, with many species exhibiting a lactational anoestrous period, where oestrous cycling is suppressed during early lactation. This anoestrous period can be used by livestock producers to manage reproductive performance. For example, in the sow oestrus first occurs several days after removal of the piglets, so weaning is used to synchronize oestrus, and therefore to synchronize the subsequent farrowing.

Yield and composition of milk also vary considerably between and within species, and with the stage of lactation. Milk contains three main components, carbohydrate, protein and lipid, suspended in an aqueous medium. Carbohydrate is principally supplied by the disaccharide lactose consisting of a β-1,4-glycosidic linkage of D-glucose and D-galactose. Lactose is unique to mammalian milk and is common to almost all mammals, and is digested by the specific neonatal enzyme lactase, cleaved into glucose and galactose, which is subsequently converted to glucose. In addition to its nutritional role, lactose is the major osmolyte in milk responsible for drawing water into the alveolus. Water is by far the largest component in milk, forming 87% in cow's milk.

Milk contains two unique families of proteins, the caseins (caesins) and the whey proteins. Caseins are synthesized by mammary epithelial cells and are easily digestible proteins providing a good amino acid source for the neonate. Some caseins are at least in part responsible for the soporific effect of milk in the young, but the change in gastric digestive enzyme (increase in gastric trypsin) with age eliminates this effect in the adult. None the less, gastric-protected formulations can be made, which can be used to calm animals. The whey proteins consist of the lactalbumins and lactoglobulins, which are again specific to mammalian milk. Triglycerides make up the major lipid component of milk. Milk fat is utilized by the neonate to form adipose tissue, but is also metabolized as an energy source. The lipid component is the most variable component between species, with marine mammals having fat contents of up to 60%, while fat content is negligible in, for example, early lactation in marsupials. Lipids are secreted as fat globules bound by a lipid bi-layer, and are present as an emulsion in the aqueous phase of the milk.

Composition of the milk changes throughout the lactation period, adapting to meet the specific needs of the developing neonate. Of particular importance is the milk produced in the first few hours postpartum, which is termed **colostrum**. The mammalian neonate is generally immunologically competent but naive. Humoral immune protection, in the form of immunoglobulin, is supplied by a process of transfer of passive immunity from the mother to the neonate. This generally occurs by transfer of maternal serum IgG from the mother to the offspring either *in utero* or, after birth, by ingestion of immunoglobulin-rich colostrum by the neonate. These maternal immunoglobulins offer immediate immune protection prior to the establishment of a fully functioning **immune system** in the neonate. The immunoglobulin content of the colostrum rapidly declines in the first 12 h following parturition and, at the same time, the neonate's ability to take up the immunoglobulin proteins also reduces in the process of gut closure.

The mammalian newborn exhibits a well-developed suckling behaviour, using thermal, olfactory and textural gradients to locate the teats on the mother's ventral surface with its nose and tongue. Contact with the teats triggers a strong suckling response. Suckling stimulates the release of oxytocin in the maternal system which, in turn, promotes milk letdown. Release of milk may be further stimulated by behaviours such as butting at the udder seen in some ruminants or by the treading response that kittens may perform while suckling. Species with few, precocious young such as ruminants and horses will suckle standing, while those with large litter sizes and more immature young will lie down to suckle. (MRC)

See also: Maternal behaviour

Lairage

Lairage generally refers to the period of time and/or accommodation in which animals are held, usually off-vehicle, during a period of transportation, or at an abattoir prior to slaughter (**see: Slaughter; Slaughterhouse; Transport**). Lairage is important because it provides the animals with an opportunity to recover from the physical and psychological **stressors** associated with transportation, whether this is prior to slaughter or part-way through a journey. In many countries there is legislation governing maximum journey durations and, thus, the timing and duration of any lairage during transportation. At the abattoir some lairage time is inevitable, given the logistics associated with receiving and scheduling animals for slaughter, and in some countries a rest period before slaughter is mandatory. However, despite these regulations, there is still uncertainty and debate about the appropriate timing and duration of lairage, and the ways in which animals should be managed during it.

From a purely meat quality perspective, the evidence suggests that, except when animals are severely stressed, lairage prior to slaughter should be kept to a few hours (**see: DFD meat**). The effects of pre-slaughter lairage will vary depending upon pre-slaughter muscle glycogen levels, which, in turn, are influenced by a variety of factors, such as the gender and character (**temperament**) of the animals, nutritional history, the length of time off food and water, the duration of transportation and the ways in which the animals have been handled. If glycogen concentrations are above the critical threshold (40–57 μmol/g for **cattle**), then there may be little benefit in having a long lairage period, although there is some evidence that improved meat quality may be obtained with supra-threshold glycogen levels.

A rest period of less than 12 h in lairage reduces DFD meat in bulls and, perhaps, also in cattle that have been minimally stressed pre-slaughter. Lairage times of 3–4 h may also improve other aspects of beef quality as well as eating quality (texture and flavour) of beef from animals that have been minimally stressed pre-slaughter. There is some evidence that these improvements may be related to the shorter than usual fasting periods. It must be remembered that any reduction in fasting and lairage time may have adverse consequences for food safety; animals will have increased ruminal volume at slaughter and this may increase the risk of rupture and carcass contamination during processing.

It may be argued that immediate slaughter is desirable from an animal **welfare** perspective, but given that this may not be achievable and that delays will occur, an extended pre-slaughter lairage period may allow animals more time to rest and recover from the various stressors associated with transportation to the abattoir. Studies with **sheep** have shown that plasma cortisol levels decline during lairage of a few hours' duration. However, the possibility for recovery is likely to be highly dependent upon the design and management of facilities. Animals in abattoir lairage are in a novel environment, may be mixed with unfamiliar animals and are exposed to different handlers and feeding and watering regimes. Such factors could add to the stress resulting from transportation. Studies have shown that animals are less likely to rest with increased human and other animal activity in the lairage facility, and when stocked at lower **space** allocations and/or larger group sizes. There is some evidence that it takes 2 or 3 h before some animals (sheep and cattle) will settle down and begin to rest in pre-slaughter lairage, and a lairage period of about 10–15 h is required to allow animals to recover from the effects of food restriction and transportation.

Evidently, the latency to rest and the time to recover will depend upon transportation duration and the degrees of fatigue, thirst and hunger it may induce and, thus, motivation to rest, drink and eat, if water and food are available. Work has shown that cattle penned without feed after a short period of transportation do not drink immediately, but drink during the subsequent 6 h. However, cattle transported for long periods (about 24 h) all drank immediately after they were unloaded, and the longer the period without water, the more they drank when given the opportunity. Thus, depending upon total time that the animals have been deprived of food and water, and how long they will be held in lairage pre-slaughter, food and water may need to be provided. However, the issue of gut-fill and potential for carcass contamination needs to be considered.

There are reports that increased time in lairage can result in increased bruising, which has detrimental effects on both carcass quality and animal welfare. The incidence of bruising will also depend upon the quality of the facilities and management – for example, whether animals are repeatedly disturbed and if groups of unfamiliar animals are mixed, resulting in increased levels of agonistic interactions.

In some countries there are requirements to break journeys that extend beyond a specific duration. These breaks can be for as little as 1 h, during which time the animals must be given food and water. Studies with sheep have indicated that it is inadvisable to provide feed and water on the vehicle during a stop because of aggressive interactions between the animals, such as riding and trampling. Also, the mechanics of providing food and water on board the vehicle make it impractical. Consequently animals must normally be unloaded into yards or pens and then re-loaded after feeding and watering. Short rest breaks of a few hours are likely to be insufficient to allow any real recovery from transportation, particularly as the animals will be entering a novel environment. Furthermore, studies on sheep have shown that they are primarily interested in feeding following journeys and are not motivated to drink until 2–4 h after feeding. Consequently, if provided with only a short break during a journey, they do not have time to drink and this could result in the development of negative water balance during the remainder of the journey, with the outcome that animals are dehydrated at the end of their journey. It has been recommended that sheep require a journey break of about 6 h if they are to have sufficient time to eat, drink and rest.

The implications for welfare of the additional handling associated with unloading and loading animals is a key consideration for determining breaks during transportation. Research with sheep has shown that loading leads to elevated cortisol levels and **heart rates**, which are sustained for the first few hours of travel, indicating that the procedure is stressful for the animals. When the sheep were loaded a second time, the cortisol response was attenuated, suggesting that, perhaps, the initial response was to novelty rather than the loading procedure per se. The degree of **stress** associated with unloading and loading will depend upon the quality of **handling** and the facilities and, in some instances, it may be better for animal welfare to continue the journey to completion than to break it for a short period if animals have to be unloaded and reloaded. (JCP)

Further reading

Grandin, T. (1990) Design of loading facilities and holding pens. *Applied Animal Behaviour Science* 28, 187–201.

Small, A., James, C., James, S., Davies, R., Howell, R., Hutchison, M. and Buncic, S. (2007) Construction, management and cleanliness of red meat abattoir lairages in the UK. *Meat Science* 75, 523–532.

Lameness

Lameness describes a multitude of conditions, including foot problems, musculoskeletal problems and joint problems, that can hinder or inhibit normal **locomotion**. Lameness occurs in all species and, in some species, causes significant economic and **welfare** effects. In farm animal species, lameness and/or musculoskeletal problems is the second most common cause for **culling** in **pig**s and the third most common cause for culling in dairy cows. The causes of lameness can be traced to four main factors: **housing**, hygiene, genetic selection and nutrition. Different housing systems can influence the amount of physical trauma to the body and the feet – either directly by perhaps including physical aspects such as rough flooring or poorly maintained slats, or indirectly by perhaps increasing the amount of time that animals spend standing. Clean, hygienic facilities help in inhibiting the spread of infectious bacterial and viral diseases such as **footrot** and infectious arthritis. Genetic selection for increased mature body size or for growth/production rate is believed to be a primary culprit for the increased prevalence of lameness in recent history, particularly in broiler **chicken** and pig production. Nutritionally unbalanced diets can cause deficiencies of nutrients such as calcium, which will affect **bone strength**, and overfeeding may result in overweight animals, which are more susceptible to **laminitis** and have an increased risk of overloading musculoskeletal tissue.

Lameness is considered to always be painful, and can inhibit normal functioning and decrease welfare. Lameness is known to activate **stress** responses in animals, increasing concentrations of hormones such as cortisol in the blood. Cortisol is a **glucocorticoid** that is a primary stress response hormone that acts to counteract the stress, but it can have both positive effects (blocking inflammatory reactions) and negative effects (decreasing immune cell populations, which can lead to infection).

As mentioned, genetic selection for increased production is believed to have increased the incidence of lameness. Studies have shown that the fastest-growing animals and animals that are overweight have a higher incidence of lameness than animals of normal weight and animals with a slow **growth rate**. Increased production output, specifically in reproducing animals, has also shown an increase in lameness when compared with animals of lower production rates. Animals bred for a specific conformational trait have also exhibited higher incidences of lameness, particularly in the form of hip dysplasia and other related conditions. It has been suggested that selection for high growth and production rates has negatively affected bone strength. Decreased bone strength can lead to conditions such as arthritis, osteoarthritis and fractures, all resulting in lameness.

In some cases, early identification of lameness may increase the chances of recovery. Identification of lameness primarily depends on observations of **posture** and **gait**. In general, identification of a change in gait characteristics (length, fluidity) or in postural equality (all limbs weighted evenly when standing) is the first sign of lameness. Behavioural measures can also indicate problems, e.g. when eating behaviour and willingness to stand or move have decreased.

As mentioned, all species are susceptible to lameness and, although it occurs in different forms, there are two general categories into which lameness falls.

Foot lesions and problems

Laminitis
Laminitis is a common disease of the foot or hoof that is typically a result of unbalanced feeding, excessive weight, and environmental or production stress. Small blood vessels between the bone and the laminar structures in the foot become blocked by **inflammation** or rupture, causing discoloration to the hoof. This inflammation or rupture of the vessel causes the horn to become soft, allowing ulcers of the sole to emerge in the horn. In severe cases, the bone and hoof can separate.

Foot-and-mouth disease
Foot-and-mouth disease is a highly contagious viral disease that affects a wide range of species. Symptoms include blisters and sores around the foot and mouth, and lesions on the teats or udders.

Footrot
Footrot is an infectious disease caused by the intestinal bacterium *Fusobacterium necrophorum*, which is shed in the faeces. The bacteria cause necrotic lesions in the soft tissue between the claws, resulting in swelling. If footrot goes untreated, the infection can spread to the upper leg.

Trauma
Trauma to the feet from the environment can also cause bruising and lesions of the horn. **Pain** is thought to be associated with these conditions and can cause a noticeable change in gait, posture or behaviour of the animal.

Bone and joint problems

Osteoarthrosis
Osteoarthritis is a form of arthritis that is common from age-related wear and tear or as a result of injuries. It mainly affects synovial joints, specifically in the limbs. The articular cartilage degenerates over time and the bone becomes thickened and distorted, restricting movement of the joint. Inflammation of the joint and pain are associated with the condition, and symptoms can be aggravated by excessive weight. Osteoarthrosis is more common in older animals but can occur due to injury in young animals as well.

Osteochondrosis
Osteochondrosis has multiple forms, is prevalent in young, growing animals and is a primary cause of lameness in several species. In osteochondrosis, there is a failure of epiphyseal growth cartilage to form into bone because of necrotic tissue in the area. When lesions are large enough, the necrotic tissue will protrude into subchondral bone (*osteochondrosis manifesta*). In severe cases the articular cartilage may collapse, forming a cleft into the subchondral bone (*osteochondrosis dissecans*) causing inflammation and osteoarthrosis. Epiphysiolysis, another form of osteochondrosis, is the loosening or separation of the epiphysis from the shaft of a bone.

Mycoplasma
Mycoplasma arthritis occurs in young animals and is caused by a respiratory infection with *Mycoplasma hyosynoviae*. It is prevalent in the majority of herds, even though there are no clinical signs of infection. The organisms can be found in the upper respiratory tract in animals of all ages, but the infection affects only young animals. The infection primarily affects the joints and tendon sheaths around the joints, but it can also cause pneumonia, **mastitis**, vaginitis and abortion. Clinical signs include a reduction in normal behaviour, including feeding behaviour, and a slightly elevated temperature. Mycoplasma arthritis is more common in heavier animals. *Mycoplasma* species lack cell walls, making treatment difficult due to antibiotic resistance.

Osteoporosis
Osteoporosis is characterized as a loss of integrity of trabecular bone and thinning of cortical bone. Osteoporosis is common in postmenopausal women and is associated with calcium deficiency and decreased **oestrogen** concentration. Osteoporosis can affect older animals of other species as well. After maturity, bone strength starts to decline and, if optimal bone strength was not reached during adolescence, the risk of osteoporosis increases. Osteoporosis may also occur in animals under high reproductive stress. **Gestation** and **lactation** place high calcium demands on the animal, sometimes depleting bone calcium reserves. Over time, if these reserves are not replaced due to chronic stress, osteoporosis may develop. The effects of osteoporosis are typically irreversible, and the onset of the disease causes an increase in fracture rate.

Fractures
Fractures can be a result of physical trauma or from one of the conditions listed above, such as osteoporosis or osteochondrosis. Presence of fractures will cause the animal to be unable to bear weight on the affected limb and have a reduction in activity. (ELS)

See also: Flooring

Further reading

Anil, S.S., Anil, L., Deen, J., Baidoo, S.K. and Walker, R.D. (2007) Factors associated with claw lesions in gestating sows. *Journal of Swine Health and Production* 15, 78–83.

Bradshaw, R.H., Kirkden, R.D. and Broom, D.M. (2002) A review of the aetiology and pathology of leg weakness in broilers in relation to welfare. *Avian and Poultry Biology Reviews* 13, 45–103.

Laminitis

Laminitis is a non-infectious disease of the hoof caused by an **inflammation** of the laminae. Between the hard outer horn of the hoof, the sole and the pedal bone is a cushioning layer of tissue known as the corium. Within the corium are the laminae, which are layers of sensitive tissue that are arranged vertically from the top of the hoof to its base, both acting as shock absorbers and producing the hard, keratinized tissue of the wall and sole. The inner, or dermal, laminae are stationary and, along with other connective tissue, hold the pedal bone in place within the hoof. The outer, or epidermal, laminae are connected to the hoof wall and migrate past the dermal laminae as the hoof grows from the coronary band. Connections between these two layers of laminae maintain the general structure of the hoof by acting as a bond between bone and hoof wall. These connections are continually broken and renewed as the epidermal laminae migrate downwards with the growing hoof, the whole process being controlled by locally produced enzymes, enzyme activators and enzyme inhibitors. Damage to the blood supply to the laminae or an imbalance of enzyme activators and inhibitors can result in laminitis.

Different phases of the disease are characterized by specific clinical signs. The most severe phase is acute laminitis, and this is often associated with a nutritional imbalance known as acidosis. Acidosis occurs when carbohydrate intake is rapidly increased compared with fibre intake, e.g. with unrestricted access to spring grass, high-concentrate postpartum rations or if food particles are so small that they inhibit chewing behaviour. For example, food particles shorter than 1.7 cm inhibit cud chewing in dairy cows. This, in turn, reduces the amounts of saliva and, more crucially, bicarbonate that are naturally produced during the chewing process. The bicarbonate plays a vital role in controlling ruminal acidity by acting as a buffer and, to maintain a normal ruminal pH above 6, the cow will need to chew for around 10 h each day.

When acidosis occurs, toxins are released into the bloodstream and become concentrated in the blood vessels within the laminae of the hoof, disrupting the blood supply and causing inflammation. The dermal–epidermal interface becomes damaged and cell death (necrosis), along with blood leakage (haemorrhage), occurs, and the affected area becomes swollen (oedema), further reducing blood supply. This oedema causes severe pressure between the hard hoof wall and pedal bone, resulting in intense pain. Immediate clinical signs of acute laminitis are sudden, severe **lameness** with no visual changes in hoof structure, though the affected hoof may be warmer than usual. The haemorrhaging of the corium during acute laminitis can reduce or even halt new horn growth.

Subclinical laminitis is a phase of the disease that can result from structural damage caused by either acute laminitis or physical injury. Prior to the onset of lameness, subclinical laminitis can be detected by the yellow discoloration of weak horn growth. Hoof tissue that is damaged during acute laminitis produces poor-quality, abnormal horn growth and, as a result, the wall and sole of the hoof become softer and more easily damaged. As the hoof becomes weaker, an animal is more likely to sustain secondary foot problems such as bruised soles, abscesses, ulcers, white line disease, heel erosion and cracks in the hoof wall. Indeed, lameness is often caused by these secondary problems rather than by the subclinical laminitis itself. In advanced cases, the pedal bone may begin to separate from the hoof wall as the laminae and connective tissue break down. As a consequence of the separation of the laminae, the pedal bone may rotate downwards, compressing the base of the corium and causing severe pain. Hoof horn surrounding inflamed or infected parts of the hoof tends to grow more rapidly during the subacute phase, leading to misshapen feet and, consequently, altered conformation. An animal suffering from lameness due to laminitis will place more of its body weight upon healthy hooves, and so displays a characteristic hunched standing position as an attempt is made to keep as much body weight as possible off the affected feet.

The third form of the disease is chronic laminitis, in which the hoof horn is also produced at a greater rate than healthy horn. If the disease remains untreated the internal structure of the hoof and its blood supply will continue to break down until the damage sustained is irreversible. Over time the pedal bone may penetrate the sole of the hoof.

Many factors have been implicated as contributing causes of laminitis, and these include poor nutritional management (e.g. acidosis) and excessively hard ground (**see: Flooring**). **Horses** performing fast or prolonged work on hard surfaces may develop concussive laminitis, particularly if their hoof horn quality is already poor. Similarly, hard lying areas discourage dairy cows from lying down, and excessive standing may predispose these cows to laminitis.

Many internal and environmental factors related to laminitis are associated with **parturition**. In dairy cows, most incidences of lameness occur within the first 3 months after parturition, and contributing factors can include **stress** during parturition, metabolic disorders, too little exercise and **mastitis**. Metritis, or retained placenta, can also cause laminitis in both mares and cows.

Prevention of laminitis includes: (i) avoiding excessive working or housing of animals on hard surfaces; (ii) providing comfortable bedded areas for all animals to be able to lie down together; (iii) avoiding malnutrition, including obesity; (iv) allowing animals to exercise; and (v) keeping stress levels to a minimum, especially during the period surrounding parturition. Further measures include introducing animals to elements of a new environment before it becomes permanent. For example, dairy heifers are often maintained in loose-straw yards but housed in cubicle yards when brought into the milking herd. Exposure to the cubicle yard prior to entering the herd will allow the heifers to learn how to enter and lie down in the more regimented cubicle system, encouraging optimum lying behaviour when they subsequently enter the herd. (LW)

Further reading

Bailey, S.R., Marr, C.M. and Elliott, J. (2004) Current research and theories on the pathogenesis of acute laminitis in the horse. *The Veterinary Journal* 167, 129–142.

Krause, K.M. and Oetzel, G.R. (2006) Understanding and preventing subacute ruminal acidosis in dairy herds: a review. *Animal Science Feed and Technology* 126, 215–236.

Language

The term 'language' is generally used to refer to systems that play a role in **communication**, such as computer language, dolphin language and human language. In the 17th century, **René Descartes** reasoned – contra Montaigne – that even though parrots may produce the same sounds that we make, such sounds cannot be called language. According to Descartes, parrot sounds are not produced to express thoughts, but are based on 'passions', such as a desire to obtain food or to satisfy other needs. The issue of whether or not animals have language is still debated today. For example, Savage-Rumbaugh *et al.* (1998) claim that Kanzi, a laboratory-reared bonobo (*Pan paniscus*), has language skills that are comparable to a 2.5-year-old child. However, as pointed out by, for example, Kirby (2000) in a review of this work, much of such a claim hinges on how key aspects of language are defined.

In order to evaluate whether animals have language, whether human and animals systems of language have similar or unique features, or whether there is any evolutionary overlap between human language and animal systems of communication, it is necessary to define the concept of language more precisely. Hockett (1960) lists 13 features that characterize human language, for example that human language is made up of discrete units, that its **signals** are arbitrary (there is no relation between the form of the signal and its meaning), productive (novel meanings can be created), rapidly fading (signals do not persist after being produced) and show dual patterning (small units can be combined into different possible orders to form larger units).

Even though Hockett's features have been used to compare human language and animal communication, these features have no formal definition, and many interpretations of, for example, creativity in language have been possible. Chomsky (1986) defines language more formally as the internal representational system of knowledge that makes it possible for humans to map signals to meaning by using syntax. In its broadest possible scope, this system of knowledge involves knowledge of the sound structure of language (phonology) or, in the case of sign language, knowledge of the sign patterns of language, knowledge of the ordering of signals in relation to each other (syntax), knowledge of the internal structure of words or gestures (morphology) and knowledge of the semantic-conceptual structure of language (meaning) (Jackendoff, 1996, 2002). Each of these knowledge systems is characterized by a finite set of primitives (for example nouns, verbs and adjectives in the case of syntax) and sets of rules or constraints that define how such primitives can be combined. This allows for a more precise definition of, for example, the concept of creativity in language (see below). This entry will not focus on sign language any further, but it will examine the properties of each knowledge system described above in human language, and briefly consider whether animal systems of communication share any of its features. Subserving human knowledge of sound structure are the mechanisms of speech perception and speech production. These mechanisms are considered first.

Liberman's Speech is Special (SiS) hypothesis argues that human speech perception is different from sound perception in animals (e.g. Liberman, 1996). Pinker and Jackendoff (2005) argue that this hypothesis is supported by, among other things, research on the categorization of sounds. Kuhl (1991) shows that adults and 6-month-old babies judge a particular 'i' sound to be more similar to a prototypical 'i' sound compared with a sound at the periphery of the 'i' sound category, even though the physical distance is the same in both cases. Rhesus **monkeys** do not display this perceptual magnet effect – a bias towards the prototype of the vowel category. However, Hauser *et al.* (2002) and Fitch *et al.* (2005) advocate a null hypothesis: that, until there is clear evidence to the contrary, there should be considered no differences between sound perception in humans and animals. In order to support their view, the above authors cite evidence that shows parallels in speech perception between humans and closely related primates. For example, both newborn humans and cotton-top tamarin monkeys appear to be able to discriminate between two unfamiliar languages (Dutch and Japanese) (Ramus *et al.*, 2000); these authors attribute this ability in both species to general processes of the primate auditory system. Hauser *et al.* (2002) and Fitch *et al.* (2005) claim that any differences between humans and animals regarding speech perception – such as those presented by Pinker and Jackendoff (2005) – are not relevant to the language faculty.

With respect to speech production there seems to be agreement that there are qualitative differences between humans and non-human animals. One of the most characteristic properties of human speech is the period of babbling that infants go through when they are around 5–7 months of age. During this phase children show a strong bias to imitate the sound patterns of their mother tongue, although they do not do this for other environmental sounds. In fact, deaf children appear to babble with their hands. The innate drive – but also the capacity to imitate sounds that are made by **conspecifics** – is absent in animals such as **chimpanzees**, which are evolutionarily close to us. An imitative capacity has been found in unrelated species, such as in certain dolphins and songbirds. However, despite many parallels between humans and songbirds in how sound production is learned, it seems that more is prespecified for songbirds when they learn their repertoire compared with humans, as evident in learning preferences and isolate songs (Doupe and Kuhl, 1999).

Human knowledge of sound structure consists of knowledge of a finite set of the smallest building blocks of sounds (phonemes) for each language (for example, 44 for English), in addition to a rule system that combines phonemes into syllables and then words. Pinker and Jackendoff (2005) argue that phonology is uniquely human; that the combinatorial properties of phonology have no analogue in evolutionarily related primates, although there may be such analogues for dolphins and songbirds (for the latter of which there is, for example, evidence of the use of syllables; see Doupe and Kuhl, 1999). Unfortunately, this area of language is not yet well enough researched even to approach the beginnings of an overview.

Knowledge of syntax consists of many different elements that together result in the ordering of words in relation to each other. Pinker and Jackendoff (2005) list four elements. The first element is the hierarchical or recursive structure of the strings of words and phrases that together form sentences. Recursion or recursive embedding is the insertion of a pattern (such as a sentence) into the same pattern (another sentence) (see, e.g. Jackendoff, 1993). For example, inserting the sentence 'John fears Peter' into the sentence 'Mary thinks X', where X stands for another sentence, thus results in 'Mary thinks John fears Peter'. This process is recursive in that there is no limit to the number of insertions: for example, 'Carl hopes that Mary thinks John fears Peter', 'Zoe assumes that Carl hopes that Mary thinks John fears Peter', etc. Recursivity formally defines creativity in human language.

The second way in which syntax manifests itself is in the form of a parameter that specifies which words in both small and large composite strings (phrases and sentences) in a particular language generally come first. For example, if a phrase has its main element (such as a preposition in a prepositional phrase) first, then this is also likely to be the case for verb phrases, adjectival phrases, etc. in that language. The third element that constitutes syntax is agreement, the way in which verbs and adjectives are marked with inflections corresponding to number, person, grammatical gender or other features of related nouns. The fourth element that impacts on word order is case: case marks noun phrases with inflections (nominative, accusative, dative, genitive, elative, ablative, etc.) depending on the grammatical role of the phrase in relation to a verb, preposition or another noun.

There seems to be general agreement that animals do not possess the ability to form strings that have recursive or hierarchical properties. However, Jackendoff and Pinker (2005) and Pinker and Jackendoff (2005) stress that human languages can differ in the relative contribution of each of the four elements mentioned above, with, for example, Pirahã lacking recursivity (see Everett, 2005). Comparative research in relation to the other three elements that are part of human syntactic knowledge is currently not available.

It has been widely observed that there are qualitative and quantitative differences in word learning between humans and animals (Hauser *et al.*, 2002; Fitch *et al.*, 2005; Pinker and Jackendorff, 2005). For example, animals do not have words for abstract entities such as 'yesterday', and are not able to acquire the same number of words (50,000–100,000) in the same effortless manner as do humans. Comparative research relating to knowledge of the internal structure of words (morphology) is rare. An exception is a study by Endress *et al.* (2009), who found that cotton-top tamarin monkeys were able to distinguish between the past tense morpheme '-ed' as either a prefix or a suffix, just as human beings do. More research in this area of language is also required.

Another crucial aspect of human knowledge of language is the semantic-conceptual structure of language, the meaning that we can derive from or put into words. There is no doubt that several concepts related to the properties of language discussed above are absent in other species, such as the concepts of 'recursion' and 'verb phrase'. However, other species have basic knowledge of cause and effect, navigation and – if part of a social species – knowledge of social structure, even though such knowledge may be structured differently (Povinelli, 2000). If a species has a well-developed system of communication, it should be no surprise to find some of these elements in their messages. However, Pinker and Jackendoff (2005) claim that such abstract concepts as essence (for example, a component of folk chemistry), ownership, multi-part tools, fatherhood, romantic love and **ethics** are absent, rudimentarily present or present in a different fashion in other primate species. These predictions have not yet been put to the test.

In conclusion, there is no simple answer to the questions of whether animals have language, whether they have similar or unique features in their systems of communication compared with us or whether their systems of communication evolutionarily overlap with human language. The evidence is dispersed over many areas of knowledge that make up language, and in many cases the evidence is not (yet) there. The search for an answer to these questions, however, is important, since the answers may reveal something about our own evolutionary history, and about the uniqueness of both our and other animals' systems of communication (Hauser, 1996). (EvdZ)

References and further reading

Bickerton, D. (2007) Language evolution: a brief guide for linguists. *Lingua* 117, 510–526.

Chomsky, N. (1986) *Knowledge of Language: its Nature, Origin, and Use*. Praeger, New York.

Doupe, A.J. and Kuhl, P.K. (1999) Birdsong and human speech: common themes and mechanisms. *Annual Review of Neuroscience* 22, 567–631.

Endress, A.D., Cahill, D., Block, S., Watumull, J. and Hauser, M.D. (2009) Evidence of an evolutionary precursor to human language affixation in a nonhuman primate. *Biology Letters*, doi: 10.1098/rsbl.2009.0445.

Everett, D.L. (2005) Cultural constraints on grammar and cognition in Pirahã. *Current Anthropology* 46, 621–646.

Fitch, W.T., Hauser, M.D. and Chomsky, N. (2005) The evolution of the language faculty: clarifications and implications. *Cognition* 97, 179–210.

Hauser, M.D. (1996) *The Evolution of Communication*. MIT Press, Cambridge, Massachusetts.

Hauser, M.D., Chomsky, N. and Fitch, W.T. (2002) The faculty of language science: what it is, who has it, and how did it evolve? *Science* 298, 1569–1579.

Hockett, C.F. (1960) The origin of speech. *Scientific American* 203, 88–96.

Jackendoff, R. (1993) *Patterns in the Mind: Language and Human Nature*. Harvester-Wheatsheaf, Hemel Hempstead, UK.

Jackendoff, R. (1996) *The Architecture of the Language Faculty*. MIT Press, Cambridge, Massachusetts.

Jackendoff, R. (2002) *Foundations of Language: Brain, Meaning, Grammar, Evolution*. Oxford University Press, Oxford, UK.

Jackendoff, R. and Pinker, S. (2005) The nature of the language faculty and its implications for evolution of language (reply to Fitch, Hauser & Chomsky). *Cognition* 97, 211–225.

Kirby, S. (2000) Review of *Apes, Language and the Human Mind*, by Savage-Rumbaugh, Shanker and Taylor. *Journal of Linguistics* 36, 190–195.

Kuhl, P.K. (1991) Human adults and human infants show a 'perceptual magnet effect' for the prototypes of speech categories, monkeys do not. *Perception and Psychophysics* 50, 93–107.

Liberman, A.M. (1996) *Speech: a Special Code*. MIT Press, Cambridge, Massachusetts.

Pinker, S. and Jackendoff, R. (2005) The faculty of language: what's special about it? *Cognition* 95, 201–236.

Povinelli, D.J. (2000) *Folk Physics for Apes: the Chimpanzee's Theory of how the World Works*. Oxford University Press, Oxford, UK.

Ramus, F., Hauser, M.D., Miller, C., Morris, D. and Mehler, J. (2000) Language discrimination by human newborns and by cotton-top tamarin monkeys. *Science* 288, 349–351.

Savage-Rumbaugh, E.S., Shanker, S.G. and Taylor, T.J. (1998) *Apes, Language, and the Human Mind*. Oxford University Press, Oxford, UK.

Lateralization and handedness

Lateralization of the **brain** refers to structural and/or functional differences between the left and right sides of the brain. As it usually refers to differences between the left and right hemispheres, it is also known as hemispheric specialization. However, there are known forms of lateralization (e.g. control of **sexual behaviour**) that occur at subcortical levels of the brain. Although hemispheric specialization was thought to be a unique characteristic of humans, underlying our abilities of **language** and higher **cognition**, we now know that it is common among vertebrates. Chickens, for example, have highly lateralized brains; feeding responses that require the chick to discriminate grains from a distracting background are controlled by the left hemisphere, and attack responses are controlled by the right hemisphere. The right hemisphere's role in controlling attack responses has also been shown in fish, toads, lizards and gelada baboons. In animals with their eyes placed laterally on the sides of their head, attack responses are manifested as a leftward side bias, since input from the left eye is processed primarily by the right hemisphere. Prey-catching responses are biased rightward, as shown in chicks and toads.

Hemispheric specialization can also be reflected in hand/limb preferences (e.g. many species of parrots use the left foot to hold food), but hand preferences do not always indicate the direction of brain lateralization. For example, rats have a population bias to use the right hemisphere to express emotional behaviour, as do other species, and they also show a population bias in a number of other types of lateralization, but 50% of rats prefer to use the left forepaw to reach for food and 50% the right. The same applies to marmoset monkeys. In these species, left-paw-preferring individuals are more emotional (more fearful) than are right-pawed ones. Hence, the limb preference may reflect which hemisphere is dominant and indicate the **temperament** of the animal. This has implications for the **welfare** of these species, since hand preference could be a useful indicator of responsiveness to **stress**. In fact, there is some evidence in primates that left-handers show higher levels of cortisol secretion when stressed than do right-handers.

When we consider the range of vertebrate species investigated in studies of lateralization, we can see that there is a general pattern of the left hemisphere being used for considered responses, which may require the animal to inhibit responding while it decides on the appropriate action to perform, whereas the right hemisphere is used when the animal must make a rapid response. In fact, animals are more responsive to an advancing predator when it is on their left side. The right hemisphere is also used in the expression of **emotions** and for processing spatial information. The ubiquity of lateralization among vertebrates, together with some recent evidence for lateralization in invertebrates, leads one to assume that it must be beneficial. Recent evidence shows that lateralized individuals are able to perform two tasks at once, whereas non-lateralized individuals cannot, and that lateralization assists in the formation of long-term memory.

(LJR)

Further reading

Hopkins, W.D. (ed.) (2007) *The Evolution of Hemispheric Specialization in Primates*. Elsevier, London.

Rogers, L.J. (2002) Lateralization in vertebrates: its early evolution, general pattern and development. In: Slater, P.J.B., Rosenblatt, J., Snowdon, C. and Roper, T. (eds) *Advances in the Study of Behaviour*, Vol. 31. Academic Press, San Diego, California, pp. 107–162.

Rogers, L.J. and Andrew, R.J. (eds) (2002) *Comparative Vertebrate Lateralization*. Cambridge University Press, Cambridge, UK.

Laying hen housing

Virtually all chicks that are destined to become laying hens are housed, as they need extra warmth and protection from predators. Most adult birds are also housed for at least some of the time. Free-range flocks are invariably confined to houses at night to protect them from predators, but also to encourage them to lay their eggs in nestboxes in the house. Egg laying is a morning activity. Hens will seek the shelter of their house in cold, wet weather or to avoid bright sunlight. The majority of laying hens are housed intensively in climate-controlled housing to increase output and productivity per unit area and to reduce overhead costs. Intensification requires more sophisticated management and housing systems in order to control most of the variables that affect productivity, including temperature, humidity and air quality, plus feed and water quality and availability. See Weeks (2002) for more details.

Enclosure and artificial ventilation enable control of light. The intensity, duration, spectral composition and photoperiod of light profoundly affect the maturation of pullets, onset of lay, egg size and numbers and bird behaviour, and are thus extremely important considerations in hen housing (see Manser, 1996 for a review). The spectral sensitivity of poultry differs from that of humans, and hens have four rather than three different types of cones within their retinas, thus perceiving light quite differently from us (Prescott *et al.*, 2003). Recommended minimum levels of light intensity are 10–20 lux at bird's eye level. Insufficient levels of light cause abnormal development of the eye. Reducing light levels makes birds easier to catch for depopulation, and lower intensities are often used to reduce **cannibalism** and **feather pecking**.

It is particularly important to remember that housing restricts the ability of the hen to choose her own environment, and thus humans must assume responsibility for providing an environment that satisfies the **health** and **welfare** needs of the birds. There is increasing realization that poultry are sentient beings and, as such, require more than the basics for survival, as expressed in the **five freedoms**. More consumers in the 'developed' world now demand 'welfare-friendly' produce from less intensive housing systems.

None of the current housing and husbandry systems for commercial egg production is ideal in terms of hen welfare. Each has both advantages and disadvantages in terms of health and behavioural needs, which are equally important considerations for animal welfare. Abnormal pathology and behaviour are reported from all systems. Whereas housing and husbandry systems traditionally evolved to be practical, economic and productive, a comparatively recent change of emphasis has emerged. Applied ethologists have used economic models to determine how animals prioritize different behaviours and features of their environment (Dawkins, 1990; Weeks and Nicol, 2006) (**see: Economics of behaviour**). This fundamentally different approach to design aims first to make provision for the most important facilities and resources from the animal's point of view, and then to develop practical and economic housing systems to incorporate these. Aviaries and modified cages have been developed this way, and their designs are still evolving.

Battery cages

The vast majority (about 75% worldwide) of laying hens kept for commercial egg production are housed in battery cage systems, where 'battery' refers to an array of cages. These are commonly three to nine tiers high and completely made from wire, including a sloping floor (up to 8% is recommended), which allows droppings to fall through and eggs to roll out on to collection conveyor belts. The number of hens kept in battery cages may reduce, as Switzerland outlawed them in 1992, Sweden in 1999 and the EC is also proposing to do so from 2012.

Conventional battery cages have been criticized for inadequate hen welfare by many sources, particularly in terms of the extremely restricted space allowance and associated disuse osteoporosis, plus the lack of facilities such as nestboxes, dust baths and perches causing **frustration** and preventing the expression of normal behaviour. Wire floors may be uncomfortable and lead to foot problems, although EU law now requires that claw shorteners should be provided to prevent overgrowth. Claw shorteners are generally strips of abrasive material fitted on the inside of the front of the cage, near the feed trough that is usually fitted just outside the front of the cage. While hens feed they tend to attempt to scratch as they would if foraging, and thus file their own claws. Abrasion of the feathers against the cage sides is common.

To save costs, cage sizes have decreased and number of hens per cage increased to the point where it is usual for hens not to be able even to stand fully upright in a cage, and for the floor allowance to be so restricted that not all birds can sit down at any one time. Typical inadequate allowances are around 340 cm^2 per hen, the most common allowance in the USA for example, although restaurant chains and retailers may specify more generous allocations. In the EU, a space allowance of 450 cm^2 was taken to be insufficient and has been increased to 550 cm^2, which will increase when the specifications for modified cages come in (see below). It is difficult and time-consuming to inspect individual cages in high tiers. Walkways are not a legal requirement in most countries but these, plus uniform lighting at all levels, should be provided to aid inspection and removal of sick and dead birds.

The battery cage system is, however, not only the easiest to manage and automate, but also provides the cheapest eggs. Hygiene is usually superior because the birds have minimal contact with their droppings, and the lack of litter reduces dust and **parasites** considerably. There is a reduced risk of humans contracting disease from the hens or their eggs. Birds in battery cages are, on average, the most productive and healthy with relatively low levels of mortality. The preference of hens for small group size is satisfied, but not their preference for a large space. Feather pecking and cannibalism are rare in battery cages, partly because of small group size and the very restricted space allowance.

Modified cages

Various designs of modified (or furnished, or enriched) cages have been developed with the aim of meeting more of the hens' behavioural needs while retaining the health benefits and economic advantages of conventional battery cages. The effect of cage design on health and performance was considered by Tauson (1998). Priorities for hen welfare appear to be relatively small group size, more space, provision of nestboxes and perches and a substrate for foraging and **dust bathing**. Elevating the feed trough improves feather cover by discouraging activities such as stepping off perches on to the backs of cage mates, which causes feather damage.

Modified cages that are being used commercially include: (i) the Victorsson model for eight birds, with perches and litter box on top of a nestbox based on the Edinburgh modified cage design; and (ii) the Big Dutchman reversed furnished cage for ten hens, with curtained nesting area and litter provided on a belt (based on a design by Tauson). Cages that will accommodate large groups (20–60) are also being examined. These are essentially small aviaries, the definition of which relates to whether a human caretaker has to enter the enclosure to check the birds (aviary) or not (cage).

Providing perches costs little, but advantages include normal roosting and perching behaviour, reduced foot problems and increased **bone strength**. Perches can be provided in conventional cages, but the space restriction is not ideal and can slightly increase the number of cracked and dirty eggs and make it more difficult for hens to move around the cage. There can be strain differences in these parameters and in perch use.

Further research is needed to ascertain whether sham dust bathing is reasonably satisfactory for the birds. Artificial turf may be acceptable to birds in place of a litter substrate that has so many practical problems associated with it. Both the Victorsson and Big Dutchman furnished cages provide 0.06 m^2 (600 cm^2) of floor area per bird. The EC has specified 0.55 m^2 as a minimum for all cages installed after 2003 (or 750 cm^2 per bird, including the nestbox and litter area). It is proposed that in 2012 this will apply to all cages. In addition, the proposed 'enriched' cages must be 45 cm high over most of the cage, and have a nest, 15 cm of perch space per hen, 'litter such that pecking and scratching are possible' and a claw-shortening device. There is no explicit provision for a dust bath. In the rest of the world, cage design and stocking density may change to meet the welfare specifications of the large retail customers.

Aviaries

There are several variants of aviaries. Single-tier systems are increasingly common in the UK. These have a one-tiered wire floor containing feeders and drinkers and banks of nestboxes in the centre or at the side, a litter area and a stocking density of about 9 birds/m^2. The tiered wire floor system was invented in the Netherlands to make similar use of the vertical space in a house to battery cages. It resembles battery cages, but with partitions removed. Hens may move freely between rows of narrow tiers, three high, with feed and water provided on each level (see Fig. L.1).

Other commercial aviary systems include Boleg, Laco-Voletage, Natura and Marielund, all floor-based systems that provide tiers with slatted or wire floors to make more use of vertical space, with litter for scratching and dust bathing at ground level. There may be ladders for hens to use to move between tiers. Often the top tier is reserved for roosting perches. Drinkers tend to be placed over slats. Feed is usually distributed to each tier via flat-chain trough feeders. Tiers of nestboxes with egg-collection belts are placed throughout the house. Manure belts need to be provided to prevent hens from defecating on each other. Most commercial stocking densities are around 17–20 birds/m^2.

Aviaries are complex systems that place ergonomic demands on both birds and humans, particularly at depopulation. Outbreaks of severe feather pecking and cannibalism can occur

Fig. L.1. It is common in housing systems for groups of layers to find a raised, perforated floor to collect droppings beneath roosting perches, feeders and drinkers.

Fig. L.2. Tiers of nestboxes in the 'Natura' perchery system. Note the perches in front that hens can fly on to, giving easy access.

in aviaries, with mortality as high as 27%. Effects of stocking density on feather pecking are complex. Different strains vary in propensity to feather peck and plumage condition as well as in mortality and performance. Reported proportions of floor eggs vary between 2.5 and 11.5%. None the less, with experience and good management, layer performance can be similar to that in battery cages, and equally good production has been recorded for the 65% of laying hens in Switzerland that are kept in aviaries. Aggression is generally low in large flock systems. Rearing pullets in an aviary-type environment and providing a covered outdoor area seem to be important for success.

Percheries

Percheries use the vertical space in a house by having several rows and tiers of perches arranged on frames (see Fig. L.2). There is some overlap in terminology and design with aviaries. At each level, hens can access feed and water and may also move up and down or fly across to another set of perches or to a tier of nestboxes.

Hens are particularly likely to sustain bone breaks in this alternative system (although recent work suggests that birds may be getting more susceptible to fractures in all systems). They need to have access to perches and learn to fly during rearing. The spacing between stacks of perches and nestboxes is also very important. Design and management should also keep perches and birds' feet clean, otherwise problems such as bumblefoot may result. Fewer problems are encountered where litter is provided on the floor, but EC rules do not specify it, and also permit up to 25 birds/m^2, with a minimum of 150 mm of perch space per bird. Cannibalism and other problems are often associated with such high stocking densities. Percheries are less commonly found now for various reasons, not least the unpleasant working environment for humans.

Housing for free-range hens

Hens must have daytime access to pasture in order to be described as free range. Stocking density can be as high as 1000 hens/ha if there is good vegetation cover on the ground, but lower densities are recommended to avoid health problems. Most producers restrict ranging within an enclosure (electric) fenced against predators and encourage birds to lay in nestboxes within the house. All houses provide perches for night-time roosting. Most houses also contain the drinkers and feeders to reduce losses to, and contamination by, wild birds as well to minimize problems with rain and freezing (see Fig. L.3).

Some provision of water outside is recommended. Houses may be fixed or movable. Fixed houses should be similar to deep litter or perchery houses to meet EC regulations. Movable housing can vary from small, wooden arcs or houses with wheels that are moved frequently to semi-permanent, large houses that tend to be moved for each new flock. Typical house stocking density is 10 hens/m^2.

Coccidiosis is a universal disease challenge in free-range flocks. Vaccination, good hygiene – including pasture rotation and frequent manure removal – and reduced stocking density all help to control coccidiosis. Because the birds are more active and, in most climates, exposed to cooler temperatures, they eat more. Thus, especially in developed countries, it is the most expensive system. Eggs must be sold for a premium for the system to be viable. However, in rural undeveloped countries, backyard flocks provide high-quality nutrition for humans at negligible cost.

There are some welfare problems associated with the free-range system, particularly in larger flocks. Both the health and the behavioural needs of the hens can be compromised by problems of bullying around pop-holes, feather pecking and cannibalism. Debeaking (**beak trimming**) is a widespread preventive measure that, in itself, is a welfare issue and is also practised to some extent in all other housing systems. Research is ongoing to find ways of encouraging hens to make full use of the range area provided. Greater use of range is made by both individuals and a larger proportion of the flock when tall vegetation and trees are provided for cover. Providing wide pop-holes tends to encourage use of range, although in some

Fig. L.3. Free-range layer house.

climates this can cause wet litter and hence problems such as ammonia and foot lesions.

Minority systems

These include straw yards (barns) where birds are housed on deep litter with nestboxes and perches, similar to free-range housing. Slatted or wire floor systems are similar, but with perforated flooring that reduces contact between birds and their faeces. Welfare tends to be worse because possible health benefits are traded against loss of a medium in which to forage and dust bathe. Additionally, behavioural problems such as floor egg laying, feather pecking and cannibalism are common, thus adversely affecting the economics. (CW)

References and further reading

Dawkins, M.S. (1990) From an animal's point of view: motivation, fitness, and animal welfare. *Behavioural and Brain Sciences* 13, 1–61.

Manser, C.E. (1996) Effects of lighting on the welfare of domestic poultry: a review. *Animal Welfare* 5, 341–360.

Prescott, N.B., Jarvis, J.R. and Wathes, C.M. (2003) Light, vision and the welfare of poultry. *Animal Welfare* 12, 269–288.

Tauson, R. (1998) Health and production in improved cage designs. *Poultry Science* 77, 1820–1827.

Weeks, C.A. (2002) Poultry housing. In: *Animal Health and Production Compendium*. CAB International, Wallingford, UK.

Weeks, C.A. and Nicol, C.J. (2006) Behavioural needs, priorities and preferences of laying hens. *World's Poultry Science Journal* 62, 297–308.

Website

See http://www.laywel.eu for results of a European comparison of housing systems.

Learned helplessness

Learned helplessness describes the apparent failure to respond to a change in a situation after experiencing a lack of control in a similar, previous situation. The phenomenon was described originally by Seligman (1975), following an experiment in which dogs that had experienced random, electric shocks (from which they could not escape) failed to learn how to avoid being shocked by jumping barriers into another compartment within their experimental environment. It appears that the individual might learn that the unpleasant event occurs independently of their behaviour and generalizes this learning. Learned helplessness may explain apparent **apathy** in individuals in situations over which they perceive that they have no control as a consequence of their previous experiences. (KT)

Reference and further reading

Maier, S.F. and Seligman, M.E.P. (1976) Learned helplessness: theory and evidence. *Journal of Experimental Psychology: General* 105, 3–46.

Schwartz, B. and Robbins, S.J. (1995) *Psychology of Learning and Behaviour*, 4th edn. W.W. Norton and Co., New York.

Seligman, M.E.P. (1975) *Helplessness*. Freeman, San Francisco, California.

Legislation to protect animals

Legislation to protect animals first developed in Europe and North America during the 19th century. The world's first law to protect animals was passed in the British parliament in 1822. It said 'that if any person or persons having the charge, care or custody of any horse, cow, ox, heifer, steer, sheep or other cattle, the property of any other person or persons, shall wantonly beat, abuse or ill-treat any such animal, such individuals shall be brought before a Justice of the Peace or other magistrate'.

There are three striking restrictions in this law:

1. Only some animals are covered. Many (e.g. **dogs**, **cats**, **pigs** and poultry) are not mentioned at all and, among the species that are covered, some, like bulls, are not indicated explicitly. One reason for this is that, at the time, there was a custom of arranging fights between animals. **Bull baiting** and other blood sports were popular, and the politicians were reluctant to make unpopular laws.
2. The law applies only to people who do not own the relevant animals. The Bill's advocates seem to have assumed that animal owners would want to protect and make good use of their property. Bad animal treatment is regarded as irrational or pointless – the kind of thing that would only be done by someone who does not share the owner's interest in protecting the value the animal represents.
3. Only what is described as *wanton* cruelty is covered. The adjective 'wanton' means undisciplined, random or motiveless, so those who passed this Bill do not seem to have been aiming to regulate established uses of animals.

There is clearly a link between **2** and **3**: the need to protect animals by law arose from concerns about pointless cruelties perpetrated by third parties. This equation underpinned most legislation designed to protect animals until at least the 1950s. It was only after this that attention turned in a serious way to legislation intended to protect animals used, by their owners, in a *purposeful* way in intensive animal production, laboratories and elsewhere.

In intensive farm animal production of the kind practised today, conflicts between productivity and the interests of the animals are familiar (see: **Intensification of animal**

production). Modern farm animals are typically allowed less space than they once had, and many live in barren environments in which they cannot exercise their normal range of behaviours. Some kinds of genetic selection and **breeding** goals are associated with production-related diseases. These problems have given rise, particularly in Europe, to a new kind of animal welfare legislation. Broadly speaking, the point of this legislation is to prevent farmers from doing what, in fact, it is most profitable to do. For example, in intensive egg production it is economically sound to keep hens in small cages with very high stocking densities. In places without animal welfare regulation it is not uncommon for hens to be housed with as little as 300 cm^2 per animal. If there is no regulation or other mechanism in place, egg producers will be incentivized by market forces to keep their hens in very small cages. The alternative is to produce at a higher cost than their competitors, and this would not be feasible in the long term.

Recent growth in international markets has ensured that national legislation is both less effective in securing its welfarist purpose and harder to justify to domestic producers. Both consequences flow more or less automatically from the introduction of competition from producers in countries with less stringent laws. In response to this kind of problem there has been a drive in EU countries, as well as in countries that are members of the Council of Europe, towards harmonized European legislation. New rules on animal production designed to create a 'level playing field' (so far as legislation is concerned) are now regularly introduced. Since animal products are increasingly imported to the EU from countries beyond the reach of European jurisdiction, there is a limit to how far this kind of policy can succeed – unless, of course, it can be translated to the global level. Some global initiatives are on the way, but they remain at a very early stage.

It can be seen, then, that the rationale of animal protection legislation has changed dramatically over time. Recent legislation aims to protect or improve animal welfare that is threatened by market forces and intensive production. Earlier 'anti-cruelty' legislation, by contrast, was designed to protect livestock against irrational forms of misuse.

Today, the need for animal protection legislation goes beyond farm animal production, as is readily apparent when we consider the use of animals as tools in biomedical research. The benefits of this kind of animal use are clear, even if the question of whether those benefits could be obtained without using animals is contested; and biomedical animal research is certainly approved by a large percentage of the population. But many people worry about animals having to suffer for our sake. Such people want reassurance that there is a good case for doing the research: that its goals are definitely worth pursuing, and that those goals oblige the researcher to use animals. To address this kind of attitude, legislation has been put in place in many countries to protect **laboratory animals** from unnecessary **suffering**. The legislation is designed to ensure that live animals are used for experiments only when no other suitable method is available, that no more animals are used than is necessary and that, when they are used, animals are not caused more **stress** or suffering than is required by the experimental aims (**see: Three Rs principle**). (PS)

Further reading

Rollin, B.E. (2005) Animal agriculture and social ethics for animals. In: Pond, W.G. and Bell, A. (eds) *Encyclopedia of Animal Science.* Marcel Dekker, New York.

Ryder, R.D. (1989) *Animal Revolution*. Basil Blackwell, Oxford, UK.

Sandøe, P. and Christiansen, S.B. (2008) *Ethics of Animal Use*. Blackwell, Oxford, UK, Chapter 1.

Webster, J. (2005) *Animal Welfare: Limping Towards Eden*. UFAW, Blackwell Publishing Ltd, Oxford, UK.

Lek

A lek is a relatively small area where males congregate and exhibit competitive displays to females. The lek does not contain resources of importance to females. Typically, 10–20% of males at a given lek secure more than half of the matings. Why males aggregate in a lek is hotly debated, although it is generally accepted that males form leks because they are unable to monopolize either females directly (female defence polygyny) or resources required for females (resource defence polygyny). Two key hypotheses are the 'hotspot' and the 'hotshot' hypotheses. According to the hotspot hypothesis, males gather at areas that have a high female encounter rate (several female territories may overlap in a particular area and/or females may travel along certain routes). In contrast, the hotshot hypothesis proposes that females are attracted to highly attractive males, and the less attractive subordinate males cluster around these successful males. (PE)

Further reading

Hoglund, J. and Alatalo, R. (2006) *Leks*. Princeton University Press, Princeton, New Jersey.

Lever pressing

An **operant response** (a response controlled by its consequences (**see: Conditioning – types of**)), commonly used in behavioural studies of learning. Lever pressing in this context typically results in the delivery of a food pellet to a hungry **rodent** and so allows clear quantification of the effects of other variables on the behaviour. (SuH)

Libido

Libido can be defined as the manifestation of the sexual motivation in both males and females. In applied animal behaviour, the term is most often used to describe the willingness and eagerness of male animals (especially farm animals such as boars, bulls, stallions and rams) to mount and attempt to mate with the females. A number of factors have been shown to affect libido, including: genetics, hormonal factors, post-weaning management, nutrition and season.

Libido has a genetic component, with differences within and between breeds. In boars, crossbred boars have greater libido than purebred boars and, within purebreds, it has been shown that Duroc boars have lower libido than Yorkshire boars. With cattle, *Bos indicus* bulls have generally reduced libido relative to *Bos taurus* bulls and, within *B. taurus*, beef breeds have lower libido than dairy breeds. In horses it has been reported that Arab stallions have greater libido than Quarterhorse stallions.

The major hormonal influence on libido is exerted by **androgens**, especially **testosterone**. Low-libido animals generally have lower circulating testosterone levels than their high- or normal-libido **conspecifics**. The whole picture, however, is not this simple, and other hormones such as **oestrogen, luteinizing hormone (LH), prostaglandins** and **thyroid hormones** have also been implicated in the hormonal control of libido.

Sexual inhibition can occur in management systems that keep young, post-pubertal males in male-only groups, and this has been reported in bulls, boars and rams. Ideally, males should have some contact with females during rearing but, at sexual maturity, they should be allowed only supervised contact with females to prevent any **aggression** from them that may suppress libido in the future.

There is some evidence to suggest that libido is diminished by rapid weight gain during the post-pubertal period. For ruminants, high levels of concentrate diet and high feeding levels appear to reduce libido relative to forage-fed animals. However, protein levels need to be maintained. Boars on low-energy and low-protein diets show low libido.

Many animals appear to show reduced libido in relation to climatic extremes, with libido particularly reduced in extreme heat. Some **sheep** breeds and *B. indicus* bulls are reported to have reduced libido in cold weather. (JNM-F)

See also: Sexual behaviour

Light and behaviour

Light is essential for many animals' lives and can have profound influences on their behaviour and **welfare**. These are discussed below with respect to the frequency, intensity, timing and duration of light.

Frequency of light

The frequency (or wavelength) of light reflected or emitted by objects is the characteristic that humans perceive as colour. There is a wide range of sensitivities to light frequencies within the animal kingdom, and animals are often categorized according to the number of types of retinal cone (colour-sensitive) cells they possess, i.e. monochromatic (e.g. marine mammals), dichromatic (e.g. most terrestrial mammals, including **dogs**, **horses**, **pigs**, **goats**, **cattle**, **sheep**, **deer** and many primates), trichromatic (e.g. humans and other primates and several marsupials) or tetrachromatic (e.g. some arachnids, insects, **amphibians**, **reptiles**, fish and birds). Some animals have five (flies and some butterflies) or even 12 (stomatopods – 'mantis shrimps') types of receptor for colour vision (**see also: Visual perception**).

Colour vision is usually dependent on differing signals from different cone types, and therefore it is believed that most dichromats have only limited colour perception. It has been argued that, despite having only one type of cone cell, monochromatic marine mammals do perceive colour, registered as a balance between signals from the rods and cones. Even within individuals, colour sensitivity can differ. For most animals, as the intensity of light decreases, the cone cells become non-functional and vision becomes more reliant on the rods, which are insensitive to colour, sometimes called scotovision. Some nocturnal geckos and moths are able to use colour vision in dim moonlight situations when humans and most other animals are colour-blind. Colour perception mechanisms are highly dependent on evolutionary factors, of which the most prominent is thought to be recognition of food sources, but it is also used in many other aspects of behaviour such as **mate choice**, **crypsis** and navigation.

Some animals (e.g. many insects, birds, lizards, frogs, turtles, fish and some mammals) are visually sensitive to frequencies higher than are most humans, and can see ultraviolet (UV) light (this is prevented in most adult humans, as the cornea and lens block ultraviolet frequencies). The functions of UV vision are not yet well understood, but possibilities include urine-mark visibility in degus, voles and mice, social communication in degus and birds (including domestic poultry) and twilight UV vision in **rats**. Beyond the lower end of frequencies visually perceived by humans, some pit-vipers are sensitive to infrared wavelengths and therefore 'see' heat emitted from potential prey.

In the sky and underwater, scattering of incoming light produces partial polarization that varies with the position of the sun and the direction of view. Some animals (e.g. insects, cephalopods, birds and fish) possess the ability to see polarized light, and may use this in a similar way to colour vision, e.g. signalling, species recognition, mate choice, orientation and navigation. Cuttlefishes and other cephalopods have only one spectral photoreceptor class and are therefore probably incapable of colour vision; however, some possess the ability to change colour rapidly in what appears to be signalling behaviour; the replacement of colour with polarization vision seems a plausible suggestion for how this might occur.

Many animals are housed under artificial light sources, the most common being incandescent and fluorescent lamps. These are designed primarily for the human visual sensory system, and as a consequence have characteristics that can influence animal behaviour and, potentially, welfare. Both incandescent and fluorescent lamps emit little or no UV wavelengths. As discussed above, UV vision is used widely by some species; however, UV-deficient environments do not appear to be **aversive** and there appear to be few, if any, consequences for welfare. Some artificial lamps are designed to emit UV frequencies, and these have been shown to be beneficial in environmental **enrichment** for **turkeys** (Lewis *et al.*, 2000) and breeding in reptiles. Furthermore, commonly available fluorescent lights flicker on and off at a frequency so great that most humans perceive this as a constant light, because it is above our critical flicker fusion frequency (CFFF). However, other animals have a higher CFFF and therefore perceive these lights as flashing on and off.

Despite this, studies on light preferences have shown that animals (primarily birds) do not find this aversive, although it could explain why, in vision studies, some species appear to be unable to recognize images on standard computer monitors and other screens that also have refresh rates below the species' CFFF (Railton *et al.*, 2009). Most animals are insensitive to infrared light and it is therefore often recommended that, if their behaviour during the dark phase is to be recorded, infrared lights and suitable recording equipment can be used to avoid disturbing them.

Intensity of light

Light intensity can have demonstrable effects on behaviour; for example, low intensities reduce **preening** and foraging, but increase sleeping in broiler **chickens** (Alvino *et al.*, 2009). The intensity of light from a point source decreases as the inverse square of the distance from the light source. This means that light intensity decreases rapidly as the viewer moves away from the source. In many animal housing scenarios, if light intensities for animals in tiers or blocks are maintained by measuring intensity at the top levels, the intensity is likely to be too low for animals on lower levels or, if intensity is maintained for those on the bottom levels, the intensity will be too high for animals on the top. Animals close to the lights (e.g. animals on the top levels of banks of cages) may need some protection from these intensities. It should also be remembered that albino strains of animals often prevalent in laboratories are more susceptible to light-induced blindness than are pigmented strains.

Low light intensities are sometimes used to calm animals, for example when pigs are being mixed or when broiler chickens are being caught for taking to the abattoir. Similarly, poultry are often housed at low light intensities (e.g. <10 lux) to reduce injurious **feather pecking**. However, these intensities (especially in combination with elongated photoperiods) can lead to bupthalmia (an elongation of the eye), retinal degeneration and even retinal detachment. Preference studies have indicated that turkeys prefer greater light intensities than are usually provided commercially (Sherwin, 1998). In some laying hen systems, light intensity is gradually decreased at the end of the light phase to give the hens the opportunity to move to the perches while there is still light and thereby avoid risking injury. Different intensities of light are preferred for some behaviours (e.g. Sherwin, 1998; Taylor *et al.*, 2006), indicating that provision of different intensities within an enclosure could be a suitable form of practicable and inexpensive environmental enrichment for captive animals.

Timing of the light phase

If there are no external influences, many behavioural, biochemical and physiological processes will follow an endogenously generated cycle of approximately 24 h (nycthemeral or **circadian rhythms**). These rhythms can sometimes be entrained by external cues (Zeitgebers), a primary one being light. The timing of the light phase is therefore fundamental to animals, as it influences behaviours such as **sleep**, social interaction, **elimination**, learning ability and regular feeding patterns. The primary circadian 'clock' in mammals is located in the suprachiasmatic nucleus in the **hypothalamus**. This receives information about illumination through specialized ganglion cells in the retina, and communicates it to the pineal gland. In response, the pineal secretes the hormone melatonin, which peaks at night and ebbs during the day. Patterns of animal activity are often expressed according to light and dark patterns, e.g. **nocturnal** (mice, rats, **hamsters**), **diurnal** (most avian species, farm, **zoo** and **companion animals**) or **crepuscular** (owls), i.e. active at dawn and dusk. Some animals are extremely flexible and adapt their activity patterns to the activity of humans (**guinea pigs**).

Many laboratory animals are nocturnal, and there is growing acceptance that gathering data from animals during times when they would normally be asleep (i.e. during our human light phase) is not best practice for scientific validity or animal welfare. In studies of nocturnal animals in the dark, red lights are often used to allow humans to observe the animals while avoiding disturbance and gaining more representative data. Such considerations should be extended to the chronobiology of other species. In studies requiring animals to respond normally in the dark, infrared lamps and suitable recording equipment can be used.

Duration of the light phase

In nature, changes in the duration of the light phase can predict seasonal periods of weather conditions, food availability, predator activity, etc. This in turn can affect the seasonal timing of behaviours such as migration, hibernation and **reproduction**. Most laboratory and intensively reared farm animals are totally under the influence of their caregivers with respect to the duration of light exposure, and sometimes this is manipulated to mimic seasonal effects, e.g. layer hens are placed on a regime of increasing daylength to induce egg laying.

Many different patterns of intermittent artificial light have been extensively researched, mainly in efforts to increase poultry production (e.g. Lewis *et al.*, 1996). It has been shown that a simple pulse of light can be used to indicate 'dawn'. If this flash occurs 1 h before an 8 h period of light, it may have the same stimulating effect as a 9 h period of continuous light. The effects of different **photoperiod**s have also been extensively researched, again mainly in relation to poultry production. In some countries, some farmed species are housed under almost continuous light, as this keeps the animals active and eating, thereby achieving **slaughter** weight at an earlier age. However, longer photoperiods have been associated with increased mortality in broiler chickens (Lewis *et al.*, 1996). It is widely believed that interruptions to the dark phase are aversive or perhaps even stressful to animals; some codes of recommendations indicate that animals should be provided with an uninterrupted period of darkness. (CS)

References

Alvino, G.M., Archer, G.S. and Mench, J.A. (2009) Behavioural time budgets of broiler chickens reared in varying light intensities. *Applied Animal Behaviour Science* 118, 54–61.

Lewis, P.D., Morris, T.R. and Perry, G.C. (1996) Lighting and mortality rates in domestic fowl. *British Poultry Science* 37, 295–300.

Lewis, P.D., Perry, G.C., Sherwin, C.M. and Moinard, C. (2000) Effect of ultraviolet radiation on the performance of intact male turkeys. *Poultry Science* 79, 850–855.

Railton, R.C.R., Foster, T.M. and Temple, W. (2009) A comparison of two methods for assessing critical flicker fusion frequency in hens. *Behavioural Processes* 80, 196–200.

Sherwin, C.M. (1998) Light intensity preferences of male domestic turkeys. *Applied Animal Behaviour Science* 58, 121–130.

Taylor, N., Prescott, N., Perry, G., Potter, M., Le Sueur, C. and Wathes, C. (2006) Preference of growing pigs for illuminance. *Applied Animal Behaviour Science* 96, 19–31.

Light/dark test

The light/dark test is used to measure anxiolytic-like or anxiogenic-like activity, especially in mice, as an animal model of **anxiety**. This test is based on the fact that **rodents** display spontaneous **exploratory behaviour** in response to mild

stressors (e.g. novel environment and light) and show an innate aversion to brightly illuminated areas. The test apparatus consists of a small, dark, safe compartment and a large, illuminated, **aversive** compartment. Transitions between light and dark compartments and exploratory behaviours are part of the scoring of behaviour. Anxiolytic drugs have been found to increase **locomotion** and time spent in the light zone, whereas anxiogenics decrease them. (SL)

Further reading

Bourin, M. and Hascoet, M. (2003) The mouse light/dark box test. *European Journal of Pharmacology* 453, 55–65.

Loading animals for transport

Livestock are transported from rearing units to farms, between farms, to and from livestock auction markets and to **slaughter**. Loading of animals on to transport vehicles can cause a significant amount of the overall **stress** associated with **transport**, as evidenced by increases in **heart rate** and the blood levels of stress hormones. The animals may need to be caught, restrained and confined in small spaces, encouraged to move using coercion and may be subjected to the physical exertion caused by, for example, movement, running and climbing loading ramps. Forcing animals to move quickly may cause them to slip and fall, or bump into pen and raceway walls, causing bruises.

Animals such as **horses**, **cattle**, **pigs** and **sheep** are loaded as groups or individuals, while small animals such as poultry and **rabbits** are initially collected into transport crates, which are then loaded on to the transport vehicles using a fork-lift truck. With broiler **chickens**, which are often reared intensively in sheds of 30,000 or more birds, and spent (culled) hens, catching (depopulation) involves hard, physical labour under often unpleasant working conditions. This, together with the low economic value of each bird, may result in reduced care when handling and, thus, compromised **welfare**. Mechanical catching of broilers is used in some countries in an attempt to overcome these problems, but with varied success. Particularly with intensively reared species like pigs and poultry, which have a relatively small amount of contact with humans, loading may be the first time that animals come into close contact with people and experience the **fear** associated with this.

In poultry, catching birds can cause trauma, particularly bruising and haemorrhages, and dislocated or broken bones. Birds caught and carried by a single leg are more susceptible than those caught by both legs. The reported prevalence of bruising ranges widely, and may be up to 20%. A vehicle carrying 6000 broilers may take 1 h to load. The transport crates restrict ventilation and, in hot weather, this can lead to heat stress for the birds until the movement of the vehicle promotes air circulation. So-called day-old chicks are handled at high speed in largely automated systems in modern hatcheries. These move the chicks using rubber belts that operate at progressively increasing speeds, with deflector plates to sort the chicks, before packing them into specially designed cardboard transport boxes that accommodate about 100 chicks. There is considerable scope for stress and damage to the chicks because of the speed of operation.

Large animals are usually walked on to the vehicle. This may involve the use of loading ramps. In many vehicles these are too steep (>20° to the horizontal), particularly the internal ramps giving access to the upper decks, making it very difficult for animals to climb and descend them. In modern vehicles ramps are often partly avoided by the use of raised loading decks, or replaced by hydraulic lift systems, which considerably reduce stress and the danger of injuries to the animals. In all loading systems, raceways and passages should: (i) be appropriate in width; (ii) not have sharp bends; (iii) not have protruding fixtures on which animals could damage themselves; (iv) not have distractions that might cause animals to baulk; and (v) have floors that are non-slip and free-draining. All systems must be secure and proofed against escape of animals.

The point of loading is important in that the decision must be taken as to whether the animal is legally fit to travel. In Europe, animals that are ill, infirm or fatigued (unless only slightly affected and where the intended journey is unlikely to cause them unnecessary suffering) are unfit to travel (EU Council Regulation (EC) No 1/2005). Each animal must therefore be able to be loaded without force and be able to bear weight on all four legs. (PDW)

Local enhancement

The term 'local enhancement' was first introduced into the **social learning** literature by Thorpe (1956/1963). Social learning refers to a group of mechanisms in which the observation of other individuals, or the products of their activities, facilitates or enables the acquisition of novel behaviour. However, as with most social learning terminology, there has been considerable confusion and disagreement with regard to the precise meaning of local enhancement.

Thorpe's (1956/1963) first priority was to distinguish local enhancement from **imitation**. He considered imitation a cognitively complex ability. He defined imitation in terms of 'copying a novel or otherwise improbable act or utterance, or some act for which there is clearly no instinctive tendency'. Hence, when one imitates one learns to perform a novel behaviour directly from watching a demonstrator.

Thorpe argued that there are several different processes that can, on the surface, resemble true imitation, but that these are, in fact, much more cognitively simple. He identified local enhancement as one of these more simple processes. He defined it as 'an apparent imitation resulting from directing the animal's attention to a particular object or a particular part of the environment'. Once the animal's attention has been brought to a particular object or location by the actions of a demonstrator, it might be more likely to engage in individual exploration and, due to being exposed to similar physical and environmental variables, may end up producing matching behaviour. However, unlike with imitation, the animal has not learned the matching behaviour directly from watching the model; instead, the behavioural match is a by-product of individual learning shaped by exposure to similar physical constraints.

Since Thorpe was primarily concerned with defining local enhancement in contrast to imitation, he emphasized the circumstances in which it might lead to organisms producing topographically matching behaviours. Although it might sometimes lead to animals producing physically similar behaviours, this need not necessarily be the case. An animal could have its attention drawn to a particular object or part of the environment and then use quite different behaviours from the demonstrator when it comes to interact with it. For example, a flock of parrots might come across a site where elephants have been excavating for minerals in a mudbank. If, as a direct result of exploring these excavations, the parrots discover the nutritional benefits of the minerals, this would constitute local enhancement. Even if the parrots never directly observed the elephants at the site, or if they used a vastly different behavioural technique, the mediating process would still qualify as local enhancement. All that is required for local enhancement is that an animal has its attention drawn to a particular object or part of the environment in a way that would not have occurred without exposure to the demonstrator or its products, e.g. scent cues or excavations (Heyes, 1994).

Local enhancement is very closely related to Spence's (1937) concept of stimulus enhancement. Spence defined stimulus enhancement in terms of a demonstrator's behaviour increasing the probability that an observer will be exposed to and respond more vigorously towards a certain class of **stimulus**. Thorpe had defined local enhancement in terms of an observer paying attention to the particular object or part of the environment to which the demonstrator had been oriented. Spence's concept of stimulus enhancement extends the effect of the demonstrator so that an observer generalizes its activity to all objects of the same physical appearance.

Heyes *et al.* (2000) point out that, as Thorpe defines it, although local enhancement may be the product of learning, and it may give rise to learning, it does not constitute a learning phenomenon in itself. Naturally, it follows that if it does not qualify as a learning phenomenon, it can not be considered a type of social learning. As they put it:

> For example, a rat may be attracted to odors deposited at a food site by a conspecific (Galef and Beck, 1985) because these odors have been associated with reward, and as a result of consuming the food the rat may acquire a flavor preference. However, the rat's approach to the food site, the phenomenon of local enhancement itself, is a response to current stimulation, and not necessarily either a precursor or a product of learning.
>
> (Heyes *et al.*, 2000)

Upon identifying the limitations of Thorpe's concept, some researchers have suggested discarding the category of local enhancement altogether. However, there are circumstances in which the concept of local enhancement is important, especially if one wishes to establish whether social learning has occurred or not. For example, Heyes and Dawson (1990) found that rats would compensate for a 180° difference in perspective from a demonstrator and push a vertical bar in the same direction as shown (i.e. left to right or vice versa). Although originally the rats' behaviour was interpreted as imitation, it was later shown that the results were likely to be due to the subjects responding to scent cues deposited by the demonstrator on the bar. If the rats had, in fact, pushed the bar in the direction demonstrated in the absence of scent cues, this would constitute social learning. As the results stand, local enhancement remains a likely explanation.

Many researchers have tended to use 'stimulus enhancement' and 'local enhancement' as if they were synonymous, distinguishing them only in terms of whether an observer has its attention drawn to an object (referred to as stimulus enhancement) or to a part of the environment (referred to as local enhancement). However, if one maintains the subtle differences inherent in Spence's and Thorpe's original definitions, one can establish with greater certainty whether social learning has actually occurred. (DC)

References and further reading

Galef, G.B. and Beck, M. (1985) Aversive and attractive marking of toxic and safe foods by Norway rats. *Behavioural and Neural Biology* 43, 298–310.

Heyes, C.M. (1994) Social learning in animals: categories and mechanisms. *Biological Reviews* 69, 207–231.

Heyes, C.M. and Dawson, G.R. (1990) A demonstration of observational learning using bi-directional control. *Quarterly Journal of Experimental Psychology B* 45, 229–240.

Heyes, C.M., Ray, E.D., Mitchell, C.J. and Nokes, T. (2000) Stimulus enhancement: controls for social facilitation and local enhancement. *Learning and Motivation* 31, 83–98.

Spence, K.W. (1937) Experimental studies of learning and higher mental processes in infra-human primates. *Psychological Bulletin* 34, 806–850.

Thorpe, W.H. (1956/1963) *Learning and Instinct in Animals.* Methuen, London.

Locomotion

Locomotion refers to voluntary movements that displace the whole body. This is usually confined to walking, trotting and galloping/running, but many species jump, swim and canter. Other limb movements such as kicking or pawing are performed but do not involve whole-body movement.

Locomotion in mammals is primarily by forward motion; sideways and backward motion can be performed, but muscles are designed for forward propulsion, and any other motion is much less efficient. During forward motion the centre of gravity is moved towards the front limb by the propulsive efforts of the hind limb, and the front limb is raised and repositioned to maintain the animal's balance. The positioning of the limbs may be in either a symmetrical or asymmetrical pattern, and quadrupeds may be supported by three, two, one or no limbs at any particular time. The faster the **gait**, the fewer the supporting limbs. The head may be held high to place the centre of gravity as far back as possible. Animals change their gait according to their velocity requirements. There is a velocity above which it is energetically more efficient to trot than to walk, and another where galloping is preferable to trotting.

The walk

Walking is a gait where each foot is on the ground for at least 50% of the stride. Each limb is lifted by shortening the leg through flexion of the joints, using especially the hip, knee, hock and digital flexor muscles. The limb then enters the swing phase and is placed on the ground through slow extension of the joint. Once the limb is on the ground, it checks and supports the load by tensing all the extensor muscles, particularly in the digital, hock and stifle regions. The sole of the foot is then pushed hard against the ground by contracting the digital flexors, thus enabling the pushing phase to begin, followed by the hanging-limb and swing phases. The limb motion is therefore a cyclical, not an intermittent, process. The quadruped walk can be considered as two people walking out of phase, one behind the other, so that four distinct sounds of the feet contacting the floor are heard. As the walking speed increases, efficiency decreases as the thrust phase (when the limb is supporting) is reduced at the expense of the hanging-limb phase. The major thrust is provided by the hind legs. The fore limbs, although typically supporting about 55–60% of the load of the animal, act mainly to position the limbs to enable them to act as support agents and steer the load.

During turning the limbs on the outside of the turning circle are abducted or rotated outwards, and those on the inside are adducted or rotated inwards, resulting in a change in direction. The main pivots for the fore and hind limbs are the shoulder and hip, respectively. All limb movements contain an accelerative and a decelerative force. During turning the outer limbs increase their accelerative force while the inner limbs increase their decelerative force.

Walking behaviour changes with the degree of confidence that animals have in being able to stay upright. For example, if the floor is slippery or the building is poorly lit, many species will shorten their stride and slow down the rate of walking.

The trot

The trot is used by many quadrupeds for long-distance motion that is more rapid than the walk. It is a symmetrical gait providing an even motion. For this reason it is used frequently by horses in regular motion and cows with full udders, the pendulous nature of which makes the cow reluctant to increase her speed to that of a gallop. When stimulated to move quickly over long periods, as for example if a herdsman hurries a cow down a track for **milking**, she will trot, as this provides a fast, even motion so that the forces on the udder can be absorbed in a rhythmical, swinging motion. In the trot, diagonally opposite limbs are used synchronously: left hind and right fore are followed by right hind and left fore. In a fast trot there is a period between limb changes when there are no limbs on the ground.

The gallop

The gallop is the fastest gait, and involves an asymmetrical step pattern and a lengthened, free-gliding phase. In small mammals this is accompanied by flexion of the back, but in large mammals this does not contribute to the propulsive force, which is provided mainly by the hind limbs. In the gallop there is a leading front limb, followed by the other front limb, then a pause after which the two hind limbs are placed on the ground, one fractionally before the other.

The canter

The canter is a gait of intermediate speed between a trot and gallop. It is a three-beat gait, leading on one hind limb, then the other hind limb together with the diagonal fore limb, and finally the other fore limb.

Motivation for locomotion

Animals are motivated to move in response to demand for food, water, companionship, shelter, safety, **grooming**, a sexual partner, more **space** and many other resources. This motivation increases with the duration and severity of resource restriction, especially of space. Many animals, particularly those that evolved as free-roaming animals that had to travel considerable distances in search of food, appear to need a certain amount of exercise to stay healthy and productive. Regular exercise in the form of supervised walking for tethered horses and cattle will increase muscle and bone growth in growing animals, prevent limb disorders – especially arthritis – and improve behaviour. Recommended allowances are for about 1 h walking per day at about 3–4 km/h. As these animals typically walk for 2–4 km/day, it is likely that this constitutes a requirement. Range conditions may necessitate that cattle walk long distances to water every 2 or 3 days, sometimes up to 40 km, but this will reduce feed intake and milk production.

Effects of the environment on locomotion

The environmental features with the greatest effect on locomotion are the spatial structure and the type of accommodation, including lying and walking areas. These environmental effects interact with each other, and the animal's perception of the environment is modified by its experience. There is a pronounced **circadian rhythm** of activity in most animals.

In both housed and **free-range animals**, the available space is one of the main determinants of locomotory activity. In the **grazing** animal an increase in area and reduction in food availability forces it to walk further to search for food. For example, cattle on rangeland conditions have been recorded as walking three times as far as cattle on intensively managed pasture. Increases in locomotion can be caused by external **parasites** such as flies, which interrupt grazing activity. This encourages grazing animals to move about and search for open, windy areas where they can lose the attention of the parasites, which normally attack when they are stationary, especially lying down. This may lead some animals to move to altitude – or even out to sea – to avoid being bitten by flies.

An important consequence of excessive treading activity by hoofed animals on poorly drained and clay soils is the loss of soil stability and eventual poaching damage. With housed cattle and horses the provision of adequate space for locomotion is a complex issue. Although a considerable proportion of locomotion is still associated with feeding, a significant amount is associated with social and other activities. This is because animals eating conserved food spend less time eating than grazing animals, they are stationary while they eat and the greater stocking density in a building than at pasture encourages more social interaction, especially grooming.

Space for locomotion is often restricted in animal **housing**. Caged animals most obviously suffer restriction of locomotion, but domestic animals in loose housing may also suffer restriction of movement. For example, in a cubicle house for dairy cows the space for walking is provided in the passageways between cubicles, in the feeding passage and occasionally in a separate exercise area. If the amount of space provided for locomotion is decreased below 4–5 m^2 per cow, it is likely that locomotory activity will initially increase for a few days because of **aggression** caused by competition for space. After this, however, locomotion decreases due to the restriction in space availability.

The floor (**see: Flooring**) is the most important part of the house in its influence on locomotion. Floor properties that influence locomotion include: (i) friction, which determines the slipperiness of the floor; (ii) hardness, which determines the loading on the limb; and (iii) the surface profile, which determines the stress loading on the foot, as well as interacting with frictional properties of the floor. Friction can be measured as the force required to start the motion of an animal's foot across the floor (static friction) or that required to maintain this motion (sliding friction). There are both horizontal and vertical components to this force, and the ratio is known as the coefficient of friction. Although it has been found with cattle that the number of slip movements by the leg increases rapidly when the coefficient of static friction decreases below 0.40, in practice cattle probably show a consistent increase in floor preference as the coefficient of friction increases. If animals also lie on the floor, as in stalls, they may show more injuries to the legs as hairless patches, swellings and wounds, and fewer slipping injuries as the coefficient of friction increases.

For loose-housing systems where minimizing slip is the main concern, the minimum coefficient of friction depends on the surface profile. For solid floors it should be at least 0.40, for slatted floors 0.35 and for perforated floors 0.25. The frictional properties of a solid floor may be improved by cutting grooves in a squared pattern; for example, with cattle it seems the optimal size should be 4 cm sides to the squares. This disrupts the floor profile and gives the animal's foot a greater vertical area against which it can apply pressure during the pushing phase of each stride. The repeated passage of animals over a solid floor wears away the deformations in the floor profile created by initial tamping or grooving. In cattle, slips in excess of 500 mm are common on slurry-covered smooth concrete, especially during mounting behaviour. During the slip the hooves, especially in the front feet, turn outwards from the direction of travel.

A comparison of different floor surfaces shows that soil is the most slip resistant, that a covering of slurry increases slip by 56% and that tamping increases slip resistance more than grooving. On slippery floors animals modify their behaviour to reduce the incidence of slipping. They reduce their speed of walking and reduce the angles of extension of the limbs to the floor. Slip is most likely when the foot first touches or leaves the floor and the vertical loading is least. The further the foot is placed in front of or behind the centre of the stride, the less the vertical force and the greater the chance of slip. Hence stride length is reduced as the animal attempts to compensate for the floor conditions.

Slatted floors present an obstacle to normal locomotion in domestic farm animals, in particular cattle, but also in pigs and sheep. They cannot avoid the slots, but their claws often slip into them, causing contusion of the sole and exungulation. To avoid excessive pressure on the sole, slot width should be controlled and, to prevent faeces building up on the slats, slat width should be controlled. Slats cause animals to slip more often and they alter their locomotion posture, orienting their heads more towards the floor.

Younger animals, particularly those at **play**, are more likely to engage in the faster forms of locomotion, like gambolling and galloping, than mature animals, who spend less time in play and for whom it may be dangerous. The oestrous state also has a major influence on locomotion, often increasing it by a factor of 3–4 on the day of **oestrus**. Increased locomotion is also observed just before **parturition**, whereas it will often decrease after both oestrus and parturition.

Lameness is the clinical exhibition of an abnormality of the musculoskeletal system in one or more limbs. This abnormality can arise from many different things: congenital defect, infectious or metabolic diseases or trauma induced by environmental factors. During lameness, the supporting- or swinging-limb periods, or both, are shortened, so that the stride length is reduced. Reduction in the supporting limb period is most common with acutely painful lesions such as a sole abscess or chip fracture, as the pain is accentuated by pressure on the limb. Relief is also gained by relieving pressure on the diseased foot while standing. When the swinging-limb period is contracted, this is often associated with increased abduction (bringing the legs closer together) or adduction (the opposite movement), as the animals may attempt to decrease load bearing by a particular part of the foot. Some unevenness of gait is evident before and after each lameness incident. Systems for scoring animals for the severity of lameness based on the associated behavioural changes have been devised, but their reliability may be variable.

In conclusion, locomotion represents a major need in most animals, but particularly for domesticated animals kept in restrictive environments. Keeping animals on unsuitable floors or with inadequate space will lead to behaviour problems and limb disorders, in particular lameness. This represents a common cause of welfare problems in domesticated animals. (CJCP)

Logic

Logic is a key element in all scientific reasoning. By combining observations and assumptions it is possible, by means of it, to derive new insights. For example, some people have argued that dogs do not have **self-consciousness** in the following way: if dogs are self-conscious, they will be able to recognize themselves in a mirror, but experiments show that dogs are not able to recognize themselves in a mirror, therefore dogs are not self-conscious. Such derived insights serve as hypotheses to be tested through further scientific enquiry.

In the philosophy of science it has been claimed that science proceeds by combining observations, previous hypotheses and background theories, and then pressing logic into service to derive new hypotheses that, in turn, are challenged by observations (from which, once again by means of logic, new hypotheses are derived). This is called the hypothetico-deductive method of science. The successful attempt to challenge a hypothesis by means of observations is called falsification.

Logic allows one to safely derive new insights from old ideas precisely because, in a way, the derived insights are *not* new, although they may be new in the sense that one has not thought of them before. However, they are old in the sense that they are already contained in the body of **knowledge** from which they are derived.

Take the example of the dogs mentioned above; this is a logically valid argument, which means that it is impossible, without contradicting oneself, to accept the premises and reject the conclusion. This is so because of the logical structure of the argument:

1. If p, then q.
2. Not q.
3. Therefore not p.

Whatever sentences one substitutes for 'p' and 'q', it remains certain that you cannot consistently accept **1** and **2** and reject **3**. This is so in virtue of the way the concepts 'If ... then ...', 'not' and 'therefore' are defined. Notice that the fact that the argument is valid does not mean that the conclusion is true. One may, for example, doubt the conclusion of the argument about dogs because one doubts that the ability to recognize oneself in a mirror really is a criterion of self-consciousness.

The study of the various forms of valid logical inference is called formal logic. (PS)

Longevity

Longevity refers to the length of the lifespan of an animal. Environmental, genetic and production factors play important roles in determining the longevity of an individual. Longevity between animals raised for meat and those raised for reproductive purposes have different factors that determine lifespan. Animals raised for meat have, on average, a significantly shorter lifespan than reproductive animals. Most meat animals are slaughtered at either the tail end of adolescence or the beginning of maturity, when meat quality is still desirable. This short lifespan decreases **culling** of these animals due to reasons such as **disease**, **lameness**, inefficient production and age. Short longevity is desirable for meat-rearing animal producers because this gives producers maximal profit for minimal input (i.e. time and resources). To achieve this goal of a short life for meat animals, selection for both high **growth rates** of muscle deposition and larger body structure has been successful. Management practices play an important role in controlling environmental factors to encourage maximal growth rate and decrease disease.

Longevity goals for **breeding** animals are the opposite of those for meat-reared animals. Increased longevity will increase the number of productive years and most likely increase the number of offspring produced, therefore increasing profit. Many factors play a role in determining the longevity of these animals. Environmental factors such as **housing** type and management practices can cause either acute or chronic **stress** that have an impact on longevity. In cases of acute stress, the animal may succumb to sudden death, usually from localized regions within the body – e.g. cardiovascular failure, respiratory disease, abscess rupture, GI tract torsion, etc.

In cases of chronic stress, the animal as a whole can suffer detrimental effects (decreased immune function, reproductive failure and decreased growth rate) that can give reason for the animal to be culled. Culling a reproductive animal depends on factors such as stage of **gestation** or **lactation** and severity of problem. Longevity and production rate of the parental lines play a role in determining the potential for their offspring. High offspring-yielding parents have a greater potential to have high offspring-yielding progeny, thus increasing the potential longevity of their offspring. If an animal becomes unable to reproduce or has low reproductive rates, it is typically culled from the herd, flock or group. Reproductive failure is the primary factor in determining breeding animal longevity. Reproductive failure can be due to genetic, environmental or management issues. Other reasons for decreased longevity or reasons for culling are age, lameness and disease. (ELS)

Lordosis

Lordosis is the characteristic position that a female animal in **oestrus** adopts to signify a willingness to mate with a male counterpart. The female may lift her tail sideways and arch or stiffen her back, thereby exposing the perineum and allowing her to adopt a secure position before being mounted by the male.

The sex hormone **oestrogen** is required for lordosis to occur, and is released in greater levels during the periovulatory period. (LW)

See also: Sexual behaviour

Lordosis quotient

The lordosis quotient is a measure of female sexual receptivity/motivation and is the proportion of mounts by a male in which the female shows a clear **lordosis** response. It can form part of a standard test in a laboratory setting; for example, a female rat may be presented with a male on ten occasions in order to calculate her lordosis quotient. (DSM)

See also: Sexual behaviour

Further reading

Kuehn, R.E. and Beech, F.A. (1963) Quantitative measure of receptivity in female rats. *Behaviour* 21, 282–299.

Lorenz, Konrad

Konrad Lorenz (1903–1989) was born on 7 November 1903 in Vienna, Austria, graduated in 1928 as a doctor of medicine from the University of Vienna and was awarded his second doctorate (PhD) in 1933. He was appointed to the Chair of Psychology at the University of Konigsberg in 1939, but was drafted into the army medical service in 1941 and taken as a prisoner of war in 1942 by the Russian army. Following his release in 1948 he worked at the University of Munster before moving to the Max Planck Institute for Behavioural Physiology, where he stayed until his retirement in 1973. A key feature of his work was his attempt to understand behaviour from both functional and causal (mechanistic) perspectives. Lorenz's best known popular works include *King Solomon's Ring* and *On Aggression*, which continue to inspire many into a career in biology. In 1973 Lorenz shared the Nobel Prize for physiology or medicine with **Niko Tinbergen** and **Karl von Frisch** for 'discoveries in individual and social behaviour patterns'. (PE)

Further reading

Konrad Lorenz Autobiography, available at: http://nobelprize.org/nobel_prizes/medicine/laureates/1973/lorenz-autobio.html (accessed 2 November 2009).

Luteinizing hormone (LH)

Luteinizing hormone (also called lutropin) is a **gonadotropin** produced by the anterior **pituitary gland** in both males and females that stimulates the **gonad**s of each sex. Its name is derived from the early identification of its role in the final maturation of the ovarian follicle, resulting in ovulation and the formation of the corpus luteum as a result. In the male it stimulates the Leydig cells to produce the androgen **testosterone**, which results in spermatogenesis. LH is therefore also known as interstitial cell-stimulating hormone (ICSH). (DSM)

Lying

Lying allows animals to rest, hide and **sleep**, which are important for the recovery of body systems and **health** and for safety. It can also be used for maintaining associations with **conspecifics**. It is an important behaviour for which animals demonstrate a strong **motivation**. Indeed, some scientists have argued that maintaining adequate lying time is at least as important as maintaining adequate feeding time, because feeding can be accelerated by eating faster, whereas lying cannot (**see: Needs**). Animals that may be deprived of lying include those that are transported long distances over several days in crowded conditions (**see: Transport**) and those that are kept in conditions with unsuitable **floors** to lie on, such as concrete floors without any bedding. Some animals may physically have sufficient space to lie down but be afraid to if other animals are standing close to them in a moving vehicle. Indeed, if an animal lies down in a moving lorry, it may be unable to get up because other animals close over around it and occupy its space.

Lying deprivation can have negative effects on animal **welfare**, as demonstrated by behavioural and physiological measures. For example, when deprived of lying, cattle stamp their legs, regularly reposition themselves, shift their weight from side to side, swing their heads and occasionally butt each other, all of which appear to demonstrate **frustration**. They interact more with their **housing**, possibly redirecting their attention to alleviate the stress of deprivation. There is also physiological evidence of increased **stress** – increased **ACTH (adrenocorticotropic hormone)** concentrations and a greater cortisol response to an ACTH challenge. **Growth hormone (GH)** concentrations may be reduced and the **immune system** adversely affected. Physically, animals may be subjected to foot disorders, such as sole haemorrhage in cattle (**see: Lameness**).

Both physical and physiological factors can compromise the amount of time available for lying. It is suspected that some high-producing animals – for example dairy cows – need extra rest, but also that they often do not have time for it because of the longer periods required for **feeding**, **rumination**, etc. Hence there is a negative relationship between lying time and milk yield in **grazing** or indoor dairy cows. There has therefore been extensive research into lying behaviour in this species. Cows in early **lactation** spend less time lying than cows in late lactation and dry cows, which is probably due to an increased milk yield at this time. The length of time spent lying varies with age and sex: for example, it averages 13 h/day for calves, 12 h/day for bulls and 7–10 h/day for lactating dairy cows (in approximately five periods of 1.5 h each).

The housing conditions in which intensively managed farm animals, such as dairy cows and pigs, are kept can also affect lying behaviour. Lying time of cows in cubicles declines in proportion to the degree of overcrowding, and cows in straw yards lie for longer than those in cubicles. Also, routine farm procedures, such as **artificial insemination**, can increase standing time. In New Zealand, farmers remove cows from pasture during periods of heavy rainfall to avoid damage to the swards, and confine them in concrete yards, where they are deterred from lying on the hard surface and spend most of the time standing.

In recent years, there has been economic pressure for livestock farmers to reduce the amount of space available for their animals (**see: Intensification of animal production**). For example, cubicle (free-stall) systems have been developed for cattle and require each animal to align closely with its neighbours rather than lying in a more random pattern, which occurs in straw yards or at pasture. A cubicle contains a solid bed raised off the floor with partitions to create individual lying areas for the cows. If the cubicle bed is too small for the size of cow, which is increasingly common with the increased size of modern dairy cows, then she will have difficulty lying down or getting up. This leads to leg abrasions and swellings, and cows often resort to lying either half in the cubicle or completely in the passageway. Some improvements in the use of cubicles can be achieved by training calves and heifers to accept a smaller version of the cubicle. When a cow lies down in a cubicle she normally kneels down with one foreleg, then both, and then tucks one hind limb under her abdomen as she lowers her rear end. Thus when lying down the cow eases her centre of gravity forward along a longitudinal axis to minimize the stress on the limbs. Adequate space for forward motion is essential for cow comfort, both when lying down and when rising. An estimate of cow comfort while lying can be made by seeing how much intervention is needed to get a cow up when she is lying. When rising, the cow first raises her forequarters slightly and then her hindquarters, using the outer hind leg primarily for vertical propulsion, then the inner hind leg that was tucked under, the fore leg on the same side and finally the outer front leg.

The style of lying may be affected by a range of circumstances. For example, pregnant cows often prefer to lie on their left side, because the **fetus** dwells in the right side of the abdominal cavity, at least in the latter stages of **pregnancy**. On average, about one-third of cows show a clear preference for right- or left-side laterality. As the cows get older they show less preference for left-side laterality, probably because there is more room in the abdominal cavity and less chance of damage to the fetus by lying on the right side. If there is a slope on the cubicle the cows prefer to lie with their dorsal side uphill. If a cubicle to one side is occupied, a cow prefers to lie with her back next to the cow's back in the adjoining cubicle. If cubicles are positioned end to end with open fronts, as is sometimes advocated to allow adequate lunging space, the cows are reluctant to lie facing each other. In fact they seem to prefer a cubicle with a solid front, which limits lunging space but perhaps provides psychological benefits.

Animals lie either in the sternally recumbent position, that is, on the sternum or breast bone, when they retain some awareness of their surroundings, or in the laterally recumbent position, with limbs and head outstretched. Some animals, such as ruminants, cannot spend long periods in lateral recumbancy because of their need to eructate gases from the rumen at regular intervals. Gravity plays an important part in reticulo-ruminal function and in releasing the gas bubble (normally present in the dorsal sac of the rumen) to the atmosphere via the oesophagus (**see: Bloat**). Calves without a fully developed rumen often lie for some time in the laterally recumbent position. Beef cattle that are very fat occasionally have difficulty in maintaining a sternally recumbent position, and may roll over on to their side. Sheep may roll on to their backs and be unable to get up if their wool and conformation are such that they have relatively flat backs – a feature favoured by many breeders.

When animals lie down, they should be prevented from lying in their own excreta, otherwise they will suffer from problems associated with the acidity and constant wetting of the foot, such as hock burns in poultry and 'slurry heel' in cattle, respectively. One way of removing them from their excreta for at least part of the day is to build raised beds for the animals to lie on. The development of cubicles or free stalls has had a major impact on reducing **mastitis** in dairy cows, caused by contact between the cows' teats and faeces. There are many different types of cubicle divisions available, all attempting to effectively separate the cows while minimizing the hindrance to their movements caused by the hardware.

One modern design that allows some space sharing by virtue of the absence of a bottom rail at the back of the cubicle is the Dutch Comfort design. If there is a bottom rail it should be about 50 cm from the cubicle base to avoid cows becoming trapped underneath. Front rails, which prevent the cow moving too far into the cubicle and becoming trapped, should be either near the floor (<25 cm from the floor) or on the top rail; otherwise the rail will impede a cow's rising. In Dutch Comfort cubicles the front hoop should not be wider than 35 cm, or small cows can get their shoulders stuck. Some cubicles should always be able to accommodate the largest cows in the herd.

A variety of bedding materials can be used to increase animal comfort on raised beds, which should absorb the force of the animals lying down and also absorb any urine deposited in the bedded area. Dairy cow cubicles may have a neck rail positioned perpendicular to the divisions, attached to the top rail 45 cm from the wall. This forces the cow to reverse out of the cubicle as she stands up and prevents the bedding being soiled. If cows are not able to stand fully in the cubicle, they will often stand half in and half out, which may increase stress to the hind feet. This is common in subordinate cows, which use the cubicle as a place of escape.

Bedding material should also have insulatory properties for animals, particularly young, non-ruminant animals kept in low temperatures. For example, the lower **critical temperature** can increase from 8°C for a calf on dry straw to 17°C for the same calf on dry concrete. When the temperature falls outdoors, cattle lie down for longer in order to reduce heat loss from the underside of the body.

Tethering animals often restricts their lying, at least for the first few weeks, when they display more intention of lying but fewer actual bouts of lying, and when stress levels are increased. Lame animals spend more time lying (to release the pressure from standing) and less time feeding, possibly to avoid competitive interactions with others.

Although species like the horse have anatomical adaptations that allow them to conserve energy and doze while standing up, they must lie down in order to enter REM (rapid eye movement) sleep. Lying behaviour in this species has been shown to be affected by factors such as the bedding substrate and the novelty of the environment. (CJCP)

M

Macaque

Macaques are what are typically known as 'Old World' **monkeys**. There are 15 species including the Rhesus (*Macaca mulatta*), crab-eating (*Macaca fascicularis*), stump-tailed (*Macaca arctoides*), Japanese (*Macaca fuscata*), Bonnet (*Macaca radiata*) and Barbary (*Macaca sylvanus*). They are naturally found almost exclusively in Asia. Primates belonging to the family Cercopithecinae (which includes macaques and vervet monkeys) are gregarious, inquisitive, noisy and often live alongside humans, scavenging on their crops and waste.

Macaques are commonly used in research to study normal complex neurophysiological processes, e.g. **vision** and **hearing**, and **disease**. Macaques can be trained to carry out various tasks, and this has led to them being trained to cooperate in experimental work and provide valuable insights into nerve function. Training can also reduce the effect of experimental **stressors** on them, such as the taking of a blood or urine sample. In **captivity**, cramped conditions predisposed macaques to fight and so single housing them in barren cages became commonplace. However, particularly since the late 1990s, efforts have been made to house them in harmonious groups in enriched surroundings, and fighting has been markedly reduced, though not always eliminated, as with any grouped social animal. (DBM)

Magnetoception (magnetoreception)

Magnetoception is the ability to detect magnetic fields and often involves cellular collections of magnetite, possibly in association with other specialized structures. The capacity for magnetoception has been identified in a wide range of species, ranging from invertebrates such as honeybees, fruit flies and lobsters, to some species of cartilaginous and bony fish (e.g. sharks and salmon), **reptile** (turtles and crocodiles) and bird (pigeons), but is less well documented in mammals, although it is suggested that **cattle** tend to align themselves towards the magnetic poles as a result of magnetoception. Magnetoception appears to be associated with the navigation required for migratory or **foraging behaviour** over relatively large distances for the species concerned, although it is often only one of a combination of cues used to aid orientation. (DSM)

Further reading

Rozhok, A. (2008) *Orientation and Navigation in Vertebrates*. Springer-Verlag, Berlin.

Maladaptive

The term 'maladaptive behaviour' is often used in relation to **problem behaviour**, but the two should not be considered synonymous, and a further useful distinction can be made between them.

Maladaptive behaviour is behaviour that is not well adapted (biologically appropriate) to the circumstances, and within the context of the captive animal may arise as a result of attempts to behave in an adaptive way in an environment to which complete adaptation may not be possible. For example, an animal may continue to perform **appetitive behaviour** for prolonged periods if its consummation is frustrated and there is insufficient **motivation** to undertake an alternative activity. The resulting behavioural display may seem inappropriate or '**abnormal**', but it cannot be assumed in these instances that the underlying mechanism is afflicted by a pathological process and therefore not functioning as it should. The animal may be operating according to its **phylogeny**, using evolutionary 'rules of thumb' that have resulted in the selection of behaviour strategies that are appropriate given the normal environment for the species. However, when the animal is taken out of that environment, these processes may no longer be 'fit for purpose' or result in appropriate adaptation.

By contrast, the term 'malfunctional behaviour' suggests that the underlying processes are not working as they should. By way of analogy, when a computer malfunctions we tend to think of something going wrong inside it, however a maladaptive computer may be malfunctional or it may alternatively be doing what it is supposed to do (but not necessarily what we hoped it would do when we bought it!). In this case the term 'malfunctional behaviour' should be limited to a description of those actions that are expressions of direct disruption of the nervous system or its normal regulation; i.e. the neural basis to the behaviour has no functional value in any context, e.g. seizure activity or age-related decline in function (**see: Cognitive dysfunction**). Such behaviours are the result of a genuine pathology.

With these definitions it follows that malfunctional behaviour is also, by definition, maladaptive. Most problem behaviours appear to be adaptive responses or maladaptive behaviours that are not malfunctional, but are considered inconvenient to the carer or other interested party. (DSM)

Further reading

Mills, D.S. (2003) Medical paradigms for the study of problem behaviour: a critical review. *Applied Animal Behaviour Science* 81, 265–277.

Malnutrition

At its most chronic, malnutrition is commonly due to an inadequate feed supply leading to the classic symptoms of **starvation** in humans, leading in turn to **kwashiorkor** and **marasmus**, or deficiency of one or more essential nutrients. Examples include: (i) lack of vitamin C, leading to scurvy in guinea pigs; (ii) lack of lysine and methionine, leading to reduced growth in pigs; and (iii) phosphorus deficiency in livestock, leading to chewing of wood, bones, soil and other items, a phenomenon known as **pica**. In addition to these chronic effects in farm and **laboratory animals**, other examples of malnutrition include acutely inappropriate feed, leading to illness or death, and short-term feed deprivation resulting in hunger. There are also diet-related risks in high-producing animals: for example, **culling** due to **lameness** is higher in more productive sows, associated with their increased susceptibility to conditions such as osteomalacia (related to calcium deficiency) arising from the physical limits on the amount of feed a sow can eat to replenish minerals lost during **lactation**.

Lack of feed

Natural disasters such as fire, flood and drought are an integral part of pasture-based livestock production around the world, and their impact on animal **welfare** can be catastrophic. For example, in a drought in the early 1900s in Australia, cattle numbers fell from 4.0 to 2.5 million and sheep numbers fell from 10 to 7 million and, while some animals would have been deliberately de-stocked, many would have starved to death. The main welfare issue in assessment of drought-affected animals is the subjectivity involved in deciding how much weight an animal should be allowed to lose and whether to provide supplementary feed or sell or agist stock. At what point, and at what condition score, does a hungry animal, losing weight, and maintained on a sub-maintenance ration, become an animal whose welfare is at serious risk? The interactions between economic viability, stocking rate, land degradation and animal welfare are complex and conflicting. Providing conserved fodder or supplements rather than reducing stocking rates or selling animals has the potential to have the greatest adverse impact on welfare if a drought persists.

Flood and fire can both directly kill animals and have major adverse indirect effects. Crops can be washed away or burned, remaining pastures can become contaminated or inaccessible and conserved fodder can be destroyed. The most serious welfare issues are likely to be feed shortages and an increased incidence of **disease**.

While lack of feed is commonly associated with natural disasters, it can also take place in conventional farm production. For example, chronic underfeeding occurs in flocks of both broilers, due to competition for limited feed, and laying hens, due to feed intake being restricted by **aggression**, and results in stunted birds with long-lasting effects on growth and **egg production**.

Inappropriate feed: toxic plants

Many plants and fungi biosynthesize chemicals that are toxic to animals. In ruminants, fungal (myco)toxins in the soil or pasture are suspected of causing seasonal 'ill-thrift' and reduced weight gain, despite an abundance of feed. An example is perennial ryegrass (*Lolium perenne*) staggers, which occurs in sheep, cattle, horses, alpaca and deer due to alkaloids produced by endophytes, including the neurotoxin lolitrem-B and the vasoconstrictor ergovaline. Stock **grazing** affected pastures develop a fine head tremor that progresses to nodding, trembling, incoordination and jerky, exaggerated limb movements with a stiff-legged gait. Affected stock can go down on one side with their legs extended. Left alone they can recover, regain their coordination and get up and resume grazing. When **death** occurs, it is usually by either misadventure (drowning in troughs, dams and gullies or attacked by predators) or wasting due to recumbency, starvation and/or dehydration, with many downed stock requiring **euthanasia**. Some sheep may be found dead overnight after grazing toxic pastures for prolonged periods. Drowning can occur due to the effects of ergovaline on increasing body temperature; this encourages affected animals to seek water. When removed from the toxic pasture animals return to normal movement in 1 to 3 weeks.

Feed restriction

The major effects of short-term feed deprivation, as can occur during **transport**, are dehydration and tissue catabolism. Sheep and cattle appear to tolerate such feed and water deprivation extremely well. Typical live weight losses in sheep range from 6 to 10% of body weight during road transport up to 24 h, and from about 3 to 11% in fasted cattle. Deprivation of food and water for 48 h over a wide temperature range did not induce cortisol or **prolactin** release in sheep, and fasting of cattle for 72 h did not affect plasma cortisol, **catecholamines** or haematocrit, suggesting a lack of **stress** effects. However, pigs that have been without feed for 3 h appear to suffer from travel sickness, as indicated by retching and vomiting. In contrast to this often single or occasional period(s) of short-term feed deprivation, for some farm animals – notably breeding sows and breeding chicken meat birds – there is an ongoing and deliberate feed restriction.

It is generally agreed that pigs should be restrictively fed during **gestation** at a level of about 1.5 × maintenance, which is about 0.6 × *ad libitum* intake. One reason for this restriction is that high levels of feeding during pregnancy can compromise piglet growth during lactation. While the level of feeding is sufficient for maintenance, some growth and fetal development, it also results in pigs being highly motivated to feed. **Operant conditioning** studies in which sows work for a feed reward suggest that they are hungry for a considerable period of the day. Stereotypic behaviours that can occur in these restrictively fed pigs may be due to a frustration of feeding **motivation**, as they are reduced by additional feed.

Similarly, restricted feeding (about 0.5 × *ad libitum* intake) is the usual practice in broiler (meat chicken) breeder management to maximize reproductive performance. The breeding bird is a similar weight at about 18 weeks of age to a meat bird at slaughter (less than 7 weeks of age). Welfare concerns arise from changes in bird behaviour, including abnormal oral behaviours such as pecking at non-food objects and increased drinking activity. Nevertheless, slower growth has welfare benefits such as improved **immune system** responsiveness and disease resistance, fewer skeletal problems and considerably lower **mortality**. While it is generally agreed that *ad libitum* feeding adversely affects the welfare of both pigs and broilers, there is no agreement on what level of feed restriction is unacceptable. (JB)

See also: Abnormal/abnormality; Body condition score; Feed restriction; Flocking behaviour; Hunger; Operant behaviour; Stereotypies

Further reading

Alexander, G.I. (1985) Animal welfare and natural disasters in the grazing industry. In: Moore, B.L. and Chenoweth, P.J. (eds) *Grazing Animal Welfare*. Australian Veterinary Association, Queensland Branch, Indooroopilly, Queensland, Australia, pp. 57–65.

Broom, D.M., Goode, J.A., Hall, S.J.G., Lloyd, D.M. and Parrott, R.F. (1996) Hormonal and physiological effects of a 15 hour road journey in sheep: comparison with the responses to loading, handling and penning in the absence of transport. *British Veterinary Journal* 152, 593–604.

Bruneton, J.J. (1999) *Toxic Plants: Dangerous to Humans and Animals*. Lavoisier Publishing Inc., France.

Farm Animal Welfare Council (1994) *Report on the Welfare of Sheep*. FAWC, London.

Knowles, T.G., Brown, S.N., Warriss, P.D., Phillips, A.J., Dolan, S.K., Hunt, P., Ford, J.E., Edwards, J.E. and Watkins, P.E. (1995) Effects on sheep of transport by road for up to 24 hours. *Veterinary Record* 136, 431–438.

Lawrence, A.B. and Terlouw, E.M.C. (1993) A review of behaviour factors involved in the development and continued performance of sterotypic behaviours in pigs. *Journal of Animal Science* 71, 2815–2825.

Marasmus

Marasmus is a form of combined protein and energy **malnutrition** resulting in emaciation and less extensive oedema, which is more typical of the more life-threatening **kwashiorkor** (arising as a result of just a protein deficiency). Marasmus is more typical of the unweaned animal, arising as a result of inadequate milk production from the mother due to her own malnutrition. (DSM)

Marine mammal

The history of the relationship between marine mammals and humans goes back as long as people have lived near water. Marine mammals exist in a web of cultural, religious, legal, recreational and economic relationships with humans. The varying nature of human impacts on and interactions with marine mammals, from **hunting** to rehabilitation, requires multifaceted approaches for effective management and coexistence.

Three orders represent the 119 taxa of present day marine mammals. Pinnipeds, meaning 'fin-footed', are aquatic carnivores including seals, sea lions and walruses. Taxonomic classification is further broken down into Phocidae or 'true seals' (the majority of seal species), characterized by the lack of external ears, Otariidae or 'eared seals' (sea lions and fur seals) and Odobenidae (walruses). Though pinnipeds are four-limbed, they rely mostly on their front limbs for steering while in the water. Although all species of pinnipeds rely on dry land to some degree, such as for **breeding** and bearing of young, their fat-insulated and streamlined bodies are largely adapted to life in the water.

Cetaceans, including the largest mammal on the planet, the blue whale, consist of whales, dolphins and porpoises. All cetaceans are fully aquatic and have special adaptations for their marine life history, such as blowholes to assist in fast gas exchange, vestigial back limbs and internal sex organs creating a fusiform body shape and, for many species, **communication** systems such as **echolocation** that take advantage of their fluid environment. Cetaceans are classified depending upon their method of feeding. Species possessing teeth, including toothed whales (sperm whales), dolphins and porpoises, are classified as Odontoceti, while species that rely on keratin plates of baleen (such as humpback whales) are classified as Mysticeti.

Also fully aquatic are the Sirenians, consisting of all manatee and dugong species. Slow moving, and relying on dense layers of fat insulation, these herbivorous species are confined mostly to coastal marine areas and inland river and estuary systems, and are not able to survive in cold water temperatures. Two species from the order Carnivora are considered marine mammals because they have crucial aspects of their life histories that rely on an aquatic environment, such as for **feeding** or **locomotion**: the polar bear (*Ursus maritimus*) and the sea otter (*Enhydra lutris*). These two species are the most terrestrial of all of the marine mammals. The common thread among all marine mammals is that they have evolved both anatomically and behaviourally to the constraints of an aquatic life. Dependence on water varies, with pinnipeds having both a terrestrial and marine life history strategy, while cetaceans and sirenians are constrained to a solely aquatic ecosystem. Many species, particularly cetaceans, are characterized by long life histories and slow reproductive rates.

Marine mammals have been kept in captive facilities in North America and Europe since the 1860s. Their charismatic image, coupled with their ability to be easily trained for a multitude of behaviours, made them a hugely popular attraction at recreational marine animal parks. Modern behaviour modification techniques became standard within training regimes by the 1970s, influenced largely by the positive reinforcement techniques popularized by Karen Pryor's work with captive bottlenose dolphins. While there are a few species of predominantly solitary marine mammals commonly maintained in **captivity**, such as polar bears, the majority of captive species are characterized by complex social structures.

The dynamic interactions between varying age groups, sexes and related individuals could be considered one of the primary concerns in the maintenance of suitable captive environments. This concern is a growing topic of debate in discussion of **welfare** issues surrounding captive marine mammals, especially as more information is obtained from field studies on the **ecology** and behaviour of wild populations. On the most extreme side of the debate over welfare issues is the argument that the social structure, environmental needs and behavioural repertoire of most species make them unsuitable for a captive environment. While particular myths of harmful effects of captivity have been debunked, such as the myth that echolocation in a tank is potentially harmful or a droopy dorsal fin signals depression, there are still valid questions to be raised when designing and maintaining an environment for species that often posses dynamic and complex social needs. Perhaps more so than for most other vertebrates in captivity, technical design of habitats is a crucial issue to be considered in captive welfare, given the risk of various forms of infection and **disease** stemming from water filtration systems.

Along with increasing awareness of social requirements of marine mammals came a greater understanding of the psychological needs of these species. A number of scientific studies in captive animals have shown that many cetaceans have cognitive abilities comparable to those thought to be unique to primates. Pinnipeds, sirenians and cetaceans have 'advanced' cognitive abilities, suggesting the importance of a stimulating captive environment.

Among concerns in all captive species, **stereotypies** and other behaviours associated with negative consequences of captivity are observed in marine mammals. However, positive **reinforcement** techniques – which were developed largely with marine mammals – are a source of **enrichment** and stimulation that meets many of the **behavioural needs** of captive species. These applied behaviour techniques have been used effectively to reduce stereotypic behaviour and improve health monitoring and reproductive function. Environmental enrichment programmes are widely used to enhance the well-being of captive marine mammals. Many marine mammals held in captivity, such as dolphins, spend much of their time in the wild searching for food and socializing with **conspecifics**, and are therefore exposed to a variety of constantly changing stimuli. In captivity, needs for stimulation can be addressed in part through social housing and the application of behaviour modification techniques, as well as the creative use of environmental enrichment.

The 1972 Marine Mammal Protection Act in the USA was among the most notable advancements for marine mammal welfare. Before this piece of legislation, there was little monitoring of human impact on wild populations. In addition to regulating hunting and the capture and transport of wild species, the Marine Mammal Protection Act also identified the need for scientific research on marine mammal populations. Increasing scientific knowledge on the ecology, behaviour, physiology and health of marine mammals has applications for increasing standards of care for captive individuals, as well as for the development of effective management guidelines for the **conservation** of wild marine ecosystems.

Escalating human impacts on coastal ecosystems stemming from pollution and recreation intensify the need for judicious management plans for wild marine mammal populations. Behavioural research on wild and captive animals is necessary to increase public awareness, develop wild and captive management strategies and provide guidance to regulatory agencies responsible for developing and enforcing legal aspects of human interactions with marine mammals. (CPL)

See also: Wildlife management

Further reading

Bean, M.J. and Rowland, M.J. (1997) *The Evolution of National Wildlife Law*, 3rd edn. Praeger Publishers, Westport, Connecticut, Chapter 5.

Defran, R.H. and Pryor, K. (1980) The behaviour and training of cetaceans in captivity. In: Herman, L.M. (ed.) *Cetacean Behaviour: Mechanisms and Functions*. Wiley and Sons, New York, pp. 247–305.

Gales, N., Hindell, M. and Kirkwood, R. (2003) *Marine Mammals: Fisheries, Tourism, and Management Issues*. Collingwood, Victoria, Australia.

Reynolds, J. and Rommel, S.A. (1999) *The Biology of Marine Mammals*. Smithsonian Institution Press, Washington, DC.

Markers, animal

Individual animals may be identified for legal, practical or scientific reasons. Many identification techniques are available, and the appropriate method for each situation will depend on the species and strain, number of animals in each group, reason for identification and the skill of (and resources available to) the person applying the technique. Some methods can cause discomfort, pain or distress, so the least invasive system should always be chosen (see Table M.1). (PH)

See also: Branding

Market

A market is a centralized area where producers (farmers) or dealers bring livestock for sale. These may be young, breeding or milking stock that will be sold to other farmers (sometimes via dealers) or animals that are ready for **slaughter**. Traditionally, and in undeveloped countries, there are many small markets, often simple street markets, that serve a local area so that animals travel for short distances and times. This reduces **transport stress** and the risk of **disease** transfer. Owing to the economies of scale and centralized processing and distribution of meat for supermarkets, the trend in the developed world is for fewer, larger markets. These subject the animals to long transit times. For example, a dealer might pick up a few animals from several farms and deliver them to a market, where they may wait for several hours to be sold and then have to travel to other farms or have a long journey to a centralized abattoir. Thus 1- or 2-day transits are not uncommon, particularly when animals are traded as commodities between countries and across continents.

Annual fairs are a common way of selling many species, including feral ponies, (race) horses, camels and, in the autumn, ruminants raised on marginal land such as hills and moors. These fairs may be relatively informal, with individual negotiations between buyers and sellers, or they may involve auctioneers and make use of more permanent penning and handling facilities. Live poultry tend to be traded in only very small local (street) markets. The majority of poultry nowadays is kept intensively in flocks of thousands and travel direct to slaughter.

Even in informal street markets and annual fairs where animals are not auctioned, there is generally some form of regulation or overseeing of vendors, who usually need to register and/or pay for their pitch. Animals for sale are often individually recorded and given a sale number. In many countries all animals presented need to be fit for sale, and this is overseen by a veterinary surgeon. Further disease control measures may require owners of animals to have a licence to move them both to and from the market or gathering. The UK has introduced passports for ruminants and equines that need to accompany the animals on journeys. All movements and changes of ownership have to be notified to a central agency. This is in a bid to contain diseases like foot-and-mouth and to safeguard the **welfare** of the animals by limiting the number of journeys they can make, particularly within a short period of time.

In a typical permanent auction market site there are defined areas for unloading and loading of the vehicles that transport the animals, with raised platforms or ramps to reduce the slope of the tailgate and enable the animals to transfer easily to and

Table M.1. Examples of common identification techniques in animals.

Technique	Comment(s)	Welfare issue(s)
Non-invasive		
Noting physical differences	Preferable on welfare grounds but may be unrealistic, e.g. if marking is a legal requirement	None, unless handling required
Identifying enclosure, e.g. by cage label	Legally required in some contexts, e.g. experimental animals	May necessitate single housing; undesirable for social species
Minimally invasive (should cause no more than momentary, slight stress or discomfort, if competently performed)		
Staining or dyeing of skin, pelage or plumage; clipping of fur, claws or feathers	Temporary identification only	Stains or dyes must be non-toxic and non-irritant; some colours may influence behaviour (e.g. red can be mistaken for blood)
Microchipping	Widely used in many species, including birds, rodents and larger companion or farmed animals	Inappropriate siting can cause discomfort; transponders can migrate
Leg rings, e.g. birds, rabbits; collars, e.g. carnivores, primates, farmed animals	Wide variety of rings and bands is commercially available	Rings and collars must be of correct size (and frequently checked in growing animals); colours can influence behaviour in some species
Heat-branding of horns or hooves	Semi-permanent; cannot be used until horns fully grown in some species	Can be stressful, but should not hurt if carefully done
Invasive (will cause pain and distress that should be minimized)		
Tagging, e.g. ear, wing	Tagged animals may be identified at a distance	Tags should be as small as possible and sited to avoid catching or tearing
Tattooing	Pigment is deposited in the dermal layer of the skin; marks can fade in some species	Causes pain and soreness
Freeze branding	Destroys melanocytes, producing white fur	Causes acute discomfort
Ear notching/punching	May be combined with biopsy in transgenic animals, instead of tail tipping	Painful, risk of tearing in small animals
Seriously traumatic (will cause severe pain and/or affect normal behaviour or survival)		
Toe clipping	Mainly rodents and amphibians	Very painful, reduces survival in field
Web punching	Aquatic animals	Painful mutilation; risk of infection
Heat-branding of skin	Produces permanent scar tissue	Extremely painful; illegal in many cases, e.g. cattle in the UK

from holding pens. Frequently there are races and restraining devices such as yokes so that animals can safely be identified individually (e.g. to read ear-tag numbers and check whether they correspond with the paperwork). This area requires good lighting, well-designed facilities and skilled **handling** of the animals to ensure good welfare. Equines and ruminants may have an auction (catalogue) identifying number glued to their rumps – for other species this number is more often on the pen to which they are then moved.

At a livestock market, the animals may be sold either via private treaty, which is common in small markets and in the Netherlands, or by auction. The most common practice is to auction groups of animals such as pigs, calves and sheep with potential purchasers moving to the pen containing the animals. Unless they need grading by size, quality or appearance, the animals will move directly to and from this one pen when arriving and departing the market, and this method is comparatively less stressful and also reduces the risk of disease transfer. Larger animals such as finished cattle, dairy cows or even groups of store beef cattle may be moved several times and then driven around a central static auction ring during bidding by potential purchasers.

This method of sale is more convenient for humans and often quicker, but with welfare implications for the animals, which often experience social stress from isolation if handled singly and from mixing with strangers. Additionally, they will experience the novel sights and sounds of the auction ring, which they may find frightening. The number of movements is increased along with the potential for injury either from the hardware (gates, posts, etc.) or from being hit by **goads**. Calves may be penned in groups but moved singly into a mobile auction ring that is moved to be adjacent to their pen. After sale, animals that are sold in groups usually remain with their cohorts when moved to the purchaser's vehicle. Other animals such as finished beef cattle and cull cows are often grouped in pens with strangers before being loaded for transport. Animals for export may need formal examination and certification of **fitness** by a veterinary surgeon. To reduce disease risk, many countries require transport vehicles to have litter material disposed of and to be washed and disinfected on the market premises between each load of animals. An area with suitable facilities is provided for this purpose.

Cattle have been shown to become more tired and thirsty (Jarvis *et al.*, 1996) and to sustain more bruises (e.g. Weeks *et al.*, 2002) when they have been through markets. Meat quality problems are often associated with the welfare issues of stress, **hunger**, thirst and poor handling, and tend to increase with total transit time. Thus passing through a market does not inevitably cause problems but, by adding to transit time and increasing the number of handling events, this usually increases the risk of poor welfare and disease. For this reason, as well as consumer perception of animal welfare, meat retailers and fast food outlets are increasingly wishing to avoid markets in their sourcing of meat (Murray *et al.*, 2000).

Murray *et al.* (2000) point out that a wide range of journey structures and complexity is found with all marketing methods. They also note the potential of electronic auction systems for reducing marketing costs, but that their uptake by industry has been limited and potential welfare benefits little researched. Electronic auction systems enable all potential buyers to simultaneously log on remotely and view images (usually real time) of the animals for sale. They can bid electronically, with an auctioneer controlling the sale. Purchasers still tend to prefer to examine an animal in the flesh in three dimensions and so continue to visit markets or farms even though this is more time consuming.

Most animals find handling stressful, unless, like the dairy cow, they are habituated to daily handling. The number of times an animal is handled and the nature of the handling are both important. To reduce stress:

- Handling should be minimized.
- Facilities should be well-designed (e.g. races should curve and have solid walls to prevent distraction and to encourage animals to move forward easily, and there should be no sharp edges, in order to reduce injuries).
- Human handlers should be sympathetic, experienced and trained to consider animal well-being (Grandin, 2000; Weeks *et al.*, 2002).

If handling is poor, animals may baulk, slip and fall, experience **fear** and stress and be more likely to be goaded and hit, leading to **pain** and injury. The nature of the animal also affects its reaction to handling. Certain breeds of pig such as the Pietran have the so-called halothane gene, which is associated with excitability and a stress response to handling and transport. Slaughter animals from these breeds tend to produce meat with quality problems such as pale, soft exudative flesh (**PSE meat**), and thus the live pigs are not sold via markets. Similarly, bulls are more excitable and stressed by handling and thus are comparatively rarely sold via markets, in part to avoid stressing them and the associated chance of their meat being dark cutting. Several studies have found gender differences in response to marketing (see Weeks *et al.*, 2002).

A further issue concerning animal welfare at markets is that of disease. The stress associated with passing through a market can reduce immune function. Concurrently, animals are exposed to a greater variety and number of infective agents than they would be if transferred directly from one farm to another or to the abattoir. It is well known that calves that have come from a market are less healthy during rearing than those transferred directly between farms. The practice of hawking sheep from market to market all over the country to make incremental profits was a major factor in the rapid spread of foot-and-mouth disease in the UK in 2000. Quite aside from the ethical implication of treating sentient beings like commodities, the practical consequences were to have far-reaching impacts on both human and animal welfare, as thousands of healthy as well as sick animals were slaughtered in a bid to contain the disease. Markets also ceased trading. Following on from this, a standstill period was introduced whereby animals could not be moved again for a specified number of days (initially 20 and now 6).

Worldwide there is considerable variation in the amount of regulation associated with the marketing of animals. Some countries have unrestricted trade, and a few have detailed legislation and codes covering facilities, handling and **health** and welfare issues. Many multiple retailers and restaurant chains have codes that they apply across countries, and these may have an impact on the marketing of animals to some degree. (CW)

See also: Transport

References and further reading

FAWC (1986) *Report on the Welfare of Livestock at Markets.* The Farm Animal Welfare Council, London.

Grandin, T. (ed.) (2000) *Livestock Handling and Transport*, 2nd edn. CAB International, Wallingford, UK.

Jarvis, A.M., Harrington, D.W.J. and Cockram, M.S. (1996) Effect of source and lairage on some behavioural and biochemical measurements of feed restriction and dehydration in cattle at a slaughterhouse. *Applied Animal Behaviour Science* 50, 83–94.

Murray, K.C., Davies, D.H., Cullinane, S.L., Eddison, J.C. and Kirk, J.A. (2000) Taking lambs to the slaughter: marketing channels, journey structures and possible consequences for welfare. *Animal Welfare* 9, 111–122.

Weeks, C.A., McNally, P.W. and Warriss, P.D. (2002) Influence of design and handling at auction markets on carcase bruising of cattle. *Veterinary Record* 150, 743–748.

Marmoset

The New World **monkeys**, marmosets, **tamarins** and Goeldi's monkey belong to the family Callitrichidae, subfamily Callitrichinae. There are 22 species of marmosets in four genera: six species of *Callithrix* from the Atlantic forest and, from Amazonia, 14 species of *Mico*, one species of pygmy marmoset (*Cebuella*) and one species of dwarf marmoset (*Callibella*). *Cebuella* is the smallest higher primate, weighing around 130 g, *Callibella* weighs around 150–185 g, while other marmosets weigh 300–450 g. The common marmoset (*Callithrix jacchus*) is used extensively in biomedical research and is the focus here.

Marmosets have little sexual dimorphism, are diurnal and arboreal and can live to over 15 years. Sexual maturity is reached at 15 months. In the wild, group size is usually around eight to ten. The basic unit is the family, with one **breeding** female (occasionally two). The father and offspring help to rear the young, usually twins. Offspring destined to breed in captivity should remain in their natal groups until they have had experience with at least two sets of offspring, as parental care is learned (**see: Altruism; Parental behaviour**). Marmosets scent-mark territories and are aggressive to neighbouring groups, so physical or close visual contact between captive groups should be avoided. Behaviours such as tail-raised present are displayed when threatened, and a variety of other postures, facial expressions and **vocalizations** can be used to monitor **welfare** (Stevenson and Poole, 1976).

As gummivore–**insectivores**, marmosets have specialized teeth to gnaw trees for induction of exudate production. Given the dental and intestinal adaptation to gum feeding, it is recommended that marmosets are given gum in such a way as to encourage gnawing. The diet should also contain fruit, vegetables, eggs and some animal matter, together with vitamin D_3 if there is no natural light, to maintain healthy animals.

Marmosets are quadrupedal, but are also adapted to vertical clinging and leaping, using their claw-like nails to grip supports. Enclosure furnishings should allow marmosets the opportunity to grip (e.g. rough surfaces) and include a variety of orientations. Platforms or ledges on which the marmosets can sprawl out for **grooming** interactions are important. In the wild, marmosets **sleep** huddled together in tree forks or in dense tangles of vines and leaves, and a secure place for sleeping is important in captivity. (HMB-S)

Reference and further reading

Buchanan-Smith, H.M. (2010) Marmosets and tamarins. In: Hubrecht, R. and Kirkwood, J. (eds) *The Eighth Edition of the UFAW Handbook on the Care and Management of Laboratory Animals and Other Animals Used in Scientific Procedures*. Wiley-Blackwell, Oxford, UK (in press).

Stevenson, M.F. and Poole, T.B. (1976) An ethogram of the common marmoset (*Callithrix jacchus jacchus*): general behavioural repertoire. *Animal Behaviour* 24, 428–451.

Mastitis

Mastitis is an inflammation of the mammary glands, with the most common cause of infection being bacteria that enter the mammary gland through the teat canal. Mastitis is a problem for the dairy industry in particular (**see: Cattle; Lactation; Milking**), and the control of bacteria that cause mastitis is an important component of herd management. The bacterial pathogens that are most frequently involved in causing mastitis can be placed into two categories – contagious pathogens and environmental pathogens.

The major contagious pathogens are *Streptococcus agalactiae* and *Staphylococcus aureus*. These pathogens are generally found on the skin of animals, colonizing the teat and teat canal. Infected mammary glands are the major source of contagious pathogens within a herd and transference is mainly during the milking process. Contagious pathogens tend to cause subclinical mastitis (i.e. no obvious signs of infection), though a high somatic cell count (SCC) in the milk and a decrease in milk yield are indicators of the presence of subclinical infections within a herd.

Environmental pathogens are found within an animal's surroundings and are transmitted through direct contact of the teats with infected areas. Clean and dry beds and passageways remain the most important management factors in controlling levels of environmental mastitis, as dust, water, manure and feedstuffs can all be sources of infection. Coliform bacteria (including *Escherichia coli* and *Klebsiella* spp.) and species of streptococci (including *Streptococcus uberis*) are the primary environmental pathogens. Environmental mastitis generally produces clinical signs such as a hot, swollen udder and clots in the milk, as well as a high SCC.

Cases of clinical mastitis tend to be of shorter duration than subclinical infections, when treated promptly. Unsuccessfully treated cases of clinical mastitis may become chronic, and the infected quarter is then no longer viable. The rate of intramammary infections increases during the dry period, particularly in the days leading up to calving. However, infection does not necessarily spread throughout the whole quarter, and only a few of the milk ducts may be affected. (LW)

Mate choice

Mate choice, in the most general sense, can be defined as any behaviour that favours the success of one set of an individual's potential reproductive partners over another. Mate choice can evolve in a system when the total **fitness** of an individual is limited by access to mates of a certain quality or to resources provided by particular mates.

In most multicellular organisms, there exists an asymmetry in gamete size: females produce relatively few large eggs, and males produce many small sperm. Typically, females invest more resources into each offspring through gamete production, and also often greater parental care. Males, therefore, can potentially fertilize eggs faster than they can be produced. Thus males are limited by access to females, and females are limited by access to **resources**. This asymmetry in gamete investment means that males are under strong selection to be proficient at searching and competing for mates, and that females should exert mate choice (Clutton-Brock and Vincent, 1991). There are, however, some exceptions to this rule. For example, when both male and female invest equally in feeding, guarding or brooding of eggs or young, both sexes may exercise mate choice. Moreover, in species with high male and low female investment, like seahorses, or when the sex ratio favours males, females may compete for mates and males may reject low-quality females. Here we will assume that the female is the choosing sex, and males are the competing sex.

The sensory and behavioural properties that influence the propensity of individuals to mate with a particular phenotype are considered mating preferences. Mating preference consists of two components, preference function and choosiness. The order in which an individual ranks potential males represents the preference function, and the energy that an individual invests in assessing mates represents choosiness. Mating preferences may be based on a variety of factors. Individuals may select mates based on direct and/or genetic benefits they or their offspring receive from mating with a particular partner. Individuals may also choose mates on the basis of latent perceptual biases: many animals may prefer mates that are louder, bigger or brighter simply because they provide more sensory stimulation (**see: Intersexual selection**).

Because the choosing sex is typically the sex that is limited by resources, potential mates may offer direct benefits or resources in exchange for a mating opportunity. In many animal species males defend territories, which can provide females and/or their offspring with protection from predators and competing males, as well as access to food. The female will often assess not only the male, but also his resource holdings, before making a mating decision. Males may also provide nuptial gifts to potential mates during courtship. Such gifts may help to nourish the female or provide her with protection. Furthermore, in species where the male is the primary caregiver, females may choose mates that can provide adequate food and protection for her young.

In addition to direct benefits, some species may choose their mates based on genetic benefits. The good genes theory predicts that individuals will prefer mates with costly and conspicuous traits because they indicate high-quality genes that will be passed on to their offspring. The handicap hypothesis (Zahavi, 1975) predicts that females will prefer males that acquire a trait with a high survivorship cost, since low-quality males cannot support the cost of the handicap and males with the trait will be of high quality. Some traits are thought to be condition-dependent: males vary their expression of a preferred trait to optimize their mating success and survival based on their condition, and thus the expression of the trait is an indicator of male quality. Traits may also be revealing indicators, where all males in the population attempt to express the trait and pay the same cost for the trait, but the condition of the trait is reduced in low-quality males.

In contrast to the good genes theory, the Fisherian runaway selection hypothesis (Fisher, 1930) predicts the evolution of heritable, arbitrary male traits that may provide no information about intrinsic male quality but enable females to produce sons that are in turn selectively chosen by females. Males with the preferred trait obtain more mates, and because their mates tend to be females with a preference for the trait, the genes for the female's preference and the male's trait become linked in their offspring. Females with the preference for the trait benefit because their sons will also possess the trait and will be more successful themselves. This process can be initiated by several mechanisms. For example, females may prefer male traits that are initially favoured by natural selection and become highly exaggerated, losing their natural selection advantage, or traits that mimic stimuli that females are attracted to, thus exploiting pre-existing sensory biases or feature detector for food items.

In addition to selecting a mating partner, in some species females are able to exert mate choice after copulation. By adjusting the size, chemical composition and sex of their eggs after copulation, some females are able to maximize the benefit of mating with a certain male. In species where females mate with multiple males, some can selectively fertilize their eggs with sperm from a particular mate or mates. Even in some seemingly monogamous systems, where females pair with males for one or more **breeding** seasons, females commonly seek extra-pair copulations, and selectively use the sperm from such matings to fertilize their eggs. Furthermore, females of some insect species will store sperm after copulation and will only fertilize their eggs after they have received a nuptial gift. Although such cryptic mate choice occurs after copulation, it is important to recognize that females may choose the sperm donor based on the same models of pre-copulatory mate choice.

The outcome of mate choice depends on the mating preference of an individual and the extent to which it can be expressed. Availability of suitable mates, **intra-** and **inter-sexual competition** and environment conditions can influence how much choice an individual is able to exert. For example, in some systems, male–male competition may be so fierce that female mate choice is hindered or suppressed by such contests; in other systems, however, females exert complete control over mate selection. In many species, mate choice can be a reflection of the social conditions that an individual is exposed to. Often animals will avoid mating with close relatives or group members, while also avoiding mates that are members of different species or ecotypes. The success of mating depends on the genetic compatibility of the partners, which can influence the number and quality of offspring.

In applied studies, an understanding of mate choice is critical for maximizing success in a managed system. Understanding how, why and when a particular mate is chosen can have important consequences for the success of captive programmes; if proper conditions are not met, animals may either refuse to mate or make poor mating decisions. Similarly, management programmes designed around wild populations must consider breeding cycles and resources required for mate choice for success to be reached. (HSF)

References and further reading

Andersson, M. (1994) *Sexual Selection*. Princeton University Press, Princeton, New Jersey.

Clutton-Brock, T.H. and Vincent, A. (1991) Sexual selection and the potential reproductive rates of males and females. *Nature* 351, 58–60.

Fisher, R.A. (1930) *The Genetical Theory of Natural Selection*. Clarendon Press, Oxford, UK.

Zahavi, A. (1975) Mate selection – a selection for a handicap. *Journal of Theoretical Biology* 53, 205–214.

Maternal behaviour

Maternal behaviour is defined as the pattern of behaviours given by the mother to her dependent offspring or young. These behaviours are often of extreme importance for the offspring's postnatal development and are directed towards their well-being and care. The evolutionary **fitness** of all species depends not only on the reproductive success of parents, represented by the rate of offspring generated over a lifetime, but also on the offspring's survivability rate. Thus the propagation of a species relies on the reproductive competence and successes of both parents but, especially, on the quality of maternal behaviours in mammals – behaviours that are evolutionarily well conserved throughout species.

Because maternal behaviour so often plays such a major role in the physiological and behavioural development of the offspring, the mother has been entitled by others as a 'hidden regulator'. Some of the distinct behaviours that may be shown by the mother include **grooming** or licking, **nursing**, **feeding** and cleaning, in addition to protection and comfort of the offspring. The amount of maternal care and thus the frequency of maternal behaviours performed towards the offspring vary with species. As an example, female primates spend a large part of their lifetime performing maternal behaviour as opposed to the minimal attention provided by rabbit females, which spend only a very short period of the day caring for their offspring within the nest; by contrast, in seahorses the male is responsible for the care of the offspring. This entry will focus on the maternal behaviour of mammals, as this is the area of most common **welfare** concern, but the reader is referred to Groothuis and Schwabl (2008) for a recent review of the regulation of this behaviour in birds.

Maternal behaviour is widely understood in rodents, and comprises a complex collection of activities such as nest building, licking, grooming, nursing and manipulation of pups, e.g. sniffing, exploration and retrieval. Female rodents also perform **placentophagia** – that is the consumption of the placenta immediately after giving birth – although this behaviour is not limited to rodents. **Carnivores** (e.g. dogs, wolves) and **ungulates** (e.g. cattle, sheep), along with animals such as hedgehogs and bats, also perform placentophagia. The most relevant explanation for this behaviour is that the placenta contains chemicals involved in involution and recovery of the uterus to its normal state (**prostaglandin**), and facilitation of milk ejection from the mammary gland (**oxytocin**). In addition, it contains molecules that modify endogenous opioid activity, helping to reduce the pain that arises pre- and post-delivery.

Maternal behaviour is absent during most of pregnancy and is suddenly triggered with oncoming **parturition**. It may be observed initially for a few days preceding labour in some species. Prepartum nest building observed in female birds, pigs and some rodents is a good example of a maternal behaviour that precedes parturition. Therefore, it is suggested that a complex connection and interaction exists between the **central nervous system** and the endocrine system during that final period of **gestation** and parturition. Their association with sensory stimuli generated by the offspring and environmental conditions both play an important role in triggering and maintaining maternal behaviours in the females.

From studies performed in rodents, specific regions of the **hypothalamus**, *stria terminalis* and lateral septum have been implicated as fundamental **brain** areas for the regulation of pup-directed maternal behaviours. Examples of these behaviours include **grouping**, retrieving and providing warmth to the pups and licking their anogenital regions, along with other tactile stimulation. The medial preoptic area of the hypothalamus receives input from various brain areas and is thought to play a role in identification of mainly olfactory cues from the pup, changing the female's motor activity and behaviour performance. Olfactory cues are crucial for the onset of maternal receptiveness and **recognition** of the newborn, in addition to recognition and establishment of mother–offspring bonding following birth.

The fact that the onset of maternal behaviour coincides with birth makes pregnancy and **lactation** endocrine hormones strong candidates for the initiation and persistence of maternal behaviour. During the last few days of pregnancy, endogenous levels of hormones such as **progesterone** begin to decline, while **oestrogen** and **prolactin** levels increase. These factors are believed to contribute to the enhancement of neural activity and beginning of maternal behaviour pre- and immediately postpartum. Oxytocin, an essential hormone for smooth muscle contraction at the time of parturition and milk ejection during lactation, has also been implicated as one of the central mediators of maternal care. It induces investment behaviours and maternal responsiveness when administered to **rats**, mice and ewes. The role of neuropeptide opioids, such as β-endorphin (**beta-endorphin**), during and after birth is controversial, as they may assist in **pain** reduction during labour and play inhibitory and stimulatory roles in maternal behaviour presentation.

The quality of maternal behaviour displayed may also rely on maternal experience. In general, experienced mothers (multiparous) show better maternal care and faster responsiveness to the offspring after birth than does a first-time, inexperienced mother (primiparous). Maternal behaviour in most species is terminated as the mother makes herself less available for the offspring by rejecting them. The actual time point in the offspring's life for terminating maternal care and nursing is determined by cues detected by the mother. For instance, female rats use **thermoregulation** to modulate mother–pup contact. Other factors, such as the offspring's weight and size related to birth, and eruption of teeth or tusks, also play a role in determining the time to separation and **weaning** in the young animal.

Maternal investment

Maternal investment has been defined as any investment or expenditure (e.g. time, energy) generated by the mother for an individual offspring, thus increasing its chance of survival and improving its well-being and reproductive success (Trivers, 1972). This investment relies on the cost of the mother's (or parent's) ability to invest in other offspring, future reproduction and inclusive fitness. The age of the offspring is also a factor in maternal investment. Growing offspring become more independent from the mother and thus benefit less from her; however, maternal demands intensify as the offspring become older. Offspring will try to acquire for themselves the maximum gain from maternal investment. However, at the same time, the mother must restrict maternal investment to no more than that absolutely necessary in order to successfully rear offspring and yet maximize her own future reproductive outcome.

This scenario can sometimes generate conflicts between mother and offspring interests. For example, the maternal care of **cats** and dogs undergoes deviations in the **reciprocity** of investment until the offspring are fully weaned. After giving birth, the mother initiates contact with her offspring. As the young become older, both offspring and mother concurrently initiate contact, primarily through nursing. As weaning approaches, the offspring mostly initiate nurturing from the mother, who may react evasively or aggressively, eventually leading to separation and full weaning.

There is also evidence that investment behaviour promotes reproductive success and increases offspring production. An interesting example is the female lion pride, where if two or more females in the group are nursing cubs, the mothers effectively form a 'nursery' in which they provide cubs with access to one another's milk. This has been described as leading to superior per capita reproduction in adult lionesses grouped in prides compared with paired or solitary females.

Moreover, maternal investment in offspring is expected to vary according to the offspring sex. This is the case when the reproductive success of the progeny is a function of differential levels of parental expenditure. Sex-biased maternal investment is observed in several species and happens mostly towards male offspring. In elephants, there is a greater early maternal investment in male calves than in females, because in the highly competitive seasonal mating systems, the body size of adult male elephants is an important factor for reproductive success. Also, seal females invest more in sons than in daughters until weaning and it is unlikely that higher post-weaning investment in daughters would compensate for the higher pre-weaning investment in sons. Mothers provide sons with extra resources because sons have greater implications in offspring lineage, since one male can produce enough sperm to fertilize many more eggs than one female can produce. Thus, parents tend to invest more in the gender that gives them the greatest reproductive pay-off (greater number of grandchildren) with increasing investment.

However, several theories predict that sex-biased maternal investment may vary under specific environmental or maternal conditions, as predicted by the Trivers–Willard hypothesis (Trivers and Willard, 1973). This states that, in some polygynous species (species in which the male mates with more than one female in a single breeding season, such as red deer), maternal investment is biased toward daughters, instead of sons, when the mother is under poor conditions. Social rank or female dominance status can also affect the sex of offspring generated, with dominant females birthing more sons and non-dominant females birthing more daughters. This is found in captive Japanese macaques (*Macaca fuscata*), where high-ranking mothers show a significant male-biased sex ratio with greater investment towards male infants, spending more time in contact and carrying them for longer than female infants.

Maternal–filial bond

Filial **bonding** or **attachment** is defined and mediated by the propensity of offspring to seek proximity and contact with their mother in preference to other objects or **conspecifics**. Neonates search for stimulation, such as physical contact, from the mother, which fulfils their expectations and contributes to their psychobiological balance. They also seek out resources from the mother to meet their physiological needs, such as warmth or nourishment. Early interactions between mother and newborn provide the basis for all subsequent interactions and bonding formation, as they have the ability to associate positive experiences with particular stimuli. In mammals, most stimuli are provided while animals are suckling, which is the foundation of maternal care. A robust indication of a proper bonding between offspring and mother is the presence of behaviours involving maternal leadership, assistance and training and, ultimately, cooperation towards offspring.

The establishment of filial bonding in sheep relies on the success of the first suckling bouts, suggesting that the **colostrum** (the first milk secreted by an animal coming into lactation, especially rich in maternal antibodies) ingested by the young immediately after birth facilitates the development of a preference for the mother. It has been shown that lambs receiving colostrum within the first 12 h after birth show a preference for their mother, while those ingesting saline do not. This confirms that maternal contact, effectively intermediated by feeding, is necessary for early associative learning and subsequent establishment of the mother–offspring affiliation. Offspring of group-living species are required to rapidly discriminate and bond to their own mother to maximize chances of survival. Olfactory cues are one of the most important mediators for mother–offspring recognition, facilitating maternal responsiveness, and they play a main role in establishing mother–infant bonding. Maternal recognition is reciprocal and, in general, alien mothers in the group will reject nursing and caring for young that are not their own. Under some circumstances, females may even be aggressive towards unfamiliar young animals.

The natural weaning process generally involves gradual separation of the mother and offspring; however, this event does not necessarily lead to the disruption of the maternal–filial bond. Calves of a semi-wild cattle herd, naturally weaned at the age of 7 to 14 months, do not show signs of impairment in the affectionate bonds with their mothers post-weaning. Cows, while nursing newborn calves, affectionately groomed older offspring that had already been weaned some time previously. In fact, the pre-weaning bonding between mother and offspring plays a major role in the foundation of the herd's cohesive social structure. Evidence of filial attachment post-weaning has also been found in guinea pigs. Physiologically, affiliative behaviours such as maternal care, grooming, protection and formation of a social bond with offspring are believed to be mediated by hormones such as oxytocin, **vasopressin** and endogenous opioids.

Maternal deprivation

Early adverse experiences can cause long-term psycho-physiological and neuroendocrine alterations in young animals that are carried on into adulthood. The magnitude of these effects is proportional and depends on the time (in life) of exposure (including stage of central nervous system development), duration of **stimulus**, and type and severity of the experience. Therefore, converse to maternal presence, early maternal deprivation, an **aversive** experience, negatively interferes with the proper development of both psychobiological and neuroendocrine regulatory mechanisms of a young animal. It is also expected to affect the young's behavioural responses under stressful situations.

The development and responsivity of the **hypothalamic–pituitary–adrenal (HPA) axis**, the main regulatory mechanism of **stress** response, is modulated by maternal behaviour. As previously reported in rodents, the amount of time the dam spends performing behaviours such as grooming and licking is inversely correlated with stress responsiveness in adult animals. Therefore, the ability of animals to cope with stress in life has been shown to be correlated to the amount of maternal care received early on in life. Animals subjected to early maternal deprivation show neuronal deficits and behavioural changes when exposed to stressful situations like social **isolation**. In addition, animals present **neophobia** when subjected to **fear** tests (e.g. novel open field) and evidence of impairment during cognitive performance (e.g. Morris water maze) or social recognition tests.

Maternal deprivation affects the individual's behavioural responses by eradicating the environmental predictability and controllability provided by a maternal presence. **Abnormal behaviours** are frequently observed in an animal subjected to artificial weaning, which, in other words, represents premature maternal deprivation. Young pigs perform abnormal behaviours such as excessive chewing, sucking other animals and behaviours resembling massaging of the maternal udder (such as **belly nosing**), reflecting distress caused by maternal separation. Pigs also show greater levels of aggressive behaviour compared with pigs left for longer periods with their dams. Moreover, maternally deprived young primates show **stereotypies**, which have no apparent biological function for the animal, such as digital or tail sucking and clasping, and rodents present **anxiety** and fear-like behaviours. Higher levels of **vocalization** and restlessness are observed in several mammal species when abruptly deprived of their mothers very early in life.

Maternal rejection

Maternal **rejection** naturally occurs with older animals when the mother separates or hides herself from the offspring in an attempt to wean them. However, it becomes a concern for the well-being of the neonatal offspring when rejection happens shortly following birth. This event typically develops when normal **socialization** or bonding between mother and offspring does not take place. Cleaning of offspring by the mother immediately after delivery is often crucial for maternal acceptance and bonding with the newborn, but human intervention with the dam or the infant, either during delivery or postpartum, can interfere with maternal acceptance. As observed in dogs, newborn pups when removed from the nest, washed and returned to the mother may no longer receive maternal care and be rejected by the mother.

Rejection by the mother occurs in all species, including farm animals, and is most frequently seen in primiparous dams (females that deliver for the first time). In ewes, the incidence of rejection behaviours, such as withdrawal, **aggression** and lack of cooperation with lambs during sucking attempts, declines with maternal experience, thus predicting the behaviour of primiparous ewes in future pregnancies. Additionally, dams with infected or injured udders (e.g. **mastitis**) show greater tendency to reject offspring. Agalactia (absence or failure to produce milk) or hypogalactia (lower than normal milk production) are examples of maternal inadequacy and may be associated with rejection. Maternal inadequacy is characterized by the incapability of a dam to provide sufficient maternal care to the offspring and may result from the dam's undernourishment, poor health, excessive excitability or exhaustion due to dystocia (difficult birth delivery).

Rejection by the mother causes psychological and physiological distress for young animals, especially when still dependent on maternal care and protection for survival. However, there may be some secondary and long-term advantages for maternally rejected animals. Once maternal support and help are lost, animals are forced to learn by their own means to handle undesired or stressful conditions more efficiently and succeed in searching for resources, such as food and shelter. Captive Japanese macaque infants whose mothers show high rates of rejection are less fearful and cope better with stressful situations in adulthood, suggesting that maternal rejection promotes offspring independence and the development of a less anxious personality. (RP)

See also: Dominance; Reproduction; Stress

References and further reading

Groothuis, T.G.G. and Schwabl, H. (2008) Hormone-mediated maternal effects in birds: mechanisms matter but what do we know of them? *Philosophical Transactions of the Royal Society B* 363, 1647–1661.

Nelson, R.J. (2000) Parental behaviour. In: Nelson, R.J. (ed.) *An Introduction to Behavioural Endocrinology*, 2nd edn. Sinauer Associates, Inc., Sunderland, Massachusetts.

Schino, G., Cozzolino, R. and Troisi, A. (1999) Social rank and sex-biased maternal investment in captive Japanese macaques: behavioural and reproductive data. *Folia Primatologica* 70, 254–263.

Schino, G., Speranza, L. and Troisi, A. (2001) Early maternal rejection and later social anxiety in juvenile and adult Japanese macaques. *Developmental Psychobiology* 38, 186–190.

Trivers, R.L. (1972) Parental investment and sexual selection. In: Campbell, B. (ed.) *Sexual Selection and the Descent of Man, 1871–1971*. Aldine, Chicago, Illinois, pp. 136–179.

Trivers, R.L. and Willard, D.E. (1973) Natural selection of parental ability to vary the sex ratio of offspring. *Science* 179, 90–92.

Mating behaviour and systems

The union of sexually reproducing animals during mating provides the opportunity for genetic mixing. This increases the adaptive potential as compared with asexual reproduction. Females usually produce a small number of gametes that are relatively expensively supplied with a life-support package to nourish the fertilized gamete. Males, on the other hand, produce large numbers of gametes with much more limited survival capacity. Given the increased investment in gamete production, it follows that in most situations greater investment will also be provided by females to increase the chances of survival of the fertilized egg than by males, and so females usually provide most of the parental care (notable exceptions include seahorses). Variation in male mating behaviour appears to be particularly related to the role of the male in rearing young and the defence of females.

For example, many prey species exhibit precocious development, and the amount of parental care is relatively small, most of which can be provided by the female. Prey animals are normally precocious to limit the **predation** of juveniles. For example, in cattle the only parental investment provided by the bull is to guard the cow during **oestrus** and prevent her from being inseminated by a rival bull. This minimal investment strategy encourages polygyny (individual males mating with more than one female) whereas in humans, for example, the need for greater parental investment encourages monogamy (an individual male mating with one female). Polygyny in turn causes competition between males for females, which results in **sexual selection** in the males and sexual dimorphism.

Sexual selection in many species is responsible for the increased size and strength of males, particularly in the shoulders, neck muscle and size of horns, antlers or other offensive body parts, which is advantageous for effective combat to secure access to females. The **hierarchy** of **dominance** so created ensures almost exclusive access to receptive females by the dominant males. However, despite the intense competition between the males to secure access to females, the dominant male does not usually attempt to force females to copulate. Males will guard the females, with attempted mounting, and they may show other **appetitive behaviour** such as partial erection and dribbling of accessory fluid. This is because it is in the male's interest to copulate only when the female ovum is receptive, and this information is readily imparted to the male by the female's behavioural signals, unlike humans where it is believed that the primeval female may have attempted to hide her menstrual condition to secure greater rewards from the male in the form of food and protection.

Some mammals also exhibit bisexual behaviour, for example cattle, with both sexes frequently exhibiting hetero- and **homosexual behaviours**. In the case of cows, homosexual behaviour was used in **feral** herds as the main signal to visually indicate to distant **grazing** bulls that the herd contained receptive cows. It is likely that there has also been human selection for this trait for as long as the sexes were kept apart, to indicate when a cow was ready to be inseminated (naturally or artificially). However, homosexual behaviour in cows is probably more than just a signal to bulls or man. It persists even in the presence of the bull and may serve to satiate a cow's sexual motivation over and above the attention paid to her by the bull. (CJCP)

See also: Sexual behaviour

Further reading

Clutton-Brock, T.H. (1989) Review lecture: mammalian mating systems. *Proceedings of the Royal Society of London B* 236, 339–372.

Emlen, S.T. and Oring, L.W. (1977) Ecology, sexual selection, and the evolution of mating systems. *Science* 197, 215–223.

Shuster, S.M. and Wade, M.J. (2003) *Mating Systems and Strategies*. Princeton University Press, Princeton, New Jersey.

Matriline

The matriline refers to the maternal female line of ancestors of an individual. Since mitochondrial DNA is inherited only from the ovum, all mitochondrial DNA is matrilinear and so the study of mitochondrial DNA can give an insight into the number of founding females within a population. The paternal line of inheritance is referred to as the patriline. (DSM)

Maturation of behaviour

New behaviour often arises as a result of learning (**see: Conditioning – types of**), but it may also occur as a result of the growth and development of the mechanistic processes regulating its expression, i.e. as a result of its maturation. This development reflects a largely genetic regulation of the underlying maturational process, since it occurs independently of experience, i.e. experience is not a necessary condition for its expression, although experience may affect its exact timing. Thus males of several species reared in the presence of sexually mature females express sexual behaviour at an earlier age than **conspecifics** reared in **isolation** from females, even without direct access to these females. This phenomenon is thought to be mediated largely through primer **pheromones**, but other stimuli may also play a role. Perhaps the most obvious examples of behaviour that develop as a result of maturation are **sexual behaviour**, **reproductive behaviour**, **maternal behaviour** and paternal behaviour; however, it is now recognized that many behaviours mature at specific times during development, according to their adaptive value at a given age.

In some species behavioural patterns may be seasonal due to the development and atrophy of supporting physiological structures, such as the **gonads**, which regulate the production of sex steroids and sexual behaviour as a result. Some consider this cyclical change in the underlying processes that facilitate the behaviour to be a form of maturation while others do not.

Some argue that concepts like maturation tend to bias the thought and study of animal behaviour towards the idea that animals take time to become fully functional beings (i.e. are initially immature), rather than viewing them, in normal circumstances, as fully functional beings according to the **ecology** of their life stage at any given time. For example, from a biological perspective, it is highly **adaptive** for a young animal not to engage in sexual behaviour and to direct available resources to growth and social development.

Other issues relating to the concept of behavioural maturation also relate to the idea that behaviour – rather than the underlying systems – matures. Thus, although **sensitive phases** tend to be thought of as times of particular sensitivity to certain types of learning (and so not related to behavioural maturation), they can also be considered times at which the maturation of particular types of behavioural response occur (**see: Development – behavioural**), since this is a time of maturation of the relevant parts of the nervous system. Thus, in the absence of exposure to certain stimuli at these times, a particular behavioural pattern becomes established that is relatively enduring due to the mechanistic organization (maturation) that occurs in the **central nervous system** at this time in relation to this particular function. For example, during the **socialization period**, **avoidance behaviour** towards unfamiliar individuals may mature as a result of lack of exposure at this time.

However, the process of **socialization** towards others does not occur as a result of maturation at this time, but rather as a result of experience during this phase of maturation of the nervous system. From this perspective, behavioural maturation is simply the default response that develops over time in the absence of experience. The concept of behavioural maturation may therefore be seen by some as a relic of the nature–nurture debate or the ethological versus psychological approach to the study of behaviour, rather than as a more progressive approach, which is a synthesis of these paradigms. (DSM)

Measures of productivity

Many modern, intensive farming methods aim to maximize production by increasing the numbers of animals kept while reducing the costs associated with this task. Productivity is measured in relation to herd or flock production, while **welfare** is assessed in terms of the individual animals.

There is evidence that poor farm animal welfare may adversely affect production in a number of species. For example, stockmanship (**see: Stockperson**) is a variable that is known to be associated with production. **Fear** of humans can provoke a **stress** response in many species, which may lead to poor welfare and performance. In contrast, there are also examples where poor animal welfare does not decrease overall production. A broiler unit may maintain high levels of production by using fast-growing strains of birds. However, the welfare of the individual birds may be poor, as reflected in the levels of **lameness** and breast blisters.

Commonly used measures of productivity include **growth rate**, disease incidence (**morbidity**), **culling** and **mortality** rates. Fertility parameters and data pertaining to the final product can also provide insight into the efficiency of a farming unit. Good production, when measured in relation to individual animals rather than a population, may be a necessary, but not a sufficient, indicator of good welfare. For example, **slaughter** data and finishing times may provide a pig farmer with useful information relating to the average rate of growth within the herd. Mortality and cull percentages may reveal the overall efficiency of a broiler unit, as well as indicating the level of **disease**, such as coccidiosis, within the population itself. In a dairy herd, an example of fertility data would be the number of cows returning to service and an example of product data would be milk yield produced by the herd. While these parameters may measure productivity on a farming unit, they may say little about the **quality of life** of the animals. Additional information relating to the management and husbandry of individuals within a population may also be required to ensure good animal welfare. (AK)

Measuring behaviour

The scientific study of animal behaviour is based on the principle that what an animal does can be quantified objectively. In this way it is possible to increase our understanding of the proximate and ultimate causes of behaviour by using these objective measurements to test hypotheses.

When undertaking a study of the behaviour of an individual or group of individuals, preliminary observations are important. This enables the identification of the behaviours that are relevant to the questions to be asked and the manner in which they will be measured. An **ethogram** may be produced as part of this preliminary work. In addition, the **reliability** (the extent to which they can be consistently measured by the same or different observers) and **validity** (the extent to which they measure what is intended) of the measures can be assessed at this point.

The choice of the measure will depend on whether the behaviour of interest occurs in states, events or bouts. State-type behaviours, such as eating or sitting, occur for a relatively prolonged or meaningful length of time. One can measure the frequency (number of times), latency (time taken to onset) and/or duration (length of time) for which the behaviour occurs during the period of observation. Event-type behaviours, such as discrete **vocalizations** or distinct body movements, occur for a relatively brief period or have durations that are less meaningful, and can be measured at specific points in time. One will usually record the frequency and/or latency with which these behaviours occur during the period of observation.

State- or event-type behaviours can both occur as a bout, which is a period of time during which the same behaviour occurs, whether this is a period of the same state (e.g. a bout of eating) or a period in which the same event is repeated (e.g. a bout of discrete vocalizations). One may measure the frequency of these bouts during the period of observation or the duration of time within or between bouts. Another measure of potential interest is the intensity of the behaviour, which is often less well defined but may have some objective **quantitative measure** such as the speed of the movement or the volume of the vocalization. In some instances a **qualitative measure**, e.g. a categorical scale describing the intensity of the behaviour, may be constructed. (KT)

See also: Video image recording and analysis

Further reading

Martin, P. and Bateson, P. (1993) *Measuring Behaviour*, 2nd edn. Cambridge University Press, Cambridge, UK.

McFarland, D. (1981) *The Oxford Companion to Animal Behaviour*. Oxford University Press, Oxford, UK.

Measuring welfare

Measures that reliably reflect the welfare of animals in different situations are crucial as a fundament for decision making in animal welfare legislation, guidelines, etc. Such measurements should translate the state of the animal (whether it is both doing well and feeling well; **see also: Welfare**) into an objective description that can be agreed upon and used by different observers. The commonly used measures of welfare are usually divided into two broad groups: behavioural and physiological.

In some cases, negative mental states are reflected in overt changes of behaviour, such as **avoidance behaviour** that indicates **pain** or **fear**. When using such measures it is, however, important to consider that, due to their evolutionary history, prey species are less likely to show many behavioural signs of poor welfare than predator species. Changes in overall behavioural **time budget** or the disappearance or appearance of behavioural patterns may also be informative about mental states. An animal that is sick often changes its time budget, spending more time inactive and less time on self-maintenance activities such as **grooming** and eating/drinking. **Abnormal** behaviours such as **stereotypies** and **self-mutilation** are generally considered to be associated with **frustration** of highly motivated behaviours. But since such abnormal behaviours may develop into habits and persist outside the frustrating situation that originally motivated their appearance, it has been argued that the presence of such abnormal behaviours should be considered evidence of poor welfare experienced at some time in the individual's life, and not necessarily an indicator of its present welfare.

Physiological measures of welfare are generally of two main kinds: those indicative of physical **health** and those associated with the reaction to **stress**. Health measures are relevant for two reasons: first, **disease** itself directly affects welfare and, secondly, disease may be the outcome of chronic stress and therefore indicate that the animals have been exposed to a situation of compromised welfare. Stress-related measures reflect the activation of the two main stress reaction systems: the **hypothalamic–pituitary–adrenal (HPA) axis** and the sympatho–adrenal axis. Indicators of their activation include plasma levels of **glucocorticoids** (cortisol and corticosterone) and **catecholamines** (adrenaline and noradrenaline), as well as cardiac output (**heart rate** and **blood pressure**). **Corticosteroid** level is the most commonly used measure of acute physiological stress. Measuring the plasma level of the hormone requires blood sampling, which may in itself be a potent **stressor** and so potentially influence the results. With a delay of minutes or hours, circulating corticosteroids can be found in body fluids that can be collected through less invasive methods, such as saliva or urine. For research purposes, telemetry is increasingly used to measure cardiac output and temperature non-invasively in animals that have been equipped with a microtransponder.

When using these physiological measures it is important to note that they are indicators of acute stress or **arousal**, i.e. the body's reaction to an environmental challenge, and that they do not indicate whether that arousal is experienced as **aversive** or as positive. Whereas the levels of circulating hormones and cardiac output are effective measures of acute stress, they are less useful in situations of chronic stress, as feedback regulation systems often return to near-normal levels. Indicators of chronic stress include changed responsiveness of the HPA axis, immune suppression, cardiovascular diseases and reduced growth.

More recently, interest has grown in the behavioural and physiological indicators of **positive welfare** states such as 'happiness', which include behavioural variety, the occurrence of **play** and the levels of biochemicals such as **oxytocin** and **endorphins**.

Statements about the welfare of animals may also be based on information about the system or situation the animals are in, rather than on measurements carried out on the actual animal. Studies of animals' **preferences** for different environments or resources and how much the animals are prepared to work for these are often used in animal welfare research. If an animal persists in working for access to a resource, even though the workload is increased, it can be assumed that this resource is important for the animal and that its welfare will be strongly affected by whether or not the resource in question is available.

The methods chosen to measure welfare depend on the applications for which the measurement is intended, which in turn may determine the amount of resources available and who will carry out the measurement. In research, there will usually be only one person to carry out the measurement and there is usually demand for a detailed picture of welfare that can be obtained through repeated assessments. In on-site auditing, many different observers will have to carry out assessments that should be reliably comparable, and often the assessment can be carried out only once at each site.

The welfare of an animal depends on many different aspects of its situation (whether or not it experiences pain, **hunger**, thirst, **fear**, **boredom**, etc.), and it is generally agreed that a correct measurement of welfare must account for this complexity and include parameters of different types. It is also important that the different views of animal welfare are accounted for, especially when welfare measurements are used for official purposes such as legislation. However, when welfare measurements are to form the basis for decision making, it is often difficult to handle a complex description that includes many different measures. One way of making this description more operational is to integrate the different measures into a single value, such as an index of welfare. In such an index, different aspects can be given different weights to reflect their assumed importance for animal welfare, and each constituent can be attributed a threshold value. The latter may be important in avoiding a situation that involves grave deficiencies in some aspects, but that is nevertheless given an acceptable overall score because these deficiencies are counterbalanced by positive results in other aspects. (AO)

See also: Eye white response; Measuring behaviour

Further reading

Dawkins, M.S. (1990) From an animal's point of view: motivation, fitness and animal welfare. *Behavioural and Brain Sciences* 13, 1–161.

Dawkins, M.S. (1998) Evolution and animal welfare. *The Quarterly Review of Biology* 73, 305–328.

Fraser, D. (2003) Assessing animal welfare at the farm and group level: the interplay of science and values. *Animal Welfare* 12, 433–443.

Hughes, B.O. and Curtis, P.E. (1997) Health and disease. In: Appleby, M.C. and Hughes, B.O. (eds) *Animal Welfare*. CAB International, Wallingford, UK.

Mench, J.A. and Mason, G.J. (1997) Behaviour. In: Appleby, M.C. and Hughes, B.O. (eds) *Animal Welfare*. CAB International, Wallingford, UK.

Rushen, J. (2000) Some issues in the interpretation of behavioural responses to stress. In: Moberg, G.P. and Mench, J.A. (eds) *Biology of Animal Stress: Implications for Animal Welfare*. CAB International, Wallingford, UK, pp. 23–42.

Rushen, J. and de Passillé, A.M.B. (1992) The scientific assessment of the impact of housing on animal welfare: a critical review. *Canadian Journal of Animal Science* 72, 721–743.

Sandøe, P. and Simonsen, H.B. (1992) Assessing animal welfare: where does science end and philosophy begin? *Animal Welfare* 1, 257–267.

Terlouw, E.M.C., Schouten, W.G.P. and Ladewig, L. (1997) Physiology. In: Appleby, M.C. and Hughes, B.O. (eds) *Animal Welfare*. CAB International, Wallingford, UK.

Medicines – traditional, complementary and alternative

Increasingly, animal owners are seeking alternative and complementary options for **health** and behavioural problems. Alternatives include traditional Chinese medicine (TCM), acupuncture, homeopathy, herbal medicine, T touch, flower essences, nutraceutical medicine and others. Like many of the psychotropic medications prescribed but not currently labelled for use in animals, little research has been conducted in veterinary medicine on the efficacy, safety and therapeutic benefit of alternative medicines in general. Research and education are warranted because, along with behavioural modification, they may offer an alternative or complement to conventional medicine, particularly if the latter is unsuccessful; however, until their efficacy can be quantified, their value relative to other interventions remains unknown.

Traditional Chinese medicine and acupuncture

According to the Western perception, TCM refers mainly to acupuncture. In fact, acupuncture is only a small part of Chinese medicine: the main part is clinical diagnostics and herbal medicine. TCM has a systematic approach to diagnosis and treatment and is a medical system in its own right. The temptation to understand TCM by fitting it into Western models is strong; however, reconciling both models is made difficult because terminology used in TCM is not directly translatable to Western medicine. For example, a blood deficiency is not anaemia per se, but a pattern recognition of poor hair and skin condition, disturbed dreams and sleep, weak pulse and pale tongue. Energy in Western medicine has a different meaning from energy, referred to as *qi*, in TCM, the former measurable, the latter esoteric. The spleen in TCM is recognized as a major organ for transforming food into *qi* energy, where this function does not exist in Western medicine.

TCM recognizes pathology as arising from complex interrelationships between organs and organ systems. An animal's behaviour can direct us to an underlying pathophysiology and thereby to treatment. For example, phobias and **anxiety** are considered to be a disturbance of *shen* or **consciousness**, and can be caused by a 'blood deficiency', '*qi* deficiency' or 'heart fire'. In veterinary medicine, acupuncture and Chinese herbs are used to treat not only physical problems but also behavioural problems. For example, *dan shen* (*Salvia miltiorrhiza*) and jujube (*Ziziphus jujube*) are used to 'nourish the heart and calm the spirit'. *Dan shen* contains compounds that bind to the same sites as **benzodiazepines** on the **GABA** receptor complex, and is orally active as a tranquillizer in animals. Jujube contains saponins and flavones that have demonstrated sedative and anxiolytic activity *in vivo*. Acupuncture has been demonstrated to affect **endorphin**, **enkephalin** and **catecholamine** release, to inhibit sympathetic activation during mental **stress** and to invoke **analgesia** in animals.

Homeopathy

Homeopathic preparations are manufactured from specific plant, animal or mineral parts according to pharmacopoeia standards, where they undergo a series of dilutions and processing. These preparations are extremely dilute and, while their proposed mechanism of action is considered controversial, they are widely used by veterinary homeopaths. There are two fundamental tenets of homeopathy. The first is the principle of 'similars', which states that a patient's symptoms can be cured by a drug that produces the same symptoms of the ailment in a healthy individual. Homeopathic preparations have been theorized to stimulate recovery in a similar way to physical vaccines. The second principle is the repeated dilutions of a remedy, called the principle of 'potencies'. Despite multiple dilutions, the remedies are supposed to retain their biological activity.

A prescription is based on a diagnostic approach called repertorization. This means taking the case and matching symptoms of the patient to the remedy's toxic effects or known effects when used undiluted – hence the action 'like cures like'. Research is contradictory, with the placebo effect thought to explain the efficacy of homeopathy in some cases. The therapeutic value of homeopathic preparations in veterinary medicine remains largely anecdotal. For example, **fear** in animals is treated with a remedy selected on the 'totality' of symptoms and history. The more unusual or peculiar the symptoms, the more specifically a remedy can be selected. Remedies that are often selected for 'fear' include phosphorus, aconitum, belladonna, rhododendron, gelsemium or natrum mur. Reported success in the treatment of long-standing problems by trained veterinary homeopaths suggests that traditional scientific methods might be used to assess the validity of the therapeutic properties of homeopathic remedies.

Herbal medicine (phytotherapy)

'Nervines' are herbs traditionally used for 'nervous' complaints in people. Frequently, the traditional use of a herb is supported by scientific research, as biomedical actions are understood. Herbs have similar effects to conventional medicines and lend themselves well to clinical research. Notwithstanding the issues of variable potency of different plant crops and parts, and quality of manufactured herbs, the patient is treated according to herbal principles based on the phytotherapeutic principles of actions. Nervines may be sedating, activating or balancing and are traditionally used in formulas that take into account the whole patient symptoms, not just the diagnosis.

From a modern phytotherapeutic perspective *Valeriana officinalis*, *Ginkgo biloba*, *Piper methysticum*, *Hypericum perforatum*, *Passiflora incarnata* and *Matricaria chamomilla* are used for conditions ranging from anxiety and **depression** to **sleep** disorders. Like conventional medicines, herbal remedies have contraindications: some adverse events and drug interactions have been reported, so these medicines should be applied only with appropriate knowledge and caution.

T touch

Tellington touch, or T touch, is a 'body work' method based on circular movements of the fingers and hands on the body. It was developed originally for behavioural modification and training of **horses**, but has now been applied in **companion animals**, horses and in human nursing. Statistically and clinically significant differences in T touch groups have been reported in the literature for both humans and horses.

Flower essences

Bach Flower Essences® and numerous other flower essences (derived from imparting the essences of particular flowers in water) are readily available and very popular for the treatment of animals with a variety of behavioural problems ranging from stress, fear, anxiety and **aggression** to grief. Rescue Remedy® is claimed to be a 'natural reliever of everyday stress' in the company literature. An unpublished study (Cram, 2004) assessed the effect of two flower essences on intense environmental stimulation in people. Using a placebo control group design, the two flower essences were reported to reduce physiological activation and stress in subjects. There are claims for flower essences to 'strengthen emotional equilibrium and equanimity in the face of stresses and environmental impacts', thus reducing the typical 'fight or flight' stress response.

Nutraceuticals

Nutraceuticals are nutritional elements that can be used in a therapeutic way for the treatment of particular conditions. For example, phosphatidylserine, acetyl L carnitine and antioxidants may be useful in **cognitive dysfunction**. The use of melatonin to improve sleep and some cognitive measures in people has been studied, and it may also be useful in animals with cognitive dysfunction and obsessive–compulsive disorders and, possibly, thunder phobia in dogs. Melatonin may alter serum sex hormone concentrations in male and female dogs. Female dogs treated with melatonin have been reported to exhibit significant decreases in serum oestradiol, **testosterone** and DHEAS (dehydroepiandrosterone) concentrations, and male dogs showed significant decreases in serum oestradiol and 17-HP (17-hydroxyprogesterone) concentrations. Melatonin may therefore have a role in **sexual behaviour** problems associated with hyperoestrogenism. An increasing number of nutraceuticals are available for treatment, or as an aid in the treatment of a range of physical and psychological conditions in animals. (BF)

See also: Pheromonatherapy

Reference and further reading

Ashley, P.F., Frank, L.A., Schmeitzel, L.P., Bailey, E.M. and Oliver, J.W. (1999) Effect of oral melatonin administration on sex hormone, prolactin, and thyroid hormone concentrations in adult dogs. *Journal of the American Veterinary Medical Association* 215, 1111–1115.

Cram, J.R. (2004) Flower essences reduce stress reaction to intense environmental stimulus. Bach Flower Research Programme. Available at: www.edwardbach.org (accessed 25 August 2009).

Janssens, L.A.A., Rogers, P.A.M. and Schoen, A.M. (1988) Acupuncture analgesia: a review. *Veterinary Record* 122, 355–358.

Marsden, S. and Wynn, S.M. (eds) (2003) *Manual of Natural Veterinary Medicine Science and Tradition*. Mosby, Philadelphia, Pennsylvania.

Saxton, J. and Gregory, P. (2004) *Textbook of Veterinary Homeopathy*. Beaconsfield Publishers, Beaconsfield, UK.

Schoen, A. and Wynn, S. (eds) (2001) *Complementary and Alternative Veterinary Medicine: Principles and Practice.* Mosby, Philadelphia, Pennsylvania.

Shanahan, S. (2003) Trailer-loading stress in horses: behavioural and physiological effects of non-aversive training (TTEAM). *Journal of Applied Animal Welfare Science* 6, 263–274.

Memory

Memory is the ability to store, retain and retrieve learned information over time. Without this ability to remember information, learning could not take place because, in the absence of memory, each encounter with the same **stimulus** would be perceived as novel. Similarly, without learning there would be no information available to be remembered – this demonstrates the closeness of the link between learning and memory. Memory, even its simplest form, is therefore necessary to allow animals, both wild and captive, to learn from their experiences, whether this concerns the spatial position of a recently located food source, reacquaintance with a previously familiar **conspecific**, the location of a preferred nestbox, the correct way to use an automated food dispenser successfully or **habituation** to the presence of a **stockperson**.

There are two main types of memory store: short-term (working) memory (STM) and long-term (reference) memory (LTM). In non-human animal memory research, STM tends to refer to temporarily stored information that lasts for seconds/minutes and is of limited capacity. In contrast, LTM refers to the ability to store information for a greater duration and has a huge capacity. LTM can be subdivided between explicit (declarative) and implicit (non-declarative/procedural) memory, with explicit memory being further subdivided between semantic and episodic memory. Explicit memory is a memory for facts and events (i.e. knowing that …); semantic memory is the memory for facts and general knowledge, whereas episodic memory is the memory for specific events – knowledge that is connected to a particular time or place. Episodic memory therefore implies some degree of **self-awareness** that has been thought to restrict it to humans. However, recent research has suggested that some animals may be capable of demonstrating episodic-like memory, with scrub jays (a food-storing bird) appearing to remember what food they have stored, where and when. Implicit memory is memory for skills and habits (i.e. knowing how …), and is often used to describe the memory processes behind simple learning phenomena such as habituation and associative learning.

The first stage of memory formation is acquisition – the process by which newly learned information (environmental input) is converted or encoded into a memory trace in the STM. The cellular mechanism thought to underlie STM involves the temporary facilitation of synaptic transmission between **neuron(e)s** in the **central nervous system** via increased neurotransmitter release. Some, but not all, memory traces may then undergo the second stage of memory formation, consolidation. This process incorporates both the storage and the retention of the memory trace in the LTM via rehearsal. In contrast to STM, the cellular mechanism thought to underlie consolidation to LTM involves the formation of structural changes to neurons (e.g. growth of additional synaptic connections) via protein synthesis. However, just because a memory trace has been consolidated does not mean that it will be retained or remain accessible indefinitely. The final memory process is retrieval – the way in which individuals can actually refer to stored memory traces. This applies specifically to the retrieval of memory traces from the LTM back into the STM.

The flow of information through memory is thus as follows: an animal perceives environmental input via its sensory organs (e.g. visual, olfactory, auditory, etc.). This information is encoded in the STM and can influence response output (i.e. behaviour) directly or, following consolidation strategies such as rehearsal, can be stored and retained in the LTM. It is important to emphasize that the STM does not only hold new memory traces, but also those memory traces retrieved from the LTM. This is why the term 'working' memory is a useful description of short-term memory, because the short-term memory store is where the 'work' takes place, and where the information directing the response output is controlled.

None the less, not all that is perceived ends up in memory. **Attention** plays a critical role in filtering out the information that is 'unnecessary', so that only 'necessary' information is encoded into a memory trace in the STM. Only the information that catches attention is first attended to, then perceived and encoded – and only a fraction of this information is then consolidated for future reference. Some may be insufficiently rehearsed, others may not have the appropriate conditions for memory (see below). Even those memories stored, retained and retrieved from the LTM may not necessarily go on to influence response output, because motivation plays a key role in controlling response output. For instance, pigs may ignore the presence of a stockperson after feeding time, but this is unlikely to be the case before feeding when they are hungry.

In humans we can assess memory by testing either **recognition** or recall of information. Recognition, for example 'Do you recognize this individual?', implies a verification of perceivable (present) facts. Whereas recall, for example 'Do you recall what a particular individual looks like?', implies a generation of non-perceivable (absent) facts. As a result, performance in recognition tasks is usually better, because of the presence of retrieval cues. When assessing animal memory, the majority of tasks used are tests of recognition – this is because recognition tests are easier to test non-verbally. Without verbal language it would be difficult for an animal to 'answer' a question such as 'What do you remember?' (a test of recall). In contrast, by displaying a behavioural change in response towards meeting something for the first time and then subsequent encounters, animals can non-verbally 'answer' questions such as 'Do you remember this?' (a test of recognition). Some of the recognition tests most frequently used to assess animal memory include habituation tests, delayed response tasks, the radial arm maze and discrimination training.

Before discussing some of these tests in more detail, it is important to emphasize that in memory research we must be particularly cautious in the interpretation of such memory assessments. While interpretations involving memory may adequately explain test results, it is vitally important to also consider other more parsimonious, non-memory, explanations for such results, e.g. **motivation**, satiation and error, before we move on to more complex explanations involving memory.

Habituation tests involve exposing an animal to a novel stimulus (e.g. a noise), upon which the animal shows a large response (e.g. ears pricking). If the same stimulus is then reintroduced after a subsequent time interval and the animal shows a decreased response to it, then this suggests habituation, and thus memory for the stimulus retained over that specific time interval. If, however, upon the reintroduction of the stimulus the animal shows no such decrease in the level of response (no habituation), then this may suggest memory failure. However, as mentioned previously, before we accept a memory interpretation of these results we should ensure that we have discounted alternative, non-memory explanations. For example, animals habituate only to functionally irrelevant signals – animals should never habituate to the sudden appearance of predator – so lack of habituation could also mean that the stimulus was salient on every exposure.

Habituation tests can be used to assess STM, e.g. seconds/minutes, but they can also be used to assess LTM. For instance, some species of migratory birds have been shown to recognize the songs of previously familiar conspecifics after 8 months of separation. Habituation tests are popular because they rely on a spontaneous 'natural' untrained response and, as a consequence, they are simple and often quick to use when compared with some other techniques. However, this is also a disadvantage in that any fluctuation in motivation may significantly affect performance, and so habituation tests are only effective for events that evoke a well-defined response.

Delayed response tasks are tests in which a delay, the 'retention interval', is interposed between a stimulus and the opportunity for a subject to respond to it. By altering the duration of the retention interval we can assess the duration of memory for the stimulus. Common examples of delayed response tasks are the delayed matching to sample test (DMTS), in which animals are rewarded for choosing matching stimuli, and the delayed non-matching to sample test (DNMTS), in which animals are rewarded for choosing non-matching stimuli. These tests are usually conducted using operant training procedures. The animal is initially shown a single stimulus on a screen, which it must peck/press to indicate that it is paying attention. Once pecked/pressed, this original stimulus disappears and there is a retention interval of several seconds/minutes. Following the retention interval, two other stimuli are presented simultaneously, and the animal is rewarded for pecking/pressing either the one that matches (DMTS) or the one that does not match (DNMTS) the original stimulus.

Delayed response tasks deal only with STM, with accuracy (the proportion of correct to incorrect choices) falling rapidly as the retention interval is increased from just a few seconds to 1 or 2 min. The difficulty of remembering this type of information over longer periods is thought to be due to both the limited time between the different trials and the sheer number of trials received each day (often 100 trials per day).

Again we need to account and control for other possible factors influencing behaviour besides memory itself. For example, the experimental design should ensure that the stimuli are presented randomly on both sides of the screen, so that the task cannot be solved purely by responding to stimulus position. The benefits of this type of test include the amount of control the researcher has, and the increased attention and motivation to complete the task present due to **reinforcement**. However, because it requires training, it can be laborious for both the subject and the researcher. It can also be thought of as 'unnatural', given that there is frequently little ecological relevance for the subject. Although the experimental subjects in DMTS tasks are most commonly pigeons and primates, dolphins have also been used, and a spatial version of the DMTS task has been devised and used successfully in dogs to investigate the impact of **ageing** on memory performance.

The radial arm maze typically consists of eight identical arms radiating from a central area. The ends of the arms can be baited with small amounts of food, such that a hungry animal introduced into the maze will visit each arm and eat all the food. A typical test will involve introducing the animal into the central area of the maze with four of the eight arms closed (a 'forced' trial), so that the animal can only enter and clear the four open arms of food. It is then removed from the maze for a retention interval of a specific duration, and then reintroduced into the central area of the maze that now has all eight arms open. Animals can be trained either to return back to the same four arms, now re-baited, that it previously visited (a 'win–stay' approach), or to visit the four previously closed arms (a 'win–shift' approach). Making errors (entry to 'wrong' arms) suggests memory failure over the retention interval, whereas making few errors suggests successful spatial memory for food/arm location.

Provided that we have adequately controlled for extraneous non-memory factors such as olfactory trails or search 'rules', we find that animals such as pigs can successfully learn to either 'win–stay' or 'win–shift' following retention intervals of at least

10 min. This suggests that the radial arm maze can be used for assessing both STM and LTM. While it does require training, this type of spatial memory test benefits from the fact that it is relatively 'natural' due to its use of **foraging behaviour**, and the animal is also well motivated to perform the task.

Discrimination training traditionally refers to procedures in which animals are reinforced for performing different, arbitrary responses to each of two or more stimuli, and is generally best used for assessing LTM. For instance, a pigeon can be trained to peck a disc to gain a reward when presented with one stimulus, but to refrain from pecking the disc and gain a reward when presented with a different stimulus. Once the pigeon has reached a specified level of accuracy during training, the training is stopped and a retention interval (e.g. days/months/years) begins. Following this retention interval the pigeon can then be retested; if it performs the discrimination at a level of accuracy above that expected by chance (even if the level of performance, as might be expected, has dropped since its initial training), then this suggests that the pigeon has retained a memory for that discrimination over the duration of the retention interval. Discrimination training can inform us about both the capacity and the duration of LTM – pigeons can remember several hundred images for a retention interval of several years.

The advantages and disadvantages of this type of test are very similar to those of the delayed response tasks (see above); however, unlike delayed response tasks, where the overall technique takes a long time to train but the individual trials themselves require no training, discrimination training requires the often arduous process of training an animal to respond in a particular way to each specific stimulus. When the time taken to train the subject is combined with the length of the retention interval and subsequent retesting, it can be a very long process.

Why is some information remembered for longer, and other information forgotten faster, than others? What are the conditions that influence memory during the three different memory processes? First, if we look at how memory can be improved, it appears that the more distinct, 'valued', emotionally loaded, long-lasting or frequent an event, then the better it is remembered. During acquisition there is a clear effect on subsequent performance of changing the duration of exposure to the to-be-remembered information. As presentation time is increased, so the memory improves – perhaps because it allows for more complete encoding. A similar effect occurs if, rather than increasing the duration of a single exposure to information, you can expose an animal to repeated presentations of the same information – especially the greater the length of the first exposure and the shorter the gap between the presentations. This process is known as retroactive facilitation.

When a particularly surprising and/or distinctive item appears in a sequence of information, then it is remembered especially well. This is known as the von Restorff effect. For instance, if pigeons are trained to recognize a sequence of colours and a symbol is positioned within that sequence of colours, it is remembered better than if that same symbol had appeared in a sequence consisting only of other symbols. More information can also be remembered if it is 'chunked', that is arranged into meaningful blocks of information, rather than memorized as single units, although there is still a limit to the number of chunks that can be remembered. For example, pigeons learn faster if the training of colour and pattern matching is chunked by the experimenter. Events that are emotionally loaded may be prioritized in memory storage when compared with emotionally 'neutral' events, and the relative 'value' of the information to be learned may also affect its memorability.

For example, information about an **aversive** experience such as being placed in a weigh crate, the location of a preferred food or the identity of key social companions may be more memorable than emotionally 'neutral' events or less valued information. In one study, sheep required differing amounts of encouragement to run down a **raceway** according to the aversiveness of the event that they had previously experienced at the end of the race, either no handling (no encouragement needed), restraint (some encouragement) or simulated shearing (lots of encouragement). This suggests that the sheep retained information about the experience they had, and adjusted their behaviour according to how aversive they remembered it to be.

Memories, nevertheless, can be forgotten (either temporarily or permanently) as well as improved. Memory can be disrupted if there is interference during the consolidation process. If this interference occurs prior to the presentation of the to-be-remembered stimulus, it is known as proactive interference and, if it occurs after the presentation of the to-be-remembered stimulus, it is known as retroactive interference. This active disruption of memory consolidation by both proactive and retroactive interference results in permanent loss of a memory trace, and is one way in which memories can be forgotten. Examples of potential causes of interference include elements of common husbandry procedures, such as moving animals to a new cage, **handling** or introducing a novel conspecific. Such procedures may act as stressors, and high **stress** levels have been shown to disrupt memory consolidation. But memory consolidation can also be disrupted by non-stressful procedures, either because they shift attention away from the to-be-remembered information or, if the procedure includes information that is very similar to the previously learned information, by replacing it.

Such examples of interference can also result, post-consolidation, in temporary inaccessibility of an existing memory (retrieval failure). High stress levels influence memory retrieval because context – both internal (e.g. state of **arousal** and/or motivation) and external (e.g. test chamber colour and/or smell) – plays an important role in determining which memories are retrieved. It is thought that the presence of contextual cues aids the retrieval of the appropriate memory from the LTM. The absence of specific contextual cues may therefore lead to retrieval failure and thus apparent forgetting, even though the memory is still present and may only be temporarily inaccessible. Thus, forgetting can mean either temporary (e.g. retrieval failure) or permanent (e.g. disruption) loss of a memory trace. In addition to this 'active' process of forgetting, memory traces can also decay passively. This results in permanent forgetting, and occurs if there is insufficient rehearsal during consolidation to move the memory from the STM to the LTM. If there is insufficient space in the STM then a memory trace may be forgotten before it even has time to be consolidated.

Such apparent memory failure – even if only temporary – may lead to inappropriate behaviour being displayed and, in some situations, this may result in a reduction in **welfare**. For example, if, following a stressful handling procedure, a subordinate animal fails to display appropriate submissive behaviour towards a dominant individual that it encounters soon after handling due to (temporary or permanent) recognition failure, then this may result in the dominant animal becoming aggressive towards the subordinate – behaviour that might otherwise have been avoided. Care should also be taken to avoid exposing animals in **captivity** to cues that may result in the retrieval of negatively emotionally loaded memories, because such memories of negative events and their associated emotional state may result in a reduction in welfare – even in the absence of the actual aversive event itself. (OHPB)

Further reading

Davey, G. (1981) *Animal Learning and Conditioning*. The Macmillan Press Ltd, London and Basingstoke, UK.

Mendl, M., Burman O., Laughlin, K. and Paul, E. (2001) Animal memory and animal welfare. *Animal Welfare* 10, S141–S159.

Pearce, J.M. (1997) *Animal Learning and Cognition*. Psychology Press Ltd, Hove, UK.

Shettleworth, S.J. (1998) *Cognition, Evolution and Behaviour*. Oxford University Press, Oxford, UK.

Mentation

Mentation refers to mental activity; the process of thinking. (PS)

Merkwelt

Derived from the German *merken* meaning 'to notice' and *welt* meaning 'world'. The merkwelt, as defined by Jacob Von Uexküll (1934), is the perceptual world of the animal, which, along with the **wirkwelt** (i.e. the operational world), constitutes the animal's **umwelt**, that is, its subjective world. The merkwelt of an animal thus depends on its sensory abilities and is composed of all the stimuli the animal is able to perceive. (SL)

Mesopredator

A mesopredator is a medium-sized predator, such as a domestic **cat** or fox. Mesopredators are of particular interest because their numbers in the wild will tend to increase when a larger predator is displaced from the environment (mesopredator release theory). As a result, there is increased **predation** of smaller prey species and less predation of larger ones, resulting in less **biodiversity** and potentially a major shift in the **ecology** of the environment as a result. For this reason, it is argued that it is important to conserve large top predators (e.g. wolves, etc.), since they suppress the numbers of mesopredators and so reduce predation of smaller prey species and increase biodiversity as a consequence. (DSM)

Metacognition

Research on animal metacognition – **cognition** about cognition – is primarily concerned with the question: Do animals know when they know and when they do not know?

The study of animal metacognition is rooted in research on human cognitive psychology. The keys to human metacognition are beliefs and judgements about **memory**. For example, the question 'What is the name of the frog in *The Muppet Movie*?' might elicit a *feeling of knowing* – a judgement that one could select the name on a multiple choice test, even if one cannot recall it now. Such judgements, remarkably, can be made accurately even when the answer cannot be recalled. Metacognitive judgements are subject to some consistent biases, such as overconfidence, but they can also be quite accurate, especially when people can test themselves.

Animal cognition refers to animals' ability to process external information, store and retrieve that information from memory and produce resulting thoughts and behaviours. Animal cognition is often distinguished from other kinds of animal learning, such as classical conditioning and operant conditioning (**see: Conditioning – types of**). Different animals have different cognitive specialties; for example, many primates are social in ways that parallel human **social cognition**; some caching birds, for example, have spatial memories for the location of hidden food that far exceed those of humans.

The study of animal metacognition revolves around the concept of uncertainty – that is, discriminating between situations in which one is certain versus uncertain of what to do. Animals frequently appear to be uncertain – as when a horse hesitates before jumping a hurdle that it may not be able to clear – but, by itself, appearing uncertain is not evidence of metacognition. The question that has been addressed over recent years is: Can animals report that they are uncertain? And can they report uncertainty about their memories?

Most of the evidence concerning animal metacognition is based on studies of rhesus macaque monkeys. Some of that evidence, as well as evidence from other species, is reviewed briefly below.

In seminal studies on animal metacognition performed in the mid-1990s, animals were given a task, such as identifying whether a tone was of high or low pitch. Crucially, they were also given a third response option, an 'escape' response, which allowed them to skip the trial and move on. Monkeys and dolphins both tended to escape difficult trials more than easy trials, providing the first evidence that they could make uncertain responses (see Smith and Washburn, 2005, for a review).

In another innovative study, rhesus monkeys were allowed to opt out of taking a test before it had occurred (Hampton, 2001). The subjects were shown an image on a touch-sensitive computer monitor; their task was to remember the image so that, later, they would be able to select it when it was presented along with other images in a multiple-choice test. The animals could also opt out of the test and settle for a food reward that, while less attractive than the reward for a correct response, was better than responding incorrectly and receiving no reward. The monkeys appeared to opt out of the most difficult trials, and thus they did better when they chose to take the test than when it was forced on them. This suggests they could monitor uncertainty in their memories. When pigeons were faced with the same task, however, they did not demonstrate an ability to monitor uncertainty.

A third kind of evidence for animal metacognition comes from a task in which, essentially, rhesus monkeys were asked to gamble on their memories (Kornell *et al.*, 2007). The monkeys were shown six pictures, one at a time; they were then shown nine pictures and had to touch the one that had been presented previously. After they responded, the monkeys were asked how many 'tokens' they wanted to wager on their response. The tokens were icons, displayed on the screen, that were automatically exchanged for food rewards. If a subject touched the 'high risk' icon, they would either gain or lose three tokens, depending on whether their previous response had been right or wrong. If they touched the 'low risk' icon, they gained one token, regardless of their previous accuracy. The monkeys' bets demonstrated that they were confident following correct responses and uncertain following errors. More importantly, because the monkeys had previous experience with the betting paradigm, they were able to respond metacognitively beginning on their first day of the task. This finding implies that they had learned a general metacognitive ability that was not task-specific.

Self-reflection

Do animals experience metacognition the way humans do? In the study of animal behaviour, it is always critical to recognize that similar behaviours do not necessarily result from similar underlying processes and/or experiences. Many human metacognitive experiences, such as a tip of the tongue – when one feels tantalizingly close to recalling an answer (e.g. Kermit) that just will not come – are intensely conscious and involve an element of self-reflection. Thus far, the evidence suggests that some animals, including primates and dolphins, can distinguish between what they know and what they do not know. The degree to which they are aware of, or consciously processing, their metacognitive judgements remains uncertain. (NK)

References and further reading

Hampton, R.R. (2001) Rhesus monkeys know when they remember. *Proceedings of the National Academy of Sciences* 98, 5359–5362.

Kornell, N. (2009) Metacognition in humans and animals. *Current Directions in Psychological Science* 18, 11–15.

Kornell, N., Son, L.K. and Terrace, H.S. (2007) Transfer of metacognitive skills and hint seeking in monkeys. *Psychological Science* 18, 64–71.

Smith, J.D. and Washburn, D.A. (2005) Uncertainty monitoring and metacognition by animals. *Current Directions in Psychological Science* 14, 19–24.

Terrace, H.S. and Metcalfe, J. (eds) (2005) *The Missing Link in Cognition: Origins of Self-reflective Consciousness*. Oxford University Press, New York.

Meyerson index

The Meyerson index is a laboratory measure of female sexual receptivity/**motivation** that scores the occurrence (or not) of **lordosis** in the female. (DSM)

Microenvironment

The microenvironment is the environment immediately surrounding an animal, normally within a confined space in which the animal may live, e.g. inside a cage, or be temporarily placed, e.g. a **transport** vehicle, and, for the young, within a nest. (DBM)

Milking

Milking is the removal of milk from the udders of lactating animals, most commonly dairy **cattle**, **goats** and **sheep**. Dairy cows of the *Bos taurus* species release their milk without the presence of their calf, but cows of the *Bos indicus* species usually need the psychological stimulus of the presence of the calf. The development of automatic machinery for milk extraction in the 20th century allowed many more animals to be milked by one herdsperson than previously, when cows were milked by hand. Milking units are now usually collected together into a milking parlour, where cows can be brought as a group and several animals can be milked at one time, under the supervision of a single herdsperson.

There is a wide variety of options for removal of the milk from cows, with the main considerations being the speed of throughput. Entry of cows into the parlour is often facilitated by the provision of a food reward at the same time as milking. The removal of milk in heavily lactating cows may offer reward in itself, and they are usually the first to enter the parlour if given the choice. Milking can be a stressful experience for cows if they are rapidly collected from pasture by a person with a dog or motorbike, if they are controlled in the collecting yard by an electric fence, if they are treated aggressively by the herdsperson in the collecting yard or parlour, if they are likely to slip on wet concrete floors or if they experience **pain** resulting from the slow removal of teat cups or an unsuitable vacuum level.

The best parlours allow the cow to enter her stall at will rather than being forced to enter by the herdsperson, such as autotandem parlours. Cows develop preferences to enter a specific stall at a specific time, and preventing them from doing this can increase their **heart rate** but it will not reduce their milk production. Recently, technology has been developed to enable teat cups to be automatically attached to cows, which has led to the commercial production of fully automated milking units, or milking robots. There are positive benefits to milking cows with robots – the vacuum and pulsation characteristics, which are standardized for all cows in conventional milking parlours, can potentially be tailored to the needs of individual cows in robotic milking systems. Inappropriate machine settings can be painful or induce haemorrhaging in the teats.

Stockpersons should have extra time to look after their cows if automatic milking units are employed, but are still required to fetch cows that do not want to be milked and to monitor milk storage and cooling. Robotic milking systems will increase the need for stockpeople to be better trained technically. An inadequate number can lead to long queues of cows, and cows may be reluctant to travel a long distance to be milked. Cows need to be trained and many are reluctant to be milked in these systems, even if concentrate feeds are used to entice them. Also, a milking robot may not be able to adequately clean cows that come in from the fields in a muddy condition, leading to increased levels of **mastitis**. (CJCP)

Further reading

EU Project: Animal Milking. Available at: http://www.automatic milking.nl/ (accessed May 2009).

Mind

The mind has puzzled people for centuries; both Plato and Aristotle, for example, presented theoretical accounts of it. Clearly, the mind cannot be literally observed under a microscope or dissected. It has long been known, however, that there is a close relationship between, on the one hand, mind and **brain** and, on the other, mind and behaviour. These relationships have given rise to two branches of science that come as close as we get to empirical study of the mind: *neuroscience* and ***psychology***. Beyond this, a third field, the *philosophy of mind*, involves a priori investigation of the mental. Each field informs, and is informed by, the other two.

Psychologists investigate (among other things) mental development in children, sensation and **perception**, learning and **memory**, **emotion**, **personality**, thought and **language**, and various disorders and therapies. In neuroscience the main question about the mind is: How does the **central nervous system**, and particularly the brain, function to sustain mentality?

The philosophy of mind tackles a range of questions. First, what is distinctive about the mind? We know it differs from physical objects, but how exactly? One influential answer is that the mind involves **consciousness**. Secondly, what is the relationship between the mind and the body – and, more particularly, the brain? And given that mind and body are so very different, can we explain how bodily events cause mental events (burns hurt) and mental events cause bodily events (love leads you to the florist)? Thirdly, the mind is home to **knowledge**, beliefs and desires that represent things in the world. How does my belief that a cuckoo has arrived from Africa manage to represent, or be about, a bird species and continent? For example, I need not picture these things in order to have the belief. Fourthly, we all claim to know at least a little about the minds of others, but how can we know anything of the sort if all we ever observe is bodily stimuli (e.g. a punch) and behaviour (e.g. someone crying 'ouch!')? How do I know that you feel what I feel when you act as if you are in pain? This problem becomes more difficult when one turns from humans to other species (**see also: Anthropomorphism; Mentation**). (PAR)

Miniature breed

Miniature breeds are distinguished by having a markedly smaller body size compared with other breeds. Miniature animals can be produced in two ways: selection for small size over consecutive generations, or utilization of one of many mutations for dwarfism.

The Lowline breed of **cattle** arose from a classic two-way selection experiment for yearling growth rate conducted in a population of Angus cattle by the New South Wales Department of Primary Industries, Australia. Fifteen years of selection for low growth rate resulted in a line with a mature body size 60% of the size of normal (control) cattle. When the experiment was completed, the existing animals from this line formed the foundation of the new breed. This is an excellent example of miniaturization as a result of recurrent selection. The miniature animals are proportionate, meaning that they have been equally miniaturized in all dimensions. It is likely that most of the differences in size among **dog** breeds arose from this type of selection.

Some dog and **cat** breeds exhibit disproportionate dwarfism, typically involving shortness of legs far in excess of what would be expected for their body size. Although the scientific evidence is very scant, it is likely that at least some of these breeds are reflecting homozygosity for a dwarfism mutation that has become a 'trademark' of the breed. If there are problems associated with the dwarfism in such breeds, it may be possible to overcome these by outcrossing to another breed, followed by selection during subsequent backcrossing.

Dexter cattle are a miniature breed that originated in Ireland. Within this breed, a gene for disproportionate dwarfism is segregating. Homozygotes for the normal version of the gene are proportionate miniature animals known as long-legged Dexters; heterozygotes show disproportionate dwarfism (short-legged Dexters); and homozygotes for the mutation are grossly deformed **fetuses** (bulldog calves). In principle, the latter can be avoided by ensuring that every Dexter mating involves at least one long-legged Dexter. In practice, there is a substantial overlap in leg length between heterozygotes and homozygotes. A **DNA analysis** test is now available for distinguishing the two **genotypes**. (JA, FWN)

Mink

Mink are semi-aquatic mustelids that show adaptations for a hunting lifestyle on land and in water. Two species exist: the European mink (*Mustela lutreola*), which is rare or extinct in most of its original range; and the American mink (*Mustela vison*), which is native to North America and a resident alien in Europe. The American mink is used in the production of fur for the mink industry. Farming of mink originated in North America in the 1860s, but expanded to become a worldwide industry in the 20th century. As mink only breed once per year, they have experienced relatively few generations of selection in **captivity**. Therefore, while there has been some selection for production traits such as fur colour and ease of husbandry, the behavioural traits of farmed mink are little different from their wild counterparts. Indeed, farmed mink that have escaped or been released as a result of **animal rights** activism adapt rapidly to local environments and establish stable **feral** populations. American mink are considered a pest in European countries, and may be a serious threat to native wildlife, particularly for vulnerable species such as the water vole (*Arvicola amphibius*).

In common with wild mink, farmed mink mate in early spring and give birth to their young (kits) in early to mid-May. In the wild the young would be weaned between 8 and 11 weeks of age, with the faster-growing males usually weaned earlier than the females and leaving the natal territory. Farmed mink are normally weaned at about 5 weeks of age and then housed in male–female pairs (for ease of grading, as males are normally twice the weight of females) and fed *ad libitum* to **slaughter** at about 6 months of age. Farmed mink are traditionally fed a high-protein, high-fat diet usually made up of waste from the fish, meat, dairy and fur industries. After 6 months, those males and females kept for breeding are singly housed and fed a restricted ration to ensure that they are ideal weight for fertility in the spring.

Concerns for behavioural **welfare** of farmed mink centre on the incidence of apparently abnormal behavioural responses such as **tail biting** and locomotor **stereotypies**, and the restriction of behaviours in conventional cages. Tail biting may occur in pairs or between mink housed in neighbouring cages, and it is believed this is a form of redirected hunting behaviour. Stereotypic activities such as **weaving** and somersaulting are rare in young, socially housed mink but normally develop during the winter period of feed restriction. Typically, these activities occur around feeding time and they may represent the perseverance of preprandial activities, coupled with conditioning due to predictable feeding regimes.

Restriction of behavioural **needs** in mink has centred on the value of swimming water. Mink are found to work as hard for access to swimming water as they would for food (**see: Economics of behaviour**), and to show elevated **corticosteroid** levels consistent with **distress** when prevented from swimming. This suggests that swimming water is a highly valued resource and essential for their well-being. The counter-argument is that mink need to have had experience of swimming in water before they develop a strong requirement to use it. As a consequence, farmed mink with no prior experience may not be deprived of the resource, whose value is context-specific as opposed to a universal need. This debate continues while, at a practical level, provision of water on commercial farms is being investigated.

The farming of mink for their fur has been banned in some countries (e.g. the UK), because of concerns regarding the welfare and **morality** of the trade. Despite continued campaigns by anti-fur organizations (e.g. Respect, PETA), worldwide production of mink fur has increased, with the largest industries in Denmark and China (about 25% of production each), while only about 10% of mink are still reared in North America, with the remainder in northern or eastern European nations. (JC)

Mirror neuron(e)

Mirror **neuron(e)**s are **brain** cells that become active when an animal both observes and undertakes equivalent actions, i.e. the observation mirrors the actual performance of a given behaviour. There is much speculation about their potential importance in a range of forms of **social learning** and **cognition**, although current direct evidence in support of these is lacking. (DSM)

Further reading

Rizzolatti, G. and Craighero, L. (2004) The mirror-neuron system. *Annual Reviews in Neuroscience* 27, 169–192.

Mis-mothering

Mis-mothering describes a failure or deficit in **maternal behaviour** that places the offspring at increased risk of **suffering** and/or harm. Mis-mothering can occur as a result of endogenous or exogenous factors. For example, common reasons for mis-mothering include: (i) failure of the mother–offspring bond to occur (**see: Bonding – parent–offspring**); (ii) poor maternal skills due to a lack of experience or exposure during an important developmental period; (iii) delayed **lactation** (often as a result of hormonal disorganization arising from a prolonged, complicated or artificial, such as Caesarean section, birthing process); (iv) inhibition of **nursing** as a result of a painful **mastitis** condition, possibly induced by inappropriate suckling behaviour by the neonate; and (v) active rejection of sickly or weak individuals. In some cases young may be lost due to deficits in the offspring (e.g. weakness/illness), or accidents such as smothering or crushing by the mother when she lies down, and the loss of these offspring may be attributed to 'mis-mothering' in a broad sense, even though there is no deficit in maternal behaviour. (DSM)

Mixed-species stocking

Many species are kept together in order to exploit resources more effectively. **Cattle** eat relatively long grass, scooping it into the mouth with the tongue, and this opens up a sward so that **sheep**, with their very selective biting action, can choose the highest-quality leaves and shoots. Generally, problems with **parasites** do not arise when the species are grazed together, at least in temperate climates.

In contrast, keeping donkeys and **horses** together can pose problems because donkeys can tolerate quite large infestations of lungworms without obvious symptoms. Horses are much less tolerant, so the donkey is effectively a reservoir of infection.

In African pastoral systems, **goats** are often kept in parallel with sheep or cattle. This enables households to spread risks, as goats can survive drought better than the other species.

There are other advantages to mixed-species stocking. A special case is the herding of sheep with cattle on ranches in the USA. If they have been imprinted upon each other when young, the cattle may defend the sheep from attacks by coyotes and other predators. Llamas have also been kept with sheep with the same intention.

Integrated systems with ducks, **chickens** and **pigs** being kept in pens above tanks of farmed fish are widespread in South-east Asia. Manure, urine and waste feed drop into the fish ponds, thus fertilizing algal growth, which is then eaten by herbivorous fish. This is highly efficient in terms of energy and nutrient cycles, but these systems have been implicated as a source of novel virus **diseases** that can threaten human life.

Mixed-species stocking enables resources to be used efficiently. There are disadvantages generally relating to human and animal **health**, especially when the systems are intensive, but substantial economic benefits are possible. (SJGH)

Modal action patterns (MAP)

Modal action patterns (MAP) are relatively fixed, species-typical behavioural patterns that are performed in response to specific stimuli, similar to fixed action patterns (**FAP**). However, unlike FAP, MAP take into account variations both within and between individuals in performance of the behaviour by referring to the most frequent measurement value (mode) shown in the distribution of the behaviour pattern. (LMD)

Modality

The modality of a **stimulus** refers to both its physical properties and the medium through which it is communicated by another or transmitted. Stimulus modality is not a unique property of stimuli involved in **communication**, but also relates to stimuli within the physical environment, and so determines the perceptual world in which an animal exists (**see: Perception**). (DSM)

Mollusc (mollusk) conservation and protection

Human-caused animal **extinctions** may be the ultimate of all human–animal interactions. Extinction is forever eliminating any other associations with humans. Since AD 1500, 643 animal species have been declared extinct, almost all from human activities such as **hunting** and harvesting, pollution and habitat degradation.

When we think of extinct animals, we tend to think of charismatic megafauna, passenger pigeons, dodo birds, Tasmanian tigers or Steller's sea cows. When we think of threatened or endangered species, we perhaps think of chimpanzees, gorillas, Siberian tigers, rhinoceroses, pandas, whooping cranes or California condors, each of which has campaigns and dedicated advocates devoted to saving them from that ultimate death. Yet 87% of all animal species are invertebrates, those animals without backbones. These are animals that are not necessarily cute and fuzzy and they have suffered the most sweeping extinctions.

Molluscs – notably freshwater snails and mussels – comprise 42% of the extinct animals today, more than all the extinct mammals and birds combined. In addition, 1930 non-marine molluscs are listed as threatened (only 41 marine species are so listed).

Freshwater mussels are particularly threatened. The south-eastern USA is the centre of diversity of the group, with over 300 species described from that area. Freshwater mussels were living in the streams there when the Appalachians were being formed, and are there now that the mountains have been extensively eroded. Endangered mussels with colourful names like purple wartyback, fluted kidneyshell, elktoe, orange nacre mucket and pink heelsplitter grow in their own particular stretches of the Tennessee and Mississippi River systems and, because of human activities, are losing the battle of survival.

First, Native Americans ate these mussels and they are well represented in ancient middens, the trash piles of the tribes. Secondly, European settlers discovered that these mussels made pearls inside themselves and countless millions were harvested solely for the pearl trade. Some mussel pearls were so highly valued that one was sold in 1902 for US$65,000. Thirdly, mussel shells were used to make mother-of-pearl buttons in the early 1900s, and many more millions were harvested for this industry. At the peak of production, over 9700 harvesters collected mussels for this industry. Fourthly, 125 dams have been built since the 1920s on the waterways of the American South-east, isolating stocks of mussels behind dams, degrading their habitats with silt and slowing down river flows. Some 100-year-old mussels are isolated in between dams with no way to reproduce themselves, since the fish they need to disperse their young no longer live there.

They were again harvested, beginning in the 1970s, for seeds in the Japanese pearl oyster trade. A tiny piece of a freshwater mussel shell is placed inside the mantle of a pearl oyster, which the oyster then treats as an irritant. It lays down layers of nacre around the irritant to make it smoother, and the result is a cultivated pearl. Japanese pearl oysters recently suffered a 60% death rate due to a virus, dramatically cutting down the demand for freshwater mussel shells, and so helping their conservation.

In addition to the above threats to their existence and from the increasing pollution from human cohabitation, freshwater mussels are now facing another threat, from two of their relatives. The European zebra mussel was first found in the Great Lakes of North America in 1985, probably first arriving there in ships' ballast water, and has since spread dramatically into the Tennessee River. It is so prolific that it coats all substrates layers thick with its tiny shells, siphoning food from native bivalve species and smothering them to death. Another introduced invader from China, the tiny Corbicula clam, also competes for food in lakes and streams.

A few people are working at saving freshwater mussels by transplanting them to clean waters, raising them in laboratories, educating farmers not to fertilize fields close to rivers or to graze cows near streams where they could crush mussel shells with their hooves. Some efforts are being made to legislate protection of the threatened species, much like those used to save the Banff hotsprings snail, which is only found in five hot springs near Calgary in Canada; three of the five are now protected, and snails have been successfully transplanted into hot springs where they had formerly been eradicated by human bathing activity.

But the future of many of the still-living freshwater mussels looks bleak. Unlike the millions of dollars donated for preserving condors, whooping cranes or pandas, little is generated for saving freshwater mussels or other threatened molluscs. When they become extinct, who will cry for the lost pink heelsplitters? Who will mourn what Lydeard and his colleagues called the 'silence of the clams'? (RCA)

See also: Conservation; Endangered species; Exotic species invasion

Further reading

Lydeard, C., Cowie, R.H., Ponder, W.F., Bogan, A.E., Bouchet, P., Clark, S.A. *et al.* (2004) The global decline of nonmarine mollusks. *Bioscience* 54, 321–330.

Monkey

Monkeys belong to the taxonomic order of primates; there are currently about 252 extant species of monkeys, in 461 taxa (see Groves, 2001). Primates are divided into two suborders, the strepsirrhines (consisting of the lorises and the lemurs) and the haplorrhines, which includes the tarsiers, monkeys and apes. The geographical distribution of monkeys is primarily restricted to the tropics: New World monkeys (NWM, platyrrhines) live in South and Central America and southern Mexico, and Old World monkeys (OWM) and apes (catarrhines) across Africa and Asia. Old and New World monkeys differ in several morphological features. NWM have nostrils that are wide open and far apart, while the nostrils of OWM are close together and comparatively narrow. Some NWM have a prehensile tail that can be used as an additional limb to grip; and some OWM, but no NWM, have ischial callosities, also known as sitting pads, on their backsides.

Most, but not all, monkeys differ from other mammals in several ways. They have forward-facing eyes allowing binocular vision, nails (as opposed to claws) on their fingers and toes, an opposable thumb to allow a pincer-like grip, a relatively large **brain** and a longer period of dependency of young. Some exceptions are the colobus monkeys, which do not have thumbs (their name comes from the Greek word *kolobus* meaning 'cut short'), and **marmosets** and **tamarins**, which have claws on all digits except their big toes.

These features are adaptations to their lifestyles. The majority of monkeys are arboreal (tree dwelling), and they play a critical role in their tropical forest ecosystem as seed dispersers and pollinators. Other monkeys are ground dwelling and live on mountain plateaux (e.g. gelada baboons, which feed on grass), in deserts (e.g. hamadryas baboons) or savannahs (e.g. patas monkeys, which can run faster, at 55km/h, than any other primate). The geographical range of the Japanese **macaque** is further north and east than any other non-human primate; they have thick coats to endure the harsh, snowy winter conditions and some individuals sit up to their necks in naturally occurring hot springs to escape the freezing weather. All species of monkey are diurnal, with one exception, the aptly named night monkey (also known as the owl monkey) from South America.

There is considerable variation in the diet of monkeys. Understanding natural dietary adaptations and providing appetizing food with appropriate nutritional requirements is essential for good **welfare**. While fruit, leaves, flowers, seeds and other plant parts are eaten by most monkeys in the wild, many also eat gum and other plant exudates, insects, reptiles, birds and their eggs. Some monkeys are known to eat amphibians and crustaceans and are good swimmers (e.g. the crab-eating macaque, also known as the cynomolgus or long-tailed macaque). Others hunt and eat other mammals; for example, baboons have been known to hunt gazelle kids, capuchins eat baby squirrels and blue monkeys have been seen to eat another primate, the bush baby. Challenging the intelligence of primates to forage for their food in **captivity** may be critical for welfare and in preventing the development of **abnormal** behaviours. Capuchin monkeys use tools in the wild to crack open nuts, but they are the only monkeys reported to use tools naturally in the wild (some other monkey species use tools under provisioned conditions and in captivity). Many primates will raid crops if their habitat is close to humans.

Monkeys vary greatly in size and weight; the smallest is the pygmy marmoset (adult head + body length is ~13.5 cm and they weigh 112–160 g), while the largest is the mandrill, with males measuring ~91 cm and weighing 27 kg – over twice the weight of females). Sexual dimorphism primarily depends upon the degree of competition among males for mates. In primarily monogamous species like marmosets, which rear their young cooperatively, males and females are similar in size. In polygynous mating systems, where males attempt to mate with several females (sometimes in harems), males are larger than females. One of the defining characteristics of monkeys is their social nature. All of the monkeys live in **groups**, ranging from small groups consisting of a mated pair and their immediate family to hordes of over 800 (in the case of mandrills in Gabon). There is a complex network of relationships within monkey groups – some cooperative and friendly and some competitive, as resources are scarce. The social bonds are often strengthened by **grooming**, and the enforced **dominance hierarchies** present in many species ensure individuals know their place. A series of ritualized threats minimizes the risks associated with physical fighting over resources. Grooming releases **endorphins** and appears to be pleasurable for the recipient. It is used as a 'currency' to trade for other desirable resources, such as mating. Understanding the natural social structure is critical for successful husbandry, to reduce the possibilities of fighting and in promoting affiliative relationships. Introducing unfamiliar individuals and forming new groups should be planned and managed carefully (see Jennings and Prescott, 2009) to avoid **injury** and prolonged **stress** or disturbance.

Monkeys have larger brains relative to body size than most other non-primate mammals. Their **intelligence** may have evolved to deal with challenges that they face in their social world, recognizing rank, friends and enemies, and responding appropriately to the behaviour of others. However, unlike apes, they are not able to take another individual's perspective, nor to deceive intentionally, nor do they recognize themselves in mirrors (they do not appear to have **self-awareness**).

The dominant sensory modality of monkeys is vision; they have both high visual acuity (i.e. ability to distinguish between closely spaced visual stimuli) and binocular vision. Many species of monkeys have elaborate pelage – manes, crests, moustaches, etc. – and often use their high-contrast markings in threat displays. All OWM, but only some NWP, have trichromatic colour vision similar to our own, which is important for selecting ripe fruits from unripe and semi-ripe ones, and may also confer advantages in detecting insect prey and predators. Colour vision also plays a role in **communication**; for example, during the mating season the facial and anogenital skin of rhesus macaques becomes redder and females prefer redder male faces over paler ones, possibly because it provides a cue as to the male's quality (Waitt *et al.*, 2003). Most of the NWM are polymorphic in their colour vision; all males and some females are dichromatic (based on two classes of cone receptors, like humans with red–green 'colour blindness'), while some females are trichromatic. Dichromats have difficulties distinguishing yellows, greens, browns and reds, which may influence the design of behavioural experiments and the choice of colour cues for positive **reinforcement** training.

All monkeys have a well-developed sense of smell, which is important for food identification and selection, as well as in social behaviours. Many NWM possess odour-producing skin glands and demonstrate conspicuous marking behaviours; individuals use these olfactory cues to obtain social information such as individual identity, rank and reproductive status. Removing all scent from enclosures may be stressful and result in abnormally high levels of **scent marking**. Cleaning and sanitation of enclosures and **enrichment** devices alternately can help to retain familiarity. Monkeys can hear higher-frequency sounds than humans and have good low-frequency hearing. Noise (e.g. construction noise and ultrasonic frequencies) can have an adverse effect on welfare. **Vocalizations** are also an important mode of communication for monkeys and can be useful welfare indicators.

Assessing the welfare of captive monkeys should combine a range of measures. These include: (i) coat and body condition; (ii) physical values such as longevity, **growth rate**, fluctuations in body weight, susceptibility to **disease**, reproduction and infant care; (iii) clinical measurements, such as **heart rate**, **blood pressure**, body temperature, serum concentrations of various **stress** hormones – e.g. cortisol – and immunological functions; and (iv) structured behavioural assessment. Behaviour is often the most accessible expression of welfare – monkeys that are well will appear relaxed (with **conspecifics** and in human presence), and spend much time grooming and foraging. Behavioural changes that may indicate poor welfare resulting from the present (or past) environment and/or experiences of the animal include a restricted behavioural repertoire, an abnormal **time budget**, inappropriate **social behaviour** and other abnormal behaviour patterns (Jennings and Prescott, 2009).

Non-human primates are used in biomedical research and in the production and testing of products and devices in laboratories. Ethical questions (**see: Ethics**) are raised due to their genetic proximity to humans, and the fact that they are neurophysiologically highly developed. Their potential for **suffering** may be greater than that of other animals used in laboratories, because of the difficulties in providing an appropriate environment to meet their social, physical and psychological needs. The use of **great apes** (only the **chimpanzee** has been used in laboratory testing) has now been banned in many countries. Animal welfare and anti-**vivisection** organizations campaign to phase out the use of any non-human primates in laboratory tests and experiments. An estimated 100,000–200,000 primates are used in research worldwide (Carlsson *et al*., 2004). In Europe around 10,000 monkeys are currently used in regulated procedures each year, approximately 0.1% of the total number of all animals used.

Worldwide, monkeys are used in the study of microbiology (including HIV/AIDS), neuroscience (including Parkinson's, Alzheimer's), biochemistry and pharmacology/physiology. In Europe the majority of monkeys used are in toxicity tests (safety and efficacy tests) during the development of new medicines. The most frequently used OWM are macaques (the cynomolgus and rhesus macaque), but other OWM used include baboons and vervet monkeys (also known as green monkeys). The common marmoset is the most frequently used NWM, but tamarins, squirrel monkeys and night monkeys are also used. Strict legislation surrounds their use in most countries. The use of wild-caught monkeys in laboratories is now banned in the UK, except under exceptional cases, and the rest of Europe may follow suit.

The housing and husbandry of monkeys in laboratories has improved in many countries since the late 1980s. In the best laboratories, cages are of sufficient size to permit monkeys to live in socially harmonious groups and to provide adequate enrichment to offer choice and control and to allow the monkeys to engage in species-specific behaviours and **locomotion**. Access to outdoor areas may also be given. In the worst laboratories, monkeys are housed singly in small, barren cages with little opportunity for social interaction. They are bored and fearful. They are often weaned too early and develop serious abnormal behaviours, such as **stereotypies** and self-injurious behaviours (**see: Self-mutilation**). Rennie and Buchanan-Smith (2006a,b,c) review refinements to primate use in laboratories (**see: Three Rs principle**). Positive reinforcement training (reward-based training) is a proven method of improving monkeys' perception of care staff, and may promote welfare as the monkeys learn to cooperate with many routine housing and husbandry events and, if required, in some experimental procedures such as venipuncture.

Monkeys are frequently kept in zoological collections; they are generally active, playful and inquisitive, making them popular with the visiting public (although, unfortunately, high visitor density may be a **stressor** for the monkeys if they are unable to retreat and hide). Many species are subject to **conservation** breeding programmes in **zoos** and private animal collections. In the wild, nearly half of all monkey species are threatened (in the vulnerable, endangered or critically endangered categories (IUCN, 2008)).

The threats they face are primarily from habitat destruction, **hunting** and pressure from the human population. This, combined with often small population sizes, means that many species are threatened with **extinction**. The rate of deforestation is much greater than regrowth, and encroachment by humans for farming brings monkeys and humans into close contact, and often into conflict if monkeys raid crops. Some **reintroduction** programmes, combined with conservation of forest fragments and reforesting to create corridors that connect them, have proved successful. There is a growing tourist trade to view monkeys in the wild and, while this may provide an incentive for locals to value and protect species, it has its associated problems. For example, monkeys that receive food from tourists lose their **fear**; they can become dangerous, and sometimes are shot as a means of control.

Hunting is mainly for food but, especially in Asia, also for traditional medicines, bait and body parts for ornamentation and amulets. People have hunted monkeys on a subsistence basis to feed their families for many years, but in recent years a market for monkey meat among the urban population has emerged. This **bushmeat** trade poses a very serious threat to the survival of many species. Eating monkeys is relatively common in Africa and Latin America, but is not so common in parts of Asia, where the religions of **Islam**, **Hinduism** and **Buddhism** forbid it. Monkeys are often caught in traps or shot. Live monkeys and fresh carcasses are sold at roadside stalls and in markets; the meat is often smoked to prevent it rotting. The fur of some species is prized – for example, the tails of colobus monkeys are used to make rugs.

In other areas humans worship monkeys, use them for assistance or keep them as **pets**. In India, the Hanuman langur is considered to be sacred by Hindus and will often be fed in temples. If a Hanuman langur dies before its time, Hindu worshippers may give the monkey a full funeral. In some parts of Malaysia, humans train macaques to help harvest fruit and nuts. Capuchin monkeys have been trained to assist severely disabled humans to perform basic tasks such as washing their face, combing their hair and microwaving food. Although some people keep monkeys as pets, a monkey's special needs can rarely be met, leading to health, psychological and behavioural problems and suffering. (HMB-S)

References and further reading

Carlsson, H., Schapiro, S.J., Farah, I. and Hau, J. (2004) Use of primates in research: a global overview. *American Journal of Primatology* 63, 225–237.

Groves, C.P. (2001) *Primate Taxonomy*. Smithsonian Institution Press, Washington, DC.

Hosey, G.R. (2005) How does the zoo environment affect the behaviour of captive primates? *Applied Animal Behaviour Science* 90, 107–129.

IUCN (2008) 2008 IUCN Red List of Threatened Species. Available at: http://www.iucnredlist.org (accessed March 2009).
Jennings, M. and Prescott, M.J. (eds) (2009) Joint Working Group on Refinement. Refinements in husbandry, care and common procedures for non-human primates. Ninth report of the BVAAWF/FRAME/RSPCA/UFAW Joint Working Group on Refinement. *Laboratory Animals* 43(Suppl. 1), S1:1–S1:47.
Rennie, A.E. and Buchanan-Smith, H.M. (2006a) Refinement of the use of non-human primates in scientific research. Part I: the influence of humans. *Animal Welfare* 15, 203–213.
Rennie, A.E. and Buchanan-Smith, H.M. (2006b) Refinement of the use of non-human primates in scientific research. Part II: housing, husbandry and acquisition. *Animal Welfare* 15, 215–238.
Rennie, A.E. and Buchanan-Smith, H.M. (2006c) Refinement of the use of non-human primates in scientific research. Part III: refinement of procedures. *Animal Welfare* 15, 239–261.
Waitt, C., Little, A.C., Wolfensohn, S., Honess, P., Brown, A.P., Buchanan-Smith, H.M. and Perrett, D.I. (2003) Evidence from rhesus macaques suggests coloration plays a role in primate mate choice. *Proceedings of the Royal Society B* (Suppl.) 270, S144–S146.
Wolfe-Coote, S. (ed.) (2005) *The Laboratory Primate*. Elsevier Academic Press, San Diego, California.
Wolfensohn, S. and Honess, P. (2005) *Handbook of Primate Husbandry and Welfare*. Blackwell, Oxford, UK.

Morality

Every society, cultural and social group tends to accept a number of beliefs about good and bad conduct (**see: Ethics**). A sense of values informs people's lives in terms of what they do, what they decide to do and their comments and judgements about themselves and others. People try, in various ways, to shape their lives according to moral values; they think that some kinds of life are more worth living than others, and have an outlook on life, and generally try to bring their children up to share this outlook. We might say that elementary beliefs about right and wrong are instilled into the young during the earliest stages of **language** acquisition and social interaction.

Language is impregnated with moral values and, at an early stage of social awareness, moral notions about justice and fairness, honesty and truthfulness will find expression; human children will utter moral judgements such as 'It is not fair', 'He is cheating/telling lies', and so on. Various rules requiring moral duties towards other beings may be taught by means of analogy and appeals to self-interest. A parent or guardian might instil respect for sentient animals by confronting a child with the question 'How would you like it if someone hurt you, as you are obviously hurting the cat?'

With maturity and reflection, many elementary moral notions and the arguments supporting them can become increasingly sophisticated, showing an awareness of universal moral principles. During the course of one's life, ingrained values may be challenged and new moral principles and outlook may replace the old. This process, in many ways, resembles the development of science, where new circumstances challenge traditional and established beliefs, and new theories emerge to replace the old. An example of this process is the changing attitudes towards **laboratory, farmed** and **companion animals** since the mid- to late 1970s.

A popular, and largely correct, account of morality is that it is a means of restraining certain forms of conduct. Thus restrictions on cruelty and causing unnecessary **suffering** to animals are widely recognized moral constraints on behaviour. As a guide to conduct, morality can place constraints on both the ends and means of people's actions. Even if the end is universal happiness, morality may intervene, saying 'Not by this means'. An objective, like finding a cure for a **disease**, may be regarded as morally commendable, whereas the means might not be morally commendable if certain unnecessary and painful procedures are involved. From the moral standpoint, not only are the results of an activity to be assessed, but the procedures by which these results are achieved are also subject to scrutiny.

Although morality is sometimes described as a subjective or personal phenomenon, this is not strictly correct. There are moral principles that are universally acknowledged: in general, people disapprove of lying and cheating, and even more strongly disapprove of cruelty, causing suffering and seriously injuring or killing others. Exceptions may be allowed, but they are justified with reference to a higher moral priority: we may lie and cheat and harm others in order to defend ourselves or others from wrongfully inflicted harm or **death**.

The study of morality, and of the nature of moral argument, is called 'moral philosophy' or 'ethics'. For most purposes these terms can be used interchangeably, but they do have slightly different meanings. 'Ethical' is derived from the Greek word for 'personal character' and has a broader meaning, which includes concern with the value attributed to different kinds of lives and activities. The term 'morality', which is derived from the Latin term for 'social custom', is frequently restricted to its interest in rules and duties but it is often used when considering whether certain feelings or attitudes are of a moral kind. For example, 'remorse' (in response to, for example, a failure to carry out a particular duty) is considered to be a moral feeling, whereas 'regret' might not be. To experience remorse one must have made a moral judgement about one's action or inactions. The experience of regret need not involve any such judgement. A criminal may regret leaving his fingerprints at the scene of the crime, but this would not count as an experience of remorse. These considerations provide an insight into the nature of morality. It is a common, but nevertheless mistaken, belief that morality is reducible to **feelings** or **emotions**. It is not, although feelings and emotions may have a moral component. Consequently, it is not a sub-field of psychological enquiry.

Morality (and ethics) stands in contrast to other concerns. Self-interest may conflict with morality when an obligation to others is contrasted with selfish desires. In some respects a moral obligation may be contrasted with legal, political, economic and religious considerations. This does not mean that the latter are without moral content, but that the moral interest is different. From the moral standpoint it is always possible to question whether this law, political directive, economic reform or divine command is good.

This independent and critical feature of morality is non-reducible to other disciplines. Morality is a very powerful social phenomenon. Moral claims may conflict with self-interest, physical well-being and various forms of happiness. For this reason, it is mistaken to reduce morality to self-interest or even a set of rules that cement the social order. Theories of morality attributed to animals frequently display morality in this narrow schema. It is sometimes held that animals in the wild exhibit various forms of moral behaviour. Studies of animal behaviour may very well corroborate this claim, but inferences drawn from the alleged social order of animal groups that carry the assumption that their moral system is functional in promoting social harmony will inevitably fail to depict morality. In contrast, pursuance of a moral imperative may result in self-destruction or social disintegration. The 19th-century German philosopher G.W.F. Hegel notoriously drew attention to the fact that the great moral figures of history and literature, such as Socrates and Antigone, not only brought harm to themselves, but were very disruptive of the social order. The complexity of morality reflects the complexity of human social interaction. Whether or not non-human animals have a moral outlook is a controversial but, nevertheless, challenging avenue for research. (DL)

Morbidity

Morbidity refers to the incidence of **disease** or sickness (from the Latin *morbidus*, meaning 'sick'), and can be interpreted as the number of animals showing signs of disease at any one time (prevalence) or the rate at which animals become sick (incidence). It is used as an indicator of the level of infection. It can also be used as an indictor of animal **welfare**, as diseased animals commonly have poor welfare because disease can cause **pain**, **suffering**, **distress** and lasting harm. Some diseases cause poorer welfare than others, particularly those that affect bodily functions. For example, a disease that causes severe diarrhoea or pneumonia is likely to cause poorer welfare than those causing a slow-growing subcutaneous cyst or skin wart. The impact of a disease on an animal will directly affect its physiological well-being. It will also affect its psychological well-being if an animal is unable to carry out normal activities such as **grooming**, **playing** or **hunting** – that is, it may cause mental distress as well as physical impairment. (DBM)

Morgan's Canon (Lloyd Morgan's Canon)

Morgan's Canon (1894) describes an application within the discipline of **psychology** of the scientific principle often referred to variously as the Law of Parsimony, the Principle of Parsimony ('seek the simplest explanation first'), Ockham's (or Occam's) Razor ('entities should not be multiplied unnecessarily'), in which it is stated that when there is more than one sound explanation for a phenomenon it is preferable to choose the simplest one.

Literally, the Canon states that 'In no case may we interpret an action as the outcome of the exercise of a higher psychical faculty, if it can be interpreted as the exercise of one which stands lower on the psychological scale'. This declaration was made in response to what Morgan saw as the anthropomorphic anecdotal approach to studying animal behaviour and the attribution of mental states to animals with regard to behaviours that might also be explained by simpler processes such as reflexes.

The Canon was used as one of the underpinning scientific principles fundamental to the development of **behaviourism**, although, contrary to the impression given in some texts, Lloyd Morgan himself appeared not to subscribe to a purely behaviourist school of thought, as he also states in the same text that 'The kitten has an impression of the ball with which it is playing, and the hungry dog may have an idea of a nice meaty bone', implying a belief in more complex cognitive concepts in these species. This highlights one of several problems with the application of the Canon, namely the level of evidence required before one should conclude that competing explanations are of equal merit. A further difficulty arises from the definition of a **hierarchy** of 'psychical faculties', since these may depend on the **behavioural ecology** and **phylogeny** of the species as well as its neuropsychological development. (KT, DSM)

See also: Anthropomorphism; Cognition; Comparative psychology; Emotion

Further reading

Hayes, N. (1994) *Principles of Comparative Psychology*. Psychology Press, Hove, UK.

Morgan, C.L. (1894) *An Introduction to Comparative Psychology*. W. Scott, London.

Mortality

Mortality literally means 'able to die', but more commonly it means the death rate in relation to the population at risk, i.e. it is a frequency measure of **death**. The reasons why animals die are surprisingly few from a physiological viewpoint, although there are many indirect causes. The direct causes of death are cessation of the heart, leading to a lack of oxygenation of the heart and **brain**, or destruction in some way of the vital centres of the brain in the brain stem that control respiration and **heart rate**. In humans it can be more complex, as a person may have had the brain stem destroyed (i.e. be brain stem dead), but the vital functions of breathing and oxygenation are taken over, e.g. by a ventilator and forced respiration. In such cases the mid- and forebrain may still function, but such treatment may become futile as there is no chance of recovery and so, at some point, the decision has to be made to stop such rescue 'treatment'. (DBM)

Motivation

Motivation is the process within the **brain** controlling what behavioural changes occur and when. It can be thought of as the summing of various causal factors that contribute to an animal's decision making, resulting in it behaving in a particular way. These factors combine together and, based upon the animal's assessment of its total internal state by the brain, incorporation of likely future demands and of new demands to be created by any given course of action, the animal chooses what to do next.

Internal and external causal factors

In the process of decision making, an animal is likely to receive a wide number and variety of inputs. For example, consider an animal searching for food. A number of factors could influence how this behaviour is expressed. There might be sensory input to the brain from the external environment, e.g. when a potential food item is detected. There will be internal input from body monitors such as blood nutrient levels or gut distension. There could be internal input from oscillators within the body, producing an output after a particular time and indicating normal feeding time or interval since the last feed. Nevertheless, there are likely to be other factors involved in the motivation process. The animal might have an injury, resulting in licking rather then food searching. There might be input about the presence of a potential mate or predator, which again leads to some other activity being given priority over food searching. Fluctuations in hormonal status will also change the likelihood of occurrence of various behaviours.

Each of the above factors will also be influenced by the animal's previous experience. For example, the smell of a food that the animal has experienced previously but found distasteful might reduce the probability of the animal initiating food searching or, the gut might be empty but experience shows that the gut has to be empty for several hours before food will arrive. Each input to the brain must be interpreted in relation to previous experience. Some inputs will never reach the decision-making centre in the brain because the interpretation results in their relevance being assessed as zero. The actual inputs to the decision-making centre, which are the interpretations of a wide variety of external changes and internal states of the body, are called causal factors.

Motivational state

The traditional view of motivation is that a change in the animal's internal state is sensed by the brain and leads to a build-up of drive to perform the appropriate behaviour. The drive gives rise to **appetitive** and **consummatory behaviour**. Appetitive behaviour involves searching for suitable external stimuli; when these are encountered consummatory activity, such as eating or drinking, occurs. The term '**drive**' was introduced as an alternative to the concept of instinct and was intended to distinguish between the energizing (drive) and the directing aspects of motivation, although for a range of reasons (e.g. see 'feed-forward' below), the concept has tended to be dropped over recent years. In the traditional view, behaviour was driven from within, and there is a direct causal relationship between the strength of drive and the properties of the ensuing behaviour.

In the modern view, the relationship between motivation and behaviour is not necessarily direct. Although particular combinations of factors can give rise to a given tendency, there is not necessarily a direct relationship between the tendency and the observed behaviour. For example, an animal might have a tendency to feed that is stronger than any other tendency but, in the interests of some long-term strategy, it may postpone feeding; or an animal may feed even though its feeding tendency is not the strongest tendency (e.g. as a **displacement behaviour**). This modern view has given rise to the concept of motivational state.

At any particular time the animal is in a specific physiological state, which is monitored by the brain. The behaviour that is expressed is determined by the brain as a result of monitoring these states, in combination with the animal's perception of environmental stimuli. The combined physiological and perceptual state, as represented in the brain, has been called the motivational state of the animal (McFarland, 1993, p. 306). It includes factors relevant to developing activities, as well as to the animal's current behaviour. It is thus fundamentally different from the old concept of drive that it has come to replace. The animal's motivational state can be represented by a point in a multi-dimensional motivational space. The multiple axes of the space are the important motivational stimuli, such as the degree of thirst or the strength of some external stimulus. This state–space approach is useful in describing complex motivations that the more 'generalized drive' concept is not. For example, animals might not be simply hungry, they might be hungry for a particular food type. In the state–space approach, the *x*-, *y*- and *z*-axes might represent the strength of motivation for fat, protein and carbohydrate, respectively. Any point within the state–space will represent the animal's overall tendency to eat food of the particular type.

Models of motivation (Manning and Stamp-Dawkins, 1992)

Perhaps the most famous model of motivation is **Konrad Lorenz**'s 'psychohydraulic' model. This draws an analogy between the way a water tank with a spring valve operates (e.g. immediately after flushing it cannot work until the water has had time to build up) and animal behaviour (e.g. immediately after feeding, it may be more difficult to elicit more feeding behaviour). This model was inspirational in its time, but is now considered to be unsatisfactory. For example, as long as the tank has not been recently emptied, it should be able to discharge water and should only cease to be able to do so when all its water has been used up. If animals operated in this way, they would stop behaving in certain ways only when their motivational equivalent of water has been used, i.e. through previous performance of behaviour.

However, there are a number of cases where behaviour ceases without the full sequence having been performed. For example, if food is placed directly into the stomach of a hungry dog, thus by-passing the act of eating, the dog does not eat nearly so much when subsequently offered food. It is the stimuli from the distended stomach, not the act of eating, that is the most important in switching off feeding behaviour. The major problem with Lorenz's 'psychohydraulic' model is that it does not incorporate a feedback function. Many behaviours such as **feeding**, drinking and nest building are affected by what the animal has done earlier, through the animal's own actions on its environment. These feedback effects have now been incorporated into a number of motivational models under the general heading of homeostatic models.

The body is constantly adjusting itself to cope with changes in the external environment. It maintains relative stability despite sometimes massive fluctuations in the outside world. This stability of the internal environment is termed **homeostasis**. The starting and stopping of behaviour can be seen as part of the homeostatic process. On a hot day, body fluids decrease; this stimulates the animal to drink, which helps to correct fluid loss, returning the body fluids to some 'ideal' or normal value. Homeostatic models assume that there is an ideal state or set point for the animal. If there is a difference between the set point, e.g. for body fluids, and the actual state of the animal, this difference is said to provide the motivation for drinking. This is described as a 'negative feedback' loop because error is 'fed back' into the system.

In many instances, negative feedback models of motivation seem to provide a good analogy to the behaviour of a real animal. Rats injected with salt, which dehydrates them, drink just enough water to restore their fluid balance. The effects of placing food directly into the stomach, which Lorenz's model does not explain, are also understandable as a homeostatic model: loading the stomach with food is enough to 'turn off' eating, because a full stomach is normally part of the negative feedback loop of the homeostatic feeding system.

We should not think, however, of an animal's behaviour as being controlled by something analogous to a simple thermostat. Factors influencing motivation can be considerably more sophisticated. If humans are deprived of water and then allowed to drink, they will drink almost all that is needed to restore fluid balance within 2.5 min, although changes in plasma dilution cannot be detected for 7.5 min and are not back to normal for about 12.5 min. Studies have shown that water is detected in the mouth and throat during the course of normal drinking and reduces the subsequent tendency to drink. Distension of the stomach and stimulation of the intestine also play a part in terminating drinking. The exact stimuli influencing even the same behaviour differ between species. For example, in dogs gastric distension is relatively unimportant for drinking, but it is very important in monkeys. Drinking can also occur because of an anticipated fluid deficit. When given dry food that makes them thirsty, rats will drink in anticipation of this, a phenomenon called 'feed-forward'.

Animals use a variety of cues and previous experience to predict that an internal state will depart from the tolerable range, and they behave in a way that prevents this from happening. When feed-forward control is very efficient, an observer may be unaware that any change in state would have occurred, because the action compensates for it exactly.

Simple homeostatic models cannot explain all aspects of behaviour. Animals fed to satiety on one type of food will often resume feeding if given a wider range of foods. Under experimental conditions this can be so excessive that animals will eat until they are grossly obese.

It has been argued that there should be a distinction between behaviour that is homeostatic and behaviour that is not. This separates behaviours such as feeding and drinking, where there is an identifiable set point, and other behaviours such as **aggression** and sex, where it is difficult to see what the set point would be. Others have argued that there is not really a fundamental difference between the two categories, because all behaviour has consequences for the internal state of the animal performing it.

Two behaviours that have been widely studied and that have allowed us to develop understanding about causal factors are **dust bathing** by chickens and nest building by sows. If hens are allowed free access to a suitable substrate, they will dust bathe with a clear **diurnal rhythm** approximately once every 2 days. If they are deprived of a suitable substrate, the intensity of dust bathing is greater when access is reinstated (**see: Rebound behaviour**), suggesting that there is an increase in the motivation to perform dust bathing and therefore an influence of internal causal factors. In addition, hens will explore more in the absence of dust-bathing substrate, and can learn an association between a colour cue and a suitable substrate, both indicating that chickens can be motivated to perform dust bathing in the absence of substrate.

External causal factors also affect dust bathing. Chickens provided with a finely structured substrate, such as peat or sand, perform more dust bathing than if they are provided with, for example, wood shavings or feathers, although experience can influence preferences. Heat, light and the sight of other hens performing dust bathing are other external factors that can increase the probability of dust bathing occurring. One model of dust bathing behaviour is that internal factors build up with time since the previous performance, whereas the **circadian rhythm** sets the threshold for the behaviour to be initiated. External factors (heat, light, **social facilitation** and substrate) affect only the threshold.

An alternative model based on the theory that behaviours are regulated through motivational affective states is more consistent with an 'opportunity' model. This predicts that, given the opportunity, hens will dust bathe and feel pleasure from doing so but, in the absence of external stimuli, dust bathing motivation will be low. The reality probably lies somewhere between the two models, and the relevance of the internal and external factors, respectively, may reflect an individual's situation in that specific moment in time (Olsson and Keeling, 2005).

Measuring motivation

It is important to measure motivation for two reasons. First, motivations for different behaviours often interact with each other. If we are to understand these interactions and attempt to predict when a particular behaviour might occur, we need to measure not just the strength of its own causal factors, but also the relative strengths of the motivations for other behaviours. Secondly, measuring motivation can have a practical application. Many people are concerned that the **housing** and husbandry systems we have for **captive animals** deprive them of performing much of the behaviour they would be able to do in the wild. If we can measure the strength of motivation to perform the behaviours the animal is 'deprived' of, we might be better able to understand the **welfare** implications. It is usually impossible to directly measure the internal state of an animal, but we can measure responses that are indicative of this.

Amount of activity performed

A straightforward way of measuring motivation is to give the animal the opportunity to perform a response and record how much or for how long the behaviour is performed. We can record the number of drinking movements or, alternatively, record the amount of water drunk by the animal. We can measure the motivation for dust bathing by recording the length of dust bathing or the number of dust-bathing bouts. This can be recorded at various times of the day to measure rhythmicity of motivation, or after periods of deprivation. Many complex behaviours, when they have not been elicited for some time, have a lower threshold and are performed at a high intensity when they are finally elicited. This is sometimes called '**rebound behaviour**'.

Aversiveness of stimuli

If we attempt to prevent the animal from performing a behaviour, we can measure how much the animal will persist despite this. One example would be to taint food or water so that it tastes unpleasant. The concentration of the taint can be gradually increased until the animal eats or drinks less, or stops altogether, thereby measuring the motivation to eat or drink. A variation on this is to place an **aversive** obstacle between the animal and an aspect of the environment with which the animal is motivated to interact. This obstacle might be a traverse of water, an electric grid, a blast of air, a narrow gap, etc.

Operant behaviour

Animals can be trained to press on bars, peck at keys, pull on chains, etc. to receive a reward or gain access to a section of environment that allows them to perform a particular behaviour. The animal does not usually receive the reward or gain access for each operation on the manipulandum (alterable characteristic), but might receive it after a pre-set or variable number of operations, or after a period of time has elapsed subsequent to the operation. The rate at which animals operate the manipulandum can be used as a measure of motivation. For example, under some **reinforcement** schedules (i.e. variable interval), the rate at which water-deprived rats will bar press is highly reliably related to the length of time that the animal has been deprived of water. Consumer demand studies **(see: Economics of behaviour; Elasticity of behaviour)** take this technique further by increasing the number of operations the animal must perform to receive the reward or gain access to the desired environment. The number of required operations is plotted against the number of reinforcements that the animal receives.

Vacuum activities and stereotypies

When animals are highly motivated to perform a behaviour, this behaviour is sometimes expressed even when the appropriate stimuli are not present. One example of this is dust bathing in hens. Hens kept in wire-floored cages sometimes perform all the movements of dust bathing, even though there is no substrate for them to dust bathe in. This has been termed vacuum or sham dust bathing. The term 'vacuum' is used to indicate that the behaviour is performed in the absence of suitable stimuli, or at least with minimal stimuli, and it suggests that the animal is highly motivated to perform the 'deprived' behaviour.

Many animals confined over periods of time can develop **stereotypies**. The presence, frequency and intensity of stereotypies in the behavioural repertoire of captive animals has been taken by many people to suggest that the animals are highly motivated to perform some behaviour they are unable to carry out in the confines of a cage, but the aetiology of stereotypies is complex.

In summary, there are several ways of measuring motivation. We might think that they are measuring the same aspect of behaviour and in a way that we might expect would change in the same way. However, studies measuring three aspects of feeding motivation show that this is not necessarily the case. In hungry rats, the amount of food eaten, the amount of quinine accepted and the rate of bar pressing do not all rise together. Over the range 0–54 h of food deprivation, the amount of quinine that rats will accept steadily rises, as does their rate of bar pressing, but their food intake reaches a maximum after only 30 h and actually declines slightly thereafter. Thus, a rat appears to be becoming 'hungrier' in that it will accept food that is increasingly bitter, and yet it eats less. 'Motivation' is a term that attempts to integrate a range of causal factors, including the animal's internal state (hormones, nervous inputs, etc.), external changes and previous experience, so it is hardly surprising that these various components do not always change in exactly the same way. (CS)

See also: Motivational state

References and further reading

Fraser, A.F. and Broom, D.M. (1997) *Farm Animal Behaviour and Welfare*, 3rd edn. CAB International, Wallingford, UK.

Manning, A. and Stamp-Dawkins, M. (1992) *An Introduction to Animal Behaviour*, 4th edn. Cambridge University Press, Cambridge, UK.

McFarland, D. (1993) *Animal Behaviour: Psychobiology, Ethology and Evolution*, 2nd edn. Longman Scientific and Technical, Harlow, UK.

Olsson, I.A.S. and Keeling, L.L. (2005) Why in earth? Dust bathing behaviour in jungle and domestic fowl reviewed from a Tinbergian and animal welfare perspective. *Applied Animal Behaviour Science* 93, 259–282.

Motivational state

Motivation is the product of an assortment of stimuli involved in leading an animal to perform certain behaviour patterns towards a goal. An animal's motivational state determines the likelihood of an animal performing one behaviour pattern rather than another at any point in time. Motivational state is influenced by both internal factors, such as gut fill, and external factors, such as the sight of food, although the relative importance of internal and external factors varies between different types of motivated behaviour. The behaviour pattern with the strongest motivation is the one that the animal will direct the most time and effort towards trying to achieve.

Animals in **captivity** have surroundings quite different from their wild counterparts, which can be quite stressful and may only allow a limited number of behaviour patterns to be performed. The **frustration** or thwarting of strongly motivated behaviour patterns is **aversive** and can lead to behavioural changes and, ultimately, **stereotypies**. There is evidence for two types of motivational explanations for stereotypic behaviour: (i) that they derive from **redirected** or **vacuum behaviour** performed by frustrated or thwarted animals; and (ii) that they derive from escape attempts shown by frustrated or thwarted animals. To decrease the performance of such behaviour patterns, it is important to house animals in environments that allow them to fulfil the behaviour patterns that they are most strongly motivated to perform. (LMD)

Mouse

Mouse is the common name that is generally, but imprecisely, applied to **rodents**, found throughout the world, with bodies less than approximately 12 cm long. In a scientific context, mouse refers to any of the 38 species in the genus *Mus*, which is the Latin word for mouse. The **welfare** of mice and range of actions deemed acceptable towards them vary enormously according to whether they are considered a **pet**, **laboratory animal** or pest species. (CS)

See also: Vole

Mule

A mule is a cross between a male ass, *Equus asinus* (jack), and a female **horse**, *Equus caballus*. The much less common offspring of male *E. caballus* and female *E. asinus* (jenny) is called a hinny. Although genetically identical, these animals may differ in size at maturity as a result of maternal effects that include early nutrition and, to a lesser extent, uterine volume.

Mules have an appeal as working animals because they are generally larger than regular donkeys and so can haul heavier loads. Their donkey heritage means that they are more resistant to dehydration than horses. From a **welfare** perspective, this may mean that they are exposed to workloads that would overwhelm a horse of equivalent size, and the behavioural characteristic often labelled 'stoicism' – the absence of resistance – is a trait that may predispose the mule (and indeed donkeys) to abuse.

Although irregular, female mules commonly have **oestrus** cycles and often have follicles that ovulate and luteinize, but they generally lack oocytes. The degree of functional infertility in mules varies considerably (Zong and Fan, 1989) and, very occasionally, they even bear live foals (Ryder *et al*., 1985).

The differences between the **ethograms** of donkeys and horses should be considered when **handling** and training donkeys and mules. For example, donkeys tend to have different responses to **fear**-inducing stimuli, tending to freeze more readily than horses (a distinct advantage if they become stuck in barbed wire). Other differences relate to the amount of training and human contact experienced by the individual – donkeys and mules have often been exposed to fewer controlled learning experiences than horses.

Hoof trimming is critical in many donkeys and their hybrids on good pasture, because their arid origins prepare them for substrates that abrade the solar surface of the hoof. In the absence of this sort of wear and tear, hooves can rapidly become overgrown to the extent that they disturb the hoof–pastern axis and affect **gait**, thus compromising the animal's ability to move about. To compound the problem, many handlers find difficulty in administering pedicures since these animals have tremendously good balance and are often reluctant to raise their limbs on demand. Early and sustained training to pick up the feet is therefore desirable. (PDM)

References

Ryder, O.A., Chemnick, L.G., Bowling, A.T. and Benirschke, K. (1985) Male mule foal qualifies as the offspring of a female mule and jack donkey. *The Journal of Heredity* 76, 379–381.

Zong, E. and Fan, G. (1989) The variety of sterility and gradual progression to fertility in hybrids of the horse and the donkey. *Heredity* 62, 393–406.

Mulesing

Mulesing is the removal of flaps of skin around the hindquarters of a young **sheep**, particularly of the Merino breed, which can become soiled and predispose them to **fly strike**. It is a common practice in several parts of Australasia as a result of the prevalence of this breed in these areas. Although the consequence of the procedure is an effective deterrent, it is a cause for concern as it is undoubtedly painful and not normally undertaken with any form of **anaesthesia** or **analgesia**, nor necessarily antiseptic, and may be undertaken by trained non-veterinarians. It is being phased out under animal welfare legislation in many regions, but is maintained in others on the basis of a cost–benefit risk and financial analysis (**see: Ethics**). Some countries have also banned the import of wool from mulesed individuals as a result of **welfare** concerns arising from the practice. (DSM)

Mutation

A mutation is a change in a segment of DNA. Many **mutations** result from mistakes in the replication of DNA. Some mistakes are quickly corrected by the cell, but those that are not are passed on to the cell's descendants. If a mutation occurs in a cell that forms sex cells (germ line mutation), it can be passed on to the next generation. If it occurs in other cells (somatic mutation), the stage of the animal's development when the mutation occurs will dictate how much the animal is affected.

The simplest type of mutation involves the substitution of one base for another. If this occurs within a coding region of a gene, it can change a functional codon to a stop codon (nonsense mutation), result in the substitution of one amino acid for another (mis-sense mutation) or, because of the redundancy in the genetic code, it can result in a different codon that still signifies the same amino acid (silent mutation). Insertions or deletions of other-than-three bases in a coding region, which result in a change in the reading frame, cause the ribosome to read a completely different set of codons downstream (frameshift mutation), and hence typically result in a non-functional peptide. Mutations occurring outside coding regions can also have adverse effects (e.g. if a mutation occurs at an exon–intron junction, it can 'derail' the splicing of introns, resulting in a very abnormal peptide).

If a mutation exerts its effect in heterozygotes, it is said to be dominant; if it has an effect only in homozygotes, it is said to be recessive.

It is very important to realize that mutations are natural phenomena, and, even though they occur very rarely (with a frequency of around one in a million), they have been occurring for so long that they are ubiquitous: almost every animal (even the purest of pedigreed animals) is carrying several lethal recessive mutations. It is, therefore, futile to think of eliminating all recessive mutations. (JA, FWN)

Mutations

Behaviour

Normal behaviour is the result of the action of many physiological processes. Although knowledge of these processes is far from complete, it is evident that many peptides are involved in the regulation of these processes. Each peptide is encoded by a gene, and is under the control of other genes that determine when and where that peptide is transcribed and translated. A **mutation** in the encoding gene can result in an imperfect form of the peptide or the absence of the peptide; a mutation in any of the controlling genes can result in the peptide being produced at inappropriate times and/or places; or not being produced at all. As a consequence, one or more physiological processes may be disrupted. If a disrupted physiological process affects behaviour, then **abnormal** behaviour may result. Because the mutated gene may be passed on from generation to generation, the resultant abnormal behaviour can also be inherited. In this way, mutations can give rise to inherited behavioural abnormalities.

Many such mutations have been documented in **laboratory animals**. For example, there are more than 100 behavioural mutants documented in the mouse; these have been given names such as Bronx waltzer, dancer, disoriented, dervish, Jackson circler, pirouette, pivoter, twirler and waddler. In non-laboratory animals, there are far fewer reports, reflecting lack of research rather than absence of mutations. This entry focuses on behavioural mutants that have been reported in non-laboratory animals.

Welfare

Most mutations are deleterious in the sense that they result in abnormal biochemical/physiological processes, which in turn can result in 'sickness', often leading to premature **death**. Hence, many mutations have adverse effects on **welfare**. However, mutations are natural phenomena and are *very* widespread; in fact, almost all animals (including the purest of pedigreed animals) are carrying several recessive lethal mutations. Consequently, it is not realistic to aim to remove all deleterious mutations from a population, since the only way to do this is to eliminate the entire population. Fortunately, most mutations are rare, due largely to natural selection against them. When they appear in a family that comprises otherwise highly favoured animals, it is possible to eliminate recessive deleterious **phenotypes** from a population without eliminating the mutations themselves: all that is required is for each mating to involve at least one parent that is homozygous for the normal allele. Some typical examples of these types of inherited disorders are described later.

One particular type of welfare problem exists where a deleterious mutant is maintained in a population because of **selection** favouring heterozygotes for that mutation. In such cases, society at large has to decide whether, on balance, the advantages inherent in the heterozygote are sufficient to justify deliberate selection to maintain the deleterious mutation in the population. In the interests of animal welfare, breeders must be prepared to accept that their favourite mutation may have sufficiently severe welfare implications to warrant its elimination.

Behavioural mutation

One common mutation having a major effect on behaviour in animals involves roller and tumbler pigeons, which have been recognized (and bred) for many hundreds of years. Pigeons with this trait perform backward somersaults either in flight or on the ground, either once or twice (tumblers) or repeatedly (rollers). This behaviour is most frequent in parlour tumblers or rollers, in which any attempt to fly results in tumbling or rolling on the ground. Neither the gene nor the causative mutation has yet been identified.

Structural mutation

Munchkin cats

Munchkin **cats** are characterized by short legs. This short-leggedness is caused by a natural mutation that has been kept in the population by selective breeding for the short-legged phenotype, which is a form of disproportionate dwarfism (**see: Miniature breed**). The gene causing this phenotype appears to be an autosomal dominant, but reliable data have not yet been published. The short legs of these cats do not affect their climbing ability, but do seem to affect their ability to jump. The limited evidence available suggests that Munchkin cats are similar to Dachshund and Corgi **dogs** and, if this is confirmed, the welfare implications for Munchkins should be similar to those for short-legged dog breeds.

Twisty cats

Twisty cats have a phenotype characterized by radial hypoplasia: the forelimbs are severely crippled due to the absence of all or part of the long bone from elbow to wrist, leaving either no paws or only vestigial paws. Affected animals hop around much like a kangaroo. Their powerful hind legs give them the ability to jump, but they struggle to land. Cats with this disorder develop sores on their legs and have problems with ingrown toenails. In the absence of published inheritance data, it is not possible to say whether this phenotype is the result of a mutation, but information from breeders suggests that it is.

Deleterious mutations maintained by selection and favouring heterozygotes

Scottish fold cats

Scottish fold cats are characterized by ears that fold forward. This phenotype is due to a dominant mutation that has been selectively bred into a population of cats. In the 1970s, it was recognized that homozygotes for the mutation show severe clinical signs of osteochondrodysplasia, reflecting abnormalities in cartilage maturation, and resulting in an abnormal **gait** that deteriorates until affected animals can no longer walk. More recently it has become evident that a large proportion of heterozygotes are also adversely affected, albeit less severely than homozygotes, but sufficiently to raise welfare concerns. The molecular basis of the mutation has not yet been discovered, but affected homozygotes can be avoided by ensuring that each mating involves at least one non-fold parent. Also, it should be possible to select against osteochondro-dysplasia among the offspring of fold × non-fold matings.

Manx cats

The Manx cat is easily recognized because of its very short or absent tail, which is due to an autosomal-dominant mutation. The tailless phenotype is the heterozygous expression of the mutant gene, which causes disruptions in embryological development, preventing normal formation of the caudal neural tube. Tail length varies from slightly shortened (longie) to the complete lack of tail vertebrae (rumpy). Manx cats also show a range of anomalies of the lower vertebrae and anal region. Abnormal anal and perineal denervation has been shown to cause megacolon and incontinence. Differing degrees of spina bifida have also been noted within the breed, as have pelvic defects, which cause a stilted or stiff-legged gait. Homozygotes die before birth: the moribund **fetuses** can easily be recognized during **gestation** as being abnormally small and globoid, with gross malformations of the **central nervous system**. As with fold cats, it may be possible to select against deleterious aspects of the phenotype among the offspring of Manx × non-Manx cats.

Bob-tails

A mutation giving rise to naturally short tails has been reported in various dog breeds, including the Pembroke Welsh Corgi. The molecular basis has been shown to be a mis-sense mutation in a gene for a transcription factor. In a notable **breeding** programme, Boxer breeder Dr Bruce Cattanach back-crossed this mutation into his pedigreed Boxers, obviating the need for tail-docking. Like the Manx mutation, this mutation is lethal when homozygous. However, unlike the Manx mutation, homozygous embryos fail to develop at a very early stage following fertilization, to the extent that all available implantation sites are taken up by normal embryos. Thus, mating bob-tail × bob-tail produces normal litter sizes comprising normal and bob-tailed dogs. In the absence of any apparent deleterious aspect of the bob-tail phenotype, Dr Cattanach has mounted a compelling argument that this is one case where breeding a lethal mutation has no adverse welfare implications.

Hyperkalaemic periodic paralysis

Hyperkalaemic periodic paralysis (HYPP) is a muscle disorder that occurs in Quarter **horses**. It is characterized by reduced tolerance to exercise and intermittent muscle spasms, which can result in difficulty in standing. The disorder is caused by a mis-sense mutation in the gene for the alpha subunit of the muscle sodium channel: a base substitution of cytosine to guanine in the gene causes an amino acid change from phenylalanine to leucine in the transmembrane region of the resultant peptide. This change increases the permeability of sodium across the skeletal muscle cell membrane. Homozygotes show severe episodes of muscle spasm, sufficient to raise welfare concerns. Heterozygotes show just sufficient muscle tone to be attractive in the show ring. Knowledge of the molecular basis of the disorder means that a DNA test is available to distinguish heterozygotes from homozygotes (normals). The welfare concerns can be eliminated by using this test to ensure that each mating involves at least one homozygote-normal parent.

Deleterious mutations that occur naturally at low frequencies

The following represents a small sample of the many deleterious mutations that have been documented in domesticated animals. In each case, the causative mutation has been identified at the DNA level, thereby providing a DNA test for heterozygotes, which in turn provides a simple means of eliminating the disorder – simply ensure that each mating involves at least one animal that has no copies of the deleterious mutation.

Gangliosidoses

Gangliosidoses are a group of autosomal-recessive lysosomal storage diseases. They are due to defective degradation and hence accumulation of gangliosides in the central nervous system. The disorders have been noted in many different species, including cats, dogs, **sheep**, **cattle** and emus. The mutations responsible for these disorders have been characterized at the molecular level in a few of these species. There are two main groups of gangliosides: GM1 and GM2. GM1 gangliosidosis is caused by the deficient activity of lysosomal acid β-galactosidase. Affected dogs have progressive neurological dysfunction and stunted growth and die prematurely. In the Portuguese water dog, the causative mutation is a base substitution of guanine to adenine in the β-galactosidase gene, causing an amino acid change from arginine to histidine in the resultant peptide. GM2 gangliosidosis results from a deficiency of the enzyme β-hexosaminidase. Clinical signs are similar to those for GM1 gangliosidosis, namely progressive neuromuscular dysfunction and impaired growth from an early age. There are two main isozymes of β-hexosaminidase, namely hexA, composed of an α and a β subunit, and hexB, a dimer of β subunits. In cats, GM2 gangliosidosis is caused by a frameshift mutation in the gene encoding the β subunit. As expected, this results in the loss of both the hexA and hexB isozymes.

Glanzmann's Thrombasthenia

Glanzmann's Thrombasthenia is an autosomal-recessive bleeding disorder. It is caused by deficiency of a platelet membrane glycoprotein, which comprises two subunits, αII_b and β_3. A mutation in the gene for the αII_b subunit is the cause of the disorder in two breeds of dogs. In the Otterhound, a mis-sense mutation in the calcium-binding domain causes the substitution of histidine for aspartic acid within the highly conserved calcium-binding domain, resulting in severe destabilization of the glycoprotein complex. In Great Pyrenean dogs, the disorder is caused by a 14-base insertion within exon 13 of the αII_b subunit gene. This insertion also disrupts the calcium-binding domain; it causes a shift in the reading frame and a premature stop codon.

Factor IX deficiency: haemophilia B

Haemophilia B is an X-linked bleeding disorder resulting from deficiency of the coagulation factor IX. Many different mutations in the factor IX gene have been identified. In Lhaso Apso dogs, the mutation is a deletion of four bases plus a base transition of cytosine to thymine. An affected Labrador retriever was found to lack the gene altogether. A five-kilobase insertion in the exon encoding the catalytic region causes the disorder in Airedale terriers. In pit bull terrier mixed-breed dogs, large regions of the gene have been deleted (approximately 30 kilobases). Yet another mutation in mixed breeds of dogs has also been discovered: a base substitution of adenine to guanine causes an amino acid change, namely glycine to glutamic acid. There are also some cases in other dog breeds where the causative mutation has yet to be identified.

Goitre

Goitre is an enlargement of the thyroid gland that indicates impaired thyroid function. It is seen as a lump at the base of the throat and can become very large. There are two different forms of goitre, one caused by iodine deficiency and the other an inherited form. The latter is found in many species, but the causative mutation has been discovered only in cattle and **goats**. In Afrikander cattle, a nonsense mutation in the thyroglobulin gene causes recessively inherited goitre. The causative mutation is a base substitution of cytosine to thymine, which causes a codon for arginine to change to a stop codon. This truncates the peptide, rendering it useless in its role as precursor to **thyroid hormone**. In Dutch goats, a nonsense mutation in a different exon of the thyroglobulin gene is the cause of the disorder. Just as in cattle, it also involves a base substitution of cytosine to guanine, causing a codon for arginine to be replaced by a stop codon.

Narcolepsy

Narcolepsy is a disabling **sleep** disorder that affects many animals. The clinical signs include daytime sleepiness, sleep fragmentation, abnormal rapid eye movement (REM) sleep, cataplexy, sleep paralysis and hypnagogic hallucinations. A narcoleptic dog is also likely to collapse from excitement when awake, such as when offered a treat. Canine narcolepsy is caused by mutations in the gene encoding a receptor for hypocretin (otherwise known as orexin). Hypocretins are major sleep-modulating neurotransmitters. In Dobermanns, an insertion of a short interspersed nucleotide element (SINE) upstream from a splice site prevents proper mRNA formation, thereby preventing the formation of the essential peptide. Another mutation has been characterized: a guanine-to-adenine base substitution in a splice junction causes an entire exon to be deleted. Both mutations prevent the proper functioning of the peptide, leading to clinical signs of the disorder.

Citrullinaemia

Citrullinaemia is caused by a deficiency of the enzyme argininosuccinate synthetase (ASS), a vital component of the urea cycle. Lack of the enzyme results in failure of the urea cycle, and hence a build-up of excess nitrogen in the form of ammonia. An affected calf is normal at birth, but signs of **depression** occur within a few hours. The animal has an unsteady gait, its tongue protrudes, it wanders aimlessly, froths at the mouth and usually dies within 3–5 days from ammonia poisoning. The disorder is recessively inherited, and is caused by a nonsense mutation in the ASS gene. The mutation is a base substitution of cytosine for thymine in the fifth exon of the gene. This substitution causes the codon for arginine to be changed to a stop codon, prematurely ending the peptide and rendering the enzyme inactive. (JA, FWN)

Further reading

Listing of Inherited Disorders in Animals (LIDA). Faculty of Veterinary Science, University of Sydney, Australia. Available at: http://www.vetsci.usyd.edu.au/lida (accessed May 2009).

Online Mendelian Inheritance in Animals (OMIA). Reprogen, Faculty of Veterinary Science, University of Sydney and Australian National Genomic Information Service (ANGIS). Available at: http://www.omia.angis.org.au (accessed May 2009).

Mutilate

In its broadest sense, the term 'mutilate' can mean to alter part of an animal to achieve a defined goal, such as: (i) perceived improvement in appearance (e.g. ear cropping in dogs); (ii) to modify individual behaviour (e.g. **castration** of male cats); (iii) to obtain a certain production quality (e.g. castration of farm animals for leaner meat or avoidance of taint); (iv) to promote better **welfare** for the individual or its group (e.g. castration of lambs); (v) to prevent disease (e.g. docking of sheep and **fly strike**); (vi) to promote harmonious **pet**–owner interactions (e.g. declawing of cats); (vii) for population control (e.g. spaying of female dogs and cats, especially **feral** cats); (viii) to make better sport (e.g. removal of one eye of birds so they fly round in circles in a shoot; breaking the leg or jaw of badgers when baited by dogs); (ix) to prevent escape (e.g. pinioning of captive birds); or (x) to prevent damage to humans (removal of the canine teeth in captive primates). Other disfigurements that could also be classified as a mutilation may occur when marking animals (e.g. **branding**, ear notching, toe clipping). Others may use the term to refer to changes to an animal that are not justified or justifiable (in the user's opinion) by any benefit gained. (DBM)

See also: Organic farming

Mutualism

Mutualism is the interaction of two individuals from different species where both individuals gain a **fitness** benefit from the interaction. Mutualistic interactions generally involve the exchange of substances or services that the individuals would find difficult to obtain themselves; thus, behaving mutualistically is advantageous to both individuals. Similar beneficial interactions between individuals of the same species are referred to as expressions of **cooperation**. Mutualism is often confused with symbiosis and, while some mutualistic relationships may be symbiotic, not all symbiotic relationships are mutualistic. (LMD)

See also: Altruism

N

Narcotic

Narcotics (or **opioid** analgesics) are most commonly used in the treatment of acute and chronic **pain** (**analgesia**). The pharmaceutical class of opioids consists of many different chemical formulations; however, all are chemically related to morphine. Narcotic medications are highly addictive, and cessation of prolonged dosing of opioid analgesics results in often painful withdrawal symptoms. In order to prevent addiction, narcotics are usually under strict governmental legislative control (e.g. the Controlled Substances Act (CSA) 1970 in the USA or the Misuse of Drugs Act 1971 in the UK) to limit use and availability.

In animal research, narcotics are most commonly delivered via injection, while in human medicine the preferred route is by oral administration. Some narcotics are also available in patch delivery systems or as a rectal suppository. Side effects of narcotics can include drowsiness, nausea, disorientation, dizziness and suppression of respiration and digestion; these side effects must be taken into account when behavioural observations are taken during treatment with narcotics. Tolerance is also another problem of narcotic regimens, occurring over time with prolonged dosing. Tolerance means that the system adapts to the drug and therefore responds less well, requiring increased dosage to achieve a similar affect. The rate at which tolerance occurs, if at all, is specific to the individual, making it difficult to dose consistently in long-term studies.

In animal research, narcotics are most commonly used during anaesthetization and for treatment of pain in animals during and after surgery. Narcotics allow researchers to perform surgical and other procedures on animals while causing them the least amount of pain possible. Narcotics also allow terminal measures and procedures to be taken on animals concurrent with the animals' **euthanasia** following humane protocols. The use of narcotics for the prevention or treatment of pain is generally preferred by institutional animal care and use committees (**see: IACUC**). As narcotic abuse is a common problem with personnel working in close proximity to narcotics, care must be taken to become familiar with and identify the signs of narcotic abuse. (RD)

Needs of animals

Animals have needs to the extent that there are resources on which the animals depend, and the lack of such resources will be detrimental for the welfare or even the survival of the animals. Reference is usually made to both physiological and behavioural needs. Physical needs include appropriate nutrition, water, shelter or other means of maintaining the body temperature within a homeostatic range, and an environment that does not result in serious injuries or infections.

Whereas there is agreement over the physical needs of animals that need to be fulfilled to guarantee survival, the definition of behavioural needs has been more controversial. The initial notion that captive animals need to perform all behaviours in their natural behaviour repertoire was soon replaced by an argument based on **motivation**: that animals need to perform those behaviours that they are strongly motivated for. Whether or not an animal is motivated to perform a behaviour depends on an interplay between the different internal and external factors regulating motivation, making it impossible to produce a catalogue of behaviours that are needs and those that are not.

The evolutionary approach favoured by modern applied ethologists assumes that behaviours that have been crucial for survival in the natural history of a species will have been hardwired into the neural system of the animals, and so also remain strongly motivated in modern domestic animals. Failure to satisfy such motivations is assumed to compromise animal **welfare**, thus these motivations are stronger than mere **preferences** or likings. In addition to the satisfaction of needs in the sense specified here, some authors also refer to a wider notion of needs, sometimes called 'comfort needs'. What they have in mind is the possibility that animals' well-being may be further improved by providing them with resources or opportunities that are valued, although the absence of these does not necessarily result in poor welfare. (AO)

See also: Behavioural need

Negative apperception

Apperception refers to the processes involved in the integration and assimilation of one's own experiences into ongoing cognitive processes (**see also: Metacognition**) – for example, how experience might alter one's beliefs or **perceptions**. Apperception is divided into two processes: positive apperception, in which this process occurs successfully, and negative apperception, in which the process is delayed or unsuccessful. For example, the failure to recognize something for what it is or to learn from experience represents a form of negative apperception. This may be harmful or potentially beneficial to an individual, depending on the particular case. For example, the failure to perceive something as a threat following an **aversive** experience may obviously put an animal at increased risk of harm but, by the same measure, in the case of an individual who is otherwise important to the animal (such as its normal carer), negative apperception after such an experience may help to maintain a functional relationship. Such negative apperception may be explained in terms of prior **reinforcement** schedules or through a more fundamental bias in related cognitive processes, such as **attachment**. (DSM)

Neobehaviourism

The original **behaviourism** proposed by Watson rejected mental constructs in animals, and argued that learned behaviour arose only as a result of the direct **reinforcement** of its components, i.e. this suggests that there was no guiding structure in the organization of behavioural acts. Neobehaviourists allow mental states as possible intervening variables shaping the goal of behaviour, but maintain a focus on the relationships between stimuli and behaviour as the only observable data available for scientific study. (DSM)

Further reading

Amsel, A. (1989) *Behaviourism, Neobehaviourism, and Cognitivism in Learning Theory: Historical and Contemporary Perspectives.* Lawrence Erlbaum Associates, New York.

Neonatal period

Generally speaking, the neonatal period is the period immediately after birth, although it cannot properly be defined and its length significantly differs between various taxa. The neonatal period is characterized by the highest **mortality** rate in most if not all species. It is sometimes used to refer to a **sensitive phase** during which parent–offspring **bonding** and filial **imprinting** are established. (RV)

Neophobia

Neophobia is the instinctive **fear** response to novel stimuli. Specifically, it manifests itself as avoidance of new food and/or environmental conditions (including food). Widely observed in rats, neophobia appears to facilitate the ability to associate eating novel food with consequent adverse effects (**see: Aversion; Aversion learning**). Rats are opportunistic foragers and, given the wide range of food qualities they may encounter and the intensity of poisoning pressure to which they are subjected, neophobia is generally assumed to be of adaptive significance. In fact, neophobia was found to be absent among rats of the Hawea and Breaksea Islands, New Zealand, which were historically uninhabited by humans (Taylor and Thomas, 1993). To overcome neophobia, pre-baiting (exposure to non-toxic forms of the food vehicle) is now employed as an integral part of most rat-poisoning operations.

The transition from feeding from familiar to novel food after a period of neophobia is usually a gradual process. Wild rats may initially sample only a small amount of food, then gradually increase their intake over time. The interplay of conditioned aversion and innate neophobia is complex. Indeed, neophobia may actually assist in the development of conditioned aversion (**see: Avoidance behaviour**), by making it more likely that a rat will remember eating novel food, and thereby associate it with adverse consequences (Macdonald *et al.*, 1999). The gradual attenuation of neophobic behaviour is known as '**habituation**'. A neophobic animal can be habituated to novel food or environmental conditions by repeatedly exposing it to that food or environment, as long as the process is not associated with aversive effects. (RD)

References

Macdonald, D.W., Mathews, F. and Berdoy, M. (1999) The behaviour and ecology of *Rattus norvegicus*: from opportunism to kamikaze tendencies. In: Singleton, G.R., Hinds, L.A., Leirs, H. and Zhang, Z. (eds) *Ecology-based Rodent Management*. Brown Prior Anderson, Melbourne, Australia.

Taylor, R.H. and Thomas, B.W. (1993) Rat eradication from Breaksea Island (170HA), Fiordland, New Zealand. *Biological Conservation* 65, 191–198.

Nesting behaviour

Nesting behaviour is the behaviour carried out by an animal that is directed at building and maintaining a nest – either for its own use as a resting place or as an environment in which to give birth to live young or to lay eggs. Nesting behaviour has been described in insects, fish, mammals, **reptiles** and birds. Within applied animal behaviour and welfare, nesting behaviour is most often discussed in relation to sows around **parturition**, in hens prior to egg laying and in **rodents** in relation to housing as **laboratory animals**.

Sows

When given the opportunity, sows build farrowing nests prior to giving birth. Prepartum nesting behaviour occurs in wild boar sows and also in domesticated sows in semi-natural environments – as well as indoors in housing conditions ranging from restrictive **farrowing crates** to very environmentally enriched pens. The function of nesting behaviour is to create a secure environment for the piglets.

Studies in semi-natural environments (large enclosures with mixed vegetation) have given insight into the natural pre-farrowing behaviour of domesticated sows, which has turned out to be very similar to that of wild boar and **feral** sows. Within the last 1 to 2 days prior to parturition, sows leave their herd: whether the primary motivation is to wander or to leave the herd is unknown. However, **aggression** towards herd members encountered after the herd is left behind suggests that the sows are no longer motivated towards social contact.

The sow often chooses a nest site located just below a slope that is horizontally protected by, for example, a large rock or tree and vertically protected by overhanging branches. Thus, the site is sheltered, while still offering a view of the surroundings. The sow builds the nest by excavating a shallow patch, lining it by scraping grass from the close surroundings into it with her forelegs, collecting nesting materials (e.g. straw, grass, twigs, branches) further from the nest and carrying them in her mouth back to the nesting site. The sow performs many of these gathering excursions and, between excursions, she arranges the materials at the nesting site with nodding, rooting and pawing movements. Nesting can last 2–7 h and is usually complete 1–7 h prior to farrowing. The nest occupation phase lasts until 7–10 days after the birth of the piglets, at which time the sow and the piglets rejoin the herd.

Sows housed in production environments are also motivated to nest build. In outdoor systems and extensive indoor pens with large space and appropriate nesting materials, the behaviour is almost identical to that of sows in semi-natural environments and wild boar and feral sows living in the wild. Even in farrowing crates, which impose severe locomotory restrictions, most of the behavioural elements known from nesting in unrestrictive environments can be observed when the sow is given nesting materials. In crates, as well as in pens, nesting behaviours can also be seen when sows are given no nesting materials at all. The behaviour is then referred to as redirected or in-vacuum nest building.

The control of nesting seems to involve both internal and external regulation, but the exact mechanisms are still unknown. It has been suggested that the hormone **prolactin** is responsible for initiating nest building, because there usually is a concurrent rise in prepartum plasma prolactin and nesting behaviour. However, recent studies have shown that nesting can occur without a rise in prolactin. Other hormones that have been suggested to be implicated in the regulation are **prostaglandin** F2-alpha and **oxytocin**. Several studies have shown that environmental feedback from the nest acts in concert with the hormonal control, e.g. nesting starts sooner when nesting materials are readily available, and may end sooner in relation to parturition when appropriate nesting materials can be used.

The fact that sows perform nesting behaviour in very different environments shows that it is a highly motivated behaviour. This, in combination with the element of internal regulation, has led to the contention that nesting in sows is a **behavioural need**. Hence, the lack of opportunity for prepartum nest building due to locomotory restrictions and/or lack of nesting materials is widely recognized as a **welfare** problem in sows.

Hens

A hen will search for a suitable nest site before laying an egg. In nature, this nest site will usually be some distance from where the other hens are but, under commercial conditions (**see: Laying hen housing**), hens will actively choose to nest close to other individuals. The results from **preference** tests show that breeds, and even individual hens, have specific preferences for colour, types of substrate and light intensity of the nest site. However, in general, hens will prefer an enclosed nest site rather than a more natural nest, and one with deep substrate over one without. Once a preference has been established, the hen will often return to the same location time and time again. There is also a well-known preference for hens in loose-housing systems (in which the nests are placed in rows) to lay eggs in the end boxes. The reason for this preference is not clearly understood.

Experiments have shown that hens that do not gain access to an appropriate nestbox will show more gakel calls (usually thought to be a sign of **frustration**), more pacing and a higher proportion of abnormal eggs (a sign of **stress**) than hens with a nest site. Hens are also prepared to work hard to gain access to a good nestbox. Taken together the evidence suggests that access to a suitable nestbox is a strong need in the hen.

The nest-building behaviour of **chickens** is less well investigated than that, for example, in sows. Nest building in chickens is performed by the hen first glancing at and examining the nest site (this phase might take a long time); she will then sit down and perform a series of movements by which she moves material towards her (bill raking and litter gathering), and shapes the surface under her by rotating her body and scraping her feet sideways.

What little evidence there is suggests that hens will attempt to build a nest irrespective of whether it has a preformed nest or not, i.e. it is the behaviour of building the nest rather than the consequence of the behaviour that is important for the hen. However, in experiments in which hens have had the possibility of choosing a mouldable surface with and without litter, no preference was detected; nor was there a preference for mouldable over non-mouldable nests. The scant evidence, therefore, points to a need for a mouldable or pre-moulded nest, but not necessarily for access to nest-building material.

Laboratory rodents

The common laboratory rodents – **rats** and mice – will both build nests if given the opportunity. Traditionally, laboratory housing has not provided bedding material due to concerns about hygiene and the ability to observe the animals at all times. More recently, however, welfare concerns have driven research into the provision of nesting materials. Mice in the wild build dome-shaped, complex, multilayered nests, but this behaviour is rarely seen in the laboratory, perhaps due to inappropriate nesting material rather than the nest-building ability of the **mouse**. If mice are given shredded paper strips, they can build complex nests. Nests are of poorer quality if nesting material is in the form of tissue paper or compressed cotton squares.

Unlike adult mice, adult laboratory rats without prior experience of nesting materials will not nest build spontaneously when materials are provided. However, wild rats are known to build complex nests from a variety of materials. If rats are given access to materials from birth, compared with receiving their first experience later in life, they will build more elaborate nests and waste less material, indicating that nesting behaviour in rats has a component that is acquired through experience. Overall, provision of bedding for nesting has been linked to decreased variation in measured parameters during experimental studies, which may thereby contribute to a reduction in the number of animals needed for quantifying a given scientific procedure. (BID)

See also: Behavioural need; Enrichment; Redirected behaviour; Vacuum behaviour

Further reading

Arey, D.S. (1997) Behavioural observations of peri-parturient sows and the development of alternative farrowing accommodation: a review. *Animal Welfare* 6, 217–229.

Duncan, I.J.H. and Kite, V.G. (1989) Nest site selection and nest-building behaviour in domestic fowl. *Animal Behaviour* 37, 215–231.

Hess, S.E., Rohr, S., Dufour, B.D., Gaskill, B.N., Pajor, E.A. and Garner, J.P. (2008) Home improvement: C57BL/6J mice given more naturalistic nesting materials build better nests. *Journal of the American Association for Laboratory Animal Science* 47, 25–31.

Hughes, B.O., Gilbert, A.B. and Brown, M.F. (1986) Categorisation and causes of abnormal egg shells: relationship with stress. *British Poultry Science* 27, 325–337.

Jensen, P. (1986) Observations on the maternal behaviour of free-ranging domestic pigs. *Applied Animal Behaviour Science* 16, 131–142.

Neurectomy

A neurectomy is the removal of a section of a nerve in order to destroy the ability of the nerve to transmit sensation from below the operative area (**see: Denerving**). Neurectomy is not often performed in modern veterinary medicine, because non-steroidal **analgesia** makes it possible to relieve **pain** without recourse to surgery. It was formerly one treatment for diseases of the lower limb. Neurectomy was sometimes performed to prolong the usefulness of a **horse** by relieving the pain of a chronic incurable condition. Perhaps more importantly, from a **welfare** point of view, it relieved the pain. The worst sequela to neurectomy is development of an infection that will go unnoticed because the horse does not feel the pain and, therefore, will not limp (i.e. be lame).

The nerve sectioned depends on the **lameness**. For example, a bilateral posterior digital neurectomy is performed when a horse has navicular disease (disintegration of the navicular bone, a sesamoid bone found in the posterior hoof). Sixty per cent of horses are sound (not lame) 1 year after neurectomy. Volar neurectomy is used when a horse has osteoarthritis of the interphalangeal joints (ringbone) or ossification of the lateral cartilages of the third phalanx (sidebone). The volar nerve can be cut just before it divides into its anterior and posterior branches; this will interrupt sensory information from the hoof. Alternatively, the median nerve can be incised, because a branch of the ulnar nerve is spared when a volar neurectomy is performed. Most neurectomies are performed on the forelimb, because that is the site of most chronic lameness; however, the tibial nerve can be incised to prolong the usefulness of a horse with a chronic incurable lameness of the hock or the distal portion of the limb. Infra-orbital neurectomies have been recommended for head shaking in the horse, but the response to treatment is generally unreliable. (KAH)

Neuroethology

Neuroethology refers to the mechanistic study of the biological processes regulating the expression of behaviour, and tends to emphasize the phylogenetic and comparative evolutionary aspects of the neurological systems controlling behaviour. The discipline has its own academic society, the International Society for Neuroethology (see http://www.neuroethology.org for details). (DSM)

See also: Comparative psychology; Ethology

Neuron(e)

A neuron(e) is the basic cellular communication and information-processing unit of the nervous system. Anatomically, a neuron consists of three parts: (i) the cell body (perikaryon or soma), with a nucleus and various organelles; (ii) the axon (nerve fibre), with its ramiform terminals (synaptic boutons); and (iii) the dendrites, with their branches (dendritic tree) and spines. Functionally, a neuron detects and integrates stimuli from the internal and external environments, and transmits the signals from one location to another along its axon, resulting in sensory and motor reactions that permit an animal to appraise, enter and exploit its environment. (H-WC)

Neuronal plasticity

Neuronal plasticity is the capability of **neuron(e)**s to adapt cellular characteristics anatomically and/or functionally in response to changes in the internal and external environments. The plasticity is maintained throughout life, which is a prerequisite for recovery of physiological functions following **injury** or **disease** in the **central nervous system** (CNS) affecting animal behavioural and physiological **homeostasis** and well-being.

Environment plays an important role in the development of the nerve system: for example, in morphology, extending the dendritic trees and/or formation of axon collaterals with boutons and synapses; and in physiology, changing the expression of receptor genes and receptor densities as well as synthesis of neurotransmitters and their release. Compared with tame animals of the same species, wild animals have larger **brains**. This phenomenon is related to differences in the amount of learning: life in the wild requires a much broader repertoire of **adaptive** behaviour.

In **laboratory animals**, more increases in both dendritic ramifications (arbors) and the number of synapses are found in the cerebral cortex of rats raised in a simulated natural environment with ample space and access to toys than in rats raised in standard laboratory cages. Following damage to the CNS, the survival neurons sprout their axon terminals, extend their dendritic tree and/or increase the number of spines, resulting in an increase in conducting or receiving signals per neuron. Through these changes, the lost neurons' functions are replaced by the survival neurons, and the pathophysiological balance is maintained. (H-WC)

See also: Enrichment; Housing

Nicking – tail

Tail nicking is a procedure in which the ventral muscles of a **horse**'s tail are cut (myotomy). The purpose of this is to decrease the ability of the horse to lower or abduct its tail. It is performed on some **breed**s of gaited horse – horses that perform a rack or running walk in addition to the natural **gaits** of walk, trot and canter. After the myotomy is performed, the tail is held vertically by a harness so that it will remain in that position during competition. The procedure is illegal in some countries, there being two **welfare** issues: (i) the **pain** of the initial incision; and (ii) the discomfort of constantly wearing a harness. Apparently, horses learn to roll wearing the harness. The breed to which tail nicking and harnessing is most often applied is the American saddlebred. This breed was developed by crossing Thoroughbreds with Narraganset pacers, a breed that paced (moving both hind and fore limbs forward at the same time). The harnesses reduce comfort, and tail nicking is an unnecessary **mutilation**. (KAH)

See also: Docking – tail

Nociception

Nociception is the underlying mechanism by which an animal feels **pain**. The nociceptive system and the feeling of pain are normally inextricably linked. The nociceptive system comprises three types of peripheral nociceptors that respond to excess pressure (mechano-nociceptors), excessive temperatures (thermo-nociceptors) or chemical release (chemo-nociceptors). They connect to the **central nervous system** by both myelinated (A delta) and non-myelinated (C) fibres.

There are three **neuron(e)**s involved in the transfer of impulses from the peripheral nociceptors to the cerebral cortex, where any pain is actually felt by an individual as a noxious event (**see: Feelings**). First-order neurons synapse (connect with another neuron) in the dorsal horn of the spinal cord with second-order neurons. The nerve axons from the second neuron then cross over to the opposite side of the cord and pass up to the thalamus in the **brain** (forming the spino-thalamic tracts). In the thalamus they synapse with third-order neurons that connect to the cerebral cortex (thalamo-cortical tracts), where pain is felt and located to the originating part or region of the body where the nociceptors were activated. Neurons in the dorsal horn also synapse with descending inhibitory fibres from the brain that modify the passage of nerve impulses to the brain.

These synapses may mitigate the level of pain experienced by permitting (so-called gating) only impulses of a certain frequency to pass on upwards. These descending fibres grow down the spinal cord in the first week or two after birth (this has been demonstrated in **altricial** animals but not in **precocial**). Consequently, as these descending inhibitory fibres have not yet grown down and synapsed in very young animals, it is likely that they are more sensitive to noxious stimuli than older animals and adults in whom the nervous system has matured. Synapses in the thalamus also modify the cerebral response through learning and **memory** and may heighten pain felt in anticipation – e.g. going to the dentist. In the abdominal organs there are stretch receptors that, when stimulated, are able to induce a feeling of pain known as colic. These are routed to the brain by afferent sympathetic and parasympathetic components of the **autonomic nervous system**. (DBM)

See also: Endorphin; Enkephalin

Nocturnal

Nocturnal animals sleep during the day and are active at night, the opposite being **diurnal**. Nocturnal animals generally have well-developed senses of smell and **hearing** and the ability to see in low levels of lighting. Animals may adapt to following a nocturnal schedule for such reasons as to decrease the risk of **predation** or to avoid the heat, and resulting water loss, of the daylight hours. In some cases animals may adopt a nocturnal pattern of activity as either a lifetime or cultural adaptation; for example, some populations of elephants appear to have developed a more nocturnal activity in response to historical **hunting** activity/pressure. (LMD)

Noise disturbance and hearing

The term 'noise disturbance' typically refers to loud anthropogenic sound that is not a natural part of an animal's environment. Noise disturbance is of growing concern among wildlife managers, as the potential impact on the social and ecological needs of both captive and **free-range animals** is integrated into **conservation** and management strategies. Anthropogenic noises that are a by-product of industrial, military, construction or vehicular activity are typical sources that prompt concern, promote research and drive the development of protective regulatory measures.

Chronic or severe noise may negatively impact wildlife in various ways. Richardson *et al.* (1995) defined 'zones of noise influence' as (listed from lowest to highest impact):

- Zone of audibility, wherein an animal hears the noise. This is limited by both normal ambient noise and audibility.
- Zone of responsiveness, whereby an animal responds to the noise either behaviourally (e.g. withdraws from source or changes current behaviour), physiologically (e.g. a change in **heart rate**, breathing pattern or hormonal **stress** response) or is displaced from a location.
- Zone of masking, which encompasses noises that may cover up important sounds such as social **communication**, predator/prey or environmental cues.
- Zone of **hearing** loss, discomfort and **injury**, which includes injuries such as temporary threshold shifts or permanent loss or impairment of hearing. This last zone typically requires close proximity to high-amplitude sound, such as the shock waves produced from explosives.

The zones of noise influence listed above (especially the first three) are limited by audibility of the noise, and so an animal's hearing ability fundamentally determines whether a noise is in fact a disturbance. Hearing is a perceptual phenomenon. Each vertebrate species has a unique capacity to hear sounds that are informative as well as those that are not. The latter type of sound is regarded as *noise*. Hearing is a component of the acoustic communication system that includes: signal propagation (**vocalizations**), signal pathway (habitat) and signal reception (hearing). Hearing allows an individual to receive social signals from **conspecifics**, location information from predator or prey or environmental cues that indicate weather change, topographic features or impending danger. This balance of social, resource and environmental cues is unique from species to species and fosters the development of species-specific perceptual abilities.

Sound is propagated through the environment in the form of pressure waves. These waves may be described in terms of amplitude, perceived by humans as loudness, and frequency, perceived as pitch. An animal's hearing sensitivity is defined as the ability to detect sounds at the lowest possible amplitude over a range of frequencies. Signals that are important to a species must have a frequency content within the hearing range, which in mammals varies considerably by species. Hearing sensitivity is illustrated by a function called an 'audiogram'. The mammalian audiogram is usually shaped like a 'U', with the greatest sensitivity (i.e. best hearing) in the middle of the range.

Audiometric measurements can also be used to develop frequency-dependent weightings of environmental noise. With this information, estimates of exposure can be tailored for particular species and predictions can be made regarding the noise impacts. For example, audiograms for humans have long been established, and these parameters have been incorporated into what is called 'A-weighting'. A-weighting is the standard that organizations such as health and safety bodies use when developing safety guidelines for noise levels encountered in the workplace. In humans, A-weighting reduces the variability in correlations between exposure and responses, improving predictions significantly. An estimation of equal-loudness perception is derived by defining hearing thresholds across frequencies (the audiogram) and then measuring the time it takes for an animal to respond to a sound (response latency).

In nature, animals are exposed to a wide variety of sound sources. The distribution of sound levels at various frequencies is specific to different environments, but a great deal of variability may be expected over the course of an animal's lifetime. From a theoretical point of view, animals are assumed to be adapted to cope with levels of noise typical in their natural environment over a lifetime. Noise stimuli that are not a 'natural' part of an animal's environment may constitute a noise disturbance, especially if the noise **stimulus** is in the range of best hearing or maximum sensitivity for that species.

At extreme exposure levels, noise can damage hearing and compromise intraspecific communication (see the last two zones of influence, above). Based on laboratory studies, chronic exposure to elevated noise may promote a stress response that can inhibit normal behavioural and physiological functions, although the literature is inconsistent and the conditions that give rise to stress are very poorly understood. The results of studies done *in situ* in wildlife habitat have been mixed as well. *In situ* studies have typically assessed the impact of relatively short-term exposure to aircraft, vehicle traffic or construction disturbances on wildlife. As a general rule, the literature suggests that short-term behavioural and physiological responses occur, but does not provide consistent evidence for long-term physiological effects. In cases where population-level effects have been suspected, appropriate controls were not available, so that habitat effects could not be eliminated as an explanation for the correlation between noise and population-level effects. At present, it is fair to say that very little is known about the long-term impact of noise disturbance on populations.

Predictive mathematical models illustrating the possible impacts of anthropogenic noise are useful in assessing the potential negative impacts of noisy activities on animal populations. However, the value of these models is limited by the dearth of knowledge on the hearing capacities of individual species. Additionally, a thorough knowledge of the role of acoustic communication is needed to address the potential impacts of noise on social and reproductive processes. (MAO)

Further reading

Richardson, J.R., Greene Jr, C.R., Malme, C.I. and Thomson, D.H. (1995) *Marine Mammals and Noise.* Academic Press, San Diego, California.

Norepinephrine

Norepinephrine (noradrenaline) is a **catecholamine** and is classified within the amine neurotransmitter group. Upon stimulation of the **sympathetic nervous system**, norepinephrine is synthesized and released from post-ganglionic nerve terminals to bind to α_1-, α_2-, β_1-, β_2- and β_3-adrenergic receptors. Activation of these receptors on or near effector tissues causes smooth muscle constriction or dilation, dependent upon the specific tissue. Sympathetic nerve activation occurs in response to stimuli such as **fear**, **pain**, **anxiety** and **stress**.

In a stress response, norepinephrine released from nerve terminals binds to α-receptors on smooth muscle surrounding the vasculature to cause vasoconstriction in a matter of seconds. Dependent upon the region of visualization, blood flow can be decreased from 'non-essential' tissues, such as the skin or gastrointestinal tract, or blood flow is increased by increasing **blood pressure**. An increase in blood flow is critical in the stress response and, by sympathetic activation, norepinephrine released from **neuron(e)s** in the heart binds to β_1-receptors to cause increased force of contraction, increased conduction velocity and increased **heart rate**. Not all norepinephrine is released from post-ganglionic sympathetic neurons. Upon stimulation of the sympathetic nervous system, the medulla of the **adrenal gland** is stimulated to produce predominantly **epinephrine**; however, epinephrine and norepinephrine are produced in a ratio of approximately 8:1. This norepinephrine is then released into the general vascular circulation, where it can affect all cells containing adrenergic receptors.

Like epinephrine, norepinephrine can be a useful measurement for researchers when assessing animal **welfare**. This is due to the fact that norepinephrine is consistently elevated in circulating blood when animals are exposed to **aversive** stimuli such as electric shock, **restraint** and **injury**. In contrast to epinephrine, which is released directly into the peripheral vasculature, only a minor proportion of norepinephrine is released into the blood, and the total norepinephrine measured in the blood is largely influenced by 'spill-over' from nerve terminals. The measurement of norepinephrine is faced with the same challenges as that for epinephrine. (DCL)

Novel object test

The novel object test is an object **recognition** test used to study learning and **memory**, commonly in rats and mice. The procedure involves exposing the animal to a distinct environment containing two identical objects (such as plastic, glass or ceramic items) for a few minutes. The time spent exploring the sample objects is recorded. Following sample-object exposure, the animal is returned to the home cage for a retention period (usually for between 1 min and 4 h). The animal is then returned to the test environment, in which one of the familiar objects has been replaced with a novel object (or has been displaced in space), and is allowed to explore for 3 min, during which time exploratory and locomotor activity is recorded. Rats and mice have a tendency to approach and explore novelty (neophilia), and the ability to discriminate between a familiar and a novel object (indexed by the time spent exploring the novel object) indicates recognition of the previously experienced object. This test has been used to examine memory function after drug challenges, brain lesions or targeted gene **mutations**, and as a function of **ageing** and development. (SuH)

Further reading

Besheer, J. and Bevins, R.A. (2003) The impact of nicotine withdrawal on novelty reward and related behaviours. *Behavioural Neuroscience* 117, 327–340.

Ennaceur, A. and Delacour, J. (1988) A new one-trial test for neurobiological studies of memory in rats. 1: Behavioural data. *Behavioural Brain Research* 31, 47–59.

Ennaceur, A. and Meliani, K. (1992) Effects of physostigmine and scopolamine on rats' performances in object recognition and radial-maze tests. *Psychopharmacology* 109, 321–330.

Novelty

Novelty means being new or having newness. A **stimulus** is only novel if the animal is not habituated to it. Once a novel stimulus is presented to an animal, the stimulus begins to lose its novelty. It is therefore not an objectively measurable or quantifiable feature of a stimulus in the way we might describe shape, colour or loudness, but instead it relies on the 'interaction between stimulus and perceiver'. A yellow ball will be yellow and spherical to every animal, but only 'novel' to those that have not had exposure to it. Additionally, the degree of novelty may be influenced by the animal's previous experience with similar stimuli, the time since exposure and the duration of that exposure. For example, a yellow ball may be less novel to animals with experience of, say, a blue ball compared with animals that have not been previously exposed to any ball.

Although the desire to seek novelty is a fundamental behavioural tendency in many species, exposure to novelty can induce both negative and positive emotional and behavioural reactions and either avoidance or approach. It has been proposed that the response is dependent on the degree of novelty. First, intense novelty can induce a **fear** response and **avoidance behaviour** (**neophobia**). Mild novelty may cause interest, trigger approach (neophilia) and give positive reward. Low novelty may generate no interest and likewise result in avoidance. Thus, a curvilinear relationship probably exists (see Fig. N.1).

This picture is further complicated not only by the degree of novelty within the environment as a whole, but also by species differences. For example, a novel object encountered in a familiar environment may elicit a different response compared with the same novel object encountered in an unfamiliar environment. Also, depending on the species, the response to a novel object that induces high levels of fear may be quite different. The animal may engage in active defence and approach to attack the novel object; it may engage in passive avoidance and become immobilized, neither approaching nor moving away from the object; or it may engage in active avoidance and actively move away from the object.

Experimentally, in applied animal behaviour, response to novelty has been used extensively to assess fear and also to determine '**personality**' or '**temperament**'. Exposure to novelty has been achieved by either placing the animal alone into a novel arena (an **open-field test**) or presenting a novel object to the animal in a test arena, with the test arena often being the open-field arena but also sometimes being the home pen.

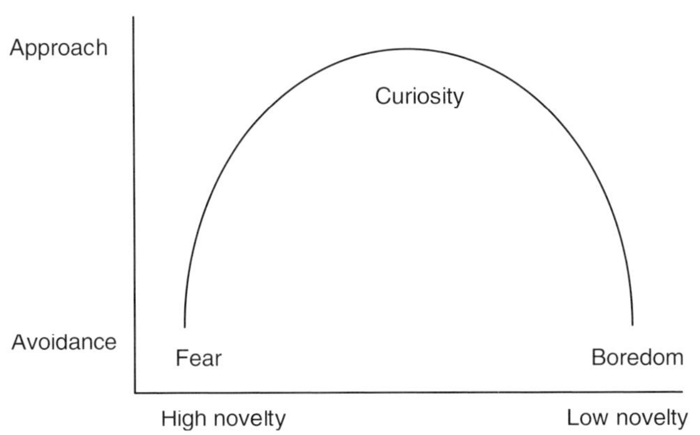

Fig. N.1. Hypothetical relationship between the novelty of a stimulus and approach/avoidance behaviour (from Hughes, 2007).

With open-field or novel-arena tests, the types of measures that are often taken include activity (e.g. number of grid squares crossed), amount of time spent in the pen periphery, amount of time spent in the pen centre, time spent immobile, number of **defecations**/urinations, total distance travelled, number of escape attempts and the number and type of **vocalizations** performed. The behavioural measures may be combined with physiological measures such as **heart rate** and circulating cortisol, **prolactin** and **oxytocin** levels. For novel object tests, all these measures can also be taken and used in combination with measures of direct response to the object – i.e. latency to approach within a certain distance, latency to make contact with the object, number of contacts or total time spent in contact with the object.

In general, novelty tests have been adapted from those used with rodents and applied to other species without validation as to what exactly they are measuring. Behaviourally, farm animals are quite different from rats and mice, and thus the conclusions drawn from farm animals' behavioural responses to novelty cannot be extrapolated from similar responses of rodents. The novel arena is usually an open enclosure, which may be square, rectangular or circular in shape.

The dimensions of pens used experimentally vary quite considerably, being perhaps as small as 1–2 m^2 and up to >200 m^2, largely depending on species but not always. Thus, the arena may be seen by poultry as an environment that could lead to increased **predation** risk, and by ungulates as a confined space. Novel objects that have been used are equally varied, including buckets, balls, traffic cones, umbrellas, etc., but, almost exclusively, objects that have no evolutionary context. Not surprisingly, the reliability and validity of these tests is open to question, and often there has been poor repeatability and low or no correlation between measures both within and between tests. (JNM-F)

Further reading

Forkman, B., Boissy, A., Meunier-Salaun, M.C., Canali, E. and Jones, R.B. (2007) A critical review of fear tests used on cattle, pigs, sheep, poultry and horses. *Physiology and Behaviour* 92, 340–374.

Hughes, R.N. (2007) Neotic preferences in laboratory rodents: issues, assessment and substrates. *Neuroscience and Biobehavioural Reviews* 31, 441–464.

Noyau social organization

Noyau social organization is characterized by a primarily solitary lifestyle, with overlapping home ranges between the sexes. Adult males typically have larger home ranges that encompass the ranges of several adult females. Temporary social groups are formed between a female and her offspring. Synonymous with dispersed mating, noyau is considered one of the simplest social structures and is observed among prosimians and **orang-utans**. (LAN)

See also: Social behaviour

Nursing

The terms 'nursing', 'suckling' and 'sucking' are used inconsistently in applied **ethology**. Quite often, nursing refers to what the mother does when she gives milk to her young, and suckling refers to the behaviour of the offspring. Some authors use the word suckle for the maternal aspect of the interaction, and the word suck for the offspring behaviour. In this article, nursing will be used to refer to the whole of the interaction, i.e. the complex of closely intertwined physiological and behavioural processes in both the mother and the offspring.

Nursing is the natural way for mammalian mothers to provide nutrition for their offspring during the first period of life. In each species, anatomical, physiological and behavioural adaptations have evolved so that the specific nursing pattern is fitted to the life history, the reproduction strategy and the social and physical environment in which the species naturally lives.

Anatomy of nursing

In all placental mammals, milk is produced in milk alveoli, from where it is transported to the teat openings via milk ducts. In relation to nursing pattern, the most important difference in milk gland anatomy is between ruminant species that possess milk cisterns – i.e. caverns that store part of the milk produced between the nursings – and other mammals that do not have cisterns. Usually, the milk stored in milk cisterns can be withdrawn just by stripping (compression of the teat cistern from its base to its teat), combined with sucking (underpressure applied at the opening of the teat canal or milk ducts), whereas the milk accumulated in milk alveoli needs to be expulsed through a contraction of the myo-epithelial cells surrounding the alveoli (milk ejection).

Female mammals have one to eight pairs of mammary glands, and the number of teats normally equals or exceeds the number of young in a litter. This adaptation secures milk supply for all surviving offspring, although competition can still occur for the most productive teats. Artificial **selection** for high litter size can result in a situation (such as the one in the domestic pig) where the number of offspring surpasses the number of teats, leading to more intense competition and increased **mortality** among the young.

The position of the teats corresponds to the usual nursing position of the species, e.g. primates that nurse in a sitting position while their young cling to the ventrum of the mother have sternal glands, whereas ruminants have glands in an inguinal position. In those dairy cattle breeds that underwent an intense period of breeding for high milk yield, the teats of the unnaturally voluminous udder are positioned so low that neonatal calves have difficulty in finding the teats by themselves and often fail to drink the **colostrum** if not assisted.

Physiology of nursing

A specific hormonal state is necessary in the female mammal in order for **lactation** and nursing behaviour to be initiated. **Prolactin** is the hormone needed for lactation initiation and maintenance in many mammalian species. However, once nursing behaviour is initiated, it does not seem to depend directly on the female's hormonal state; rather, it seems to be regulated by chronobiological, motivational, reflexive and cognitive processes.

The release of milk (milk ejection) is caused by the contraction of milk alveoli in reaction to an increase of **oxytocin** in blood. The whole causal chain involves several steps. First, unconditional stimuli (such as tactile stimuli from the pre-ejection teat massage) or conditional stimuli (such as seeing the baby or hearing its voice) are transmitted via the sympathetic nervous connections to the **hypothalamus**. In reaction to this stimulation, oxytocin is secreted from the hypothalamic cells and released to the bloodstream from the rear part of the **pituitary gland**, called the neurohypophysis. Some species such as the pig have a refractory period after each milk ejection, i.e. another milk ejection cannot occur before a certain interval elapses since the previous ejection (about 15 min in the pig). The mechanism of this block resides somewhere in the transmission of the ascending stimuli to the descending hypothalamic command because, during the refractory period, the oxytocin release and milk ejection can be induced by direct hypothalamic stimulation but not by tactile stimulation of the glands.

After emptying of the glands during a nursing bout, milk production in the glands starts anew at the full speed. As the gland gradually refills, the speed of milk secretion slows down. This has the consequence that if intervals between nursings become prolonged, the total milk production (per day) is reduced and the young obtain less milk. When a teat is not used for a certain period of time (usually several days), the milk secretion in the glands connected to it stops altogether – the teat dries up and the process of mammary involution starts. The negative feedback process is mediated through inhibitory substances contained in the milk, reduced blood flow to mammary tissue and the pressure of the accumulated milk on epithelial cells. Nursing behaviour thus has a crucial feedback effect on the physiology of milk synthesis and secretion.

The act of nursing is accompanied by dynamic hormonal release. In the dam, oxytocin, prolactin, **insulin** and somatostatin concentrations increase, peak and then decrease in the blood during or within 60 min of nursing. In the sucking offspring, **glucocorticoids**, **endorphins**, **cholecystokinin**, oxytocin and insulin, among other hormones, are released in connection with nursing. Often it is not (just) the release or ingestion of milk but rather the performance of the nursing/sucking behaviour that triggers the hormonal surge. The hormones released during and after nursing modify the catabolic/anabolic balance of the mother's metabolism, affecting her further milk production; for the young the hormones enhance the digestion of milk; and, in both the dam and the offspring, they provide internal reward, thus increasing further the motivation to engage in nursing.

The mother is alert or even aroused during the nursing in some species like the pig; or she may behave indifferently, e.g. a late-lactation cow that grazes while nursing a calf; or nursing may occur during a somnolent state such as in the rat.

Time structure of nursing

Three hierarchical levels of time structuring of the nursing behaviour can be distinguished. First, a nursing bout has a specific inner time structure related to the timing of milk flow. Secondly, nursing bouts occur at predictable intervals, creating a species-typical rhythm of nursing. Thirdly, nursing behaviour changes during the course of lactation.

Time structure of a nursing bout

The structure of a nursing bout typically consists of: nursing initiation, an introductory pre-ejection phase, a period of milk flow, a post-ejection non-nutritive phase and the termination of nursing. The nursing can be initiated by the young, when they approach the mother and solicit nursing through **vocalizations** and/or attempts to nuzzle the teats; alternatively, the dam can start the nursing through calling the young (e.g. grunting in the sow, bellowing in the cow, seeking them out in **hider** species like the red deer or the rabbit, and assuming the nursing position (lying on the side by the sow; crouching over the young in the rat).

After making contact with the nipples, the young start to stimulate them by nosing, sucking, butting (cattle) and/or pawing (rat). This stimulation usually lasts several tens of seconds to several minutes before it triggers milk ejection. The young then switch to intensive sucking and swallowing of the milk. This nutritive phase of the nursing bout can be as brief as 15 s (in the pig) or as long as 5 min or more in cattle.

In some species such as the rabbit, nursing is terminated by the dam immediately after milk flow ceases; in most species, the young remain in contact with the nipples for a further period of up to several minutes. This post-ejection non-nutritive phase is often characterized by similar types of behaviours to those seen in the introductory phase. The termination of nursing may be gradual, such as when piglets fall asleep at the udder, or abrupt, such as when the sow rolls over on her belly or stands up.

Nursing in some species does not contain all the phases: for instance, in the rabbit the mother enters the burrow and crouches over the pups, which orient to the nipples (using pheromonal and tactile cues) within a few seconds. The milk flow phase starts immediately and is ended abruptly after 3–4 min by the doe jumping out of the burrow. In other species, such as rats or humans, a nursing bout may contain more than one milk flow phase.

The switch from one phase to the other is usually triggered by a specific **stimulus**, e.g. oxytocin is released after there has been enough stimulation by the young, or the intense sucking phase in the calf starts when the first milk is tasted. In some cases, though, the duration of a phase seems to be internally fixed. For instance, once milk ingestion has started in a calf, the very intense **motivation** to suck lasts for a pre-set period of 5–10 min.

The duration of a nursing bout is not a good indicator of the amount of ingested milk. Sometimes the relation can even be negative, such as when low milk intake stimulates the young to prolong the post-ejection phase of the bout.

Nursing intervals and synchronization

The frequency of nursing varies widely among species. Rabbit mothers visit their young once per day to nurse them, whereas a sow gives milk about 30 times a day at the beginning of lactation. The timing of nursing may have a more or less pronounced **diurnal pattern**. For example, lactating cows have a prominent peak in nursing activity early in the morning at the end of the night rest, and otherwise nurse about six times a day at quite regular intervals; a rabbit doe nurses at an individually specific time in the night and the pups prepare for her coming 1 h in advance. In the pig, there is little difference between the inter-nursing intervals during the day and in the night.

In group-living mammals, females tend to synchronize their nursing. For instance, if two lactating sows are housed in one room, they nurse within 1 min of each other in about 90% of cases and, even in larger groups of sows, the nursing synchronization is prominent and much stronger than the synchronization in general activity (**see: Social facilitation**). This synchronization probably evolved to enable group cohesion during alternating of foraging and nursing, but may also be a maternal strategy to reduce allosucking by alien piglets.

Changes in nursing behaviour during lactation

From the biological fitness point of view, nursing is crucially important for both the mother and the young. Maternal investment is very high during lactation (substantially higher than during **gestation**), and this investment decreases her prospect for further survival and reproduction. For the offspring, milk supply is absolutely critical for survival during the early suckling period, but less so later in lactation when they combine foraging for solid food with milk intake. The changing balance between the accumulating maternal investment and decreasing importance of milk for the young is reflected in the nursing behaviour. The frequency of nursing declines during the lactation, and signs of behavioural conflict become more common: the nursings are increasingly initiated by the young, and terminated by the mother; the young attempt sucking even when the mother is busy with another activity; and the mother may nurse in a position that gives less access to the teats for the young, such as the standing posture in the pig.

Communication during nursing

Since nursing is an interactive process, **communication** between the mother and the young is of paramount importance. All sensory channels could be used during the nursing, but acoustic, tactile and olfactory communication prevails. In the mother's acoustic signals, information about her identity, location and timing of nursing phases is often included. For instance, the nursing sow produces typical rhythmical grunts. The inter-grunt intervals and spectral composition of the individual calls contain information about the dam's identity; the abrupt start and end of each call enable an easy localization; and a typical increase in the grunting rate announces an imminent milk ejection. The young most often vocally signal their identity (through frequency-modulated calls that function like 'signatures') or a disruption in the milk intake process ('protest' screams of a piglet that has been displaced from its teat by an intruder). Tactile stimuli (prolonged/intensive teat nuzzling and/or butting) are typically used for the communication of need.

Olfactory communication primarily serves to identify the progeny to the mother. Calves, foals, lambs and goat kids typically suckle in the anti-parallel position, i.e. access the udder from the front while having lateral contact with the mother's body and swinging the tail to increase odour dispersion.

Behaviours that enable identification of the young by the mother are often part of the introductory and milk ingestion phase. For example, calves often pass under the chin of their mothers before going to the udder; piglets wander to the snout of the nursing sow during the introductory phase – and especially just after milk ejection – and make close naso-nasal contacts with the mother (sometimes even putting the nose into the mother's open mouth) while producing specific croaking calls, and mothers repeatedly sniff their suckling young during the nursing bout, concentrating particularly on the anogenital region.

Non-nutritive nursing

As has been already described in the section above on the time structure of nursing, there are non-nutritive phases included in each nursing bout. However, some nursing bouts do not contain any milk let-down and the young are not able to obtain any alveolar milk during those bouts. These non-nutritive bouts may be due to external disturbance in some cases, such as with the **epinephrine**-caused blocking of milk ejection in the cow. In the pig, non-nutritive nursings are part of the normal behavioural repertoire and are probably a way of bargaining the level of milk investment between the sow and the litter.

Allonursing

As a rule, a lactating female would only nurse her own offspring. However, allonursing (giving milk to young that are not the biological offspring of the dam) can occur due to a number of mechanisms. A mother may make a misdirected attachment (sometimes labelled '**mis-mothering**') to an alien young during the sensitive postpartum period and then nurse it as her own offspring. Alternatively, mothers may tolerate the sucking by alien offspring even though they recognize them correctly. For instance, some beef cows allow alien calves to suck together with their own progeny. Thirdly, mothers may aggressively repel alien young, but do not succeed in avoiding all allonursing because it is not possible to check the identity of all young at each nursing (such as is the case in the pig) or because the allosuckers break through the defence from time to time. The latter occurs, for instance, when a calf 'steals' the milk from an alien dam – it is sucking together with the biological offspring, reaching the udder from the side or from behind so that the cow cannot access it for olfactory identification.

One specific form of allonursing is communal nursing. Two female house mice, for instance, give birth in the same nest and then nurse the two mixed litters indiscriminately.

Some forms of allonursing (such as mis-mothering) can be maladaptive, i.e. they decrease the **fitness** of both the dam and the young. Sometimes allonursing is beneficial for the young but not for the female, such as when piglets from a very numerous litter succeed in allosucking at another sow. Finally, allonursing may be an adaptive strategy also for the 'foster' mother due to **reciprocity** (e.g. communal nursing in mice) or due to **kin selection** (e.g. nursing orphaned or weak offspring of a relative female).

Nursing behaviour and welfare

Species-specific patterns of nursing have changed relatively little during the process of **domestication**. In the natural environment, both the female and the young use the available space to assume appropriate nursing/sucking position and to bargain the initiation and termination of nursings through moving towards or away from each other. When movement is restricted in **captivity**, abnormalities can occur, leading to **welfare** problems. Improper design of crates for lactating sows (e.g. horizontal bars low above the floor) prevents full access to the udder and thus disturbs the nursing order among the piglets. Crated lactating sows are not only heavily confined (**see: Confinement**), but cannot walk away when the litter demands over-frequent nursing later in lactation. This results in chronic stress in the sows and forces them to use alternative strategies such as spending a lot of time lying on their belly.

(MS)

Nutritional wisdom

The concept of 'nutritional wisdom' historically has been taken to mean that animals possess the ability to detect and identify internal dietary deficiencies and seek out appropriate food items to correct the internal imbalance. This led to the original hypothesis that animals possessed specific hungers for nutrients, e.g. a hunger for salt or protein.

Influential evidence for this initial hypothesis came from Richter's studies, carried out in the 1930s/40s, on cafeteria feeding with rats. Richter demonstrated that rats immediately altered their food selection patterns in order to rectify a response to an artificially induced nutrient deficiency or a requirement for nutrients through changes in physiological state. Richter's work was later criticized on the basis that animals would require a specific appetite for each of the numerous micro- and macronutrients. However, in partial support for Richter's assumptions, there is evidence to suggest that certain animals are able to correct a sodium deficiency, and moose have been observed to skew foraging strategies in order to obtain adequate amounts of sodium in their diet. However, even though sodium hunger may be recognized by the animal, it is debatable whether generalist feeders are genetically programmed to recognize foods that will correct this specific deficiency. Thus an element of learning is involved in identifying the sodium source.

The contemporary view of nutritional wisdom discounts the possibility that animals are genetically programmed to directly taste or smell nutrients. Instead, this view recognizes that animals learn about the foods they ingest. Thus the selection of a diet occurs through learned associations that are modified by past experience. Learned associations are acquired through the pairing of the food's oro-sensory cues (smell, flavour, texture, temperature) with post-ingestive feedback from the body. This more complex view also takes into account that learned associations can be further modified by physiological status and environmental conditions such as the presence of **conspecifics**. (SR)

See also: Diet selection; Self-medication

Further reading

Provenza, F.D. and Villalba, J.J. (2006) Foraging in domestic vertebrates: linking the internal and external milieu. In: Bels, V.L. (ed.) *Feeding in Domestic Vertebrates: From Structure to Function*. CAB International, Wallingford, UK, pp. 210–240.

O

Observational learning

Observational learning is a subcategory of **social learning**. Many animals can learn by observing the behaviour of another animal, or the consequences of the other animal's actions (Heyes and Galef, 1996). One of the most widely studied examples is feeding among domestic chickens. When chickens eat a distasteful food, they shake their heads, gape and wipe their beaks, i.e. the disgust reaction. Chicks observing others giving this reaction after pecking at coloured beads coated with a distasteful substance subsequently avoid these colours. However, adult hens do not avoid coloured food that another hen has found to be distasteful (Sherwin *et al.*, 2002), indicating that observational learning might be more important for juvenile animals in some circumstances. Observer hens seeing demonstrator hens not feeding, feeding on normal food or feeding on highly palatable food did not change the amount of food they ate; however, there was a positive correlation between the pecking rate of the demonstrators and the proportion of food of the demonstrated colour eaten by the respective observers. There was also an interaction with the colour of the food: the effect of observing a hen eating highly palatable food induced the birds to eat more red food, which they otherwise avoided.

Natural populations of black rats also show observational learning. Young rats learn from their mothers how to strip seeds efficiently from cones. Observational learning can sometimes be as simple as directing another animal's attention to an area of the environment (**local enhancement**), or a feature of the environment (**stimulus** enhancement). Careful laboratory studies have addressed the question of whether more complex observational learning occurs, e.g. whether animals can imitate opening an 'artificial fruit' (Whiten and Custance, 1996) or whether parrots can imitate human movements (Moore, 1996). (CS)

References

Heyes, C.M. and Galef, B.G. (1996) *Social Learning in Animals: the Roots of Culture.* Academic Press, San Diego, California, pp. 17–47.

Moore, B.R. (1996) The evolution of imitative learning. In: Heyes, C.M. and Galef, B.G. (eds) *Social Learning in Animals: the Roots of Culture.* Academic Press, San Diego, California, pp. 245–265.

Sherwin, C.M., Heyes, C.M. and Nicol, C.J. (2002) Social learning influences the preferences of domestic hens for novel food. *Animal Behaviour* 63, 933–942.

Whiten, A. and Custance, D. (1996) Studies of imitation in chimpanzees and children. In: Heyes, C.M. and Galef, B.G. (eds) *Social Learning in Animals: the Roots of Culture.* Academic Press, San Diego, California, pp. 291–318.

Observer effect

Observer effect refers to the possible changes that the act of observing will have on the phenomenon being observed. For example, animals may be aware when human observers are present and may consequently alter their behaviour when being observed. This may lead to inaccurate representations of the animals' normal behaviour patterns. (LMD)

Odour

Odours are received through both the nose, the chemosensitive trigeminal pathway, and the **vomeronasal organ**. The vomeronosal organ is particularly involved in receiving **pheromones** that control **aggressive behaviour** and the oestrous cycle of many mammalian species. Some authors dismiss the need for distinction between odours and pheromones, and some refer to the latter as social odours.

Pheromones are a specialized group of chemical attractants produced by animals to stimulate other animals, especially in intra-species social **communication**, but also some inter-species communication. They are secreted by interdigital, infraorbital, inguinal, sebaceous and salivary glands and are particularly concentrated in the perineal region. They are also present in milk, urine and the other body secretions, including sweat. In ungulates there are many different types, including alcohols, diols, alkanes, ethers, diethers, ketones, primary amines and aromatic alkanes. The oestrous pheromones are mainly released from the body surface, particularly the hindquarters and genital region, rather than in urine, faeces or vaginal mucus.

For example, cows in oestrous display spend much time sniffing and licking each others' ano-vaginal areas. Bulls respond to odours produced by **teaser** bulls even more than to pheromones identified in the blood of oestrous cows. Bulls can also detect a change in the pheromonal secretions of cows up to 4 days before the day of **oestrus**. On the day of oestrus the main olfactory system of nasal detection is operative, and the **flehmen** expression is not always used. The ability of bulls to predict oestrus in cows relates to their tendency in wild herds to 'guard' cows as oestrus approaches. In domestic herds in which there is no competition between bulls this does not occur. To test for cows' odours, bulls sample the urine of potential oestrous cows and mark the ground with pheromones by rubbing their head and neck on the ground. **Homosexual** cow relationships appear to be stimulated by sniffing and licking, particularly in the perineal regions.

Other animals, such as dogs, cats and rats, can also detect the oestrous odours of cows, but they are not able to distinguish pre-oestrous odours from those produced at other stages in the cycle in the way that bulls can.

Pheromones are also used to convey **fear**, and cattle respond to the pheromones produced by **conspecifics** that have been mistreated, with increased cortisol production and a reluctance to engage in activities where they might expose themselves to perceived dangers, such as **feeding**. Fear pheromones are present in the blood and also in other body fluids, such as urine. (CJCP)

Oestrogen(s)

This is a group of sex steroid hormones found in females and males; it is secreted by the ovarian follicular cells during the oestrous cycle and from the Sertoli cells in the testis. The **adrenal gland** cortex secretes a precursor (dehydroepiandrosterone) that may be converted into oestradiol by tissues. These hormones are responsible for the secondary sex characteristics (e.g. mammary gland development, cornification of the vagina in preparation for mating) and are closely involved in a feedback loop to the **pituitary gland** for the regulation of the ovarian cycle (**see: Oestrus**), eventually leading to ovulation and the formation of a corpus luteum.

Oestrogens also affect the behaviour of animals preparing the females for mating and acceptance of the male(s) at the time of oestrus (a period of 'desire'), through behaviours such as standing still for the male to mount (most species) as well as flicking the tail to one side (bitches) or **lordosis** (rabbits). In male animals, the role of oestrogens may be in synchronizing sperm development, whereas the administration of oestrogens to males (for tumours of the Sertoli cells in dogs) leads them to develop female characteristics. (DBM)

See also: Sexual behaviour

Oestrus

Mammalian females show a cyclical reproductive pattern, with the majority of species exhibiting distinct periods of increased sexual activity (**see: Sexual behaviour**) termed oestrus, or heat (also estrus, while oestrous/estrous is used in the adjectival form). Equivalent sexual periods in other taxonomic groups are generally not referred to as oestrus. The term oestrus is derived from the Greek word *oistros*, meaning gadfly, or frenzy, as it was thought that gadflies around cattle drove them into a frenzy of activity and noise like that shown by cows in heat. The primates are an exception among the mammals, showing little or no overt behavioural cycle, and reproductive activity is regulated by a menstrual cycle.

There are two important behavioural components to the oestrus cycle, exhibited as increases in both receptivity and in **proceptivity**. Females are receptive to males during the oestrous period in that they will allow mounting and intromission; they frequently adopt a specific stance characterized by immobility, arching of the back and deflection of the tail – in rodents this is termed **lordosis**, but many other mammals exhibit species-specific variations of this behaviour.

As well as a passive increase in receptivity towards males, females also show proceptive behaviours during oestrus, behaviour patterns designed to advertise the female's reproductive status in order to attract males and to initiate copulation. This often involves an increase in general affiliative behaviour (**see: Affiliation**), with an increase in proximity accompanied by other solicitous responses such as butting or rubbing against males.

Oestrous females in herd situations may also seek to drive away other females or even to disrupt copulation, particularly where a strong **social hierarchy** exists and dominant females ensure their precedence over subordinates. In addition to behavioural signals, females may utilize other auditory, visual or olfactory cues to advertise their oestrous state, such as: (i) calling; (ii) adopting specific postures; and (iii) colour changes to draw attention to their sexual parts, through either the release of **pheromones** from specific glands or by an increased urination rate. During the anoestrous period, females may show contraindications, actively avoiding males, striking out or showing other aggressive responses, vocalizing and moving away if approached by males.

The cyclical behaviour patterns observed during oestrus are coordinated with the reproductive cycles, which regulate ovarian events, particularly follicular development and oogenesis culminating in ovulation, and uterine preparation for fertilization and the receipt and implantation of the embryo. This coordination is dependent on the hypothalamic–pituitary–ovarian endocrine axis. The **hypothalamus** secretes **gonadatropin-releasing hormone (GnRH)**, which regulates the release of the pituitary gondotropic hormones, **follicle-stimulating hormone (FSH)** and **luteinizing hormone (LH)**, which, in turn, indirectly lead to production of the ovarian steroids **oestrogen** and **progesterone**.

The oestrous phase of the cycle is predominantly under the control of oestrogen produced by the growing and maturing ovarian (Graafian) follicles. The earliest events of oogenesis and folliculogenesis are independent of pituitary hormones; however, the later stages of follicular recruitment, establishment of dominance, completion of meiotic metaphase I and ovulation are all critically dependent on FSH and LH levels. Growing follicles actively produce oestrogen, which acts as an important regulator for the oestrous behaviour patterns. Raised levels of oestrogen also prepare the female tract for sperm transport to the site of fertilization in the oviduct.

Following ovulation, the formation of a corpus luteum producing progesterone results in dioestrous behaviour patterns (sexual quiescence). The expression of proceptive and receptive behaviour patterns by females during the oestrous period helps to ensure that mating will take place at an appropriate time during the reproductive cycle. In species where spontaneous ovulation occurs unrelated to copulatory behaviour (unlike the induced ovulators, such as the domestic cat), ovulated eggs have a limited fertile lifespan within the oviduct, during which they must be fertilized by the male's sperm before becoming atretic. Expression of oestrus ensures that males and females are drawn together at the right point of the cycle to maximize the chances of fertilization occurring.

The oestrous cycle is an endogenous cycle exhibited by mammalian females, but its expression may be entrained by environmental cues – principally by light, but temperature and nutritional plane may also affect the cycle, although both of these tend to be closely correlated with light levels. The environmental effects of light are regulated via the pineal gland, which secretes the hormone melatonin. Melatonin secretion is inhibited by impulses from the optic nerve, in turn initiated by light falling on the retina, and is increased during periods of darkness. Varying melatonin levels affect pituitary gonadotropin release in different ways, according to the species. Species sensitive to these environmental cues are termed seasonally polyoestrous, as they show repeated oestrous cycles until mating results in pregnancy, but only during certain periods of the year, while at other times they exhibit no oestrous behaviour and are termed seasonally anoestrous.

The seasonal timing of oestrous behaviour and response to melatonin is closely linked with the **gestation** period of different species so that offspring can be produced – usually in late spring or early summer, when conditions are advantageous to the young and they have a maximal growth and development period before the onset of more rigorous winter conditions. Hence sheep and goats, which have a 6-month gestation period, are short-day breeders with oestrous behaviour occurring in response to declining day length in autumn and early winter, while horses are termed long-day breeders as they respond to increasing day length in the spring, reflecting their longer 11-month gestation period. Other species such as pigs are relatively unaffected by day length and will cycle at all times of the year, while others still may only show single, isolated oestrous periods with prolonged anoestrous phases in between, and these are termed monoestrous – an example being the wolf.

As well as responding to environmental cues, the timing and incidence of oestrus may also be influenced by the presence or absence of male animals. In a range of species the introduction of a male has been shown to advance the appearance of oestrus (Whitten effect) and to cause synchronization of oestrus in groups of females (Bruce effect). In a number of cases these effects have been demonstrated by exposing females to male secretions such as urine, saliva or other specific scent gland products.

Familiarity with the overt signs of oestrus has traditionally been important in livestock husbandry, allowing for efficient and safe mating of livestock species – for example, the observation of 'bulling' in cattle, female–female mounting behaviour, as an indicator that cows are in heat and ready to be introduced to the bull. Similarly, sows may be tested by applying pressure to their backs in order to determine whether they will stand for a boar. A range of semi-automated systems has been devised for oestrus detection so, for example, general activity levels increase in the oestrous period and pedometer systems set to automatically flag increased movement levels in cattle have been used to indicate oestrous cows within a herd. Pressure sensors have also been used to indicate when female mounting has occurred, or vasectomized males may be used to mark oestrous females when they mount.

In horse **breeding**, mares are traditionally exposed to a **teaser** stallion, whose job is to elicit signs of oestrous behaviour. Teasing may be performed by the covering stallion but, with more valuable stallions covering large numbers of mares, frequently a separate teaser stallion that has high **libido**, but is easily and safely managed, may be used to test the mares before they are covered (mated) by the breeding stallion, which is thus protected from the adverse responses of an anoestrous mare.

Increased availability of on-farm ultrasound scanning has allowed the extent of ovarian follicular development to be monitored directly in many cases without having to consider the rather more indirect behaviour signs of oestrus. However, observation of the correct behavioural responses that females exhibit during the oestrous period remains one of the best indicators of an animal's readiness to mate. (MRC)

Olfaction

Most higher animals are able detect **odours** from inanimate and animate objects and utilize the information to modify their reproductive, ingestive or **social behaviour**. However, despite the existence of two nostrils, there is only limited evidence that mammals can localize odours efficiently. The detection of odours has been studied less than the other primary senses, but they are clearly of major significance, especially in animals such as cattle, which evolved as prey, that needed durable, covert signalling. Odours are used particularly in intra-species **communication**, where they synchronize **reproduction**, act as territorial markers and **signal** the presence of predators. Other possible odour signals include **aggression**, **hunger** and **anxiety**.

The first stage in odour perception is the odour entering the oronasal cavity by the mouth or nasal orifices. This chemically stimulates olfactory receptor **neuron(e)s** (ORNs), which terminate in cilia that are bathed in mucus. Acting in parallel is the chemosensitive trigeminal pathway, the neurons of which also terminate in the nasal cavity. Neurons in this pathway are activated by many of the same chemical stimuli that activate ORNs, but the trigeminal system is particularly specialized in detecting irritants.

The third major site of odour reception, in addition to the olfactory and nasal trigeminal free nerve endings, is the **vomeronasal organ**, which is not well developed in adult humans but is well developed in most mammalian species. It is situated in the roof of the mouth and consists of blind, tubular diverticula, lined with olfactory epithelium. Vomeronasal organ neurons are not ciliated but end in microvilli, which are particularly sensitive to low-volatile compounds in water. These are common in urine and are used for social and sexual communication, including the detection of oestrous females by the male. The reception of odours by the vomeronasal organ is also used for **reinforcement** and the maintenance of sexual interest during copulation.

Air is sampled via the buccal cavity and the nose, although the nostrils are partially closed by the curling of the lip in this behaviour. The characteristic **flehmen** expression, whereby the head is directed upwards with the mouth ajar, the tongue flat and the upper lips curled back, is thought to aid odour sampling in ruminants by allowing air to be inhaled so that it passes over the roof of the mouth. Flehmen is also believed to act for appeasement, being the antithesis of the threat display. Different forms of flehmen exist in other species, for example in the cat, where it resembles a facial grimace.

Territorial animals use **scent marking** extensively, particularly **ungulates** – the dikdik builds dung heaps and rubs its forehead on them to increase its scent. The pronghorn urinates before defecating to amplify the scent. Deer have tarsal glands and also sudoriferous scent glands in their forehead that they use for marking territory. Sheep are reputed to be able to track **conspecifics** in mountain territory, presumably by detecting tarsal gland secretions. Cattle are not territorial in this way and defecate more or less at random. However, bulls do paw and dig their horns into the ground, then throw the soil over their withers, which may amplify their scent. In deer, smell is also important in **diet selection**. Although specific tests have not been conducted with cattle, it is likely that smell is unimportant for cattle **grazing** a uniform pasture with a limited range of botanical species, because repeated exposure to odour stimuli reduces sensitivity as the receptor sites become locked up.

In humans smell perception may be impaired, not so much by the sensory apparatus as by the lack of detailed information processing by the **brain**. Much of this is due to lack of training, because we focus on developing other sensory abilities, but human perception is also less acute. Humans can actually identify their own odours, synchronize menstrual cycles using body odours and recognize their mothers' odour at 5 days of age.

Many animals are highly sensitive to the odours of potential predators, and increase their investigatory behaviour, spending more time sniffing the air and in cautious movement. Most domesticated animals are far more sensitive to chemical odours than are humans. For example, the chemical sensitivity of cattle ranges from being able to detect very weak solutions of sodium salts (10^{-4} M solutions of sodium bicarbonate) up to large hydrocarbon molecules and steroids. In comparison with man, a microsmatic animal, ungulates are macrosmatic. Humans can usually detect an intensity difference about one-third that of the actual intensity, whereas rodents can detect a difference of one-twenty-fifth. (CJCP)

Further reading

Sommerville, B.A. and Broom, D.M. (1998) Olfactory awareness. *Applied Animal Behaviour Science* 57, 269–286.

Ontogeny

Ontogeny refers to the development of a behaviour pattern in an individual through the integration of learned and innate aspects of the behaviour, and is one of four groups that **Niko Tinbergen** used to divide the study of behaviour (**see: Tinbergen's four questions**). Ontogeny can also refer to the history and development of an organism from the fertilized egg to maturity. (LMD)

See also: Development of behaviour; Emancipation; Maturation of behaviour

Open-field test

The open-field test involves placing an experimental animal into a large, unfamiliar arena in order to measure its behavioural responses. The open-field test was originally developed by Hall (1941) as a method of measuring fearfulness in rats, and subsequently used by Hall and Broadhurst (Broadhurst, 1975) as a basis for performing divergent selection for defecation (Maudsley reactive and non-reactive rats). A large number of responses to threatening and potentially threatening stimuli have been measured in later experiments comparing the Maudsley strains. Because many behavioural differences indicate a higher level of fearfulness in the reactive strain and a lower level of fearfulness in the non-reactive strain, these studies provide some support for the hypothesis that open-field defecation in rats reflects fearfulness and that fearfulness is partly under genetic control. However, a number of studies using the open-field test for measuring fearfulness have produced very ambiguous results. Caution should therefore be used when interpreting open-field behaviour, and alternative tests are often preferable.

In addition to its use for laboratory rodents, the open-field has also been used in modified forms on a large number of other species, including cattle, goats, sheep, poultry, foxes, pigs and rabbits. The aim of the experimenter is often to quantify the balance between an animal's anxiety-related and exploratory tendencies as a measure of fearfulness, or simply to quantify the animal's level of activity shown in response to this novel test situation. Interpretation of open-field behaviour in terms of activity in a novel environment has some empirical support, but there is often little support for interpretation of open-field behaviour in terms of fearfulness.

There are several **motivations** and **emotions** that may influence an animal's behaviour in the open field, and limited possibilities for controlling the direction of their effects on an animal's behaviour. The most obvious emotion that may be observed is novelty-induced **anxiety**, which results in **avoidance behaviour**, arousal and observation of the surroundings. Curiosity is another fundamental emotion elicited by novelty, and leads to approach, investigation and exploratory responses.

A major problem with the open-field, as opposed to the **approach test**, is that it is difficult to define approach (reflecting curiosity) and avoidance (reflecting anxiety) in the open field. Whereas some herbivores and ungulates may prefer open spaces, many other species may be attracted to the relative safety that the walls of an open-field test arena represent. However, when measuring relative levels of approach to and avoidance of the central areas of an open-field arena, one normally presupposes that the apparatus is large enough for the animal to distinguish between open and protected areas. One also presupposes that the species does in fact perceive open space as aversive. In cases where the animal has no aversion to open space, or the open-field is too small, responses often give little information besides indicating the animal's activity in a novel environment.

In addition to the above, socially reared animals may be highly motivated to reinstate social contact with **conspecifics** when tested singly in the open field. This may be observed as high levels of activity aimed at escaping. **Fear** of humans in aversively handled animals, or attraction to humans in hand-reared animals, may also influence their behaviour in the open field. Whereas livestock treated poorly by **stockpersons** may avoid the handler and flee into the centre of the open field, animals handled gently may be distressed by the absence of the handler and seek to reinstate contact. In some experiments, blood sampling is performed prior to testing. Although such **aversive** handling may strongly affect subsequent behaviour during testing, this has seldom been investigated systematically. Whereas fearful rats may defecate and show signs of behavioural suppression when exposed to the open field, this may not apply to other species, which may behave quite differently when exposed to a novel enclosure.

Spaciotemporal measures that may be recorded in the open-field include total distance walked (locomotive activity) and **locomotion** in the centre relative to locomotion at the periphery of the apparatus. However, locomotion in the periphery may be difficult to interpret, as discussed above. Total locomotion is also difficult to interpret in terms of anxiety, because changes in activity may have several causes. Changes in activity may reflect **disinhibition** of locomotion following a period of deprivation in a small home cage or pen, or reflect the animal's **coping style** during forced exposure to **novelty**. Locomotive activity is also difficult to interpret because immobility (the lack of locomotion) is frequently interpreted as indicating increased anxiety in a novel environment. This latter interpretation is based on the fact that immobility is often associated with increased avoidance of potentially dangerous objects.

Species-specific ethological measures recorded in the open field may include different postures, defecation, **vocalization**, rearing on the hind feet, sniffing or rubbing components of the room and apparent attempts to escape. Sometimes, physiological measures such as **glucocorticoid** secretion and **heart rate** are also recorded. When several measures are simultaneously recorded in the open-field test, scores generated by principal component analysis may aid in interpreting behaviour.

To summarize, interpretation of open-field behaviour in terms of anxiety or fearfulness (emotionality or emotional reactivity) is often unclear. The open field has seldom been subjected to satisfactory validation (but see Gray, 1987; Salvatierra and Arce, 2001), and some attempts at validating it have failed (Andersen *et al.*, 2000). Because of this, it is probably best to avoid using the open-field test for measuring these emotional reactions or **personality/temperament** traits. Despite its general weaknesses as a test of anxiety or fearfulness, the open-field test may, however, be useful for assessing the level of activity displayed in a novel enclosure.

(AMJ)

References

Andersen, I.L., Færevik, G., Bøe, K.E., Janczak, A.M. and Bakken, M. (2000) Behavioural evaluation of methods for assessing fear responses in weaned pigs. *Applied Animal Behaviour Science* 69, 227–240.

Broadhurst, P.L. (1975) The Maudsley reactive and non-reactive strains of rats: a survey. *Behaviour Genetics* 5, 299–319.

Gray, J.A. (1987) *The Psychology of Fear and Stress*. Cambridge University Press, New York.

Hall, K. (1941) Temperament: a survey of animal studies. *Psychological Bulletin* 38, 909–943.

Salvatierra, N.A. and Arce, A. (2001) Day-old chicks categorised on latency to peck, exhibit a stable fear pattern until 15 days of age. *Applied Animal Behaviour Science* 73, 103–116.

Operant response

An operant response is one that can be modified by its consequences, an element of which may be referred to as an operant. A free operant response is an operant behaviour that can occur at any time, because it is not dependent on any time or resource-limiting event. Running may be considered a free operant response, whereas hunting depends on the availability of prey and so is not a free operant response. In an experimental situation evaluating operant conditioning (**see: Conditioning – types of; Instrumental conditioning**) it is usually preferable to use free operant responses as the substrate for the experiment. The level at which an operant response occurs prior to any significant reinforcement is referred to as the operant level, and is the baseline against which an assessment of the effects of potential reinforcement or punishment can be assessed.

(DSM)

Operant test

An operant test is an experiment in which the delivery of a resource or event is contingent on an animal's behaviour (operant response). Because approach and avoidance are operant responses, even simple **choice tests** are, strictly speaking, operant tests. However, common usage of the term 'operant tests' excludes choice tests. An operant test may be used to investigate an animal's **motivation** and preference. The subject may be required to perform a response, or refrain from performing a response, in order to either obtain a **positive reinforcer** or avoid an aversive (**see: Aversion and aversives; Reinforcement – types of**). It is usual for only one reinforcer to be contingent upon responding at a given time (unlike choice tests). Reinforcers or aversives may be compared retrospectively by reference to a common measure of responding. Another typical feature of an operant test is that the response is specifically trained and sometimes manipulated to vary the cost of access to a reinforcer, although some procedures remain difficult to classify (**see also: Economics of behaviour** for an explanation of some of the terms used in this entry).

Research questions addressed by operant tests

The research questions addressed by operant tests can be broadly divided into four categories:

(A) To ascertain whether an animal is motivated to obtain or avoid a resource or event (**see: Choice test**). These work on the principle that an animal will learn to perform, or refrain from performing, an operant response to obtain a resource or event that is rewarding, or to avoid an event that is aversive.

(B) To ascertain whether an animal has a preference among alternative reinforcers (**see: Choice test**). These operate on the principle that the more strongly motivated an animal is to obtain or avoid a reinforcer, the greater its response (e.g. intensity, frequency, rate) should be. There are two approaches: (i) in the first, used mainly to compare negative reinforcers, subjects respond to obtain/avoid the alternative reinforcers sequentially. The validity of this procedure is limited when comparing positive reinforcers. (ii) The second approach, used to compare positive reinforcers, is a hybrid operant × choice test, in which reinforcers are made available concurrently. In practice though, simple choice tests are most frequently used to ascertain whether a preference exists.

(C) To measure the strength of an animal's motivation for a reinforcer (**see: Choice test**). These are often based upon those designed to ascertain whether a preference exists. However, for positive reinforcers, an important difference is that the reinforcer of interest is compared to one for which the animal has an independent motivation. Strength of preference (between the reinforcer of interest and an alternative) can be assessed if the animal has free access to an alternative during the test; strength of motivation is assessed if the animal has no access to an alternative during the test. These tests work on the principle that the stronger an animal's motivation to interact with one reinforcer compared to another, the greater its response should be for one compared to the other, so long as they satisfy the same motivation. If this is not the case, a second approach may be used in which the degree to which an animal is prepared to give up or work for one reinforcer is compared to another.

(D) To ascertain whether the strength of an animal's motivation for a reinforcer, or the strength of its preference between two reinforcers, is altered by changes in its internal or external environment. For example, we might ask: 'is feeding motivation in restricted-fed animals reduced by bulking out the diet with roughage?', 'does nesting motivation increase as oviposition time approaches?', or 'is handling rendered less aversive by previous experience?' These are similar to those designed to ascertain whether a preference exists (B) or to measure the strength of motivation or preference (C). The only difference is that a single reinforcer is evaluated and the procedure is repeated before and after changes in the subject's internal or external environment.

Specific procedures

Listed below are procedures that have been used by ethologists, together with the type of research questions that might be addressed using the given approach, using the letters (A) to (D) from above:

1. Response acquisition (A). An attempt is made to train the subject to perform an operant response, either to obtain the hypothesized reinforcer, or to avoid or terminate exposure to it. Limitations include the fact that motivation may be confounded with the degree of compatibility between response and reinforcer (salience), because animals have a biological propensity to associate certain behaviours with certain stimuli or outcomes. For example, it may be easier for a chicken to learn to scratch for access to a dust bath than to peck for one. However, this can be avoided by either training a response that resembles natural aversive or appetitive behaviour for the motivation in question or training several different responses.
2. Response frequency, or number of reinforcements earned (also called the 'y-intercept of the demand curve') (B, C, D). The subject is required to respond on a fixed ratio schedule (**see: Reinforcement – types of**) during a session of fixed duration to obtain or avoid something. This procedure is carried out for each reinforcer of interest, or comparator, in turn. The greater number of responses performed or reinforcements obtained is used to determine a stronger motivation for that outcome (e.g. number of times an animal accesses food versus a companion). However, when comparing positive reinforcers that satisfy different motivations, motivational strength is confounded with the quantity or interaction time required to reach satiation.
3. Passive avoidance – dual schedules, with multiple reinforcements per session (A, B, C, D). The subject is trained to respond on two ratio or interval schedules, either concurrently or alternately, for access to a standard positive reinforcer (e.g. food). Then, in a series of treatments, response X is followed by exposure to an aversive of interest (e.g. the type of vibration associated with **transport**) plus the standard positive reinforcer (e.g. food), while response Y is followed by access to the standard positive reinforcer only. A decline in the magnitude of response X, relative to response Y, reflects the level of motivation to avoid the aversive.
4. Passive avoidance – single schedule, with multiple reinforcements per session; aversive versus positive comparator (C). The subject is required to respond on a single ratio or interval schedule during a set time for access to the comparator. Responding on this schedule is followed by exposure to the aversive of interest plus the comparator. If the response magnitude declines substantially, then motivation to avoid the aversive is comparable to motivation to obtain the comparator.
5. Runway test/chamber entry test – latency to approach/leave (A, B, C, D). The subject may initially be trained to move between location X (e.g. a chamber or one end of a runway), containing no reinforcer, and one or more adjoining locations Y (e.g. an adjoining chamber, the other end of a runway or the arms of a Y-maze), each containing either a reinforcer of interest, a comparator, or no reinforcer (control). Then, in a series of treatments, the subject is placed in location X (for latency to enter) or Y (for latency to leave) and allowed to move at will. If the latency to enter/leave location Y when it contains the hypothesized reinforcer is less than the latency to enter/leave this location when it contains no reinforcer, then the subject is motivated to obtain/avoid the reinforcer; a smaller latency to enter/leave location Y in one treatment than in another indicates a stronger motivation. However, it is important to note if there is a ceiling effect on approach/withdrawal rate.
6. Runway test/chamber entry test – passive avoidance (A, B, C, D). The subject is trained to move along a runway or enter a chamber to reach a standard positive reinforcer (e.g. food). Then, in a series of treatments, traversing the runway or chamber entry is followed by exposure to the aversive of interest or a comparator, plus the standard positive reinforcer. In a common variant, subjects are forced to traverse a runway without any positive incentive. (However, this is not an operant test according to common usage, because subjects do not learn to perform a response.) An increased latency to enter/traverse the chamber/runway in the test indicates that a motivation to avoid the hypothesized reinforcer exists and a greater latency in one treatment than in another indicates that motivation to avoid the reinforcer in the former is stronger.
7. Runway test/chamber entry test – passive avoidance versus positive comparator (C). The subject is required to traverse a runway or enter a chamber to reach the comparator. Traversing the runway or entering the chamber is then followed by exposure to the aversive of interest plus the comparator. For example, this has been used to investigate the perception of noxious gases used in the euthanasia of mink. If latency to enter/traverse the chamber/runway declines substantially, then motivation to avoid the aversive is comparable to motivation to obtain the comparator. If the subject refuses to enter/traverse the chamber/runway, then motivation to avoid the reinforcer of interest exceeds motivation to obtain the comparator.
8. Cross point of two concurrent demand curves (B) (**see also: Economics of behaviour**). The subject is required to respond during a series of concurrent fixed ratio schedules to obtain a fixed quantity of reinforcer X on one schedule and a fixed quantity of reinforcer Y on the other. The schedule (relative 'price') of reinforcer Y is steadily increased, while the schedule for reinforcer X is held constant or decreased. The cross point is the price at which the curves overlap, i.e. the price at which demands for the two reinforcers are equal. From

this the relative motivation for X compared to Y can be assessed.
9. Matching law (B). The subject is required to respond on a series of concurrent variable-interval schedules to obtain different reinforcers. The session ends either after a fixed period of time or when a given number of reinforcements have been earned. The schedules are changed once a criterion for stable responding has been met and the process repeated. According to the matching law, the relative number of responses performed on the two schedules (P_X/P_Y) and the relative amount of time spent responding on the two schedules (T_X/T_Y) both increase as the relative rate of reinforcement on the two schedules (r_X/r_Y) increases. From this, the relative strength of motivation for the two alternative reinforcers can be calculated. This approach has been used, for example, to assess the strength of preference for different rooting substrates by pigs.
10. Work rate (B, C, D). In a series of treatments, the subject is required to respond by pushing on a door or operating a switch until a fixed amount of work (force × time) has been accumulated, to obtain a single exposure to either a reinforcer of interest or a comparator, for example access to a nestbox or food for chickens. More rapid completion of the task or greater average effort in one treatment over another indicates that motivation to obtain the reinforcer in the former is stronger.
11. Break point (B, C, D). In a series of treatments, the subject is required to respond on a single progressive ratio schedule (an ascending series of fixed ratio schedules, increased after every reinforcement), to obtain a reinforcer of interest or a comparator. The session ends when the subject has ceased responding for a specified period. A greater number of responses performed, or a higher schedule reached, in one treatment compared to another indicates that the motivation to obtain the reinforcer in the former is stronger; for example, this has been used to assess the effect of food deprivation on sows. When comparing positive reinforcers that satisfy different motivations, motivational strength is confounded with the quantity/interaction time required to reach satiation. This may be ameliorated by ensuring that the size of each reward is small.
12. Highest price paid (also called reservation price) (B, C, D). In a series of treatments, the subject is required to respond on an ascending series of fixed ratio schedules, or on a schedule with increasing response difficulty or aversiveness ('prices'), to obtain either a reinforcer of interest or a comparator. The schedule is gradually increased until the subject fails to earn the reinforcer. The reservation price is the highest price reached, and this was used to determine the value of group housing to sows (Kirkden and Pajor, 2006). It estimates motivation for a single access to a reinforcer, or for the quantity obtained in the time allowed. A higher schedule reached indicates that the motivation to obtain a single access to the reinforcer is stronger in the given circumstances.
13. Price elasticity of demand or slope of the demand curve (B, C, D). This involves a similar methodology to that used to determine highest price paid, except that the reinforcer is made freely available for a set time (rather than being a single event). The schedule is gradually increased, sometimes until the number of rewards earned ('demand') falls to zero. Demand is plotted against price to obtain a demand curve. If the absolute slope of this curve is less in one treatment compared to another, then motivation to obtain the reinforcer in the former is stronger. Alternatively, if the slope of the curve for the reinforcer of interest is not significantly different from zero, or if its price elasticity of demand is less than 1 ('inelastic demand'), then motivation for this reinforcer is strong. Empirical findings suggest that, for two reasons, it is not valid to infer a strong motivation from inelastic demand. First, price elasticity of demand is influenced by procedural variables such as reward size and response type and for a given reinforcer it can vary widely between studies. Secondly, practically every reinforcer that has been evaluated in an operant test exhibits price inelastic demand. This is likely to be an artefact of the methodology of operant tests. Therefore a comparator should be used to assess the relative, not absolute, strength of motivation.
14. P_{max} (B, C, D). This procedure is a variant on the price elasticity of demand approach and is used when there are non-linear demand curves. Demand is measured at a series of prices, in the same way as in the price elasticity of demand approach. P_{max} is defined as the price at which the price elasticity of demand is equal to 1. At this price, the subject performs a maximal number of responses ('work') for the reinforcer. A greater P_{max} in one treatment compared to another indicates that motivation to obtain the reinforcer in the former is stronger. This approach has been used, for example, to help to determine preferred stocking density by laboratory rats.
15. Consumer surplus (also called travel cost consumer surplus) (B, C, D). Demand is measured at a series of prices, in the same way as in the price elasticity of demand approach (13). Price is plotted against demand. The consumer surplus estimates motivation for access to a specific quantity of a reinforcer (e.g. to a single reinforcement during the course of a day, the first ten reinforcements or a satiating quantity), or motivation for access after a particular prior level of consumption (e.g the tenth reinforcement of the day). It corresponds to an area under the curve, bounded by two points on the demand axis. For example: the consumer surplus for the first reinforcement is an area between demand = 0 and demand = 1 (equal to the reservation price); the consumer surplus of a satiating quantity is the entire area under the curve. A greater consumer surplus in one treatment than in another indicates that motivation to obtain a specified quantity of the reinforcer in the former is stronger. Although claims for using this parameter have been made in the animal welfare literature (e.g. Mason *et al.*, 2001), this has not been done in practice, because the surplus can only be measured if the price of each commodity is raised relative to the price of others rather than all raised together.

Some limitations affecting several of the procedures, which should be considered

i. The five general limitations listed for **choice tests** (i.e. motivation depends on context experience, availability of alternatives and other cues and does not reveal what is good for welfare in the long run) also apply to all operant testing procedures, with the exception that an open economy is integral to the matching law approach (9).

ii. When alternative positive reinforcers are evaluated sequentially, preference may be masked if a need to satisfy the motivation overrides a preference for how to satisfy it.
iii. When comparing positive reinforcers that satisfy different motivations, with access to multiple reinforcements per session, reinforcer magnitude (quantity/duration) may be controlled either by the experimenter or by the subject, but either approach can be problematic. If the experimenter has control, restriction/interruption of bouts may devalue some reinforcers to an extent that depends on the motivation in question. However, if the subject has control and cost is manipulated, cost and reinforcer use may not co-vary (Mason *et al.*, 1998). However, for some reinforcers, reward magnitude may be controlled by the experimenter without devaluing them. For other reinforcers, the subject may be given control, but if cost and resource use do not co-vary, the approach will not yield valid measures of motivation, although it may be possible to usefully determine the highest price paid.
iv. When comparing positive reinforcers that satisfy different motivations or negative reinforcers that involve different mental states, motivational strength may be confounded with the degree of compatibility between response and reinforcer, because animals have a biological propensity to associate certain behaviours with certain stimuli/outcomes. This problem can be solved by training responses that are not specific to one of the motivations being measured.
v. When using consumer demand approaches in which income is constant, price changes have income effects (Kirkden *et al.*, 2003). The problem can be ameliorated by either ensuring that income is high; or evaluating a small quantity of the reinforcer of interest where this is possible.
vi. When comparing positive reinforcers that satisfy different motivations, there are several important considerations: motivational strength depends on deprivation level, so when a single measure is extracted from a demand curve, the measure is of limited practical value. Also, motivational strength is confounded by the rate at which subjects satiate for a resource as they earn more. Finally, motivational strength is also confounded with the quantity consumed or interaction time when price is low. However, in this latter instance it is possible to factor out this quantity or time in the analysis.

(RDK)

See also: Behavioural elasticity; Operant response; Reinforcement

References and further reading

Dawkins, M.S. (1990) From an animal's point of view: motivation, fitness, and animal welfare. *Behavioral and Brain Sciences* 13, 1–61.

Kirkden, R.D. and Pajor, E.A. (2006) Motivation for group housing in gestating sows. *Animal Welfare* 15, 119–130.

Kirkden, R.D., Edwards, J.S.S. and Broom, D.M. (2003) A theoretical comparison of the consumer surplus and the elasticities of demand as measures of motivational strength. *Animal Behaviour* 65, 157–178.

Mason, G.J., McFarland, D. and Garner, J. (1998) A demanding task: using economic techniques to assess animal priorities. *Animal Behaviour* 55, 1071–1075.

Mason, G.J., Cooper, J. and Clarebrough, C. (2001) Frustrations of fur-farmed mink. *Nature* 410, 35–36.

Rushen, J. (1990). Use of aversion-learning techniques to measure distress in sheep. *Applied Animal Behaviour Science* 28, 3–14.

Operational definition

An operational definition of something defines it in terms of how it can be recognized, rather than its intrinsic properties. For example, **stress** might theoretically be defined in terms of an individual's cortisol level, and thus cortisol levels might be used to compare the stress resulting from different procedures. However, this approach does not tell us about the nature of stress, beyond its measurement. Thus operational definitions allow scientific investigation at one level, but caution is required in their extrapolation when it comes to defining the nature of a potentially subjective phenomenon that is defined in this way – i.e. in the example above, the nature of stress when defined by cortisol level is not clear, and it cannot be assumed to be directly related to suffering. (DSM)

See also: Ostensive definition

Opioid

Any chemical (natural or synthetic) that has an effect like the **narcotic** (sensation-numbing) alkaloids derived from opium, such as morphine, may be referred to as an opioid. This includes natural **brain** chemicals, such as the **endorphins** and **enkephalins**. Strictly speaking the term opiate should be applied only to the natural chemical derivatives of opium and not to the synthetic drugs like heroin and hydrocodone that are related to these compounds. (DSM)

Optimal fitness

The term 'optimal fitness' may be considered a misnomer for either of the terms 'maximal fitness' (**see: Fitness; Inclusive fitness**) or 'optimal strategy for fitness' (**see: Optimal strategy**), because natural selection favours strategies that maximize, not optimize, fitness, i.e. a strategy may be optimized to maximize fitness (**see: Optimality**). (PE)

Optimal foraging

Foraging means 'searching for food'. Within behavioural **ecology**, the study of **foraging behaviour** was combined very early on with **optimality** theory to make models that predict the behaviour of animals. This connection between optimality theory and foraging behaviour is so strong that it is used to describe almost all studies of foraging, and hence the term 'optimal foraging'.

There are two classical models of optimal foraging: the marginal value theorem and the prey-choice model (Stephens and Krebs, 1986).

The marginal value theorem

The marginal value model predicts the behaviour of an animal that is searching for patchily distributed food. The food in the patches is depleted as the animal forages in them. The question the model answers is: how long should the animal stay in one patch before changing to another patch? The most important assumptions that are made are:

- The animal has perfect knowledge of the content of a patch once it visits it.
- The animal does not know how far it is to the next patch.
- The cost of travelling from one patch to the next is the same as foraging in the patch.

What the model maximizes is the intake rate of the animal. These assumptions might seem to make the model totally unrealistic. Experiments have, however, shown that the predictions of the model fit surprisingly well, at least qualitatively. The two main predictions from the model are that you should stay longer in better than average patches and, as the mean travel time increases, the time spent in a patch – the giving up time – should increase.

Picking mushrooms is a good example of the kind of situation that the model makes predictions about. Mushrooms are not distributed evenly in the landscape, but in patches, and we usually cannot see the next patch and therefore do not know exactly how far it is to the next patch. We also want to pick as many mushrooms as possible, as fast as possible.

The prey-choice theory

The second classical model is the prey-choice model. It models a situation in which the animal encounters prey, with two different types of prey of differing benefits (in the model, energy contents) and costs (in the model, handling times). The assumptions of the model are:

- The prey types are encountered sequentially and randomly (the probability of encountering a prey is independent of when the prey has previously been encountered).
- Searching and handling of prey are mutually exclusive activities.
- The energy content, cost and encounter rate are fixed.
- Encounter without attack is instantaneous and incurs no cost.
- The animal has complete information concerning the encounter rates, costs and benefits of the model.

The strongest and most controversial prediction from the model is that the bad prey type (i.e. the one with the lower energy content/handling time value) should either always be eaten or never be eaten. This is irrespective of the encounter rate of the two prey types. This all-or-none **preference** is not generally found in animals – they instead show partial preferences (Krebs *et al.*, 1977). An example of the type of situation modelled is a fish that stays in a fixed position in a stream encountering different types of prey floating by.

Both of these models are rate-maximizing models, i.e. what is maximized is intake over time. This has very strong similarities with models of learning, and many situations investigated by experimental psychologists can also be understood in terms of optimal foraging.

The assumption of complete information is often criticized. Sampling alternative food sources, and acquiring information concerning their availability and handling, is obviously an important part of the foraging behaviour of many species. This has been extensively studied in, e.g. bumblebees, which have one flower species on which they specialize and one or two alternative species that are continually sampled (Heinrich, 1979). Attempts have also been made to model the relation between exploration and exploitation in foraging situations (McNamara, 1982; Inglis *et al.*, 2001).

There are also models of optimal foraging that take other aspects into account. In models of risk-sensitive foraging, for example, the variability of the food sources is considered. What is maximized in these models is not the rate of intake but the probability of intake rate staying over a certain threshold (in effect, maximizing the likelihood of survival; Caraco *et al.*, 1980). The digestibility and nutritional content of the prey/food are other aspects that are often modelled.

(BAF)

References

Caraco, T., Martindale, S. and Whittam, T.S. (1980) An empirical demonstration of risk-sensitive foraging preferences. *Animal Behaviour* 28, 820–830.

Heinrich, B. (1979) 'Majoring' and 'minoring' by foraging bumblebees, *Bombus vagans*: an experimental analysis. *Ecology* 60, 245–255.

Inglis, I.R., Langton, S., Forkman, B. and Lazarus, J. (2001) An information–primacy model of exploratory and foraging behaviour. *Animal Behaviour* 62, 543–557.

Krebs, J.R., Erichsen, J.T. and Webber, M.I. (1977) Optimal prey selection in the great tit. *Animal Behaviour* 25, 30–38.

McNamara, J.M. (1982) Optimal patch use in a stochastic environment. *Theoretical Population Biology* 21, 269–288.

Stephens, D.W. and Krebs, J.R. (1986) *Foraging Theory*. Princeton University Press, Princeton, New Jersey.

Optimal strategy

Organisms have to allocate their limited resources between competing life history traits. The suite of behavioural, physiological and developmental trade-offs that results in the greatest propagation of the relevant genes (i.e. evolutionary pay-off) can be considered the optimal strategy. Ethologists have used the concept to analyse the day-to-day activities of animals, such as **foraging behaviour**. The logic here is that individuals that optimize components of their behaviour will leave the most descendents (i.e. will have maximized their **fitness**). Although this concept has been applied with great success in so much as aspects of behaviour can be predicted very accurately given knowledge about the costs and benefits of engaging in an activity, there are good theoretical grounds as to why studies of behavioural optimization should be interpreted with caution. For example, the optimal strategy may depend on what other strategies are being employed within the population (**see: Game theory**). (PE)

Further reading

Stamp-Dawkins, M. (1995) *Unravelling Animal Behaviour*, 2nd edn. Longman, London.

Optimality

Optimization, as applied to animal behaviour, is the behavioural repertoire that yields the greatest benefit to cost differential. It is built on the premise that animals perform a suite of behaviours in order to survive and reproduce, and those that perform these behaviours efficiently leave the most surviving offspring, who in-turn inherit these characteristics (**see: Optimal strategy**). Thus, over evolutionary time, the behaviours are predicted to become more and more efficient and, ultimately, optimal. The basic concept – that natural selection favours those individuals that behave optimally – has enabled behavioural ecologists to make very precise predictions about how animals will behave under particular ecological conditions, given knowledge of the costs and benefits associated with the particular activity.

In essence, the theory proposes that individual reproductive success is maximized through the optimization of day-to-day activities, such as foraging. However, the link between short-term optimization and lifetime reproduction success is not straightforward. For example, a less-than-optimal forager that is more alert to predators may leave more surviving offspring than the optimal forager that is unaware of the approach of danger (**see: Optimal foraging**). This also highlights another problem with the theory: it is difficult to categorize behaviours into discrete, independent units. For example, while collecting food, a bird may be classed as foraging, but it is also likely to be watching out for predators and/or potential mates, thus it is difficult to justify the categorization of this behaviour as foraging. Optimality theory has also been criticized on the grounds that animals are adapted to past environments, not the present, and that phylogenetic inertia will often mean that animals are less than optimal. There are also methodological difficulties: to 'prove' that animals are behaving optimally, there needs to be a close fit between observed and predicted results, yet the scientific method is primarily based on the refuting of a null hypothesis. Thus, the demonstration of a statistically non-significant difference between what is observed and what is predicted might arise from poor scientific methodology.

The above criticisms have led some to reject optimality theory. However, the fact remains that the theory does enable behavioural ecologists to make very accurate predictions about how animals will behave under certain circumstances. It is of no surprise that some of the best examples of optimality theory involve cases in which short-term gains are almost certainly linked to long-term reproductive success. For example, a parent bird that optimizes the rate of delivery of food to a nest full of chicks is likely to maximize its reproductive success. In a similar vein, male dung flies that optimize the rate at which they fertilize the ova of females will sire the most offspring. Thus, the fact that animals appear to behave optimally under some circumstances would indicate that natural selection has operated in the past to produce very efficient (optimal) behaviours under certain circumstances. (PE, MLW)

Further reading

Alcock, J. (2005) *Animal Behaviour: an Evolutionary Approach*, 8th edn. Sinauer Associates, Sunderland, Massachusetts.

Orang-utan

Meaning 'forest people' in Malay, orang-utans are the largest arboreal mammal inhabiting the rainforests in Asia. There are two extant species of orang-utans, Bornean (*Pongo pygmaeus*) and Sumatran (*Pongo abelii*), found on the islands of Borneo and Sumatra, respectively, and both are at risk of **extinction**. Characterized by their reddish coat and long arms, orang-utans are primarily frugivores, eating other items opportunistically. As adults, orang-utans are sexually dimorphic, with males being larger and possessing enormous cheekpads. They are members of the **great apes**, but are unique in their semi-solitary behaviour and **noyau social organization**. (LAN)

See also: Endangered species

Orchidectomy

Orchidectomy is the removal of the male **gonad**, also known as **castration**, though that term can also apply to removal of the ovaries (ovariectomy) of females. This is a surgical procedure usually carried out in immature animals (**see: Mutilation**). It is done to make animals more tractable and easier to handle, to stop them being aggressive and to stop them mating when animals are kept in mixed-sex groups. In male cats, benefits may be claimed for the cats themselves, as their roaming and fighting can lead to them having recurrent fights and infections, and hence a reduced **quality of life**.

It is normally performed with an anaesthetic in animals with a high economic value (dogs, cats and horses), but not in young farmed animals such as piglets, lambs and calves. However, recent research has shown that young animals are also able to feel **pain**, and methods to provide some form of **analgesia** and/or **anaesthesia** are being researched and developed for farm animal species. (DBM)

Organic farming

Organic farming takes an holistic approach to farming and its principles are based upon the maintenance of the natural equilibrium that exists between animals, plants and the soil. Emphasis is given to the concept of naturalness (**telos**) within the definition of animal **welfare** of many organic systems. In 2005 the International Federation of Organic Agricultural Movements (IFOAM) outlined the ethical principles for organic farming. These are the principles of **health**, **ecology**, fairness and care. However, official standards for organic certification vary between different areas. Organic farming works with and supports natural ecological systems, and practices that may cause an imbalance between and within these systems are avoided, including high stocking densities, artificial fertilizers, chemical pest control and genetically modified organisms.

Claims over the production of a high-quality product is one of organic farming's main principles, and feed rations are designed to maintain animal health rather than optimize production. In support of this, minimum **slaughter** ages are set for each type of animal to prevent excessive **growth rates**, and overproduction is further avoided by limiting stocking densities to a farm's ability to recycle manure and to produce most of its own feed requirements.

Maintaining animal health is claimed to be a high priority for organic systems, and pharmaceutical therapeutics (drugs) may be used only as a last resort if alternative products have little or no effect on a **disease**. Though sick animals must be treated promptly, it is the prevention of disease that is one of the basic principles of organic farming. Breeds or strains of animals that have a naturally high resistance to disease are selected, and are fed nutritionally balanced feed rations that further promote their ability to ward off disease.

The health and welfare of stock are also given high priority in **housing** systems, and environmental diseases are controlled by cleanliness and appropriate stocking densities. The design of housing must allow for free movement, access to food and water and must have sufficient **lying** area, which should also be littered with a natural substance such as straw. Additionally, housing systems should also cater for **behavioural needs** at different stages of life and allow for the expression of natural behaviours such as socializing with **conspecifics**, rooting behaviour and avoidance of dung.

Castration of males is permitted to avoid tainted meat, but other systematic **mutilation** of body parts, such as tail-docking (**see: Docking – tail**), teeth trimming and **dehorning**, is not permitted unless these actions can be justified in terms of health and safety. (LW)

Further reading

Lund, V. (2006) Natural living – a precondition for animal welfare in organic farming. *Livestock Science* 100, 71–83.

Vaarst, M., Roderick, S., Lund, V. and Lockeretz, W. (eds) (2004) *Animal Health and Welfare in Organic Agriculture*. CAB International, Wallingford, UK.

Orientation response

The orientation response is the behaviour shown by an individual when a novel, unpredicted or potentially threatening **stimulus** is presented to it. The response is associated with increased attention towards the stimulus to allow its evaluation. At the same time, there is often an increase in **arousal** and activation of physiological systems in preparation for the 'fight or flight' response, depending on the outcome of this evaluation. Repeat presentation of the stimulus may result in **habituation**, as evidenced by a loss of the orientation response, and so it may be used to investigate this form of learning. (DSM)

See also: Alarm reaction

Ostensive definition

An ostensive definition defines something by way of examples. Implicit within the use of ostensive definitions is an assumption that this information can be generalized or used to create a reliable concept. Thus a stressor might be defined ostensively as something like a painful procedure, a scary event or something that makes the individual feel sad. In this case, the definition is likely to lead to the assumption that stressors are detrimental to **welfare**, and this may or may not be the intention of the individual providing this definition. Thus the **reliability** and **validity** of the definition as a form of communication are dependent on the quality of the examples offered, as well as the level of understanding of the receiver of the information. Ostensive definitions are often used when it is difficult to precisely define or interpret the nature of the subject of interest. (DSM)

See also: Operational definition

Ostrich

The ostrich (*Struthio camelus*) is the world's largest living bird, standing around 2 m tall and weighing 100–150 kg. Native to Africa, four subspecies, generally geographically separated, have been recognized largely on the basis of the colour of the skin and wing and tail feathers. Males have black body feathers, unlike the brown–grey body feathers of the female; both genders have white (or brown) primary wing and tail feathers. In sexually active males, the skin on the face, abdomen and legs flushes red. Young chicks are mottled brown, white and black, moulting out to brown–grey plumage in juveniles. The feathers are symmetrical, lacking interlocking barbules typical of other birds. The adult ostrich is a neotenous bird retaining chick-like characteristics (e.g. large eyes and soft, downy feathers) when becoming sexually mature.

The ostrich exhibits a flightless, cursorial lifestyle in open savannah, scrub and desert areas. Although capable of running at speeds of 40 km/h, this is usually only a response to a threat. The main diet of the bird is formed by vegetation, which is processed by an alimentary tract characterized by a large, expandable proventriculus, a large, muscular gizzard (which relies on swallowed stones to assist in grinding food) and a relatively small intestine compared with the very long colon. Digestion is assisted by microbial fermentation in large, paired caeca and proximal colon, and the short-chain fatty acids produced are absorbed in the distal colon. Micturition and defecation are separate events.

The ostrich is a herbivore feeding on a wide variety of plants according to the prevailing flora. Green annual forbs and grasses are preferred, although leaves, flowers and fruit from succulent plants will also be consumed. Selection of food items is by sight and the beak is used to strip the leaves. Up to one-third of the bird's time is spent **feeding**, a behaviour that occurs in bouts, often while walking, interspersed with periods of vigilance.

Breeding is relatively opportunistic and generally reliant on environmental conditions (especially rainfall). Male ostriches establish and defend territories around which the females wander before settling on one male with which to establish a nest. Courtship involves an elaborate display by the male, including a booming call and showing the female potential nest sites. Mating can occur outside of established pairs, and many females will lay some of their eggs in other nests. Egg laying takes place every 48 h and the clutch of roughly ten eggs can take 3 weeks to complete. The number of eggs in a nest can be much higher because of egg dumping by other females, but, once incubation is established, many of these eggs are pushed out (although it is unclear how the incubating female recognizes her own eggs).

The egg is the largest laid by any living bird (average of 1500 g, measuring 15 × 12 cm) and has a gross composition similar to that of the domestic fowl egg. Incubation takes 42 days and is continuous with the male sitting at night and the female sitting by day. Hatched chicks are ~30 cm tall and fully **precocial**, leaving the nest after 24–48 h to follow their parents in looking for food and water. Typically, groups of chicks form crèches looked after by one pair of birds until they are well-grown juveniles (**see: Cooperation**). Adult height is reached by 12 months of age and sexual maturity is at 2–3 years.

For many centuries ostriches were hunted for their feathers, which were used as human adornment. During the 19th century it became apparent that **slaughter** of wild ostriches was not sustainable, and in South Africa some birds were brought into **captivity**. Nevertheless, the ostrich remains threatened in much of its natural range. Captive breeding was fully established in the late 19th century by development of an artificial incubator, although captive adult birds were used to foster-rear most chicks. Farming was largely based around providing grazing on irrigated lucerne (alfalfa) supplemented by maize. Breeding birds were either kept in 'free-range' large groups or as pairs in small enclosures. **Captive breeding programmes** were also established to develop a bird more suited to a farming environment and for the regular plucking of mature feathers from living birds. To a large extent, this farming system still applies in South Africa.

Farming brought a rapid expansion of the worldwide market for feathers, and vast fortunes were made, particularly by farmers around Oudtshoorn, South Africa, which became (and remains) a major centre for ostrich farming. At this time ostriches were exported to North America, Europe and Australia, where farming operations were also established. In the early 20th century demand for feathers fell dramatically. By the 1930s only remnant farming operations remained in South Africa, and ostrich had disappeared from other parts of the world. By the middle of the century, a marketing cooperative was established in South Africa that developed the market for leather tanned from the skin of ostrich slaughtered at around 12–14 months. Characterized by patches of raised quills, ostrich leather was marketed as a high-quality, high-value product. The leather market led to a renaissance of ostrich farming in South Africa. More recently, the product range was extended by sale of the high-quality leg meat, often erroneously marketed as being low in cholesterol.

International sanctions against South Africa during the 1980s produced a shortage of ostrich hides, and ostrich farming became more attractive for other parts of the world. Aggressive marketing caused an extremely rapid spread of ostrich farming in North America, which led to an increasing awareness around the world. By the mid-1990s ostrich farming operations were established in Israel, Australia, Europe and other African countries.

Outside of South Africa, ostrich farming has been attempted in a wide range of climates and geographical conditions, although the farms are usually small in size. Often breeding birds are kept in 'trios' (one male with two females) in small enclosures. Eggs are artificially incubated and the chicks are intensively reared. Common problems have involved maintaining the health of birds, getting good **egg production**, fertility, hatchability, and poor survivability and **growth rates** of chicks. Commercial production and marketing of the products has been a major restriction to development of an ostrich industry to match that of South Africa. As a result, many farming operations in North America, Europe and Australasia closed down by the end of the 1990s. The few remaining ostrich farms in the UK are small-scale and service a local market they have developed.

Despite this decline in interest, ostrich farming continues to capture the imagination of farmers in many developing countries. At the start of the 21st century the market for ostrich products has recovered to a large extent. Ostrich farming remains strong within South Africa but, although more widespread geographically, elsewhere its commercial significance is limited. (DCD)

Further reading

Deeming, D.C. (1999) *The Ostrich: Biology, Production and Health.* CAB International, Wallingford, UK.

Overshadowing

If two potential conditional stimuli are repeatedly presented simultaneously in a classical conditioning procedure, it is possible that only one of them will be associated with the conditional response. This is because more intense or relevant stimuli may overshadow less intense or relevant ones. At a practical level, it is particularly important to be aware of overshadowing when training animals, because much training is done with verbal commands, but these may be overshadowed by the unintentional visual cues provided by the trainer at the same time, such as hand gestures or body posture. Thus the animal may initially appear to have learned to respond to the command, when in fact it has not. (DSM)

See also: Blocking

Oxytocin

Oxytocin is a nonapeptide (consisting of nine amino acids). It acts systemically as a hormone and within the **central nervous system** as a neurotransmitter or neuromodulator. The molecular structure is the same in all mammals, and the amino acid sequence of the oxytocin-like nonapeptide in birds and reptiles differs by only one or two amino acids.

Oxytocin is produced in the supraoptic and paraventricular nuclei of the **hypothalamus** and is secreted from the posterior **pituitary gland**. Oxytocin is also found in other locations because of projections of **neuron(e)**s from the paraventricular nuclei to sites in the **brain**, the brain stem and the spinal cord. The release into the bloodstream is highly pulsatile. The nature and frequency of pulses vary according to the species, sex and reproductive condition, but the pulsatility is an important functional aspect. Effects are often produced on pulsatile release that cannot be obtained via general increase in basal level, or only at much higher concentrations. **Opioid**s are of importance in the regulation of oxytocin, both at the level of the posterior pituitary and within the hypothalamus.

Oxytocin is well known for its involvement in **reproduction**, but it also seems to have general anti-**stress** properties. The types of behaviour that are induced are dependent on the prevalent steroid pattern and on the environmental cues.

Parturition

Parturition is associated with a cascade of hormonal changes, including the release of plasma oxytocin. Oxytocin causes uterine contractions when the uterus is responsive to it (due to various hormonal events, oxytocin receptors are up-regulated prior to parturition). Experiments on rodents and pigs have shown that environmental disturbances, i.e. movement to a less attractive location early in parturition, can disrupt parturition through an opioid-mediated inhibition of oxytocin secretion. When an opioid inhibitor was given prior to movement, parturition proceeded irrespective of movement and, when oxytocin was administered after movement, parturition was reinstated. Oxytocin, or an artificial version of it, is often used to promote parturition in domestic animals and humans. In the chicken, the hypophyseal nonapeptides arginine vasotocin and mesotocin (which is the avian form of oxytocin) are responsible for oviposition.

Lactation

During **lactation**, oxytocin is involved in the nursing process. Oxytocin is released into the blood in response to sensory manipulation of the teats. Via the circulation, it reaches the mammary glands, where it causes milk ejection by contracting the myoepithelial cells. In oxytocin-knockout mice, transfer of milk failed but the deficit could be repaired by administration of oxytocin. Milk ejection in dairy cows has been studied in detail because the efficiency of milk letdown has great practical and economic importance. Natural suckling by a calf has been found to cause greater oxytocin release compared with machine **milking**. This is probably related to the greater sensory input from manual stimulation or suckling by a calf than from the milking machine. Manual pre-stimulation of the teats before attaching a milking machine, as well as hand-milking during the entire milking process, enhances the levels of oxytocin. The reflex can become associated with environmental stimuli, so that oxytocin is released in anticipation of the sensory stimulation. Hence, the sound of the milking machine may evoke letdown in cows, and women may respond to the cry of a hungry baby in the same way, because of prior associations between these sights and sounds and nipple stimulation.

Mother–offspring bonding

In several species, oxytocin has been shown to promote mother–offspring **bonding**. In rats, the oxytocin released within the brain at parturition is involved in stimulating the rapid onset of **maternal behaviours** acting on up-regulated oxytocin receptors. Similar actions of oxytocin have been demonstrated in sheep, in which the central release of oxytocin is triggered by stimulation of the uterine cervix and vagina. In sows, oxytocin seems to be involved in timing of maternal responsiveness to piglets, i.e. to the change from being relatively responsive during the initial phase of parturition to being passive and non-responsive during parturition and the early postparturient period – which allows piglets to access the udder and reduces the risk of piglet crushing – to a later increase in responsiveness. In humans and non-human primates, correlation studies suggest that a link between oxytocin and maternal behaviour also exists in these species.

Sexual behaviour

Oxytocin is involved in **sexual behaviour**, but there is large interspecies variation in the distribution of oxytocin receptors in the brain. This may be one of the reasons why species vary in the role of oxytocin in sexual behaviour. For example, it plays a significant role in inducing sexual behaviour in rats (e.g. penile erection in the male and **lordosis** in the female), but not in mice. In general, in relation to mating, oxytocin is thought to reduce aggressiveness between individuals and to facilitate receptive behaviour, sexual arousal, orgasm and sexual satiety.

Social behaviour and memory

Oxytocin also has more general effects on the **social behaviour** and **memory** of several species. In monogamous female voles, it enhances selective social behaviours and the formation of pair bonding. Unmated females develop partner **preferences** when given oxytocin centrally or peripherally, providing this is done in a manner in which the pulsatility of oxytocin is mimicked. Data on mice have also indicated that oxytocin is involved in social memory, i.e. the development of social familiarity. In contrast, other studies implicated oxytocin as an amnestetic peptide. Apparently an intracerebral oxytocin stimulus may enhance forgetting, whereas systemic oxytocin may enforce memory.

Anti-stress system

Oxytocin seems to be implicated in a general anti-stress system. In conscious rats, massage-like stroking of the abdomen is followed by increased oxytocin levels. At the same time, there is a lowering of **blood pressure** and release of vagally regulated gastrointestinal hormones, indicating a shift from sympathetic to parasympathetic autonomic dominance. The massage seems to sedate the rats, as reflected by reduced locomotor behaviour, and they have prolonged withdrawal latency to heat and mechanical stimuli. Peripherally administered oxytocin produced similar acute effects. The effects of a 5-day treatment with oxytocin were pronounced and long-lasting (e.g. decreased blood pressure for 10 days in males and 20 days in females) and, when rat pups were treated immediately after birth, these effects were lifelong.

An anti-stress system involving oxytocin (or mesotocin, which is the avian equivalent) does not seem to exist in birds. In broiler cockerels, stepping and wing flapping increased dose-dependently with intracerebral oxytocin administration, but **preening** decreased. The experimenters interpreted the results as suggestive of a role for oxytocin in inducing a state of **arousal** in chickens that resembles **fear/anxiety**. This seems to contrast with the effects found in mammals. (BID)

Further reading

Kendrick, K.M. (2000) Oxytocin, motherhood and bonding. *Experimental Physiology* 85, 111–124.

Nelson, L. (2000) *An Introduction to Behavioural Endocrinology*. Sinauer Associates, Inc., Sunderland, Massachussetts.

Russel, J.A. and Leng, G. (1998) Review. Sex, parturition and motherhood without oxytocin? *Journal of Endocrinology* 157, 343–359.

Uvnäs-Moberg, K. (1998) Oxytocin may mediate the benefits of positive social interactions and emotions. *Psychoneuroendocrinology* 28(8), 819–835.

Uvnäs-Moberg, K., Johansson, B., Lupoli, B. and Svennersten-Sjaunja, K. (2001) Oxytocin facilitates behavioural, metabolic and physiological adaptations during lactation. *Applied Animal Behaviour Science* 72, 225–234.

P

Paedomorphosis

Paedomorphosis describes the evolutionary retention of juvenile characteristics into adulthood, i.e. it is the phenotypic outcome of the exclusion or reduction of some ancestral developmental stages during the **ontogeny** of an individual. A classic example of paedomorphosis is seen in the axolotl, (*Ambystoma mexicanum*), which is a form of salamander that, as an adult, has the external gills that are characteristic of the young of other species of salamander, but which are normally replaced by lungs in these related species as the animal matures. The opposite to paedomorphosis is peramorphosis, which reflects the outcome of the inclusion of all relevant normal developmental processes seen in the ancestor as only part of the related ontogeny of the feature of an individual – i.e. the animal appears to develop in certain ways beyond the stages seen as the norm in close relatives or its presumptive ancestors. The exaggeration of certain traits as a result of **sexual selection** and gigantism within a taxon, may arise as a result of peramorphosis, i.e. the animal may continue to develop these features beyond the normal limit of adulthood seen in related species – for example, the extensive antlers of some species of deer.

Such phylogenetic changes in the relative timing and rate of developmental processes are collectively referred to as **heterochrony**. Historically, within the field of applied animal behaviour, emphasis has been given to the consideration of the importance of paedomorphosis arising as a result of **domestication.** For example, juvenification may be associated with greater (but still limited) developmental adaptability in response to the man-made environment and a reduced risk from **aggression**, which are favourable traits; however, many domestic species appear to show a mix of paedomorphic and peramorphic change. The occurrence of such changes in these species can also be used as a basis for criticism of assessments of the welfare impact of domestic management regimes (**see: Intensification of animal production**), which are based on simple analogy with the wild type in similar circumstances. Some texts appear to confuse the term 'paedomorphosis' (a phenotypic outcome) with 'neotenization' (a slowing of development), which describes one of several mechanisms leading to paedomorphosis (see heterochrony for a more extensive description of the mechanisms underpinning this phenomenon). (DSM)

Further reading

Goodwin, D., Bradshaw, J.W.S. and Wickens, S.M. (1997) Paedomorphosis affects agonistic signals of domestic dogs. *Animal Behaviour* 53, 297–304.

Price, E.O. (1999) Behavioural development in animals undergoing domestication. *Applied Animal Behaviour Science* 65, 245–271.

Pain

Pain, in humans, is defined as an unpleasant sensory and emotional experience associated with actual or potential tissue damage. A similar definition of pain also applies to animals: pain is an **aversive** sensory and emotional experience by the animal in response to damage or a threat to the integrity of its tissues. In animals, as in humans, pain has both a physiological sensory and a psychological or emotional component. Physiological components are the conscious experiences of pain resulting from activated nociceptors (from the Latin *nocere*, 'to hurt') following tissue damage, while subjective contents are an unpleasant sensory and emotional experience that occurs with actual or threatening tissue damage.

However, pain in animals is relatively difficult to recognize and assess, since the exact nature of the emotional component remains uncertain, and an animal is unable to report the qualitative and quantitative dimensions of the sensory (pain) and emotional experience. The level of pain experienced can be affected by the maturation of the nervous system and by other factors such as **memory**.

Nociception that results in a feeling of pain will result in altered behaviour. Within a few milliseconds there may be a very rapid withdrawal motor reflex – for example, when touching something very hot. This is mediated through synapses between the sensory nerve and the relevant motor nerves in the spinal cord with the first-order **neuron(e)**. However, after that time, any motor movement or altered behaviour is most likely to be voluntary, e.g. to run away and escape, to limp, to vocalize. These behaviours are a result of the pain that is felt in the cerebral cortex, and the motor response will depend on the area affected.

Pain is classified as being acute or chronic, according to its characteristics. Acute pain is related to nociceptor activation following **injury**, and ends when the stimulus ends. The chronic pain phase is defined as an intense pain lasting weeks or months beyond the expected injury healing time, or as pain involving an alteration in the nervous system that is capable of maintaining a painful state after the stimulus is removed and the injury is healed. Chronic pain is characterized by an abnormally lowered pain threshold, so that normal, innocuous stimuli elicit pain (allodynia). The experience of pain upon nociceptor stimulation is more intense than normal (hyperalgesia).

Following tissue inflammation and injury, specific nociceptors and/or 'silent' nociceptors are elicited, from where arisen nerve signals are transferred to the surface layers (lamina I and II) and the neck (lamina V) of the dorsal horn of the spinal cord by lightly myelinated (Aδ) and unmyelinated (C) fibres; the signals then ascend to the thalamus through the spinothalamic tract, mediating the discriminative component of pain sensation and, together with the spinoreticular tract, produce the emotional-affective component of pain sensation.

Based on the 'gate control theory' of pain, the nociceptive system undergoes modulation at the dorsal horn level of the spinal cord. A pathological activity of sensitized nociceptors leads to plastic alterations at both peripheral and central levels following an injury or **inflammation**. These changes increase peripheral sensitization and central hyperexcitability, such as abnormal sensation in the somatosensory system (neuropathic pain), and reduce activities of pain inhibitory systems, such as the functions of the periaqueductal grey matter, Raphe magnus nucleus and dorsal horn (laminae I and II) in the spinal cord, leading to a decrease in the pain threshold and hyperalgesia (increase in the intensity of pain), as well as allodynia (pain elicited via activation of normally non-painful Aβ fibres by an innoxious stimulus such as touch). (H-WC, DBM)

Further reading

Flecknell, P. and Waterman-Pearson, A. (2000) *Pain Management in Animals*. Saunders, London.

Wall, P.D. and Melzack, R. (1999) *Textbook of Pain*, 4th edn. Churchill Livingston, London.

Palatability

Palatability refers to the hedonic quality of orosensory stimuli or the experienced pleasantness of a **taste**. Palatability may be indexed by changes in observable behaviours such as taste reactivity responses, which have been categorized in a number of species. The term is also used to refer to food properties that promote ingestion, or the immediate effect of a food on ingestion. (SuH)

Parasite

A parasite is an organism living within or on the body of another species (i.e. the host) and obtaining from the host food and/or protection. Parasitism is thus beneficial for the parasite and costly for the host. Parasites may spend their whole life on one host, or spend different developmental stages in hosts of different species. Their impact on animal welfare is variable, being dependent on load and virulence, but particularly a cause for concern in organic farming systems, where the use of agents to control parasite loads is limited. (BM)

Further reading

Ebert, D. (1998) Experimental evolution of parasites. *Science* 282, 1432–1436.

Hovi, M., Sundrumm, A. and Thamsborg, S.M. (2003) Animal health and welfare in organic livestock production in Europe: current state and future challenges. *Livestock Production Science* 80, 41–53.

Parasitism

Parasitism, or parasitic symbiosis, can be defined as the symbiosis between animal host and infecting organism, whereby the latter has a perceived benefit and the animal host suffers detrimental effect or at least where the infecting organism lives at the expense of the animal host. The infectious organism is called the **parasite** and, based on the definition, can be a virus, bacterium, fungus, protozoan, helminth or arthropod. However, classic parasitology usually excludes viruses, bacteria and fungi, to be examined within the discipline of microbiology. It distinguishes between endoparasitism and ectoparasitism. Endoparasites are internal, and include the gut and blood parasites, while ectoparasites are mainly involved in cutaneous infestations. In some cases parasites themselves will contain parasites, i.e. hyperparasitism, for example the *Plasmodium* parasite within the *Anopheles* mosquito, or the *Dipylidium* tapeworm cyst within the *Ctenocephalides* flea larva.

The forms of parasitic coexistence are too numerous to cover in this context but, in reference to behaviour, parasites have been found to have unique abilities to influence animals' behaviour. To illustrate, some parasites affect the ability of their host to increase the chances of the host becoming prey and so enabling the completion of the parasite's lifecycle (*Leucochloridium* in snails). In other cases the parasite may manipulate the male host into diverting all its energy to nurturing the parasite, the former having been rendered sterile and displaying female behaviour. The parasites will now be treated as his brood to be protected (*Sacculina* in crabs). (FLR)

Parasympathetic nervous system

The parasympathetic nervous system (PNS) arises in the craniosacral division of the spinal cord. The preganglionic nerve fibres originate from the motor nuclei of certain cranial nerves in the brain stem, as well as the 2nd to 4th sacral segments of the spinal cord. With few exceptions, most organs in the body of vertebrates are innervated by the PNS. The cranial nerves primarily innervate effectors within the head region (**see: Central nervous system**), such as salivary glands and papillary muscles. An exception to this is the vagus nerve (cranial nerve X), which innervates the heart and lungs, stomach and upper intestine and components of the urinary system. PNS nerve fibres (also called cholinergic fibres) terminate in varicosities, which are the sites of release of the neurotransmitters. The principal neurotransmitter in the PNS is **acetylcholine**, which is released at both the pre- and postganglionic levels. Acetylcholine is synthesized in the nerve terminals from choline and acetate, and stored in vesicles until it is released directly into the space surrounding the target cells. The activity of acetylcholine is efficiently inhibited by hydrolysis with an enzyme, acetyl cholinesterase, which is stored in the synaptic cell membranes.

The central function of the parasympathetic nervous system is to maintain metabolic equilibrium and replenish energy reserves when the body is in a state of rest, repair and restoration. The effects of dominant parasympathetic activity include decreased **heart rate**, increased visceral activity (e.g. digestion) and an overall decrease in metabolic rate, all of which contribute to the maintenance of **homeostasis**. Effects of activation within the PNS are much more refined than those exhibited by the **sympathetic nervous system** (SNS). There is no mass simultaneous activation of parasympathetic nerves, such as the general excitation seen in the SNS. Rather, individual nerves can be independently activated, which is facilitated by the typical 1:1 synaptic arrangement of pre- to postganglionic synapses. This configuration allows for very specific activation of target effectors. For example, the vagus nerve can at any point in time function to regulate cardiac activity without having to exert any regulatory influence on the digestive system.

The PNS and SNS were traditionally viewed as exhibiting only coupled reciprocal control, where an increase in one branch lead to a decrease in the other. In many instances the branches do behave in a rigid bipolar or continuum fashion, with PNS activity at one end of the plane and SNS at the other. According to this model, the state of autonomic control at any point in time is characterized by a point on the continuum. In reality, however, the interrelationship between PNS and SNS activity is better represented by a bivariate model that encompasses reciprocal regulation as well as co-active and independent regulation. Co-active regulation, or coupled non-reciprocal control, describes positively correlated activity within the two branches. For instance, an increase in PNS activity is coupled with a similar increase in SNS activity. Independent, or uncoupled, activity occurs when a change in one branch is not accompanied by a change in the other branch.

There is increasing interest in the neural mediation of stress responses in farm animals, in particular in measuring activity within the subdivisions of the **autonomic nervous system**. In general, however, very little attention has been paid to the role that the PNS plays in instrumenting stress responses in any species. Of particular interest is the vagal nerve, which has both afferent and efferent capabilities. Two structures in the brainstem, the dorsal and ventral vagal complexes, are responsible for the regulation of PNS activity. The dorsal complex regulates parasympathetic activity beyond the level of the diaphragm, whereas the ventral complex controls activity above the diaphragm. Emerging research has shown that the ventral vagal complex is capable of expressing both inhibitory and excitatory control over cardiac activity, vocalizations and some respiratory functions. For instance, certain stressors can initiate PNS induced changes in heart and respiratory rate that do not require any activation of the sympathetic nervous system. The PNS, in particular the vagus, has also recently been identified as playing an important role in the neural regulation of the immune response. Afferent signals in the vagus nerve are involved in communicating with the CNS, via cholinergic anti-inflammatory pathways, to initiate early and rapid immune system responses to inflammation. This anti-inflammatory role is brought about by vagal inhibition of the release of TNF, HMGB1 and other pro-inflammatory cytokines through the activation of certain nicotinic receptors. (RM-F)

Further reading

Berntson, G.G., Cacioppo, J.T. and Quigley, K.S. (1991) Autonomic determinism: the modes of autonomic control, the doctrine of autonomic space, and the laws of autonomic constraint. *Psychological Reviews* 98, 459–487.

Berntson, G.G., Cacioppo, J.T. and Quigley, K.S. (1993) Respiratory sinus arrhythmia: autonomic origins, physiological mechanisms, and psycho-physiological implications. *Psychophysiology* 30, 183–196.

Goehler, L.E., Gaykema, R.P.A., Hansen, M.K., Anderson, K., Maier, S.F. and Watkins, L. (2000) Vagal immune-to-brain communication: a visceral chemosensory pathway. *Autonomic Neuroscience* 85, 49–59.

Wang, H., Yu, M., Ochani, M., Amella, C.A., Tanovic, M., Susarla., Li, J.H., Wang, H.C., Yang, H., Ulloa, L., Al-Abed, Y., Czura, C.J. and Tracey, K.J. (2003) Nicotinic acetylcholine receptor alpha 7 subunit is an essential regulator of inflammation. *Nature* 421, 384–388.

Parathyroid hormone

Parathyroid hormone is a peptide hormone active in calcium **homeostasis**. Release is stimulated by a decrease in extracellular calcium concentrations, from a number of small parathyroid glands adjacent to or embedded within the thyroid gland. Parathyroid hormone acts by stimulation of osteoclastic activity in promoting calcium release from bone, to both enhance calcium uptake in the intestine and reduce calcium excretion from the kidney. (MRC)

Parental investment

Parental investment is 'investment by the parent in an individual offspring that increases the offspring's chances of surviving (and hence reproductive success) at the cost of the parent's ability to invest in other offspring' (Trivers, 1972). The evolutionary premise underlying parental investment is the concept of life history trade-offs, i.e. current reproductive effort diminishes residual reproductive value, or prospects for future reproduction, either by reduced survivorship or fertility. Parental investment among mammals is usually assessed in the context of maternal investment because of the energetic and time costs involved in **gestation** and **lactation**. Sex-biased parental investment is a process that involves differential allocation of resources and risks to producing and rearing daughters compared with sons, as a function of their anticipated reproductive output.

Sex-biased parental investment includes both differential partitioning of resources to sons or daughters, as well as facultative adjustment of secondary sex ratios. Sex-biased allocation of resources devoted to rearing offspring, rather than modifications in secondary sex ratio, is more likely among species with lengthy periods of infant dependency, while variation in sex ratio at conception is more likely among species that spend little time in rearing offspring. According to the Trivers–Willard model (Trivers and Willard, 1973), the difference in maternal timing of sex-biased investment is because sex-biased maternal investment is regulated by three life history requirements: (i) maternal condition is correlated with offspring condition at independence; (ii) offspring condition at independence is correlated with their condition as a reproductively mature individual; and (iii) adult condition has a direct impact on variation in reproductive output.

Since the concept of parental investment was formalized, and the conditions under which sex-biased investment were identified, over 1000 papers have been published that have attempted to determine the degree to which differential allocation of resources to sons and daughters has reproductive consequences for the mother. Antithetical conclusions have been derived from the published literature because of a number of brambles in the trail:

- Over a dozen ways to assess 'maternal condition', ranging from kidney fat index to body mass to chronological age to visual estimates, have been adopted by scientists.
- A minimum of three generations is required to test the model, because the maternal pay-off to sex-biased investment arises from offspring reproductive output.

- An unbiased secondary sex ratio or pattern of investment can be interpreted as either stochastic variation or support for the model, depending upon the relative costs and benefits of rearing daughters and sons.
- Both the production and rearing of progeny result from a myriad of complex factors, with animals residing in a variety of mating systems, so the search for a universal Holy Grail within labyrinthine biological systems creates nearly insurmountable odds.
- Ecological, demographic, social, and phylogenetic factors mould life history strategies, so female flexibility in response to local circumstances could increase fitness more than an inflexible strategy dependent only upon 'condition'.
- If females have evolved mechanisms to respond to local environmental perturbations, then adjustments in maternal investment over the short term could reflect either adaptive modifications or small sample size effects.
- Whether or not female mammals actually possess the ability to modify sex ratio at conception remains unknown.
- Females have evolved age-specific reproductive strategies, so the costs and benefits of adjusting maternal investment patterns might depend more upon chronological age or parity than upon body condition.

In sum, testing models of parental investment by collecting solid empirical data is extraordinarily difficult.

Evaluating parental investment presents an enormous academic challenge, but also provides a solid evolutionary basis for assessing trends and colony management programmes in zoological institutions. For a perspective on the topic, consider a few examples. In red **deer**, *Cervus elaphus*, variance in male reproductive success is a function of body size, which is strongly dependent upon levels of maternal expenditure. In bighorn **sheep**, *Ovis canadensis*, only a weak correlation connects maternal and lamb mass, with neonatal mass having an impact on adult male body mass and reproductive success. In rhesus **macaques**, *Macaca mulatta*, maternal mass accounted for about 8% of the variance in infant mass and about 2% of the variance in sons' progeny production. In neither African **elephants** nor Cape **buffalo** is sex ratio of progeny related to maternal condition, but heavier reindeer are more likely to have sons than are lighter reindeer.

One might reason that captivity could favour a male-biased sex ratio if body condition is enhanced under food provisioning and if maternal condition has a direct impact on the sex ratio of progeny. Suppose that life in a zoological institution resulted in increased body condition among females or in accelerated reproductive maturation. If animals have evolved physiological mechanisms to modify secondary sex ratio, and females in good condition tend to produce males, one might then obtain a male-biased sex ratio in captive settings. One might find that first-time mothers have more difficulty giving birth (if the advanced physiological state favouring reproduction is not accompanied by an increased rate of skeletal maturation). **Zoos** might also experience sampling error in sex ratio due to small sample sizes.

Confinement in zoological settings might be associated with increased **stress** levels that hinder reproductive output and thereby favour production of daughters, because female ability to partition extra resources into sons will be hampered by stress-related endocrine disruption. Comparative studies of both ungulates and primates in the wild and in captivity have not found any evidence that enhanced food availability influences secondary sex ratio. Furthermore, a survey of 66 mammalian species in North American zoos found little evidence that secondary sex ratio was male biased among polygynous species. In sum, although a logical argument can be made that maternal investment in mammals directly depends upon maternal condition, evidence that maternal condition impacts sex-biased investment is inconsistent and has been interpreted as both strong, as well as weak.

One might expect parental investment to be prominent in long-lived mammals where females need to allocate resources to dependent young that increase the likelihood of offspring survival, but that also hinder prospects for future reproduction by delaying subsequent conception. Females have probably evolved a context-dependent life history strategy whereby maternal effort is dictated not by sex of offspring, but by size of offspring, condition of mother and resource availability. Individuals have probably evolved an adaptive physiological flexibility that enables them to adjust maternal expenditure in offspring according to demographic, social and individual context. For example, among non-human primates, sexually monomorphic, dual-sex dispersing species, such as the white-handed gibbon, *Hylobates lar*, should pursue different investment strategies from either sexually dimorphic, male dispersal species, such as savannah baboons, *Papio cynocephalus*, or sexually dimorphic, female dispersal species, such as **chimpanzees**, *Pan troglodytes.*

Since the phylogenetic history of mammals, birds, **reptiles**, amphibians and insects has produced a plethora of mechanisms for producing and rearing progeny, factors contributing to variation in secondary sex ratio or levels of maternal expenditure are not identical across taxa. Many alligators and turtles have temperature-dependent sex determination, whereby offspring sex is not due to sex chromosomes, but is a function of nest incubation temperature. Females can control birth sex ratios within litters by adjusting nest temperature. A number of insect species are haplodiploid, in which case the queen of the colony directly controls sex of offspring by either fertilizing or not fertilizing eggs. Among avian species, females have two different sex chromosomes, while males have a duplicate copy of the same sex chromosome, which is opposite to the case in mammals, so mechanisms of sex ratio adjustment at conception have to differ from those in mammals, where females are the homogametic sex.

The potential for sex-biased parental investment among females creates a golden opportunity for exploration among zoological institutions because of their long-term database. A productive research approach for the future is to identity factors that can, and do, modify parental investment among a variety of creatures under specific conditions. Uncovering proximate mechanisms that influence both secondary sex ratio and levels of maternal investment could enhance captive management programmes. (FBB)

References and further reading

Bercovitch, F.B. (2002) Sex-biased parental investment in primates. *International Journal of Primatology* 23, 905–921.

Cameron, E.Z. (2004) Facultative adjustment of mammalian sex ratios in support of the Trivers–Willard hypothesis: evidence for a mechanism. *Proceedings of the Royal Society of London, Series B* 271, 1723–1728.

Faust, L.J. and Thompson, S.D. (2000) Birth sex ratio in captive mammals: patterns, biases, and the implications for management and conservation. *Zoo Biology* 19, 11–25.

Hewison, A.J.M. and Gaillard, J.-M. (1999) Successful sons or advantaged daughters? The Trivers–Willard model and sex-biased maternal investment in ungulates. *Trends in Ecology and Evolution* 14, 229–234.

Trivers, R.L. (1972) Parental investment and sexual selection. In: Campbell, B.G. (ed.) *Sexual Selection and the Descent of Man*. Aldine, Chicago, Ilinois, pp. 136–179.

Trivers, R.L. and Willard, D.E. (1973) Natural selection of parental ability to vary the sex ratio of offspring. *Science* 179, 90–92.

Parturition

Parturition is the act of birth (**see: Gestation**). This process may take a few hours to several days depending on the species, the number of pregnancies and whether the animal is polytocous, i.e. has several young at one time. It frequently occurs at night in some species, e.g. in rodents, but at any time in other species (cattle, sheep, dog). It comprises three stages: the first stage ends when the cervix is fully dilated due to hormonal changes (relaxin release) and uterine contractions push the **fetus** up against the closed cervix, gradually dilating it. The second stage involves the expulsion and delivery of the **fetus**. The third stage is when the placenta is expelled. Some animals like the sow, cat and dog and many rodents, have a strong desire to nest build at this time and will work hard to access nest-building materials, such as straw (sows) or pluck their own fur out (rabbits). Assistance may be required when difficulties threaten the welfare of the mother and offspring during parturition. Such difficulties are collectively termed dystocia. Caesarean section is an example of assisted parturition, where an incision is made through the abdominal and uterine walls in order to deliver the fetus. (LW, DBM)

Paternal behaviour

Paternal behaviour patterns are associated with, or characteristic of, a father. Female reproductive success is limited by time and energy constraints and benefits from ensuring the survival of offspring. Commonly, the males of a species maximize their reproductive success by mating with multiple females (polygyny) and leaving them to care for the offspring. However, paternal behaviour is found in about 6% of species where two parents are necessary to provide for and defend the young, and is generally associated with monogamy (a single partnership between one parent of each sex). In mammals, much paternal behaviour shown is qualitatively similar to **maternal behaviour** with the exception of **lactation**, although there is often a quantitative difference in the behaviours expressed. In some species (e.g. sea horses), paternal behaviour involves greater investment and input to the rearing of young than that given by the mother, with the father taking on the more typical maternal role. (LMD)

See also: Parental investment

Pavlov, Ivan

Ivan Petrovich Pavlov (1849–1936) was a Russian physiologist and psychologist who was awarded the Nobel Prize in 1904 for his research on the digestive system. Pavlov is widely known for his studies on classical conditioning of the salivation response in **dogs**. (BM)

Pecking behaviour

The pecking behaviour of the domestic **chicken**, *Gallus gallus domesticus*, has been extensively researched, in large part because of the adverse welfare and economic impacts of damaging forms of pecking in this species. These include **feather pecking**, vent pecking and **cannibalism**. Pecking behaviours are often described by the substrate that is pecked; for example, feed pecking is distinguished from litter pecking. The morphology of the peck may also be used to describe it; for example, an aggressive peck is fast and directed at the head of another bird. The response of the recipient is also often used in descriptions; for example, gentle feather pecks ignored by the recipient are sometimes distinguished from severe feather pecks, in which the recipient attempts to escape, on this basis. However, the behavioural motivation and function underlying the particular pecking behaviour may be obscured by such descriptions: pecks directed at feed may result in feed particle ingestion or be part of exploratory foraging and not result in any feed intake. The more commonly used descriptions of types of pecking behaviour used in the behaviour research literature are used here.

The ancestral species of the domestic fowl spends about 60% of its active day foraging for feed. Feral domestic fowl spend about 50% of the active day foraging. Feed pecks consist of the bird lunging at the food with an open beak, which then grasps the feed item. The head then withdraws before the item is swallowed. During foraging behaviour the bird pecks, pulls, tears and strikes substrates, pausing frequently to scratch alternatively with both feet and explore the scratched area with the beak. Newly hatched chicks have innate preferences to peck at small, spherical objects. These preferences are modified in the light of experience. The nutritional consequences of ingestion and **social learning** from **conspecifics** ensure that chicks peck less at inedible particles and more at edible particles. However, fowl will continue to peck at inedible substrates throughout life, to some extent. A **diurnal rhythm** of feeding develops by about 8 weeks of age, so that there are peaks in the morning and towards dusk. Feeding behaviour tends to be synchronized between individual hens due partly to 'social facilitation'.

Aggressive pecking commonly takes the form of a single, rapid and hard peck directed at the head of another bird. The recipient is often feeding, and the peck results in the cessation of feeding and an attempt to escape or show submission. The behaviour is rare before 3 weeks of age and becomes more common after 6 weeks. Fowl use these pecks, together with other aggressive behaviours, to establish a **dominance hierarchy** or 'pecking order'. Once formed, usually by about the tenth week of life, aggressive pecking becomes less frequent. Males perform more aggressive pecking than females. Increased competition for scarce resources can increase aggressive pecking. Aggressive pecking increases with group size up to about 100 birds, but thereafter increasing flock size seems to result in less aggressive behaviour. This is possibly because hens in very large flocks are unable to distinguish between individuals, and hence adopt a strategy to avoid unnecessary agonistic interactions. Within some flocks, certain birds may attract more aggressive pecking than others. These 'pariah' birds have difficulty in feeding and suffer from poor body condition and low **egg production**.

Aggressive behaviour and aggressive pecking are often confused with feather pecking behaviour and cannibalism. Flocks in which cannibalism or feather pecking have occurred are frequently described as 'aggressive', and flocks without these problems as 'docile'. However, aggressive pecking can be distinguished from feather and vent pecking, in that it tends not to cause feather or tissue damage, tends to be associated with feeding and results in a behavioural response from the recipient. Feather pecking, on the other hand, does not seem to be related to competition for resources, is often ignored by the recipient unless it is severe and is unaffected by dominance position. Feather pecking may result in **feather** removal, feather damage or tissue damage.

Feather pecking involves a hen pecking at the plumage of another bird. The behaviour is common among laying chickens, both caged and those housed in alternative systems such as percheries or free-range. The behaviour also occurs in other species, such as **turkeys** raised for meat, or pheasants. It is much less common in broiler chickens. The behaviour has not been observed in the ancestral species of the modern chicken, the red jungle fowl (*Gallus gallus spadiceus*), when in its natural environment. Gentle feather pecking is defined in the behaviour literature as pecks to a conspecific's feathers which elicit little reaction in the recipient and do not involve feather pulling. Severe feather pecking is usually considered as either more forceful pecks with the recipient attempting to escape, or attempts at feather pulling which sometimes result in feather removal. The behaviour is important, since feather removal is likely to be painful and the resultant plumage damage reduces food conversion ratios, making flocks less economically efficient. Furthermore, measures taken to reduce this behaviour, such as **beak trimming**, also have an impact on chicken welfare and economics.

Feather pecking behaviour has been seen in chicks less than 2 weeks old. It is unclear whether this early pecking is under the control of the same behavioural systems as the more damaging pecking seen later in life. Peaks of feather pecking are observed at 4–6 weeks and 11–12 weeks of age, but the behaviour increases greatly after the onset of lay at around 18–20 weeks of age. This increase in damaging pecking may be associated with the increase in circulating gonadal hormones at this time. There are two main theories on the underlying motivational basis of feather-pecking behaviour. Both theories suggest that feather pecking behaviour results from the redirection of other forms of pecking behaviour. In the one case, feather pecking is considered to be a form of redirected foraging or ground-pecking behaviour. Various studies have shown a reciprocal relationship between foraging behaviour and feather pecking: improved quantity and quality of foraging opportunities reduces feather pecking. In the other case, feather pecking is said to result from chicks inappropriately **imprinting** on feathers as a substrate during the development of **dust-bathing** behaviour. Such imprinting is less likely in the presence of a suitable litter substrate.

It is also possible that feather pecking is not derived from just one behavioural system. Feather pecking can 'spread' from bird to bird, which is one reason why it can be especially problematic in the large flocks housed in non-cage systems. There may be more than one mechanism responsible for this: there is evidence that damaged plumage can attract more pecking and that the behaviour may be socially transmitted when birds observe the behaviour in others.

Many factors increase the risk of feather pecking. Genetic influences are important: there are strain differences in propensity to feather peck and it is possible to selectively breed both for and against the behaviour. It is possible that selection for improved production may have inadvertently resulted in strains more likely to feather peck. This may be remedied by future selection for lower feather-pecking strains. Various factors are known to reduce the likelihood of feather pecking: (i) provision of good-quality litter in rearing and laying environments; (ii) dim lighting; (iii) feeding of an adequate diet; (iv) feeding mash rather than pellets; (v) the provision of perches in rearing and laying period; (vi) small group size; and (vii) lower stocking densities.

An epidemiological survey of alternatively housed UK flocks produced evidence that the following risk factors were associated with a greater likelihood of an outbreak of feather pecking: (i) less use of range on sunny days; (ii) frequent diet change; (iii) flock inspections by more than one person; (iv) the use of bell rather than nipple drinkers; and (v) the use of nestbox lighting. The causal relations of these risk factors with feather pecking are not yet ascertained. The same survey found that nearly 57% of responding farms reported outbreaks of feather pecking in their most recent flock.

Object pecking may be another form of redirected ground pecking. Hens will peck at inedible objects throughout their life, probably as a form of environmental exploration. They will **contra-freeload**, i.e. work for food by performing foraging pecks in the presence of a freely available food source. Object pecking is generally viewed as pecking at non-food, non-litter objects within the birds' environment, such as cage walls or furniture within pens such as perches. Increased object pecking has been noted experimentally in chickens undergoing fasts to mimic induced moulting. Object pecking has been reduced experimentally where hens have been encouraged to forage in litter enriched with grains. It is possible that there may be a relationship between feather pecking and object pecking.

Because of the interest in selectively breeding low-feather-pecking lines of laying hens, there is a requirement to find a suitable measure of the propensity of a bird to feather peck. Pecking at inanimate objects has been explored as one such measure, with equivocal results. In one experiment, birds identified as feather peckers were more likely to peck at a variety of objects, including buttons and feather bunches, than birds characterized as non-peckers. Birds from a high-feather-pecking line pecked more at inanimate objects than those from a low-feather-pecking line in one experiment, whereas in another similar experiment no such relationship was found. In another study, no correlations were found between pecking at feather bunches and cannibalism. Inedible objects have also been used as a form of environmental enrichment in poultry, in an attempt to reduce injurious pecking behaviour. For example, provision of white string elicits much exploratory activity, and there is some anecdotal evidence that it can reduce feather pecking in laying chickens. The provision of destructible objects, such as polystyrene blocks, may be most effective at reducing feather pecking.

Vent pecking is directed at the cloaca of conspecific birds, and usually results in tissue damage that can progress to evisceration and death, causing very high mortality in affected flocks. It is the commonest cause of cannibalism, when chickens peck and consume the flesh of their conspecifics. In the UK nearly 37% of farmers reported that the problem had occurred in their last flock. The behaviour often starts at the onset of lay. The reasons for its occurrence remain obscure, but the exposure of the cloacal mucosa after oviposition seems to be the stimulus that attracts pecking. The relationship between feather and vent pecking is unclear: many researchers consider them to be separate phenomena, whereas some suggest that vent pecking may develop from severe feather pecking. Providing perches during rearing may help prevent this vent pecking, by allowing birds escape from pecking flock mates, provided that perches are at a sufficient relative height. Perching within pecking distance of birds below may actually increase the risk of vent pecking. Risk factors for vent pecking identified in one UK survey include: (i) the use of nestbox lighting; (ii) changing the diet three or more times in lay; and (iii) the use of hanging bell drinkers.

Due to the tissue damage and painful death that can result from cannibalism, the welfare impact can be severe. Cannibalism frequently follows vent pecking, but may also be an indirect consequence of severe feather pecking which exposes and damages skin elsewhere on the body, such as the tail base. Experiments have shown that deliberately damaging birds' feathers increases the risk of cannibalistic attack. The phenomenon is affected by many of the same factors that feather pecking is affected by. For example, providing a diet with adequate protein and a good-quality litter can reduce cannibalism. It is more common at higher light intensities, in birds being fed pelleted food rather than mash and in larger group sizes. Beak trimming reduces deaths from cannibalism.

As with feather pecking, outbreaks of cannibalism are unpredictable. Outbreaks can result in large mortality rates. Certain genetic strains of birds show more cannibalism than others. There is a possibility than cannibalistic birds have different individual characteristics than other birds: there is some evidence that they have larger body mass than others and are more active. It may be that a small number of such birds are responsible for starting outbreaks, and that other birds may then contribute secondarily. Following the first cannibalistic attack, hens in the group are more likely to perpetrate another attack. The mechanism for this is uncertain. Blood is known to be a potent stimulus for pecking in birds, but there is also some evidence that social learning of cannibalism can occur. The causal relations of these risk factors with vent pecking are not yet ascertained. (GD)

See also: Laying hen housing; Measuring behaviour

Further reading

Blokhuis, H.J. (1986) Feather pecking in poultry: its relation with ground pecking. *Applied Animal Behaviour Science* 16, 63–67.

Green, L.E., Lewis, K., Kimpton, A. and Nicol, C.J. (2000) Cross-sectional study of the prevalence of feather pecking in laying hens in alternative systems and its associations with management and disease. *Veterinary Record* 147, 233–238.

Leonard, M.L., Horn, A.G. and Fairfull, R.W. (1995) Correlates and consequences of allopecking in White Leghorn chickens. *Applied Animal Behaviour Science* 43, 17–26.

Savory, C.J. (1995) Feather pecking and cannibalism. *World's Poultry Science Journal* 51, 215–219.

Vestergaard, K. and Lisborg, L. (1993) A model of feather pecking development which relates to dustbathing in the fowl. *Behaviour* 126, 89–106.

Perception

Perception is the process by which individuals register and evaluate information detected from the internal or external environment, consciously or unconsciously. For a **stimulus** to be perceived, information about it must first be detected and then undergo processing. An important aspect of perception is that the resulting representation of reality differs from reality itself; individuals sample information, and then form representations about the world and their situation in it. An individual's subjective representation of all perceived stimuli (i.e. its perceived world) is termed its '**umwelt**'.

It could be said that the senses provide the material for perception, while the processing system (usually a nervous system) provides the tools to shape it. The quality and quantity of stimuli that can be perceived depends on the individual's sensory apparatus, while the complexity of perception will depend on the individual's cognitive abilities and previous experiences.

Perception is a very broad term, being used to describe the process by which, for example, a plant *detects and orchestrates a response* to seasonal change, or at the other extreme, to distinguish between whether an animal *sentiently experiences* nociceptive stimuli as **pain** or merely responds behaviourally to the stimuli. The term therefore has a hierarchy of meanings depending on the organism in question and that organism's processing capabilities. In animals with a degree of behavioural flexibility, the perceptual process becomes much more complex, but it also retains some of the simpler aspects of perception, such as reflexes. Moreover, in humans and other sentient animals, the meaning of perception depends on what aspect of the process is under study. For example, in research fields focusing on sensory detection, the term is defined as 'information pick-up', while definitions within fields focusing on the later evaluation of information include perception as a mechanism for 'hypothesis generation'.

Information processing usually takes place via the nervous system, but even organisms that do not possess nervous systems, such as plants or *Paramecium* spp., are sometimes said to perceive stimuli in order to respond to them. Perception in these cases would refer to registering stimuli, presumably without **qualia**, leading to an inflexible evaluation of their biological relevance to the organism; these organisms can generally only respond in a hard-wired manner to the stimuli they register. Yet perception in these organisms can still differ from reality, leading to incorrect behavioural decisions. For example, a summer-flowering plant that uses temperature to induce flowering may flower too early if it experiences unusually warm springtime weather. This is a very broad meaning of perception, and is not usually relevant to animal behaviour or **welfare** studies.

In sentient or conscious animals, registering stimuli can produce qualia, which are the sensory **feelings** generated early on in the perceptual process. In these animals, the evaluation stage of perception can involve integrating information from different sensory modalities, categorizing and recognizing stimuli and determining their relevance. These processes may call upon learned information and memories, as well as genetically constrained, reflexive mechanisms. Some examples of this perceptual level come from work by Kendrick and colleagues showing that **neuron(e)**s in sheep's **brain**s fire in response to specific categories of **conspecifics** and humans, whether the eliciting stimuli are olfactory or visual.

It is perception at this cognitive level that has been described as a process that generates hypotheses about reality – an individual develops moment-to-moment beliefs about the real world on the basis of the cues it detects and, through its behavioural responses, it might test those beliefs. For example, if walking through a tropical forest, we glance ahead and think we see a dangerous snake in our path, we might prepare ourselves for avoiding it, but also cautiously inspect it to get a better idea of what it really is. If on closer scrutiny we see that it is merely a piece of vine, the hypothesis that it was a snake is refuted, and we can continue along the path.

Perception can also be used as a term to differentiate between stimuli that are subconsciously processed versus those of which a human or animal is aware; perceived stimuli would be those that the individual is aware of, compared with all the other sensory information the individual would also have detected. In conscious or sentient animals, recognition of which stimuli are behaviourally relevant helps to direct attention toward those stimuli. Any irrelevant stimuli can then be ignored even if they have been detected by the sensory apparatus. Examples of such 'inattentional blindness' include a study by Simons and Chabris (1999), who asked human volunteers to monitor particular aspects of a crowd scene; the volunteers were so intent on the task that most failed to notice the irrelevant but surprising presence of a man in a gorilla suit in prominent positions within the crowd.

Perceptual deficits in brain-damaged individuals have also contributed to our understanding of this level of perception. A classic example is work by Weiskrantz and colleagues on 'blind-sight' in humans and non-human primates with damaged visual cortices. These individuals are unable to see stimuli on the affected side, even though their eyes are undamaged. However, if asked to guess characteristics of the stimuli, they are correct more often than chance would predict, suggesting that they can subconsciously see the stimuli but that they do not experience any visual qualia on that side. Therefore, there appears to be a distinction between conscious and sub-conscious perception, at least in the primate species studied thus far.

From an applied animal behaviour and welfare perspective, an individual's sensory biases will be important, as well as the way the individual evaluates stimuli, i.e. the emotions or motivations that relevant stimuli provoke.

Sensory perception

The type of stimuli that can be detected and perceived will depend on the animal's sensory apparatus, and this can differ markedly between species and even between conspecifics. Therefore, each individual might detect different aspects of the world, giving it a distinct perception of stimuli from other individuals. Moreover, natural selection means that the type and range of senses that a species possesses are tuned to ethologically relevant stimuli. For example, **nocturnal** animals either have extremely sensitive vision or rely much less on that sense than their diurnal relatives would; they communicate not with visual displays, but with sound or scent, and their hearing and olfactory apparatus will be highly tuned towards detecting signals from conspecifics and predators or prey.

In humans, we traditionally divide the senses into five modalities: vision, audition, **olfaction**, gustation and somato-sensation. Sometimes differences between humans' and other animals' sensory abilities take the form of shifted sensitivity ranges within a sensory dimension, e.g. elephants hear lower frequencies of sound than humans (infrasound), while many small animals hear higher frequencies (ultrasound). Similarly, while many non-human primates have similar trichromatic colour vision to humans, most other mammals have dichro-matic vision, which extends into the ultraviolet in some species. As for birds, most have tetrachromatic vision, allowing extremely fine colour discrimination and often extending into the ultraviolet. Bees and octopuses can see polarized light, and pit vipers and some boas have infrared detectors in their snouts for 'seeing' warm-blooded prey. Most vertebrates are much more reliant on olfaction than are humans, using scents to discriminate between food items, predators and familiar and unfamiliar places, and to communicate with conspecifics.

The vast majority of these communications are ones that we cannot currently understand or monitor directly, and we may even be unaware that they occur. However, we have harnessed dogs' superior olfactory abilities for various applications, including training them as 'sniffer dogs' for detecting illegal substances, finding individual humans or identifying certain cancers. We can also influence animals' behaviour through chemical sensory means, either deliberately (e.g. using anxiolytic pheromones to help reduce behavioural problems in cats and dogs) or inadvertently (e.g. stressing laboratory rodents can cause them to release an alarm odour that increases **stress** in nearby conspecifics in an uncontrolled manner).

Some animals use senses that have no real equivalent in humans. Examples include **echolocation** in cetaceans, bats, swiftlets and oilbirds, which locate objects by emitting sounds and monitoring the delays in echoes deflected by the objects. Platypus, some fish, sharks and rays can detect weak **electric fields**, including those generated by the movements of other animals, predators or prey; this sense is known as **electroception**. A final example is **magnetoception**, by which some birds, reptiles and crustaceans can detect spatial changes in the earth's magnetic field; indeed, despite thousands of years of domestication, chickens have retained their magnetoceptive abilities (Freire *et al.*, 2005). This sense is also one that, along with vision and olfaction, allows homing pigeons to navigate long distances, an ability that has long been recognized by humans using pigeons for delivering messages and in competitive racing.

For any animal behaviour experiment, it is important to be aware of the sensory abilities and limitations of the study species and even the individuals themselves (e.g. in case of impaired vision or audition), to ensure that they can perceive the relevant stimuli. Animals are slower to learn tasks when non-salient cues are used, e.g. rats learn significantly faster with olfactory or tactile cues compared with classic operant tasks that use coloured lights. Understanding an animal's sensory biases can also help ensure that the experiment is not confounded by cues we cannot perceive. The well-developed olfactory abilities and **communication** of rats and mice have hampered many experiments in this way. For example, rats observing a conspecific demonstrator complete a task to obtain a reward appeared to be able to imitate the demonstrator's actions, but Mitchell and colleagues (1999) later discovered that the rats were actually responding to attractive scents that the demonstrator left in key locations on the apparatus.

Where the sensory capabilities of captive animals differ from those of humans, welfare problems can be caused by stimuli that we cannot easily detect, making the problem particularly difficult for us to understand and to solve. For example, rats have very sensitive vision for low light conditions, which means that captive individuals can suffer retinal damage from lighting that is comfortable for humans. They also have ultrasonic hearing and communicate using ultrasonic **vocalizations**, and some apparently silent laboratory equipment, such as computer monitors, produces loud ultrasonic sounds that might harm their welfare. The ultraviolet vision of birds can also mean that their behaviour is affected, and their welfare possibly decreased, by the ultraviolet-deficient lighting in captivity, as shown by Cuthill and colleagues' research.

Evaluative perception

From an applied animal behaviour perspective, the way animals evaluate stimuli to potentially generate an appropriate behavioural response is one of the most important aspects of perception. Many of these evaluative processes can be unconscious but others may involve some form of awareness (attention, sentience or consciousness). In sentient animals, perception may generate certain emotions and motivations in relation to a stimulus, e.g. a perceived predator might generate fear, while a familiar playmate might generate pleasure or excitement. Equally, the same stimuli might be evaluated very differently by different individuals. For example, the same male mouse scent might be perceived as threatening by a subordinate male conspecific, attractive by a female conspecific, unhygienic by a human and appetizing by a cat.

The way stimuli are evaluated can be determined genetically (i.e. through natural selection), especially when the stimuli have important ethological significance. For example, rabbits, rats and mice, even those bred for generations within laboratories, respond fearfully to predator odours. This rapid, innate evaluation of the odour as dangerous would enable the animals to behave appropriately even on their first encounter with a predator. Genetically constrained responses are also found in reproductive contexts and in very young animals, where perception of crucially relevant cues and rapid appropriate responses to them are very strongly selected for; examples might include sign stimuli, like the red spot on a herring gull's beak that induces its chicks to peck it for food.

Within an individual's lifespan, experience and learning can also affect the way stimuli are evaluated. Association of a conditional stimulus with an unconditional one usually describes a situation when an animal learns to perceive a detectable but previously irrelevant cue as relevant, because it predicts a relevant event. If the conditional cue predicts a food reward, an animal might learn to perceive it as having a positive meaning, and may even start to show feeding **appetitive behaviour** in response to it. **Ivan Pavlov**'s dogs are an example, with the sound of a bell attaining relevance as a predictor of food, even inducing salivation. At the other extreme, some animals learn to fear humans wearing particular clothing, because of associations between the uniforms and past **aversive** events.

Many behaviour problems in **companion animals** can be because of the way that the animals evaluate stimuli, e.g. dogs perceiving other domestic animals as prey or perceiving human visitors as a threat. Therefore, some corrective training techniques aim to change the way in which the animals evaluate relevant stimuli. To illustrate, a horse might shy at the sound of a squeaky gate, evaluating it as some kind of threat, but careful systematic desensitization (**see: Desensitization – systematic**) might eventually mean that the horse would start to perceive the sound as irrelevant, requiring no behavioural response.

The animal's current physiological and psychological state can also affect the way it evaluates sensory information. The hormonal state of an individual can determine its attraction or aversion to relevant stimuli, such as female mammals in **oestrus** becoming more attracted to males, or birds that usually have no apparent interest in sticks collecting them in springtime to build nests. On a different physiological scale, when animals are deficient in certain nutrients, their appetite for those specific nutrients becomes exaggerated and they perceive foods containing them as being particularly attractive ('state-specific appetite'). Then, once they have consumed enough, or too much, of those nutrients, those foods become unattractive or even aversive ('state-specific satiety').

Examples of animals' psychological states influencing their evaluative perception include findings from **cognitive bias** studies in humans and other animals, showing that when individuals have recently experienced negative events, in extreme cases becoming depressed, they evaluate ambiguous cues as being more negative (Mendl and Paul, 2004). In this way negative experiences can lead to animals behaving in ways that reduce the risk of experiencing further negative events. Similarly, when animals are trained to expect a large reward and are instead given a moderate reward (a situation known as successive negative contrast), they show a much less positive response towards that reward than animals which were originally only trained with the moderate reward.

In summary, perception is a hierarchical process that generates a subjective representation of an individual's situation in the world, involving varying levels of sentience and consciousness. It is relevant to applied animal behaviour science in terms of the kinds of sensory stimuli that individuals can detect and the ways in which they evaluate them. Understanding an individual's perceptual biases can help us to refine experimental designs, to solve behavioural problems and to improve the individual's welfare. (CCB)

See also: Sentience

References and further reading

Burn, C.C. (2008) What is it like to be a rat? Rat sensory perception and its implications for experimental design and rat welfare. *Applied Animal Behaviour Science* 112, 1–32.

Freire, R., Munro, U.H., Rogers, L.J., Wiltschko, R. and Wiltschko, W. (2005) Chickens orient using a magnetic compass. *Current Biology* 15, R620–R621.

Kendrick, K.M., Leigh, A. and Peirce, J. (2001) Behavioural and neural correlates of mental imagery in sheep using face recognition paradigms. *Animal Welfare* 10, 89–101.

Maddocks, S.A., Cuthill, I.C., Goldsmith, A.R. and Sherwin, C.M. (2001) Behavioural and physiological effects of absence of ultraviolet wavelengths for domestic chicks. *Animal Behaviour* 62, 1013–1019.

Mendl, M. and Paul, E.S. (2004) Consciousness, emotion and animal welfare: insights from cognitive science. *Animal Welfare* 13, 17–25.

Mitchell, C.J., Heyes, C.M., Gardner, M.R. and Dawson, G.R. (1999) Limitations of a bidirectional control procedure for the investigation of imitation in rats: odour cues on the manipulandum. *Quarterly Journal of Experimental Psychology: Section B* 52, 193–202.

Sales, G.D., Milligan, S.R. and Khirnykh, K. (1999) Sources of sound in the laboratory animal environment: a survey of the sounds produced by procedures and equipment. *Animal Welfare* 8, 97–115.

Simons, D.J. and Chabris, C.F. (1999) Gorillas in our midst: sustained inattentional blindness for dynamic events. *Perception* 28, 1059–1074.

Weiskrantz, L. (1990) *Blindsight: a Case Study and its Implications*. Oxford University Press, Oxford, UK.

Perching behaviour

Many birds will settle either standing or sitting on branches of trees to rest, a behaviour referred to as perching. However, not all birds are able to perch, as this behaviour requires suitable foot anatomy. Perching birds usually have three toes facing forward and one toe pointing back, which enable the bird to grip around the perch. Passeriforme birds, which make up over 50% of all bird species, are all able to perch but many species of Galliforme birds are also able to perch.

Perching is a highly motivated behaviour that is considered to be one of the most important behaviours to laying hens, and it has been shown that hens will work for access to a perch. Under commercial conditions, if there is limited perching space, hens will compete with each other to obtain a place to perch, particularly at night. Perches enable birds to move around at different levels and perch for rest. While perching, birds will also often perform other behaviours such as **preening**.

Perching behaviour is particularly common at night, when many birds will perch sitting in branches of trees throughout the dark period (**see: Roosting**). Birds seek out places to roost well before dark, even on short winter days, and it has been shown that the cue to go to roost is a reduction in the light intensity below a certain level. As dark approaches, birds become very active as they search for a suitable roosting site. Many species of bird, including species of parrot and groups of free-living hens, have preferred roosting sites within their **home range**, and flock members will move to these sites at night. Birds that perch in elevated locations with other flock members at night reduce the risk from ground predators and also benefit through heat conservation.

Even when there is enough space, hens have been observed to perch tightly together, suggesting they prefer to roost together. Under commercial conditions, hens will spend as much as 85% of the dark period roosting on perches, if provided, and if a perch is unavailable at night or access to their usual perch is prevented, hens will walk around more, attempt to perch at alternative sites and take longer to settle. During daylight hours, the amount of time spent perching may depend on a number of factors, including availability of perches, opportunity to perform other behaviours and breed. Hens in wire-floored cages that provide perches will spend between 25 and 50% of the light period perching, although there is much individual variation. This is less in pens with a litter floor.

Hens tend to prefer high perches and, when roosting at night, hens will all choose the highest available perch. Selecting higher perches may also be associated with **dominance** in some birds, and this may commonly result in problems with owners of parrots who allow their birds access to high perches. Experimental evidence has shown that hens do not show any preference for different perch shapes and designs. However, different perch design can result in differing foot damage. Evidence suggests that, in general, perches with a rectangular cross-section cause less damage than those with a circular cross section. Rectangular perches result in less keel bone damage than circular ones, but round ones with flat upper and lower surfaces appear to result in even less keel bone damage and fewer bumblefoot lesions and allow good grip. For budgerigars, irregular perches are recommended for this reason, and it is recommended that various irregularly sized perches are used in order to exercise the feet and prevent foot problems.

The development of perching behaviour is affected by early experience. Chicks must learn to roost when young. Under natural conditions, junglefowl and feral hens will roost on the ground while their chicks are young and, when they are about 6 weeks of age, the hen will begin to take chicks on to lower branches at night for roosting, encouraging them to fly up with a 'roosting call'. Chicks reared under commercial conditions without access to perches are less adept at using them later in life. Learning to use perches at an early age increases the muscle mass and **bone strength** of the birds and enables them to use the perches better as adults. It also helps birds to develop the cognitive skills necessary for moving around in three-dimensional space. Rearing chicks without access to perches before 4 weeks of age can result in a number of additional problems, such as an increase in the incidence of cloacal cannibalism and aggression, as well as an increase in eggs laid on the floor rather than in the nestboxes. In addition, rearing without access to perches seems to impair the spatial skills of the domestic hen, and its effect may be both pronounced and long lasting.

As well as providing the opportunity to perform an important highly motivated behaviour, making perches available to birds may reduce welfare problems through improved muscle and bone strength. As birds step on to and off a perch, they exercise their muscles and increase dynamic forces on the tibia, resulting in increased muscle and bone mass and strength. When a bird sits on a perch, it has to constantly contract its muscles to grip and support its weight. Despite improved bone strength, the provision of perches within commercial **laying hen housing** can result in an increase in injuries, particularly to the keel bone in hens. It is likely that this is due to collisions when jumping on to or between perches. Studies have demonstrated that jumping downwards between perches at steep angles causes difficulties for hens, and these difficulties increase with greater inter-perch distances, which may increase the risk of **injury**.

In conventional battery cages, birds are unable to perform this important behaviour – one of the reasons conventional cages are being banned within the EU. However, in enriched or furnished cages suitable perches must be provided for the hens. Another of the most common currently used alternative systems of **egg production** is the perchery system. The use of perchery or aviary systems for laying hens, and the provision of perches for other species of captively housed birds, may improve welfare through providing a more complex environment and improving bone strength, but care must be taken in the design and positioning of perches in order to prevent injuries. (RJNM)

Performance testing

Performance testing involves recording the **phenotype** of animals for one or more traits, with a view to identifying those animals most likely: (i) to be useful within a generation; and/or (ii) to produce the best offspring. The traits can be anything that can be objectively measured or subjectively assessed. In **companion animals**, particularly dogs, various aspects of behaviour are typical traits that are performance tested. This applies most commonly to organized breeding programmes for guide dogs and sniffer dogs. Performance testing is also widely used in breeding schemes for the control of multifactorial disorders, such as hip and elbow dysplasia. It is also used in a wide range of farm animals to select for increased productivity, e.g. milk yield, and reduced **disease**, e.g. **lameness** in dairy cattle. It is also used in some countries in the selection and grading of stallions. There is the potential for performance testing to be used for the comparison of different first-crosses among dog breeds for suitability to a range of environments in which dogs are kept, ranging from flats to backyards to 2 ha blocks.

In typical cases, the traits being performance tested are multifactorial, meaning that their variation results from the combined effect of a range of genes and non-genetic factors. For such traits, the procedures of quantitative genetics developed and utilized successfully in production animals can be applied equally to other species. First, estimates of genetic and phenotypic variances and covariances must be obtained for the relevant set of traits; then selection indexes can be calculated, providing the best possible predictions of: (i) future performance within that generation; and (ii) performance of progeny in the next generation. In the case of multifactorial disorders, this latter prediction is couched in terms of the expected incidence of affected offspring from each potential parent or the expected score for the disorder in the offspring of a potential parent. Both types of prediction are based on a multi-point scoring scheme reflecting subjective assessment of a phenotype, such as a radiograph in the case of hip or elbow dysplasia. (JA, FWN)

Peripheral nervous system (PNS)

The PNS is that part of the nervous system that extends beyond the **brain** and spinal cord of vertebrates and is typically not specifically protected by hard structures. The peripheral nervous system is divided into different systems in a variety of ways, which can lead to some confusion. The afferent nervous system (or sensory nervous system) consists of those **neuron(e)s** within the PNS that carry signals from sensory receptors towards the **central nervous system** (CNS), while the efferent nervous system (or motor nervous system) is the part that conveys signals away from the CNS to effector structures like muscle and endocrine glands.

However, the PNS may also be divided on the basis of the types of structures served; thus the somatic nervous system may be considered to consist of the part involved in the maintenance of posture and movement (i.e. the supply to and from those structures that primarily effect skeletal muscle tone). The visceral nervous system may then be considered that part which serves other structures, such as the gut, skin, etc. However, this distinction is not necessarily reliable, with some including part or all of the nerve supply from the exteroreceptors (sense organs detecting external stimuli such as the eye, nose, tongue and skin) within the somatic nervous system. For some, the visceral nervous system is synonymous with the **autonomic nervous system**, while for others the latter is one component of the former, being limited to the peripheral innervations of those structures associated with the **stress** response. In this case, the nervous supply associated with the neurological coordination of visceral organs involved in the process of digestion may be considered to belong to a separate enteric nervous system.

The autonomic component of the peripheral nervous system may be further divided into the **sympathetic nervous system** (which broadly prepares the body for action) and **parasympathetic nervous system**, which largely antagonizes the effects of the former. Further confusion may arise because both the peripheral and parasympathetic nervous systems may be abbreviated to PNS. (DSM)

Personality

There is no universally accepted definition of personality, even within the human psychological literature, but the term 'personality' in animals is often used to describe the biologically based behavioural tendencies of an individual and refers to the relatively stable phenotype arising from interaction between the **genotype** and environment that produces consistency in behavioural style, but which also differentiates the individual from other members of the same species. In general, consistent behaviours (habits) that are correlated form traits (although it should be noted that certain behaviours may be related to different traits depending on the context in which they are elicited), and it is the traits that define the personality of the individual. Traits and habits are not synonymous. For example, if an animal runs away from an umbrella when it is opened, it has a habit of avoiding opening umbrellas, but if it tends to avoid a range of other unusual circumstances then we may say it is neophobic, or fearful of new things, which would be a trait. It is also important to distinguish personality from ability. The latter focuses on maximal performance capacity, while the former is more focused on typical performance.

The term is used synonymously by some authors with '**temperament**' and '**behavioural syndrome**', although some differentiate these terms. The term 'personality' is avoided by some, in order to avoid accusations of anthropomorphic bias (**see: Anthropomorphism**). In the human literature, 'temperament' is often used to describe inherited, relatively stable behavioural tendencies that are apparent early in life and form the basis of personality. However, in the animal literature the term temperament may be used to focus attention on relatively stable, emotionally dependent behavioural dispositions. Thus, from this perspective, fearfulness may be considered an aspect of temperament, but independent problem-solving ability would not, although it might be considered an aspect of personality.

The term 'behavioural syndrome' is frequently used within the context of **behavioural ecology** to describe collections of correlated behaviours expressed within a given context or across different contexts. The focus in this context is often the adaptive value of such predispositions, with different biases assumed to be superior in different contexts. This ecological approach also logically leads to the inference that certain expressions may be **maladaptive** in certain contexts and give rise to problems in certain environments, or that disorder in the underlying mechanism of a trait is possible giving rise to pathology or dysfunction. These concepts have not been given much attention in the animal literature, but deserve careful consideration. While it is possible to study single/narrow traits in isolation (e.g. by developing a specific scale), this approach does not necessarily relate traits to a more general framework of personality, which can lead to confusion and, potentially, be problematic.

Other major focuses of personality research include: the role of environmental and genetic influences on the development of personality, its physiological and neurological basis and the relationship between cognitive styles and personality (i.e. how personality affects the way things are perceived and evaluated).

Gosling *et al.* (2003) have suggested there are three necessary criteria for the establishment of personality in animals: (i) ratings should relate to real attributes of the animal; (ii) assessments by different independent observers must demonstrate reliability; and (iii) assessments must have predictive validity. Personality may be inferred from direct measurements of a range of behaviours that may be related in some way (such as behaviour in social situations) or subjective ratings of more holistic traits (e.g. gregariousness) and, in general, the latter are not only easier to undertake, but may also show greater reliability.

Although different personality traits may be identified from different behavioural focuses and there are potentially thousands of ways to describe the personality of an individual, one review across species, including octopuses and a range of vertebrates (Gosling and John, 1999), suggests that three traits appear quite consistently: (i) extraversion (arousal and sensitivity to reward, possibly linked to Gary's behavioural activation system (**see: Emotion**), sometimes described as boldness, curiosity, energy, exploratory tendency, neophilia, novelty seeking, sociability, surgency or vivacity); (ii) neuroticism (sensitivity to novelty or potential punishment, possibly linked to Gray's behavioural inhibition system (**see: Emotion**), sometimes referred to as emotionality, reactivity, fearfulness or nerve stability); and (iii) agreeableness (tendency to bond, sometimes referred to as affection, affability, affiliative tendency, affinity, altruistic tendency, aggressivity, fighting tendency or hostility, trust).

However, the equivalence of these traits cannot be assumed in different species, especially when there is no evidence of contextual and physiological similarity. It also needs to be considered that human perceptions of personality may largely reflect their own biases in perception rather than any consistent differences in the subjects being observed – for example, if a report-based psychometric instrument developed for use in people is applied to a non-human species, it should come as no surprise that the same structure is revealed in the non-human species.

The recognition of the potential importance of personality in animals has significant implications for animal science. Not least, it provides another dimension against which the suitability of experimental subjects in a given context can be assessed (**see: Individual differences**). Gosling also argues that the recognition of animal personality will facilitate a greater understanding of the biological basis of human personality, by facilitating the exploration of biological continuities through the exploitation of relevant animal models. Animal models potentially allow personality to be studied in a more controlled environment, with the manipulation of specific factors of potential interest with fewer confounding variables.

Personality in animals is also of practical importance, as animals with a particular personality may be more or less suited to certain environments and thus be at different risk of **welfare** and management problems in the captive environment or survival if being reintroduced as part of a **conservation** programme. They may also differ in their suitability to undertake different tasks, e.g. working/competing horses and dogs.

The study of human personality is also of practical importance in animal science – for example, the personality of those responsible for the care of animals (e.g. **stockpersons** and owners) has recently become an area of interest for those studying animal welfare and behaviour problems (e.g. Podberscek and Serpell, 1997; Hemsworth and Coleman, 1998). (DSM)

References and further reading

Gosling, S.D. (2008) Personality in non human animals. *Social and Personality Psychology Compass* 2, 985–1001.

Gosling, S.D. and John, O.P. (1999) Personality dimensions in non-human animals: a cross-species review. *Current Directions in Psychological Science* 8, 69–75.

Gosling, S.D., Lilienfeld, S.O. and Marino, L. (2003) Personality. In: Maestripieri, D. (ed.) *Primate Psychology: the Mind and Behaviour of Human and Non-human Primates.* Harvard University Press, Cambridge, Massachusetts, pp. 254–288.

Hemsworth, P.H. and Coleman, G.J. (1998) *Human–Livestock Interactions*. CAB International, Wallingford, UK, 151 pp.

Podberscek, A.L. and Serpell, J.A. (1997) Aggressive behaviour in English cocker spaniels and the personality of their owners. *Veterinary Record* 141, 73–76.

Sih, A., Bell, A.M., Johnson, J.C. and Ziemba, R.E. (2004) Behavioural syndromes: an integrative overview. *Quarterly Review of Biology* 79, 241–277.

Pest control – ethics of

Animals are considered pests when they are found in locations where they are not desired and where they cause damage to human property. Of the vertebrate pests, **rodents** probably constitute the largest problem. A considerable proportion of the world's food supply is lost to rodents, and rodents also spread some serious **diseases**; thus there is an obvious need for rodent pest control.

Whereas the **euthanasia** of laboratory **rats and mice** may only be carried out by trained personnel following detailed guidelines, a wide range of methods for killing these animals when considered pests are available to the general public. The most widely applied method is through bait poisons with anticoagulants, which when ingested cause death from internal bleeding after a prolonged period of disease, presumably accompanied by considerable **distress**. Many other commonly used methods also raise serious **welfare** concerns (Mason and Littin, 2003). The negative image of 'vermin' and the fact that the presumably distressful deaths occur out of human sight probably contribute to this discrepancy in which methods are accepted to kill animals of the same species. Given the detrimental consequences if rodent populations are not controlled, it can be argued that there is a much stronger human interest in such pest control than in most other cases when animals are killed.

Irrespective of the view on how to balance human and animal interests, there seems to be consensus that animals should not be made to suffer unnecessarily. There are methods for rodent control that cause little or no welfare problems, and Mason and Littin (2003) suggest rodent-proofing buildings, the use of well-designed and monitored snap traps or electrocution traps, cyanide gas and alphachloralose bait poison. However, these methods may be less convenient or more costly than currently predominant methods.

The view as to which animals are pests is not constant. For example, large carnivores, such as wolf and lynx, have been heavily hunted and continue to be disliked by the owners of potentially predated domestic animals, but are now protected in some countries where wild populations still remain. (AO)

See also: Free-range animals; Pest species – welfare of

Reference

Mason, G. and Littin, K.E. (2003) The humaneness of rodent pest control. *Animal Welfare* 12, 1–37.

Pest species – welfare of

Innumerable unwanted, free-living animals are routinely trapped, snared, poisoned, shot, gassed and otherwise killed or managed as part of strategies relating to environmental protection, **conservation** (including **wildlife management** and the control of **urban wildlife**) and the protection of human and animal health, safety and comfort. Pest control carries inherent risks to animal welfare. The risks are to 'target' animals that are killed by the control agent, target animals that are subject to control but are not killed (e.g. those that consume a sublethal dose of poison or that escape from a trap), non-target animals exposed directly to the control agent (e.g. primary poisoning, by-catch in traps), and – when pesticides are used – non-target animals exposed to poisoned carcasses. Target and non-target animals that are not exposed to the control agent can also be indirectly affected positively and negatively by impacts on their environment and on animals with which they interact or on which they depend.

Historically, there has been little concern for the welfare of animals deemed to be pests, but growing attention from the public and animal welfare non-governmental organizations is driving more research and regulatory change to benefit pest **welfare**.

Vertebrate pest animals are no less capable of experiencing **pain** and **distress** than other vertebrate animals. Indeed, they may experience more than other animals in some cases, owing to the actions inflicted upon them and to the types and degrees of **suffering** of which they are capable.

There are several explanations as to why this pain and distress might have been ignored:

- It normally happens unseen and we can therefore ignore it, remain ignorant of it or convince ourselves that we are not to blame for it.
- Pests are not valued as highly, and their suffering is considered easily justified on the basis of the needs of the **health** and **well-being** of humans, domestic animals and the environment being more important than those of pests (i.e. the harm that pests do is considered to outweigh the harm done to them) (**see also: Ethics**).
- Wild animals are subject to similar levels of suffering and a less than ideal manner of death in their daily lives, so what we do is considered less of a concern.
- Some common definitions of animal welfare do not fit with pest control, where the aim is to curtail normal freedoms, happiness and so on or to cause ill health or **death**. The lack of a concept to grasp, or definition with which to work, can be used as a reason to delay progress.
- Cases of domestic **animal cruelty** or animal mistreatment, where the human cause is clearer or mitigation easier to effect, are more easily tackled.

Why should we be concerned about the welfare of vertebrate pest animals? First, they can experience pain and distress and it has been argued that all animal suffering should be minimized. Secondly, it has also been argued that people are responsible for solving problems that they create: in many cases, it is introduced wildlife that causes particular problems. If we introduced the wildlife, we must control it and we must limit the amount of suffering we cause when we do so. Thirdly, there may be a legislative requirement to do so (for instance, to assess the animal welfare impacts of a pesticide when applying to register it for use in some countries). Fourthly, growing public awareness and concern means that the use of some more inhumane pest control methods is becoming less and less acceptable in some countries (for instance, the use of snares, leg-hold traps, strychnine, drowning, burrow fumigation with toxic 'tear' gas and glueboard traps is increasingly being debated).

A wide range of vertebrate animals are controlled, from mammals, birds and fish to **reptiles** and **amphibians**. Accordingly, control methods are wide ranging. Lethal methods include: shooting, **hunting** with bows (e.g. big game), hunting with dogs (e.g. **pigs**, game birds) and other animals (e.g. **rabbits**), explosives (e.g. for **roosting** animals), electrocution (e.g. **rodents**, roosting animals), traps, nets and snares, poison baits, injections, fumigants and gases, encouraging **disease**, the introduction of predators, fishing, burrow collapse, drowning (including drowning of trapped aquatic and terrestrial mammals) and methods of preventing **lactation** so that milk-dependent young die. Non-lethal methods include: live capture traps from which animals are later released, repellents and deterrents, physical barriers such as exclusion fences, and control of fertility or **breeding**.

All pest control methods have some impact on animal welfare. An ideal control method would be one that is completely effective but does not cause any suffering. There is currently no such ideal method. Hence, it has been suggested that we should use only the most humane control method available (i.e. one that causes the least amount of pain and distress), improve the humaneness of existing methods and conduct research to find more humane tools.

Reducing the pain and distress caused by existing tools can be as easy as following best practice when using it (e.g. ensuring that a lethal dose of poisonous bait is consumed, or using appropriate firearms and ammunition when shooting). It is easier with physical methods than with poisons, due to the extensive research required, for instance, on drug–poison interactions, appropriate doses, methods and routes of delivery, impacts on the safety of human users (including the possibility of intentional misuse) and domestic animals. There is thus a growing drive to find alternatives to existing methods.

Common definitions of animal welfare do not fit with pest control, where the focus is not on 'happiness', health or natural behaviour but the aim is to curtail normal freedoms or to cause ill health or death. Instead, a notion of welfare as the absence of suffering (in terms of pain, distress and so on), caused by compromise in each of five domains relating broadly to the **five freedoms**, seems the most applicable (see Sharp and Saunders, 2008). Considering the total animal welfare impact of pest control means thinking about all of the effects of the pest control in each of these domains, both intended and unintended, and on all sentient animals impacted. *All* animals includes target and non-target animals, such as dependent young or members or those of the same or another species or group.

There are published guidelines for assessing the impacts of pest control tools on animal welfare. There are also several reviews of animal welfare impacts of vertebrate pest control tools that outline methods for their assessment. Sharp and Saunders (2008) proposed a model for a 'systematic, comprehensive and transparent process that helps to generate consensus among diverse stakeholders regarding the humaneness of control methods'. The model also allows the relative animal welfare impact of different control tools to be compared. The animal welfare impact of most control methods has not been assessed. Some reviews are listed in 'References and further reading' below.

How can pest control be justified, given its impact on animal welfare? From a utilitarian perspective, pest control might be justified if the positive impacts outweighed the negative impacts. However, because pest control affects people, animals and the environment, the interests of all three should be considered when considering positive and negative impacts. This can lead to a conflict because of differing values placed on each of these. For instance, is it justified to kill several introduced predators to protect a few rare native species? Is it justified to kill sentient animals to protect insentient plants? Pest control should also be necessary: the necessity of control itself and the necessity of killing versus non-lethal alternatives should be considered. Consideration of a list of principles, such as the following, could help to justify pest control and ensure that harm to animal welfare is minimized.

- The aims or benefits and the harms of each pest control programme must be clear. Control should only be undertaken if the benefits outweigh the harms.
- Control should only be undertaken if the benefits outweigh the harms and if it is likely that the aims or benefits will be achieved.
- The methods that most effectively and feasibly achieve the aims of the control programme must be used.
- The method that causes the least amount of pain and distress while still achieving the aims must be used (this requires an assessment of the humaneness of all existing methods).
- The methods must be applied in the best possible way.
- Whether or not each control programme actually achieved its aim must be assessed.
- Once the desired aims or benefits have been achieved, steps must be taken to maintain the beneficial state.
- There should be research to reduce the negative animal welfare impacts of existing control methods and to develop novel methods that cause less pain and distress. (KL)

See also: Conservation; Exotic species invasion; Pest control – ethics of

References and further reading

Humane Vertebrate Pest Control Working Group (2004) A national approach towards humane vertebrate pest control. Discussion paper arising from the proceedings of an *RSPCA Australia/AWC/VPC joint workshop*, 4–5 August, Melbourne, Australia. RSPCA Australia, Canberra (http://www.rspca.org.au/events/seminar2003.asp).

Iossa, G., Soulsbury, C.D. and Harris, S. (2007) Mammal trapping: a review of animal welfare standards of killing and restraining traps. *Animal Welfare* 16, 335–352.

Littin, K.E. and Mellor, D.J. (2005) Strategic animal welfare issues: welfare and ethical issues arising from the destruction of free-living wildlife for disease control and environmental reasons. *OIE Scientific Technical and Reviews Series* 24(2), 767–782.

Littin, K.E., Mellor, D.J., Warburton, B. and Eason, C.T. (2004) Animal welfare and ethical issues relevant to the humane control of vertebrate pests. *New Zealand Veterinary Journal* 52(1), 1–10.

Mason, G. and Littin, K.E. (2003) The humaneness of rodent pest control. *Animal Welfare* 12, 1-37.

Sharp, T. and Saunders, G. (2008) *A Model for Assessing the Relative Humaneness of Pest Animal Control Methods*. Australian Government Department of Agriculture, Fisheries and Forestry Canberra.

Pet

Any animal kept primarily for companionship. Originally this word was applied to humans, implying inferior size, rank or importance, and was only more recently used to describe animals. Because of the inferiority implied, and also because of possible confusion with its use to describe people rather than animals, the term **companion animal** is more usually used in the scientific literature. (JWSB)

Petting

Petting is an interaction or display of affection and touching or stroking by a human. Many people enjoy a 'bond' or relationship with animals, especially with animals such as **cats** and **dogs**, and will frequently stroke or pet their animal. Small mammals, such as mice, **rabbits** and **guinea pigs**, are also touched or stroked regularly as are some domesticated bird species such as budgerigars. Larger animals, such as **horses**, and even cold-blooded animals (poikilotherms) such as **reptiles**, are also frequently petted (**see: Companion animal**).

Petting behaviour may have been one of the reasons dogs came to live alongside humans. The human tendency to nurture and a need for warmth and protection may have been a primary incentive to domesticate wolves some 12,000–80,000 YBP. Animals that have evolved to live within groups may enjoy being petted more than do members of species that are solitary. Most dogs, for instance, enjoy being petted, reflecting the social nature of the dog pack, where dogs may exhibit **allogrooming** or rest in proximity to one another. Cats, too, although solitary hunters, interact within their **social groups**, and human petting may replace the mutual **grooming** commonly observed among felines.

Petting may have significant health benefits for humans. Stroking an animal may have a calmative effect. Human **blood pressure** and **heart rates** decrease when a dog is petted, suggesting that a positive relationship with a companion animal may result in a reduction in cardiovascular **disease**. In human–dog contact, concentrations of β-**endorphin**, **oxytocin**, **prolactin** and β-phenylethylamine increase in both species during positive interspecies interactions such as gentle stroking. For instance, oxytocin, the hormone thought responsible for **bonding**, doubles its plasma concentrations in both humans and dogs after positive interactions.

Petting an animal gives humans an opportunity for close physical contact with another creature, which may be of particular benefit to those people who live alone or those who have less frequent contact with other people (the elderly or teenagers). As people pet an animal, they also talk to them in a language variant known as 'motherese', the same type of language used with babies. Petting an animal may also be used by humans as a secondary reinforcer for good behaviour in some species such as dogs and horses, an alternative to offering food.

Individual animals may differ in their acceptance of being petted and in the areas of their bodies that they prefer to be touched. This may be related to both the physiology and the **behavioural ecology** of the species. The vibrissae (whiskers) around the mouth of animals such as horses and cats, for instance, may be extremely sensitive, as they have a rich afferent nerve supply and are thought to be important in judging distances and detecting vibrations. Few animals appear to enjoy these areas being petted. Cats demonstrate more negative responses when being stroked in the caudal (tail) region, while more positive responses are shown when they are stroked in the temporal region (between the eyes and ears). The perioral region (chin and lips) is intermediate. Horses are sensitive in areas around the withers, mouth, flank and elbow regions and may dislike certain regions being touched – for instance the ears, eyes or groin. Grooming of horses at the withers, however, has a calming effect.

Traditionally, dogs have been petted along their dorsal region (back), but more recent educational programmes encourage chest and chin stroking, as this is thought to be less challenging to an unknown canine. No observable differences were found, however, in heart rate measurements based on the regions of the dog's body touched when kennelled greyhounds and guide dogs were studied, although heart rates decreased when stroking continued for a significant time.

Acceptance of petting may depend on the individual animal's experience when young. Dogs may become fearful, assertive or even aggressive when touched, especially those that were inadequately socialized when young. Some cats also dislike being petted and will resort to biting or attacking. Cats frequently signal the end of social exchange in this way, which can be a problem (petting **aggression**). Education of animal carers to familiarize their companion animal with being petted all over the body is important (**see: Habituation**). Education of children, who are most likely to suffer bites that require hospitalization, and of other people in the correct way of approaching, handling and petting animals may be influential in enhancing human–pet relationships (**see: Bite prevention programmes**).

Further unwelcome outcomes from petting include disease transmission via skin and saliva. Rabies, salmonellosis, toxocariasis, ringworm and sarcoptic mange are some of the diseases that can be spread from animal to human by petting. Routine precautions, such as washing hands after petting animals, may lower the risk of **zoonotic disease** transfer. For the majority of people, nevertheless, the benefits of petting animals are usually considered to outweigh the disadvantages. (JR)

Further reading

Katcher, A.H., Friedmann, E., Beck, A.M. and Lynch, J. (1983) Looking, talking, and blood pressure: the physiological consequences of interaction with the living environment. In: Katcher, A.H. and Beck, A.M. (eds) *New Perspectives on Our Lives with Companion Animals*. University of Pennsylvania Press, Philadelphia, Pennsylvania, pp. 351–359.

McGreevy, P., Righetti, J. and Thomson, P. (2005) The reinforcing value of physical contact and the effect on canine heart rate of grooming in different anatomical areas. *Anthrozoos* 18, 236–244.

Odendaal, J.S.J. and Meintjes, R.A. (2003) Neurophysiological correlates of affiliative behaviour between humans and dogs. *The Veterinary Journal* 165, 296–301.

Soennichsen, S. and Chamove, A.S. (2002) Responses of cats to petting by humans. *Anthrozoos* 15, 258–265.

Phenotype

Phenotype is any observable trait or characteristic of an individual, such as its morphology, development or behaviour. The expressed phenotype results from an individual's **genotype**, environmental factors and an interaction between these. (LMD)

Phenotypic cline

A phenotypic cline is a gradual change in the **phenotype** of a species across a geographical range (sometimes referred to as character gradients). For example, in the common guillemot (*Uria aalge*), a polymorphism characterized by the presence (bridled) or not of a white ring around the eye and a white line running from this ring to the nape changes in frequency from less than 1% in the south of its Atlantic range to over 50% in the north. (PE)

Pheromonatherapy

Pheromonatherapy refers to the use of **pheromone** or social **odour** signals in a clinical context to manage behaviour problems (often related to some form of **stressor**) in mammals and birds. In this context, the message usually consists of an artificially synthesized mixture of relatively simple chemicals in a particular combination delivered into the immediate environment of the animal to be treated, e.g. by spray or diffuser. These mixtures are inevitably simpler combinations than those emitted by animals in their natural state, but they are used at higher concentrations. Patrick Pageat, a French veterinary behaviour specialist, is largely credited with the initial isolation of the mixtures that underlie the formulation of many of these products, which are now specifically available for the management of a range of behaviour problems in cats, dogs, rabbits, horses and pigs. (DSM)

Further reading

Mills, D.S. (2005) Pheromonatherapy: theory and applications. *In Practice* 27, 368–373

Pheromone

The term 'pheromone' was initially used in 1959 by Karlson and Lüscher, and was derived from the Greek word *pherein*, meaning 'to carry', and the suffix *mone* was borrowed from hormone or the Greek word *hormon*, meaning 'to stimulate'. They defined the term pheromone as 'substances secreted to the outside of an individual and received by a second individual of the same species in which they release a specific reaction, for example, a definite behaviour or developmental process'. This definition held true in insect research on pheromones, but however it fell short in light of research in the area of mammalian pheromones. The main point of contention to this definition was the term 'specific reaction'. Mammalian behavioural responses are more complex than those of insects and are affected by a large number of influences and stimuli. Although higher animals' responses are biased by the presence of chemosignals, they do not usually respond in a uniform manner with one specific reaction, and so some prefer the term 'social odour' in these species.

Pheromones are most thoroughly studied in insects; however, more recent research has begun to characterize the functional role of pheromones in mammalian species. Pheromones, like hormones, are chemicals or groups of chemicals secreted by an organism which transmit a message that is detected by receptors. Unlike hormones, however, pheromones must travel outside of the body in order to transmit their message to another organism of the same species that have receptors along and near the nasal and oral cavities. The **vomeronasal organ** (or Jacobson's/Jacobsen's organ), which lies between the nasal and oral cavities, is the mammalian chemosensory organ most responsible for detecting pheromones. Some pheromones are thought to be detected by receptors in the olfactory membrane (**see: Olfaction**).

Releaser versus primer pheromones

Mammalian pheromones can be categorized as either releaser or primer pheromones. Primer pheromones, also called biostimulating pheromones, elicit long-term effects on the animal's developmental processes or their neuroendocrine/endocrine systems. Releaser pheromones, sometimes referred to as signalling pheromones, on the other hand, are noted by more immediate and shorter-lasting changes in behaviour. A common example of primer pheromones is those found in adult male **urine**. Prepubertal female mice have been shown to increase production of inositol-triphosphate (IP_3) in response to adult male urine. This increase in IP_3 production has been shown to hasten their maturation to puberty. Sexual pheromones that elicit a standing-to-be-mounted behavioural response are a great example of releaser pheromones, eliciting a short-term behavioural response in the receptive female.

Pheromone secretion

Pheromones can be secreted through any bodily secretion, including urine, saliva, sweat, milk and faeces. In addition, many species have specialized scent glands that are specifically used to deposit species-specific scents. Examples of such glands are those found in male **cats** or lemurs, which spray their scent from glands located near the rectum. Chemosignalling through the urine is one of the most widely studied routes effecting social behaviour. Pheromone signals in wolf urine are used to mark territory and warn unrelated wolves to stay away. Similarly, mice and other social animals secrete a cocktail of molecules through their urine which, together, communicate the individual's social and reproductive status. It has also been shown that young mammals in the presence of same-sex sexual pheromones will enter into puberty sooner than those that have no exposure. Many sexual pheromones have also been shown to increase territorial and social aggressiveness in males and maternal aggressiveness in females. Male mice, following removal of the vomeronasal organ, exhibited a significant reduction in aggressive and **sexual behaviours** and showed no sexual preference in mounting **conspecifics**.

Vomeronasal organ and pheromone transport

The vomeronasal organ is located within the nasal septum, and is tube-like in shape and lined with mucus. It is connected to either the nasal and/or oral cavity via the vomeronasal or nasopalatine duct, depending on the species. Running along the length of the lumen of the tube-like vomeronasal organ is a large blood vessel along with the vomeronasal sensory **neuron(e)**s. The lumen is crescent-shaped and microvilliated, the large blood vessel lining the inside of the crescent, while the outer side is lined with sensory neurons. The entire tube is protected within a hard capsule of cartilage. The vomeronasal nerves run from the vomeronasal organ to the accessory olfactory bulb of the **brain**. The protection of the cartilage maintains the sensory neurons of the vomeronasal organ away from the airflow within the nasal cavity. This system assures that stimulation of the chemoreceptor system is limited to the mechanism of vascular pumping. In this way the vomeronasal organ can actually take in non-volatile pheromones such as those often secreted in urine and saliva; many of the pheromones secreted from scent glands and from vaginal secretions are also non-volatile. In order for volatile secretions to be efficiently used as chemosignals they must be accompanied by a transporter protein.

Lipocalins are elicited by the body and bind to volatile pheromones so that they can be efficiently transported in body secretions. They are large proteins, approximately 17–30 kDa, produced by the chemosensory epithelia. Lipocalins bind to volatile pheromones via an odourant-binding site. It has been suggested that lipocalins may have alternative functions in pheromonal signalling aside from their transporter role. Some evidence suggests that lipocalins may also act as chemosignals, binding directly to pheromone receptors. Male mice mark territory using urine deposits, containing non-volatile and volatile pheromones. Volatile pheromones secreted in **mouse** urine bind to the lipocalins – called major urinary proteins, or MUPs. A mouse may produce up to 15 or more different MUPs in its urine in different concentrations. MUPs are some of the most widely studied lipocalins. They have been shown to continue binding to volatile pheromones following secretion, producing a slow release of volatile pheromones well after the urine is dry. The slow release of volatile pheromones attracts attention from competing mice to the mark. However, it has been suggested that the unique mixture of MUPs in the urine acts as an identifier **signal**, providing a virtual fingerprint of the individual that left it. This identifier signal may be the stimulus that elicits a countermarking response from other males.

The vomeronasal regulation of chemosignalling was originally thought to be a small subset of the main olfactory system. Contrarily, the current projected sensory map of the vomeronasal system is considerably different from that of the main olfactory system. In fact, the target centre for the axonal projections of the vomeronasal sensory neurons is the accessory olfactory bulb. The neural organizations of the olfactory and vomeronasal systems are quite different. The most pointed difference between these two is the axonal relationship with the glomerulus and mitral cells. Olfactory sensory neurons expressing a specific receptor project their axons to a single glomerulus located near the outermost region of the main olfactory bulb, while the axonal projections of vomeronasal neurons, expressing the same receptor, target one of 15–30 smaller glomeruli located within the accessory olfactory bulb. Similarly, mitral cells within the main olfactory bulb send out one dendritic process to receive information from the glomerulus, whereas mitral cells of the vomeronasal system send out several dendritic processes to gain information from multiple glomeruli, suggesting a role of the mitral cell as the processor for all vomeronasal input of a particular receptor.

Pheromone receptors

Recent research has identified approximately 250 pheromone receptors from three structurally distinct families, referred to as V1Rs, V2Rs and V3Rs. All three classes of pheromone receptors are G protein-linked receptors. These classes of receptors are distinct in the G protein α subunit expressed in the neuron. V1Rs are found in neurons expressing Gαi2, V2Rs are found in Gαo-expressing neurons while V3Rs are exclusive to neurons expressing Gαq/11. Olfactory neurons also generally express mostly Gαo. The significance of this difference is largely unclear; however, it is apparent that the neurons of the vomeronasal system express in large quantities only one of the two G protein subunits.

A mutant strain of mice deficient in Gαi2 displayed only half the number of neurons carrying V2Rs compared with their non-mutant counterparts. Aggressive and sexual behaviours were investigated in these mice, as these behaviours are known to be altered by pheromonal signalling. These mutant mice exhibited a severe reduction in both maternal **aggression** and male territorial aggression. The mice, however, showed no differences in sexual behaviour or partner preference. These results suggest a distinct separation of function between the vomeronasal sensory neuron types. Further studies were conducted to investigate the role of the receptor types in the prepubertal murine female increase of IP_3 response to male urine, a response which hastens their maturation into puberty. Pertussis, a bacterial toxin that blocks the activity of V1Rs and V2Rs, was shown to have no effect on this primer response. Further investigation of the prepubertal vomeronasal membrane *in vitro* determined that, when cultured with male urine and a V3R antagonist, the IP_3 response was prevented, suggesting a role of Gαq/11-containing neurons in this response in mice.

Response to pheromone

Pheromones have been shown to elicit a number of behaviour responses in mammals. Regulation of sexual behaviour is the most widely known function of pheromone **communication**. Certain female pheromones involved in sexual attraction are secreted from sebaceous glands located in or near the mons pubis of the female. In response to pheromones of females in heat, males commonly respond with **flehmen**, a behavioural response involving the lifting of the head and curling of the upper lip. Boars secrete a combination of two pheromones in their saliva, which elicits a wide array of behavioural responses from oestrous sows, including attracting them to the boar's location and initiation of the standing response, which is a stance adopted by oestrous sows signalling that they are ready to mate. Boar urine has been shown to elicit a similar response. This behavioural response is used in production in order to identify sows in **oestrus**. Studies have determined that synthetic boar pheromones used *in lieu* of a live boar or urine are just as effective in detecting oestrus in sows.

Similar to the response found in mice, puberty in female swine has also been shown to be accelerated by the presence of a boar. This hastening to sexual maturation can reduce the time to puberty by as much as 30 days. In addition, lactating sows exhibit a shorter postpartum anoestrus period in the presence of boar. Similar priming pheromones from rams and bucks have been shown to increase the age of puberty in **sheep** and **goats**. Rams are often used in production to induce oestrus in a flock or in order to synchronize oestrous cycling among the ewes, and this is often referred to as the 'ram effect'. Cattle research has revealed that as much as a 3.3-month hastening in puberty in heifers can be achieved following exposure to vasectomized bulls. Bull-exposed cows also experienced a shorter postpartum period, entering oestrus as rapidly as 60 days following calving, with a mean difference of 5–10 days faster than non-exposed cows.

One of the few well-characterized single-molecule pheromones is the **rabbit** nipple-search pheromone, a chemosignal found in rabbit milk aiding rabbit pups in locating the milk-producing teats. A similar maternal pheromone has been identified in sows that stimulates a feeding response in piglets. Piglets have been shown to refuse to feed if the odourant is washed off the sow. Post-weaned pigs with this maternal pheromone applied to either their feeder or directly to their snout exhibited increased activity over a 48 h period, and increased weight gain and improved feed:gain ratios over a 28-day period when compared with pigs not exposed to this pheromone post-**weaning**.

More recently, simple fatty acid mixtures, based on the cutaneous secretions of a range of species, have become commercially available for the treatment of a range of behavioural problems in **companion animal** species (**see: Pheromonatherapy**).

Pheromonal regulation of behaviour and physiology is an important consideration in animal production settings. Many production animals are maintained in large group sizes and often at high densities. These practices maintain an environment in which high doses of pheromones are constantly being taken up by the animals. The flooring substrate and waste removal practices utilized may also maintain high levels of pheromones in the environment. The consequences of the continuously high degree of chemosignalling stimuli are not yet known. It is well known that aggressiveness can be drastically altered by the presence of certain pheromones. Reduction or removal of certain pheromones from the housing environment might suppress antagonistic behaviours, thereby improving the welfare of the animals. Reproductive, developmental and neuroendocrine responses may also be altered greatly by the high doses of pheromone maintained in production housing.

Animals as a model for understanding human pheromones

Animal models may not present an accurate representation for the study of pheromone signalling in humans. The human vomeronasal organ degenerates well before birth, and continues to exist only in vestigial form. Similarly, the majority of the pheromone receptor genes identified through animal research are found only as non-functional **pseudogenes** in humans. Only a few functional pheromone receptor genes are identified in humans, and their current functions are unclear. One such receptor, the hV1RL1, has been found to be expressed in the nasal mucosa in humans; however, its role as a pheromone receptor is, as yet, uncertain. It has been suggested that, with complex **language** and communication, pheromone signalling lost its adaptive advantage in the species. However, compelling evidence does exist to support a possible role for pheromones in menstrual cycling. Synchonization of female menstrual cycles (the 'Whitten effect') has been well documented in women housed in close proximity for a prolonged period of time. Primer pheromones present a likely explanation for this response, suggesting the possibility of a role for pheromones in human physiology. (RD)

Further reading

Brennan, P.A. and Keverne, E.B. (2004) Something in the air? New insights into mammalian pheromones. *Current Biology* 14, R81–R89.

Rekwot, P.I., Ogwu, D., Oyedipe, E.O. and Sekoni, V.O. (2001) The role of pheromones and biostimulation in animal reproduction. *Animal Reproduction Science* 65, 157–170.

Wyatt, T.D. (2003) *Pheromones and Animal Behaviour*. Cambridge University Press, Cambridge, UK.

Philopatry

Philopatry is the tendency of an individual to return to or stay in a particular home area. Species that breed at or near their birthplace exhibit natal philopatry, while others that return to the same breeding site exhibit breeding philopatry (or site fidelity). Natal philopatry is often sex-biased, with one sex (usually females in birds and males in mammals) more likely to disperse then the other. It is generally thought that sex-biased dispersal is a mechanism to avoid **inbreeding**. (PE)

Photoperiod

Photoperiod is the duration of light during a standard 24 h day. It is of ultimate (adaptive, evolutionary) significance to animals because it dictates the windows of time during which certain activities are possible (e.g. **sleep** versus activity). It is also of proximate (mechanistic) significance because it provides predictive information about critical matters such as seasonal changes in environmental conditions (weather, food supply). It is practically relevant to those interested in husbandry of both domestic and captive wild animals, because it plays a critical role in coordinating the different elements of animals' lives on a day-to-day basis as well as on longer time scales, such as seasonally. Timing of extremely important processes such as **reproduction**, migration, hibernation/aestivation and replacement of body covering (e.g. plumage, pelage) is controlled fundamentally by seasonal changes in photoperiod in many kinds of animals.

Activity cycles of almost all animals, as well as many temporal patterns of physiological activity, such as metabolic rate, hormone secretion and digestive system activity, are controlled fundamentally by endogenous **circadian rhythms**. The only known exceptions are species that live in essentially constant conditions, such as the fossorial naked mole rats. These endogenous rhythms persist in the absence of light/dark cycles, but gradually drift away from the natural night/day cycle (i.e. free-run) if deprived of the entraining effects of light/dark cycles. Aberrant activity patterns (e.g. arrhythmicity) can result if animals evolved to occupy environments typified by night/day changes in light encounter certain unnatural conditions. For instance, many types of bird become arrhythmic if held in constant bright light, whereas very dim light permits these endogenous rhythms to persist.

Circadian rhythms are susceptible to modification by environmental inputs, the most obvious being the entraining effects of the onset of the light period, but also the modulating effects of such factors as food supply and moonlight. Circadian rhythms are regulated by endogenous clocks in the **central nervous system**, particularly in the suprachiasmatic nucleus (SCN) of the **brain**, and these clocks receive modulating inputs from various sources, most notably the eyes of mammals (via the retino-hypothalamic tract, a separate pathway connecting certain populations of retinal ganglion cells directly to the SCN, and thence via the brain stem to the pineal). Photoreception for regulation of circadian rhythms in non-mammals is more complex and less dependent on the eyes.

Except for exactly on the equator, photoperiod varies in a highly predictable manner on a seasonal basis. Consequently, it provides extremely reliable information to animals regarding impending changes in environmental conditions that are relevant to survival and reproductive success. Many kinds of animals from all vertebrate classes have been shown to display behavioural and physiological responses to seasonal changes in photoperiod, which are also integrated with inputs from other kinds of cues, such as temperature, food supply/nutritional plane and social interactions with **conspecifics**. Temperature has a particularly strong modulatory effect on ectotherms.

Seasonal changes in photoperiod may act either as drivers of behavioural, physiological and morphological changes, or entrain endogenous circannual rhythms, or both. In the temperate zone, where changes in photoperiod are pronounced, photoperiod clearly plays a dominant role in orchestrating the transitions animals make between different life cycle stages. For instance many temperate-zone birds and mammals are long-day breeders, initiating reproductive development as days lengthen in spring in anticipation of improving environmental conditions, and then terminating breeding prior to deterioration of those conditions in autumn. Examples of long-day breeders include most temperate-zone birds and many rodents. In other species, preparation for **breeding** does not commence until well after the start of the decline in day-length after the summer solstice. Many kinds of large wild animals such as deer, and domestic animals such as sheep, fall into this category. A small but significant number of bird species, and many kinds of fish (e.g. autumn- and winter-breeding salmonids) also breed on declining day-length or even on or near the shortest days of the year.

Both birds and mammals are known to become 'photorefractory', in that at certain stages of the annual cycle individuals cease to display a reproductive response to photoperiod. The terminology used with reference to birds and mammals is reversed, however: long-day breeding rodents are said to be photorefractory when their reproductive axis ceases to be inhibited by short days (i.e. they recrudesce the reproductive system and resume breeding despite continued short days), whereas long-day breeding birds are said to be photorefractory when they cease to maintain reproductive competence despite continued long days (i.e. the reproductive system regresses to an inactive state, sometimes before the summer solstice). Despite this difference in terminology, the neuroendocrine processes underlying the reproductive cycles of birds and mammals are fundamentally similar in many respects.

Although some kinds of animals will display fairly normal seasonal changes in reproductive condition, hibernatory or migratory activity and body covering (plumage/pelage) replacement in the absence of any changes in photoperiod (i.e. circannual rhythms), natural coordination of the full suite of elements to animals' annual cycles is facilitated by natural changes in photoperiod. Some species will proceed from one life cycle stage (e.g. breeding) to the next (e.g. plumage moult and preparation for autumn migration) without any change in photoperiod (e.g. many kinds of temperate-zone and Arctic-breeding birds); however, in the absence of further photoperiodic changes (e.g. a period of short days typical of autumn), many of these will then fail indefinitely to proceed to the next stage after that (e.g. redevelopment of the **gonads** and preparation for spring migration). In fact, expression of endogenous circannual rhythms can be very photoperiod-dependent, even in those species that display them.

Sensitivity to changes in photoperiod can be fairly exquisite. Taxa living within the tropics but far enough from the equator for there to be small seasonal changes in photoperiod are capable of sensing and responding to changes in photoperiod of as little as 15 min. However, this degree of sensitivity to small changes in photoperiod is not unique to tropical taxa; temperate-zone taxa are affected significantly by changes of as little as 30 min, and perhaps less. Response patterns to particular seasonal photo-regimes can be sufficiently specialized as to constrain species' abilities to cope effectively with rapid environmental changes that alter the optimal timing for reproduction. Effects of the photocycles should be considered very carefully when maintaining captive populations of animals, particularly if breeding is desired. (RS, DSM)

Further reading

Coppack, T. and Both, C. (2002) Predicting life-cycle adaptation of migratory birds to global climate change. *Ardea* 90, 369–378.

Gwinner, E. (1986) *Circannual Rhythms. Endogenous Annual Clocks in the Organization of Seasonal Processes.* Springer, Berlin.

Hahn, T.P. and MacDougall-Shackleton, S.A. (2008) Adaptive specialization, conditional plasticity, and phylogenetic history in the reproductive cue response systems of birds. *Philosophical Transactions of the Royal Society B* 363, 267–286.

Hau, M., Wikelski, M. and Wingfield, J.C. (1998) A neotropical forest bird can measure the slight changes in tropical photoperiod. *Proceedings of the Royal Society of London B* 265, 89–95.

Nicholls, T.J., Goldsmith, A.R. and Dawson, A. (1988) Photorefractoriness in birds and comparison with mammals. *Physiological Reviews* 68, 133–176.

Phylogeny

Why does the family dog often walk in tight circles before lying down? Can conservationists predict which species will be most sensitive to environmental change? The way animals behave today is dependent to a great extent on their evolutionary history (**see also: Evolution**). While processes such as natural selection can lead to changes in behaviour over very short evolutionary timescales, animals also exhibit traits 'held over' or retained from their evolutionary ancestors. Behaviour can therefore be specific to particular species because of recent and unique selective pressures, or common across groups of closely related species through shared ancestry. Such similarities and differences between species are the raw material of phylogeny.

Broadly defined, phylogeny refers to the evolutionary history of species. This can mean the specific evolutionary relationships among species, or how an animal's appearance, physiology or ecology has been shaped by evolution. Species relationships are depicted graphically as a phylogenetic tree. Positioned at the tips of the tree are the names of present-day species, and these are connected by branches indicating how related species are to one another. The fundamental relationships between species are typically inferred through comparisons of DNA sequences (e.g. mitochondrial DNA, or mtDNA), but morphological and behavioural traits are also sometimes used. Once a phylogenetic tree has been developed, the current distribution of a given trait across species can be used to reconstruct the evolutionary origin of that trait. This technique is called the **comparative method** (as distinct from '**comparative psychology**') and is a powerful – yet often underused – tool in animal behaviour. Knowing the phylogeny of an animal and using the comparative method can provide unique insights into the origin of behaviour. Phylogeny can also help predict how animals might react under different conditions.

Thinking phylogenetically

'Even in the psychological field, living organisms are phylogenetically derived entities whose specific origin and form can only be interpreted in the light of their phylogenetic history' (Lorenz, 1974).

Two influential pioneers in **ethology**, **Konrad Lorenz** and **Niko Tinbergen** (**see also: Tinbergen's four questions**), emphasized the importance of phylogeny in the study of animal behaviour. Despite this, incorporating phylogeny into behavioural research is seldom done. More commonly, function is interpreted exclusively from the context in which behaviour is observed today. Yet, as Lorenz points out above, only by understanding the phylogenetic history of an animal is a clear picture of its behaviour obtained. For example, the apparently functionless circling behaviour of a dog before **lying** down is puzzling when considered only in its present-day context. We should remember, however, that the domestic dog has its origin with the ancestors of the grey wolf. In the wild, wolves walk repeatedly in circles to flatten down areas of grass or snow before bedding down. We can reconstruct this behaviour back to the common ancestor of dogs, wolves and other canids. From this we can conclude that dogs are expressing a behaviour that has been retained from their evolutionary past.

Understanding the phylogeny of behaviour can serve more practical purposes than deciphering pet psychology. Phylogeny can also provide clues about the behaviour of animals in the wild. The successful preservation of rare and endangered species is greatly facilitated by knowing an animal's behavioural ecology (**see also: Conservation; Wildlife management**). Such information takes time and resources to obtain. Data collection is further complicated if species are located in areas geographically inaccessible to biologists, or when population numbers have declined so low that animals are rarely observed in the wild. However, because closely related species often share traits through common ancestry, we can get at least some idea about the ecology of rare species by observing the behaviour of their better-known relatives. Experiments on more common sister species could also help to determine the potential success of an ambitious conservation strategy (e.g. translocation of animals), without having to disturb the already vulnerable species in question.

Using behavioural phylogenies

'[The naturalist's] main source of inspiration is comparison. Through comparison he notices both similarities between species and differences between them' (Tinbergen, 1963).

Faced with budget cuts, lack of personnel and/or time before species decline into **extinction**, conservationists are forced to make difficult decisions about where to allocate scarce resources. In an effort to help prioritize, biologists have started to develop large datasets of life history, **ecology** and behaviour, representing decades' worth of work on hundreds of species. With these databases, researchers have begun to use phylogenetic comparative methods to identify the sorts of factors that species vulnerable to extinction typically share. Chief among these seems to be the extent to which species have evolved specializations to particular habitats. In the event that these habitats disappear (through climate change or human activities such as logging), species are forced into habitats for which they are poorly adapted. The result is often population decline and, ultimately, extinction if evolution is unable to catch up with rapid changes in the environment.

On the other hand, the comparative method has also revealed that some traits can evolve remarkably quickly. This can pose a different set of problems for conservationists, because recovery efforts must take into account the potential for rapid specialization. For instance, many island populations of the Caribbean rock iguana (genus *Cyclura*) have become critically endangered through human development and the introduction of predators (cats and dogs). Iguanas use elaborate head-bobbing displays to acquire territories that are vital for reproduction (**see also: Territoriality**). Researchers have found that different populations perform different types of head-bobbing patterns (see Fig. P.1).

This means that lizards brought together in captivity from different islands may have difficulty communicating effectively with one another, and this will in turn jeopardize the success of **breeding** programmes. Furthermore, the phylogenetic analysis of displays reveals dramatic changes in head-bobbing patterns over extremely short time periods, as rapidly as six generations (~30 years). Animals raised from consecutive generations in **captivity** and then reintroduced back into the wild could find it difficult to acquire territories (and subsequent mating opportunities) because they possess a display of the 'wrong' type.

The imprint of evolutionary history is apparent in all types of behaviour, from the way domestic pets behave in our home to how animals interact with each other in the wild. Placing animal behaviour in a phylogenetic context is essential for a complete understanding of its origin, but also its present-day function. By comparing closely related species, we can obtain a window into the past and a means to piece together the steps leading to the evolution of different forms of behaviour. While this information has obvious relevance to the interpretation of behaviour generally, it has practical implications for applied science as well (e.g. conservation biology). (TJO)

References and further reading

Brooks, D.R., Mayden, R.L. and McLennan, D.A. (1992) Phylogeny and biodiversity: conserving our evolutionary legacy. *Trends in Ecology and Evolution* 7, 55–59.

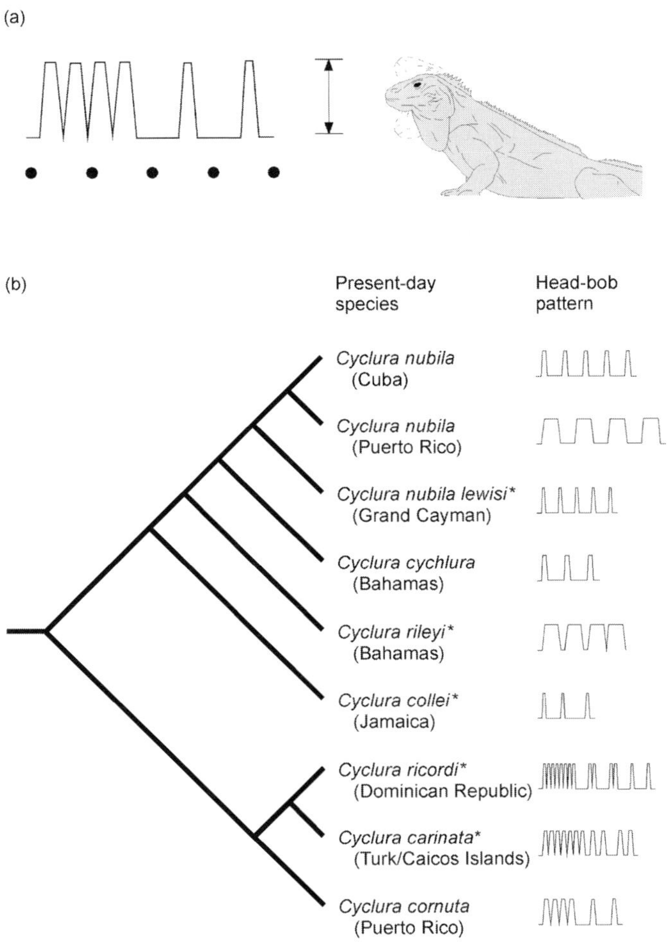

Fig. P.1. The phylogeny of Caribbean rock iguanas (genus *Cyclura*). (a) Territorial displays in these lizards consist of rapid up/down movements of the head, called 'head-bobs'. Dots represent 1 s intervals; (b) a phylogenetic tree illustrating the evolutionary relationships between different groups of living rock iguanas, followed by the head-bob pattern that animals use to resolve territorial disputes. Asterisks indicate endangered or critically endangered species, while the remainder are considered vulnerable to extinction. This figure is compiled from information presented in Martins and Lamont (1998).

Fisher, D.O. and Owens, I.P.F. (2004) The comparative method in conservation biology. *Trends in Ecology and Evolution* 19, 391–398.

Lorenz, K. (1974) *Studies in Animal and Human Behaviour*. Harvard University Press, Cambridge, Massachusetts.

Martins, E.P. (1996) *Phylogenies and the Comparative Method in Animal Behaviour*. Oxford University Press, New York.

Martins, E.P. and Lamont, J. (1998) Estimating ancestral states of a communicative display: a comparative study of Cyclura rock iguanas. *Animal Behaviour* 55, 1685–1706.

Tinbergen, N. (1963) On aims and methods of ethology. *Zeistschrift für Tierpsychologie* 20, 410–433.

Pica

The tendency to ingest non-nutritive items is called pica. Pica has been documented to occur most notably in the domestic **dog**, **cat** and **pig** (see: **Stone chewing**), although it is seen in other species. A variety of different objects have been noted to be ingested, including rocks, rubber, plastic, paper, card, wool, cotton, silk and synthetics.

The behaviour can occur at any age, although the occurrence of the behaviour in juveniles has been suggested to be part of normal investigative behaviour, particularly with puppies. Onset of the behaviour in the adult has been described as an **abnormal** behaviour. In certain cases the behaviour has been classified as a **compulsive disorder** – for example, when objects are persistently selected as seen in wool sucking or chewing in cats.

The cause is not fully understood, and may be related to an underlying clinical condition or an environmental trigger such as early weaning, stress or separation anxiety. The occurrence of pica in the domestic cat is likely also to have a genetic component, as it has been noted to occur proportionally more in the oriental breeds. (SR)

See also: Malnutrition; Self-medication; Urine drinking

Further reading

Bradshaw, J.W.S., Neville, P.F. and Sawyer, D. (1997) Factors affecting pica in the domestic cat. *Applied Animal Behaviour Science* 52, 373–379.

Pig

The pig is a domesticated descendent of the Eurasian wild boar (*Sus scrofa*). The wild boar is a member of the even-toed **ungulates** (artiodactyla), which first evolved in the Eocene epoch (56–34 million YBP), and it has essentially retained the mode of life of the Eocene artiodactyls: an unspecialized diet, absence of horns and antlers and the **carnivore**-like habits of using dens, making nests and bearing litters. Originating in South-east Asia, the wild boar is the most widely distributed member of the nine species of the *Suidae* family. It is found in subarctic, temperate and tropical regions of the Old World, including most areas of Europe, the Mediterranean basin and North Africa, the Russian taiga, India, South-east Asia, Sri Lanka and Japan.

A recent morphological analysis has proposed that four subspecies exist. On the basis of archaeological records, the domestic pig appears to have been independently domesticated on at least two occasions – first in the Near East at around 9000 YBP, and secondly, in the Far East (China) around 4900–7000 YBP. However, recent mitochondrial DNA evidence suggests that up to six separate domestication events occurred, and that modern European breeds are derived from local wild boar populations in central Europe.

Historically, pigs have been kept on a small scale. In early **domestication**, pigs were allowed to roam relatively freely in woodlands surrounding human settlements. These pigs differed little from wild boar except that they were smaller in size, and were more amenable to human contact. As woodland gave way to agricultural land during the Middle Ages, pigs' capacity to damage crops meant that they were increasingly confined, and access to woodland was restricted. In rural areas the world over, and even within urban settings, individual families would rear a few pigs to supplement their diet, in effect 'recycling' waste food by feeding it to the pig. During the course of the dramatic changes in agriculture over the last 200 years, deliberate artificial **selection** began to be carried out and pigs grew more rapidly, became fatter and had a larger mature size. Asian pigs, introduced to Europe in the 18th and early 19th centuries, were crossed with local breeds, and the genome of the modern European Large White and Landrace breeds reflect this hybridization. There remain some behavioural and phenotypic variations between breeds, reflecting the differences in selection pressure placed on individual traits, but modern domestic pigs are now much larger than wild boar – mature individuals weighing in excess of 300 kg compared with 35–230 kg for wild boar (see Fig. P.2).

Current farming systems

Pig farming itself in the industrialized world is now one of the most intensive of all livestock production systems. It has moved away from small-scale production, and the ever-increasing scales of commercial livestock operations, influenced by advances in mechanization, economic pressure and tightening legislation regarding **feeding**, **transport**, **slaughter** and meat hygiene have changed the way our animals are kept (**see: Intensification of animal production**). On commercial pig units, animals may now be kept in large buildings, with a high degree of automation to supply feed, water, heat and ventilation and to remove waste. Intensification has been a commercial success because it has reduced costs of production, by reducing feed and labour costs and improving **disease** control. While the trend for fewer, larger commercial units is spreading globally, consumers are now questioning intensification and its effect on the **welfare** of pigs. All aspects of pig production are attracting welfare concerns – **breeding** herd **housing**, slaughter herd housing, transport and slaughter – with major concerns associated with close confinement and housing in barren, frustrating environments, although outdoor systems also have potential welfare problems, even though they offer a more naturalistic environment (see Fig. P.3).

There are essentially three different types of commercial pig unit: (i) a breeding-only unit, which houses only the breeding herd and piglets commonly up to about 25 kg; (ii) a rearing-/finishing-only unit, which houses the pigs from 25 kg up to age of slaughter; and (iii) the combined unit, which houses both the breeding herd and the rearing/finishing herd up to

Fig. P.2. Wild boar sow with her piglets (image courtesy of Fritha Langford).

Fig. P.3. Outdoor sow and piglets rooting and foraging (image courtesy of Marianne Farish).

slaughter. Early-weaning practices in some countries have resulted in greater separation of the various phases of slaughter pig rearing but, in terms of housing and welfare considerations, pig production is essentially split into three distinct phases: (i) the dry sow, meaning all gestating sows, sows awaiting service and barren sows within the herd; (ii) the farrowing sow/lactating sow and her litter; and (iii) the weaned pig being reared for slaughter.

The dry sow

The breeding cycle of the sow already in the herd begins at **weaning** from the farrowing house. After weaning, the sow is usually moved to a special service area with boar contact, which encourages **oestrus**, optimally – from a production viewpoint – at around 4–5 days after weaning. Service itself can either be natural, using a boar, or by **artificial insemination** (AI) or, commonly, a mixture of both. After serving, the sow will typically remain in the service area for 3–4 weeks, where the **stockperson** will watch for any signs that she is not pregnant – i.e. signs of oestrous behaviour at 21 days after service. Typically, between 10 and 20% of sows do not hold their pregnancy, and those that consistently fail will be culled from the herd and replaced by new breeding females, which are called gilts until after the birth of their first litter. Most units will also cull sows at weaning on the basis of age or number of litters, irrespective of whether or not the sow is still producing and rearing good piglet numbers. The mean **culling** rate for sows is often as high as 40–50%, the majority due to fertility and age reasons, but also including culling due to other causes such as **lameness** and various ailments. Another 5–10% of sows may die annually, thus resulting in an annual replacement rate of 50–60%.

Once the sow does not obviously come back into oestrus, or after pregnancy is confirmed by ultrasound scanning, she will then be returned to the gestating sow system, where she will spend approximately 12 weeks prior to moving back to the farrowing house about a week before the predicted date of birth. There is a great diversity of dry sow housing systems currently in use worldwide, ranging from pasture-based systems to close-confinement stall systems. Although the last few decades have seen sow housing moving increasingly away from extensive to intensive, recent legislation in some countries, notably those within the European Union (EU), will reverse this trend as close confinement systems have already become (in e.g. the UK and Sweden) or will become illegal. There are signs that the movement away from close confinement during gestation may spread further than Europe, with major producers in North America (e.g. Smithfields and Maple Leaf Foods) bringing in their own voluntary change.

Until recently, the major factor behind sow housing systems was that of economics of production. Since World War II there has been a steady reduction in the number of individual pig producers, but increases in the number of pigs being kept. The trend away from extensive systems with small herd numbers towards larger, intensive units was initially fuelled in some countries by government-backed price incentives in the 1940s and accelerated during the 1950s as new, intensive, system-based technology was applied.

Troughs and peaks in production and total sow numbers in response to pig prices are a notorious feature of pig production in many countries, and this has also accelerated the decline of the small producer. It made economic sense to increase herd sizes, stocking densities and turnover and decrease labour costs by mechanization wherever possible. The ultimate developments in terms of dry sow housing were those of stalls and tethers. Keeping sows in permanent confinement gave the farmer a number of advantages over less intensive systems – for example: (i) increased stocking density; (ii) increased cost effectiveness, as sows on concrete slats with a mechanised slurry handling system have reduced straw and labour costs; (iii) increased ease of management, as the stockperson is able to monitor individual sows easily and adjust husbandry regime where necessary.

Arguably, the degree of skill required to manage confinement systems is also less than for loose-based systems. However, it has since become apparent that such intensive systems, although perhaps conferring some welfare advantages over loose-housed systems, have a number of serious welfare disadvantages and project a negative image of the industry to the consumer.

The farrowing sow

Sows generally move to the farrowing house around 5–7 days prior to expected farrowing date, which itself is usually based on 115 days after service. Just as with gestating sow housing, there is a great diversity of housing types for the farrowing sow, but worldwide (including those countries with little or no confinement housing of gestating sows) the majority of sows give birth in a confinement system of a **farrowing crate** (see Fig. P.4), although this may be only temporarily in place immediately around farrowing itself.

Prior to intensification, sows most often farrowed in open pens with large amounts of straw. Again, economics forced change. As straw use and pen size decreased and litter size increased, piglet mortality – primarily due to crushing – was found to rise. Introduction of crates into this situation was found to reduce mortality back towards pre-intensification levels. Coupled with this effect on piglet mortality, which is the main reason given by farmers for continued usage, crates were also found to confer management benefits, such as reduced labour costs and ease of piglet handling.

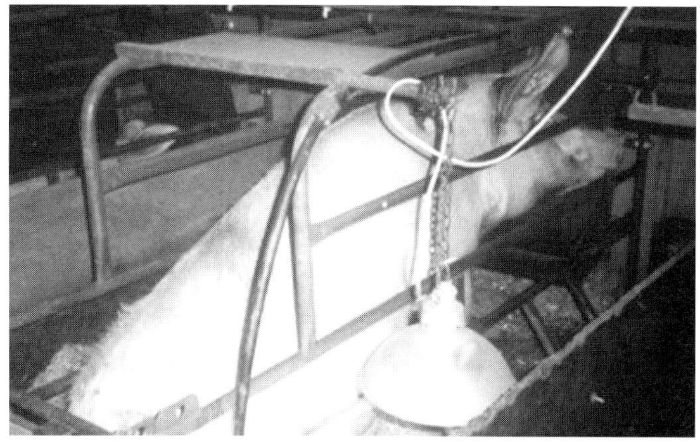

Fig. P.4. Conventional tubular metal farrowing crate, with forward creep area and supplementary piglet heating lamp (image courtesy of JNM-F).

The great amount of research on modification of farrowing crate design has yet to improve liveborn piglet mortality much below 10%, and this fact, together with increasing concerns for the welfare of the sow and a desire for alternative systems that require minimal financial input, has led to a re-examination of the factors necessary for defining the optimum farrowing conditions. As with gestating sow housing, there has been recent development of both old and new farrowing systems. A return to basic individual straw pens with farrowing rails has often given increased mortality from crushing, as the sow is given greater freedom of movement without necessarily having either the maternal behavioural skills or the physical ability to carry out careful posture changing. Other designs include circular or oval crates (see Fig. P.5) or the incorporation of farrowing areas into communal systems (see Fig. P.6) and, as with alternative gestating sow housing, these alternative farrowing sow systems present their own set of potential welfare challenges, not just for the sow, but also for the litter.

Fig. P.5. Oval farrowing crate (image courtesy of J.J. McGlone).

Fig. P.6. Experimental communal pen farrowing system (image courtesy of JNM-F).

The slaughter pig

Once weaned, piglets usually pass through a number of different housing systems during their growth, up to the age they reach slaughter weight. The number of changes in accommodation varies, but intensive units typically move piglets at weaning to specialist nursery accommodation and thereafter move pigs only once more, into finishing accommodation. It is not uncommon, however, for pigs to move through four different systems from weaning, each specific for particular stages of the growing phase. The age at which pigs are slaughtered also varies from country to country and is also dependent on the type of product required, e.g. pork, bacon, ham or processed products. Slaughter weights can therefore be anywhere from 60 kg up to 125 kg, with a slaughter age of between 18 and 26 weeks.

As with sow housing, there is a very large variety of housing types for slaughter pigs, but the majority of pigs, including those bred from outdoor-housed sows, are reared indoors. The goal of the producer is to grow the pig from weaning to slaughter in as short a time possible and using as little feed as possible. This is best achieved indoors, where the thermal environment can be controlled and feed wastage minimized. There is, however, much variety, ranging from fully enclosed, fully slatted systems to open-fronted, kennel-type, deep-straw systems with access to outdoor pens. In general, the slaughter pig has been subject to less welfare research than the sow, but this has been a result of political expediency rather than any particular lack of welfare problems.

Transport and slaughter

At some stage in their lifetime, the vast majority of pigs will be transported, except those that are born and then die or are euthanized on the same farm. The number of times and distances over which they are transported will vary depending on whether they are in the finishing or the breeding herd, and on the nature of the pig industry within the country or region in which they are being reared. For example, it is not uncommon for herds in North America to employ multi-site production, meaning that, at weaning, pigs may be transported to a nursery unit and, later on, transported again to a growing and finishing unit. Over the last few decades, the average distances and journey times that pigs have to be transported to reach the slaughter plant have increased as the number of slaughter plants has reduced. The maximum journey time in particular has been the focus of much recent research, with resulting recommendations and legislation in some markets, such as the EU.

If the process of transportation is too stressful for a pig, the ultimate result can be the death of that pig or the pig becoming non-ambulatory or 'fatigued'. In North America, the current in-transit + post-transit mortality and non-ambulatory rates are around 0.15 and 0.40%, respectively, representing in total around 11 pigs in every 2000 transported. It is therefore in the interests of the producer (and of course the pig) that the stress of transportation is minimized. The transportation and slaughter process can be thought of in terms of its constituent parts, namely loading, transport, unloading, **lairage** and slaughter. All of these parts have welfare concerns and have been subject to a fair amount of research.

Ultimately, the end point of most transportation events is the slaughter of the pig. Once lairage time is complete, the pig is usually removed from the pen and driven towards the **slaughterhouse** manually. The pigs will enter a forcing pen and, from here, they will enter the **raceway**, which may be double or single file, until they enter the restraint system in single file. The restrainer will minimize body movements and make electrical stunning, in particular, easier to achieve. Stunning may be carried out automatically or manually by passing an electric current across the **brain**, or across brain and heart. Alternative stunning methods are CO_2 or argon gassing, which can be carried out with pigs kept as groups. Regardless of stunning method, the aim is to render the pig insensible, and the pig is then exsanguinated by severing the carotid arteries and jugular veins, leading to death. (JNM-F)

Further reading

Genov, P.V. (1999) A review of the cranial characteristics of the Wild Boar (*Sus scrofa* Linnaeus 1758), with systematic conclusions. *Mammal Review* 29, 205–238.

Larson, G., Dobney, K., Albarella, U., Fang, M.Y., Matisoo-Smith, E., Robins, J. *et al.* (2005) Worldwide phylogeography of wild boar reveals multiple centers of pig domestication. *Science* 307, 1618–1621.

Marchant-Forde, J.N. (2009) *The Welfare of Pigs*. Springer Science + Business Media B.V., Dordrecht, the Netherlands, 348 pp.

Rodriguez-Zas, S.L., Southey, B.R., Knox, R.V., Connor, J.F., Lowe, J.F. and Roskamp, B.J. (2003) Bioeconomic evaluation of sow longevity and profitability. *Journal of Animal Science* 81, 2915–2922.

Piscivore

A piscivore is a form of carnivorous animal that feeds predominantly or totally on fish. The balance of piscivorous species within a given aquatic environment, like any other type of feeder, may have important implications for the ecology and behavioural patterns of other animals of that area. (DSM)

Further reading

Diehl, S. and Eklov, P. (1995) Effects of piscivore-mediated habitat use on resources, diet, and growth of perch. *Ecology* 76, 1712–1726.

Pituitary gland

The pituitary gland is divided into two major sections of distinct embryonic origin: the neurohypophysis of neural origin and the adenohypophysis derived from the ectoderm of the roof of the embryonic mouth. The neurohypophysis (posterior pituitary) is comprised of the median eminence (ME), infundibular stem and the *pars nervosa*. The infundibular stem spans the distance from the intra-hypothalamic ME and the lobes of the hypophysis shallowly embedded within the skull. The adenohypophysis (anterior pituitary) is made up of the *pars distalis*, the *pars tuberalis* and the *pars intermedia*.

Projections of hypothalamic **neuron(e)s** project into the median eminence, where they release hypothalamic hormones, such as **corticotropin-releasing hormone** (CRH), into the hypothalamo–hypophysial portal vessels, thereby stimulating or inhibiting secretion and synthesis of trophic hormones from the anterior pituitary gland. These trophic hormones include **adrenocorticotropic hormone (ACTH)**, **follicle-stimulating hormone (FSH)**, **growth hormone (GH)**, **luteinizing hormone** (LH), **prolactin** (PRL) and thyroid-stimulating hormone (TSH). Hormones synthesized from the precursor **pro-opio-melanocortin** (POMC) are also produced in the *pars intermedia* of the anterior pituitary, and these include β-**endorphin** and met-**enkephalin**. The *pars nervosa* of the posterior pituitary secretes the non-peptide hormones **oxytocin** (OT) and arginine vasopressin (AVP, also called **antidiuretic hormone** or ADH), which are stored here after manufacture in the **hypothalamus**. Maintaining water balance is a primary action of the posterior pituitary.

Animal behaviour and **well-being** research focuses mainly on the hormones of the anterior pituitary. ACTH, a major regulator of **glucocorticoid** release in the **adrenal gland**, is often used as a measure of physiological **stress**. Adverse or noxious stimuli increase the concentration of ACTH secreted by the anterior pituitary. The gonadotrophins LH and FSH are used as measures of reproductive competence. PRL has many physiological effects, including its effects on lactation in mammals and brooding in birds. PRL responsiveness is also directly and negatively related to the neuronal concentration of **dopamine**, a neurotransmitter directly related to an animal's stress level and stress-coping ability. Animal production research also focuses on GH, TSH and β-endorphin for the regulation of growth, metabolism, feed conversion and appetite. (RD)

Placentophagia

Placentophagia describes the act of eating the placenta, normally by the mother shortly after giving birth in eutherian placental mammals. In marsupials, the placenta is resorbed and so ingestion does occur. Placentophagia occurs in both carnivorous and herbivorous species as a normal behaviour, with the notable exception of the camel and many aquatic species. Placentophagia may result in enhanced opiate sensitivity that has analgesic benefits for the mother postpartum, but may also affect **parental behaviour** and **bonding**, with some males of species showing a high level of **paternal behaviour** also engaging in the activity. (DSM)

Further reading

Kristal, M.B. (1980) Placentophagia: a biobehavioural enigma (or *de gustibus non disputandum est*). *Neuroscience and Biobehavioural Reviews* 4, 141–150.

Play

Play is one of those topics that everyone seems to know about but is not quite able to define or explain. Recent attempts to develop rigorous, multifaceted criteria by which to identify play have led to the fascinating conclusion that not only do most species of mammals and many species of birds qualify as exhibiting play, but so do 'lower' vertebrates and even an assortment of invertebrates (Burghardt, 2005). However, despite this, play is a relatively rare phenomenon in the animal kingdom, with the most clear-cut cases being among mammals and some birds. Typically, it is behavioural episodes that seem to be performed for their own sake without fulfilling any obvious biological need that are likely to be labelled as play.

To avoid the risk of labelling any immature act by an infant or, for that matter, a cage stereotypy by an adult as play, care must be taken, for if all puzzling acts are thrown together and labelled as play, then the label becomes meaningless. Again, criteria capable of distinguishing between playful and non-playful actions are critical. Burghardt proposes five such criteria:

- It is not completely functional in either its form or context.
- It is spontaneous, seemingly performed for its own sake.
- It is modified compared with the functional behaviour that it imitates.
- It is repeated in the same form for extended periods of the life cycle.
- It is such that it primarily occurs when the animal is fed, healthy and free of **stress**.

Even detailed criteria such as these need to be applied with caution. For example, many instances of play in adults can be linked to clear biological benefits at the time at which the play occurs. Thus, criteria need to be viewed as a guide, not as a means to pigeonhole behavioural phenomena. A major reason for this definitional difficulty is that play may have differing levels of complexity and may be co-opted for different functional purposes in different lineages of organisms that engage in play. Among **rodents**, for example, play among immature animals varies markedly across a range of species, with some not playing at all, some having rudimentary forms of play and others being extraordinarily exuberant in the complexity of their playful performances. Only some of these species appear to derive developmental benefits from the performance of play. Similarly, while playful behaviour similar to that seen in juveniles may be present in the adults of some species, not all adults engage in such play with any clear connection to functional outcomes.

When considering non-human animals, play is generally placed into one of three categories: social play, object play or solitary locomotor play. Running, jumping and turning without reference to another subject is an example of locomotor play, manipulating objects qualifies as object play and social play is when two or more animals engage in play together. These categories are limited, of course, because the same actions (e.g. running, leaping and turning) may occur in locomotor play and during social play. None the less, they do provide a crude classification that appears to reflect something important about the biology of play. After all, some species may only exhibit one or two of these three types of play, and developmental studies suggest that they may have different times of onset and offset. Indeed, there has been considerable debate as to whether there is one type of playful **motivation** that can be expressed in multiple overt patterns, or multiple motivational systems with their own characteristic type of play. Although the matter is still unresolved, the evidence for a multiple systems view is building at the expense of the single-system view.

Detailed studies of these forms of play in various species reveal that, generally, object play involves the behavioural repertoire typical of food procurement, with the most elaborate examples being seen in carnivores with their prey-killing behavioural tool kit. Predatory play, which involves using actual prey as play objects, may be considered a version of object play. A **cat** playing with a **mouse** or a ball of yarn applies the predatory repertoire to a non-functional, playful context. Locomotor play most typically involves the runs, leaps and erratic movements of anti-predator behaviour, with the most elaborate forms being seen in species most subject to predation (e.g. **goats**, **deer**, **rabbits**, mice). Social play, the most widely studied form of play, can involve the behaviour patterns from a variety of functional behavioural systems, such as **conspecific aggression**, sex and **parental behaviour**. The form of social play most often studied is play fighting, which most commonly involves the behavioural repertoire of conspecific aggression although, for some lineages, this competitive form of play involves **sexual behaviour**, especially the pre-copulatory elements (e.g. murid rodents).

While these categories have some coherence in terms of the functional behavioural repertoire that is used, cases where these categories seem to overlap diminish their descriptive utility – for example, tug-of-war games and chasing incorporate object play and locomotor play with social play. At a finer level of behavioural analysis, such apparent cases of overlap may disappear. For example, it has been commonly thought that in play, animals combine the behavioural repertoire from several functional systems, such as aggression and sex in play fighting. However, detailed, descriptive studies have shown that at a micro-level, the sexual and aggressive elements do not combine but, rather, the sex sequence is not commenced until the attack and defence sequence of aggression is completed, and vice versa (Pellis and Iwaniuk, 2004). Therefore, there may be a greater degree of discreteness in the categories of play than may appear at a more molar level of observation. However, this is an empirical issue still requiring resolution.

Play fighting is the most commonly studied form of play, with more known about its form and consequences, and so forms the basis of this entry. When two animals are engaged in a play fight, they try to gain some advantage over one another, such as striking, biting or nuzzling the partner, usually on some specific body target. The subjects then engage in a sequence of attack, defence and counter-attack, leading to the wresting and jostling for position so apparent in play fighting. These targets can be derived from sexual or aggressive behaviour, or indeed, from predatory behaviour. Note that, in the latter, it is really a form of predatory play, with the partner acting as the surrogate prey. Whatever the origin of the target, the play fighting sequence involves a pattern of usage of attack and defence that is not the same as that seen in functional aggression; indeed, it has a cooperative appearance, in that the interaction leads to frequent role reversals.

Play is most often associated with immature animals, with juveniles providing its most exuberant examples. In addition, play is generally thought to occur in a stress-free environment where the basic necessities of life have been met. When directing the play socially, subjects most typically engage in play with partners of the same age and often of the same sex. These are broad trends and they should not mask the fact that there are important exceptions. For example, some species have high rates of adult play, especially in the form of social play that may often be directed at other adults, including partners of both sexes. Furthermore, in some species, social play by immature animals may be directed preferentially at adults rather than at other immature animals. Thus, the general trends should be viewed as guides to the phenomenon, not as substitutes for detailed knowledge of the species under investigation. Indeed, when trying to understand the functions of play and how play may be used in welfare contexts, the peculiarities of the play present in the species of concern need to be carefully characterized.

For a range of species, including birds and mammals, social play among sub-adults and adults often occurs in contexts of social tension or ambiguity, such as when potential sex partners meet or when there is some turmoil in existing **dominance** relationships. In these contexts, playful modes of interaction appear to be useful in either establishing or reinforcing social bonds, or in testing and manipulating social relationships. For a variety of species, especially those that engage in playful interactions as adults, the organization of social play encounters as juveniles is modified in a manner consistent with such interactions that serve to refine cognitive and emotional skills needed for subtle and complex social interactions as adults.

In **rats**, social play mostly involves play fighting. In play fighting, a partner attempts to contact the nape with its snout, which, if contacted, is gently rubbed. The defending rat responds to such an attack with one of a variety of tactics in order to block access to its nape. Juvenile rats engage in defensive actions most likely to prolong bodily contact with their partners – typically by rolling over on to their backs when attacked. The attacker may then hold the defender down with its forepaws while attempting to make further contact with its partner's nape but, more often than not, will stand on the supine partner's belly with its hindpaws. Standing with all four paws on its wriggling partner leads to the rat losing its stability and, indeed, facilitates the defender's ability to counter-attack, thus enabling it to gain access to the original attacker's nape. This configuration of attack and defence behaviour is not present either before or after the juvenile phase – rather, when juvenile, when play is at its most frequent, young rats adopt modes of interaction that increase their experience of instability during social interactions.

Such instability may be excellent training for the physical and emotional vicissitudes to be expected in the rough and tumble of the social world. Learning what to do in such situations makes for an adult with a greater range of options in dealing with the unexpected. There are data from a variety of species that suggest that play in general serves to provide such training for this.

Play in adults and in immature animals thus reflects the presence of complex social exchanges and the development of important social, emotional and cognitive skills. As such, the presence of play in captive animals of species that have play as part of their normal repertoire represents an important indicator of their **well-being**. Indeed, as play is most often associated with a stress-free environment, it has been used to monitor the well-being of both free-living and captive animals. Generally, it has been found that animals that engage in play are healthier, better nourished and more socially adjusted than ones that play infrequently. This relationship implies that play is simply a measure of the absence of stress, but there is increasing evidence for a more complex relationship between the two. First, play experience in the juvenile phase may be important in promoting the development of strategies for **coping** with stress as adults. Secondly, there are laboratory data suggesting that engaging in play can reduce stress. It is well known that social **grooming** and huddling can reduce stress, and social play may tap into the same mechanisms. The added advantage of the playful solution to stress reduction is that it combines the benefits of amicable social contact with that of rigorous exercise.

Therefore, under a number of husbandry conditions, in the research laboratory, in the **zoo** and in the farmyard, the presence of play may well indicate the unfolding of species-typical development, well-adjusted adults and animals capable of dealing with the day-to-day vicissitudes of life. There are, however, some important caveats to bear in mind given that our knowledge of play is still in its infancy. First, in any given case, direct measures of stress, behavioural or physiological, need to be made to ensure that the play being observed is indeed a marker for the absence of stress or stress reduction. For example, in a study of captive squirrel **monkeys**, it was found that while play led to the reduction of stress in the juveniles, the stress levels in the cohabiting adults actually rose. Thus, it cannot be assumed that, if the average level of play is relatively high, then the stress level is uniformly low. In fact, there is some evidence suggesting that rougher forms of social play can be used as a form of 'punishment'; this may lead to subordinate members of the group being stressed when they are the targets of playful interactions with dominant animals.

Secondly, the presence of play may indicate that the animals are in a state of **distress**. For example, in gibbons, there appears to be an elevation of social play when new troops are formed. A reduction in play may thus indicate that a more stable set of relationships within the troop has been achieved, and so, presumably, a reduction in the level of stress. Thirdly, we know that the frequency of play can vary with environmental context, but we do not know what the minimum, baseline level of play may be for any species. Without such a baseline, it is hard to determine what may represent too little or too much play in any particular species in a given context.

In the absence of a baseline for the frequency of play, information on the context and form of the play fighting occurring may be useful, as would information on which individuals are playing with whom. For example, in observations of a troop of captive **gorillas** (Calgary Zoo, Alberta, Canada), the silverback responded to the play overtures of one of his sons, and would sometimes even initiate play with that son. The other son, of the same age, was typically left out and, if he attempted to join in, the silverback would strike him forcefully, inhibiting for a while any further attempts to join the fray. Thus, while all three appeared to be engaged in social play, the experiences, positive and negative, for each individual would probably have been very different. Again, it seems important to examine the details of the play performed.

Even with these notes of caution, it would still seem prudent to use play as a marker for general well-being, bearing in mind that additional data may be needed in order to help evaluate whether the state of play should be cause for intervention. Given the seeming importance of play for many species of mammals and some species of birds, the potential value of this behaviour for animal husbandry and welfare concerns should not be underestimated. (SMP, VCP)

See also: Aggression; Agonistic behaviour; Cooperation

References and further reading

Burghardt, G.M. (2005) *The Genesis of Animal Play*. MIT Press, Cambridge, Massachusetts.

Pellis, S.M. (2002) Keeping in touch: play fighting and social knowledge. In: Bekoff, M., Allen, C. and Burghardt, G.M. (eds) *The Cognitive Animal: Empirical and Theoretical Perspectives on Animal Cognition*. MIT Press, Cambridge, Massachusetts, pp. 421–427.

Pellis, S.M. and Iwaniuk, A.N. (2004) Evolving a playful brain: a levels of control approach. *International Journal of Comparative Psychology* 17, 92–118.

Pellis, S.M., Pellis, V.C. and Foroud, A. (2005) Play fighting: aggression, affiliation, and the development of nuanced social skills. In: Tremblay, R.E., Hartup, W.W. and Archer, J. (eds) *Developmental Origins of Aggression*. Guilford Press, New York, pp. 47–62.

Soltis, J., Wegner, F.H. and Newman, J.D. (2003) Adult cortisol response to immature offspring play in captive squirrel monkeys. *Physiology and Behaviour* 80, 217–223.

Polydipsia – stress-induced

Polydipsia describes excessive **drinking behaviour**. Polydipsia may arise as a result of physical **disease**, e.g. renal failure, diabetes, etc. or from psychological factors, often referred to as stress-induced polydipsia. This may arise incidentally in animals under chronic **stress** or be induced deliberately. The most commonly induced form is a schedule-induced behaviour, in which the animal is incidentally or deliberately placed on a fixed-interval reinforcement schedule (**see**: **Reinforcement – types of**) for drinking: i.e. the animals may be rewarded with food for drinking after a set time. This tends to increase the amount of drinking that occurs around the time of the interval associated with reinforcement. This procedure in the **rat** has been proposed as model of obsessive–**compulsive disorder** in people, but its validity is questionable; however, it may explain some forms of repetitive/stereotypic behaviour in captive animals.

The phenomenon was first described by Skinner, who referred to it as superstitious behaviour. It has since been shown that not only may a terminal or **consummatory act** that is reinforced increase, but also that other interim behaviours incidentally associated with this act (adjunctive behaviours) may become more common, even if they are not contiguous with the reinforcement (**see: Contiguity**). A range of other schedule-induced behaviours have also been demonstrated, including **aggression**, air-licking, eating and wheel running. Studies of schedule-induced polydipsia indicate that a range of other factors affect the intensity of the behaviour: for example, it is increased when a preferred flavour is used or the animals are underweight and there is a peak interval reinforcement time effect. (DSM)

Porcine stress syndrome

Porcine stress syndrome (PSS) is a non-infectious genetic **disease** that develops in (homozygous) pigs expressing the mutated autosomal recessive ryanodine-receptor (*ryr-1*) gene, also called the halothane gene. The individuals expressing this gene are greatly sensitive to the anaesthetic halothane, hence the name of the gene. Genetically predisposed pigs are highly susceptible to acute **stressors**, such as **handling** and **transport**, that lead to a complex of conditions. Malignant hyperthermia, back muscle necrosis (BMN) and pale soft exudative pork muscle (**PSE meat**), followed by sudden death, can develop. These conditions result from a substitution of the nucleotide cytosine (C) for thymidine (T) on the *ryr-1* gene, which ultimately alters the function of the protein ryanodine. The mutated protein facilitates the opening of the calcium channels and inhibits their closure, leading to greater flow of calcium ions and excessive stimulation of muscle contracture.

The expression of this gene is linked with heavily muscled and lean (low fat deposition) **breeds** (i.e. Pietrain). Genetic selection for meatiness and carcass conformation, both economically beneficial traits, facilitated the spread of the disease, characterized by stress susceptibility and predisposition to developing an inferior meat quality. The mutated *ryr-1* gene is also a matter of animal **welfare**, since PSS includes degeneration of the muscle and **pain**, and can cause public concern that diseased pigs may be used for meat production.

Pathological changes in the muscle occur when the cellular metabolism changes from aerobic to anaerobic, developing biochemical abnormalities. The switch of energy utilization results from a prolonged increase in calcium flux through the calcium channels of the sarcoplasmic reticulum. The deficit in the intracellular removal of calcium causes persistent muscle contraction. This state leads to excessive production of lactic acid (acidosis), primarily in IIB white muscle fibres, and increased body heat with poor production of adenosine triphosphate (ATP). An increase in body temperature is an outcome of an exacerbation of muscle metabolism that leads to acidosis and causes constriction of the peripheral blood system. **Norepinephrine** and cortisol concentrations are high. Both smooth and skeletal muscles can be affected.

Homozygous individuals can be visually identified by their shorter body and bulging oval hams, with a thin layer of fat. Back muscle necrosis is a particular manifestation of PSS and can be seen in pigs over 50 kg of body weight. Unilateral BMN is characterized by a curvature of the body towards the affected side. Symptoms occur during or following a stressful situation and are initially visible as rapid tremor of the tail and skeletal muscle occur. Pigs may present lesions of erythema (blotched red and pale skin), twitching of the face, tachypnoea (fast respiration), tachycardia (fast heartbeat) and an increased body temperature (>41.5°C). Death is usually preceded by collapse and muscle rigidity, and occurs within 15–20 min from the start of symptoms. Some of these clinical signs may be induced by anaesthesia with halothane. Environmental temperatures above 22°C predispose the condition.

Generalized oedema, congestion of lungs and frothy exudate from the trachea are among the few pathological changes seen. Hepatic congestion and hydrothorax may be observed. There is a rapid onset of rigor mortis, and dark blood caused by oxygen desaturation may occur. The muscle's pH drops to 6.0 or less within 45 min post-mortem. Muscle with a pale, soft exudative appearance can be seen in 60–70% of PSS-susceptible pigs within 15 min of post-mortem. Back muscle necrosis, which is characterized by pale areas of necrosis in the multifidi and longissimus dorsi muscles, is seen when sliced in sections, as the pale areas contrast with the red intact muscle bundles. Histological observation of degenerating and regenerating, in addition to hypercontracted, muscle fibres may be present. Cardiac muscle may show lesions of acute heart failure, such as degeneration and fragmentation.

The clinical diagnosis is established by the development of clinical signs, such as skin blanching and rigidity after the animal is subjected to a stressful stimulus or to a volatile anaesthetic (i.e. halothane). A differential diagnosis of PSS from other fatal diseases (i.e. anthrax, twisted bowel, bloody gut, gastric torsion, electrocution, vegetative endocarditis) is based on body conformation, breed and the acute nature of the disease in relationship to a stressor like handling, restraint, transport and high environmental temperature.

Laboratory tests are used to aid in the identification of live PSS-susceptible animals, but the procedures are not rapid enough to diagnose the disease in time to prevent death. Polymerase chain reaction (PCR), a DNA-based assay, is used for the detection of the *ryr-1* gene in the blood or tissue and is the most common screening tool, although it may lack sensitivity and specificity to identify PSS-susceptible animals or carriers. It allows identification of homozygous PSS-resistant and PSS-susceptible animals, as well as heterozygous carriers. This technique can also be used in the meat to identify genetically susceptible animals post-mortem. Other tests include the caffeine contracture test, which involves *in vitro* exposure of isolated biopsied muscle tissue to caffeine and halothane, measurement of enzymes and clinical chemistry of the serum. Blood typing also assists in tracing PSS-susceptible animals, along with the *in vivo* halothane challenge test, with 3–6% halothane in oxygen for 3–5 min, which causes susceptible pigs to become rigid and rectal temperature to rise. **Anaesthesia** should be discontinued upon development of these clinical signs to prevent death. Measurement of muscle pH from the carcass at **slaughter**, and pale soft exudative muscle consistency, are also indicative of the condition. Due to its acute nature, PSS episodes are usually not able to be treated in time to prevent death. However, pigs that develop PSS under anaesthesia may be successfully treated if detected early.

Animals sensitive to **stress** must be housed in a cool environment with watering areas available (i.e. dripping system) and receive adequate ventilation. It is advisable to starve animals 12 h prior to transport, which should be carried out at temperatures below 22°C. Mixing of unfamiliar susceptible animals should be prevented to reduce exposure to social stress, and medication may be used to minimize stress prior to handling, movement or transport.

Genetic selection against the PSS trait is the best tool in preventing or reducing the prevalence of the disease within the swine population. A policy of testing and eliminating PSS-susceptible and carrier animals has been adopted, especially with the advent of DNA-based assays, and has played a major role in controlling the spread of the condition. Currently, several breeds and lines that no longer express the *ryr-1* gene are commercially available. However, breeding of homozygous, stress-susceptible sire lines with resistant dam lines can yield heterozygous animals that have higher carcass quality and prolificacy. Prevention of PSS episodes can be assisted by proper management practices that minimize exposure of animals to stressful events. (RP)

Further reading

O'Brien, P.J. and Ball, R.O. (2006) Porcine stress syndrome. In: Straw, B.E., Zimmerman, J.J., D'Allaire, S. and Taylor, D.J. (eds) *Diseases of Swine*, 9th edn. Blackwell Publishing, Oxford, UK.

Smith, W.J., Taylor, D.J. and Penny, R.H.C. (1998) *A Colour Atlas of Diseases and Disorders of the Pig*, 2nd edn. Elsevier Health Sciences, UK.

Taylor, D.J. (1999) The porcine stress syndrome (PSS), pale soft exudative (PSE) muscle and malignant hyperthermia. In: *Pig Diseases*. St Edmundsbury Press, Suffolk, UK.

Positive welfare

Historically, the scientific assessment of animal **welfare** focused primarily on the **health** of animals and so was dominated by the physical (and physiological) evidential approach (positivism) of the veterinary profession, with recognition of the importance of behavioural parameters emerging after the publication of **Ruth Harrison**'s text *Animal Machines* in 1964, which highlighted the biological frustrations associated with intensive farming. This led to a new generation of animal welfare scientists focused on the goal of how to establish the presence or absence of **suffering** in captive animals, as encapsulated in the **five freedoms**. The perspective that good welfare was associated with the absence of suffering (absence of indicators of negative welfare) probably arose as a result of the historical focus on **disease** and the traditional acceptance that animals can feel **pain,** with little consideration given to the relevance of positive **emotions** in veterinary health care (although it is an interesting historical note that the **Brambell Report**, which led to the production of the five freedoms, acknowledges the importance of pleasure to animals). The extent of animal **sentience** remained very uncertain and, in order to avoid accusations of poor science and **anthropomorphism** in this emerging field, many scientists avoided the subject of specific positive emotional states such as 'happiness'.

However, in recent years, with an increase in our understanding of the neurophysiological and potentially functional basis of these emotions, there has been growing scientific interest in their assessment in non-human animals, which has started to inform thinking within the field of animal welfare science. This approach has been referred to as a 'focus on positive welfare'. This focus does not seek to produce methodologies that will replace previous approaches to the assessment of animal welfare but, rather, as with previous innovations, to complement it. In this respect, it has been highlighted that there is an important distinction to be made between the consideration of positive feelings associated with the presence of certain resources or occurrence of certain events (what an animal likes) and the resources or events that it is motivated to achieve (what an animal wants) (Yeates and Main, 2008). Three main approaches to this subject have emerged: one focused on neurophysiological correlates of positive emotional states, one focused on the evaluation of resources that may be desired by an animal and one focused on the behavioural assessment of the relative difference in cognitive or emotional state between individuals in different circumstances or with different histories (**see: Cognitive bias**). However, it must be recognized that the immediate preferences of animals are not necessarily concordant with their long-term interests. Thus indices of positive welfare are not necessarily associated with a high quality of life in the longer term. (DSM)

See also: Measuring welfare; Welfare measurement

Reference and further reading

Boissy, A., Manteuffel, G., Bak Jensen, M., Oppermann Moe, R., Spruijt, B., Keeling, L.J. *et al.* (2007) Assessment of positive emotions in animals to improve their welfare. *Physiology and Behaviour* 92, 375–397.

Mendl, M., Burman, O.H.P., Parker, R.M.A. and Paul, E.S. (2009) Cognitive bias as an indicator of animal emotion and welfare: emerging evidence and underlying mechanisms. *Applied Animal Behaviour Science* 118, 161–181.

Yeates, J.W. and Main, D.C.J. (2008) Assessment of positive welfare: a review *The Veterinary Journal* 175, 293–300.

Posture

Posture is the arrangement or position of an animal's body and limbs. The basic, normal postures for quadrupeds can be termed 'standing', 'sitting', 'kneeling', 'lying laterally' and 'lying sternally'. Standing and **lying** postures are often held for relatively long periods of time. Sitting and kneeling are more often transitional postures between standing and lying, and are often held for relatively short periods of time. These postures are defined in Table P.1.

Postures can yield useful information either through relative amounts of time spent in normal postures, the number of changes between postures or the presence of abnormal postures.

Normal posture is often used in behavioural observations to give a general indication of the animal's state of **arousal** and/or level of comfort. Many **ethograms**, especially those used for the collection of **time budget** data, will include the separate postures either as behavioural classes or as modifiers (i.e. extra information that is added on to the original behaviour classification). Including posture in this way allows the time budget information to be analysed in greater detail if required. For example, instead of just having information on, say, time spent alert, we would now be able to distinguish between time spent standing alert and time spent lying alert.

Table P.1. Posture definitions.

Posture	Definition
Standing	Standing upright with soles of feet in contact with ground
Sitting	Sitting on buttocks with soles of front feet in contact with ground
Kneeling	Standing on rear feet but resting on carpus of front limbs
Lying sternally	Lying on sternum with belly in contact with ground
Lying laterally	Lying on side with belly exposed and limbs extended

The general classifications of normal posture are often used in comparison of systems, to determine treatment effects on such things as general activity and **resting behaviour**. For example, comparison of time budgets of pigs in a straw-bedded system versus a fully slatted, non-bedded system might show differences in all five posture classifications. Straw-bedded pigs may have increased time spent standing and lying laterally and decreased time spent sitting, kneeling and lying sternally. These differences may be due to the straw-based pigs spending greater amounts of time in active behaviours such as rooting and **exploratory behaviour**, and also more time sleeping in a position that demonstrates greater comfort. Of course, for these interpretations to be valid, additional behavioural elements would need to be collected together with the posture data.

Within a given system, the time spent in the various postures would fall within a fairly limited, 'normal' range. Within this range, there may be some degree of variation between individuals, but this variation would be small. Taking population means, however, may mask individuals within the system showing abnormal variation from that mean. An animal showing abnormal variation from the mean may well represent an animal that is encountering a **welfare** problem and, thus, posture data can be used to identify animals at risk. For example, an animal that is showing greater than normal time spent lying may be suffering from a clinical or subclinical illness. This may be a whole-body illness, causing lethargy or lassitude or it may be a localized infection or **injury**, such as a hoof or limb problem that is causing localized **pain** and **lameness**. In both examples, the animal is likely to spend less time carrying out active behaviour and thus spend a greater amount of time lying. Similarly, any animal spending different amounts of time to normal in other postures will also be worthy of investigation.

Posture changing

Other information that can be contained within time budget data, but may also be investigated separately, is that of posture changing or the number of transitions between postures. With scan sample data, the data contain only information on the posture that the animal is in at the instant of recording (**see: Measuring behaviour**). If an animal is, say, standing on two consecutive scans, it may mean that the animal has been standing for that whole time, but the animal may also have lain down and stood up again. To get the real incidence of posture-changing, continuous behavioural observation techniques would need to be used. However, counting the number of transitions in posture across scans can give a rough estimation of how much posture-changing the animal is carrying out. For example, say scan data show the following sequence:

Stand–stand–stand–lie laterally–
lie laterally–lie sternally–sit–stand–stand

Looking for times when the posture is different between scans would show that in this sequence, there are a total of four postural transitions:

Stand–stand–stand–lie laterally–

lie laterally–lie sternally–sit–stand–stand

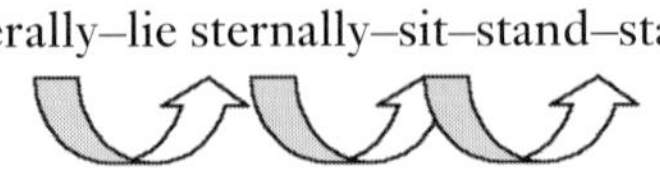

Comparisons of scan data and continuous data show that the number of postural transitions contained within scan data are very highly correlated to the actual number of posture changes carried out by the animal. Posture changing in itself is used as a measure of comfort. Ordinarily, many domestic animals may spend long periods of time in a single posture. Certainly, domestic carnivores and pigs show long periods of inactivity and/or **sleep**, and these periods contain long bouts within a single posture.

Domestic ruminants have much shorter bouts of activity and inactivity and therefore, usually, much shorter periods in a single posture. However, this difference notwithstanding, increased posture changing may well be indicative of increased restlessness, which may itself be caused by various external or internal factors. An example of an external factor is the example used earlier: comparison of pigs' behaviour in bedded and non-bedded pens may show that non-bedded pigs have an increased propensity to change posture more frequently, as a concrete **floor** is less comfortable than a straw-bedded floor. Conversely, sows in crates and cows in tie-stalls may show a decreased level of posture-changing behaviour relative to sows and cows in open pens. This is due to the physical constraints placed upon them, making transitions from standing to lying and lying to standing more difficult to achieve.

An example of an internal factor affecting posture changing is that of approaching **parturition**. As parturition nears, posture changing often increases markedly. In pigs, posture changing of the sow can also impact the survival of the young. Piglets are susceptible to being crushed by the sow when she lies down. Also, sows that carry out particularly high levels of posture changing during the immediate pre-parturition phase appear to be more likely to savage their piglets.

Abnormal posture can be used to describe modifications of the normal postures that are usually seen when the animal is undergoing challenge. The challenge may be such factors as illness, **disease**, pain or **fear**, and thus, the manifestation of abnormal behaviour can be an indicator that the animal is having its welfare compromised. As already noted, abnormal amounts of time spent in postures may indicate problems. Modification of that posture may help to pinpoint the cause of the problem. An animal that is standing but only bearing weight on three limbs is an obvious example of an animal showing an abnormal posture – probably indicating a **lameness** issue with the raised limb that should be investigated.

Husbandry procedures such as tail docking (**see: Docking – tail**), **castration** and **mulesing** often induce abnormal postures indicative of the animal trying to decrease or cope with pain, such as 'statue standing' or lying on the side with limbs rigidly extended. In situations that induce fear, animals may become immobile and crouched or hunched, or adopt a posture that enables them to carry out defensive aggression. (ELS)

Further reading

Grant, C. (2004) Behavioural responses of lambs to common painful husbandry procedures. *Applied Animal Behaviour Science* 87, 255–273.

Marchant, J.N., Broom, DM. and Corning, S. (2001) The influence of sow behaviour on piglet mortality due to crushing in an open farrowing system. *Animal Science* 72, 19–28.

O'Callaghan, K.A., Cripps, P.J., Downham, D.Y. and Murray, R.D. (2003) Subjective and objective assessment of pain and discomfort due to lameness in dairy cattle. *Animal Welfare* 12, 605–610.

Tuyttens, F.A.M., Wouters, F., Struelens, E., Sonck, B. and Duchateau, L. (2008) Synthetic lying mats may improve lying comfort of gestating sows. *Applied Animal Behaviour Science* 114, 76–85.

Precautionary principle

The precautionary principle states that if an action might theoretically or logically cause harm, then those who wish to undertake this action have a greater moral responsibility to demonstrate good evidence that the action does not cause harm – i.e. action is not justified because of a lack of evidence of harm. For example, the precautionary principle would suggest that the wide-scale planting of genetically modified crops should not be undertaken until there is good evidence that they will not have a significant incidental impact on wild ecosystems. (DSM)

Further reading

Foster, K.R., Vecchia, P. and Repacholi, H. (2000) Science and the precautionary principle. *Science* 288, 979–981.

Precocial

A state of neonatal maturity characterized by the organism being free-living and adopting behaviours that enable it to be in part independent of its parents. Generally considered to be the opposite of **altricial** development. Precocial bird and mammal neonates are fully covered by **feathers** or fur and are often able to thermoregulate effectively. In birds, hatchlings are self-feeding, albeit with parental supervision. In mammals, precocial offspring are active from birth but are reliant on maternal milk. (DCD)

Predation

Predation is the act of an individual or group of individuals of one species (the predator) killing an animal of another species (the prey), with the intent of subsequently ingesting the killed animal. Within all ecosystems, predation is one of the most prominent interactions between species. The term predation is mostly used for an act of killing an animal, but it is also sometimes used to describe herbivory or the killing of seeds (seed predation). More generally, predation is a specific type of a broader set of interactions in which one species, in this case the predator, is positively affected and the other species, in this case the prey, is negatively affected. There are also cases where individuals kill and ingest members of the same species – specifically termed **cannibalism**.

The stages of predation

Predation can be described as occurring across a series of stages: encounter, detection, recognition, attack, capture and consumption. Predators benefit from increasing the likelihood of each stage while the prey benefit from decreasing this probability. For example, wolves hunting elk search for (encounter, detection and recognition), stalk (attack) then run down (capture) an elk, whereas the elk seeks to avoid being seen and attempts to evade attack and capture by escaping. Predation is often studied from either the predator perspective or the prey perspective, and studies usually focus on one or a few stages.

Predator strategies

There is a wide array of strategies employed by predators to obtain prey. Predators can actively hunt, as individuals or groups, tracking prey (e.g. an owl hunting mice or lions hunting wildebeest) or can use ambush, sit-and-wait strategies or 'surprise attacks' (e.g. rattlesnakes hunting rodents by striking out from a hidden location). There are also other sit-and-wait strategies that use deceptive, trapping methods: spiders build webs to trap insects, deep-sea fish lure prey with bioluminescence appendages and several insect species mimic female cues of a prey species to lure males (aggressive mimicry). The sensory systems of predators are often well adapted to detect and recognize prey, especially prey specialists (those that feed on one or few prey species).

Prey strategies

There are equally as many strategies employed by prey to avoid predators. Prey can seek out areas that limit their exposure to predators either in terms of space (e.g. dense cover, rock crevices) or time (e.g. be active when their predator is not), become harder to detect or recognize as prey (e.g. cryptic coloration) or become harder to capture (e.g. speed, armour).

Predator and prey population dynamics

Predation can have a large effect on both predator and prey population dynamics. The simplest case is when one predator species feeds on one prey species. This generally results in both the predator and prey populations cycling relative to each other, with the predator population 'lagging' behind the prey population. The explanation for this centres on the time lag between predators capturing food and being able to use these resources to produce offspring. For example, when the prey is at high densities the predator's population size (via increased survival and birth rate) should increase. This should cause the prey population to decline, which in turn makes it harder for all predatory individuals to survive and reproduce. As the predator population declines, this should favour conditions for the prey population to increase, and so forth (the Lotka–Volterra model).

Thus the peak in predator population size lags slightly behind that of the prey population size. In real ecosystems, food webs are usually more complicated (more predators and prey and thus alternative options), and data showing this tight predator–prey relationship are scarce. However, a classic example of such data is the Canadian lynx and snowshoe hare, where research shows a cyclic relationship.

Predation shapes phenotypes

For many prey species, predation is a major source of mortality and therefore is a strong agent of natural selection on prey phenotypes (a **phenotype** comprises the physical characteristics of an organism). Individuals of a prey species that are able to evade predation can pass heritable traits to the next generation, while successful predators can also pass heritable traits themselves to the next generation.

Over generations, predation can lead to the evolution of many forms of anti-predator mechanisms (behavioural, morphological and/or physiological) and, in turn, the precision of predator strategies and these traits are said to be 'co-evolved' traits. One example is a co-evolutionary arms race where successful predators and prey are both selected to improve survival, but the adaptations of the other species tend to offset any change in survival. For example, cheetahs that tend to be faster survive and reproduce better; however, because the cheetahs tend to be faster, the gazelles that tend to survive and reproduce are also faster. This tends to lead to both the cheetah and the gazelle becoming faster. Another example is toxins in prey western newts and resistance to these toxins in predator garter snakes, where over evolutionary time the newts have become more toxic, while the snakes have better resistance.

Predation and domestic animals

Domestic animals (from livestock to dogs) are not considered part of most ecosystems, but they are not immune to predation. Wolves, coyotes, wild cats and birds of prey are very capable of predating on domestic livestock, fowl, etc. In some cases, domestic animals have essentially lost the skills needed to evade predation (or ancestors of these domestic animals may never have encountered such predators), so they are 'easy' prey. In addition, some species of domestic animals (especially **cats**) can be significant predators in their local ecosystem. This may be viewed as beneficial by people when it is directed towards pest species (**see also: Pest control – ethics of** and **Pest species – welfare of**), but in other situations it is of greater concern because it may devastate indigenous fauna, or at least have a significant impact on local levels of wildlife. (BC)

See also: Anti-predator training; Predator avoidance

Further reading

Caro, T. (2005) *Antipredator Defenses in Birds and Mammals.* University of Chicago Press, Chicago, Illinois.

Lepczyk, C.A., Mertig, A.G. and Liu, J. (2004) Landowners and cat predation across rural to urban landscapes. *Biological Conservation* 115, 191–201.

Lima, S.L. and Dill, L.M. (1990) Behavioural decisions made under the risk of predation: a review and prospectus. *Canadian Journal of Zoology* 68, 619–640.

Predator avoidance

Predator avoidance involves behaviour, morphology or physiology that reduces an individual's probability of encountering, being detected, recognized, attacked, captured and/or consumed by a predator – collectively reducing predation risk. Most animal species are prey to other animals, and this predation is a major source of natural selection. This has led to the evolution of a vast array of anti-predator mechanisms.

Stages of predator–prey interactions

Predation

Predation be described as occurring across a series of stages: encounter, detection, recognition, attack, capture and consumption. Predators benefit from increasing the likelihood of each stage, while the prey benefit from decreasing this probability. Depending on the particular stage of predation, there are many different forms of anti-predator mechanisms prey can use to evade their predators.

Reducing encounter probability

Prey can reduce the probability of encountering a predator by adjusting their spatial and temporal activity patterns. Thus, they can avoid dangerous areas or times of the day (or night). For example, certain prey species may have become **nocturnal** to avoid diurnal predators, or vice versa.

Reducing detection

To decrease the chance of being detected by a predator, prey species can be cryptic: morphologically by blending into their background using camouflage (e.g. cryptic patterns, colours), or behaviourally by becoming motionless or moving with the background envronment. For example, walking stick insects not only blend into their background but some species also move to match vegetation movement.

Reducing recognition

Mimicry can be used by prey species to 'dupe' predators by reducing the prey's probability of being recognized. Prey can mimic organisms that are noxious, toxic and/or unpalatable (Batesian mimicry) to prevent predation, such as the classic example of the Viceroy butterfly mimicking the wing colour patterns of the unpalatable Monarch butterfly. In addition to visual cues, mimicry also occurs in other sensory modalities: the hiss of burrowing owls sounds like the rattle of a rattlesnake, and may keep potential mammalian predators out of their burrows (acoustic mimicry).

Reducing attack

Prey can deter an attacking predator. Morphologically, prey may threaten predators with spines, horns or other external weapons (e.g. rhinoceros' horns). Prey can also use startle behaviours (flashing wings, making a loud noise, releasing pungent odours) that confuse, distract or deter predators. Predators may also decide not to attack a prey animal that has performed a 'pursuit deterrence' behaviour. Pursuit deterrence involves a prey animal communicating to the predator that it has been spotted and/or that the prey will probably outrun the attack (escape ability). For example, several ungulate species 'stot' (jump straight up into the air) when a predator is spotted and it is suggested that stotting deters the predator from continuing or beginning an attack.

Evading capture
To evade capture prey can simply attempt escaping and/or take refuge in a structure the predator cannot enter or climb (e.g. rock crevices or trees). Prey can also avoid capture using their bodies, such as turtles and crustaceans retreating into their shells or prey that can roll up into a ball, protecting themselves with armour (e.g. armadillos). Prey can also use defensive behaviour such as clawing, biting and/or stinging to injure the predator. Some species can 'drop' their limbs (termed tail autotomy or limb autotomy) to escape capture (e.g. some lizard, rodent, insect and crustacean species).

Evading consumption
Consumption by a predator can be avoided by several mechanisms, including toxicity/unpalatability. Other potential mechanisms are death feigning and **fear** screams. Although not completely proved to be anti-predator behaviours, animals that feign death (e.g. **tonic immobility**) or scream during the capture stage may avoid being consumed by confusing or distracting the predator.

Grouping to avoid predators

Prey species can avoid predation by forming permanent or temporary groups with **conspecifics** (members of the same species) and/or heterospecifics (members of other species). Individuals benefit from joining groups for several reasons (**see: Grouping**).

Increased vigilance
Prey animals can increase the probability of detecting a predator by being in a group, sometimes called the 'many eyes' hypothesis. For example, individual sparrows in flocks spend less time being vigilant (head up) because, with more flock members, there is a greater probability that some other bird(s) is looking up to detect predators.

'Selfish herd' or 'Dilution effect'
These hypotheses assume that an individual in a group has a decreased probability of being captured, because the predator has more prey animals from which to select in the group. Schools of fish may make it difficult for predators to 'lock into' one individual to capture, and schooling reduces the chance of an individual fish being caught.

Confusion effect
When a predator attacks a group, the general confusion of the prey animals escaping may interfere with the predator's ability to track and capture an individual.

Warning signals

Prey species can benefit from perceiving and responding to warning signals (acoustic alarm calls, visual or chemical signals – **see also: Signal**) produced by conspecifics or heterospecifics. Individuals in close enough proximity to hear, see or smell the warning signal are often related (kin) to the signaller and can benefit from a seemingly altruistic behaviour (**see also: Altruism; Kinship**).

Mobbing

Animals in groups or individuals can mob a predator and decrease the predator's chances of a successful kill. Mobbing occurs when prey animals closely approach a predator (most often when it is not hunting) and engage it by chasing it, striking at it or performing other antagonizing behaviour. Predation risk may be low in mobbing situations, as most of the predators being mobbed rely on surprise attacks to capture prey. Several examples include monkey troops mobbing snakes, birds mobbing hawks and seals mobbing sharks. An impressive example is that of California ground squirrels that will closely approach and kick dust at rattlesnakes.

Evolution and development of anti-predator behaviour

Innate versus learned
Although a false dichotomy, anti-predator behaviour is sometimes categorized as innate or learned. Some behaviour can appear 'innate' because individuals successfully evade predation prior to extensive experience with the predators; however, it is difficult to prove that no experience was necessary for those behaviours to develop. Learned behaviour may be acquired through individual experience or can be obtained through social learning. In a classic study of **social learning**, researchers were able to make birds react with mobbing behaviour towards a plastic jug because it was paired with the mobbing call of a conspecific.

Predator naivety
Predator-naive prey (i.e. prey that do not recognize and respond appropriately to predators) may or may not be capable of learning about novel or ancient predators. Novel predators can arise by invasion or introduction to an area where the predator has never existed (e.g. **feral** cats and foxes in Australia), and ancient predators are those that the prey species have not been in contact with for many generations (e.g. wolves reintroduced in North America). For example, although moose have not been in contact with wolves for hundreds of years, they are able to respond appropriately after experiencing predation and predator cues. However, wallabies are sometimes incapable of escaping or learning about novel prey such as feral cats. Domesticated animals also sometimes lose their anti-predator behaviours, making them susceptible to predators. This may have occurred through artificial selection or due to lack of a persistent exposure to predators over time. (BC)

See also: Anti-predator training; Exotic species invasion; Vigilance

Further reading

Caro, T. (2005) *Antipredator Defenses in Birds and Mammals*. University of Chicago Press, Chicago, Illinois.

Lima, S.L. and Dill, L.M. (1990) Behavioural decisions made under the risk of predation: a review and prospectus. *Canadian Journal of Zoology* 68, 619–640.

Predator control

Predator control is perhaps one of the most controversial issues in **wildlife management**, involving the active management and limiting of predator populations through human intervention. In natural systems, predators play an important role in ecosystems and can regulate prey populations. However, predator–prey dynamics can be unstable in disturbed systems, with increased predation driving prey species to **extinction** in certain situations. For example, exotic predators such as the black rat introduced on to islands have caused the extinction of many rare avian species. Thus predator control may be deemed necessary to protect threatened and endangered species. Habitat changes can also lead to increases in the number of native predators.

Agricultural practices, including sheep/cattle ranching and poultry farms, along with wildlife refuges, may inadvertently increase the carrying capacity for predators and create higher mortality losses for prey populations. In fact, preventing economic losses is one of the primary reasons for predator control. Besides predation on livestock, predators may also compete with humans for game species or pose a threat to human health or safety. Because of these human–wildlife conflicts, predator control programmes have been created to reduce agricultural losses, increase game species numbers, minimize risk to humans and/or assist in the recovery of **endangered species**.

The theory of predator control is based on the negative relationship between predator and prey abundance, with increased prey survivorship with fewer predators. Several different lethal and non-lethal techniques have been employed to remove predators, including trapping, translocation, ground and aerial shooting, poisons, snares and sterilization (**see: Pest control – ethics of**). However, depredation management is not always effective due to the complexity of predator behaviour and ecological interactions among predators and prey. Predator control has threatened the persistence of some predator species, while others have been extremely resilient to control efforts. Removing top predators also can create **trophic cascades** and **mesopredator** release, impacting non-target species. Consequently understanding the **ecology** and behaviour of predators is as critical as knowing the benefits and costs of predator control. (LAN)

See also: Free-range animals; Predation

Preening

Feathers are critical not only to bird flight but also for thermoregulation, and can be important in other aspects of avian biology, e.g. reproductive success. As a result, feathers require regular maintenance to ensure that they are in good condition and so serve their various functions efficiently. Preening with the beak, or sometimes the feet, is one of a range of behaviours that serves to maintain the condition of the feathers; others include **dust bathing**. In that it involves care of the integument it is analogous to **grooming** behaviour. Flight feathers only function efficiently if they form a surface to resist the air, and so preening ensures that the mutual attachment of barbules is as continuous as possible. Care of down and other feathers covering the body is important in trapping a layer of air close to the skin and so maximizing the efficiency of **thermoregulation**. Preening is most commonly performed by individuals upon their own feathers, but mutual preening can be a key component in the formation of a pair bond in some species, e.g. parrots.

Preening is seen as an important behaviour of birds, albeit forming a small percentage of their **time budgets**. Depriving a bird of the ability to preen normally is considered by many as being a **welfare** issue. The act of tending feathers is seen as a precursor of other more deleterious behaviours. Self-plucking, or chewing of feathers, is interpreted as an extension of normal preening behaviour, but taken to an extreme when the individual bird is in a stressful situation. Typically, birds maintained within poor captive conditions, or deprived of performing some other key component of their behavioural repertoire, develop behaviours that lead to damage of the feathers either on their own body or on their companions, which are resistant to behaviour modification even after the **stressors** are removed. (DCD)

Preference

Animals show preferences in their daily decision making, choosing between available resources (such as food sources, potential territories or mating partners) and behaviours. **Ethology** and **behaviour ecology** studies (**see: Optimal foraging**) support the assumption that such decision-making mechanisms are the evolutionary product of natural selection, i.e. through **evolution**, animals' preferences have been shaped so that the animals will look for and prefer situations and resources which promote their survival. Equally, they would be motivated to avoid environments detrimental to **welfare**.

Preference studies as an instrument in animal welfare research were first proposed by the ethologist W.H. Thorpe in his appendix to the 1965 Brambell Committee Report. The first preference experiment also addressed a question discussed in the **Brambell Report** – **flooring** material for laying hens in cages. By placing hens in a pen with two compartments, each with a different flooring and studying how much time the birds spent in each of these compartments, Hughes and Black (1973) could demonstrate that hens did not show any strong preference away from the 'chicken wire' thought unsuitable by the Brambell committee.

The hen cage floor example illustrates the essential value of preference studies and the consumer demand studies (**see: Economics of behaviour**) that have developed from them: that these tests offer a way of asking the animals directly rather than relying on human judgement of what might or might not be good for animal welfare. However, as experience has shown, the matter is somewhat less straightforward than was initially believed.

The most common design of a preference test is similar to that described above: an experimental set-up with several compartments containing different resources, where animals can choose where to stay and where their choice can be monitored. Because preference can depend on time of day, internal state and what activities the animal is carrying out, it is important that the set-up permits monitoring the behaviour of the animal over extended periods of several days. Such monitoring can be automated with the help of video equipment and automatic event recorders such as a photo beam being broken each time an animal passes from one compartment to the other.

Even the most carefully designed preference experiment will, however, only give information about whether one resource is preferred over another, and tell us nothing about how strong this preference is. Experiments measuring preference strength have come to play an important role in animal welfare research. The underlying assumption is that the more an animal prefers a resource, the more its welfare will be affected by frustrating access to this resource. However, some caution is needed in assuming a direct link between motivation strength and welfare, as there may be cases when animals (including humans) are strongly motivated to access resources that are not good for their long-term welfare.

Strength of preference can be measured by making access to a resource costly. The cost can consist of a task that the animal has to perform, such as making a number of lever pushes or button presses during a given time. This method is described as operant testing (**see: Operant test**). Because the operant task is usually artificial, some researchers advocate the obstruction method, where the animal has to make its way through an obstruction, such as a narrow gateway, a water bath or a weighted door. Independent of the type of cost, in the simplest approach the animal is asked to pay a single predetermined cost to obtain access to the resource and, if it is found willing to pay that cost, it is assumed to be motivated to use the resource. It is possible to refine this measure if the cost can be increased in a stepwise or continuous manner so that the motivation can be quantified.

The British biologist Marian Dawkins has been an important driving force behind the introduction and development of operant testing as an instrument in animal research. She originally suggested that an unknown **motivation** could be quantified by titrating it against the motivation for a resource for which motivation was well known and could be easily manipulated, such as feeding motivation. This approach was further refined by bringing in consumer demand theory from economics research. In consumer demand studies, the demand for a resource is studied when the cost for gaining access to it is increased. The demand (the amount of the resource obtained) is plotted on the y-axis against the cost on the x-axis. The slope of the resulting curve can be used as a measure of the demand strength.

> Commodities for which a given percentage increase in price results in a decrease in the quantity demanded are said to have *elastic demand* and are sometimes called luxuries; those for which a given percentage increase in price results in little change in the quantity demanded are said to have *inelastic demand* and may be called necessities … Elasticity of demand is a key concept for the study of animal welfare … because it shows how important different environments or commodities are to the animals themselves.
>
> (Dawkins, 1990)

An alternative approach to make resource access more costly is through restricting the time an animal has available by changing the light schedule so that the active period is shortened. The assumption is that animals will persist in performing the most important behaviour patterns, whereas less essential behaviours are strongly reduced or disappear completely when the animal has less time.

There are a number of factors that may affect the outcome and the possibility to extrapolate outside the test situation, which ought to be taken into account in designing the study. These include differences in preferences between animals (depending on sex, breed, age, reproductive state, etc.), previous experience with any of the resources, cues from resources, duration of access to resources as well as overall time or energy budgets. Moreover, animal choice and preferences may be affected by how many options are available and whether they are independent or whether some of these can substitute for or complement each other. There has been much discussion of how these issues impact on the conclusions that can be drawn when using preference tests for welfare recommendations. (AO)

References and further reading

Bateson, M. (2004) Mechanisms of decision-making and the interpretation of choice tests. *Animal Welfare* 13, S115–S120.

Dawkins, M.S. (1990) From an animal's point of view: motivation, fitness, and animal welfare. *Behavioural Brain Science* 13, 1–61.

Fraser, D. and Nicol, C.J. (2010) Preference and motivation testing. In: Appleby, M.C. (ed.) *Animal Welfare*, 2nd edn. CAB International, Wallingford, UK (in press).

Hughes, B.O. and Black, A.J. (1973) The preference of domestic hens for different types of battery cage floor. *British Poultry Sciences* 14, 615–619.

Mason, G., McFarland, D. and Garner, J. (1998) A demanding task: using economic techniques to assess animal priorities. *Animal Behaviour* 55, 1071–1075.

Matthews, L.R. (1998) Using economic techniques to assess animal priorities: repays the investment. *Animal Behaviour* 55, 1076–1078.

Nicol, C.J., Caplen, G., Edgar, J. and Browne, W.J. (2009) Associations between welfare indicators and environmental choice in laying hens. *Animal Behaviour* 78, 413–424.

Preparedness

The concept of biological preparedness proposes that different species (or possibly populations within a species that have been selected for a given trait) have differing propensities to learn certain types of association or that certain types of association are biologically easier to form. It is suggested that natural and human **selection** have favoured biases in the types of association that can be formed. As such, the concept of biological preparedness opposes the universal generality of the Laws of Association, proposed by early behaviourists, who suggested that external relationships are all important in the learning and that all associations are equipotent, i.e. have equal probability of being formed on the basis of the equivalence in relation to the Laws of Association (**see: Conditioning – types of**).

Biological preparedness suggests that different species may be predisposed to develop **fear** in relation to stimuli that are of evolutionary significance, or perhaps different breeds of animal selected for different functions. For example, noise fears are less common in the typical gun dog breeds. However, biological preparedness also implies that a given species may find it easier to learn one type of association over another – for example, an **aversion** to a given food may be more readily made with the flavour of a food than with a similarly contingent visual stimulus. It might be argued that the concept of biological preparedness extends also to the predisposition to learn certain associations at certain times (**sensitive phases**), which may be of phylogenetic adaptive significance – for example, children appear much more sensitive to learning to fear animals around the age of three, when historically they would be much more mobile and independent.

Other phenomena that it has been suggested can be explained by biological preparedness include: (i) a greater resistance to **extinction** of certain associations over others; and (ii) the ability to form certain associations without manifest attention to them. However, some argue that other explanations are preferable in understanding these phenomena. (DSM)

Further reading

McNally, R.J. (1987) Preparedness and phobias: a review. *Psychological Bulletin* 101, 283–303.

Problem behaviour

Problem behaviour is a personal rather than a physical construct, as it describes any behaviour that is problematic for the individual reporting it. The problem is not the animal's behaviour per se, but rather the problem it poses to the reporter. Thus the problem may be resolved by altering the understanding, attitude or opinion of the reporter without necessarily effecting any change in the behaviour of the performer. For example, the behaviour of a cat in **oestrus**, which wails and appears to a naive owner to be in **pain** may be a problem until it is explained that this is the normal behaviour of the species. The term does not imply that the behaviour is **maladaptive**, nor necessarily associated with poor **welfare**. Although some consider the term to be synonymous with behaviour disorder or **abnormal** behaviour, the former term, in particular, implies that the behaviour is not structured appropriately and reflects a medicalization of the construct (as has been applied in human psychiatry in relation to problem human behaviour), which has given rise to numerous misconceptions in the field (see Mills, 2003).

While some behaviour problems may be disorders, such as panic disorders where the response is out of proportion to the potential threat and interferes with normal functioning, this does not apply to perhaps the majority of problem behaviours seen in clinical practice, such as **urine marking** and scratching in the cat, most causes of **aggression** and barking in the dog or bucking in response to back pain in the horse.

Broadly speaking problem behaviour may be derived from one of three types of action: (i) those that have adaptive value within the species but are inconvenient for the keeper (these are adaptive behaviours); (ii) attempts to behave in an adaptive way in an environment to which complete adaptation is not possible (these may be considered maladaptive behaviours); and, finally, (iii) expressions of direct disruption of the nervous system (i.e. the neural basis has no functional value in any context and so is truly malfunctional). Accordingly, there is no objectively consistent or defining feature of a problem behaviour but, once its broader social context is recognized, so too are the full range of approaches to its scientific study and resolution. This refocuses attention away from just the individual performing the behaviour on to a broader perspective that extends to the physical and social environment to which the animal is responding. The study of problem behaviour is therefore a multidisciplinary challenge. (DSM)

Reference

Mills, D.S. (2003) Medical paradigms for the study of problem behaviour: a critical review. *Applied Animal Behaviour Science* 81, 265–277.

Procedural knowledge

Procedural **knowledge** refers to the state of knowing 'how' to do something. Procedural knowledge is often contrasted with declarative knowledge (also referred to as descriptive knowledge, factual knowledge or propositional knowledge), which describes factual knowledge. Thus a trainer may know how to get a dog to sit for a treat (procedural knowledge), but may not know much about the reasons why this works (declarative knowledge). Declarative knowledge forms the basis of thinking about something, and may be further divided into episodic knowledge (knowing about the context of something) and semantic knowledge (knowing the meaning of something, but not in a personal way). The extent to which non-human animals are capable of having different forms of knowledge is debated widely; however, the possession of knowledge does not equate to either **awareness** of that knowledge or **consciousness**. (DSM)

See also: Cognition; Epistemology; Metacognition

Proceptivity

Proceptivity refers to the behaviour carried out by the female that enhances the creation, maintenance and escalation of a sexual interaction (**see: Sexual behaviour**). It specifically refers to the pre-consummatory, that is, pre-ejaculatory, phase of a sexual interaction. There is evidence to suggest that **oestrogen** and adrenal **androgens** play an important part in increasing proceptivity. (JNM-F)

Production animal diseases

Diseases can be defined as conditions of organisms that impair body functions and are associated with certain physical signs. Animal production systems tend to show a higher prevalence of certain diseases associated with increased productivity. The term production disease is often used to refer to non-infectious conditions such as metabolic and management problems, typically associated directly with the high productivity demands of these systems, while the broader term 'production-related disease' may be used to include secondary problems, such as infertility and a range of infectious diseases to which these animals are particularly prone because of their circumstances.

There are a number of additional factors that contribute to the high occurrence of different forms of disease in these animals. For example, animals in production systems tend to have similar **genotypes**, and since many of these conditions have a substantial genetic component, they are more likely to be widespread due to the widespread prevalence of genetic risk factors in the less diverse population. Prevention may therefore require selective breeding, but this may compromise productivity. These animals are usually fed similar concentrate feed rations and are often kept in close contact with their faeces. Additionally, different production stages usually happen at different sites and the animals are moved together in large groups. These factors can all contribute to large outbreaks of certain diseases and the ease with which disease can spread through groups of animals. The multifactorial nature of these problems means that their management should also be multifactorial and potentially multidisciplinary, with consideration given to strategies aimed at both the individual and population level of the problem. Examples of some of the more common production diseases are discussed briefly below (see also specific individual entries where appropriate).

Ascites

This condition is often found in broiler **chicken**s and is a major cause of **mortality** in these animals. It is an accumulation of fluid in the peritoneal cavity. The most common cause is increased back pressure in the venous system (pulmonary hypertension) associated with right ventricular cardiac failure. This problem arises because selection for high growth rate in these birds means that the physical growth of some individuals exceeds the capacity of the lungs to provide sufficient oxygenated blood. Higher growth rate demands a higher level of nutrients and oxygen, which is provided in part by increasing **blood pressure**, which results in chronic hypertension.

Challenges that increase cardiac demand may trigger a bout of ascites and associated mortality as cardiac capacity is exceeded. For example, exposing animals to a lower environmental temperature (**see: Critical temperature**) for a short period may increase metabolic demand for thermoregulatory processes and trigger the problem. Similarly, any problem that reduces the efficiency of oxygenation, ranging from mild infectious respiratory diseases to poor air quality, may result in a compensatory cardiac response that serves to try to increase blood flow through the lungs, but which may ultimately result in cardiac failure. Although the condition can be treated theoretically with drugs, this is uneconomic and often impractical for animals destined for human consumption. Reducing the birds' metabolic oxygen requirement by slowing growth or reducing feed can prevent ascites.

Mastitis

Mastitis is very common in lactating production animals, such as cows, sows, mares, goats and ewes. It is the **inflammation** of the mammary gland(s) associated with bacterial infection. The milk will often be visibly abnormal, e.g. in colour, or have clots, and the udder may swell, be red and hot and be painful to the touch. Risk factors for developing mastitis include poor sanitary conditions, trauma inflicted by offspring and systemic infection.

Lameness

Lameness is the third most economically important problem in dairy cattle (mastitis and infertility being more economically important). Lameness is also commonly found in **sheep**, **goats**, **pig**s and broiler chickens. The incidence of lameness has increased since the late 1980s. Production animals housed on concrete floors or which are unable to get appropriate exercise often show some degree of lameness. Management aimed at reducing abrasion and contact with faeces, or prolonged soaking of the feet, foot trimming and foot bathing are all important treatment measures for managing the condition.

Ruminal acidosis

Also known as metabolic acidosis or milk acidosis, this is a common disease found in ruminants since grain and concentrate feeding became a widespread practice associated with a demand for higher productivity. The pH of the rumen becomes more acidic due to increased starch intake, lack of forage and associated chewing and rumination. It consistently causes anorexia, and movement of feed though the rumen slows and may even stop. Faeces will be loose and dehydration will occur. Ruminal acidosis is a predisposing factor for many other problems in cattle fed high-grain diets, such as **laminitis**, sudden death syndrome and liver abscesses. Its exact relationship with another production disease, ketosis (acetonaemia), is unclear.

Ketosis arises as a result of the mobilization of fat reserves to provide glucose due to a negative energy balance with the diet. The condition is most commonly seen in early **lactation** when demand is at its highest. Fatty acids are converted by the liver to ketones, which, if allowed to accumulate, become toxic, suppressing appetite and exacerbating the problem further. It is possible that the anorexia associated with ruminal acidosis may precipitate ketosis, or vice versa.

Hypocalcaemia

Another production condition of lactation that is particularly prevalent in mature cows is hypocalcaemia, resulting from the initial demand for calcium in the milk at the onset of lactation depleting the mobile reserves in the body. Hence the condition is also known as milk fever, or parturient paresis, in dairy cows. Given the importance of calcium to the activity of muscle cells, it presents as an afebrile flaccid paralysis around the time of **parturition** that can ultimately lead to heart muscle failure and death.

Hypomagnesaemia

Also known as grass tetany or staggers, this condition may occur in association with hypocalcaemia or independently, and arises as a result of a lack of magnesium uptake, particularly when much is lost in the milk during lactation. It is important to identify or eliminate its co-occurrence, as treatment resulting in excessive magnesium may be equally dangerous. Typically, the animal is more excitable and twitchy when hypomagnesaemia is involved. The problem can arise because the animal is feeding on species of herbage that are low in magnesium, or on lush herbage that is naturally lower in this mineral. Alternatively, the condition may arise secondarily to some cause of anorexia or reduced appetite. The use of nitrogen and potassium fertilizers on pasture may increase the risk by both encouraging a flush of lush grass and because these minerals reduce magnesium uptake from the rumen.

Diarrhoea (scours)

Diarrhoea is a condition commonly found in young production animals and can have a variety of different causes. In calves, those fed large amounts of milk or inappropriately formulated milk replacers produce a large volume of loose faeces. Piglets may contract a bacterial infection characterized by watery or greasy, yellow–white diarrhoea. Piglets may appear weak, dehydrated and undersized, weight gains may be depressed and individuals sometimes die. A contributing factor to mortality is that piglets become covered with diarrhoeic faeces and stay damp.

Other conditions

There is also a wide range of highly infectious diseases of relevance to production animals, which, while not associated directly with productivity (and so not typically considered production-related disease), because of certain factors associated with productivity such as stocking density, especially in the case of poultry, or movement policies, may be of particular concern to those working with production animals. These include conditions such as Marek's disease, a common paralysing and often fatal viral disease of domestic poultry. The virus matures in the feather follicle and is released into the environment with dust or dander. Newcastle disease is another common viral disease of domestic poultry and many other bird species. It usually presents as a respiratory disease, but infected birds can also show depression, nervous manifestations or diarrhoea. The virus is shed and easily transferred through respiration and in the faeces. Infection may be introduced by wild birds, and control for both these measures and similar conditions is through biosecurity, an all-in/all-out population policy and vaccination.

While production diseases tend to be thought of in relation to high-productivity, intensive systems, it is important to realize that less intensive systems, such as **organic farming** systems, may be associated with their own range of production diseases associated with the particular husbandry practices, e.g. endoparasitic problems if anthelmintics are not used and pasture management is inappropriate. (LMD, DSM)

See also: Farm assurance scheme; Intensification of animal production

Progesterone

Progesterone is a C-21 **steroid hormone** belonging to the hormone class progestogens, which have hydrophobic properties. Progesterone is involved in the female oestrous cycle (**see: Oestrus**) and during pregnancy, its main function being to maintain gestation.

Progesterone is mainly produced in the **adrenal** cortex. Acetyl-Coenzyme A is a side-chain cleaving enzyme that cleaves the initial substrate cholesterol to form pregnenalone, from which progesterone is synthesized. Progesterone can then take part in two different pathways. It can either be used in the synthesis of **corticosteroids** through the synthesis of deoxycorticosterone, corticosterone, cortisol or aldosterone; if the progesterone takes part in the sex steroid pathway, it is then a precursor to 17-α hydroxyprogesterone, androstenedione, **testosterone**, estradiol-17β and estrone (**see: Oestrogen**). Progesterone is synthesized in the corpus luteum after ovulation, the placenta during pregnancy, and the **brain**. The placenta uses maternal cholesterol to produce progesterone. The majority of placental progesterone enters the circulation of the mother, and the rest enters the fetal circulation and is utilized as a substrate for fetal corticosteroids.

During the early follicular phase (early growth phase of the follicle) of the oestrous cycle, **lutenizing hormone** (LH) and **follicle-stimulating hormone (FSH)** are at medium levels and oestrogen and progesterone levels are low. In the late follicular stage, there is further growth of the follicle. Concentrations of LH and FSH remain similar to that of the early follicular phase; during this time oestrogen levels increase and progesterone levels remain low. The rise in oestrogen causes a rapid increase in LH and FSH, which stimulate ovulation to occur. During the luteal phase, the corpus luteum develops and, with this, oestrogen and progesterone levels increase, these high levels of progesterone and oestrogen having a negative feedback on the **pituitary gland**, thus inhibiting the release of LH and FSH.

If fertilization does not occur, the corpus luteum regresses sooner or later, depending on the species, and there is a reduction in the levels of oestrogen and progesterone. There is no longer a negative feedback effect on the pituitary, and LH and FSH increase to the levels they were at during the early follicular phase. However, if fertilization occurs, progesterone aids implantation of the blastocyst by reducing the maternal immune response so that there is acceptance of the pregnancy. Progesterone helps to maintain pregnancy and remains at high levels until the placenta is expelled during **parturition**. The rapid decrease in progesterone after parturition is followed by a rapid increase in **prolactin**, which is involved in milk synthesis.

Progesterone receptors are present in mammary tissue in many different species. It has been suggested that progesterone stimulates ductal growth and lobulalveolar development in prepubertal mice. In dairy cows, a mechanism has been suggested that, during late pregnancy, progesterone has a role in mammary development by promoting the survival of mammary epithelial cells by inhibiting apoptosis. Progesterone's inhibitory effect of apoptosis of murine mammary epithelial cells help to prolong lactation in these mice. The mechanism of how progesterone inhibits apoptosis is not clear but it is thought that progesterone inhibits matrix proteinases which, in turn, would prevent the degradation of basement membrane and, as a result, prevent mammary apoptosis.

In lambs, it has been suggested that progesterone and its metabolites suppress fetal arousal behaviour. In studies where progesterone or its metabolites were injected into fetal circulation, it was found that fetal electrocorticogram, electrooculogram and **electroencephalogram** activities, behaviour **arousal** and breathing movements were reduced. During the last days of pregnancy, there is a decrease in progesterone in placental and fetal circulations. This reduces the inhibitory effects that progesterone has on behaviour arousal, thus facilitating a vigorous behavioural phenotype following parturition of the lamb and increasing the survivability of the newborn.

It has been suggested that adrenal progesterone may be a sensitive indicator of **pain** in some contexts. Bulls given **anaesthesia** before **electroejaculation** have lower levels of progesterone than bulls that were not given anesthesia. Progesterone and its metabolites are thought to act as a natural anaesthetic in animals and humans. A mechanism suggested for this is that progesterone interacts with $GABA_A$ receptors, thus increasing the binding of the neurotransmitter **GABA**. One of GABA's roles is to inhibit pathways of the **central nervous system** and, as a result, help desensitize that system to painful stimuli.

Progesterone is also a neurosteroid. As a neurosteroid, progesterone plays a neuroprotective role and is implicated in myelination and synaptic function. In this capacity it is suggested that progesterone may be involved in **memory** and **cognition**. Progesterone plays a vital role in keeping individuals healthy. It is a thermogenic compound involved in raising core temperature. Progesterone is also involved in reducing muscle spasm, and the widening of bronchi and mucus regulation. (AGF)

See also: Sexual behaviour

Further reading

Crossley, K.J., Nicol, M.B., Hurst, J.J., Walker, D.W. and Thorburn, G.D. (1997) Suppression of arousal by progesterone in fetal sheep. *Reproduction Fertility and Development* 9, 767–773.

Falk, A.J., Waldner, C.L., Cotter, B.S., Gudmundson, J. and Barth, A.D. (2001) Effects of epidural lidocaine anesthesia on bulls during electroejaculation. *Canadian Veterinary Journal* 42, 116–120.

Feng, Z., Marti, A., Jehn, B., Altermatt, H.J., Chicaiza, G. and Jaggi, R. (1995) Glucocorticoid and progesterone inhibit involution and programmed cell death in the mouse mammary gland. *Journal of Cell Biology* 131, 1095–1103.

Mellor, D.J. and Gregory, N.G. (2003) Responsiveness, behavioural arousal and awareness in fetal and newborn lambs: experimental, practical and therapeutic implications. *New Zealand Veterinary Journal* 51, 2–13.

Skarda, J., Fremova, V. and Bezecny, I. (1989) Progesterone alone is responsible for stimulation of the growth of ducts and of mammary alveolar structures in mice. *Endocrinologia Experimentalis* 23, 17–27.

Stefnanon, B., Colitti, M., Gabi, G., Knight, C.H. and Wilde, C.J. (2002) Mammary apoptosis and lactation persistency in dairy animals. *Journal of Dairy Research* 69, 37–52.

Prolactin

Prolactin is a phylogenetically old hormone. It is a peptide consisting of about 190 amino acids (varies between species) and it is secreted from the anterior **pituitary gland**, and during gestation also from the placenta. It is controlled by a complex array of excitatory and inhibitory factors involving **serotonin** as a stimulatory factor and **dopamine**, **epinephrine** and **norepinephrine** as inhibitory factors (although the latter two substances can also be stimulatory in some circumstances). There is a circannual and **diurnal rhythm** in many species, with longer days stimulating release and release levels being higher at night than during the day. Prolactin is named after its promotion of **lactation** in female mammals, but it has many additional functions. In relation to behaviour, its primary involvement is in **sexual behaviour** and **parental behaviour**. In relation to animal welfare, prolactin is relevant as a **stress** measure.

Prolactin plays an important role in lactation of mammals. It is involved in the development of the mammary gland along with **steroid** and **growth hormones**, and in the initiation of milk production along with **oestrogen**, **glucocorticoids** and **insulin**. Prolactin is also very important in the maintenance of milk production. Studies have shown that milk production is reduced when a prolactin inhibitor is administered either prepartum or during established lactation. Suckling causes a dramatic increase in prolactin, and in cows there is a correlation between the prolactin level and milk yield.

In some species, prolactin is implicated in prepartum nest building in the female (**see: Nesting behaviour**). An example of this is the rabbit doe, which builds a nest of grass that she lines with her own fur. The fur loosens through pregnancy so that it can easily be removed at the time of nesting. This loosening depends on hormonal changes. The primary change is release of prolactin in response to an increased oestrogen/**progesterone** ratio at the end of pregnancy. It has also been suggested that prolactin is responsible for initiating prepartum nest building in sows, because there usually is a concurrent rise in nesting activity and prolactin as parturition approaches. However, studies have shown that nesting can occur without the rise in prolactin, and the exact hormonal regulation of nesting in pigs remains to be determined.

Maternal behaviour and care for the neonates involves prolactin. In rats, it has been found that infusion of prolactin in steroid-primed rats induces maternal behaviour very quickly. In hamsters, decreased prolactin levels during pregnancy induced by a prolactin inhibitor also impaired maternal care. There was reduced maternal **aggression** toward male intruders, increased aggression toward pups, disturbed pup retrieval behaviour and increased maternal **cannibalism** (**see: Infanticide**). Similarly, in mice with the prolactin receptor gene deleted, maternal behaviour was disturbed, indicating that the prolactin receptor is important in mediating maternal behaviour.

Prolactin also seems to be related to paternal behaviour in species that display this behaviour. In the males, prolactin is elevated through pregnancy and remains elevated after birth. In mice, comparison of unmated males and expectant fathers showed that most expectant fathers performed paternal behaviour, whereas unmated males did not. In expectant fathers, prolactin levels were significantly higher than in unmated males – they were in fact comparable to those of the expectant mothers. In gerbils there seem to be two male strategies. Prolactin and **testosterone** levels rise during pregnancy, and testosterone then either drops at the birth of the pups or it remains elevated. In the former case the males behave paternally and, in the latter case, they pursue additional copulations. Similar results have been found for common marmosets and dwarf hamsters. It should be noted, however, that these studies are correlational and therefore merely suggestive of a causal relationship.

The role of prolactin in parental behaviour in birds, i.e. broodiness, incubation of eggs and feeding of young, has been studied in many species of birds. The initiation of the breeding cycle is caused by annual change in daylight stimulating prolactin production. Further increase in centrally acting prolactin after oestrogen and progesterone priming induces nesting and incubation behaviour. Prolactin rises to a peak during the incubation phase, where the pituitary also shows enhanced responsiveness. In some species, a brood patch develops on the breast to improve heat transfer from the female to the eggs. Prolactin acts together with ovarian steroids to increase vascularization and **feather** loss in the patch. Stimulation of the brood patch by the eggs or a familiar nest site maintains the high levels of prolactin release. In photorefractory bird species, prolactin is involved in autumnal gonadal regression. Prolactin is not considered to cause photorefractoriness, but the seasonal peak in prolactin may serve to fine-tune the timing of the end of the breeding season.

In males, sexual stimulation (e.g. exposure of a stallion to an oestrous mare or ejaculation in the bull) produces a rapid prolactin release, with fast decline after mating. Sexual behaviour can also be induced by short-term, moderate prolactin administration. In contrast, prolonged treatment or higher dosages decrease sexual behaviour in males. Hence, rats receiving a single, centrally administered prolactin dose showed increased sexual activity, but a long-term treatment had the opposite effect. Similarly, in rams given a prolactin inhibitor, frequencies of mounting and mounting attempts were reduced. An association between low **libido** and increased prolactin levels has been found in several other species (e.g. rats, mice and bulls). The inhibition may occur through either dopaminergic **neuron(e)**s and through inhibition of spinal reflexes and neuromuscular connections controlling erection and ejaculation or effects on release of steroid hormones from the **gonads**. Similar paradoxical effects of prolactin have been found with female rats, with low, short-term doses increasing **lordosis** and higher, longer-term doses inhibiting it.

Prolactin can be used as a stress measure. Plasma levels increase within a few minutes in response to stressful stimuli in many species, although the relationship is not straightforward. For example, prolactin levels in sheep increased in response to **restraint**, **transport**, **isolation** and/or **novelty**. Similar findings have been made in pigs. However in rescue dogs, following neutering, a reduction in prolactin has been recorded, with no effect on cortisol with adequate analgesia, and it has been suggested that this may reflect an interaction with prior sensitization of dopamine pathways as a result of chronic **frustration**. Dogs with generalized **anxiety** are reported to have raised levels of prolactin, but those with phobias or mild anxiety appear to have normal levels. However, in pigs and horses, prolactin levels rose in response to a repeated painful **stressor**, i.e. nose snaring in the pig and twitching in the horse. The rise in prolactin may be due to release of **β-endorphin**, which decreases levels of dopamine acting to inhibit prolactin release, as well as to hormonally induced prolactin release. The role of prolactin in the stress response is not clear, but it has been suggested as modulating immune function and suppressing **reproduction** temporarily. Prolactin enhances the responsiveness of the **immune system** by stimulating **cytokine** release from helper T-cells. It inhibits reproduction by interaction with the hypothalamic–pituitary–gonadal axis.

(BID)

Further reading

Dawson, A. and Charturvedi, C.M. (2001) *Avian Endocrinology*. Alpha Science International Ltd, Pangbourne, UK.

Nelson, L. (2000) *An Introduction to Behavioural Endocrinology*. Sinauer Assour Associated, Inc., Sunderland, Massachusetts.

Prolactin-inhibiting factor (PIF)

Also known as prolactin-inhibiting hormone (PIH) and prolactostatin. Unlike other pituitary hormones, **prolactin** release is largely under inhibitory control, so if the connection between the **hypothalamus** and **pituitary gland** is severed, circulating levels of prolactin will increase, while the levels of other pituitary hormones fall. This phenomenon was observed before prolactin-inhibiting factor was identified, but was used to postulate its existence. It is now apparent that prolactin-inhibiting factor is the neurotransmitter **dopamine**. (DSM)

Prolactin-releasing factor (PRF)

Also known as prolactin-releasing hormone (PRH, PRLH)/ and prolactin-releasing peptide (PrRP). Although the regulation of prolactin release appears to be largely under inhibitory control, a number of stimulating factors exist, such as **oestrogen**, **gonadotropin-relasing hormone** and **thyrotropin-releasing hormone**. Prolactin-releasing factor is a specific peptide produced by the **hypothalamus** that is released in a pulsatile way and that stimulates the production of **prolactin** by the **pituitary gland**. This peptide is not hormone-specific and also appears to increase the release of both **corticotropin-releasing hormone** and gonadotropin-releasing hormone. (DSM)

Pro-opio-melanocortin (POMC)

Pro-opio-melanocortin is an important biochemical precursor synthesized primarily by the anterior **pituitary gland**, **hypothalamus** and melanocytes in the skin; it has the potential to give rise to a number of biologically important substances (see Fig. P.7). Initial cleavage may be to adrenocorticotropic hormone (**ACTH**, which leads to cortisol secretion), γ-melanocyte-stimulating hormone (γ-MSH, which stimulates melanin pigment production in melanocytes) and β-lipotropin (which is involved in fat and lipid metabolism). Depending on the cell type, ACTH may be further metabolized into α-melanocyte-stimulating hormone and corticotropin-like intermediate peptide (CLIP, or corticotropin-like intermediate lobe peptide, the normal function of which is unknown but which may be used as a biomarker in certain forms of cancer). β-lipotropin may also be further metabolized into **β-endorphin** and γ-lipotropin (which may itself be further metabolized to α-lipotropin).

(DSM)

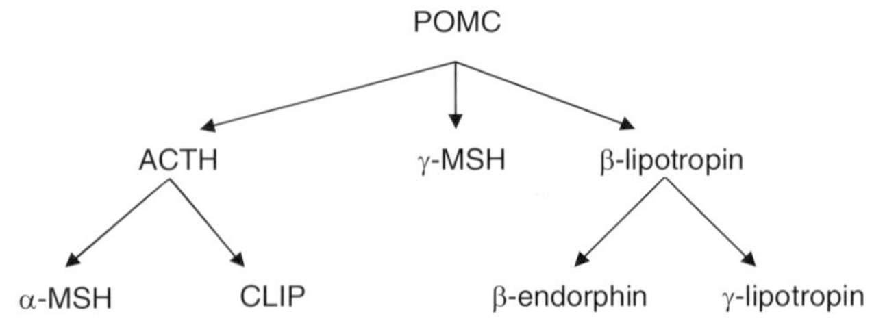

Fig. P.7. Pro-opio-melanocortin (POMC) as a precursor to other biologically important substances.

Prostaglandin

Prostaglandins are a member of the eicosanoid family and are found in a wide range of species such as insects, fish, corals, birds and mammals. Prostaglandins are lipid compounds synthesized from essential fatty acids in any cell in the body, with the exception of red blood cells; prostaglandins do not travel in the blood like hormones but are used within the cell to regulate many different cellular mechanisms. Prostaglandins carry out a wide range of functions that are vital for

homeostasis of the individual, such as maintaining **blood pressure**, **parturition** and associated events such as muscle contraction and nest building. Prostaglandins are also vital in a variety of male sexual behaviours such as **libido**, mounting and ejaculation. Prostaglandins also help maintain skin integrity and sleep–wake cycles. Integrity of the **central nervous system** is partially dependent on prostaglandins, as they are a feature of processes such as myelination.

Prostaglandins can be synthesized by two pathways. The first pathway is initiated with double-unsaturated omega-6 linoleic acid (linoleic acid), while the second begins with triple-unsaturated omega-3 α-linoleic acid (α-linoleic acid). The linoleic acid pathway gives rise to series 1 and series 2 prostaglandins. Series 1 prostaglandins (TXA_1, PGE_1, $PGF_{1\alpha}$ and PGD_1) are formed from the precursor dihomo-γ-linolenic acid and are involved in clotting, dilation and **inflammation**. Series 2 prostaglandins (TXA_2, PGE_2, $PGF_{2\alpha}$, PGD_2, PGH_2 and PGI_2) are produced from the precursor arachidonic acid and function in parturition, reducing blood pressure, temperature modulation and regulation of clotting. The α-linoleic pathway produces Series 3 prostaglandins (PGE_3, PGH_3, PGI_3 and TXA_3) from the precursor eicosapentaenoic acid, which is found in fish liver oils. Series 3 prostaglandins have a protective role in the body and it is suggested that they help reduce incidences of asthma, heart attacks and arthritis.

Sexual behaviour

Prostaglandin $F_{2\alpha}$ has been used in the swine industry to improve the sexual performance of boars by increasing boar libido and increasing successful mounting behaviours. The mechanism of how prostaglandin optimizes male **sexual behaviour** in swine is not clear; however, it has been suggested that $PGF_{2\alpha}$ results in the testes releasing higher levels of the **steroid hormones testosterone** and **oestradiol**. Data supporting this hypothesis come from a study where it was found that boars with higher blood concentrations of oestradiol required less time to mount and begin ejaculation in the presence of an artificial sow (Louis *et al.*, 1994). Sexually inexperienced boars given intramuscular $PGF_{2\alpha}$ have higher success rates when trained to an artificial sow compared with boars given a placebo. Sexually experienced boars injected with prostaglandin also showed a marked improvement in successful semen collection.

Prostaglandin influences other sexual behaviours in boars besides libido and mounting behaviours. Boars given a low dosage of prostaglandin execute mild scratching behaviour around the face and neck using the hind legs. High doses of prostaglandin result in more vigorous scratching behaviour around the facial area. There is no evidence that administration of prostaglandin effects semen quality; however, it does increase ejaculation duration. Beef heifers undergoing prostaglandin immunization do not display oestrous behaviours for an extended period of time, due to the formation of, and extended life of, the corpus luteum. Prostaglandin causes the corpus luteum to regress; however, after immunization, the corpus luteum does not regress and continues to produce progesterone. Other behaviours such as **lying**, standing and eating do not appear to be affected by this process. Female garter snakes receiving a large dose of prostaglandin have inhibited mating and a large decrease in sexual activity, while garter snakes given a small dose of prostaglandin take up to three times longer to mate even though there are no differences in other steroid hormones (Whittier and Crews, 1986).

Nesting behaviour

Nest building is a natural behaviour carried out by female members of many species prior to parturition as one of the processes of preparation for the arrival of the newborn. This powerful maternal instinct is controlled by both internal and external factors. In swine, internal factors encourage the seeking out and preparation of the nest site; the sow begins nosing, rooting and pawing the intended area for the nest. External factors affecting **nesting behaviour** include environmental temperature, the availability of substrates such as straw, earth, sawdust or other substrates deemed suitable and sensory feedback.

Before parturition can begin, prostaglandin $F_{2\alpha}$ ($PGF_{2\alpha}$) is required, its function being to degrade the corpus luteum. As the corpus luteum regresses, **oestrogen** and **progesterone** concentrations decrease, which are required for the maintenance of parturition. As these hormones decrease, **oxytocin** and **prolactin** levels increase, allowing parturition to take place. Prostaglandin $F_{2\alpha}$ can be used to induce parturition. Pregnant sows injected with $PGF_{2\alpha}$ display nest-building behaviour shortly after. At about 12 h after $PGF_{2\alpha}$ is administered to the sow, a second, more intense bout of nest-building behaviours is evident, if the substrates are available, and parturition takes place approximately 22–36 h after $PGF_{2\alpha}$ is given. Sows in a pen with straw or a similar pen without straw both displayed pawing and rooting behaviours when administered peripheral $PGF_{2\alpha}$; however, rooting behaviours were more intense with sows in pens with straw. Space requirement is also important for $PGF_{2\alpha}$ nest-building behaviours to be displayed. Dosage levels of $PGF_{2\alpha}$ affect the type of behaviour carried out within the range of nest-building behaviours. Low dosages of $PGF_{2\alpha}$ were highly correlated to gathering behaviours of nest-building substrates, while at high dosages of $PGF_{2\alpha}$ these showed a higher incidence of pawing behaviours in pregnant sows (Burne *et al.*, 2001).

These effects of $PGF_{2\alpha}$ are specific to post-pubertal females; post-pubescent boars increase sexual behaviours such as mounting, scratching and libido, and juvenile pigs show signs of **pain** and changes in activity.

Stress

Prostaglandin E_2 (PGE_2) – a metabolite of arachidonic acid – is involved in the regulation of impulsive behaviours. This mechanism is mediated through the EP1 receptor, which is encoded by the *Ptger1* gene. Studies in rodents found that the PGE receptors EP1 and EP3 are effective through the **hypothalamic–pituitary–adrenal (HPA) axis**. Mice that lack the EP1 receptor display abnormal behaviours such as abnormal social interaction, impulsive behaviours and **aggression** towards **conspecifics**. These mice also become highly sensitive to sound and are startled easily. The EP1 receptors have been found to be expressed presynaptically in the amygdala and the paraventricular nucleus of the **hypothalamus**.

In a study to determine the function of PGE_2 and its receptors, **stress** was defined as either stress caused by sickness or as a psychological sickness. During sickness stress, there is an increase in **cytokines** in blood, this increasing the expression of two enzymes, cyclooxygenase (COX) and PGE synthase (PGES), required for the synthesis of PGE_2 in **brain** microvessels. During psychological stress PGE_2 is produced in pyramidal neurons that have COX. Sickness and psychological stress stimulate PGE_2 production in structures in the brain, and this binds to EP1 receptors in either the amygdala or the hypothalamus. This pathway reduces the frequency of impulsive behaviours during times of stress.

Pain

Prostaglandin E_2 is also a mediator for inflammation, fever and spinal nociceptive processing. Arachidonic acid is converted to prostaglandin H_2 by the enzymes COX-1 and COX-2. Prostaglandin H_2 is then catalysed to PGE_2 by PGES. COX-1 is coupled to cytosolic PGES (cPGES), and COX-2 is coupled to microsomal PGES-1 (mPGES-1). COX-1 has been suggested as being involved in both spontaneous pain and slow-developing, diffuse pain. It has been suggested that COX-2 and mPGES-1 are involved in the processes of pain, inflammation and fever. In rats, it was found that inhibiting COX-1 and COX-2 decreased **nociception**. In addition, down-regulation of cPGES and mPGES-1 has also decreased nociception.

General health

In the cardiovascular system, prostaglandins of the E series are vasodilators of the vascular bed, causing a decrease in blood pressure. Prostaglandin F_2 also reduces peripheral blood resistance, but without any effect on blood pressure. Thromboxane is a prostaglandin that is primarily produced in platelets, and it functions as a vasoconstrictor and it inhibits platelet aggregation. Another prostaglandin, prostacyclin, is also an inhibitor of platelet aggregation and inhibits cellular proliferation of vascular smooth muscle. Endothelial cells in the lung produce prostacyclin and, alongside PGE_2, functions in maintaining the lung in a dilated state, modulating vascular tone and as a bronchodilator. Prostacyclin has also been implicated in respiratory distress syndrome.

In the gastrointestinal tract, prostaglandins have three main functions: motility, cytoprotection and secretion. Prostaglandins E_2 and F_2 are involved in the contraction of the gastrointestinal longitudinal muscle; PGF_2 is involved in the contraction of the circular muscle; and PGE_2 relaxes the circular muscle. Prostaglandins of the E series promote secretion of mucus and bicarbonate and, acting with prostacylcin, increase mucosal blood flow, promoting cytoprotection. The prostaglandin E series reduce gastric acid secretion but promote the secretion of water and electrolytes into the jejunum.

In the cortex of the kidney, prostaglandins are produced in the cortical collecting tubes, glomeruli and the arterioles. Prostacylcin is the major prostaglandin in the kidney cortex and it uses its vasodilatation capabilities to regulate blood flow and control both the filtration rate of the glomerulus and renin secretion. In the kidney medulla PGE_2 is the main prostaglandin; it is synthesized in the loop of Henle, collecting ducts and interstitial cells. Prostaglandin E_2 increases renal blood flow and decreases sodium transport from the loop of Henle to the interstitial cells. It also decreases urea and sodium reabsorption from collecting ducts and, finally, acts as a vasopressin antagonist to decrease water permeability. In doing this PGE_2 controls water and salt secretion from the kidney.

In skin, PGD_2 is the major prostaglandin produced and plays an important role in the integrity of the skin. Prostaglandin D_2 has a number of functions, including reducing platelet aggregation, decreasing nitric oxide release, regulating body temperature, increasing bronchoconstriction and modulating **sleep** patterns. In mice, it was found that decreased levels of PGD_2 were implicated in an increase in scratching behaviour. During the scratching behaviour, the toenails of the mice disrupt the subcutaneous layer, thereby damaging the skin; this precedes the onset of spontaneous dermatitis. (AGF)

References and further reading

Burne, T.H.J., Murfitt, P.J.E. and Gilbert, C.L. (2000) Deprivation of straw bedding alters $PGF_{2\alpha}$-induced nesting behaviour in female pigs. *Applied Animal Behaviour Science* 69, 215–225.

Crowe, M.A., Enright, W.J., Swift, P. and Roche, J.F. (1995) Growth and estrous behaviour of heifers actively immunized against prostaglandin F2α. *Journal of Animal Science* 73, 345–352.

Estienne, M.J. and Harper, A.F. (2004) Semen characteristics and libido in boars treated repeatedly with $PGF_{2\alpha}$. *Journal of Animal Science* 82, 1494–1498.

Louis, G.F., Lewis, A.J., Weldon, W.C., Miller, P.S., Kittok, R.J. and Stroup, W.W. (1994) The effect of protein intake on boar libido, semen characteristics, and plasma hormone concentrations. *Journal of Animal Science* 72, 2038–2050.

Matsuoka, Y., Furuyashiki, T., Yamada, K., Nagasi, T., Bito, H., Tanaka, Y. *et al.* (2005) Prostaglandin E receptor EP1 controls impulsive behaviour under stress. *Proceedings of the National Academy of Sciences of the United States of America* 102, 16066–16071.

Miller, S.B. (2006) Prostaglandin in health and disease: an overview. *Seminar in Arthritis and Rheumatism* 36, 37–49.

Whittier, J.M. and Crews, D. (1986) Effects of prostaglandin F2 alpha on sexual behaviour and ovarian function in female garter snakes (*Thamnophis sirtalis parietalis*). *Endocrinology* 119, 787–792.

Proximate factor

A proximate factor is an internal or external **stimulus** that produces a particular response in an animal. Proximate causation focuses on the immediate or short-term events that are responsible for a behavioural or physiological response. For example, a proximate causal explanation for **feeding** is **hunger** or lack of feeding for some time. (BM)

See also: Tinbergen's four questions; Ultimate factor

PSE meat

Pale, soft, exudative, or PSE, meat is a classification of meat that is characterized by pale colour, lack of firmness and fluid seepage. The meat is generally inferior in taste and texture, and there is nutrient loss in the fluid seepage. Such meat is rarely used as a raw product, but is occasionally used as a portion of processed meat products. This condition is a result of a rapid decrease in the pH, along with high temperatures in the muscle after **slaughter**, resulting in an increased rate of glycolysis and a build-up of lactic acid. These combined conditions cause a disruption in the protein structure of the muscle tissue due to variable sizes of the sarcomeres, a component of the muscle fibres.

Acute stressors prior to slaughter and the condition known as **porcine stress syndrome** (PSS) are considered to contribute to this condition. Genetics has also been attributed as having an effect on the occurrence of PSE meat, with selection for increased leanness and feed efficiency resulting in a marked increase in the incidence of PSE meat. It has been suggested that the swine industry selected pigs against the occurrence of PSE meat while maintaining the selection for optimum leanness and feed efficiency. Rapid chilling of the meat immediately after slaughter and alterations to the pre-slaughter environment, including during transport and lairage, to reduce stress have been suggested as reducing the occurrence of PSE meat. If transporters and workers in the slaughter plants are properly trained in handling procedures that minimize animal welfare problems immediately prior to slaughter, then the risk of PSE meat production is also reduced. (RP)

Pseudogenes

A pseudogene is a DNA sequence that is not functional, i.e. it does not code for a protein, and that has a functional counterpart in the same organism (Hartl and Jones, 2005). It is presumed that pseudogenes are mutated relics of gene duplications in the evolutionary history of organisms. (SJGH)

Reference

Hartl, D.L. and Jones, E.W. (2005) *Genetics. Analysis of Genes and Genomes*, 6th edn. Jones and Bartlett Publishers, Sudbury, Massachusetts.

Psychogenic

An adjective, meaning of psychological origin. It is applied to conditions in which the usual organic cause cannot be established, but a likely psychological one exists. For example, in psychogenic paralysis, there is no identifiable physical cause, and thus the condition is assumed to be a reaction to psychological trauma (where evidence for trauma or **stress** exists). (JH)

Psychology

Psychology as a discipline cannot be defined in any clear or straightforward way. Individuals who call themselves psychologists may study phenomena as varied as mental illness, the learning abilities of cephalopods, the nature of dreams, the neuroanatomy that underlies **fear** or the biochemistry of **stress**. Thus it would appear that subject matter cannot be a defining characteristic of the field.

The kinds of questions that psychologists ask might be another way to define the discipline. These include such ill-defined problems as: (i) determining the characteristics of human and non-human animal nature; (ii) understanding how the mind and body are related; (iii) investigating the extent to which genes versus environment versus interactions between the two determine behaviour; and (iv) understanding exactly what humans and other animals have in common with one another, mentally and behaviourally. Finding the answers to even one question of such a scope is a tall order for any discipline – and psychology in one way or another endeavours to take on all of these, plus more. The vague parameters of such questions, their diversity and their seemingly endless iterations prevent one from using these elements as a means by which to characterize the discipline.

Nor does methodology or approach suffice to define psychology. Psychology is not so much characterized by a single approach or suite of approaches as it is by recurring argument regarding which approach is most appropriate for its variable subject matter. Is the nature of the mind a topic suitable to empirical analysis, or is a qualitative approach more appropriate? Can behaviour be best understood by applying a molecular approach, breaking it into its smallest constituent elements? Or are behavioural phenomena best evaluated by taking a more holistic, molar approach that focuses on purpose or function? Debates such as these about the ideal methodology continue in the psychology literature.

An examination of psychology's history can in part explain the divisiveness that seems to characterize the field. Psychology is the product of an uneasy marriage between philosophy and physiology – two disciplines that are fundamentally different in their preferred method of enquiry. The dialectic between these two historical influences has shaped much of psychology's development, as well as the paradigms currently dominating the field. In fact, a brief review of that history is perhaps the best way to try to describe the field to non-psychologists.

In most introductory textbooks, psychology is defined as the scientific study of **brain**, mind and behaviour. That study is undertaken from many different levels of analysis – from the molecular level of genomics and brain biochemistry to the more molar level of family and social systems. Considering this discipline's unlikely parentage, however, it should come as no surprise that the textbook definition is not without controversy. There are practitioners in psychology, for instance, who argue that the empirical method cannot adequately be applied to some of the kinds of questions addressed by the field. There are others who assert that the mind, for instance, is not a topic suitable to the science of psychology, because it is impossible to subject it to empirical analysis. Indeed, there may be no single aspect of the field on which all psychologists agree. But the diversity of opinion, approach and focus in the field is also part of its richness. After all, we humans would hardly be satisfied by an explanation of ourselves as individuals that was too tidy and concise.

Although one can trace its disciplinary roots, it is hard to say precisely when psychology as a field began. If indeed psychology is a field in which the relationship between mind and body is considered and evaluated, then perhaps the very first humans who ever pondered the nature of their nature were the first psychologists. In this way, psychology may well be one of the oldest academic disciplines.

The word 'psychology' derives from the Greek word *psykhe*, meaning 'spirit' or 'soul' and the Greek word *logia*, meaning 'study of.' The etymology of the word clearly reflects this oldest and perhaps most fundamental of questions posed by psychologists: what is the relationship, if any, between the mind and body? Many early investigations in the field can ultimately be described as efforts to answer this question (**see: Dualism, Cartesian**).

For some philosophers, the mind was simply not something that could be subject to empirical analysis. The mind of an individual is a private, idiosyncratic thing, to which only the owner has access (and in some cases, as theorists like Freud would argue, even that access is limited). Nevertheless, there were those who insisted that the mind was indeed a subject suitable for scientific investigation, and their efforts to pioneer means by which to ask empirical questions about personal experience paved the way for the new discipline's development.

Among these persons was the German physician and physiologist Wilhelm Wundt. Wundt was perhaps the first person to call himself a psychologist. In 1879 at the University of Leipzig (where he was teaching as a Professor of Philosophy!), Wundt established what many consider to be the first laboratory entirely devoted to psychological investigation, and it is this moment that is generally regarded as the formal establishment of the field. It did not take long after Wundt's initial work for others attracted to the field to begin to critique his methods and to develop their own brand of psychology.

Wundt's preferred method for investigating the nature of the mind was **introspection**, a technique he applied primarily to the study of sensory experience. In the mid-19th century, it was widely believed that psychological phenomena (such as **consciousness** and sensory experience) were not suitable for empirical study, since they could not be directly measured or manipulated. Wundt disagreed. He had studied under the physiologist Hermann Helmholtz, who had shown that a number of elements of human **perception** were measurable, and that sensory experiences of many kinds could be tied directly to the nervous system. Wundt believed that, with appropriate training, he could develop human subjects who could accurately and objectively report on their own experiences, using a common vocabulary that made data obtained in this way replicable and testable. While today it is difficult to imagine a science based on subjective self-report, Wundt was nevertheless a powerful advocate for the necessity of an empirical approach to the developing field of psychology. Paradoxically, his own preferred method of introspection is today considered quite the opposite of scientific rigour.

Intellectuals attracted to the new field of psychology had to travel to Germany to learn about the discipline from its new master, Professor Wundt. Among these pilgrims were the British psychologist Edward Titchner and the American psychologist William James. Upon his return home to the UK, Titchner found no ready home for his new ideas about a science of the mind, and thus he moved to the USA to take up an appointment at Cornell University. There, Titchner extended the ideas he had learned about in Leipzig into an approach to psychology called *structuralism*. Much as contemporary physicists believe that there is a fundamental particle from which all other matter is derived, structuralist psychology assumes that it is possible to break conscious experience into its smallest underlying components. By understanding its fundamental structure, Titchner believed, one can truly come to understand the nature of mental experience.

In vehement opposition to this approach was William James. James asserted that the mind was much more than the sum of its elements. Much like the Heisenberg uncertainty principle's assertion about the limits of our abilities to study atomic particles, James argued that consciousness was not something that could be stopped at a given moment in time so that it could be subjected to analysis. In so doing, he noted, one is no longer observing the phenomenon in which one was originally interested.

The structuralist approach, in James' view, was a bit like someone trying to explain a house by describing each of its bricks and boards individually. To James, the more important aspect of these components was how they all worked together to form a house. By analogy, the functions served by the mind are the important thing, and not the elements that come together to generate those functions. This emphasis on the adaptive utility of different mental elements became known as a *functionalist* approach. The functionalists stressed the application of data gathered in the name of psychology to everyday, real-world problems. Contemporary functionalists today include evolutionary psychologists who focus on the phylogeny and adaptive value of a wide range of behaviours, as well as neuroscientists who answer the mind–body question with the argument that 'mind' is nothing more than a function of the brain. This latter emphasis on immediate or proximate function is characteristic of most of the kinds of questions contemporary psychologists attempt to answer.

Other opponents of the structuralist approach of Wundt and Titchner included the German psychologists Max Wertheimer and Wolfgang Köhler. Both the functionalists and these German thinkers rejected the elementism of the structuralist approach. While the American functionalists emphasized the *purpose* of mental and behavioural elements, however, Wertheimer and his colleagues focused upon what they considered to be the wrong-headedness of the elemental approach itself. To Wertheimer, the mind could not be reduced to elements without distorting its true nature. From his perspective, organisms do not experience the world in small, discrete pieces, but rather in whole, meaningful configurations. The assertion that the whole of an experience was more than the sum of its parts became known as a *gestaltist* principle, from the German word for 'configuration' or 'whole'. The gestaltists argued that psychology should take a more molar approach to its subjects. For instance, the study of behaviour from a more molar approach would mean focusing on a suite of behavioural elements that together might comprise a single goal-directed, purposive act.

An example of this approach put into practice can be seen in the work of Wolfgang Köhler during the early 1900s. Visiting the Prussian Academy of Science's anthropoid research station in the Canary Islands, Köhler became marooned there by the unfortunate and coincident start of World War I. Köhler did not make idle use of his time, however. He spent the next few years investigating the nature of intelligence and learning in chimpanzees, summarizing his research in *The Mentality of Apes* (1917). In this volume, Köhler described the holistic nature of chimpanzee problem-solving; in many cases, it did not seem to him as though the animals solved a problem through combining elemental experiences or through trial and error, but rather through the more global application of **insight**.

The gestaltist and structuralist perspectives thriving in Europe emphasized the power of descriptive analysis, and also the value of studying behaviour and mental experience as it occurred naturally – that is, outside of the artificial environment of the laboratory. These values set the stage for the coming field of **ethology** – the study of animal behaviour as it occurs in the natural environment. The legacy of the structuralists and gestaltists can also be seen today in the works of phenomenologist psychologists such as John Compton and Elisabeth Ströker.

Yet another school of thought in psychology that developed largely out of a rejection of structuralism is the *behaviourist* school (**see: Behaviourism**). Behaviourist thinkers, such as the Americans John Watson and **B.F. Skinner** and the Russians Ivan Sechenov and **Ivan Pavlov**, argued that psychology could *not* be a science as long as it continued to include in its scope of investigations the study of consciousness and mental phenomena. The mind, it was argued, was inherently unsuitable to empirical analysis. Mental experience was personal and therefore inaccessible to the kind of public testability demanded by the empirical method. Behaviour, however, was a different thing entirely. Behaviour could be observed by others and quantified; the conditions under which it occurred could be replicated and the causes of behaviour empirically tested. To the behaviourists, the only valid subject of study in a psychology that wished to call itself a science was behaviour (**see: Experimental analysis of behaviour**).

Focusing upon behaviour, the physiologist Ivan Pavlov documented the kinds of 'automatic' external triggers that invariably invoked the same response. He called these triggers 'unconditioned stimuli', and described the means by which other, previously neutral, stimuli ('conditioned stimuli') could come to elicit the same responses. The form of learning in which an organism appears to use the presence of one **stimulus** as a **signal** for another has come to be called *classical conditioning*, and it was Pavlov who first described this form of learning in detail. Principles of classical conditioning are employed today in many arenas of applied animal behaviour – from conditioning wolves and coyotes to avoid sheep to using milk barn stimuli to trigger milk letdown in dairy cattle (**see: Conditioning – types of**).

In America, psychologists such as Watson directed their attention to experience with the consequences of behaviour that could modify the frequency of that behaviour in the future. Operant conditioning (so called because the organism must operate on its environment in order for learning to occur) was the name given to this kind of learning in which animals learn about the consequences of their actions. Two primary kinds of consequence were described by the behaviourists: **reinforcement** (any consequence that results in a behaviour being more likely to recur) and **punishment** (any consequence that decreases the likelihood that a behaviour will occur again). The principles of reinforcement and punishment objectively described and empirically tested by early behaviourists are employed today in nearly every aspect of domestic animal handling.

Investigations by behaviourist psychologists into the proximate causes of behaviour also resulted in a compilation of rules or 'laws' for explaining behaviour, such as the assertion that all behaviour is the consequence of some environmental trigger or stimulation, and that no behaviour changes unless it is reinforced or punished in some way. Influenced by the Darwinian revolution in science that was taking place at the turn of the 19th century, behaviourists argued that all organisms learned in the same way, and that rules governing the shaping of behaviour applied equally to all living things – from amoebae to humans. Thus, if one wanted to understand learning in wild elephants, one could study learning in domestic rats maintained in a university laboratory, and any information gleaned from the rat studies was believed to apply equally to wild elephants.

The assumption of universal laws of learning that applied to all organisms, and the reliance on laboratory work rather than on observations of animals in nature, were among the elements of behaviourism attacked by ethologists such as **Konrad Lorenz**. Influenced by the same European schools of thought that had shaped gestaltist psychology, the ethologists argued that behaviour could not be understood separate from the environment in which it evolved: natural history and behaviour had to be taken as one. Necessarily, then, little if anything useful could be learned about behaviour from examining it under laboratory conditions. The feud between classical ethologists and classical behaviourists was to influence the development of **comparative psychology** for years to come.

Comparative psychology focuses on the comparative study of human and non-human animal behaviour – traditionally in laboratory or other controlled settings. Perhaps the first comparative psychologist (though not by training) was **Charles Darwin**'s disciple, **George J. Romanes**. Romanes considered that, if all living things were descendants of a common ancestor, then at least some components of intellect and behaviour should be similar across species. To support his point he collected anecdotes of non-human animals engaging in behaviour that suggested problem-solving and reasoning capacities much like those of humans. Though Romanes' underlying assumptions seem reasonable, his casual methods and anthropomorphic interpretations were the subject of an aggressive attack by the American psychologist C. Lloyd Morgan. Morgan's argument that parsimony should be the rule applied in interpretation of behavioural data, and that data should be based on empirical observation, became known as **Morgan's Canon**, and this principle dominates the field of comparative psychology today.

Still other schools of thought that can be found under the umbrella of psychology include the philosophy of Sigmund Freud and his followers – a philosophy that emphasized the influence of unconscious motivations in the shaping of behaviour. Additional perspectives include those of Kurt Lewin (who emphasized how social situations affect behaviour) and George A. Miller (who is credited with launching the cognitive revolution in psychology). Each of these perspectives (like the others reviewed above) endorses a different preferred theoretical explanation for behaviour, and often a different preferred methodology for pursuing hypotheses derived from those explanations.

In summary, psychology as a field is nearly impossible to define – it cannot be reliably identified by its subject matter, by its methods, nor by its preferred levels of analysis. Nevertheless, the study of brain, mind and behaviour reaches across a great many areas of interest to students of applied animal behaviour, and thus it behoves us to try to understand the complexity of this many-faceted discipline. (KNM)

Further reading

Freedman, D.K. and Weiner, I.B. (eds) (2003) *Handbook of Psychology:* Vol. 1, *History of Psychology*. John Wiley and Sons, Inc., Hoboken, New Jersey.

Gibson, E.J. (2003) What psychology is about: ruminations of an opinionated aged psychologist. *Ecological Psychology* 15, 289–295.

Koch, S. and Leary, D.E. (eds) (1985) *A Century of Psychology as a Science*. McGraw-Hill Book Company, New York.

Punishment – types of

If the consequence of an action reduces the likelihood of the action being repeated, this consequence is said to have punished the action. In other words, punishment is anything that reduces the likelihood of the preceding action being repeated in a similar situation.

Punishment can occur in one of two ways, either through the presentation of something **aversive**, in which case it is referred to as positive punishment, or through the removal or denial of something pleasant, which is then referred to as negative punishment, or 'response cost'. For example, an electric shock applied to a rat when it presses a lever is an example of positive punishment, if lever-pressing behaviour then reduces in the long term. On the other hand, putting a dog on a lead (removal of freedom) to reduce over-exuberant off-lead behaviour would be considered a negative punisher if the boisterous off-lead behaviour is then reduced in future.

The consequence can only be considered punishing if it has a suppressing action on the behaviour at which it is aimed. Aversive consequences without long-term effect on the preceding action are not punishment, although they may disrupt the behaviour in the short term. If the aim was to punish the behaviour, this inappropriate use of an aversive stimulus raises ethical concerns over the management of the animal. Disruptive stimuli (interrupters) may be used in training to distract the animal (giving the trainer time to redirect the animal or take control of the situation in some way) – for example, clapping one's hands to startle a dog that is barking so that the dog can be directed towards expressing a more acceptable behaviour. However, what may serve as a disruptive stimulus for one individual may be a punisher for another, more sensitive, individual.

The effect of punishment is assessed by its long-term impact and not by its immediate effect on the behaviour, since a goal-directed behaviour (**appetitive behaviour**) that is followed by a variant of its goal will cease in the short term. Thus an attention-seeking behaviour that is followed by some form of **attention** (such as being reprimanded) may not actually be punished by this consequence, even though behaviour stops in the short term. To the contrary, it may actually be reinforced by the giving of attention, and so more likely to occur the next time the animal seeks attention.

The fact that whether or not an event constitutes punishment is dependent on the subject is important to bear in mind at all times when interacting with animals, as it allows us to determine whether actions the trainer considers not to be aversive are perceived as such by the animal. If a behaviour is reducing in frequency (being suppressed), one of the possible explanations may be that something is punishing the behaviour and thus something the trainer considers innocuous is perceived as aversive by the animal. There are also examples of where the event punishing the action may not be a deliberately administered stimulus or even a direct consequence of the animal's behaviour. Something that occurs as a random event but coincides with the action of the animal and is perceived as aversive by the animal may punish the prior behaviour. For example, if a cat is using the litter tray as a boisterous dog runs into the room, litter tray use may be punished and a **house-soiling** problem begin.

Just as conditioned reinforcers are used in training (**see: Reinforcement**), conditioned punishers can also be created and utilized. If a neutral stimulus is paired sufficiently frequently with an aversive event (be it negative punishment or positive punishment), this stimulus will take on the quality of a punisher and can then influence behaviour. Conditioned punishers can be very useful, especially with regard to the timing of delivery of punishment in practice (see later); however, it is important to realize that the potential negative consequences of the primary punisher will also be seen with the use of conditioned punishers. It is also important to realize that conditioned punishers can be created inadvertently. It has been demonstrated that the environment and people present at the time of punishment delivery can also take on punishing characteristics (Schilder and van der Borg, 2004), which has potentially serious implications for the relationship between trainer and animal, as well as between the subject and other animals present at the time punishment occurs.

In addition, from a declarative knowledge perspective (i.e. the information an individual has), the use of punishment does not increase certainty about the environment and so it is not conducive to good **welfare**. This is because punishment informs the individual that a given action was not appropriate, but does not, on its own, provide information about what would be appropriate in such circumstances.

Therefore, significant reliance upon punishment as a means of training an animal or teaching people is highly controversial. If it is to be used it is important that it is used correctly to ensure its success, as well as to safeguard the welfare of the trainee. The following must be considered when utilizing punishment in training.

The contiguity of punishment and behaviour to be punished is important for the punisher to be effective. Timing of punisher delivery is critical, as an association needs to be made by the animal between the action and the consequence and, if there is time for an additional action to intervene between the undesirable behaviour and the aversive event, then the latter behaviour may be the one associated with punishment rather than the target behaviour. A delay between action and punishment delivery is less effective in suppressing behaviour in the longer term.

Secondly, the intensity of the punisher needs to be sufficiently aversive to suppress the undesirable behaviour, but not so aversive that it causes substantial emotional interference. **Fear** and **anxiety** are not only undesirable from a welfare perspective, but they may also interfere with the learning that is trying to be achieved; for example, the animal may become more inhibited and therefore less likely to exhibit the desired behaviour. Aversive stimuli of insufficient intensity are not only ineffective, but may result in **habituation** if frequently repeated, which may mean that a more intense event will need to be applied to achieve the desired effect. Should the intensity be gradually increased over time, the intensity may be escalated to a point where serious welfare concerns are manifest without the desired effect being achieved. Thus the appropriate application of punishment requires considerable skill, as it will often vary for a given behaviour and between different individuals.

The behavioural consequences of punishment are often situation dependent and thus, for it to generalize across contexts, repetitions in different contexts will normally be necessary. Single-trial learning can occur when event intensity is sufficiently great. However, if this is not the case, then punishment needs to be delivered at every occurrence of the undesirable behaviour, or else negative reinforcement of the behaviour in the circumstances that avoid punishment may occur.

Another important consideration associated with the potential use of punishment relates to its impact on motivation towards a goal rather than on the behaviour. The motivation for the goal is not removed by suppressing the initial behaviour associated with it, and thus the animal may start to perform a new behaviour in place of the old behaviour directed towards achieving the same goal. This behaviour too may be inappropriate. Thus, whenever a punishment is utilized as a means of reducing behaviour, a new appropriate behaviour needs to be reinforced to guide the animal towards appropriate behaviour. (HZ)

Reference and further reading

Chance, P. (2003) *Learning and Behaviour*. Brooks/Cole, Pacific Grove, California.

Lieberman, D.A. (1999) *Learning: Behaviour and Cognition*. Wadsworth Publishing Co. Inc., London.

Schilder, M.B.H. and van der Borg, J.A.M. (2004) Training dogs with the help of the shock collar: short- and long-term behavioural effects. *Applied Animal Behaviour Science* 85, 319–334.

Q

Qualia

Qualia (singular: quale) are the subjective **feelings** accompanying sensory experiences, sometimes referred to as 'raw feels'. Classic examples include what it feels like to see red, smell roses, taste coffee or experience **pain**. The term was first coined by Lewis (1929), and it is a key concept in phenomenal consciousness. (CCB)

See also: Consciousness; Perception

Reference and further reading

Lewis, C.I. (1929) *Mind and the World Order*. C. Scribner & Sons, New York.

Ramachandran, V.S. and Hirstein, W. (1997) Three laws of qualia: what neurology tells us about the biological functions of consciousness. *Journal of Consciousness Studies* 4, 429–457.

Qualitative measure

A measure of something that has a discrete meaning or quality that cannot necessarily be described in terms of quantities (i.e. numerically). For example, eye colour can be described in terms of categories of colours: blue, brown, green, etc. Qualitative measures of animal behaviour might be whether an individual is performing one of the following categories: running, walking or standing. (KT)

See also: Measuring behaviour; Quantitative measure

Quality of life

In discussions focusing on human treatment of animals, the term 'quality of life' is sometimes used as a synonym for '**welfare**' to refer to animals' **well-being**. The term may serve to remind the reader that the study of animal welfare is analogous to the study of human quality of life. (PS)

Further reading

UFAW (2007) Quality of life: heart of the matter. *Animal Welfare* 16(Suppl.).

Quantitative measure

A quantitative measure describes something in terms of quantities (i.e. numerically) – for example, the number or weight of individuals. Quantitative measures of animal behaviour might include the volume of a **vocalization** in decibels or the length of time for which an individual is vocalizing. Quantitative measures may be either objective or subjective. (KT)

See also: Measuring behaviour; Qualitative measure

R

Rabbit

The family Leporidae of the order Lagomorpha comprises 11 genera, which are commonly known as rabbits and hares. As a highly adaptable species, rabbits are found throughout the world, with individual species having their own origin. The European rabbit (*Oryctolagus cuniculus*), native to south-west Europe and north-west Africa, is the only species of this genera to be domesticated and has become a popular laboratory, farm and **companion animal** both within and outside its native range.

At the beginning of the 19th century, selection of rabbits for colour, shape and size began. As farm animals, Californian and New Zealand rabbit breeds are most commonly used for meat, due to their ability to grow quickly. Angora rabbits, with their long, silky coats, are most commonly used for fur production, and New Zealand rabbits are usually used for laboratory purposes, mainly medical and safety product research. The most recent domestic role for rabbits is within the companion animal sector, where over 2.8 million rabbits are kept in 0.7 million households as pets and/or exhibition animals (conformation or show jumping) in the UK alone. In some countries rabbits are considered a significant pest species, especially where they are not native (**see: Exotic species invasion**).

There are many differences in the conformation of modern domesticated rabbits and European wild rabbits. The wild European rabbit is small and grey-brown in colour, growing up to 45 cm in length and weighing an average of 1.3–2.2 kg. There are over 50 breeds and 500 varieties of domestic rabbit recognized, with a large range in size and weight. The largest of these breeds, the Continental Giant, is a minimum of 65 cm in length and 5.5 kg in weight, while the smallest breed, the Netherland Dwarf, weighs a maximum of about 1.2 kg. In addition to the variety of sizes, domestic rabbits also come in lop breeds, meaning that the usually upright, alert ears (see Fig. R.1) are drooped by the sides of the head (see Fig. R.2).

Fig. R.1. Erect-eared rabbit.

Fig. R.2. Lop-eared rabbit.

In fully lopped rabbits, the ears are incapable of upright movement. Similar to other domestic species, the breeding of extreme phenotypes can be a concern. Some of the giant breeds may grow too quickly or be too heavy for their skeletal structures, some smaller breeds have blunted noses that may change the rabbit's field of vision or interfere with dental function, and rabbits with lop ears are more prone to ear infections as humidity within the ear changes.

Rabbits are herbivorous, with diets that are naturally very high in fibre. In the wild the diet would be made up of a variety of grasses and hays. However, this type of diet is comprised of large amounts of relatively indigestible cellulose. A unique normal feeding process involving **coprophagy**, or re-ingestion of faeces, allows nutrients to be obtained from food during a second passage through the digestive tract. Thus rabbits have two types of faeces: soft, mucus-covered faeces that the rabbit eats directly from its anus; and the commonly seen hard, dry faeces that are the waste left over from the second passage through the digestive tract. Domesticated rabbits have similar digestive processes to their wild ancestors;

however, they often get a very different diet. This diet usually includes supplementation or even replacement of this natural diet with dry/pelleted food. Many of these commercially available feeds are often high in carbohydrate (particularly sugar) and protein but low in fibre, a combination of which often results in health problems. The diet of the rabbit can affect their behaviour; for example, overweight rabbits may no longer engage in coprophagy or the behaviour may become diminished. In the wild, rabbits spend a large part of their active time finding and eating suitable grasses and hays. Pelleted feed provided readily in bowls for domesticated rabbits is consumed more quickly than a high-fibre diet and does not require any effort to find, leaving the rabbit with more free time and perhaps an unfulfilled **need** to forage.

European rabbits (unlike many other species of rabbit) are gregarious in the wild, living in groups of up to 20 individuals in underground tunnels known as warrens that cover vast amounts of space. In general, when domestic rabbits are housed in large group enclosures, the behavioural repertoire of their wild counterparts is retained. However, this housing system is rarely found within modern-day society, with rabbits mainly being kept in singles, pairs or small groups, and different behaviour or revision of behaviour is often seen. In addition, rabbits in the wild are **crepuscular**, that is, they are most active at dawn and dusk, while domestic rabbits tend to have their active periods at times that fit with their human caregiver's schedule. For example, a pet rabbit may get turned out in the garden during the afternoon for a run or may be fed mid-morning. It is not yet known what effects, if any, this may have on domestic rabbit **welfare**.

In many countries, there is a dichotomy in the statutory and legal protection provided for rabbits kept for commercial versus companion purposes. Laboratory and farm rabbits are typically housed in cages or pens, individually or in groups. In the UK, housing guidelines and **codes of practice** have been produced by the Home Office, Department for Environmental Food and Rural Affairs (Defra) and the Laboratory Animal Breeders Association (LABA), to provide basic welfare requirements, but there is less protection for companion rabbits within the UK, where housing is usually either outdoors in hutches or indoors as house rabbits. The **RSPCA** and Rabbit Welfare Fund (RWF) provide housing recommendations based on the **five freedoms**, although there are no specific minimum housing requirements. Codes are being developed in light of recent UK legislation that should afford pet animals the same protection as farmed ones. There is a general consensus that rabbit housing, regardless of the purpose of the rabbits, should be large enough to allow hopping and sitting with ears alert, that there should also be an area with a minimum height of 45 cm and that a shelter be provided.

As with many domestic species, rabbits kept in conditions very different from their wild counterparts are prone to development of **abnormal behaviour**, including stereotypic behaviour patterns. These are especially prevalent in laboratory, fur- and meat-farming conditions, where the welfare of the rabbits has to be balanced with cost of the end product. The most commonly found abnormal behaviour patterns in rabbits are circling, fur pulling/plucking and water bottle 'playing' or **polydipsia**. (JRH, LMD)

Further reading

Defra (2007) Available from: http://www.defra.gov.uk/foodfarm/farmanimal/welfare/onfarm/othersps/rabcode.htm (accessed 2 December 2009).

McBride, A. (1998) *Why Does My Rabbit…?* Souvenir Press, London.

McBride, E.A., Magnus, E. and Hearne, G. (2004) Behaviour problems in the domestic rabbit. In: Appleby, D. (ed.) *The APBC Book of Companion Animal Behaviour.* Souvenir Press, London.

Morton, D.B., Jennings, M., Batchelor, G.R., Bell, D., Birke, L. *et al.* (1993) Refinements in rabbit husbandry. Second report of the BVAAWF/FRAME/RSPCA/UFAW joint working group on refinement. *Laboratory Animals* 27, 301–309.

Raceway

Otherwise known as a chute, a raceway is used to move animals in a controlled manner from one place to another, usually single file. Sometimes raceways are used to administer drugs to animals, as when tightly packed they are unable to move. Attempts to encourage animals to move down a raceway may be made by people shouting at, hitting or goading them with sticks or electronic sticks (**goads** or prodders). Using excessive pressure to force animals to move in a raceway will lead to them becoming frightened, difficult to handle, unwilling to enter the raceway a second time and, on occasions, dangerous. Animals usually prefer to move from a dark area to light. Solid walls and curved raceways are best, but moveable gates are an acceptable alternative. V-shaped races support the animal and prevent them from lying down. If they do lie down, animals may be reluctant to get up.

Raceways may lead into a loading ramp and then into a vehicle, in which case the loading ramp should have side-gates and a slope of suitable incline for the type of animal being moved (e.g. less than 4-in-7 for cattle and a step-up of less than 20 cm). Sloped raceways or ramps should have battens for the animals to get a grip, and the floor should be non-slip. Adequate staffing is essential for moving cattle through raceways and the animals should be moved slowly and preferably without the use of goads. Staff should be calm, confident and able to predict the animals' behaviour. Sticks can provide a useful means of extending the arm to accelerate movement, but should not be used to hit the animals with unnecessary force, using persuasion rather than aggression. Raceways may also lead into a crush, where individual animals are managed, often for veterinary procedures, weighing or pregnancy diagnosis. The ideal crush has solid sides to make the animal feel secure, but frequently this is not possible as access is needed, for example to hooves when the animals are treated for **lameness**.

Animals are put into a raceway either for one-by-one treatment such as drenching, or to move them in single file into an area where they can be handled individually, as in a crush or dip. Although they move more easily in the race if it has solid sides, they will accustom to one with open sides, which should preferably be V-shaped to support the animal and prevent it from lying down. Cattle prefer to move in a circular rather than a straight direction and, if they are stressed, bunching may occur. Often an animal may then lie down, and it may be difficult to make it stand again. (CJCP)

Further reading

Grandin, T. (2007) *Livestock Handling and Transport*, 3rd edn. CAB International, Wallingford, UK.

Rat

Laboratory rats (*Rattus norvegicus*) in the zoological order of Rodentia (**rodents**) are descended from the wild rat. Albino rats, caught on the trading ships, were used for showing and research (most notably at the Wistar Research Institute in the USA). Hooded rats, so called because of the dark fur over the back of the neck, were probably assigned to the sport of rat catching by dogs. The long and close association of rats with humans probably facilitated taming them, as well as causing **disease** transmission. (DBM)

Further reading

Suckow, M.A., Weisbroth, S.H. and Franklin, C.L. (2005) *The Laboratory Rat*, 2nd edn. Academic Press, London.

Whishaw, I.Q. and Kolb, B. (2005) *The Behavior of the Laboratory Rat: a Handbook with Tests*. Oxford University Press, New York.

Ratites

Ratite birds are cursorial, flightless birds typically of very large size. Their primitive palate defines them as part of the Palaeognathae (along with the Galliformes and Anseriformes) but, although being closely related, the various species are recognized as a distinct taxonomic grouping. Living ratites include the **ostrich** (Struthionidae) of Africa, the emu (Dromaiidae) and cassowaries (Casuariidae) of Australasia, and the rheas (Rheidae) of South America. Extinct families included the elephant birds (Aepyornithidae) of Madagascar and the moas (Dinornithidae) of New Zealand. Although the kiwis (Apterygidae) of New Zealand are very much smaller they are also classed as ratites, and share a close taxonomic relationship with the moas. Penguins (Spheniscidae) are of the Neognathae and so are not ratites.

All ratite birds lack well-developed wings and so are well adapted to rely on their legs for locomotion. They lack the flight **feathers** with interlocking barbules seen in other birds and are covered by long, soft feathers that are more reminiscent of down feathers seen on other birds. With the exception of the kiwis, ratites are **diurnal** and have herbivorous diets of foliage or, in the case of the cassowaries, fruit. They lay the largest of all bird eggs (400–1500 g), which hatch into **precocial** offspring.

Ratites have gained prominence in recent years due to their commercial exploitation. Farming of the ostrich for feathers was started in the mid-1860s in South Africa, and soon spread to other parts of the world including Europe, North America and Australia. Around the same time rheas were exploited for their feathers. However, due to the collapse of the feather market after World War I, farming of ostriches went into severe decline, surviving only in South Africa. After World War II, ostrich farming revived in South Africa with the development of a leather market, and then a market for leg meat. During the 1980s economic sanctions against South Africa curtailed the supply of ostrich hides to the world market. The gap in the market was recognized in the USA, where interest in farming ostriches increased rapidly, largely fed by unrealistic prices for birds and aggressive marketing. Against this backdrop, interest spread rapidly around the world, but it soon became clear that the prices were unsustainable and that maintaining ostriches in many countries of the world was very difficult and unprofitable. Interest in ostrich farming outside of South Africa had all but disappeared by 2000.

Farming of emus had been practised in Australia since the early 1970s, but there was considerable interest in emu farming in the USA, largely because the birds were smaller and easier to manage. For the same reason rhea farming was also promoted in the USA and adopted around the world. However, unlike the situation for the ostrich, markets for leather, meat and feathers from these species did not exist and so farming enterprises have been slow to develop and interest typically waned around the world. In its native Argentina the rhea is being investigated as a species that could be used commercially to exploit the environmentally sensitive grassland that is so easily destroyed by cattle.

Ratite farming led to considerable welfare issues because of the inexperience of many of the people attempting to keep the birds. Poor husbandry and a lack of understanding of the basic needs of ratites in captivity led to problems of high mortality, poor growth rates and chronic health problems in many birds. In particular, farming of ostriches and emus at northern latitudes promoted grave concern over the ability of birds to survive colder temperatures and wetter conditions. Despite such concerns, research into the **welfare** of ostriches, emus or rheas was not funded by producers or other agencies and so reliable reports of either good or bad practice, or how the birds respond to novel environments, are rare. Even in South Africa, studies into the welfare of ostriches appear to be a low priority. Although the popularity of ratite farming has declined on a global scale it remains locally of interest, in several countries around the world where environmental conditions are favourable. How welfare issues are being addressed in these locations is largely unknown at a scientific level. (DCD)

Rats and mice

The terms 'rats' and 'mice' can refer to many different species, most of which are in the family Muridae, but the species with the closest relationships with humans and which have been extensively domesticated are the Norway or Brown **rat** (*Rattus norvegicus*) and the house **mouse** (*Mus musculus*). Because of their relationships with humans, these two species have the most diverse and complex man-made welfare issues of all rat and mouse species, so this article focuses on them.

Rats and mice are **rodents**, an order that makes up about 40% of all mammal species, and that includes squirrels, **hamsters**, gerbils, **guinea pigs**, voles, mice and rats, among many other species. Rodents are characterized mainly by their teeth, which are specialized for gnawing through hard substrates; these grow continually throughout the rodent's lifetime and, as the rodent gnaws, the teeth are worn down, retaining their sharpness. In **captivity**, **welfare** problems can result if gnawing substrates are not provided, because the teeth can overgrow, causing difficulties with chewing food, and causing physical injuries to the animal. Rat and mouse teeth are also one characteristic that has led to them becoming major pest species. Rats and mice have three main relationships with humans, and their welfare issues differ depending on these contexts: they exist as pest species, as **laboratory animals** and as **companion animals**. Each of these contexts is discussed in turn, after a description of their characteristics as wild, undomesticated animals.

Wild rats and mice

R. norvegicus and *M. musculus* are generalist species, living in diverse habitats across all continents except Antarctica. They are opportunistic omnivores, consuming a highly varied and flexible diet, and they are highly social animals, with social structures that can vary depending on the resource availability of the habitat they occupy (Berdoy and Drickamer, 2007). Their general biological characteristics in the wild are shown in Table R.1. Despite generations in captivity, even domesticated rats and mice retain many of the characteristics of their wild ancestors, so the information on their wild biology can explain some of the welfare issues they face in captivity.

Sensory perception and physical abilities

Rats and mice usually show **nocturnal** or **crepuscular activity**, being most active at dawn and dusk, but their **circadian rhythms** are flexible to some extent, responding to food availability, weather conditions and **predation** risk. Because they usually occupy dimly lit environments, their eyes are extremely sensitive to small differences in light intensity (Burn, 2008). They can see colours, having two cone types peaking in the green and ultraviolet wavelengths, but their acuity is very poor, so their vision would appear very blurred to a human.

Their hearing overlaps with humans' but also extends into the ultrasound (humans hear from 0.02 to 20 kHz; rats hear from 0.25 to 80 kHz (Burn, 2008); and mice hear from 10 to >100 kHz (Latham and Mason, 2004)). Correspondingly, they vocalize audibly using a variety of 'squeaks', and also ultrasonically. Their ultrasonic **vocalizations** include: (i) a male courtship 'song' in mice and rats; (ii) long calls that occur in negative situations and after male ejaculation; and (iii) a highly variable 'laugh' that mainly occurs in positive situations. Rats and mice have extremely well-developed olfactory senses, including a well-developed **vomeronasal organ** specialized primarily for detecting non-volatile **pheromones**. Olfactory cues allow them to discern each others' characteristics including sex, kinship, health, reproductive state, familiarity and **dominance** status. Both species produce an alarm **odour** in negative situations, and rats also produce a reward odour that attracts colony members to **resource**-rich areas.

Being generalist omnivores, rats and mice have similar gustatory senses to humans, tasting sweetness, saltiness, sourness, bitterness and umami (the 'fifth taste'). Like us, they tend to favour foods high in sugar and fat, and to avoid bitter foods, many of which contain natural poisons (**see: Diet selection**). Rats and mice have highly sensitive whiskers (vibrissae) capable of making fine tactile discriminations. They are also thigmotaxic, meaning that they usually choose to maintain physical contact with vertical surfaces, so they tend to move alongside walls or other landmarks, rather than across open spaces.

Both rats and mice are adept at climbing (although not compared with some other closely related species such as the black rat, *Rattus rattus*), and also at digging. They dig burrows, with tunnels that connect chambers for storing food or for sleeping in. Both species build nests inside the sleeping chambers, especially when raising young, but mice build more elaborate cup-shaped nests than do rats; rats often just **sleep** on flat pads of material. The nesting materials depend on what is locally available and can include dry leaves, grass, feathers or paper. Rats also have a propensity for swimming, capable of swimming up to about 400 m across open water (Russell *et al.*, 2005), but mice usually avoid immersion in water.

Foraging

Both species are central-place foragers, so they tend to forage around the burrow site and, when they find food, they carry it back to the burrow or to the nearest harbourage site to consume it. Rats and mice eat seeds, fruit, nuts, molluscs, eggs, small vertebrates and any food made available by humans (Berdoy and Drickamer, 2007; Wurbel *et al.*, 2009). Rats have also been known to dive for molluscs or for eels caught in fishing nets. They are more predatory than mice, and can kill small mammals, birds or reptiles, sometimes leading to conservation issues when they colonize islands and prey on endemic species (**see: Exotic species invasion**).

Table R.1. Some general biological characteristics of rats and mice in the wild (mostly taken from Berdoy and Drickamer, 2007; Wurbel *et al.*, 2009).

Biological characteristic	Rat (*Rattus norvegicus*)	Mouse (*Mus musculus*)
Adult female body weight (g)	250–300	15–21
Adult male body weight (g)	250–500	17–23
Age at weaning (days)	16–34	21–30
Age at puberty (weeks)	9–15	6–11
Gestation duration (days)	20–23	19–21
Oestrus frequency if not mated (days)	4–5	4–5
Litters per year (per female)	4–6	Limited data
Pups born per litter (*n*)	7–8	5–9
Pups weaned per year (*n*)	20–30	Up to about 50
Home-range size (m^2)	30–45 or more rarely up to about 150	2–5 but can be orders of magnitude higher if nomadic
Life expectancy (weeks)	Limited data	11–15
Maximum lifespan (months)	12–18	12–18

The diversity of rat and mouse diets enables them to spread into diverse habitats, but it comes at a cost, because every unfamiliar substance eaten could contain toxins or pathogens. Both species therefore initially sample very small quantities of unfamiliar food, and learn to avoid it if they become unwell afterwards. They also learn socially (**see: Social learning**) what each other have eaten through olfactory cues (Burn, 2008); for example, as well as being attracted to conspecifics' feeding sites through reward odours, carbon disulfide present in rats' breath causes **conspecifics** to prefer foods that they smell on each other. They also avoid new foods they have sampled if they subsequently smell an ill conspecific, known as the 'poisoned partner effect'. Rats have two additional mechanisms that protect them from poisoning: (i) they can be highly neophobic (avoiding novelty), depending on genetic and environmental factors; (ii) and they have very high diversity in the genes regulating enzymes for denaturing toxins (Berdoy and Drickamer, 2007).

Population structure and social and reproductive behaviour

In large resource-rich areas, rats and mice live in large dynamic colonies with polygynandrous or promiscuous mating systems, respectively (Berdoy and Drickamer, 2007; Wurbel *et al.*, 2009). However, when resources are more scarce or widely distributed, they exist in small polygynous groups, where one dominant male usually cohabits with one or more females and their offspring. Dominance and territorial boundaries are marked using **urine**, mainly deposited by the dominant male, which will be overmarked by challenging males and refreshed by the dominant male, who patrols the territory many times daily. In mice, urine marking can lead to visible pillars of dried proteins from the urine deposits. **Aggression** is more common in males than females, but is rarely or never directed towards pre-pubertal individuals.

The dominance structures differ between the two species: in mice, males frequently fight for dominance and the larger mouse usually wins, but in rats the hierarchy is more stable. This means that rats fight more rarely, and the hierarchy is predicted more by age than by size in stable environments (rats and mice grow continually throughout their lifetime). The lower incidence of fighting in rats also goes hand-in-hand with their more diverse repertoire of postural **signals**, meaning that overt fights are usually unnecessary. In fact, rats have the most diverse forms of **play** of all rodent species studied to date, and even play as adults (Pellis and Iwaniuk, 2004). In both species adolescent, subordinate males are the first to disperse from the natal territory if resources (including mating opportunities) become scare, but in mice females are almost as likely to disperse as are males. However, dispersing individuals have a very low chance of survival.

Mice and rats exercise mate choice, with pairs that choose each other having fitter offspring than pairs that have had no opportunity to select their mate (Berdoy and Drickamer, 2007). Courtship involves the females giving olfactory and postural signals (hopping, darting and ear wiggling) that they are ready to mate, and males in larger colonies competitively chasing the females. The females will often nest cooperatively, helping to raise each other's young. In small colonies where there is a dominant male, the dominant male may kill any offspring not fathered by him, but in larger colonies this is rare, perhaps because it is difficult for the males to discern the paternity of litters. Litters with multiple paternity are fairly common, at least in mice. Also, pregnant female rats deposit a pheromone in the nest that discourages cohabiting males from killing the females' offspring (Burn, 2008). Females of both species can reproduce all year round but, in reality, they only breed when resources are fairly plentiful and the temperature is warm enough, so this usually limits breeding to the spring and summer.

Mortality

In the wild, fewer than 50% of rats and mice survive beyond puberty, and still fewer survive to reproduce. Deaths are often from lack of resources, fighting injuries or disease, but can also be due to predation. Their main predators are carnivores, such as foxes, cats and mustelids (e.g. mink, weasels or polecats), and birds of prey. Their circadian rhythm is one mechanism of **predator avoidance**, with their nocturnality making them cryptic from visual hunters but, when nocturnal hunters are prevalent, rats become diurnal (Berdoy and Drickamer, 2007). Rats and mice, even after generations of domestication, are also innately fearful of fox, cat and mustelid odours. Another major cause of death for commensal rats and mice (those sharing resources with humans) is, of course, pest control by humans.

Rats and mice as pests

Rats and mice cause billions of dollars of damage annually, consuming stored food, chewing through electrical cables and they can carry diseases. Rats are blamed for carrying more diseases than mice, and several studies have found that rats do harbour many pathogens and parasites that can harm humans (Berdoy and Drickamer, 2007) – **zooonotic disease**. However, it was *R. rattus*, not *R. norvegicus*, that carried the flea that caused the bubonic plague, and the few studies that have investigated mice as disease vectors have shown that they too can carry diseases, such as salmonella. Phobias of mice, and particularly of rats, are fairly common, and the media's portrayal of rats is especially negative, with them often being used to convey scenes of squalor or horror.

Perhaps for these reasons, the welfare of rats and mice in pest control is of low priority on the public agenda. However, some of the methods used to kill the many millions of rats and mice would be considered inhumane in other circumstances (**see: Pest control – ethics of; Pest species – welfare of**). Methods for controlling rodent pests include ingested poisons, contact poisons, fumigant poisons, live traps, lethal traps, sticky boards, repellents and preventing rodent entry to human habitations (Mason and Littin, 2003). Anticoagulant poisons are the most common form of control, being used in over 90% of rodent control situations in the UK and the USA. Some rats and mice have become resistant to anticoagulants but, in affected animals, anticoagulants can cause symptoms including inactivity, anorexia, bloody diarrhoea, breathing difficulties, paralysis and multiple haemorrhaging (painful in humans). These symptoms first appear 1–3 days after ingestion, and death occurs after 4–8 days. If sublethal quantities are ingested, symptoms can still occur in a milder form, and can persist for many weeks. Other methods are also reviewed by Mason and Littin (2003), who conclude that 'cyanide, alpha-chloralose, electrocution traps and well-designed snap traps all usually kill swiftly and with little distress' and that preventative methods such as rodent-proofing are also humane and are under-used at present.

Rats and mice as laboratory animals

Mice are the most commonly used laboratory animal, followed by rats, and together they comprise about 85% of all research animals. There are now several thousand strains of domesticated mice and rats, including outbred, inbred, transgenic, knock-out and mutant strains. Most mice used in research are now genetically modified in some way. Welfare issues in laboratory rats and mice can arise from:

- genetic predispositions to certain defects, diseases or other disabilities;
- experimental procedures; or
- housing and husbandry.

In the European Union (EU), laboratory rats and mice are legally protected by Council Directive 86/609/EEC, using the **three Rs principle** as a framework for preventing unnecessary **suffering**. Rats and mice are protected in many other countries also but, at the time of going to press, they remain legally unprotected in the USA.

The specific welfare issues associated with genetics or experimental procedures vary enormously according to the research in question, so cannot be covered in this entry, but their welfare impacts can be assessed on a case-by-case basis. Flecknell and colleagues have described many behaviours to help researchers identify when rats and mice may be in pain (e.g. Roughan and Flecknell, 2003); analgesics are not routinely given to rodents after potentially painful procedures. For example, in studies published between 2000 and 2002, analgesics were reported to have been given only after 19.8% of potentially painful procedures (Richardson and Flecknell, 2005). Other species-specific welfare indicators can include vocalizations, stereotypic behaviours in mice (rats rarely perform these) and chromodacryorrhoea (a Harderian gland secretion) in rats (Mason *et al.*, 2004).

Some welfare problems associated with rat and mouse housing and husbandry include isolation of these social animals (but male mice sometimes show high levels of aggression if group housed); small cages; aversive handling methods; and lack of shelters, nesting material or any other environmental **enrichment** to create 'standardized' barren environments (Lawlor, 2002; Sherwin, 2002). The sensory environment can also be harmful to the animals (Burn, 2008): for example, normal lighting conditions can rapidly cause permanent retinal damage to the animals' eyes, being adapted for dim light conditions; procedures are often carried out in the daytime when nocturnal animals would normally sleep; some laboratories (and homes) contain equipment that emits loud ultrasonic noise; and the homogenous diets provided in most laboratories provide no opportunity for the animals to nutritionally regulate or to gain positive experiences from gustatory diversity.

Rats and mice as companion animals

Despite the negative media portrayal of these animals, rats and mice are popular **pets**. Many websites and fan clubs exist extolling the virtues of these 'fun' and easily cared-for companions. So-called 'fancy' rats and mice are available with various coat colours and, for example, curly fur or 'dumbo' ears (positioned much lower than other rats' ears).

Yet, despite rats and mice seeming to be fairly low-maintenance pets, their robust survival can conceal some common welfare problems. A major issue is that it is still very common to keep just one individual in isolation. Allowing rats or mice to breed can be a problem because it is often difficult to find homes for the pups, but keeping single-sex animals together can fulfil their social **needs**, such as playing and sleeping huddled together. Because rats and mice are nocturnal, social housing is particularly important for their long active periods at night, when their human owners cannot provide stimulation for them. However, aggression must be closely monitored, especially between male mice. The way that the individual animals are introduced is important for establishing a harmonious relationship and, if aggression is serious or persistent, the animals may need to be separated.

Cages should be provided with safe environmental enrichment and, particularly for rats, games and positively rewarded training can provide extra stimulation for the owner and the pet when interacting outside the cage. Rats and mice appreciate dietary diversity and can be fed kitchen scraps, but these should be chosen carefully; obesity is common in these pets, so high-calorie foods should be avoided, as should chocolate and dairy products because rats and mice lack the enzymes necessary to digest them properly.

Respiratory problems are common in these animals, and they should be checked regularly for behaviours indicating **pain**, **distress** or illness, including lethargy, stereotypic behaviour, sneezing or a hunched posture with raised (pilo-erected) fur. As with laboratory rats and mice, it is fairly rare for veterinary clinicians to prescribe analgesics for pet rats and mice suffering from potentially painful conditions (Roughan and Flecknell, 2003). (CCB)

References

Berdoy, M. and Drickamer, L.C. (2007) Comparative social organization and life history of *Rattus* and *Mus*. In: Wolff, J.O. and Sherman, P.W. (eds) *Rodent Societies: an Ecological and Evolutionary Perspective*. University of Chicago Press, Chicago, Illinois, pp. 380–392.

Burn, C.C. (2008) What is it like to be a rat? Rat sensory perception and its implications for experimental design and rat welfare. *Applied Animal Behaviour Science* 112, 1–32.

Latham, N. and Mason, G. (2004) From house mouse to mouse house: the behavioural biology of free-living *Mus musculus* and its implications in the laboratory. *Applied Animal Behaviour Science* 86, 261–289.

Lawlor, M.M. (2002) Comfortable quarters for rats in research institutions. In: Reinhardt, V. and Reinhardt, A. (eds) *Comfortable Quarters for Laboratory Animals*, 9th edn. Animal Welfare Institute, Washington, DC, pp. 26–32.

Mason, G. and Littin, K.E. (2003) The humaneness of rodent pest control. *Animal Welfare* 12, 1–37.

Mason, G., Wilson, D., Hampton, C. and Wurbel, H. (2004) Non-invasively assessing disturbance and stress in laboratory rats by scoring chromodacryorrhoea. *Alternatives to Laboratory Animals* 32, 153–159.

Pellis, S.M. and Iwaniuk, A.N. (2004) Evolving a playful brain: a levels of control approach. *International Journal of Comparative Psychology* 17, 90–116.

Richardson, C.A. and Flecknell, P.A. (2005) Anaesthesia and post-operative analgesia following experimental surgery in laboratory rodents: are we making progress? *Alternatives to Laboratory Animals* 33, 119–127.

Roughan, J.V. and Flecknell, P.A. (2003) Evaluation of a short duration behaviour-based post-operative pain scoring system in rats. *European Journal of Pain* 7, 397–406.

Russell, J.C., Towns, D.R., Anderson, S.H. and Clout, M.N. (2005) Intercepting the first rat ashore. *Nature* 437, 1107.

Sherwin, C.M. (2002) Comfortable quarters for mice in research institutions. In: Reinhardt, V. and Reinhardt, A. (eds) *Comfortable Quarters for Laboratory Animals*, 9th edn. Animal Welfare Institute, Washington, DC, pp. 6–17.

Wurbel, H., Burn, C.C. and Latham, N.R. (2009) The behaviour of laboratory mice and rats. In: Jensen, P. (ed.) *The Ethology of Domestic Animals: an Introductory Text*, 2nd edn. CAB International, Wallingford, UK, pp. 217–233.

Rearing environment

The rearing environment is the environment an animal experiences up until maturity and includes all the stimuli, both social and non-social, that animals will be exposed to during this time. In natural environments, the rearing environment typically includes a mother and peers and a variety of changing environmental stimuli. In most cases, it is the mother that directs the offspring's attention to important or relevant environmental stimuli. However, in animal production there is often little or no contact with the mother and the environments tend to be relatively static with few opportunities to interact with environmental stimuli. Many domestic rearing environments are geared towards preparing the young physically for life in the later environments or reaching productivity criteria (such as **slaughter** weight) as efficiently as possible, with less attention paid to behavioural and psychological requirements.

Early-life experiences are very important to the development of normal behaviour patterns, and these experiences can have lasting effects throughout the life of the animal. **Imprinting** to different stimuli in the rearing environment occurs during early life. For example, many **precocial** species imprint on slowly moving objects that elicit a 'following response', and in natural environments this would encourage the young to follow their mother. There are **critical periods** or **sensitive phases** that occur in early life, where exposure to certain stimuli can result in lasting changes. Preferences for mates, foods, substrates, colours and so on are started in the rearing environment. For example, there is a period of time in the early life of songbirds during which the young birds can learn songs from their parents, even though the young do not start singing themselves until several months later. In some cases the preferences formed in early life can be altered by learning and experience, but in other cases they persist throughout life.

In intensive production environments, the rearing environment is dramatically different from that found in the wild (**see: Intensification of animal production**). These discrepancies between wild and human-made environments can lead to a number of behavioural problems and **welfare** concerns. One of the main problems in animal production rearing environments is the absence of a mother and associated early or artificial **weaning**. A well-known example of this is the experiments of Harlow and colleagues in which infant rhesus macaques were reared in environments with varying degrees of maternal and peer **isolation**. Animals kept in more isolated environments developed higher levels of stereotypic behaviour and **self-mutilation** behaviours than those reared with peers. These stereotypic and **abnormal** behaviours continued as the macaques aged and gave way to additional stereotypic and abnormal behaviours. Even **hand rearing** of animals alone can produce negative life-long changes in behaviour, such as heightened fear and aggression and poor parenting and social skills. This is a common problem in hand-reared exotic birds, which often imprint on humans as their sexual partners. Maternal deprivation can also affect the way an animal deals with stressors. For example, maternally deprived rats have been shown to adopt a reactive **coping** strategy compared with rats reared with their mothers that adopted a proactive coping strategy. The presence of a mother also aids in behavioural synchronization compared with those reared without a mother.

In addition to the stress of being separated from their mother, some mammals also experience nutritional stress as they are changed from their mother's milk to a milk replacer. For example, lambs removed from their dams at 2 days of age had slower growth rates, higher cortisol levels and lower immune responses than lambs kept with the ewes and lambs fed a mixture of ewe's milk and milk replacer.

Wild-caught animals placed in captivity may display less stereotypic and abnormal behaviour than captive-born animals. However, the offspring of the former show similar levels of abnormal behaviour to the latter. This indicates the importance of rearing environment and parental care in the development of normal behaviour, and well-managed early experiences can help protect the animal from developing abnormal behaviours and a range of welfare problems later in life. (LMD)

Further reading

Cooper, J.J. and Nicol, C.J. (1996) Stereotypic behaviour in wild-caught and laboratory-bred bank voles (*Clethrionymus glareolus*). *Animal Welfare* 5, 245–257.

Latham, N.R. and Mason, G.J. (2008) Maternal deprivation and the development of stereotypic behaviour. *Applied Animal Behaviour Science* 110, 84–108.

Novak, M.A., Meyer, J.S., Lutz, C. and Tiefenbacher, S. (2006) Deprived environments: developmental insights from primatology. In: Mason, G.J. and Rushen, J. (eds) *Stereotypic Animal Behaviour: Fundamentals and Applications to Welfare*, 2nd edn. CAB International, Wallingford, UK, pp. 153–189.

Rebound behaviour

Rebound behaviour refers to the increase in performance of a behaviour pattern after prevention of performance of that behaviour for a period of time. While some behaviour patterns may only need to be performed when elicited by external stimuli, such as anti-predator behaviour, others may need to be performed at regular intervals or certain rates (**see: Behavioural need**). It has been suggested that if the rate of performance of these latter behaviours falls below a certain preset level, there will be an increase in the amount of the behaviour performed at the next opportunity to compensate for the difference.

Animals in confined conditions can be limited or prevented from performing a number of normal behaviour patterns. However, the mere lack of occurrence of a behaviour pattern is not sufficient to determine whether there is a decrease in the animal's **welfare**. The demonstration of a rebound of the prevented behaviour patterns when the animal is removed from confinement can be used as an indication that the animals' needs may not be met in the management system and that their welfare may be compromised. (LMD)

Further reading

Kennedy, J.S. (1985) Displacement activities and postinhibitory rebound. *Animal Behaviour* 34, 1375–1377.

Nicol, C.J. (1987) Behavioural responses of laying hens following a period of spatial restriction. *Animal Behaviour* 35, 1709–1719.

Reciprocity

Reciprocity in animal behaviour refers to the tendency to respond to social actions with similar social acts. Thus positive gestures are reciprocated as may negative ones. It is also referred to as a 'tit-for-tat' or 'equivalent retaliation' behaviour strategy. A misunderstanding of the need for reciprocity of **appeasement** and greeting gestures or the failure of carers to recognize such behaviours (e.g. moving slowly, looking away, yawning and indirect eye contact in dogs) can result in **agonistic behaviour**, ultimately culminating in overt **aggression** and an associated problem behaviour as a consequence.

Within **ethics**, reciprocity is sometimes referred to as the 'golden rule' of society and may be expressed as 'do to others as you would like them to do for you', and is a common code within social communities globally. (DSM)

Recognition among animals

Recognition is such an extensive concept it is difficult to define. Researchers often define it operationally in terms of what their subject matter is, such as 'discriminative behaviour related to social categories' (Colgan, 1983), or defined 'by the nature of the objects being discriminated' (Sherman *et al*., 1997). Recognition can be broadly thought of as *a unique response to something in the environment based on some form of previous experience*. Here, 'response' could be physiological and not externally manifest; 'something' could be as simple as a cell or as complex as a group of animals; and 'previous experience' could be at the level of the **genotype** (i.e. not necessarily during an organism's lifespan). An important point is that recognition is a fundamental and ubiquitous biological process that plays a key role in interactions at various levels of biological organization, from cells to whole organisms. At an organismal level, recognition in some form plays a central role in essentially all processes of assortment among animals (e.g. all social interactions including **mate choice** and **cooperation**). Recognition among organisms (e.g. self, kin, group, sex), particularly social interactions, is the focus of this item.

Social recognition behaviour is inextricably linked to Jacob Von Uexküll's original (1934) concept of **umwelt**, one of the founding concepts of **ethology**. In order to understand an animal one must understand how the animal perceives its surroundings: its umwelt. The animal's umwelt includes its perceptual world (**merkwelt**) and that within this world that is relevant (**wirkwelt**). The process of recognition defines the wirkwelt. Many aspects of heterospecific recognition concerning an animal's behavioural ecology (e.g. predator and prey recognition) occur at this level. An animal's social world may be understood in exactly the same manner, as nested within the larger perceptual framework defined by its umwelt. We can think of this as an animal's 'social umwelt': the social information that is: (i) available; and (ii) relevant. The resolution of **conspecific** identity (i.e. intraspecific social recognition) is among the most basic of such information. Although intraspecific social recognition by no means comprises all aspects of recognition behaviour by animals, it is recognition at this level that is most commonly addressed when the term recognition is used, usually in reference to **communication** about identity. A final important point here is that although all communication is not 'about' recognition (e.g. identity of the signaller), recognition necessarily defines and underlies all communication. Recognition can thus be thought of as the 'carrier wave' of all communication behaviour among animals.

Approaches to the study of recognition

For the purpose of study, recognition behaviour can be broken into three essential components: production, perception and action (Sherman *et al*., 1997). Although all studies necessarily focus on one or a combination of these aspects, few do so explicitly. The result is that distinctiveness at whatever level of organization is being studied (e.g. individual, group, etc.) is often confused with recognition ability. That an animal has a distinctive trait (e.g. vocal individuality) does not confer recognition ability. The trait must also be perceived and acted upon. As a result, many investigations of recognition behaviour use **experimental 'playback'** paradigms or procedures in order to invoke reactions.

In order to compare the array of approaches used in studies of recognition behaviour it is important to distinguish between those that take evolutionary or 'ultimate' perspectives as compared with mechanistic or 'proximate' perspectives. Studies focusing on proximate forces acting on recognition behaviour are likely to be testing for the existence of recognition, the cues used or the level of resolution achieved. Cues that animals use for recognition of conspecifics can be grouped into four general categories: spatial location, familiarity, **phenotype** matching and allele/genotype matching (although beware, as many researchers use these terms in their 'ultimate' perspectives on recognition: see below). The salient sensory cues used for recognition vary considerably among taxa. Although the process is often multi-modal, most studies have focused on the acoustic/auditory channel using a limited number of taxa. A substantial amount of research has also focused on olfactory recognition, mostly with social insects and mammals. In comparison, there has been considerably less research on electrical and visual recognition, and very little in the area of cross-modal comparisons of recognition.

Different organisms are capable of different degrees of recognition, referred to here as *recognition resolution*. The ability to resolve individuals, as compared with groups (e.g. castes of social insects, breeding colonies, etc.) or species, appears to be mostly limited to birds and mammals. Tests of individual recognition often focus on parent–offspring recognition or neighbour–stranger (Dear Enemy) recognition in territorial animals. Often, however, true individual recognition in these circumstances is unnecessary. Beecher and colleagues (1989) made the point that true individual recognition is the ability to discriminate each individual in a group from every other individual. Functional individual recognition, in contrast, is the differential treatment of an individual based on individually distinctive cues. Individual recognition capabilities can be further thought of in terms of lateral and longitudinal resolution. *Lateral resolution* refers to the number of individuals one is able to resolve at a given point in time (e.g. immediate kin, more distant kin, cohort members, etc.). *Longitudinal resolution* refers to the durability of recognition over time (e.g. how long it lasts).

Identifying selective pressures acting on recognition behaviour invokes an ultimate perspective (i.e. an evolutionary timeframe) on the subject. Sherman and colleagues (1997) developed such a theoretical framework based on the nature of the objects being discriminated. Two basic and non-exclusive categories are whether the 'objects' are potential mates or whether they are kin. **Kin recognition** is the necessary ingredient for nepotism, where genetic relatedness is used to explain cooperation through kin selection. Recognition between parent and offspring – maternal recognition in most mammals – is a special subset of kin recognition. Mate recognition, on the other hand, is key to assortative mating and possible sympatric speciation, and to the process of sexual selection via mate choice. The two types of recognition overlap in the area of optimal outbreeding: obtaining a balance between **inbreeding** and outbreeding when selecting a mate.

In summary, recognition processes are fundamental to every aspect of biology. The patterns and mechanisms of recognition behaviour in animals can be approached from many different and often interrelated perspectives. Taken together, they are a critical part of understanding the behavioural ecology and consequently the conservation of any species or species complex. (SJI)

See also: **Altruism; Conspecific attraction; Cooperation; Kinship; Social behaviour; Territoriality**

References and further reading

Beecher, M.D., Loesche, P., Stoddard, P.K. and Medvin, M.V. (1989) Individual recognition by voice in swallows: signal or perceptual adaptation? In: Dooling, R.J. and Hulse, S.H. (eds) *The Comparative Psychology of Audition*. Erlbaum, New Jersey, pp. 277–294.

Colgan, P. (1983) *Comparative Social Recognition*. J. Wiley, New York.

Fletcher, D.J.C. and Michener, C.D. (1987) *Kin Recognition in Animals*. Wiley and Sons, Chichester, UK.

Hepper, P.G. (1991) *Kin Recognition*. Cambridge University Press, Cambridge, UK.

Sherman, P.W., Hudson, H.K. and Pfennig, D.W. (1997) Recognition systems. In: Krebs, J.R. and Davis, N. (eds) *Behavioural Ecology: an Evolutionary Approach*, 4th edn. Blackwell, Oxford, UK, pp. 69–96.

Redirected behaviour

A redirected behaviour is one that is directed towards an apparently inappropriate target. This may occur as a response to a stimulus where the primary target for the behaviour is unavailable (**see: Frustration**). For example, **wood chewing** in horses may represent redirected grazing behaviour. Redirected **aggression** is aggressive behaviour directed towards a target other than those engaged in the confrontation, possibly because it is not possible to reach those. Redirected behaviour differs from **displacement behaviour** in that, although both seem to occur when the animal is thwarted in its original activity, in redirected behaviour the target for the behaviour is substituted rather than the behaviour itself. (KT)

See also: Emancipation

Reflex

Reflexes are simple reactions to stimulation involving automatic, involuntary and fixed movement patterns in response to external stimuli, such as blinking or sneezing, and can also be classified as fixed-action patterns. The sense organs that detect the stimulation send messages to the **central nervous system**, which sends responses directly to the effector organs. Reflexive reactions do not require higher cognitive processes and thus their measurement is not considered a good indicator of **sentience** in non-human animals. (LMD)

Regan, Tom

Tom Regan (b. 1938), Professor of Philosophy at North Carolina State University, USA, is a leading exponent of the **animal rights** movement in the USA. He denies that human beings are unilaterally significant from a moral point of view by arguing against the belief that 'animals are here for us' as mere resources. In contrast to **utilitarianism**, Regan applies a *strong* rights-based perspective to our treatment of animals and concludes with an abolitionist stance towards the use of animals in agriculture, biomedical research and sport **hunting**.

Regan rejects the idea that acts are morally right or wrong as a function of their consequences. For example, even though the use of animals in biomedical research has the potential to produce good consequences for future patients (animal and human alike), Regan denies that this is adequate justification for harming animals' interest in continued existence.

Regan sets out in *The Case for Animal Rights* (1983) and *Defending Animal Rights* (2001) the nature and content of moral respect. His view of respect accords with the philosophical rights view espoused by rights theorist Joel Feinberg (1984), who contends that 'To have a right is to have a claim to something and against someone, the recognition of which is called for by legal rules or, in the case of moral rights, by the principles of enlightened conscience'. A moral right serves as a *valid claim* that makes demands of others. Rights serve to protect or promote individual integrity and vital interests in self-determination, physical security and life. As a valid claim, a right serves as a *protective barrier* or 'trump card', shielding the rights-bearer from unjustified violation or infringement of their person or interests by others.

Where animals are concerned, Regan understands these demands to include, at minimum, *negative* or *non-interference* rights, whereby moral agents are obligated not to thwart animals' interests. Claim rights may also encompass *positive* rights, which are understood as obligations to provide conditions that *benefit* the rights-holder. Regan is less clear whether animals can demand that we actively take steps to benefit them. One consequence of Regan's view is that if an animal has the right to life, for example, then it would be morally egregious for us to kill it for our benefit, especially if there are viable vegetarian options available and it is not a matter of survival that we have to eat animal protein. Animals may also not be used as organ and tissue donors just because people would benefit from their deaths.

Regan melds the concept of inherent value with certain psychological properties of the beings in question, and the notions of moral respect and rights. He argues that certain animals, *subjects-of-a-life*, possess certain moral status-making features that make them inherently valuable. They are individuals who possess

> beliefs and desires, perceptions, memory, and sense of the future, including their own future: an emotional life together with feelings; an ability to initiate actions in pursuit of their desires and goals; a psychophysical identity over time; and an individual welfare in the sense that their experiential life fares well or ill for them, logically independently of their utility for others, and logically independently of their being the subject of anyone else's interest. (Regan, 1983, p. 243)

Regan contends that *moral agency* is not a prerequisite for membership in the moral community. By 'the respect principle' outlined in his discussion of rights, he concludes that all subjects-of-a-life have equal rights to respectful treatment and to be protected from harm that cannot be overridden in order to promote the good of some other beings.

His focus on interspecific psychological similarities also makes use of the argument from *marginal cases*. If we think that human moral subjects, like infants and 'nonparadigm' or 'misfortunate' human beings, deserve equal moral respect and equal moral consideration just like adult 'paradigm case' human beings, then there ought to be similar obligations to respect the lives and inherent value of all subjects-of-a-life who have similar moral status-making features, regardless of species. Regan thinks that there is no non-bias, human-centric ground for attributing rights to 'marginal' humans that would not also apply to many animals. Moral rights are something you either have or you don't: they are 'all or nothing', otherwise the notion of equal consideration would be subject to arbitrariness. Members of the class of subjects-of-a-life are (at least) all normal mammals greater than a year of age. He leaves the line between those animals that are subjects-of-a-life and those that are not intentionally vague, awaiting better insight into animals' minds to help provide guidance.

Regan acknowledges that conflicts of rights between rights-holders do occur. He offers two ways in which rights may be prioritized:

1. The *mini-ride principle*: in the case of comparable harms, we ought to override the rights of the fewest number of individuals. This appears to have a utilitarian flavour to it in practice.
2. The *worse-off principle*: in the case of non-comparable harms, we ought to avoid harming the individual who will be worse off, i.e. more basic rights trump trivial or peripheral ones, irrespective of the species of the rights-holder. (RA)

References

Feinberg, J. (1984) *Harms to Others.* Oxford University Press, New York.

Regan, T. (1983) *The Case for Animal Rights.* University of California Press, Berkeley, California.

Regan, T. (2001) *Defending Animal Rights.* University of Illinois Press, Champaign-Urbana, Illinois.

Reinforcement – types of

The actions an animal performs can be associated with various consequences. There can be no significant consequence to the action, which means there is no feedback that alters the likelihood of the behaviour in similar circumstances. An **aversive** consequence will reduce the likelihood of the action being repeated, while the opposite will increase the likelihood of the action being repeated. In this context, the term 'reinforcement' is used in different ways by different authors and this can lead to some confusion. In its broadest sense, the term is used to describe any outcome of the behaviour that has motivational significance, and thus the term 'aversive reinforcement' is synonymous with **punishment** (reinforcing aversion towards performing the behaviour), and the term 'appetitive reinforcement' synonymous with the narrow-sense definition of reinforcement, which is the focus of this entry. This narrow-sense definition restricts the term to any associated event that increases the likelihood of the preceding action being repeated.

Reinforcement in this narrower sense can occur when an event or object is presented to the animal (e.g. food is acquired, access to a desired activity is gained). This is known as positive reinforcement. Alternatively, it can occur when an undesirable event or object is removed from the animal (e.g. some level of restriction is lifted). This is known as negative reinforcement. By definition, these events can only be considered reinforcers if the action that precedes their presentation increases in frequency or probability. It is for this reason that the term positive reinforcement is preferable to reward, since the latter could imply compensation for the action, without necessarily altering its probability in future, i.e. in human terms we might think of a reward as a 'thank you' message, while a positive reinforcer is a message that says 'if you do it again, then you might get more'.

For an event to serve as a reinforcer for an action it needs to be associated with the animal's behaviour, and there are two important considerations in relation to this: **contingency** (the predictability of the association) and **contiguity** (the closeness of the two events in time or space). These rules of association can be used in a number of ways to the benefit of behaviour modification, as well as for the teaching of very specific actions. For example, during the procedure known as differential reinforcement of other, any action except the unwanted action may be reinforced, which leads to an increase in a variety of actions perceived as desirable or appropriate and a simultaneous reduction in the unwanted action without the need for reinforcing a single action incompatible with the unwanted action.

Reinforcement can be presented according to different schedules (response contingencies), and these will influence the preceding behaviour in different ways. If a reinforcer is presented every time an action occurs, this is referred to as a continuous schedule of reinforcement and will typically lead to the fastest acquisition of new behaviour. However, when a continuous schedule of reinforcement is used and reinforcement is then withdrawn, **extinction** of the behaviour will occur rapidly. Intermittent schedules of reinforcement on the other hand, where not every response is reinforced, may lead to a reduced rate of acquisition of the behaviour but, in most cases, will increase the behaviour's resistance to extinction (a phenomenon known as the partial reinforcement effect or PRE). A number of different intermittent reinforcement schedules are described:

- Fixed ratio schedule: a set number of actions need to be performed for each reinforcer. For example, every third response is reinforced.
- Variable ratio schedule: a variable number of actions need to be performed for each reinforcer; however, they average at a given ratio. For example, on average every third response is reinforced.
- Fixed interval schedule: a response is only reinforced a set time after the previous reinforcement has been delivered.
- Variable interval schedule: a response is only reinforced on average a set time after the previous reinforcement; however, this time is variable.

Each of these schedules will see response curves that differ in key aspects. Typically, there is a decline in the rate of performance (post-reinforcement pause) after reinforcement, and this is greater on a fixed compared with a variable schedule, although the rate increases markedly towards the time of expected reinforcement in a fixed schedule.

Outside the laboratory in a training context, it is most common to find the use of a continuous schedule used for the acquisition of a response and then a variable ratio schedule used for maintenance. For the maintenance of high-level performance differential reinforcement may used, in which only behaviours with specific characteristics are reinforced – for example, behaviours with a particularly short latency or those that are particularly satisfactory for some other reason. This is a form of intermittent reinforcement in which the physical characteristics of the preceding behaviour determine the probability of reinforcement and, as the animal approaches a continuous schedule, the standard is raised higher. For most behaviours a degree of extrinsic reinforcement is required (such as the provision of food or a toy) for their maintenance, although some behaviours are associated with intrinsic reinforcement (such as **play**). There is some debate over whether the acquisition of information can have intrinsic reinforcing value (**see: Contra-freeloading**).

Elimination of reinforcement, in the form of its complete omission or a significant decline in the rate of reinforcement, induces ratio strain, which is characterized by a disruption in the normal pattern of responding and may ultimately lead to extinction of the response. Prior to extinction, the behaviour may intensify for a short while during this period of non-reward, before it rapidly declines. This increase in the behaviour is called an extinction burst and is believed to arise as a consequence of the frustration associated with the decline in reinforcement. The absence of an expected reward may be considered a form of negative punishment.

In most cases where behaviours are taught to animals it is necessary that they be performed at the request of a person. This process is popularly referred to as cueing a behaviour or putting a behaviour on cue, and is an example of **stimulus** control. Stimulus control exists when behaviours, which have been taught through reinforcement, are performed differentially according to the stimuli present. The cue is simply a stimulus that tells the animal that reinforcement is available should a specific action be performed. In the absence of the cue, no reinforcement is available, i.e. reinforcement is not only a response-contingent event but also a stimulus-contingent event.

Reinforcement may be delivered by either primary or secondary (or tertiary) reinforcers. A primary reinforcer is something that is typically reinforcing for the species (a natural reinforcer) – for example food, water or copulation. The significance of a secondary reinforcer on the other hand needs to be learned by association with a primary reinforcer (and that of a tertiary reinforcer by association with a secondary reinforcer, and so on). A phrase such as 'good dog' has no intrinsic value to a dog, neither does the sound of a clicker (commonly used in pet training (**see: Clicker training**)) or a whistle – the secondary reinforcer commonly used in marine mammal training. The value of a secondary reinforcer arises as a result of its classical conditioning with a primary reinforcer (**see: Conditioning – types of**).

A practical advantage gained by using a secondary reinforcer is the ability to increase the temporal contiguity between the action and the delivery of reinforcement. Some actions are taught where the practicality of delivering a reinforcer immediately after the desired response or close to the point of the desired response is difficult. For example, when teaching an assistance dog to cross the room to turn on a switch, the desired action takes place at a distance from the trainer. When using a secondary reinforcer, such as a clicker, the clicker can be sounded at the moment the action is performed (high temporal contiguity) and the secondary reinforcer can then be delivered later in time and in a different place, if a delay has been built into the association between the secondary and primary reinforcers.

Not all reinforcement is deliberate, and any action that is maintained in an animal's behavioural repertoire may be being reinforced in some way. This is important to consider when behavioural change is desired and behaviour modification is implemented. For example, for some dogs even eye contact with a person can reinforce a desire for social contact and can be the reason an attention-seeking behaviour is maintained.

(HZ)

Further reading

Chance, P. (2003) *Learning and Behavior*. Brooks/Cole, Pacific Grove, California.

Lieberman, D.A. (1999) *Learning: Behaviour and Cognition*. Wadsworth Publishing Co. Inc., London.

McGreevy, P.D. and Boakes, R.A. (2007) *Carrots and Sticks: Principles of Animal Training*. Cambridge University Press, Cambridge, UK.

Reintroduction

Reintroduction is an attempt to establish a species in an area that was once part of its historical range, but from which it has since been extirpated or become extinct (IUCN, 1998). The concept has been with us long before it became associated with **endangered species**. For example, pheasants and waterfowl have been raised in captivity for hundreds of years to restock wetlands and other areas for sport **hunting**. For species

conservation, reintroduction is often seen as a last resort for establishing or reinforcing a wild population. Reintroduction programmes may involve translocating wild animals from one area to another, often from an area where they are plentiful to an area where they formerly ranged (**see also: Translocation**).

It may involve removing animals when they are particularly vulnerable to natural threats such as **predation** or flooding, maintaining them in **captivity** for a period of time and returning them to the capture area when their chances of survival are greater. For example, caiman and sea turtle eggs are removed from nests in the wild, incubated in protected areas and the young are then released near the nesting site when they are large enough to have a higher probability of survival (**see also: Headstarting**). In this way, the young reinforce wild populations. Reintroduction programmes may also include breeding and rearing animals in captivity over several generations. This may be due to: maintaining genetic diversity in the wild population by creating a genetic reservoir in captivity; the lack of adequate conditions in the former range to release the animals; or the wild population being so small that the only way to protect it from **extinction** is to bring the remaining individuals into captivity for **captive breeding programmes** (e.g. the California condor, *Gymnogyps californianus*).

Reintroduction programmes are often expensive, time-consuming and unsuccessful. They depend on scientific research to determine whether reintroduction is feasible, the cooperation of stakeholders (i.e. governments, local communities, non-governmental organizations), public relations and conservation education, a commitment to long-term funding and post-release monitoring. To a large extent, successful reintroduction depends on reducing or removing whatever threats had reduced or eliminated the wild population. Threats can include habitat loss, poaching, predation, **disease**, competition with introduced non-native species and human–animal conflicts. However, if the threats can be ameliorated and the impacts of reintroduction on the release animals and the ecosystem in which they will be released are evaluated scientifically, the programme stands a much better chance of success. The behaviour of release candidates is critical to ensure survival. This entry will focus on the importance of behaviour in reintroduction.

Effects of captivity

From a historical perspective, two generalities can be made. First, more behaviourally 'hard-wired' taxa such as fish, amphibians and reptiles tend to be reintroduced more successfully than birds and mammals if anthropogenic causes of decline are managed (e.g. collection of specimens for food or the pet market). Generally, these animals reproduce early in life, produce many young at a time and do not require parental care. Thus, it does not take many years for these animals to establish themselves. The second generalization is that wild-born animals tend to survive better when reintroduced than those that are bred in captivity. This is particularly true for long-lived mammals and birds. These k-selected species take several years to become reproductively mature, and offspring per reproductive season are few. Such long-lived animals also have long periods of parental care. Thus, these animals may spend many years in captivity before sufficient numbers are available for reintroduction. Behavioural variability from long periods in captivity or generations of offspring produced from captive breeding are important in reintroduction programmes because traits necessary for survival may be under-represented.

In **zoos** and aquariums, reintroduction species are increasingly provided captive environments that stimulate the full range of species-appropriate behaviour. Opportunities can be created through environmental enrichment such as group housing for social species, providing a variety of foraging options, topography and manipulanda (e.g. twigs, stones) (**see also: Enrichment**). Although the full behavioural repertoire may be represented in captivity, the proportion of time spent performing the behaviours and the frequency of occurrence may not be what is required to survive in the wild. Behavioural research, including the development of **ethograms** in captivity and the field, can assist in preparing release candidates.

Captivity can also affect **tameness**, migration (no knowledge of traditional migratory routes), philopatry (tendency to return to the same breeding area or colony), social skills for reproduction and establishing territory. Release candidates must be able to avoid predators, acquire and process food, interact socially with **conspecifics**, find or construct shelters or nests, move on complex terrain and orient and navigate in a complex environment. The Universities Federation for Animal Welfare (**UFAW**) reintroduction guidelines (1992) suggest that the captive environment should mimic the wild as much as possible to allow the individual animals to learn and exhibit their full range of behaviours necessary for survival.

Reintroductions of captive-bred animals should not proceed without first reviewing the scientific literature and conducting field studies to determine which behaviours observed in the wild should be encouraged in captivity. Another way to identify potential behavioural challenges that may impact survival is to conduct experimental reintroductions with similar, but more common, surrogate species to develop the expertise and methodology needed for the release programme species. For example, the Siberian ferret (*Mustela eversmanni*) was used as a surrogate for the endangered black-footed ferret (*Mustela nigripes*), the Andean condor (*Vultur gryphus*) for the California condor and the greater sandhill crane (*Grus canadensis tabida*) for the whooping crane (*Grus americana*).

Rearing and release methods

Reintroduction programmes often manage small populations of the endangered species in captivity. In order to produce a large number of offspring (usually from a small founder stock) for the reintroduction programme as well as to manage the genetic diversity of the captive population, it is not always possible to have all offspring raised by their parents. In the whooping crane recovery programme, in order to increase the number of chicks produced per year, parents are encouraged to lay a second egg clutch (double clutching) by removing offspring from the first clutch. Other chicks are removed because parents may not provide appropriate care or may be too aggressive towards the young. Thus, some chicks are hand-reared by caretakers. Caretakers wear costumes and use puppets that resemble the parent. This also serves to keep the chick from **imprinting** on humans.

In the wild, **parental investment** is important to the survival of **precocial** young. Depending on the species, parents feed the young, identify appropriate food items and demonstrate how to handle them, and protect the young from predators. Young also learn predator recognition, nest-building skills, offspring rearing, migratory routes and appropriate **vocalizations** from parents. However, the importance of parental rearing in reintroduction programmes is variable. A potential drawback of parental rearing in captivity is that the parents may behave differently from their wild counterparts, particularly if they were not reared by wild founder stock. Some offspring may learn inappropriate behaviours such as stereotypic pacing. With hand-rearing, humans decide what behaviours are important to survival in the wild and encourage these behaviours by becoming the surrogate parent. Puppets, mannequins, costumed caretakers and adult conspecific models are often used to teach young, hand-reared animals to locate food and water, encourage flocking and social behaviour and teach parenting skills.

Young may also be cross-fostered or reared by adults of a similar, but more common, species. Exposure to predators, environmental enrichment and rearing with adult models in view are husbandry techniques designed to provide exposure of the young to novel stimuli that may be encountered in the wild.

Regardless of rearing method, for some species, captive rearing methods do not affect post-release survival as much as the animal's adaptability to challenges at the release site. Finally, animals are individuals. Some may be unusually inquisitive and others may be very aggressive. Generalizations may be made by species, release group or rearing method, but some animals will inevitably be better release candidates than others.

The elimination of tameness and appropriate responses to predators are important factors in behavioural modification in captivity for many species. Tameness is the absence of aggression towards humans and tolerance to **handling**. The effects of tameness on survival are mixed. It has caused human–animal conflicts such as released California condors perching on suburban homes and destroying roof shingles or puncturing car tyres. Conversely, tameness allows for animals to be easily recaptured after release, led to safer areas or for more effective manipulation for minor veterinary procedures such as inoculations.

Release candidates must sometimes be taught how to behave to evade predation (**see also: Anti-predator training**). **Predator avoidance** behaviour may involve finding habitat that is inaccessible to the predator or adopting a body position such as crouching. In captivity, predator avoidance can be encouraged by providing the individual with a psychologically traumatic experience such as controlled exposure to a real predator. Techniques have involved the use of dogs, taxidermy models, falconry birds and costumed or uncostumed humans. Caution must be used when including anti-predator training in a reintroduction programme so the individual does not habituate to the predator or become injured or killed, thus removing it from the pool of release candidates.

Typically, animals are 'hard released' or 'soft released'. Hard-released animals are simply transported to the release site and released. This works well with r-selected species like insects, amphibians, fish and many reptiles. Most mammals and birds are soft released. Soft-released animals are placed in pens at the release site for a period of time before being released. The soft-release method gives the release animals time to adapt to their new surroundings. The pen provides the animals with opportunities to forage for local food items, navigate on new substrate and interact with resident animals without risking predation or injury from aggressive conspecifics at the site. It also gives managers the ability to observe the animals to determine whether any final intervention is needed before release.

Behaviour is also considered in the timing of release. The season of release is important especially if the resident wild animals are territorial (e.g. breeding season) or the food supply is low (e.g. dry season). Another consideration is age. Birds are usually not released until they are at least fledged and able to fly with a flock. Mammals are not released before they are weaned. A further behavioural consideration is the composition of the release group. Some animals are socially solitary and others must be in groups of a certain size, with the appropriate age structure and sex ratio.

The role of applied ethology studies

While science has been used extensively to develop population genetic management strategies, veterinary care and health monitoring techniques and propagation methods, applied ethological studies are far less prominent. This is understandable because endangered species are often on the brink of extinction, sample sizes are small and recovery teams often operate in a crisis mode. Behavioural studies of reintroduction species are time-consuming, require regimented experimental conditions and may involve some degree of risk to the release candidate (i.e. **stress**, **pain**, suppressed reproduction). Using more common surrogate species, however, can minimize risk to the reintroduction species (but animal **welfare** risks to the surrogate must still be considered). Rigorous systematic studies involving behavioural observations pre- and post-release have helped to determine the optimal rearing conditions and release strategy. Behavioural research has also helped identify the best release candidates. Wildlife managers and biologists have collaborated in determining which behaviours are most essential to the programme. If there are advantageous behaviours that are being selected against in captivity (such as vigilance), they can be identified, studied and encouraged. Important areas of study include predator recognition and evasion, development of parental behaviour or **foraging behaviour** and **reproduction** and **social behaviour**.

A study of behaviour in captivity cannot predict optimal behavioural strategies in the wild without comparison to field studies. Ideally, studies should be conducted at rearing and release sites. If there is a lack of field studies or if it is difficult to observe animals in the wild, it may be necessary to remove individuals from the wild for observation in a naturalistic enclosure. This can be done *in situ* or at a remote facility.

The following are considerations when planning an applied ethology study with a reintroduction species:

1. Commitment of decision makers. Funding, personnel and material resources must be assured for the duration of the study. In addition, there must be commitment to the agreed upon sample sizes and study design.

2. Statistics. The experimental design and statistical analysis should be developed a priori. This includes appropriate sample size and replication. In some instances, it is difficult to propagate large numbers of some species in captivity necessary to have a statistically useful sample size without involving multiple captive-breeding institutions. In addition, study results may be affected by year (i.e. drought, change of release site). Without repeating a study over the course of years, it is difficult to determine whether results are due to year effects or whether they represent typical behaviour of the study species.
3. Wildlife management. Release site selection, method of release, timing of release and post-release monitoring may vary from site to site or year to year based on decisions made by wildlife managers. Such variability may need to be reduced to produce sound study results.
4. Captive management. Conditions in captivity can vary due to the number of pens available, removal of animals from the study for veterinary purposes, social behaviour of penmates and other factors. This too should be considered to control for variability in the experimental design and interpretation of results.
5. Animal welfare. With small sample sizes, the survival of every individual is important. However, to determine best release methods or release candidates, some studies may need to involve physical risks to some animals. Humane endpoints need to be considered.
6. Peer review. Many proposed studies are reviewed and approved by outside scientists, as well as those knowledgeable about the study species (such as Recovery Team members), and a statistician.
7. Publication of results. There are far too few reintroduction-related studies in the scientific literature. Regardless of what the results are, it is important for researchers to know, for example, what rearing and release techniques improved post-release survival and what did not. Descriptions of methods, including rearing and release, statistical design and results, should be as detailed as possible so that the study can be replicated or comparisons can be made with other studies.

A note on animal welfare and ethical dilemmas

There are important animal welfare and ethical considerations with reintroduction. In captive population management, decisions with welfare implications include: (i) deciding which animals should be removed from a group and relocated to another facility for **breeding** (**see: Transport**); (ii) separating mothers from young; (iii) how and where to house offspring that are surplus to the genetically managed population; and (iv) what to do with post-reproductive animals. Some animals may be artificially inseminated, and may be deprived of the experience of natural breeding. As discussed above, training avoidance behaviours such as predator evasion or deterring the animal from interacting with the human environment may involve aversive conditioning such as mild shock, loud noises or pursuit. The welfare of prey animals should be considered in preparing predators for reintroduction. If a captive-bred predator will need to kill live prey post-release, it must be taught to recognize, capture and feed on live animals. Prey animals will defend themselves and can pose significant dangers to predators (e.g. through biting, scratching, kicking, etc.).

From an animal welfare perspective, the **well-being** of the prey animal is just as important as the well-being of the predator. This is a critical difference between conservation and welfare philosophies. From the conservationist's perspective, the more endangered species is accorded a comparatively greater value than that of the more common species (**see: Ethics**). Conservationists value naturally occurring biological diversity and understand the interdependency of species; the loss of one species can result in the loss of others. This is in contrast to animal welfare (and **animal rights**) philosophy, which focuses exclusively on individual animals. Thus, ethical considerations must balance the welfare of individual animals versus their value to a reintroduction programme and conservation.

Defining reintroduction success

The success of a reintroduction project may be measured by a series of benchmarks: length of time the animal survives in the wild, breeding success, survival of offspring or when a population becomes independent from human support. Some view a programme as being successful either when the wild population reaches 500 individuals that do not depend on human support or if demographic analysis predicts the population to be self-sustaining. (MK)

References and further reading

AZA (Association of Zoos and Aquariums) (1992) *Guidelines for Reintroduction of Animals Born or Held in Captivity*. Prepared by the AZA Reintroduction Advisory Group, Silver Spring, Maryland.

Beck, B.B. (1995) Reintroduction, zoos, conservation and animal welfare. In: Norton, B., Hutchins, M., Stevens, E. and Maple, T. (eds) *Ethics on the Ark: Zoos, Animal Welfare and Wildlife Conservation*. Smithsonian Institution Press, Washington, DC, pp. 155–163.

Clark, T.W., Reading, R.P. and Clarke, A.L. (1994) Introduction. In: Clark, T.W., Reading, R.P. and Clarke, A.L. (eds) *Endangered Species Recovery: Finding the Lessons, Improving the Process*. Island Press, Washington, DC, pp. 3–17.

IUCN (1998) *IUCN Guidelines for Re-introductions*. Prepared by the IUCN/SSC Re-introduction Specialist Group. The World Conservation Union Re-Introduction Specialist Group. Gland, Switzerland.

Kleiman, D.G. (1996) Reintroduction programs. In: Kleiman, D.G., Allen, M.E., Thompson, K.V. and Lumpkin, S. (eds) *Wild Mammals in Captivity: Principles and Techniques*. University of Chicago Press, Chicago, Illinois, pp. 297–305.

UFAW (1992) *Welfare Guidelines for the Reintroduction of Captive-bred Mammals to the Wild*. Universities Federation for Animal Welfare, Wheathampstead, UK.

Websites

Alexander Center for Applied Population Biology Reintroduction Program, http://www.lpzoo.com/conservation/Alexander_Center/reintroduction/index.html (accessed 2 December 2009).

IUCN/SSC Re-introduction Specialist Group, http://www.iucnsscrsg.org

Rejection

Rejection of the young animal by its mother is a natural process, by which **weaning** is achieved, but in most if not all species of farmed mammals, individual mothers may reject the neonate. This is most common with the first **parturition**, and farmers can often overcome it by confining the mother with her newborn, or restraining her. If this fails, young can often be fostered (**see: Fostering**); this often happens spontaneously in **pigs**, which are particularly accepting of the offspring of other females. Fostering lambs and calves is very commonly done by farmers, though seldom achieved with **horses**, where foal rejection can be a troublesome problem. **Breeds** differ in this aspect of maternal behaviour – this is evident in horses, and some breeds of **sheep** are more likely to reject their lambs than others. Among British breeds, the Suffolk (a breed associated with relatively intensive, lowland sheep farming) is more likely to reject her lambs than the Scottish Blackface (a hill breed, developed in a harsher environment). A cow is more likely to reject her calf if it is her first, and some of the behaviours she may show suggest that there is an element of **fear** of the newborn. Another cause could be a failure of the **bonding** process. In all species so far studied, there is evidence for a **sensitive phase** soon after birth, when the mother and neonate bond together, and this could easily be disturbed in crowded or **stress**-inducing conditions. (SJGH)

Further reading

Crowell-Davies, S. and Weeks, J.W. (2005) Maternal behaviour and mare–foal interaction. In: Mills, D.S. and McDonnell, S.M. (eds) *The Domestic Horse: The Origins, Development, and Management of its Behaviour*. Cambridge University Press, Cambridge, UK, pp. 126–138.

Relational ethics of care

This is the ethical position according to which special emphasis is given to responsibilities incurred due to relationships formed with others. Examples include spousal relationships, parent–child relationships and the **human–animal bond**. In the latter case, acceptable animal use is contingent on there being a sustained effort to promote mutually beneficial relationships and husbandry conditions that foster attitudes of caring. (RA)

See also: Ethics

Further reading

Midgley, M. (1983) *Animals and Why They Matter*. University of Georgia Press, Athens, Georgia.

Relationship

The term 'relationship' describes the connection between two or more individuals or things. It can describe the extent of the association and influence of one situation on another, for example the relationship between body size and heat loss. Relationships between variables may be measured and represented as an equation, e.g. the simple linear regression equation: $y = ax + b$, with one variable (y) being dependent on the other (x).

The term 'relationship' may also be used to describe the connections between individuals or groups of animals. These may be defined in terms of genetic similarity between individuals (e.g. **kinship**) or the influence of one group of individuals on the population of another (e.g. predator–prey relationships). 'Relationship' is commonly used to describe the quality and quantity of the social interactions between individuals or groups of individuals.

Relationships between individuals of a social species

The relationships between individuals of social species have been explored in detail by zoologists such as Goodall, Harlow, Hinde and Stevenson-Hinde in the 1960s and onwards. Their primary interest was the social relationships of primates, how these could be described and how they compared with relationships among humans. Their observations were based on individuals within natural social groups in **field studies** (Goodall) and on separation experiments in the laboratory (Harlow). The results of their research underpin the basis of our understanding of animal relationships today.

Hinde defined relationships as 'a series of interactions that occur over time between two individuals that are known to each other' (Hinde, 1987). An interaction is a series of interchanges that occur over a shorter period of time; for example, an aggressive interaction may involve posturing and threatening interchanges. These interactions can be described in terms of their type (e.g. biting or kicking), their quality (e.g. intensity or ferocity) and duration and can be used to form a picture of the kind of relationship between the two individuals. Within a social group or network there will be a complex of relationships, with each individual having a relationship with other individuals – possibly everyone – in the group.

Early attempts to describe the relationships between individuals of a group tended to be oversimplistic. For example, individuals in a group would be placed in a linear **hierarchy** from the most dominant to the least (e.g. pecking orders in horses). Group dynamics in many species are now thought to be more subtle, with dominance of one individual over another depending on more fluid variables such as the context or resource, in addition to more static ones such as the individual's age or size. It is also recognized that relationships between individuals occur over many dimensions and, for a more complete picture, a number of these need to be measured. For example, looking at the 'nearest neighbour', i.e. the individual in the closest physical proximity to the individual in question at any given time, within groups of horses for example, has shown that individuals have preferences for certain individuals and these 'friendships' may more accurately portray the relationships within a herd than interactions relating to dominance. In addition, looking at the relative frequency of interactions, e.g. how many times A helps individual B out or the number of times B solicits help from A, may more accurately describe the strength of the relationship between individuals. This tool has also been used to look at relationships between individuals that are based on reciprocity, for example cooperative alliances between male baboons (reciprocal **altruism**).

The relationships between individuals may also be defined by the role each individual plays within the group. A mature female within a band of horses may, for example, have a leadership role that is characterized by being the individual that initiates the moving on of the group towards richer pasture or water. A mature male may have the role of primary defender of the group. Roles may not be static, and aggressive displays may signify a change in the roles between individuals.

Human–animal relationships

The relationship between humans and animals varies enormously depending on the species in question and the context. Perhaps the closest desirable relationship is that between a **pet** animal and its human 'owner'. Several aspects of this relationship have been investigated, including: (i) how it evolved (**see: Domestication**); (ii) the ways in which it can be described (in terms of love, trust, obedience, etc.); (iii) the benefits to the human of this relationship (in terms of health and mental wellbeing); and (iv) the influence of the human's characteristics and attitudes on the pet (particularly in terms of welfare and behaviour problems).

The relationship between pet and owner seems most closely to resemble a surrogate parent–child relationship, but the two types of relationship do differ. Infantile characteristics such as friendliness, dependency and playfulness are often among those favoured by the owner in their pet. Some **companion animals** retain specific infantile behaviours, for example, kneading by cats when being held. However, a degree of **reciprocity** is also apparent in the human–cat relationship: for example, the cat may terminate the relationship if the human does not fulfil their role, e.g. by providing food. The relationship between a dog and its human owner is one of the closest **human–animal bond**s and the most long-standing, since dogs were the first animals to be domesticated several thousand years ago.

Attachment

A particularly strong relationship may be described as '**attachment**'. This can be defined as 'an affectional tie that one person or animal forms between itself and another specific individual, which binds them together in space and endures over time' (Ainsworth and Bell, 1970). Behaviours that indicate that one individual has an attachment bond with another include proximity seeking and maintenance towards the attachment figure, protestation and following when they are leaving their sight and greeting upon their return. Such attachment behaviours are adaptive in that they ensure that the individual remains in familiar company and is therefore more likely to be protected from danger. Attachment typically describes the parent–child relationship but may be applicable to other relationships, i.e. between mating pairs, and occur in other non-human species.

Features of attachment are preference for the attachment figure, which assumes the ability to discriminate and respond differentially to the attachment figure and response to separation from and reunion with the attachment figure that is distinct from responses to others. Attachment has an internal structure that endures over time, even during periods when attachment behaviours are not being activated. Ainsworth devised the 'strange situation test' to activate these behaviours by subjecting infants to periods of separation from their attachment figure (their mother) in a strange environment (Ainsworth and Bell, 1970). This technique has since been used with dogs and results suggest that attachment may also describe the relationship between a dog and its owner (Topal *et al.*, 1998). (KT)

See also: Affiliation; Altruism; Anthrozoology; Companion animal; Measuring behaviour; Social behaviour

References and further reading

Ainsworth, M.D.S. and Bell, S.M. (1970) Attachment, exploration and separation illustrated by the behaviour of 1 year olds in a strange situation. *Child Development* 41, 49–67.

Hinde, R.A. (1987) *Individuals, Relationships and Culture: Links between Ethology and the Social Sciences.* Cambridge University Press, Cambridge, UK.

Podberscek, A.L., Paul, E.S. and Serpell, J.A. (2000) *Companion Animals and Us: Exploring the Relationships between People and Pets.* Cambridge University Press, Cambridge, UK.

Serpell, J.A. (1996) *In the Company of Animals: a Study of Human–Animal Relationships.* Cambridge University Press, Cambridge, UK.

Topal, J., Miklosi, A., Csanyi, V. and Doka, A. (1998) Attachment behaviour in dogs (*Canis familiaris*): a new application of Ainsworth's 1969 strange situation test. *Journal of Comparative Psychology* 112, 219–229.

Releasing factor

A stimulus that triggers an **innate releasing mechanism** resulting in the performance of a fixed action behaviour pattern (**FAP**). 'Releasers' are specific events or observations (also known as sign stimuli – although reserve the term for releaser for social cues, and use sign stimuli to refer to physical cues) within a more complicated context that can be identified as the initiator of the FAP. For example, **Niko Tinbergen** found the presence of a red belly on an approaching individual to be the releaser for aggressive behaviour in male stickleback fish. (KT)

Further reading

Tinbergen, N. (1951) *The Study of Instinct.* Oxford University Press, Oxford, UK.

Reliability

Reliability refers to the quality of a test or measuring procedure. A test or a procedure is considered reliable to the extent that it yields the same result on repeated trials. Reliability may require one observer to obtain the same results over time (longitudinal reliability, or repeat reliability) or may focus on the ability of different observers to acquire the same results (**inter-observer reliability**). (PS)

See also: Validity

Religious slaughter

Some religions require food animals to be killed by a specific, traditional method, the protocol for which may be both rigid and immutable. These protocols were described in ancient religious texts and, given the requirement that an animal is unblemished at the time of **slaughter**, had the aim of maximizing food safety and moral respect for the animal while minimizing suffering in the context of the standards of the time.

The traditional Jewish method of slaughter, *shekita* (or *shochetim*), requires the animal to be conscious at slaughter, restrained in a specialized pen with the head held fast. A transverse, reciprocal cut is made in the neck using a long, very sharp and undamaged blade with an uninterrupted motion. The animal is then restrained until it has bled out. **Sheep** may have the wool shaved from the neck to facilitate the cut and are restrained on their back in a cradle. All followers of the Jewish religion are required to eat meat killed by only this method. Some premises may administer an immediate post-cut stun for **cattle** where rabbinical approval has been given.

The traditional Muslim method of slaughter, *halal*, is in many respects similar to *shekita*. However, in *halal* **slaughterhouses** in some countries, where consumers are concerned about animal **welfare**, pre-stun methods that do not kill the animal and where the heart continues to beat have been deemed acceptable by religious authorities.

In many Western countries, slaughter without prior stunning is illegal, but there may be specific exemption for religious slaughter. In the European Union (EU), for example, Directive 93/119, Article 3 requires that no animal should be caused or permitted to sustain any avoidable excitement, **pain** or **suffering** during restraint, stunning, slaughter or killing. Directive 74/577, Article 2, provides for religious slaughter to be exempt from this requirement. However, there is no such exemption in certain European countries, including Sweden and Switzerland.

It has been argued, in a highly regarded scientific journal, that the behavioural responses of animals slaughtered using religious methods, as well as neurophysiological studies, suggest that these methods are painless and humane. Many other scientists strongly disagree with these conclusions. However, such methods would undoubtedly have been more humane than those likely to have been used previously, where more than a single cut was permitted and where there was no requirement for the operator to be proficient, for the knife to be razor sharp and free from blemishes or for all vessels supplying blood to the brain to be severed. Due to the difficulties in restraining animals in an appropriate position, throughput rates on religious slaughter lines may be slower than on conventional kill lines.

The principal welfare problems associated with religious slaughter are considered to be:

- **fear** and **distress** during restraint;
- severe pain and distress due to the throat being cut;
- distress resulting from inhalation of blood through the severed trachea;
- prolongation of pain and distress in cattle due to their unique pattern of blood supply to the brain, and thus a prolonged period of sensibility; and
- prolongation of fear and distress due to reduced rate of blood loss resulting from the neck folding and/or occlusion of arteries.

There is evidence that pre-slaughter restraint of animals in boxes designed for religious slaughter may cause higher levels of stress than conventional boxes and that the period of restraint may be prolonged. There is also evidence that manual inversion of sheep may be physically demanding and may lead to animals being lifted by the fleece.

The Farm Animal Welfare Council (**FAWC**) in the UK, in their report (FAWC, 2003), concluded that, during the transverse neck cut at slaughter there is massive injury to tissues, including skin, muscle, trachea, oesophagus, carotid artery, jugular vein and major nerves. They considered that this results in the conveyance of very high levels of painful sensory information to the brain of the conscious animal, causing significant pain and distress. This conclusion differs from that of other scientists, who conclude that the sharpness of the knife prevents stimulation of pain receptors.

During slaughter, blood from the severed blood vessels is inhaled through the cut end of the trachea, which enters the lungs causing further distress through an experience of asphyxiation, with associated choking reflexes.

The course of the vessels supplying blood to the brain, for sheep and for cattle, is presented diagrammatically in Figs R.3 and R.4. The principal supply for most species is via the two carotid arteries, which are located on the ventral, or lower, aspect of the neck, either side of the midline; the vertebral arteries do not join directly with the cerebral vessels. In cattle,

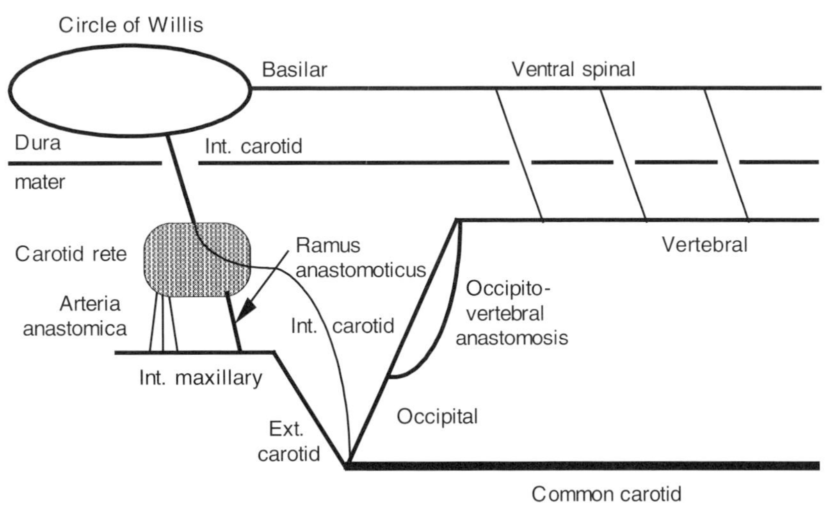

Fig. R.3. Arterial arrangement in the sheep (from Baldwin and Bell, 1963).

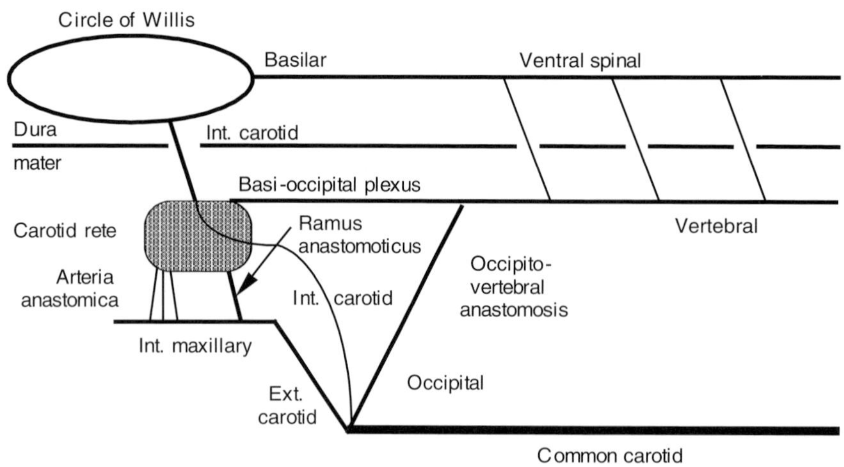

Fig. R.4. Arterial arrangement in the calf (from Baldwin and Bell, 1963).

additionally, the two vertebral arteries, which run either side of the spinal column in the dorsal, or upper, neck, join directly with cerebral vessels and supply a significant proportion of blood directly to the brain. The vertebral arteries are not severed by the transverse ventral neck cut used in *shekita* and *halal* slaughter protocols, with the result that blood supply to the brain may be prolonged.

Furthermore, the carotid arteries may be occluded following neck cut on occasions when the head slips free of the restraint and the neck folds. Occlusion may also result from the retraction of the connective tissue coat of the artery, and collapse of the elastic walls; blood may then collect in the connective tissue sealing the cut ends. This results in an increase in blood pressure in the vertebral artery, which therefore continues to supply the brain, thereby delaying insensibility. The practice by some religious slaughtermen, of placing a hand into the neck wound immediately after neck cut, in an attempt to prevent occlusion of the arteries due to elastic collapse of the walls of the carotid, causes additional pain and distress and may be ineffective. Supporters of religious slaughter argue that ballooning of vessels does not occur due to the sharpness of the knife but the FAWC found, during visits to observe religious slaughter, that ballooning did occur.

Scientific studies, using encephalography, show that the duration of consciousness after a knife cut may be up to 60 s in adult cattle, while in calves, consciousness has been shown to remain for up to 120 s, depending on the degree of occlusion of the blood vessels. Sheep and **goats** become insensible in less than 7 s. This contrasts with an immediate loss of consciousness that has been shown to occur following captive-bolt stunning.

A significant volume of meat from animals killed by religious methods is unsuitable for use as *halal* or *kosher* product due to religious considerations, but is otherwise fit for human consumption, including hindquarters, as religious texts require that they not be eaten unless certain blood vessels are removed. This meat may be sold on the open market, without being labelled as slaughtered by religious methods. This would be of some considerable ethical concern to non-Jewish and non-Muslim consumers, for whom animal welfare is an important aspect of meat quality, were they to be aware of this practice. The societal consensus in European countries, that pain and distress arising from slaughter without stunning is ethically unacceptable, has resulted in legislation being enacted to ensure that animals are stunned prior to killing. The FAWC in the UK has recently recommended that the exemption for religious slaughter of the legal requirement for pre-slaughter stunning, under the Welfare of Animals (Slaughter and Stunning) Regulations 1995, should be repealed. However, this recommendation was not accepted by the UK Government. (SH)

See also: Buffalo; Consciousness; Islam; Judaism

References and further reading

Baldwin, B.A. and Bell, F.R. (1963) Blood flow in the carotid and vertebral arteries of the sheep and calf. *Journal of Physiology* 167, 448–462.

FAWC (2003) *Report on the Welfare of Farmed Animals at Slaughter and Killing*. Farm Animal Welfare Council, PB8347, London.

Gregory, N. (2000) *Animal Welfare and Meat Science*. CAB International, Wallingford, UK.

HSA (1998) *Slaughter by Religious Methods*. Humane Slaughter Association, Wheathampstead, UK.

Repetitive behaviour

Behaviour is repetitive when the same behaviour pattern is repeated over a short period of time. Bouts of the repetitive behaviour may also be repeated. The behaviour is usually similar each time it is performed. It has been suggested that the performance of rhythmical, repetitive behaviours may be have comforting properties (**see: Comfort behaviour**) and that it may become addictive, through the associated release of endogenous **opioids** (Cronin *et al*., 1986). Examples of repetitive behaviour include normal adaptive behaviours such as ingestive behaviour of herbivores and blinking, as well as maladaptive **abnormal** behaviour patterns such as the **stereotypies** of **weaving**, **crib-biting** and **stall-walking** in **horses**.

Repetitive behaviours may be learned or variations of **instinctive behaviour** patterns, e.g. **vocalization**, chewing, walking, that are necessary for the survival or reproduction of the animal. Once developed, some repetitive behaviours may be more easily elicited and become less dependent on external factors for their reinforcement or development. A repetitive behaviour may however be dependent upon the continued presence of a **stimulus**, for example food in the mouth stimulates chewing and swallowing until the mouth is empty. Other repetitive behaviours may be **reflexes** that have been conditioned by the association of a reward with the performance of the behaviour, e.g. attention-seeking spinning by a **dog**.

Some repetitive behaviours present as behaviour problems in animals, and so understanding the **aetiology** of a given case is important, and an assumption should not be made that all repetitive behaviour problems are stereotypies. For example, horses that repeatedly shake their heads may simply be responding normally to a stimulus that is repeatedly being presented (e.g. poorly fitting tack or biting insects). However, for some, the performance of the behaviour appears to be related to neuralgic pain and in others it may have become a **habit** or be a dysfunctional behaviour and become emancipated from the stimulus that originally elicited its performance. In these cases the bouts of repetitive behaviour may be quite sustained and the animal may appear to have difficulty ceasing it.

Just as the term 'stereotypy' is inappropriate for all of these behaviours, so is the term 'obsessive–compulsive disorder' or 'compulsive disorder', since these terms imply a common mechanistic basis. However, it has been suggested that when a repetitive behaviour appears **abnormal**, it may be more useful to refer to it as abnormal repetitive behaviour or stereotypic behaviour (with the term stereotypic used to convey that it is like a stereotypy, but it may not have the precise mechanistic basis of a stereotypy as implied in the human medical literature) until its precise mechanism is more clearly elucidated, when it can be more precisely labelled (see Fig. R.5). However, these conventions have yet to be widely adopted within the literature. (KT)

Further reading

Cronin, G.M., Wiepkema, P.R. and van Ree, J.M. (1986) Endorphins implicated in stereotypies of tethered sows. *Experientia* 42, 198–199.

Garner, J.P. and Mason, G.J. (2002) Evidence for a relationship between cage stereotypies and behavioural disinhibition in laboratory rodents. *Behaviour and Brain Sciences* 136, 83–92.

Mason, G.J. and Rushen, J. (2006) *Stereotypic Animal Behaviour: Fundamentals and Applications to Welfare*, 2nd edn. CAB International, Wallingford, UK.

Reproduction

Reproduction refers to the propagation of a new individual, and may arise sexually or asexually. While sexual reproduction involving the fusing of gametes from different individuals is the norm in the animal kingdom, sexual reproduction by hermaphrodite individuals (in which a single individual produces both the eggs and sperm of a mating) and asexual reproduction also occur in both invertebrates and vertebrates. For example, the mangrove killifish (*Kryptolebias marmoratus*) has a mixed mating strategy, with hermaphrodites engaging in both self- and cross-fertilization, while the female hammerhead shark (*Sphyrna*) may reproduce asexually through parthenogenesis. Some species can alternate between sexual and asexual reproduction (heterogamy) according to the prevailing conditions, in order to capitalize on the advantages associated with each strategy. Sexual reproduction has the advantage of allowing genetic mixing through the fusion of gametes produced by meiosis (even in the case of intra-individual sexual reproduction). This can facilitate adaptation in a changing environment and may result in change in the gene frequency within the population. By contrast, asexual reproduction facilitates rapid colonization of new areas, and may be a particularly successful strategy in stable environments. (DSM)

See also: K and r selection

Reproductive behaviour

Reproductive behaviour is poorly defined as an entity and can be used to refer to all behaviour associated with successful reproduction or a subset of these that are more directly associated with mating. Thus it may refer to behaviours associated with intrasexual competition and **intersexual selection, courtship behaviour, mating behaviour (see: Sexual behaviour), nesting behaviour, maternal**

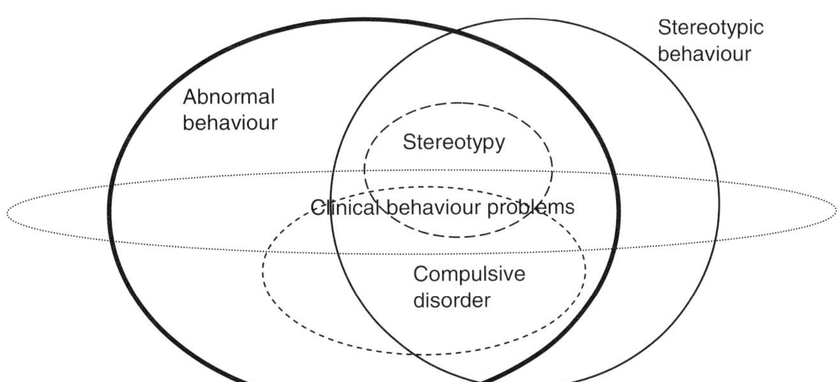

Fig. R.5. Venn diagram illustrating the possible relationship between different types of repetitive behaviour, as proposed by Mills and Luescher (2006) in Mason and Rushen (2006). See text for details of the various terms. Note: the relative areas of the ellipses do not represent the relative magnitudes of the respective sets.

behaviour – including brooding and **nursing** – **paternal behaviour** and other aspects of parental behaviour, as well as behaviours associated with maximizing reproductive opportunity such as **territoriality**. (DSM)

See also: Reproduction

Reproductive suppression

Reproductive suppression may be defined broadly as inhibition of reproductive physiology and/or **reproductive behaviour** in an otherwise fertile individual in response to specific environmental or physiological conditions. By this definition, reproductive suppression involves disruption of normal reproductive processes by inhibitory influences, rather than simply the absence of stimulatory influences. Reproductive suppression has been documented in a broad range of taxa and occurs in both sexes. It may be triggered by a variety of factors from the internal and external environments, and may involve inhibition at a number of different stages of the reproductive process.

Reproduction can be highly risky and energetically demanding, especially for females. Moreover, the likelihood that any particular **breeding** attempt will succeed – that is, that it will produce surviving offspring – may depend upon prevailing conditions in the physical, biotic and social environment, as well as upon the potential breeder's physical and physiological condition. Thus, to the extent that current reproductive attempts decrease prospects for survival or future reproduction, individuals of many species may increase their lifetime reproductive success by timing their reproductive attempts to occur during favourable conditions and, correspondingly, curtailing their reproductive activity at other times. Natural **selection** may therefore be expected to favour those individuals that undergo reproductive suppression under inauspicious conditions.

Reproductive suppression can occur in response to variation in either physiological or environmental factors that are otherwise associated with, or predictive of, poor reproductive outcomes. Physiologically, the most important of these variables is energy balance. Reproduction in many (e.g. mammals) but not all (e.g. reptiles) taxa is suppressed when animals enter negative energy balance as a consequence of such factors as low food availability, high workload or activity levels, and low ambient temperature, or combinations of these factors. The metabolic signal(s) detected by the body and transduced into reproductive suppression are unknown. A critical role of body fat, once thought to be the key signal, has largely been discounted, and more recent attention has focused on potential roles of other metabolic fuels and metabolic hormones. In addition to energy balance, reproduction may be suppressed by such physiological states as **lactation** and illness.

The external environmental factors that influence reproductive outcomes can generally be categorized as dietary, physical or social in origin. In some cases, these same factors are used as proximate cues to trigger suppression (or activation) of reproduction. In other cases, animals respond more directly to factors that are not inherently crucial for reproduction but are predictive of critical changes in climatic or dietary conditions. For example, many species become reproductively quiescent on a seasonal basis, thereby avoiding breeding during periods of low food or water availability or adverse climatic conditions. The environmental cues that may be used to synchronize reproductive activity with seasonal changes in the environment are quite diverse, however, including such variables of direct reproductive significance as rainfall (e.g. zebra finch, *Taeniopygia guttata castanotis*), ambient temperature (e.g. green anole lizard, *Anolis carolinensis*) and food availability (e.g. Columbian ground squirrel, *Spermophilus columbianus*), as well as such predictive variables as **photoperiod** (e.g. sheep, *Ovis aries*) and availability of non-nutritive plant compounds (e.g. montane vole, *Microtus montanus*).

In many species, reproductive physiology and/or reproductive behaviour may be partly or completely suppressed by cues from the social environment. Social suppression of reproduction often involves agonistic or dominance-related interactions between same-sex adults. Among cooperatively breeding naked mole-rats (*Heterocephalus glaber*) and common marmoset monkeys (*Callithrix jacchus*), for example, behaviourally subordinate females typically exhibit low reproductive hormone levels and atrophied reproductive tracts, fail to ovulate and engage in little or no sexual behaviour while living with a dominant female, but commence normal reproductive function shortly after separation from the dominant female. In other species, **pheromones** may trigger reproductive suppression independently of direct behavioural interactions. For example, ovulatory or lactating rats (*Rattus norvegicus*), as well as female house mice (*Mus musculus*) living in large groups or dense populations, release chemosignals that extend or suppress oestrous cycles and reduce fertility in other females.

Reproductive suppression is less common and, usually, less pronounced in males than in females, presumably reflecting the lower costs of reproduction and/or higher maximum potential reproductive success in males. Suppression in males may involve either a delay in the initial onset of reproductive function or a subsequent inhibition of reproductive activity in mature, potentially fertile adults. Male orang-utans (*Pongo pygmaeus*), for example, may undergo a prolonged adolescent period, characterized by reduced levels of **androgens** and gonadotropic hormones, arrested development of secondary sexual characteristics and low attractiveness to females, if living in proximity to an adult male. In other species, fully mature, socially subordinate males may undergo inhibition of **sexual behaviour** (e.g. dwarf mongoose, *Helogale parvula*), impairment of spermatogenesis (e.g. naked mole-rat, *Heterocephalus glaber*) and/or suppression of androgen secretion (e.g. Alpine marmot, *Marmota marmota*).

Females have considerably more complex reproductive physiology than males and, not surprisingly, exhibit a greater variety of mechanisms by which reproduction can be suppressed. As in males, organismic or environmental factors may inhibit or delay maturation in young females or may reversibly impair reproductive function in otherwise fertile adults. Interactions with or cues from dominant females, for example, can delay puberty (e.g. Mongolian gerbil, *Meriones unguiculatus*), suppress ovulation (e.g. pine vole, *Microtus pinetorum*), inhibit sexual behaviour (e.g. rhesus macaque, *Macaca mulatta*), delay conception (e.g. yellow baboon, *Papio cynocephalus*), block implantation (e.g. white-footed mouse, *Peromyscus leucopus*), induce spontaneous abortion or prenatal litter reduction (e.g. golden hamster, *Mesocricetus auratus*) or impair maternal care (e.g. ring-tailed lemur, *Lemur catta*) in subordinate females.

The environmental and organismic factors that determine poor reproductive outcomes may be considered **stressors**, and reproductive suppression is frequently assumed to be mediated by a **stress** response. Although reproduction can certainly be impaired by aspects of the stress response (e.g. elevated circulating concentrations of **glucocorticoids** or endogenous **opioids**), it is becoming increasingly clear that reproductive suppression frequently occurs in the absence of generalized stress and may instead be mediated by more specialized neuroendocrine mechanisms. The final common pathway for many of these mechanisms may be disrupted hypothalamic secretion of **gonadotropin-releasing hormone**, leading to inhibited release of **gonadotropin**s from the **pituitary gland** and, consequently, to impaired endocrine, gametogenic and/or ovulatory function in the **gonad**s.

Reproductive suppression may have profound implications for **captive breeding** and **conservation** programmes. As a consequence of the reproductive system's sensitivity to environmental factors, a particular species may show contrasting reproductive patterns in captivity and in the field, or even in different captive or natural environments, and may fail to breed under non-natural conditions. Clearly, the success of breeding and conservation programmes may be critically dependent upon an understanding of the organismic, environmental and social influences on reproduction. (WS)

Further reading

Bronson, F.H. (1989) *Mammalian Reproductive Biology*. University of Chicago Press, Chicago, Illinois.

Ferin, M. (2006) Stress and the reproductive system. In: Neill, J.D. (ed.) *Knobil and Neill's Physiology of Reproduction*, 3rd edn. Elsevier, St Louis, Missouri, pp. 2627–2696.

Vandenbergh, J.G. (2006) Pheromones and mammalian reproduction. In: Neill, J.D. (ed.) *Knobil and Neill's Physiology of Reproduction*, 3rd edn. Elsevier, St Louis, Missouri, pp. 2041–2058.

Wallen, K. and Schneider, J.E. (1999) *Reproduction in Context*. MIT Press, Cambridge, Massachusetts.

Wasser, S.K. and Barash, D.P. (1983) Reproductive suppression among female mammals: implications for biomedicine and sexual selection theory. *Quarterly Review of Biology* 58, 513–538.

Reptile

Traditionally, reptiles are the class of vertebrates that include squamates (lizards, snakes and amphisbaenids), turtles, crocodilians and the two species of the tuatara (found only on a few islands in New Zealand). Current taxonomic practice argues that if all these animals are in a single class, Reptilia, then so are their descendants – dinosaurs and birds. Others argue that the traditional Reptilia is not a monophyletic group, and thus should be broken up with, perhaps, only lizards and snakes as 'true' reptiles. Staying with traditional views, reptiles are ectothermic amniotes comprising roughly 3000 snakes, 4500 lizards, 160 amphisbaenians, 300 turtles and 23 crocodilians.

Reproductively, reptiles have traits found in both birds (oviparity) and mammals (viviparity). Post-hatching/-natal parental care, long thought absent in reptiles, is now known to be both complex and lengthy in all crocodilians studied, absent in virtually all turtles and the tuatara and highly variable in lizards and snakes. In fact, in some lizards, cohorts of offspring live with their parents and younger siblings for some years. Many crocodilians not only guard their nests for many months, but aid in the hatching of their babies when they hear them vocalize in the nest and transport them, very gently, to water. Probably the largest group, lizards, presents the most extreme diversity in terms of size, diet, reproductive mode, age at maturity, diversity of sensory mechanisms, parental care, sociality, social organization and habitat use. All this diversity makes any generalizations on biology, behaviour or husbandry impossible. Thus, it is necessary to consult references that provide such basic information, and some starting points are cited below.

Recognizing this diversity is also important because reptiles are increasingly featured in conservation plans, maintained and bred in **captivity** and even reintroduced to former native habitat. Conversely, many reptiles have been introduced into environments where they can be serious pests and do biological harm. The brown tree snake (*Boiga irregularis*), introduced from New Guinea into Guam, is perhaps the most egregious example, but in Florida alone the following are just a few of the introduced species that may need control: green iguanas (*Iguana iguana*), Nile monitors (*Varanus niloticus*), brown anoles (*Anolis sagrei*) and Burmese pythons (*Python molurus*). In Japan and Europe, North American emydid turtles such as red-eared sliders (*Trachemys scripta elegans*) are disrupting native species. Studies of natural history of these animals in native habitats, as well as in captivity, are needed if effective controls can be instituted. Unlike some invasive large mammals, the secretive nature and high reproductive potential of introduced reptiles makes control through hunting and trapping impractical and inefficient.

On the other hand, some species, especially edible reptiles, are being ranched and becoming domesticated. Alligator, iguana and turtle farms are prominent in several countries. In addition, many species of lizards and snakes are being bred for the **pet** trade, with aberrant colour morphs being developed. Such captive breeding is altering the behaviour and morphology of reptiles through both deliberate and unconscious **selection (domestication)**, as well as through phenotypic plasticity.

The literature on captive reptile maintenance, **breeding** and rearing has greatly expanded since the late 1980s. There are extensive works on veterinary care and volumes on captive care of almost all major taxa. The most commonly maintained species have been the subjects of many reports, many developed and written by amateurs or professional breeders. Academic and **zoo** biologists have also contributed much. European and American herpetoculture magazines abound. These often contain useful hints as well as leads for suppliers of equipment, diets, food supplements, innovative caging and other supplies.

A misconception about reptiles, still widespread, is that their behaviour is relatively unaffected by their rearing conditions and environment. In other words, unlike birds and mammals, reptiles are often viewed as more 'hard-wired', learning little if at all, and are psychologically not affected by the nature of their environment. Thus, reptiles not on exhibit in zoos and laboratories are frequently reared in relatively sterile and non-stimulating environments. What is increasingly recognized is that reptiles, being ectothermic and ecologically specialized, often have a very narrow range of environmental conditions in which they can thrive. Chief among these are temperature, humidity, appropriate ultraviolet lighting and dietary nutrients. Discrimination studies on the sensory abilities of reptiles have documented keen thermal, chemical, visual, tactile and even auditory perception. But all these vary markedly across reptilian groups.

While the view that the learning ability of reptiles is limited was challenged years ago, the extent of the abilities of reptiles and their need for appropriate environments for psychological and behavioural **well-being** is of even more recent concern and only beginning to be explored. Environmental enrichment has been shown to be beneficial in promoting health and behavioural performance in turtles, snakes and lizards, and experimental studies are urgently needed. Some reptiles even **play** in the mammalian way.

These considerations are important in learning about the many members of the second largest vertebrate class. Reptiles possess many traits that make them useful in basic research as well as for exploring solutions to behavioural, medical, ecological and conservation issues important to their survival and the habitats in which they are often an integral part.

(GMB)

See also: Development of behaviour; Enrichment

Further reading

Almli, L. and Burghardt, G.M. (2006) Environmental enrichment alters the behavioral profile of ratsnakes (*Elaphe*). *Journal of Applied Animal Welfare Science* 9, 89–109.

Burghardt, G.M. (1977) Learning processes in reptiles. In: Gans, C. and Tinkle, D. (eds) *The Biology of the Reptilia* (Vol. 7) *Ecology and Behavior*. Academic Press, New York, pp. 555–681.

Greene, H.W. (1997) *Snakes: the Evolution of Mystery in Nature*. University of California, Berkeley, California.

Pianka, E.R. and Vitt, L.J. (2003) *Lizards: Windows to the Evolution of Diversity*. University of California, Berkeley, California.

Warwick, C., Frye, F.L. and Murphy, J.B. (1995) *Health and Welfare of Captive Reptiles*. Chapman and Hall, London.

Rescue shelter

A rescue shelter is an establishment that provides protection and care for displaced animals, usually **companion animals** but sometimes **horses**, farm animals and wildlife, or is restricted to looking after specific groups or species such as **reptiles** or donkeys. Animals enter rescue shelters for various reasons, including: (i) being accidentally lost after escaping or straying; (ii) being deliberately abandoned; (iii) being handed in by their owner; or (iv) having been seized by an animal **welfare** inspector through relevant legal powers in cases of suspected cruelty or neglect. The most frequently cited reasons for companion animals being taken by their owners to a rescue shelter are behavioural and accommodation problems.

The first official rescue shelter was founded in London in 1861 and was known as 'The Home for Lost and Starving Dogs', subsequently to become the Battersea Dogs' Home. Although most are run without government support, rescue shelters have subsequently become one of the key elements of the protection of animal welfare in many communities. They have four main functions: (i) to provide for the direct **needs of animals** in their care, including the alleviation of immediate **suffering** through appropriate veterinary treatment or **euthanasia**; (ii) to provide for the long-term care of animals not immediately reunited with their owner or placed with a new owner; (iii) to seek to reunite accidentally displaced animals with their owners; and (iv) to attempt to find new places or homes for displaced animals (rehoming).

Rehoming of companion animals may be dependent upon successful **temperament** and **health** evaluation. Where animals are not considered suitable for rehoming they may be committed to long-term care within the shelter, transferred to another shelter or euthanized. Some shelters have a policy of providing lifetime care of displaced animals that have not been rehomed ('no-kill' shelters), and in some countries such a policy is supported by legislation, e.g. Italy. In other cases the emphasis is on providing short-term shelter to all displaced animals, with those animals that cannot be rehomed being euthanized.

(BJ)

See also: Abandoned animal

Further reading

Companion Animal Welfare Council (2003) *Report on Companion Animal Establishments: Sanctuaries, Shelters and Rehoming Centres*. CAWC, Cambridge, UK.

Marston, L.C. and Bennett, P.C. (2003) Reforging the bond – towards successful canine adoption. *Applied Animal Behaviour Science* 83, 227–245.

Marston, L.C., Bennett, P.C. and Coleman, G.J. (2004) What happens to shelter dogs? An analysis of data for one year from three Australian shelters. *Journal of Applied Animal Welfare Science* 7, 27–47.

Resource

Any substance or organism that is valuable for a given animal may be referred to as a resource. Typical examples of resources are food or mating partners, but the term may also refer to **enrichment** devices for captive animals. Access to resources is usually supposed to give some benefits for **individual fitness**, so the less available a resource is the more competition there may be to access/monopolize it.

(BM)

See also: Resource-holding potential

Resource-holding potential

Game theory models, when applied to animal contests, predicted that three main asymmetries will influence how animals fight and thus determine the winner: (i) prior ownership of the **resource**; (ii) the value that each contestant places on the resource; and (iii) the overall fighting ability of each contestant, originally called resource-holding power and later resource-holding potential (RHP). Prior ownership was viewed as an asymmetry that is uncorrelated with the expected pay-off and used as a convention to settle the contest with minimum costs. Resource value is correlated with the potential pay-off in terms of expected **fitness** gain from victory. The animal that places the higher value on the resource should have a higher motivation to win and thus: (i) be prepared to accept higher costs; and (ii) be more likely to win if both opponents have equal RHP.

RHP includes any feature that might influence the ability to win at a particular time and thus comprises intrinsic aspects such as body size, weapon size, previous experience and physiological states such as energy reserves or lactic acid, which induces fatigue. In addition, extrinsic aspects such as position relative to that of the opponent may influence RHP; for example, male amphipods that hold females are much more successful than are males attempting to displace them. Body or weapon sizes are easily measured and thus, in empirical work, are the most frequently used estimates of RHP. All other things being equal, the contestant of greater RHP is expected to be the eventual winner and this has been shown for a wide range of organisms, particularly with respect to body size. However, there are different theoretical models that explain the process by which this occurs, and these have only recently been tested in an effective manner.

Early models, such as the asymmetric war of attrition, presumed that each individual would have information about its own abilities and gather information about those of the opponent. That is, opponents would engage in mutual RHP assessment and the animal perceiving it was the weaker one would quit immediately and avoid wasted costs of engaging in a contest that it would probably lose. This idea was developed in the 'sequential assessment model', in which it was presumed that contestants would engage in bouts of fighting, gathering increasingly accurate information, with the weaker animal quitting as soon as an inevitable lost cause was perceived. Typically, interactions would start with displays before escalating to more costly activities. These displays should not easily be faked, otherwise that trait would spread in the population and then offer no way of assessing RHP (**see: Honest signals**). Displays should thus correlate with true RHP, and this has been shown for vocal displays such as the croaks of toads, with large toads being able to produce calls of low auditory frequency, or red **deer** stags, which can only roar vigorously if in good condition. Visual displays may also demonstrate the size of the animal or the size of the weapons that it may possess, such as with claw-waving displays of crabs. For the most part these displays are expected to provide honest information, but bluff, exaggeration or dishonesty have been shown, particularly in stomatopods, snapping shrimps and hermit crabs.

Alternative models of how RHP asymmetry determines the winner presume that contestants do not gather information about their opponent but, instead, the surrender decision is based on the accumulated costs of staying in the contest. The decision will be triggered once a threshold level of accumulated costs, that the contestant is willing to accept, has been crossed. Possible costs have been modelled in terms of time, as in the 'war of attrition without assessment' (WOA-WA), and energy as in the 'energetic war of attrition' (E-WOA). In the 'cumulative assessment model' (CAM), the costs can accrue as a result of the contestant's own agonistic activities, but they can also be influenced by the actions of the opponent. How each contestant competes is determined by its own changing state or RHP, even though the relative RHP may influence the costs per unit time of being in the fight. In these models the smaller of two contestants would typically surrender first, so these models clearly predict that RHP asymmetry determines the winner. However, animals will fight for a period and thus will always be prone to incur costs prior to quitting. Indeed, the WOA-WA and E-WOA predict that fight costs to the loser should equal the value of the resource to that animal.

In contrast, the sequential assessment model predicts that animals will fight only until the asymmetry in RHP becomes apparent, with minor adjustment for resource value but without incurring costs up to the resource value. Own size assessment (OSA) is thus an inferior strategy to mutual assessment (MA) because, in the former, contestants typically pay higher costs for the same outcome. However, MA may not be possible because some species may lack the ability to gather information about the opponent due to inadequate sensory acuity or experience, high costs of assessment or a lack of reliable cues. Even with escalated fighting, in which opponents inflict wounds and induce fatigue in each other, contestants may only have accurate information about their own state and may have relatively little information about the condition of their opponent.

A prediction of the MA model is a negative relationship between contest duration and the relative size of the opponents. This supposedly diagnostic finding for mutual RHP assessment has been found in numerous studies, but these rarely considered other models. However, it has been demonstrated that both OSA and MA models produce the same relationship. The models can, however, be distinguished by the relationship between the winner RHP and contest duration. For mutual assessment this is negative whereas for own size assessment it is weakly positive. Furthermore, with OSA there will be an increasing duration of contests with increasing size of size-matched opponents, but this is not predicted by MA models. Tests using these criteria have shown that species of chameleons, spiders, crabs and amphipods cannot gather information about opponents. Other studies, however, clearly show that both contestants gather information about the RHP of the opponent, as is seen by the calls of toads and roars of red deer. Thus, it is clear that RHP is a major determinant of contest outcome, but the manner by which that occurs may vary considerably between species. (RWE)

See also: Aggression; Intrasexual selection; Intraspecific behaviour

Further reading

Bradbury, J.W. and Vehrencamp, S.L. (1998) *Principles of Animal Communication*. Sinauer Associates, Sunderland, Massachusetts.

Elwood, R.W., Pothanikat, R.M.E. and Briffa, M. (2006) Honest and dishonest displays, motivational state and subsequent decisions in hermit crab shell fights. *Animal Behaviour* 72, 853–859.

Maynard Smith, J. (1982) *Evolution and the Theory of Games*. Cambridge University Press, Cambridge, UK.

Stuart-Fox, D. (2006) Testing game theory models: fighting ability and decision rules in chameleon contests. *Proceedings of the Royal Society B* 273, 1555–1561.

Taylor, P.W. and Elwood, R.W. (2003) The mismeasure of animal contests. *Animal Behaviour* 65, 1195–1202.

Respondent conditioning

Also known as either Pavlovian or classical conditioning, respondent conditioning refers to a form of associative learning that is assessed by evaluating the animal's change in response to stimuli within the environment as a result of the association of these stimuli. In other words the range of stimuli to which the animal will respond is changed as a result of learning (contrast this with **instrumental conditioning**, in which a response changes in response to its performance). The phenomenon was first studied and described by **Ivan Pavlov** (hence the alternative name). Typically, in respondent conditioning, an unconditional stimulus that elicits an '**innate**' or unconditional response is paired with a neutral **stimulus** that, prior to conditioning, does not trigger the response, in such a way that it eventually comes to trigger the response when not paired with the unconditional stimulus. It is then referred to as a conditioned stimulus (or conditional stimulus), and the response that it triggers a conditioned response (or conditional response). Careful behaviour assessment often reveals that the conditioned response is not identical in form to the unconditional response, and may contain greater **appetitive behaviour** elements. (DSM)

See also: Conditioning – types of; Contiguity; Contingency

Resting behaviour

Resting is a **comfort behaviour** that is important for **health**, in conserving energy and aiding both **thermoregulation** and maintenance of physiological efficiency. Many social species synchronize resting behaviour to reduce risk of **predation**. There may be competition for resting places, even if there is enough physical space for all animals to lie down, as some spots may be more comfortable or provide more protection from the elements or the **space** between them may be insufficient (**see: Flight zone**). This may lead to **agonistic behaviour** and monopolization of some spots by some animals. Some intensive production environments do not provide enough space for all animals to rest simultaneously, and thus behavioural synchronization may not be possible. This may also increase agonistic events and restlessness, leaving some individuals with less time to rest and **sleep** than others.

The ability to find space away from neighbours while resting may be important for thermoregulation. Different postures will also be performed in different amounts, depending on the amount of space available, with animals tending to spread out more or lie in a recumbent position when given more space. The optimum amount of resting time is not known for most species. As a result, the duration of resting behaviour is not a clear welfare indicator. If animals are stressed they tend to decrease resting behaviours and increase **vigilance**. However, animals that spend a lot of their time resting may be sick, diseased, in a depressed or anhedonic state and so have decreased **welfare**. (LMD)

See also: Flooring

Restraint

Restraint is used to limit some or all of an animal's movement. In order to minimize injury to the animal and provide operator safety, restraint is used for transportation (**see: Transport**), examination, convalescence, surgery, **breeding** and handling of animals. The type of restraint used can depend on the animals' anatomy, physiology, behaviour, **temperament** and previous experiences. The procedure being conducted, location and available facilities can also influence the restraint method, as can the experience of the restrainer and the risk to the safety of the animal and persons involved.

Various restraint techniques have been developed, ranging from those that allow the animal relative freedom of movement within a confined area to those severely restricting or even stopping all movement. One of the most basic restraint techniques is psychological restraint. Psychological restraint involves counteracting or incorporating an animal's own behaviour, such as its flight distance (**see: Flight zone**), into its restraint. A skilled restrainer can approach an animal to the edge of its flight zone and observe it without that animal attempting to escape. Animals can also be trained to perform various acts or allow certain procedures to be carried out without the use of any means of physical, mechanical or chemical restraint. For example, chimpanzees involved in biomedical research can be trained to present their arm and hold it steady while receiving an injection.

Restraint can be achieved in some instances by diminishing certain sensory perceptions of the animal, such as by blindfolding it or placing it in a darkened room. Darkness can have a strong calming effect on animals, although placing a **nocturnal** species into a darkened room may be counter-productive. Restraint can also be achieved by stimulating particular senses, as can happen when particular body areas of some animals are stroked in the appropriate direction. Horses, for example, are particularly responsive to stroking and massaging above the shoulder (withers). Cooling can be used to reduce the ability of ectothermic species to respond to handling (thermal restraint); however, prolonged cooling can be deleterious to their health.

Trained, experienced personnel can use their hands, and sometimes other body parts, as a physical restraint tool. For example, a traditional method for restraining sheep being sheared involves the sheep being held slightly off vertical with its back being cradled by the shearer's legs and its inside front leg placed under the shearer's rib cage. Gloves can be used to protect the hands during restraint; however, these tend to be cumbersome and reduce the restrainer's ability to determine how tightly they are grasping the animal.

An alternative to using the hands or other body parts is usage of hand-held devices such as ropes, nets, snares and tongs. Such devices have proved particularly useful for restraining, or at least applying an initial hold, on free-ranging wild animals (**see also: Free-range animals**). One such example of hand-held restraint used primarily on horses is the **twitch**. The twitch, consisting of a loop of rope or chain on a pole, is slipped over the horse's upper lip and twisted, applying pressure and presumably **pain**. The pain sensed when the twitch is first applied leads to the release of opiates, which sedate the horse.

Confinement is a restraint technique often associated with livestock production, research laboratories and **zoos**. Animals are typically confined to a specific area such as a cage, crush, **raceway** or stockyard. Mobile barriers, such as sheets of perspex or plywood, are also used further to confine an animal to a particular area. Close confinement can be achieved mechanically by using apparatus such as squeeze cages or squeeze chutes. These devices are essentially a solid, bar-sided or mesh-sided box designed and constructed to hold one animal. Slings are another close confinement device and can be used to restrain certain animals for an extended period. Close confinement of some mammals, birds and reptiles can be achieved using special bags or by wrapping them in a towel. Adjustable plastic tubes are another alternative for some reptiles and small mammals. Complete restraint of some species, such as birds, snakes and primates, can be achieved with a restraint board. Typically under sedation, the animal is fastened with tape or Velcro straps to a board. Another technique used completely to confine laboratory primates is the restraint chair; animals are fastened to the chair with neck, waist and limb restraints.

By placing animals into specific positions, some forms of physical and mechanical restraint can keep animals relatively still and, in some species, induce a lowered state of awareness. Rabbits, for example, appear to enter a torpid state if placed on their back and held for a short period. An important consideration when applying physical or mechanical restraint is the amount of pressure applied. Animals will struggle if insufficient pressure is applied, while excessive pressure causes pain or discomfort.

Drugs, such as diazepam and ketamine, have revolutionized the practice of animal restraint. Many species of animals that were previously difficult to restrain using other techniques, or requiring a major manipulative procedure, can now be handled with less stress and trauma using chemical restraint. Chemical restraints, used to sedate, immobilize or anaesthetize, can be administered either orally or injected, with immobilization typically occurring after 2–20 min. Chemical restraints can be used to restrain animals briefly or for several hours.

Electroimmobilization, involving a mild electric current passing through the animal's body, has been used to restrain some species by tetanizing their muscles with electricity. However, this procedure has been shown to be aversive to the animal and has no analgesic or anaesthetic effect.

Restraint techniques vary in aversiveness, with most causing some stress to the animal. When using any means of restraint, it is necessary to weigh the benefits of a controlled animal against the distress it causes the animal. (MKP)

Retroduction

Retroductive inference was described by the 19th-century American philosopher and logician C.S. Peirce, who spoke of a kind of inference to explanatory hypotheses that humans share with the problem-solving abilities of many other animals. It differs from concepts of deductive and inductive inference focusing on inferences which take place when humans and other animals respond to anomalies. Deductive inference, argued Peirce, cannot yield new facts or truth (and is inapplicable to models of animal thought processes) and involves a generalization that must reach beyond the supporting evidence of its premises. Animal trainers are familiar with the difficulties animals face when attempting to generalize a sequence of actions. Retroductive inference, however, draws from perceived certainties to the solution of a problem.

For example, a dog displays signs of burning and blistering around the lips and in the mouth, accompanied by abdominal pain, vomiting and diarrhoea. A veterinary surgeon may reason from these clinical signs to a plausible hypothesis regarding the intake of a toxic substance. This hypothesis, if true, would explain the physical signs, and is consequently worthy of further investigation, even if on detailed examination the condition was caused by something other than poisoning. But in another culture, in another period in time, it might be plausible to infer that the dog had been bewitched. This inference is not merely unacceptable to anyone educated in a Western scientific tradition, it would not even arise.

Retroductive inference is selective with regard to the generation of hypotheses, whereby certain wild hypotheses – such as witchcraft – are restricted. On the other hand, some certainties enjoy the protection of the relevant community. Retroduction may involve a search through a hierarchy of hypotheses to find the most plausible, while at the same time limiting the scope of potential solutions to a problem. Accordingly, it is a valuable tool in the economy of research. Peirce spoke of special aptitudes for guessing correctly which ones humans share with other animals. Just as the chicken that successfully pecks its way out of the eggshell does not waste time on other activities, argued Peirce, so a scientist confronting a pressing problem cannot afford to waste time on endless hypotheses.

Although widespread in human thinking, retroductive inference – generating and selecting the most plausible solutions – can be observed in many forms of animal behaviour involving defensive strategies, tool making and predatory and reproductive activities. (DL)

See also: Logic

Further reading

Hartshorne, C., Weiss, P. and Burks, A. (eds) *Collected Papers of Charles Sanders Peirce* (1931–1958), 8 vols. Harvard University Press, Cambridge, Massachusetts.

Rhinoceros

Rhinoceros, along with their relatives the **horses**, are members of the order Perissodactyla. There are five extant species of rhinoceros: black (*Diceros bicornis*), white (*Ceratotherium simum*), Javan (*Rhinoceros sondaicus*), Sumatran (*Dicerorhinus sumatrensis*) and Asian greater one-horned or 'Indian' (*Rhinoceros unicornis*). Two species, the black and the white rhinoceros, are endemic to Africa, with the remaining three residing in Asia. Second to the elephant, rhinoceros are the largest terrestrial mammals and are included in the heterogeneous group of animals referred to as megaherbivores. White rhinos, the largest rhino, can weigh over 2300 kg, while the Sumatran rhino does not exceed 800 kg in weight. The black, white and Sumatran species have two horns, whereas the Javan and Indian have only one. After a gestation period of 15–17 months, females give birth to a single calf, weighing 40–60 kg depending on species. Females generally mature at age 5–6 and males at 7–8 years of age.

Scientific studies are beginning to document the behaviour and **ecology** of the two African species and the Indian rhinoceros, but little is known about the Sumatran and Javan species. Javan rhino are the least numerous and are rarely seen in the dense tropical jungles of Vietnam and Indonesia, where they are found. The primary means of monitoring this species is by installing 'camera traps' to document their presence. By contrast, the use of radiotelemetry to track and study white and black rhino in the more open habitats where they reside has led to a much better understanding and improved conservation management of these species, especially the white rhino.

All rhino species are strictly herbivorous, but possess different foraging strategies reflecting the preferred habitat in which they are found. White rhinos occupy grassland habitat and have wide lips adapted for grazing on grass. In contrast, while sympatric with white rhinos, black rhinos prefer thick, bushy habitat (such as the bushveldt of southern Africa) and have narrow, prehensile lips for browsing on leaves and branches, including acacia trees with 5 cm thorns. The Javan and Sumatran rhinos are also primarily browsers. The Indian rhino is a grazer, feeding on grasses in lush alluvial floodplains. Rarely far from water, the Indian rhino is the most amphibious rhino.

All rhino species are relatively solitary, although some – especially white rhinos – can be found in small groups. Most social groups are short-term aggregations, but adult white rhinoceros females will form stable and enduring associations with subadults or, on occasion, other adult females. However, most rhino **social behaviour** occurs between mother and offspring and short-lived consort relations between males and females. Female rhinos have long inter-birth intervals – generally 2–4 years – and invest heavily in their offspring. Studies show that white rhino mothers in good condition invest more in male than in female calves. This finding is predicted by evolutionary models of **parental investment**. In polygynous species competitive males can produce many more offspring than females, but poor-quality males may have reproductive success much lower than the average female. Thus, mothers in good condition can maximize **fitness** by producing high-quality sons (when resources allow) that will fare better in reproductive competition.

Male–male competition for access to females, whether through territoriality or direct fighting in the presence of an oestrous female, can be fierce. Male white rhinos are highly territorial and maintain well-defined boundaries. They exclude other males, but may allow a submissive bull to share their territory. These beta-bulls typically do not breed and do not display dominant behaviour, such as urine spraying and dung-scraping. Although less well studied, black and Indian males appear to be less territorial. Courtship behaviour can include significant aggression between the male and female, especially among Indian rhinos. Females are not territorial and have substantial home range overlap with other females as well as with males. White rhino females' ranges are large and encompass several male territories. Because of their preference for grassland habitat, males defending territories with more grassland have more opportunities for mating. Females may select mates by remaining on the territory of a preferred male during their fertile period.

As relatively solitary species, the lack of direct contact limits most forms of **communication**. Rhinos use a number of **vocalizations**, mostly during brief aggressive or sexual encounters, but rely on olfactory signals for day-to-day communication. A great deal of information is transmitted in communal dung-heaps, which function like a community bulletin board. Rhinos visit dung-heaps, investigate the dung left by other rhinos and deposit their own dung. From sniffing dung they can determine the sex, age, reproductive condition, **dominance** status and the age of the dung.

All rhinoceros species are dependent on **conservation** measures and all but one are classified as **endangered** or **critically endangered** by the International Union for the Conservation of Nature (IUCN). Only the white rhino has moved up the list to 'near threatened'. Unlike most endangered species, many rhinos still have sufficient habitat to survive and, in some cases, thrive. Poaching is by far the primary threat to rhinoceroses. Their horns are valued in the Middle East, where they are used as dagger handles, and in China, where the horn is believed to have medicinal properties as a fever reducer. These markets create incredible demand for rhino horn, and locals can earn more than a year's income by slaughtering one rhino. Although the northern subspecies of white rhino is almost extinct, the southern white rhino is one of the great conservation success stories, coming back from about 20 individuals 100 years ago to more than 12,000 today. This remarkable turnaround can be attributed to the political willpower in several southern African countries, particularly South Africa, in establishing secure reserves and anti-poaching patrols. The Indian and black rhinos are also possible success stories in the making, both with growing but precarious populations. The Javan and Sumatran rhinos are critically endangered, with numbers so low – and still declining – that they are at immediate risk of **extinction**. The northern subspecies of white rhinoceros numbers only a few individuals and will probably become extinct in the near future.

Followed by anti-poaching measures and habitat protection, **translocation** is the most important conservation measure for white, black and Indian rhino. This strategy entails removals of surplus rhino from reserves with growing populations to repopulate protected areas where rhinos are rare or have been extirpated. This action has been highly effective for southern white rhino, once confined to a single reserve but now widespread throughout southern Africa. In the past decade translocations have figured prominently in expanding the geographic range of black rhino populations. While this programme is successful, the **temperament** of black rhino poses challenges to successful translocation. Stress following translocation can be debilitating, and panicked rhino can run into fences and sustain injuries. In the unstable social environment created by translocation, black rhinos often fight and even kill one another. Behavioural research is addressing ways to mitigate these problems by managing the post-release environment. Releasing many rhino at once can put all rhinos on an equal footing and reduce attacks by residents on newcomers. Dispersal and territory establishment can be managed by 'seeding' an area with rhino dung, as black rhinos prefer to settle adjacent to areas where these olfactory signals indicate prior occupancy by other rhinos (called **conspecific attraction**). Aggression is also mitigated by releasing rhinos into larger reserves at lower rhino densities, giving the rhinos better opportunities to avoid direct encounters with unfamiliar **conspecifics**. A key finding is that black rhino **aggression** is related to lack of familiarity and, although solitary by nature, once social relationships are established aggression is less common.

Four of the five species of rhino are maintained in captivity. More than 75% of captive Sumatran rhinos died before reproducing and only one pair has produced offspring, making the successful establishment of a viable population unlikely. Captive black rhino populations are marginally self-sustaining and appear to be limited mostly by disease and nutrition issues. White rhinos are also marginally self-sustaining, but demographically are headed for a population crash because most breeders are **ageing** females. Captive white rhinos have been the subject of more behavioural research because reproduction is compromised by enigmatic breeding problems. Many of the founding population (F_0 generation), given appropriate husbandry and management, reproduced well, but reproduction among captive-born (F_1) females has been extremely sluggish. Behavioural studies demonstrated that F_1 females show normal signs of behavioral oestrus and reproductive behaviour and that males are just as interested in courting and mating F_1 females. The prevailing hypothesis espoused by managers – that mothers or the older F_0 females behaviourally or physiologically suppress reproduction in younger F_1 females – has been contradicted by behavioural and reproductive data. Research indicates that the problem arises during the development of female calves in captivity, not as adults. Several researchers still actively pursue this intractable problem, now focusing on the developmental period. This research programme provides an example for how using behavioural research to test hypotheses can be directly applied to management of conservation breeding programmes. (RS)

See also: Captive breeding programmes; Conservation behaviour; Endangered species; Mate choice; Reintroduction

Further reading

Emslie, R. and Brooks, M. (1999) *African Rhino Status Survey and Conservation Action Plan*. IUCN, Gland, Switzerland.

Linklater, W.L. and Swaisgood, R.R. (2008) Reserve size, conspecific density, and translocation success for black rhinoceros. *Journal of Wildlife Management* 72, 1059–1068.

Owen-Smith, N. (1988) *Megaherbivores: the Influence of Very Large Body Size on Ecology*. Cambridge University Press, Cambridge, UK.

Swaisgood, R.R., Dickman, D.M. and White, A.M. (2006) A captive population in crisis: testing hypotheses for reproductive failure in captive-born southern white rhinoceros females. *Biological Conservation* 129, 468–476.

Risk factor

A factor that increases the chance within a population of a particular condition occurring. Risk factors may be causal or non-causal (e.g. associated with the occurrence of the condition but do not contribute to its development). Potential risk factors for a particular condition can be identified by comparing the incidence of the factors between similar populations of affected and non-affected individuals. Risk factors can include genetically determined tendencies or (lack of) previous exposure to environmental factors including diet, pathogens or social interactions during key points in the individual's life. (KT)

See also: Causal factor

Ritualization

This is the evolutionary process through which a behaviour becomes highly stereotyped with limited inter-individual variation. Ritualization is usually observed in behaviours having a communicative function and where the **signal** has to be unequivocally interpreted by the receiver. Aggressive and submissive displays are typical examples of ritualized behaviour. (BM)

See also: Emancipation

Further reading

Zahavi, A. (1980) Ritualization and the evolution of movement signals. *Behaviour* 72, 77–80.

Robotic milking

Robotic **milking** is a type of automated milking system that is available for dairy cows. The cows present themselves at a gate when they are ready to be milked, where their individual electronic tags are recorded. The cows are then allowed access to the milking parlour where a robot cleans the udder and attaches the milking cluster, based on individual cow data stored in the computer and using proximity sensors, while the cow is fed a grain ration. There are a number of advantages to this type of system, as the computer records daily milk yield, allows detection of **disease** – such as **mastitis** – and compares these data with past records to determine any variation from an individual's expected values. In addition, the cows are given more control over their behaviour and can choose when and how often they are milked, and there is no need to dock the cows' tails (**see: Docking – tail**) in this system. (LMD)

Rodent

Rodents are the commonest and most diverse of mammals, comprising about 40% of all mammalian species. Their name is derived from the Latin *rodere*, meaning 'to gnaw'. These animals are seen more as pests in most countries in that they compete for the same food as humans (it has been calculated that they eat 42.5 million t of food each year, i.e. 5–10% of total production). They are important vectors of **disease** and it is said that they have killed more humans than all wars put together. On the other hand, they have contributed enormously to our understanding and treatment of various diseases in humans. They are also increasingly kept as **pets**. Rodents are defined by a continually growing (several millimetres a week) pair of incisors on upper and lower jaws, which differentially wear, ensuring a sharp edge to the teeth. Their well-developed head muscles enable them to crack open various seeds to obtain food, as well as to gnaw through relatively hard materials such as wood and even concrete. They tend to be highly social animals, live in large groups with a **hierarchy** and are able to reproduce rapidly and in large numbers (the young develop to sexual maturity early and they have relatively short **gestation** periods). (DBM)

See also: Hamster; Mouse; Rat; Vole

Rollin, Bernard

Bernard Rollin (b. 1943), Professor of Philosophy, Physiology and Bioethics at Colorado State University, USA, has helped to expose and call into question the underlying values in animal and veterinary science, respectively. Rollin has written extensively on animal ethics, animal biotechnology, veterinary ethics, animal agriculture and genetic engineering. His (1995a) contemporary animal *husbandry ethic* attempts to foster a new consensus animal ethics by offering alternative solutions that focus on meeting the evolutionarily inherited adaptations of farmed animals. Rollin's ethic incorporates elements of interspecific equity, human virtues of restraint and diligence and animal **welfare** science. Rollin rejects the 'one-size-fits-all' application of a single moral principle (like the principle of utility) to the animals' issue and opts to rejuvenate a mutually advantageous social contract with animals.

In animal agriculture and biomedical research, we should revisit the tenets of good husbandry practices grounded in an 'interspecific contract' that binds us to certain commonly accepted obligations for resource animals under our charge. The social ethic regarding animals demands that those that benefit from animal production and use, provide animals with proper care and **housing** and management practices, be solidly grounded in an appreciation of the animals' points of view.

Presently, the industrialized paradigm of animal use has left this contract in a state of disequilibrium: lopsided in favour of human interests at the cost of genuine commitment to animal welfare. Producers and agribusinesses, for example, can today employ technologies that undercut practices that once underscored the contract: 'If one's animals did not fare well, then you did not either'. While not intentionally perpetuating cruelty to animals, the aim of efficiency and abundance of cheap food under the industrial paradigm have, through the deployment of technological fixes and layers of agribusiness, masked the tenets of a mutually advantageous relationship. The industrial paradigm has shifted focus from standards and values commensurate with the status of animals as biological and biographical beings to the terminus state of commodities.

Rollin's critique reflects to two major philosophical underpinnings, namely, the notions of **telos** or 'natural' or 'intended' lives, and the principle of welfare conservation. In the first instance, Rollin appropriates the Aristotelian notion of telos as an heuristic device to articulate the content of our duties towards animals. For Aristotle, telos or 'end' captures the 'purpose' and hence, 'good' of the being in question. The telos of a living being, for example, is to live what amounts to a life that is suited for things of that kind. For humans, our species is constituted towards a life suited for rationality and sociability.

A good life for any specific animal is one in which everything is functioning as evolution has adapted it to function. Telos is 'a nature, a function, a set of activities intrinsic to [an individual of a particular species], evolutionarily determined and genetically imprinted'. From our ordinary phenomenal experiences of animals, we know that '[they] have natures – the pigness of the pig, the cowness of the cow, "fish gotta swim, birds gotta fly" – which are essential to their well-being. Common sense tells us that animals that are built to move need to move to feel good; there is no point in trying to prove that they are fine if kept immobile.' Hence, '[t]o promote welfare of animals, we need to raise them in ways that respect their natures' (Rollin, 1995a).

These telic ends impose ethical limits on our treatment of animals and prescribe responsibilities as part of a fair exchange: 'use for care'. The peculiar species constitution of individual animals serves as a window to the functional needs of animals and reflects how animals can be harmed or benefited if their living conditions do not 'jive with their natures'. Animals fare well to the extent that they are exercising their natural capacities unencumbered, and they fail to do so when their lives do not accord with how they are designed.

According to Rollin, under today's industrialized conditions of raising and using animals, animals no longer live in a manner that corresponds to their adaptations and repertoire of natural behaviours. Animals do not have the requisite adaptations to meet the challenges of their human-made abodes. These conditions parallel 'putting round pegs into square holes'. What is needed is a concerted, system-wide effort to reshape housing systems and management practices to accord with certain optimal conditions for which animals are adapted.

The idea of central species defining properties is closely related to the principle of welfare conservation (Rollin, 1995b). This principle reflects his view of the morality of genetically modifying agricultural animals in order to fit their environments. Genetic alteration is wrong if it would worsen an animal's well-being in relation to **conspecifics** that do not receive the treatment or if it would have had a better life without the intervention.

In earlier works, Rollin uses the language of interests and rights. Rollin understands rights less stringently than strong animal rights proponent **Tom Regan** (Rollin, 1989, 1992). Deploying the language of rights (broadly construed) has utility, because as a tactic it affords protections to animals that ought not be subordinated to the general good. Rollin holds that while it may be acceptable to change the natures of animals through genetic means if doing so will on balance promote its welfare, it would be wrong to intervene in the human genome this way. Adding Nussbaum's capabilities view and Thompson's agrarian ideal view are nice departure points to Rollin's telic and husbandry ethic views, respectively. (RA)

See also: Ethics

References and further reading

Rollin, B.E. (1989) *Unheeded Cry: Animal Consciousness, Animal Pain and Science.* Oxford University Press, Oxford, UK.

Rollin, B.E. (1992) *Animal Rights and Human Morality*, revised edn. Prometheus Books, Buffalo, New York.

Rollin, B.E. (1995a) *Farm Animal Welfare: Social, Bioethical and Research I*ssues. Iowa State University Press, Ames, Iowa.

Rollin, B.E. (1995b) *The Frankenstein Syndrome: Ethical and Social Issues in the Genetic Engineering of Animals.* Cambridge University Press, New York.

Romanes, George J.

George J. Romanes (1848–1894) was perhaps the closest academic colleague of **Charles Darwin**, and is now best known for his interest in comparative **psychology**, although he was originally recognized scientifically for his work on the nervous system of invertebrates. He used anecdotal case studies to support his suppositions on the mental ability of non-human animals, and while some argue that there is much **anthropomorphism** in his work, others highlight the rigour with which he investigated cases and his attempt to frame them within an evolutionary context.

It is suggested that Lloyd Morgan formulated his canon (**see: Morgan's Canon**) in direct response to the publications of Romanes. Romanes was not a poor-quality scientist, as suggested by some of his critics, but was critical of Darwinian natural **selection** as originally proposed, highlighting several problems with the concept, and coined the term 'neo-Darwinism' to distinguish his refinements. This term has since been adopted by biologists to describe current views of evolutionary theory, which go beyond the original refinements proposed. Of particular significance was his insight into the importance of selection working at the level of reproductive performance and the necessary reproductive **isolation** required for speciation to occur in many contexts. (DSM)

Further reading

Romanes, G.J. (1881) *Animal Intelligence*. Kegan Paul, Trench and Co., London.

Roosting

At night, many birds **sleep** either standing or, more usually, in a sitting position, perching on tree branches – a behaviour known as roosting. Birds usually seek out places to roost well before dark, even in short winter days, and the cue to go to roost has been shown to be the light dropping below a certain intensity.

Many species of bird have preferred roosting sites within a **home range** to which group members move at night. It is likely that there has been strong **selection** for roosting behaviour, as birds that choose to rest at night in elevated locations run a significantly reduced risk from predators. In addition, by roosting together, birds may benefit from conservation of heat. The formation of large roosting sites in urban areas can pose a serious environmental problem. Starlings sometimes flock into cities, and up to a million birds may assemble at a roosting site, producing huge quantities of droppings. In addition to the **health** and noise problems this may cause, it can result in damage to the structures upon which they roost by the weight of the birds and their droppings, as well as through damage from chemicals in the droppings.

Chicks must learn to roost when young. Junglefowl and feral hens will roost on the ground with their chicks until they are about 6 weeks of age, at which time the hen will begin to take chicks on to lower branches at night for roosting, encouraging them to fly up with a 'roosting call'.

Roosting appears to be a strongly motivated behaviour in domestic fowl. When perches are absent or inaccessible, birds will display signs of unrest, probably due to **frustration** or **exploratory behaviour** as they search for an alternative site to roost. In commercial systems, in which perches are provided, the majority of hens will use perches to roost at night, with most on the highest perches available. It is likely that hens housed in conditions that lack a perch to roost on experience reduced **welfare** due to reduced **bone strength** and inability to avoid pecking by **conspecifics**, as well as frustration to perform the behaviour itself. (RJNM)

Route tracing

Route tracing involves the relatively invariant following of a track within an enclosure. In addition, the repetitive and apparently functionless characteristics of route tracing mark it out as a stereotypy. Recognized in incarcerated humans, route tracing may reflect an individual's **behavioural need** to move within its habitat even though it may have been provided with key resources for survival. Such movement for its own sake may be driven by the need to explore and locate fresh resources. The prevalence of this response in isolated animals is strongly suggestive of the value placed on contact with **conspecifics** and the deleterious effect isolation can have on the welfare of both domestic and **captive animals**, especially social species.

Repetitively pacing animals in enclosures often cause concern among lay people, who most often label this and other locomotory stereotypic behaviours as a sign of **boredom**. The problem with the term boredom lies in its tendency to evoke human mental states that relate to a lack of cognitive challenge. A more conservative approach is to diagnose animals that begin to show stereotypic behaviours as being subject to thwarted **motivation** and then to identify the motivation that is being thwarted. To do so, applied ethologists may add substrates and reinforcers to measure their effect on the tendency to stereotype.

Although it is sometimes seen as a means of filling vacuum periods, route tracing can affect **welfare** by reducing the time available for maintenance behaviours, such as eating and resting. Additional effects on animal welfare relate to asymmetrical wear and tear, e.g. of pads. Blocking the path with obstacles and distracters is virtually fruitless as a management response, but radical changes in the enclosure and the institution of a new **time budget** may disturb the pattern shown and the frequency of stereotypic episodes. There is current debate as to whether the incomplete disappearance of the stereotypy constitutes evidence for so-called **emancipation** (the liberation of stereotypic behaviours from their initiating causes). On one hand it is argued that the resistance of route tracing to change means that the behaviour can become less relevant to the animal's immediate needs than it was when it first emerged. On the other hand, the (albeit incompletely) reduced frequency and intensity of the stereotypy in environments that have been, however transiently, enriched is seen as an indication that its function has become less critical. (PDM)

See also: Stall-walking

R–R interval

The R–R interval, alternatively referred to as the inter-beat interval, normal-to-normal interval or heart period, can be defined as the time difference (in ms) between two successive R waves in a **electrocardiogram** QRS complex. Although any two points in adjacent cycles may be used to calculate the interval, the R–R interval is most commonly reported. (RM-F)

RSPCA and Inspectorate

The Royal Society for the Prevention of Cruelty to Animals, now generally known as the RSPCA, was founded in England in 1824. Its formation followed the development in the early 1800s of the humanist movement, which advocated the protection of certain basic rights for the most vulnerable in society. From this arose the first animal **welfare** legislation, presented to the British House of Commons by Richard Martin MP and passed in 1822. This bill acted 'to prevent the cruel and improper treatment of any horse, cow, ox, heifer, steer, sheep or other cattle in the charge, care or custody of any person or persons'. It applied only to farm animals or beasts of burden. Further attempts were made by Martin to extend this legislation to **dogs**, **cats** and other animals, but these failed. It was some years before **animal cruelty** legislation was extended to cover the majority of vertebrate animals.

The RSPCA began with a meeting on 16 June 1824 in Old Slaughter's Coffee House in London. The meeting sought to form an organization 'instituted for the purpose of preventing cruelty to animals' that would seek both to prosecute offenders under the new legislation, but also to 'alter the moral feelings of the country' towards the welfare of animals. Twenty-two people attended the meeting, including the inaugural Secretary of the society, the Reverend Arthur Broome, Richard Martin MP and William Wilberforce MP, the anti-slavery campaigner. In 1835 the Duchess of Kent and Princess Victoria became patrons of the Society, and Victoria was active in her encouragement of the Society's activities. When Victoria became Queen she permitted the prefix 'Royal' to be assigned to the Society.

The RSPCA movement quickly spread internationally. In 1845, the French Societé Protectrice des Animaux was formed. This was followed in 1866 by the formation in New York of the American Society for the Prevention of Cruelty to Animals. By 1869, the SPCA had also been founded in the states of Massachusetts, Pennsylvania, New Jersey, California, Illinois and Minnesota. Prevention of cruelty to animals laws, similar to those enacted in England, were introduced in each of these states. In 1877, the leading US societies met to form an alliance that became known as the American Humane Association. Similarly, SPCAs were also formed in Canada in 1869, in Australia in 1871 and in New Zealand in 1882. Those Societies based in Commonwealth countries individually gained the right to the use of the term 'Royal' and most are still referred to as 'the RSPCA'.

An integral part of the foundation of the RSPCA was prosecution of prevention of cruelty to animals legislation. At first this involved the formation of a committee inspecting markets, **slaughterhouse**s and working horses in and around London. The Reverend Broome also used his own money to employ two men to carry out such inspections. In 1825, 63 offenders were brought before the courts. By the late 1830s an Inspectorate of five men had been established, based in London.

Then, as now, the Inspectorate worked both to prosecute animal welfare legislation and to educate the community about the laws regarding animal welfare. The UK RSPCA is the most well-resourced Society of the movement, with more than 320 RSPCA inspectors operating across England and Wales. The legal powers of RSPCA Inspectors vary, as they are dependent on the enabling legislation of the state, but they can include the right to enter property and seize animals where animal welfare has been compromised to the point of breaking the law. These rights can also be seen as limitations in terms of the possible actions of Inspectors. While RSPCA policies may not support practices such as the keeping of hens in battery cages, Inspectors cannot act to prevent these practices where they are legally permitted. Inspectors also perform a valuable role in educating individuals about the care and protection of animals and promoting the policies of the RSPCA.

The main objectives of the RSPCA today are fundamentally the same across the world and have changed little from the original intent of its founders: (i) to prevent cruelty to animals by enforcing existing legislation; (ii) to work towards improving such legislation for the protection of animals; (iii) to educate the community about the humane treatment of animals; and (iv) to encourage and sustain public debate on animal welfare.

The RSPCA works to improve animal welfare across the whole spectrum of human–animal interactions, including **farmed animals**, **companion animals**, animals used for sport or entertainment, **wild animals** and animals kept for research purposes. As well as its law enforcement role, the RSPCA's activities include the practical application of animal welfare, such as running **animal shelters**, clinics, hospitals and specialist wildlife centres, as well as high-profile campaigning, government lobbying and education. Market research consistently indicates that the name 'RSPCA' has an extremely high level of recognition and is well respected by the general public. While there is no formal affiliation between RSPCAs internationally, most have established connections and associations within their region and internationally to promote animal welfare and to work cooperatively on international issues.

In practice the RSPCA's work is limited to protecting the welfare of *sentient* animals. In most cases this will reflect the definition of animals contained in the organization's enabling legislation, which is generally restricted to vertebrates. More recently the RSPCA has argued for the inclusion of other species (e.g. cephalopods) where the level of neural development is such that pain perception has been demonstrated, or is likely to occur.

The original philosophy of the RSPCA movement reflected that of the first animal welfare legislation: that unnecessary cruelty to animals was unacceptable. The Society sought to improve the treatment of animals without challenging the inherent right of humans to use animals as sources of food, clothing, labour or entertainment. Since its foundation, the RSPCA and the general animal welfare movement that it inspired has continued to expand, improve and enforce animal welfare legislation around the world.

The RSPCA and like-minded animal welfare organizations continue to accept some use of animals by humans, provided that it is justified and humane. In contrast, many **animal rights** organizations advocate an end to the use of animals by humans altogether. Acceptance of the use of animals by humans is at the core of the difference between the RSPCA and the animal rights movement. This is made clear in their respective positions on the killing of animals for food. Many animal rights groups have a stated policy opposing the killing of animals for food. The RSPCA movement does not oppose this but has worked progressively to improve standards for farm animals. In recent years this has extended to the establishment of **farm assurance schemes** that guarantee a high standard for the welfare of animals on accredited farms.

(BJ)

See also: Five freedoms; Humane societies; Humane Society of the United States

Further reading

Brown, A. (1974) *Who Cares for Animals? 150 Years of the RSPCA.* William Heinemann, London.

Budd, W.B. (1988) *Hear the Other Side. A History of the Royal Society for the Prevention of Cruelty to Animals in South Australia 1875–1988.* Investigator Press Pty Ltd, Hawthorndene, South Australia, 237 pp.

Rumination

A challenge facing many species of herbivorous mammal is how to access the nutrients contained within the cellulose cell walls of plant cells in the absence of the ability to produce a cellulose enzyme. Rumination (colloquially referred to as chewing the cud) is the process used to maximize the digestibility of plant material by ruminants (i.e. animals with a multi-chambered stomach including a rumen). This includes species such as the cow, **sheep**, **goat**, **deer** and **camelids**. Typically these animals graze areas quite rapidly (**see: Feeding**; **Foraging behaviour**; **Grazing**) when the opportunity is presented, with herbage initially entering the large rumen. Later this material is repeatedly regurgitated, masticated and swallowed. As the material is broken down into smaller parts, and cell walls fragmented as a consequence, it is able more readily to leave the rumen and pass down the digestive tract for digestion to occur. Thus ruminants do not need to eat as frequently as some other herbivorous species such as **horses** (referred to as trickle feeders), and their feeding behaviour is interspersed with prolonged periods during which ingestion does not occur, but chewing behaviour is prevalent.

While rumination allows the release of nutrients high up in the digestive tract and so is very efficient when feeding on higher-quality forage, when faced with low-quality forage the rumen may become full of material of little value and the extraction of sufficient nutrients may be more difficult than in a trickle feeder. Also, if the gas associated with fermentation in the rumen is not released efficiently then the animal may develop the potentially fatal condition **bloat**. (DSM)

Running wheel

Voluntary wheel running is one of the most widely reported behaviours performed by captive animals (Sherwin, 1998). In addition to pet and laboratory **rodents**, a wide range of animals will use running wheels, including hens, foxes, bobcats, domestic **cats**, **macaques**, **rabbits**, **ferrets**, dormice, flying squirrels, ground squirrels, bandicoots and Tasmanian devils. Wheel running can be a prodigious activity. In 24 h, the distance run can be as great as 43 km for **rats**, 31 km for wild mice, 19 km for lemmings, 17 km for the red fox and 16 km for laboratory mice.

It is a highly motivated, plastic behaviour, influenced by a wide variety of internal and external causal factors. It occurs spontaneously, usually within a few minutes or hours of access to a wheel becoming available. Some animals show preferences for square or triangular wheels (it has been calculated that mice running in square wheels jump at the corners at a rate of 15/s!), or wheels containing hurdles. Running wheels are generally preferred to other apparatus providing the opportunity for extended locomotion or exercise, e.g. treadmills.

There are several hypotheses regarding the function and causality of wheel running. One is that caged animals run in wheels as a means of regulating energy balance (Collier and Leshner, 1967). Another is that wheel running is motivated by information gathering, searching and exploration (Mather, 1981). A third suggests that wheel running is a self-reinforcing '**abnormal**' behaviour that is a positive experience for the animal caused by feedback dysfunction, rather like humans enjoying a roller-coaster ride (Sherwin, 1998). It is sometimes suggested that wheel running is a stereotypy; however, the immediacy with which it occurs, the plasticity of its performance, its demonstrably high **motivation**, its performance in complex environments and its various purported functions all indicate that it is not a stereotypy – at least not according to widely accepted definitions. (CS)

Further reading and references

Collier, G. and Leshner, A.I. (1967) An invariant in mouse running wheel behavior. *Psychonomic Science* 8, 9–10.

Mather, J.G. (1981) Wheel-running activity: a new interpretation. *Mammal Review* 11, 41–51.

Sherwin, C.M. (1998) Voluntary wheel-running: a review and novel interpretation. *Animal Behaviour* 56, 11–27.

Runway test

This is a test used to measure the **motivation** to approach a reinforcing **stimulus** (**see: Reinforcement**). The stimulus is usually a distinctively flavoured food presented to a food-deprived animal, usually a **rat** or **mouse**; other reinforcers may be presented, such as: water to a thirsty rat; access to intravenous injections of drugs of abuse; or access to a sexually receptive partner. The procedure has also been used to measure motivated behaviour in **pigs** and fowl. The apparatus typically comprises a start box, straight-alley runway and goal box. The animal is confined to the start box before the trial begins. At the beginning of a trial, the door of the start box is opened and the animal is free to move down the pathway into the goal box. The dependent measure is running time, which is the time from leaving the start box to entering the goal box. The acquisition of runway responding can be used to measure learning of an association between the reinforcing stimulus and the goal box (**see: Conditioning – types of**). This paradigm can also be used to measure aversive responding because reductions in reinforcement as well as omission of reinforcers have been shown to be aversive (**see: Aversion learning**). (SuH)

Further reading

Ettenberg, A. and Geist, T.D. (1993) Qualitative and quantitative differences in the operant runway behavior of rats working for cocaine and heroin reinforcement. *Pharmacology, Biochemistry and Behavior* 44, 191–198.

McFarland, K. and Ettenberg, A. (1998) Haloperidol does not affect motivational processes in an operant runway model of food-seeking behavior. *Behavioural Neuroscience* 112, 630–635.

Russell and Burch

Bill Russell and Rex Burch were commissioned in the mid-1950s by **UFAW** (the Universities Federation for Animal Welfare, Sir Peter Medawar) to write a book on minimizing animal **suffering** during experiments. It was eventually finished in 1959 under the title *The Principles of Humane Experimental Technique*. Within this text the concept of the **three Rs principle** (replacement, reduction and refinement) of laboratory practice was described, which has now been widely supported as an ethical obligation of research scientists. The three Rs also form a fundamental ethical framework and principle in legislation regulating research. (DBM)

Further reading

Russell, W.M.S. and Burch, R.L. (1959) *The Principles of Humane Experimental Technique*. Methuen, London.

Ryder, Richard D.

Richard D. Ryder (b. 1940) is a British psychologist, writer and former chairman of the **RSPCA**.

The pejorative term '**speciesism**', which is employed by analogy with 'racism' and 'sexism' to deplore human beliefs and practices presupposing that *Homo sapiens* is morally superior to other animal species, is today associated with the philosopher **Peter Singer**. In fact, however, it was coined by Ryder, who, in his book *Victims of Science* (1975), says that he uses the word 'to describe the widespread discrimination that is practised by man against other species [...]. Speciesism and racism both overlook or underestimate the similarities between the discriminator and those discriminated against.'

It was in the early 1970s that Ryder became concerned about animal use in experimentation (he had himself undertaken animal experiments previously). At this time he contributed to the influential *Animals, Men and Morals: an Inquiry into the Maltreatment of Non-humans* (1972). It was in a widely noticed review of the former that Singer set out the anti-speciesist arguments he would later present in more detail in *Animal Liberation* (1975).

Despite their shared denunciation of speciesism, Singer and Ryder hold radically different views. Since 1985 Ryder has advocated a theory he calls 'painism'. In painism the central utilitarian idea that harms and benefits falling to several individuals can be added together, and that the morally right thing to do is whatever will bring about the greatest net benefit, is rejected. Troubled by the idea that large enough collective benefits can justify individual sacrifice, Ryder ascribes something like moral rights to each individual, human or animal. He also urges us to care especially about the plight of the 'maximum sufferer'. But, although he defends a prohibition on merely using others, including animals, for our own purposes, he sees his own position as one that steers a course between **utilitarianism** and the **animal rights** view (**see: Ethics**). (PAR, PS)

References

Godlovitch, S., Godlovitch, R. and Harris, J. (1972) *Animals, Men and Morals: an Enquiry into the Maltreatment of Non-humans.* Taplinger Publishing Co., New York.

Ryder, R.D. (1975) *Victims of Science: the Use of Animals in Research.* Davis-Poynter, London.

Singer, P. (1975) *Animal Liberation.* New York Review/Random House, New York.

S

Sacred animals

Sacred animals are those offered some degree of veneration, respect or reverence. In many **cultures**, individual species or groups of animals may be given special 'revered' or 'consecrated' status, and this may affect not only the lives of the animals themselves, but also the lives of the people who interact with them, and the condition of the environment in which they live. Venerated status often derives from longstanding religious dictats; below are some examples of the religious views that different cultural groups have of animals:

- Buddhism: One of the five precepts of the *Dhammapada* – 'he who destroys life – digs up the very roots of his life'.
- Hindu verse: He who kills a cow will 'rot in hell for as many years as there are hairs on the cow'.
- Islam: 'There is not an animal on earth, nor a flying creature on two wings, but they are people like unto you.'
- St Francis of Assissi: 'Not to hurt our humble brethren is our first duty to them, but to stop there is not enough. We have a higher mission – to be of service to them whenever they require it.'

These dictats express deep intentions to respect and protect animals, but these aspirations are not necessarily achieved in practice. For example, to illustrate the **welfare** issues that veneration can cause, in Bali and Thailand long-tailed **macaques** are treated as sacred (*druwe*). In villages such as Ubud, Bukit Sari and Karangasem in Bali, thousands of macaques collect to be fed. These animals sometimes come into abrasive contact with man through damage to buildings and vehicles, by stealing and begging for food. Conversely, many **monkeys** are injured by vehicles, in fights with the dog population and as a result of fighting in competitive **groups** of monkeys in artificially tight groupings caused by the ready food source and lack of **predation** or control (**see: Social stress**).

These are not 'natural' behaviours, and the animals may become reliant on human supplies of food and lose natural foraging abilities. Wild monkeys from the forest of the Gumang Mountains, Bali, join these 'temple groups' to receive food from tourists – is their religious status adversely altering wild behaviours? If so, does this matter? These animals are very adaptable, and are simply using a new resource provided by man. Studies in American brown **bear** have shown that removal of feeding by tourists (and careful control of rubbish) eventually led the bears to leave tourist locations and to return to natural foraging – this reduced both the number of road-killed bears and the number of dangerous interactions between the bears and man, and led to a more sustainable bear population. Local feeding of macaques may lead to unsustainable populations whose long-term welfare is threatened by complete dependence on human food and protection from predation.

In India, there are about 500,000 rhesus monkeys, of which nearly 60% live among humans. The monkeys have a consecrated status in Hinduism, where they are viewed as gods. Originating from the temples where they were sacred, monkeys have dispersed among the human population in a number of cities, from India to Indonesia. The rhesus monkeys, while sacred, are viewed by many as a problem because they move freely through and around public buildings, and are injured by cars, electrocuted in the tangles of electrical wiring on telegraph poles, defecate and urinate in buildings, may act as vectors for **disease** – including rabies – and fight with the local dog population.

In Hinduism, the cow is venerated. Verses of the *Rgveda* refer to the cow as Devi (goddess), identified with Aditi (mother of the gods) herself. Milk, curd, butter, **urine** and dung are used in rites of purification and penance. Cows are used to pull ploughs in the arid areas of India, but usually only **buffalo** or bullocks are used as **draught animals**. While protecting the cow from **slaughter** (except at the end of its 'natural' life), this veneration does not prevent some cattle from being subject to very hard living conditions on the street, or in very arid areas where feed is very poor – **starvation**, disease and neglect are still potential welfare threats to these animals, despite their being given religious status. The slaughter of **milking** cows is legal in only a few Indian states, so cattle may have to travel long distances by road for slaughter, and India exports large amounts of leather derived from cattle despite the cows' sacred position. **Transport** of these cattle from state to state subjects them to journeys that would be 'unnecessary' if slaughter was permitted locally – and it is the (sometime illogical) imposition of requirements such as these that, on a daily basis, affect the welfare of animals and do not provide any apparent immediate 'benefit' to man. (AB)

Sacrifice

Sacrifice is the offering of something precious to a deity and, through history, living animals have been considered as 'precious' and suitable as offerings. Animals are universally killed for food, but the killing of animals as an 'offering' raises the possibility that lack of 'control' of the killing process, often carried out by priests or preachers, or by the public, can result in unusually long, or even artificially protracted, 'time to **death**', and the potential for significant suffering during the process of killing. Some examples of sacrifice still common in parts of the world include santeria, or *La Regla Lucumi*, which originated in West Africa, and is practised in many countries, but most commonly in Cuba, Haiti, Brazil, Trinidad, Puerto Rico and the USA. Sacrifices or offerings (*ebó*) of **chickens**, **goats**, doves and pigeons are used to mark significant events – the initiation of priests, births, deaths and marriages. The animals are killed by decapitation, or by severing the blood vessels in the neck. This method of killing is very poorly controlled in comparison with the requirements for slaughter of farmed livestock.

In many parts of the world, sacrifices of **sheep**, **goats**, **buffalo**, **cattle** and birds are carried out to seek blessing for festivals, to atone gods or to mark visits by important guests. During the *Eid-El Kabir* festival, Muslims kill sheep and goats by public neck cutting. In countries where this is unlawful, local legislators have experienced difficulty in enforcing **slaughter** legislation in the face of claims that this interferes with religious rights. In Indian villages, animals are sacrificed to the gods Kali, Durga and Rudra to atone for sin. However, even within countries, different groups, religions and geographical areas have very different views on ritual slaughter, and this poses real difficulties for legislators who may wish to protect animals via national law, or via local orders, but find that inconsistencies from area to area make this complex.

The term 'sacrifice' may also be used within the context of killing during the process of **religious slaughter** for meat, and also as a euphemism for the killing of a subject as part of an experimental procedure. (AB)

Further reading

O'Brien, D.M. (2004) *Animal Sacrifice and Religious Freedom: Church of the Lukumi Babalu Aye vs. City of Hialeah.* University of Kansas Press, Lawrence, Kansas.

Scully, M. (2003) *Dominion: the Power of Man, the Suffering of Animals and the Call to Mercy.* St Martins Griffin Publishers, New York.

Scent marking

Scent marking describes the release of chemical substances (e.g. **pheromones**) by an animal, for example by rubbing a specific gland on a surface, which acts as a **signal** for other animals. Scent marking is often used to signal **territory** ownership or to attract potential mating partners. (BM)

Further reading

Hurst, J.L. and Beynon, R.J. (2004) Scent wars: the chemobiology of competitive signalling in mice. *Bioessays* 26, 1288–1298.

Zala, S.M., Potts, W.K. and Penn, D.J. (2004) Scent-marking displays provide honest signals of health and infection. *Behavioural Ecology* 15, 338–344.

Seasonality

Seasonality refers to fluctuations in growth, development and reproduction seen with respect to changing seasons. **Photoperiod** – i.e. the relative lengths of day and night – can reliably regulate seasonality because, unlike other 'seasonal' attributes such as temperature, rainfall and predominant weather patterns, photoperiod is consistent in its changing pattern over the course of a year.

From an evolutionary standpoint, it is therefore unsurprising that animals have used the predictability of changing day length to prepare for and optimize the timing of **reproduction**, migration and dormancy. The annual variation in day length between winter and summer will depend on latitude. Near the equator, variation is minimal; at the poles, variation is maximal. Within the polar circles at midsummer, there are 24 h of daylight and, at midwinter, there are 0 h of daylight.

A wide variety of animals from diverse taxa (for example, annelids, molluscs, arthropods, bony fish, frogs, turtles, lizards, birds and mammals) use the day length as an anticipatory cue to make seasonal preparations. From a biological **fitness** viewpoint, it is essential to reproduce at a time and place when food and environmental conditions will be best suited to successfully rearing the offspring.

Animals therefore may need to migrate, establish territory, build up fat reserves, moult, etc. to ensure fitness. The major controlling pathway of seasonality involves the pineal gland and its release of melatonin. Melatonin is released maximally during darkness and, as daylight increases, light **signals** received by the retina and transmitted to the pineal gland via the suprachiasmatic nucleus suppress melatonin synthesis. Melatonin then acts on the release of **gonadotropin-releasing hormone (GnRH)** and subsequently affects **luteinizing hormone (LH)** and **follicle-stimulating hormone (FSH)** release, thereby exerting effects on reproductive function.

From an applied animal behaviour aspect, domestication of farm animal species, in particular, has sought to try to minimize the effects of seasonality. In the more intensive commercial livestock industries, the producer is aiming to control nutrition, photoperiod and thermal environment, so that there is little variation for the animal over the course of its life (**see also: Intensification of animal production**). For a breeding herd, the producer is trying to maintain continuous, predictable output in terms of number of offspring produced and subsequent growth rate. Eradication of seasonality has perhaps been best achieved in **pigs** and **cattle**. Feral pigs and cattle exhibit specific breeding seasons, with the aim of producing offspring from March through to June. Under commercial conditions, this seasonal breeding has all but disappeared, although there is evidence of reduced fertility if environmental conditions around the time of conception are suboptimal. **Sheep**, **goats** and **horses** continue to be seasonal breeders, but are most likely to be maintained in extensive farming systems. (JNM-F)

Further reading

Chemineau, P., Guillaume, D., Miguad, M., Thiery, J.C., Pellicer-Rubio, M.T. and Malpaux, B. (2008) Seasonality of reproduction in mammals: intimate regulatory mechanisms and practical implications. *Reproduction of Domestic Animals* 43 (Suppl. 2), 40–47.

Secure base hypothesis

The secure base hypothesis is derived from a development by Mary Ainsworth of John Bowlby's **attachment** theory. Attachment theory proposes that the young of many species have a need for a reliable psychological relationship with a caregiver, who provides support and acts as a psychological buffer at times of **social stress**, and that this relationship has important influences on the future development of the young individual. It is suggested that the form of the attachment biases the emotional response and expectation of future close social relationships, and that this may account for a range of social disorders. Ainsworth developed a test, commonly referred to as the Ainsworth Strange Situation Test (ASST), which sought experimentally to describe different types of attachment in young infants by subjecting them to a range of mild social **stressors**. The secure base hypothesis specifically proposes that the attachment characterized by a healthy 'secure base' facilitates exploration and response to challenges within the surrounding environment with more confidence. Some suggest that the 'secure base' also facilitates the development of a wider range of social cognitive abilities and skills.

The ASST has since been adapted for use in animals to evaluate the utility of the concept of attachment to describe the relationship between a **pet** and its owner. **Dog**s and **cat**s appear to show some evidence of attachment behaviour in modifications of the test, but this is less evident in hand-reared wolves, suggesting that the capacity for attachment to humans may, in this species, be a product of **domestication**. (DSM)

Further reading

Palmer, R. and Custance, D. (2008) A counterbalanced version of Ainsworth's Strange Situation Procedure reveals secure-base effects in dog–human relationships. *Applied Animal Behaviour Science* 109, 306–319.

Topal, J., Miklosi, A., Csanyi, V. and Doka, A. (1998) Attachment behaviour in dogs: a new application of Ainsworth's Strange Situation Test. *Journal of Comparative Psychology* 112, 219–229.

Selection

By **Charles Darwin**'s (1859) own description, two pieces of information led him to the theory of natural selection:

1. The effects of selective breeding (artificial selection). Animal and plant breeders, even before Darwin's time, were aware that offspring tended to resemble their parents. Selecting only the parental stock with the best characteristics to breed in each generation led to steady improvements: maize with larger ears, fish with fancier fins, cows that produced more milk, and so on (**see also: Breeding**).

2. The 'struggle for existence'. This was a concept Darwin picked up from Thomas R. Malthus' (1798) *An Essay on the Principle of Population*. Malthus presented calculations showing that all species produce far more offspring than can possibly survive and reproduce. Individual females of some insect species, for example, produce hundreds of eggs in their lifetime. Obviously not all of the eggs of all of the females live to become egg-laying adults. Most species are more or less in a state of equilibrium. This means that, on average, each parent (in a sexual species) produces two surviving offspring. Malthus thought that population numbers were controlled primarily by competition for **resources**. (He predicted that the exponential population growth of humans would eventually be checked by starvation.)

Darwin's crucial insight was that, because not all organisms can live to reproduce, any trait that gives the individual bearing it an edge in the struggle to survive and reproduce would be more likely to be transmitted to future generations. This is natural selection. In artificial selection, the breeder, wittingly or unwittingly, allows only animals with particular traits to breed. Natural selection has the same effect, but through natural variation in survival and **reproduction**.

Selection in general can be defined as: 'nonrandom differential survival or reproduction of classes of phenotypically different entities' (Futuyma, 1998). This definition has several facets: 'differential survival or reproduction' means that not all individuals leave the same number of descendants; 'nonrandom' means there is some reason that some individuals do better than others; 'phenotypically different entities' means that there are measurable differences between individuals (the **phenotype** includes all characteristics of organisms except their genes). The reason this definition refers to 'entities' rather than individuals is that, theoretically, selection can also operate either below the level of the individual (at the level of genes) or above (at the level of groups, species or higher taxa). Selection is artificial when humans do the selecting; otherwise it is natural. Note that human activities also cause natural selection: selection for bacteria resistant to antibiotics is a good example. Selection leads to evolutionary change – that is, change across generations – only when the phenotypic differences are at least partly genetic; otherwise, it affects only the distribution of traits in the parental generation (**see also: Evolution**).

The classic example of natural selection in action: industrial melanism

Two centuries ago, before the Industrial Revolution, pale lichen covered the trunks of trees throughout the UK. Many insect species rested on these lichens during the day, protected by cryptic colour patterns closely resembling the colour of the lichens. (Cryptic means camouflaged, i.e. the opposite of conspicuous.) But lichens are sensitive to pollution, and were unable to survive on tree trunks near major industrial areas. In the late 1840s, the frequency of very dark (melanic) individuals began to increase in moth populations near big cities, where soot had killed the lichen and turned the tree trunks black. One species, the peppered moth (*Biston betularia*), went from about 1% melanic to over 90% melanic in less than 50 years.

In this species, the colour difference between light and dark morphs is controlled by a single gene. What caused the increase in dark forms probably was not the pollution itself but the effect that the colour change of the trees had on the crypsis of the moths. By pinning dead moths to trees, researchers showed that light moths were more likely to be eaten by birds when against dark backgrounds, and dark moths were more likely to be eaten against light backgrounds. They then released large numbers of marked moths in both polluted and unpolluted areas, and showed that light moths had higher survival in unpolluted areas and dark ones had higher survival in polluted areas. People watching from blinds observed that birds were more likely to eat moths that perched on the 'wrong' background, i.e. light moths that perched on dark trunks and dark moths that perched on light trunks. Today, in some parts of the UK, pollution has decreased, the lichens have made a comeback and so have the light-coloured moths.

This classic example illustrates several things. First, the researchers clearly detected selection – non-random differential survival of phenotypically different entities – and the colour trait is known to be genetically heritable. So this satisfies all the criteria for **evolution** by natural selection. We will never know with absolute certainty, however, whether differential predation by birds caused the increased frequency of melanic morphs in the late 1800s. Second, because peppered moths have a 1-year life cycle, the dramatic change in morph frequency must have occurred in fewer than 50 generations. This is rapid for evolution by natural selection – it would be nothing, however, for artificial selection. By selectively breeding only dark moths, we could produce the same effect in a few generations. Finally, although the colour change was dramatic, on an evolutionary scale the change was quite minor. It did not result, for example, in the origin of new species. (GFG)

Reference and further reading

Cook, L.M. (2000) Changing views on melanic moths. *Biological Journal of the Linnean Society* 69, 431–441.

Endler, J.A. (1986) *Natural Selection in the Wild*. Princeton University Press, Princeton, New Jersey.

Futuyma, D.J. (1998) *Evolutionary Biology*, 3rd edn. Sinauer, Sunderland, Massachusetts.

Self-awareness and self-recognition

The study of self-awareness and self-recognition in animals is part of the wider field of animal **cognition**. A topic with many ramifications, it is of considerable interest in developmental, comparative and evolutionary **psychology**, cognitive science, the philosophy of mind, animal **welfare** and animal **ethics**. It is also inextricably tied up with the highly complex question of whether higher animals can recognize other animals as having minds, since self-awareness and other-awareness seem to be two sides of a single coin.

Following **Charles Darwin**'s publication of *The Descent of Man* (1871), the idea that there is psychological continuity between animals and humans became a routine background assumption in much research on mentation in non-human species. Clearly, the assumption was supported by Darwinism. An argument from analogy also lent it credibility: where animals displayed behaviour in response to stimuli in ways similar to the way we do, it was reasonable to assume that they, like us, experienced mediating mental states.

However, an animal with self-awareness would not merely be aware of something that *happens to be* itself, as is the case when a cat cleans its paws. It would be aware of itself *as* itself. Self-awareness seems, therefore, to represent a step-change in psychological evolution.

Self-recognition – often equated with the ability to recognize oneself in a mirror (mirror self-recognition, MSR) – is a relatively primitive aspect of self-awareness. It emerges in humans around 18–24 months, with more abstract forms of self-awareness involving introspection and mental state attribution (**see: Theory of mind**) developing later. Even so, MSR is a highly sophisticated cognitive capacity and seems to be extremely rare in the animal kingdom: other than in humans it has been identified in our closest kin, the **great apes**; and lately it has been identified in a number of other species, including some dolphins and European magpies. This more recent discovery is of great interest, because dolphins, birds and primates differ so profoundly in their **brain** organization and evolutionary lineage. (PAR)

Further reading

De Waal, F.B.M. (2008) The thief in the mirror. *PLoS Biology* 6, e201 doi:10.1371/journal.pbio.0060201, available from: http://biology.plosjournals.org/perlserv/?request=get-document&doi=10.1371/journal.pbio.0060201&ct=1 (accessed 2 December 2009).

Self-consciousness

Self-consciousness is the awareness of oneself as distinct from one's environment (**see: Self-awareness and self-recognition**). Sometimes, particularly in philosophical contexts, the term 'self-consciousness' is used to refer to various forms of awareness of oneself that are supposed to accompany conscious thought, sensation and other forms of experience.

In **René Descartes**' philosophy, it is argued that every conscious experience is connected with an infallible **knowledge** of oneself as an experiencing subject. Later philosophers have mainly been critical of Descartes' notion of self-consciousness. While some, like the Scottish philosopher David Hume, have wanted to jettison self-consciousness altogether, reducing it to a bundle of normal, object-related experiences, other philosophers, notably from the so-called phenomenological tradition in philosophy, have argued that a pre-reflective awareness of oneself is a precondition of any experience. (PS)

Self-medication

The selection and consumption of substances that have little nutritional value but are either pharmacologically active or act in some form to improve symptoms of ill **health**.

There are numerous anecdotal reports of self-medication from the natural world; **elephants**, for example, and other herbivores will ingest soil (**geophagia**), which assists digestion by helping to absorb toxic, plant-derived alkaloid compounds. Similarly, laboratory **rats** will eat clay in response to gastro-intestinal malaise. Detailed observational studies have also shown a correlation between **parasite** infestation and ingestion of certain plants. For example, the plant *Vernonia amygdalina* is not routinely ingested by **chimpanzees** despite it being widely available within the local habitat. However, individuals have been observed to ingest the bitter pith of the plant when showing signs of parasite infestation and after consumption symptoms have reduced. Worms have been noted with the leaves, suggesting that either the plant acts to enhance mechanical expulsion in some way or contains anti-parasitic compounds.

Whether animals actually actively seek out and consume certain foods when sick still remains to be resolved. However, when provided with substances with medicinal properties or drugs and the opportunity to learn about their effects, there is compelling support for the hypothesis that animals can learn to self-administer medication to alleviate **pain** or other symptoms of ill health. Under controlled conditions, one study found that sound and lame broiler **chickens** were trained to discriminate between two differently coloured feeds, one of which contained carprofen (a non-steroidal anti-inflammatory drug with analgesic properties). The lame birds ingested more of the medicated feed than the sound birds, and the amount of carprofen consumed increased with the severity of **lameness**, suggesting that these birds had learned to self-medicate to reduce the pain of lameness. (SR)

See also: Nutritional wisdom

Further reading

Danbury, T.C., Weeks, C.A., Chambers, J.P., Waterman-Pearson, A.E. and Kestin, S.C. (2000) Self-selection of the analgesic drug carprofen by lame broiler chickens. *Veterinary Record* 146, 307–311.

Villalba, J.J. and Provenza, F.D. (2007) Self-medication and homeostatic behaviour in herbivores: learning about the benefits of nature's pharmacy. *Animal* 1, 1360–1370.

Self-mutilation

Self-mutilation is apparently directed, repetitive behaviour resulting in self-induced **injury**, often referred to in the literature as a stereotypy, an obsessive–compulsive behaviour, or simply a compulsive behaviour. Some form of self-mutilation has been described in many domestic species, ranging from excessive self-barbering in laboratory **rodents**, incessant licking or biting in **dogs**, self-directed biting or kicking in **horses** and **feather plucking** in birds. The behaviour and resulting injury not only affect the **health** and behavioural **well-being** of the animal, but can be detrimental to the **human–animal bond**.

Self-mutilation may be traced to an original physical or environmental **aetiology**. Often, however, the behaviour continues long beyond, and independent of, the inciting **stimulus**. The neurochemical basis for development or maintenance of self-mutilation has been traced to the dopaminergic, serotonergic and opiate neurotransmitter systems in various species studied, resulting in an associated pharmacologic component to the overall therapeutic plan.

Horses

In horses, self-mutilation has been observed in association with **pain** emanating from the gastrointestinal tract, urogenital tract or other anatomical site. Removal of the primary source of pain usually eliminates the self-mutilation behaviour. It is most commonly seen, however, in intact male horses, both intensively managed and in rather extensive turnout situations, so clearly not the product of any specific confinement situation. These horses may bite themselves in the chest or pectoral area, forelimbs, thorax, flanks and tail or may strike or kick at walls or other solid objects, resulting in injury to their fore or hind limbs. This form of self-mutilation has been described as self-directed inter-male **aggression** (Dodman *et al.*, 1994). That is, the behaviours involved appear to be similar to those observed in natural inter-male agonistic encounters. Exposure to the sight or scent of other stallions (such as at turnout, in a stall or trailer, or associated with a multi-user faecal pile in a pasture or lot) or exposure to mares might trigger an episode of self-mutilation. In other cases it might be noted that the stallion or gelding has been thwarted in normal **sexual behaviour** opportunities.

Provision of environmental changes to reduce exposure to triggering stimuli or to establish more frequent normal sexual behaviour opportunities may reduce the behaviour, but rarely abolishes it. **Castration** is a consideration, though it may not be a preferred option for the owner or manager due to economics; castration is not always reliable in reducing or eliminating the behaviour. Attempting to simply reduce the horse's ability to reach the affected areas on its body, such as placing a neck cradle, may result in the horse redirecting the behaviour towards injuring another part of the body, or incurring secondary injuries related to the device or barrier itself. Single-subject reports have cited the success of management changes such as decreased concentrate and increased roughage diets, exercise and provision of a **goat** as a companion, and a pharmacologic approach via the use of a specific **opioid** antagonist.

Psittacines

Companion psittacine birds are prone to the self-mutilation behaviour known as feather plucking. In this behaviour, birds will pick and remove all accessible **feathers** and may then begin to self-injure their skin, resulting in chronic wounds. The increased incidence of feather plucking in certain species suggests a possible genetic basis. There are various pain-related, dietary and systemic medical aetiologies. Commonly, though, the management and care of the caged bird results in social deprivation, **social group** (in which humans may be included) disorder or lack of other appropriate environmental stimuli failing to meet the bird's behavioural needs (Jenkins, 2001).

Treatment for feather plucking includes managerial changes such as providing appropriate environmental **enrichment** devices, activities and behaviour modification techniques to affect the bird's response to stressful situations. Pharmacologic therapy and physical means to prevent the behaviour are largely ineffective without appropriate management changes.

Canids

In domesticated dogs, self-mutilation is often presented as lick granulomas or acral lick dermatitis, but many other forms of self-injurious biting and scratching are also seen. Some of these behaviours seem to be breed or group specific. As in other species, the aetiology may be found in an inciting pain or allergic stimulus, such as in the case of a dog that begins to lick at an allergic dermatitis or surgical wound, and the behaviour continues beyond physical resolution of the problem or on to unaffected areas of the body. In other cases the inciting stimulus is conflict, **frustration** or **fear** associated with **confinement**, **housing**, **restraint**, inconsistency in management or **handling** or disorders of social interactions with other dogs or humans (Luescher, 2003). Owners may inadvertantly reinforce the behaviour, and thus it may become a conditioned response. Attempted distraction may interrupt or aggravate the performance of the behaviour.

Management and behaviour modification techniques to resolve this include: (i) removal of an offending environmental stimulus; (ii) desensitizing the dog to the stimulus; (iii) increasing structured activity and exercise; and (iv) counter-conditioning training. Clomipramine, a tricyclic antidepressant drug, has been used in several **companion animal** species with behaviours classified as **compulsive disorders**. With this or any pharmacologic treatment, the duration of use must be adequate to observe an effect, behaviour modification and management techniques should be instituted and the animal should be weaned slowly off the medication, with a return to effective levels should the behaviour recur in the interim.

Humans

For the animal specialist, it is instructive to note that researchers and clinical therapists have described medical/physiological, psychological and social aetiologies for self-mutilation or self-injurious behaviour in humans. Along with behaviour modification therapy, use of pharmacologic treatments for self-mutilation in humans, based upon the suspected aetiology or other accompanying disorders, includes endogenous opioid antagonists, selective serotonin reuptake inhibitors and several classes of anti-depressants. (NKD)

See also: Compulsive disorder; Endorphin; Enkephalin; Stereotypies

References

Dodman, N.H., Normile, J.A., Shuster, L. and Rand, W. (1994) Equine self-mutilation syndrome (57 cases). *Journal of the American Veterinary Medical Association* 204, 1219–1223.

Jenkins, J.R. (2001) Feather picking and self mutilation in psittacine birds. *Veterinary Clinics of North America: Exotic Animal Practice* 4, 651–667.

Luescher, A.U. (2003) Diagnosis and management of compulsive disorders in dogs and cats. *Veterinary Clinics of North America: Small Animal Practice* 33, 253–267.

Selye, Hans

Hans H.B. Selye (1907–1982) is best known for coining the term '**general adaptation syndrome**' to describe the similar physiological **stress** response mounted by the body in relation to a range of **stressors**, and his extensive work in the role of a range of stressors in physical **disease** manifestations led to him being called the 'father of stress'. Indeed, the word 'stress' was introduced into both the French and German language as a direct result of his work, since no native equivalent existed at the time of his initial presentations in these countries. As a medical endocrinologist, his work has been instrumental in the recognition of the influence of psychological factors on human **health**, but it has also been very influential on the perspectives taken by many on **welfare measurement** in animals. His initial experimental evidence reputedly came from an initial problem associated with the poor **handling** and husbandry of animals involved in an endocrinological experiment, in which both treatment and control animals showed similar adrenal enlargement, atrophy of the thymus and gastrointestinal ulceration. (DSM)

Semiochemical

A semiochemical is a chemical substance secreted by an organism with the purpose of carrying messages that may be interpreted by either members of the same species (intraspecific) or different species (interspecific). **Pheromones** are examples of intraspecific semiochemicals that can carry information regarding sexual receptiveness, locations of food sources, **territory** boundaries and dangerous situations. (LJ)

See also: Communication; Odour

Sensitive phase

Sensitive phases refer to the stages of an animal's life in which the animal is particularly sensitive to certain types of experience or stimuli, and the concept is gradually replacing the older idea of **critical periods**. Sensitive phases have gradual beginnings and ends (hence the preference for the term 'phase' over period, which implies a more abrupt start and finish) and, while they are the period of maximum sensitivity in which the appropriate experience may occur, the correct behaviour can still be learned at a later stage (thus they are not as 'critical' in development as originally believed). (LMD)

See also: Development of behaviour; Preparedness

Sensitization

The process by which an individual reaches a period of increased reactivity following repeated experience of a **stimulus**. If an individual experiences a series of arousing positive or negative stimuli, sensitization describes the likelihood that they will respond more quickly or with more intensity to this or another stimulus that is presented soon after. For example, rats that are repeatedly given small electric shocks will show a heightened response to this or any other arousing stimulus for a period of time. (KT)

See also: Arousal; Habituation

Sentience

Sentience is the capacity to perceive or feel things, one of several aspects of **consciousness**. Sentience refers to the response of the **central nervous system** to activation of the peripheral sensory system. A sentient being is one that has its own experience of life, meaning that 'there is something it is like' (Nagel, 1974) to be that being. Sentience is sometimes termed 'phenomenal consciousness and awareness of (the quality of) sensory input is an important aspect of sentience'. The mainstream view in Western society is that at least vertebrate animals are sentient. Most countries' animal protection legislation includes all vertebrates, but no (or only exceptional cases of) invertebrates. In the European Community, the Treaty of Amsterdam explicitly refers to sentience in its statement that the member states desire to 'ensure improved protection and respect for the welfare of animals as sentient beings'.

The question of sentience is central for bioethics and animal **welfare** research, because it provides the fundament for important considerations over how individuals should be treated – viewing animals as sentient is a major reason for many people to be concerned about how animals are treated. However, the scientific understanding of sentience (human and animal) is still limited. Present knowledge of neurobiology is not able to explain consciousness in terms of material mechanisms of the nervous system. The discrepancy between the subjective consciousness experienced (and assumed to exist in other beings) and what is known about the mechanisms potentially is sometimes termed the 'explanatory gap'. As long as the explanatory gap persists, there will be considerable philosophical disagreement and uncertainty about sentience.

Philosophical theories that consider any kind of consciousness dependent on capacity for **language** or capacity for second-order thought will determine animals as non-sentient. According to this view, defended by philosophers such as Peter Carruthers, whether animals are sentient cannot be determined. The mainstream view in biology is less sceptical. Although it views sentience and other aspects of consciousness as essentially private – that is, they can only be experienced from within – it is possible to assume that other individuals are sentient, based on their behavioural and physical similarity to us. While the assumption is uncontroversial for adult human beings, when aiming to extend the assumption to non-human animals the issue becomes more complicated, as verbal account is not available, and the behavioural and physical similarities are smaller.

Even though common sense may suggest sentience in many species, a scientific argument for attributing sentience must be based on systematically collected evidence. Such a systematic approach is suggested by Smith and Boyd (1991) – to find out whether animals of a particular kind are sentient we can ask whether these animals have anatomical, physical and biochemical mechanisms similar to those we know are essential for human subjective experiences, and whether the animals behave in similar ways to experiencing humans. The authors synthesize this into a list of neuroanatomical/physiological and behavioural criteria for determining the capacity to experience **pain**, **stress** and **anxiety** in non-human animals. For any of these experiences, this list includes the possession of higher **brain** centres and evidence of behavioural reaction towards potentially nociceptive, anxiogenic or stressful experiences. Further evidence is added if these behavioural reactions are modulated by drugs that have a known anxiolytic or analgesic effect in humans. In addition, one should look for peripheral nervous structures (including receptors, **signal** substances and hormones) for each specific type of reaction. The more of these criteria that are fulfilled, the more evidence there is to support the view that an animal is indeed sentient.

Sentience is the fundamental criterion for moral consideration in utilitarian **ethics** (**see: Utilitarianism**), and is also central for the subjectivist or hedonist view of animal welfare. It also plays an important indirect role in the animal rights ethics proposed by **Tom Regan**, as only sentient beings are considered 'subjects of a life' and hence worthy of consideration. Even though he has strongly argued against attributing sentience to animals, in later texts Peter Carruthers has affirmed that animals nevertheless should be given moral consideration.

The other important aspect of sentience for animal ethics and animal welfare is the *quality* of the subjective experience. Because animal welfare science and animal ethics have been predominantly preoccupied with identifying and preventing negative experiences, the question of sentience has often been reduced to a question of capacity to feel pain, stress and anxiety. It is not immediately obvious that the interest in avoiding negative experiences is always greater than the interest in having positive experiences, as humans may be prepared to go through unpleasant and often painful experiences (the training required for international sports performance being one example) to gain later positive experiences, and few humans would probably opt for a life without pain if it meant a life without gain.

Nevertheless, given that animal welfare legislation at this point is primarily concerned with reducing animal **suffering**, it seems valid to maintain focus on negative experiences. Much animal welfare research addresses the quality of subjective experiences only indirectly, assuming a relation between absence of negative experiences and good health, possibility to express motivated behaviours and the absence of stress reactions. Later approaches have attempted to come closer to a direct measure of the sentient experience, by studying behaviours as expressions of mental states (see Wemelsfelder, 1997) and through the adaptation of tests from human clinical **psychology** into non-linguistic cognitive measures of animal **emotion** (see Paul *et al.*, 2005). (AO)

References and further reading

Allen, C. (2005) Animal consciousness. In: Zalta, E.N. (ed.) *The Stanford Encyclopedia of Philosophy*, Winter 2005 edn. Available at: http://plato.stanford.edu/archives/win2005/entries/consciousness-animal (accessed 2 December 2009).

Boissy, A., Manteuffel, G., Jensen, M.B., Moe, R.O., Spruijt, B., Keeling, L. *et al.* (2007) Assessment of positive emotions in animals to improve their welfare. *Physiology and Behaviour* 92, 375–397.

Nagel, T. (1974) What is it like to be a bat? *Philosophical Review* 83, 435–450.

Paul, E.S., Harding, E.J. and Mendl, M. (2005) Measuring emotional processes in animals: the utility of a cognitive approach. *Neuroscience and Biobehavioural Reviews* 29, 469–491.

Smith, J.A. and Boyd, K.M. (1991) *Lives in the Balance: the Ethics of Using Animals in Biomedical Research*. Oxford University Press, Oxford, UK.

Webster, J. (guest ed.) (2006) Sentience in Animals (Special Issue). *Applied Animal Behaviour Science* 100 (1–2), 1–152.

Wemelsfelder, F. (1997) The scientific validity of subjective concepts in models of animal welfare. *Applied Animal Behaviour Science* 53, 75–88.

Separation anxiety

Separation anxiety refers to the feelings of **anxiety** generated by the impending or actual separation of an individual from one to which it has a close relationship. It is classified as a disorder when the response interferes with normal functioning, e.g. the individual cannot be left alone for short periods of time or shows signs of panic or self-harm. The term 'separation anxiety' was widely used in the first English language texts in the field of veterinary behavioural medicine to describe a range of **problem behaviours** in **dogs** relating to destructiveness, **vocalization**, **elimination** and hyperactivity in the owner's absence. It has been suggested that the condition may affect up to 50% of dogs in the UK at some time. The condition has since been reported in a range of species, including **cats**, which might not typically be considered gregarious.

More recently, it has been recognized that there are a range of aetiological and potentially mechanistic processes that might play a part in the development of such distress, such as: (i) the level of **attachment** the animal exhibits towards a particular person; (ii) the association of owner departure at a particular time with a previously traumatic event; and (iii) the capacity of the animal to cope with isolation per se and its level of trait anxiety (**see also: Temperament**). This has led some to abandon the term 'separation anxiety' for the general presentation of **distress** when left alone in favour of the terms 'separation-related problems' or 'separation-related disorders', and less commonly terms like 'separation distress', 'separation-related anxiety' and 'separation stress syndrome'. The rationale here is that the term 'separation anxiety' implies a single emotional element dimension to the presenting problem, when in fact the presenting **sign** may have diverse emotional foundations.

In France, there is a distinctive approach to veterinary behavioural medicine encapsulated largely by Pageat's (1995) text, *The Behavioural Pathology of the Dog*. In this text, a distinction is made between 'separation anxiety', a condition arising from a failure of the puppy to full development detachment from its maternal figure, and 'hyperattachment', an attachment that usually develops against a background of **depression**. Thus the French school approach tends to emphasize signs relating to an absence of adult behaviours and an early age of onset in separation anxiety, while others tend to refer to signs of over-attachment. However, there are no peer-reviewed scientific publications based on empirical studies to support the distinction made in France, or any difference in the general treatment strategies employed.

One recent study using a modified human test of attachment (**see: Secure base hypothesis**) concluded that separation anxiety is not based on hyper-attachment per se, although no attempt was made to subcategorize affected subjects on the basis of potentially clinically relevant aetiological factors such as those discussed above. However, it is acknowledged that the problem may be related to a different style of attachment, rather than hyper-attachment. This hypothesis is supported by retrospective data on behaviours associated with separation anxiety, which suggest that dogs with separation anxiety tend to follow their owner excessively and excitedly greet the owner upon his/her return for a longer period. This might indicate that the nature of the interactions between owner and dog, rather than the level of attachment per se, is important to the expression of the problem. (DSM)

Reference and further reading

Flannigan, G. and Dodman, N.H. (2001) Risk factors and behaviours associated with separation anxiety in dogs. *Journal of the American Veterinary Medical Association* 219, 460–466.

Pageat, P. (1995) *Pathologie du Comportement du Chien*. Editions du Point Vétérinaire, Maisons Alfort, Cedex, France.

Parthasarathy, V. and Crowell-Davis, S.L. (2006) Relationship between attachment to owners and separation anxiety in pet dogs (*Canis lupus familiaris*). *Journal of Veterinary Behaviour* 1, 109–120.

Sepsis

Strictly speaking, sepsis describes a generalized immune reaction to infection and is an often fatal condition. The term is often erroneously applied to infectious conditions in which the pathogen is circulating via the blood. This latter condition is more accurately referred to as a septicaemia (blood poisoning), which may lead to sepsis but not necessarily so. (DSM)

Serotonin

Serotonin (5-hydroxytryptamine, 5-HT) acts as a monoamine transmitter as well as a neuromodulator in the **central nervous system** (CNS). It is synthesized from the aromatic amino acid L-tryptophan by the enzyme tryptophan hydroxylase (the rate-limiting step) to 5-hydroxytryptophan (5-HTP), and by an amino acid decarboxylase (non-specific enzyme) to 5-HT. 5-HT is metabolized by monoamine oxidase (MAO) to 5-hydroxyindoleacetic acid (5-HIAA).

Following synthesis, 5-HT is transported to the nerve terminals and stored in synaptic vesicles. It is released into the synaptic cleft by exocytosis in response to depolarization of the axon terminal (action potentials). Released 5-HT is coupled to the 5-HT receptors to stimulate postsynaptic **neuron(e)**s. The deactivation of 5-HT is achieved by its re-uptake into the presynaptic terminals, uptake by astrocytes and metabolism by MAO to 5-HIAA. The release of 5-HT from neurons is regulated by axon traffic, while maintenance of the content of 5-HT in the CNS is regulated by multiple factors, including the availability of precursor L-trypotophan, auto- and hetero-receptors, activity of enzymes and functions of astrocytes.

The distribution of 5-HT in the CNS forms a diffuse network. The cell-bodies of the serotonin neurons are mainly found in the raphe nucleus of the brainstem. Axons of the 5-HT neurons arising from the dorsal raphe nucleus innervate the cerebral cortex and the neostriatum; the median raphe nucleus innervates the limbic system; the rostral raphe nucleus innervates the forebrain regions; whereas the caudal raphe nucleus projects within the brainstem and the spinal cord with some overlap.

Serotonin mediates biological processes by interacting with its receptors. There are seven subtypes of 5-HT receptors, 5-HT_{1-7}, which have been recognized through various studies using selective agonists and antagonists, radio-ligand-binding analysis and identification of receptor genes. Several subpopulations for some of these receptors, such as 5-HT_{1A-1F} and 5-HT_{2A-2C}, have been grouped according to operational, structural and transductional properties. Each type and subtype of receptors has been implicated as playing important roles in physiological, pathological and psychopathological conditions.

Based on the broad distribution of 5-HT and its receptors in the CNS, 5-HT has a wide range of biological functions in controlling physiological homeostasis, including **pain** reaction, **thermoregulation**, **sexual behaviour**, motor activity and appetite (**feeding**, **hunger** and energy **homeostasis**). Serotonin also plays important roles in controlling states of **consciousness**, such as **sleep**, **memory** and mood, as well as various neuropsychiatric disorders such as **aggression**, violence, **anxiety**, **depression**, migraine and drug abuse. Through regulation of the **hypothalamic–pituitary–adrenal (HPA) axis**, 5-HT regulates the neuroendocrine control mechanisms in the HPA axis in releasing **prolactin** and **growth hormone (GH)** during **reproduction**, and **ACTH** and **corticosteroid**s in response to **stressor**s, such as environmental danger or threat.

Serotonin regulates the transmission of painful stimuli at both the dorsal horn and the brainstem sites, especially in the periaqueductal grey matter. Exogenous intracerebroventricular administration of 5-HT exhibits an analgesic effect (**see: Analgesia**). The analgesic effects are confirmed following administration of drugs that enhance 5-HT transmission – for example, blockers of retake such as zimelidine and clomipramine.

Depletion or decrease of 5-HT concentrations, ratio of 5-HIAA:5-HT or metabolism of tryptophan have all been implicated in neuropsychiatric abnormalities such as cognitive decline, mood disorders and depressive symptoms, especially in long-lasting and chronic **diseases**. Depression can be worsened by experimental depletion of 5-HT via a tryptophan-depleted diet. In contrast, selective 5-HT re-uptake inhibitors, such as fluoxetine, have proved to be effective antidepressants.

Brain serotonin functions as an inhibitory factor of aggression. Depletion or decrease of 5-HT concentrations and ratio of 5-HIAA:5-HT have been implicated in aggressiveness, and violence in human and animals and **cannibalism** in rats.

Experimental studies in animals have shown that **domestication** of rats selected for tame behaviour towards humans has resulted in increased levels of 5-HT and 5-HIAA in the midbrain and **hypothalamus**. Experimental increase of 5-HT and/or 5-HIAA in the brain regions, such as in the lateral hypothalamus and amygdala, blocks or retracts killing behaviour in rodents. 5-HT has also been highlighted as an important factor in the **stress** response, and low levels of endogenous 5-HT (or its metabolites) have been associated with depression and suicidal behaviour.

There are correlations between 5-HT activity and impaired impulse control and stress-related aggression in adolescents who suffer from mental retardation. Administrations of compounds that specifically elevate 5-HT levels in the brain (e.g. fluoxetine) have been used effectively to treat depression and associated aggressiveness. In addition, previous studies have shown that functions of 5-HT_1 and 5-HT_2 receptors are linked to aggression. Aggressiveness can be increased due to impaired receptor functions or by knockout of 5-HT_1 receptor genes.

Serotonin has a tonic, inhibitory effect on sexual behaviour and reproduction, such as inhibition of **luteinizing hormone (LH)** secretion and ovulation. The effects of 5-HT on sexual behaviour are positively correlated to the stimulation of the pre-optic area (POA) and median eminence of the hypothalamus, and differently regulated by 5-HT receptors in rodents. Recently, the roles of 5-HT in modulating sexual behaviour have been reevaluated, and it is concluded that 5-HT can either inhibit or facilitate secretion of **gonadotropins**, sexual hormone release and sexual behaviour, depending on which brain regions are involved and what subtypes of 5-HT receptors are activated. Effects of 5-HT on sexual behaviour are positively correlated to the stimulation of the pre-optic area (POA) and median eminence, but negatively correlated to the stimulation of the ventromedial nucleus (VMN), zona incerta and arcuate nucleus. The 5-HT_2 receptors in the POA are involved in the stimulatory function of 5-HT, while 5-HT_{1A} receptors mediate the inhibitory role in the VMN. With systemic injection of the selective 5-HT_2 and 5-HT_1 agonists, MK 212 and Ru 24969, sexual behaviour is either facilitated or inhibited, respectively. In addition, inhibited sexual behaviour is effected by peripherally administered 5-HT2 antagonist, pirenperone, and its effect is attenuated by the 5-HT agonist, quipazine.

Serotonin cannot pass the brain–blood barrier and is regulated differently in the CNS, the serotonergic neurons, peripheral tissues and the platelets. The brain and blood could be assumed to be two compartments. Previous studies have found that there are no correlations of either MAO activity or 5-HT levels between brain and blood.

Although supplemental tryptophan, a precursor of 5-HT, decreases aggression in feed-restricted male chickens, the effect of tryptophan on modification of aggression is related to centrally enhanced neuronal firing rather than peripherally increased 5-HT resulting from conversion of tryptophan. However, blood 5-HT concentration could functionally be a monitor of the changes of the central 5-HT system, as the 5-HT systems between blood platelets and serotonergic neurons share some of anatomical and functional similarities, including the active carrier mechanism, intracellular storage, receptors and binding sites. In addition, 90% of cerebral 5-HIAA is released directly into the blood.

Analysis of blood levels of 5-HIAA may directly or indirectly provide clues of the changes of the CSF 5-HIAA and brain 5-HT. Platelets have been proposed as a peripheral tool for observation of a central 5-HT system. Further studies are needed to examine the cellular mechanisms of regulating 5-HT systems between the periphery and the brain, and to determine how each functions in affecting animal behaviour and adaptation, and their correlations. (H-WC)

Service animal

A service animal is defined by the American Disabilities Act as 'any guide dog, signal dog or other animal individually trained to provide assistance with a disability', and this definition may be usefully employed in science. Service animals perform some of the functions and tasks that disabled human individuals cannot perform themselves. Examples include guide **dogs** for the blind, hearing dogs for the deaf, seizure alert dogs and mobility assistance dogs. In some contexts the term may also be applied to 'working animals' such as military dogs. (EP)

See also: Animal-assisted therapy

Sex differences

For all animals, including humans, sex differences exist in the behaviour, morphology (sexual dimorphism) and physiology of males and females. For example, the primary sex organs of males and females differ, and the same can be said about their **brains** and behaviour. The physiological mechanisms that mediate most sex differences in brain function are attributed to differences in the gonadal hormone secretion and action among males and females. However, recent work has shown that sex differences can be induced by external sources of gonadal steroids and by sex chromosome genes directly affecting the brain. The evolutionary processes that led to differences between males and females are ultimately a consequence of anisogamy (differences in the gametes). Anisogamy has led to sex differences in morphology, behaviours and possibly **cognition**. Differences in these phenotypic characteristics have provided the substrate for natural selection and **sexual selection** to bring about differences between the sexes.

Sexual differentiation

Sex differences are usually attributed to genetic differences between males and females. In mammals, females have two X chromosomes and males have one X and one Y chromosome. Many of the cells in neural substrates, especially those associated with the brain, also differ in genetic sex. Genetic sex, however, is only one factor that will influence sex differences. Several studies have shown that hormones constitute an important internal causal factor that can affect the behaviour, morphology, brain and physiology of males and females. The sex differences may be expressed early in development, during the prenatal or early postnatal periods, when hormones organize neural and endocrine substrates or sex differences may be manifested late in development, during puberty, when hormones activate mediation of neural and endocrine substrates.

Most sex differences in brain function are attributed to differences in the gonadal hormone secretion and action among males and females. In mammals, for example, the presence of a Y chromosome induces the formation of the testes. The testes produce a surge of **testosterone** during fetal development and neonatal life. This testosterone surge causes masculine neural development, at least in mammals, and creates a phenotypic male. Testosterone and other **androgens** also cause the transformation of the genitals into male organs and the organization of male-like behaviours. If a male rodent pup is castrated or treated with an anti-androgen shortly after birth, as an adult it will not display, or have, suppressed male-typical behaviours such as mounting, but will display female-like behaviours such as **lordosis**.

The absence of a Y chromosome results in the formation of ovaries and a phenotypic female being produced. In mammals, morphology, physiology and behaviours typically associated with females develop in the absence of testosterone. If androgens are administered to female rodent pups shortly after their birth they will display, as an adult, more male-like behaviours and fewer female-like behaviours. In birds, the situation is more complicated and reversed. Males are the default sex and a surge of estradiol around hatching feminizes the neural development of female birds.

External steroids – those not produced by the animal's own sex organs – can affect sexual differentiation in mammals and birds. Birds still in the egg can be exposed to large quantities of testosterone deposited in the egg yolk by the mother. It has been suggested that the mother can actively regulate the amount of testosterone in the egg yolk, which in turn affects the behaviour of the young after they hatch. In rats, mice and probably other rodents, intrauterine position can affect how females will behave as adults. Females that are surrounded on either side by brothers during development are more likely to be exposed to the testosterone produced by them. These females, known as 2M (M = male) females, are more likely as adults to display more male-typical behaviour and fewer female-typical behaviours as compared with females that were surrounded *in utero* by sisters (0M female) or by a sister and a brother (1M female).

This picture of sexual differentiation, however, is more complex than originally thought. Recent work has shown that, in a finch and some transgenic mice, the sexual **phenotype** may be insensitive to the effects of sex hormones during development, or that sex differences occur prior to the onset of sex-specific patterns of gonadal hormone activity. These findings suggest that sex chromosome genes acting within particular brain cells could contribute to their sexual differentiation.

Anisogamy

Differences between males and females are ultimately a consequence of sexual reproduction and anisogamy. Theoretical models suggest that, soon after the beginning of sexual **reproduction**, there was a process of disruptive **selection** that ended in the differentiation of gametes, or anisogamy. Briefly, larger gametes produced **zygotes** with higher survival as compared with zygotes produced by smaller gametes. The reason is that larger gametes have more nutrients for the development of the zygote. Thus, there was a selective pressure for the production of progressively larger gametes, even though this implied the production of fewer gametes, assuming a relatively fixed amount of resources. At the same time, there was another selective pressure favouring small gametes that were able to swim fast and find a larger than average gamete before other competing gametes did. These small and fast gametes, sperm, were basically parasitizing larger gametes, including their DNA in the zygote without offering any nutrients. Because these small gametes were cheaper to produce, individuals producing them were able to do so in large quantities.

The final result of disruptive selection for both small and large gametes favoured the extreme anisogamy that we observe in animal species, with small and mobile gametes (sperm) produced in large quantities, and large and sessile gametes, eggs, produced in small quantities. Males, producers of sperm, are the sex that seeks mates more actively, whereas females, producers of eggs, are the receptive sex. Moreover, males are often the limited sex, whereas females are often the limiting sex with regard to mating opportunities.

Sex differences in behaviour

The fact that males produce numerous and cheap sperm, while females produce few and expensive eggs, has two important consequences. First, the potential reproductive success of males and females is entirely different. A male may potentially fertilize many females, whereas females are limited by the number of eggs, and thus offspring, that they can produce and raise. Males, thus, will benefit by trying to fertilize as many females as possible, and will compete among them for access to reproductive females, a process called male–male competition. In contrast, females will benefit by mating with the best mate possible, i.e. the mate that will maximize their fitness. Both mate choice and male–male competition are strong evolutionary forces responsible for much of the observed sexual dimorphism. Such characters as weaponry, large body size and heightened **aggression**, generally male characteristics, are a result of male–male competition (**see: Intrasexual selection**). Additionally, females may amplify the development of male characters via **mate choice**, because such characters as bright plumage, complex **vocalizations**, reckless behaviour or large body size may be indicators of male quality (**see: Intersexual selection**).

When females show a preference for elaborate and extreme traits, there may be directional selection favouring those extravagant traits in males, which will also result in high sexual dimorphism. A confirmation of the importance of sexual selection (mate choice and male–male competition) in the production of sexual dimorphism is the fact that in monogamous systems, in which sexual selection is relatively weak, sexual dimorphism is less apparent than in non-monogamous systems.

A second consequence of the initial difference between males and females (i.e. production of either sperm or eggs) is a sexual difference in parental investment. Because males' initial investment is relatively low and their reproductive success may be improved by further mating, they will generally opt for not providing any other parental investment than their gametes.

Females, because of their initial larger investment, are therefore forced to provide much of the parental care to their offspring (**see: Parental investment**). This is the reason why maternal care is generally much higher than paternal care, and is an illustrative example of sexual conflict, which occurs when benefits and costs of reproduction are asymmetric for males and females, and thus each sex will behave in such a way that maximizes its own **fitness** to the detriment of the other sex's fitness.

Sexual differences in behaviour are also dependent on phylogenetical constraints. Whereas in mammals and birds the chromosomes of an individual will determine the sex, in many reptiles incubation temperature determines an individual's sex, and some fish species may change sex in response to their social environment. Furthermore, females are the heterogametic sex in birds, whereas males are the heterogametic sex in mammals. Similar differences are found in relation to **philopatry** and dispersal. Whereas females are the philopatric sex in mammals, female birds are the dispersive sex.

Sex differences in cognition – a case study in voles

Several studies have reported sex differences in cognition in humans, but little is known about sex differences in cognition among non-human animals. Recently, a study to determine whether meadow voles (*Microtus pennsylvanicus*) were able to distinguish between areas containing varying numbers of over-marks from two scent donors, suggesting a capacity for relative numerousness, has been reported (Ferkin *et al.*, 2005). Over-marks are formed when two scent marks overlap, and is a common feature of **scent marking** in voles. It is possible to test for relative numerousness by allowing voles to explore an area that contains a set of over-marks by one individual and a smaller or larger set of over-marks by another individual. If voles displayed the capacity for distinguishing more from less, they would respond preferentially to the donor which was the top-scent donor on more of the over-marks as compared with the donor that was the bottom-scent on most of the over-marks.

Female voles were better than males at spontaneous discrimination of more frequent versus less frequent top-scent donors. Females were capable of spontaneously discriminating between areas in which donor A's marks were on top of donor B's marks four times relative to the three times donor B's marks were on top of donor A's marks. Males could make a similar discrimination if donor A's marks were on top of donor B's marks six times relative to the one time donor B's mark was on top of one of donor A's marks.

This type of asymmetry in relative numerousness by voles is interesting, in that the literature is replete with studies suggesting that sex differences exist in mathematical skills that favour males over females, particularly in primates and humans. One argument is that males have better spatial and navigational abilities than females, which provides them with a greater capacity to solve problems in geometry and other mathematics-related activities. It is intriguing that male meadow voles have better spatial ability but worse prenumerical ability than female meadow voles. The findings suggest that sex differences in distinguishing more over-marks from fewer over-marks are somehow separate from the spatial ability of meadow voles. Moreover, for voles it appears that spatial ability may be a poor predictor of relative numerousness, and vice versa. Female biases in relative numerousness may be the developmental link for more complex numerical processes, such as subitizing, estimation, counting and arithmetic reasoning in this species.

Recency in areas and dominance status may both be features of the top-scent donor and attributes of males that sire offspring in voles. In that male voles wander through large **home ranges** that encompass the territories of one or more females, resident females may be able to select potential mates based on asymmetries in the number of times that a particular male was the top- or bottom-scent donor in their **territory**. Females may need to distinguish between and remember the individuals that may enter their territories in order to keep track of them. In contrast, males are not territorial and may not need to respond to over-marks, as do female voles. Thus, female voles may have a better 'memory for scent marks' relative to male voles. Alternatively, female voles may be more responsive than male voles to asymmetries in the relative number of top- and bottom-scent marks in an area, because the top-scent mark conveys greater or more salient information to females relative to that of the bottom-scent marks. At this point in time, it is not clear why male voles cannot distinguish small asymmetries in the relative number of over-marks of female **conspecifics**. (MHF, JdB-T)

Reference and further reading

Darwin, C. (1871) *The Descent of Man, and Selection in Relation to Sex*. John Murray, London.

Ferkin, M.H., Pierce, A.A., Sealand, R.O. and delBarco-Trillo, J. (2005) Meadow voles, *Microtus pennsylvanicus*, can distinguish more over-marks from fewer over-marks. *Animal Cognition* 8, 82–89.

Kimura, D. (2002) Sex differences in the brain. *Scientific American* Special issue 'The hidden mind' 12, 32–37.

Parker, G.A., Baker, R.R. and Smith, V.G.F. (1972) The origin and evolution of gamete dimorphism and the male–female phenomenon. *Journal of Theoretical Biology* 36, 181–198.

Phoenix, C.H., Goy, R.W., Gerall, A.A. and Young, W.C. (1959) Organizing action of prenatally administered testosterone propionate on the tissues mediating sexual behaviour in the female guinea pig. *Endocrinology* 65, 369–382.

Trivers, R.L. (1972) Parental investment and sexual selection. In: Campbell, B. (ed.) *Sexual Selection and the Descent of Man*. Aldine, Chicago, Illinois, pp. 136–179.

Sexual behaviour

Sexual behaviours are designed to ultimately result in the deposition of sufficient viable spermatozoa in the reproductive tract of the female at the optimum time of the oestrous cycle to ensure conception. Male farm animals are often required to mate frequently (to reduce costs, maximize genetic gain, etc.) and, thus, level of sexual behaviour is especially important to the value of these animals. When males are mating frequently, the reserves of sperm in the epididymides are depleted, resulting in the need for females to be mated more than once to ensure conception. Furthermore, males are often required to mate a large number of females in a short period. Sexual behaviour therefore is clearly important from the perspective of both the species and the farmers or owners of domestic animals. This commentary will consider the characteristics and regulation of the sexual behaviour of our common domestic animals.

Male sexual behaviour

There are two major characteristics of males that affect the likelihood of achieving copulation, sexual **motivation** and mating competency. Sexual motivation, sometimes referred to as **libido** or sexual drive, refers to the tendency to copulate. Mating competency, sometimes referred to as mating dexterity, is the ability to copulate, that is the ability to mount, gain an erection of the penis, thrust and achieve intromission.

Male sexual behaviour is obviously necessary to achieve impregnation in natural mating systems. Furthermore, the level of male sexual behaviour may affect the level of reproductive performance such as the number and timing of offspring produced, which has clear economic implications. The level of sexual behaviour of rams and bulls in competitive and uncompetitive natural mating systems will affect numbers of females mated and pregnancy rate. Furthermore, some studies have shown that a high level of sexual behaviour in bulls can lead to greater weaning or marketing weights, because the offspring are born earlier.

Male sexual behaviour is quite varied, although the essential features of intromission, pelvic thrusting and ejaculation are characteristic of all male mammals. One of the behavioural components of male sexual display in all hoofed livestock, except the **pig**, is the 'olfactory reflex' known as the **flehmen**. The animal fully extends its head and neck, contracts its nares and raises its upper lip while taking shallow respirations. Flehmen normally occurs following the smelling of **urine** and nosing the ano-genital region of the female, and is considered a form of urine testing. Olfactory and gustatory stimuli are used in these species to assist in detecting oestrous females.

A number of components of male sexual behaviour are common to a number of species, such as nosing the perineum, nudging, flehmen, flicking the tongue, striking out the fore limb and low-pitched bleats in **sheep** and **goat**s. Other components of male sexual behaviour are reasonably specific to an individual species. For example, boars generally utter a short series of characteristic grunts and champ their jaws, producing saliva, while their heads are held in close proximity to that of the female. Boars will also vigorously nose the flanks of females during courtship.

Hormonal and neural control of male sexual behaviour

The hormonal and neural control of sexual behaviour has been most studied in laboratory species, particularly the **rat**. Some elegant research by Frank Beach in the 1940s and 1950s with rats and **guinea pig**s clearly demonstrated the critical role of the gonadal hormone **testosterone**. Testosterone is necessary to elicit sexual behaviour: **castration** diminished sexual behaviour, with the extent dependent on age and/or sexual experience. This research also indicated that a threshold concentration of testosterone is required to elicit sexual behaviour and, above this threshold, the concentration of the hormone is ineffectual in increasing copulatory behaviour. In addition to maintaining sexual behaviour, androgens stimulate the latter stages of spermatogenesis and prolong the lifespan of epididymal sperm. They also promote the growth, development and secretory activity of the sexual accessory glands of the male.

Most males have sufficient testosterone, and thus most male sexual behaviour problems are not due to low testosterone concentrations. However, in those species where there is a seasonal fluctuation in sexual behaviour, such as **horse**s and sheep, there is a corresponding fluctuation in testosterone secretion. Studies by Beach showed that pre-castration differences in copulatory behaviour still existed after testosterone therapy, indicating the impact of genetics and experience on the control of male sexual behaviour.

Some sexual reflexes, such as erection and ejaculation, are organized by neural circuits in the spinal cord. In some laboratory studies, it has been shown that the ejaculatory response is normally inhibited by **neuron(e)**s in the **brain**. The medial preoptic area is the forebrain area most critical for male sexual behaviour. Stimulation of this region produces copulatory behaviour, while lesions abolish it. Implantation of testosterone directly into this area reinstates copulatory behaviour that had previously been abolished by castration in adulthood. Autoradiography studies show that neurons in the medial preoptic area contain testosterone receptors.

Factors affecting male sexual behaviour

A number of factors can affect male sexual behaviour, such as genetic, seasonal, social, sexual and psychological factors. **Breed** effects are apparent in many farm species. For example, dairy bulls generally display higher levels of sexual behaviour than beef bulls. Research on **cattle** and several laboratory species indicates that level of sexual behaviour is moderately heritable. The considerable variability in male sexual performance, with many individuals exhibiting low levels of sexual behaviour and mating competency, may be in part due to relaxed genetic selection for sexual performance in non-competitive captive populations.

Seasonal effects are evident in some species, such as sheep, cattle, horses, goats and **dog**s, with reproductive activity intense during the breeding season and subdued or absent during the remainder of the year. **Photoperiod** is the principal external agent controlling sexual activity in seasonal breeders. Adverse temperature effects, particularly high temperatures, on sexual behaviour are apparent even in constant breeders such as pigs and **rabbit**s.

Social and sexual effects on male sexual behaviour can be profound. The social environment during rearing can have marked effects on the development of the sexual behaviour of boars. Social contact, particularly tactile contact with other pigs, has been shown to be influential in the development of high levels of sexual behaviour of boars. Both heterosexual sexual experience and age influence the sexual behaviour of rams. Cockerels isolated from pullets and individually reared from an early age show diminished sexual behaviour in adulthood in comparison with males reared in all-male **group**s.

Depriving mature boars of female contact, possibly through reduced olfactory stimulation, will depress but not eliminate the sexual behaviour of mature boars; however, this effect is not permanent since the sexual behaviour of mature boars can be restored within 4 weeks of housing near female pigs. In contrast to the effects on mature boars, deprivation of female contact at the time of puberty may have adverse long-term effects on the sexual behaviour of boars. The oestrous status of the female does not influence the effectiveness of females in stimulating the sexual behaviour of pubertal and mature boars, and olfactory – and perhaps auditory – stimuli from the female are most likely implicated.

Sexual stimulation, through observation of other animals mating, can increase short-term levels of sexual behaviour in males, but the presence of dominant males, even when physically separated, can inhibit the sexual behaviour of rams and bulls. Males generally show a refractory period or quiescence following copulation, in which the sexual stimulus properties of the female are less effective. However, successive introduction of 'new' or unfamiliar females can often re-initiate and maintain sexual activity for long periods in roosters, boars, rams, bulls and buck goats, a phenomenon called the 'Coolidge effect' (**see: Biostimulation**). This phenomenon is clearly valuable for species in which a single male inseminates all the members of his harem.

The incidence of '**buller-steer syndrome**', in which a steer is repeatedly mounted by other steers to the extent where it may be injured or even killed, is estimated at 1–3% with a resulting mortality of 1% in North American feedlots. The animals ridden ('bullers') behave much like a female in **oestrus** by standing to be ridden. The causes of buller-steer syndrome are not fully understood, are conflicting and are controversial. Buller behaviour may be involved in the establishment and maintenance of a social hierarchy. However, additional contributing factors are the submissive behaviour of the buller, the use of various anabolic implants to stimulate growth, seasonal factors and the number of animals per pen.

Locomotor and penile injuries may not physically allow the achievement of copulation or may inhibit copulation because of **pain**, while **injury** sustained during copulation in many male farm animals may produce a psychological effect for some time after physical recovery has occurred, again inhibiting copulation. Obesity and skeletal defects – for example, in bulls and boars – affect sexual behaviour, often by affecting mobility and thus mating competency. Commercial **turkey**s, because of their broad-breasted breeding, are unable naturally to mate and so **artificial insemination** is a routine practice.

Poor orientation of the mounting response, such as side or head mounting, is often seen in young males. However, proper orientation is probably a learned response and, if the male is of satisfactory sexual motivation, mating competency should improve with the positive reinforcement of copulation.

Female sexual behaviour

Sexual behaviour of female farm animals is cyclic. By definition, females will mate only during oestrus, with the exception being induced ovulators, such as rabbits and **cat**s, where mating induces ovulation. Three characteristics of female sexual behaviour that affect the likelihood of successful mating are attractivity, **proceptivity** and receptivity.

Attractivity refers to the extent to which the female evokes a sexual response in the male. Odours, visual characteristics and behaviour may affect attractiveness. Oestrous females generally are more attractive to males, although boars and bulls will mate restrained, non-oestrous (unreceptive) females. Rams prefer woolly ewes to shorn ewes, and older ewes are usually mated more frequently in competitive mating situations than are maidens.

Proceptivity refers to invitation or soliciting behaviour, and consists of affiliative behaviours such as male-seeking and physical appetitive behaviours such as signalling. Oestrous sows will vigorously seek boars, as well as mounting other females. Mounting cows indicate a sexually receptive female to bulls, although the presence of bulls can suppress mounting behaviour. About 90% of all mounted cows are in oestrus, but only 70% of all mounting cows. Sexually receptive ewes and nanny goats show increased activity and show an increase in non-specific bleats. Nanny goats tend to mount and be mounted by other females, but this is the exception in receptive ewes. Harems of ewes indicate sexual receptivity to rams. Oestrous mares approach stallions frequently, urinating and rhythmically everting their vulvas ('winking'), thus signalling their readiness to copulate. Female fowl may at times solicit mating by approaching the male, and female fowl prefer males with characteristics that indicate high fitness, such as symmetrical features or sexual secondary characteristics, e.g. large spurs or expansive decorative **feather**s.

Receptivity refers to the behavioural responses sufficient to enable copulation, such as the immobility response. Receptive sows show what is often called the 'standing response', which is the primary stimulus eliciting mounting in boars. Oestrous mares show a characteristic stance, with the tail lifted off to the side of the perineum, the head turned back to the stallion and one fore leg flexed. Oestrus is defined as the state during which the female accepts the male's copulation attempts. The behavioural features are synchronized with various physiological changes to ensure appropriate timing of sperm and eggs at the site of fertilization. The signs of oestrus are characteristic for each species.

Hormonal and neural control of female sexual behaviour

Oestrogens (oestradiol) are an absolute requirement in ensuring oestrus. They act on the **central nervous system** to elicit sexual behaviour, as well as stimulating the pre-ovulatory **luteinizing hormone** surge and, in turn, ovulation. There are some differences between species in that ewes and cows require progesterone prior to oestrogens to optimize their sexual behaviour, ovulation and other physiological events associated with successful copulation.

As in the male, some sexual reflexes in the female, such as **lordosis**, are organized by neural circuits in the spinal cord. The most important forebrain region controlling female sexual behaviour is the ventromedial nucleus of the **hypothalamus**. Its destruction abolishes copulatory behaviour and its stimulation facilitates the behaviour. Both oestradiol and progesterone exert their stimulatory effects on female sexual behaviour in this region: implantation of these two hormones directly into ventromedial nucleus of ovariectomized females reinstates both proceptive and receptive behaviour. Autoradiographical studies show that neurons in this region contain oestrogen and progesterone receptors.

Factors affecting female sexual behaviour

Low levels of sexual behaviour in females will result in problems with oestrus detection and sexual receptivity, and consequently depress reproductive success in both natural and artificial matings. As with the male, season, sexual stimulation and sexual experience can affect the sexual behaviour of female farm animals. Sheep and cattle are seasonal breeders.

Sexual stimulation affects the sexual behaviour of females. In cattle, it appears that both natural matings and matings with vasectomized bulls reduce the duration of sexual receptivity or oestrus.

Exposure to unfamiliar females, regardless of their sexual status, hastens the onset of puberty in gilts. However, the effects on puberty attainment of exposure to unfamiliar females are relatively minor in comparison with the effects of boar contact.

The impact of male stimulation on sexual behaviour is well illustrated by the behavioural response of oestrous sows to boar stimulation. About 60% of oestrous sows will show either the immobility or standing response to firm hand pressure on their back in the absence of boars, but this can be increased to 90% by providing auditory and olfactory contact with boars, and further increased to 100% with the addition of visual and tactile contact with boars.

While it appears that male contact has an important role in stimulating the female's sexual behaviour, there are situations where continuous stimulation may adversely affect female sexual behaviour. Housing sows adjacent to boars, which allows continuous boar stimulation, over several weeks has been shown to create difficulties with oestrus detection, particularly in gilts or maiden sows. Habituation by females to the important boar stimuli (e.g. auditory and olfactory stimuli) that facilitate the standing response of the oestrous female may be responsible. Furthermore, prolonged boar stimulation at the time of detection may also reduce sexual receptivity. There are contradictory reports of the effects of exposure to rams during the non-breeding season on the onset of breeding activity in ewes.

Stress can affect sexual behaviour. Overcrowding of female pigs around the time of mating may depress their sexual behaviour. Stress during the follicular phase of the oestrous cycle is known to disrupt oestrous patterns and fertility.

The behaviour of domestic animals, including their sexual behaviour, is dependent upon a complex interaction between the organism's internal and external environment. The internal environment includes the central nervous system, the metabolic state of the organism and the endocrine milieu, while the external environment includes the organism's social, climatic and physical features. An understanding of the regulation of sexual behaviour is important in optimizing the reproductive performance and welfare of domestic animals. (PHH)

See also: Breeding; Homosexual behaviour; Mating behaviour and systems

Further reading

Carlson, N.R. (1986) *Physiology of Behaviour*, 3rd edn. Allyn and Bacon, Boston, Massachusetts.

Fraser, A.F. and Broom, D.M. (1990) *Farm Animal Behaviour and Welfare*. CAB International, Wallingford, UK.

Jensen, P. (ed.) (2009) *The Ethology of Domestic Animals*, 2nd edn. CAB International, Wallingford, UK.

Sexual selection

Sexual selection was identified by **Charles Darwin** as an explanation for the evolution of traits that appear to hinder survival (e.g. the tail of the peacock). It was traditionally considered as **selection** for traits that enhance mating success via direct male–male combat (**intrasexual selection**) or female choice (**intersexual selection**), but subsequently modified to cover any traits that enhance **breeding** success, in order to include post-copulatory sexual selection for fertilizations, in the form of sperm competition and cryptic female choice, and post-zygotic sexual selection in the form of **infanticide**.

However, the distinction between the effects of natural and sexual selection are sometimes ambiguous, as survival may be considered a component part of an individual's life-history that has evolved to maximize lifetime reproductive success. Thus, sexual selection might be defined as: those natural selection forces that operate differently in males and females because of the strategies of the sexes. With this definition, sexual selection becomes a general term to describe selection acting differently in each sex, favouring adaptations that facilitate the sex-specific differentiation of phenotypes. (PE)

Further reading

Carranza, J. (2009) Defining sexual selection as sex-dependent selection. *Animal Behaviour* 77, 749–751.

Shade-seeking behaviour

Providing shade is one of the primary methods used to protect livestock and outdoor captive animals from high heat load (direct solar radiation). The temperature can be up to 20°C lower in the shade than in the sun. In hot conditions black *Bos taurus* cattle can have rectal temperatures over 1°C lower (38.8 versus 40.1°C), and respiration rates some 50 breaths per minute lower (70 breaths per minute versus 130 breaths per minute) if they have access to shade. The **welfare** of animals is, in most cases, improved by the provision of shade, because it not only facilitates more effective temperature regulation and so reduces the risk of potentially fatal conditions such as heat exhaustion and heat stroke, but also reduces the risk of sunburn and certain forms of skin cancer, especially in animals with pale skin (notably Hereford **cattle** and **cats** with white ear tips). More effective body temperature regulation also improves productivity. However, shade may lead to a decrease in productivity if its use reduces the amount of time spent grazing, in the case of herbivores.

If natural shade or shade structures are not available, then animals exposed to hot climatic conditions may use whatever source of shade they can find. This may be shade cast by fence posts, water troughs and feed bunks – and even other animals. It is common to see hot cattle or **horses**, without access to shade, with their heads held low in the shadow of another. This behaviour may lead to other problems because bunching may decrease heat loss. Bunching is more common in dark-coated cattle than in white cattle, probably because they absorb more solar heat than white cattle. Temperatures on the poll of black cattle of up to 65°C have been recorded, and it has been observed that black cattle move to shade before white cattle.

Shade-seeking behaviour is not simply a function of ambient temperature. Related factors such as solar radiation levels (solar load), relative humidity, **stocking density**, micro-climate under the shade (temperature and relative humidity), wind speed, the type of shade (e.g. natural, shade cloth or iron roof), time of eating and **genotype** will influence when an animal will seek shade and for how long it will use it. In some circumstances animals may move out from under shade on extremely hot days if there is little air movement under the structure and/or the area is crowded. They may also leave shade to drink and eat; however, if space is limited they may restrict visits to water because of the risk of losing their space in the shade. Solar load plays a major role in shade usage; however, its importance has not been fully documented. Many animals will move out from under the shade as the sun sets, even when ambient conditions are still hot.

European (*Bos taurus*) breeds of cattle will start to seek shade when the ambient temperature exceeds a temperature threshold that typically lies between 20 and 27°C. At temperatures greater than 27°C, 70–100% of cattle will be under shade if it is available. The amount of time that *Bos taurus* cattle spend in shade increases by approximately 23 min/day for every 1°C increase in ambient temperature between 15 and 36°C. Above 36°C there will usually be 95–100% usage of shade.

The provision of shelter for captive animals should be considered essential. Natural shade in the form of trees, etc. is usually the first preference of many species, and trees have been shown to be particularly effective in relieving thermal radiation. The evaporation of water from the leaves of trees also cools the surrounding air, unlike purely physical shelters. Manmade shelters with a solid roof normally provide the highest level of protection and may be preferred, but the advantages of shade may be lost if the design of the structure is poor. For example, if the roof of the structure is low (e.g. 2 m rather then 5 m), the radiant heat from a galvanized metal roof may drive animals from the shade. Other common problems with artificial shelters concern the build-up of moisture from urine and faeces and a lack of air movement.

(JG)

See also: Free-range animals; Heat stress

Further reading

Gregory, N.G. (1995) The role of shelter belts in protecting livestock: a review. *New Zealand Journal of Agricultural Research* 38, 423–450.

Shaping (operant)

Shaping (also known as successive approximation) describes the training process whereby differential **reinforcement** or **punishment** is used to encourage an animal to perform increasingly stringent approximations of a predetermined behaviour. It is typically undertaken using positive reinforcement, with any approximation initially reinforced and then only those responses that more closely resemble the goal reinforced in order to guide the form of the behaviour towards the end goal. For example, if we want a dog to sit quickly on command, we might start by rewarding any sit response to the command. Once this response has become reliable, i.e. the animal has learned the association between the command and sitting, we might only reward those responses that occur within a certain time following the command. In this way the animal learns that it has to sit more quickly for the reward. In practice, it is often helpful to focus on shaping only one aspect of the learned response at a time when trying to improve performance, i.e. the topography, latency, duration, frequency or error rate of the behaviour. (DSM)

Shearing

Shearing, also known as clipping, refers to the practice of removing the fleece (wool) of an animal, typically a **sheep** but also some breeds of **goat** such as Angora. In the most primitive systems knives are used to remove the fleece from the animal, but more usually hand-shears (similar to heavy-duty scissors) and electric clippers are used, the latter predominating in the more developed areas of the world. The aim is not to remove all of the wool completely, but to leave the latest growth on the animal. Thus shearing is usually carried out when the fleece has 'started to rise', i.e. the tightly packed wool fibres move away from the skin, pushed up by the new wool growth. In temperate climates such as Northern Europe, south-eastern Australia and New Zealand, this occurs at the start of summer – after the last frost but before the hottest weather.

In recent times the predominant trend in developed countries has been a move to the use of contract shearing. Specialized teams of shearers and fleece-wrappers travel to farms as and when required, often shearing in Europe during early summer (May and June) and in the southern hemisphere during the European winter months, which is summertime in the southern hemisphere.

Experienced shearers maintain that the secret of shearing is to get the sheep in a comfortable position. Thus the animal is not struggling against the shearer and its **stress** levels are lower. One advantage of removing the fleece is that it decreases the risk of blowfly strike – the flies being attracted to the faecally soiled wool and laying their eggs there that hatch into maggots. Care of the animal is paramount when shearing. The skin must be kept tight at all times in order that no flaps develop that can easily be cut by mistake. Pulling the wool, however, is inadvisable, as this pulls the skin out into flaps, which means that a cut is almost certain. The fleece should be dry at shearing, as this aids cutting and helps to prevent mechanical clippers from becoming blocked. (MM, HO)

Sheep

Sheep are animals belonging to the family Bovidae, subfamily Caprinae and genus *Ovis*.

Ovis

Sheep were one of the first species to be domesticated by humans, and archaeological data suggest that this process began in the Levant region of the Middle East and South-west Asia around 10,000 years ago. There is some doubt surrounding the wild origins of domesticated sheep (*Ovis aries*) and, until recently, it was believed that there were two separate groups of the wild mouflon, one in Europe (*Ovis musimon*) and one in Asia (*Ovis orientalis*) and that modern domestic breeds had largely developed from the European mouflon. However, the European mouflon is now considered to be a relic of the first domestic sheep that were brought into Europe by migrating groups of people, rather than the remnants of a separate species.

As newly migrated communities settled in different regions the sheep they had taken with them became isolated from other groups of sheep and they began to develop characteristics peculiar to their region. Distinct breeds emerged, often taking their name from the geographical area in which they developed. Sheep provided these early farmers with meat, wool, skins and milk, and humans have since continued to harvest these products, although geographical region and the economic climate determine how each of these commodities is valued.

During the Mediaeval period sheeps' wool was extremely valuable as the principal cloth material, and England (in particular, the great monasteries) gained great wealth by exporting wool to Europe. Some counties of England profited enormously from the wool trade, and this was largely determined by the type and quality of the fleece produced by the local sheep. For example, the Ryeland sheep of Herefordshire, which derived its name from the vast areas of rye grass grown in that county, produced the finest wool in the British Isles, fetching such high prices that it became known as 'lemster-ore' (referring to an area of Herefordshire known today as Leominster). The county of Lincolnshire was also well established as a wool-producing area by the 13th century, and its importance continued throughout the Middle Ages and up to the Industrial Revolution. The large-framed Lincoln Longwool sheep were kept on the rich grass of the Lincoln Fens and produced four to five times as much fleece as the Ryeland. The coarse Lincoln wool was exported in considerable quantities to neighbouring European countries, making Lincoln one of the richest cities of Mediaeval Britain.

As the cotton trade grew, wool was replaced as the preferred material for clothing and meat, or mutton, became the primary commodity, helping to sustain the growing urban populations during the Industrial Revolution. During this transitional period farmers, such as the pioneer Robert Bakewell, began to practise artificial **selection**, a process where animals were selectively bred for specific attributes and which allowed for the development of specialist breeds designed to maximize production of one commodity, i.e. either wool, meat or milk. The selection for economically important traits was highly successful and, in a few generations, lambs could be fattened to slaughter weight up to 12 months earlier than previously.

Worldwide there are now over 200 distinct **breeds** of sheep, of which 70 are British and, even today, the influence of British sheep breeds is considerable throughout the world. Many countries now farm sheep that are descended from British stock, the most notable exception being the fine-woolled Spanish Merino that is the favoured breed in Australia. More recently, breeds such as the Dutch Texel and the French Charolais and Rouge de l'Ouest have also become popular, particularly for crossing with cross-bred ewes to produce robust, fast-maturing lambs.

Milk sheep can be found in many European countries, including Greece, Italy and Turkey, as well as the Middle East and Saudi Arabia. There are around 25 different dairy sheep breeds, including Assaf, Friesland and British Milksheep, although any animal that has a placid temperament and can rear twins or triplets would be suitable. Sheep milk is particularly high in solids and is used for making cheese, yoghurt and butter. The fat in sheep milk also contains a high level of medium-chain triglycerides (MCT), which are highly beneficial to humans both medically, by inhibiting cholesterol deposition, and nutritionally. Sheep milk is used to feed premature human infants, as well as providing a source of dairy products for those people unable to digest cow's milk.

The genus *Ovis* is characterized by the presence of small glands in the groin area and the tear duct, as well as interdigital glands. These glands all secrete a clear fluid, which gives sheep a characteristic scent. The size, shape and colour patterns of domestic sheep vary considerably between breeds, though most domestic breeds are now principally black and/or white, with a few older-type breeds of a reddish (tanay or moorit) colour. Modern domestic sheep also have long tails, are no longer able to shed their fleece and the female (ewe) is generally smaller than the male (ram). The presence of horns in males and females differs between breeds. For example, both genders of the Swaledale breed are horned, whereas the Herdwick ram is horned but the ewe is polled. Both genders of the Kerry Hill breed are polled.

Ovis aries

Ovis aries breeds on a seasonal basis and, although rams are capable of mating all year round, their **libido** is diminished and semen is of poorer quality outside the **breeding** season. Ewes are most fertile during September, October and November in the northern hemisphere, because their **oestrus** cycle is triggered by a decrease in the amount of daylight. Sheep are, therefore, classed as short-day breeders. A ewe is seasonally polyoestrus, so will continue to cycle every 15–17 days for an average of 30 h, either until mating is successful or to the end of the season (the anoestrus period). Breeds such as the Mouflon that originate from areas closer to the equator, where there is less variation in seasonal temperature, have a longer breeding season. Onset of oestrus can be artificially synchronized in a flock with the use of intravaginal sponges containing either **prostaglandin** to induce oestrus or progestagen to prolong the length of an oestrus cycle. Artificial synchronization of oestrus allows greater control over lambing dates and shortens the lambing period.

During oestrus the ewe is more restless and tends to bleat more frequently. To detect oestrus the ram smells the ewe's **urine**, using both smell and taste, showing the '**flehmen** response' of curling his upper lip and raising his head. The ewe reciprocates by smelling the body and genital area of the ram and the pair begin to circle each other. The ram then faces the rear end of the ewe, licking her genitalia and nudging her while grunting. If the ewe is ready to mate she will adopt a characteristic position (**lordosis**) that enables the male to mount, and copulation takes place.

Embryonic implantation occurs approximately 23 days after fertilization and the **gestation** period lasts for approximately 148 days, when ewes produce lambs to coincide with the warmer weather and rich spring grass. Prolificacy varies between breeds, with those from harsher climates tending to produce fewer lambs than breeds from milder climates. Some breeds such as the Finnsheep can produce up to five lambs, although single or twin lambs are preferred as the ewe can generally raise one or two offspring with little or no intervention. The birthweight of a lamb can be affected by litter size, sex, the age and breed of its dam, along with the level of nutrition, particularly during the latter stages of pregnancy.

Optimum flock health is maintained by breeding from healthy sheep of good conformation and ensuring that ewes have a full set of teeth to optimize grass intake. Lambs are born with a set of eight milk incisors, which are replaced with permanent incisors at the rate of two each year so that a set of eight mature incisors would mean a minimum age of 4 years, and the sheep would then be said to be 'full-mouthed'. The teeth gradually spread as the sheep ages, and either fall out or break off. When this occurs, a sheep is said to be 'broken-mouthed' and, as it becomes increasingly difficult to ingest sufficient grass to maintain body condition, affected individuals are culled from the flock. Ensuring that fluctuations in nutritional requirements are met throughout the different stages of the **reproduction** cycle is vital when maintaining a healthy flock.

There are four major production stages: breeding (30–45 days), late gestation (30–45 days), **lactation** (30–45 days) and maintenance (210–240 days, the latter including the first 100 days of gestation). At each of the four phases the diet should provide sufficient nutrients for the maintenance of the ewe and lamb, both before **parturition** and up to **weaning**. In general terms, maintenance should be achieved if a ewe consumes 3% of its own body weight daily during the breeding phase and late gestation phases, 4% during lactation and 2% during the maintenance phase. An increase in the level of nutrition at mating time increases the likelihood of twins, a practice known as 'flushing'.

Requirements for specific nutrients also fluctuate between the four phases. During the late gestation phase the fetal lamb has a high requirement for carbohydrate and is able to maintain its own blood sugar concentration at the expense of the dam's. Too little carbohydrate can cause pregnancy toxaemia in the ewe, resulting in neurological dysfunction. This is often termed 'twin lamb disease', as it is more likely to occur when a ewe is carrying two or more lambs. Nutritional requirements for calcium, phosphorus and water increase markedly during the lactation phase, to ensure adequate milk supply for growing lambs.

A copper deficiency during late gestation inhibits nerve development and limb coordination in the lamb, causing a condition known as 'swayback'. Copper metabolism is different in sheep to other farmed species because it is not excreted very efficiently from the liver, which acts as the main storage organ. Excessive intake of copper will result in copper toxicosis, where the liver becomes saturated and damaged. Copper is then released into the bloodstream, killing off red blood cells, and the condition is fatal if left untreated. Copper also plays an important role in the natural pigmentation of hair and wool and in the crimp of wool. A copper-deficient fleece will have reduced elasticity and affinity for dye, reducing both its thermal and water-shedding properties for the sheep and affecting its value as a commodity.

Non-nutritional **diseases** to which sheep are susceptible include scrapie, **footrot**, **fly strike**, bluetongue and orf. Orf is a viral infection that causes severe blistering and scabbing around the muzzle and lower leg, and can be prevented through vaccination. Bluetongue is also a viral infection, which is transmitted by *Culicoides* spp. (midges), and can again be prevented through vaccination. Bluetongue causes fever, **lameness**, abortion, emaciation and can end in **death**, though the **mortality** rate is around 10%. Ulceration of the mouth with a swollen, cyanotic tongue gives the disease its name. Scrapie is a degenerative disease of the **brain** caused by a form of protein known as a prion, and is untreatable. Footrot is the most common disease in sheep, and is caused by bacteria found in the soil and faeces. When conditions are moist and warm, bacteria thrive and colonize the join between the hooves, from where they penetrate the skin. Symptoms of footrot include poor growth, cracked hooves, lameness and necrosis of the tissue beneath the horn, causing a foul smell. Vaccination can help prevent footrot, along with regular hoof trimming and the use of foot baths with antiseptic agents.

Fly strike is a particular problem in warm, muggy weather, when the green-bottle fly lays its eggs in the wool around the tail region. When hatched, the maggots feed on skin and flesh. Sheep with dirty hindquarters are particularly susceptible to fly strike, and prevention includes control of scouring, keeping the hindquarters clean and dry by cutting away any dirty fleece (dagging) and tail-docking of lambs (**see: Docking – tail**) to reduce the likelihood of dags forming. In countries such as Australia the prevention of fly strike through dagging is more difficult, as sheep roam over extensive areas of land and are both harder to identify and catch for treatment. The traditional method of controlling this disease was to remove the skin from around the perianal region of lambs (**mulesing**), and the resulting scar tissue helped to form a barrier against maggot infestation.

Although sheep are largely managed in extensive farming systems, a number of welfare problems relating to **confinement** occur, especially during **transport**, which can be over enormous distances as live export is preferred by some countries to coincide with religious festivals.

The British sheep industry has a stratified breeding system that incorporates both pure-bred and cross-bred sheep adapted to different regional environments, which perhaps explains the diversity of breed that still exists in this nation. Hardy hill sheep such as the Welsh Mountain, the Scottish Blackface and Swaledale are kept on poor-quality hill grazing for the first four breeding seasons, when they produce pure-bred lambs for flock replacement. For the fifth breeding season the ewes are brought off the hills to milder conditions on upland farms (a process that is known as 'drafting', and the ewes called 'draft ewes'). Here they are crossed with upland breeds, such as the Bluefaced Leicester or Border Leicester, to produce strong, prolific and hardy ewes for commercial flocks. There are about 12 recognized crosses, including the Mule, a Bluefaced Leicester ram crossed with a Swaledale ewe.

Many of these cross-bred, or half-bred, ewes produced on upland farms are sold on to lowland farms, where they are then crossed with a lowland meat breed such as the Suffolk or Texel to produce meat lambs. A ram that is crossed with a half-bred to produce meat lambs is known as a 'terminal sire'. Cross-bred ewes tend to mature early and can be bred from in their first year. They generally produce twin lambs and make good mothers.

Once a lamb is born the dam will lick it thoroughly clean and dry, and it is during this phase that the bond between a lamb and its dam is developed, initially using olfactory cues (**see: Maternal behaviour**). As the lamb and ewe begin to recognize each other, visual cues become increasingly important in reuniting them as the lamb grows and strays further afield. As a lamb develops and becomes more nutritionally independent, the ewe will initiate the weaning process by refusing to stand or butting the lamb away if suckling is attempted.

Flocking behaviour in sheep is a strong anti-predator response that can cause confusion in a potential predator, which is then unable to focus on one animal as a target. A flock continually assesses its environment, not only by looking out for potential predators but also by watching each other (**see: Vigilance**). Each individual tends to graze with at least two other flock members within view, enabling a whole flock to react very quickly should one animal respond to any perceived threats. An alarmed sheep will hold its head up and walk forward with rapid, but short, steps. A ewe with young lambs is more likely to stand her ground for longer against potential predators, stamping the ground with one front foot. The tendency for sheep to flock together when confronted allows farmers to handle and move large groups over extensive areas with relative ease, particularly when trained sheepdogs are used. (LW)

Sign

An objectively measurable phenomenon that is associated with a particular condition to such an extent that it is indicative of its presence. A sign must be able to be observed by someone other than the patient (i.e. not something about which the patient complains but cannot be observed – a **symptom**). In animals one can deal only with what is observable, and therefore all evidence of conditions are signs. For example, excessive barking can be a sign of **separation anxiety**; a high rectal temperature may be a sign of infection. (KT)

See also: Signal

Signal

Signals are a major component of animal **communication** – behaviour that is integral to the most important aspects of their lives. The other major components are the animal producing the signal (variously referred to as the signaller, sender, actor or generator) and the animals perceiving the signal (receivers, reactors or detectors). Signals are best thought of as the specialized information carriers linking communicating animals (i.e. signallers and receivers) in a communication network. Signals can be defined as adaptations to transfer information. This biological definition is close to the everyday use of the term signal as something that is intended to convey information. While signals always contain information, not everything containing information is a signal. This important distinction is illustrated by the difference between a weathervane and trees swaying in the breeze. The weathervane is designed to show wind direction and has specializations to achieve this, such as an easily interpreted direction indicator (the vane) and letters indicating the cardinal compass points. On the other hand, trees bend because of the force of wind and, in doing so, they incidentally give relatively imprecise information on wind direction; they do not show any specializations to indicate direction.

While signals are a diagnostic characteristic of animal communication (so much so that communication is sometimes defined as behaviour involving signals), they are only one source of information available to animals. However, signals' roles as specialized information carriers and the nature of communication mean that such information is more complex, more diverse and possibly more important than information from other sources.

Indeed, diversity is a striking characteristic of animal signals. Part of this diversity is explained by signals that use the different sensory capabilities of receivers. The term signal modality (or channel) refers to the sense used to detect the signal; for example, animals see visual signals and hear acoustic signals. Multimodal signals require two or more of the receiver's senses. There are six signal modalities:

- *Visual signals* most commonly use reflected light (although some species can generate light for signalling) and often involve movement (such as the display of brightly coloured plumage).
- *Acoustic signals* transmit information through pressure differences in the transmission medium (i.e. air or water). In regions very close to the signaller, termed the near field, information is transmitted by displacement of the transmission medium.
- *Vibrational signals* (or seismic signals) transmit information through vibrations in the substrate.
- *Chemical signals* are detected by the sense of smell/taste. They differ from other signal modalities in that signals such as scent marks are deposited (and persist) on the ground or vegetation and can be approached by the receiver, often several days after the signaller produced the signal (**see: Chemical communication**).
- *Tactile signals* transmit information through the sense of touch, and therefore the signaller and receiver must be in physical contact.
- *Electrical signals* are used by two groups of freshwater fishes (South American gymnotids and African mormyrids) that have an electrical sense and specializations of muscle or nervous tissue to generate electrical signals (**see: Electric field**).

Each of these six modalities differs in transmission speed, persistence, inherent directionality and the effect of obstacles. These characteristics influence signal form and how signals are affected by the transmission environment, thereby contributing to signal diversity. Another aspect that increases signal diversity is that animals often produce graded signals, in which signals vary in form as factors such as **arousal** and **motivation** change.

Conflicts of interests between signallers and receivers generate considerable signal diversity. Such conflicts arise because it is rare for information transmission to mutually benefit both signaller and receiver. This asymmetry in pay-off can help to explain otherwise puzzling features of signals such as differences in signal conspicuousness (less conspicuous signals if the pay-off asymmetry is small and more conspicuous signals if the asymmetry is large). Conflicts of interest and the realization that signallers and receivers exert a selection pressure on each other (resulting in co-evolution and arms races) are central to evolutionary explanations of signal features.

A closely related question is whether signals are honest. When potential prey are signalling to deter predators from attacking, or females are using male signals to choose the best mate, we might expect selection to favour signals that best achieve signallers' interests (even at the expense of receivers), that is, signals are unlikely to be honest. However, we would also expect selection to favour receivers that are adept at extracting the information that best furthers their survival and reproduction. This co-evolutionary arms race between signaller 'salesmanship' and receiver 'sales resistance' is most likely to result in signals that are 'honest on average'.

Another factor that contributes to signal diversity is that signals may be graded. That is, the form of the signal varies with internal factors such as motivation.

Signals can originate in two ways – through either **ritualization** or sensory exploitation. Signals originate by ritualization through **selection** for increased efficiency of information transfer, so that a behaviour (termed the signal precursor) that contained some information becomes specialized to transmit that information – through exaggeration, repetition, simplification and increased stereotypy. Many signal precursors fall into one of three categories: (i) intention movements that contain information on what the animal is likely to do next; (ii) autonomic responses (often involuntary) that change the body's state; and (iii) displacement activities that often indicate motivational conflict. By contrast, signals originate by sensory exploitation through selection to exploit receivers' 'pre-existing sensory biases'. For example, signals used in mate attraction may be selected to contain features to which the receiver is already sensitive in a different context, such as prey detection.

The question of whether signals contain honest information has been a major focus of research on communication, particularly in relation to mate choice, resource defence and predator–prey interactions. A rather risky example is the use by male spiders of signals for mate attraction that stimulate females' prey detection receptors. (PKM)

Further reading

Bradbury, J.W. and Vehrencamp, S. (1998) *The Principles of Animal Communication*. Sinauer, Sunderland, Massachusetts.

Greenfield, M.D. (2002) *Signalers and Receivers: Mechanisms and Evolution of Arthropod Communication*. Oxford University Press, Oxford, UK.

Maynard-Smith, J. and Harper, D. (2003) *Animal Signals*. Oxford University Press, Oxford, UK.

McGregor, P.K. (ed.) (2005) *Animal Communication Networks*. Cambridge University Press, Cambridge, UK.

Searcy, W.A. and Nowicki, S. (2005) *The Evolution of Animal Communication: Reliability and Deception in Signaling Systems*. Princeton University Press, Princeton, New Jersey.

Singer, Peter

Australian philosopher Peter Singer (b. 1946) is presently Ira DeCamp Professor of Bioethics, University Center for Human Values at Princeton University, New Jersey. Singer is also co-founder of the Centre for Bioethics at Monash University, Victoria, Australia. He is a prolific philosopher and has written on biomedical **ethics**, moral theory, human rights and the ethics of globalization. Singer's *Animal Liberation* (1975) is often credited with thrusting concern for treatment of animals used in agriculture and research into mainstream discourse. *Animal Liberation* gave philosophical teeth to other contemporaneous works, most notably, **Ruth Harrison**'s *Animal Machines* (1964) and Stanley Godlovitch *et al.*'s *Animals, Men and Morals* (1972). *Animal Liberation* inspired and spawned a generation of philosophically savvy animal protectionists and **animal rights** advocates (broadly speaking).

Singer's championing of animal protection or animal rights (in a broad sense) is underscored by his commitment to **utilitarianism**. Singer's brand of utilitarianism is *preference utilitarianism*. Unlike hedonistic utilitarianism, which focuses on maximizing happiness or minimizing unhappiness as to what is intrinsically valuable, Singer takes satisfaction of preferences and interests (and their frustration) as constitutive of what is valuable (Singer, 1979a). For Singer, morality requires that we maximize preference satisfaction (including consciously pursuing desires and projects and goals) or minimize preference frustration.

Singer begins with the view that an animal's ability to experience **pain** and pleasure, i.e. **sentience**, is an interest-making property, and should be the basis of our ethical assessment of our treatment of animals. He contends that, '[t]here can be no moral justification for regarding the pain (or pleasure) that animals feel as less important than the same amount of pain (or pleasure) felt by humans' (Singer, 1990). Elsewhere, he argues that '[a]s long as a sentient being is conscious, it has an interest in experiencing as much pleasure and as little pain as possible. Sentience suffices to place a being within the sphere of equal consideration [of like] interests' (Singer, 1979b). On his **preference**-based analysis of what is intrinsically valuable, Singer contends that animals that can conceive of their own futures and have the capacity for forward-looking desire satisfaction are harmed morally when they are not allowed to exercise their preferences.

According to Singer, sentient animals are *sufficiently* like us in morally relevant ways – they have desires, preferences about their future and can experience pain and pleasure. He undermines the weight given to human **language** and rationality by contending that 'pain is pain', no matter the identity of the experiencing subject, and that equal consideration of similar interests does not hinge on seemingly unique human qualities. The combination of the ideal of interspecific equality, coupled with the sentience criteria and his brand of utilitarianism, and the fact that large scale animal farms and research facilities allegedly are bound to cause immense suffering, leads him to the conclusion that we should boycott these industries. Singer concludes that we ought to become moral vegetarians and we should cease unnecessary use of animals in medical research, since the overall benefit to humans and animals does not justify the harm done to animals (Singer, 1975, 1990).

Singer's prescription for moral vegetarianism is not based on the thesis that killing certain animals is wrong in itself. Since Singer is committed to utilitarianism, his objections to killing emphasize the gross harmful effects that killing animals has on reducing the amount of preference for a pleasant life. If husbandry conditions are improved such that the preference for a pleasant life is maximized, then Singer's utilitarian calculus might produce a different conclusion to the moral obligation to be vegetarian. Singer's approach as discussed here is consistent with efforts to *reform*, not eradicate the various industries of animal use.

In terms of reform, an effective strategy in some cases has been to boycott industries of animal use, for example commercial animal agriculture. However, if removing oneself from the decision-making process minimizes utility, then reform may be better brokered through active participation in public policy formation or conscientious consumerism. Because the utilitarian calculus is contingent on shifting empirical facts, **vigilance** as regards structural changes and improvements in animal **welfare** on an ongoing basis should be observed in order to promote the best possible balance of benefits over harms for all considered.

Singer has helped to popularize two concepts that get used as fuel in the animal liberation debate. The first, '**speciesism**', coined by **Richard Ryder** (1970, privately printed leaflet), criticizes discrimination against animals (or privileging human interests) *simply* on the grounds of species identity. If we lack a morally relevant difference for being partial to human interests, then we should abandon species boundaries as a way of excluding animals from the moral community; otherwise, we embrace an indefensible form of prejudice akin to racism or sexism (Singer, 1979a).

Singer also made famous the 'argument from marginal cases'. Exemplifying the virtues of logical consistency, the argument contends that we must provide a morally relevant difference to justify differential treatment of animals and human beings. However, the qualities that are often given, such as rational agency, language, self-awareness and self-governance, fail to apply to all and only human beings. 'Marginal' human beings – for example, profoundly mentally disabled individuals such as those with hydrocephaly, Tay-Sachs or anencephalic newborns – may not have these above-mentioned qualities and may never demonstrate them. If the permissibility of animal agriculture or animal research is justified on the basis of the former qualities, then we would justify using 'marginal' humans.

Fellow utilitarian Raymond Frey holds that, because the value of a life is a function of its quality, and its quality corresponds to the enrichment opportunities open to the subject, Singer's argument does not go far enough with respect to equal consideration. Frey is less hesitant than Singer about using marginal human beings to promote the overall good of other beings, in place of non-defective adult animals. (RA)

See also: Ethics

References

Godlovitch, S., Godlovitch, R. and Harris, J. (1972) *Animals, Men and Morals: An Inquiry into the Maltreatment of Non-humans*. Victor Gollancz, London.

Harrison, R. (1964) *Animal Machines: The New Factory Farming Industry*. Vincent Stuart, London.

Singer, P. (1975) *Animal Liberation: A New Ethic for Our Treatment of Animals*. Jonathan Cape, London.

Singer, P. (1979a) *Practical Ethics*. Cambridge University Press, Cambridge, UK.

Singer, P. (1979b) Killing humans and killing animals. *Inquiry* 22, 145–156.

Singer, P. (1990) *Animal Liberation*, 2nd edn. Avon Books, New York.

Skinner, Burrhus Frederic (Fred)

Burrhus Skinner (1904–1990) was one of the early champions of **behaviourism** and the experimental laboratory approach to **psychology** and learning (referring to the mind as a 'mysterious and dangerous metaphor'). His enthusiastic work on **instrumental conditioning** has led many to refer to it as Skinnerian learning, in recognition of his enormous contribution to the field. This includes the development of a controlled experimental environment that allows the automatic recording of the behaviour of animals in response to different contingencies, and which bears his name (Skinner box). An additional possible reason why Skinner achieved such notoriety within the discipline relates to his philosophical skill and appreciation of its global influence. He was not afraid to challenge conventional thinking in relation to the implication of his work and was quoted as saying 'All human beings are controlled, but the ideal of behaviourism is to eliminate coercion, to apply controls by changing the environment in such a way as to reinforce the kind of behaviour that benefits everyone'.

Although some of his work was sponsored by the military, Skinner objected to the harm caused by war. He believed that behaviour should be shaped with positive **reinforcement**, and often rejected the use of **punishment**. Although some Skinner boxes have electrified floors for delivering **aversives**, this was not a focus of his work, and others were more willing to use much more powerful currents than he, challenging his results that punishment had only temporary suppressive effects on behaviour.

Skinner originally aspired to be an author and poet before turning to a career in psychology, and published one notable piece of popular fiction, *Walden Two*, a story about a community living a Utopian existence according to behaviourist principles, which challenges many of the principles conventionally associated with a free society. The book has sold over 1 million copies globally. (DSM)

Slats

Slats are made from a number of parallel beams (slats) with gaps or slots in between. They are used for floors in cattle, pig and sheep **housing**, so that faeces and urine can fall between the slats into a pit, without the floor having to be cleaned. A high **stocking density** of animals is often required so that the animals tread the excreta through the slots. Slats may be made from concrete, wood or plastic. Poorly constructed slatted floors cause animals to walk with their heads down, fixing their gaze on the floor ahead of them. This enables them to position their hooves carefully; there is less walking activity in total and it is slower than if the animals are on solid floors. Animals may even spend less time grooming their hindquarters because of the risk or fear of overbalancing.

However, well-designed slatted floors allow animals to be kept at high stocking densities without bedding and to remain reasonably clean. If slats are too narrow, there is an unacceptable strain on them, particularly if they are very long; but, if they are too wide, there is inadequate disposal of faeces between the slats. Optimum slat width is about 150 mm for cattle, and the slats should be T-shaped in end-on profile to encourage dung to fall into the pit below. The gaps between the slats should be about 40 mm wide. If they are less than this the faeces do not pass through easily. Great care should be taken that slats have been manufactured from high-grade materials to the required loadings for the class of stock to be kept on them, including the necessary reinforcement. (CJCP)

Slaughter

In comparison with killing an animal for sport or pest control or the humane destruction of **companion animals**, the process of slaughtering an animal for meat is constrained by requirements for food safety, in terms of minimizing food contamination and drug use, and economical considerations. Consumers demand cheap meat, which drives producers to develop faster and more cost-effective production systems.

Animal slaughter for meat is almost always conducted in a **slaughterhouse**. To slaughter an animal is to cause **death** by loss of blood; sticking is the act of making an incision into an animal to allow bleeding. In the European Union (EU) all animals are required, under Directive 93/119, to be stunned prior to slaughter. Under this Directive, Article 3 requires that no animal should be caused or permitted to sustain any avoidable excitement, **pain** or **suffering** during **restraint**, **stunning**, slaughter or killing. There is a derogation for **religious slaughter**. Equivalent legislation or guidelines have been passed in most developed countries, including the US Code of Regulations Title 9, Chapter 3, Part 313 Humane Slaughter of Livestock, the Canadian Meat Inspection Act and the Australian NACCAW guidelines on The Welfare of Animals at Abattoirs 1994 (as amended).

To stun an animal is to render it unconscious and insensible to external stimuli of any sort. The principal **welfare** issues arising from the stunning and slaughter of animals for food relate to:

- **fear** and **distress** due to poor pre-slaughter **handling**;
- severe pain, fear and distress due to failure to stun; and
- severe pain, fear and distress due to recovery of sensibility prior to death during bleeding.

Although all of these problems may occur during the slaughter of any animal, specific welfare issues tend to be associated with particular handling, stunning and killing systems and to certain species. Many handling, stunning and slaughter systems that worked well and with minimal welfare problems, when throughput was low and labour plentiful, have inherent problems at higher throughput speeds or where tasks are automated. Some welfare problems specific to certain commonly used systems and to species commonly slaughtered for meat are examined in some detail below. The methods used to monitor each problem and actions which may be taken to address it are then discussed.

Fear and distress due to pre-slaughter handling

Cattle, sheep and pigs

Injury, pain, fear and distress may arise in poorly designed premises and handling systems for **cattle**, **sheep** and **pigs**. The mixing of different groups of animals from different farms or pens on arrival at the **lairage**, the pre-slaughter holding area, may cause fighting, resulting in fear and injury. This is a particular problem for pigs and uncastrated male cattle. Hot or humid lairage conditions may cause discomfort in ruminants, but **heat stress** is a particular problem for pigs, which have a narrow thermoneutral zone and are less able to control their body temperature. In pre-slaughter handling systems, including unloading ramps and **raceways**, animals may be reluctant to move forward due to noise, draughts, poorly positioned personnel, or poorly designed races with right-angled turns and open sides. This may result in lairage personnel beating, prodding or dragging animals, or twisting sensitive parts of their anatomy in order to persuade them to move forward, especially where throughput speeds are high. At some premises, electric goads are routinely used for unloading and moving pigs through the lairage and into stunning pens or gas-stunning equipment.

Of particular note are high-throughput pig slaughterhouses where, typically, approximately 600 pigs may be killed per hour, through 90% carbon dioxide stunner/killers. For controlled-atmosphere stunning, pigs are loaded into suspended cages, or gondolas, and lowered into pits full of carbon dioxide. Carbon dioxide is very **aversive** to animals, and required throughput rates for some stunners that are currently widely in use may be achieved only by using teams of lairage personnel equipped with electric goads. The use of a **goad** to load a gas stunner is illustrated in Fig. S.1.

Driving practices may be monitored by routine recording of the number of carcasses affected by bruising and stick marks and the extent of the lesions; stick marks in cattle are shown in Fig. S.2.

There are a number of systems available for the assessment of carcass damage in pigs. Such systems use a point-scoring

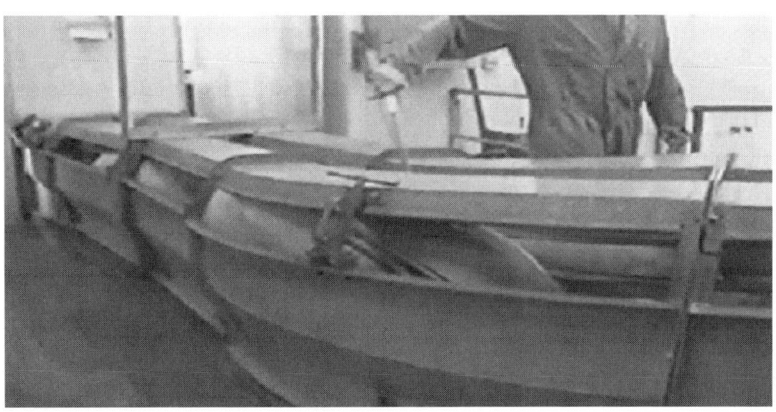

Fig. S.1. The use of electric goads is often necessary when loading animals into controlled-atmosphere stunners.

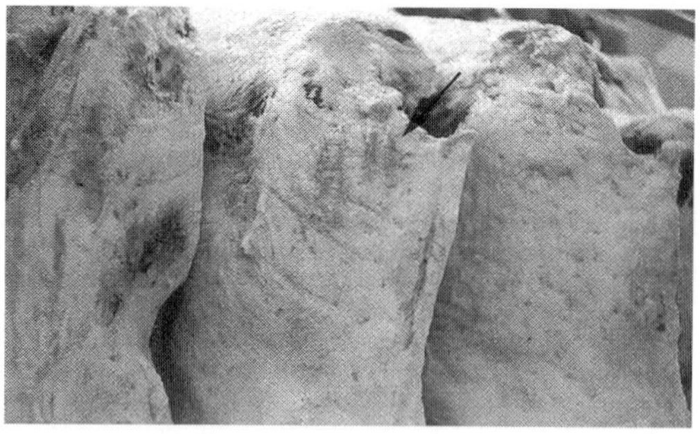

Fig. S.2. Stick bruising on cattle carcasses due to poor pre-slaughter handling.

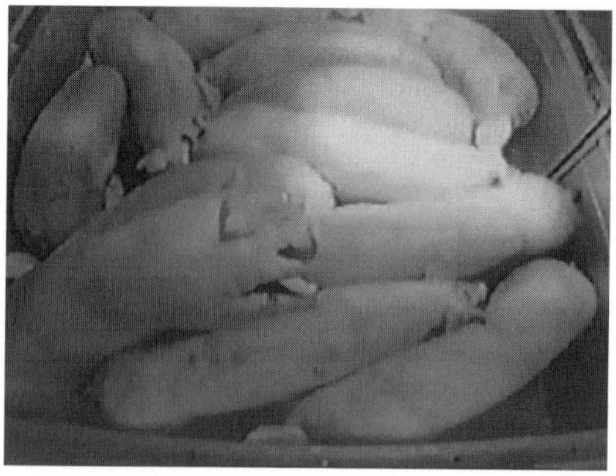

Fig. S.3. Pigs fighting in the lairage.

system, usually from 1 to 4 or 5, to record the extent of damage in a number of defined areas of the carcass, usually four or five areas. The sum of scores may then be used to compare damage between carcasses. A sample of several carcasses from each herd killed allows comparison of handling between herds. However, this will give a measure of overall damage, including damage to animals on the farm and during loading. Handling in the lairage may be assessed by routine measures of the frequency of goad use, beatings, tail twisting, animals slipping or falling, **vocalization**, etc. in a given time period.

The cumulative amount of pre-slaughter handling stress may be roughly estimated by measuring the pH of carcasses. When animals experience high levels of **stress** during fasting and loading on the farm, in **transport** and in handling in the lairage, their carcasses may have either low pH, causing pale, soft, exudative (**PSE**) **meat** or high pH, causing dry, firm, dark (**DFD**) **meat**. pH meters are available, but must be used at specific anatomical locations on the carcass and at specific times post-slaughter in order to allow valid comparison between both carcasses and slaughterhouses.

Fighting in the lairage may be avoided by preventing the practice of mixing animals on arrival. Extension of lairage area or reduction in throughput rate may be considered if mixing of groups occurs frequently (see Fig. S.3.).

Heat stress may be avoided by installation of suitable ventilation systems in the lairage. Sprinkler systems for pigs may reduce the chance of heat stress and will also reduce fighting. However, sprinkler systems may increase the risk of heat stress through increasing humidity in very hot weather, and may cause **cold stress** when used to prevent fighting in cold weather. Pigs suffering from cold stress are clearly identified, as they shiver. Temperature, relative humidity and ammonia levels may be continuously monitored in the lairage, and action may be taken to reduce them when excessive. As a guide, temperature of over approximately 26°C and a relative humidity of roughly >90% are likely to cause problems for pigs, although higher temperatures are tolerated where relative humidity is low. Levels of ammonia of over 25 ppm are aversive.

An adequate supply of clean water should be available at all times to all lairaged animals. There should be a sufficient number of drinkers or troughs, in good working order and at animal height, in each lairage pen. Pressure on throughput may cause overstocking of pens or inappropriate classes of livestock to be held in them, with the result that the drinker:animal ratio may be too small or drinkers may be at an inappropriate height. Poor lairage management, due to poor personnel training or overstretch, may result in non-serviceable drinkers.

Pre-slaughter handling problems may be prevented by improving unloading facilities and raceways to eliminate steep ramps (those over 20°), slippery floors, sharp corners, noises caused by machinery or lairage personnel, open-sided raceways and those going from light to dark areas. Scientific research, which includes a considerable body of work by Temple Grandin based in the USA and Canada (see Further reading, below), has identified a number of problem areas in this respect. In practice, problem areas may be identified by observation of animals being moved through the lairage to identify locations at which animals refuse to move forward, which physical coercion is routinely used, where animals slip or where there is excessive vocalization.

Expert advice may be sought when designing the layout of, and materials used in, new lairage facilities or when existing facilities are being extended, in order to ensure that they cause minimal welfare problems during unloading and pre-slaughter handling. In many countries cattle must be identified at the slaughter plant for the purpose of disease control, including that for new variant Creutzfeldt–Jakob disease. This causes additional stress if restraint is required in order to read ear tags. Electronic systems of identification may help to avoid this additional **stressor**.

Poultry: broiler chickens, end-of-lay hens, turkeys and ducks

All poultry are 'harvested', either manually or by machine, into crates that may be stacked into a module. Heat stress frequently causes severe discomfort, distress and death in hot weather. This is a particular problem for meat birds, due to their high metabolic rate and narrow thermoneutral zone and to the high stocking densities at which they are transported. Hens may suffer from cold stress in cold weather due to poor **feather** cover at the end of lay.

At the plant, birds are removed from the crates, either manually or by tipping on to a conveyer belt, and inverted to be placed in shackles, i.e. metal frames shaped to allow the legs to be slotted downwards into them to secure the bird on the kill line. This causes fear and distress, and frequently pain, which arise from the handling and shackling of birds with leg

disease or fractured bones, from which conditions poultry frequently suffer. Typical line speeds of up to 200 birds/min for broilers and 50 birds/min for **turkeys** make speed of shackling imperative and so preclude careful handling.

Heat stress in the lairage may be monitored by estimating the proportion of birds that are panting. The problem may be addressed by reducing the **stocking density** of birds in the crates, careful planning of 'harvesting' and journey to the plant to reduce lairage times and, if necessary, installation of fans in the lairage.

Pain at shackling is inevitable in plants using water bath stunners. However, levels of excitement may be reduced by using blue light in live bird areas, installing breast comforters on the line and reducing noise, draughts and personnel in live bird areas. Scientific research has shown that blue light has a calming effect on poultry, and a breast comforter, consisting of a continuous rubber strip fixed at breast level along the kill line, prevents the bird from swinging and so reduces distress. Sharp bends in the line cause birds to flap and are therefore usually avoided at the plant design stage. The use of controlled-atmosphere stunners avoids the need to shackle live birds.

However, there is some controversy over current, commercially available gas stunners that use carbon dioxide, which is aversive to birds but is inexpensive and produces good carcass quality. The alternatives include inert gases, such as argon, which are not detectable by the birds. However, inert gases may be more expensive and cause anoxic convulsions, which may be aesthetically unpleasant to operatives and give poor carcass quality due to blood splash.

Severe pain and fear due to failure to stun or recovery of sensibility

Cattle, sheep and pigs

Cattle are most commonly stunned through use of a **captive bolt** stunner, as illustrated in Fig. S.4, which disrupts the **brain** structure by accelerating the brain against the skull.

The positioning of the stunner is critical and requires considerable skill. UK slaughterhouses generally use cartridge stunners, while in the USA hydraulic instruments are commonly employed. Re-stun rates may be up to 5% for hydraulic stunners but are lower for cartridge instruments. The effectiveness of head restraints to aid stunner positioning is controversial. An electrical stunning box is available for cattle, widely used in Australasia. There is some debate concerning the effectiveness of electrical stun boxes for cattle. The electrical stun box delivers a head stun, which causes immediate insensibility, followed by a nose to heart stun that stops the heart, eventually leading to death, and finally a spinal depolarization to reduce convulsions and protect the slaughterman. The final depolarization may mask signs of an effective stun. The boxes involve complex electric and computer systems, which must be serviced by only trained personnel. The currently available box does not incorporate instrumentation indicating currents applied under load. Dirty cattle or electrodes or inadequate water supply to electrodes will impair the effectiveness of stun.

While captive bolt stunners usually kill an animal outright, cattle may recover from an electric stun where there is an

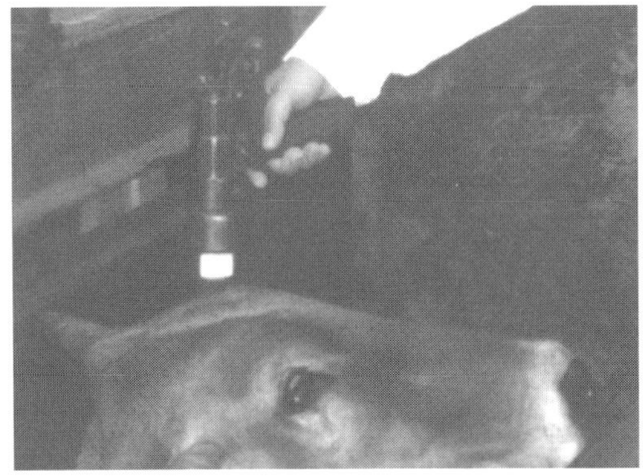

Fig. S.4. The use of a captive bolt stunner for cattle.

inadequate heart stun; stun to stick time is therefore critical, and should be less than 40 s to ensure that a stunned animal does not recover sensibility. Clonic convulsions, i.e. rapid, repeated and uncontrolled paddling limb movements, occur in animals for which there is inadequate spinal depolarization, which may delay sticking.

Sheep are stunned by a head-only electrical stun, using a hand-held electrical stunner, or by captive bolt stunner. Many sheep stunned using an electrical stunner show signs of recovery of sensibility on the bleed line, i.e. that part of the kill line incorporated to allow the legally required, minimum duration of bleeding before further dressing of the carcass may begin. The maximum voltage applied by hand-held stunners is limited due to health and safely considerations, and there may be variation in electrical impedance of sheep heads, due to either wool on the head for some breeds or extensive skin contamination from muddy conditions in the field or during transport.

There exists, therefore, a risk that insufficient current may be delivered, causing inadequate stunning. Sheep are usually stunned while manually restrained, with the resultant possibility that positioning of stunning tongs may be poor, leading to failure of stunning. Stunning for 3 s, at a constant voltage of 250 V, has been shown to produce an adequate stun if tongs are correctly positioned.

Pigs may be stunned using either a hand-held electrical stunner, to give a head-only stun, an automatic stunner that applies both a head and heart stun, or a controlled-atmosphere stunner, usually carbon dioxide. Where head-only stunning is used, the duration of application of the head electrodes is critical; this should be 3 s for high-voltage units and 7 s for low-voltage stunners. The stun to stick time for these pigs should be less than 15 s, to ensure loss of sensibility until death. The positioning of head tongs is critical, and should be immediately below the ears to ensure that the shock is applied across the brain. Poor positioning of head tongs or poor electrode contact may cause some pigs to be non-stunned or to show signs of recovery on the bleed line. Signs of recovery in pigs after carbon dioxide stunning may be due to inadequate duration of exposure to the gas, air pockets in the stunner or inadequate carbon dioxide concentration.

Effectiveness of stun may be monitored by recording signs of recovery of animals, both immediately after stunning and then on the bleed rail at a point 50 s from stun point in cattle and 35 s for pigs. These may include the return of eye and righting reflexes and rhythmic respiratory movements. Legislation and guidelines for duration and voltage required for stunning and for stun to stick times, as well as other aspects of handling, slaughter and killing, have been informed to a large extent by scientific research. This work has included analysis of spontaneous electrocorticograms (ECoGs) and visually evoked responses (VERs) or somatosensory evoked potentials (SEPs) taken during stunning and slaughtering performed under various conditions. ECoGs measure electrical activity in the brain; an electrical pattern similar to an epileptic fit is thought to signify insensibility of the animal, as people with epilepsy have no memory of events occurring while experiencing a fit. VERs are measured by implanting an electrode in the visual cortex of an animal and then shining a light source into the animal's eye. Where no electrical response to the stimulus is recorded, the animal is considered to be insensible. SEPs are measured by a similar method, but using a mechanical stimulus, such as a pin-prick.

If signs of recovery of animals on the kill line are seen, stunning tongue positioning may be observed and, if necessary, corrected; duration of stunning may also be recorded and corrected if inadequate. Additionally, stun to stick times may be recorded and sticking techniques monitored. Poor bleed-out due to small stick holes in pigs may delay brain death. Although large stick holes may increase risk of contamination in the scald tank, trimming of stick holes is a viable option, which avoids risk of poor bleed-out and recovery of sensibility prior to death. In cattle a thoracic cut, where all vessels leaving the heart are severed, is essential for rapid bleed-out.

Concentration of carbon dioxide and duration of exposure of pigs may be checked where signs of ineffective stunning are seen after controlled atmosphere stunning. A gondola or pig may be marked and timed from entry to exit of the stunner. As a guideline, pigs should be exposed to a concentration of carbon dioxide of at least 90% for 60–100 s in order to produce an effective stun under commercial conditions. There is experimental evidence that inert gases, such as argon, are less aversive to pigs than is carbon dioxide, and research into appropriate gas concentrations and mixtures and effective duration of immersion in these atmospheres is being carried out; effective exposure duration for a 95% argon atmosphere is thought to be >180 s. Gas stunners incorporate a gas concentration monitor. However, the concentration of gas in the stunner varies with position, and the monitor will indicate the gas concentration only at the location of the monitor. If it is placed near the bottom of the stunner, pigs may be exposed only to the display concentration for a part of the time that it is in the stunner – the remainder of the time they will be exposed to lower concentrations. An independent measure of carbon dioxide concentration, at several sites in the stunner, may be necessary. Where signs of recovery of pigs are seen on the line, line speeds may be reduced until no further recovery events are recorded during monitoring.

Poultry

Poultry are usually stunned in an electrically charged water bath, followed by automatic or manual neck cutting, but may also be killed outright in a controlled-atmosphere chamber. In parts of Europe 90% carbon dioxide stunning is used, which is currently only possible under a legal derogation in the UK, where either 90% argon or inert gas and 2% oxygen or 25–30% carbon dioxide and 60% argon or inert gas, in atmospheric air, is required for controlled-atmosphere stunning and killing. Research into the use of alternative gases and gas mixtures is being carried out. Some low-frequency electrical stunners may kill the bird outright, through cardiac arrest, but high-frequency equipment causes only a stun.

An effective stun causes birds to vibrate gently with wings held away from the body; rhythmic breathing and eye movements are absent. Birds may miss the water bath and so be non-stunned due either to head arching or incorrect water bath height. Additionally, where there has been uneven growth of birds in a flock, smaller birds may miss a stunner that has been adjusted to be correct for the average size of bird. However, even where all birds enter the water bath, some may not be stunned. This is because variable impedance of birds, or the leg shackle contact point, may result in birds receiving different currents, which, where a bath is set at the minimum 'effective' voltage in order to reduce carcass damage, may result in some birds receiving insufficient current to cause a stun.

Recovery of birds on shackle lines may also be due to failure to neck cut by either manual or automatic cutters. During manual cutting, commonly used to kill turkeys, birds may be missed due to operator fatigue and/or fast line speeds, while automatic cutters may miss birds due to incorrect height adjustment, uneven bird size or neck arching by non-stunned birds. Failure to cut birds may result in conscious birds entering the scald tank.

Where birds are being inadequately stunned, or stunned but not neck cut, the line speed may be reduced and additional back-up personnel placed at the stun bath, neck cutter and at the bleed line to re-stun or cut affected birds. Water bath and neck-cutter heights may be adjusted and water bath voltage increased. Birds may be re-stunned with a hand-held electric stunner or may be decapitated. A recently developed commercial percussion stun/killer, as illustrated in Fig. S.5, is more effective and humane than decapitation.

Fig. S.5. Percussion stunner/killer for emergency poultry slaughter.

Welfare issues in minority species

The slaughter and killing of each minority species, such as **rabbits**, **ostriches**, **deer**, wild boar and **horses**, have specific welfare problems. Common to most of them, however, is the fact that slaughter facilities may be widely spread geographically, causing prolonged transport times and associated feed and water deprivation and fatigue. Species-specific handling and slaughter equipment that has been tested for efficacy may not be commercially available, with the result that equipment designed for more common species is modified without any appropriate testing. Additionally, there may be little scientific evidence available to determine correct handling, stunning and slaughter techniques, including correct voltages and stun durations. Restraint may be problematic where animals have only relatively recently been domesticated, such as ostriches, deer and wild boar.

There is some ethically based controversy about horse slaughter for meat in developed countries, as these are regarded as companion animals by many people; horse slaughter in these countries may therefore be carried out with maximum secrecy and welfare issues may remain hidden. Horses are usually killed in a similar way to cattle, using a captive bolt and neck cut, although a rifle may be used in some countries. The most significant welfare issues in horse slaughter relate to very long journey times, water and food deprivation, injury and discomfort due to cramped and poorly designed transport vehicles and fighting resulting from fear and the mixing of animals. Double-decked horse transporters are to be phased out in the USA.

A specific problem for ostrich slaughter is identification of rhythmic breathing, which may be difficult due to spinal reflexes that produce contraction of limb muscles and may be confused with respiratory movement. An effective stun has been found for most birds, with an electrical stunning current of >400 mA at 50 Hz applied to the head only for at least 3 s, provided that they are bled within 60 s from start of stun. Exact current and duration of stun required vary with design of head electrodes, which is a subject currently under scientific investigation. Mechanical stunning using a captive-bolt stunner is possible, but hoisting the bird may be difficult as there is no tonic, or passive, phase to the epileptic fit produced. A captive needle pistol, using air pressure, has also been used to stun ostriches.

Due to difficulty in handling and transporting deer, they are usually killed in the field, either by rifle at a distance or by captive bolt. For stunning by captive bolt, deer may be restrained in an electrically operated crush, which lifts the animal's feet off the ground in order to reduce struggling and injury. Rifle shooting is less reliable than captive bolt stunning, and animals may suffer injury and distress for a significant period before being killed. A modified, hand-held electrical stunner, applied for 0.5 s using 130 mA, with a stun to stick time of <10 s, has been shown to be effective in deer provided that there is adequate head restraint.

Wild boar may be stunned using hand-held electrodes or killed with a rifle, either in the field or in a pen, or by using a shotgun. A minimum stunning current of 140 mA, achieved by applying 100 V, has been shown effectively to stun rabbits.

With an increased consumer awareness of food production methods and greater public concern for animal welfare, there is a risk that pre-slaughter handling, stunning and slaughter practices that are currently widely in use may become unacceptable. It is essential that these practices be carefully monitored and deficiencies immediately and effectively rectified. Active research to develop more humane and more reliable methods of handling, stunning and killing animals is imperative. (SH)

See also: Dark cutting

Further reading

FAWC (2003) *Report on the Welfare of Farmed Animals at Slaughter and Killing*. Farm Animal Welfare Council, PB8347, London.

Grandin, T. (2007) *Livestock Handling and Transport*, 3rd edn. CAB International, Wallingford, UK.

Gregory, N.G. (2007) *Animal Welfare and Meat Production*, 2nd edn. CAB International, Wallingford, UK.

Warriss, P.D. (2009) *Meat Science: an Introductory Text*, 2nd edn. CAB International, Wallingford, UK, Chapter 10.

Slaughterhouse

A slaughterhouse, or abattoir, is a premise where animals are killed and slaughtered to produce food. A large slaughterhouse may also be described as a processing plant, especially where there is further processing of the meat produced. Slaughter is specifically the causing of **death** through loss of blood. (SH)

See also: Lairage; Slaughter

Sleep

Sleep is characterized by a temporary period of inactivity and a raised response threshold. It is exhibited in a distinct **diurnal rhythm** and with a characteristic posture, usually with neck recumbency. For example, in cattle this is usually in the sternally recumbent position with the head either resting on the ground or tucked round and held against the thorax.

Four levels of alertness can be distinguished:

- Alert wakefulness (AW): eyes fully open, characterized by a low-voltage, fast-activity **electroencephalogram** (EEG) output by the **brain**.
- Drowsiness (DR): upper eyelids relaxed, increased **arousal** threshold and reduced alertness. Some high-voltage, slow-activity electroencephalogram (EEG) output.
- Quiet sleep (QS): eyes almost closed, increased arousal threshold and all EEG output of the high-voltage, slow-activity type.
- Active sleep (AS): eyes fully closed, all EEG output of the low-voltage, fast-activity type, some rapid eye movement (REM), heightened arousal threshold. Otherwise known as paradoxical sleep.

In diurnal species AW tends to predominate during the day and DR at night; for example, in cattle 52, 31, 13 and 3% of the 24-h day are spent in AW, DR, OS and AS, respectively. AS occurs only at night and often after a bout of **rumination**. Rumination often accompanies DR at night and may also accompany AW and occasionally QS. The state of QS in many animals is preceded by regular, monotonous actions like ruminating. Reticulo-ruminal motility is decreased at night, probably because of reduced basal metabolic rate during DR and QS. The normal transition is from AW to DR to QS to AS. Bouts of AS are short in cattle and horses in comparison with other mammals, typically only 5 (cattle) to 15 (horses) min, but they are numerous. Hence, sleep in these species can be considered as polyphasic in contrast to humans, in whom it is monophasic.

Sleep is also characterized by its consistency, as patterns vary little from day to day when health is good but are specific to individuals. Occasionally during AS, animals demonstrate tachycardia, rapid breathing and/or limb movements, which are thought to be indicative of dreaming. The amounts of both AS and QS often decline with age.

There are two probable functions of sleep: immobilization and recuperation. In some animals regular, temporary immobilization is beneficial in guarding against **predation**, providing that the sleeping site is secure. In large, **grazing** prey animals this is often not the case, and hence most grazing ungulates have short sleep periods. In contrast, for most forest-dwelling prey such as the dormouse, safe sleeping sites are available and sleeping times are long, even extending to hibernation in many species. Thus immobilization offers two evolutionary advantages – energy conservation, which is of use to predator and prey alike, and security from predation, since most predators search for moving prey.

Recuperation is likely to be of importance, since prolonged deprivation leads to a state of exhaustion, the exhibition of **abnormal** behaviours and even hysterias. However, there is no evidence that longer sleep patterns per se occur more frequently in more complex animals, suggesting that recuperation of brain function in general is not directly related to sleep duration. Indeed AS may specifically function to maintain the brain in a state of readiness during the subconscious state. Also, animals that are physically more active do not use sleep more as a method of recuperation. On the contrary, many inactive or sporadically active animals such as sloths, bears and lions sleep for long portions of the day. The choice of whether to become an active feeder for most of the day and minimize sleep, or whether to save energy by maximizing sleep, is determined by the security of the sleeping site and the rate at which the ingested food can be digested.

Sleep is therefore an important component in an animal's behaviour repertoire. Its inclusion in welfare indices is advocated by some scientists, although the duration of sleep per se is not directly related to **welfare** – disturbances in sleep pattern may be a more useful measure. Some modern housing systems, with high density of animals and low levels of comfort, may preclude adequate sleep in some animals, reducing their welfare significantly. (CJCP)

See also: Flooring

Further reading

Smith, C. (1985) Sleep states and learning: a review of the animal literature. *Neuroscience and Biobehavioural Reviews* 9, 157–168.

Van Reeth, O., Weibel, L., Spiegel, K., Leproult, R., Dugovic, C. and Maccari, S. (2000) Physiology of sleep (review). Interactions between stress and sleep: from basic research to clinical situations. *Sleep Medicine Reviews* 4, 201–219.

Social behaviour

Social behaviour refers to the behavioural interactions occurring between individuals of the same species or between different species that share some sort of structured relationship, which might be cooperative (**see: Cooperation; Social group**) or competitive. Social behaviours which promote social inclusion or cohesion within the group, e.g. reciprocal **altruism**, may be referred to as prosocial behaviours.

For a social interaction to exist one individual must first elicit a response from another (this may be termed a reaction), which feeds back to the first individual, i.e. there is a bidirectional exchange of information between the animals involved (**see: Communication**). Some species capable of interspecific social behaviour (**see: Interspecific behaviour**) may adapt their intraspecific behavioural repertoire in order to improve communication with another species. For example, meowing by the **cat** is rarely used intraspecifically but is commonly used during vocal exchanges with humans. However, interspecific social behaviour does not require that they view heterospecifics as **conspecifics**, as is sometimes implied. For example, **dogs** adapt similar social **play** sequences according to whether the play partner is another dog or a human and use different perceptual processes when interacting with the two species. None the less, many trainers and animal carers now recognize the potential value of using 'ethologically relevant' signals to improve social interaction with another species.

The potential for cooperative social behaviour often arises from the need or desire to achieve common goals efficiently, e.g. **reproduction**, distribution of **resources**, etc., and in this case a social **bonding** may exist between the two animals. This may be identified by a tendency to associate with the individual and **distress** when separated. Thus the patterning of social behaviours can be used to define the relationship between and role of individuals within the **social group** (e.g. least preferred and most preferred associates). In addition to behaviours associated with cohesion between individuals (including acts of **affiliation**, such as **courtship behaviour**, as well as **maternal behaviour** and **paternal behaviour**), social behaviour also includes interactions associated with the defence of resources against others (e.g. **aggressive behaviour**, **agonistic behaviour** and acts of **territoriality**), which, if persistent or of high intensity, can lead to **social stress**.

Although **social releasers** may result in the exhibition of fixed action patterns or **modal action patterns** from another individual, much social behaviour is typically quite variable, since social stimuli are less predictable than physical environmental stimuli to which behaviour may be directed. The response of one individual is dependent not only on the stimulus presented by the other individual, but also by its internal **motivational state**. Thus at one time a threat may result in withdrawal by an individual but at another time the same act may result in a reciprocal threat, for example if the individual is more hungry. Therefore, it can be useful to conceptualize social behaviour as collective expressions of individual, goal-directed strategies (which may be compatible or not) by more than one individual in a group, which affect other individuals.

However, this is not to say that genetic factors are not important in the expression of social behaviour, and sociality is a heritable trait (**see: Personality**). The relative importance of genetic factors to the evolution of social behaviour is debated and a focus of **sociobiology**. Clearly, genetic factors are important to the evolution of sociality, which, with its associated advantages and disadvantages, affects the evolution of social behaviour in any given context. Disadvantages include: (i) the potential costs associated with increased competition for breeding- and feeding-related resources; (ii) increased **disease** and **parasite** burden risk; (iii) increased exposure to **predation** as a result of a group being more conspicuous; and (iv) increased risk of exploitation by others, any or all of which can result in particular types of social behaviour and related social **cognition**. For example, it has been suggested that the risk of exploitation by others within a social group (itself a potential form of social behaviour) may have been a key driver in the evolution of a range of complex problem-solving abilities.

Advantages to sociality include: (i) increased defence capacity against predators and protection of young; (ii) greater potential to identify and exploit important resources in the environment (such as food and mates, although competition following identification may be increased) and the ability to exploit different ecological niches; (iii) improved population stability; and (iv) the ability to modify the environment (rather than be modified by it) – all of which depend on the development of related social behaviours. Psychological benefits and associated health benefits may also arise from certain social behaviours – for example, mutual **grooming** in some species has been shown to result in lowered **heart rate**, as well as **oxytocin** and **endorphin** release, which may be associated with relaxation and affiliation. The relative importance of these potentially beneficial and detrimental factors to any given species can be expected to affect the nature of social behaviours expressed in relation to the given context.

Historically, the study of the social behaviour of animals in **captivity** has given great emphasis to competitive interactions between individuals over resources and the significance of **hierarchy** as a consequence of which there may be reduced physical harm and social stress. This emphasis may in part be an artefact of spatial restrictions and the limited distribution of high-value resources within many **captive animal** systems, which may lead to increased competitive behaviour (see Lockwood, 1979). Another challenge relates to the interpretation of competitive social interactions, with some authors emphasizing the importance of the assertion of **dominance** and eliciting **submissiveness** in other individuals as a result. Submissiveness is thus defined as the response associated with yielding to a display of threat by another, and so is a response that follows threat. However, such a sequence of events might imply social instability or uncertainty, since aggression is a divisive and potentially risky behaviour used when in response to a threat to resources. By contrast, yielding or avoidance before any threat is more conducive to maintaining social relationships. It is perhaps useful to distinguish this from submissiveness as defined above, and refer to it as deferential behaviour, since it pre-empts unnecessary stress and reflects a more healthy social situation.

A third related term, **appeasement**, might also be usefully distinguished. This refers to behaviours expressed after a social conflict aimed at restoring good social relations. After a dispute, there is a risk that the loser may be rejected from the social group, which would greatly reduce the individual's biological **fitness**. Appeasement gestures (also referred to as post-conflict resolution behaviours) may serve to reduce this risk. Whereas the primary function of deference and submissiveness is to communicate a willingness to yield in different circumstances (avoidance versus termination of threat), the function of appeasement is primarily to maintain social cohesion.

Unfortunately, some species appear to use at least some of the same behaviours (e.g. head turning away in the dog) in different circumstances and, if the context is not clearly defined, this can lead to confusion about the relative importance of different social behaviours. While increased competition with **conspecifics** is undoubtedly an important cost to social living, it is worth remembering that social living and its associated behaviour arises from an intrinsic attraction between individuals. Thus communication, coordination and cohesion should perhaps be considered the fundamental basis of social behaviour more than competition. In cases of social tension involving **companion animals**, it is increasingly recognized that it is preferable to encourage deference rather than to try to assert dominance, since the former strategy is logically less stressful, more socially harmonious and therefore more conducive to good welfare.

Early exposure at particular times (**see: Sensitive phase**) and social play appear to be very important in shaping the social behavioural repertoire and skills of an individual, as well as defining the range and type of individuals to which social attraction may be exhibited (**see: Socialization**). Inappropriate or limited opportunity can give rise to a range of behaviour problems, such as **aggression** to unfamiliar types, which also reflect a welfare problem associated with social **anxiety**. Other problems can arise as a result of unmet social behavioural goals, and include problems such as excessive **vocalization** and **separation anxiety** by dogs when isolated from their social attachment figures. (DSM)

Reference and further reading

Keeling, L.J. and Gonyou, H.W. (eds) (2001) *Social Behaviour in Farm Animals*. CAB International, Wallingford, UK.

Lockwood, R. (1979) Dominance in wolves – useful construct or bad habit. In: Klinghammer, E. (ed.) *Symposium on the Behavior and Ecology of Wolves*. Garland STPM Press, New York, pp. 225–245.

McDonald, D.W. (1983) The ecology of carnivore social behaviour. *Nature* 301, 379–384.

Robinson, G.E., Grozinger, C.M. and Whitfield, C.W. (2005) Sociogenomics: social life in molecular terms. *Nature Reviews Genetics* 6, 257–270.

Social cognition

Social cognition is the sum of the processes that deal with information coming from, or relevant to, other animals. It integrates a variety of social phenomena, such as **communication**, social recognition, **social learning**, **cooperation**, social bonding, inference of relationships between other individuals and, sometimes, **theory of mind**. Research on social cognition first focused on primates, but has since extended to many other species. There is now a growing body of data on fish (e.g. guppies) and invertebrates (e.g. octopus) that benefits the field of comparative social cognition and thus the understanding of the mechanisms, function and evolution of social cognition.

Depending on their sociality level and thus on the complexity of their social environment, different species may have evolved different socio-cognitive abilities. It is the basis of the social complexity hypothesis that suggests that living in stable, socially complex groups favours the **evolution** of social intelligence. This idea has been applied and validated, for instance, in primates but also recently in birds (corvids), by comparing the performance of highly social and relatively asocial species in socio-cognitive tasks.

Research on social cognition has also been stimulated, especially since the late 1980s, by the **welfare** implications of socio-cognitive abilities in domestic species. Information gained from social partners through their behaviour, or from other channels of communication, is indeed a major factor affecting how animals adapt to their social and non-social environment, especially when this environment is complex and artificially manipulated by humans. Domestic animals do not cope (**see: Coping**) equally with the various challenges of their rearing environment (e.g. crowding, social instability, foraging, potentially stressful human interventions, instrumental learning) due to differences in their socio-cognitive abilities (**see also: Social stress**). Knowing more about the cognitive capacities of animals and about how they use social information according to the quality of social bonds between individuals and their social motivation can thus facilitate interventions aimed at lowering **fear** responses or favouring the acquisition of appropriate instrumental responses through better-adapted management practice (e.g. use of knowledgeable conspecific demonstrators), hence enhancing the animals' welfare. (SL)

Further reading

Miklósi, Á., Topál, J. and Csányi, V. (2004) Comparative social cognition: what can dogs teach us? *Animal Behaviour* 67, 995–1004.

Reznikova, Z. (2007) *Animal Intelligence. From Individual to Social Cognition*. Cambridge University Press, Cambridge, UK.

Social facilitation

Social facilitation occurs when the presence of a **conspecific** changes the motivation of the observer and its behaviour, resulting in the tendency of individual animals to do what other individuals are doing. This phenomenon can be considered as social influence but not as **social learning**, as it leads only to an increase in performance of a behaviour already known and stops when no other individual is present. Social facilitation takes part in synchronization of various behaviours, such as **feeding** behaviours, for instance. (SL)

See also: Social behaviour; Social cognition

Social flux

Social flux describes inconsistent **group** membership in social animals. Groups with a high social flux have higher intra-group **aggression** rates compared with groups of low social flux. They can form temporary intra-group social bonds (often with kin) and have stable numbers, but nevertheless depend on immigration to maintain their numbers. Changes in group membership can be due to high **mortality** or dispersal rates or movements among groups because of overlapping **home ranges**. They can also be due to human manipulation of certain sex/age classes, resulting in a skewed sex ratio and thus relatively high inter-group transfer of one particular sex. For example, many **feral** horse populations have skewed sex ratios due to removal of males by humans, and this leads to increased social flux (Linklater, 2000).

Social flux in feral dogs has been shown to result from widespread **resource** distribution and, particularly, from variation in the predictability, dispersion and richness of food sources, as well as cooperative hunting techniques and fluctuating density of predators (MacDonald and Carr, 1995).

In the domesticated situation, high social flux can lead to considerable welfare concerns in socially kept animals. Higher aggression rates can lead to increased instability or injuries, and thus to **stress**. To prevent stress, the animals must first be raised and kept in **social groups**, preferably with peers as well as older animals present, provided that this is part of the species' natural social organization (**social behaviour** is partly a learned process). Secondly, sufficient knowledge of species-specific natural behavioural requirements can lead to **welfare**-friendly enclosure designs and appropriate **grouping**. (MVD)

References

Linklater, W.L. (2000) Adaptive explanation in socio-ecology: lessons from the equidae. *Biological Reviews of the Cambridge Philosophical Society* 75, 1–20.

MacDonald, D.W. and Carr, G.M. (1995) Variation in dog society: between resource dispersion and social flux. In: Serpell, J. (ed.) *The Domestic Dog: its Evolution, Behaviour and Interactions with People*. Cambridge University Press, Cambridge, UK, pp. 199–216.

Social group

Social groups are more than simple collections of individuals: the animals make an effort to stay together. A social group may be defined in terms of its composition (characteristics of individuals making up the **group**), social structure (type and quality of relationships and interactions occurring between members of the group) and dynamics (rate of change in membership of the group). Groups are often recognized by a consistent set of social relationships and interactions between members of the group. Their members share similar goals and interests and may include genetic relations.

Within a group of social animals there tends to be a leader or leaders and followers, and the leading animal can change depending on time and circumstance. This is sometimes associated with the presence of a **dominance hierarchy**. However, dominance hierarchies can also be observed within non-social groups if forced into close contact. Social groups of animals are also identified by their affiliative preferences, coordination of activity, **communication** between individuals within the group and division of labour. For example, some groups, called eusocial (**see: Eusociality**), are highly organized and are characterized by reproductive division of labour. Some members have lost their reproductive capability and become workers, while others retain **reproduction** and are breeders. All members of the eusocial group work to increase the reproductive output of the colony (**see: Inclusive fitness**). This type of social group is often found in ants, bees, wasps and termites, but can occasionally be shown by vertebrates, like the naked mole rat.

In contrast to social insects, social groupings of vertebrates are not fixed or species specific. They depend on numerous factors, such as the available food supply, population density, ages of individuals and male:female ratios. There are benefits and costs to living in a social group that affect the size, structure and dynamics found. Some social groups are made up of monogamous pairs of animals, while others only form social groups at certain times of year. For example, blue tits vigorously defend nesting sites from other birds during the **breeding** season, but flock together in groups during the winter months for increased predator protection and food location. (LMD)

Social interaction test

The social interaction test was developed in rats by File and Hyde in the late 1970s to measure anxiogenic and anxiolytic drug effects. It relies on a natural spontaneous behavioural response (i.e. unconditioned response) like other animal models of **anxiety**, such as the **elevated plus-maze** or the **light/dark test**. The social interaction test is based on the observation of active **social behaviours** in a pair of unfamiliar animals. The time they spend performing social investigation (i.e. sniffing a partner), social contact (e.g. **allogrooming**) or play-fighting is the dependent variable measured. A decrease in social interaction is then indicative of an anxiogenic effect, whereas an anxiolytic effect is revealed by an increase in social interaction and motor activity. As the novelty of the environment and light are mild **stressors** for rodents, different levels of anxiety can be generated by manipulating those factors that will result in the test being more sensitive to anxiolytic or anxiogenic effect detection.

From a **welfare** point of view, the social interaction test has the advantage over some other models in its avoidance of the use of water or food deprivation and electric shock. From a practical point of view, it is a quick test to perform as it does not require extensive training of the animals. Recently, a modified version of the social interaction test that also allows measurement of social **motivation**, by providing the test subject with an opportunity to escape from its partner, has been developed.

While this test was first validated in adult male rats, recent studies showed that it seems to provide sensitive and consistent results across age groups, and can also be used successfully in female rats and other rodents such as mice and gerbils. The social interaction test has played a major role in giving further insight into the neural basis of anxiety, and is also used to help determine the genetic basis of anxiety disorders. Recent work also suggests that this model could be used for studying the neurobiology of social behaviour. The social interaction test paradigm is not only sensitive to a number of physiological factors that can affect anxiety, but can also be used to assess the effect of environmental factors (e.g. **handling**, **housing**). (SL)

See also: Choice test

Further reading

File, S.E. and Seth, P. (2003) A review of 25 years of the social interaction test. *European Journal of Pharmacology* 463, 35–53.

Varlinskaya, E.I. and Spear, L.P. (2008) Social interactions in adolescent and adult Sprague–Dawley rats: impact of social deprivation and test context familiarity. *Behavioural Brain Research* 188, 398–405.

Social learning

Social learning occurs when an individual learns by observation of another individual (i.e. demonstrator). Such learning has been demonstrated in various species, from fish to birds and mammals, and seems to be potentially influenced by sex, age, social rank and the social relationship between the observer and demonstrator. Social learning can take place between **conspecifics** or heterospecifics, including humans as shown in dogs, dolphins and primates. The mere presence of a demonstrator and the subsequent modification of the observer's behaviour do not necessarily imply that social learning has occurred. Social influence such as **social facilitation** (i.e. increase of performance of a behaviour already known in the presence of other individuals) could indeed then be at **play**. Social learning can be assumed to take place when the observer performs better only after he/she has seen the demonstrator performing the behaviour and if he/she still subsequently displays this behaviour when the demonstrator is absent. It can imply a variety of behavioural processes:

- **Local enhancement** (observer attracted toward one particular object) and stimulus enhancement (observer attracted toward any similar object): the acquisition of the behaviour to be learned is facilitated by increasing the frequency of the observer's behaviours directed toward the location or object in question following the demonstrator's presence near or interaction with the object.
- Social **goal emulation**: the observer is thought to realize what the demonstrator's goal is and then displays its own pattern of action to achieve the same goal.
- **Imitation**: defined by Thorpe as the copying of a novel or otherwise improbable act for which there is obviously no instinctive tendency; the existence of imitation in animals has been the subject of a lot of debate, especially because some researchers claim that imitation is a prerequisite for the existence of animal **culture**, but recent results argue for imitative behaviour in some species.

Social learning has been showed to play a role in the acquisition of various behaviours such as, for instance, food **preferences**, **foraging behaviour**, **mate choice** or **predator avoidance**. It has an adaptive value when individual learning by trial and error would be too costly (e.g. learning to avoid eating toxic plants), and can be taken advantage of in the context of animal welfare to facilitate the adaptation of domestic species to their rearing environment. (SL)

See also: Social cognition; Social stress

Further reading

Galef, B.G. and Laland, K.N. (2005) Social learning in animals: empirical studies and theoretical models. *BioScience* 55, 489–499.

Shier, D.M. and Owings, D.H. (2007) Effects of social learning on predator training and postrelease survival in juvenile black-tailed prairie dogs, *Cynomys ludovicianus*. *Animal Behaviour* 73, 567–577.

Social parasite

Social parasites are individuals that increase their **fitness** by non-fatally exploiting the **social behaviour** of others. A true social parasite should exhibit some dependence or adaptation associated with the behaviour, which can be inherited, in order to distinguish from incidental exploitation of available resources. It is often thought of in terms of interspecific interactions; for example, cuckoos exploit the parental behaviour of other species (brood **parasitism**), and many insects exploit chemical **signals** to obtain food from eusocial species (**see: Eusociality**); however, it may also be intraspecific, with some considering certain forms of exploitation of one sex by another as a form of social parasitism (**see: Sexual selection**). Social parasitism is also sometimes known as involuntary **altruism**. (DSM)

Social releasers

Social releasers are significant biological stimuli that play an important role in the display of **ritualized behaviour** of many animals. These releasers can be chemical, visual or auditory, and elicit various types of behaviour such as, for instance, **sexual behaviour**, anti-predator behaviour or **maternal behaviour**. Social releasers are generally considered as being more complex in their nature and in the response they provoke than **sign** stimuli, which are based on simple structural characters or movements that set off a single, stereotyped response pattern. (SL)

See also: Releasing factor; Social behaviour

Social stress

If **stress** is an organism's attempt to maintain **homeostasis** in response to environmental challenge, then social stress refers specifically to challenge originating from the social environment, in particular that associated with the presence and absence of **conspecifics** or their cues, since the potential contribution of non-conspecifics to the social environment is as yet unclear. It therefore has the potential to influence, at least to some extent, the lives of all those higher animal species that engage in social contact. Social stress can result from direct interactions with conspecifics, either familiar or unfamiliar, or in response to indirect encounters with conspecifics – for example, exposure to olfactory or auditory cues. In addition, social stress can be used to refer to situations where (preferred) social contact is absent, such as is the case during social **isolation** or separation from a peer **group**.

Examples of situations when social stress may occur for some species include: (i) exposure to unfamiliar conspecifics; (ii) dominant/subordinate relationships; (iii) maternal/infant separation; (iv) peer group separation; (v) isolation; (vi) mixing or introducing new individuals; (vii) crowding; and (viii) **sexual behaviour**. These situations may be especially relevant for animals that are held in captivity, because normal behavioural responses to social stress, such as escape, may be impossible.

'Acute' social stress describes a response to either a mild or severe social challenge that is short-lived, and can be considered to be **adaptive**. It may therefore be expected to have little long-term effect, if any, on an animal's **welfare** unless it is particularly severe (e.g. an acute stress with a chronic (long-term) consequence). 'Chronic' social stress, on the other hand, describes a response to either a mild or severe long-term social challenge. It may result in significant behavioural and neuroendocrine changes and, ultimately, in pathological consequences (see later) reflecting an animal's inability to cope with/adapt to the challenge. It can therefore have a major impact on animal welfare. In addition to the physical effect of social stress, there is also likely to be an emotional component to social stress with, for example, animals attaining negative states of **fear** and/or **anxiety** when reacquainted with individuals that have previously defeated them in an aggressive encounter. This combined effect of the emotional and physical response to the challenge of social stress makes it a key component of animal welfare research.

Although the use of the term 'social stress' is intended to separate it from other potential sources of stress, in reality there is often great difficulty in differentiating between the specific causes of stress, be they either social or originating from some other aspect of the environment – and different sources of stress may result in qualitatively different physiological, immunological and behavioural responses. If we were to study the effect of increased housing density on broiler **chicken** welfare, for example, can we differentiate between potential **stressors** from the physical environment (e.g. enclosure size, room temperature, ammonia levels, etc.) and those from the social environment (e.g. crowding) – especially if the different stressors induce different, potentially contradictory, responses? Even within the potential social stressor of crowding we find that there are many different potential components of social stress, such as proximity of individuals, group size, activity and competition for resources. It can therefore be concluded that the response to social stress is generally non-specific, in that the outcome may be influenced by many different variables.

The resulting response of animals to social stress is also subject to great variability. It can depend, for instance, on the species, strain, sex, age, status and previous social experience of the particular individual involved. For example, if two **pigs** are about to fight each other, both may respond to the stress of the encounter, but they may both respond differently according to the outcome of previous aggressive interactions that they have experienced. Thus, the threat of further aggression could be less stressful for a previously successful pig than for one that has been repeatedly defeated. In addition, individuals with different **coping** strategies (active/proactive and passive/reactive) may react differently to stressful situations, and/or the way that each individual animal perceives and evaluates a given social situation may result in further variation. In other words, we might expect different patterns of response to social stress in different groups of animals, with behavioural, physiological and immunological responses to stress varying between groups, and also between individuals within those groups.

While the social environment is a potential source of stress, it is also the reason why individuals group together – even if it may occasionally appear against their own immediate interest. Emphasis on the negative aspects of the social environment often blocks appreciation of the positive aspects of living in a social group, such as anti-predator effects, foraging benefits, reproductive benefits, thermal regulation, positive affiliative relationships and social support. Living in a group can actually help to reduce the impact of environmental stressors by influencing an animal's psychological state; for example, the presence of unafraid companions decreases the stress response to the threatening presence of a human. There are therefore many advantages to living in groups and, particularly in free-living animals, these may outweigh the potential costs of social stress. Besides, in a stable social group, much of the social stress observed during the formation of a social **hierarchy** (e.g. dominant/subordinate interactions) is quickly over and remains at low levels (i.e. acute social stress) – so this apparently 'negative' stress is adaptive, and may not necessarily indicate poor welfare in the longer term. Once the social hierarchy has been established, any other potential causes of social stress (e.g. the arrival of interlopers or separation) are rare. Problems arise when there is either a threat to the stability of the group or the group composition is unbalanced, and this is frequently the case for captive animals due to the constraints of **captivity**.

Under captivity, formerly acute and tolerable levels of social stress may become chronic due to human influences, such as mixing (e.g. sows coming out of **farrowing crates**), crowding (e.g. high stocking densities for broiler chickens), spatial restrictions (e.g. laying hens in battery cages) and the formation of an unbalanced or unrepresentative group structure (e.g. **zoo animals**). This can result in increased behavioural, neuroendocrine and possibly emotional changes that may compromise welfare. Despite this apparent reduction in welfare, singly housed individuals (e.g. laboratory **rats**) will still work for access to social contact – thus it appears that the 'benefits' of living alone, i.e. avoiding the potential stress of direct social interactions, are outweighed by the costs, i.e. no direct interaction at all, of isolation. Although this is likely to be a species-specific and context-specific response as, for example, chickens in a safe and familiar environment will often show preferences for being alone.

Perhaps we should therefore accept that, as a result of their genetic heritage, the priority of most social animals will be to seek social interaction, despite the consequences. However, this priority for social interaction has now been compromised by human influence to such an extent that the benefits of group living may become redundant, and the costs of group living may be far higher than for free-living groups, resulting in the development of chronic social stress and hence poor welfare. For example, the constraints of captivity may induce a lack of both controllability and predictability to the social interaction. Interactions may be inescapable due to confinement (e.g. within cages/pens), with escape attempts by subordinates frustrated. Indeed, it may be as stressful, or even more so, being dominant as it is being subordinate. In captivity, dominant individuals (e.g. laboratory mice) may become as stressed defending their social position in the continued presence of rival animals as their subordinate rivals become from being unable to escape the presence of their social superiors.

Selective **inbreeding** of domestic species/strains may have resulted in the production of increasingly homogeneous individuals. For example, inbred strains of mice are derived from 20 or more consecutive generations of brother/sister (full sibling) matings. This may not only cause **communication** problems between animals, resulting in unnecessary or inappropriate aggression rather than non-fighting assessment, but also problems with the formation of a stable social hierarchy. If, for instance, all the animals in one group have very similar social strategies, e.g. all aim to be dominant, then this serves only to increase conflict within the group. This issue of group composition may be as important for determining the level of social stress in a group as other, more obvious, social stressors such as personal space allowance. Another effect of captivity is that, even in groups that are potentially able to form stable social hierarchies, such an attempt may be continuously disrupted by the introduction or removal of individuals from the group. An example of this is seen on pig farms, where individual sows are removed for, and subsequently reintroduced after, farrowing.

In an attempt to try and reduce the influence of multiple variables, several models of social stress have been devised, and these are studied predominately under laboratory conditions and using **rodents**. The most commonly used models include the social defeat model, the social instability model and the social disruption model. These three models focus on agonism as a source of social stress. The social defeat model involves one or more direct aggressive encounters between the naive subject and an 'attacker'. By concentrating on two individuals with known social backgrounds, this model can be fairly reliable about interpretations of the data in terms of an interaction + defeat (or interaction + victory). Social instability models involve the formation of stable groups of individuals that are then mixed together whereas, for social disruption models, high-aggression individuals are introduced into previously stable groups. In contrast to the social defeat model, neither the social instability model nor the social disruption model can be so reliable about the specific source of any social stress expressed by the subjects – because the subjects may have experienced victories or defeats due to the group housing.

But how can we identify when an individual or group is experiencing social stress, and how can we then identify when that level of social stress is threatening welfare? Some 'indicators' of social stress may be more likely than others to demonstrate that an animal is suffering a decrease in welfare due to social stress. Such indicators will only result either from prolonged exposure to social stress, e.g. pathology and **abnormal** behaviour, from exposure to extreme stress (e.g. gastric ulceration) or that attempt to reveal the subjective state of an animal (i.e. what it feels), e.g. anxiety tests.

The majority of indicators, e.g. aggressive behaviour, indicate the presence of social stress in general – without allowing us to determine whether welfare is threatened. In addition, the absence of any of these indicators does not necessarily indicate an absence of social stress. Our predictions for what behavioural and endocrine responses to social stress we might expect will also depend strongly on the time at which the measure is taken after the social 'challenge', e.g. immediately after mixing or 24 h later, and the precise time of day/night/year that the indicator is measured if the indicator is affected by **diurnal**/seasonal **rhythms** or motivational influences.

Common behavioural 'indicators' of social stress include (among others): (i) agonistic (i.e. aggression + defence) behaviour; (ii) general activity; (iii) social investigation; (iv) self-directed behaviour (e.g. tail manipulation in rats); and (v) abnormal behaviour, including **stereotypies**. Abnormal behaviour may develop when the natural adaptive response to a social stressor is frustrated due to the constraints of captivity, and an animal no longer has control over its environment. For example, bar-biting in laboratory rodents may represent frustrated attempts by individuals to escape from social conflict. Such behaviours may therefore suggest the presence of welfare-threatening chronic social stress.

Further behavioural changes that can be associated with social stress include: (i) emotionality-linked behaviours such as anxiety and performance in anxiety tests, e.g. the **elevated plus-maze**; (ii) alcohol/sucrose consumption; (iii) sexual behaviour; and (iv) **preference** and consumer demand tests (although it should be noted that the effect of social stress itself may act to disrupt an animal's ability to make an accurate decision, because of its effects on learning and **memory**).

Several different measures of immunocompetence can be measured, all based on the premise that stressed animals have increased activation of the **hypothalamic–pituitary–adrenal (HPA) axis** and therefore increased levels of corticosterone/cortisol that can be immunosuppressive, i.e. can suppress immunity. Few measures are specifically related to the measurement of social stress, but are used to assess a general, non-specific stress response. These include (among others): (i) peripheral lymphocyte counts; (ii) lymphocyte proliferation; (iii) salivary/faecal immunoglobulin A (IgA); and (iv) serum total immunoglobulin G (IgG).

Physiological 'indicators' include (among others): (i) weight loss (or reduced growth); (ii) an increase in HPA activity (especially an increase in the level of cortisol/corticosterone in blood plasma and/or saliva; (iii) **glucocorticoid** metabolites in urine and/or faeces); (iv) an increase in **sympathetic nervous system** (SNS) activity (e.g. increase in the levels of adrenaline in blood plasma and tyrosine hydroxylase in the **adrenal glands**); and (v) an increase in **heart rate**, reproductive success, **blood pressure** and body temperature. Examples of pathologies that can occur as a consequence of chronic social stress include: (i) an increase in adrenal weight (adrenal hypertrophy/hyperplasia); (ii) a decrease in thymus weight (thymus involution); and (iii) an increase in gastric ulceration.

Social stress can also generate changes in neurochemical transmission in the **brain** with, for example, increased serotonergic activity in response to social stress that may relate to the role that **serotonin** (5HT) plays in mediating social behaviours, including aggressive and sexual behaviour. Social stress can also influence **corticotropin-releasing hormone** (CRH) and mRNA levels in the paraventricular nucleus region of the **hypothalamus** – where most CRH is synthesized. CRH is critically involved in the instigation and control of HPA axis activity.

In all cases, whatever the 'indicator', the prediction that we make about how a group/individual might respond to the challenge of social stress is specific to each particular instance, and therefore cannot be generalized. Care should also be taken when interpreting the results of potential 'indicators' of social stress. Commonly used physiological and behavioural indicators of social stress – for example, heart rate and general behavioural activity – may increase during an **aversive** agonistic encounter with an unfamiliar conspecific, but a similar increase may also be observed during, presumably positive, sexual behaviour. In addition, an increase in heart rate might also be expected with any increase in general activity levels – even in the absence of any stressor, social or otherwise – i.e. we should be sure of the functional significance of the indicator.

There are several different approaches to the welfare problems that can be posed by social stress. One approach is to decrease aggression (e.g. during the mixing of **weaner** pigs) by either the use of drugs (e.g. tranquillizers) or altering the environment in some way (e.g. mixing groups in the dark). However, these may only serve to hide/delay the problem rather than address the initial cause of the issue itself. Olfactory cues such as boar taint may reduce aggression during the mixing of weaner pigs, because the smell of the dominant individual may suppress aggression in groups of subordinate animals; however, this may also cause a state of fear or anxiety in the other animals if they associate the smell of the dominant individual with previous experience of defeat.

Another approach is to attempt less interference by humans, i.e. reduce the constraints imposed by captivity. This can be done by identifying those situations/procedures most likely to result in social stress and either reducing their frequency or avoiding them entirely, in order to both encourage stability and promote the beneficial effects of affiliative relationships and social support, e.g. reduce the amount of mixing and separation of groups.

However, we should not attempt to avoid stress entirely, as this would create problems of its own. Indeed, in some instances (e.g. **zoos**) the introduction of controlled levels of stress may be used to improve welfare. Good husbandry practice should ensure correct provision of resources in order to reduce competition. If a food trough is placed at the end of an enclosed corridor and there is not enough room for all the animals in the group to feed at the same time, then this will increase the chances of aggressive competition or **resource** monopolization by particular individuals, to the exclusion of the others. Attention can also be directed towards trying to create the 'best' possible group composition, e.g. by mixing active and passive coping strategists, higher- or lower-aggression animals or individuals with a range of sizes. In this way we might facilitate the use of non-fighting assessment as a means for individuals to discriminate between one another. Finally, the creation of suitable space(s)/barriers within the social environment may allow room for escape or shelter – however, as for all environmental modifications, you have to be sure that these are actually beneficial, as in some species alterations to the environment may lead to an increase in aggression rather than a decrease. (OHPB)

See also: DFD meat; Dominance; Feedlot; Frustration

Further reading

Blanchard, R.J., McKittrick, C.R. and Blanchard, D.C. (2001) Animal models of social stress: effects on behaviour and brain neurochemical systems. *Physiology and Behaviour* 73, 261–271.

Mason, G. and Mendl, M. (1993) Why is there no simple way of measuring animal welfare? *Animal Welfare* 2, 301–319.

Mendl, M. (2001) How do animals cope with social problems? In: Broom, D.M. (ed.) *Coping with Challenge: Welfare in Animals Including Humans*. Dahlem University Press, Germany, pp. 211–228.

Moberg, G.P. and Mench, J.A. (2000) *The Biology of Animal Stress – Basic Principles and Implications for Animal Welfare*. CAB International, Wallingford, UK.

Zayan, R. and Dantzer, R. (eds) (1990) *Social Stress in Domestic Animals*. Kluwer Academic Publishers, Dordrecht, The Netherlands.

Socialization

Socialization is the process by which animals adopt the behaviour patterns appropriate to the social environment in which they live, allowing them to coexist/interact with other individuals. It is through this process that social preferences are formed and the species identity of a young animal is established. An animal is said to be 'socialized' when interacting with another animal if it displays a recognition and acceptance of it, and behaves accordingly towards it.

The adaptability of behavioural tendencies is most rapid during the early stages of their lives and declines with age. Four consecutive, interrelated phases of behavioural development have been suggested. These are: the **neonatal period**, the **transitional phase**, the **socialization phase** and the **juvenile period**. The duration of these stages varies between species. Experiences during the early stages of life appear to be more important than those occurring during later life, with the socialization period playing a major role in determining social behaviour in adulthood, and a disruption of this process resulting in profound influences on adult behaviour.

The process of socialization does not necessarily cease at the end of the socialization phase. While social behaviour is most amenable to modification at this time, if the animal has not been socialized during this period the formation of social relationships can still occur. Remedial socialization has been attempted in a number of species, such as the **cat**, **dog** and **pig**, with some success. Socialization also appears to be reversible to some extent and requires reinforcement into adulthood. For example, puppies that are initially well socialized to people have been found to become less friendly if they are subsequently housed in kennels and receive minimal human contact. This may be a consequence of stress-induced **dishabituation**.

The basis for socialization and the formation of social bonds in mammals has generally been attributed to the concept of **attachment**, with the formation of inter- and, in some cases, intraspecific relationships during socialization being influenced by the social organization of the species. The formation of social bonds has been studied in many species, with arguably some of the most dramatic attachments demonstrated by the **imprinting** process in **precocial** birds such as geese and **chickens**. Precocial animals, whose locomotor and sensory abilities are well developed at birth, have a shorter sensitive phase during which socialization occurs, with imprinting playing an important role in the formation of social behaviour (Belyaev *et al.*, 1985). Filial imprinting is the process whereby the newly hatched young become rapidly attached to parents or surrogates, independently of the consequences of the behaviour.

Konrad Lorenz's studies of filial imprinting in waterfowl led to the idea of a fixed period in a rigid sequence of development, during which environmental influences have a profound and irreversible effect on subsequent social behaviour. Although the term 'imprinting' is usually associated with precocial birds, the concept of a period of sensitivity can also be applied to the development of social behaviour in **altricial** species. In mammals, the concept of a **'critical period'** in behavioural development is more usefully replaced with that of a **'sensitive phase'**, during which the formation of social attachments is facilitated. The end of this phase coincides with the appearance of fear in many species, possibly related to the maturation of the pituitary–adrenal system. In most animals the first interactions occur with the mother, particularly in altricial species, which tend to form strong mother-infant attachments. Secondary socialization appears to involve social play, with littermates or peers.

Species identity is not strictly genetically pre-programmed. Social behaviour may therefore be directed more broadly towards humans and other species in a domestic situation. The ease with which **wild animals** become domesticated is also undoubtedly influenced by their species-typical patterns of social development. While species-specific sexual behaviour is essential (this appears to be a separate process, possibly dependent upon other innate stimuli), social attachment has a learned component.

Humans have been able to exploit this process, thereby enabling the animal to at least tolerate, and often value, the presence of people. Indeed, the process of intentional socialization is central to producing animals to fulfil the role of human companions and in the training of animals for certain working roles – for example, livestock-guarding dogs. The sensitive phase for socialization seems to be longer, with less sharply defined boundaries in domestic animals compared with wild animals (hence the use of the term phase rather than period), and this phenomenon has been demonstrated experimentally through the selection of silver foxes for domesticated traits (Belyaev *et al.*, 1985). While the ability to socialize with people may, in part, have resulted as a consequence of **domestication**, this phenomenon is also present, to some extent, in many wild species, with hand-reared individuals often displaying affiliative behaviours towards people.

In order to understand adult behaviour, it is necessary to understand the **ontogeny** of that behaviour. However, because of the speed with which it occurs, it has proved difficult to study the development of social behaviour and it is not clear what types of interaction are required for normal social development. Several different experimental approaches have been used to examine the process of socialization. These include deprivation experiments, in which the animals are subject to varying levels of opportunity to socialize to other individuals (Harlow and Harlow, 1962). In many of these experiments extreme social isolation is accompanied by spatial or sensory deprivation, and individuals are also subjected to the stress of being taken out of a restricted environment for testing, giving rise to the criticism that the **abnormal** behaviours observed may be the result of any of these factors.

Cross-fostering experiments, in which an individual of one species is removed from its **conspecifics** at birth and reared with individuals of another species, is another method that has been employed in the study of socialization. While some behavioural differences have been observed, these must be treated with caution. However, cross-fostering, or cross-rearing, experiments as they are sometimes called, may provide a useful method for studying **behavioural plasticity** in the development of social behaviour. The main shortfall with both of these methods is that they do not provide an understanding of the mechanisms by which these social behaviours are acquired under 'normal' conditions (Bekoff, 1977). (EB)

References

Bekoff, M. (1977) Socialization in mammals with an emphasis on nonprimates. In: Chevalier-Skolnikoff, S. *et al.* (eds) *Primate Biosocial Development*. Garland Publishing, UK, pp. 603–636.

Belyaev, D.K., Plyusnina, I.Z. and Trut, L.N. (1985) Domestication in the silver fox (*Vulpes fulvus* Desm): changes in physiological boundaries of the sensitive period of primary socialization. *Applied Animal Behaviour Science* 13, 359–370.

Harlow, H.F. and Harlow, M.K. (1962) Social deprivation in monkeys. *Scientific American* 207, 136.

Socialization period/phase

The socialization period refers to a **sensitive phase** during which animals appear most responsive to forming social **attachments** (**see: Socialization**), and its definition is a function of the natural **ecology** of the species. The temporal boundaries of the socialization phase have often been defined on the basis of **isolation** and **handling** experiments, which are not without criticism. These have led to the suggestion that it occurs between 3 and 12 weeks in the **dog** (peaking at around 8 weeks of age) but is much shorter, between 2 and 7 weeks, in the **cat**. This is typically followed by a **juvenile period** during which, it has been suggested, social referencing is used to a greater extent to shape perceptions of, and interactions with, the social environment, although some include this in a more extended definition of the socialization period (i.e. it is a time during which social norms are established). In addition, some define the time during which initial maternal **bonding** occurs as a period of primary socialization.

Although the bond formed with a maternal figure is in many ways unique, there is evidence to suggest that this, including the form of attachment developed at this time, plays an important role in later social development (often referred to as secondary socialization) (**see: Secure base hypothesis**). There is sometimes confusion between the socialization phase and a more extended period during which there is a greater predisposition to habituate to a range of physical stimuli in the environment (typically extending up to about 14 weeks in the dog), although the process of social relationship formation should be considered distinct from habituating to the presence of others. (DSM)

See also: Habituation; Handling

Societies and animal welfare

A society is a population of human beings who share a distinctive **culture** and/or certain political institutions. More broadly, a society is an economic, social and cultural structure through which a group of people is united. One can speak about societies at different levels: compare Danish society, Western society and modern society. The word 'society' may also refer to an organized voluntary association of people for a certain cause, as, for example, in the Royal Society for the Prevention of Cruelty to Animals (**RSPCA**).

There is a growing awareness of the need to understand human relations to animals in the context of culture and other aspects of society. For example, the ranking of animals according to status, the so-called socio-zoological scale, is clearly shaped by societal values. Thus in the 19th century, when the first animal **welfare** laws were passed, the highest-ranking animals to be protected by the law were **horses** and **cattle**; **cats** and **dogs** were not covered at all. Today, the importance of cats and dogs kept as **pets** has grown and so they have, as a result, risen on the socio-zoological scale. To describe and explain this process, philosophical, sociological and historical studies are called for.

It has become clear that the notion of animal welfare, and norms governing acceptable treatment of animals, must be understood in a wider social context. This is not only so when it comes to societies and contexts differing from one's own: it is also true that one's own notion of animal welfare, and one's own standards describing good care of animals, may be enlightened by being understood in the context of one's own culture and society.

Organized voluntary associations of people who work to improve conditions for animals in human care have in the past played a very significant role in putting animal welfare on the political agenda in Northern European and in other Western countries. Over the last few decades new organizations have developed; these differ from traditional animal welfare societies in their tendency to focus on specific causes, in being run more like corporations than societies, in working at an international level and in using modern, highly professional campaigning and fundraising methods. (PS)

Sociobiology

Sociobiology is a term popularized by and often associated with **E.O. Wilson**, through his classic text *Sociobiology: the New Synthesis*, although the term was apparently first coined by the developmental geneticist John Paul Scott, who undertook the classic experiments on the **sensitive phases** for **socialization** in dogs. A key tenet of sociobiology is that **social behaviour**, along with other behaviours, has some genetic basis that can be selected for or against. The relative importance of genetic factors in different aspects of behaviour undoubtedly varies and is widely debated. At the extremes, some proponents of sociobiology lean towards genetic determinism, while some critics suggest that the complexity or nature of some traits preclude meaningful genetic **selection**. An early challenge to sociobiology focused on the development of **altruism** as a trait, whereby one individual incurs a cost to the benefit of another. It was initially thought that such a behavioural tendency would be selected against because donors would be at a disadvantage to recipients. However, these acts can be explained in terms of **inclusive fitness**, reciprocal altruism and other longer-term gains for the donor. Historically, the development of **eusociality** in insects has been used by sociobiologists to provide a particularly strong example of the implications of natural selection and inclusive fitness to explain social behaviour. In humans, sociobiological concepts have also been used to explain a range of cross-cultural behavioural tendencies, such as the taboo associated with incest and, perhaps most controversially, in the defence of human **eugenics**. (DSM)

Further reading

Alcock, J. (2001) *The Triumph of Sociobiology*. Oxford University Press, Oxford, UK.

Dawkins, R. (1977) *The Selfish Gene*. Oxford University Press, Oxford, UK.

Hamilton, W.D. (1964) The genetical evolution of social behaviour I & II. *Journal of Theoretical Biology* 7, 1–52.

Wilson, E.O. (1975) *Sociobiology: the New Synthesis*. Harvard University Press, Cambridge, Massachusetts.

Sollwert

The term 'sollwert' is used in applied animal behaviour to refer to the desired state of an animal at a given time. Thus potential welfare concerns arise when an animal's current perceived state (**istwert**) differs from its sollwert, especially when the animal is restricted in its ability to adapt or achieve its sollwert. (DSM)

See also: Merkwelt; Umwelt; Wirkwelt

Further reading

Wiepkema, P.R. and Van Adrichem, P.W.M. (eds) *Biology of Stress in Farm Animals: an Integrative Approach*. Oosterbeek Nijhoff, Dordrecht, the Netherlands.

Soring

Soring is the practice of applying caustic material, such as mustard oil, to the pasterns of **horses**, in particular Tennessee Walking Horses, to irritate the skin. Then chains, rollers or a series of large wooden beads are placed around the pastern so that they will abrade the burned area in order to make the lesion even more painful. The purpose is twofold: (i) to encourage the horse to shift his weight to the hindquarters; and (ii) to induce the horse to lift the fore legs higher and more quickly when executing the running walk. The hooves are allowed to grow very long, and the horse wears heavy shoes and 1 lb (450 g) bell boots.

Soring is illegal in the USA, according to the Horse Protection Act of 1970, Public law 91–540, but enforcement is weak because there are insufficient inspectors. The law is enforced by the US Department of Agriculture, which is also charged with inspection of other much larger animal populations, such as commercially bred dogs and animals in research facilities. It is not surprising that the relatively small number of Tennessee Walkers is not inspected or that lay inspectors are hired. The lay inspectors are often Tennessee Walker owners themselves and are reluctant to accuse their friends and colleagues. The citation rate for soring is much higher by USDA inspectors than by lay inspectors. Soring can be detected by the use of thermography, but this is probably not necessary because the pastern lesions are usually quite obvious.

The Tennessee Walker, originally known as the Southern Plantation Horse, is an American breed. Its original purpose was to carry a rider across a plantation in comfort while traversing long distances quickly. It was only when the horses began to appear in shows that an exaggerated running walk became desirable and soring was found to get results faster than training. Soring has also been used on jumpers to encourage them to lift their legs. (KAH)

Sow behaviour

The European wild boar, *Sus scrofa*, is the closest living relative of today's domestic **pig** and, where free-living domestic pigs have been compared, their behaviour is essentially the same. The usual **social group** of wild sows consists of up to five adult females, probably related, and their offspring of the previous 1–2 years. Sows sleep in a simple communal nest, but may forage separately. Their natural habitat is forest, which provides both feeding opportunities and their preferred dense cover. Pigs are omnivorous, eating a wide and varied diet of both vegetable and animal matter. Dietary staples are wild fruits, nuts, acorns, roots, tubers, bulbs and seeds, as well as grasses, insects (e.g. worms, beetles and ants), small animals such as rodents, frogs, lizards, snakes, birds and birds' eggs. Pigs will also eat carcasses when given the opportunity, as well as cultivated foods such as wheat and maize. The majority of each day is spent investigating the environment and foraging. Boars are either solitary or live in small bachelor groups; they join the sow groups temporarily in autumn or early winter for mating purposes. **Oestrus** in sow groups is synchronized and **breeding** is seasonal, with piglets usually being born in the spring.

Wild boar are not **territorial**; instead, they form **home ranges** without well-defined boundaries, which can shift from time to time in response to food availability. The size of the home range may vary with the habitat (increasing in less plentiful areas) and the sex of the animal (being smaller in females). Home ranges calculated from radio-tracking studies have varied from 120 to over 1000 ha. Wild boar habitats are likely always to feature resting places and nests, wallows and watering places, rubbing and scratching places, rooting areas and a network of regularly used interconnecting paths. While several studies have observed wild boar to be **nocturnal**, it is probable that, in the absence of human influence, they would be diurnal, but switch to a nocturnal pattern for safety in response to disturbance and predation by humans. Farmed sows will often give birth at night, perhaps because it is quieter than during the day.

Pigs have been domesticated for at least 5000 years. Steady increases in levels of control exerted by humans over their porcine companions have culminated, in many regions of the world, in highly intensive, **confinement** agriculture, with large numbers of pigs kept in close proximity and minimal space allowance per individual. While pigs are in many ways highly adaptable, the conditions in which we keep them today are far from those they have evolved to inhabit, and this has multiple effects on their behaviour. In a farmed environment, artificial groups will be abruptly imposed, comprising single-sex, single-age, single-weight, often unrelated animals. Growing gilts are unlikely to meet a piglet until they give birth to their first litter. Mating is imposed, with no choice of sexual partner, and **parturition** may be artificially induced or augmented through injection of hormones if it is progressing too slowly.

The average duration of pregnancy is 114 days, or about four months, the majority of which the wild boar, **feral** or free-living sow will spend with her social group. Intensively farmed sows may be kept in groups that are small or large (anything from a handful of individuals to over a hundred) and static or dynamic (meaning that the group composition stays the same or changes). **Housing** may be indoors or outdoors, and enclosure or paddock designs can vary widely. They may be bedded or unbedded, barren or enriched. Pigs are selected for rapid growth and, if allowed to eat *ad libitum*, sows may gain weight excessively, which can negatively affect milk production and impair their fertility, as well as costing farmers money. Therefore, most farmed pregnant sows are subject to severe food restriction, limiting them to a once- or twice-daily intake of concentrated feed, equivalent to approximately 60% of *ad libitum* intake, which takes only a few minutes to eat. While this precisely calculated diet contains sufficient nutrients for growth and production, it is insufficient to induce feelings of satiety. Hence it is probable that pregnant sows feel hungry most, if not all, of the time.

Feeding systems for group-housed sows have been devised to minimize **aggression** and control of **resources** by dominant individuals, ensuring that all individuals receive adequate food: these include individual feeding stalls, electronic sow feeders (which recognize individuals and allocate feed accordingly), trickle feeders (which 'tie' each sow to a single feeder by delivering food at the feeding rate of the slowest sow) and dump or spray systems that distribute feed widely on to the floor. Despite these measures, group-housed sows will fight on occasion over resources or after being mixed with unknown individuals. Gestation stalls, rectangular metal enclosures similar to the **farrowing crate**, and tethers, ties around the sow's neck or girth that confine her in one place, have all been banned in the UK since 1999, and stall use will be illegal after the fourth week of pregnancy throughout the European Union from 2013. However, stalls remain a common housing method in other regions such as North America. Gestation stalls severely restrict a sow's movement, preventing exercise and turning around, and thus violating at least one of the Farm Animal Welfare Council's **five freedoms**, the freedom to express normal behaviour. However, they also prevent fighting and simplify feeding and management, allowing large numbers of sows to be kept easily in small amounts of space.

The combination of feed restriction, lack of foraging opportunity and restrictive housing in dry sows predisposes them to develop oral **stereotypies**, which can include **bar biting**, sham chewing and licking. These behaviours may be displayed at high levels, particularly immediately after feeding. Oral stereotypies are more common in sows housed in gestation stalls, though group-housed sows will also display them, especially in barren environments. Environmental complexity – for example, the provision of space, **enrichment** or straw bedding (which can be ingested and manipulated) – can reduce the exhibition of oral stereotypy, as can the inclusion of high-bulk, low-nutrient diet additives to promote feelings of satiety.

In the free-living sow behavioural changes, including seclusion seeking and reduction in excessive movement and conflict, begin around 1 month before parturition. A few days before giving birth, the free-ranging sow leaves her group and walks up to 6.5 km to find a suitable nest site. Simple 'dummy' farrowing nests may be constructed on more than one site before the final farrowing nest is built. Nest building starts within 24 h of birth, peaks at 6–12 h before and decreases as parturition approaches, ceasing within 4 h of birth.

Nesting occurs in two phases: the exploratory phase and the nest-building phase. During the exploratory phase the sow seeks a suitable nest site. The nest-building phase consists of the sow carrying mouthfuls of bedding materials to the nest location and rooting and pawing the collected pile into a nest. In semi-natural conditions, domestic sows build farrowing nests consisting of shallow holes in the ground, rooted out and filled with grass and small branches. Survival of piglets has been noted to increase with greater distance of the farrowing nest from the group nest. The nest provides thermal and physical protection, and the sow will occupy it during birth and, with her litter, for 7–10 days afterwards. During this period she will leave the nest briefly to forage. Recognition between sow and piglets does not develop immediately after birth, but happens gradually during the period of nest occupation, equipping them to rejoin the extended family group, which they do within about 10 days of birth, but remain as an intact social unit until weaning. **Nursing**/suckling continues for at least 12–16 weeks, and **weaning** is a gradual process during which growing piglets start to forage and eat increasing amounts of solid food while still regularly, but less frequently, suckling from their mother.

Farmed sows will be transferred from gestation accommodation to farrowing quarters a few days before they are due to give birth. Outdoor sows may farrow in arcs or huts, while indoor sows are confined to pens or, commonly, a farrowing crate. In farrowing crates, domestic sows show behaviour changes (restlessness, posture changing, stereotyped bar gnawing, pawing behaviours, attempts to escape from the crate, frothing at the mouth and pupil dilation) starting about 1 day before farrowing. These behaviours share many elements and follow the same course as nest building in unrestrained sows, and may be interpreted as nest-building behaviour, frustrated and displaced by restraint. Sows may be more restless in crates than when loose-housed in pens, and their response to crating may be mediated by their accommodation during gestation, with an increased response to confinement in a crate following loose-housed gestation. Physical signs of impending parturition observed in domestic sows include swelling of the vulval lips, enlargement of the udder, **colostrum** secretion from the teats and an increase in respiratory rate.

Few published data are available on the birth process in free-living wild sows, because of the shy and cryptic nature of the wild boar and the difficulties of observing a behaviour that occurs in an enclosed nest and is performed by an animal that actively seeks seclusion. Domestic females will usually settle on their sides soon before parturition starts, a position they will remain in for the majority of the birth period. Over 90% of piglets are born with their mother lying on her side, although sows occasionally give birth lying sternally or standing up. A little straining, possibly some **vocalization**, the passage of some fluid and rapid tail swishing will immediately precede the birth of the first piglet. In between births, the sow may stand briefly and, if able to do so, turn to inspect and make nose-to-nose contact with her newborns.

Sows do not lick or groom piglets, either for cleaning, stimulation or identification, and perform little overt maternal care other than exposing their udder. Nose-to-nose contact between sow and offspring appears to be important for the establishment and maintenance of **bonding**, and free-ranging sows maintain contact with their newborns through sniffing and grunting. Some sows, especially when giving birth for the first time, will attack newborn piglets and may kill them (**see also: Infanticide**). Even if this 'savaging' behaviour is not displayed, sows sometimes become agitated during farrowing, repeatedly alternating between standing and lying. This may be related to the discomfort of giving birth.

In domestic sows the second stage of parturition (expulsion of piglets) lasts 2.5–4.0 h, with mean inter-birth intervals of 15–16 min. The duration of parturition is similar in wild boar, although litter sizes are much smaller than in domestic sows. Although it seems logical that the length of farrowing will vary with the number of piglets, no clear relationship has been reported between litter size and duration of farrowing. There is considerable variation around this average duration, and in the length of inter-birth intervals. Piglets may be born head or rear-end first, with few complications resulting from either presentation. Piglets born towards the end of farrowing have an increased chance of being stillborn due to oxygen deprivation during the birth process. Confinement in a restrictive farrowing crate, which prevents normal pre-farrowing behaviour and constrains behaviour during farrowing, is almost certainly stressful. This **stress**, combined with lack of opportunity for exercise, may prolong the duration of farrowing, leading to increased levels of stillbirths.

Nursing (which the sow does) and suckling (which the piglets do) is much more complex in the litter-bearing pig than in species that give birth to one or two offspring at a time. During farrowing and for a few hours afterwards, colostrum is readily available from the sow's udder. Piglets locate a teat and suck soon after birth, and the nutrient- and antibody-rich colostrum is important for their survival. Once the whole litter has been born piglets will start to form a **teat order**, in which piglets consistently nurse from 'their' teat, recognizing it by smell and sight, and returning to it at each suckling bout. In order to maximize milk intake and survival, most of the litter must be present and working at each bout to elicit milk letdown.

Suckling has been characterized as comprising five phases: (i) gathering and finding teat positions (before and during which the sow may 'call' her piglets to the udder by grunting); (ii) massaging the udder; (iii) slow and steady sucking; (iv) a period of 15–20 s of rapid sucking that coincides with the sow's brief ejection of around 20 ml of milk; and (v) a return to alternating periods of massage and slow sucking after letdown (which is thought to stimulate subsequent milk production). These phases appear indistinguishable in wild boar and domestic pigs. Once a teat order has been established and colostrum has turned into mature milk, piglets will gather at their mother's udder to suckle approximately once an hour, day and night. 'Weaning' of domestic piglets consists of an abrupt separation of sow and litter, and generally takes place 2–6 weeks after birth. As little as 4–5 days afterwards the sow will be mated again.

Good **maternal behaviour** makes an important contribution to the survival of piglets in a free-ranging situation, and this is also the case in domesticated sows kept outdoors or in pens. However, sows in farrowing crates are unable to display much maternal behaviour, and production in these systems does not rely on good mothering. Thus, modern pigs are not selected for their mothering abilities, but rather for prolificacy, rapid, lean growth and docility.

In intensive European production systems, an average sow may produce 2.2 litters and up to 25 piglets per year. Litters can average 10–15 piglets (with some sows giving birth to up to 20 piglets at a time), compared with the wild boar, which will produce three to seven 'boarlets'. Farmed sows spend almost their entire adult lives either pregnant or lactating, with short breaks of only a few days after they have been separated from their piglets. Once their productivity declines, they will be sent to **slaughter**. Around 40% of sows in the average breeding herd are replaced each year, and a sow may produce four to six litters of piglets in total before being sent to slaughter around her third birthday. (MH)

Further reading

Fraser, D. (1980) A review of the behavioural mechanism of milk ejection of the domestic pig. *Applied Animal Ethology* 10, 301–317.

Jensen, P. (1986) Observations on the maternal behaviour of free-ranging domestic pigs. *Applied Animal Behaviour Science* 16, 131–142.

Space

In terms of animal **housing**, space can be defined as that property of the environment that accommodates the interaction of the animal with the environment, that is, that accommodates its behaviour. It is a multifaceted concept that can be described either physically or functionally, and can include qualitative aspects as to how well it provides for the animal.

As a physical resource, space exists in three dimensions: length, width and height. Together, these allow space to be described in terms of a single dimension (distance), two dimensions (area) and three dimensions (volume). Examples of these uses in animal husbandry include the length of a **lying** stall (one dimension), the area of the **floor** (two dimensions) and the volume of an aviary system (three dimensions). Allometric relationships, expressing space in terms of body size, are based on body weight (BW) raised to the powers of 1/3 (distance), 2/3 (area) and 3/3 (volume). Provided the body shape and behaviour of the animals remain consistent over a weight range, the coefficients used to describe space requirements remain constant throughout that range. For example, the floor space requirement of an animal can be expressed as Area = $k^*BW^{2/3}$, and k will remain constant over a range of body weights.

A 'fourth' dimension of space is time. If the behaviour of animals does not need to be synchronized, then behaviour-related space can be shared over time. A large number of animals may eat sequentially from a single feeding space if feed is continuously available. However, if feed is limited, sufficient feeding spaces may be required to accommodate all of the animals simultaneously. Because most animals in a group will be resting simultaneously at some point of the day, lying space must be sufficient to accommodate the entire group, either as individual lying stalls for each animal, or a loafing area sufficient to accommodate them all.

Space requirements may be expressed as either space allowances (space/animal unit) or its reciprocal, **stocking density** (animal unit/space). For linear resources, based on distance, such as the width of a feeding space, space allowance is expressed as length/$BW^{1/3}$. For two-dimensional space, such as floor area, space allowance is expressed as area/BW$^{2/3}$ and for volumetric space, e.g. tank volume for fish, as volume/BW. If a resource is shared by several animals over time, its allocation is usually referred to by its stocking density, animals/unit (e.g. cattle/drinker).

The total space available to an animal or group of animals can be subdivided by its behavioural use. The primary uses include ingestive behaviour (eating and drinking, either together or separate locations), **elimination behaviour** and **resting behaviour.** In some conditions a portion of the space is dedicated to thermoregulatory behaviour, such as when a wallow is provided (**see: Thermoregulation**). The space required for **feeding** behaviour and **drinking behaviour** will vary with the type of feeder or drinker provided, whether feed is provided continuously or at intervals, and the time required for the animal to consume the feed and water. The time required for eating varies with the nature of the diet and the size (age) of the animal, resulting in different amounts of feeding space being required at different stages of production. Eating space may overlap completely with resting space if animals are fed on the floor.

Elimination space is somewhat determined by manure-handling systems, with bedded systems requiring additional space to accommodate sufficient bedding to absorb moisture. The remainder of the space in the pen or lot is sometimes referred to as free space. This space is used for active behaviours, including social interactions among the animals (**see: Social behaviour**). In some situations, particularly those of limited duration such as holding pens or **transport** vehicles, space is limited to the amount needed for standing or lying only.

In situations in which an animal is confined to an individual stall or uses a defined area for a particular behaviour (e.g. a free stall), the space can be seen as static and dynamic. Static space is that required physically to accommodate an animal in a particular posture. It consists of the space taken up by the body of the animal. However, animals in these situations will change postures and must have additional space to do so. Typically, an animal changing from a lying to a standing posture requires more length than that of its body to accommodate this movement. The space required to accommodate this movement is referred to as dynamic space and requires that stall dimensions be greater than the size of the animal's body in either the lying or standing posture.

Surrounding an animal is a region that the animal seeks to keep clear of other animals, called the individual distance zone or individual (personal) space (**see also: Inter-animal distance**). An animal approaching this zone may be discouraged by means of a threat, or allowed to enter if they show some type of **appeasement** behaviour. The size of this individual space will vary with the behaviour the individual is engaged in. While at rest, individual space may be non-existent, even though the same animal would enforce a zone while foraging. Some behaviours, such as grooming, require an animal to enter another's space. Social relationships may also affect the size of individual space. Dominant animals have larger zones than do subordinate, and close affiliates may be allowed to enter the zone more freely than others.

Similar to the individual distance zone is the **flight zone**, or the distance an animal allows another to approach before the animal moves away. The distance that it travels before ending its flight is the escape distance. For **group** situations, social tension will be reduced if animals are able to move about without having to enter each other's flight zone or, if they do, if they are able to flee beyond their escape distance to end the encounter.

Groups of animals, in a free-range environment, may maintain their space as either **home ranges** or territories (**see: Free-range animals; Territory**). Territories are defended against non-group members and have no overlap with other groups. Home ranges are not defended and may overlap, although the same space is rarely occupied simultaneously by both groups. Most domestic animals maintain free ranges, facilitating the management of multiple groups in a large range. However, under conditions of defendable resources, both horses and pigs have been reported to be territorial.

The quality of space refers to its ability to accommodate or enhance an animal's behaviour or welfare. Higher-quality space should better accommodate an animal's behaviour in the same amount of space, or maintain its ability to perform the behaviour in less space. Partitions or dividers in a feeder reduce **aggression** and facilitate uninterrupted eating, increasing the quality of the feeding space. Appropriate slopes and grooving in the floor can increase traction while an animal stands or lies down, increasing the quality of the space without changing its dimensions. Comfort-enhancing features such as insulated floors, freedom from draughts and walls for visual isolation also improve the quality of the space. Quality of space can also be enhanced using environmental **enrichment**.

Space is said to be adequate if animals are able to maintain normal behaviour and welfare status. Crowding refers to the impact of low space allowance (or high stocking density) on the animals, and is first evident in behavioural changes indicative of space restriction. Very low space allowance, or very high stocking densities, result in overcrowding in which the animal's welfare is impacted, as evident in severe behavioural adaptation, physiological responses, lower productivity or poor **health**. (HWG)

Speciesism

A form of prejudice, which discriminates against members of other species and ignores their **suffering** and interests simply because they are not human. Privileging human interests for no justifiable reason other than on species lines is, according to those who use this term, tantamount to sexism and racism. The term originates from a privately printed leaflet produced by British psychologist and animal advocate **Richard Ryder** (1970). (RA)

Spontaneous recovery

Spontaneous recovery refers to the reappearance following **extinction** of a previously learned association. Some limit the term to associations learned through **respondent conditioning** (referring to the equivalent phenomenon in relation to associations learned through **instrumental conditioning** as resurgence). An important implication of spontaneous recovery is that the process of extinction does not result in the permanent breaking or 'forgetting' of the previous association, but rather that it generates new learning, which may itself be lost or suppressed. In training, the phenomenon of spontaneous recovery is important as it highlights that it is not possible to guarantee that **problem behaviour** will not re-emerge if it has been eliminated through the training of the animal. (DSM)

Sport – use of animals

Animals can provide food, clothing, companionship and can also provide entertainment and be used for 'sport', i.e. as a source of diversion or recreation. The definition of sport is very wide, and this article refers to sport activities the reader in the northern hemisphere may not immediately associate as sport – things like bear **baiting** and **cock fighting**. Sporting use of animals can be divided into: (i) non-lethal – racing, chasing, fighting, recreational; and (ii) lethal – sport **hunting** and fishing.

In general, sport is not directly essential to human life, but *may* give added 'value' to life. The 'products of sport' – income from betting and keeping animals and animal breeding, the economic value of the animals themselves and food produced after sport hunting and fishing – can be considered as outcomes of the sporting use of animals.

If we accept the premise (and some people may not) that humans wish to be entertained, but also that animals have some basic needs – we can use this as a basis for discussions on situations where the sporting (recreation/diversion) wishes of people impact on the lives of animals. Some people may consider it a principle of animal use that the wishes of humans should not override the basic needs of the animals, and the 'benefit' to the people should 'justify' the 'cost' to the animals. (**see: Ethics**).

In many parts of the world the 'sporting' use of animals falls under general animal **welfare** legislation – but for large areas of the world there is no statutory protection for animals used in racing, fighting, baiting or as a hunted species in sport hunting. Many societies have created 'standards' by which a group or community decides to regulate its behaviour with regard to the sporting use of animals. These standards are sometimes formal – for example, in **horse** and camel racing, animals are often categorized by weight and age so that sporting competition is 'fair'. Sporting standards may also be 'cultural', or based on **seasonality** – for example, the creation of hunting seasons for sport-hunted animals in many **cultures** is based on protection of animals at mating or nesting times, or the restriction of sporting activity to a cultural group (falconry, stag hunting – once the preserve of the monarchy) reflects long-held history and distinctions of privilege. Additionally, in many cultures there are strong rituals and taboos associated with the use of animals in sport, and to prevent excess (and perhaps to prevent cruelty) some countries have made certain sporting uses of animals unlawful – cock fighting, **dog** fighting, bull and **bear** baiting, for example – but in other parts of the world these activities are still very actively pursued and carried out in public. The following examples illustrate some of the complex issues relating to the sporting use of animals.

1. Greyhound racing, a common sport in the UK, Ireland, the USA, South Africa and Australia, and variations of dog racing found in many developing countries. The dog is expected to perform, and will be trained, fed and kept with this aim in mind. The dog may be rewarded by the physical challenges and by the bond created with the owner/trainer, it may be required to travel and be subject to intensive training, and it may also be injured during training or racing and, if the animal does not perform, or when it becomes aged, then welfare concerns – neglect, poor **housing**, poor levels of care – may develop. Rehoming or destruction of unwanted greyhounds has become a high-profile welfare issue in some countries. The owner has social links, belongs to a sporting community, and is entertained and diverted by racing, and may also derive personal reward from the contact with the dogs and may (sometimes) derive financial reward. Spectators are entertained and may make (or lose) money by placing bets on the animals, but the public is not directly exposed to the animals that do not succeed, those past their prime or those that have been injured. The regulatory authority wishes to protect human and animal welfare, and will do so only so far as the law requires and resources allow. Racehorse ownership has some parallels with sporting dog ownership and is widespread across the world, and many horses are carefully tended, often into old age. Others are kept to perform at a higher sporting level but still may become beloved companions once they retire from sport – horse-riding is a popular pastime, and owners of these recreational and **companion animals** are likely to consider the benefit (human enjoyment) versus cost (cost to horse) balance to be at least equal, or even to pan out in favour of the horse.

2. Bear baiting, particularly in Pakistan. Adult bears may have been taken from the wild, but it is more common for young animals to be taken into captivity (as they can be trained) and sometimes adults are killed during the capture of their young. The bear will experience a life of restricted **space**, with limited scope to perform 'natural' hunting, social and reproductive behaviours, and will be subjected to **injury**, **pain**, **fear** and **distress** during the fighting. The bear is usually tethered, and its teeth and claws may have been blunted. Spectators see this activity as entertainment, and there is a strong social element to attendance at bear baiting as there is to dog fighting. Money will change hands in betting over which animals will win and, additionally, the public sees an animal interaction that is likely to reinforce their view that bears are dangerous **wild animals**, and that this justifies this activity. The local authority may have the powers to prevent bear baiting – the Prevention of Cruelty to Animals Act of 1890 and the Pakistan Wildlife Act prohibit bear baiting – but a lack of political will and inadequate resourcing of policing of this activity means that it continues today in some countries.

3. Rodeos, principally in the USA, Spain, Mexico, Brazil, Australia and Canada, fiestas and bullfighting in Spain, Portugal, southern France and several countries in Latin America, roping, riding, throwing of bulls, steers and calves, and **bullfighting** all exemplify a deep-seated part of the cultural use of **cattle** in many parts of the world. The animals are usually domesticated cattle that have not been made familiar with close contact with people. In some of these activities it appears to be necessary to goad or taunt the animals to get them to 'perform', and this is carried out with hooters, gunfire or fireworks, or by the use of spears, swords or darts. In the first stages of a bullfight (*toreo*, *corrida de toros*, *corrida de touros* or *tauromaquia*) the picador stabs the muscles on the bull's neck to enrage it (to draw first blood). This is followed by the placing of the *banderillas* (sharp, short spears) on the bull's

flanks to damage the neck and flank muscles and to incite the bull to charge. During the last stages of the bullfight, the *tercio de muerte* ('the third of death'), the matador uses a cape and a sword. The matador attempts *estocada* – to stab the sword between the shoulder blades, through the thoracic inlet and into the heart. The reliability and duration of this method of **killing** is in sharp contrast to that which would be permitted under local law (in most countries) in a **slaughterhouse**.

4. Sport hunting, shooting with guns or bows, or hunting of animals using dogs, hawks or **ferrets** is widely carried out in almost all areas of the globe. The origin of these activities is clear – as a means of gathering animal food and skins. However, the scale and focus of sport hunting has shifted away from subsistence food collection in many areas and towards trophy-type hunting, where the major motivation is the killing of the animal as a demonstration of tracking, technical and armaments capability. These activities can produce food, and some proponents argue that an animal shot in the field is not subject to the **confinement** and **transport** experienced by farmed animals. **Conservation** and population number issues are widely debated in relation to hunted species, but the predominant animal welfare concerns with sport hunting are that: (i) the methods used may result in injury or distress; (ii) there is the potential for a poorly controlled means of killing (for example, when hunting with dogs); and (iii) the action of chasing or tracking may induce periods of considerable distress and fatigue in the hunted animal. There has been only a small amount of objective scientific work on the impacts of hunting on wild animal species when compared with the literature available on, for example, farmed or **laboratory animals**.

In countries where animal welfare legislation is present and is applied in the area of animals in sport, the implementation of the law can be ineffective due to poorly designed legislation that is difficult for courts to interpret – who defines what is 'unnecessary' suffering in relation to sport? Limited resources, variable enforcement and the common finding that responsibility for enforcement is spread between government departments, so reducing coordination, may all significantly reduce the effectiveness of legislation aimed to protect animals used in sport. (AB)

Further reading

Bateson, P. and Bradshaw, E.L. (1997) The physiological effects of hunting red deer (*Cervas elephas*). *Proceedings of the Royal Society B* 264, 1–8.

Lewis, A.R., Pinchin, A.M. and Kestin, S.C. (1997) Welfare implications of the night shooting of wild impala (*Aepyceros melampus*). *Animal Welfare* 6, 123–131.

Websites

http://www.aht.org.uk/ahtequine.html
http://www.gra-america.org
http://www.andalucia.com/bullfight/home.htm

Stall-walking

Stall-walking (USA), also known as box walking (UK), is the pacing of a fixed route around the stable. Typically, a circular route is traced but, in larger stables or even at pasture, **horses** may trace a 'figure of eight'-shaped route. Affected horses may be prompted to begin a bout of stall-walking when separated from **conspecifics**, so horses in sale yards are more likely to be detected with this behaviour than with any **feeding**-related **ethopathy** (behaviours not found within the **ethogram** of free-ranging members of the species). Clearly, it is not a behaviour that is possible in standing stalls (tie stalls, USA) or when a horse is in cross-ties. Evidence of this activity may take the form of a track in the bedding of a stable. Occasional stall-walkers may conduct the activity at the trot, and these are among the more likely to sweat and become exhausted as a result of the stereotypy (**see: Stereotypies**). Stall-walking must be declared at certain types of auction and tends to lower the value of affected animals. It is reported to occur in 2.2% of stabled horses.

After **weaving**, stall-walking is the most common locomotory stereotypic behaviour in horses. Both are considered similar in causation and reflect a strong association with social needs. Stall-walking is often accompanied by **vocalization** as further evidence of its being a form of barrier **frustration**. Stall-walking is reported with higher than expected frequency in horses used for endurance trials. It is possible that persistence in apparently redundant locomotory activity reflects a shift in the **endorphin** activity of horses being conditioned for stamina and resistance to dehydration. In some cases the behaviour can be eliminated by incorporating a mirror in the stable.

Concerns about the behaviour relate to the effect on the horse's ability to maintain bodyweight, as it consumes time that would otherwise be spent feeding and resting. Furthermore, stall-walking in a single direction may cause asymmetric hoof or shoe wear, and sometimes lateralized atrophy and hypertrophy of the lumbar musculature. That said, stall walkers are less likely to be isolated than horses with other stereotypies, such as **crib-biting**, because the emergence of stall-walking is not generally attributed to observational learning. (PDM)

Star-gazing

Star-gazing describes several different behaviours: (i) the posture of **horses** that extend the neck while being ridden; (ii) an **innate** response that can reflect head, neck or chest pain; or (iii) a learned evasion of the bit. Mechanical devices such as martingales (tie-downs, USA) are frequently used to correct the posture but are undesirable as isolated remedies because they fail to address the proximal causes. Star-gazing also occurs as a **compulsive disorder** in **dogs** and as a neurological **sign**, involving extension and elevation of the neck (opisthotonus), in herbivores, most notably **sheep**. (PDM)

Startle response

A startle response is a rapid behavioural reaction to a sudden **stimulus**. Usually, it includes immediate cessation of the previous activity and a motor reflex such as a twitch, postural tension (freezing behaviour) change, jump or **vocalization**. Startle responses may occur when the animal is exposed to a sudden noise, touch, electrical or visual stimulus and act to prepare it for fight or flight. The particular motor reflex that occurs varies between species and circumstance. The behaviour is associated with a range of neuro-hormonal changes related to **stress**. The extent of the response, such as the length of time spent freezing, may indicate the severity of the reaction. Repeated or chronic stimulation of the startle response will have detrimental effects not only on an animal's **welfare**, but also lack of stimulation for a prolonged period, especially during development, may result in an exaggerated response or inability to cope efficiently when startled later in life (**see: Coping**). (KT)

Further reading

Broom, D.M. and Johnson, K.G. (1993) *Stress and Animal Welfare*. Chapman and Hall, London.

Mackenzie, S.A. and Thiboutot, E. (1997) Stimulus reactivity tests for the domestic horse. *Equine Practice* 19, 21–22.

Starvation

Starvation refers to a severe/prolonged lack of nutrients to the point of potentially threatening life. Adult birds and mammals of many species may need to lose more than 50% of normal adult weight before they die, if water is still available, although starvation is recognized as occurring long before this. Wild animals often starve as a result of prolonged adverse weather conditions, but in **captivity** animals should be buffered against these by appropriate care. It may occur in animals as a result of **disease** (causing a failure to feed, process, digest or absorb nutrients) or lack of access to nutrients in the environment. In captive animals, the latter may be a result of inappropriate management, such as having too many animals within a given system (excessive **stocking density**) or inadequate provision (failure of stockmanship), either by withholding food or providing an inadequate diet. Occasionally, starvation is a deliberate management strategy used in laying hens, which is undertaken to extend the productivity of the flock, as it induces moulting and subsequently increases **egg production** (**see: Forced moulting**).

Starvation may be associated with specific nutrient deficiency diseases (**see: Malnutrition**), but often results in more general effects, such as: (i) stunted growth in the young; (ii) loss of **libido** and shrinkage of the **gonads**; (iii) muscle wastage (as protein is used to provide energy); (iv) a fall in core body temperature and weakness (as energy is conserved); (v) anaemia; (vi) diarrhoea; or (vii) immunodeficiency.

Starving **wild animals** may not only be more willing to be approached by humans but may also more readily approach them. Supplemental **feeding** is often used to prevent wildlife from dying during periods of difficult weather, but may actually disturb the local **ecology** by skewing survival – for example, by altering the balance between species, as some are provided with supplementation and others not (this may have knock-on effects for higher predators too), or by preserving individuals that may be less well adapted to a wild existence (more tame) at the expense of others within the species, which may have otherwise had greater reproductive potential. (DSM)

Stem cells

Stem cells are the cells that arise from the fertilized egg, or **zygote**, and which develop into an **embryo** and later into a **fetus**. At about the eight-cell stage the embryonic cells have the ability to differentiate into all tissues of the body (totipotent), but they gradually lose this ability as they further divide (pluripotent) until they differentiate into specific cell types. Research is currently aimed at trying to grow and differentiate these cells *in vitro* to replace defective cells – and even organs – in the body, and to better understand some **disease** processes. (DBM)

Stereotypies

Stereotypies are behaviour patterns that are repetitive, invariant and have no obvious goal or function. They may consist of a single recurring behaviour, such as **bar biting** in tethered sows or **weaving** in stabled **horses**, or may involve a complex sequence of behaviour, as in one route-tracing polar **bear**, which performed a partial rearing up with an idiosyncratic head-swing at a particular spot in the enclosure on each turn. Stereotypies are sometimes confused with impulsive/compulsive behaviours. However, **compulsive disorders** involve the repetition of an inappropriate goal, whereas the **goal-directed behaviour** remains variable.

Stereotypies seem restricted to captive animals, mentally ill or disabled humans and subjects given stimulant drugs or suffering from **brain** damage. In this respect, they can be considered as **abnormal** behaviours. With respect to captive animals, however, there is controversy as to whether stereotypies represent the normal response of a normal animal to an abnormal environment (i.e. **maladaptive**) or whether they are abnormal in the sense of lacking in function and/or being the expression of an underlying pathology (i.e. malfunctional). Similarly, stereotypies in captive animals are generally considered as a sign of impaired **welfare**. However, the nature, duration and extent of the impairment have remained elusive. In the best case, stereotypies are reminiscent of earlier **frustration**, like mental scars, or may even help animals cope with adverse conditions. In the worst case, they reflect acquired brain dysfunction and/or chronic **suffering**. The answers to these questions have to be sought in the causation, the mechanisms underlying development and the effects of stereotypies on the animals themselves.

Stereotypies in captive animals do not appear suddenly in full form, but instead originate from particular behaviour patterns that become stereotypic over time. These source behaviour patterns include adjunctive behaviour, **appetitive behaviour**, **displacement behaviour**, intention movements and **redirected behaviour** (including **vacuum behaviour**); behaviour that indicates frustration, motivational conflict or thwarting. This suggests that stereotypies develop when animals are chronically or intermittently prevented from performing highly motivated behaviours or from attaining essential functional goals. Food-restricted **pigs**, for instance, develop oral stereotypies from appetitive and redirected **foraging behaviours**; gerbils denied access to a shelter develop stereotypic digging from attempts to dig a burrow; and mice develop various stereotypies from escape attempts when trapped in a barren cage. In the wild, these responses may be crucial for survival and reproductive potential. Therefore, they may be particularly resistant to **habituation** and **extinction**, which may favour stereotypy development.

However, the motivational problem underlying stereotypy development can rarely be inferred directly from the behavioural content or form of a stereotypy. Different **motivations** may induce the same behavioural response, leading to the same stereotypy, while the same motivation may trigger alternative behavioural strategies, leading to different stereotypies. In mice, for instance, attempts to escape barren laboratory cages were found to be influenced by the motivation to return to the mother after **weaning**, to seek shelter and to explore external **odour** cues. However, while some mice attempted to escape by squeezing or gnawing through the bars of the cage top (leading to stereotypic bar-mouthing), others attempted to jump out of the cage (leading to stereotypic jumping). Furthermore, some species seem to have a very low propensity to develop stereotypies or may not develop them at all.

This is particularly puzzling in closely related species with similar **behavioural needs**, such as house mice and Norway **rats**. Their general behavioural biology is largely similar, and rats, like mice, show high levels of bar-related escape behaviours when housed in barren laboratory cages. Thus, they both seem to be faced with similar thwarting of highly motivated behaviours, in response to which they perform similar behavioural responses. Unlike in mice, however, with rats these responses do not seem to develop into stereotypies. This suggests that species differences (as well as individual differences within species) may reflect, at least in part, differences related to the mechanisms by which behavioural responses to thwarting develop into stereotypies.

Therefore, the **causal factors** eliciting performance of potential source behaviour need to be distinguished from the mechanisms by which these behaviours develop into stereotypies. Due to the specificity of animal–environment interactions, the causal factors eliciting the source behaviour patterns from which stereotypies develop can be highly heterogeneous. In contrast, developmental changes and formal aspects of established stereotypies suggest similarities, rather than differences, in the mechanisms underlying stereotypy development.

In order for behaviour to become stereotypic, it needs to be performed repeatedly in the first place. Several factors may lead to initial repetition of behaviour. For example, constant or intermittent exposure to a stimulus will frequently elicit the corresponding response. This is particularly true when the motivation to perform that response is high and when there is little behavioural competition, due to the barren nature of the environment. For example, constant exposure of a gerbil in a barren laboratory cage to cage corners that resemble structures in the wild that would normally orient the beginning of a digging sequence readily elicits digging behaviour. When animals are frequently exposed to the same **stimulus**, the response to that stimulus often habituates. However, if a response is crucial for survival and reproductive potential in the wild, it may be highly resistant to **habituation**. In mice, for instance, habituation of escape attempts might be maladaptive because the reproductive potential of a wild **mouse** that gave up attempts to escape when trapped would be severely compromised. The same applies to hungry sows that stopped seeking food.

Once a sequence of behaviour has been initiated, repetition may be facilitated by a lack of consummation. This applies to many food-restricted farm animals, such as pregnant sows and broiler breeders. They may get stuck in an appetitive sequence of behaviour (e.g. foraging), when the behavioural response does not lead to the **perception** of a situation that correlates with the animal's expectations (e.g. satiety). And, in gerbils, a simple chamber with a tunnel-shaped entrance completely prevented the development of stereotypic digging, while the same chamber without the tunnel did not. Gerbils simply may not perceive a dark chamber as a shelter unless they experience moving into it through a narrow tunnel. This underscores just how specific the environmental **needs of animals** can be.

While specific causal factors may often explain the occurrence of particular source behaviour patterns and their repetition, they usually fail to account for the changes in form and performance typically observed in the course of stereotypy development. Three major developmental changes can be distinguished: increasing performance, increasing invariance and **emancipation**. Thus, performance of stereotypies normally increases over time, both in frequency and duration, leading to excessive performance (also referred to as over-performance) when fully developed. Excessive performance is a common feature of stereotypies, and may be the main reason for most stereotypies being detected in the first place. This may reach a point where animals do little else but **feeding**, drinking, **grooming** and resting, apart from performing their stereotypies. For example, laboratory mice may spend up to 50% of their total active time performing stereotypic bar mouthing; and polar bears were observed to perform up to 80% of their active time performing stereotypic route-tracing.

Besides the increase in performance, stereotypies will normally become increasingly *invariant*, i.e. fixed in form and orientation (also referred to as **ritualization**). In some cases, this decrease in variability of performance goes together with the loss of behavioural elements while, in other cases, additional, sometimes idiosyncratic, motor patterns are added (e.g. the head-swing in the route-tracing of the polar bear). Thus, stereotypies may become more or less complex with increasing invariance. It has been proposed that this process may also reflect response degradation, by which attempts at functional behaviour degrade into insufficient components or dysfunctional forms of that response. Together with increasing invariance, performance will normally *emancipate* from the originally eliciting circumstances. Stereotypies will thus become elicited by a wider set of stimuli or motivational states than early in development.

Stereotypies may eventually become so emancipated that they persist even in the absence of the original eliciting stimuli. This is sometimes referred to as the stereotypy becoming established or irreversible. Some consider establishment as a separate developmental change, although it might be fully explained by emancipation. Thus, with increasing emancipation, new eliciting stimuli may come to substitute entirely for the originally eliciting circumstances. In bank **voles**, for example, stereotypy performance was fully abolished by environmental **enrichment** introduced 3 months, but not 9 months, after the onset of stereotypy development, and partially abolished by enrichment introduced after 6 months.

These developmental changes indicate that there must be additional mechanisms that contribute to stereotypy development. Two alternative views have been proposed. First, stereotypies might reflect acquired behavioural strategies to cope with adverse circumstances (*coping hypothesis*). Alternatively, they might result from pathological changes at the neural level, leading to a disruption of normal brain functioning (*pathology hypothesis*).

According to the **coping** hypothesis, stereotypies develop because of rewarding properties associated with their performance in the initial situation of conflict or thwarting. Various forms of reward could be associated with stereotypy performance, on both a physiological and a psychological level. Thus, stereotypies would act to maintain or restore, at least partially, an optimal physiological or psychological state under conditions from which the animals are unable to alter or escape by behavioural means. The coping hypothesis explicitly considers stereotypies to be responses that the animals acquire through associative learning. The rewarding consequences would act as reinforcers, thereby increasing the probability (i.e. motivation) of the same behavioural pattern being performed on subsequent occasions.

Two variants of the coping hypotheses have been examined in some detail: *self-narcotization* and *coping with stress*. The first idea was inferred from the finding that endogenous **opioids** (**see: Endorphin**) were involved in stereotypies of tethered sows. In analogy to the 'runners-high', it was speculated that animals might stereotype to narcotize themselves as a means of avoiding perception of the frustrating situation to which they are exposed. However, experimental work failed to provide supporting evidence. Alternatively, performance of stereotypies might be reinforcing for having **stress**-reducing effects. The adaptive value of such a strategy would lie in its potency to attenuate the deleterious consequences of chronic stress. Despite some correlational evidence, selective prevention of stereotypy performance in pigs and horses did not lead to the increase in stress levels predicted by the hypothesis. Furthermore, a short term increase of corticosterone levels following prevention of stereotypic bar-mouthing in mice could be explained by the disruption of familiar behavioural routines.

However, stereotypies might develop for having beneficial effects on some other, as yet unexplored, physiological or psychological system. Attempts have been made to test general predictions that are independent of the system on which stereotypy performance might act. For example, if stereotypies help animals to cope with adverse environments, animals with established stereotypies should find such environments less aversive than non-stereotypers. In bank voles, a negative relationship between stereotypy level and preference for an enriched environment seemed to support this prediction, although alternative explanations have been proposed. Alternatively, increased stereotypy performance following selective prevention, i.e. post-inhibitory rebound (**see: Rebound behaviour**), would also speak for a coping effect. Post-inhibitory rebound was found in **crib-biting** horses following prevention of crib-biting by a commercial cribbing collar. This makes sense in the light of recent evidence suggesting that crib-biting might serve to stimulate saliva release to buffer gastric acidity induced by concentrate feed, thereby preventing ulceration. However, selective prevention of stereotypic bar mouthing in laboratory mice did not lead to post-inhibitory rebound, indicating that coping may not be a general characteristic of stereotypy performance.

Adaptive models of stereotypy development compete with explanations based on maladaptive or pathological processes. Again, two slightly different hypotheses are currently being discussed: *behavioural sensitization* and *behavioural disinhibition*.

The first is based on the common assumption that stereotypies are caused by chronic conflict or thwarting of highly motivated behaviour. In response, the animals are thought to perform **appetitive behaviours** that cannot become consummatory (**see: Consummatory act**), due to the lack of a suitable goal object. The positive sensory feedback of the appetitive behaviour on the neural systems controlling it would lead to an increased probability of this behaviour being performed on subsequent occasions. In addition, progressive sensitization of the activated neural pathways would lead to emancipation of the behaviour. While the same process under normal circumstances might facilitate habit formation, it would, by coincidence, act on behaviour that would not normally be repeated over and over again, thereby leading to fixation of malfunctional behaviour.

Indirect evidence for this hypothesis stems from studies on drug-induced stereotypies. Psychomotor stimulant drugs, such as amphetamine and apomorphine, have the potential to induce stereotypic behaviour through activation of dopaminergic systems in the brain, with the behaviour becoming more stereotyped with increasing doses of the drug. An increase in form and intensity, similar to that observed in many environmentally induced stereotypies, can also be induced by repeated administration of a constant dose, indicating behavioural sensitization to the drug. Furthermore, behavioural sensitization to the drug is also obtained by repeated exposure to various **stressors** such as mild tail pinch, inescapable foot shock, food deprivation or immobilization. During stressful experiences endogenous opioid peptides are massively released, and these have been shown to exert modulatory effects on dopaminergic pathways. Thus, stress induced by conflict or thwarting might play an important role in stereotypy development, via the sensitizing effect of endogenous opioids on **dopamine** systems in the basal ganglia. In line with this, pharmacological experiments in bank voles suggest that their stereotypies are mediated by dopamine, and that endogenous opioids might be critically involved in the early stages of stereotypy development.

Alternatively, stereotypies might reflect impaired mechanisms of inhibitory control of behaviour (**see: Disinhibition**). Stereotypies in human mental disorder (e.g. schizophrenia, autism) and stereotypies induced by drugs (e.g. amphetamine) or brain lesions depend on reduced behavioural inhibition, possibly caused by impaired dopaminergic regulation in the dorsal basal ganglia. Impaired basal ganglia function is associated with a series of other characteristic behavioural changes, including enhanced rates of behavioural initiation, impulsivity, impaired response suppression and perseveration (e.g. in extinction tasks). In bank voles, it was recently found that stereotypic bar-mouthing correlated with increased rates of behavioural activation, hyperactivity, impairments of response timing and slower extinction learning, and that all of these signs intercorrelated. These findings suggest a single, underlying deficit that is consistent with impaired response selection caused by dorsal basal ganglia dysfunction. Thus, stereotypies might be the most apparent expression of an acquired brain pathology. On the other hand, individual variation in stereotypy performance and correlated measures might simply reflect the normal range of variation of a population of healthy individuals. To exclude this possibility, it remains to be demonstrated that these signs change together as stereotypy develops, and back again when the stereotypy is cured, e.g. by environmental enrichment.

Despite decades of extensive research, our understanding of the causal factors and mechanisms underlying stereotypy development has remained elusive. Because the welfare implications of stereotypies clearly depend on these aspects, no definitive assessment can be made. Present evidence suggests that stereotypies are unlikely to be learned responses for coping with environmental adversity. Although they all seem to develop from attempts to cope with behavioural thwarting, they are likely to reflect true behavioural disorders based on some form of **central nervous system** (CNS) dysfunction, at least in their fully established form. Stereotypies are clearly heterogeneous in causation, yet the similarities in development indicate common underlying neural changes. Even if they reflect CNS dysfunction, the question still remains as to how such dysfunction is perceived by animals. It was proposed that stereotypies become 'hard-wired' with establishment, thereby losing emotional significance.

However, in the absence of a coping effect, the problems that gave rise to stereotypy development might persist to be perceived as **aversive**. Others consider stereotypy performance to be a direct sign of animal suffering caused by the '**boredom**' induced by a lack of control over the environment. While this hypothesis has remained vague and has not produced testable predictions, others have stressed the possibility that stereotyping animals might become frustrated about their increasing inability to put knowledge into action, based on evidence for an action–knowledge dissociation associated with impaired inhibitory control, as also observed in human patients. Since all of these hypotheses are essentially speculative, identifying the emotional correlate of stereotypy performance is likely to remain a major challenge in the near future. (HW)

See also: Corticosteroids; Disorders of behaviour; Feather; Feedlot; Malnutrition; Needs of animals; Repetitive behaviour

Further reading

Mason, G. and Rushen, J. (eds) (2006) *Stereotypic Animal Behaviour: Fundamentals and Applications to Welfare*, 2nd edn. CAB International, Wallingford, UK.

Steroid hormones

The steroid hormones are characterized by a series of three joined 6-carbon rings fused with a fourth 5-carbon ring. They constitute a large endocrine family synthesized by the **gonads** and by the **adrenal glands**, with their function varying according to the functional groups attached to these rings. Steroids include the sex hormones: **androgens** (**see also: Testosterone**), **oestrogens** and progestagens (**see: Progesterone**), and two important homeostatic regulatory hormones, the **glucocorticoids** and mineralocorticoids (**see: Corticosteroids**).

The lipid basis of the molecules means that they are able to pass freely across membranes and, as a consequence, cannot be stored within vesicles in their producing cells, and so are not available for instantaneous release in the same way that the peptide hormones are, but must be synthesized on demand. Membrane permeability also means that they are not limited to activating cell surface receptors, but can bind to intracellular receptors and frequently have direct effects on their target cell's DNA via nuclear receptors. Their hydrophobic nature results in steroid hormones often being associated with specific binding proteins for transport in the aqueous plasma, and thus active hormone levels can be influenced not only by the level of release but also by the amount of available binding protein, which affects the proportions of bound and free hormone within any system.

Breakdown and removal of steroid hormones is also a more complex process than for peptide hormones, involving catabolism in the liver. Taken together, the biophysical properties of the steroid hormones mean that they are frequently involved in regulation of long-term and slower-acting development and behavioural effects with a relatively slow response time and a long half-life, although they are also capable of initiating rapid, acute responses in some circumstances, e.g. in relation to the **stress** response. (MRC)

Stimulus

In broad terms, a stimulus is any factor or event that elicits a response. A stimulus–response system is observable at any biological level, from genes and cells to organisms, **social groups** and **ecosystems**. (BM)

See also: Stimulus–response behaviour

Stimulus–response behaviour

Early ethologists and behaviourists devoted much attention to stimulus–response systems. These systems are expected to be relatively stable and fixed for **instinctive behaviours** and/or for basic survival needs (e.g. avoidance of danger). The factors affecting stimulus–response systems, and how the link between a particular **stimulus** and a behavioural response may be modified by experience, have been the topic of various learning theories (**see: Conditioning – types of**). (BM)

Further reading

Inglis, I.R. and Langton, S. (2006) How an animal's behavioural repertoire changes in response to a changing environment: a stochastic model. *Behaviour* 143, 1563–1596.

Stocking density

Stocking density is the ratio of animal units to **space**. The units are often based on body weight (BW) and should be expressed in appropriate terms for the type of space provided. For space measured in distance, the most appropriate term would be $BW^{1/3}$/length; for area, $BW^{2/3}$/area; and for volume, BW/volume. In the case of a **resource** used by one animal at a time (point source), such as a defined feeder space or a drinker, stocking density can be expressed as animals/unit (e.g. pigs/feeder space). Some literature uses 'stocking density' to refer to animals/pen, but this confounds **group** size and true stocking density. The appropriate term would be $BW^{2/3}$/area. Stocking density is the reciprocal of space allowances. Crowding refers to the impact of high stocking density on animals, and is first evident in behavioural changes indicative of space restriction. Very high stocking densities result in overcrowding in which the animals' **welfare** is impacted, as evident in severe behavioural adaptation, physiological responses, lower productivity or poor **health**. (HWG)

See also: Feedlot

Stockperson

With increasing **intensification of animal production (see: Farming of animals)**, the dependency of livestock on the stockperson to effectively care for and manage them correspondingly increases. While all forms of livestock production rely to varying degrees on key inputs from the stockperson, this reliance is arguably greater with increased **confinement** of the animal.

The key feature of any employee, irrespective of the industry in which he or she works, is how well he or she does the job. Delving further, work performance is predominantly affected by three interrelated factors: capacity, opportunity and willingness to do the job. Therefore, in considering the work performance of the stockperson in a livestock industry, capacity includes characteristics of the person such as technical skills and **knowledge**, and attitude and behaviour towards the animals, while willingness includes **motivation**, job satisfaction and work attitude, and opportunity includes working conditions and facilities, actions of co-workers and organizational policies and rules. In order to optimize the role of the stockperson in animal productivity and **welfare**, an understanding of the impact and regulation of the work performance of the stockperson is necessary.

Capacity

Technical skills and knowledge

Knowing and being skilled at the techniques that must be used to accomplish a task are clearly prerequisites to being able to perform that task. Thus, these job-related characteristics will be the most limiting factors to job performance in situations where specific technical skills and knowledge are required to perform the tasks. While there are few data in agricultural industries, this basic premise is nevertheless universally accepted.

Some of the key characteristics of stockpeople include a basic knowledge of both the behaviour of the animal and its nutritional, climatic, **housing**, health, social and sexual requirements, together with a range of well-developed husbandry skills and knowledge to effectively care for and manage farm animals. For instance, stockpeople may have knowledge and skills in a number of diverse management and husbandry tasks such as: (i) **oestrus** detection and mating assistance; (ii) semen collection and preparation and **artificial insemination**; (iii) pregnancy diagnosis with ultrasonagraphy; (iv) artificial rearing of early-weaned animals; (v) milk harvesting; (vi) controlling and monitoring feed intake for optimizing growth, body composition, milk production and reproductive performance; (vii) pasture management for optimizing pasture production; (viii) routine health checks; (ix) monitoring and adjusting climatic conditions in indoor units; (x) administering antibiotics and vaccines; (xi) **shearing** and crutching **sheep**; (xii) teeth and tail clipping of piglets; (xiii) **castration** of males; and (xiv) effective and safe animal **handling**. These are all skilled tasks, and stockpeople are required to be competent in many of these tasks.

Attitudes and behaviour towards animals

Fear is generally considered an undesirable emotional state of **suffering** in both humans and animals. While fear thresholds have been reduced by **domestication**, fear responses have not been eliminated from domestic animals. There is substantial variation between commercial farms in animals' fear of humans and, furthermore, there are significant sequential relationships between the stockperson's attitudes and behaviour towards their animals and the fear of humans and productivity of farm animals.

Studies in both the dairy and pork industries have shown that negative attitudes of stockpeople towards handling their animals are correlated with a high percentage of negative behaviours by stockpeople in handling their animals, which in turn is correlated with high fear levels and low productivity in these animals. Negative attitudes of stockpeople towards handling include beliefs that considerable verbal and physical effort is required to move their animals and that these animals do not require **petting** and stroking. These are correlated with a high percentage of negative behaviours used by stockpeople in handling their animals. Negative behaviours used by stockpeople include not only forceful hits and slaps, but also audible slaps, pushes, shouting, fast speed of movement and sudden movement.

In contrast, positive behaviours include pats, strokes, the hand of the stockperson resting on the back or flank of the animal, talking and slow deliberate movement. The sensitivity of dairy cows and **pigs** to human behaviour is highlighted by the impact of moderate negative interactions, such as audible slaps, pushes, shouting, fast speed of movement and sudden movement by humans, something that is not intuitively obvious to most of us. Similar effects of human–animal interactions have been found in the poultry industries. Laboratory studies indicate that the likely mechanism responsible for the adverse effects of high fear on animal productivity is via a chronic **stress** response and its consequences. There is also evidence in the dairy and pork industries that human–animal interactions may directly affect a number of job-related variables, such as job satisfaction.

The attitude of the stockperson towards interacting with his or her animals is obviously a key antecedent of the behaviour of the stockperson, which in turn is an important determinant of the animal's fear of humans. The existence and importance of these sequential relationships are demonstrated in intervention studies in both the dairy and pork industries, in which cognitive–behavioural training of stockpeople, designed specifically to improve these key attitudes and behaviours of stockpeople, decreased animal fear and improved both animal productivity and welfare.

The development of fear responses in livestock is not surprising, considering that stockperson interactions may be biased towards negative ones. Opportunities for positive human contact are probably reduced in modern production units, and many routine husbandry tasks undertaken by stockpeople may contain **aversive** elements. This bias towards negative interactions, together with the sensitivity of farm animals to even moderate negative interactions, highlights the challenge confronting stockpeople in improving their interactions with farm animals and emphasizes the importance of training.

Willingness

While the impact of these other characteristics may be obvious, they have been less researched in the livestock industries than the impact of the stockperson's attitudes and behaviour.

Job motivation and commitment

It is convenient to group these two attributes together, as they together refer to the extent to which a person applies his or her skills and knowledge to the management of the animals under his or her care (e.g. how reliable, thorough, conscientious, etc. a person is). Factors such as job satisfaction, meaningfulness of work, utilization of skills, etc. will affect work motivation and commitment. High job performance in any industry relies on a combination of motivation, technical knowledge and skills and an opportunity to perform the job. Clearly, low motivation will limit job performance regardless of technical skills and knowledge of the individual. For example, while an ability to identify any departures in the behaviour, health or performance of the animal is an important technical knowledge and skill, the willingness to regularly inspect animals and to promptly provide or seek appropriate support to address any departures is obviously critical in achieving high animal productivity and welfare.

There are various theories on how work conditions may affect job satisfaction and work motivation. For example, motivation to perform tasks will improve with rewards such as feelings of pride or accomplishment. In fact, a hierarchy of needs may influence motivation, with esteem needs (recognition by others) and self-actualization (self-fulfilment) motivating workers once the physical, safety and social needs of the individual have been met.

Job satisfaction

This factor is influential because of its indirect effects on work performance: job satisfaction will directly affect other job-related characteristics such as job motivation and commitment, motivation to learn new skills and knowledge and thus, in turn, technical skills and knowledge, etc. Job satisfaction refers to the extent to which a person reacts favourably or unfavourably to his or her work and is considered to derive from the extent to which a person's needs or expectations are being met by the job.

As mentioned above, it is generally considered that job satisfaction is influenced by rewards (personal and financial), job design and enrichment (e.g. involvement in decision-making processes), work performance, animal comfort and **health** and the working environment. As with some of the above job-related characteristics, there is little evidence in the agricultural industries that job satisfaction affects animal productivity. However, it is generally recognized that job performance in any industry is influenced by job satisfaction via its effects on job motivation and commitment, motivation to learn new skills and knowledge, etc. Several authors have suggested that a decline in job satisfaction is associated with staff turnover and absenteeism.

Personality

Although there is some disagreement among psychologists, a **personality** trait is generally considered to be a relatively enduring characteristic that exerts a general effect on the individual's behaviour and, while personality cannot be observed directly, it can be inferred from the person's behaviour. Most researchers agree that personality can be characterized in terms of the 'big five' personality traits: 'extraversion/introversion', 'emotional stability', 'agreeableness', 'conscientiousness' and 'intellect'. Furthermore, it is also generally well accepted that personality factors may be useful in matching people to some kinds of jobs. For example, discipline and conformity may be important factors in some jobs in which routine tasks are performed by teams of people, while independence, introversion and self-motivation may be important in others in which the tasks are more problematic and where the individual may at times work alone. Indeed, there is some limited evidence to support the importance of personality factors in the livestock industries.

Research on stockpeople in the pork industry has shown that the importance of personality factors on piglet survival may vary according to the working place, with the relative importance of the traits depending on the type of farm. Self-discipline was a trait that appeared to be important at all farms studied; however, high insecurity and low sensitivity were favourable traits in relation to piglet survival at small, independent owner-operated farms, while stockpeople that were highly reserved and bold, suspicious, tense and changeable were associated with higher piglet mortality at large, integrated farms.

While an early study in the dairy industry indicated that the highest-yielding herds were associated with stockpeople who were introverted and confident, recent research found that personality factors, based on similar measures used in the earlier study, were not significantly correlated with cow productivity. However, this recent study found that some personality factors were significantly correlated with the attitudes of stockpeople. Agreeableness, which is one of the 'big five' personality traits and is generally considered to be associated with cooperation, good nature and tolerance in a person, correlated positively with a positive attitude towards cows. Agreeableness also correlated positively with the use of positive behaviours and tended to correlate negatively with the percentage of negative behaviours.

These results support the general view in the current literature that personality factors may influence stockperson performance in ways that are relevant to animal productivity and welfare.

Empathy

Several authors have suggested that the degree of empathy may predispose people to be good stockpeople. Empathy can be described as the capacity to put oneself in the place of another, and stockpeople may therefore perform better if they have a good insight into the emotional responses of the animals under their care. There is limited evidence in the dairy and pork industries to suggest that empathy may be a factor underlying the development of positive attitudes towards pigs: empathy towards animals was positively associated with positive attitudes both about animals and towards interacting with animals.

Opportunities

There is a clear need to identify the attributes that best allow a person to meet the job requirements of a stockperson. These characteristics will be a combination of learned factors, such as attitudes, handling and observational skills and technical knowledge, and dispositions, such as personality and empathy.

The sequential relationships between stockperson attitude and behaviour and animal fear, welfare and productivity demonstrate the opportunities to improve animal welfare and productivity by appropriate selection and training of stock-people. Other job-related characteristics such as technical skills and knowledge and job motivation and commitment should also be targeted in the selection and training of stockpeople. A number of recent studies in the dairy and pork industries demonstrate excellent opportunities to improve animal welfare and productivity through training and selection of stockpeople.

While there is little evidence in the livestock industries relating personality and empathy directly to work performance of the stockperson, these characteristics may have indirect effects on animal welfare and productivity. For example, the antecedents of attitudes are many and varied. Experience, demographic variables, various general attitudes and personality traits may indirectly affect behaviour through their influence on attitudes and thus these characteristics, such as personality and empathy, may operate indirectly through attitudes. Furthermore, personality traits may be useful in matching people with specific jobs in the livestock industries. For instance, as stated above, independence, introversion and self-motivation may be important factors in which the tasks are more problematic and where the individual may at times work alone.

Therefore, there is a clear need to reduce the limitations that human–animal interactions impose on the welfare and productivity of commercial farm animals. While our understanding of the regulation and impact of human–animal interactions has improved considerably over the last decade or so, recognition of the role of stockpeople in the welfare and productivity of livestock has only recently occurred.

Traditionally, stockpeople in modern animal agriculture have been regarded as unskilled labourers. The role of stockpeople as the key people responsible for the day-to-day welfare and productivity of the animals under their care has not received due acknowledgment, even though the welfare and productivity of large numbers of animals are entrusted to the care of these stockpeople. Much has been done to improve genetics, nutrition, health and housing of livestock, but efforts to target stockpeople, who perform such a key function, have just begun.

It is likely in the near future that the general public, consumers, governments and the livestock industries will place an increasing emphasis on ensuring the competency of stockpeople to manage the welfare of farm animals; this is likely to occur in both intensive and extensive animal industries. Appropriate strategies to recruit and train stockpeople in the livestock industries will be integral in safeguarding the welfare and productivity of commercial farm animals. (PHH)

See also: Memory; Transport

Further reading

Coleman, G.J. (2004) Personnel management in agricultural systems. In: Rollin, B.E. and Benson, J. (eds) *Maximizing Well-being and Minimizing Suffering in Farm Animals*. Iowa State University Press, Ames, Iowa, pp. 167–181.

Hemsworth, P.H. (2004) Human–livestock interaction. In: Benson, J. and Rollin, B.E. (eds) *The Well-being of Farm Animals: Challenges and Solutions*. Blackwell Publishing, Oxford, UK, pp. 21–38.

Hemsworth, P.H. and Coleman, G.J. (1998) *Human–Livestock Interactions: the Stockperson and the Productivity and Welfare of Intensively Farmed Animals*. CAB International, Wallingford, UK.

Stone-chewing

Picking stones up into the mouth and chewing them has been commonly reported in **pigs**, and also in **dogs**. It has variously been described as **play** behaviour, redirected **foraging behaviour** or a stereotypic behaviour (**see: Stereotypies**). In pigs, stone-chewing is often observed in sows housed on paddocks, and most frequently on paddocks where plant cover has been removed. The major issues surrounding stone-chewing involve speculation as to its cause and/or function and potential **health** problems that it may cause the animal in chewing or swallowing the stones.

Horrell and A'Ness (1999) observed wild boar in a semi-natural environment and commercial pigs in six different environments. They reported that commercial sows housed on arable land and housed indoors in small straw yards with stones available spent 46.5 and 40% of observation time, respectively, chewing stones, which was 10–20% more than sows in three grass-based paddock treatments. Wild boar spent less than 1% of their time chewing stones in their enriched environment. Dailey and McGlone (1997) reported that sows on pasture chewed grass, sows in crates chewed the metal bars of the crates and sows on a soil-covered paddock chewed stones to a similar extent. Both sets of authors conclude that stone-chewing represents thwarted foraging behaviour and may be a form of stereotypic behaviour, as it is most prevalent in relatively barren environments of bare-earth paddocks. Horrell and A'Ness (1999) hypothesize that it is a function of **boredom** or a **coping** response to **stress**.

In terms of health concerns, Davies *et al.* (2001) examined teeth and stomach contents of indoor- and outdoor-housed sows and found that tooth damage (30 and 28%, respectively) and wear (91 and 88%, respectively) were in fact very similar in both populations and thus independent of stone-chewing, which was only available to outdoor-housed sows. Even though nearly 40% of outdoor-housed sows had stones in their stomachs, there was no indication that stone-chewing affected the physical health of the sows. (JNM-F)

References

Dailey, J.W. and McGlone, J.J. (1997) Oral/nasal/facial and other behaviours of sows kept individually outdoors on pasture, soil or indoors in gestation crates. *Applied Animal Behaviour Science* 52, 25–43.

Davies, Z.E., Guise, H.J., Penny, R.H.C. and Sibly, R.M. (2001) Effects of stone chewing by outdoor sows on their teeth and stomachs. *Veterinary Record* 149, 9–11.

Horrell, R.I. and A'Ness, P. (1999) Stone-chewing in outdoor pigs. In: Boe, K.E., Bakken, M. and Brastad, B.O. *Proceedings of the 33rd International Congress of the ISAE*, p. 88.

Stray voltage

Animals can be affected by stray voltage in their housed environment, particularly if it is a wet environment and the building is inadequately earthed. This has been detected in **cattle**, which can readily detect the low-level electric currents that often exist in parlours where the wet environment and connection of machinery to their udder makes them particularly susceptible. Most cows will respond behaviourally to a 3 mA current of 0.7 V, and some will respond to a 1 mA current of 0.2 V. The resistance to the passage of electricity is only about 250–400 Ω, as a result of their direct contact with the floor at four points. The resistance provided by humans is two to ten times greater, depending on footwear, etc. At low levels of current, typically **heart rate** is slightly elevated, and this increases with the current together with the appearance of overt behavioural responses. Mild behavioural responses include flinching and **vocalization**, progressing to a startled response and **avoidance behaviour**. In terms of voltage, about 1–2 V will elicit a response from the majority of cows, and it is recommended that stray voltage should be less than 0.35 V, which will be perceived by less than 10% of cows.

Stray electricity may also derive from off-farm sources, in particular where there are three-phase supplies nearby. It can be caused by: (i) large voltage drops on the farm, which cause the resultant supply to be out of phase with the central supply; or (ii) faulty equipment grounding, especially in a corrosive environment. Electricity can be used on the farm to control the movement of animals, for example into the milking parlour by electronic crowd gates, in front of self-feed silage where an electrified bar may be used, by electric fences or by using electrified training devices. All of these may produce stray electricity if not properly earthed. Intermittent, irregular shocks are more harmful than continuous stimulation, due to greater difficulty in **habituation**. If the stray electricity is present in a water trough, animals may lap at the water like a dog rather than immersing their muzzle in the water. (CJCP)

Stress

The word 'stress' brings with it the connotation of a negative state; however, stress is a normal biological state that every living being experiences. Not all stress is negative or **aversive**. For this reason the term has been broadened to include stress, **distress**, which is negative stress, and **eustress**, which is positive stress. Under these categories, stress can be used to designate everyday challenges to **homeostasis** with which an animal is able to effectively cope, without causing negative **welfare**. Eustress is stress that is positive, such as stress due to wilful exercise, **play** behaviour, **reproductive behaviour** and 'thrill-seeking' behaviour, as seen in humans. Distress then is used to refer to stress that does have a negative impact on animal welfare. The challenge provided to scientists and producers is being able to distinguish when an animal is being stressed to the point of impaired welfare and determining how to eliminate this stress.

The concept of stress has been studied for around 100 years. Much has been learned but many more questions remain to be answered. The general definition of stress itself is often debated. Moberg (2000) defines stress as 'the biological response elicited when an individual perceives a threat to its homeostasis'. He goes on to state that 'When the stress response truly threatens the animal's **well-being**, then the animal experiences "distress"'. Similarly, Charmandari *et al.* (2005) define stress as 'a state of threatened or perceived as threatened homeostasis'. In contrast, Broom (2001) defines stress as 'an environmental effect on an individual which overtaxes its control systems and results in adverse consequences, eventually reduced **fitness**'. Biological fitness is commonly measured by the ability of an animal to produce offspring that make up future generations. Thus, Broom's definition judges welfare to be poor if fitness is reduced, as opposed to Moberg's and Charmandari's definitions, which allow for less profound experiences, which may not reduce reproductive ability, to be classified as stress.

The idea that is common in almost all definitions of stress is that an animal must make changes, behavioural and/or physiological, to respond to the **stressor**. Furthermore, that if the changes are profound enough, the stress is considered deleterious, often termed 'distress'. It may be useful to think of the concept of stress relative to engineering standards. A load-bearing beam in a home has a considerable amount of weight challenging its integrity. However, it is not until the weight becomes so great that the beam may break that we are concerned about the stress placed on the beam. Animals are, of course, not inanimate objects, and our concern extends prior to the point of 'breaking' the animal. It is this challenge, deciding when the stress is merely 'stimulation' or 'distress', which occupies a great deal of research effort in the study of stress. The complexities are further detailed by the realization that animals are not identical beings and that each can experience stress in its unique manner. Previous experience, social setting, mood and **health** can all alter the level of stress an animal experiences.

The duration of time over which an animal experiences stress is important as to whether the stress causes deleterious effects. Often, an animal can experience short-term (acute) stress and suffer few, if any, long-term effects to its health or productivity. For example, animals are often restrained during which time they make enormous attempts to escape and exhibit elevations of **heart rate**, **blood pressure**, **epinephrine**, **norepinephrine** and cortisol – all classic indicators of a stress response. Almost all stress researchers would consider such an animal to be experiencing 'distress'. However, once the animal is released, behaviour will quickly return to normal, along with physiological indicators of stress. In this situation, there is little chance that a long-term deleterious effect will be experienced by the animal. In contrast, chronic stress (long term) can have profound deleterious effects. For example, animals housed at very high **stocking density** or with very aggressive **conspecifics** have been shown to have reduced growth and/or reproductive ability. It is also well established that chronic stress impairs normal immune function. Thus, the ability of an animal to escape a stressor, and thus relieve its state of stress, is critical to maintenance of its well-being. This concept is important, because many situations in animal husbandry do not allow animals to control their exposure to stressors.

In order to maintain homeostasis and respond to stressors, animals make both behavioural and physiological adjustments. To respond to stressors that pose both a psychological and physical threat, such as a predator, animals respond by fighting or fleeing. Obviously, the behavioural component of the response is to fight or flee. The physiological response is characterized by elevations in heart rate, stroke volume, blood pressure and mobilization of glucose. These changes in physiology allow the animal to meet its needs of moving oxygen to the muscles and **brain**, and supplying the body with energy in the form of glucose to mount the behavioural response.

All of these responses are adaptive in the sense that they provide the opportunity for the animal to successfully cope with the predator. This brings up an important point: the state of being stressed is a normal, purposeful condition that allows species to survive. Common usage of the term 'stress' conveys a negative connotation and implication that all stress should be eliminated. This is not the case. The stress response is a normal function of the body, just like sweating in response to excessive heat. Sweating allows the body to return its temperature to normal levels. Activation of the stress response also allows the body to return to its normal state. The problem arises when the stress response is so profound or of such a long duration that the health of the animal is threatened.

When such profound activation of the stress response occurs, deleterious effects to both physical and psychological health can occur. Physically, animals subjected to unrelieved stress can have an impaired **immune system**. This impairment permits infections from bacteria and viruses to make the animal sick. Furthermore, typical hormones often highlighted in the stress response, such as **glucocorticoids**, can have negative effects on many other physiological systems. For instance, glucocorticoids are known to suppress **luteinizing hormone (LH)**, which results in suppression of reproductive function. Glucocorticoids are also known to suppress **insulin-like growth factor**, which results in suppressed growth. See Lay and Wilson (2004) and Charmandari *et al.* (2005) for a more comprehensive review of the negative impact of stress on health and productivity. Psychologically, animals subjected to unrelieved stress can experience states similar to **anxiety** and **depression**, as observed in humans. However, due to the subjectivity of these states, very little is understood about their intricacies. Dantzer (1994) provides a comprehensive review of subjective states and their importance to the animal's experience of stress.

An animal's reaction to exposure to a stressor is also dependent upon the type of stressor. Stressors can be physical, environmental, nutritional, metabolic or psychological. In general, almost all physical stressors cause the animal to experience some type of psychological stress. It is often difficult or impossible to separate which reactions are due to the physical stressor and which are due to the psychological stressor. For instance, an animal that is being chased by a predator has physiological responses indicative of the physical stress of running; however, these responses are also indicative of a fear response. In cases such as these, some physiological responses could be maximum, such as heart rate, while others could be additive, such as an increase in cortisol, which is further increased by the feeling of fear.

Some stressors are of an entirely psychological nature. This is often the case when an animal is fearful. Poultry can exhibit a phenomenon termed '**fowl hysteria**', which is characterized by a stimulus, such as a door slamming, causing the flock to act as if panicked. Research has consistently shown that a psychological stressor is more potent in causing harm as compared with a physical stressor. Some stressors are due to environmental effects. Extremes in temperature and air quality are common environmental stressors. These types of stressors typically do not invoke a psychological stress response. However, these are significant stressors in terms of animal welfare with chronic exposure to either able to cause **death** or **disease**.

Nutritional and metabolic stressors also exist that push the animal past its ability to maintain homeostasis. In production agriculture, demands for high productivity require that nutrients be obtained and utilized at a rate in order to maintain a positive net energy balance. However, situations of high productivity, such as a dairy cow that produces great quantities of milk, a hen that produces many eggs or a sow that produces very large litters, can stress some individual animals to the point that their welfare is poor. Measures of poor welfare in these situations are illustrated by dairy cows that do not conceive, hens that have sequestered so much calcium from their bones to lay eggs that their bones easily break, and sows that emerge from **lactation** with little body fat and develop body sores from lying on hard surfaces.

Measures of productivity are often used to help determine whether an animal is being subjected to significant stress. In livestock production systems, productivity for an animal is characterized by measures of growth and function of the reproductive system, such as the production of eggs, offspring and milk. In many ways, these measures are very useful. Exposure to a chronic stressor activates physiological systems that have the ability to impair, disrupt and suppress the above measures of productivity. Thus, if an animal is being exposed to a stressor and these measures of productivity are negatively affected, the animal is in a state of distress.

However, there are several problems with using productivity as an indication of distress. First, productivity that is lower than expected can be due to causes other than stress. For instance, genetics play a large role in all aspects of productivity, and an unstressed animal may have low productivity simply due to its genetic make-up. Non-stressful environmental factors also play a role in productivity. For instance, the length of light exposure can decrease egg production as well as changes in diet, neither of which are necessarily stressful to a hen. Secondly, many stressors are not so severe as to decrease productivity but are still a welfare concern. Transportation stress is of significant concern, yet this stress does little to affect reproductive status or growth of most animals (**see: Transport**).

If a stressor does decrease productivity, it can be considered a severe stressor. Another problem with using productivity as a measure of stress is in determining the timespan to consider. Often, animals produce at a very high level, but have shorter life expectancy due to their high level of productivity. In this situation, life productivity is low, while weekly productivity may be high. Also of consideration is the problem of determining what is 'normal' productivity. Without a normal, or 'gold standard', we cannot say whether productivity is high or low. Using the herd average as the standard represents flawed logic, because the entire herd could be stressed. Thus, lacking a legitimate standard, productivity has limited value as a measure of stress.

The question then remains: what measures should be used to quantify stress? Given the current state of knowledge, the best approach is to collect a wide array of information, including: physiology, behaviour, productivity, cognitive state and measures of subjective states. None of this information will be a perfect measure by itself but, viewed together, can provide insight as to the animal's state of stress. Although we have learned a great deal about stress from previous research, much more needs to be learned before simple quantitative assessments can be recommended. (DCL)

See also: Corticosteroids; DFD meat; Measuring welfare

References

Broom, D.M. (2001) Coping, stress, and welfare. In: Broom, D.M. (ed.) *Coping with Challenge: Welfare in Animals, Including Humans*. Dahlem University Press, Germany, pp. 1–9.

Charmandari, E., Tsigos, C. and Chrousos, G. (2005) Endocrinology of the stress response. *Annual Revision of Physiology* 67, 259–284.

Dantzer, R. (1994) Animal welfare methodology and criteria. *Reviews in Science and Technology* 13, 277–302.

Lay, D.C. Jr and Wilson, M.E. (2004) Considerations when using physiological data in assessing animal well-being. *Journal of Animal and Veterinary Advances* 3, 614–626.

Moberg, G.P. (2000) Biological response to stress: implications for animal welfare. In: Moberg, G.P. and Mench, J.A. (eds) *The Biology of Animal Stress: Basic Principles and Implications for Animal Welfare*. CAB International, Wallingford, UK, pp. 1–21.

Stress hormones

Stress hormones are those hormones whose concentrations are considered to be altered, either increased or decreased, when an animal is exposed to a **stressor**. An animal's physiological response to **stress** concentrates on meeting its needs for the 'fight or flight' response. Toward this end, hormones are altered in order to: (i) change blood flow so that blood is directed toward those tissues such as skeletal and cardiac muscle and away from 'non-essential' tissues such as the gastrointestinal tract; (ii) increase energy by creating and liberating glucose; and (iii) cause lypolysis to liberate free fatty acids from fat as an energy source to be used in the response. Thus, the main stress hormones have been considered to be **glucocorticoids** (predominantly cortisol and corticosterone), **epinephrine** (a neurohormone, also known as adrenaline) and **norepinephrine** (a neurotransmitter, also known as noradrenaline).

The challenge in defining a single or class of 'stress hormones' is that the hormones and neurotransmitters currently used are all non-specific to the stress response. Each of those listed have base functions in the body to maintain **homeostasis** and, thus, are present and measurable even when an animal is not stressed. The reason they have been commonly referred to as stress hormones is because, when an animal is challenged with an aversive stimulus, the levels of these hormones and neurotransmitters are increased. However, these same compounds' levels are also increased when animals experience states not considered to be stressful, such as exercise and copulation.

Along with these traditional hormones listed above, stress research has also measured a variety of other hormones when assessing stress in animals. Traditional stress hormones cause a wide variety of effects in the body, and thus they also alter many other hormones and compounds, which have been used to assess stress in animals. For instance, it is well known that glucocorticoids inhibit **luteinizing hormone (LH)**, which has allowed researchers to use LH as a measure of stress in animals. Similarly, glucocorticoids also inhibit **insulin-like growth factor-1** (IGF-1), which has allowed its successful use in measuring stress as well. (DCL)

See also: Measuring welfare

Stressor

A stressor is anything that induces a **stress** response. Use of the term has been advocated as it helps to differentiate linguistically the **stimulus** causing the response from the actual response, both of which are sometimes (somewhat confusingly) referred to as 'stress', e.g. the mixing of unfamiliar animals is a stress for them (stimulus) and the mixing of unfamiliar animals is a cause of stress (response). (DSM)

Stroke volume

The stroke volume is the amount of blood pumped from the left ventricle in a single heartbeat, and is measured in ml/beat. The stroke volume can be calculated using the formula: SV = EDV – ESV, where EDV is end-diastolic volume and ESV is end-systolic volume. It is not, therefore, the total volume of blood that the ventricle can hold, but just that portion that is ejected during ventricular contraction. The stroke volume can be influenced by the physical size of the heart, the contractility of the cardiac muscle (i.e. the intrinsic ability of a cardiac muscle fibre to contract at a given fibre length), duration of contraction, the preload (i.e. the amount of ventricular stretching prior to contraction) and the afterload (i.e. the pressure that the ventricular cardiac muscle has to generate to eject blood into the aorta).

In general, stroke volume will increase with increasing heart size, increasing duration of contraction, increased contractility in response to positive inotropic compounds such as **catecholamines**, increasing preload and descreasing afterload. Stroke volume will decrease with decreasing heart size, decreasing duration of contraction, decreased contractility in response to negative inotropic compounds such as **beta-blockers**, decreasing preload and increasing afterload. Responses to acute challenges often result in linear increases in **heart rate**, but the changes in stroke volume may be quite different and dependent on the characteristics of the individual. For example, exposure to a sudden **stressor** may stimulate the sympathetic–adrenal–medullary axis, resulting in catecholamine release, which will increase contractility – i.e. have a positive effect on SV. However, catecholamines also increase **blood pressure**, which will increase afterload – i.e. have a negative effect on SV. (RM-F)

Stunning

Stunning refers to the process whereby an animal is rendered insensible, which is normally required by law in **slaughterhouses** prior to **slaughter**, in order to reduce the **suffering** caused by the slower exsanguination **killing** procedure that leads to the **death** of the animal. Stunning may be undertaken in one of a variety of ways, depending on the species. Electrical stunning is commonly used in **pigs** and poultry (with **cattle** being stunned in electric boxes in some countries), but the **restraint** and **handling** prior to stunning can result in **distress**. Gaseous stunning is also used commercially for both pigs and poultry, and usually involves high levels of carbon dioxide (which many animals appear to find **aversive**) or inert gas such as argon (which appears to be less detectable by the animals). In many countries, cattle are normally stunned by physical percussion (**captive bolt**).

Some **religious slaughter** procedures have been criticized on animal **welfare** grounds, as they are often interpreted as requiring the animal to be fully conscious during the procedure. It has been argued that the original aim of these customs was to minimize suffering and provide a 'respectful' death, and therefore they need to be revised to ensure that suffering is minimized rather than the procedure adhered to for its own sake. For this reason, some religious leaders allow stunning prior to slaughter. (DSM)

Further reading

Becerril-Herrera, M., Alonso-Spilsbury, M., Lemus-Flores, C., Guerrero-Legarreta, I., Olmos-Hernández, A., Ramírez-Necoechea, R. and Mota-Rojas, D. (2009) CO_2 stunning may compromise swine welfare compared with electrical stunning. *Meat Science* 81, 233–237.

Gregory, N.G. (2008) Animal welfare at markets and during transport and slaughter. *Meat Science* 80, 2–11.

Mota-Rojas, D., Maldonado, M.J., Becerril, M.H., Flores, S.C.P., Gonzalez-Lozano, M., Alonso-Spilsbury, M. *et al.* (2008) Welfare at slaughter of broiler chickens: a review. *International Journal of Poultry Science* 7, 1–5.

Subclinical

A subclinical condition is one that is present within an animal but not overtly manifested or detectable by routine physical or laboratory examination procedures. As such, it is not in itself normally an immediate **welfare** problem, but is of importance to animal welfare. A subclinical condition may make an animal more prone to a clinical manifestation of the problem that does cause suffering, e.g. a racehorse with a subclinical **lameness** may be at greater risk of a more catastrophic **injury** if it is raced, or predispose an animal to other problems, e.g. an animal with a subclinical infection may be immuno-compromised and so prone to a range of other infections.

By definition, subclinical conditions are difficult to detect and so their true significance in animal welfare often remains speculative, e.g. the role of subclinical **pain** in problems of **aggression** in **companion animals**. Many infections are subclinical in the early stages and towards the end of full recovery, and this may be a time when there is a high risk of spread; similarly, some carriers post-infection continue to shed the pathogen but show no clinical signs and are otherwise healthy. For this reason quarantine facilities may be imposed when the **health** status of a new individual to an area remains unknown. Quarantine involves **isolation** of the individual for a predetermined period and can be a significant **stressor**, as individuals are often subjected to restricted **space**, close **confinement** and may be deprived of their normal social contacts and interactions for a prolonged period (**see: Social stress**) in the interests of health control. Environmental **enrichment** can be particularly important to these animals.

Vaccination may be used as an alternative precaution to quarantine but, in some cases, it can be difficult to distinguish between antibodies mounted to the vaccination and those relating to a subclinical infection (which might interfere with the efficacy of the vaccine). Therefore, blood tests pre- and post-vaccination may be required, not only to establish the efficacy of the vaccination procedure but also to ensure that a subclinical infection is not present. (DSM)

Subjective experience

Subjective experience is sometimes used as another term for conscious mental states of humans or animals (**see: Consciousness**). These states are taken to be subjective in the sense that they exist only by virtue of being the object of some kind of awareness. Whereas it makes sense to say that I have a wound on my arm but that I am not aware of it, it does not in normal circumstances seem to make sense to say that I have a certain feeling, such as a feeling of **pain**, but am not aware of it.

The objective–subjective distinction is sometimes taken to relate primarily to a difference between the way things *appear* (e.g. there is an oasis ahead) and the way things *are* (it is a mirage). Things can appear differently to different observers, depending on, among other things, ambient conditions (e.g. light levels), the perceptual apparatus of the observer (some animals see in black and white) and salience (parent birds attend to the chick's gape). The term 'subjective experience' is often used to mark this fact. From one view, science is an attempt to understand the world as it really is, free of subjectivity. The scientist moves beyond subjective experience and examines objective reality. (PS)

See also: Feelings

Submissiveness

Submissiveness describes the behavioural **signal**s and actions made by an individual that demonstrate its subordinance to another individual(s) within its **social group** (**see also: Dominance**; **Hierarchy**). Such behaviours serve to appease more dominant individuals, prevent escalation of **aggression** and promote the successful integration of the individual into the **group**. Some individuals prefer the term 'appeasement gesture', as this emphasizes their proximate function, rather than the potentially less reliable predictive value of general social status. Submissive signals aim to dissipate aggressive threats by showing a lack of threat on the part of the submissive individual, and include avoiding eye contact, lowering of the ears and/or tail, grinning, crouching or even rolling over to expose the abdomen. Such signals are often the opposite (antithesis) of the signals displayed by the dominant individual. (KT)

See also: Appeasement

Subspecies

A subspecies is a taxonomic rank, subordinate to species. In accordance with the International Code of Zoological Nomenclature, a subspecies is identified by the Latin binomial of the species followed by the name of the subspecies (resulting in a Latin trinomial). For example, tigers are known by their Latin name *Panthera tigris* (the Latin binomial); the Sumatran and Bengal subspecies follow the nomenclature: *Panthera tigris sumatrae* and *P. t. tigris*, respectively. Subspecies are characterized by differences in morphology or through different DNA sequences. The distinction between a species and subspecies tends to revolve around why the groups do not interbreed. If two groups do not interbreed because of intrinsic genetic differences, they tend to be classified as separate species. If they do not interbreed due to some geographical barrier, but would freely interbreed should the barrier be removed, then they are classified as subspecies.

(PE)

Suffering

Suffering refers to the presence of highly unpleasant or **aversive** mental states in animals. A paradigmatic case of suffering is **pain**, which has been defined by the International Association for the Study of Pain (IASP) Task Force on Taxonomy as 'an unpleasant sensory and emotional experience associated with actual or potential tissue damage, or described in terms of such damage' (IASP, 1994). However, it is important not to assume that all suffering involves pain. First, the word 'suffering' indicates simply that the states in question are unpleasant, or aversive, to a degree where it becomes a problem. Pain, on the other hand, can come in degrees, ranging from very mild to very strong; and there is no reason to claim that mild and controllable pain amounts to suffering – although, clearly, people will disagree about where to draw the line on what is acceptable. Secondly, there seem to be forms of suffering, not involving tissue damage, that typically feel very different from pain. Examples include **fear**, **anxiety** and nausea.

There is ongoing discussion, particularly in connection with fish, about whether certain animals can be said to suffer in the full sense of the word, or whether they merely have the ability to react to physical and other stimuli without actually feeling pain and enduring other forms of suffering (**see: Fish welfare**). Those who contend that fish and similar animals cannot feel pain or suffer in other ways point out that these animals lack the neural structures involved in human pain and suffering. Those maintaining that a wide range of animals, including fish, are able to suffer argue, first, that other neural structures in these animals may do the job; and, secondly, that behavioural data seem to support the hypothesis that these animals are able to feel pain and suffer. Given the scientific uncertainty, some people conclude that it is best to give animals the benefit of doubt and assume that they are able to suffer (**see: Precautionary principle**). (PS)

Reference

IASP (1994) Part III: Pain terms: A current list with definitions and notes on usage. In: Merskey, H. and Bogduk, N. (eds) *Classification of Chronic Pain*, 2nd edn. IASP Press, Seattle, Washington, pp. 209–214.

Suffering – its recognition and assessment

René Descartes, in his writings, considered that animals could *feel* **pain** but the question for him was whether they could actually *suffer* pain. From his viewpoint, animals were analogous to machines and, unlike humans, lacked **consciousness**, as they could not reason, reflect or even speak and, because animals lacked those abilities, they could not suffer. **Jeremy Bentham** questioned this view: 'The question is not, can they reason? Nor can they talk? But, can they suffer?'

This controversy over whether animals can suffer still exists today, not so much in relation to mammals, where it is generally accepted that they can suffer, but more when considering other chordata (phyla such as birds, **reptiles**, amphibia and fish) (see EFSA, 2009). Some think that only humans can truly suffer, whereas others are of the view that all those with a nervous system are able to suffer, but to varying degrees according to the development of their nervous systems. In addition to vertebrates, this includes some invertebrates such as cephalopods (e.g. octopus) and some crustaceans (e.g. lobsters), which do not have a **central nervous system** as such, but have a series of ganglia that may assume the same function.

Suffering is intimately linked with consciousness, **self-consciousness**, **awareness** and **self-awareness**, but it is always associated with unpleasant mental experiences that encompass different **feelings**, such as pain and **distress**. These mental states are brought about by responses to stimuli that may be external (e.g. injury, cold, barren or small cages) or internal (hunger, thirst, social isolation). It seems likely that animals that are aware of themselves in relation to others may potentially suffer more than those that are just sentient, i.e. have the ability to experience and respond to 'pain and pleasure' (**see: Sentience**).

Suffering is used as a generic word to cover those adverse mental states and would include, in addition to those above, feelings such as **fear**, **frustration**, **boredom**, grief and discomfort. Suffering is used in an holistic (all-embracing) sense to denote the breadth of mental experiences an animal might be feeling. For example, an animal that has a broken leg is likely to be in pain (suffering in the present), but may also associate that pain with someone who caused it (suffering due to fear in the presence of that person, or **anxiety** when anticipating their presence). Furthermore, that animal may be unable to **play**, or seek shelter, or find food and water, and so it will experience other feelings such as distress, discomfort, **hunger**, thirst (**see: Drinking behaviour**) and frustration. It may also think that it will soon be hungry and that it will be unable to forage or kill for food, or escape in the presence of a predator (thoughts about suffering in the future). It is not clear to what degree mammals other than the higher primates are able to reflect in this way (so-called '**mentation**'; see DeGrazia, 1996).

As knowledge increases about the sentience of animals, the difference between non-human animals and humans seems less clear-cut, as observed by **Charles Darwin** (1871) in *The Descent of Man* over 100 years ago: 'There is no fundamental difference between man and the higher animals in their mental faculties'. The **Brambell Report** (1965) noted that farm animals were sentient, i.e. 'Animals show signs of suffering from pain, exhaustion, fright, frustration and can experience emotions … and pleasure' and considered 'that it is morally incumbent upon us to give the animal the benefit of doubt and to protect it so far as is possible from conditions that may be reasonably supposed to cause it suffering, though this cannot be proved'.

The study of animal suffering is important as it underpins public policy on how we ought to treat animals (**see: Ethics**) and provides evidence for the suitability, or otherwise, of the methods we use to farm, keep and care for animals (FAWC, 2009). In any system of husbandry some suffering may be necessary and justifiable, but to cause more than the minimum level is unacceptable and so there is a notion of 'avoidable suffering'. For example, farm animals are subjected to uncomfortable procedures such as mixing with other animals that come from different farms at markets, or the discomfort of vaccination to prevent disease, testing for **diseases** such as tuberculosis, and **transport**: all these will cause some degree of suffering but are seen as necessary and unavoidable for farming. However, there are also approved surgical mutilations such as **docking** and **castration** that are carried out with no **anaesthesia** nor long-term **analgesia**, and the suffering caused therein could have been avoided had these measures been taken.

Recognition and assessment of suffering

Not all animals will suffer negative emotional states to the same extent or in the same way as others. It will vary according to: (i) the adverse physical, physiological and psychological circumstances in which an animal finds itself; and (ii) the cognitive capacity of the species, as well as the individual being and its past experiences in life. Furthermore, *recognition* of suffering will depend on the anatomical and physiological characteristics of the species or the individual being studied. In addition, the environmental circumstances in which the observations are being made will play an important part; for example, one is more likely to be able to observe a greater variety of behaviours in complex rather than in simple environments.

While it is difficult, it is not impossible to assess the level of suffering. Research into animal **welfare** is an evolving science, and many new ideas of **welfare measurement** are being developed. They include observing an animal's behaviour, measuring physiological changes, monitoring **brain** function and associated neuropeptide levels and their receptors, and even trying to devise experiments that ask an animal what 'it thinks' about a particular resource or environment (see below). At a basic level, an assessment of **mortality**, **morbidity** (**health** records) and measuring of productivity in some way (e.g. in farmed animals **growth rate**, milk production, fertility, fecundity) of animals kept under certain **housing** and husbandry conditions are all valuable, but not complete, measures. All such observations have to be compared with the levels (prevalence, i.e. the rate at which a factor is observed at any one point in time) or frequency of occurrence (incidence, i.e. the rate at which the factor occurs in unit time) of the same measures in other animals considered to be 'normal' in those circumstances. It is perhaps worth emphasizing that it is not to do with what is 'natural' that relates more to its wild state.

The assessment of the degree of suffering can then be determined by how far those measures have deviated from normality and, in general, the greater the deviation, the greater is the suffering. Suffering, or severity as it is sometimes referred to, is a combination of intensity and duration: the greater the deviation and the greater the time it lasts, the greater will be the severity. Obviously, this is only true when the behavioural or clinical **sign** reflects an adverse effect, such as loss of body weight, **lameness**, increases or decreases in body temperature, and so on. But it is also applied to other measures such as blood **corticosteroid** levels and autonomic responses such as **heart rate** and **blood pressure**. In these latter cases it is important that these are seen in context, as many of these measures are often elevated in circumstances relating to good as well as to poor welfare, e.g. **play**, exercise. Furthermore, even inquisitiveness, which is associated with a mild form of **arousal**, will show several identical types of change, although they may be quantitatively different, e.g. heart rate, rise in blood pressure, increased alertness. In addition, there may well be other caveats and concerns in interpreting these signs; for example, playing with companions for longer than normal is unlikely to be a cause for concern, but excessive weight gain through overfeeding/-eating is a predictable indicator for future suffering (arthritis), even though an animal might enjoy eating (**see: Positive welfare**).

Behavioural measures

When assessing an animal for **signs** of suffering it is important to minimize any interaction between the observer and the animal. Undisturbed or unprovoked 'natural' behaviour can reveal much about how an animal is feeling. Play would indicate happiness, **lying** and sleeping or chewing the cud could reveal a lack of fear and contentment, whereas an animal that is lying on its side, kicking at its abdomen and groaning would indicate pain, especially in the context of it just having been docked or castrated or had some other painful operation, or was suffering from colic. Excessive self-licking or stereotypic behaviour (**see: Stereotypies**), such as bar-chewing or licking water bottles, could indicate boredom. Frustration again may be indicated by animals trying to carry out natural behaviours but doing them in a modified form and for longer periods of time. For example, **rabbits** in small, barren cages repeatedly scratch the floor of the cage in an attempt to burrow; they try to rear up to scan the environment for danger, but they are unable to do so because the height of the cage will not permit it, so they end up doing a half-rear and then sit down again.

Immobility may indicate **apathy** and boredom or an animal being in pain as it hurts to move. By comparing the times spent carrying out these **abnormal** behaviours, or normal behaviours carried out more frequently or for longer, one can make a judgement on the individual and whether it falls within a 'normal' range. Moreover, it is possible to compare two different systems of husbandry looking at the prevalence of normal and abnormal behaviours, e.g. stereotypic behaviour. One can also measure the behavioural diversity of an animal or **group** of animals. For **companion animals**, where owners get to know their animals extremely well, a **dog** not getting out of bed to greet you in the morning, not barking when the doorbell sounds or not wanting to come out for a walk, or a **cat** not asking for food, are all signs that indicate a change from normal for that individual animal, and have to be explained in some way.

It is also possible to use behavioural measures to deliberately provoke an animal to respond and observe if it responds in a normal way. For example, by changing the environment one can assess inquisitiveness and other sorts of response times. In

rats, turning the lighting colour from white to red to mimic night-time (rats cannot see red and so it appears dark to them) should stimulate their night-time behaviour (when they are usually most active) within a few minutes (**see: Light and behaviour**). If they do not do this then something is wrong.

Pressing on a site on the body is often used to make a veterinary diagnosis, e.g. to determine if there is a fracture or deep wound, as it will cause an animal to react. The exact response will depend on the species and the temperament of the individual animal, and it may react by trying to bite, by vocalizing or by making vigorous escape attempts. Again, a failure to respond would normally mean that the animal is not in pain but, if it does respond, then that is an important sign.

Asking the animal

The ultimate expression of how an animal feels about its environment is how it behaves, i.e. behavioural measures are arguably the most relevant measures of good and poor animal welfare. Giving animals a choice of environments (putatively good or bad), making them work for a resource (**see: Economics of behaviour**) and measuring their **cognitive bias** (i.e. are they feeling optimistic or pessimistic) provide additional tools for discriminating whether an animal feels it is in a good or poor environment, or prefers one thing to another. By observing what environment animals choose to be in when given a choice, and how hard they will work to gain a resource of some sort, or how hard they will work to stay or leave a particular environment, can be interpreted as reflecting a preference. How closely that is correlated with long-term welfare is debatable (animals might choose to 'overeat'), or whether it is an ideal environment as opposed to the best on offer.

Other signs

Animals that do not eat lose body weight, and so measurement of body weight can be a helpful guide. An animal that has lost 20% of its body weight is likely to be suffering more than one that has lost just 5% over the same period. Animals with abnormal **postures** (hunched, holding a limb off the ground, as in lameness) or appearance (starey coat, ruffled **feathers** – both indicating a lack of **grooming**) may also be suffering. In farmed animals a decreased productivity such as a reduction in growth rate or milk yield may indicate a problem.

Another important point to note is that all clinical signs, and physiological and behavioural measures, can usually be scored objectively, i.e. in a repeatable and reliable manner, and it is the interpretation of those signs where the subjective element of welfare assessment comes in (validation of what an increase in a particular behaviour or blood hormone actually means in terms of **well-being**). Any clinical signs of disease, e.g. spasms, fits, breathing difficulties, discharges, etc. all indicate a change from normal, and so have to be examined carefully.

Interpretation of 'abnormalities'

Several criteria are important: one is whether the sign can be recorded accurately by different observers and, secondly, how one interprets those abnormalities in terms of suffering. Inter- and intra-observer variation can be minimized through training and ensuring that the sign to be scored or evaluated is simple. Whether it is present or absent would be the best evaluation, but to ask whether an animal is in pain presupposes an assessment of multiple signs, and is bound to cause more difficulties and more variation. Secondly, it may be possible to record a sign on a numerical basis, e.g. body weight loss, body temperature, cortisol levels, heart rate.

Alternatively, quantifying lameness, diarrhoea or breathing difficulties is far more difficult, but can still be objectively recorded as with numerical data. This may be by banding – e.g. for lameness it can be from normal to just limping to marked lameness to hopping lame, being unable to weight bear at all. This can give us 3 or 4 bands with which to categorize animals. Having made objective observations of both quantitative (metric) and semi-quantitative (bands or parametric) signs, they then have to be interpreted in terms of animal well-being, and this is where the subjective step comes in. It is not the observation of the signs that is subjective (given the narrow choice of a sign being only present or absent), it is how that sign is 'subjectively' interpreted in terms of animal suffering that is key (i.e. does the sign really reflect suffering?). In another sense, deviation from normality, whether it is because of pain, distress, frustration, boredom or whatever, does not matter, except from the point of view of making a diagnosis on the cause, as it may be possible to treat by avoiding or alleviating an animal's condition in some way, e.g. through the use of analgesics. (DBM)

References and further reading

Darwin, C. (1871) *The Descent of Man, and Selection in Relation to Sex*. John Murray, London.

Dawkins, M.S. (2008) The science of animal suffering. *Ethology* 114, 937–945.

DeGrazia, D. (1996) *Taking Animals Seriously: Mental Life and Moral Status*. Cambridge University Press, New York.

EFSA (2009) *Report on Sentience in Fish*, available from: http://www.efsa.europa.eu/EFSA/efsa_locale-1178620753812_1211902344910.htm?WT.mc_id=EFSAHL01 (accessed 2 December 2009).

FAWC (2009) *Animal Welfare in Great Britain: Past, Present and Future*. Farm Animal Welfare Council, London.

Supplementation

Supplementary feeds are usually based on processed feeds, or concentrates, and are most likely to consist of cereals or other high-energy and protein feeds (**see: Concentrate feeding**). They are usually made into a pellet, or compound, with the addition of a binding agent. They are brought on to the animal unit either in bags or they may be delivered by an auger into a feed bin. The cost of these processes makes compound pellets an expensive form of food, unless the raw ingredients are inexpensive. In deciding how much supplementary feed to offer animals, an animal manager will normally estimate the anticipated increase in productivity and determine a break-even point where additional feed would provide no additional financial return.

However, feeding additional concentrates will often do more than simply increase **growth rate** or milk yield, and consideration should be given to changes in product composition and savings in forage when concentrate intake is increased. For example, with dairy cows, the provision of additional supplementary concentrated feed will reduce milk fat content and, to a lesser extent, protein content will increase, as the amount of concentrate and the ratio of glucogenic to lipogenic precursors increases. At most levels of concentrate intake, milk fat and protein yields increase with concentrate intake but, at very high intakes, in excess of approximately 12 kg/day, the milk fat content declines rapidly with additional concentrate, so that milk fat yield declines (the low milk fat syndrome). Under virtually all circumstances this will be an uneconomical level of supplementary concentrate to feed to cows.

The saving in forage when concentrate supplements are fed should enable the stocking rate to be increased. The manager will decide how to maximize income, and this will entail looking at responses to the different resources available, in particular the most limiting ones. Feeding additional concentrates may increase the gross margin per hectare on a dairy farm (largely because **stocking density** increases) but reduce the gross margin per litre of milk (because feed cost per litre increases). Under strict quota limitations, therefore, farmers may prefer to feed low levels of concentrates but, if the land area of a farm is restricted, the farmer may prefer to feed more. The output per animal is likely to be particularly important to farmers keeping pedigree stock, who may feed high supplementary concentrate levels so that the animals' performance is high and others are encouraged to buy them.

The allocation of supplementary concentrates should take account of the animals' physiological state – lactating or non-lactating, pregnant or non-pregnant. As nutrient intake increases, particularly energy, productivity may increase up to a certain level and then no further increase can be obtained from greater energy intakes. The extra energy may be stored as body tissue, to be used at a later date, but other nutrient excesses may be excreted or become toxic. The response of high-yielding cows to increased energy supply from concentrates may be offset by a similar reduction in low-yielding cows, if they receive fewer concentrates.

However, it may be worthwhile allocating different forms of supplementation to differentially productive animals, e.g. rumen-protected protein and mineral/vitamin supplements may be reserved for high-producing animals. Mineral supplements should be provided in accordance with local deficiencies in forage and/or soil. For example, phosphorus deficiency is prevalent in much of northern Australia because of weathered soils and low-quality forage. (CJCP)

Sympathetic nervous system

The sympathetic nervous system (SNS) arises from the thoracic and lumbar spinal cord (spinal segments T1–L3) and is regulated through the **hypothalamus**. Myelinated preganglionic sympathetic nerve fibres give rise to unmyelinated postganglionic nerves that are generously dispersed throughout the body. Most organs in the body, as well as somatic structures such as the skin, muscles, mucous membranes and certain bones, are innervated by this subdivision of the **autonomic nervous system**.

Activation of the SNS results in what may be described as a general excitation of effectors, triggered primarily by the release of **catecholamines**, such as **norepinephrine** from postganglionic sympathetic **neuron(e)**s, and **epinephrine** from the **adrenal gland** medulla. The terminal branches of SNS postganglionic fibres end in varicosities that synapse with effectors. Norepinephrine is produced and stored within these varicosities, and can later be released into the synaptic cleft as and when required. The SNS is characterized by a very high ratio of post- to preganglionic synapses (termed divergence), which enables it to excite many effectors simultaneously. These effects are magnified in effectors that also boast adrenergic receptors through the generalized release of epinephrine into the bloodstream.

The SNS controls those activities associated with the mobilization of energy necessary for an organism to respond to a challenge, such as a **stressor**. Responses governed by the SNS include modulations in cardio-pulmonary, respiratory and pupillary activity, besides the mobilization of fat and glycogen from liver hepatocytes and adipose cells. The activation of the SNS, for example, increases **heart rate** and respiration rate, redirects blood from skin and splanchnic vessels to voluntary muscles, regulates the secretion of hormones (e.g. epinephrine), increases blood glucose levels and causes piloerection in preparation for a fight-or-flight reaction. Aside from times of **stress**, in healthy animals the SNS is always tonically active, meaning there is a continuous low level of discharge of nerves facilitating their effectors. (RM-F)

See also: Parasympathetic nervous system

Symptom

Symptoms are what a patient reports to the doctor that normally cannot be seen, e.g. feeling nervous, having a headache. Clinical signs are those that the doctor can observe for her/himself. It follows, therefore, that animals can only be assessed by the use of clinical signs and not symptoms. (DBM)

See also: Sign

Synchronization of behaviour

Synchronization of behaviour literally refers to acts that occur together, although the term may also be used to refer to behaviours that are related in time in some other way. While all synchronized behaviours share this common temporal characteristic, there are diverse reasons why and how certain behaviours may become synchronized. Synchronization may occur between animals with similar objectives, e.g. synchronization of **mating behaviour**, or opposing objectives, e.g. **hunting** versus hiding in predator and prey.

Some factors resulting in synchronization relate to the physical environment and others to the social environment, or to a combination of these. **Proximate factors** resulting in synchronization include the appearance of a resource that stimulates the behaviour, e.g. the arrival at a water hole by migrating animals that results in synchronized **drinking behaviour**. In the latter example the synchronization is incidental, while in other situations it may be intentional – for example, the synchronization of **feeding** and vigilance behaviour within a **social group** in the presence of a potential threat. The behaviour of another may be important in triggering synchronization, e.g. behaviours affected by **social facilitation** or behaviours that are being mimicked. In some cases, specific physiological factors are important in helping to maximize the fitness benefit of synchronization, e.g. the hormonal effects on breeding and migratory behaviour in relation to lunar or diurnal factors.

In **captivity**, behaviours that are normally synchronized may be prevented, and this may give rise to **welfare** concerns, through the deprivation and/or **frustration** of adequate amounts of motivated behaviour. For example, the provision of inadequate **space** may prevent synchronous feeding, drinking or **resting behaviour**, and physical barriers between members of the opposite sex may prevent mating. The importance of behavioural synchronization in the design of animal facilities should therefore be considered (**see also: Confinement**).

(DSM)

Further reading

Engel, J. and Lamprecht, J. (1997) Doing what everybody does? A procedure for investigating behavioural synchronization. *Journal of Theoretical Biology* 185, 255–262.

Pays, O., Jarman, P.J., Loisel, P. and Gerard, J.F. (2007) Coordination, independence or synchronization of individual vigilance in the eastern grey kangaroo? *Animal Behaviour* 73, 595–604.

Systemic measures

Systemic measures are those taken from the animal that are a result of the additive and subtractive contributions of the entire variable being measured. For instance, systemic **blood pressure** is the result of the pressure in all the vessels of the body, from the aorta to the vena cava. Similarly, the concentration of a hormone measured in the circulating blood is a result of all of the hormone being cleared and produced in the entire body at the time at which it is measured. (DCL)

T

Tail-biting

Tail-biting is an abnormal injurious behaviour that occurs predominantly, but not exclusively, in intensively housed pigs. In essence, a **pig** will take the tail of a pen-mate in its mouth and chew or bite on it. Initially this may appear to cause little or no damage to the tail and is tolerated by the other pig; however, the biting may intensify over time until the tail bleeds and becomes painful. It is unlikely that all outbreaks follow this pattern; however, at some point a tail is damaged somehow and this presence of blood results in a sudden escalation of the behaviour and thus of damage. Without intervention the biting can become generalized and lead to dramatic damage and eventually **cannibalism**. Even small amounts of tail damage can have a major impact on the health of the bitten pig, causing abscesses in the hindquarters, lungs, spine and pelvic region, as well as an increase in the spread of **diseases**.

Tail-biting is described as a multifactorial problem involving both internal and external risk factors. These include genetics, **health**, high **stocking density**, poor nutrition, insufficient temperature control and a lack of rooting material. Experiments attempting to induce outbreaks by modifying one factor alone have often failed, suggesting that outbreaks occur only when enough factors combine to make the pigs sufficiently uncomfortable. This association between stressful factors and biting implies that it may be indicative of reduced **welfare** of both the biter and the bitten individuals.

Estimating the extent of the tail-biting problem is difficult as the levels vary according to the production system, whether tails are docked and how the estimate is produced. Records from abattoirs place the level at around 3% in the EU (EFSA, 2007); however, the same report stated that abattoir data may record fewer than half the lesions found on farms. A survey of farms in the UK found that 66% had experienced tail-biting in the previous year (Chambers *et al.*, 1995), and thus it is a considerable problem. It is rarely seen in extensive production systems, even when docking is not used.

Tail-docking (**see: Docking – tail**), where either the tip or most of the tail is clipped off shortly after birth, has been the most common technique used by farmers to prevent tail-biting. There are several welfare and ethical concerns regarding this procedure and, although it is thought to reduce the incidence of tail-biting, it does not eliminate the problem. The search for viable alternatives such as provision of rooting materials is continuing. (PoS)

See also: Tail tip necrosis

References and further reading

Chambers, C., Powell, L., Wilson, E. and Green, L.E. (1995) A postal survey of tail biting in pigs in south west England. *Veterinary Record* 136, 147–148.

EFSA (2007) Scientific report on the risks associated with tail biting in pigs and possible means to reduce the need for tail docking considering the different housing and husbandry systems. *The EFSA Journal* 611, 1–98.

Moinard, C., Mendl, M., Nicol, C.J. and Green, L.E. (2003) A case control study of on-farm risk factors for tail biting in pigs. *Applied Animal Behaviour Science* 81, 333–355.

Schrøder-Petersen, D.L. and Simonsen, H.B. (2001) Tail biting in pigs. *The Veterinary Journal* 162, 196–210.

Tail tip necrosis

Tail tip necrosis can affect many species, but is commonly used to describe a condition in **cattle** that begins by the tail being stepped on, causing a lesion to form. It is more prevalently found in cattle kept on slatted floors, at high stocking densities, in warmer weather and in cattle above 200 kg in weight. The lesion at the tail tip starts off swollen, followed by **inflammation** and suppuration. In severe cases of infection, metastasis to other parts of the body can occur and result in abscesses and osteomyelitis. Treatment includes antimicrobial therapy, tail amputation (**see: Docking – tail**) and, in worst cases, **slaughter** of the animal. (LMD)

Tamarin

Together with the Goeldi's **monkey** and the **marmosets**, tamarins belong to the family Callitrichidae, subfamily Callitrichinae and are distributed throughout the Atlantic and Amazonian rainforests of South America. There are four species of lion tamarins (*Leontopithecus*) and 15 species of tamarins (*Saguinus*; Rylands *et al.*, 2000), of which *Saguinus fuscicollis*, *Saguinus labiatus*, *Saguinus mystax* and *Saguinus oedipus* are the most frequently used in laboratory experimentation. *S. oedipus*, despite being endangered, was used to study colonic adenocarcinoma, as it is the only primate model that, like humans, spontaneously develops colitis and/or colonic cancer. Several species of callitrichines are either endangered or critically endangered, and **reintroduction** programmes are under way for some.

Tamarins are similar to marmosets in many respects. They are **diurnal**, arboreal, give birth to twins as the norm and all group members assist with infant care. Olfactory **communication** is highly developed. Tamarins can live up to 15–20 years. *Saguinus* become sexually mature at 18 months, *Leontopithecus* at 24 months.

However, tamarins are different from marmosets in some critical ways that impact on their **welfare** and captive conditions. They are slightly larger (average adult body weight for both males and females is 400–500 g for *Saguinus* and up to 700 g for *Leontopithecus*), have larger **home ranges** (up to

50–100 ha) and longer daily path lengths, related to a more frugivorous diet and less dependence on gum-feeding.

Tamarins are **insectivore**–frugivores and their dentition is not adapted for gnawing. The long, slender hands and fingers of *Leontopithecus* are used for probing for concealed prey in specific microhabitats. A considerable proportion of prey items are located by touch rather than by sight, and the most important foraging site is epiphytic bromeliads. There are also differences in foraging strategies among *Saguinus*. Three distinct insect-foraging patterns have been described. The first pattern is shown by *Saguinus geoffroyi*, which hunts for insects on thin, flexible branches in the low shrub layer of the forest understorey. The second is shown by *S. labiatus*, *S. mystax*, *Saguinus imperator* and possibly *Saguinus midas*, which have a similar insect-foraging style to the marmosets, exploiting insects on leaves and branches in the lower and middle levels of the forest.

Visual scanning plays an important role in the detection of their prey. The third pattern is shown by *S. fuscicollis* and possibly *Saguinus nigricollis* and *Saguinus bicolor*. These species are predominantly manipulative specific-site foragers, concentrating their feeding efforts on relatively large cryptic prey. In those species of tamarin studied in the wild, ripe fruits account for 20–65% of total feeding time; nectar is also consumed, principally in the dry season. Like marmosets, without natural light tamarins need a vitamin D_3 supplement added to their diet of fruits, vegetables and animal protein. Feeding enrichment should take their natural feeding adaptations into account (e.g. hiding food in bromeliads for *Leontopithecus*).

Like marmosets, tamarins are arboreal but have an even greater tendency to avoid the ground, where they behave nervously. It is therefore important to provide appropriate cage furnishing to increase spatial complexity and allow better utilization of the vertical, that is, the arboreal dimension. The type of cage furnishing, its dimension, orientation and its placement have to vary to take into consideration specific adaptations, needs and **preferences**.

Generally, breeding of *Saguinus* is less successful in **captivity** than for *Callithrix*, and prior experience with offspring is even more critical for parenthood. Infants are provisioned and carried for longer periods. In general, tamarins are even more aggressive towards unfamiliar **conspecifics** than are marmosets, and visual barriers must be present to stop threats and inter-group aggressive behaviour that may lead to redirected **aggression** towards **group** mates. It is argued that, because of these differences in weight, home range size, daily path length, **reproduction** and arboreality between marmosets and tamarins, in the laboratory, minimum cage sizes should be larger for tamarins than for marmosets (Prescott and Buchanan-Smith, 2004).

All tamarins have claw-like nails to facilitate grip, and all cage furnishings should take grip into account. **Locomotion** is primarily quadrupedal, but vertical clinging and leaping are seen in several species. Positioning of vertical supports is particularly important for *S. fuscicollis*, which locomotes extensively by vertical clinging and leaping. *Saguinus* usually sleep in vine tangles and forks of trees, and in captivity secure nesting places such as a nestbox some distance from the **floor** should be provided. *Leontopithecus* are unusual in that they primarily use holes in tree trunks and branches as sleeping sites, and therefore providing tree holes in captivity is appropriate.

Some tamarins form stable mixed-species associations in the wild. Those between *S. fuscicollis* and either *S. labiatus*, *S. mystax* or *S. imperator* are the most stable. Mixed-species groups of *Saguinus* have proved a successful method of housing, which has a number of advantages such as greater activity and increased social encounters. However, careful monitoring is necessary around the time of the birth of offspring, as individuals may get protective of young. Other important considerations are to provide food and water in multiple locations, as *S. fuscicollis* is dominated by its associating species. Multiple nestboxes should also be provided.

All callitrichines possess specialized scenting organs and have a rich repertoire of chemosignalling behaviours. The most conspicuous of these are scent-marking patterns, involving the circumgenital and subrapubic glands. Scent-marking itself and the chemical **signals** deposited are important in many areas of their behavioural biology, such as reproduction, and possibly territorial defence and inter-group spacing. Tamarins also have a wide range of **vocalizations** and a variety of other species-specific **postures** and facial expressions that can be used to monitor behaviour and welfare.

Marmosets and tamarins generally do not like being handled. However, they can be trained using positive reinforcement techniques (**see: Reinforcement**) to cooperate with routine husbandry procedures such as capture, weighing, veterinary procedures such as oral administration or palpation, or to provide samples (such as saliva or urine). Marmosets and tamarins are increasingly producing triplets in captivity, probably due to their rich diet. **Hand-rearing** of the third infant, or of rejected infants, is feasible but the infants must be reintroduced into a conspecific group as soon as possible to minimize the serious adverse effects of separation from their natal groups. (HMB-S)

References and further reading

Buchanan-Smith, H.M. (2001) Species-specific housing and husbandry for marmosets and tamarins (Callitrichinae). *Proceedings of the 4th International Conference on Environmental Enrichment, The Shape of Enrichment Inc.*, pp. 95–105.

Buchanan-Smith, H.M. (2010) Marmosets and tamarins. In: Hubrecht, R. and Kirkwood, J. (eds) *The Eighth Edition of the UFAW Handbook on the Care and Management of Laboratory Animals and Other Animals Used in Scientific Procedures*. Wiley-Blackwell, Oxford, UK (in press).

Carroll, J.B. (ed.) (2002) *EAZA Husbandry Guidelines for the Callitrichidae*. Bristol Zoo Gardens, Bristol, UK.

Prescott, M.J. and Buchanan-Smith, H.M. (2004) Cage sizes for tamarins in the laboratory. *Animal Welfare* 13, 151–158.

Rylands, A.B., Schneider, H., Langguth, A., Mittermeier, R.A., Groves, C.P. and Rodríguez-Luna, E. (2000) An assessment of the diversity of New World primates. *Neotropical Primates* 8, 61–93.

Tameness

A standard dictionary definition of 'to become tame' would be to become gentle, tractable or docile, and 'to tame' a **wild animal** would be to bring it from the wild state into submission and under human control. But this does not explain what it means to be wild or tame. A wild animal is usually thought of as one that is fearful of humans and runs away if it can. However, this **fear** of humans is in itself a behavioural pattern that has been learned from experience of human **predation** over countless generations. A species of animal that has had no contact with humans as predators over its evolutionary history has no fear of them and can be quickly exterminated, like the dodo, or like the fox on the South American island of Chiloe, which was killed by **Charles Darwin** in 1834 and described by him thus in his description of the voyage of the HMS *Beagle*:

> A fox (*Canis fulvipes*) of a kind said to be peculiar to the island, and very rare in it, and which is a new species, was sitting on the rocks. He was so intently absorbed in watching the work of the officers, that I was able, by quietly walking up behind, to knock him on the head with my geological hammer. This fox, more curious or more scientific, but less wise than the generality of his brethren, is now mounted in the museum of the Zoological Society.

In order to tame an animal that has a perception of humans as predators, either potential or real, its fear of danger must be annulled. This can be most easily achieved with very young animals that can be persuaded that the tamer is a substitute parent. This can be done either by feeding the baby or by changing its **imprinting** from the mother to the tamer, as was described by **Konrad Lorenz** in his well-known and memorable account of the goslings that followed him everywhere. A crucial part of taming is **habituation**, the process in which new and unusual stimuli are slowly and continuously presented to a wild animal, so that it becomes accustomed to the presence of a person, or another animal, or a vehicle, and over time loses its fear and its flight response.

Lack of fear may be seen as the essence of tameness, but the extent of this fearlessness depends on the integration of the many facets that make up the complex characteristics of the species. An animal will be more or less fearful and therefore more or less easy to tame depending on whether it belongs to a social or solitary species, whether it is normally a predator or prey and, if it is a prey, whether it escapes by fleeing or hiding. A burrowing species like a **rabbit** can be easily tamed, while a superficially similar species like a hare that depends on speed for escape cannot be tamed. Similarly, a wild **goat** that is by nature a mountain animal and depends on agility for escape has less instinctive fear than a gazelle, which is a plains animal depending on speed for escape from predators. This means that the natural flight distance of a goat is much shorter than that of a gazelle, and partly explains why goats could be easily tamed and then domesticated by prehistoric people, while gazelle remained wild (**see also: Domestication**).

There is one form of taming in which loss of fear is not paramount, this being in the physical and mental subjugation of individual animals for their use by humans: the **horse**, the **elephant** and **camelid**s have traditionally been 'broken in', often by cruel methods rather than by the gentler method of habituation. However, it has been shown that horses can become accustomed to a harness and rider more quickly and with less trauma than by physical subjugation through a process referred to as '**horse whispering**'.

The state of wildness or tameness of a species of animal is closely connected to the culture of both the human and the animal populations. **Culture** can be defined and seen as a way of life imposed over successive generations on a society of humans or animals by its elders. Where the society includes both humans and animals then the humans act as the elders.

The process of taming a wild animal, whether it is a wolf or a wild goat, can be seen as changing its culture. The animal is removed from the environment in which it learns from birth either to hunt or to flee on sight from any potential predator, and it is brought into a protected place where it has to learn a whole new set of social relationships as well as new feeding and reproductive strategies.

It is common knowledge that in any region of the English countryside, where there is a culture of **hunting** foxes with hounds or shooting them as vermin, the foxes will be wild and fearful and will be rarely seen, but in towns where they can scavenge or are even fed by householders, foxes have little fear and are even known to come into people's houses. These animals have not been actively tamed but they have lost their fear of humans and developed a culture in which they exploit, albeit unconsciously, the urban dwellers' desire for contact with the natural world.

Other common species in which there is a dichotomy between the wild and the tame are the rabbit, the brown **rat** and the house **mouse**, which are all treated as vermin in the wild but, as tamed, domestic animals, they become much-loved animal companions (**see: Companion animal**).

Without hunting, animal populations will become increasingly tame, as can be seen in the wildlife parks of East Africa where herds of large **ungulates** and even the large **carnivores** are seemingly unaffected by the presence of crowds of people in vehicles all round them. It will be one of the great problems of the future how to manage the wild places and their faunas, because it will not matter whether these faunas include African elephants, Asian lions or giant tortoises – they will become increasingly hedged in and the wild will merge with the tame.

(JC-B)

See also: Flight behaviour

Further reading

Clutton-Brock, J. (1999) *A Natural History of Domesticated Mammals*, 2nd edn. Cambridge University Press, Cambridge, UK/Natural History Museum, London.

Taming

Taming refers to the process whereby an animal is made more approachable and tractable to humans. As such, it may be an important element of the **socialization** process for animals living within human society, but the two are not synonymous. Socialization is a broader process, in which an animal learns appropriate behaviour towards all members of its **social group**. A tame animal may still show inappropriate behaviour to other species, e.g. a **dog** may be tame but not well socialized towards **cats**, and it may merely tolerate humans (not flee when approached) without being socialized towards them (positively engage in a social context).

Taming may be established by a combination of gradual **habituation** (**see: Desensitization – systematic**) and positive reinforcement (**see: Reinforcement – types of**) and is a lifetime trait, although the ease with which an animal may become tame has a genetic element. Taming should not be confused with **domestication**, which arises as a result of interaction with people (taming) and **selection** over generations. (DSM)

Taste

Chemoreception by taste (gustation) has been extensively studied in both man and animals. Generally, four primary tastes are identifiable and can be correlated to physiological requirements: sweet for nutrient supply, salt to control electrolyte balance, bitter to avoid toxins and sour to regulate the body's pH. Bitter compounds, such as tannins, also reduce the digestibility of plants, providing another reason for herbivores to avoid them. There may also be separate metallic and monosodium glutamate tastes, but probably all tastes derive from combinations of **signals** from these four primary taste receptors. The receptors are located in specific areas of the tongue, and differ in their taste discrimination, sensitivity and positioning on the tongue between species. For example, in man the tip of the tongue is sensitive to sweet and salt taste, but in cattle sweet taste is mainly at the base and salt at the tip. This may derive from the longer period of time that food is masticated in the middle to rear section of the buccal cavity in cattle compared with man.

Different animals derive different sensations from food items, in relation to the benefit that they derive from them (**see: Euphagia; Hedyphagia**). For example, several compounds that taste sweet to man, such as monellin and thaumartin, do not elicit a similar response in cattle. Cattle generally experience a greater sweet taste from monosaccharides than from disaccharides, with the most potent sweeteners being glycine, sodium-saccharin and xylintol. When presented with sodium chloride, xylintol enhances the salt taste, acting as a flavour enhancer. Tastes can also act antagonistically; for example, the taste of the potent sweetener thaumatin can be reduced by salt, a phenomenon known as mixture suppression. Flavours also function to elicit avoidance and many food components, such as shea nut meal, cherko (dried coffee grounds) and cocoa bean meal, are unpalatable to domestic animals because of their bitter flavour.

Both strong salty and bitter tastes have the potential to reduce the rate of intake of foods, whereas a sweet taste is likely to enhance it. However, at low concentrations animals increase their appetite for feeds if they have enhanced salt taste, with differences in concentration of as little as 1 g Na/kg DM (dry matter) being readily perceived by cattle, demonstrating that for important nutrients there is an optimum concentration in food items that they will recognize and select. Animals may self-adapt to a salty taste over time, i.e. any reduction in palatability diminishes. In ruminants, the regurgitated bolus provides a mechanism for chemoreception of ruminal conditions, especially acidity.

Taste perception develops at an early age. For example, it has been demonstrated in the sheep **fetus** in mid-pregnancy, and maternal diet can affect dietary preferences of juveniles through *in utero* influences. Taste perception and discrimination thresholds change with advancing age. In cattle, the discrimination threshold to sucrose is less for calves than for adult cows. In sheep, the intensity of salt flavour ascribed to a standard sodium chloride (NaCl) solution increases fourfold from the fetus to adulthood, whereas that to potassium chloride (KCl) decreases by about 30%. Adult cattle prefer pure water to a 0.09 M NaCl solution but calves do not, suggesting that a similar increase in NaCl perception with age occurs in cattle as in sheep. KCl is a less potent salt flavour than NaCl for cattle. The appetite for sodium is particularly acute in many species (salt wisdom), which may show excessive licking behaviour in an apparent attempt to obtain supplementary sodium. (CJCP)

Taste aversion

A learned rejection of a **taste** stimulus due to previous pairing of that taste with **aversive** consequences such as sickness. Conditioned taste aversions are **adaptive** because they promote avoidance of harmful or poisonous substances. Protocols to study taste aversion learning generally involve presentation of a bottle containing a novel-flavoured solution (such as a dilute saccharin solution) to a thirsty **laboratory animal**, such as a rat. Upon cessation of drinking, an illness-inducing drug such as lithium chloride is administered via injection. A number of days later, the strength of the learning is assessed by giving the rats the opportunity to consume the illness-paired flavour from a bottle for a specified time. If the rat has learned the flavour–illness association, it will drink less of the solution than will a rat that had received saline during the learning phase instead of the lithium chloride. A two-bottle choice procedure may also be used to assess learning in which intake of the toxin-paired solution is compared with intake of a solution that has not been associated with the toxin (usually water). Taste aversion can occur after just one pairing of the novel solution with the effects of a toxin, and the response may be generalized to other, similar flavours. (SuH)

See also: Aversion learning; Conditioning; Generalization

Further reading

Garcia, J., Kimeldorf, D.J. and Koelling, R.A. (1955) Conditioned aversion to saccharin resulting from exposure to gamma radiation. *Science* 122, 157–158.

Teaser

A teaser is a male animal used to assist in domestic **breeding** contexts for the detection of **oestrus** in females. In a natural situation, reproductive males exhibit species-specific responses towards the female (such as sniffing, nudging and nibbling) as a form of **courtship behaviour** prior to copulation. This courtship functions to determine the receptiveness of the female as well as to prepare physiologically both the male (e.g. erection) and female (e.g. **lordosis**) for copulation.

The use of teasers in **horse** breeding is a product of market demands because using a teaser to detect oestrus protects stallions, which are often extremely valuable, from injuries that may be sustained from courting a mare that is not in oestrus (dioestrus or anoestrus).

It is important that the teaser does not copulate successfully. Thus the teaser may be entire (non-castrated) but restrained and/or too short to mate easily with the females (e.g. use of a small pony with horses) or may be physically altered so that he retains his **libido** but not the ability to inseminate. This may involve vasectomy, radical **circumcision** of the penis itself, deviation of the shaft or **castration** and subsequent treatment with **testosterone**. Depending on factors such as facilities available and personal preference, teasing may be conducted in several ways. Usually, the teaser is controlled by a handler who will direct it to approach females on an individual basis. Alternatively, groups of females may be allowed access to the teaser, which may be confined to a small yard.

The nature of the teaser's job prevents him from being able to consummate normal copulatory behaviours. Teasers are commonly kept in **isolation** or semi-isolation, brought out for teasing (i.e. only during the spring and summer months), often handled inappropriately (e.g. repeated jerking of the bit in the mouth of a horse) and rarely, if at all, get to mate with the females they are teasing.

In racehorse and performance horse breeding, teasers are conventionally ponies with high libido. They are often considered difficult to handle and, as a consequence, are usually handled roughly. Subsequently, this inappropriate handling contributes to the teaser becoming difficult to handle and the decline in its compliance is self-perpetuating. Collectively, the **welfare** of such animals is therefore a cause of concern. (AW-S)

Teat order

Within the first few days of life, the offspring of some mammalian species develop a teat order. That is, each individual limits their suckling to one or maybe two teats or nipples, thus reducing competition and allowing **nursing** of all offspring during suckling bouts. Teat orders are reliably established in piglets but can occasionally be found in other animals, such as kittens. With the **intensification of animal production**, many species can produce a larger number of offspring than they have teats to feed with, and thus cross-fostering needs to be performed to ensure all individuals can nurse during a suckling bout. (LMD)

Telos

The ancient Greek concept of telos corresponds to the modern-day notion of an aim, end or purpose and is the root of the word 'teleological' (meaning having to do with a thing's purpose). To understand the concept of telos properly it is necessary to separate two questions about purposes. In the first we ask: What is the purpose of that (e.g. dorsal fin)? In the second we ask: What is the purpose of doing that, or what is that animal aiming to do by doing that (e.g. scratching the ground)? The first question invites speculation about functions. The second is about agency and points towards goals.

Enquiries about purpose often appear in ancient Greek writings, especially in Aristotle's work. The notion that life, or being human, has a purpose is critical in Aristotle's ethical thought and re-emerges in Christian theology. It lies behind the familiar question 'What is the meaning of life?' Turning from human goals to functional teleology, Darwinism asks what role an adaptation plays in making a species fitter for survival.

The notion of a thing's telos focuses attention on the final, or ultimate, purpose of the thing. For this reason the telos is sometimes described as completion, or perfection. There is a direct connection here with the concept of nature, and with the ethical outlook known as perfectionism. The idea that something is natural (and hence the worry that something is unnatural) is both highly flexible and always in need of clarification. Naturalness is generally thought to be a good thing, presumably because we are, somewhat distantly, prone to take an ancient Greek perspective. Perceived departures from what is natural (bald broilers, genetic technology as a whole) are often condemned as unnatural (**see also: Abnormal**).

Perfectionism in animal **ethics** is the belief that animals are best off when they are able to live in the way they naturally do, expressing normal behaviours. One version of this view states that it is unethical to alter farm animals using modern techniques of genetic modification (though gradually bred change is often thought less worrying). Another, associated with the American philosopher **Bernard Rollin**, suggests that the important issue is to let animals live according to the nature that they actually possess, i.e. their telos. In the latter view, it is not enough to ensure that an animal does not suffer or is not frustrated: it should also be allowed to express its nature. (PAR)

See also: Welfare

Temperament

Temperament has often been broadly defined as a set of relatively stable and genetically based characters that differentiate individuals of the same species in terms of internal dispositions and reactivity to external stimuli. This definition highlights the supposed stability over time and the innate nature of temperament, which, according to some scientists, differs from **personality** in the fact that personality refers to more dynamic and modifiable traits – that is, traits that may change during the course of an individual's lifetime. As a consequence of this difference between the two terms, temperament is usually referred to describing individual differences in non-human animals and, to some extent, in human children, while personality is more commonly applied to adult humans.

Most recent research on the topic, however, opposes this presumed difference between the two terms. Indeed, personality and temperament seem to be influenced by the same factors (e.g. level of particular hormones in the blood, age and sex), share the same level of genetic inheritance across generations (i.e. between 20 and 40% of the inter-individual variation in temperament and personality seems to be genetically determined) and a similar degree of variation within individuals. Inter-individual differences in temperament may have important effects on an animal's behaviour and response to social and non-social stimuli. As such, temperament should be taken into account when, for example, designing **enrichment** devices for **captive animals** or when analyses are performed at the species level. (BM, RV)

See also: Corticosteroid; Individual differences; Lairage; Welfare measurement

Further reading

Réale, D., Reader, S.M., Sol, D., McDougall, P.T. and Dingemanse, N.J. (2007) Integrating animal temperament within ecology and evolution. *Biological Reviews* 82, 291–318.

Temperament test

The aim of a temperament test is to improve our knowledge of an animal and its likely future behaviour above that of chance alone. Temperament tests have been created by a range of organizations and individuals in order to assess useful, predictable behavioural and putative emotional tendencies in a range of species. Temperament tests may assess a single behavioural tendency/trait or a range of factors to give a greater overview of the **temperament** or **personality** of animals assessed (e.g. flightiness and boldness), or of their compatibility with a particular environment (e.g. pet–owner matching).

For example, temperament tests in **pet** dogs have been used to select suitable pets from rescue centres or to identify those already in the population that are, or are likely to be, unsuitable for this particular role (e.g. those with aggressive behaviour problems). Unfortunately, many of these tests seem to have been developed without a systematic scientific approach and often excessive claims of **reliability** and **validity** are made. This is unfortunate, since the outcome of tests may be used as the basis for important welfare-relevant and even survival-related decisions. Even those that are more robust may not be feasible and so of limited value.

Reliability, validity and feasibility determine whether a given test is a good measure, the right measure and a useful measure. Reliability concerns the degree to which the test scores are free from errors of measurement. This involves identifying the consistency of the results across sub-tests, tests, observers, assessment centres, etc. Measures of reliability include consistency within the observer of the test (**intra-observer reliability**), between observers (**inter-observer reliability**), within the animal (test–retest reliability) and within components of measures designed to assess the same behaviour (internal consistency) (see Fig. T.1).

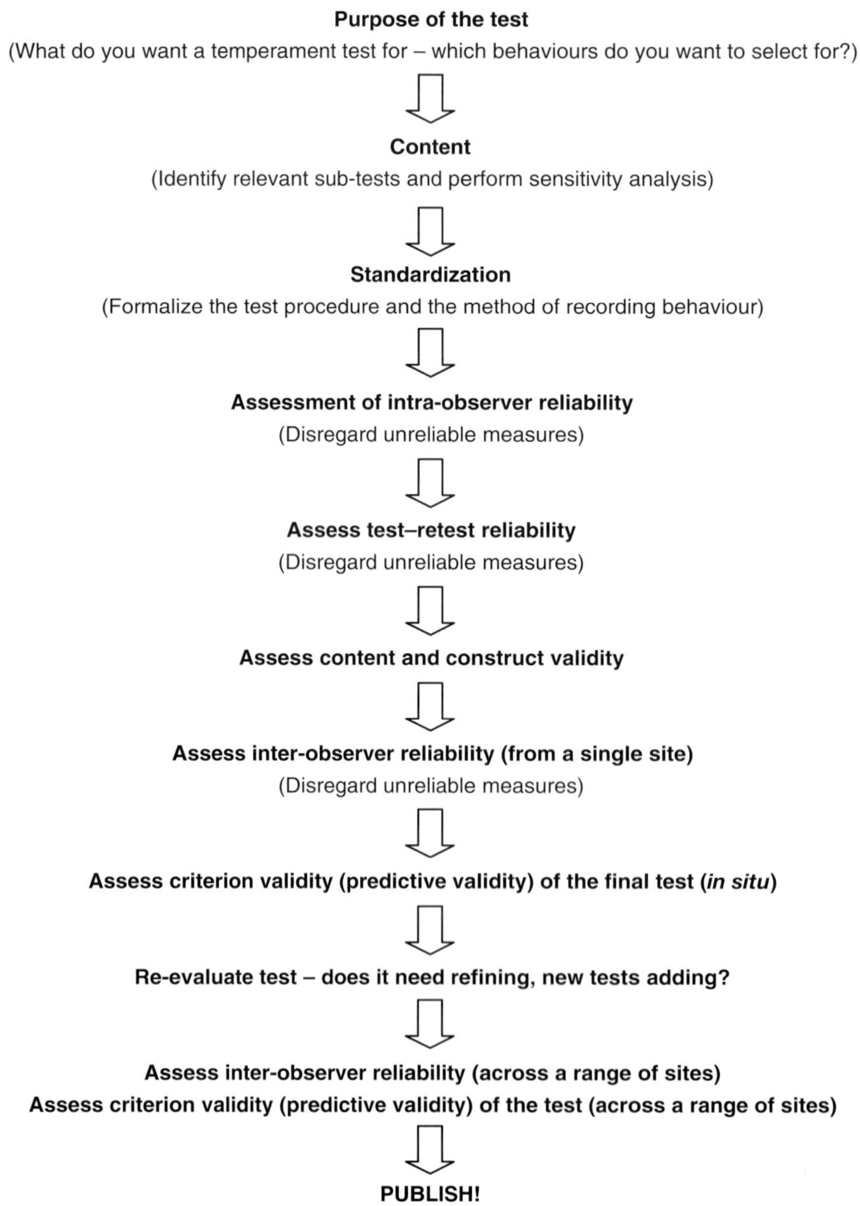

Fig. T.1. Diagrammatic representation of the process involved in creating and evaluating a temperament test (adapted from Taylor and Mills, 2006).

Evidence of the consistency and hence predictability of an animal's behaviour is considered by some to be what differentiates a temperament assessment from a **behavioural test**. Reliability is one, but not the only, requirement for validity. Validity concerns the appropriateness, meaningfulness and usefulness of the specific inferences made from the test results. It needs to be demonstrated that temperament tests are actually assessing what they claim, i.e. the trait(s) of interest, e.g. fearfulness, if they are to be valid. Validity assessments for temperament tests are fraught with difficulty, as it is unlikely that any test will be wholly predictive of an animal's behavioural reaction in every given circumstance. Finally, the quality of temperament tests must also address issues of practicality and appropriateness for academic or commercial use. Tests that are impractical, overly long and difficult to assess are unlikely to be performed accurately or reliably, if at all, and so a scientifically developed test will often require refinement for practical use.

For test developers two additional considerations need to be made in order that a test is reliable, valid and feasible. That is, consideration of the purpose of the test and standardization of the test procedure. If the goals of a temperament test are not clearly identified, i.e. the aspects of temperament that the testers wish to identify, then it is unlikely that the test will be valid. The next step in the development process is the selection of appropriate tests and corresponding scores for the animal's behaviour. If this stage is not standardized (formalized) then it is unlikely that the test will be reliable. It is advisable that these two additional prior requirements are fulfilled before the test developers can proceed to assessment of reliability and validity. (KT, DSM)

See also: Individual differences

Reference

Taylor, K.D. and Mills, D.S. (2006) The development and assessment of temperament tests for adult companion dogs. *Journal of Veterinary Behaviour – Clinical Applications and Research* 1, 94–108.

Teratogen

A substance that, when administered during pregnancy (particularly during the early stages, when the organs and limbs are forming), causes embryonic and subsequently fetal abnormal development. The most famous case was that of thalidomide, used in the 1960s to treat morning sickness in pregnant women, which led to a failure of the limbs to develop normally, leaving them shortened. As the limbs appeared similar in structure to those of seals, the name of *phocomelia*, literally meaning 'seal-limbed', was given to this condition. Interestingly, the thalidomide case is used by both opponents and proponents of animal testing to support their case, since some argue that it demonstrates the futility of testing while others argue that it demonstrates the need for more testing to increase confidence in safety. (DBM)

Territoriality

One of the most useful general definitions of animal territoriality was developed in the middle of the 19th century, when Alfred Nobel defined a **territory** as 'any defended area'. Since then, this definition has been refined and elaborated, but not by much. These days, most scientists focus on **space** that is defended for a long period relative to an animal's lifespan, to distinguish the defence of a territory from the defence of more ephemeral **resources**, such as individual food items or basking sites. Hence, a critical component of animal territoriality is that an individual must use a particular area for an extended period of time, a pattern of behaviour called 'site fidelity'.

The second component of Nobel's definition focuses on defence. Territory defence includes a wide range of **aggressive behaviour** patterns that tend to discourage other individuals from remaining in, or returning to, a particular area; these include chases, fights and the use of jaws, claws, horns or other weaponry to inflict costs on other individuals. However, many territorial animals are able to discourage other individuals from entering space using behaviour patterns that do not have such an obvious and immediate effect on the recipients. Thus, experimental studies using tape-recordings have shown that the song of a territory owner reduces the chances that other members of the species will enter the territory, relative to their entry rates when control sounds are played from the same location. Familiar examples of territorial advertisement include the songs of birds or crickets, visual **signals** such as the push-up displays of lizards, or the scent marks deposited around the territory by wolves or rabbits.

Nobel's definition is silent with respect to the types of individuals against which a territory is defended. Years of study have shown that territory owners often make fine discriminations among different categories of intruders when defending their territories. For instance, male house mice defend exclusive territories against other dominant males, but permit adult females, subordinate males and juveniles to share their territory. In many species, territory owners behave differently when interacting with individuals with whom they share a common border (neighbours) than when interacting with otherwise comparable strangers. In some species, owners are less aggressive to neighbours than to strangers (the so-called 'dear enemy' effect), while in other species the reverse is true, and owners react more aggressively to intrusions by neighbours than to intrusions by strangers. In some animals, discrimination among different categories of intruders may even extend to selected members of other species, as in algae-eating reef fish that defend their territories against other species with similar feeding habits, but ignore species with different feeding habits.

Because the **fitness** consequences of territorial behaviour are not immediately obvious, biologists have devoted considerable attention to studying the **adaptive** significance of territoriality. One important approach to questions about territory function developed in the middle of the 19th century, when concepts borrowed from economics were applied to studies of territory function. The economic approach to territoriality assumes that individuals defend territories when the benefits of territory defence exceed the costs, where benefits and costs are both measured in terms of lifetime reproductive success.

Because, in most animals, it is difficult if not impossible to measure lifetime reproductive success, most theoreticians and empiricists take short cuts, and rely instead on indirect indices of the benefits and costs of territorial behaviour. For instance, the benefits of a **breeding** territory might be estimated by counting the number of independent young produced in a single breeding season. Along the same lines, estimates of the cost of territory defence might rely on measurements of the number or intensity of attacks directed at intruders or neighbours, or the rate of production of territorial advertisement signals by territory owners.

One shortcoming of the classical economic approach to territoriality is that, by focusing on benefits and costs over the short term, it tends to ignore the effects of territorial behaviour on the long-term social and spatial behaviour of individuals. Most of the individuals living in the same area, because of their spatial proximity, are likely to encounter one another repeatedly over an extended period of time. This is not only true of individuals that defend territories adjacent to one another (neighbours), but also of territory owners and other types of individuals that inhabit the same area (e.g. floaters (that live in large, undefended home ranges) or intruders (that live nearby and repeatedly visit particular territories in the neighbourhood)). Whenever pairs or groups of animals interact with one another over a period of time, the behaviour of one individual at a given time can affect the behaviour of other individuals at a later time. In this situation, the behaviour of territorial animals is best analysed using the same theoretical tools that are used to study any other sort of ongoing social interactions in animals, e.g. learning-based and game-theoretical models (**see: Game theory**).

This recent theoretical framework helps explain much of the behaviour that is characteristic of territorial animals. For instance, most territorial animals exhibit the 'prior residency effect', in which residents retain possession of familiar space when competing with intruders, even if the intruders are larger, older or otherwise more competitive than the residents. Prior residency effects are predicted by a learning-based approach, which assumes that familiar space is more valuable to an individual than novel space, and that animals inflict punishment when they interact aggressively with one another. Assuming that intruders have somewhere else to go (i.e. that space is divisible), theory suggests that as long as residents can inflict some sort of costs on intruders, residents will eventually win contests over space with intruders, because the latter will eventually switch to using other areas that do not incur the ongoing costs of social interactions with residents. This prediction has been supported by empirical studies of territorial animals, showing that individuals can 'lose every battle' but eventually 'win the war', by persistently using familiar areas and repeatedly engaging their opponent in costly social interactions in those areas.

These recent studies have interesting implications for many species of concern to applied biologists. Many species studied by conservation biologists or pest managers are territorial, and many of the species domesticated by humans are derived from ancestors with sophisticated territorial behaviour; examples include **dogs**, mice, **rats** or **rabbits**. As a result, all of these animals come equipped with the ability to respond to punishment by avoiding the individual and/or the area in which that punishment occurred and, conversely, are likely to increase their level of aggressiveness towards intruders after they have become familiar with an area suitable for long-term residency. Thus, a basic knowledge of territorial principles can be useful for managing the behaviour of these and any other animals with an evolutionary history of territoriality. (JAS)

See also: Domestication; Dominance

Further reading

Adams, E.S. (2001) Approaches to the study of territory size and shape. *Annual Review of Ecology and Systematics* 32, 277–303.

Morrell, L.J. and Kokko, H. (2005) Bridging the gap between mechanistic and adaptive explanations of territory formation. *Behavioural Ecology and Sociobiology* 57, 381–390.

Stamps, J.A. (1994) Territorial behaviour: testing the assumptions. *Advances in the Study of Behaviour* 23, 173–232.

Territory

A 'territory' can be defined as an area that is defended (physically, vocally, e.g. howling or birdsong, or with **scent marking**, etc.) against a particular category of members of the same species (**conspecifics**) or the members of other species (heterospecifics) for an extended period of time. The geographic distribution of food (or some other defendable **resource**, e.g. potential mates) is an important determinant of whether an animal will defend a territory or not. (JAS)

See also: Home range; Territorial behaviour

Further reading

Adams, E.S. (2001) Approaches to the study of territory size and shape. *Annual Review of Ecology and Systematics* 32, 277–303.

Morrell, L.J. and Kokko, H. (2005) Bridging the gap between mechanistic and adaptive explanations of territory formation. *Behavioural Ecology and Sociobiology* 57, 381–390.

Testosterone

Testosterone is a member of the **androgen** family of hormones and, as a 19-carbon steroid, is derived from cholesterol. It is produced in relatively large quantities by the Leydig cells within the testes in males and in small amounts by the ovaries and placenta in females, and by the **adrenal gland** cortex in both sexes. Other members of the androgen family include androsterone, androstenedione and dihydrotestosterone (DHT – a testosterone metabolite). The major functions of testosterone can be described as being androgenic and anabolic. Testosterone is key for the development of secondary sexual characteristics during puberty and the control of **libido** and certain elements of **reproductive behaviour**. It also has anabolic effects on muscle and bone.

The control of testosterone production is under strong and complex regulation. Luteinizing hormone-releasing hormone (LNRH), **luteinizing hormone (LH)** and **follicle-stimulating hormone (FSH)** are known to play the crucial role here, but there are also feedback interactions with growth hormone, inhibin B and **oestrogens** that remain not clearly understood. Nitric oxide can stimulate testosterone release at low concentrations, but will inhibit release at high concentrations. There is also evidence of a neural pathway between the paraventricular nucleus (PVN) in the **hypothalamus** and the testes, which may act much more rapidly to influence testosterone release than the LH pathway.

Of particular note in the field of applied animal behaviour and **welfare** are the relationships between testosterone and reproductive behaviour, **social behaviour** (including **social stress**) and cognitive function (**see: Cognition**). Comparing species, mean testosterone levels are lower in species that are monogamous rather than polygynous. Primate species that practise uni-male mating systems have lower testosterone levels than those that practice multi-male–multi-female mating systems. Within species testosterone levels appear to affect libido, but reproductive behaviour is also closely linked to social behaviour and the social rank of the individual and thus, social stress.

In general, a number of **stressors** have been shown to act on the reproductive system by inhibiting testosterone production. These stressors include physical and mental **stress**, **depression** and (in humans) endurance-type sport. Species studied include rats, mice, hamsters, non-human primates and humans. However, the picture is not a simple one and different types of stressors impact circulating testosterone levels to different degrees. Whereas it has been shown that chronic stress provokes a decrease in blood testosterone, the initial stages of acute stress may actually cause an increase. This is particularly the case in individuals that have a greater experience in winning agonistic social interactions and in individuals having high **dominance** in a social **hierarchy**. The type of stressor appears to be important, with agonistic interactions and sporting competition having a particularly noticeable effect. It may therefore be that these types of stressors actually induce a type of anticipation or excitement response in the individual and do not constitute a negative stress, but rather may be a positive one.

Although it is generally thought that LH is the primary driver for testosterone release, the relationships between stress, LH and testosterone are mixed. Stress can decrease, increase or leave unaltered LH levels. At the same time, studies have shown that stress may decrease testosterone without altering LH levels, or may increase testosterone yet decrease or not alter LH levels. These apparently conflicting results perhaps need closer attention in relation to the mechanisms that might exist for either lowering or elevating testosterone levels and, also, the social dominance of individuals and the effects that this might have on hormonal responses.

In general, many stressors act by inhibiting LNRH in the hypothalamus, thereby decreasing both LH and testosterone levels. It is proposed that these effects or stress are mediated by **ACTH** and **glucocorticoids** such as cortisol. These hormones are known to directly inhibit testosterone secretion by Leydig cells and also to promote apoptosis (cell death) of those cells. Stress may also initiate oxidative stress, which itself reduces steroidogenesis in the testes and can reduce blood flow through the testes.

However, also during stress, **catecholamine** release is activated and we see increasing levels of **epinephrine** (adrenaline) and **norepinephrine** (noradrenaline). Norepinephrine stimulates LH, thereby potentially stimulating the release of testosterone, but it has also been hypothesized that sensitivity to LH may be increased in the testes and thus testosterone secretion can be stimulated without a rise in LH. Regardless, 'stress' can cause physiological changes that may both stimulate and inhibit testosterone release simultaneously. Superimposed on this are the resting levels of cortisol and catecholamines in the animal prior to the stressor being applied, and the magnitude of the responses once the stressor is applied. Certainly, in the case of cortisol, resting levels can differ between animals of different social status, and the magnitude of responses of both the **hypothalamic–pituitary–adrenal (HPA) axis** and the SAM axis may well differ between high- versus low-ranking animals and winners versus losers. Overall, factors likely to contribute to an increase in testosterone levels as stress is applied include: (i) an absence of chronic stress; (ii) dominant status within a relatively stable social hierarchy; (iii) experience of winning; and (iv) a stress reactivity that favours norepinephrine release.

Sex hormones are known to play a crucial role in the development of the **brain**, and testosterone is also known to be present throughout life in the cortical regions of the brain that are important for cognitive processes. Androgen receptors are found in the hippocampus, prefrontal cortex and amygdale – areas that are crucial for learning and **memory** – and also, in primates, throughout the cortical regions of the brain. In rodents, gonadectomy impairs subsequent performance in maze tests, but can be normalized by testosterone dosing. In humans, testosterone supplementation has been shown to improve spatial cognition and working memory in healthy older men, but not in younger men. In young women, a single testosterone injection can improve mental rotation and object location memory. However, there remains a great deal more research needed to clarify testosterone effects on cognition.

(JNM-F)

Further reading

Chichinadze, K. and Chichinadze, N. (2008) Stress-induced increase of testosterone: contributions of social status and sympathetic reactivity. *Physiology and Behaviour* 94, 595–603.

Janowsky, J.S. (2006) Thinking with your gonads: testosterone and cognition. *Trends in Cognitive Sciences* 10, 77–82.

Tethering

Tethering is a traditional method for restraining domestic and **companion animals** such as **dogs**, **horses**, **goats** and pregnant sows (**see: Restraint**). A flexible cable, typically a rope or chain, is used to link a collar or girdle on the animal to a fixed point. This arrangement restricts the movement of the animal to a defined area.

A range of **welfare** problems can arise from tethering. These include: (i) exposure to the elements (if no shade or shelter is provided); (ii) **hunger** and **starvation** (where no food is supplied, or where the available area becomes overgrazed); (iii) thirst and dehydration (if water is unavailable); (iv) **frustration** resulting from limitations to natural behaviours and an inability to socialize with other animals (this may result in the development of **stereotypies**); and (v) strangulation due to tangling (the risk of this increases with the length of tether used). The conditions for a tethered animal can change quickly: shade moves with the sun and water buckets can easily be upturned, and the accumulation of **urine** and faeces within the range of the tether can make conditions unhygienic. The period for which an animal is tethered under inappropriate circumstances can be critical.

In response to these recognized welfare problems, the use of tethers is becoming increasingly regulated. Common aspects of legislation include: (i) the maximum time an individual animal may be tethered; (ii) provision of shelter; (iii) provision of food and water; (iv) specifications for the design of the tether (e.g. a double swivel attachment to the collar can reduce the risk of strangulation); and (v) a ban on the use of tethers for pregnant sows. A more modern derivative of the tether uses a swivel attachment on a raised length of wire rather than a fixed point (sometimes called a running line). This increases the area available to the animal while minimizing the risk of strangulation. (BJ)

See also: Confinement

Theory of mind

Theory of mind (often referred to by the abbreviation ToM) is one of the central concepts in the study of animal (**social**) **cognition** as well as in the cognitive development of children. To have a theory of mind means to understand that others have **knowledge** and a **perception** of the world which may be different from one's own, to attribute beliefs, **motivations** and desires to others and to be able to make use of that knowledge in guiding one's own actions. Knowing about the knowledge of others can be used intentionally to deceive others or to gain information from understanding and making use of information that other individuals may possess. Having a theory of mind also enables individuals to communicate with the intention to inform, and thus the sender can adapt the **communication** according to his/her knowledge about the receiver's knowledge.

The concept of theory of mind was first introduced by Premack and Woodruff in 1978. In their paper entitled 'Does the chimpanzee have a theory of mind?', they described an experiment involving Sarah, a **chimpanzee** that had been trained for more than 10 years in the animal cognition laboratory and so having long experience of interacting with humans. In this experiment, Sarah watched videos showing short sequences in which human actors were confronting different types of problems (such as reaching for a banana outside the cage or washing a floor with a hose). Putting the video on hold before the problem was solved, the experimenters offered Sarah two different photographs: one showing the actor taking the right action to solve the problem, the other showing an unsuccessful action. More often than not, Sarah chose the photograph depicting the appropriate action to solve the problem. The experimenters argued that, in order to make such correct choices, Sarah must have been able to understand the actor's motivation.

After Premack and Woodruff's initial experiment, a large body of literature including both experimental and observational studies has been published addressing the question of whether non-human animals possess a theory of mind. There are several anecdotes of **deceit behaviour** from field observation of primates, suggesting that (at least some of) these animals are capable of using **grooming** as a way of distracting a dominant individual in order to get access to food, as well as to hide their actions from a dominant individual in order to avoid **aggression** (see Shettleworth, 1998). However, because the researchers making the observations have no information about the animals' previous experiences, it is not possible to know whether animals did use their knowledge about the knowledge and/or intentions of the other (theory of mind) or whether they behaved in a way that had been rewarded previously (associative learning), or even acted simply through chance. Carefully designed experiments are therefore essential to study theory of mind. Initial optimistic interpretations of field observations and early experiments were received by critics asking for more parsimonious interpretation, and over the course of the process, research methods have become increasingly sophisticated.

The false belief paradigm, originally introduced for studying cognitive development in children, has been widely used and adapted for use in animals. In the original set-up, a child is first introduced to a puppet. While the child and the puppet are both present, the experimenter hides a toy. Then the puppet is removed and the experimenter moves the toy to a different hiding place, still with the child observing. The child is then asked where the puppet will look for the toy. Until children are about 4 years of age, they will expect the puppet to look for the toy where they themselves know it is to be found. Young children are obviously not able to distinguish between their own knowledge and that of the puppet that has not seen the same events, and so they do not possess a theory of mind. Older children will correctly respond that the puppet will look for the toy in the first hiding place (see Shettleworth, 1998).

In different adaptations of the false belief paradigm, known as the guesser and knower paradigm, Povinelli and co-workers have tested whether primates are able to detect others' knowledge or ignorance. In one such experiment, chimpanzees were trained to gesture toward an experimenter through a hole in a transparent wall that held a food treat on the other side. When the animals had learned the task, they were tested in a series of sessions with a food treat positioned in front of two experimenters, one 'non-attending' experimenter who could not see (wore a blindfold, or held a hand or a cardboard screen in front of their eyes) and one 'attending' experimenter who could see (blindfold/hand/screen not covering the eyes but placed elsewhere near the face). A correct response for an animal that understands that seeing means knowing where the food treat is would be to reach a begging hand towards the 'attending' experimenter; however, in the experiment chimpanzees responded as often to either one of the two experimenters (Povinelli and Eddy, cited in Shettleworth, 1998).

So far, no absolutely conclusive evidence able to convince the most critical minds has been presented of theory of mind in non-human animals, although others claim that the sophisticated laboratory experiments are too artificial and distant from the animals' normal environment and communication methods and/or that all available evidence for ToM in humans is dependent on verbal affirmations that cannot be obtained in non-human animals (see Heyes, 1998 for original article and corresponding open peer commentary). Recognizing that absence of evidence is not evidence of absence, researchers are continuing their efforts to refine experiments that might be able to answer a reformulation of Premack and Woodruff's original question – 'Do non-human animals have a theory of mind?', with primates and cetaceans being the most popular candidates.

The capacities for **imitation** (the ability to learn a new behaviour through copying the behaviour of another individual) or self-recognition/self-perception (as often tested in the mirror recognition test) are sometimes associated with theory of mind. Although being able to learn through imitation as well having a concept of one's own body – as reflected in the mirror recognition test – suggests a certain level of cognition, these capacities do not require (and are therefore not evidence of) having a theory of mind. (AO)

References

Heyes, C.M. (1998) Theory of mind in nonhuman primates. *Behavioural and Brain Sciences* 21, 101–148.

Premack, D. and Woodruff, G. (1978) Does the chimpanzee have a theory of mind? *Behavioural and Brain Sciences* 4, 515–526.

Shettleworth, S.J. (1998) *Cognition, Evolution and Behaviour*. Oxford University Press, New York.

Thermoregulation

Animals thermoregulate in various ways depending on whether they need to lose, retain or gain heat, and depending on the species. In order to *retain* heat, the circulation in the arterioles to the outer surfaces of the body is constricted. In order to *gain* heat, shivering may occur, which is a rapid constriction of the motor muscles of the body, thus generating heat from the biochemical activity of muscular contraction. Some animals may also seek shelter and huddle together, or lie on warm surfaces. In species such as rats and mice, they eat more and may build nests. To *lose* heat most animals vasodilate the arterioles to the outer surfaces and, in addition, some animals sweat and/or salivate and pant – the evaporation of the sweat and saliva decreases temperature due to the latent heat of evaporation. Many animals may seek cold surfaces and lie out flat on them. Thermoregulatory behaviour can form an important part of the assessment of the **welfare** of an animal. (DBM)

See also: Floors; Lying

Thigmotaxis

Strictly speaking, thigmotaxis refers to the movement of an organism in response to touch, but some species, e.g. rodents, are described as being thigmotactic because they appear to prefer to maintain contact with the walls of an open area when introduced into a new environment. A reduction in thigmotactic behaviour (e.g. walking along a wall) may be taken as a measure of **anxiety** reduction in the screening of drugs with this putative effect. (DSM)

Threat postures

Threat **postures** are the behavioural displays that are typical of aggressive interactions before physical contact is made. Threat postures may be sufficient to solve a conflict or may escalate into overt **aggression**. In many species, threat postures are characterized by attempts apparently to increase body size (e.g. through fur bristling), by display of dangerous anatomical features (e.g. antlers) or by bright colorations. (BM)

Three Rs (3Rs) principle

The principle of the three Rs (3Rs) was created by scientists William Russell and Rex Burch and published in 1959 in a book by the Universities Federation for Animal Welfare (**UFAW**) entitled *The Principles of Humane Experimental Technique* (reprinted 1992). Their aim was to increase the humanity of scientific research that used animals by providing a framework for scientists to use when planning their research.

It embodies the notions of *replacing* animals in research with non-sentient material, *reducing* the numbers of animals within an experiment to the minimum consistent with good statistical analysis and *refining* the experimental protocols so that the minimum of **suffering** is caused to achieve the scientific objective. These principles can be applied to any research project that involves animals where there is the potential that the animals involved may experience **pain**, suffering, **distress** or lasting harm (including **death**) as a consequence.

Although the 3Rs principle has been incorporated into some of the laws controlling the use of animals in research, including the European Directive on the Use of Laboratory Animals (86/609/EEC) and the UK Animals (Scientific Procedures) Act, both written in 1986, neither of those states the principle explicitly. This is perhaps because the principle of the 3Rs, and the book that described it, lay largely ignored until the early 1990s, when it started to become heavily promoted by animal protection organizations such as the Humane Society, FRAME and UFAW. The book was reprinted by UFAW in 1992, and the 3Rs have now gained mainstream acceptance within laboratory animal science, both within industry and academic research. It is now expected that prior to and during ethical review of proposed harmful animal-based research, consideration of the 3Rs should be made and projects adjusted accordingly.

Centres of excellence in the 3Rs were initially created within university departments, e.g. the Netherlands Centre for Alternatives to animal use (NCA at Utrecht University) and the Center for Alternatives to Animal Testing (CAAT at Johns Hopkins University in the USA). However, national governments have now responded with the creation of national centres of excellence in the 3Rs (e.g. the National Centre for the 3Rs in the UK – NC3Rs). More recently, the European Commission has partnered with industry in the form of the European Partnership for Alternative Approaches to Animal Testing (EPAA) to help accelerate the development and acceptance of new 3Rs methods in regulatory toxicology.

Replacement

Russell and Burch defined replacement as 'any scientific method employing non-sentient material which may, in the history of animal experimentation, replace methods which use conscious living vertebrates'.

Current important replacement techniques include cell- or tissue-based (*in vitro*) systems that use bacteria, primary human or animal cells or permanent cell lines. Examples of *in vitro* replacement tests include: (i) the Ames test to detect genotoxicity; (ii) the embryonic stem cell test to detect reprotoxicity; (iii) Hep2G human liver cells used to detect hepatoxicity; (iv) human, blood-derived monocytes to detect bacterial contamination of biological products; (v) reconstituted human epidermis to test for likely skin irritation; and (vi) DNA microarrays to assess the direct effects of chemicals on gene expression.

Future replacement techniques are likely to involve the combination of individual, cell-based assays, even within the same system inside a microchip to build up a picture of the entire body, or the creation of miniaturized tissue organs. Other important replacement techniques include computer systems (*in silico*) and chemical-based assays (*in chemico*). *In silico* techniques include computer models that predict the toxicity of a chemical based on its similarity to others (QSARs) or computer models of physiological systems such as the human heart. *In chemico* techniques are assays based purely on the chemistry of the chemical of interest, and can include assessment of the chemical 'reactivity' as a measure of allergenic potential or the use of high-tech chromatography to identify known toxins within a mixture. The latter technique is now a replacement for the detection of toxins within shellfish, a test that currently constitutes an LD_{50} test in mice (requiring death as an endpoint for 50% of the mice injected with the shellfish mixture).

Absolute replacement includes methods that do not depend on any animal material, while relative replacement includes methods that depend on either animals of lower neurophysical **sentience** (such as insects and worms) or animal tissues. Currently, many *in vitro* cell-based replacement systems rely on primary animal tissues or animal-derived proteins, such as: (i) the Ames test, which uses proteins experimentally derived from rats; (ii) fetal bovine serum, which is obtained directly from the still-beating hearts of calves excised from their mothers at the **slaughterhouse**; and (iii) the bovine corneal opacity and permeability test for eye irritation, which uses the eyes of slaughtered food cows. There are increasing efforts to move away from these relative replacements in the interests of sound science and animal **welfare**.

Replacement techniques feature most heavily for the prediction of the toxicity of chemicals, including cosmetics and pharmaceutical drugs. This is perhaps because industry can directly appreciate the cost and time savings of replacement techniques and has been under the most public pressure to avoid the use of animals. The ban on the use of animals to test the safety of cosmetics, which entered force on 11 March 2009 in the European Union, arguably further encouraged industry to invest in alternatives. Similar incentives have yet to appear for academic research, which tends to be more variable and therefore harder to replace than regulatory toxicology. However, some important replacements have recently appeared, such as the use of non-invasive imaging scanners, e.g. functional magnetic resonance imagers (MRI) that enable researchers to 'see' activity in the human **brain**, rather than invasively using non-human primates.

Reduction

Russell and Burch defined reduction as the 'reduction in the number of animals used to obtain information of a given amount and precision'.

Reduction is generally seen as something to be considered within a single experiment or procedure, known as intra-experimental reduction (de Boo and Hendriksen, 2005). However, reduction can also be an aim across an entire research project (consisting of many procedures; supra-experimental reduction) or on a much wider scale, such as something for governments and agencies to consider across entire research programmes (extra-experimental reduction).

Examples of reduction techniques include: (i) prior assessment of the required numbers of animals needed, by using pilot studies or statistical power calculations (Festing and Altman, 2002); (ii) reducing the number of concurrent control animals; (iii) usage of historical control animals; and (iv) usage of imaging machines so that disease progression and effects of treatment can be assessed non-destructively in the same animal. Harmonization of testing requirements for chemicals and medicines across countries can also be seen as a reduction exercise, reducing the numbers of animals used in what would be otherwise essentially duplicative tests.

Refinement

Russell and Burch defined refinement as 'simply to reduce to an absolute minimum the amount of stress on those animals that are still used'.

Refinement considerations can and should be concerned with the entire lifetime experience of the animal involved, beginning with their capture or **breeding**, **housing** and **transport**, **handling** and cage-cleaning routines, routine veterinary care, food and **enrichment** of their cage environment, in addition to attempts to reduce the suffering involved in the procedures themselves.

Perhaps, not surprisingly, refinement efforts fall within the interest of animal welfare scientists and, with this in mind, Buchanan-Smith *et al.* (2005) recently elaborated refinement to 'any approach which avoids or minimises the actual or potential pain, distress and other adverse effects experienced at any time during the life of the animals involved, and which enhances their **well-being**'.

Examples of procedural refinements include the use of less severe techniques, such as: (i) taking ear notches instead of amputating toes for identification; (ii) the use of **analgesia** and **anaesthesia**; (iii) training animals to cooperate with procedures; and (iv) using positive **reinforcement** such as food treats when handling.

Examples of housing-related refinements include the use of food hoppers and foraging devices, toys and platforms to encourage natural behaviour, larger enclosures including outdoor access and access to natural foraging and bedding materials. Social housing is one of the greatest contributors to positive welfare in social animals and is increasingly being considered, even if only prior to entry into a research protocol. Other positive refinements include the use of solid rather than wire **flooring**, provision of nestboxes for mice and positive human handling and **socialization**. All have been shown in studies significantly to improve the welfare of caged animals. A typically enriched non-human primate enclosure may include social housing, platforms and swing ropes, foraging devices, a varied diet, pools of water, positive handling and outdoor access.

Implementation of the 3Rs

Recent studies on the impact of the 3Rs on experimental animal research have shown mixed results. The use of animals is thought to have decreased significantly from the 1970s but is still very high, having been estimated at 115 million annually (Taylor *et al.*, 2008) and is currently, in Europe at least, no longer decreasing. This suggests that, on a macro scale, the 3Rs have yet to make any significant impact. Looking within reports of animal use within scientific journals, some authors have found that the use of pain relief has increased (refinement) and intra-experimental animal use has decreased (reduction) (Carlsson *et al.*, 2004; Richardson and Flecknell, 2005). However, specific mention of the 3Rs or 3Rs-related parameters within animal research scientific papers remains low (with the exception of animal welfare science papers), and there are suggestions coming from the animal welfare community that not only should scientific journals have ethical policies that promote the 3Rs, but that papers themselves should have sections describing the positive 3Rs changes made (Olsson *et al.*, 2007; Wurbel, 2007).

Current issues in relation to the 3Rs

Validation and acceptance of 3Rs methods

While the general aims of promoting humane research within the framework of the 3Rs are often applauded, a number of practical problems with the principle have arisen in more recent years. These include problems with the acceptance of alternative methods, competing Rs within the 3Rs and a tendency to focus on the 'easiest' R – refinement.

Significant changes to standard test procedures based on the 3Rs are often referred to as 'alternative methods'. As such, they need to be formally assessed to ensure they are reliable and provide the same results as the original test method. This process is called validation and requires inter-laboratory, if not international, collaboration. To facilitate this, Centres for the Validation of Alternative Methods (VAMS) have been created in Europe (ECVAM), the USA (ICCVAM) and Japan (JAVAM). Following positive validation statements from these bodies, new alternative tests can and (indeed, under 86/609/EEC) should be used. Current problems with this process include: (i) the lack of funding for the development of new tests; (ii) the speed at which validation projects proceed; (iii) reluctant regulatory acceptance following validation; and (iv) the obvious problems with comparing a potentially superior method with an outdated, highly variable gold standard (the old animal test), which would in all likelihood fail itself if it were to go through the same validation process (Balls and Combes, 2005).

Competing Rs

Unfortunately, as experience in the 3Rs has increased so has knowledge that each R is not mutually exclusive, and sometimes Rs can be in conflict depending on the nature of the improvement being considered (de Boo and Hendriksen, 2005). For example, the use of imaging machines such as fluorescence or infrared imagers can lead to the reduction in the numbers of animals used (positive reduction); however, it may adversely affect the welfare of those that are used in that they may be used for longer periods (across the entire disease progression) and may incur repeated handling and exposure to anaesthesia, **confinement** and pharmaceutical agents in order to be imaged (negative refinement).

Similarly, the reuse of animals across different procedures is often considered a positive reduction technique, but can have such severe refinement consequences that it may not be advocated. For example, neurophysiological experiments on non-human primates require the implantation of electrophysical recording devices in their skulls. Once implanted these can allow the animal to be used in many different research projects, avoiding the use of new animals. However, such primates are often kept confined for many years and are repeatedly exposed to behavioural experiments that none the less can cause distress.

The 'easiest' Rs

Of the 3Rs, refinement and reduction may be seen as the easiest Rs to tackle and replacement as the hardest of the 3Rs to achieve. None the less, animal protection groups would prefer animal users to focus on the replacement of animals, since this would negate the use of any animals in the future, an ultimate aim also advocated by Russell and Burch. This has led animal protection groups to criticize the 3Rs, because it allows animal users to claim discharge of their ethical responsibilities by attending to only one part of one of the 3Rs (usually refinement). This concern is reflected in the use of the word 'alternative', which has crept into 3Rs literature, confusing replacement with essentially the same animal test but with improved animal welfare. For example, the fixed-dose procedure is seen as an alternative to the notorious LD_{50} test for acute toxicity. There is no doubt that the fixed dose represents a significant improvement in animal welfare, not requiring the death of 50% of the animals, but neveretheless it remains a very serious animal procedure.

There is a danger in considering that implementation of the 3Rs is all that is required when considering the use of animals in harmful research. Assuming full replacement is not possible, the harm versus benefit of the research should also be assessed. This assessment should weigh up the harms done to the animals (after the 3Rs have been applied) against the potential benefits of the resulting research. The benefits of the research should be considered, not just in terms of knowledge gained but also for the quality of the research and the likelihood of it producing positive improvement to human or animal health. For example, even though researchers may make positive 3Rs improvements to these tests, it could be argued that using animals in cosmetics or household product testing should not be allowed because the impact on human health is trivial. Similarly, testing of chemicals for carcinogenicity should arguably not be permitted given the known limitations of the rodent model of cancer. To this end the 3Rs can be seen as a 'reactive' solution to be applied *after* the genuine need to perform research on animals has already been decided.

Therefore, while the 3Rs principle provides a useful framework for seeking to improve animal-based research, applying these simple rules should not negate other important ethical and scientific considerations that should be made before the harmful use of animals is even considered. (KT)

See also: Humane Society of the United States; Russell and Burch

References and further reading

Balls, M. and Combes, R. (2005) The need for a formal invalidation process for animal and non-animal tests. *Alternatives to Laboratory Animals* 33, 299–308.

Buchanan-Smith, H.M., Rennie, A.E., Vitale, A., Pollo, S., Prescott, M.J. and Morton, D.B. (2005) Harmonising the definition of refinement. *Animal Welfare 14* (*Special Issue on the 3Rs*)(4), 379–384.

Carlsson, H.E., Hagelin, J. and Hau, J. (2004). Implementation of the 'three Rs' in biomedical research. *Veterinary Record* 154, 467–470.

de Boo, J. and Hendriksen, C. (2005) Reduction strategies in animal research: a review of scientific approaches at the intra-experimental, supra-experimental and extra-experimental levels. *Alternatives to Laboratory Animals* 33, 369–377.

de Boo, M.J., Rennie, A.E., Buchanan-Smith, H.M. and Hendrikson, C.F.M. (2005) The interplay between replacement, reduction and refinement: considerations where the Three Rs interact. *Animal Welfare 14* (*Special Issue on the 3Rs*)(4), 327–332.

Festing, M.F. and Altman, D.G. (2002) Guidelines for the design and statistical analysis of experiments using laboratory animals. *Institute for Laboratory Animal Research Journal* 43, 244–258.

Olsson, I.A.S., Hansen, A.K. and Sandoe, P. (2007) Ethics and refinement in animal research. *Science* 317, 1680.

Reinhardt, V. and Reinhardt, A. (2002) *Comfortable Quarters for Laboratory Animals*, 9th edn. Animal Welfare Institute, Washington, DC.

Richardson, C.A. and Flecknell, P.A. (2005) Anaesthesia and post-operative analgesia following experimental surgery in laboratory rodents: are we making progress? *Alternatives to Laboratory Animals* 33, 119–127.

Russell, W.M.S. and Burch, R.L. (1959) *The Principles of Humane Experimental Technique*. Methuen, London.

Stephens, M.L., Goldberg, A.M. and Rowan, A.N. (2001) The first forty years of the alternatives approach: refining, reducing and replacing the use of laboratory animals. In: Salem, D.J. and Rowan, A.N. (eds) *The State of the Animals*. Humane Society Press, Washington, DC, pp. 121–135.

Taylor, K., Gordon, N., Langley, G. and Higgins, W. (2008) Estimates of laboratory animal use in 2005. *Alternatives to Laboratory Animals* 36, 327–342.

Van Zutphen, L.F.M., Baumans, V. and Beynen, A.C. (2001) *Principles of Laboratory Animal Science*. Elsevier, London, 428 pp.

Wurbel, H. (2007) Publications should include an animal-welfare section. *Nature* 446, 257.

Thyroid hormone

There are two thyroid hormones produced and secreted by the thyroid gland, these being thyroxine (T_4) and triiodothyronine (T_3). Thyroxine and triiodothyronine are produced from tyrosine, having either four (T_4) or three (T_3) iodine atoms on the thyronine ring. Low dietary intake of iodine will lower the production of these two hormones.

The production and secretion of these hormones are initiated by the **hypothalamus**, which secretes thyrotropin-releasing hormone (TRH), which enters the hypophyseal portal system to reach the anterior **pituitary gland** to bind to membrane receptors on thyrotrophs – cells in the anterior pituitary that produce thyroid-stimulating hormone (TSH); the anterior pituitary gland then produces and secretes TSH. Upon thyroid stimulation, thyroglobulin – which is the storage form of T_3 and T_4 – is transported from the lumen of the follicle, by pinocytosis, into the follicular cells of the thyroid gland. In these cells the large protein, thyroglobulin, is hydrolyzed into T_3 and T_4 and secreted into the peripheral circulation. The majority of thyroid hormone produced and in circulation is T_4. The majority of T_3 and T_4 concentrations are then bound by thyroxine-binding globulin, while the remaining 30% is bound to albumin or thyroxin-binding pre-albumin. Only a fraction of 1% remains unbound in the peripheral circulation. The large percentage of bound hormone supplies the body with a large pool from which to draw and prevents its degradation until it reaches its target tissue.

Of the two thyroid hormones it is T_3 that is considered the biologically active hormone, although T_4 to a lesser extent can bind to thyroid receptors. After entering the peripheral circulation, T_4 is converted into T_3 in organs such as the anterior pituitary gland, the kidney and the liver. Most tissues in the body have nuclear receptors for thyroid hormone, and by activating these receptors thyroid hormones set the basal rate of heat production and oxygen consumption for an animal. Thyroid hormones also influence the rate of lipid, protein and carbohydrate metabolism, increasing these rates when thyroid hormone concentrations are elevated and decreasing them when thyroid concentrations are low.

Overall secretion of thyroid hormone is regulated by a negative **feedback** system, in which high concentrations of circulating hormone cause the anterior pituitary gland to decrease production of TSH and thus decrease the production of thyroid hormone. Thyroid hormone is particularly important in the processes of physical and behavioural maturation, as well as in cognitive development (**see: Cognition**), and differences in thyroid hormone effects have been suggested as one of the factors controlling neoteny. (DCL)

Further reading

Smith, J.W., Tudor Evans, A., Costall, B. and Smythe, J.W. (2002) Thyroid hormones, brain function and cognition: a brief review. *Neuroscience and Biobehavioural Reviews* 26, 45–60.

Thyrotropin-releasing hormone

Also known as either TRH, thyrotropin-releasing factor or TRF, this is a peptide hormone produced by the **hypothalamus** that stimulates the **pituitary gland** to produce the peptide thyroid-stimulating hormone (TSH, or thyrotropin), which itself regulates the production of thyroid hormones. Its action is antagonized by growth hormone-inhibiting hormone. The relationship between TRH and welfare is complex and depends on the specific **stressor** and species concerned. TRH can also stimulate **prolactin** production. (DSM)

Further reading

Moberg, G.P. and Mench, J.A. (2000) *The Biology of Animal Stress – Basic Principles and Implications for Animal Welfare.* CAB International, Wallingford, UK.

Tid-bitting

Tid-bitting is a behaviour performed during courtship displays in many species of galliformes, in which the male pecks repeatedly at the ground to attract females. During this the cock may also scratch and give food calls. The behaviour may also be performed as a threat during **competition** between males. Tid-bitting is also performed by female birds, which will peck at a food source to attract their chicks to it. (RJNM)

Time budget

Time budgets measure the amount of time animals spend performing various behaviour patterns, and are often measured over 24 h periods. Some behaviours may occupy a large amount of an animal's available time while others may take very little time to perform, but this does not necessarily reflect their relative importance. Time can be thought of as a commodity, as only a finite number of activities are possible during a set period of time. Thus an animal may modify the amount of time spent performing different behaviour patterns depending on the time and behaviours available. For example, when the milk supply of deer fawns was reduced, their time spent **nursing** also decreased but their time spent **grazing** increased to compensate for the change in nutrition.

The way animals behave throughout a period of time can be viewed as the result of a series of **preferences** or choices between different behavioural options. In **wild animals**, the daily time budgets followed can be seen as strategies for **coping** with environmental changes, such as increases or decreases in food availability or variations in temperature. Allocating the appropriate amount of time to different behaviours and performing these behaviours at appropriate times of day will affect an animal's ability to survive and reproduce. Factors that affect the amount of time spent on different behaviour patterns include food reserves, risk of **predation**, time available to perform the behaviour and the results of behaviours performed earlier.

Time budget measurement can be used as a tool to assess animal behaviour and evaluate **welfare**. Resources or environments can be manipulated and the effects on different behaviour patterns can be recorded (**see: Economics of behaviour**). Time budgets of wild and captive animals can also be compared to determine the potential effects **captivity** may have on behaviour (Veasey *et al.*, 1996). As measuring time budgets requires only recording of the start and stop times of certain behaviour patterns, it is an easy technique to implement. It can be done through live observation or through video-recording and measuring time budgets from tapes. Thus, it does not require any expensive or elaborate equipment.

However, there are problems with using only time budget measurements. While measuring time budgets can determine how behaviour is affected by different factors, a difference in the amount of time spent performing a behaviour pattern does not necessarily indicate deprivation or **suffering**. The behaviour may be strongly dependent on external cues and is not shown in situations where the cue is not present. For example, anti-predator behaviour is commonly found in wild species and is important to survival and **reproduction**. However, in captive environments, where predators are not present, anti-predator behaviour is not important and thus rarely shown. In addition, some behaviour patterns may be performed when possible but the animal does not suffer if the behaviour is prevented, i.e. the animal *wants* to perform the behaviour but does not *need* to perform the behaviour.

One way around these constraints has been devised by placing costs on the performance of certain behaviour patterns and measuring what happens as the costs increase. In other words, the animal's **motivation** to perform certain behaviour patterns is measured. These methods have been adapted from human economics and are generally referred to as consumer demand experiments. As part of these experiments, the time devoted to performing certain behaviour patterns can be measured as it becomes harder and harder to perform the behaviour. If the amount of time devoted to a behaviour pattern decreases as costs increase, the behaviour is thought to be less important to the animal. However, if the time devoted to performing a behaviour pattern remains fairly constant as the cost increases, the behaviour is thought to be important to the animal. Alternatively, the amount of time available to perform a range of behaviours can be decreased (cost) and the animals will reschedule their behaviour patterns to allow more time to be spent on the more important behaviours.

Consumer demand techniques are not without their own problems (**see: Preference test**). Thus, care should be taken when designing these types of experiments to avoid or compensate for as many problems as possible. (LMD)

Further reading

Veasey, J.S., Waran, N.K. and Young, R.J. (1996) On comparing the behaviour of zoo housed animals with wild conspecifics as a welfare indicator, using the giraffe (*Giraffa camelopardalis*) as a model. *Animal Welfare* 5, 139–153.

Tinbergen, Nikolaas (Niko)

Niko Tinbergen (1907–1988) was born in The Hague, Netherlands, on 15 April 1907 and studied at the University of Leiden, where he was awarded his PhD in 1932. Together with **Konrad Lorenz** and **Karl von Frisch**, he was awarded the Nobel Prize for Physiology or Medicine in 1973. He first met Lorenz in 1936 when he was invited to lecture at Leiden, and the two soon became close friends and colleagues, with Lorenz traditionally being seen as the scientific speculator and Tinbergen the more analytical scientist.

The two became separated by World War II and Tinbergen, like Lorenz, was held hostage during some of this time. After the war he was invited to set up a behaviour research group at Oxford, UK, and this ultimately led to his reunion with Lorenz. Perhaps his most enduring legacy is what has become known as **Tinbergen's Four Questions**, which outline four approaches to the study of animal behaviour relating to proximate and ultimate factors: namely its physical mechanism, lifetime development, **evolution** and **adaptive** function. Tinbergen was notoriously modest about his abilities and, in addition to being a formidable scientist, was also an accomplished illustrator. (DSM)

Further reading

Nikolaas Tinbergen autobiography, available at: http://nobelprize.org/nobel_prizes/medicine/laureates/1973/tinbergen-autobio.html (accessed 24 November 2009).

Tinbergen's Four Questions

In 1963, **Niko Tinbergen** wrote 'On the aims and methods of ethology', and proposed what are now referred to as Tinbergen's Four Questions. These questions, based in part on previous suggestions by Ernst Mayr, have helped guide behavioural research for the past four decades. These four logically distinctive and mutually exclusive types of questions about causation, development, **adaptive** utility and evolutionary history can be profitably applied to any bio-behavioural phenomenon. Importantly, by asking questions at multiple levels of analysis, our knowledge about behaviour is enriched. Broadly, a bio-behavioural question could focus on how something works, or why it is as it is.

Proximate questions are those employed to explain how something works or how it develops. For instance, studies of functional morphology tell us how behaviour is patterned and its structural basis. Studies of behavioural genetics identify the degree to which genes are responsible for behaviour, and the field of genomics identifies those genes. Studies of behavioural endocrinology tell us about the hormonal control or regulation of behaviour. These three examples illustrate causal questions. By answering them, we learn about how behaviour functions. A logically distinct type of proximate question focuses on the development (or **ontogeny**) of behaviour. Ontogenetic questions might ask about the degree to which a particular behaviour requires specific experiences to be properly performed, and address the time course of development.

Ultimate questions are those employed to explain why we see the diversity of behaviour. For instance, studies that focus on the **evolution** of behaviour tell us how or when a particular behaviour evolved. They might also tell us how many times a behaviour evolved. To do so, evolutionary biologists construct phylogenetic trees (hypotheses about the relationships between species) and then 'optimize' (i.e. map) behavioural traits on these trees. A logically distinct type of ultimate question focuses on the current adaptive utility of a trait. Only traits that increase the **fitness** of individuals who possess them will evolve or be maintained by natural **selection**. For instance, if long legs aid in escaping predators, we will expect natural selection for leg length and running speed to evolve.

Importantly, these four types of questions (or levels of analysis) produce questions that are mutually exclusive only within a level. Consider birdsong: we can ask about the evolutionary history of song learning. Song learning has evolved in parrots, hummingbirds and passerine birds. Among passerines, it is seen in a broad group called the oscine birds. These are questions about the evolutionary history of birdsong. We can also ask about the current adaptive utility, or function, of birdsong. Male birds may sing to attract females or to defend their **territory** from other males. In some species, males that sing more songs have more mates and therefore have higher fitness. It would be illogical to use evidence that, because male birds sing to defend territories, song learning has evolved only once. Questions within a level of analysis are mutually exclusive only with other questions within that level.

We can ask proximate questions about birdsong as well. For instance, recent work on the genetics of song has discovered that humans and birds both express the *FoxP1* and *FoxP2* genes. In birds, these genes are specifically expressed during song learning. A set of **neuron(e)**s, called the higher vocal centre (HVC), seems to be responsible for the neural control of song learning. In some species, the size of the HVC is correlated with the number of songs they produce, while in other species the size of the HVC changes seasonally and becomes largest when song learning is required. Finding evidence that the HVC does not change seasonally has no direct bearing on whether or not *Fox* genes are expressed during song learning; nor does it directly bear on hypotheses about the evolution of song-learning abilities, or about whether or not males that exhibit larger repertoires have higher fitness. Again, these questions are mutually exclusive.

By taking a Tinbergenian approach to studying behaviour, it forces us to examine qualitatively different sorts of questions. By doing so, we generate considerable knowledge about the diversity of behaviour. Recognizing that these are qualitatively different questions is essential, as well as ensuring that arguments about explanation are contrasting different hypotheses at the same level of analysis.

A fifth question to help extend the scope of behavioural research might also be proposed: What is the applied value of a trait or phenomenon? Formal recognition of a fifth question has several benefits and no real costs. Virtually all questions can be applied in some way and, by recognizing the applied value of a phenomenon, we expand the scope of inquisition. Much, but not all, applied research asks proximate questions, but these questions are targeted to address a larger question. For instance, researchers who study animal **welfare**, **conservation** biology and **comparative psychology** routinely apply the results of their studies to non-humans or to humans to increase welfare, manage wild or captive populations or to understand the biological basis of human behaviour. In many cases, the applied benefits of a study are the central question.

Not all researchers will desire to apply behavioural knowledge, just as not all evolutionary biologists are interested in studying proximate questions about causation. This, in itself, is no reason not to recognize the study of applied behaviour as a formal, logically distinct field of enquiry. It is essential to value high-quality research at whatever level of analysis it is conducted and not to dismiss the importance of focused studies operating only at a single level. By formally testing hypotheses about application, we can clearly identify applied benefits and this may help 'market' behavioural research to the public. Formal acknowledgement of the applied value of behaviour should make the future of bio-behavioural study even richer. (DTB)

Reference and further reading

Mayr, E. (1961) Cause and effect in biology: kinds of causes, predictability, and teleology are viewed by a practicing biologist. *Science* 134, 1501–1506.

Tinbergen, N. (1963) On the aims and methods of ethology. *Zeitschrift für Tierpsychologie* 20, 410–433.

T-maze

A T-maze is an apparatus employed to examine spatial memory in animals. The animal is placed at the bottom of the T-shaped apparatus in a start box, is required to move along the apparatus and then to turn either left or right to reach a reward. After that the animal is taken back to the start box, where it is retained for a period of time (termed the retention interval). At the end of the retention interval the animal is released and allowed to search for the reward. There are several types of tests. For example, in a match-to-sample procedure, the animal has to choose the same arm that it visited recently in order to receive a reward. In a non-match-to-sample procedure, the animal has to choose the alternative arm to the one last visited.

Knowledge of animals' spatial **memory** abilities is necessary in many species and, in particular, for farm animals in extensive production systems, since such animals would need to locate and exploit distant resources. For example, laying hens in large systems are not always in sight of resources such as food, water, litter and nestboxes, and may need to remember the location of these resources, as well as to navigate to them. Clearly, a failure to find resources would severely compromise the **welfare** of animals in production systems.

Research involving T-mazes, and also other apparatuses such as radial-arm mazes, has proved invaluable in investigating spatial memory in animals, and hence identifying potential cognitive deficits which would impinge on welfare (**see: Cognition**). For example, Mendl *et al.* (1997) investigated the effects of rearing pigs in both a substrate-enriched and a substrate-impoverished environment on their performance in a T-maze, and concluded that substrate-impoverished environments provide inadequate stimulation for the expression of their natural behaviour. (RF)

Reference

Mendl, M., Erhard, H.W., Haskell, M., Wemelsfelder, F. and Lawrence, A.B. (1997) Experience in substrate-enriched and substrate-impoverished environments affects behaviour of pigs in a T-Maze task. *Behaviour* 134, 643–659.

Tongue rolling

Although it occurs rarely, the most common stereotypy among adult **cattle** is tongue rolling; it is also found in many other captive species, notably **horses** and **pigs**, when they are subject to marked **confinement**. The behaviour involves repetitive, circular movements of the tongue inside or outside of the mouth. Among adult dairy cows, tongue rolling occurs more frequently when the animals are tethered, rather than being loose housed, but the method of feeding is more important than the way cows are housed.

Tongue rolling in cows occurs most when they are fed a small amount of food. Adding straw to cows' diets reduces tongue rolling, even though the energy content of the diet is not significantly altered. By transferring ruminal content between cows (using ruminal fistulas), researchers have shown that tongue rolling in cows occurs most when they can eat only a small amount of feed, during a short period of time and have low ruminal content after the meal. Tongue rolling occurs also in young calves and fattening bulls. Providing calves with more water or increased fibre, or with objects that they can suck or chew, such as a piece of rubber tyre or a chain, reduces tongue rolling. There has been little scientific investigation of tongue rolling in horses, but it may be that its occurrence relates to similar factors.

There are large differences between animals in the extent to which they show tongue rolling; most individuals show little or no tongue rolling, while a small number perform the behaviour frequently. Little is known about the individual differences that underlie this. There are few clear relations between the occurrence of tongue rolling and signs of **stress** or production levels in farm animals, and the relation between tongue rolling and animal **welfare** is uncertain. (JRu)

See also: Bar biting; Feedlot

Further reading

Lindström, T. and Redbo, I. (2000) Effect of feeding duration and rumen fill on behaviour in dairy cows. *Applied Animal Behaviour Science* 70, 84–97.

Redbo, I. (1998) Relations between oral stereotypies, open-field behaviour, and pituitary–adrenal system in growing dairy cattle. *Physiology and Behaviour* 64, 273–278.

Tonic immobility

Tonic immobility is an unlearned response to a brief period of physical **restraint** and is characterized by a reduced-reactivity, 'catatonic-like' state (Jones, 1986). Tonic immobility can also refer to the seemingly immobile state, or 'freezing', response to the presence of a predator or other dangerous stimuli. It was initially characterized in 1636 by Daniel Schwenter, and has since been referred to by several different names including **death** feigning, animal hypnosis, catalepsy and fright paralysis. Tonic immobility response has been characterized in many invertebrate and vertebrate species. It is most commonly used and best defined in bird species, including domestic fowl. Although there is no loss of **consciousness**, animals in a state of tonic immobility appear to be in a hypnotic state, remaining virtually immobile with suppressed respiration and body temperature. Altered **heart rates**, decreased movement and **vocalization**, and muscle rigidity are also common. The animal is not entirely immobile, however, because tonic immobility is also marked by muscle tremors in the extremities, spontaneous eye movements and jerking or bobbing movements of the head and neck.

There are two well-accepted definitions of tonic immobility: the first includes two stages of tonic immobility (Jones and Faure, 1981) and the second defines this phenomenon as having three levels (Rovee and Luciano, 1973). The two-stage definition includes the initial stage that lasts from the induction of tonic immobility until the first vigilant movement of the head. This first stage is marked by severe suppression of alert activity; however, it may include muscle tremors or jerking movements of the extremities and spontaneous movement of the eyes and head. The second stage continues from the first attentive head movement until the animal rights itself. This stage is marked by vigilant head movements and may consist of increasing frequency of leg movements and vocalizations.

The three-level definition describes tonic immobility by the level or degree of immobility or awareness. An animal in level one of tonic immobility has open eyes and generally makes loud vocalizations. This level is reached immediately at the onset of immobility and again just before its cessation. At level two, vocalizations are reduced and the eyelids flutter open and shut. The third level is marked by cessation of all vocalizations and movements, with the exception of jerking movements of the body and extremities and characteristic head bobbing. An animal in level three of tonic immobility also maintains closed eyelids.

In a clinical setting, tonic immobility is generally induced with brief physical restraint, often lasting only between 10 and 30 s, being applied to the animal in order to induce tonic immobility. The protocol for physical restraint varies within and across species. In avian research, the bird is typically laid upside down in a U- or V-shaped cradle and restrained by light pressure on the sternum for a brief period (approximately 15 s). Piglets are also often laid upside down in a U- or V-shaped cradle while the experimenter places a cloth bag filled with sand on the piglet's chin, to stretch its head back, and the piglet's legs are stretched gently back to induce the immobility response. However, animals may also be laid on a flat surface with or without cushioning, or in an induction box, and may be laid laterally, ventrally or dorsally. After this initial restraint, the animal voluntarily remains in the characteristic catatonic-like state for a varied period of time afterward.

Measures taken generally include: (i) number of inductions required to achieve the immobile state; (ii) time to first involuntary head movement; (iii) number of involuntary head movements; and (iv) duration time of tonic immobility or the latency time before the animal rights itself. Tonic immobility responsiveness, including duration of immobility, may vary with the induction protocol used, including the orientation of the animal during induction. As the majority of tonic immobility experiments are performed by laying the animal dorsally, it had been suggested that the disorientation associated with tonic immobility may be due to inner ear involvement. However, vestibular involvement seems unlikely as tonic immobility may also be induced laterally and ventrally. Additionally, research performed on frogs following surgical removal of the ears showed that altered frogs performed similarly to intact frogs on tonic immobility tests.

Tonic immobility responsiveness, including ease or ability to induce and duration of immobility, is used to assess an animal's fearfulness, as tonic immobility has been shown to be positively correlated with an animal's fearfulness. Animals that showed prolonged tonic immobility durations or lesser inductions required to induce tonic immobility also scored higher in **fear** response to such fearfulness tests as **novel object tests** and environment tests. Tonic immobility has been suggested as being a superior measure of fearfulness that overcomes some of the pitfalls of the more traditional fear tests. The uniformity that tonic immobility affords, both between investigators and between animals, provides reduced individual variability. Specific tonic immobility protocols, like that published by Jones (1986), allow for the test to be performed similarly by multiple experimenters. Compared with many fearfulness tests, **memory** and past experience have less of an effect on the test results. However, the degree of fear that the protocol for induction and the tonic immobility itself elicits is currently unknown. Alternatively, fear alone does not elicit an immobile response.

Physiological studies have been performed to assess the relationship between tonic immobility responsiveness and the fear elicited by the test. Fear-eliciting stimuli increase respiration and heart rate. It has been shown that respiration rate increased with induction of tonic immobility, suggesting a possible increase in fear responsiveness. Following induction of tonic immobility, respiration rate decreased steadily for the duration of the immobility response. Heart rate, however, has been studied in more depth and lends less convincing evidence of the fear-inducing quality of tonic immobility. Numerous studies of the effect of tonic immobility on heart rate have yielded inconclusive results. However, the most recent studies suggest that the test provokes an increase in heart rate, reflective of increased plasma concentrations of **catecholamine**s in response to **stress** or fear, immediately following the initial restraint of the tonic immobility protocol. These studies also found a fall in heart rate, significantly below the pre-induction rate, following the initiation of immobility. It has been suggested that the inconclusive results of previous studies may be due, in part, to the technological inability at the time to obtain heart rate measures without adding additional fearful stimuli to the immobility protocol.

The process of inducing tonic immobility has been shown to elicit a stress response in several species. Induction of tonic immobility in Japanese quail has also been shown to be marked by an increase in plasma **corticosteroid** levels compared with pre-induction levels, suggestive of increased physiological stress. Similarly, strains of Japanese quail bred for high and low stress, as measured by high and low corticosterone response to a brief mechanical restraint, exhibited greater and reduced fearfulness, respectively, as measured by the tonic immobility test. Domestic fowl showed increased tonic immobility duration following continuous infusion of corticosterone by mini-pump (15 mcg/h) for 4 and 11 days. Hens from a strain of white Leghorn bred for long duration of tonic immobility were also shown to have higher heterophil/lymphocyte ratios than birds bred for a short duration of tonic immobility. Increased heterophil/lymphocyte ratios suggest an increase in stress and reduction in immunocompetence. Studies have shown a positive correlation between tonic immobility responsiveness and the degree of fluctuating asymmetry, a measure of chronic stress. Similarly, strains of hens with reduced stress-coping ability have also been shown to have increased latency to the first head bob and overall tonic immobility duration, as well as reduced inductions required to elicit immobility.

Tonic immobility is apparently similar to a hypnotic or sleeping state. Reduction in both respiration and heart rate suggests a possible physiological similarity between the two states. Further evidence, however, has shown a marked difference between tonic immobility and **sleep**. Electroencephalographic (EEG) activity was measured from the hippocampus of **rabbit**s during periods of movement, non-movement, tonic immobility and sleep, results showing that rhythmic, slow activity was identical during tonic immobility to that during periods of non-movement and was increased only during periods of movement and active sleep.

Tonic immobility responsiveness has also been linked to other behavioural strategies. It has been suggested that reduced tonic immobility responsiveness may indicate a more active or bolder behavioural strategy. Piglets that did not become immobile after experimental attempts to induce tonic immobility also exhibited more active behavioural strategies, such as shorter emergence times from a test box. Contrarily, white Leghorns exhibiting high aggressiveness also maintained longer tonic immobility durations compared with less aggressive birds. Strains of Japanese quail bred for long tonic immobility duration were found to struggle more than birds bred for short tonic immobility duration (**see: Coping**).

The evolutionary origin of the tonic immobility responsiveness is not completely known. However, it appears to be an adaptation in surviving predatory attack. When restrained by a predator tonic immobility, or feigning dead, increases the chance of survival. It has been shown that tonic immobility duration increases in the presence of a predator simulation. It has also been shown that birds and mammals in the wild increase their chance of surviving an attack when they succumb to a state of tonic immobility following the initial attack of a predator. It is common that predators, such as **cat** and **dog** species, will temporarily release their hold on their prey once they become immobile and, in some instances, even walk away from their prey completely. Immobility or freezing in the sight of a predator might also lessen the chance of initiating an attack. Although freezing might reduce the chances of attracting the attention of a predator, it is also possible for an animal successfully to flee, thereby escaping attack, especially if the animal is a considerable distance from the predator. It has been shown that the distance to the predator is negatively correlated with the ability to induce tonic immobility and the duration once the animal is induced.

Tonic immobility responsiveness can be affected by many environmental factors, including time of day and age of individual. Extensive research on the diurnal periodicity of the immobility duration in domestic fowl showed that the shortest tonic immobility duration occurred immediately before and after the onset of the light cycle. The longest duration however, was most often observed at mid-day, 5–7 h after lights on. The evolutionary explanation for this **diurnal rhythm** is not entirely clear; however, it has been suggested that it may reflect the periodicity of predatory attacks. The evidence for age effects on tonic immobility response is less clear. Some studies have found that animals, including white Leghorn chicks, do not acquire the tonic immobility response until 1 week or so of age, while contrary evidence has shown that white Leghorn chicks can be induced less than 24 h post-hatch.

Researchers have suggested that animals may reduce their tonic immobility responsiveness with age. However, it has also been suggested that repeated inductions or taming by regular handling were confounded with age and may have had a greater impact on reducing tonic immobility responsiveness. Further studies have confirmed a **habituation** to repeated inductions. Consistent handling has also been shown to reduce tonic immobility duration, as well as to suppress fearfulness to humans and novel environments.

Social environment has also been shown to influence the immobility response. White Leghorns housed in isolation have been shown to have increased fearfulness and tonic immobility duration compared with **group**-housed hens. However, in group-housed hens, **dominance** position has been shown to affect tonic immobility responsiveness, with subordinate birds exhibiting shorter durations compared with their dominant counterparts. Birds maintained at high stocking densities (**see: Stocking density**) showed increased fearfulness, as evidenced by increased immobility durations compared with birds maintained at lower stocking densities. Birds housed in groups with 50% of birds marked for identification showed reduced inductions required to achieve tonic immobility compared with birds from groups where all birds were marked. Tonic immobility response is also affected by **housing** and physical environment. Domestic fowl have been shown to have longer duration of immobility in tonic immobility tests when housed in cages compared with those housed in more spacious aviaries. Similarly, birds housed in floor pens exhibited significantly shorter tonic immobility durations compared with birds housed in conventional cages.

The presence of familiar objects, **conspecifics** or people during testing has been shown to reduce the tonic immobility duration. Although the presence of humans can elicit a fear response in most animals, the degree of familiarity the animal has to a specific human is negatively correlated to the degree of fear that human elicits. Additionally, research animals in some circumstances have been found to imprint to an experimenter. The presence of a human to whom an animal has imprinted will elicit a considerably suppressed fear response (**see: Imprinting**). Use of the same experimenter for induction is ideal in order to reduce the confounding effects on tonic immobility response from familiarity with the experimenter. Different experimenters also exhibit differences in the roughness of handling, which can also alter the latency time before the animal rights itself. The presence of novel objects, predators, loud noises and increased light have been shown to increase latency time before the animal rights itself.

The tonic immobility response can also be used to immobilize an animal for other tests and measures without the use of pharmaceuticals or continuous manual restraint. Mechanical and pharmacological methods of restraint can be considered contrary to good welfare of the animal. They can also cause **injury** and **pain** and elicit chemical responses that might interfere with the experimental results. The use of the tonic immobility phenomenon to immobilize the animal may be a cost-effective and welfare-efficient animal handling technique. (RD)

References and further reading

Gallup, G.G. (1977) Tonic immobility: the role of fear and predation. *The Psychological Record* 27, 41–61.

Jones, R.B. (1986) The tonic immobility reaction of the domestic fowl: a review. *World's Poultry Science Journal* 1442, 82–96.

Jones, R.B. (1996) Fear and adaptability in poultry: insights, implications and imperatives. *World's Poultry Science Journal* 52, 131–174.

Jones, R.B. and Faure, J.M. (1981) Tonic immobility ('righting time') in laying hens housed in cages and pens. *Applied Animal Ethology* 7, 369–372.

Rovee, C.K. and Luciano, D.P. (1973) Rearing influences on tonic immobility in three-day-old chicks (*Gallus gallus*). *Journal of Comparative and Physiological Psychology* 83, 351–354.

Tool use

Tool use is first evidenced in the hominid fossil record about 2.6 million years ago, and was originally thought to be a defining feature accompanying the emergence of the genus *Homo*. *Homo habilis* (or 'handy man') possessed a larger **brain** than their predecessors, the australopithecines, and the correlation between the two traits led early physical anthropologists to postulate that an increased brain size was directly related to the development of complex behaviours such as tool use. At the time of these discoveries the humancentric view of 'man the toolmaker' was widely accepted; it was several years later in the early 1960s when Louis Leakey appointed a young Jane Goodall to conduct a study of **chimpanzees** in the wild. Goodall was one of the first people to witness tool use in non-human primates, at Gombe in Tanzania, when she saw one of her study animals dipping for ants using a twig, which had been modified by stripping it of its leaves.

Tool use can be loosely defined as the manipulation of an object to perform a specific task. It may often involve modification of existing materials in an apparently purposeful way to perform a function that cannot be accomplished without the tool for a beneficial effect, such as exploiting different foods. The significance of tools is, by extension, what they can tell us about the cognitive abilities of the toolmaker and user. For example, stone hammers may show little modification of the raw material but chimpanzees show a degree of foresight and planning when using them, carrying favoured stones hundreds of metres to the nut-cracking site. Knapping of stone tools by early hominids produced stone fragments that have proved invaluable to archaeologists trying to reconstruct early human material culture – recent interest in chimpanzee tool use behaviour has shown that their culture too leaves traces in the archaeological record (that might be mistaken for human artefacts).

Since the earliest ethological studies confirmed that chimpanzees use tools, their deployment has been witnessed in all three remaining species of **great ape**. **Gorilla**s were first seen to use water in spectacular splashing displays in 2001, but it was not until 2005 that a female gorilla was captured on film using a branch to test the depth of water in a murky pool that she later crossed, using the same branch as a walking stick. One of the first instances of tool use by **orang-utan**s was unexpected – after 9 years of fieldwork an adult male was observed to select a dead branch to scratch his rear in 1976. It was not until 1994 that Carel van Schaik recorded instances of orang-utans using thin sticks to poke into the centre of fruits covered in irritant hairs to extract the seeds and pulp, and the variety of tool use behaviours observed has increased since then. Bonobos have not been observed to use tools in the wild but, as with other species of great ape, they are prolific tool makers and users in both **captivity** and populations that have been reintroduced into the wild. Tool use is generally more prevalent in captivity rather than in the wild, even when corrected for observational bias. This is at least in part due to the increased opportunities for experimentation in captive environments, where basic biological needs are met easily and, additionally, suitable objects may be deliberately made available to the animals in question.

Tool use is not restricted to apes – capuchin **monkey**s in the Caatinga forests of north-eastern Brazil use stones to dig for tubers, to crack hard seeds and to pierce hollow branches to access food during the long dry season. The earliest anthropologists would have been even more incredulous to discover that the New Caledonian crow (a species restricted to a Pacific island archipelago) has shown expertise in the production of complex tools that are selected for different food-finding tasks, and shows a level of innovation that may exceed that of chimpanzees. A wide variety of other animal species can be regarded as tool users exhibiting varying degrees of competence in the selection, manufacture or fashioning of tools and degree of cognitive insight, including Galapagos finches, green herons, Egyptian vultures, sea otters and bottlenose dolphins.

Culture can be defined in a biological sense as inter-generational transmission of **group**-specific behaviour at least partly acquired from **social learning**. We know very little about cultural variation in tool use repertoires among species other than humans and chimpanzees. Chimpanzees in West Africa, east of the Nzo-Sassandra River in Côte d'Ivoire, habitually use stone or wooden hammers and anvils to crack open hard-shelled nuts to extract the nutritious seeds within. Chimpanzee populations to the south and east of the Sanaga river in Cameroon, however, have not been observed exhibiting this behaviour (despite all the necessary materials being present) in spite of decades of behavioural studies by researchers. Between these two regions, however, the behaviour is not consistently present or absent, and it seems that this particular behaviour may have evolved on several occasions, or else it has become extinct in some areas since its invention. It has been suggested that stone tool use in capuchins has developed to exploit different foods due to the harshness of the long dry seasons, so resorting to an ecological explanation of the invention of a culture. This explanation for the invention of a culture is not so obviously reconcilable in the chimpanzee example, however – there are few ecological differences between habitats where chimpanzees do and do not crack nuts in West-Central Africa.

That tool use is evident in a diverse array of taxonomic groups throughout the animal kingdom should not detract from its emergence as a vital factor in the evolution of humankind's behavioural repertoire. Humans have gone further than their closest living relatives in the manufacture of a complex tool assemblage – insights from the other proficient tool users are invaluable in understanding the development of our own tool-using culture. (BJM)

See also: Cognition; Foraging behaviour; Imitation; Intelligence – comparative

Further reading

Kacelnik, A., Chappell, J., Weir, A.A.S. and Kenward, B. (2004) Tool use and manufacture in birds. In: Bekoff, M. (ed.) *Encyclopedia of Animal Behaviour*, Vol. 3. Greenwood Publishing Group, Westport, Connecticut, pp. 1067–1069.

McGrew, W.C. (1992) *Chimpanzee Material Culture: Implications for Human Evolution*. Cambridge University Press, Cambridge, UK.

Mercader, J., Panger, M. and Boesch, C. (1998) Excavation of a chimpanzee stone tool site in the African rainforest. *Science* 296, 1452–1455.

Van Schaik, C.P., Ancrenaz, M., Borgen, G., Galdikas, B., Knott, C., Singleton, I. *et al.* (2003) Orangutan cultures and the evolution of material culture. *Science* 299, 102–105.

Whiten, A., Goodall, J., McGrew, W.C., Nishida, T., Reynolds, V., Sugiyama, Y. *et al.* (1999). Cultures in chimpanzees. *Nature* 399, 682–685.

Tooth

A tooth is a structure embedded in the jaws of animals that may be used for eating, **grooming** and defence. Many animals have teeth of different sizes and shapes, a condition known as heterodonty, allowing different teeth to be specialized for different tasks. These specialized teeth can be categorized into four different types, including incisors (I), canines (C), premolars (P) and molars (M), although not all animals have all four types of teeth. Incisors are used to shear or cut through food; canines are used to grip food and for defence; premolars are the transitional teeth between canines and molars and have the initial grinding function during transfer of food to the back of the mouth; molars are the principal grinding teeth that mechanically break down the food during chewing prior to swallowing.

As teeth have the primary function of securing nutrition and assisting digestion, it is not surprising that the number and structure of teeth vary greatly with species. Some animals, such as sharks, have multiple sets of teeth over the course of their lifetime whereas others, such as mammals, have only two sets – deciduous and permanent. In many animals, like **carnivores**, once fully erupted the permanent teeth stop growing and, indeed, gradually wear down over time. In other animals, especially herbivores like **rodents**, lagomorphs and equines, the permanent teeth continue to grow, with the rate of growth ideally matching the rate of wear due to diet. The dentition of any animal can be described using a dental formula, which gives details of the number of teeth in one half of the maxilla (upper jaw) over the number of teeth in one half of the mandible (lower jaw). For example a dental formula of 3.1.2.3/3.1.2.3 would mean that the animal in question has a total of six incisors, two canines, four premolars and six molars in the top jaw and the same in the bottom jaw.

In general, teeth are comprised of:

- a pulp cavity, containing blood vessels, nerves and soft connective tissue;
- dentine, a mineralized connective tissue between the pulpy cavity and the outer layers of the tooth;
- cementum, a bony substance made up of collagen and hydroxyapatite that covers the root of the tooth;
- enamel, a highly mineralized substance of hydroxyapatite and enamelins covering the outer exposed surface of the tooth.

In some species, such as **horses** and young **cattle**, the age of the animal can be estimated by either the number of permanent teeth that have erupted or their state of wear. Also in some species, extended canine teeth are used during fighting with **conspecifics** or during defence from predators. In **pigs**, these teeth may be removed soon after birth to prevent the piglets damaging each other or the sow's udder. (JNM-F)

Training of animals

Animal training describes the process involved in ensuring that learning comes about in a predictable way within an animal in response to human intervention. Therefore, training involves not only the application of learning theory (**see: Conditioning – types of**) but also interspecific **communication**, either of which may affect the **welfare** of the trainee. For example, in relation to the application of learning theory, the excessive, unnecessary or inappropriate use of **punishment** may compromise welfare, through both physical and non-physical means; poor training technique resulting in the non-contingent application of **aversives** may result in **learned helplessness** or chronic **anxiety**.

However, even training methods that focus on the delivery of positive **reinforcement** may give rise to problems. For example, the frequent use of continuous schedules of reinforcement may give rise to the expectation within the trainee of a high rate of reinforcement that, if not met, might lead to **frustration** and **distress**. Failure to appreciate how context-specific learning may initially be can lead to unrealistic expectation of the animal's behaviour and the inappropriate application of aversives as a consequence; for example, a **dog** may learn a response in the training hall, but not generalize to the wider environment. Understanding the principles of **stimulus** control in this regard is crucial. Thus a dog may respond to a command issued by one person but not by another for the same reason (and not due to stubbornness or a **preference** for one individual over another), i.e. because the command is recognized in one context due to its similarity to the one used in the training context, but not in the other due to its difference. When 'instructing' an animal, it is important to appreciate that it is really a request that is being made, which the animal may ignore deliberately, perhaps because of other, competing **motivations**, or inadvertently because it does not recognize the request due to poor communication. In either case the onus is on the trainer to remedy the situation through appropriate means, which will differ according to the cause.

Although many 'new age' or 'ethological' training methods for animals have been developed that are claimed to be more humane, these generally involve the application and communication of simple and traditional learning principles in a particular way, for example through the use of more ethologically relevant cues. Because animals differ in how they respond to others and to reinforcement, the most sound training package is one that uses a flexible approach and is based on establishing the application of declarative **knowledge** relating to animal training rather than one focused on **procedural knowledge**, as is often the case in the 'new age' training packages.

In order to train an animal efficiently, the author recognizes ten considerations for the trainer, all of which have potential welfare implications:

- Is the animal physically and psychologically fit for training?
- What is the end goal and how is this defined?
- How is the **attention** of the animal to be obtained in order to establish communication with it?
- What is going to be the cue for the behaviour (i.e. the stimulus that indicates that a reinforcement schedule is in operation)?
- What is being done to encourage the animal to make the decision the trainer desires?
- What is being done to reassure the animal that this decision is a good choice?
- For efficient training it is also worth considering which aspect of the behaviour is to be reinforced (e.g. its latency, duration, etc.).
- What is the most appropriate training set-up, e.g. differential reinforcement of a spontaneous or lured behaviour (**see: Shaping (operant)**)?
- What degree of generalization or discrimination is required in the end so that training can be arranged accordingly?
- It is important to determine how the training can be objectively assessed and monitored so that modifications can be made on a rational basis.

It has been claimed that punishment is associated with a higher risk of **problem behaviour** (Hiby *et al.*, 2004), but it is difficult to establish whether the use of punishment is causative or the consequence of misbehaviour. Any training method can be misapplied or abused and give rise to welfare problems, but certain methods of training attract particular criticism, for example the use of electronic training ('shock') **collars**. While opponents claim these devices are inevitably cruel – often focusing on their use as a form of punishment – proponents point out that they can be used as part of an integrated training system (whereby the current is used to interrupt ongoing behaviour rather than to punish it, to notify the animal that an alternative form of positive reinforcement is available – consideration three, above). Used in this way, the device would be expected to cause minimal **suffering** in the short term and its success, particularly in establishing a reliable recall, is to the long-term benefit of the animal, for example by allowing it greater freedom.

Clearly, these debates are complex and the ethical arguments articulated are not always consistent. Owners are generally more opposed to the use of electric current for the control of behaviour than to other forms of aversives – for example, a spray of citronella to control barking; however, if either method is effective, then it must be acting as a form of punishment and it is perhaps anthropomorphic (**see: Anthropomorphism**) to suggest that an electric current is necessarily more aversive than a strong odour to an animal with a long coat and highly developed sense of smell. Further research is undoubtedly required to answer such questions.

Other causes for concern, in relation to the welfare of animals undergoing training, incude the general lack of regulation of this field as a profession and the difficulty faced by the average animal owner in establishing the quality or value of a given trainer. (DSM)

Reference and further reading

Barry, J. (2008) *The Ethical Dog Trainer*. Dogwise, Wenatchee, Washington.

Hiby, E.F., Rooney, N.J. and Bradshaw, J.W.S. (2004) Dog training methods: their use, effectiveness and interaction with behaviour and welfare. *Animal Welfare* 13, 63–69.

Transgenics

Transgenics are animals that have a gene, or genes, from another animal, usually of a different species, incorporated into their **genome** in such a way that the transgene is incorporated into the germ line and so will be inherited. However, it is often applied to any animal that has had its genome modified in some way. This can be by either: (i) the addition of a specific gene, or genes; (ii) the addition of either a specific sequence of DNA (called a 'knock-in', say with a faulty gene to study its function) or of several copies of a gene, for example **growth hormone (GH)** for increased agricultural productivity; or (iii) the deletion of part of an animal's normal DNA ('knock-out').

These procedures are carried out on recently fertilized ova by micro-injection techniques. Such animals are used in research to better understand normal and **abnormal** development, as well as for the identification of specific genes and the roles that genes play in **disease**, such as causation, susceptibility and resistance. (DBM)

See also: Farmed animals; Xenograft

Transitional phase

The term transitional phase in relation to behavioural development refers to a **sensitive phase** of development between the **neonatal period**/phase and **socialization phase**, during which there is rapid development and adaptation of the external sensory organs to the prevailing environmental circumstances. (DSM)

Translocation

Translocation is the intentional capture and transfer of **wild animals** or populations from one part of their historic range to another (**see also: Reintroduction**). Translocations are used to establish, re-establish or augment a wild population to increase the viability of a species or to supplement **game animal** populations. They are also used as a control measure to remove nuisance animals from areas where they are causing damage, and thus alleviate human–animal conflict. Most translocation research has focused on mammals and birds, but translocations have also been conducted with fish, **amphibians** and invertebrates. Though translocation has become an increasingly popular **conservation** tool, most translocations fail to produce sustainable populations and involve risks such as **disease** transmission. This has led to an increased interest in determining the factors that influence the success of translocations and to design methodology to minimize transmission of disease.

Post-translocation **mortality** is highest in the first days to weeks following release, as animals make settlement decisions and modify the release-site habitat to accommodate their needs. Problems associated with this initial establishment phase include: (i) post-release dispersal (i.e. long-distance movement away from the release site); (ii) **predation**; (iii) **stress** response to the novel environment; (iv) difficulty finding food; and (v) **competition** for **resources** (e.g. **territory**) with **conspecifics** – either residents or fellow releasees.

Among the proposed explanations for the high mortality during the establishment phase, post-release dispersal and predation are thought to be important factors. Immediate rejection of a release site indicated by post-release dispersal has been documented in many species. In some cases, translocated animals travel all the way back to their natal habitat (i.e. 'homing'). Long-distance movement and, for many species, the required habitat modifications (e.g. digging burrows) leave translocated animals particularly vulnerable to predation immediately after release.

Post-release **dispersal and habitat selection** may initially be high because animals: are site faithful; are not familiar with the physical characteristics of the release site or with the individuals with whom they were released; or are at a disadvantage when competing for resources with residents. Biologists studying translocations are beginning to understand these problems and modify translocation methodology to address these issues. For example, to dampen post-release dispersal and decrease stress, biologists have used 'soft' release techniques, where the animals are provided with some form of support during the release (e.g. a period of time in an enclosure on the release site and/or supplemental food being provided after release). Compared with 'hard' release (e.g. direct release without an acclimation period), soft releases are generally thought to enhance the likelihood of translocation success, via increasing site fidelity and post-release survival. However, results of the few studies that have directly addressed this question are mixed, possibly because other factors that influence translocation success (e.g. predators) outweighed the effects of this technique.

Whether soft release methods are beneficial may also depend on the social system of the species. For example, in wild **rabbits**, soft release enhances survival for females but not for males. European wild rabbits live in highly stable **social groups** composed of a few males that defend a group of related females from unfamiliar males. Because translocated rabbits were trapped and held with unfamiliar individuals in their acclimation cage, soft release may have increased rather than decreased stress of translocation for males, thus rendering this procedure detrimental.

Newly translocated animals may also leave the release site because there are no resident conspecifics, which may indicate that the habitat is unsuitable. For example, black **rhinoceros** move less during the first 5 days after release if cues that signify conspecific presence (i.e. dung) are broadcast at the release site.

Alternatively or in addition, founder group composition may influence translocation success. Group size, age and sex class ratios, and familiarity between founder group members have all been shown to affect survival post-release. While there is a positive relationship between the size of the founder group and survival across taxa, suggested age and sex class ratios may vary by **mating system**. For example, black **bear** are polygynous. Males have large **home ranges** that overlap with as many as nine females. Thus, suggested translocation schedules include a release of males prior to release of females to allow males time to establish breeding territories into which females can settle.

Composing founder groups of intact social groups has been shown to have a significant influence on translocation success. For black-tailed prairie dogs, founder groups composed of intact families were more successful in terms of post-release survival, reproductive success (see Fig. T.2) and population viability. Maintaining family unity appeared to reduce the effects of successful predation on prairie dogs and dampen post-release dispersal.

Thus, any species dependent upon social interactions for survival and **reproduction** may benefit substantially from the maintenance of social groups during translocations. Species for which social interactions enhance individual **fitness** would be especially likely to benefit from the maintenance of social groups during translocation. Those species include species with kin-selected behaviours (e.g. Belding ground squirrels' alarm calls), those that rely on **reciprocity** (e.g. **allogrooming**; helping at a communal nest) and those that receive direct benefits from relationships with group members (e.g. coalition formation, social foraging, sexually selected cooperation, **social learning**, enhancement of immune function and reduced disease transmission and stress).

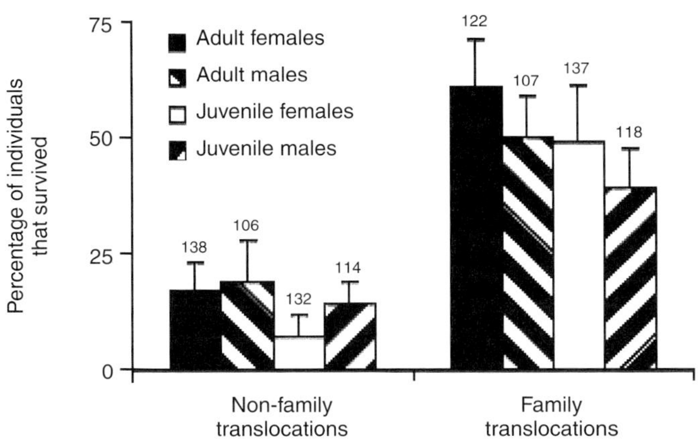

Fig. T.2. Differential survival rates of intact versus incomplete social groups.

Predation is a significant impediment to translocation success. The effects of predation are especially prevalent during the establishment phase. Most attempts to minimize predation on newly translocated animals have involved monitoring of release sites post-release and predator removal (**see also: Predator control**). In many cases, novel predators are one of the primary selective forces that caused the initial decline of the target species. In these situations, it is crucial that the novel predators are removed from the release site prior to translocation and/or releasees are trained to avoid the novel predators. (DMS)

Further reading

Fischer, J. and Lindenmayer, D.B. (2000) An assessment of the published results of animal relocations. *Biological Conservation* 96, 1–11.

Griffith, B., Scott, J.M., Carpenter, J.W. and Reed, C. (1989) Translocation as a species conservation tool: status and strategy. *Science* 245, 477–480.

Wolf, C.M., Garland, T. Jr and Griffith, B.J. (1998) Predictors of avian and mammalian translocation success: reanalysis and phylogenetically independent contrasts. *Biological Conservation* 86, 243–255.

Transmissible disease

Diseases caused by pathogens that are passed on from one animal to another are transmissible. The transmitting animal may or may not be of the same species. Transmission can occur through contact with excretions (faeces, **urine**) and secretions (venereal, saliva), by inhalation (droplets in the breath) and by inanimate objects that have been contaminated (fomites). A contagious disease is, strictly speaking, one that is spread by direct contact between animals, while an infectious disease is one that may or may not be easily transmissible. When animals are immunosuppressed infection may occur more easily. (DBM)

Transport

Animals are transported for a number of reasons. It is unfortunately the case that the reason for transport, the type of animal and its inherent value will potentially affect the amount of care accorded to the process and the likelihood that the animal's **welfare** might be compromised. For example, **pet** animals, **breeding** stock and racehorses are likely to be handled carefully and their welfare will generally be good. In contrast, the trade in animals transported to **slaughter** is subject to commercial pressures that may lead to them being handled less carefully. Very young animals, such as small calves, may also be particularly vulnerable by virtue of their physiological immaturity.

Modes of transport

The majority of animals are transported by road. Air travel is used for particularly valuable animals such as pets, racehorses and breeding stock, but also increasingly for day-old chicks. Rail travel has decreased in importance from its early historical pre-eminence, with the improvement of roads. Animals such as large **zoo** species, racehorses and **horses** used for other recreational purposes travel in relatively sophisticated vehicles, often in individual compartments. **Cats** and **dogs** are often transported in individual crates or containers. Animals destined for slaughter usually travel in simple vehicles, often multi-decked, in which the decks are subdivided into pens holding a number of individuals. The number could range from as few as four or five in the case of adult **cattle** to perhaps 30 in the case of **sheep**. Transport by sea is used for a relatively small proportion of journeys, mostly for exported animals. Sheep are exported from Australia to the Middle East in ships that hold up to 100,000 animals. The number of animals transported is therefore significant, although the number of journeys is small.

Poultry such as broilers, **turkeys** and hens are transported commercially in loose plastic crates or in the drawers of 'modules' (see Fig. T.3). These are metal frames into which the plastic drawers slide. Each crate or module drawer might hold from 15 to 25 birds, depending on live weight. An entire module might hold 12 drawers and therefore 300 birds. The modules can be handled and lifted on to a flat-bed lorry using a forklift truck. A loaded lorry might carry 5000–6000 birds.

Because ventilation is passive, and is impeded by the close stacking of the crates or modules, close environmental control is very difficult. This may lead to very high temperatures developing, particularly in the centre of the load, with the danger of the birds suffering hyperthermia. A 5°C increase over ambient temperature is common, and temperatures of ≥30°C in the centre of the load have been recorded. By contrast, the birds in the outermost containers may suffer hypothermia if the weather is cold, especially if they become wet so that their **feathers** lose their insulation properties. To reduce this possibility many poultry transporters are fitted with adjustable side curtains that protect the otherwise exposed birds. Unfortunately their use may make the situation of the innermost birds on the lorry more difficult by further reducing air circulation.

So-called day-old chicks for stocking of commercial growing units are transported from the hatcheries in specially designed cardboard boxes. The chicks are in fact usually 1–3 days old when transported and are reliant on the contents of their yolk sac for energy and water from hatching until they reach the rearing units. Chicks of modern broiler **genotypes** have higher metabolic rates than traditional genotypes, and use up these reserves more quickly. It is therefore essential to avoid delays in placement, and to ensure that metabolic and thermoregulatory demands on the chick are controlled. Chicks must be kept warm, so temperature and humidity control of their microenvironment is very important.

Fig. T.3. A module containing 12 drawers used in the commercial transport of poultry.

Fish are sometimes transported live, particularly ornamental fish or for restocking. The main factor in determining **fish welfare** seems to be water quality. To this end it is important to ensure sufficient volumes of water to prevent the build-up of dangerous waste products such as ammonia, and to have adequate oxygenation.

Major factors affecting welfare

Welfare during transport is a function of the length of the journey and the conditions prevailing. The time taken is often more important than the distance travelled (**see: Journey times**). However, short journeys under poor conditions may be worse than long journeys under good conditions. Legislation and **codes of practice** relating to the protection of the welfare of animals during commercial transport therefore often prescribe maximum journey times and the various factors that influence these conditions. These include rest, feed and watering intervals, minimum **space** allowances on the vehicle, how animals are penned, ventilation requirements as they affect the microclimate around the animals, and other aspects of vehicle design. For example, current European Union Directives specify that, using appropriate vehicles, adult sheep and cattle may be transported for a total of 29 h. After the first 14 h the vehicle must stop for 1 h and the animals must be offered liquid and, if necessary, be fed. After this they may be transported for a further 14 h. Journeys may be prolonged for a further 2 h (over the 29 h) in exceptional circumstances and if this is to the benefit of the animals. After the whole journey the animals must be rested for 24 h before any further transport. Legislation also often prescribes the level of experience or training required by drivers and the records of each journey that must be kept. Records enable the monitoring of compliance with the law.

Individual journeys for animals for slaughter are often part of a wider marketing system where the animals also pass through one or more live auction markets (**see: Market**). With these relatively complex systems it is often difficult to ensure that the animal's welfare through the whole process is being adequately protected, as each stage in the chain is treated separately rather than being considered as part of the wider whole. Animals may therefore go for very long periods without adequate rest or access to food and water. Additionally, animals that pass through live auction markets suffer a greater amount of bruising. This is probably both because they are necessarily handled more, so there is a greater chance of falls, slips and contact with sharp projections, and because handlers in some livestock markets may be less careful than they should be.

An important part of the transport process is the preparation and loading of the animals. Preparation includes making up groups of animals and, if appropriate, ensuring a period of food deprivation before transport. **Pigs** do not travel well on full stomachs and, in animals destined for slaughter, a balance needs to be struck between prevention of unnecessary **hunger** and the requirements for good hygiene at slaughter. Full guts make evisceration more difficult, and the consequences of accidental cutting or tearing of the gut wall more serious.

Loading and unloading of the larger animals can be a particularly stressful procedure. The use of level loading docks and hydraulic tail lifts, rather than ramps, will reduce **stress** and the chances of physical **injury**. For most animals the provision of appropriate bedding on the vehicle is important to cushion them from the effects of vibration and to soak up **urine** and faeces. Because of this, legislation often prescribes its provision unless alternative arrangements are in place to deal with waste material. For example, pigs are sometimes transported on perforated **floors**.

Time spent in transport

Journey times can range very widely – from an hour or two in the case of breeding animals sent by air, or meat animals sent to slaughter at local abattoirs, to several days for animals exported by road or sea for slaughter or further fattening. Journeys involving sea crossings can be of particular concern from the point of view of animal welfare, partly because they are often long, and partly because conditions at sea tend to be inherently poorer.

Stresses of transport and animals' responses to them

Even when carried out with care, the transport of animals is likely to be stressful. The stresses involve removal from their home environment and close **confinement** in unfamiliar surroundings and exposure to vibration and noise, unfamiliar animals and extremes of temperature. Animals often suffer the breakdown of social groupings, which may lead to stress or even fighting, and deprivation of food and water (**see: Social stress**). Falls may result in trauma such as bruising. The microclimate in the interior of vehicles may be poor, with the build-up of noxious gases such as ammonia. The commercial transport of poultry is of particular concern in this regard. Too high stocking densities can compromise the ability of animals to thermoregulate, and especially to lose heat, at high ambient temperatures (**see: Stocking density; Thermoregulation**).

Animals react to the stresses of transport in various ways. Psychological stress leads to increases in **heart rate** and the release into the blood of hormones such as **catecholamines** (adrenaline/**epinephine**) and **corticosteroids** (cortisol, corticosterone), and physical stress to increased circulating levels of enzymes such as creatine kinase (CK). Pigs can become travel-sick and show elevated concentrations of the hormone vasopressin. Dehydration is reflected in increased plasma osmolality, and prolonged fasting in increases in non-esterified fatty acids (NEFA) and ketone bodies. Prolonged transport often leads to significant loss of live weight, partly because of the enforced period of fasting but also as a direct consequence of the stresses associated with the journey. Animals that usually remain standing during even prolonged transport, such as adult cattle, probably suffer from greater fatigue than animals that will lie down if given suitable conditions, such as pigs.

The transport of many **wild animals**, for example to import them into zoos and game parks or relocate them to new areas or reserves, is particularly difficult because of the severe reactions they often show to the process. For these animals it may be essential to tranquillize them immediately after capture and before loading on to the vehicle, to prevent them dying. Even then, some will show 'capture myopathy'. Tranquillizers are unacceptable for meat animals because of the potential danger to human health of drug residues.

It is important that transported animals are fit to travel. Unfit animals include those that are heavily pregnant or have recently given birth, newborn mammals and animals that are lame, injured or severely ill.

Temperature, humidity and ventilation

Pigs and poultry are particularly sensitive to high temperature and humidity during transport. This reflects their relatively poor ability to thermoregulate under the confined conditions of transport. Normally, pigs rely on evaporative heat loss facilitated by wallowing. This is obviously impossible within the confines of a normal transport vehicle. Poultry lose heat by spreading their wings, again impossible in the confines of a crate, or by panting so that water is evaporated from the respiratory tract. The value of this is obviously reduced or nullified at the high relative humidity that tends to develop under these conditions. Ventilation tends to be compromised at higher stocking densities and therefore, when high ambient temperatures are expected, many hauliers will increase the space allowance per animal.

It is, however, important to be aware of the dangers of high temperatures to stock and in summer to travel whenever possible during the cooler parts of the day and to avoid stopping unnecessarily, so that ventilation is maintained. In animals exported over long distances the range of environmental temperature experienced may be large, with both very low and high temperature being potentially stressful. In northern latitudes where very low temperatures can occur, it may be necessary to have insulated vehicles. Low temperatures are also a danger to poikilothermic animals such as **reptiles**. In this case heated containers are essential.

Space allowances

For the commercial transport of slaughter animals hauliers will tend to use the maximum stocking densities possible, as this will reduce the unit cost of transport. High stocking densities make it more difficult for individuals to maintain their balance during changes in velocity of the vehicle. The idea that at high stocking densities standing animals support one another seems to be unfounded, based on observations of the number of falls occurring given different amounts of space. Generally, animals should be given at least enough space to lie down if they wish. Under these conditions, provided that bedding is adequate to ensure comfort, pigs and sheep will often lie down during longer journeys. Adult cattle may not do so if transported as **groups**, and this may make them more prone to fatigue after long journeys.

Provision of food and water

During commercial transport and the associated **handling**, many animals may become hungry and thirsty because of the impracticality of supplying food and water during the journey. This highlights the need to ensure that animals are adequately fed, watered and rested before loading. However, it is generally considered unwise to feed pigs right up to this time, and a minimum period of 4 h from the time of last access to feed and loading has been recommended. This will reduce the chances of the animals suffering travel sickness. The mortality rate during transport is also reduced. The reason for this is unclear, but may be because a full stomach can impair venous return, leading to circulatory insufficiency. Pigs appear to become quickly dehydrated during transport. To alleviate this, European Union legislation requires that pigs have continuous *ad libitum* access to drinking water in journeys longer than 8 h. Experience, however, shows that the amount of water actually consumed is relatively small compared with the animals' normal intake, but for reasons that remain obscure.

Ruminants may be more resilient than non-ruminants to the effects of food and water deprivation, because the rumen acts as a store of nutrients and water. However, long periods without food undoubtedly disrupt ruminal function and this may itself be distressing to the animal. Horses are particularly susceptible to dehydration and should therefore be offered frequent access to water.

Effects of transport of animals for slaughter on carcass and meat quality

Transported animals can die in transit, especially under adverse conditions. In Europe and North America **mortality** rates for slaughter pigs average about 0.1%, but can be much higher in very hot weather or in genotypes that are stress susceptible. Mortality rates for poultry are at least double this. Transport often results in bruising, which, as well as being a welfare issue, reduces carcass value. The fatigue and stress associated with long journeys can deplete muscle glycogen stores. If severe, this can lead to the production of so-called dark firm dry **(DFD) meat**, characterized by a reduced extent of post-mortem acidification and a high ultimate pH value. As well as its appearance being unattractive to consumers, and having poor eating quality characteristics, because of the high pH, DFD meat spoils rapidly and thus has a relatively short shelf life.

Effects of transport on disease control

Disease can compromise animal welfare as it is likely that diseased animals will suffer **pain** or **distress**. Transport has implications for disease control because it can lead to the wide geographical spread of disease between animal populations. Since exposure to stress may reduce the function of an animal's **immune system**, transport can also affect the susceptibility of individuals to infection. This effect is exemplified by the disease known as 'shipping fever', usually caused by the bacterium *Pasteurella*, which is a problem in young calves transported long distances to **feedlots**, particularly in North America.

Recovery from transport

Animals may need considerable time to recover after long transport. This will be especially so if they are dehydrated and, in the case of ruminants, have been deprived of food for a long time such that ruminal function is disrupted. Recovery times are poorly defined but are probably at least 1–2 days, even under ideal conditions.

Improving the welfare of animals during transport

In the future it is likely that there will be pressure for the commercial transport of animals to come under greater control, with more stringent monitoring of compliance with legislation. There are also likely to be greater restrictions on the maximum length of journeys before animals must be rested, and increases in the length of these rest periods. The design of livestock transport vehicles will probably improve, with greater use of air suspension systems to reduce vibration and noise, monitoring of the microclimate within the pens and automatic control of ventilation to maintain optimum temperature and humidity. To reduce the dangers of overheating, mechanical ventilation systems are likely to replace current passive systems. Loading and unloading systems are likely to be improved, with the replacement of ramps by hydraulic lift systems. Much greater attention to the adequate training and accreditation of vehicle drivers will also ensure that the people primarily responsible for the welfare of the animals during transport have adequate knowledge to fulfil this role. (PDW)

See also: Lairage; Loading animals for transport; Restraint; Slaughterhouse

Further reading

Grandin, T. (ed.) (2007) *Livestock Handling and Transport*, 3rd edn. CAB International, Wallingford, UK.

Trophic cascade

A trophic cascade describes the change within an ecosystem that arises when one level of a food web increases in success and so reduces the density of the next level down, which in turn affects the third tier. Thus a shift upwards in the success of a large top predator may result in a reduction in the level of their prey in the ecosystem, which might be a **mesopredator**. This may, in turn, result in an increase in the numbers of the normal prey of the mesopredator, which might be herbivores, and increase competition within them for **grazing**. (DSM)

Further reading

Polis, G.A., Sears, A.L.W., Huxel, G.R., Strong, D.R. and Maron, J. (2000) When is a trophic cascade a trophic cascade? *Trends in Ecology and Evolution* 15, 473–475.

Schmitz, O.J., Hanback, P.A. and Beckerman, A.P. (2000) Trophic cascades in terrestrial systems: a review of the effects of carnivore removals on plants. *American Naturalist* 155, 141–153.

Turkey

Turkey is the common name for a large gallinaceous bird with a fan-shaped tail and a bare, wattled head, which has become widely domesticated for food. The name is derived from its distinctive 'turk-turk' call. Turkeys show sexual dimorphism, with the male being approximately 1.5 times larger than the female. Other differences between the sexes can easily be identified during sexual displaying: the male's head region will change colour from pink to a mix of blue, red and purple, while the female's head region remains pink.

Natural habitat, feeding and maintenance behaviour

Wild turkeys (*Meleagris gallopavo*) range over North America from southern Mexico to the north-eastern USA. Their habitat is diverse, ranging from low, moist and hot coastal plains to swamps and mountains. *M. gallopavo* includes five subspecies of wild turkeys and all domestic varieties. Recognized subspecies are the Mexican turkey (*M. g. gallopavo*) of central Mexico, Merriam's turkey (*M. g. merriami*) in the mountainous areas of south-western USA and northern Mexico, the Rio Grande turkey (*M. g. intermedia*) of Texas and north-eastern Mexico, the Florida turkey (*M. g. osceola*) and the Eastern turkey (*M. g. silvestris*), ranging from Texas and Nebraska to the Atlantic coast and the north-eastern USA.

Wild turkeys are omnivorous, with their diet consisting of a variety of berries, seeds, leaves, grasses, insects, crustaceans and small **reptiles**. Turkeys do not show any specialized adaptations of feeding – they exhibit the basic avian feeding pattern. It is not necessary for the bird to raise its head to swallow food, which suggests that swallowing is an active process. Drinking is accomplished differently from feeding, as the head is raised, the beak is extended upward and the beak is closed several times in rapid succession. This suggests that there is a degree of active swallowing in addition to the flow by gravity to the oesophagus.

Scratching behaviour is an important component of foraging for food, and gallinaceous birds scratch with one foot while standing on the other. Wild turkeys extend the foot forward and scratch backward and outward in a sweeping motion to form an inverted 'V' pattern on the ground. In domestic turkeys scratching behaviour is rarely seen on the range or in the litter provided in houses. The reduction in scratching in domestic turkeys may reflect reduced **motivation** associated with genetic change during **domestication**. Alternatively, turkeys may still be motivated to scratch but are prevented from doing so by physical constraints or inappropriate environmental provision (Hale *et al.*, 1969).

Newly hatched poults (young birds) will peck indiscriminately at objects that contrast with the background. This response leads to the poults pecking at feed and images in the water, and so initiates feeding behaviour. **Social facilitation** of this behaviour is provided by other poults. If novel objects are presented to the birds, curiosity pecking is initiated, and there appears to be little evidence of **neophobia** as compared with **chickens**. The digestive tract of the turkey is similar to that of other poultry species. In turkeys, along with many species of galliform, **preening** and **dust bathing** are common compared with water bathing. Dust bathing is influenced by litter condition; it is facilitated when dry, friable litter is available. Preening is observed throughout the day, but especially just after lights on and just before lights off. Turkeys can develop problems of **feather pecking** and **cannibalism** in captivity.

Social behaviour

Many interactions occur between individuals and **groups** of turkeys. **Social behaviour** in turkeys includes aggressive, submissive and sexually related activities, parent–offspring interactions, synchronized behaviour and competitive activity.

The social stratification of a population of turkeys is highly developed in the wild turkey (**see: Hierarchy**). The status of each individual is determined by the end of its first year, and usually remains fixed for life. Males are usually forced into two contests around 6 or 7 months of age when they leave the brood flock: one establishes his position within his group of siblings and the other determines the status of his sibling group with respect to other groups. Fighting consists of pecking at the head and neck. The strongest fighter in the group becomes the dominant bird, and the order of rank is established among siblings and seldom changes as long as the dominant bird lives.

Females in the group also establish social dominance through contests. When males leave the brood flock, females fight for individual rank among themselves but, unlike the males, there is not fighting among flocks (Watts and Stokes, 1971).

Wild turkeys are gregarious except at **breeding** time, and are polygamous. They are non-migratory, even though they are good fliers. Their breeding behaviour is based on the **lek**. The female builds the ground nest and will lay 8 to 15 eggs per clutch, which she will brood. At hatching the poults weigh 2 ounces (55 g) and have yellow down with usually brown markings. Within 10–12 days the first flight **feathers** begin to show and, by the end of the first month, all the yellow down has been replaced with plumage.

Domestication

Domestication first occurred in the seed-planting **culture** of the Aztec Indians of Mexico. Domesticated turkeys were then brought from Mexico to Europe through Spain during the 16th century, and several European varieties have since been developed.

Present farming practices

The turkey cycle

Fertility under natural mating conditions in commercial flocks does not have a high success rate, possibly because of the natural lek-based mating system. Genetic selection for broad-breasted strains now means that commercial males may be up to three times heavier than the females to which they are mated, and **artificial insemination** is therefore usually necessary (Hocking, 1993).

The average female breeding turkey is kept for over 1 year and does not enter lay until 28–32 weeks. It is necessary to rear breeding birds under high light intensities to induce them to lay. Breeding turkeys typically produce between 45 and 60 live young per year. The incubation period for a turkey egg is approximately 28 days. Once the poults are hatched they are moved to a brooder unit, when birds are placed within surrounds where feeders and drinkers are available. The purpose of the surround is to condition the poults to locate drinkers and feeders, and to ensure the birds are maintained at the correct temperature. Day-old poults are initially raised at 35°C, and this temperature is usually reduced by 1°C every 3 days until 21°C is reached. At 6 weeks of age the surrounds are removed and the poults are either given access to the whole shed or to a range, or are moved to another farm, called 'brood and move'. Around 18 weeks female turkeys reared for meat purposes are depopulated (slaughtered) and the whole carcass is typically used. Around 21–25 weeks male turkeys used for meat purposes are depopulated, and the meat from these birds is usually cut into portions. Birds of both sexes reared for meat are called growers.

Growing and breeding turkeys reared in the UK are usually given *ad libitum* feed and water, although breeding birds receive diets with lower protein levels and grow more slowly than grower birds. Growers typically have three or four changes in diet, whereas breeder birds will have more changes in diet. For the first 8 weeks turkeys are fed turkey starter crumb feed, which is high in protein, and are then fed on turkey grower feed from around 8–12 weeks, and from 12 weeks to depopulation the birds are reared on turkey finisher feed, which is high in fibre and fat. In some countries no animal substance is permitted in turkey feed, so the main source of protein is from soybean.

Housing of meat turkeys

Commercially reared turkeys are grown in a variety of **housing**, ranging from conventional (enclosed, controlled environment) sheds to **free-range** conditions. Another popular method of rearing turkeys is in a pole barn system. This allows access to natural daylight, but there is little ventilation or temperature control in these systems. Free-range systems provide the birds with access to an outside area and subject them to natural light conditions. The majority of turkeys are reared in single-sex flocks, but smaller flocks may be mixed sex.

Welfare concerns

Due to the variety of turkey-rearing systems birds may be kept in many different environments, so some of the **welfare** issues raised below will not be relevant to all birds. For example, there is a huge range of flock sizes of commercially reared turkeys, from a few hundred in pole range or range systems to tens of thousands in conventional houses. Different systems have various minimum available **floor** area guidelines, with conventional houses having a minimum of 260 cm^2/kg and enclosed range areas with 10 m^2/kg.

Turkeys are vaccinated against numerous **diseases**, administered to the birds through a variety of methods. For example, turkey rhinotracheitis live vaccine is sprayed on to the turkeys when they are 1 day old; avian encephalomyelitis live vaccine is administered to the birds in their drinking water; and pasteurella, Newcastle disease and Paramyxovirus type 3 vaccines are all administered via intramuscular injection.

Lighting used in conventional turkey houses is either incandescent or fluorescent bulbs. Turkeys reared in conventional houses are typically kept in light intensities between 1 and 10 lux as a control strategy to reduce feather pecking and cannibalism. Intensities as low as 1 lux can cause buphthalmus (enlarged eye), and increased intraocular pressure and enlarged **adrenal glands** (Siopes *et al.*, 1984). Turkeys will preferentially select high light intensities while spending very little time in very low-light environments (Sherwin, 1998; Barber *et al.*, 2002).

Another way the industry controls feather pecking is by **beak trimming** of birds kept in open housing where the light environment cannot be controlled. These include breeders, as they require light to stimulate lay, and turkeys reared in pole barns and extensive systems, where they are subjected to high light intensities. This practice is also known as debeaking, or partial beak amputation, and most UK turkey companies will beak trim the birds before 21 days of age, and will remove up to one-third of the upper mandible. Three main methods are used to beak trim birds: electronic trimming (bio-beaking), cold cutting (with secateurs) and hot cutting (with a cautery iron). There is no evidence that beak trimming causes chronic **pain** in turkeys trimmed up to 21 days old (Grigor *et al.*, 1995). Breeding birds may be beak trimmed twice, as they are reared for over a year and it is possible that the beak may grow back.

Some turkey producers toe clip birds, either removing one, two or three toes at the junction of the first and second digit, or at the nail end. This prevents damage from scratches on the skin, which can downgrade the carcass. This may occur if the flock crowds into a small area such as a corner when startled. Back scratches may also occur during catching and transportation to the killing plant.

Males have a secondary sexual organ, the snood, which is at the base on the beak. The snood can be pulled during male fights and can be pecked by females in a mixed flock. This can cause damage and provide an area for disease entry. Some turkey producers remove the snood at 1 day old to prevent this damage from occurring. The organ is well innervated, but there is no information as to whether the snood removal process is painful or whether it leads to chronic pain.

It is important to keep the litter in a good, friable condition. Certain litter conditions can cause the development of breast blisters, which may cause discomfort to the birds and also downgrades the carcasses. Contact with ammonia or other noxious materials, including faecal enzymes in the litter, could result in breast blisters. The length of the **photoperiod** is also known to increase the chance of birds developing breast blisters. The longer the dark period the greater the chances of breast blister development, possibly because birds lie down for longer without moving.

Leg weakness can be a welfare problem for commercially reared turkeys (**see: Bone strength**). Genetic predisposition, nutrition, infection, mycotoxins, medication, husbandry practices and interactions between them have all been associated with leg weakness. Large strains usually show higher incidences of leg disorders than lighter stocks. This may be due either to weight effects or a consequence of a genetic correlation between body weight and defective bone growth (Hocking, 1993).

Sudden changes in the environment, such as the presence of a predator or rapid change in the light environment of the turkey house, can panic the birds, which may cause them to trample and damage other birds. Turkeys will 'pile up' against walls or in corners if startled, which can lead to smothering and **death**. (HM)

Further reading and references

Barber, C.L., Prescott, N.B., Wathes, C.M., Le Sueur, C. and Perry, G.C. (2002) Preferences of growing ducklings and turkey poults for illuminance. *British Poultry Science* 43, S10–S11.

Grigor, P.N., Hughes, B.O. and Gentle, M.J. (1995) An experimental investigation of the costs and benefits of beak trimming in turkeys. *Veterinary Record* 136, 257–265.

Hale, E.B., Scleidt, W.M. and Schein, M.W. (1969) The behaviour of turkeys. In: Hafez, E.S.E. (ed.) *Behaviour of Domestic Animals*. Baillière, Kidlington, UK, pp. 554–592.

Hocking, P.M. (1993) Welfare of turkeys. In: Savory, C.J. and Hughes, B.O. (eds) *Proceedings of the Fourth European Symposium on Poultry Welfare*. Universities Federation for Animal Welfare, Potters Bar, UK, pp. 125–138.

Martrenchar, A. (1999) Animal welfare and intensive production of turkey broilers. *World's Poultry Science Journal* 55, 143–152.

Schorger, A.W. (1966) *The Wild Turkey – Its History and Domestication*. University of Oklahoma Press, Norman, Oklahoma.

Sherwin, C.M. (1998) Light intensity preferences of domestic male turkeys. *Applied Animal Behaviour Science* 58, 121–130.

Siopes, T.D., Timmons, M.B., Baughman, G.R. and Parkhurst, C.R. (1984) The effects of light intensity on turkey poult performance, eye morphology, and adrenal weight. *Poultry Science* 63, 904–909.

Watts, C.R. and Stokes, A.W. (1971) Social order of turkeys. *Scientific American* 224, 112–118.

Twitch

Twitches are used to constrict the skin, usually of the upper lip of **horses**, in a bid to restrain them. Such nose twitches are simple and relatively safe for both horse and operator. For brief, mildly invasive procedures, they may help to immobilize the horse. That said, some experienced horses respond aversively to the mere sight of the twitch. This response strongly suggests that **pain** is involved. Twitch designs vary from a simple loop of rope attached to a metal ring that is twisted to tighten the rope, through to the so-called humane twitch (a pair of metal pliers, which, in the hands of a novice, may cause less tissue damage than a homemade twitch).

Pressure is a function of area (e.g. pounds per square inch or kPa), so with any piece of equipment used on horses, the broader the area that touches the animal the less discomfort it causes. Thus broader areas are covered by tack, such as the girth, to which we expect the horse to habituate. In contrast, other sources of pressure are intended to cause a response, so **sensitization**, rather than **habituation**, is required. The narrower the structure in touch with the horse, the greater the stimulation and potential for pain, but so too the likelihood of permanent damage. For this reason soft, thick rope is generally favoured in twitch designs. Because horses wearing a twitch may evade the pain by moving the head rapidly from side to side, twitches with plastic handles are generally the safest for personnel. The constriction around the nose should be loosened every 15 min or so to maintain cellular perfusion distal to it. In many cases, the effectiveness of the twitch tends to wane after this period, possibly because of neurotransmitter depletion at the level of synapses in the **opioid** pathways.

A rope gag (also known as war bridle) involves a cord passed under the upper lip and over the poll, before being threaded through a loop at the side of the face to form a running noose. It has an action similar to the twitch when pulled tight. Variations on this design, including the use of metal chains rather than cord, appear in the equestrian literature (lip chain). A piece of rubber tubing slipped over the portion of the rope that presses on the underpart of the lip reduces the risk of **injury** without appearing to reduce the effectiveness of the device. Although they are vehicles for escalated force, these devices are sometimes advocated for problems in the ridden horse. The inherent dangers with this sort of equipment are considerable. Rather than force, better timing and consistency is preferable to severe equipment that, through habituation, creates self-perpetuating reliance on force.

Among veterinary professionals, physical **restraint** of horses with twitches during painful procedures is often possible, but rarely preferable, to chemical restraint. In a cost–benefit analysis (**see: Utilitarianism**), the use of the twitch seems legitimate, if the alternative is a brutal struggle or an opportunity for the horse to practise a hazardous flight response. Fundamentally, the use of force in a fearful animal can escalate the aversiveness of the interaction and is ultimately likely to compromise the horse–human bond, making the horse less compliant for future procedures.

In very difficult horses, the ear is sometimes grasped briefly to control the horse while a twitch is put on the upper lip. However, this technique should be avoided because it is far more likely to create head-shyness in the long term than compliance in the short term. Twitching the ear is not appropriate because it may make the animal head-shy, scar the skin or paralyse the ear.

At best the twitch can be seen as a variant of acupuncture (Lagerweij *et al.*, 1984). Its effectiveness varies between individual horses, its application increases plasma concentrations of **beta-endorphin** and its action can be blocked by the opioid antagonist naloxone. Even though the twitch may work through the mediation of beta-endorphins, there is little doubt that it works because it involves pain (Webster, 1994). Current recommendations suggest that twitching a horse's lip should be undertaken only when chemical restraint is not available. Twitching should be regarded as a last resort of restraint, justified only as a means by which to inject a sedative or tranquillizer. Interestingly, the effects of a nose twitch can often be achieved by grasping and twisting a fold of skin on the side of the neck, presumably because this too causes pain and thus an **endorphin** response. Using this technique, handlers can often restrain fearful horses well enough to inject some form of chemical restraint. Hand twitches (taking the skin in the hand and twisting it) may be applied to the ear or skin at the side of the neck and, historically, tongue twitches were also used, but the consequence of **injury** in this case can be catastrophic (e.g. the tongue may be severed).

The **heart rate** of horses when twitched undergoes a transient increase (Morris, 1988) before returning to baseline values. This return to baseline is quicker in the crib-biters than in normal horses (Minero *et al.*, 1999). Although it is unclear why, **crib-biting** horses are said to be calmer during twitching and less reactive to the twitch in general than non-crib-biters (Minero *et al.*, 1999). These findings may reflect altered **dopamine** activity in the brains of these stereotypic animals.

(PDM, AM)

References

Lagerweij, E., Nelis, P.C., Wiegant, V.M. and van Ree, J.M. (1984) The twitch in horses: a variant of acupuncture. *Science* 225, 1172–1174.

Minero, M., Canali, E., Ferrante, V., Verga, M. and Ödberg, F.O. (1999) Heart rate and behavioural responses of crib-biting horses to two acute stressors. *Veterinary Record* 145, 430–433.

Morris, D. (1988) *Horsewatching*. Jonathan Cape, London.

Webster, A.J.F. (1994) *Animal Welfare: a Cool Eye towards Eden*. Blackwell Science, London.

U

UFAW (Universities Federation for Animal Welfare)

UFAW is an independent scientific and educational charity that promotes high standards of **welfare** for **companion**, **farmed**, **laboratory**, **zoo** and **wild animals**. It supports developments in the science and technology that underpin advances, holds workshops and symposia and advises regulatory authorities within and outside the UK. Its publications include the quarterly scientific journal *Animal Welfare* (see http://www.ufaw.org.uk). (PH)

Ultimate factor

Ultimate factors refer to causes that favoured the **evolution** of a particular response in an animal, and are supposed to give **fitness**-related benefits to that animal. For example, an ultimate causal explanation of feeding would refer to it as a behaviour favoured by natural **selection** as a way to maintain an animal in good physical condition and thus to increase its potential reproductive success. (BM)

See also: Proximate factor; Tinbergen's Four Questions

Ultradian rhythm

An ultradian rhythm is a biological rhythm that occurs with a frequency of less than one per day. Not surprisingly, the major biological rhythm in animals is the **sleep**/wake cycle, and many animals will have a single period of sleep/inactivity and a single period of wakefulness/activity each day. This cycle is endogenously generated, and many other biological processes are linked to this daily cycle but also have ultradian rhythmicity superimposed upon **circadian rhythm**icity.

Ultradian rhythms have been described in behaviour (activity), cardiovascular function, hormone release, body temperature, cerebral hemisphere dominance, nasal airflow and rapid-eye-movement (REM)/non-REM sleep. Activity-related ultradian rhythms are more common in neonatal and juvenile animals than in adult animals, and are characterized by bouts of vigorous activity interspersed with bouts of rest and/or sleep. In adult animals, ultradian rhythms of activity are closely related to feeding cycles and will be influenced by whether the animal is carnivorous, omnivorous or herbivorous. Environmental temperatures may also influence activity; for example, animals may show peaks in activity in the morning and evening, and avoid activity during the hottest part of the day.

Many hormones are secreted in a rhythmic, pulsatile fashion and, among those known to show ultradian rhythmicity, are **growth hormone**, **norepinephrine**, **dopamine**, **thyroid hormone** and cortisol. The rhythmicity is species dependent – for example, growth hormone peaks every 2–3 h in rats but every 5 h in calves. Similarly, cattle show cortisol peaks every 2 h, whereas rhesus monkeys have a rhythm of one every 90 min. Other physiological processes to show ultradian rhythmicity include **heart rate** and heart rate variability – indicating rhythmicity in autonomic function – and also rhythmicity in **electroencephalogram** (EEG) activity from the cerebral hemispheres. The cerebral hemispheres show alternating dominance, with activity switching from right to left and vice versa. This is accompanied by changes in the airflow through the nasal passages, known as the nasal cycle, which in humans has a periodicity of around 4 h. Throughout sleep, the body switches between REM sleep and non-REM sleep every 90–110 min.

Within the animal **welfare** field, ultradian rhythmicity has not received much attention, but some studies have looked for changes or disruption in biological rhythms in animals exposed to **stressors**. There remains scope for further research application. (RM-F)

Ultrasound

Ultrasound is sound pressure with a frequency above the threshold of human hearing – i.e. around 20 kHz and above. Within the animal sphere, ultrasound may either pertain to sounds emitted by some species for the purposes of **conspecific** communication or for **echolocation** of food, or to the medical sonography primarily used by animal caretakers to confirm pregnancy or examine specific tissues within the body. The terms 'ultrasound' and 'infrasound' are based purely on the human hearing range (20 Hz to 20 kHz), whereas many animals have normal hearing ranges that extend either below and/or above the human range. Higher-frequency sounds attenuate (lose intensity) more rapidly with distance and are susceptible to scattering and reflection by physical elements in the environment, but are more directional than lower-frequency sounds.

Ultrasound in communication

Ultrasonic **communication** has been documented throughout the animal kingdom, including insects, **amphibian**s, mammals and birds. The properties of high-frequency sounds outlined above mean that, for communication purposes, ultrasonic calls may have limited value over longer distances. Among insects, many male moths are known to produce ultrasounds from special organs located in front of the tegulae, which produce a pulse of sound at up as high as 80 kHz on each wing beat. The sounds serve the dual purpose of helping to establish **territory** by keeping other males at a distance and also attracting female moths for mating. Some ants and some katydids have also been documented as being able to produce ultrasonic pulses, with potential **distress** signals, 'reassurance' **signal**s and mating signals being emitted.

Within vertebrates, ultrasonic communication is well described in **rodents**. The ultrasonic calls of **rats** have attracted a great deal of research attention, and calls have been categorized into three distinct types. Rat pups will give characteristic 40 kHz **vocalizations** when they are separated from their mothers. Other rodent species such as **voles** and mice also emit these calls, ranging in frequency from 35 kHz to 55 kHz depending on the species, and invariably they result in the mother locating and returning the pup to the nest. Traditionally, these calls have been thought of as being emitted by the pup as a specific distress call meant to induce retrieval, but an alternative hypothesis is that they may be a by-product of an abdominal compression reaction in which abdominal muscles are contracted during expiration to help venous blood return to the heart, thereby maintaining cardiac output during **stress**. The larynx acts as a brake during exhalation and produces ultrasound. This hypothesis is still awaiting full testing.

Adult rats can also produce what are called 22 kHz vocalizations and 50 kHz vocalizations. The 50 kHz calls tend to be produced during **play**, non-aggressive social interactions and, by males, during copulation – i.e. during situations that may be emotionally positive. The 22 kHz calls are more associated with negative situations and have been recorded during aggressive interactions, in response to predators, when startled and in a model of chronic **pain**. They are also produced by the male during the post-copulatory phase. These calls have been widely used in psychological and pharmacological research as being an indicator of stress and, particularly, **anxiety**. As noted above, ultrasonic calls have properties that make them particularly useful as warning signals. Even though many predators such as canids and felids can hear up to 50 kHz, the fact that the calls rapidly attenuate means that rodents can signal to others in very close proximity to themselves without warning the predator that is further away. Producing lower-frequency calls as a predator approaches runs the risk of the predator being able to eavesdrop and obtain information about prey that may be nearby.

Other species known to produce ultrasonic calls with communicatory function are frogs, birds, toothed whales and bats. The latter two groups of animals use ultrasound primarily for echolocation, whereas the ultrasonic frog and bird species appear to have evolved their ultrasonic calls specifically for interspecific communication and under pressure from their habitat. Within any given environment, there are varying degrees of background noise. The noise may come from other animals or from abiotic sources, such as wind, rain or running water. The noise will have a certain frequency range, and the degree to which it interferes with the animal's call will depend on the overlap of this frequency range with the frequency range of the call. Much ambient noise, such as rushing water or traffic noise for example, may span the entire human audible frequency range and, thus, there may be selection pressure to shift the dominant frequency of calls above this range and into the ultrasonic range, even though the audible distance of the call may be lessened. Species of frogs and birds living by waterfalls, and even some birds in urban settings, are known to produce calls with ultrasonic components.

Toothed whales comprise the suborder Odontoceti of the order Cetacea, and include dolphins, porpoises, pilot whales, beaked whales, sperm whales and orcas. Many of the toothed whale species have a wide range of vocalizations used for communicating with each other, including clicks, squeaks and whistles, and many of these contain elements in the ultrasonic band. All toothed whales so far investigated also produce purely ultrasonic clicks, which are used for both locating prey and navigation. It has recently been proposed that some elements of ultrasonic communication may have in fact derived from non-communicatory ultrasounds, i.e. the ultrasounds were evolved for one purpose but have been 'co-opted' for a role in communication. For example, bats are widely known as being highly ultrasonically vocal, as they use ultrasound for echolocation purposes in navigation and **hunting**. Consequently, bat hearing is acutely tuned to hearing ultrasonic sounds and is maximally sensitive over the frequency band of their echolocation pulses. It is therefore proposed that bats can use not only their own calls, but also that the calls of conspecifics can convey useful information. Bat ultrasonic calls have been implicated in synchronizing roosting behaviour, attracting each other to food sources and playing a role in mother–offspring communication. There is also evidence that echolocation calls can convey information about individual identity.

The ability of animals to hear ultrasound is also exploited by humans in both training (ultrasonic whistles for **dogs**) and in scaring unwanted animals away from human resources (ultrasonic deterrents for rodents, insects, **deer**, etc.).

Ultrasound in echolocation

Echolocation (or bio-sonar) refers to the ultrasonic sounds emitted by animals to ascertain the location, range and identity of objects in their environment. The principle is that the pulse of sound emitted echoes back from objects in the environment, being received by the animal's ears. The magnitude of the returning sound, the time taken to return and any differences in the timing of the return received by the ears provide information about relative size of the object, the distance from the animal and the direction of the object. The animal can then process the information and respond accordingly.

The best-known animal echolocators are the suborder of bats known as the Microchiroptera (microbats). These bats use echolocation both to navigate in the **nocturnal** environment and to locate and home in on prey. The sound itself is produced in the larynx and emitted through the open mouth or the nose. Microbat calls range from 14–100 kHz, and each species will vocalize using a narrow range within this general range that is suited to the environment and prey of that particular species. During the search phase of foraging, bats will call at 10–20 pulses/s. Once a target is located, call rate can increase and may reach levels of 200 pulses/s.

The other famous echolocators are the toothed whales. Toothed whales have a wide variety of vocalizations used mostly for social communication, but they also have a characteristic click that is used for echolocation. The source of these sounds is the nasal air sacs, which are located behind a specialized fatty tissue structure called the melon, which appears to focus the sound into a narrow beam. The clicks are emitted in a series, called a click train, and the pulse burst can contain more than 600 clicks/s, with individual clicks lasting between 50 and 130 μs. The peak frequency of echolocation clicks is typically around 100 kHz, but varies depending on the specific echolocation task. The sound waves travel 4.5 times faster in water than in air, and the typical frequency clicks have an effective range of between 5 and 200 m for targets of 5–15 cm in length. The echoes are received back through the lower jaw and transmitted up to the inner ear, with the animal responding based on the information received.

Other animals known to use echolocation primarily for navigational purposes are some species of shrew, the nocturnal oilbird and swiftlets. The birds, however, use clicks that are audible to humans, and are not thus classed as ultrasound. The shrews' echolocation is carried out using ultrasonic squeaks.

Ultrasound in veterinary medicine

Ultrasonography is widely used in both human and veterinary medicine. It can be used both as a diagnostic tool and a therapeutic tool. In diagnostics, it can be used as a non-invasive method of visualizing soft tissue structures. As with echolocation, the methodology is based on the transmission of ultrasound and interpretation of the reflected sound back from the tissue. The sound wave is typically produced by a piezoelectric transducer encased in a probe. Electrical pulses make the transducer ring at the desired frequency, which is usually between 1 and 18 MHz. Superficial structures are scanned using higher frequencies, while deeper structures are scanned using lower frequencies. The transducer also receives the echo, which in turn creates vibration that is turned into electrical impulses for processing and digital imaging. As a technique, diagnostic ultrasonic imaging has multiple applications but, within the veterinary field, it is most commonly used to confirm pregnancy in domestic animals. It may also be used by veterinarians to diagnose disease, as in human medicine, and by animal scientists to determine production parameters such as back-fat depth and loin eye muscle area.

Pregnancy checking in domestic animals can help with the management of pregnancy – ensuring that care of the mother is tailored towards her gestational needs. The exact benefits will depend on whether the species is **companion animal** or **farmed animal**. With companion animals (such as pedigree dogs or competition horses), the mother and/or offspring may have high commercial value as well as emotional value. With farm animals, the commercial value of the individual animal is likely to be considerably less, but the collective financial well-being of the farm will be dependent on high proportions of successful conceptions. Many farms employ ultrasonic pregnancy checking as a management tool to ensure that 'open' females are not entered into the gestation herd or flock, but are either retained in the breeding system or culled due to reproductive difficulties. (JNM-F)

Further reading

Arch, V.S. and Narins, P.M. (2008) 'Silent' signals: selective forces acting on ultrasonic communication systems in terrestrial vertebrates. *Animal Behaviour* 76, 1423–1428.

Au, W. (2004) Echolocation signals of wild dolphins. *Acoustical Physics* 50, 454–462.

Bushong, S.C. (1999) *Diagnostic Ultrasound: Essentials of Medical Imaging Series*. McGraw-Hill, New York, 150 pp.

Han, C.M. and Hurd, C.D. (2004) *Practical Diagnostic Imaging for the Veterinary Technician*. Mosby, Philadelphia, Pennsylvania, 304 pp.

Litvin, Y., Blanchard, D.C. and Blanchard, R.J. (2007) Rat 22 kHz ultrasonic vocalizations as alarm cries. *Behavioural Brain Research* 182, 166–172.

Umwelt

Jakob von Uexküll (1936) termed an individual's entire subjectively perceived world its 'umwelt'. Umwelts differ between individuals, because sensory abilities differ and because stimuli have different relevance to each individual; Von Uexküll used the example of a flower stem being a path for ants, but a food morsel for cows. (CCB)

See also: Ethological relevance; Perception

Further reading

Von Uexküll, J. (2001) An introduction to Umwelt [Translated from J. von Uexküll (1936). *Niegeschaute Welten*, 11–17, S. Fischer Verlag, Berlin]. *Semiotica* 134, 107–110.

Ungulates

Ungulates are the mammals that have replaced claws with hooves during their **evolution**. The surviving primitive ungulates are **elephants**, hyraxes and the aardvark. Fairly soon after they first arose, the ungulates diverged into even-toed (order Artiodactyla) and odd-toed (order Perissodactyla) ungulates. The asses, **horses** and zebras, the tapirs, and the **rhinoceroses** are perissodactyls and the **pigs**, peccaries, hippos, **camelids**, chevrotains and musk deer, **deer**, **giraffes** and bovids are artiodactyls. Bovids comprise the pronghorn, wild **cattle** (bovines), duikers, gazelles and grazing antelopes, and goat antelopes.

All ungulates are terrestrial herbivores or omnivores. Specializations for a diet of plant material are evident in their teeth, jaws, digestive systems and behavioural ecology. A key feature is **rumination** (retention of food in a specialized stomach chamber system where it is fermented by symbiotic microorganisms), which most artiodactyls (but not the pigs, peccaries or hippos) employ to extract energy and nutrients from plant material. Perissodactyls have other adaptations – they house their symbionts in the hindgut, which is a less effective way of extracting energy and nutrients but is better able to cope with a coarser diet.

Many ungulates have social systems with strong social **bonding** and a clear **dominance hierarchy**. These, and other characteristics, suited them for **domestication** and they include the most important **farmed animals**. (SJGH)

Urban wildlife

Urban wildlife is defined as animals that make use of resources (breeding habitat, roosting substrates, food, etc.) in urbanized areas. Urban wildlife includes a broad range of animals (e.g. house sparrows, Canada geese, pigeons, bats, **deer**, opossums, raccoons, etc.) that can have different degrees of association with humans.

Species highly dependent upon humans, particularly in terms of food (e.g. pigeons, **feral cats**, house **mouse**), are known as synanthropes. Synanthropes can be full synanthropes (also called exploiters or urbanophiles), which obtain most of their resources in urban areas, or casual synanthropes (also called adapters or moderately urbanophilic), which can adapt to exploit resources in urban habitats but also use other resources in more natural environments. Full synanthropes are generally found in areas with low vegetation within cities, but casual synanthropes do use urban vegetation for food or nesting/**roosting**. Finally, species that are highly sensitive to urbanization (e.g. rare in cities) are called avoiders or urbanophobes.

Urban wildlife can be composed of native and non-native species. Native species are local species that use certain resources in urban areas, but they are also dependent on other habitats (e.g. casual synanthropes). Non-native species have been introduced locally from other ecosystems (or have colonized a local area after a regional introduction). Non-native species fare well in cities because of their high competitive ability and the high availability of foraging resources in cities. Two well-known examples are house sparrows (*Passer domesticus*) and European starlings (*Sturnus vulgaris*). The house sparrow, a native from Eurasia, was initially introduced to North America (in New York) in 1851. The species had expanded its geographic distribution to many states on the east and west coasts by the 1900s. Nowadays it is commonly seen **breeding** in cities, although it can also be present in farms and residential areas. Even more remarkable is the introduction of about 100 European starlings in New York in 1890, which expanded their distribution throughout North America with an estimated abundance of 200 million individuals. Part of the success of house sparrows and European starlings has been attributed to their ability to outcompete native species for nest cavities, which could result in the decline of native cavity-nesting species.

The role of urban wildlife is particularly important because of two worldwide trends in land use that are changing the representation of urban habitats. First, more than 50% of the human population is expected to be living in urban areas around the world by 2015. This increment in human density will be accompanied by even higher rates of urban sprawl, which will fragment and reduce the representation of natural habitats. Secondly, in some regions of the world there is an increase in the immigration rate to rural areas by people seeking lifestyles that provide closer access to natural areas (ex-urban development). Both trends are causing changes in land use that provide conditions conducive to the colonization of synanthropic species due to at least three factors. First, breeding and foraging resources increase around human settlements due to the presence of food leftovers, nesting substrates in buildings, etc. Secondly, natural predators are thought to be less abundant in cities, which would increase the survival of urban species, although further evidence is needed to support lower predation risk in cities. Thirdly, urbanized landscapes modify environmental conditions that minimize climatic fluctuations throughout the year, such as an increase in temperature (heat island effect) and water availability due to irrigation. These three conditions create similar urban environments in different regions of the world, which leads to homogenization of wildlife – an increase in the similarity of species composition across cities. Homogenization in turn can decrease species diversity at the regional level.

Variations in resource abundance, predation levels and environmental conditions along the rural–urban gradient modify species richness and abundance. Generally, the number of native species decreases, and the number of non-native species increases, with increasing levels of urbanization. Because of these opposing patterns, the overall number of species tends to peak at intermediate degrees of urbanization. This inverted U-shape pattern has been found in birds, small mammals, lizards and butterflies. In birds, a few synanthropic species tend to dominate urban communities due to their extremely high densities relative to other habitat types. These birds are usually ground-foraging seed eaters or omnivores that have adapted to nesting in human substrates. In mammals, synanthropic species that reach the highest densities are often found inside buildings.

Urban sprawl can modify ecological relationships, such as predator–prey interactions. Fragmentation of natural habitats reduces the size of the remnant patches, which can decrease the densities of top-predators, or drive them to local extinction. For instance, in coastal southern California, coyote populations have dwindled due to the high degrees of urbanization around native sage-scrub habitat. The lack of this top-predator increases the abundance of native (raccoon, grey fox) and non-native (domestic cat, opossum), low-level predators, a process known as **mesopredator** release. This intensifies the negative effects of habitat fragmentation on native scrub-breeding birds because low-level predators, especially the domestic cat, have high abundances in small-sized remnant patches, thereby increasing local bird mortality due to **predation**.

The overlap in spatial use by humans and wildlife in urbanized areas generates different types of conflicts. The overabundance of some species can damage private and public property, increase road danger and spread diseases. For instance, the accumulation of pigeon faeces can block rainwater drainage systems, decrease the aesthetic value of buildings – particularly those of cultural relevance – and increase slippery footing on walkways. Deer collision with vehicles is probably enhanced by increasing deer abundance, expanded road networks, greater traffic load and high average vehicle speeds. This results in around 1.5 million deer–vehicle collisions yearly in the USA. Finally, urban wildlife species that are hosts of infectious **diseases** (e.g. West Nile virus, avian influenza, Lyme disease, etc.) can increase the chances of transmission to humans in the presence of vectors. Wildlife managers and public health officials have developed various strategies to minimize the risks of transmission by **culling** different species or reducing the availability of resources (e.g. water for mosquito breeding).

In conclusion, urban wildlife encompasses a diverse set of species that require to be managed in different ways according to their degree of association with humans. Despite conflicts, the presence of wildlife in cities not only brings ecological but also psychological benefits by providing humans with the opportunity to experience the natural environment. To maintain these benefits, it is important that the diversity of native vegetation is enhanced in urban areas. (EF-J)

See also: Conservation; Exotic species invasion; Wildlife management

Further reading

Chace, J.F. and Walsh, J.J. (2006) Urban effects on native avifauna: a review. *Landscape and Urban Planning* 74, 46–69.

Fernández-Juricic, E. and Jokimäki, J. (2001) A habitat island approach to conserving birds in urban landscapes: case studies from southern and northern Europe. *Biodiversity and Conservation* 10, 2023–2043.

Hansen, A.J., Knight, R.L., Marzluff, J.M., Powell, S., Brown, K., Gude, P.H. and Jones, K. (2005) Effects of exurban development on biodiversity: patterns, mechanisms, and research needs. *Ecological Applications* 15, 1893–1905.

Marzluff, J.M., Bowman, R. and Donnelly, R.E. (eds) (2001) *Avian Conservation and Ecology in an Urbanizing World*. Kluwer Academic Publishers, New York.

McKinney, M.L. (2006) Urbanization as a major cause of biotic homogenization. *Biological Conservation* 127, 247–260.

Urine

Urine is an aqueous solution of waste materials from metabolic processes in the body. Blood passes through the kidneys, where it undergoes filtration. Metabolites are secreted along with water into the ureter. The urine then travels down the ureter into the bladder, where it is stored until the animal urinates – expelling the urine out of the bladder through the urethra. From a physiological aspect, the production of urine is important in **homeostasis**, to maintain the water balance within the body. With dehydration, water is conserved as blood passes through the kidneys and urine volume decreases and becomes more concentrated. With subsequent rehydration, water will be removed in larger quantities and urine volume and dilution will both increase.

From welfare and behaviour aspects, urine can have an important role in intra- and interspecific **communication**; it can contain **stress** markers that can be collected non-invasively; and it can impact **welfare** in relation to hygiene and air quality. Mares' urine is also used in the manufacture of the drug used in **oestrogen** replacement therapy, and there have been concerns about the welfare of the horses involved.

Urine and communication

Chemicals contained within urine, and sensed by the olfactory system and the **vomeronasal organ** (VMO), can convey a variety of information that can have an effect both within and between species. Urine can convey information about individual identity, social **dominance**, territorial limits, reproductive status, predator/prey proximity and even emotional states, such as **fear**. One source of information about individual identity is most probably conveyed by soluble antigens of the major histocompatability complex (MHC) reacting with bacterial flora to produce volatile compounds that give the urine a characteristic odour. Another mechanism involves major urinary proteins (MUP), which bind and release volatile compounds, and have been well described in rodents.

Experimental studies have shown individual **recognition** using urine in many species – for example mice, rats, bank voles, hamsters, lynx and pigs. For social dominance, **androgen** hormones and metabolites appear to be an important element in conveying information; and in rodents, relative dominance may be conveyed by the quantity of urine scent marks left within a **territory**. More dominant mice make more urine marks, whereas subordinate mice make fewer marks. Territorial marking by many species often includes urination around territorial boundaries, in combination with visual markers such as scratching. Within domestic species, territorial urine spraying in cats can become a problem behaviour that necessitates intervention to remedy.

Urine has long been known as a conveyor of information about reproductive status. Male urine, as noted above, can convey information about territory and dominance, and can act as an advertisement or attractant to females seeking a mate. Female urine can contain information about stage of **oestrus**, and thus a female's receptivity for mating. A number of urinary volatiles have been shown to peak during oestrus in a variety of species. A characteristic male response to female urine, particularly obvious in ungulates, is that of **flehmen**, a certain type of lip curling thought to facilitate use of the vomeronasal organ.

There are a great many studies that have shown that urine from a predator species can be detected and discriminated by prey species, which subsequently alter aspects of their behaviour, presumably to decrease risk of predation. The most common predator species used have been felids, canids and also humans. The most common prey animals tested have included rodents and lagomorphs, but also marsupials and ungulates. Behavioural strategies that may occur in response to predator odour include decreases in activity and behaviours such as **grooming** and **reproductive behaviour**, shifts to a more strategic location and increases in vigilance. Temporal patterns of behaviour may also change with alteration of feeding times.

Urine and stress markers

One of the major problems with gathering physiological data relating to responses to stress is that the collection of such data may itself be a **stressor**. For example, collection of blood for stress hormone analysis will either involve the implantation of an indwelling venous catheter under anaesthetic or **restraint** of the conscious animal and insertion of a needle to draw the blood sample. Both of these methods can be considered invasive and, thus, there has recently been an upsurge in experimental work looking at alternative body fluids, such as saliva and urine, which can be collected in a much less invasive way.

Among the compounds analysed in urine and used as stress markers are cortisol, cortisone, **epinephrine**, **norepinephrine** and immunoglobulin A (IgA). The use of urine collection has been especially popular in studies involving **companion animals**, such as cats and dogs undergoing boarding at catteries and kennels, and in **laboratory animals** undergoing experimental procedures. It has also been used in a number of studies on welfare in pigs. Ideally, samples are collected as the animal urinates freely (voluntary voiding).

Urine and welfare in animal housing

Animals other than those that are free-ranging are, to a greater or lesser extent, constrained in terms of elimination behaviour. The degree of constraint will largely depend on the degree of spatial restriction – pigs in a small pen will be more constrained than pigs in an enclosed paddock. Ordinarily, most animals will choose to separate themselves from their faeces and urine and may allocate specific areas of their habitat for elimination. For domestic species, **housing** invariably disrupts normal eliminative habits and, thus, animals may be unable to separate themselves from their own waste. This can impact the animals' welfare, either by forcing them to be exposed to substances that may be ordinarily avoided, by exposure to situations that challenge comfort and **thermoregulation** or by exposure to potential **health** hazards, either by direct contact with pathogens or through indirect effects of poor air quality, in particular ammonia.

Farmed animals and laboratory animals are perhaps the most impacted by the housing systems in which they are kept. For farm animals, the housing system will either attempt to delineate areas within the pen into **lying** areas and dunging areas by use of relatively small amounts of bedding applied to one area, or it may aim to remove the animals from their waste by utilizing slatted floors so that urine and faeces fall through into a slurry pit. Alternatively, the pen may be deep bedded, with fresh bedding added on top of existing bedding until the animals are moved out after a period of weeks or months. In all these situations, the animals will be unavoidably exposed to their own urine and faeces, and to that of their pen-mates. Once in the environment, urine will react with bedding and faeces to produce ammonia, and this can be present in harmful concentrations in farm animal housing with insufficient ventilation. Wet bedding can also have a negative impact on pen hygiene and pathogen load, affecting illness incidence such as increased respiratory disease in calves or behaviour – for example, reducing lying time in dairy cows. For laboratory rodents, there have been some concerns raised that frequent cage cleaning to remove urine may have a negative impact by disrupting the rodents' olfactory environment. However, recent studies suggest that frequent cleaning has little impact on rodents' welfare.

Urine and pregnant mares

A drug containing conjugated **oestrogens** is manufactured from the urine of pregnant mares. It is used as a form of hormone replacement therapy during menopause and by women who have undergone hysterectomy. Welfare concerns over the housing of the horses used to produce this hormone, i.e. the use of tie stalls, and the fact that many of the foals born to the mares were sent for **slaughter**, have been raised as objections to the process. In the last couple of decades, the industry has downsized as synthetic oestrogens have become more widespread, and also has adopted a welfare code of practice that must be followed for the ranch to maintain its contract, with non-pregnant mares having access to pasture for many months of the year. A few studies carried out on mares in these systems have not found any major welfare problems. (JNM-F)

Further reading

Apfelbach, R., Blanchard, C.D., Blanchard, R.J., Hayes, R.A. and McGregor, I.S. (2005) The effects of predator odors in mammalian prey species: a review of field and laboratory studies. *Neuroscience and Biobehavioral Reviews* 29, 1123–1144.

Brennan, P.A. and Zufall, F. (2006) Pheromonal communication in vertebrates. *Nature* 444, 308–315.

Burn, C.C. and Mason, G.J. (2008) Effects of cage cleaning frequency on laboratory rat reproduction, cannibalism and welfare. *Applied Animal Behaviour Science* 114, 235–247.

Kurien, B.T., Everds, N.E. and Scofield, R.H. (2004) Experimental animal urine collection: a review. *Laboratory Animals* 38, 333–361.

Urine drinking

Urine drinking, urophagia, is usually a form of **pica**, a depraved or **abnormal** appetite, and can be considered an abnormal behaviour. It is believed to usually derive from nutrient deficiencies or other **stress**es in animals, but also occurs in males testing for **oestrus pheromones** in females, e.g. in dogs and cattle. The behaviour may involve direct drinking from the anovaginal area of females or from the penis of males, or from pools on the ground. It has been most commonly observed in overwintering cattle indoors. The effects on production levels and herd health are small, but infectious agents, such as *Leptospira* and *Mycobacterium tuberculosis*, can be passed from animal to animal in this way. However, urine directly expelled contains urea, which has antibacterial and antifungal properties.

Urine-drinking behaviour has been observed in group-housed calves as an extension of cross-sucking behaviour (Lidfors, 1993) (**see: Cross-suckling**), and in piglets that have been experimentally challenged with regard to their salt balance (Benda *et al.*, 1998). **Wild animals** also show this behaviour, and it has been reported in colobus monkeys (Rode *et al.*, 2003) and African tree monkeys (Lambert, 2000). The fructivorous diet is low in minerals and it is postulated that, in this case, urine drinking is a strategy to minimize mineral losses, of sodium in particular – an idea that would appear to be further evidenced by the aforementioned work on salt-challenged piglets (Benda *et al.*, 1998). There may also be a role for urine drinking in nitrogen balance. (DA)

References

Benda, T., Salyi, G., Glavits, R. and Cseplo, A. (1998) Olaquindox toxicity in piglets. *Magyar Allatorvosok Lapja* 120, 451–457.

Lambert, J.E. (2000) Urine drinking in wild *Cercopithecus ascanius*: evidence of nitrogen balancing? *African Journal of Ecology* 38, 360–362.

Lidfors, L.M. (1993) Cross-sucking in group-housed dairy calves before and after weaning. *Applied Animal Behaviour Science* 38, 15–24.

Rode, K.D., Chapman, C.A., Chapman, L.J. and McDowell, L.R. (2003) Mineral resource availability and consumption by colobus in Kibale National Park, Uganda. *International Journal of Primatology* 24, 541–573.

Urine marking

Urine marking involves the use of urine as a scent mark to denote the individual's identity, presence or status or to mask the presence of others (for example in the marking of **territory** or **home range**). Urine marking can sometimes be discriminated from eliminative urination by the volume of the urine expelled and possibly the **posture** of the animal. In many species it can occur in either sex (and in neutered animals) but tends to occur more often in entire males. **Dogs** may mark by raising one hind leg against a vertical object ('raised leg urination', RLU), **cats** may back up against a vertical surface with an erect, quivering tail. This may not only serve as a visual display but also help to make the smell more noticeable to other individuals by placing it nearer to head height. (KT)

See also: House-soiling; Scent marking

Urine washing

Urine washing, where an animal urinates on the palm of the hand, and then wipes their hand on the soles of their feet, is a behaviour that has been observed mainly in primates. It may have thermoregulatory benefits as well as disinfecting properties for the skin and coat, especially aiding wound healing. (DA)

Utilitarianism

Utilitarianism is a consequentialist ethical theory that specifies that the rightness or wrongness of an action depends entirely on the outcome that it produces. Utilitarianism directs us to promote the greatest total amount of good or best possible balance of benefits over harms for all those affected by the action. In the case of animal welfare and ethics, this approach contends that we are justified in using animals if – and only if – the overall benefit to humans and animals is not outweighed by the harm done to animals, and there is no other more efficient 'optimific' way of achieving the same result. (RA)

Utilization distribution

The utilization distribution of an area is the probability distribution that refers to the likelihood of the location of an individual within a given area at any time. Areas of highest probability within a territory often make up the **core area**, with the **home range** being a more extensive area that is also used routinely (i.e. has a relatively high level of probability with the utilization distribution). Certain areas may vary in their probability of use depending on time-bound or physical factors, such as season and availability of relevant resources. (DSM)

V

Vacuum behaviour

An action that occurs in the absence of the external stimulus that usually initiates it. Such behaviours are usually **instinctive behaviour** patterns that the individual is highly motivated to perform, but where the appropriate outlet is absent. As such, the behaviour appears to serve no obvious purpose in its current context. For example, **dust-bathing** behaviour observed in hens when no substrate is available (Vestergaard, 1980), or in circling before **lying** down in the house by dogs. The **welfare** implications of vacuum behaviours are variable and need to be assessed on a case by case basis. (KT)

See also: Abnormal/abnormality; FAP/FMP

Reference and further reading

Manning, A. and Stamp-Dawkins, M. (1998) *An Introduction to Animal Behaviour*, 5th edn. Cambridge University Press, Cambridge, UK.

Vestergaard, K. (1980) The regulation of dust bathing and other patterns in the laying hen: a Lorenzian approach. In: Moss, R. (ed.) *The Laying Hen and its Environment*. Martinus Nijhoff, The Hague, the Netherlands, pp. 101–113.

Validity

In science 'validity' is used mostly as a term to describe the quality of a test or measuring procedure. A test or a procedure is considered valid if it succeeds in *measuring what the researchers set out to measure*. Often the validity of a specific test or procedure is assessed by comparing how well its results match the results of other tests or procedures designed to measure the same thing. In **logic** and philosophy, validity attaches strictly to *arguments*. In a valid argument the conclusion is logically *entailed* by the premises. It is important to appreciate that such entailment is no guarantee that one has made a good argument: it can easily generate a false conclusion if the relevant premises are false. Only if true premises are used will validity secure a true conclusion. A valid argument with true premises is best described as cogent, or sound. (PS)

See also: Reliability

Variation

Variation refers to the instance or magnitude of change within any biological level. Species inhabiting different or variable habitats usually have a higher degree of inter-individual or inter-population variation than species localized in a single, stable habitat. Examples of variation may be the variable degree of difference in body size between males and females of a given species or the number of alleles for a single locus. (BM)

Further reading

Ball, G.F. and Baltharart, J. (2008) Review. Individual variation and the endocrine regulation of behaviour and physiology in birds: a cellular/molecular perspective. *Philosophical Transactions of the Royal Society B* 363, 1699–1710.

McNamara, J.M., Barta, Z. and Houston, A.I. (2004) Variation in behaviour promotes cooperation in the Prisoner's Dilemma game. *Nature* 428, 745–748.

Vasopressin (antidiuretic hormone (ADH))

Vasopressin is a peptide neurohormone that is produced by the **hypothalamus** and released into the circulation from the posterior **pituitary gland**, which regulates circulatory pressure by constricting blood vessels and increasing water resorption in the kidneys. Two related forms are commonly recognized in mammals: arginine vasopressin and lysine vasopressin, the latter being the normal form found in pigs and marsupials, the former the more typical form in other mammals. The equivalent compound in birds is called vasotocin. Interest has grown in vasopressin recently as a potential mediator of certain forms of emotional **stress**. (DSM)

Veganism

Some adherents of moral veganism oppose the consumption of all animal products because they believe that, because animals experience many of the same physiological and emotional states as humans and can suffer the harm of **death**, commitment to true humanity involves extending the circle of compassion to include all living things and ending institutionalized cruelty of animals. Others may be vegans out of concern for the environment and for unjust social practices associated with animal agriculture. (RA)

Vegetarianism

The philosophical underpinnings of those committed to ethical vegetarianism are diverse. Some adherents protest against the use of animals as mere things to be exploited for human benefit in ways that make animals and their products cheaply and abundantly available. They disagree with the justification behind the infliction of **suffering** and **death** on animals for meat, eggs, milk and other by-products.

One variety of ethical vegetarianism contends that our obligation be based on respecting the moral status of animals. This variety can be divided into two overlapping but distinct sub-strands:

- Animal **welfare** utilitarians stress that we have a duty to boycott animal products as long as the way animals are raised promotes excessive suffering. Their boycott is due to the confined conditions of poultry and swine, for example, which suffer from severe restriction of movement and inhumane **slaughter**.

- **Animal rights** vegetarians contend that vegetarianism is morally required because a meat-based diet violates respectful treatment of animals, especially adult non-human mammals. These animals are rights-holders and have inherent value. Killing them is wrong because it deprives them of a future life they would have enjoyed. Giving animals pleasant lives or painless deaths does not truly amount to respect.

Other varieties have emerged, which can be linked to environmentalist or social justice movements. Under the former view we should adopt vegetarianism, or at least reduce our meat consumption of animal products, because of the catastrophic environmental impacts of large-scale animal agriculture, including the pollution by toxic animal wastes of air, streams and groundwater, and deforestation. Under the latter view a Kantian vegetarian, while discounting the independent moral worth of animals, would contend that subscribing to a meat-based diet effectively treats humanity simply as a means and not as an end in itself. They contend that there is enough scientific evidence to show that a meat-based diet high in animal fat is not healthy, and that a vegetarian diet prevents diseases. Hence, continuing such a meat-based diet conflicts with the duty to promote one's **well-being**. Others, like Kathryn Paxton George, argue to the contrary that meat consumption is one of the most efficient ways for women to prevent iron deficiency. Moderate vegetarians, namely lacto-ovo and demi-vegetarians, approve of some animal use as long as good animal welfare/care practices are fastidiously observed. (RA)

See also: Ethics

Further reading

George, K.P. (2000) *Animal, Vegetable, or Women? A Feminist Critique of Ethical Vegetarianism*. SUNY Press, New York.

Vices

The term vice is used in two distinct ways in relation to applied animal behaviour and animal **welfare**. In Aristotelian **virtue ethics**, vices result from choosing excesses with respect to emotional states and desires. Vices are distinguished from involuntary actions – actions performed under compulsion or in ignorance, for which the agent did not deliberately contribute to that act. From the perspective of virtue ethics, intentionally causing harm to animals is seen not so much as a violation of the animals' rights per se but as a character flaw of the perpetrator.

The term 'vice' is also commonly used to describe the repetitive behaviour of animals in **confinement**, most notably behaviours such as **weaving** and **crib-biting** in the **horse**. This use of the term is discouraged in the literature because of confusion with its use in virtue ethics, which may bias thinking about these behaviours as a defect primarily in the animal, rather than arising as a consequence of its husbandry (**see: Stereotypies**). (RA)

Video image recording and analysis

The use of video recording equipment has greatly increased the ability to observe, capture and analyse many aspects of animal behaviour. The ability to place a camera out of sight has enabled researchers and wildlife filmmakers to record elusive species and behavioural sequences and to minimize the impact of the observer on these. The use of time lapse and long-play videotapes has allowed researchers to record behaviours over relatively long periods without having to be physically there, saving them time and money. The video camera has also increased the detail with which behaviour can be recorded and analysed by enabling the researcher to make more careful observations of the behaviour in the laboratory, with the videotape slowed or paused to catch fast or complex behaviours. Special high speed video recording is necessary for capture of particularly rapid behaviour. Video recording has also facilitated the ability to test **inter-observer reliability** and **intra-observer reliability**. Over-reliance on the use of video cameras, however, can adversely affect a piece of research when, for example, equipment fails or behaviours are out of view or not in focus and cannot be recorded properly.

The use of computerized event recorders in conjunction with the playback of a videotape can speed up and increase the detail and reliability with which data can be recorded during an observation. Preliminary observations are required to identify the behavioural events or states of interest and to give these, and the individuals performing them, a code such as a key on the computer keyboard. The observer then watches the video and presses keys to identify an individual or the start of a behaviour. This may be done continuously as in a live observation or by using predefined intermittent time points. The use of event recorders allows the user to pause during an observation or to repeat observations if they feel they have made a mistake. The raw data can then be moved directly into a data analysis package, minimizing the risk of error at this stage. However, the more sophisticated event recorders now available can also perform preliminary analysis of the data, including giving the percentage of time spent in each behaviour and the reliability between observers. (KT)

See also: Measuring behaviour; Observer effect

Further reading

Flowers, J.H. and Leger, D.W. (1982) Personal computers and behavioural observation: an introduction. *Behaviour Research Methods, Instruments and Computers* 14, 227–230.

Martin, P. and Bateson, P. (1993) *Measuring Behaviour*, 2nd edn. Cambridge University Press, Cambridge, UK.

Noldus, L.P.J.J. (1991) The Observer: a software system for collection and analysis of observational data. *Behaviour Research Methods, Instruments and Computers* 23, 415–429.

Unwin, D.M. and Martin, P. (1987) Recording behaviour using a portable microcomputer. *Behaviour* 101, 87–100.

Vigilance

In general terms, vigilance is alert observation to detect objects or events in the surrounding environment (**see also: Arousal**). When in a highly vigilant state, animals actively scan their environment with heightened sensory perception to increase their chance of detecting a subject of interest. Once detected, this **stimulus** elicits a corresponding response by the animal.

However, the level of vigilance and reaction can vary. For example, antelope as well as other ungulates will occasionally lift their heads from the ground to look around, visually surveying the surroundings while foraging. If disturbed or danger is sensed, an extreme alert posture may be formed, with the head up, nose lifted, ears erect and oriented towards the disturbance. In both situations, vigilance is occurring but at different levels.

Animals are able to enhance their vigilance by increasing the rate and length of scans, engaging in vigilant behaviour more frequently and for longer periods. While having increased vigilance confers advantages by providing critical information useful for an individual's response to its environment, animals cannot remain at a constantly high vigilant state. There are two main reasons for this: (i) detection efficiency tends to decrease over time; and (ii) physiological constraints. Human subjects have been shown to tire under sustained vigilance, making more false alarms or fewer correct detections as time increases. Increased vigilance also can be energetically costly to an individual by reducing their food intake rate. Trade-offs between time spent being vigilant and time spent performing other necessary activities, such as feeding, must be made in order to maximize **fitness**.

Vigilant behaviour can serve several purposes for an organism: predator detection and **predator avoidance**, social interactions, mating and location of food or other resources. The role of vigilance in predator detection has received considerable attention. Animals are known to alter their behaviour in the presence of predators, including increasing vigilance. Early detection of predators is obviously advantageous to allow a quick response and avoid being eaten. Multiple factors can influence anti-predator vigilance such as predation risk, habitat, refugia, presence of young, group size, location and distance to **conspecifics**, and body size.

In areas where there are many predators and the risk of predation is high, vigilance is increased in order to lower that risk. Certain habitat types can pose greater risk to some species. While dense cover can provide refugia for some, it can also obstruct visibility, reduce mobility and provide hiding cover to predators. Large ungulates have been found to be more vigilant in dense cover than in open grassland habitat, whereas the opposite may be found for other species, depending on the proximity to refugia for escape and the particular behaviour of the predator (e.g. stalking or ambush behaviour of mountain lions, aerial dives of hawks). **Parental behaviour** also has a tremendous effect on vigilance. Adults caring for offspring typically increase vigilance to protect their investment and decrease the risk of mortality for their defenceless young. However, the time required for anti-predator vigilance can place a huge demand on a single individual's activity budget (**see: Time budget**).

One strategy to reduce the cost of vigilance is to forage in **groups**. Among social vertebrates, collective vigilance tends to be positively related to group size, while individual vigilance decreases with increasing group size. Each individual forager can devote more time to feeding and less time to vigilance when in a group because there is a greater chance that at least one individual in the group will detect the predator. This has been called the 'many-eyes' hypothesis, based on the greater number of eyes available to scan the environment as group size increases. The effectiveness of this strategy, however, depends on the maintenance of collective vigilance, despite lowered individual vigilance, and the ability of the whole group to be alerted once a predator has been detected. Alarm calls are often used to communicate a detected risk among group members, conferring the benefits of vigilance to the entire group. Similarly, herding behaviour may decrease the risk of predation due to a dilution effect. An individual's probability of being attacked is reduced in a group, since the predator can only target one prey per attack.

As distance to **conspecifics** increases and individuals become more isolated, vigilance tends to increase. Based on the selfish herd theory, the centre of the herd may have the lowest risk of predation. Correspondingly, individuals near the centre tend to be less vigilant compared with individuals located on the periphery of the group. Body size also appears to be a factor in the level of vigilance observed. A negative allometric scaling relationship between body size and vigilance was found for African antelope. This could be due to lower predation risk to larger-sized animals, greater efficiency of vigilant behaviour or the higher cost of vigilance due to time constraints on foraging needed to meet their energetic requirements.

Besides functioning as an anti-predator behaviour, vigilance also plays a role in social interactions and **communication**. For example, dominant males survey their territories to prevent intrusion from other males. In harem-based mating systems, a male may watch over a group of females to maintain and protect the harem from outside copulations. Vigilance can also help to detect and locate potential mates. In addition, vigilance may be directed at other conspecifics to help determine spacing and minimize intraspecific competition within a foraging group.

Finally, vigilance can assist with foraging by scanning the environment for suitable or preferred food items. As with anti-predator vigilance, the time spent being vigilant can incur a cost and reduce food intake rate, unless an animal can combine searching with handling food. By processing food while searching for the next bite, vigilance costs can be reduced, particularly in areas of high food biomass. Animals may also decrease search time by forming a search image or **gestalt** of their preferred items to facilitate a selective search for those items distributed across the environment.

Of course, these different functions of vigilance are not mutually exclusive. Animals can detect multiple stimuli and direct their behaviour accordingly. The primary goal of vigilant behaviour, however, remains the same – effective detection within one's environment. (LAN)

See also: Cooperation; Flight behaviour; Flocking behaviour; Sheep

Further reading

Quenette, P.-Y. (1990) Functions of vigilance behaviour in mammals: a review. *Acta Oecologica* 11, 801–818.

Underwood, R. (1982) Vigilance behaviour in grazing African antelopes. *Behaviour* 79, 81–107.

Virtue ethics

Virtue ethics holds that morality consists of the exercise of virtues – settled dispositions of character founded on reasoning well about emotional impingements and desires.

Aristotle (384–322 BC) serves as a main source of inspiration for this agent-centred view of ethics, which highlights moral agents and their lives instead of actions or obligations per se. Virtue ethics approaches encourage those who interact with animals to cultivate the virtues of compassion, generosity, care and empathy. (RA)

See also: Ethics

Visual perception

Visual perception constitutes the ability to detect light as it reaches the eye and interpret the information contained within it, the resulting perceptual construct being known as sight or vision. Visual perception relies upon light reflected from surfaces in the environment; different colours and textures absorb and reflect different wavelengths of light, these then entering the eye, often passing through a lens which focuses them upon a photosensitive area, e.g. the retina of vertebrates and the eye of the octopus being an example of convergent evolution. The retina acts as a transducer, converting light into neural impulses, which are sent via the optic nerve to the lateral geniculate nucleus and visual cortex for processing. Sampling of the optic array is achieved through rapid movements of the eye, saccades, followed by fixations upon a specific object and smooth pursuit movements that facilitate tracking.

The retinae of most diurnal species contain two types of photoreceptors: (i) rods, which respond to low levels of illumination, allowing the perception of achromatic colour (scotopic vision); and (ii) cones, which respond to levels of brighter light and enable the perception of chromatic colour (photopic vision or colour vision). The number and type of cones in the retinae vary across species, determining the visible colour spectrum. Most mammals have two types of cone, some primates and marsupials have three while birds can have as many as six. The distribution of photoreceptors across the retina also varies across species, affecting visual acuity. In primates the highest density of cones is found near the centre of the retina in the fovea, allowing objects in frontal vision to be viewed with maximum clarity; however, in many other species including many domestic mammals, photoreceptors are most densely distributed in a horizontal band across the retina, the visual streak, this providing enhanced detection of movement across the visual field. However, in some species there is variation in the distribution, which appears to correlate with muzzle length, and thus short-nosed breeds of dog (brachycephalics) may have more of a fovea, while longer-nosed breeds (dolicocephalics) may have more of a visual streak.

The positioning of the eyes in the head influences visual perception, with frontally located eyes – typically found in predators – providing a narrower field of view than that afforded to species with more laterally positioned eyes, usually in prey species, that allow greater peripheral vision. The overlap, however, between the fields of vision of frontally positioned eyes allows a larger proportion of the visual field to experience depth perception through stereopsis, based upon the correspondence between the slightly different images delivered to the **brain** from the two eyes.

As the perception of the environment is influenced by the components of the visual system and the processing performed by the brain, it is useful to distinguish between the environment surrounding the individual and its **umwelt**, the environment as it is perceived by a given individual, which is what determines responses to visual stimuli. (FJW)

See also: Hearing

Further reading

Bruce, V., Georgeson, M.A. and Green, P.R. (2003) *Visual Perception: Physiology, Psychology and Ecology*. Psychology Press, Hove, UK.

McGreevy, P.D., Grassi, T.D. and Harman, A.M. (2004) A strong correlation exists between the distribution of retinal ganglion cells and nose length in the dog. *Brain, Behaviour and Evolution* 63, 13–22.

Wynne, C.D.L. (2001) *Animal Cognition*. Palgrave Macmillan, New York.

Vivisection

Literally meaning 'cutting up alive', the term was applied to the very invasive research carried out before **anaesthesia** was first discovered in 1845. Before then, animals were literally tied down or nailed to a table or floor so they could not move during the experimental surgery. Animals were used to further knowledge of anatomy and function, to demonstrate known facts before the public or as a part of an educational process, e.g. the training of veterinarians in anatomy. Groups who describe themselves as antivivisectionists are generally opposed to the use of animals in all experiments, whether or not the work is invasive or causes **suffering**. (DBM)

See also: Animal experiments; Animal rights

Vocalization

Vocalizations are sounds produced by animals via their vocal tract. Thus, vocalizations can be considered a subset of acoustic **signal**s, different from other animal sound such as those produced by crickets when rubbing together the striated surfaces of their wings. All sound is produced as a result of vibrations. The animal may modulate these sounds – for example, filtering and resonating different frequencies as the sounds pass through the vocal tract.

Vertebrates produce vocalizations by forcing air through a tube. The tube is constrained by membranes, constricting airflow and causing the membranes to vibrate. In mammals the vocal apparatus is called the larynx, and the vocal cords (glottis) block respiratory airflow to create vibrations. In birds, the analogous structure is called the syrinx; this structure consists of two sides that can independently produce sounds.

In birds, authors typically distinguish 'songs' from 'calls', with songs typically being longer and more complex than calls. Songs are also learned, while many calls can be produced with little or no learning. Calls are typically associated with a wide range of functions, such as the 'begging' calls that nestlings use to solicit parental care, or the 'alarm' calls used to alert others to the presence of a potential predator. In contrast, songs are used for intra- and intersexual **communication**; for example, male birds sing to advertise territorial boundaries to male rivals, or their quality as a potential mate to female listeners.

Vocalizations and other acoustic signals can be described in terms of the wave properties, including amplitude (typically measured in dB) and frequency (typically measured in Hz). More complex vocalizations can be broken down into component 'notes' or 'syllables', and each of these may vary in frequency or amplitude over time. Changes in frequency in time can be visualized using a spectrogram. Although historically many vocalizations have been classified according to gross spectrographic features, e.g. distinguishing a bark from a growl, from a whine, etc. in the dog, vocalizations within a given type may vary content with meaning; for example, different types of bark in the dog, are reliably associated with different contexts when they would be expected to convey different meanings. The use of computer algorithms offers a reliable way of characterizing the consistent features of different types of vocalization, opening up the way to a deeper appreciation of the range of auditory signals used in vocalization and other means of acoustic communication. (DW)

Further reading

Pongrácz, P., Miklósi, Á., Molnár, Cs. and Csányi, V. (2005) Human listeners are able to classify dog barks recorded in different situations. *Journal of Comparative Psychology* 119, 136–144.

Vole

These animals are **rodents**, and two species commonly studied are the bank vole (*Clethrionomys glareolus*) and the field vole (*Microtus agrestis*). In the USA they may be called field mice or meadow mice, but they do not belong to the genera *Mus*, i.e. true mice. (DBM)

See also: Laboratory animal; Mouse

Vomeronasal organ

The vomeronasal organ is also known as Jacobson's organ. It consists of a small, paired, tubular structure in the nasal cavity, lined with ciliated epithelium and including chemoreceptors and a collection of **neuron(e)**s. In mammals it is located in the septum and in **reptile**s in the roof of the mouth. It is present in most vertebrate animals. It develops in the **fetus** in man and higher primates, but is thought to subsequently regress, though opinions vary. It is present in most **amphibian**s and reptiles, though not in the crocodile nor chameleon, and it is also absent in birds and fish, where the vomeronasal receptors are located in the olfactory epithelium.

It usually has a narrow range of receptors that respond to a small range of compounds (**pheromone**s and possibly some other large molecules), far fewer than the more general olfactory receptors. The vomeronasal organ is not only anatomically distinct from the olfactory receptors, at least in tetrapod vertebrates, but is also linked to a different part of the **central nervous system** (accessory olfactory bulb). This suggests not only a difference in function but also a difference in evolutionary origin – probably evolving first in the **amphibian**s.

The vomeronasal organ stimulates behavioural responses to pheromones, especially related to **sexual behaviour** and the mediation of reproductive status, but its role in other behaviour systems is increasingly being recognized. Male and female **rats**, **guinea pig**s and **hamster**s whose vomeronasal organs have been surgically isolated no longer exhibit sexual behaviour (including the defence of territories). The mechanism for this is probably through the mediation of the release of **gonadotropin-releasing hormone (GnRH)** – and therefore the release of **luteinizing hormone (LH)**.

In amphibians and some reptiles it is also related to predator and prey recognition. The amphibian vomeronasal system has further been shown to respond to chemical cues released by injured or alarmed **consepecifics**. Pheromones are detected by passing inspired air over the vomeronasal organ, exemplified by the **flehmen** response (demonstrated by the equine, bovine, ovine and feline species), licking and teeth chattering (canids), touching the organ with the tongue (snake species and the komodo dragon), nose-bobbing (salamander) or, in the frog, showing no behavioural act, the cilia simply wafting introduced water across the vomeronasal organ. (DA)

See also: Odour; Olfaction

Von Frisch, Karl Ritter

Karl von Frisch (1886–1982) was born on 20 November 1886 in Vienna, the city where he also began his university studies initially in medicine, although he later switched to zoology (studying in both Munich and Vienna) and was awarded his PhD from Vienna in 1910. Much of his academic life was spent associated with Munich. He was jointly awarded the Nobel prize for Medicine or Physiology in 1973 with **Konrad Lorenz** and **Niko Tinbergen**, in recognition of his work on honeybees. He is largely recognized for having elucidated the honeybee dance as a means of navigational **communication**, but he studied the sensory system of bees more widely. This included demonstration of their colour vision and the role of the **pheromone**s produced by the queen in the regulation of the hive. In addition to his reputation relating to bees, he also published on various aspects of marine biology. (DSM)

Further reading

Von Frisch autobiography, available from http://nobelprize.org/nobel_prizes/medicine/laureates/1973/frisch-autobio.html (accessed 24 November 2009).

W

Weaned pig (weaner)

A weaned **pig** is a pig that is permanently deprived of suckling from the sow's udder. Before weaning, pigs rely on maternal milk for nourishment, though solid food may be gradually introduced. Naturally, **weaning** is gradual as milk consumption decreases and solid food consumption increases over time until pigs are fully weaned, at between 8 and 17 weeks of age; in swine production systems pigs are weaned abruptly, usually between 2 and 5 weeks of age. (RP)

Weaner

A weaner is a young animal that has been recently weaned from its mother – i.e. it is no longer nutritionally reliant on maternal milk or, in a husbandry setting, has been artificially deprived of maternal milk by maternal separation. The term is most often used for piglets (**see: Weaned pig**), but is also used for calves and lambs. For piglets, a weaner is a pig usually between the ages of 3 and 8 weeks of age. The bottom end of the scale will vary depending on the exact age of **weaning** – it may be as early as 2 weeks of age or as late as 5–6 weeks of age, depending on the system. The top end of the scale is fairly static, however, and after this age the piglet becomes a grower and then a finisher. For calves, weaning age will differ depending on whether the calf is a dairy calf or a beef calf. For dairy calves, weaning is usually around 6–8 weeks of age, although they are separated from the cow at between 1–4 days of age. Beef calves are usually reared on the cow up to the age of 6–8 months. For lambs, the usual weaning age is around 2–3 months, but may be much earlier in dairy **sheep**. In all cases, weaning under commercial conditions occurs quite differently to natural weaning. It occurs early and it is abrupt, with a sudden change from milk to solid feed, separation from the mother and often mixing with unfamiliar animals. The process is therefore considered to be particularly stressful for the weaner, and there are a number of welfare issues that have been subject to research.

Weaner pigs

The normal weaning age for piglets is anywhere between 8 and 19 weeks old. Under natural or semi-natural conditions, weaning is a gradual event. In effect, natural weaning starts early on in **lactation**. The frequency of suckling declines gradually from the first week, and the number of suckling events terminated by the sow increases, perhaps indicating that the sows become less inclined to nurse. Piglets begin to consume solid food from around 4 weeks postpartum and, by 8 weeks, solid food constitutes a large part of the piglets' diet. The number of piglets missing from suckling also gradually increases and weaning is completed anywhere between 8 weeks and 19 weeks postpartum, and there can be quite large variation in weaning age occurring within a given litter. Piglets also mix with piglets from other litters from around 2 weeks of age. Under commercial conditions, weaning occurs anywhere from 2 to 6 weeks of age and most typically at 3–4 weeks. At this point, piglets are most likely removed from the sow and placed into a new environment – the nursery accommodation. They may be kept as a litter or mixed with other litters weaned at the same time. In North American systems in particular, weaner pigs may also be subjected to long-distance transportation at weaning, with truck trips of hundreds of miles from the breeding farm to the rearing farm.

Major factors that affect the **welfare** and behaviour of weaner pigs include: weaning age, group size, group composition, space allowance, floor type, feeding system and thermal environment. A variety of **housing** systems are used for weaned piglets. Piglets are typically housed in highly controlled environments with supplementary heating in partly or fully-slatted pens or raised in flat decks in groups of varying sizes (10–40). They may be moved from the first-stage weaner accommodation to larger, second-stage accommodation after 2–4 weeks or remain in the same pen until the age of 8–10 weeks (30–40 kg) or, in a few instances, until **slaughter**. The pen area per pig varies from 0.2 (<20 kg) to 0.3 m^2 per pig (<30 kg). Weaner pigs are typically fed *ad libitum* (dry) or restricted (liquid) with an animal:feeder space ratio of 1:1–12:1, depending on the feeding system. Within nursery accommodation, the ambient temperature recommended and generally used (non-bedded, perforated floors) is in the range of 26–30°C, e.g. a temperature of 28°C for piglets weaned at 26–28 days of age.

The age at which piglets are weaned makes them particularly susceptible to disease. At around 3 weeks of age, a piglet's immune function is at its lowest. The protection afforded by maternal antibodies is by now very low, and its own immune system is just beginning to become functional. The challenges of a new environment, new pathogens, **social stress** and poor nutrition can have a large impact on piglet health around this time, and they become particularly prone to enteric disorders. Most of the bacterial pathogens associated with post-weaning enteric disorders are strains of enterotoxin-producing *Escherichia coli*. However, those strains of *E. coli* are also found in the digestive tract of healthy pigs, and the environmental conditions the piglets are exposed to before and at weaning play the most important role. Production results are strongly related to both the degree of segregated rearing systems and the level of hygiene. In herds rearing weaners on deep-litter bedding, both segregation and degree of hygiene are better, and the use of antibiotic three to four times lower than in herds with pigs in traditional post-weaning pens with solid **flooring**. Slatted flooring in the dunging area can lead to better pen hygiene, a lower infection pressure and lower **mortality** and **morbidity**, in comparison with solid flooring. Pens with fully slatted floors do have potentially better hygiene, but this type of flooring can also be harsh on piglets' hooves and limbs. An increased risk of injury from flooring can result in an increased risk of infection.

In terms of **space** allowance in relation to **disease**, 0.25m² per pig at post-weaning stage (between 10 and 25 kg live weight) is a best compromise regarding health maintenance and production criteria. From epidemiological studies, a progressive increasing risk of disease has been found when the number of pigs per pen, stocking rate and space at the feeder are more than 13 pigs, <0.30m² per pig and <7 cm, respectively. At the farm level, the factors involved are not independent. Certain housing systems tend to induce certain practices. Hence large pens are associated with more mixing of piglets from different litters at weaning.

Another important aspect of the weaning environment is **enrichment**. Fully slatted systems with no obvious manipulable material for the piglets are associated with unwanted behaviours such as aggression and pen-mate manipulation (e.g. **belly nosing**, **tail-biting**). Straw in particular can reduce **aggression** and promote **play** behaviour, and reduce belly-nosing.

Weaner calves

Natural weaning in **cattle** occurs when the calf is around 10–12 months of age. Generally, beef calves remain with their mothers until weaning at approximately 5–9 months of age. Veal and dairy calves, however, are separated from their mothers soon after birth and weaned around 6–8 weeks of age. Veal and dairy calves continue to be fed a liquid diet of either milk or milk replacer. Dairy calves undergo weaning at 1–3 months of age, whereas veal calves, especially those used for 'white' veal production, continue on a liquid-only diet with minimal roughage up to slaughter at between 5 and 7 months. Following separation, dairy and veal calves are often housed individually. In the USA, veal calves may still be reared individually in crates, which may be too small to allow the calf to turn round. Dairy calves are often reared individually in hutches with small runs, which allow basic body movements but not free locomotion and restricted social contact. In the EU, individual housing for calves older than 8 weeks of age is banned, and thus **group** housing is the norm. Group housing is becoming increasingly popular in the USA as well. The major advantages of individual housing are that of reduced disease transmission and ease of individual observation and treatment. However, calves are denied social contact.

The effects of individual versus group housing have undergone a fair amount of research, but direct comparison is often difficult as studies often vary a number of factors between treatments. With veal and dairy calves, studies have shown that mortality may be higher in group housing and the incidence of respiratory disease may be greater. However, difficulties in interpretation arise due to confounding effects of prior transportation. Group housing has no effect on likelihood of coliform, Salmonella or Cryptosporidium infection, but group housing may increase the risk of Johne's disease. More problems are found if group size is larger than 6–8 animals.

Behaviourally, individual housing does not allow calves to show many elements of their normal behavioural repertoire. Ordinarily, group-housed calves may spend around 2% of time engaged in **social behaviour**. Individually housed animals vocalize more and will work for social contact, and thus social behaviour must be seen as having some degree of importance to the calf. Individual housing may also have short-term effects on the calf's increased fearfulness when introduced to unfamiliar animals. The biggest restriction that individual housing places on the calf is its ability to carry out its full locomotory repertoire, and crates may also prevent certain postures that the calf would normally adopt. The main behavioural disadvantage of group housing is the increases seen in **cross-suckling** – where calves will attempt to suck on pen-mates. This behaviour is greatly influenced by the method of milk presentation during this phase of life. If milk is delivered such that it is sucked from a teat, then cross-sucking incidence is greatly reduced, relative to bucket-fed calves.

Weaning ages for dairy and beef cattle differ considerably. Beef cattle are often left with their mothers on pasture-based systems until they are abruptly weaned anywhere between 5 and 9 months of age. Alternatives to this abrupt, complete separation are either to practise fenceline separation that allows the calf some social contact with the cow, or to attach a device to the calf that prevents the calf from suckling from the udder. Both these methods appear to reduce the calf's stress responses when the cow is finally removed. After weaning, many beef calves are transported to a **feedlot**. These are often open pens with dirt, concrete or slatted floors and a feed bunk. The range in sizes varies greatly, but it is not uncommon in the USA to have hundreds of animals within each pen. At this time, feedlot cattle are often subject to long-distance **transport**, mixing with unfamiliar animals and sudden change to a grain-fed diet to maximize growth. Unsurprisingly, mortality can occur and a number of health issues become apparent, such as increased respiratory disease, bacterial and protozoal infections, acidosis, **bloat** and liver abscesses. Without shelter, cattle may also be subject to both **heat** and **cold stress**.

For dairy cattle, abrupt weaning often occurs between 1 and 3 months of age, although separation has already occurred. Nevertheless, weaning is often accompanied by signs of obvious distress, with increased **vocalizations** and decreased growth. Initially, calves are weaned on to a grain-based diet, and they must undergo a further change on to a forage-based diet at a later date. Weaned dairy heifers are usually kept in groups and most have access to outside areas for at least part of the year. Obvious factors that will impact the welfare of these individuals are such things as group size, space allowance and bedding availability. However, the area of heifer housing has been somewhat neglected research-wise.

Weaner lambs

Natural weaning in sheep occurs when the lamb is around 5–6 months of age, and usually corresponds to the time that the ewe is coming into **oestrus** for the next breeding cycle. The ewe–lamb **bonding** is particularly strong during the first month, but there is evidence that it weakens over time and, by 12 months of age, relationships with peers appear to be more important. Artificial weaning in commercial production will depend on whether the flock is a meat production flock or a dairy flock. With a typical meat flock, weaning will occur at 2–3 months of age and the lamb will stay on the ewe for this time. With a dairy flock, as with dairy cattle, maternal separation may occur immediately after **parturition** (1–4 days). If not occurring then, disruption of the ewe–lamb bond during the first month appears to be particularly stressful for both parties, and the lamb usually remains on the ewe until around 4 weeks of age.

Lambs from dairy ewes will therefore undergo a period of artificial rearing, during which time they will be group-housed with other lambs of the same age and fed milk replacer from a bucket. These lambs appear to shift their social relationships towards pen-mates and can benefit from **gentling** from a **stockperson**. As well as the social aspect, artificial rearing also affects feeding and growth rates. Allowing the lambs to have a mix of bucket-fed maternal milk and milk replacer counteracts some of the negative effects of replacer-only diets, improving growth, cellular immunity and behavioural responses to **open-field tests**. However, growth remains below that of lambs reared on the ewe. Gradual separation of lambs from dairy ewes has a larger negative impact than abrupt separation. Artificial rearing of lambs can also be associated with behavioural disturbances, such as redirected sucking on pen-mates and other non-food substrates, suggesting that these systems do not meet all the **behavioural needs** of young lambs.

For meat lambs and dairy lambs, weaning is usually carried out at 3 months. For meat lambs that have stayed with the ewe for this period, the stress of weaning and maternal separation is apparent. Separation causes increases in stress hormone concentrations, increases in activity, high vocalization rates and disruption of **circadian rhythms** of activity. Although lambs can be successfully switched to solid food by 1 month of age, even at 3 months of age the lamb usually has a period of reduced growth as the rumen is not fully matured. As with piglets, susceptibility to disease increases through impaired immune responses. Gradual separation, or allowing the lamb to remain with the ewe although prevented from suckling, appears more stressful for the lamb than abrupt separation. At 3 months, weaning may also disrupt social structure. After separation from the ewe, lambs re-associate in temporary groups, and an increase in pair association between twins is observed. It may take a year before full flocking behaviour is attained. Rearing of lambs in single-sex groups can also impact sexual behaviour, particularly of males. (JNM-F)

Further reading

Marchant-Forde, J.N. (2009) (ed.) *The Welfare of Pigs*. Springer Science and Business Media B.V., Dordrecht, the Netherlands, 330 pp.

Napolitano, F., De Rosa, G. and Sevi, A. (2008) Welfare implications of artificial rearing and early weaning in sheep. *Applied Animal Behaviour Science* 110, 58–72.

Rushen, J., de Passillé, A.M., von Keyserlingk, M.A.G. and Weary, D.M. (2007) *The Welfare of Cattle*. Springer Science and Business Media B.V., Dordrecht, the Netherlands, 310 pp.

Weaning

Weaning is defined as a process of permanent deprivation of nursing milk from the mammary gland of the mother (called udder for most animals) or any other artificial source of milk. Nursing animals rely on maternal milk for physical, cognitive and emotional formation, which provides the framework for proper development. Natural weaning happens slowly as the young animal is gradually introduced to adult-like solid feed, hence adjusting the digestive tract to the new diet. The solid feed should complement and/or compensate for nutrients, until then provided to the offspring through the maternal milk, for successful growth and development. An animal (or infant) is considered to be completely weaned when it no longer receives milk as part of its diet, hence relying solely on solid feeding for nourishment.

The age at which natural weaning occurs is evolutionarily species-specific. According to the literature, the time span during which an animal should stay **nursing** on its dam can be correlated to several factors. One of the assumptions is that weaning in mammals should happen when the offspring reach triple or quadruple that of their birth weight or one-third of adult weight. Another parameter relates to **gestation** length: the age at which weaning occurs is proportionally correlated with gestation length, but this correlation may be significantly affected by the adult size of the animal. For instance, small primates nurse for a period shorter than the gestation length, whereas large-primate nursing time exceeds their gestation period. Also, dental (or tusk) eruption has been described as a potential threshold for terminating nursing in some species (e.g. primates, **elephants**). Naturally, weaning age may be anticipated as a consequence of a decrease in the availability of essential **resources**, such as food and water, thus reducing and terminating maternal milk production.

Most animals living in their natural habitats (e.g. wild animals) tend still to wean their offspring gradually and at their natural time. However, this scenario is not commonly, if ever, seen in domestic species mostly subjected to animal production practices. In species such as swine, for instance, artificial (or enforced) weaning is a common practice and usually occurs abruptly, and is accompanied by complete separation from the dam. Additionally, it generally takes place several months prior to the biological age of weaning. For instance, the natural age for **pigs** to wean can range from 8 to 12 weeks but can happen up to 17 weeks of age (see Table W.1).

In swine production, pigs are artificially weaned at an average of 21 days of age and almost universally at a range of between 2 and 6 weeks of age. Dairy calves are separated from their dams soon after birth, ranging from 1 to 4 days of age, and fed with milk supplement, while they would naturally be weaned from their mother at around 12 months of age. Beef calves are usually weaned between 6 and 8 months of age. This implies that the still-nursing animals are taken away from their dams and are subsequently reared alone or with other artificially weaned animals.

The constraints dictated by the **intensification of animal production** systems have caused dramatic changes to the weaning age of several domestic species. This occurs because agricultural production has some important advantages with weaning animals earlier than their natural age. One advantage is the early control of diseases. An example of this is called 'all-in/all-out' management – a common practice in swine production. This practice involves removing, at the same time, all piglets from their dams and placing them in another separated, sanitized facility (nursery), eliminating pathogen exposure from the sow and older pigs. Piglets, when early-

Table W.1. Age ranges (weeks) for natural and artificial weaning for a range of animal species.

Species	Natural weaning age (range)	Artificial weaning age (range)
Primates	52–78	13–26
Swine	8–17	2–4
Rats	4–5	3
Dairy cattle	43–52	26–35[a]
Beef cattle	43–52	26–35
Sheep	18–23	9–13
Horses	52	18–26
Cats	8–12	5–6
Dogs	10	3–10

[a] Permanently separated from the dam at 1–4 days postpartum.

weaned at around 2 weeks of age, are still under the immunological protection of maternal antibodies transferred in the colostral milk (**see: Colostrum**). However, to be effective, it is extremely important that newly weaned animals are housed in an uncontaminated environment for efficient **disease** control.

Artificial weaning also results in a more rapid return of the female to an active reproductive state. The separation of the offspring from the mother triggers the female hormonally to return to an active reproductive cycle, thus improving its reproductive efficiency (**see: Reproduction**). Maximizing the female's reproductive rate results in reduction of financial investments and higher profits due to a greater number of offspring produced in a shorter period of time. However, this idea is contradictory for some species (e.g. rhesus **monkey**, beef **cattle**), where non-interference in the natural weaning process may instead inhibit maternal reproductive system, by avoiding the inherent psychological **stress** associated with maternal–offspring separation. Permanent mother–infant separation prior to natural weaning is also a common husbandry practice in monkey-breeding colonies.

A better understanding of the nutritional needs of young animals has allowed improvement in feed quality and composition and nutritional management practices. Nowadays, diets are formulated to substitute all nutrients ingested from the milk and in fact taste similar to the maternal milk, thereby increasing palatability. Complex diet formulations favour higher consumption and thus better weight gain and growth rates of young animals, enhancing survivability. The earlier an animal is weaned, the greater are its nutritional requirements and more costly is the feeding programme. When abruptly performed, weaning is marked by weight loss and transitorily reduced growth rate, due to lower feed intake and stress resulting from permanent maternal deprivation. **Abnormal** behaviours are also observed in several animal species subjected to artificial weaning. These consequences may be more severe in animals weaned too young, and certainly are indicative of a poor welfare state.

Weaning – a stressful event

Weaning is a traumatic and stressful event that may be accompanied by several other **stressors** that, when summed, can affect the health and **well-being** of the weaned animal and, ultimately, its dam. Physical separation of a young animal from its mother, and in some cases from its siblings, is undoubtedly stressful for both parties. As an alternative and to reduce the negative impact of maternal separation, weaned animals may be placed in an adjacent area or in the proximity of their dam. This procedure facilitates weaning by permitting visual contact and exchange of calls between the animals. However, this procedure is not always feasible, depending on the animal species and production system under which animals are raised. Another option is the use of an anti-suckling device, commonly used in beef cattle to carry out a two-stage weaning method. This apparatus is first placed on the nostrils of the calves and prevents the animal from nursing while still being kept with the dam. This stage is then preceded by full physical separation of calves from cows, which can also experience distress post-separation.

Most animals when weaned undergo a transitory loss of weight and reduced growth, with some being more susceptible to transient diseases such as diarrhoea and mild respiratory disturbances. These effects can be caused by sudden change of diet from milk to solid feed or by simply subjecting animals to a different environment. All these factors are combined into neurophysiological stress, caused by the lack of maternal contact. In some species, permanent maternal separation is performed when the animal is still very young and its digestive system is not yet fully formed to process diets other then milk. An example of this scenario is the maternal–offspring separation in dairy calves that is performed when calves are at maximum of 4 days of age. Thus, supplementation with artificial milk coupled with solid diet is provided to allow the digestive tract gradually to adapt to a solid diet such as total mixed ration and hay. Dairy calves are fully weaned from milk supplementation at approximately 3–6 weeks depending on body weight/condition and when enough grain mix is being appropriately consumed.

There are also negative consequences for the mother when the offspring is separated from her. A sudden removal of a suckling offspring from its dam can cause milk to accumulate in the udder and the occurrence of **mastitis** (udder **inflammation** usually caused by bacterial infection). Maternal milk production peaks within days or weeks after giving birth, varying with species, as young animals increase their milk consumption. Milk production is then gradually decreased as young animals increase solid diet consumption. Additionally, behavioural changes can be observed in the dam such as increased number of calls in an attempt to communicate with the offspring, more pacing or walking (restlessness) and reduced feed intake, indicating distress. In some cases, acute reduction of milk production is seen. In cattle, and most likely in other species also, the longer period the offspring spends in contact with the dam, the more intense are the behavioural responses of the dam immediately after separation and, in cattle, the presence of calves can positively affect the **health** and productivity of cows.

Usually, artificial weaning is accompanied by environmental changes, when weaned animals are transferred to other facilities or areas to be housed. Adequate ambient conditions, such as good air quality, light intensity and suitable environmental temperatures are vital requirements for young animals in order to minimize disease occurrence and allow proper growth. For animals housed in groups, overcrowding in the environment should be avoided (**see: Social stress**). Attention and care, by trained personnel, must be provided especially to animals weaned at very young ages.

The effects of early maternal separation have been extensively studied and shown to cause long-term psycho-physiological effects on the **brain** and behaviour of animals. The process of maternal separation affects the individual's behaviour by eliminating the predictability and controllability provided by the maternal presence and its interaction with the offspring. Weaning has been associated with a range of abnormal behaviours and/or **stereotypies**, which are **repetitive behaviours** with no biological function, and are considered to be indicative of poor psychological well-being. These behaviours are more evident in artificially weaned animals and are suggestive of neurophysiological stress, therefore being of welfare concern.

Behavioural responses to and after weaning or maternal separation are at some levels, species specific. **Crib-biting** in foals has been described to be associated with concentrate feeding and has been shown to increase after weaning. Artificially weaned piglets perform abnormal behaviours, such as **belly nosing**, where the animal mimics nursing by nosing the abdominal region of another pen-mate. Young weaned pigs also show chewing behaviour towards the pen bars, which is indicative of stress. Behaviours such as nosing, chewing or sucking other animals resemble massaging the maternal udder and are contradictorily discussed as to whether they are abnormal behaviours associated with feeding or linked to social interaction. Early-weaned pigs also show reduced capacity to learn a spatial task and have an impaired ability to recognize familiar animals. Monkeys that have been separated from their mothers at birth or within the first year of life and raised in part or total social **isolation** show self-directed stereotypic behaviours, such as self-sucking of digits or tail and self-clasping in the absence of a **conspecific**.

Dam and offspring, when abruptly and permanently separated from each other, endure a traumatic experience. Increased rates of **vocalization** have been observed in several species following maternal deprivation in both dams and offspring, as both use calls for identification. In cattle, the mother and her calf vocalize intensively and show restlessness for several days, accompanied by less time lying and eating post-separation. These behavioural responses are exacerbated in beef calves suddenly weaned, at approximately 6–8 months of age, compared with behaviours of animals subjected to the two-stage weaning process, which allows the calf to gradually wean and separate from the mother.

Aggressive behaviour is another factor enhanced by early artificial weaning and has been found in piglets following maternal deprivation. **Rat**s maternally separated also become more aggressive and show depressive and anxiety-like behaviour that can be accompanied by fearful or escape behaviour. The occurrence of abnormal behaviours, increased **aggression** and cognitive deficits suggest that the stress of weaning, an early adverse experience, alters the development and function of both the neuroendocrine and **autonomic nervous system** of animals.

Negative early experiences, such as maternal separation in **rodent**s and other species, have been associated with multiple alterations in the adult brain, and are correlated with specific behavioural phenotypes depending on the timing and nature of the event. Maternal deprivation, especially if performed very early in life, is likely to interfere with the proper development of both psychobiological and neuroendocrine regulatory mechanisms in the developing brain. The stress-associated effects may be severe when associated with transitory nutritional deficit resulting from sudden alteration in the diet composition. This last factor may contribute to an impairment of neuronal mechanisms caused, for instance, by amino acid deficiency.

Early weaning of piglets can provide a useful model for studying early maternal separation effects on behaviour and neuroendocrine mechanisms. Previous studies revealed that piglets, when weaned at up to 28 days, show abnormal behaviour and intensified aggression later in life. In addition, piglets subjected to early weaning show suppression of genes encoding proteins associated with the stress response in the brain. A large body of scientific evidence indicates an age-dependent sensitivity of brain pathways to stress hormones. In pigs, for instance, the first 3 weeks of age are considered critical 'windows' of brain plasticity to alter behaviour and brain development. Weaning or maternal separation alone is a highly stressful episode that results in the release of **corticosteroids**, such as cortisol, and an increase of their concentration in plasma and **urine**. Cortisol crosses the blood–brain barrier and affects several brain areas by binding to corticosteroid receptors. One of the primary outcomes of central **glucocorticoid** action is to mediate a negative feedback system that suppresses **hypothalamic–pituitary–adrenal (HPA) axis** activity when its levels are elevated. Therefore, at early brain developmental stages, the disruption of the glucocorticoid negative feedback system may compromise the developing brain and be accountable for some aspects of cognitive impairment and abnormal behaviours.

Brain areas with abundant levels of corticosteroid receptor binding sites can suffer disruption of their functions under stressful events, such as weaning, primarily during the **sensitive phase** of brain development. High concentrations of the hormone in these brain areas cause behavioural and cognitive deficits that may not be reversible. Hence, early environmental manipulations, primarily during the first weeks of age or during the critical stage of brain development, may result in permanent changes in the HPA axis function and cause lifelong changes in behavioural responses. Currently, cerebral changes caused by stressors such as weaning can be studied in detail with the help of molecular biology techniques. These techniques can be used to characterize, for instance, changes in gene expression profile in various tissues following exposure to stress. Molecular biology techniques are viable approaches for examining underlying neuronal changes that may affect animal behaviour and welfare of animals. Thus, connecting neurophysiological changes to behavioural alterations in response to stressors may facilitate a better understanding of the mechanisms underlying neuronal and behavioural effects of stress in animals weaned early in life. (RP)

See also: Fence-line behaviour; Rejection

Further reading

Flower, F. and Weary, D.M. (2003) The effects of early separation of the dairy cow and calf. *Animal Welfare* 12, 339–348.

Plotsky, P.M., Sanchez, M.M. and Levine, S. (2001) Intrinsic and extrinsic factors modulating physiological coping systems during development. In: Broom, D.M. (ed.) *Dahlem Workshop on Coping with Challenge: Welfare in Animals Including Humans*. Dahlem University Press, Berlin, pp. 169–196.

Poletto, R., Siegford, J.M., Steibel, J.P., Coussens, P.M. and Zanella, A.J. (2006) Investigation of changes in global gene expression in the frontal cortex of early-weaned and socially isolated piglets using microarray and quantitative real-time RT-PCR. *Brain Research* 1068, 7–15.

Poletto, R., Steibel, J.P., Siegford, J.M. and Zanella, A.J. (2006) Effects of early weaning and social isolation on the expression of glucocorticoid and mineralocorticoid receptor and 11β-hydroxysteroid dehydrogenase 1 and 2 mRNAs in the frontal cortex and hippocampus of piglets. *Brain Research* 1067, 36–42.

Waters, A.J., Nicol, C.J. and French, N.P. (2002) Factors influencing the development of stereotypic and redirected behaviours in young horses: findings of a four year prospective epidemiological study. *Equine Veterinary Journal* 34, 572–579.

Worobec, E.K., Duncan, I.J.H. and Widowski, T.M. (1999) The effects of weaning at 7, 14 and 28 days on piglet behaviour. *Applied Animal Behaviour Science* 62, 173–182.

Weather – effects on behaviour

Climatic physiology of farm animals is well understood, because they will not thrive in environments that depart from their climatic optima. Thus cattle of temperate climates, whose core body temperature is 38.2°C, have a thermoneutral zone from about –20°C to +26°C (**see: Thermoregulation**). This zone is defined as the temperature range within which metabolic heat production is independent of air temperature. Below it, the animals must generate extra heat. Above it, their heat loss mechanisms based on evaporation (sweating, panting) become unable to cope and metabolic rate rises, raising body temperature and endangering the life of the animal. The yak can feed and move normally at air temperatures of –40°C, but at 13°C its respiration rate increases, and at 16°C its heart rate and body temperature rise. Yak require the annual mean temperature to be below 5°C, and the average in the hottest month must not exceed 13°C. The pig has a very narrow thermoneutral zone (+16°C to +24°C), and for this species careful climatic control in **housing** and outdoors is essential.

Responses to uncomfortable ambient temperatures can be either physiological or involuntary (shivering, increased respiratory rate, sweating, etc.) or cognitive (shelter- or **shade-seeking behaviour**). Animals respond to rain, and to precipitation generally, by cognitive processes. Whether these behaviours are performed will depend on the strength of **motivation** to perform or continue other behaviours, such as grazing. (SJGH)

Further reading

Li, C. and Wiener, G. (1995) *The Yak*. FAO Regional Office for Asia and the Pacific, Bangkok, Thailand.

Weaving

Weaving is the lateral swaying of the head. It may include swaying of the rest of the body (notably the shoulders) and picking up the front, and sometimes the hind, legs. Also recognized in exotic species, notably **elephants** and **bears**, weaving is a common **ethopathy** in **horses**. It usually occurs over the stable door or in the face of some other barrier. As with other **stereotypies**, weaving may develop over time (e.g. it may change from a stationary head-swing to a side-to-side pacing). More than one stereotypic behaviour may be performed by an individual horse, e.g. a weaver may also be seen **stall-walking** on occasions. But the risk of locomotor versus oral stereotypic behaviours seem to be largely independent within an individual.

Concern for the **welfare** of weavers relates to the effects of excessive wear and tear on the hooves and the musculoskeletal system. Weaving may have a long-term effect on the animal's ability to learn; stereotypic horses have been reported to persist in non-rewarding activities longer than control horses (Hausberger *et al*., 2007; Hemmings *et al*., 2007). Weavers may also show altered time budgeting that represents a shift toward stereotyping at the expense of resting and feeding, an outcome that may result in a failure to thrive. This is one of the chief reasons for weaving being an unwelcome behaviour that must be reported at auction in some countries.

Weaving is reported more in Thoroughbreds (3.25% ± 3.23) than in other breeds, but it is unclear whether this reflects their **breeding** or environment, as gallopers are managed more intensively than any other equine athletes. It tends to arise at a median age of 60 weeks (i.e. after youngstock have been sold from the stud to new homes) and so does not appear to be related to **weaning** so much as to environmental and social disturbances at the time of dispersal. Along with other locomotory stereotypies, it is commonly seen in stabled horses before feeding or other arousing events, such as the arrival or departure of **conspecifics**. So it seems unlikely that weavers are subject to understimulation (or what might be labelled '**boredom**'), but rather that they have a thwarted **motivation** to move, e.g. to be with other members of the group or to reach food (**see also: Frustration**). Weaving may either be the expression of an unusual form of species-typical anticipatory behaviour and/or a learned response to a desirable outcome, such as leaving confinement.

Although reported as both more and less prevalent in standing stalls than in loose boxes, weaving is observed in pasture only when horses have encountered a barrier. If being in a stable is frustrating for a horse, it is possible that the lack of free exercise plays a role in the motivation to weave. However, the relationship between exercise routine and stereotypic behaviour is unclear. Exercise routine has not been identified as a risk factor in epidemiological studies and there is little evidence that exercise has a consistent effect on the incidence of stereotypic behaviour in the stable. However, stimuli that signify turning out (to pasture) often act as catalysts for episodes of locomotory stereotypic behaviours.

When compared with the traditional solid-sided loose box, the provision of partitions that permit close tactile and visual contact with neighbouring horses reduces weaving. For example, in short-term studies, weaving was not observed when horses could socially interact with their neighbours on all four sides of their enclosure (Cooper *et al.*, 2000). This adds weight to the suggestion that weaving can be regarded as a frustrated escape response.

Found in approximately 70% of UK establishments, anti-weaving bars, V-shaped apertures through which horses can put their heads and necks while limiting lateral movement, are the most common prophylactic for, or response to, weaving. Other attempts to prevent or interrupt weaving include fixing an upright bar or suspending a heavy object to occupy the **space** above the door, or completely preventing the horse from putting its head over the stable door. In response to all of these physical measures, most weavers simply move to an area within the box and weave without having their head and neck over the door. The merits of such physical impediments to stereotyped locomotion are spurious, since they probably increase affected horses' frustration with the environment by reducing the utility of the space provided.

Modifying cues that precede feeding or feeding times may reduce pre-feeding weaving but not post-feeding stereotypic behaviour. This, and the involvement of certain stimuli that reliably precipitate episodes of weaving, suggest that the role of learning in the emergence of stereotypies can be significant. So, while changing the husbandry routines may be an effective treatment of weaving, it is unlikely to be permanent because affected horses will most probably learn new associations that predict feeding in novel routines. Such changes would have to be continual to maintain their novelty value to avoid the behaviour returning to its original level. Crucially, however, flux in routine is considered harmful to horses, not least in terms of gastrointestinal **health**.

Exercise may help to reduce time spent weaving by causing fatigue and therefore prompting rest, which is an appropriate activity for stabled horses. Additionally, a practical measure may be to provide stabled horses with mirrors. In both short-term and long-term studies, mirrors have a similar effect to social contact, although it is unclear whether horses are simply distracted by reflected visual stimuli or whether they regard the image as another horse.

At the time of its emergence, weaving may be seen as a sign of frustration and possibly an indicator of compromised welfare. The extent to which a horse that is currently weaving can be used as an indicator of compromised welfare is contentious. Many authors highlight the importance of the process of **emancipation** in liberating stereotypic behaviours from their initiating causes and argue that mature stereotypies may have lost any putative role in ameliorating distress. One reason for this being contentious relates to our poor understanding of the neural mechanisms that might reinforce weaving. For example, the role of **endorphin**s as an intrinsic reward for weaving remains equivocal since **opioid** antagonists have been shown to both decrease and increase weaving.

Husbandry systems that facilitate social housing and therefore, by implication, are associated with reduced emergence of weaving are becoming increasingly popular, even though they may remain unattractive to some owners because they may permit undesirable social interactions (e.g. biting and bite threats) and the spread of contagions. (PDM)

References and further reading

Cooper, J.J., McDonald, L. and Mills, D.S. (2000) The effect of increasing visual horizons on stereotypic weaving: implications for the social housing of stabled horses. *Applied Animal Behaviour Science* 69, 67–83.

Hausberger, M., Muller, C., Gautier, E. and Jego, P. (2007) Lower learning abilities in stereotypic horses. *Applied Animal Behaviour Science* 107(3–4), 299–306.

Hemmings, A., McBride, S.D. and Hale, C.E. (2007) Perseverative responding and the aetiology of equine oral stereotypy. *Applied Animal Behaviour Science* 104, 143–150.

Mills, D.S. and Davenport, K. (2002) The effect of a neighbouring conspecific versus the use of a mirror for the control of stereotypic weaving behaviour in the stabled horse. *Animal Science* 74, 95–101.

Weight gain

Weight gain is the increase in an individual's weight over a given amount of time. In the farm animal industries, weight gain is often measured in terms of average daily gain (ADG) in grams (g) or kilograms (kg) per day and is used an indicator of how well an animal is growing in a given husbandry system. It can be combined with other parameters for **welfare** assessment. However, as with other parameters, it should not be used in isolation, and fast and efficient growth does not necessarily indicate good welfare. Its importance as a measure somewhat depends on the purpose for which the animal is being kept. For **companion animal**s, the ideal is for the animal to grow normally, with normal weight gain until it reaches its final mature body size and thereafter to maintain its weight within a normal range for the remainder of its natural lifespan. These goals are similar for **zoo animals**. For **laboratory animal**s, depending on the species and use to which they are put, they may or may not have a truncated lifespan and thus may or may not reach mature body size. However, the goal again would be to achieve normal weight gain up to maturity, then maintenance.

For **farmed animal**s there are different management approaches to weight gain, depending on whether the animal is to be part of the **breeding** flock or herd or to be reared for meat production. For a breeding animal, growth and maturation are managed to maximize and prolong reproductive life. Often, this means controlling weight gain below that of the animal's maximal potential and then reaching a plateau of maintenance, but ensuring sufficient nutrition to support growth of the offspring during uterine development and subsequent **lactation** (for mammals) or maximize **egg production**, fertility and offspring viability (for poultry). For an animal being reared for meat, the aim is to get that animal to gain weight at its full potential – i.e. to grow maximally for minimal input and to reach **slaughter** weight in the shortest time possible.

Generally, the amount of weight that an individual will gain will depend on the relationship between the total amount of energy taken in and the total amount of energy expended.

Each animal has its own basal energy expenditure – that is the amount of energy required to maintain the body's normal basal metabolic activity, i.e. to keep the heart beating, sustain breathing, repair tissues and keep the **brain** and nerves functioning. Superimposed upon this are extra energy requirements needed for growth (if applicable), **thermoregulation** and carrying out a normal behavioural repertoire. Also, if the animal is under **stress**, extra energy is required to meet the subsequent metabolic increases, such as increased **heart rate** and respiration rate, increased **locomotion**, etc.:

- If energy in exceeds energy expended, then the animal will gain weight.
- If energy in equals energy expended, the animal's weight will be static.
- If energy in is less than energy expended, then the animal will lose weight.

For breeding animals, growth, and thus weight gain, is usually controlled so that the different systems within the animal's body develop at a similar rate.

Due to genetic selection, many animals have the potential to grow extremely fast, and it is not uncommon for growth in the muscular system to outpace growth of the skeletal system. Rapidly growing animals such as some horses, pigs, broiler chickens and certain breeds of dog are known to have an increased incidence of leg weakness and bone diseases such as osteochondrosis (**see: Lameness**). In order to maximize reproductive longevity, growth is controlled so that the breeding animal has the musculoskeletal foundation to support reproductive function. This is often achieved by not allowing the animal to have *ad libitum* access to feed, but ensuring that daily energy intake remains above energy expenditure, thereby allowing limited growth. Once the animal has reached what is considered an 'adult' size and weight, feed intake may be reduced even further, so that the animal is theoretically obtaining levels that allow maintenance of body weight and support reproductive function. Again, this is achieved by restricting feed intake.

Monogastric species, such as breeding sows and broiler breeders, are often fed a single meal of high-energy concentrate diet, which can be consumed in a relatively short period of time, when they would often spend a great amount of their daily **time budget** engaged in foraging for a relatively low-energy diet. As a consequence, these animals are usually in a state of chronic **hunger**, which can impact the animal's physiology, behaviour and welfare. Also, some animals such as pigs and cattle have their commercial reproductive lifetime during the period in which they are actually still growing and have not yet reached full adult body size. It is therefore important that the amount of feed given takes growth into account, as well as stage of gestation or lactation, so that the animal does not enter a negative energy balance and start losing body condition.

For meat animal production, a more important figure than just weight gain is the ratio between weight gain and the amount of feed eaten – the gain:feed or G:F ratio. To maximize profit, the farmer is aiming for a high G:F ratio, meaning that the animal is making efficient use of the energy within the feed to put on weight. For example, it is more cost effective for a pig to grow 500 g/day by eating 800 g/day of food than to grow 510 g/day but eat 900 g/day of food. For the first pig, the G:F is 0.625, but for the second pig it is only 0.567. However, the financial benefit of greater feed efficiency does also need to be balanced against the costs associated with the extra time taken to reach slaughter weight.

Typical or expected weight gains are available for many domestic species. Although meeting the expected weight gains cannot be taken as an indicator of good welfare by itself, falling below expected weight gains can be an indicator of a welfare challenge. The most common causes of an animal failing to thrive are clinical or subclinical **disease**, inadequate access to feed or inadequate thermal environment. Many diseases may cause lethargy and anorexia (loss of appetite), meaning that the animal may not get up and feed. In groups of animals with poorly designed, competitive feeding systems, subordinate animals may fail to gain access to their allotted feed allowance and thus fail to meet their nutritional demands. Animals being housed outside their thermoneutral zone may either expend more energy maintaining body temperature if it is too cold or may become anorexic if it becomes too hot. In both cases, the energy balance is disrupted and weight gain will decrease.

(ELS)

See also: Body weight – metabolic; Feeding systems

Welfare

'Animal welfare' is a wide-ranging, and often value-laden, term that is used with somewhat different meanings by different people. It appears to have been first used by Major C. W. Hume, who was instrumental in founding the University of London Animal Welfare Society in 1926.

In non-technical assertions like 'securing the welfare of animals in our care is vital', the term 'welfare' refers to positive **well-being** or **quality of life**. In the technical literature on animal welfare, by contrast, it makes sense to speak about welfare as a continuum running from negative to positive. In some contexts (e.g. in the notion of an animal welfare organization) the phrase 'animal welfare' carries the implication that we have an ethical obligation to treat animals well, or so as to meet certain minimum standards. However, many academics who study the living conditions of domestic and wild animals normally aim to use the term 'animal welfare' descriptively, simply to record the state in which an animal happens to be, without implying anything normative about the acceptability of the conditions in which the animal is kept. Also, in the academic study of animal welfare the term is normally used strictly to describe reactions of the individual animal being studied: welfare is a state of the animal. Here 'welfare' is used as a synonym of 'well-being' or 'quality of life', whereas in other, non-academic, contexts it can be used in statements about the environment in which the animals live.

The study of animal welfare is mainly undertaken as a part of natural science, building on disciplines such as **ethology**, **pain**- and **stress**-physiology and veterinary medicine (**See also: Measuring welfare**). However, important underpinnings of the study of animal welfare are philosophical in nature. Before studying animal welfare, we need a definition that clarifies what it is for an animal to be in conditions that are good for it and what it is for an animal to live a good life. From philosophical discussion of human well-being dating back to ancient Greece, two main views about the nature of welfare emerge: (i) *perfectionism* (and associated *objectivist* views), according to which welfare is connected with *doing well*, e.g. by realizing important species-specific potentials; and (ii) *hedonism* (and allied *subjectivist* views), in which welfare is connected with *feeling well*, e.g. in the experience of pleasure and the avoidance of pain.

Often, advocates of these rather different views will arrive at similar conclusions about what is needed for an animal to live a good life. Positive **feelings** typically follow if an animal is allowed to pursue its natural goals, and pain or feelings of **frustration** may follow if an animal is thwarted in what it is naturally disposed to do. (Hens in battery cages appear to experience frustration when they are prevented from laying eggs in a nest.) However, it is not difficult to think of cases in which the two views might well diverge. Fighting to rise in the social **hierarchy** is natural in many animals and therefore positive from a perfectionist view. Nevertheless, a hedonist may consider it desirable to prevent painful fights – perhaps by limiting the **aggression** of male individuals through **castration**.

It is fair to say that subjectivist views presently dominate academic discussion of animal welfare. When Ian Duncan (1996) states that 'sentience, in other words feelings, is what animal welfare is all about', he is presenting a mainstream view. However, there are dissenting views: **Bernard Rollin** (1993) has observed: 'It is likely that the emerging social ethic for animals ... will demand from scientists data relevant to a much increased concept of welfare. Not only will welfare mean control of pain and suffering, it will also entail nurturing and fulfilment of the animals' natures, which I call **telos**.'

Again, in the following influential definition offered by Donald Broom (1986), the emphasis is on **coping**, or functioning – a basic form of perfection – rather than on feeling, although Broom himself acknowledges that **feelings** may be an important part of functioning:

> The welfare of an individual is its state as regards its attempts to cope with its environment. Coping can sometimes be achieved with little effort and expenditure of resources, in which case the individual's welfare is satisfactory. Or it may fail to cope at all, in which case its welfare is obviously poor. Or, if the individual does cope with the conditions it encounters, this may be easy, with little expenditure of resources, or may be difficult taking much time and energy, in which case welfare is deemed to be poor.

Well-being can be measured in a variety of ways. Each method of measurement tends to suit some definitions of welfare better than others. Measures focusing on biological, psychological and social functioning are readily understood within a perfectionist framework. If, on the other hand, the researcher allows the animals to rank different outcomes (**see: Choice test**; **Economics of behaviour**), or concentrates on psychological well-being and **distress**, the measurements will make more sense in a hedonist (or, more broadly, subjectivist) framework. So researchers need to consider whether and how the data they record can be interpreted so as to say something about well-being in a specified sense. Measures of functioning often indicate little, directly at least, about pleasure, **suffering** and other subjective states of animals. It follows that the use of these measures within a subjectivist view of animal welfare requires critical discussion. (PS)

See also: Organic farming

References and further reading

Appleby, M.C. and Sandøe, P. (2002) Philosophical debate on the nature of well-being: implications for animal welfare. *Animal Welfare* 11, 283–294.

Broom, D.M. (1986) Indicators of poor welfare. *British Veterinary Journal* 142, 524–526.

Duncan, I.J.H. (1996) Animal welfare defined in terms of feelings. *Acta Agriculturae Scandinavica (Section A: Animal Science)* 27, 29–35 (Suppl.).

Duncan, I.J.H. and Fraser, D. (1997) Understanding animal welfare. In: Appleby, M.C. and Hughes, B.O. (eds) *Animal Welfare*. CAB International, Wallingford, UK, pp. 19–31.

Rollin, B.E. (1993) Animal welfare, science and value. *Journal of Agricultural and Environmental Ethics* 6, 44–50 (Suppl. 2).

Sandøe, P. and Christiansen, S.B. (2008) *Ethics of Animal Use*. Blackwell, Oxford, UK.

Welfare measurement

Ethical decision making (**see: Ethics**) and enforceable legislation derived from an interest in the physiological and psychological **well-being** of animals requires a set of well-defined measurements to assess animal **welfare** and criteria in order to distinguish between acceptable and unacceptable practices. However, the objective assessment of animal welfare is a topic of great debate due to the complexity and practical implications of the process. Measurement may focus on the physical and/or mental state of the animal, or its **'telos'**. Historically, many approaches to measuring animal welfare have emphasized the presence or absence of indicators of poor welfare (e.g. absence of **stress** or specific **aversive** situations such as **fear**, **hunger**, **injury**, etc., as espoused within four of the **five freedoms**); recently, however, greater attention has been paid to the potential to elucidate indicators of positive well-being, e.g. happiness, (see Désiré *et al.*, 2002) and the determination of the **needs of animals** (**see: Economics of behaviour**) and their provision (or not).

A hierarchy of needs has been proposed in relation to welfare. Some are life-sustaining, such as the need for food and water, others are necessary to maintain **health** and minimize the risk of ill health, such as the need for an environment that minimizes the risk from pathogens, while yet others are necessary for comfort, such as the need for appropriate bedding. Some may be more important than others, but all may be required if welfare is not to be seriously compromised (Hurnik and Lehman, 1985).

Initial problems when assessing animal welfare are its definition and the avoidance of **anthropomorphism** or **anthropocentrism** in the assessment process: the welfare and potential needs of an animal should be considered on the basis of the **ecology** and behaviour of each species and not on what we, as humans, believe is important. Indeed, the **behavioural ecology** of a species may be the starting point from which to set the criteria for measuring welfare. For example, good cage illumination is important for diurnal animals but may be detrimental for **nocturnal** species.

The measurement of welfare may focus on the individual or population. The latter may focus on factors like the incidence of disease, such as **mastitis** and **lameness** in the case of a dairy herd and the level of output from the unit (milk production, conception rate, etc.), and so be used to compare different systems as well as different populations. It may or may not include specific personal measures from a sample of individuals.

Various welfare measurements of the individual have been proposed, focusing on either physiological or behavioural measures. Behavioural signs of welfare have frequently attracted more attention, perhaps because they are often more conspicuous and their measurement is usually less invasive than physiological measurements. In relation to physiologically related measurements, attention is often focused on specific neurological, endocrinological and immunological parameters.

Behavioural approaches to the assessment of animal welfare include not only traditional approaches to **measuring behaviour** to quantify specific behaviours and determine the **ethogram** or **time budget** of individuals but also assessments of the variety and complexity of behaviours shown, and experimental approaches, e.g.: (i) consumer demand approaches to **preference** testing (**see: Economics of behaviour**) to allow the quantification of demand, which may be used to infer need (Mason *et al.*, 2001); and (ii) **cognitive bias** tests, which potentially assess emotional bias and may be used to infer subjective state (Harding *et al.*, 2004).

Another approach to the assessment of animal welfare focuses not on how animals are responding but, rather, on what is being provided; this might take the form of some kind of animal needs index (list of provisions potentially impacting on welfare), against which the management system can be rated. This approach is favoured within many regulatory frameworks, both legislative and voluntary, such as **farm assurance schemes**.

Historically, perhaps because of the traditional role of the veterinary profession in relation to the protection of animal welfare, much emphasis has been placed on measures of health to assess animal welfare; however, the importance of other measures is now universally recognized, with animal welfare scientists coming from almost every biological field, highlighting the fact that no single measure is sufficient to assess welfare but, rather, that welfare should be assessed using a range of measures and approaches. Different measures are required not only for different species, but possibly even for individuals in different circumstances. Because welfare itself cannot be measured directly (only inferred from the measures taken), a philosophical approach to assessing 'welfare' is to focus on addressing more precise questions relating to animal well-being, in order to identify whether there are any serious challenges to its well-being.

Different approaches tend to emphasize different measures. For example, the veterinary-medical approach suggests that animals should be in good physical health, free from **disease** and **injury** and not displaying any self-inflicting behaviour; the biological approach suggests that animals should breed successfully and should be able to rear their offspring; and the behavioural-ecological approach emphasizes that animals should live in physical and social environments resembling those of their wild **conspecifics**, where they can acquire and maintain the species-specific behavioural skills that they would need in their natural environment. (DSM, RV, BM)

References and further reading

Broom, D.M. and Johnston, K.G. (1993) *Stress and Animal Welfare*. Kluwer Academic Publishers, Dordrecht, the Netherlands.

Désiré, L., Boissy, A. and Veissier, I. (2002) Emotions in farm animals: a new approach to animal welfare in applied ethology. *Behavioural Processes* 60, 165–180.

Harding, E.J., Paul, E.S. and Mendl, M. (2004) Cognitive bias and affective state. *Nature* 427, 312.

Hurnik, J.F. and Lehman, H. (1985) The philosophy of farm animal welfare: a contribution to the assessment of farm animal well-being. In: Wegner, R.M. (ed.) *Second European Symposium on Poultry Welfare*, German branch of the World's Poultry Science Association, Celle, Germany, pp. 255–266.

Mason, G.J. and Mendl, M. (1993) Why is there no simple way of measuring animal welfare? *Animal Welfare* 2, 301–319.

Mason, G.J., Cooper, J.J. and Clarebrough, C. (2001) Frustrations of fur-farmed mink. *Nature* 410, 35–36.

Neville, G. (2004) *Physiology and Behaviour of Animal Suffering*. Blackwell Science, Oxford, UK.

Novak, M.A. and Petto, A.J. (1991) *Through the Looking Glass*. American Psychological Association, Washington, DC.

Segal, E.F. (1989) *Housing, Care and Psychological Well-Being of Captive and Laboratory Primates*. Noyes Publications, Park Ridge, New Jersey.

Stamp Dawkins, M. (1998) Evolution and animal welfare. *The Quarterly Review of Biology* 73, 305–328.

Stamp Dawkins, M. (1998) *Through Our Eyes Only? The Search for Animal Consciousness*. Oxford University Press, Oxford, UK.

Various (2005) Science-based assessment of animal welfare. *Revue Scientifique et Technique* Special issue 24(2), 483–547.

Well-being

Animal 'well-being' is often used as a synonym of animal '**welfare**'. There is a small difference in usage, however. 'Welfare' is sometimes used to describe the conditions in which animals live and how they are cared for, whereas 'well-being' is used only to describe the resulting state in the animal. (PS)

Whipping

Whipping refers to the use of the whip as an instrument of punishment or retribution. The most common form of whip is a long, thin, flexible instrument that tapers distally. It is designed to produce **pain** or discomfort and has a long history of use on both humans and animals, ostensibly for behaviour modification but, not infrequently, for retributive purposes. Its use on animals is often based on an anthropomorphic view of animal behaviour in which an animal is considered to be responsible for and cognizant of its actions or their consequences. There are various important **welfare** considerations surrounding the use of the whip in training animals.

When the whip is used in **punishment**, it is supposed to reduce the frequency of, and ultimately delete, a particular response. Critically, the **aversive** stimulus has to be contingent upon and contiguous with the behaviour targeted for removal. Delayed punishment has serious welfare implications and furthermore is frequently ineffective in removing a behaviour.

Even 'well-timed' punishment can be an impoverishing training tool. Haag *et al.* (1980) found that punishment in **horses** lowers their ability to offer a new behaviour in solving a problem. It can cause problems because it tells the animal what not to do, but *not what to do*. Punishment has the potential to desensitize an animal to the punishing **stimulus** if the punishment intensity is not correct, and risks deleterious emotional changes that can interfere with attention and learning. Punishment can also foster unwelcome aversive associations with the person delivering it. All of these conditions apply to the misapplication of the whip.

The animal most commonly subjected to the whip is the domestic horse. Importantly, humane use of the whip can be an integral part of **horse training**. In this regard its irritant effect (through a tapping motion) may trigger certain desirable locomotory responses, but its application should cease at the immediate onset of the desired or targeted behaviour.

Use of the whip in horse racing is ubiquitous. Here it is largely intended to be used as a tool for training and eliciting acceleration. Unfortunately, it is largely misused in this context. However, the whip does motivate some horses (that are difficult to motivate by other means) to gallop faster. Its most humane use is to train it as a light acceleration **signal**. To follow the laws of learning theory, the whip should be removed the moment the horse exhibits *any* acceleration. So, at best, humane training would demand a brief period of very mild whip use (closely spaced whip-taps until the onset of the targeted response) that is then rapidly replaced with a cue such as the jockey's body motion or squeezing leg signals to ultimately elicit the response. The use of the whip as an accelerating device in horse racing can be seen as abusive when it is excessive or applied for too long or to an animal that is unable to accelerate. 'Humane' whips have been developed that limit the force that can be applied. It should also be noted that the whip can be employed in all equestrian disciplines as a mild negatively reinforcing signal that can be as subtle as the tapping of human fingers. Therefore in itself the whip is not necessarily a severe punishing stimulus. (AM)

Reference

Haag, E.L., Rudman, R. and Houpt, K.A. (1980) Avoidance, maze learning and social dominance in ponies. *Journal of Animal Science* 50, 329–335.

White noise

White noise is a sound that contains all frequencies at equal amplitudes. This can be represented by a flat power spectrum, or a uniform sound spectrogram with all frequencies represented equally. In this sense, white noise can be seen as analogous to white light, which contains all visible frequencies. In theory, white noise may extend across an infinite range of frequencies, although in practice this could never be produced, as it would require infinite power. More practically, the frequencies should be flat (that is to say represented equally) across the range audible to the listener. For humans this range is approximately 20 Hz to 20 kHz. Noise can also be 'coloured' by emphasizing certain frequencies to achieve 'red', 'pink', 'brown' noise and other variations. White and coloured noises can be useful to mask other sounds from the environment, such as sounds that might startle livestock. This noise is also sometimes used in experiments as a control **stimulus**. In this way, the effects of sounds that may be biologically meaningful (such as animal **vocalizations**) can be compared not only with silence but also with another sound equal in energy, and perhaps even in duration, rate and other features. (DW)

Wild animals

The label 'wild animals' can have legal implications and influence animal management and **conservation**. In identifying animals as 'wild', we may consider the evolutionary histories of species as well as the current environment, or living conditions, of individual animals or populations of animals, including their relationship to humans (see Fig. W.1).

Many animals around the world that are free-ranging and not domesticated group easily together as 'wild animals', and many captive domesticated animals, such as dairy cows or seeing-eye dogs, could be excluded from 'wild animals' without debate. However, free-ranging species that have undergone **domestication** and captive non-domesticated animals present greater challenges in labelling. How would we categorize domesticated or semi-domesticated animals that have free-ranging populations, such as horses or European honey bees in North America? How would we describe captive or semi-free-ranging individuals of 'exotic' species, such as animals in **zoos** or sanctuaries, dolphins trained to assist military operations or 'sea monkeys' in a child's aquarium?

In the *Origin of Species*, Darwin (1859) uses 'organic beings in a state of nature' to refer to non-domesticated plants and animals arising through **evolution** by natural **selection**. Darwin sets up this apparent dichotomy between domesticated and non-domesticated animals but makes it clear that animals cannot be cleanly separated into these two simple groups; he explains that evolutionary history yields a continuum of variation rather than a strict dichotomy. 'Wild animals', as a faunal-oriented equivalent to 'organic beings in a state of nature', is a convenient and useful term to represent the opposite of 'domesticated'. Some legal statutes that regulate or prohibit the keeping of 'dangerous wild animals' in private homes use the domestic–non-domestic dichotomy, defining 'wild animals' as species 'other than domestic' ones. The

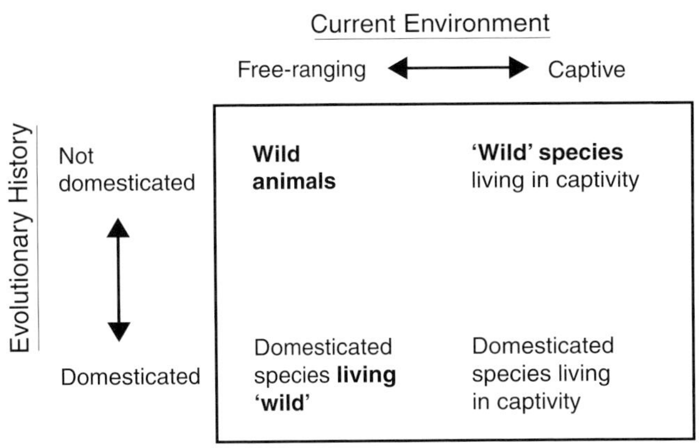

Fig. W.1. Animals live in environmental conditions that can be arranged in a continuum between extremes of captive and fully dependent on human care-givers to free-ranging and not reliant on people. The evolutionary histories of species also form a continuum based on the degree to which selective breeding by humans has influenced the species. Animals may be called 'wild animals' based solely on evolutionary history, their environment or a combination of both. Free-ranging animals that are not domesticated are always 'wild animals'. Other animals may sometimes be considered wild animals. Captive individuals of domesticated species are not 'wild animals'.

statutes may not directly discuss the evolution of wild animals, but the distinction between legally and illegally held animals relies on whether the animal is a result of selective breeding for traits favourable to people or whether the species evolved only naturally, via natural selection.

Most animals in zoos and aquariums are also 'wild animals' based on the evolutionary histories of the species, but they may not be considered truly wild compared with their free-ranging **conspecifics**. The potential for captive animals to impact their species' evolutionary future is restricted by humans and, in captivity, they may exhibit certain behaviours not seen in nature and may not exhibit the full range of their species' natural behaviours. Drawing distinctions between captive wild animals and animals living in nature, or ranking their relative levels of 'wildness', can be important for achieving conservation goals. For example, **captive breeding** and **reintroduction programmes** may characterize the behaviour of a species in the wild, examining free-ranging wild animals to establish standards by which to measure the behaviour of captive reintroduction candidates. A measure of success for a programme might be how 'wild' released individuals behave relative to the original 'wild' population.

Arranging animals according to evolutionary history and current environment might not yield groups that are meaningful or sufficiently distinct for management strategies for ecosystems or protected lands. Identifying all free-ranging, non-domesticated species as 'wild animals' encompasses non-native animal species into this group, but natural resource management plans and conservation efforts generally devise different strategies for native species than for non-native species. Thus, a species' historical presence might be part of a resource-based definition of 'wild animals' or an important characteristic for subdividing species of 'wild animals' for management planning. A recently proposed 'rewilding' strategy for ecological restoration of North America suggests intentionally introducing non-native or feral species as proxies for fauna that have gone extinct. For this strategy, ecological function is the primary consideration in identifying rewilding candidate species.

The management of non-native species and free-ranging, domesticated (i.e. **feral**) animals has long been highly controversial, drawing proponents for protection and for removal. The US Wild Free-roaming Horse and Burro Act of 1971 protected wild horses, a feral species, as an added element of animal diversity and as symbols of the 'pioneer spirit of the West', but controversy over the presence of horses on public lands continues. Similar controversy has surrounded dingoes in Australia, with arguments both for their preservation as wild animals and for their removal as potential livestock predators. Taxonomic classification of dingoes as a subspecies of dogs or a subspecies of wolves does not alter dingo behaviour or **ecology**, but it may shift the debate between the management and control of a domesticated species and strategies for conservation and coexistence with a wild species of carnivore.

Most large carnivore species range widely and come into conflict with people. Development activities increasingly fragment the landscapes that can meet the large-area requirements of carnivores and other wide-ranging species, restricting their movements or funnelling wild animals into sub-optimal habitats. Humans are therefore altering the spectrum of environmental conditions, inadvertently and intentionally fencing wild animals inside smaller areas and more highly altered and human-dominated landscapes. These changes blur the distinctions between captive and free-ranging conditions and alter the evolutionary outcomes of species, most dramatically in extinctions. Our perception and definition of 'wild animals', and our actions towards them, may shift as we alter these continua. (EEB)

See also: Feral; Free-range animal; Habituation; Reintroduction; Urban wildlife; Wildlife management

Reference and further reading

Darwin, C. (1859) *The Origin of Species*. John Murray, London. [Reprinted in Penguin Classics, 1985.]

Donlan, C.J., Berger, J., Bock, C.E., Bock, J.H., Burney, D.A., Estes, J.A., Foreman, D., Martin, P.S., Roemer, G.W., Smith, F.A., Soulé, M.E. and Greene, H.W. (2006) Pleistocene rewilding: an optimistic agenda for twenty-first century conservation. *American Naturalist* 168, 660–681.

Wildlife management

Wildlife management is the process of deliberately affecting the future of wild species and their habitats through the deployment and manipulation of various **resources**, including human, financial, material and intellectual. Originally linked to **hunting wild animals** for consumption or display, it now encompasses many human attempts to protect, conserve, limit, enhance, recover, use or eradicate wildlife. The modern professional discipline of wildlife management requires the application of scientific and technical principles to wild populations and their habitats (especially mammals, birds and fish) in order to regulate such populations for economic, recreational or scientific purposes.

Frequently, wildlife management requires the intentional manipulation (or intentional non-manipulation) of habitat rather than of the particular species directly. This requires strategic planning, managing resources, deploying necessary human and financial assets and measuring results. It also includes researching species and their physiology, behaviour and **ecology**. But wildlife managers are not necessarily ecologists; they have different professional backgrounds, including biology, veterinary medicine, agronomy and forest engineering. Wildlife management includes creating and implementing laws that regulate the use of wildlife and the kinds or amount of wildlife that people can harvest. Wildlife management techniques are used to intervene directly into the species' life history, the habitats' carrying capacity and the human effect upon them. Strategic interventions of wildlife management include:

- habitat control or management, in order to favour or eliminate some species;
- **culling**, translocating or deterring wild species, in order to control any damage caused by them;
- culling infected wild species, in order to prevent transmission to livestock, humans or other wild species;
- predator control or management, in order to prevent damage to domestic species (livestock, **pets** and ornamental animals), other wild species and humans;
- providing supplemental water, in order to increase some species populations;

- providing supplemental food, in order to increase some species populations; and
- conducting census counts and behavioural and physiological studies, in order to acquire the knowledge necessary for pursuing these strategies.

Wildlife specialists collect such information and manage such populations to achieve a variety of recreational, economic, cultural and aesthetic benefits for people while ensuring the future of wildlife and their ecosystems. They can aim to increase, maintain or decrease a species population, or choose to leave the species alone.

Conservation biology is one kind of wildlife management that aims to maintain or increase species that have been negatively affected by human action. But wildlife management encompasses much more than conservation biology, such as exploitation for food or other products that contribute to human needs. Wildlife management is also concerned with the negative effects of wildlife on human assets, such as predators near livestock, birds at airports and rats where food is processed.

Sustainable use of wild species is another objective of wildlife management. In this case the aim is to provide economic benefits to local communities by harvesting products or by-products from wild animals without negatively affecting the viability of the targeted population or its environment. The use to be sustained can range from culling for food, to shearing, to recreational observation. Animal welfare concerns vary depending on the degree and kinds of use. Examples of such concerns, in consumptive and non-consumptive uses, are inhumane methods of culling that cause much suffering before death and constant harassment for photographs in some forms of **ecotourism**.

Wildlife control involves removing, scaring or culling wild animals from certain places. Control techniques range from non-lethal repellents and fertility control to lethal traps, chemicals and shooting. Wildlife managers regularly confront the need to destroy wild animals for various reasons, such as protecting **endangered species**, protecting public **health** and controlling the population of species considered pests. Concern for animal welfare calls for methods of **euthanasia** that are humane to the target species and not dangerous to non-target species. It is difficult to satisfy both conditions when little is known about the effectiveness and response to control methods (physical or chemical) on wild species. Many control methods thus confront strong opposition from animal welfare specialists, and criteria and recommendations for euthanasia and control in the wild are still highly controversial. Wildlife managers and veterinarians are addressing this issue: the American Society of Mammalogists and the American Veterinary Medicine Association, among others, have suggested criteria for euthanasia.

The wildlife manager's concern with animal welfare also applies to **captive breeding programmes**, **reintroduction** to native habitats, **translocation**, **zoos** and rehabilitation centres. Since wildlife requires moving freely, acquiring food and shelter, and interacting with the environment while subject only to pressures of natural selection, any form of **captivity** implies a potential animal welfare problem. Even as they employ temporary captivity for the purposes of rescue and conservation, wildlife managers should work to alleviate any human-induced **stress** and **suffering** caused by that captivity.

Wildlife management in general is related to ecology as the science of animal welfare is related to individual animal behaviour and physiology. Both disciplines are forms of professional expertise that intervene in animal life and therefore involve a serious ethical dimension (**see: Ethics**). Wildlife managers must employ the science of animal welfare if they are to improve the quality of captive animals' lives and prevent or alleviate anthropogenic harm to animals in the wild, besides the more known traditional priorities of this discipline (i.e. protection, conservation, environmental protection, sustainability, control and human **well-being**). (CB)

See also: Biodiversity; Conservation behaviour; Endangered species; Free-range animals

Wilson, Edward Osbourne (E.O.)

E.O. Wilson was born in Birmingham, Alabama on 10 June 1929, and is largely credited with the development of the theory of **sociobiology**, but is also a proponent of the **biophilia** hypothesis, a secular biological basis to **ethics** and **conservation**. In addition to his scientific work, he has written extensively for the popular press on his theories and discoveries. (DSM)

Wind-sucking (aerophagia)

This term is used to describe two quite distinct conditions in the **horse**: a stereotypic behaviour and a behaviour dependent on a conformational fault of the perineum. Poorly distinguished from **crib-biting**, wind-sucking describes the stereotypic engulfing of air by horses with (Australia) or without (UK) holding on to a fixed object. The fixed objects most commonly used include the feed trough (hence the term crib-biting) and the stable door. Other substrates include fence posts, railings, other horses and even the affected horse's own limbs. Horses often lick the substrate before using it. The neck is always flexed during the consummatory phase of the behaviour. In the past, it was commonly assumed that every grunt a wind-sucker made marked a bolus of air that was being swallowed. However, when wind-sucking/crib-biting horses were radiographed, it became clear that little if any air travels as far as the stomach. Instead, each grunt accompanies transient ballooning of the cranial third of the oesophagus with air, which then departs through the horse's open mouth. Significantly, no deglutition is involved and therefore the air is described as having been engulfed rather than swallowed. Aerophagia is considered a redundant term in the light of this discovery.

The question of why horses are often highly motivated to wind-suck remains, but several theories have been advanced. For example, it may be that, in holding on to fixed objects with their teeth and then arching their necks, horses generate saliva that acts to buffer gastric acidity. This is supported by evidence that foals crib-bite less when treated for gastric ulceration. Another possibility is that, in distending the cranial oesophagus, horses somehow stimulate the release of **endorphins**, as hinted at by studies showing that **opioid** antagonists transiently eliminate crib-biting.

Since it disturbs the normal equine **time budget**, wind-sucking can compromise a horse's ability to maintain condition by occupying time that it would normally spend **feeding** and resting. It is a behaviour that must be declared prior to auction in some countries.

Wind-sucking, as a stereotypic behaviour, should not be confused with the acquired conformational fault of the perineum in mares that leads to pneumovagina (also known as wind-sucking; **see: Caslicks**). (PDM)

See also: Stereotypies

Further reading

McGreevy, P.D., Richardson, J.D., Nicol, C.J. and Lane, J.G. (1995) A radiographic and endoscopic study of horses performing an oral stereotypy. *Equine Veterinary Journal* 27, 92–95.

Wire-gnawing

When placed in cages, laboratory **rodents** and **rabbits** often chew the wire bars of the cage or lid. This behaviour can rapidly become highly persistent and prevalent. In one study on mice, all 24 animals developed wire-gnawing between 20 and 30 days of age in one strain and before 40–50 days of age in another. It occupied 11% of the total activity, although large individual differences occurred. When wire-gnawing is prevented by placing the bars so close together that the nose cannot be pushed between them, mice show an initial increase in **corticosteroid** levels indicating an acute **stress** response; however, these levels return to pre-treatment baseline values within 3 days as the animals develop new behavioural habits. Prevention of wire-gnawing has no significant effects on chronic measures of both the sympathetic–adrenal–medullary system and the **hypothalamic–pituitary–adrenal (HPA) axis**, indicating that wire-gnawing does not reduce chronic stress and therefore is not a '**coping** strategy'. When reared in cages with two sets of bars, mice interact more with a set that provides a potential escape route than one that does not, possibly indicating that the motivation for wire-gnawing lies in prolonged attempts to escape. Environmental **enrichment** in the form of hay for rabbits, a **running wheel** for **hamsters** or cardboard tubes for mice have been shown significantly to reduce wire-gnawing. (CS)

Wirkwelt

Derived from the German *wirken* meaning 'to act' and *welt* meaning 'world'. Defined by Jacob Von Uexküll (1934) in his **umwelt** theory, the wirkwelt is the operational world of an animal. The wirkwelt of an animal thus depends on its effectors and is composed of all its potential responses toward its environment. The interrelation between the wirkwelt and the **merkwelt** (i.e. the perceptual world) enables the animal to interact with its umwelt. (SL)

Reference

von Uexküll, J. (1934) *Streifzüge durch die Umwelten von Tieren und Menschen*. Springer, Berlin.

Wood-chewing

Wood-chewing (lignophagia) occurs naturally in a number of species, including some of the herbivores: for example, **deer** stripping bark from a tree. In mammals it can make up a small part of the feeding strategy and may only be observed under certain conditions. It is thought as occuring to supplement a nutritional requirement, possibly for more fibre in the diet, although the stripping of willow bark, which contains salicylic acid (aspirin), may be a form of **self-medication**.

Abnormal wood-chewing has commonly been reported in domestic **horses**, particularly when they are stabled for long periods and fed a high cereal–low forage type of ration, browsing including incidental wood-chewing is a normal behaviour in the donkey. However, the behaviour pattern is not exclusive to equids, as it has been observed in sheep in the form of slat-chewing when they are confined to metabolism crates and provided with little forage. In general, the wooden surface is grasped, stripped and then ingested. Damage can occur to any available wooden surfaces such as the top of the stable door and the walls. The chewing of wooden railings has also been observed in horses kept at pasture.

Wood-chewing has been described as an **abnormal** behaviour by some authors when it occurs repetitively and out of context. It has also been recognized as a behavioural problem by owners or keepers, particularly when the chewing occurs at their financial expense. Habitual wood-chewing can also be potentially harmful due to the risk of swallowing wood splinters.

In the horse it is important to distinguish between wood-chewing and **crib-biting**, another oral stereotypic behaviour, as in both cases a surface is grasped by the teeth, though ingestion of wood does not occur during crib-biting. Therefore, the impression of tooth marks from the incisors on wooden surfaces should not be taken as a definite sign of wood-chewing.

There is good evidence to suggest that feeding a ration high in cereal and low in fibre is implicated in the development of wood-chewing and that it may arise as a consequence of a strong motivation to forage, coupled with harmful changes in the pH of the gut. Measures such as painting surfaces with taste repellents or removing chewable surfaces from the stable should therefore be questioned on welfare grounds, as they act to curb the problem behaviour through prevention of the behaviour rather than addressing the cause or underlying motivation. (SR)

See also: Stereotypies

Further reading

Cooper, J. and McGreevy, P. (2002) Stereotypic behaviour in the stabled horse: causes, effects and prevention without compromising horse welfare. In: Waran, N. (ed.) *The Welfare of Horses*. Kluwer Academic Publishers, Dordrecht, the Netherlands, pp. 99–124.

Wool-eating

Wool-eating is behaviour involving an animal (typically a **sheep**) eating its own wool or the wool of other individuals. The behaviour can result from either of two situations. First it can be a response to insect infestation, typically blowfly strike or sheep scab. By removing the wool the animal can use its mouth to scratch the site of infestation and help to alleviate some of the **pain**. In addition, any maggots or eggs in the wool will be removed from the animal. The second situation in which wool-eating occurs is when sheep are under **stress** from a lack of high-quality feed **resources**. It is a traditionally held belief that animals eat wool in order to supplement their diet with essential micronutrients, as wool contains trace elements. This situation typically occurs when livestock are overwintered on poor-quality feeds. Akgul *et al.* (2000) showed that levels of most micronutrients were normal in affected animals, although levels of copper and zinc were lower than in unaffected animals, as were haemoglobin levels and the total protein value of blood serum.

This behaviour can be passed to animals in an affected **group** from those from a non-affected group when the groups are mixed. Thus, not only is wool-eating a sign of a problem within an individual, it could also be symptomatic of a problem within its group, even if the individual observed shows no malnutrition or insect infection. As either situation is a serious **welfare** and production problem, if wool-eating is observed in a flock the whole flock requires inspection and action to correct the problem. (MM, HO)

Reference

Akgul, Y., Agaoglu, Z.T., Kaya, A. and Sahin, T. (2000) The relationship between the syndromes of wool eating and alopecia in Akkaraman and Morkaraman sheep fed corn silage and blood changes (haematological, biochemical and trace elements). *Israel Journal of Veterinary Medicine* 56(1). Available at: http://isrvma.org/article/56_1_4.htm (accessed 24 November 2009).

X

Xenograft/xenotransplantation

Xenotransplantation is the transplantation of an organ, tissue (grafts) or cells from one species of animal to another (not to be confused with **transgenics**). The technique was developed to make up for the shortage of human donor organs for transplantation. Recently, xenotransplantation has been the subject of much research, but also concern, as transgenic pigs were developed whose organs would not be subject to the very rapid (hyperacute) rejection that would normally occur in the first few seconds after a blood supply from the recipient is connected to the transplanted organ. To date, despite their potential for saving human lives, no organ transplant procedures have become adopted as routine for fear of infection passing to humans from the donor species (xenozoonosis), as well as doubts that they will be successful for a variety of other reasons, e.g. physiological compatability. (DBM)

Xenopus

Species of *Xenopus* are fully aquatic **amphibians** occurring in Africa south of the Sahara. The best known is *Xenopus laevis*, employed intensively in laboratory research for over 70 years. Early work concerning physiology, biochemistry, embryology and endocrinology contributed significantly to understanding in these disciplines. Currently, this species is one of the most widely used **laboratory animals** in developmental, cell and molecular biology.

The common name 'African clawed toad' (or 'clawed frog') refers to the keratinized black claws on three of the toes. The terms 'toad' and 'frog' are used inconsistently for *Xenopus*, but these names do not have strict scientific relevance and *Xenopus* is actually distinct from both 'groups'. Its morphology is distinctive, with flattened body and limbs splayed laterally (see Fig. X.1); its streamlined shape, powerful legs and webbed feet contribute to specializations for life underwater. Relatives in the family Pipidae include species of *Pipa* (in South America) and *Hymenochirus* (in Africa). Fossils of extinct pipids occur in South America, Africa and the Middle East, indicating a wide distribution before the Atlantic opened up (about 90 million YBP). The earliest known fossils are from the Lower Cretaceous (120 million YBP).

There are 18 species of *Xenopus*. These have major genetic and evolutionary interest, forming a polyploid series with two lineages: one with 20 and 40 chromosomes, the other with 36, 72 and 108. The 20/40 chromosome group has a separate genus name, *Silurana*, as distinct from *Xenopus*, with multiples of 36 chromosomes. *Silurana tropicalis* is the only species retaining the ancestral diploid state ($2n = 20$); all other species evolved through allopolyploidization – hybridization between different species accompanied by genome duplication. *X. laevis*, with 36 chromosomes, is tetraploid and contains the genetic information of two (unknown) ancestral species. Six species are octoploid and two (*Xenopus ruwenzoriensis* and *Xenopus longipes*) are dodecaploid (108 chromosomes), containing six sets of genetic information. Such complexity is extremely rare among vertebrate animals, although it occurs in plants, including ferns.

The habitats of *Xenopus* include most types of water bodies in sub-Saharan Africa, especially the still waters of swamps, ponds and streams. *Xenopus* are less common, or absent, in rivers and lakes, especially those with well-established fish communities. The two species most frequently employed in laboratory research, *X. laevis* and *Xenopus tropicalis*, are ecologically distinct. *S. tropicalis* is limited to lowland tropical forest in West Africa, from Nigeria to Senegal. *X. laevis* is a generalist savannah species distributed from South Africa to Sudan, with a series of subspecies corresponding with latitudinal zones. Most research has been based on *Xenopus laevis laevis* from South Africa, but there is significant evolutionary divergence within this subspecies, corresponding with the winter rainfall area of the western Cape and the summer rainfall area further north. However, this should not confuse existing research findings, because supply of *X. l. laevis* to laboratories has been predominantly from the Cape.

Fig. X.1. *Xenopus laevis* in characteristic pose, partially hidden in a refuge, with forelimbs outstretched in readiness to capture prey organisms that may drift past in the water column or land on the water surface above. The hind legs have powerful musculature for rapid swimming and the foot has black claws on the innermost three digits. The eyes are directed upwards. The prominent bulges on the posterior dorsum are created by the large, inflated lungs in the body cavity. 'Stitches' are visible around the eyes and along the flanks: these are the lateral line organs sensitive to movement, vibrations and pressure changes. This subspecies, *X. l. victorianus* from East and Central Africa, has deep yellow pigmentation on the legs and a white belly; the dorsum is camouflaged with olive-brown pigments. Body length (from nose to vent) is about 60 mm.

The diet of *Xenopus* includes a wide variety of invertebrates. The prey range is distinct from that of other anurans because *Xenopus* feed underwater and are scavengers, detecting dead animals by **odour**. The tadpoles are filter feeders, exploiting plankton. Although normally fully aquatic, *Xenopus* can tolerate extreme conditions, including aestivation in dried-up ponds for over 6 months and survival for many months without food. *X. laevis* tolerates a wide temperature range, from 4 to 28°C, and can overwinter under ice in frozen ponds. However, physiological adjustment to this range requires gradual acclimation. The temperature optima of tropical forest species reflect their more uniform habitat conditions: e.g. 25–30°C for *S. tropicalis*.

Sensory perception, including vision and **olfaction**, shows dual adaptations for both air and water. In most amphibians, the lateral line system (present in tadpoles) disappears at metamorphosis; in *Xenopus*, the lateral line is well developed in adults and can detect changes in water pressure (depth), currents and vibrations. This provides a highly accurate system for locating unseen prey. *Xenopus* communicate by specialized **vocalizations**, maintain territories underwater and have dominant/subordinate **hierarchy**; they are capable of complex learning and long-term **memory**.

The skin of *Xenopus* produces a considerable diversity of secretions, including toxins that deter attackers, natural antibiotics protecting against infection (including 'magainins' that kill bacteria by perforating membranes) and a wide range of peptides that have pharmacological significance.

Export of *X. laevis* from South Africa for use in laboratories and the **pet** trade has led to establishment of introduced populations in many countries, including the USA, the UK, France, Italy, Portugal, Indonesia, Japan, Chile and Ascension Island.

Maintenance of *X. laevis* in laboratories began with its use in human pregnancy diagnosis based on the fact that hormones in urine of pregnant women trigger egg laying by female *Xenopus*. The current major application of *Xenopus* in molecular genetics exploits the large size and very active gene transcription of its eggs. Following micro-injection of DNA from animal or plant cells into the oocyte, all stages in gene expression occur leading to translation into protein (of the foreign organism).

For maintenance in **captivity**, semi-natural conditions can be created in aquariums reflecting the habitat conditions of *Xenopus* in the wild, including dim illumination, refuges and minimal disturbance. This approach can be successful for long-term maintenance (with lifespan exceeding 20 years). In large-scale maintenance, *Xenopus* may adapt well to simplified environments, but some laboratories with demanding research protocols experience problems with oocyte quality that may be attributable to **stress**. Animals taken from the wild are caught as adults, after at least 3 years in natural habitats where they are **nocturnal**, secretive and accustomed to the security of cloudy water. Transfer to laboratories, where *Xenopus* experience clear, shallow water, illuminated from all sides, with frequent disturbance, may cause stress. Wild-caught animals may carry parasitic and microbial infections.

Ovarian condition varies depending on season and recent **breeding** history: *X. laevis* may spawn up to three times during spring/summer at the Cape, South Africa (September to February), so some females arriving in laboratories have gravid ovaries prepared for ovulation while others have depleted ovaries following natural spawning. Adult *X. laevis* collected during April–July (late autumn/winter at the Cape) may be taken from water at 10°C, unprepared for immediate spawning and experience shock when transferred to laboratory conditions. These problems are avoided with laboratory-raised animals. For intensive laboratory 'production', semi-automated aquariums provide environmental control, including water quality, and *X. laevis* can reach maturity in 6 months post-metamorphosis.

Research based on *Xenopus* has contributed to major advances in biomedical and fundamental science, and its use seems certain to increase in the future. It has been an anomaly that very demanding research techniques should be subject to the inherent variability of wild-caught animals. It is appropriate that future research should be based primarily on purpose-bred animals. This shift in practice will simultaneously resolve other issues concerning **ethics**, legal regulations, **disease** risks and behavioural stress associated with transfer of animals adapted to natural conditions into intensive laboratory maintenance.

Research on the welfare and parasitic infections of *Xenopus* has been supported most recently by grants BB/C506272/1 and BB/D523051/1 from BBSRC. (RCT)

Further reading

Evans, B.J. (2008) Genome evolution and speciation genetics of allopolyploid clawed frogs (*Xenopus* and *Silurana*). *Frontiers of Bioscience* 13, 4687–4706.

Reed, B.T. (2005) *Guidance on the Housing and Care of the African Clawed Frog* Xenopus laevis. Research in Animals Department, RSPCA, Horsham, UK.

Tinsley, R.C. (2010) Amphibians. In: Kirkwood, J.K. and Hubrecht, R. (eds) *UFAW Handbook on the Care and Management of Laboratory and other Research Animals*, 8th edn. Blackwell Science, Oxford, UK (in press).

Tinsley, R.C. and Kobel, H.R. (1996) *The Biology of* Xenopus. Oxford University Press, Oxford, UK.

Z

Zoo

Zoo is an abbreviation of the term 'zoological garden' or 'zoological park', and is now the most commonly used term for a place where live, non-domestic animals are maintained in **captivity** for the purposes of display to the public. Zoos have existed since ancient times, the earliest known being those of the Egyptians (*c*.2500 BC), Chinese (*c*.1000 BC) and Greeks (*c*.500 BC). The first use of the term 'zoological garden' was in 1828, when the Zoological Society of London founded the London Zoological Gardens at Regent's Park; the shortened version of the name, London Zoo, was rapidly adopted.

They are many types of zoos but, according to the World Association of Zoos and Aquariums (WAZA), the key factors that all zoos have in common are: (i) they possess and manage collections that primarily consist of wild (non-domesticated) animals, of one or more species, that are housed so that they are easier to see and to study than in nature; and (ii) they display at least a portion of this collection to the public for at least a significant part of the year, if not throughout the year.

In Europe (that is member states of the European Union (EU)), zoos are defined legally as 'all permanent establishments where live animals are kept, with a view to public display for seven days per year or more, with the exception of circuses and pet shops' (Council of the European Union, 1999). EU member states are allowed to exempt certain establishments from this definition based on the number or species of animals held. Most states have exempted those only keeping domesticated species. For instance, in the UK a zoo is legally defined as 'an establishment where wild animals are kept for exhibition ... to which members of the public have access, with or without charge for admission…' (Zoo Licensing Act 1981), and **wild animals** are defined as non-domesticated species.

Although the term 'zoo' includes all establishments exhibiting wild animals to the public, in popular thinking zoos are often thought of as distinct from wildlife parks or safari parks. This distinction is usually based on the types of enclosure in which the animals are kept, with zoos being regarded as having traditional cages and wildlife parks featuring much larger and more natural enclosures. However, this distinction is totally arbitrary because there is no clear cut-off point where a zoo becomes a wildlife park, and an enormous range of enclosure styles and sizes can be present within a single establishment. Aquariums also fall into the legal definition of zoo in most countries, as do specialist **marine mammal** parks. There are also many other types of zoos, such as animal theme parks in which visitors are entertained by various rides and roller coasters in addition to live animal exhibits, as exemplified by Disney's Animal Kingdom in Florida, USA, opened in 1998. Wildlife sanctuaries that care for rescued animals, with or without an intention to rehabilitate them back into the wild, are also legally zoos if they admit the public to view the animals.

The first known zoos were those of the ancient Egyptians; Queen Hatshepsut organized the earliest recorded animal-collecting expedition (*c*.1500 BC) and Ptolemy II founded the largest ever zoo at the time in Alexandria (*c*.280 BC). Even by today's standards this was an enormous zoo and included 96 elephants and the first known captive polar bear! Zoos have been created by many cultures all over the world including the Chinese (e.g. Empress Tanki's marble house of deer, *c*.1150 BC) and the Aztecs of Mexico (Montezuma II in the 16th century had a zoo so large it reportedly required 300 staff). Not many other forms of public entertainment can claim to have been so popular throughout the world and to have persisted for so long – at least 5000 years. This universality and longevity demonstrates how powerful humans' desire to interact with wild animals must be. This desire has taken many forms over the years, primarily moving from domination and using wild animals as symbols of power and wealth in ancient and medieval menageries to scientific curiosity in the 19th-century zoos and then appreciation, respect and the desire to conserve in modern zoos.

Zoos have faced criticism from **animal rights** campaigners who fundamentally believe that animals should not be kept in captivity. They have also been criticized by animal **welfare** campaigners, who do not object to animals in captivity as long as they enjoy good welfare standards, which they claim many zoo animals do not. Continuously improving husbandry and veterinary techniques in zoos now means that most zoo animals enjoy very high standards of physical **health** and welfare, and often live longer and breed more successfully than their wild counterparts.

Most concern for zoo animal welfare now centres on their psychological well-being and the inevitable lack of opportunity, which results from being in captivity, to perform a full range of natural behaviour. The absence of the need for such behaviour, such as escaping from predators, may be beneficial for welfare. But there is good evidence that the internal **motivation** to perform some types of behaviour is very strong, and the lack of opportunity to do so can lead to poor welfare and possibly the development of **stereotypies** or other **abnormal** behaviour. The occurrence of such behaviour is usually assumed to be an indicator of suboptimal welfare. The prevalence of these behaviours across zoos in general is hard to determine, but there is no doubt that it does occur and that vast improvements have been made since the peak of anti-zoo activity in the 1980s. In many zoos new enclosures are bigger, provide greater complexity for their occupants and are designed with a greater understanding of species-specific behaviour and needs. Many zoos also place a large emphasis on environmental

enrichment, which encompasses many techniques to increase zoo animals' opportunities to perform a wider range of natural behaviour.

Criticism of animal welfare standards during the latter part of the 20th century created the need for zoos to justify the keeping of wild animals in captivity. At the same time there was increasing concern about species **extinction** and the need for wildlife **conservation**, and a growing recognition within the zoo community of the role zoos could play. As a result, in recent years zoos have become much more than simply visitor attractions; they now participate in education, conservation and scientific research. These 'new' roles have become legal requirements in many parts of the world e.g. the EU (Council of the European Union, 1999), but progressive zoos have been performing them for some time. WAZA and various regional zoo associations around the world – e.g. Europe (EAZA), North America (AZA) and Africa (PAAZAB) – have long required their members to be active in conservation, education and research, as well as having high standards of animal welfare. WAZA published their first World Zoo and Aquarium Conservation Strategy (WZACS) in 1993 and completely updated it in 2005. This document outlines what zoos around the world can and should achieve within the broad goals of conservation, education and research and is applicable to all zoos, whether publicly or privately owned and operated on a not-for-profit or a commercial basis.

Zoos and conservation

Despite conservation now being the most widely recognized role of zoos among the public, it is the most recent goal to be widely adopted by zoos. It was initiated most notably by Gerald Durrell, who founded Jersey Zoo (Durrell Wildlife Conservation Society) in 1959. Durrell and other early pioneers realized that zoos could be modern arks, saving species from otherwise certain extinction by bringing them into captivity and **breeding** them in safety, with the long-term goal of releasing them back into the wild once conditions became suitable. They began the first inter-zoo exchanges of animals for breeding purposes and established studbooks to keep breeding records, the forerunners of today's **captive breeding programmes** (see Box Z.1).

Conservation has also been the most successful PR story for zoos in recent times, and the modern ark 'breeding and **reintroduction**' scenario is now well established in the public's mind. However, there is a danger that zoos may become a victim of this PR success if the public believes that every animal in a zoo is destined for reintroduction to the wild. This is clearly not the case, and the public may be very disappointed to discover this. Most captive breeding programmes are not directly linked to reintroductions; rather, they are regarded as 'safety nets', managed with the aim of maintaining self-sufficient, genetically healthy captive populations in case the worst-case scenario happens in the wild. One criticism often made against the value of zoos for conservation is that many zoos hold large numbers of non-endangered species that take up much of the available **space**; another concerns the staff and effort required to run breeding programmes. Although this is true, zoos argue that while planning their collections they need to balance conservation needs with the ability to attract the paying public, or they cannot exist at all. In addition, many species that are not endangered today may soon become so and it is impossible to predict which species those will be.

There have been some very notable reintroduction successes, and zoos have been responsible for saving an increasing number of species from extinction and re-establishing them in the wild, e.g. Bali starling, scimitar-horned oryx and black-footed ferret. Zoos have been particularly successful when breeding and releasing native species; for instance, the hazel dormouse and British field cricket in the UK. However, considering the number of **endangered species** and the number of species held by zoos worldwide, these successes have been relatively few. It is argued by many anti-zoo campaigners that the cost of reintroducing zoo-bred animals to the wild in terms of money, time and animal welfare is not worth it and that the effort and money would be better spent protecting natural habitats. They also go so far as to argue that *ex situ* conservation (i.e. captive breeding) is detrimental as it reduces effort and finances

Box Z.1. Captive breeding programmes.

It is unfeasible (and unpopular with visitors) for a single zoo to hold enough individuals of one species to avoid **inbreeding** and gene loss and to maintain a genetically healthy population. To ensure large enough populations to do this, several zoos must cooperate in regular animal exchanges and good records must be kept of animal transfers and breeding history. Captive breeding programmes were established to do this and usually operate within a region (e.g. Europe) and are coordinated by the regional zoo association (e.g. EAZA). The various associations have established Taxon Advisory Groups (TAGs) to oversee the programmes within taxon groups, e.g. Old World monkey. Initially, paper records were used to maintain studbooks but these have been replaced by the computer databases ARKS, SPARKS and, recently, ZIMS, developed by the International Species Inventory System (ISIS).

Most regions have a higher and lower level of genetic management for species. The higher level involves specific targets for maintenance of genetic variation (often 90% of the founder diversity for 200 years) and requires the coordinator to make annual recommendations for pairing-up of certain individuals for breeding. The lower level is less intensive and involves monitoring and recording of breeding only.

EAZA's first programmes began in 1985 for six species and, in 2006, there were 40 TAGs operating 163 EEPs (higher-level management) and 153 ESBs (lower-level management); in North America the AZA has 107 SSPs (higher level) and 315 PMPs (lower level).

In addition there are currently 182 international studbooks coordinated by WAZA, which operate globally across all regions.

directed to *in situ* conservation efforts. However, there is little evidence that the money raised by and for zoos would find its way into *in situ* conservation schemes if zoos did not exist.

It is recognized within WZACS (WAZA, 2005) that simply participating in captive breeding programmes is not enough, and that zoos need to contribute more to *in situ* conservation. This can be done in several ways, including simply donating money or through cooperation and sharing of technology and expertise. Many zoo associations such as EAZA run annual fund-raising campaigns for field conservation that enable all zoos to contribute in a meaningful way without the need for any additional staff expertise. Many larger zoo organizations such as the Wildlife Conservation Society (New York zoos) have their own extensive programmes of field conservation projects.

Zoos, education and scientific study

Education of the general public and scientific investigation were both recognized as being roles of zoos by the Ancient Greeks, who were also the first known to charge entrance fees (*c*.500 BC). Although some of the early modern zoos were founded on scientific principles by learned societies (see below), public education was not generally accepted as an important function in modern zoos until comparatively recently. Paignton Zoo (Devon, UK), founded by Herbert Whitley in 1923, was an early pioneer of modern zoo education. Whitley was so determined that his zoo was a place of education and not entertainment that he twice closed its gates (in 1924 and 1937) rather than pay entertainment tax on ticket sales.

Zoos can be used as educational facilities for an enormously diverse range of subjects, including art, tourism and media studies, but their main strength is obviously zoology and related biological disciplines, and their emphasis must now be **ecology**, the environment and conservation education. Most zoos have formal education programmes that deliver a range of specialist courses to various student groups, from pre-school to university level. However, informal education of general visitors is also considered to be extremely important and has been the main driving force behind the trend towards holistic, immersion exhibits in zoos. According to this principle, every detail of the zoo, from animal enclosures to retail outlets and rubbish bins, should be designed to create the impression that visitors are in the animals' natural environment. In addition to the architecture, vegetation and visual imagery, zoos use physical props, artificial weather conditions, sounds and even smells to convince visitors that they are in a tropical rainforest or a desert or a mangrove swamp. It is believed that by doing this visitors gain a better appreciation of the animals together with their environment, and that this ultimately leads to greater concern for them and their conservation.

The impact of zoos' conservation education efforts on their 6 billion visitors worldwide each year is almost impossible to evaluate, but it is claimed by many zoos to be their most important function. Indeed, it seems intuitive that seeing actual, living wild animals close up will have an important and positive influence on many people. Zoos are making increasing numbers of attempts to evaluate their educational impact, but to date the body of hard evidence to support their claim that it is a significant factor in influencing people for wildlife conservation is still small.

Although some ancient zoos, such as Emperor Wen Wang's Garden of Intelligence (China, *c*.1000 BC), were established for the advancement of biological knowledge, most were simply menageries kept for human amusement and to demonstrate power and wealth. The first 'modern' zoo to be founded on scientific lines was at Versailles, France, in 1664 and became the Ménagerie des Jardin de Plantes (Paris) in 1793, 'a place of science for knowledge and progress'. The Zoological Society of London established London Zoo along the same principles in 1828, at which time it was only open to fellows of the society for scientific study. It did not open to the general public until 1847. Major cities around the world created zoos along similar principles shortly afterwards, e.g. New York (Central Park Zoo) and Melbourne, Australia, both in 1860.

During this heyday of European empire building and increased global exploration, these great city zoos had a seemingly never-ending supply of new species to be inspected, described and classified, and much zoo science was simply that. During the 20th century, as fewer and fewer large vertebrates remained to be documented, science within zoos became less important. Zoology was branching towards either the laboratory, where precisely controlled experimental manipulations could be performed, or the field, where results would not be affected by 'unnatural' surroundings. However, a number of prominent researchers, such as Frans de Waal, still appreciated modern zoos as useful facilities for some types of research. Good zoo exhibits with reasonably natural environments and social groupings of animals, excellent animal records and well-informed, cooperative staff can provide excellent halfway houses between the two extremes of **laboratory study** (totally unnatural but highly controlled) and field (totally natural but completely uncontrolled). This realization is growing, and progressive zoos are increasingly developing strong links with academic institutions, such as universities and natural history museums, in order to make good scientific use of their distinctive collections. (AP)

See also: Animal enclosure; Enrichment; Zoo animal

References and further reading

Council of the European Union (1999) *Act 1999/22/EC – EU Zoos Directive*. Available at: http://eur-lex.europa.eu/smartapi/cgi/sga_doc?smartapi!celexplus!prod!DocNumber&lg=en&type_doc=Directive&an_doc=1999&nu_doc=22 (accessed 14 October 2009).

Hancocks, D. (2001) *A Different Nature: the Paradoxical World of Zoos and Their Uncertain Future*. University of California Press, Berkeley, California.

Norton, B.G., Hutchins, M., Stevens, E.F. and Maple, T.L. (eds) (1995) *Ethics on the Ark: Zoos, Animal Welfare, and Wildlife Conservation*. Smithsonian Books, Washington, DC.

Tudge, C. (1991) *Last Animals in the Zoo: How Mass Extinction Can Be Stopped*. Island Press, Washington, DC.

WAZA (2005) *Building a Future for Wildlife – The World Zoo and Aquarium Conservation Strategy*. WAZA Executive Office, Bern, Switzerland. Available at http://www.waza.org/conservation/wzacs.php (accessed 14 October 2009).

Zoo animal

A zoo animal is any individual animal that lives in a **zoo**. Zoos are usually defined by the presence of captive, non-domestic animals so most zoo animals are non-domesticated species. However, many zoos also keep domestic species, usually to provide opportunities for visitors to touch animals, and these are also zoo animals (**see: Domestication**). (AP)

Zoonotic disease

A zoonotic disease is one that passes from animals to humans and vice versa (although the latter may also be referred to as an anthroponosis). Bacteria such as *Salmonella* spp. and *Campylobacter* spp. may infect other animals directly through the feed or food. Transfer may also occur from humans to their companion animals, and from farmed animals to humans (e.g. from cows' milk, rabies through bites from **cats** and **dogs**). (DBM)

Zoophilia

Zoophilia is a term derived from the Greek *zôon* (animal) and *philia* (friendship or love) that refers generally to the state of having an affinity for animals. The more common form of this affinity refers to sexual attraction by a human to a non-human animal. (DGL)

See also: Biophilia

Further reading

Beetz, A.M. and Podberscek, A.L. (2005) *Bestiality and Zoophilia: Sexual Relations with Animals*. Purdue University Press, West Lafayette, Indiana.

Zygote

A zygote is the single diploid cell resulting from fusion of two haploid gametes in the process of fertilization. Following the initiation of cleavage to produce a two-cell structure, the zygote is generally thereafter termed an **embryo**. (MRC)

Species Index

Note: terms in **bold** appear as main entries in the text. Where '*see*' is used, this refers to another entry in the index. Where '*see also*' is used, this refers to the main entry in the text.

Subject Index

Note: terms in **bold** appear as main entries in the text. Where '*see*' and '*see also*' are used, they may refer either to the main entry in the text or to another entry in the index.